AF717969

Advances in Ceramic-Matrix Composites II

Related titles published by the American Ceramic Society:

Phase Equilibria Diagrams, Phase Diagrams for Ceramists, Volume X: Borides, Carbides, and Nitrides
Edited by Anna E. McHale
© 1994, ISBN 0-944904-74-2

Silicon-Based Structural Ceramics (Ceramic Transactions Volume 42)
Edited by Brian W. Sheldon and Stephen C. Danforth
© 1994, ISBN 0-944904-76-9

Advances in Ceramic-Matrix Composites (Ceramic Transactions, Volume 38)
Edited by Narottam P. Bansal
© 1993, ISBN 0-944904-69-6

Microwaves: Theory and Application in Materials Processing II (Ceramic Transactions, Volume 36)
Edited by David E. Clark, Wayne R. Tinga, and Joseph R. Laia, Jr.
© 1993, ISBN 0-944904-66-1

Structural Ceramics Joining II (Ceramic Transactions, Volume 35)
Edited by Arthur J. Moorhead, Ronald E. Loehman, and Sylvia M. Johnson
© 1993, ISBN 0-944904-65-3

Functionally Gradient Materials (Ceramic Transactions, Volume 34)
Edited by J. Birch Holt, Mitsue Koizumi, Toshio Hirai, and Zuhair A. Munir
© 1993, ISBN 0-944904-64-5

Phase Diagrams for Ceramists, Volume V
Edited by Robert S. Roth, Taki Negas, and Lawrence P. Cook
© 1983, ISBN 0-916094-47-2

For information on ordering titles published by the American Ceramic Society, or to request a publications catalog, please call 614-890-4700, or write to Customer Service Department, 735 Ceramic Place, Westerville, OH 43081.

Narottam P. Bansal
March 10, 1995.

Advances in Ceramic-Matrix Composites II

▼▲▼

Edited by

J.P. Singh
Argonne National Laboratory

Narottam P. Bansal
NASA—Lewis Research Center

▲▼▲

Ceramic Transactions
Volume 46

Published by
The American Ceramic Society
735 Ceramic Place
Westerville, Ohio 43081

Proceedings of the Ceramic-Matrix Composites Symposium, presented at the 96th Annual Meeting of the American Ceramic Society, held in Indianapolis, IN, April 25–27, 1994.

Library of Congress Cataloging-in-Publication Data

Advances in ceramic-matrix composites II /Edited by J.P. Singh,
Narottam P. Bansal.
p. cm. -- (Ceramic transactions, ISSN 1042-1122 ; v. 46)
Includes index.
"Proceedings of the Ceramic-Matrix Composites Symposium presented
at the 96th Annual Meeting of the American Ceramic Society . . .
Indianapolis, IN, April 25–27, 1994"--CIP verso t.p.
ISBN 0-944904-80-7
1. Ceramic-matrix composites--Congresses. I. Singh, Jitendra
Prasad, 1948– . II. Bansal, Narottam P. III. American Ceramic
Society. Meeting (96th : 1994 : Indianapolis, Ind.) IV. Ceramic
-Matrix Composites Symposium (1994 : Indianapolis, Ind.) V. Title:
Advances in ceramic-matrix composites 2. VI. Series.
TA418.9.C6A288 1994
620.1'4--dc20 94-23785
CIP

Copyright © 1994, The American Ceramic Society. All rights reserved.

No part of this book may be reproduced, stored in a retrieval system, or transmitted in any form or by any means, electronic, mechanical, photocopying, microfilming, recording, or otherwise, without written permission from the publisher.

Permission to photocopy for personal or internal use beyond the limits of Sections 107 and 108 of the U.S. Copyright Law is granted by the American Ceramic Society, provided that the base fee of US$5.00 per copy, plus US$.50 per page, is paid directly to the Copyright Clearance Center, 27 Congress Street, Salem MA 01970, USA. The fee code for users of the Transactional Reporting Service for *Ceramic Transactions Volume 46* is 0-944904-80-7/94 $5.00+$.50. This consent does not extend to other kinds of copying, such as copying for general distribution, for advertising or promotional purposes, or for creating new collective works. Requests for special photocopying permission and reprint requests should be directed to the Director of Publications, The American Ceramic Society, 735 Ceramic Place, Westerville OH 43081, USA.

For information on ordering titles published by the American Ceramic Society, or to request a publications catalog, please call 614-890-4700.

Printed in the United States of America.

1 2 3 4–98 97 96 95 94

ISSN 1042-1122
ISBN 0-944904-80-7

Contents

Particulate, Platelet, and Whisker Composites

In-Situ and Nanocomposites

Layered Composites and Joining Methods

Ceramic-Metal Composites

Composite Properties Characterization and Test Methods

Micromechanical Modeling and Design

Coating and Surface Modifications of Fibers and Composite Interfaces

Environmental Effects

Preface

Ceramic composites are leading candidate materials for high-temperature structural applications. This proceedings volume contains papers given at a symposium on ceramic-matrix composites held during the 96th Annual Meeting and Exposition of the American Ceramic Society in Indianapolis, IN, April 24–28, 1994. This symposium provided an international forum for scientists and engineers to discuss all aspects of ceramic composites. A total of 181 papers, including invited talks, oral presentations, and posters, were presented from 16 countries (the United States, Germany, Sweden, Canada, the People's Republic of China, France, the Republic of Korea, Brazil, India, Israel, the United Kingdom, Japan, Russia, Taiwan, Venezuela, and Spain). The speakers represented universities, industry, and government research laboratories.

This volume contains 87 invited and contributed papers, all peer-reviewed by American Ceramic Society procedures, on various aspects of ceramic-matrix composites. The latest developments are covered in the areas of fiber and whisker reinforcements, composite (particulate-, platelet-, whisker-, and fiber-reinforced) processing and characterization, test methods, micromechanical modeling and design, coating and surface modifications, composite interfaces and environmental effects, ceramic-metal composites, in situ and nanocomposites, layered composites and joining methods, etc. The result is that all of the most important aspects necessary for understanding and further development of ceramic composites are discussed.

The organizers express their gratitude to all participants and session chairs for their time and effort, to authors for their timely submissions and revisions of the manuscripts, and to reviewers for their valuable comments and suggestions. Without the contributions of all involved, this volume would not have been possible. Financial support of the American Ceramic Society is gratefully acknowledged. Thanks are due to the staff of the meetings and publications departments of the American Ceramic Society for their tireless efforts. Especially, we greatly appreciate the helpful assistance and cooperation of Lori Kozey throughout the production process of this volume. Finally, special thanks are extended to Jinho Joo and Dileep Singh for their invaluable help during the organization of this symposium and the preparation of the proceedings.

J.P. Singh
Narottam P. Bansal

Fibers and Whiskers

MICROSTRUCTURAL AND STRENGTH STABILITY OF A DEVELOPMENTAL CVD SiC FIBER

Ramakrishna T. Bhatt, Anita Garg, and David R. Hull
NASA Lewis Research Center
21000 Brookpark Road
Cleveland, OH 44135

ABSTRACT

The effects of thermal exposure on the room temperature tensile strength and microstructure of a developmental 50 μm CVD SiC fiber have been studied. The fibers were heat treated between 600 and 2000°C in 0.1 MPa argon and air environments for up to 100 hr. In the as-fabricated condition, the fibers showed ~ 6 GPa tensile strength. After argon treatment, the fibers showed strength degradation after 1 hr exposure beyond 1000°C, but those exposed between 1600 and 2000°C retained ~ 2 GPa strength. TEM results showed microstructural changes both in the surface coating and SiC sheath. Flaws created by the rearrangement of carbon in the surface coating and growth of equiaxed SiC grain zone in the SiC sheath are the suggested mechanisms of strength degradation. After air treatment, fibers showed strength degradation after only 2 min exposure at 600°C. Strength retention after 2 min at 1500°C was ~ 2 GPa. Oxidation of the surface coating is the primary reason for strength degradation.

INTRODUCTION

SiC fibers fabricated by chemical vapor deposition methods (CVD) are being used as reinforcement for metal and ceramic matrix materials because of their chemical purity, high modulus, and high strength. SiC fibers, such as the well-known SCS-6 fibers from Textron Specialty Materials (TSM), Lowell, Massachusetts are themselves composites, consisting of a pyrolytic-carbon coated graphite core, a

To the extent authorized under the laws of the United States of America, all copyright interests in this publication are the property of The American Ceramic Society. Any duplication, reproduction, or republication of this publication or any part thereof, without the express written consent of The American Ceramic Society or fee paid to the Copyright Clearance Center, is prohibited.

SiC sheath, and a carbon-rich surface coating as shown in Fig. 1(a). The SiC sheath has two zones: the outer zone is typically silicon-rich with high aspect ratio SiC grains with preferred radial orientation; the inner zone is typically carbon-rich with lower aspect ratio SiC grains [1].

It has been reported previously [2,3] that these fibers when heat treated in argon beyond 1500°C showed recrystallization and growth of the outer silicon-rich zone of the fiber, but the inner carbon-rich zone of the showed much retarded grain growth. In addition, a good correlation was found between the thickness of recrystallization zone and the tensile strength of the fibers. These studies suggested that CVD SiC fibers containing only the inner carbon-rich SiC zone microstructure should show better strength retention for high-temperature applications than those containing the silicon-rich SiC outer zone structure. Based on these observations, a 50 μm CVD SiC fiber with carbon-rich SiC grains was developed by TSM. The objectives of this paper are to determine thermal exposure effects on strength and microstructure of this new fiber and to suggest possible the strength degradation mechanisms.

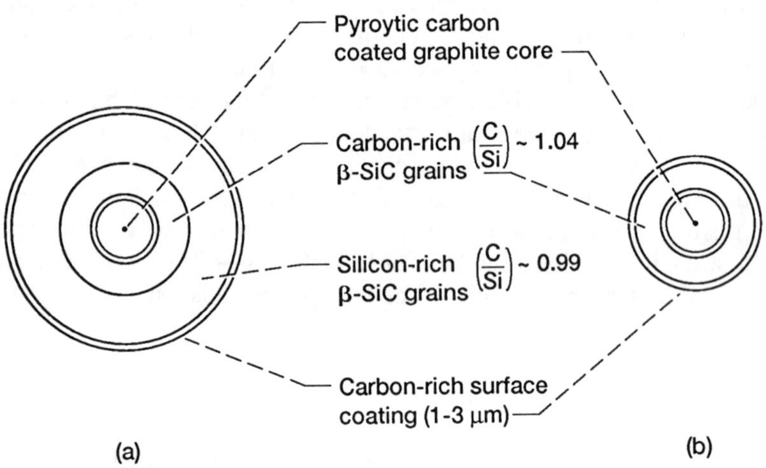

Fig. 1. Cross-section of CVD SiC fibers: (a) Textron SCS-6 fiber (~140 μm), (b) Textron developmental SiC fiber (~50 μm)

EXPERIMENTAL
Material

The 50 μm SiC fiber used in this study was manufactured by CVD of SiC onto a heated pyrolytic carbon coated ~25 μm graphite core in a single stage reactor. Microstructurally, this fiber and the standard grade of CVD SiC fibers (SCS-6) manufactured by TSM are distinctly different. Schematics of the fiber cross-sections are shown in Fig. 1. The 50 μm SiC fiber contains a single zone of carbon-rich SiC, whereas the SCS-6 fiber contains two zones of SiC having different composition. The C/Si ratio in the SiC sheath of the 50 μm SiC fiber is similar to that of the inner zone of the SCS-6 fiber as determined by electron microprobe. Both fibers had ~ 1 to 3 μm thick carbon-rich coating on the outer surface of the SiC sheath. The coating is used to improve strength and abrasion resistance of the fiber [4].

Thermal treatment and tensile testing

For heat-treatments, batches of fifteen individual fibers of length ~ 125 mm were placed in grafoil envelopes. These envelopes were heated under an argon pressure of 0.1 MPa from 600 to 2000°C for 1 to 100 hr in a graphite lined box furnace. For air exposure tests, individual fibers of length ~ 175 mm were heated in a slot furnace for 2 min. Only 25 mm of the 175 mm length was heated to the maximum furnace temperature.

For room-temperature tensile testing, individual fiber specimens, ~ 60 mm long, were placed at the center of a paper picture frame and the fiber ends were glued on to the picture frame using a fast curing epoxy leaving 25 mm as the gauge length. The picture frame was inserted in the pneumatic grips of a tensile testing machine and the sides of the picture frame were cut so that the fiber was freely supported between the grips. The fiber specimens were pulled at a constant cross-head speed of 1.3 mm/min. At least ten fibers were tested for each exposure condition.

Transmission Electron Microscopy (TEM)

For transmission electron microscopy (TEM), fibers were placed in a mixture of epoxy and 6 μm alumina powder and packed longitudinally in a 3 mm diameter glass tube. After the mixture hardened, ~ 1 mm thick disks were sectioned using a diamond saw. The disks were mechanically ground from both sides, dimpled, and ion-beam thinned. Details of specimen preparation technique are described elsewhere [3]. The microscope used for conventional and analytical TEM was a

Phillips 400T operating at 120 kV.

RESULTS

Effects of thermal exposure on the room-temperature tensile strength of the 50 μm SiC fibers after heat treatment in 0.1 MPa argon for 1, 10, and 100 hr from 600 to 2000°C are shown in Fig. 2. For comparison, the strength of the as-fabricated fibers (~6 GPa) is also shown in the figure. Each data point represents an average of 10 to 15 individual tensile tests and the error bar represents one standard deviation. After one hour heat treatment in argon up to 1000°C, the fibers displayed strength levels similar to that of the as-fabricated fibers. From 1100 to 1600°C, the tensile strength decreased with increasing temperature. In the same temperature range, a slight decrease in strength was noticed with increasing exposure time, but the temperature of exposure had a greater influence on strength degradation than the time of exposure. Beyond 1600°C, the strength approached ~ 2 GPa, and the time of exposure had no significant influence on the strength.

To identify flaws responsible for fiber fracture, attempts have been made to retain the primary fracture surface. Since fibers fractured into multiple pieces, identification of the primary fracture surface became difficult. Therefore, it is not clear whether the fractures are initiated by surface or interior flaws.

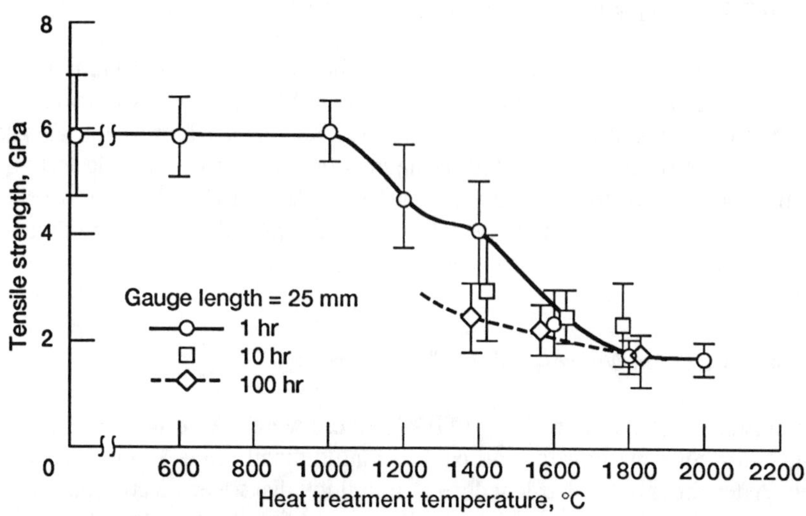

Fig. 2. Effects of temperature on the room temperature tensile strength of 50 μm SiC fibers heat-treated in 0.1 MPa argon

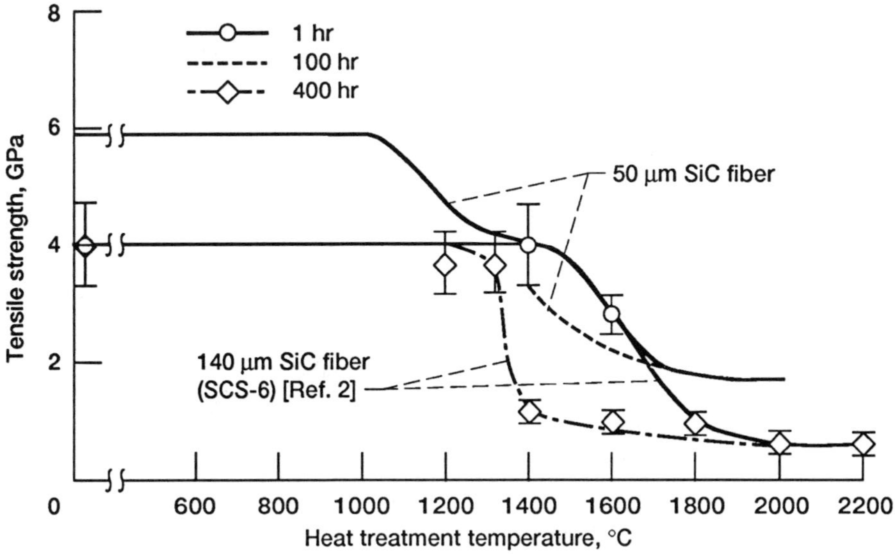

Fig. 3. Comparison of room-temperature tensile strengths of the 50 μm and 140 μm (SCS-6) SiC fibers after heat-treatment in 0.1 MPa argon

In Fig. 3, the average room-temperature tensile strength data for the 50 μm SiC fibers after 1 and 100 hr exposure in argon between 600 and 2000°C is compared with the data reported previously for the SCS-6 fibers in the same environment after 1 and 400 hr exposure in the same temperature range [2]. Comparison of the data in Fig. 3 indicates the following: First, the average as-fabricated tensile strength of the 50 μm fibers is higher than that of the SCS-6 fibers. Second, after 1 hr exposure the 50 μm SiC fibers begin to show strength degradation at a lower temperature than the SCS-6 fibers. Third, the residual strengths of the 50 μm SiC fibers after 1 hr exposure between 1400 and 1700°C are similar to those of the SCS-6 fibers. Fourth, the residual strengths of the 50 μm SiC fibers exposed between 1600 and 2000°C for up to 100 hr are generally higher than those of the SCS-6 fibers.

In Fig. 4, the room temperature tensile strength data of the 50 μm SiC fiber after 1 hr exposure in argon between 600 and 1800°C is compared with the data reported for the two types of SiC fibers derived from polymer precursor methods (Commercially available Hi-Nicalon and a developmental Hi-Nicalon) [5]. For

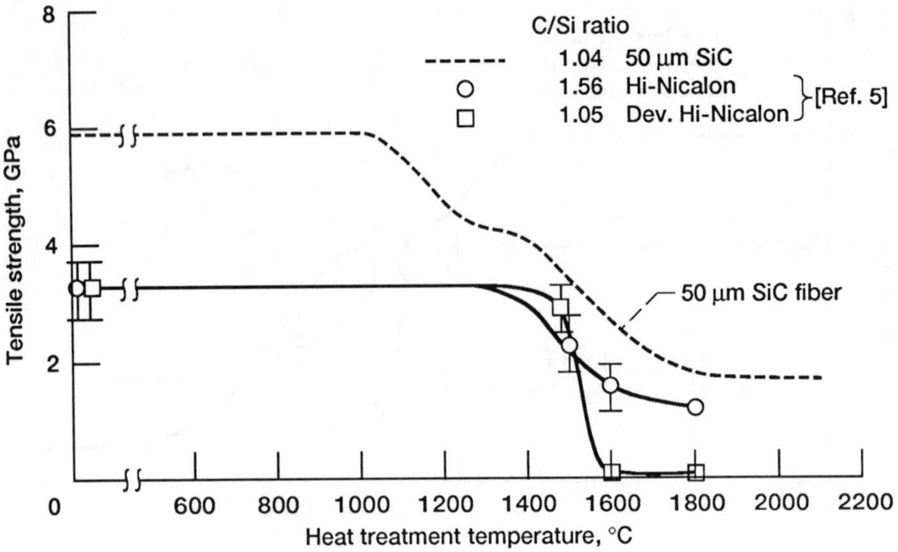

Fig. 4. Comparison of room-temperature tensile strengths of the 50 μm and 14 μm Hi-Nicalon SiC fibers after 1 hr heat-treatment in 0.1 MPa argon

comparison purposes, the carbon to silicon (C/Si) ratios of the fibers are also given. The C/Si ratio of commercially available Hi-Nicalon is significantly higher than that of the 50 μm SiC fibers. On the other hand, C/Si ratios of the developmental Hi-Nicalon and the 50 μm SiC fibers are similar. According to Fig. 4, the 50 μm SiC fiber shows strength degradation at a lower temperature (1000°C) than both types of Hi-Nicalons. Beyond 1500°C, the 50 μm SiC fiber displays higher strength retention than both types of Hi-Nicalons. In fact, the developmental Hi-Nicalons have no strength at all after 1 hr exposure in argon beyond 1600°C.

Influence of oxidation on the room-temperature tensile strength of the 50 μm SiC fibers is shown in Fig. 5. For this study, fibers were heat treated in air for 2 min between 600 and 1500°C and then tested at room temperature. Exposure temperatures above 1500°C were not used to avoid strength degradation due to rapid and thick SiO_2 growth on the fiber surface. Fig. 5 shows that with increasing exposure temperature, strength of the fiber decreased continuously and attained a level similar to that of the fibers heat-treated between 1500 and 1800°C for 1 hr in argon.

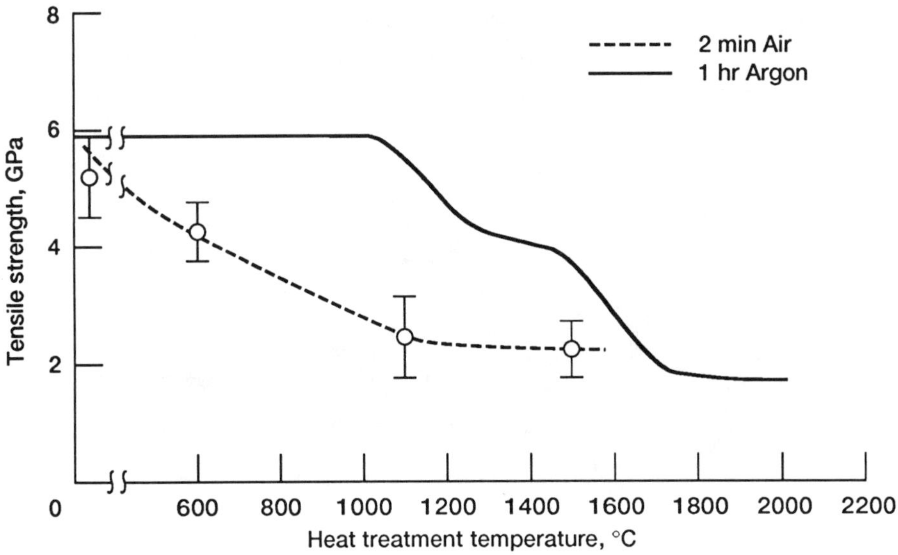

Fig. 5. Effects of temperature on the room temperature tensile strengths of the 50 μm SiC fibers heat-treated in air

To determine argon heat-treatment effects on the microstructure of fibers, TEM analysis was performed on the as-received and heat-treated fibers. In all cases, microstructure of the pyrolytic carbon coated graphite core, SiC sheath, and surface coating were analyzed. Results of this analysis for the as-received fibers are shown in Fig. 6. The core of the fiber contains randomly oriented turbostratic carbon (TC) blocks (Fig. 6(a)). The pyrolytic carbon layer on top of the graphite core consists of TC blocks having a textured orientation. This layer is ~ 0.6 μm thick (Fig. 6(b)) and contains elongated SiC grains that are oriented normal to the radius of the fiber. The aspect ratio of these SiC grains is ~ 13. The carbon-rich SiC sheath is ~ 12 μm thick and consists of elongated β-SiC grains running along the radius of the fiber (Fig. 6(c)). These β-SiC grains are highly faulted and have an aspect ratio of ~ 8. The surface coating on top of the SiC sheath has two layers (Fig. 6(d)). The first layer adjacent to the carbon-rich SiC sheath consists of textured TC blocks and small elongated SiC grains similar to that present in the pyrolytic carbon layer on the core. The aspect ratio of SiC grains is ~ 8. The outer most carbon layer consists of textured TC blocks and finely dispersed SiC particulates.

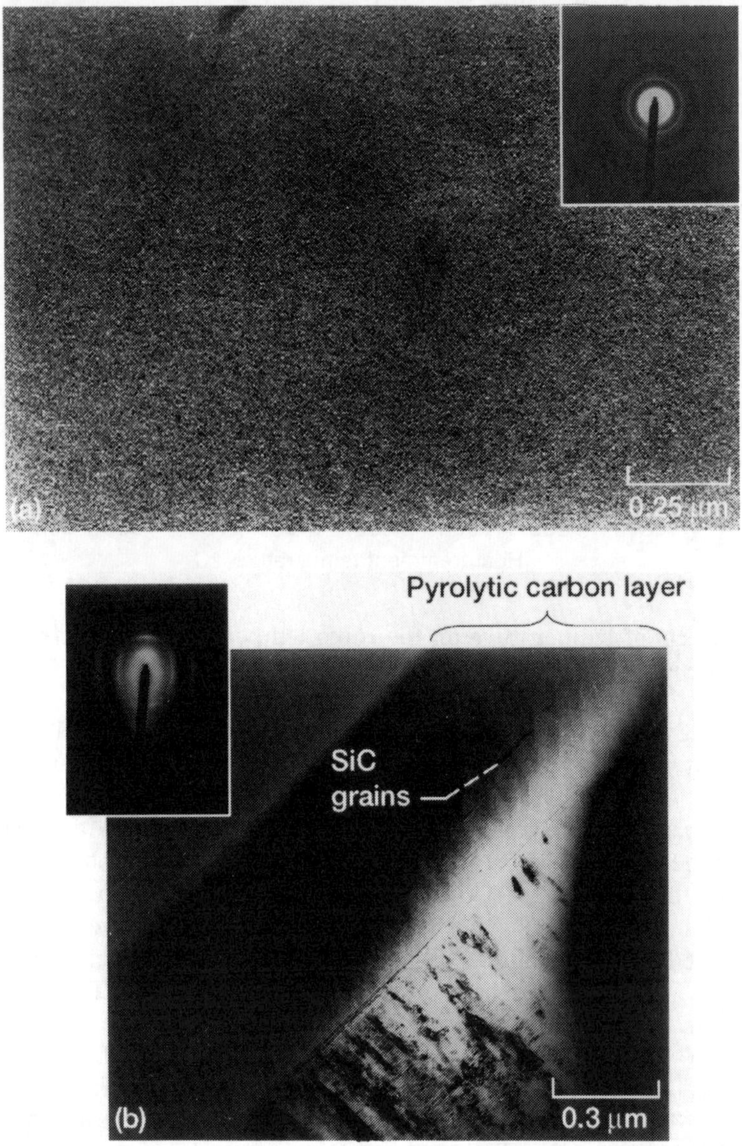

Fig. 6. TEM micrographs and corresponding SADP of different zones of an as-received 50 μm SiC fiber: (a) Graphite core (b) pyrolytic carbon layer on the carbon core (c) carbon-rich SiC sheath (d) carbon-rich surface coating

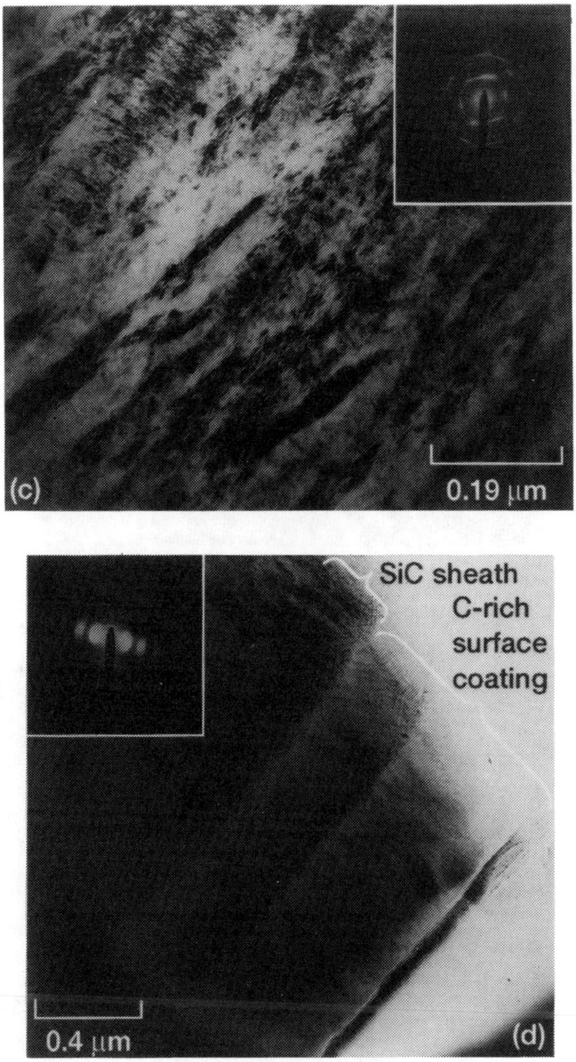

Fig. 6. Continued

Fig. 7. TEM micrographs and corresponding SADP of a 50 μm SiC fiber heat-treated in argon at 1400°C for 1 hr: (a) pyrocarbon layer on the carbon layer (b) SiC sheath and carbon-rich surface coating interface

One hour heat-treatment in argon at temperatures to 2000°C had no significant effect on the microstructure of the graphite core and it remained turbostratic, but other zones of the fiber showed microstructural changes after 1 hr heat treatment beyond 1200°C. With increasing temperature of exposure, spheroidization of SiC grains was observed in the pyrocarbon layer on top of the graphite core, and in the carbon-rich surface coating (Figs. 7 and 8). At 1400°C, a zone of equiaxed SiC grains formed at the interfaces of the SiC sheath and the pyrocarbon layer on the graphite core, and of the SiC sheath and the carbon-rich surface coating (Figs. 7(a) and (b)). This zone, however, stopped growing after 1 hr exposure beyond 1600°C. Typical thickness of the zone is ~ 100 nm. As the heat treatment temperature was increased, no change in the aspect ratio of the faulted grains in the SiC sheath was noticed, but secondary SiC grains started nucleating and growing along the original grain boundaries and at the intersection of two or more grains (Fig. 8). The rate of growth of these secondary grains was slow.

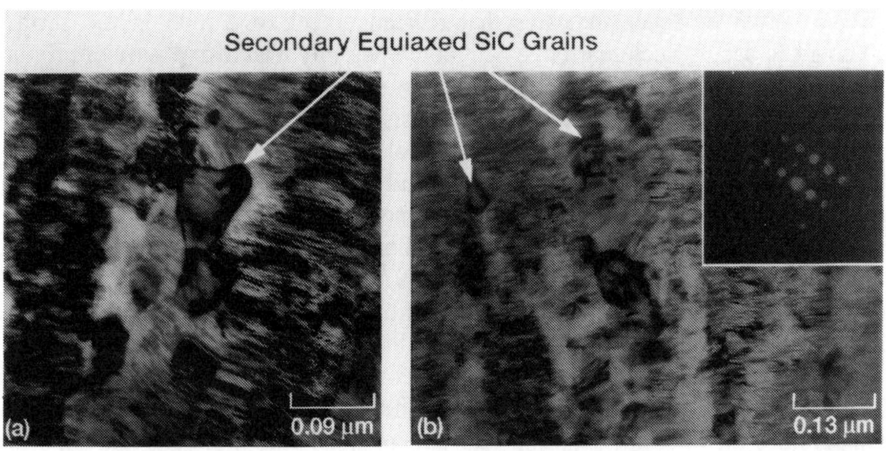

Fig. 8. TEM micrographs of the interfaces of a 50 µm SiC fiber heat-treated in argon at 1800°C for 1 hr: (a) interface between pyrocarbon layer and SiC sheath, (b) interface between SiC sheath and 2nd carbon-rich surface coating

DISCUSSION

It is evident from this study that the developmental 50 µm SiC fibers show high as-received strength, ~ 6 GPa, and that even after 1 hr argon treatment between 1600 and 2000°C, these fibers display high retained strength, ~2 GPa. This value is significantly higher than the strengths of SCS-6 or of Hi-Nicalon fibers after exposure to similar conditions. In argon environment, tensile strength degradation of the 50 μm SiC fibers starts at ~ 1000°C and occurs in two stages as displayed by the shape of the tensile strength vs heat-treatment temperature plot (Fig. 2): the first stage ranging from 1000 to 1200°C and the second stage ranging from 1300 to 1600°C.

The temperature at which strength degradation in the first stage starts also coincides with the deposition temperature of the carbon-rich surface coating [4]. Based on earlier strength characterization studies of other high strength carbon-coated CVD SiC fibers [2,6], we believe that strength degradation in this fiber in the first stage is due to changes occurring in the surface coating which exposes original surface flaws on the SiC sheath. Although in this temperature range, microstructural changes in the carbon-rich surface coating were not observed, except for a slight coarsening of SiC grains in the outer most carbon layer, circumstantial evidence indicates that rearrangement of carbon in the surface coating can occur. Such rearrangement could possibly lead to exposure of original flaws on the SiC sheath, thus nullifying the advantages of the carbon-rich coating. Strong evidence for this hypothesis comes from a previous study [7] which showed ~10 to 20% increase in the electrical conductivity after annealing the carbon-coated SiC fibers in argon beyond 1000°C. This increase in electrical conductivity was possibly due to rearrangement of carbon in the surface coating. One of the factors responsible for the carbon rearrangement can be associated with diffusion of trapped by-products of CVD reaction in the surface coating. It has been documented that during deposition of CVD or PVD SiC, H_2 and Cl_2 are formed as by-products [8]. Depending on the deposition conditions, significant amounts (3 to 15 at%) of these by-products are generally trapped in the fibers. These by-products diffuse out when the fibers are heated above the deposition temperature thus possibly causing rearrangement of carbon blocks and exposure of original flaws on the substrate.

The second stage of strength degradation takes place between 1300 and 1600°C. In this temperature range, TEM study indicates that this degradation is the result of the flaws created by nucleation and growth of the equiaxed SiC grains in the SiC sheath at the pyrocarbon layer/C-rich SiC sheath, and the C-rich SiC sheath/C-rich surface coating interfaces, and along the grain boundaries of elongated SiC grains of the SiC sheath. The growth secondary SiC grains at the boundaries of the original grains is observed only in the 50 μm SiC fibers heat-treated beyond 1300°C. Such growth was not observed in the heat-treated SiC fibers with Si-rich composition (SCS-6). The slow growth of these equiaxed SiC grains in the 50 μm SiC fibers with increasing time and temperature helps in limiting the size of the flaws. This allows the fiber to maintain the strength levels to ~ 2 GPa.

In air environment, strength loss in the 50 μm SiC fiber is seen even at 600°C. This loss is attributed to oxidation of the carbon-coating, similar to the oxidation and strength loss seen in SCS-6 fiber exposed to air [6].

SUMMARY OF RESULTS

(a) In comparison to other commercially available CVD or polymer derived SiC fibers, the 50 μm SiC fibers presently show the best room-temperature tensile strength in the as-fabricated condition and after exposure to high temperatures in argon .

(b) In argon environment, exposure of original surface flaws due to rearrangement of carbon-blocks in the coating, and the flaws created by the formation of equiaxed SiC grains are possibly some of the factors responsible for the strength degradation of the fibers.

(c) In air environment, flaws created during oxidation of the carbon-rich surface coating is the mechanism for strength degradation.

CONCLUSIONS

Based on the characterization results, we conclude that the microstructures of the as-fabricated 50 μm SiC fiber are metastable and that microstructural changes occur after high temperature exposure. However, these changes stop after a certain time interval of exposure at ~ 1600°C. Once the microstructure is stabilized, fibers still display high residual strength which may be controlled by the surface flaws.

Thus, CVD SiC fibers with slightly carbon-rich SiC composition appear to have better potential for high temperature composite applications than other CVD fibers with silicon-rich SiC composition.

REFERENCES

[1] X.J. Ning and P. Pirouz, "The Microstructure of SCS-6 SiC Fiber," J. Mater. Res., **6**, 2234-48 (1991).
[2] R.T. Bhatt and D.R. Hull,"Strength and Microstructural Stability of a SiC

fiber in Argon Environment," Ceram. Eng. Sci. Proc. pp. 1832, Vol 12 (1991).
[3] X.J. Ning, P. Pirouz, and R. Bhatt, "The Effect of High Temperature Annealing on the Microstructure of SCS-6 SiC Fiber,"Mater. Res. Soc. Symp. Proc., 250, 187-92 (1992).
[4] R. Krutenat, Textron Specialty Materials, Private Communication.
[5] M. Takeda, J. Sakamoto, Y. Imai, and T. Isikawa,"Properties of Stoichiometric Silicon Carbide Fiber Derived from Polycarbosilane," Ceram. Eng. Sci. Proc. Vol 15 (1994) In Press.
[6] R.T. Bhatt and M.D. Kraitchman, "Environmental Effects on the Tensile Strength of Chemically Vapor Deposited Silicon Carbide Fibers," NASA-TM 86981, (1985).
[7] J.A. DiCarlo, "Thermal Stability of SCS-6 SiC Fibers," 2nd Annual HITEMP Review, NASA CP-10039, 1989, pp. 69-1.
[8] B. Bhushan, A.J. Kellock, N. Cho, and J.W. Ager III, "Characterization of Chemical Bonding and Physical Characteristics of Diamond-like Amorphous Carbon and Diamond Films," J. Mater. Res., Vol 7, #2, pp 404-410, (1992)

TENSILE CREEP AND STRESS-RUPTURE BEHAVIOR OF POLYMER DERIVED SiC FIBERS

H.M. Yun, J.C. Goldsby, and J.A. DiCarlo

NASA Lewis Research Center, Cleveland, Ohio, 44135

ABSTRACT

Tensile creep and stress-rupture studies were conducted on polymer derived Nicalon, Hi-Nicalon, and SiC/BN-coated Nicalon SiC fibers. Test conditions were temperatures from 1200 to 1400 °C, stresses from 100 to 1600 MPa, stress application times up to 200 hours, and air, argon, and vacuum test environments. For all fibers, creep occurred predominantly in the primary stage. Hi-Nicalon had much higher 0.2 and 1 % creep strengths than as-produced as well as coated Nicalon fibers. The stress-rupture strength of Hi-Nicalon up to 100 hours was also higher than that of the coated and as-produced Nicalon fibers. SiC/BN coating on Nicalon increased only the short-term low-temperature rupture strength. Limited testing in argon and vacuum suggests that for all fiber types, creep and rupture resistances are reduced in comparison to the results in air. Possible mechanisms for the observed behavior are discussed.

INTRODUCTION

High temperature structural applications, such as, advanced gas turbine engines for civil transport aircraft, have generated great interest in intermetallic and ceramic composites reinforced by high strength continuous length fibers. The maximum use temperatures of these composites can depend upon key fiber properties, such as, oxidation resistance, creep resistance, and stress-rupture resistance. Because of their composition, SiC-based fibers can withstand extremely high temperatures in an oxidizing atmosphere by forming a protective silica layer [1]. However, in terms of creep resistance, polymer derived nanocrystalline SiC fibers, e.g. Nicalon and Tyranno fibers, which have high reinforcement potential because of their small diameter and high as-produced strength, have recently been shown to possess low tensile creep strengths of below 100 MPa at 1300 °C for 1 % creep strain in about 10 hours [2]. At this temperature, decomposition of their inherent Si-C-O oxycarbide phase and SiC grain growth were reported to control their creep resistance as

To the extent authorized under the laws of the United States of America, all copyright interests in this publication are the property of The American Ceramic Society. Any duplication, reproduction, or republication of this publication or any part thereof, without the express written consent of The American Ceramic Society or fee paid to the Copyright Clearance Center, is prohibited.

well as causing weak stress-rupture behavior. Recently a low oxygen-containing nanocrystalline SiC fiber, Hi-Nicalon, has been fabricated by a radiation curing method [3,4]. Because of its reduced oxygen content (~0.5 wt. %), this fiber has greater tensile strength retention at high temperatures. However, because it contains a large content of free carbon, Hi-Nicalon creep and stress-rupture as well as oxidation resistance may not be as good as stoichiometric SiC fibers.

The primary objective of this study was to measure the tensile creep and stress-rupture behavior of newly processed Hi-Nicalon and SiC/BN-coated Nicalon fibers, and to compare this behavior to that of as-produced uncoated Nicalon in air and inert environment. The purpose was to understand whether the new Hi-Nicalon fiber with low oxygen content or Nicalon with a surface coating to prevent oxycarbide decomposition offered any creep and stress-rupture strength advantages over the as-produced Nicalon for long-term stress application at high temperature. Emphasis was placed on the low creep strain region below 1 % since this value is often quoted as the upper creep limit for structural composites. In addition, an analysis was performed to identify whether rupture strength or creep strength limits the maximum use temperature of these fibers for a 1000 hour application.

EXPERIMENTAL PROCEDURE

Typical properties of the fibers in this study are listed in Table I. The primary phase of the fibers was beta-SiC. The mole compositions of beta-SiC varied from 49 to 65 % for Nicalon and Hi-Nicalon, respectively. Other major phases were about 23 % Si-C-O and about 28 to 35 % free C for Nicalon and Hi-Nicalon, respectively. These additional phases which are a consequence of the fiber fabrication process serve as matrix for the beta-SiC nanocrystals and influence the mechanical strength of the fibers. The average diameters of these fibers were nominally 14 and 15.5 μm for both Nicalon and Hi-Nicalon, and for SiC/BN-coated Nicalon, respectively. The nominal thickness of the inner layer of BN and outer layer of SiC was about 0.4 μm. Average SiC grain sizes were about 1.7 nm and 4 nm for the Nicalon and Hi-Nicalon, respectively.

For this investigation, tensile creep and stress-rupture measurements were made by dead weight loading. The test temperatures were in the range of 1200 to 1400 °C with loading up to about 200 hours. The tests were done mostly in air, but some experiments were done in argon and in vacuum of 10^{-5} Pa to determine possible environmental effects. Before starting the creep rupture experiments at constant temperature, fibers were dead-weight loaded at various stresses at room temperature, and then subjected to a constant warm-up rate to ~1500°C. Temperatures for fiber fracture were recorded for each stress level. The results of this "warm-up" test allowed determination of the maximum stresses that the fibers could withstand during the long-time constant temperature tests.

TABLE I. NOMINAL PROPERTIES OF POLYMER DERIVED SiC FIBERS

Property \ Fiber	NICALON NL202	Hi-NICALON	COATED NICALON (SiC/BN)
Manufacturer	Nippon Carbon	Nippon Carbon	Nippon Carbon (3M Coating)
# of Filaments	500	500	500
Composition (mol.%)	~49% β SiC + ~23% Si-O-C + ~28% C	~65% β SiC + ~35% C	~49% β SiC + ~23% Si-O-C + ~28% C ~0.4μm BN : ~0.2μm SiC
Avg. Grain Size (nm)	1.7	4	1.7
Avg. Diameter (μm)	14	14	15.5
Density (g/cm^3)	2.6	2.7	2.6
Elastic Modulus at RT (GPa)	~ 200	~ 270	~ 200
Tensile Strength at RT (MPa)	~2400	~2800	~2400

In preparation for the elevated temperature creep and stress-rupture tests, 250 mm long fiber specimens were first separated from as-received multifilament tows and then glued on paper tab grips for air testing. The fibers were inserted through a slit in the furnace and hooks were attached to the paper grips (Fig. 1). For argon and vacuum testing a similar rig was used. The uniform temperature regions of the furnace (hot zone) were determined to be 1 and 4 " for two types of $MoSi_2$ heating element furnaces. The hot zone length for the argon (graphite heating element) and vacuum (tungsten heating element) furnaces was 4". The creep deformation was calculated by assuming the main creep deformation occurred only in the hot zone lengths. A hot-grip method [5] was also used and the creep strains were found to be in fairly good agreement with the cold grip method. The fiber length change was monitored by a free floating cored LVDT (Schaevitz Engineering, Pennsauken, NJ). The displacement sensor was interfaced with a computer to record the data. The maximum temperature gradient along the hot zone of the furnace (1" and 4" hot zone) was within 6 °C. A Pt-Rh (for the air and vacuum furnace) and W-Re (for the argon furnace) thermocouple electronic controller provided a constant temperature at center of the hot zone (± 2 °C).

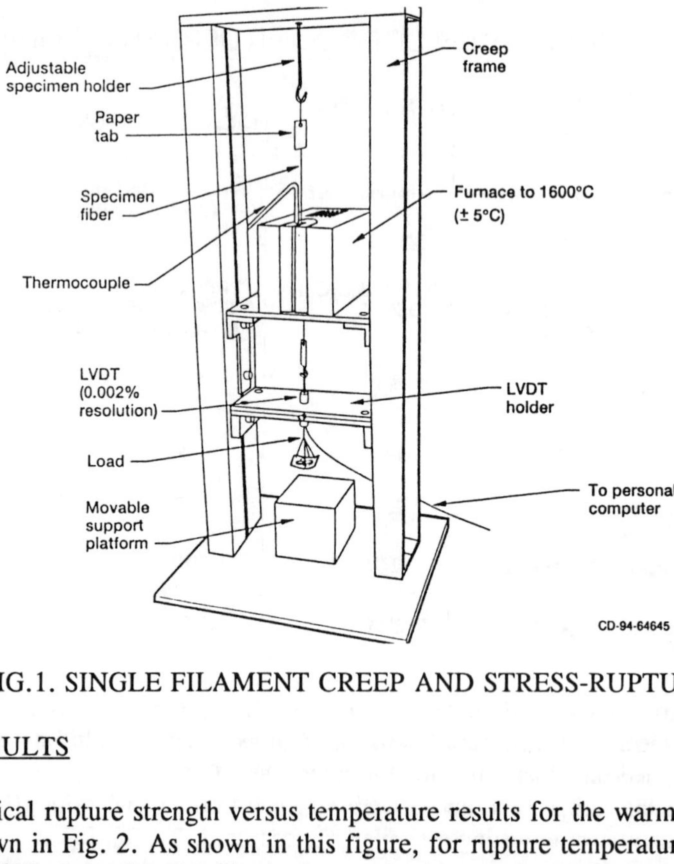

FIG.1. SINGLE FILAMENT CREEP AND STRESS-RUPTURE RIG.

RESULTS

Typical rupture strength versus temperature results for the warm-up test are shown in Fig. 2. As shown in this figure, for rupture temperatures from 1000 to 1600 °C for the SiC fibers, the required stresses were in the range from 100 to 1600 MPa. The general ranking of the fibers in terms of maximum rupture temperature at a given stress were Hi-Nicalon, coated Nicalon, and uncoated Nicalon. The coated Nicalon data indicated higher rupture temperatures in argon than the as-produced Nicalon, suggesting that the coating on Nicalon delays the decomposition of the Si-C-O phase. Generally, the rupture temperatures in air for the Hi-Nicalon were higher than those measured for the same fiber in argon and vacuum.

For the constant temperature tests, Fig. 3 illustrates the stress-rupture strength versus time for Hi-Nicalon (a), and coated & as-produced Nicalon (b) as a function of testing temperature. At 1200 to 1315 °C and stresses below 500 MPa, most fibers did not rupture in less than 100 hours. Therefore it was not possible to use these results to draw best fit lines for rupture time.

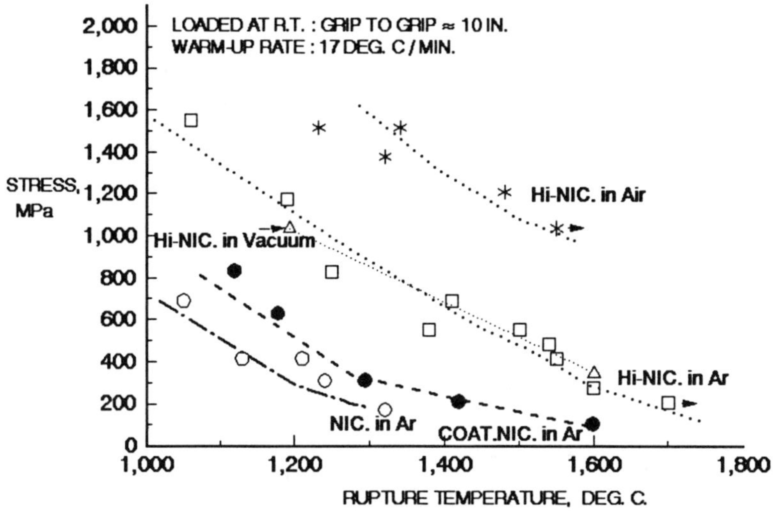

FIG.2. WARM-UP RUPTURE STRENGTH OF Hi-NICALON, NICALON, AND SiC/BN-COATED NICALON FIBERS

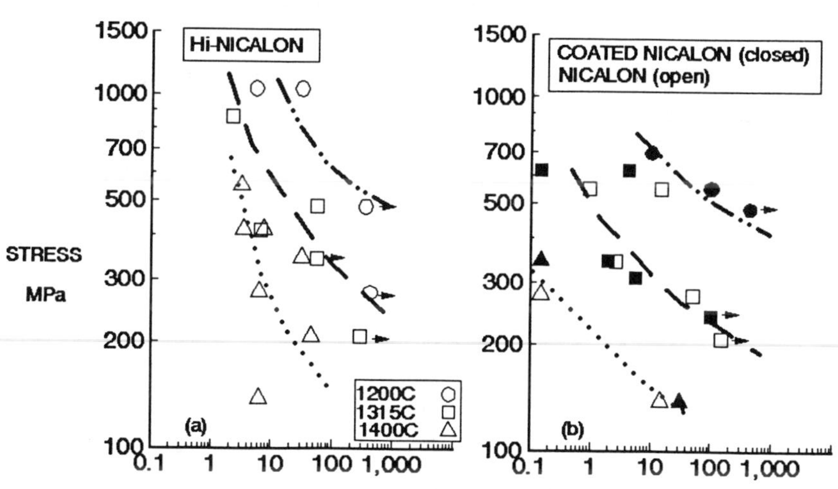

FIG.3. STRESS-RUPTURE STRENGTH OF Hi-NICALON, NICALON, AND SiC/BN-COATED NICALON IN AIR

At the moment, the initial drop with time is assumed to follow a $t^{-1/N}$ dependence with N of about 5 to 6. This is believed to be related with slow crack growth. At 1400 °C, the strength drop was more severe for all polymer derived SiC fibers, even at low stresses and long times, suggesting that fiber decomposition might also be occurring.

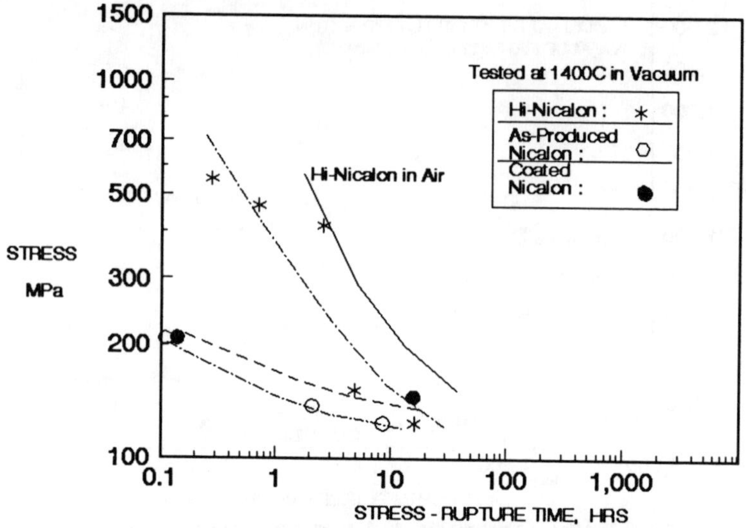

FIG.4. 1400C STRESS-RUPTURE STRENGTH OF POLYMER DERIVED SiC FIBERS IN VACUUM

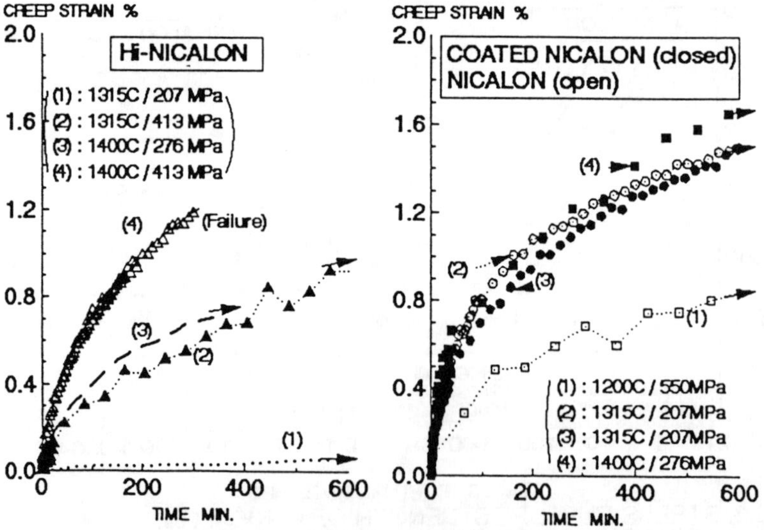

FIG.5. REPRESENTATIVE CREEP CURVES OF Hi-NICALON, NICALON, AND SiC/BN-COATED NICALON IN AIR

The stress versus rupture time data at 1400 °C in vacuum are shown in Fig. 4. In general, the rupture strengths of the fibers in vacuum were lower than those in air, especially for the Hi-Nicalon fibers. The coated Nicalon fibers indicated a slight advantage for rupture strength in vacuum in comparison to the as-received fibers.

Typical results for creep deformation versus time for the Hi-Nicalon and Nicalon are shown in Fig. 5. In this figure, creep data were taken from 1200 to 1400 °C in air. For all fibers, primary creep occurred; while for Hi-Nicalon steady-state creep may also evident. In general, the Nicalon fiber required lower stress than Hi-Nicalon to obtain the same creep strain. At 1200 and 1315 °C, the creep deformation of Hi-Nicalon was relatively small; most cases below 1 % in 100 hours. At 1400 °C, creep deformation increased to about 2 % for the same time conditions.

For estimation of the creep strength of the polymer derived SiC fibers, the times for 0.2 and 1 % total creep strain were determined from the creep curves. Fig. 6 illustrates the stress dependence of the 0.2 and 1% creep time of Hi-Nicalon and Nicalon in air. For the same stress, Hi-Nicalon had a longer creep time than Nicalon fiber at the higher temperatures, but a comparable creep time at 1200 °C. Least squares fit lines through the data in Fig. 6 suggest a power law creep time relation of $t \approx \sigma^{-m}$. From the best fit lines, the stress exponents m were determined to be about 4 for the Hi-Nicalon and about 5.1 for both coated and uncoated Nicalon at 1200 °C, and about 2.5 from 1315 to 1400 °C. It can also be seen that the data for Nicalon exhibit an increase in scatter with increasing temperature.

The inserts in Fig. 6 illustrate the effect of temperature on the 0.2 % creep time at 400 MPa and at temperatures in the range of 1200 to 1400 °C. The apparent activation energy for creep, Q_{creep}, can be obtained from the slope of the best fit line using the following equation:

$$Q_{creep} = R\,(\Delta \ln(t)/(\Delta(1/T)) \qquad (1)$$

where R is the universal gas constant, t is the creep time (hr), and T is theabsolute temperature (K). The Q_{creep} was determined to be about 610 kJ/molfor Hi-Nicalon and 430 to 660 kJ/mol for Nicalon (coated & as-produced). The 660 kJ/mol energy for the Nicalon was determined at the two lower temperatures, and the 430 kJ/mol at the two higher temperatures. The Hi-Nicalon value is somewhat higher than that of Bodet et al. [6] who reported Q_{creep} to be 340 to 420 kJ/mol. On the other hand, the 610 kJ/mol value from

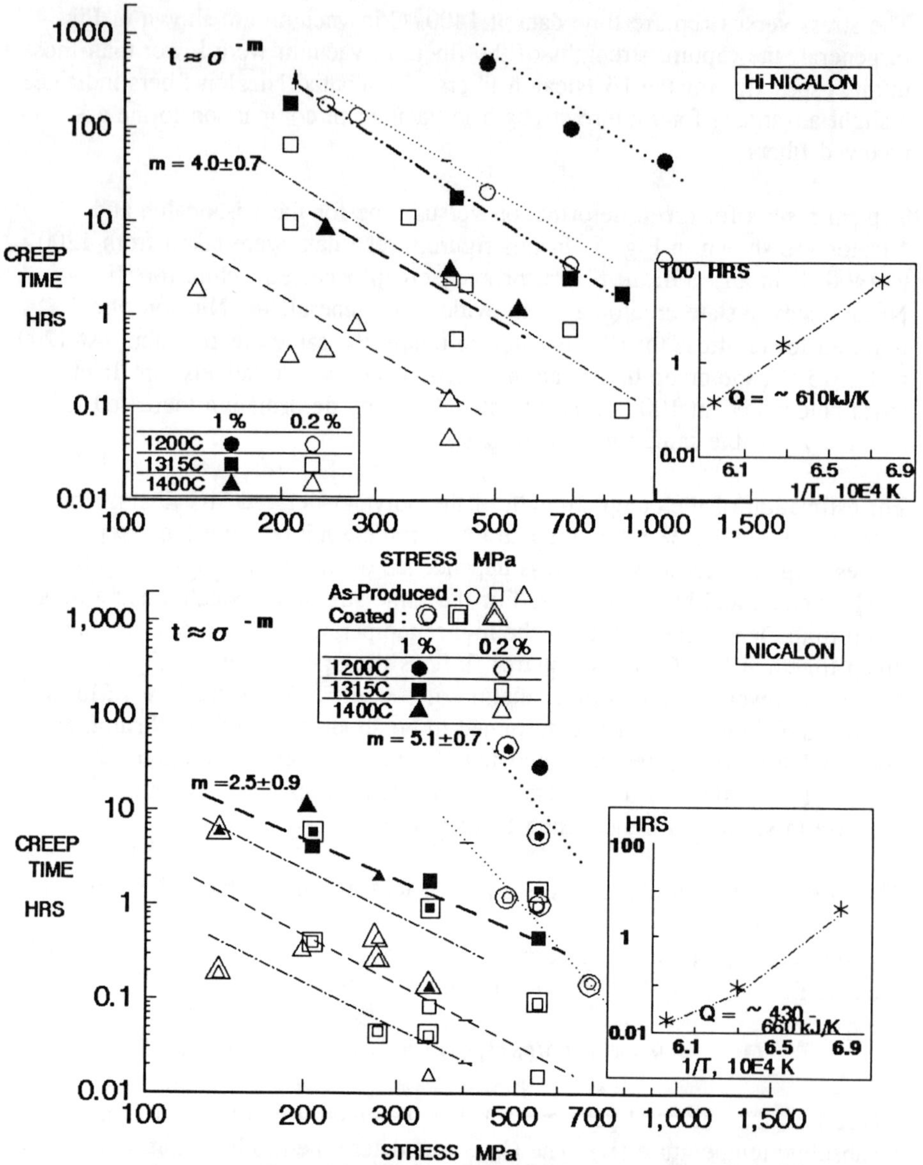

FIG.6. STRESS-EFFECTS ON 0.2 AND 1 % CREEP TIME OF POLYMER -DERIVED SiC FIBERS

this study appears to agree with the sintered bulk SiC data and is similar to that of the activation energy (560kJ/mol) for grain boundary diffusion of carbon [7].

DISCUSSION

The single filament creep and stress-rupture results of this study indicate that radiation-cured low-oxygen Hi-Nicalon fiber appears to have higher strength capabilities than as-produced Nicalon fibers. The SiC/BN coating on Nicalon improves only the short-term rupture strength, not the long-term stress-rupture strength. The creep curves of the coated Nicalon in air were similar to the uncoated fibers, suggesting the double layer SiC/BN coating did not delay decomposition of the Si-C-O phase, at least in fibers fully exposed to the oxygen environment. The abrupt change in Nicalon creep above 1200°C may be related to the fact that during fabrication these fibers did not experience temperatures greater than about 1200°C.

The influence of free carbon in Hi-Nicalon appears to be significant for creep behavior. The initial free carbon of the as-produced Hi-Nicalon may cause the measurable initial creep strain in the primary stage similar to the creep behavior of Nicalon. Nevertheless, the amount of creep was much smaller than that of Nicalon. This suggests that the radiation-curing which significantly reduces the amount of the Si-C-O phase can also minimize the creep contribution of this phase. Also the Hi-Nicalon have probably been produced at higher temperatures than Nicalon which may help to stabilize the free carbon phase and reduce its creep contribution. In terms of carbon phase stabilization at high temperatures, creep deformation should be further studied after the Hi-Nicalon fibers are pre-annealed at temperatures of 1400°C and above. In comparison to as-produced fibers, Bodet et.al. [6] reported creep deformation at 1400°C decreases after a pre-anneal at 1600°C for 1 hr.

The influence of the air environment appears to exist for the creep and rupture of the SiC fibers. Creep in vacuum of Hi-Nicalon and Nicalon at 1400°C was observed higher than that in air, and resulted in a shorter rupture life in vacuum. This higher creep in vacuum is not fully understood. However, the improved rupture strength in air could be explained by the protective SiO_2 layer which forms on the fiber during testing in air and could then inhibit release of the decomposition gases, SiO and CO.

In other studies [8], it was found that the primary creep behavior of the polymer derived SiC fibers in air can be expressed in terms of the tensile stress, σ, absolute temperature, T, and time, t, by the relation:

$$\epsilon = A \sigma^n t^p (-pQ_{creep}/RT) \quad (2)$$

Here n is the stress exponent, p the time exponent, Q_{creep} the effective activation energy for creep, and A a creep constant. This equation can be manipulated to predict creep stress versus time for a given strain, that is:

$$\sigma = \sigma_0 (t \exp (Q_{creep}/RT))^{-1/m} \quad (3)$$

where σ_0 depends on creep strain and m is a time exponent (m = n/p). Creep parameters determined here for the Hi-Nicalon and Nicalon are summarized in Table II. For Nicalon, the previously reported value of n/p = 3 [8] is in agreement with the m value measured here for 1300 - 1400°C, but lower than the m value at 1200°C.

TABLE II. APPROXIMATE CREEP STRENGTH MODEL FOR POLYMER DERIVED COMMERCIAL SiC FIBERS AT 0.2 % CREEP

	Hi-NICALON	NICALON / COATED NICALON	
σ_0	$(3.48\pm0.25) \times 10^{-2}$	$(7.3\pm0.7)\times10^{-2}$ *	$(9.9\pm0.5)\times10^{-3}$ +
m	4.0 ± 0.7	5.1 ± 0.7 *	2.5 ± 0.9 +
Q	610 kJ/mol.	660kJ/mol. *	430kJ/mol. +

* at 1200C
\+ 1300 - 1400C

Using the stress-rupture curves in Fig. 3, one can estimate the 1000-hour stress-rupture strength at three temperatures for each of these fibers. By using Eq. 3 and extrapolation from 0.2 and 1 % creep strength curves in Fig. 6, one can project the 1000-hour creep strength for the fibers. Fig. 7 illustrates the projected 1000-hr creep and stress-rupture strengths in air. In ceramic composite application, the polymer derived SiC fibers should not be expected to deform beyond 1%. Therefore the 0.2 and 1 % creep strengths will probably limit the use temperature of these fibers, rather than the fiber stress-rupture behavior. The 1000 hr creep strength of Nicalon appears to change abruptly between 1200 and 1300 °C, and the Fig. 7 shows discontinuous line. Thus, even though Hi-Nicalon has higher creep strength than the uncoated or SiC/BN coated Nicalon, the Hi-Nicalon is creep limited to a use temperature of about 1250°C at 200 MPa and 0.2 % creep strain. This can be compared to the 1350 °C stress-rupture use temperature limit for this fiber at the same 200 MPa stress level.

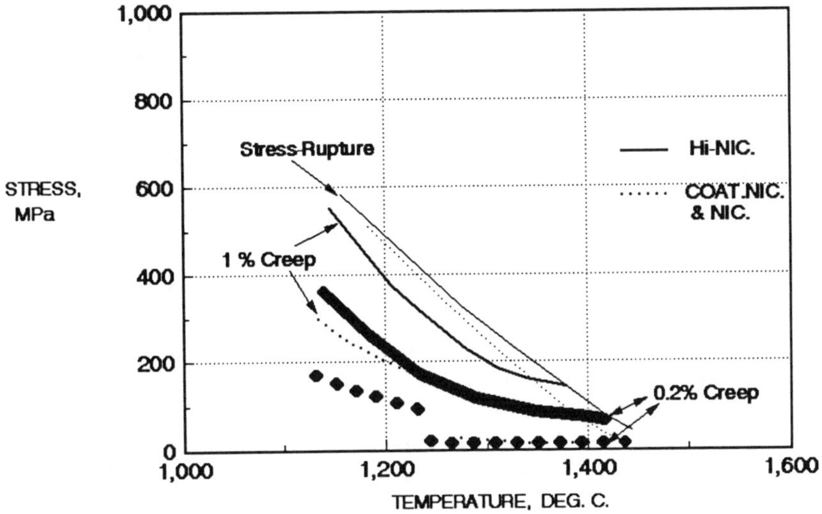

FIG.7. PROJECTED 1000 - HR CREEP AND RUPTURE STRENGTHS OF POLYMER DERIVED SiC FIBERS

CONCLUSIONS

For times up to 1000 hours at 1200 to 1400 °C, Hi-Nicalon is projected to have a greater 0.2 % creep strength and stress-rupture strength than the SiC/BN coated and as-produced Nicalon. The greater creep strength of Hi-Nicalon appears to be related to its low oxygen content. Excess free carbon beyond the stoichiometric Si/C ratio in Hi-Nicalon may reduce its creep strength, but apparently does not seriously influence its oxidation resistance as suggested by its high stress-rupture strength in air.

REFERENCES

1] T.F.Cooke: "Inorganic Fibers - A Literature Review", J. Am. Ceram. Soc., 74, 2959-78, (1991)

2] R.L.Bodet and R.E.Tressler: "Effects of Chemical Environments on the Creep Behavior of Si-C-O Fibers", "Proc. of the 6th European Conference on Composite Materials", edited by R.Naslain,J.Lamon,D.Doumeingts, Bordeaux, France, Sept. 20-24, 1993, pp. 75-83.

3] K.Okamura: "Ceramic Fibers from Polymer Precursors", Composites, 18, 2, 107-20, (1987).

4] K.Okamura and T.Seguchi: J.Inorganic and Organometallic Polymers, 2, 171-79, (1992).

5] H.M.Yun and J.C.Goldsby: "Tensile Creep Behavior of Polycrystalline Alumina Fibers", NASA Technical Memorandum 106269, 1993.

6] R.Bodet,X.Bourat,J.Lamon and R.Naslain: "Tensile Creep Behavior of a Silicon Carbide-Based Fiber with a Low Oxygen Content", PhD Thesis, Domaine Universitaire, France, 1993.

7] J.E.Lane,C.H.Carter and R.F.Davis: "Kinetics and Mechanisms of High Temperature Creep in Silicon Carbide", J. Am. Ceram. Soc., 71, 4, 281-95, (1988).

8] J.A.DiCarlo: "Creep Limitations of Current Polycrystalline Ceramic Fibers", Comp. Sci. Technol., 51, 213-22, (1994).

THE TIME DEPENDENT HIGH TEMPERATURE MECHANICAL BEHAVIOR OF POLYCRYSTALLINE α-SiC FIBERS

K. L. Rugg, L. A. Giannuzzi, and R. E. Tressler, Department of Materials Science and Engineering, The Pennsylvania State University, University Park, PA 16802

ABSTRACT

A preliminary investigation of the time dependent high temperature (> 1200°C) tensile properties of polycrystalline α-SiC fibers have been assessed utilizing dynamic and static fatigue measurements in conjunction with creep experiments. From these initial results we cannot confirm the existence of a threshold stress intensity for crack growth as has been proposed for monolithic sintered α-SiC. The fibers exhibit primary creep to strains less than 1%. The strength and creep results are used to approximate the overall time dependent stress rupture characteristics of the polycrystalline α-SiC fibers.

INTRODUCTION

An understanding of the mechanisms of creep and fracture of currently available fibers may lead to improved fibers for reinforcing ceramic matrix composites (CMC's) for high temperature uses. The advancement of the high temperature properties of CMC's has spurred interest in the development of polycrystalline SiC fibers in hopes of exploiting its strength capabilities as well as realizing similar creep properties to that of monolithic SiC. When a ceramic is utilized as a high temperature structural component, the time dependent mechanical properties should be thoroughly assessed in order to establish its reliability. In this paper, we report on the high temperature time dependent strength and creep properties of the Carborundum α-SiC fiber and compare these properties to that of monolithic α-SiC and other SiC-based fibers where applicable.

EXPERIMENTAL TECHNIQUES

The α-SiC Fiber

The fiber under investigation is a developmental 30μm nominal diameter sintered polycrystalline alpha silicon carbide (α-SiC) fiber manufactured by Carborundum [1]. Two batches of fiber were studied, Batch CL-1-210-45-1-PS received May, 1992, hereafter referred to as "old" fiber, and Batch A-176-77A received November, 1993, referred to as "new" fiber. The primary difference in the basic

To the extent authorized under the laws of the United States of America, all copyright interests in this publication are the property of The American Ceramic Society. Any duplication, reproduction, or republication of this publication or any part thereof, without the express written consent of The American Ceramic Society or fee paid to the Copyright Clearance Center, is prohibited.

microstructure between the two batches of fibers is that the newer fibers exhibit better grain size control and lower porosity. The α-SiC fiber contains additions of boron and carbon which serve to aid in the sintering process. Earlier work in our laboratory has shown that the grain size of the fiber is ~ 2μm, and the grain size remains stable even after annealing for 100 h in air at 1400°C [2]. There appears to be no additional grain boundary phases present, except for carbon-rich inclusions at the SiC triple junctions. In addition B_4C has been shown to precipitate intragranularly during heat-treatment of the fiber. The microstructure of the α-SiC fiber is similar to that of monolithic α-SiC except for grain size. Monolithic sintered α-SiC has an average grain size of ~ 5 μm [3]. Therefore, it is reasonable to suggest that the high temperature mechanical properties of the fiber would mimic the properties of the monolithic material except for grain size effects.

High Temperature Strength Experimentation

It has been shown that monolithic sintered α-SiC exhibits subcritical crack growth (SCG) at high temperatures and that there is a threshold in the stress intensity (K_{Th}) below which crack growth does not occur [4]. Therefore, it is the object of this study to determine if the same behavior occurs in Carborundum developmental α-SiC fibers. Dynamic fatigue and interrupted static fatigue have been used to investigate this behavior.

The velocity of a crack in a material being stressed can be represented by the empirical relation [5]

$$v = A\, K_I^N \tag{1}$$

where v is the crack velocity, K_I is the applied stress intensity factor, and A and N are the experimentally determined crack growth parameters for the material at a specific temperature and a specific environment. The magnitude of N indicates the susceptibility to SCG; as the value of N increases, the susceptibility for SCG decreases. When there is no threshold stress intensity for SCG, the strength of the material undergoing dynamic fatigue can be modeled by [5]

$$\sigma_f = \left[B\,(N+1)\,(d\sigma/dt)\, S_i^{N-2} \right]^{1/(N+1)} \tag{2}$$

where σ_f is the fracture stress, S_i is the initial (i.e., fast fracture) strength of the material, and B is a constant that incorporates A and the crack geometry. By plotting the log of the fracture stress versus the log of the stressing rate, the parameters B and N may be determined. In the absence of a threshold stress intensity for SCG, the time to failure for the material undergoing static fatigue can be modeled by [5]

$$t_f = B\, \sigma_A^{-N}\, S_i^{N-2} \tag{3}$$

where t_f is the time to failure and σ_A is the applied stress. Values of B and N were calculated from dynamic fatigue results and were compared with the results of interrupted static fatigue tests.

Experiments have been carried out on both fiber batches using two different testing devices. The first, designated AFT (i.e., advanced fiber tester), operates by suspending a fiber from a micropositioning motor through a furnace to the load cell [6]. This apparatus is capable of testing in controlled atmospheres. The second tester is a tabletop Instron (Model TM, Instron Corp., Canton MA). A fiber is suspended from a hook attached to the load cell through a furnace to the crosshead. Our Instron apparatus is capable of testing only in air. The total fiber gauge length for the Instron tester is more than twice as large as that of the AFT, 210 mm versus 90 mm, respectively, and the hot zone length of the furnace on the Instron is approximately twice as long as that on the AFT.

Dynamic fatigue tests were conducted at stressing rates ranging from 1×10^{-2} to 3×10^{1} MPa/s at 1400°C. Interrupted static fatigue (ISF) tests were carried out at applied stresses of 300 to 650 MPa for time periods of 4 to 24 hours after which surviving fibers were quickly pulled to failure at 80 MPa/s for AFT tests or 30 MPa/s for Instron tests.

High Temperature Creep Experimentation

Creep experimentation was performed utilizing the AFT. To determine the apparent activation energy for creep, individual fibers were tested under constant load (a nominal stress of 278 MPa was used) at temperatures ranging between 1250°C-1400°C. Similarly, in order to calculate the stress dependence of the creep behavior, individual fibers were tested at constant temperature (1400°C) under various loads (the nominal stress range was between 550-650 MPa). All creep tests were performed in UHP argon.

RESULTS AND DISCUSSION

High Temperature Strength Behavior

Some of the old fibers were pre-oxidized in air for 10 hours at 1200°C prior to testing. The room temperature strength of the old fiber was 782 ± 172 MPa for a 90 mm gauge length. The new fiber had a strength of 944.1 ± 117.3 MPa for the same gauge length and a nearly identical strength of 944.9 ± 176.8 MPa for a 210 mm gauge length.

Dynamic Fatigue: Dynamic fatigue results at 1400°C obtained for each batch of fiber using both of the testers are shown in Figs. 1 and 2. Findings for the new fiber on the AFT and the old fiber on the Instron are based on a limited number of specimens.

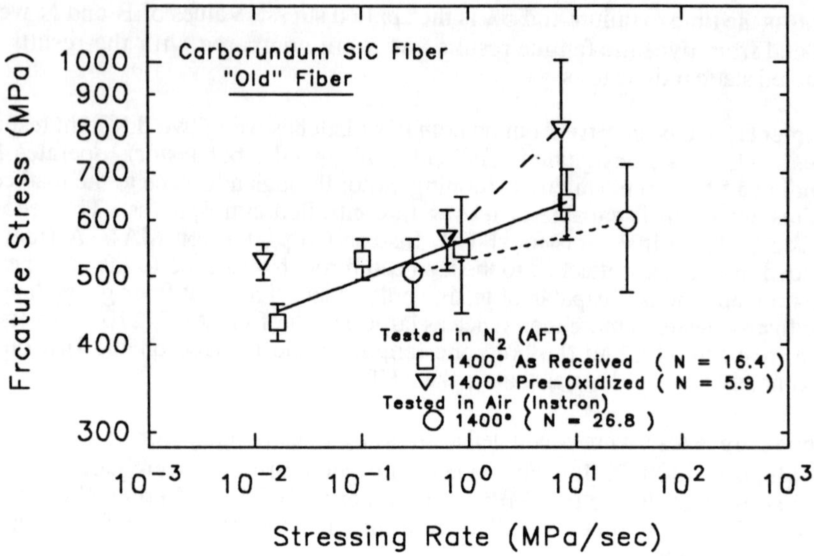

Fig. 1 Dynamic fatigue of "old" Carborundum α-SiC fibers at 1400°C.

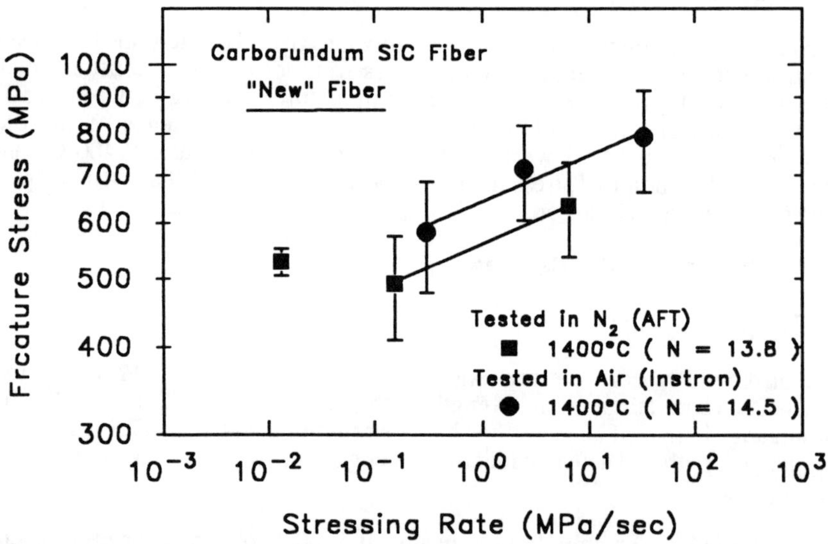

Fig. 2 Dynamic fatigue of "new" Carborundum α-SiC fibers at 1400°C.

The fast fracture strength of the old fiber at 1400°C (fig. 1) shows little dependence on the type of tester used. While it may be expected that the longer gauge length of the Instron specimens would lead to lower strengths, it is possible that oxidative blunting counteracts this trend leading to the observed correspondence. For the old fiber tested in nitrogen the crack growth exponent N was calculated to be 16.4. For the same fiber tested in air, the N value was 26.8, but is based on fewer data points, therefore there is much uncertainty in this value. There is an increase in strength at the fastest stressing rate used for the fiber pre-oxidized in air at 1200°C for 10 hours prior to testing in nitrogen, indicating the effect of oxidative blunting of intrinsic surface flaws. Using only the fastest two stressing rates, an N value of 5.9 was determined, indicating that crack growth was much easier in the pre-oxidized material. The discrepancy in behavior between the pre-oxidized fibers and the fibers tested in air may have resulted from insufficient time at temperature during the latter tests to allow significant oxidation effects on crack growth. At 1200°C in nitrogen, the old fiber was weaker (583 ± 92 MPa) at the fastest stressing rate used than it was at 1400°C (635 ± 70 MPa). The stressing rate sensitivity at 1200°C was also lower, i.e. N = 34.

The strength of the new fibers tested at 1400°C on the AFT in nitrogen (fig. 2) was surprisingly lower than that observed on the Instron in air, except at the lowest applied stressing rate. The reasons for this behavior are not yet known, but may be related to flaw blunting due to oxidation. The minimum in the dynamic fatigue curve for the tests in nitrogen was characteristic of blunting of the severe flaws below the threshold at the slowest stressing rate [7]. The N value for tests in air was 14.5. Using only the data for the two fastest stressing rates an N value of 13.8 was determined. These N values are consistent with those reported for bulk sintered α-SiC under similar conditions [8,9].

The presence of SCG during dynamic fatigue becomes evident upon inspection of fracture surfaces. Unfortunately, nearly all of the fibers tested shattered upon failure and many fracture surfaces were lost. Fig. 3 shows a fracture surface which was preserved from a new fiber tested at 1400°C in nitrogen at a stressing rate of approximately 0.1 MPa/s. The fiber failed at a stress of 408 MPa. The arrowed region of intergranular fracture is indicative of SCG. The fracture path became transgranular during catastrophic propagation after the crack had reached a critical length. There was no evidence of cavitation on grain facets in the SCG region indicating that a diffusion mechanism as proposed by Chuang [10], and not accumulation of creep damage, was the likely mechanism of crack propagation. This result is consistent with the absence of a tertiary creep region as will be discussed in the section on creep testing which follows. Using a stress intensity solution for a round bar with a semi-elliptical surface crack under tension [11] and measured flaw geometry parameters 2a = 23 μm and b = 11.5 μm, the toughness was be calculated to be 2.7 MPa√m.

Interrupted Static Fatigue: Fig. 4 shows the observed Interrupted Static Fatigue (ISF) behavior and the behavior predicted by Equation 3 of both the old and new fibers at 1400°C. The curves on fig. 4 show the predicted behavior for the strongest, weakest, and average strength fiber of the original (fast fracture)

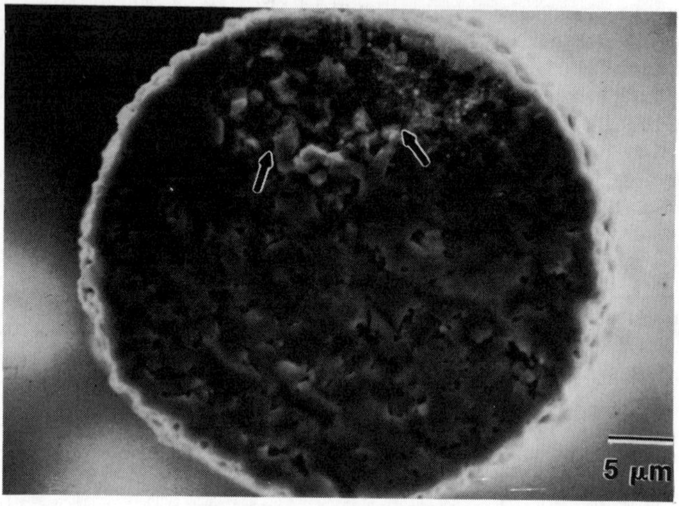

Fig. 3 Fracture surface of a new Carborundum fiber which failed at 1400°C in N_2 at an aplied stressing rate of 0.1 MPa/s. The arrowed region shows intergranular fracture indicative of the range of subcritical crack growth.

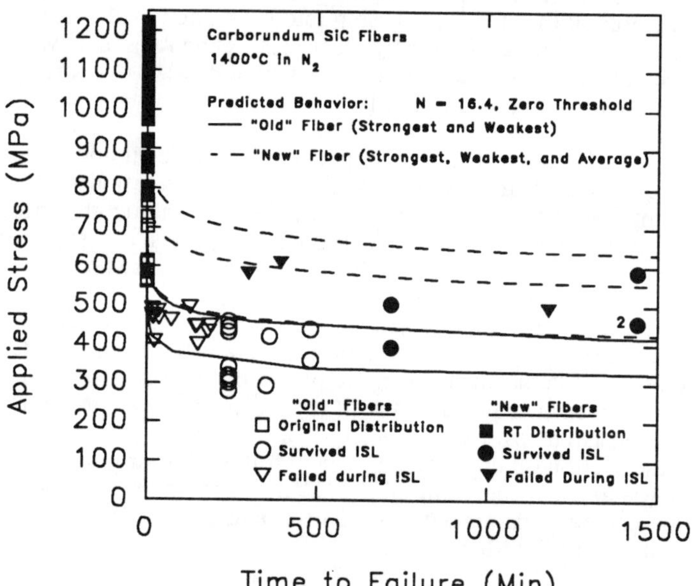

Fig. 4 Interrupted static fatigue behavior of new and old Carborundum α-SiC fibers at 1400°C in N_2. The curves indicate the expected behavior based on equation 3 for several fibers in the original strength distributions.

distributions. All data falls within the bands of expected results. If a data point had appeared above the prediction line for the strongest fiber, it would have been evidence for a threshold stress intensity for SCG. A statistical analysis can yield some further insight. With an N value of 16.4, the model predicts that 3 out of the 9 fibers in the original strength distribution (old fibers) would survive 4 hours at 400 MPa. In ISF tests conducted, 5 of 11 fibers survive at least 4 hours at stresses ranging from 400-460 MPa. Similar results have been observed for the old fiber at 1200°C in nitrogen. This greater survival probability indicates that either the N value is incorrect or there is a threshold effect. It is evident that additional tests are needed to conclusively determine whether or not a threshold stress intensity exists for this fiber. We have conducted an insufficient number of tests at this point to perform the same kind of analysis for the new fiber. Fig. 5 shows the difference in behavior for the new fiber tested in nitrogen and air environments at 1400°C. While initially stronger when tested in air, the fibers weaken more rapidly as indicated by the shorter failure times at lower applied stresses.

High Temperature Tensile Creep Behavior

Creep tests were conducted on specimens from the new batch of fibers. The temperature dependence and stress dependence of the creep behavior of the α-SiC fiber are shown in fig. 6 and fig. 7 respectively. It should be noted that the creep curves in these figures show that the fiber exhibits only primary creep (i.e., a continuously decreasing creep rate throughout the duration of the test) for the test conditions investigated. The creep tests were terminated prior to failure of the fiber.

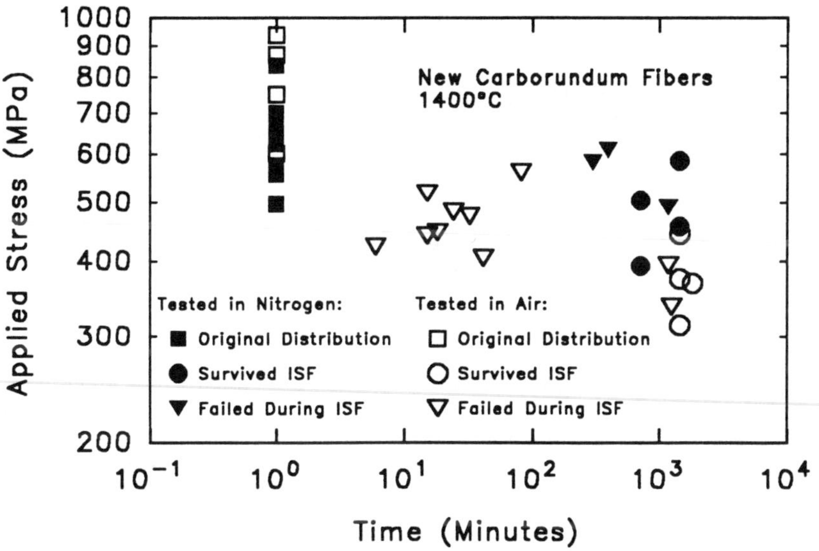

Fig. 5 Interrupted static fatigue behavior of new Carborundum α-SiC fibers at 1400°C in N_2 and air. The fibers tested in air are initially stronger, but weaken more quickly than those tested in N_2.

The model used by Lewinsohn et al. [12] to describe the primary creep behavior of CVD SiC fibers was employed in the present study. In this method it is assumed that the creep strain at constant stress is a single valued function of temperature-compensated time θ, such that

$$\varepsilon = f(\theta), \text{ and } \sigma = \text{constant.} \tag{4}$$

The temperature-compensated time is defined by the equation

$$\theta = \int_0^t \exp(-Q/RT)\, dt \tag{5}$$

Differentiating equations 4 and 5 with respect to time and combining them yields:

$$\dot{\varepsilon} \exp(-Q/RT) = F(\theta) = Z, \sigma = \text{constant} \tag{6}$$

The quantity Z is known as either the Zener-Hollomon parameter [13] or the temperature-compensated creep rate [14,15]. Therefore, if several creep curves for different temperatures and constant stress are obtained, the apparent activation energy, Q, may be determined as that value for which the dependencies of creep strain ε on temperature-compensated time θ become identical.

The stress dependence of the primary creep behavior may be described by

$$\dot{\varepsilon} = A\sigma^n \tag{7}$$

where n is known as the stress exponent. The time dependence of the primary creep may be expressed by

$$\varepsilon = At^p \tag{8}$$

where p is the time exponent [16]. Therefore, the creep strain at any given time, temperature, and stress, may be expressed by

$$\varepsilon = A\, \sigma^n\, t^p \exp(-pQ/RT) \tag{9}$$

where A is a constant and the remaining symbols have their usual meaning. The creep parameters describing the behavior of the α-SiC fiber for the conditions noted are as follows. The apparent activation energy, Q, is 405 ± 14 kJ/mole, and the stress exponent, n, is 2.31 ± 0.03. The value of p was determined to be 0.3. These creep parameters are consistent with a mechanism of creep by interface reaction controlled grain boundary diffusion [17]. The fiber creep parameters compare with compressive creep results of monolithic α-SiC with Q = 387-541 kJ/mole [18] and Q = 338-434 kJ/mole [19] under similar test conditions. An apparent activation energy of Q = 625 kJ/mole has been reported for tensile creep behavior of monolithic α-SiC [3]. Compressive creep stress exponents of lower

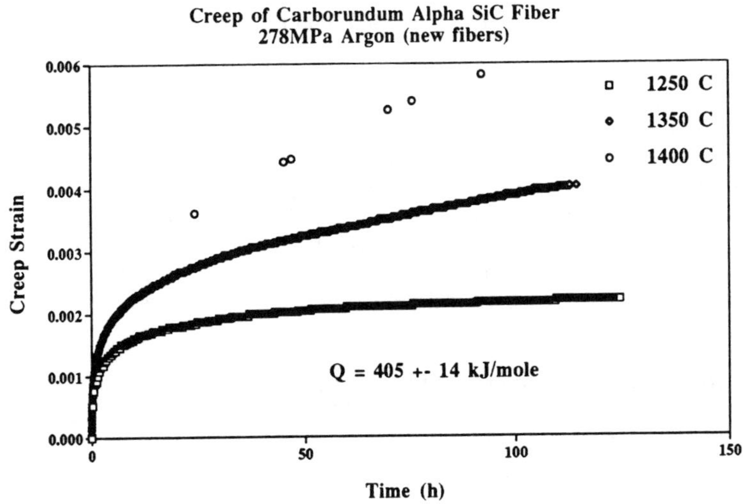

Fig. 6 Temperature dependence of the creep rate of Carborundum α-SiC fibers at 278 MPa in argon.

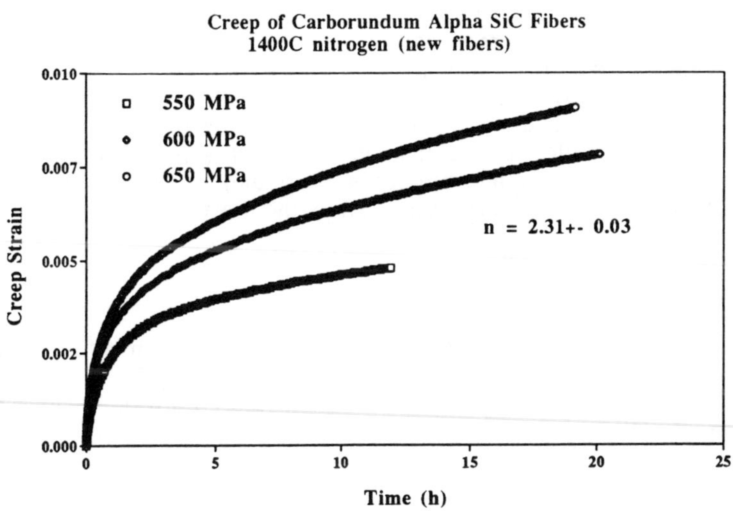

Fig. 7 Stress dependence of the creep rate of Carborundum α-SiC fibers at 1400°C in argon.

values (n = 1.0-1.4 [18] n = 1.44-1.71 [19]) than that of the present study have been reported.

Since primary creep is the only regime of creep observed, it may be presumed that microstructural transformations are occurring during the duration of the creep test. For example, stress assisted grain growth during the creep test would account for the decreasing creep rate over time. In addition, intragranular precipitation of B_4C has been observed in a crept fiber by TEM [20], however, it is not readily understood if this precipitation affects the creep behavior of the fiber. Microstructural analyses of crept fibers are currently underway to attempt to quantify the affect of microstructural development on the creep behavior of these fibers.

Fig. 8 compares the creep strain vs. time for the Carborundum α-SiC fiber and the most creep resistant CVD SCS-2mil fiber [12] tested under the same conditions (test results courtesy of C.L. Lewinsohn). The fibers were tested using the same creep testing apparatus (i.e., the AFT rig). This data would suggest that the Carborundum α-SiC fiber is the most creep resistant fiber developed to date. The α-SiC fibers were also creep tested using dead load tests on the same rig as performed with the CVD fibers. Measurable creep rates of the α-SiC fiber could not be acquired with the dead load apparatus. This is further evidence that the α-SiC fiber is more creep resistant than the CVD fibers. It should be noted that the 2mil fiber was heat-treated at 1000°C for 300h. Heat-treated 2mil fibers exhibit slower creep rates than as-received fibers [21]. Therefore, the as-received 2mil fiber would show equal or less creep resistance in the AFT rig than its heat-treated counterpart. Instantaneous creep rates of the α-SiC and 2mil fiber were calculated at a constant creep strain of 0.004, and at a constant test time of $2x10^5$ s. The creep rate of the 2mil CVD fiber is on the order of 2 times faster than the creep rate of the α-SiC fiber at constant strain, and a factor of ~ 5 times faster at the constant time value. Hence, regardless of the absolute creep rate value obtained by different methods (e.g. constant strain or minimum creep rate), it is evident that the α-SiC fiber is qualitatively more creep resistant than the most creep resistant CVD 2mil SiC fiber.

CONCLUSIONS

Subcritical crack growth was observed during static and dynamic fatigue testing of sintered α-SiC fibers. The crack growth exponent was 12-16 at 1400°C. From these initial results we cannot confirm the existence of a threshold stress intensity for crack growth as has been proposed for monolithic sintered α-SiC. Only primary creep was observed for the specified test conditions. An apparent activation energy of Q = 405±14 kJ/mole and a stress exponent of n = 2.31±0.03 was calculated between 1250-1400°C and 550-650 MPa. These creep parameters are consistent with an interface reaction control via grain boundary diffusion mechanism of creep. The α-SiC fiber is the most creep resistant SiC-based fiber to date.

ACKNOWLEDGMENTS

This work is supported by NASA Grant #NAGW-1381 and the CAM COOP Program. Helpful discussions with C.A. Lewinsohn are greatly appreciated.

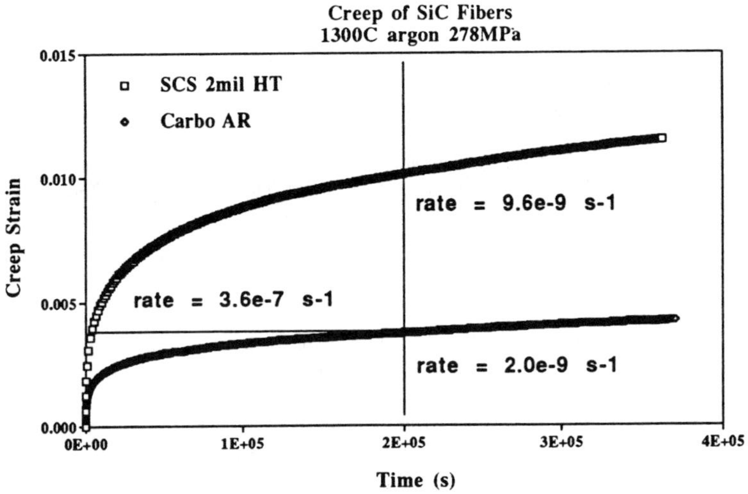

Fig. 8 Comparison of creep behavior between the Carborundum a-SiC fiber and the SCS-2mil fiber tested under the same conditions of 1400°C, 278MPa, in argon.

REFERENCES

1. V. Venkateswaran, G.V. Srinivasan, F.Frechette, and R.S. Storm, "Development of Continuously Sintered Polycrystalline SiC Fiber," to be presented at the 17th Annual Conf. on Composites, Materials and Structures (Restricted Session) Cocoa Beach, Fl, (1993).

2. D.J. Pysher, K.L. Rugg, and R.E. Tressler, High Temperature Properties of Carborundum α-SiC Fibers, HITEMP REVIEW 1992, *NASA Conf. Pub. 10104*, [1] 22-1-12 (1992).

3. J.C. Conway,Jr., J.J. Mecholsky, Jr., S.M. Weiderhorn, O.M. Jadaan, D.L. Shelleman, D.C. Cranmer, and R.F. Krause, Jr., "High Temperature Mechanical Properites and Reliability," in Center for Advanced Materials Annual Report-1990, CAM-9002, The Pennsylvania State University, 313-360 (1990).

4. E.J. Minford, D.M. Kupp, and R.E. Tressler, "Static Fatigue Limit for Sintered Silicon Carbide at Elevated Temperatures," *J. Am. Ceram. Soc.*, **66** [11] 769-773 (1983).

5. S.M. Wiederhorn, "Subcritical Crack Growth in Ceramics"; Fracture Mechanics of Ceramics, Vol. 2. Plenum, New York, 1974.

6. D.J. Pysher, L.A. Giannuzzi, and R.E. Tressler, "An Apparatus for Mechanically Testing Single Filaments at High Temperature in a Controlled Atmosphere or Vacuum," submitted to *Rev. Sci. Intruments.*

7. E. Minford, J.A. Costello, I.S.T. Tsong, and R.E. Tressler, "Oxidation Effects on Crack Growth and Blunting in SiC Ceramics"; Fracture Mechanics of Ceramics, Vol. 6. Plenum, New York, 1982.

8. R.K. Govila, "Flexure Stress Rupture Strength of Sintered a-SiC," in Time Dependent Failure Mechanisms and Assessment Methodologies, ed. J.G. Early, R. Shives, and J.H. Smith, Cambridge University Press, 1983.

9. E.J. Minford, "Flaw Behavior Near the Threshold Stress Intensity for Slow Crack Growth in Silicon Carbide Ceramics at High Temperatures," PhD Thesis, The Pennsylvania State University, University Park, 1983.

10. T.-J. Chuang, "A Diffusive Crack-Growth Model for Creep Fracture," *J. Am. Ceram. Soc.*, **65**(2) 93-103 (1982).

11. *Stress Intensity Handbook*, Vol. 2. Edited by the Committee on Fracture Mechanics, The Society of Materials Science, Japan. Pergamon Press, Oxford, U.K., 1987.

12. C.A. Lewinsohn, L.A. Giannuzzi, C.E. Bakis, and R.E. Tressler, "An Investigation of the Creep Behavior and Microstructural Changes During Creep of CVD Silicon Carbide Monofilaments," HITEMP REVIEW 1993, *NASA Conf. Pub. 19117*, [III] 86-1-11 (1993).

13. C. Zener and J.H. Hollomon, J. Appl. Phys., **15**, 22, (1944).

14. J.E. Dorn, in Creep and Fracture of Metals at High Temperatures; Proc. NPL symposium, HMSO, London, 89, (1956).

15. J.E. Dorn, Creep and Recovery, ASM, Cleveland, OH, 255, (1957).

16. E.N. da C. Andrade, Proc. R. Soc. London, **A84**, 1, (1911).

17. B. Burton, "Interface Reaction Controlled Diffusional Creep: A Consideration of Grain Boundary Dislocation Climb Sources," *Mater. Sci. Eng.*, **10** 9-14 (1972).

18. R.D. Nixon and R.F. Davis, "Diffusion-Accommodated Grain Boundary Sliding and Dislocation Glide in the Creep of Sintered Alpha Silicon Carbide," *J. A. Ceram. Soc.*, **75** [7] 1786-95 (1992).

19. J.E. Lane, C.H. Carter, Jr., and R.F. Davis, "Kinetics and Mechanisms of High-Temperature Creep in Silicon Carbide: III, Sintered α-Silicon Carbide," *J. A. Ceram. Soc.*, **71** [4] 281-95 (1988).

20. L.A. Giannuzzi, and R.E. Tressler, "The Creep Behavior of Polycrystalline Silicon Carbide Fibers," HITEMP REVIEW 1993, *NASA Conf. Pub. 19117*, [III] 87-1-13 (1993).

21. C.A. Lewinsohn, private communication, 1994.

THERMAL DEGRADATION MECHANISMS OF NICALON® AND TYRANNO® FIBERS

Constantin Vahlas, Centre National de la Recherche Scientifique, Laboratoire Marcel Mathieu, Centre Hélioparc, 2 avenue Angot, F-64000 Pau, France.
Temporarily with Materials Science Program, University of Delaware, Newark, DE 19716.

ABSTRACT

A calculational thermodynamic investigation of the thermal degradation of Nicalon® NL-202 and of Tyranno® Lox-M fibers in inert gas flow and at atmospheric pressure has been performed, based on the minimization of the Gibbs energy of the Si - C - O - H (-Ti) chemical systems. The simulation is based on a critically selected thermodynamic database of the participating compounds. The results are compared to recently reported TEM studies of as-received and of heat treated materials, which illustrate their sequential structures and nanotextures as a function of treatment temperature. The degradation mechanisms and the corresponding temperature ranges of the two fibers in the same conditions are compared and the observed differences are discussed.

INTRODUCTION

Models for the structure and the nanotexture, and mechanisms for the thermal degradation of silicon oxycarbide NL-202 Nicalon® and of silicon-titanium oxycarbide Lox-M Tyranno® fibers were recently proposed [1-4], based on previous works on these materials [5-7]. The as-received fibers were found to be microcomposites containing less than 2 nm in diameter silicon carbide (SiC) nano-grains and a free carbon network. This network consists of basic structural units (BSUs) i.e. the elemental entities common to many other carbonaceous materials which were shown to be less than 1 nm in diameter [8]. An SiC_xO_y intergranular solid solution was additionally evidenced in NL-202 and the possible existence of a corresponding $SiC_xO_yTi_z$ phase was discussed in [4] for Lox-M, in view of the similarity between the two fibers concerning their precursors, elemental composition, morphology and properties. Heat treatments reported in [1] for NL-202 and in [3] for Lox-M were realized following the same operating conditions, i.e. alumina environment, atmospheric pressure, 10 l/h Ar flow, 600°/h heating rate and 5 h isothermal exposure at the final temperature. It was thus possible, based on TEM characterization, to compare their sequential structures and nanotextures as a function of treatment temperature.

To the extent authorized under the laws of the United States of America, all copyright interests in this publication are the property of The American Ceramic Society. Any duplication, reproduction, or republication of this publication or any part thereof, without the express written consent of The American Ceramic Society or fee paid to the Copyright Clearance Center, is prohibited.

Two major overlapping steps were recognized for the thermal degradation of NL-202, following the treatment temperature. In the first step, which is still apparent at 1523K, the major phenomenon occurring is the modification of the structure of free carbon due to the release of hydrogen from the carbon network. In the second step, which is completed by 1773K, a drastic change in the fiber nanotexture is observed, due to the release of the contained oxygen in the form of carbon and silicon monoxides. This change is materialized by the destruction of the intergranular phase, by the strong degradation of the free carbon, and by the widening of the β-SiC crystals whose diameter now attains 8 nm. An intermediate third step was also evidenced at 1673K with the occurrence of a skin-and-core effect, indicating that for the adopted heating profile, the external part of the fiber has, in contrast to the internal part, already reached a structure and a composition corresponding to its degraded state.

The nanotextures of the as-received and of the 1623K treated Lox-M were found to be similar, although difficulties for the unambiguous characterization of the latter sample were reported [4]. The major evolution between the two temperatures is the improvement of the organization of the free carbon network. Lox-M was found to be degraded at 1673K, the diameter of the β-SiC crystals being increased at 60 nm.

In the light of these recent experimental contributions in understanding the two fibers a thermodynamic study of the conditions of their thermal degradation allows to define the relationship between the behavior of the material under different operating conditions and the corresponding degradation or reaction mechanisms. This thermodynamic study has been presented in [9] and in [4] for NL-202 and for the Lox-M, respectively. In this paper, the obtained results are compared and the observed differences are discussed. The principle and the different types of calculations as a function of the experimental situation to be simulated are first introduced, together with the critical selection of the thermodynamic data for the species considered. The results of calculations are presented by means of yield diagrams, indicating the number of moles obtained at equilibrium for the condensed and for the gaseous species as a function of temperature. The validity of the approach adopted is discussed and the results obtained are compared to available experimental information.

CALCULATIONS

The GEMINI2 software* was used to minimize the total Gibbs energy (G) of the chemical systems generated from combinations of the elements Si-C-H-O-Ar for NL-202 and of the elements Si-C-O-Ti for Lox-M. Twenty five condensed stoichiometric compounds, forty four gaseous species and three solution phases (liquid and bcc Ti-Si and fcc Ti-C) were considered in the calculations. It is worth noting that, since there are no thermodynamic data available for the $SiC_xOy(Ti_z)$ solid solutions, they were replaced in calculations by an equivalent mixture of $SiC+SiO_2+C$ $(+TiC_x)$. This assumption, the validity of which we have discussed in [9], does not allow the specific evolution of the fiber during the transition period of degradation to be simulated. However, it is expected to have little influence on the calculated degradation temperatures and on the fibers and the gas phase

* **G**ibbs **E**nergy **MINI**mization (GEMINI2), available from THERMODATA, BP66, F-38402 Saint Martin d'Hères, France

compositions beyond them. Indeed, as it is evidenced by the elemental analysis and the TEM investigation, there is no SiC_xOy solid solution left at the end of the degradation process of NL-202. The considered compounds are listed in Table I together with the source of their thermodynamic description.

Table 1. Species considered in the thermodynamic calculations and the origin of their thermodynamic description.

Nr	Species	Ref	Nr	Species	Ref	Nr	Species	Ref	Nr	Species	Ref
1	C-diamond	10	20	$TiSi_2$-solid	13	37	CH_4O-gas	16	56	O_2-gas	10
2	C-graphite	10	21	Ti_3Si-solid	13	38	CO-gas	16	57	O_3-gas	16
3	Si-diamond	10	22	Ti_5Si_3-solid	13	39	CO_2-gas	16	58	Si-gas	16
4	SiC-hcp, l	11	23	Ti_5Si_4-solid	13	40	C_2-gas	16	59	SiC-gas	11
5	SiC-cubic	11	24	Ti_3SiC_2-solid	14	41	C_2H-gas	12	60	SiH-gas	12
6	SiO_2-Lquartz	12	25	$Ti_5Si_3C_x$-sol.	14	42	C_2H_2-gas	12	61	SiO-gas	16
7	SiO_2-Hquartz	12				43	C_2H_4-gas	12	62	SiC_2-gas	11
8	SiO_2-tridym.	12	26	$Ti\text{-}C_x$-fcc**	15	44	C_2H_4O-gas	12	63	SiH_4-gas	12
9	SiO_2-cristob.	12	27	Ti-Si-bcc	14	45	$C_2H_4O_2$-gas	12	64	SiO_2-gas	11
10	SiO_2-liquid	12	28	Ti-Si-liquid	14	46	C_2H_6-gas	12	65	Si_2-gas	16
11	Si_2H_6-solid	12				47	C_2H_6O-gas	12	66	Si_2C-gas	11
12	Ti-hcp	11	29	C-gas	16	48	C_3-gas	16	67	Si_2H_6-gas	12
13	TiO-solid	12	30	CH-gas	16	49	H-gas	12	68	Si_3-gas	16
14	TiO_2-(C)R	12	31	CHO-gas	16	50	HO-gas	12	69	Ti-gas	12
15	TiO_2-(S)A	12	32	CH_2-gas	12	51	HO_2-gas	12	70	TiO-gas	12
16	Ti_2O_3-solid	12	33	CH_2O-gas	12	52	H_2-gas	10	71	TiO_2-gas	12
17	Ti_3O_5-solid	12	34	CH_2O_2-gas	12	53	H_2O-gas	12	72	Ar-gas	10
18	Ti_4O_7-solid	12	35	CH_3-gas	12	54	H_2O_2-gas	12			
19	TiSi-solid	13	36	CH_4-gas	12	55	O-gas	16			

Various quantities of inert gas in contact with the fibers were considered in order to differentiate the chemical environments at their skin and in their core. This approach helps to materialize, in the calculations, the available volume around the fibers and within their internal nanopores [1,3]. It is based on the modification of the global equilibrium composition when the inert gas dilution is modified in order for the equilibrium constant to remain unchanged [9]. Hence, calculations were made with 1 g-atom (skin) and 10^{-4} g-atom Ar (core) in contact with 1 g-atom of the fibers whose initial elemental compositions were experimentally determined [1,3] as: Si 33.2 (37.9), C 46.4 (43.6), O 14.4 (14.8), Ti 0.7 (0.0) and H <5.28 (3.7) at% for Lox-M (NL-202), respectively.

In figure 1 are presented the calculated results for Lox-M at the skin environment as a function of treatment temperature. The evolution of the condensed products is illustrated in figure 1a. In these calculations the fiber is represented by a mixture of SiC, C and SiO_2

** In order to avoid demixing artifacts, the excess free energy of TiC_x was modified as: $G^{xs} = -6000Y_CY_{vac}$ J/mol. Y_i are the atomic fraction of carbon and vacancies in their sublattice ($Y_C+Y_{vac}=1$).

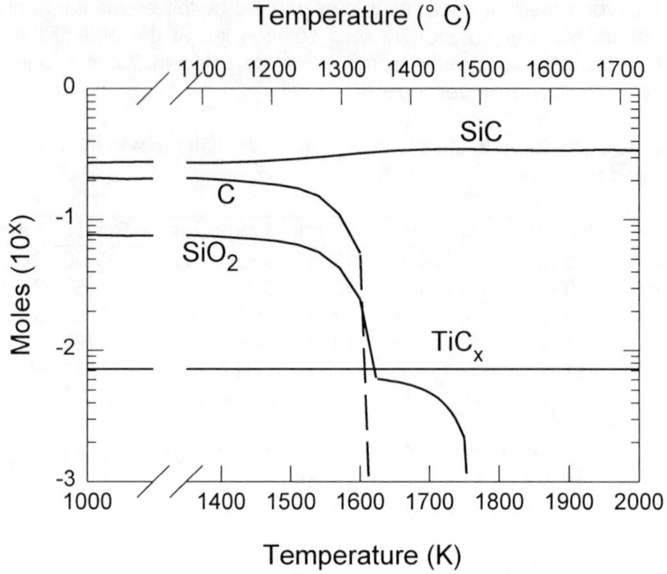

1a

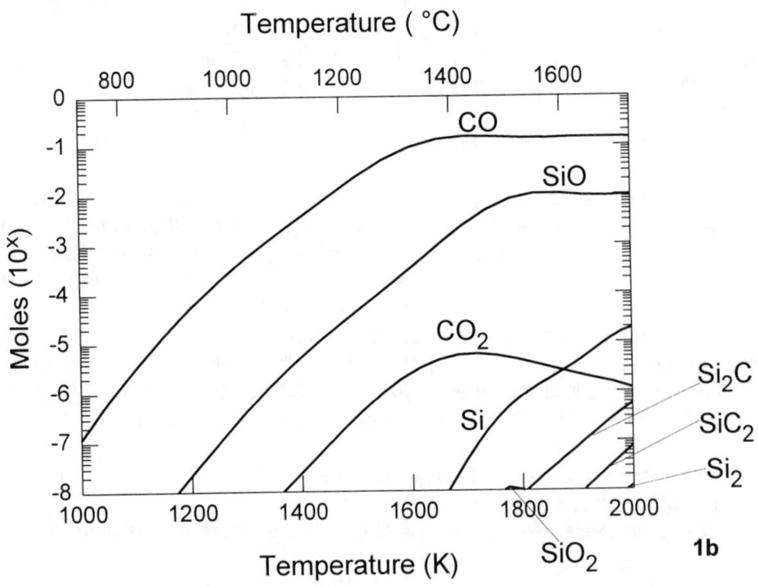

1b

Figure 1. Thermal evolution of one g-atom of Lox-M. Condensed (2a) and gaseous (2b) species in contact with one g-atom Ar as a function of treatment temperature.

which remains stable up to 1500K. Above this temperature there is a slight decrease of the oxygen and of the free carbon content of the material. At 1600K the free carbon vanishes, SiC increases significantly and the oxygen content reaches a plateau an order of magnitude below its initial concentration. At 1750K the oxygen containing condensed species also vanish. The above behavior is to be related to the evolution in the same operating conditions of the gaseous by-products of the heat treatment, presented in figure 1b. From this figure it appears that silicon and especially carbon monoxide, SiO and CO, are the main gaseous species. Their quantities continuously increase with temperature before reaching a plateau at 1600K for CO and at 1750K for SiO, corresponding to the exhaustion of the free carbon and of the oxygen reserves of the fiber, respectively. At and beyond 1750K the quantity of CO is ten times higher than that of SiO. Other species, especially CO_2 and Si are present at high temperatures but in very low quantities. They also qualitatively follow the evolution of the free carbon and of the oxygen within the fiber. Beyond 1800K their increase is delayed or stopped due to the appearance of small quantities of gaseous Si_xC_y species.

Figure 2 presents the evolution with temperature of Lox-M in contact with 10^{-4} g-atom Ar (core), for the condensed (2a), and for the gaseous (2b) species. The general shapes of the two diagrams are equivalent to diagrams of figures 1a and 1b, respectively with two main differences: The curves are now flatter with more sharp changes, and the degradation temperature is shifted 200 degrees higher, i.e. at 1800K. This is also illustrated in figure 2b where the drastic increase of CO, SiO, CO_2 and Si is stopped also later, i.e. at 1800K. The transition temperatures of 1600 and 1800K of figures 1a and 2a respectively are to be compared to the thermal degradation temperature of the fiber, the upper limit of which was experimentally determined to be 1673K [3].

From figures 1a and 2a it appears that titanium is found in Lox-M exclusively in its carbide form. A zoom on the y axis of the diagram of figure 2a is presented in figure 3, aiming to evidence any minor evolution of TiC_x with temperature. The composition of the TiC_x phase in terms of carbon content for one g-atom of titanium is also presented. The quantity of TiC_x remains unchanged indeed, up to 1650K. Beyond this temperature TiC_x definitively decreases, but not in significant amounts: At 2000K the number of moles of TiC_x is $7.1997 \cdot 10^{-3}$ to be compared to $7.2 \cdot 10^{-3}$ in the as-received material. The carbon content of the TiC_x phase follows an equivalent evolution: while up to approximately 1600K titanium carbide remains stoichiometric, beyond this temperature it loses continuously carbon but again in very weak amounts: At 2000K TiC_x is 10^{-4} g-atom understoichiometric in carbon. It thus has to be pointed out that these variations are extremely weak and, even if they evidence the upper stability limit of the material in the selected operating conditions, they are not representative of a significant change in the composition and, consequently in the physical properties of the fiber. The formation of stoichiometric titanium carbide is in agreement with the Si-Ti-C phase diagram, whose isothermal sections at 1273K and 1473K were recently calculated [14]. Indeed, the global composition point of Lox-M is located at the interior of the SiC-TiC-C part of this diagram. Since this part of the isothermal section remains stable within a large temperature domain, it can be reasonably assumed that titanium will remain present in the fiber in the form of stoichiometric TiC. No gaseous Ti-containing compounds are evidenced in figure 2b. As a matter of fact TiO and Ti appear above 1900K at extremely low quantities

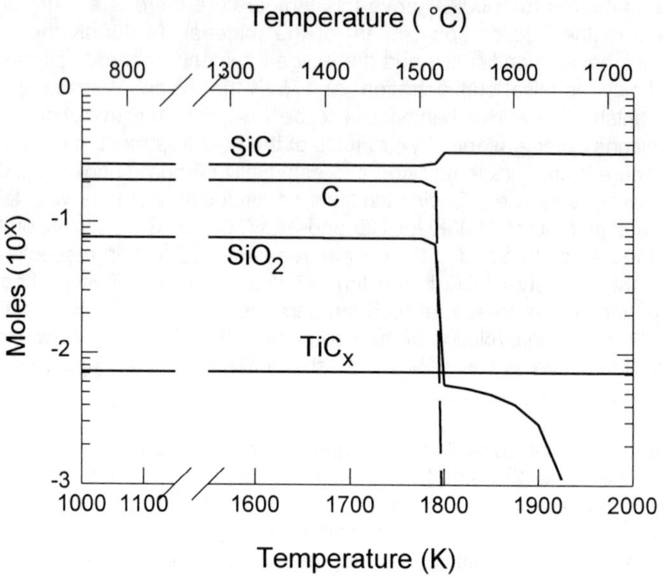

2a

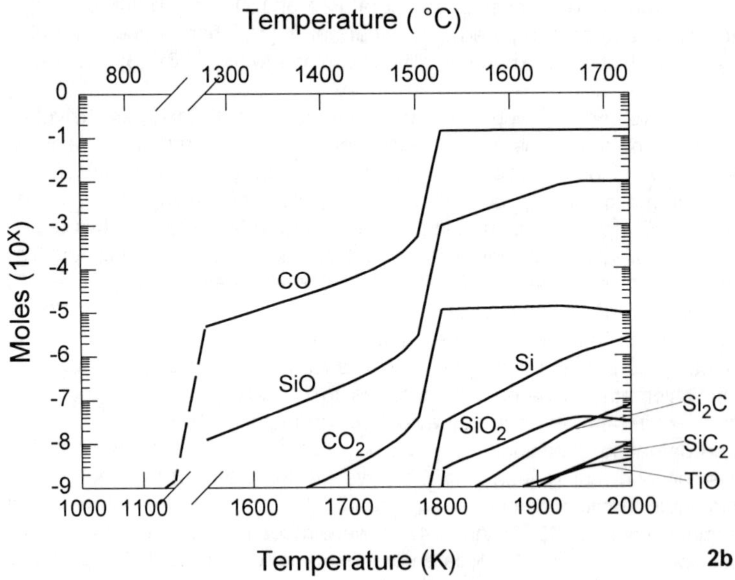

2b

Figure 2. Thermal evolution of one g-atom of Lox-M. Condensed (2a) and gaseous (2b) species in contact with 10^{-4} g-atom Ar as a function of treatment temperature.

(<2*10^{-9} g-atom) illustrating the stability of carbon and silicon oxides towards titanium compounds. Titanium carbide behaves identically in the operating conditions of figure 1.

The simulation of the first step of the NL-202, characterized by hydrogen release, evidenced a complete departure of hydrogen in the gas phase at any temperature. This result, associated to the continuous shape of the curves of the gaseous by-products indicates that the release of hydrogen from the free carbon network as it was reported in [1], was not reproduced in the calculations. This is in agreement with TEM observations [1], indicating that the release of hydrogen is more the result of the evolution of the organization of carbon, with an edge to edge association of the individual and random BSUs, performed on the depends of the C-H bonds, rather than the consequence of an abrupt modification of the chemical equilibrium in the corresponding temperature.

The simulation of the conditions of the main degradation step of NL-202 yield the same results as the above presented: Figures 1 and 2 perfectly illustrate the thermochemically predicted behavior of NL-202 concerning the nature of the condensed and of the gaseous products, the shape of the curves and the transition temperatures of 1600 and 1800K. These temperatures are now to be compared to the corresponding thermal degradation temperature, the upper limit of which was experimentally determined to be 1773K [1]. Only slight differences in the quantities of the resulting compounds were observed between the diagrams of figures 1 and 2 and those calculated for NL-202. They are due to differences in composition of the as received materials and to the absence of titanium in Nicalon® and do not influence the validity of figures 1 and 2 for the case of NL-202. Figure 4 presents the oxygen content of the two fibers as a function of the treatment temperature. Dots are experimental results from [1] concerning NL-202 and are compared to the calculational results in the core of the material. Although there is a systematic misfit of about 1 %, the agreement between the two sets of results is strong.

DISCUSSION

The evolution in composition of the condensed species as a function of temperature, which is evidenced in figures 1a and 2a, illustrates the capacity of the fibers to support without any major alterations the heat treatments in the selected experimental conditions. This is valid up to a temperature which is function of the argon dilution and which is materialized in the above diagrams by an abrupt change in composition, especially by the elimination of the free carbon, due to an endogenous oxidation mechanism leading to SiC and CO. Following the results of the thermodynamic investigation, Lox-M and NL-202 are expected to change drastically in composition, morphology and hence physical properties in the range between 1600 and 1800K (1327 and 1527°C).

The validity of these transition temperatures is studied in figure 5, where the degradation temperature is drawn as a function of the number of moles of Ar available in the calculations. From this diagram it appears that the degradation temperature does not depend on the dilution of the reactants up to approximately 10^{-2} g-atom Ar/g-atom. For higher dilution it decreases linearly with the logarithm of Ar quantity. From this figure it appears that only calculations made with 10^{-4} g-atom Ar/g-atom have a more general value and can represent the situation in the core of the fiber, while calculations made with

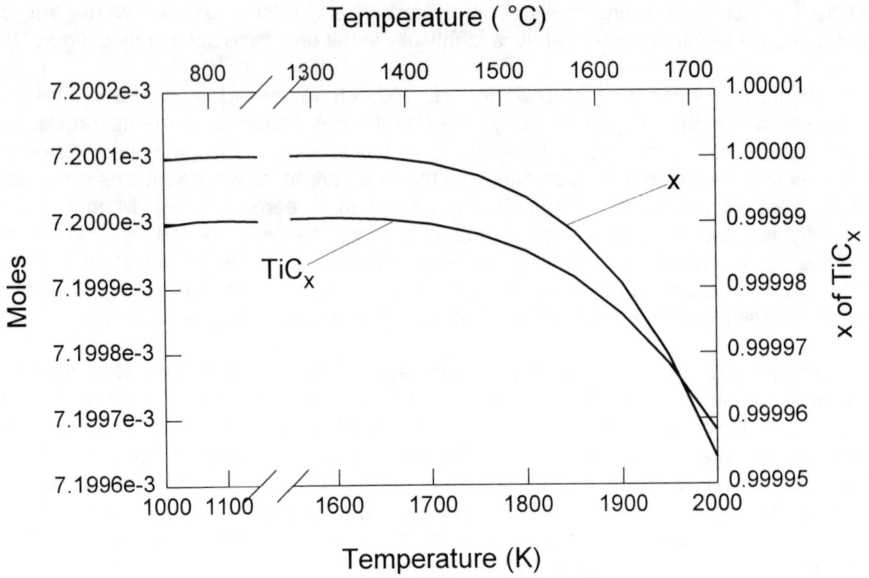

Figure 3. Thermal evolution of one g-atom of the fiber. Number of moles of TiC_x (left) and carbon content of TiC_x (right) in contact with 10^{-4} g-atom Ar as a function of treatment temperature.

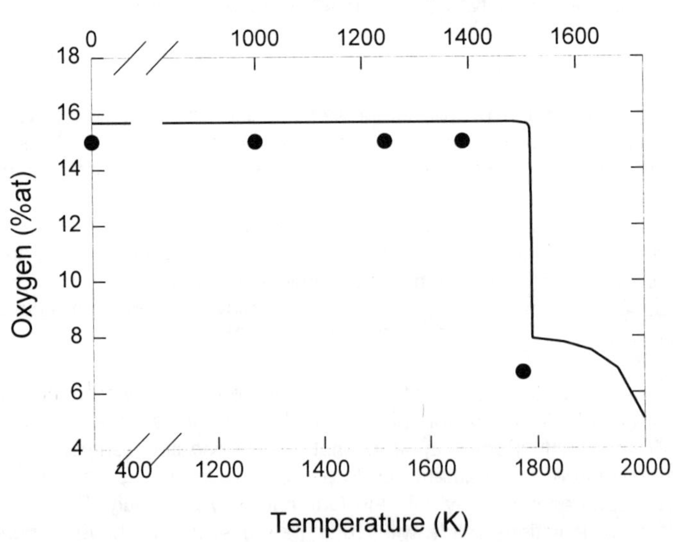

Figure 4. Oxygen content of the fibers as a function of treatment temperature. Experimental results on NL-202 [1], compared to calculation results.

1 g-atom Ar/g-atom indicate qualitatively that operating conditions are more severe at the skin part, leading to lower values for the degradation temperature. Calculations made with 10^{-4} g-atom Ar/g-atom represent in a satisfactory way the global behavior of the fibers if, in the adopted experimental conditions, their major part is found in a "core" environment. In that case they are expected to remain stable up to 1800K (1527°C).

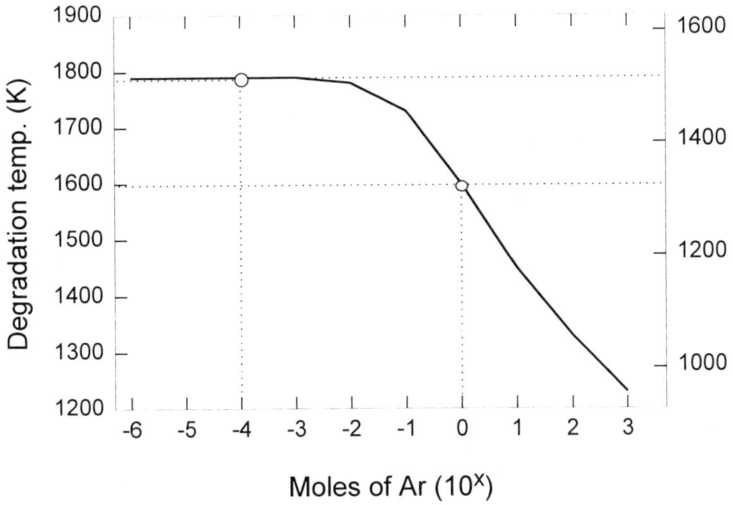

Figure 5. Thermal evolution of the fibers. Degradation temperature as a function of the number of moles of Ar available. Dotted lines correspond to the investigated conditions.

The predominant reaction in the temperature range where condensed C, SiO_2 and SiC coexist, is:

$$3\ C_{(s)} + SiO_{2(s)} \leftrightarrow SiC_{(s)} + 2\ CO_{(g)} \qquad (1)$$

while in the temperature range between the free carbon and the SiO_2 consumption the reaction which fits the best the calculated results is:

$$SiC_{(s)} + 2\ SiO_{2(s)} \leftrightarrow 3\ SiO_{(g)} + CO_{(g)} \qquad (2)$$

Equilibria (1) and (2) fit well the relative evolution of the quantities of the condensed and of the gaseous phases. For example, in the same conditions as in figure 2, at the temperature of total free carbon consumption (1785K), reaction (1) leads to 0.0327 g-atom SiO_2 and 0.3593 g-atom SiC for NL-202. These quantities as well as the resulting weight loss of the fiber are identical within 1‰ to the values calculated from the minimization of the free energy of the total chemical system.

The presented calculational results confirm the global endogenous oxidation mechanism which was proposed in [1]:

$$0.7\ SiO_{1.15}C_{0.85(s)} + 0.42\ C_{(s)} \rightarrow 0.24\ SiO_{(g)} + 0.56\ CO_{(g)} + 0.46\ SiC_{(s)} \qquad (3)$$

where $SiO_{1.15}C_{0.85(s)}$ corresponds to the proposed composition for the oxygen-containing intergranular solid solution. Indeed, the ratio of CO over SiO in equilibrium (3) is 2.3 to be compared to 2.2 in figure 2b at high temperatures, i.e. when all the oxygen available is completely consumed. At lower temperatures, i.e. when there is still oxygen left in the condensed phases, this ratio is calculated to be two orders of magnitude higher. This result is in disagreement with mass spectrometric measurements of CO and SiO above NL-202 [17], showing at the same range of temperatures a CO/SiO rate slightly higher than one. The misfit can be due to the control of the degradation by kinetic factors as has already been developed in [17], but additionally to the assumption made in calculations, of completely crystallized condensed phases. This point was discussed in [18], where the amorphous state was considered as a supercooled liquid and the activities of the constituents were calculated from estimated temperatures and entropies of melting and from the activities in the solid state. The as obtained gas phase composition has a lower CO content relatively to SiO, compared to calculations where the activities of the condensed species were considered equal to unity.

No skin-and-core effect was reported in [3] for the degradation of Lox-M. However, in view of the similarity between the two fibers concerning their precursors, their morphology and their properties, it is presumable that they present a similar, diffusion-limited degradation mechanism. In that case a skin-and-core effect may indeed occur for Lox-M between 1623 and 1673K. The thermodynamic simulation evidenced no difference between the two chemical systems. Since figures 1, 2 and 5 are similar for the two materials, it can be deduced that degradation of the two fibres is initialized in the same temperature range. In that case, the relatively narrow temperature gap where the presumed skin-and-core effect takes place for Lox-M comparatively to NL-202, indicates that the progression of the skin to the prejudice of the core is faster in the Tyranno®. A more rapid diffusion of the produced gaseous by-products may be responsible for this behavior due to a higher or more open nanoporosity as is suggested by the lower density of Lox-M relativelly to NL-202: The two densities are 2.37 and 2.55 g cm^{-3}, respectively, and only 6% of this difference is due to the difference in composition of the fibers.

The difference between experimental results and thermodynamic predictions in similar operating conditions may be also partially related to the quality of the thermodynamic data adopted to describe the compounds participating in the chemical process. Although important works have already been performed on the chemical behavior of the Si-C-O [19] and of the Si-Ti-C [20] systems, more is left to be done to improve their thermodynamic description. Additionally, solid state diffusion, or surface/interface kinetics may contribute to this misfit, the corresponding mechanisms being more complex to quantify in the presence of poorly organized constituents. An overall consideration of the fabrication process, including the spinning technology and the chemical and/or thermal treatments for the fibers stabilization can provide reasons for the difference in the thermal stability between Lox-M and the NL-202: Degradation of NL-202 is completed above

1673K (at 1773K it still contains 5 wt% oxygen indicating that a part of the Si-C-O solution phase still exists [1]) while in the same conditions at 1673K Lox-M has already been degraded. However, both of them present comparable elemental compositions, exhibit an equivalent weight loss comparatively to the as-received material when degraded and contain silicon carbide grains the size of which increases drastically at the conditions of fiber degradation as well as a poorly organized carbon network which participates in the further formation of silicon carbide and carbon monoxide with increasing temperature.

CONCLUSIONS

The thermal degradation of the NL-202 Nicalon® and of the Lox-M Tyranno® fibers in inert gas atmosphere and atmospheric pressure was theoretically investigated by performing thermochemical calculations in the Si-C-O-H-Ar-(Ti) chemical systems and was compared to recently published experimental results. The environments in the core and at the skin of the fibers were differentiated by adopting different inert gas dilution of the reactants. The stability limit of the materials was thus calculated to be at 1800K (1527 °C) or lower. At this temperature the available carbon is exhausted and the oxygen content is highly reduced. The main reaction products are silicon carbide and carbon and silicon monoxides. Titanium in Lox-M forms stoichiometric titanium carbide which exhibits a pronounced stability in terms of quantity and stoichiometry. An endogenous oxidation mechanism was proposed to be responsible of the degradation of the fibers.

From the comparison of the obtained results on the Lox-M Tyranno® and of the NL-202 Nicalon® fibers it appears that the chemical behavior of the Si-C-O-Ti and of the Si-C-O systems are identical in the investigated conditions and that similar mechanisms govern their thermal degradation. However, the degradation process is completed more rapidly in the case of the Lox-M, possibly because of a more rapid diffusion of the gaseous reaction by-products through the core of the fiber.

ACKNOWLEDGMENTS

Thanks to Marc Monthioux (CNRS, Pau), for valuable discussions, to Philippe Rocabois (LTPCM, Grenoble), for help with the critical selection of the thermodynamic data and to Evelyne Fischer and to Pierre-Yves Chevalier (THERMODATA, Grenoble), for help with the minimization software.

REFERENCES

1. P.LeCoustumer, M.Monthioux and A.Oberlin, Understanding Nicalon® fiber. *J. Europ. Ceramic Soc.*, 11, 95-103, 1993.
2. P.Schreck, C.Vix-Guterl, P.Ehrburger and J.Lahaye, Reactivity and molecular structure of silicon carbide fibres derived from polycarbosilanes. Part II XPS analusis. *J. Mater. Sci.*, 27(15), 4243-4246, 1992.
3. C.Bourgerette, P.LeCoustumer, M.Monthioux and C.Vahlas., TEM characterization of heat treated Tyranno fibers, pp. 67-74, *Proceedings of the International Conference ECCM 6th. High Temperature - Ceramic Matrix Composits*, R.Naslain, J.Lamon, D.Doumeingts, eds, 1993.

4. C.Vahlas, M.Monthioux, On the Thermal Degradation of the Lox-M Tyranno® Fibre, *J. Europ. Ceramic Soc.*, submitted.
5. L.Porte and A.Sartre, Evidence for a silicon oxycarbide phase in the Nicalon fibre, *J. Mater. Sci.*, 24, 271-276 1989.
6. C.Laffon, A.M.Flank, P.Lagarde, M.Laridjani, R.Hagege, P.Olry, J.Cotteret, J.Dixmier, J.L.Miquel, H.Hommel and A.P.Legrand, Study of Nicalon-based ceramic fibres and powders by EXAFS spectrometry, X-ray diffractmetry and some additional methods, *J. Mater. Sci.*, 24, 1503-1512, 1989.
7. V.S.R., Murthy, M.H.Lewis, M.E.Smith and R.Dupree, Structure and degradation of Tyranno fibers. *Materials Letters*, 8, 263-268, 1989.
8. A.Oberlin, High resolution TEM studies of carbonization and graphitization, pp.1-143, *Chemistry and Physics of Carbon*, 22, P.A.Thrower, ed., Marcel Dekker, NY and Bassel, 1989.
9. C.Vahlas, P.Rocabois, C.Bernard, Thermal Degradation Mechanisms of Nicalon® Fibre: A Thermodynamic Simulation, *J. Mater. Sci.*, in press.
10. A.T.Dinsdale, SGTE Data for Pure Elements, NPL report DMA(A) 195, Sept. 1989.
11. L.V.Gurvich, I.V.Veyts and C.B.Alcock, Thermodynamic Properties of Individual Substances, 4th edition vol 1 and 2, Hemisphere publishing corp., New York, 1990.
12. Scientific Group Thermodata Europe, available on line from THERMODATA, BP 66 F-38402 Saint Martin d'Hères, France and Royal Institute of Technology, S-10044, Stockholm, Sweden.
13. C.Vahlas, P.Y.Chevalier, and E.Blanquet, Thermodynamic approach to the selective deposition of refractory metal silicides, *Calphad*, 13, 275-294, 1989.
14. C.Racault, F.Langlais, and C.Bernard, On the CVD of Ti_3SiC_2 from $TiCl_4$-$SiCl_4$-CH_4-H_2 gas mixtures. 1 - A thermodynamic approach. *J.Chemical Vapor Deposition,* submitted.
15. C.Vincent, J.Dazord, J.Bouix and C.Bernard, Calcul de l'energie libre de formation de Gibbs du carbure de titane TiC_y par la methode des sous reseaux. Application a la determination de diagrammes de depot chimique en phase vapeur. *Thermochimica Acta*, 147, 1-15, 1989.
16. I.Barin, Thermochemical Data of Pure Substances, VCH, Weinheim, Germany, 1989.
17. P.Rocabois, C.Chatillon and C.Bernard, Mass spectrometry experimental investigation and thermodynamic calculations of the Si-C-O system and $Si_xC_yO_z$ fibre stability. pp.93-100, *Proceedings of the High Temperature Ceramic Matrix Composites - ECCM6 Conference*, Bordeaux, France, R.Naslain, J.Lamon and D.Doumeingts, eds, Woodhead Publishing Limited, 1993.
18. K.L.Luthra, Thermochemical analysis of the stability of continuous "SiC" fibers *J. Amer. Ceram. Soc.*, 69(10), C231-C233, 1986.
19. P.Rocabois, C.Chatillon and C.Bernard, Multiple Knudsen cell mass spectrometric investigation of the evaporation of silicon oxicarbide glass. *Surface and Coatings Technology*, 61, 86-92, 1993.
20. C.Racault, Elaboration par voie solide et gazeuse et caractérisation physico-chimique de céramiques à base de Ti_3SiC_2, *Thesis*, Université de Bordeaux I, 1993.

THE TEMPERATURE DEPENDENT FRACTURE STRENGTH OF DOPED SINGLE CRYSTAL ALUMINA FIBERS

H. Sayir, S. C. Farmer, P. Lagerlof, A. Sayir
NASA Lewis Research Center, Cleveland, OH 44135

ABSTRACT

The tensile strength of impurity doped, laser heated float zone (LHFZ) grown, single crystal sapphire (α-Al_2O_3) fibers was measured at elevated temperatures, 1100 and 1450 °C. The effects of MgO doping, TiO_2 doping and TiO_2/MgO co-doping were examined. MgO doping significantly enhanced the high temperature strength. At 1450 °C, tensile strengths of MgO doped sapphire (~1760 MPa) are almost twice the strength of undoped fibers. In contrast, the effect of TiO_2 doping and TiO_2/MgO-codoping on the tensile strength of sapphire fibers was not statistically significant at either 1100 or 1450 °C. The relationship between strength distributions and temperature is examined and possible strengthening mechanisms are discussed .

INTRODUCTION

High temperature aerospace structural applications require composite materials that are lightweight, stiff, and strong. Alumina (α-Al_2O_3) can be considered a candidate material in fiber form for composites.[1-7] The measured tensile strength

To the extent authorized under the laws of the United States of America, all copyright interests in this publication are the property of The American Ceramic Society. Any duplication, reproduction, or republication of this publication or any part thereof, without the express written consent of The American Ceramic Society or fee paid to the Copyright Clearance Center, is prohibited.

of single crystal alumina however is much less than the estimated theoretical value and this limits its application in composite bodies. The tensile strength of commercially available single crystal c-axis sapphire fibers (Saphikon™)* is less than 4 GPa at room temperature. The low strength is believed to be due to processing and handling flaws. Furthermore, the fiber strength decreases with increasing temperature reaching a minimum between 300 - 400 °C, then rises to a maximum at around 900 °C. Loss of strength in this temperature range (300 - 900°C) has been described as a stress corrosion phenomenon.[9] With further increase in temperature, fiber strength rapidly decreases. At temperatures above 1000 °C, strength values fall below 1 GPa.[8-12,17]

In previous work on the temperature dependent bend strength of undoped fibers,[11] it was demonstrated that the fracture strength strength minimum at intermediate temperatures (300 - 900 °C) could be eliminated by Mg^{2+} or Ti^{4+} doping. This raised the issue that stress corrosion is perhaps not the only strength controlling mechanism in the fracture behavior of these fibers. The effects of dopants could not be clearly explained. However, it was argued that mechanisms other than stress corrosion may contribute to the fracture behavior of sapphire fibers and that the unusual temperature dependence of fracture behavior is due to overlapping temperature regimes for more than one mechanism.

The high temperature behavior of sapphire (T > 1000 °C) has been studied by several researchers. Fracture appears to be controlled by dislocation induced flaw nucleation. The high temperature tensile strength of sapphire fibers can be improved by alloying alumina with Mg, Ti, Cr, and Fe.[3,4,14,15] These alloying elements increase the critical resolved shear stress of alumina.[15,16] However, there is no systematic study available to represent a study of the effect of different impurities on the fracture behavior of single crystal Al_2O_3 in tension, under atmosheric condition. The work presented in this paper focuses on the effects of doping (TiO_2 doping, MgO doping and TiO_2/MgO co-doping) on the high temperature tensile strength of sapphire fibers grown by the laser heated float zone technique.

---

* Saphikon,™ Inc., Milford, NH.

EXPERIMENTAL

The undoped and impurity doped single crystal alumina fibers were fabricated in two steps: 1) powder mixing and preparation of feed rods and 2) growth of single crystal c-axis fibers by laser heated float zone technique (LHFZ).[6,7] Feed rods were prepared from pure alumina powder and three doped powders (500 ppm MgO, 500 ppm TiO_2, and 500ppm TiO_2/MgO codoped). The chemical analysis, supplied by the manufacturer (Ceralox) showed that undesired impurity concentrations were less than 20 ppm. The powders were first mixed to form a slurry. Subsequently the water was extracted and the mixture extruded into cylindrical rods (green fibers). The green fibers were then used as feed rods to grow sapphire fibers by the laser heated floating zone technique. Fibers were seeded by c-axis single crystals. All sapphire fibers tested in this study had the c-axis within 3° of the fiber axis. Orientation was determined by transmission optical microscopy using polarized light.**

For the tensile test, cold grips were used with a total fiber gauge length of 23 cm.[17] The fibers were tested in air by placing them in a $MoSi_2$ furnace*** with a hot zone of 2.5 cm. All fibers fractured within this 2.5 cm length. A scanning electron microscope (JEOL 840-A) was used to identify the fracture origins of the fibers and a transmission electron microscope (Phillips 400T) was used to investigate dopant precipitation.

RESULTS AND DISCUSSION

The effects of TiO_2 doping, MgO doping and TiO_2/MgO codoping on the tensile strength of alumina measured at 1100 and 1450 °C are shown in Fig.1 with their respective room temperature tensile strengths. Commercially available c-axis sapphire fibers from Saphikon also tested under the same conditions serve as a reference for comparison. The average tensile strength of as-received

---

** Model Axiphot, Carl Zeiss, Inc., Germany.

*** CM, Inc., Bloomfield, NJ.

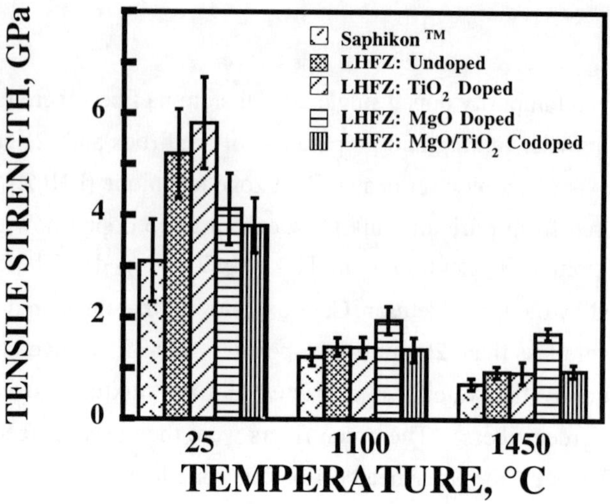

Fig. 1 The effect of the impurity doping on the tensile strength of c-axis sapphire fibers. Each histogram represents 20 data points and the error bars are ± one standard deviation.

Saphikon fibers at room temperature was ~3 GPa. The tensile strength of undoped LHFZ-grown fibers at room temperature was ~5 GPa. The dopants either did not improve the average room temperature tensile strength of the fibers (MgO doping and TiO_2/MgO codoping) or the scatter in the data masked any effect (TiO_2 doping). We believe that the room temperature tensile strength of LHFZ fibers is governed by processing related flaws rather than intrinsic material properties.

The average room temperature tensile strength of LHFZ grown fibers was ~ 50 % higher than Saphikon fibers. In contrast, the tensile strengths of LHFZ-grown fibers measured at 1100 and 1450 °C were only slightly higher than Saphikon fiber. The sharp decrease in the tensile strength of undoped sapphire with increasing temperature observed in this work agrees with results reported by

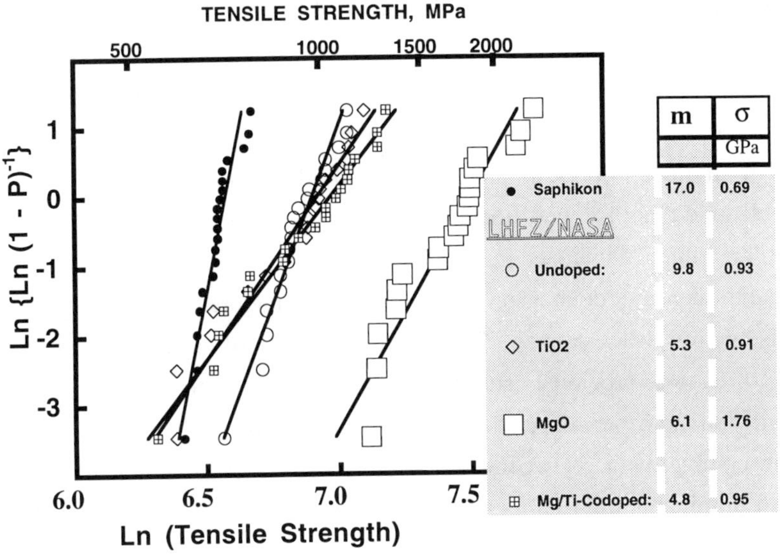

Fig. 2 Weibull distribution of the tensile strength of undoped and impurity doped sapphire fibers at 1450 °C. [m = Weibull modulus; P = estimated cumulative probability]

Shahinian,[10] and Newcomb and Tressler.[18] The effect of TiO_2 doping and TiO_2/MgO-codoping on the tensile strength of fibers was not statistically significant at either 1100 or 1450 °C (Fig. 1). However, dramatic improvements in the strength of MgO doped fibers occurred at both temperatures. The average tensile strength of MgO doped fibers was ~1760 MPa at 1450 °C (standard deviation of 246 MPa). This represents nearly a twofold increase in the tensile strength of MgO doped sapphire fibers compared to that of undoped LHFZ fiber at the 1450 °C temperature. MgO doping clearly improves the strength retention of sapphire fibers at 1100 °C and 1450 °C.

The change in average strength with temperature is accompanied by a change in strength distribution, Fig. 2. The Weibull modulus for the room temperature fracture strength of undoped fibers is 4 to 5. At elevated temperatures the strength distribution narrows. At 1450 °C, the fracture strength Weibull modulus ~10. The increase in Weibull modulus with temperature suggests that the material may be less sensitive to the initial flaw population that controls low temperature strength. In doped fibers, an increase in Weibull modulus is **not** observed. The Weibull modulus of doped fibers was 4 to 6, similar to the room temperature Weibull modulus of undoped material.

The influence of dopants on high temperature fracture behavior may relate to effects on slow crack growth. The fracture strength of sapphire fibers at 1200 and 1400 °C was previously studied by Sayir et. al.[17] The decrease in fracture strength and the dependence of stress rupture values on strain rate was attributed to slow crack growth. Newcomb and Tressler[18] also determined that slow crack growth dominates high temperature fracture behavior and suggested that the controlling parameter of fracture strength during slow crack growth above 1000 °C is the threshold stress intensity factor, K_{TH}.

The decrease in strength and narrowing of the strength distribution with temperature (Figure 2) is then explained as follows. Once a particular flaw in the undoped fiber was subjected to stress at T = 1450 °C, and started to propagate above K_{TH}, the measured strength values correspond to flaw sizes after slow crack propagation rather than to the original flaw population. The flaw sizes were thus larger and populations corresponding to K_{TH} were narrower resulting in a higher Weibull modulus for the undoped fibers. The Weibull moduli of the impurity doped fibers (Fig. 2) were around 4 to 6 at 1450 °C. Weibull moduli of this magnitude are observed for sapphire fibers at low temperatures where initial flaw population determined the tensile strength of the material. This implies that all the dopants may decrease susceptibility of sapphire fibers to slow crack growth presumably through an increase in K_{TH}.

The enhanced strengthening of c-axis sapphire with MgO doping in comparison to TiO_2 doping and TiO_2/MgO codoping as shown in Figs. 1 and 2 has yet to be addressed. MgO doping has a profound influence on high temperature strength. In contrast, TiO_2 and TiO_2/MgO codoping had limited effect. Examination of fracture revealed that the TiO_2 doped and codoped fibers

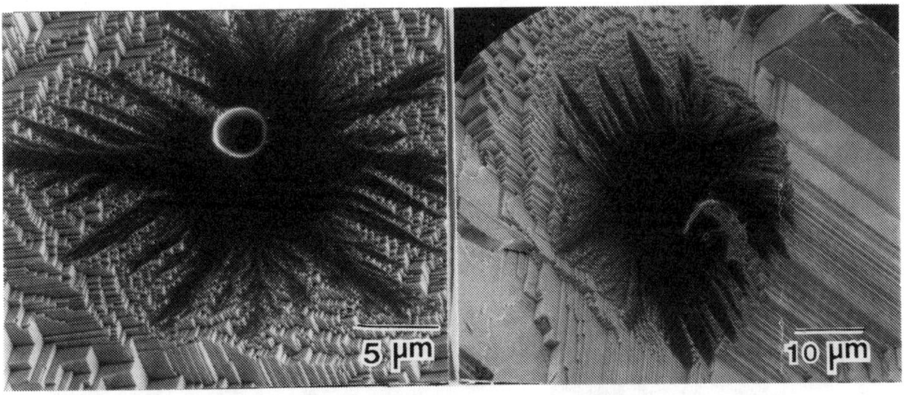

Fig. 3 The primary fracture origin of Ti^{4+}-doped sapphire fiber tested at 1450 °C.

were populated with internal voids, Fig. 3. Thus, the intrinsic potential for the increase of tensile strength with the TiO_2 doping and TiO_2/MgO-codoping could not been adequately assessed in the present work. Additional experiments are needed to study the intrinsic contribution of TiO_2 doping and TiO_2/MgO-codoping on K_{TH} and hence on the high temperature mechanical properties of Al_2O_3.

Although our results imply MgO raises K_{TH}, the exact mechanism responsible for the observed strength increase is not apparent. Both precipitation hardening and solution hardening have been considered. Large angle tilting experiments were performed on MgO doped fibers using a transmission electron microscope. There was no evidence for dopant precipitation. In the absence of precipitation, solid solution hardening alone can have a strong effect on the high temperature deformation of sapphire. The Ti- and Mg-doped crystals were grown in air and therefore were 4+ and 2+ respectively imparting a distinct color to the LHFZ fibers. The Ti^{4+} ion and Mg^{2+} ion and their associated charge compensating defects introduce an asymmetric distortion into the alumina lattice. Complex formation by addition of MgO doping may serve as an effective strengthening mechanism at high temperatures.

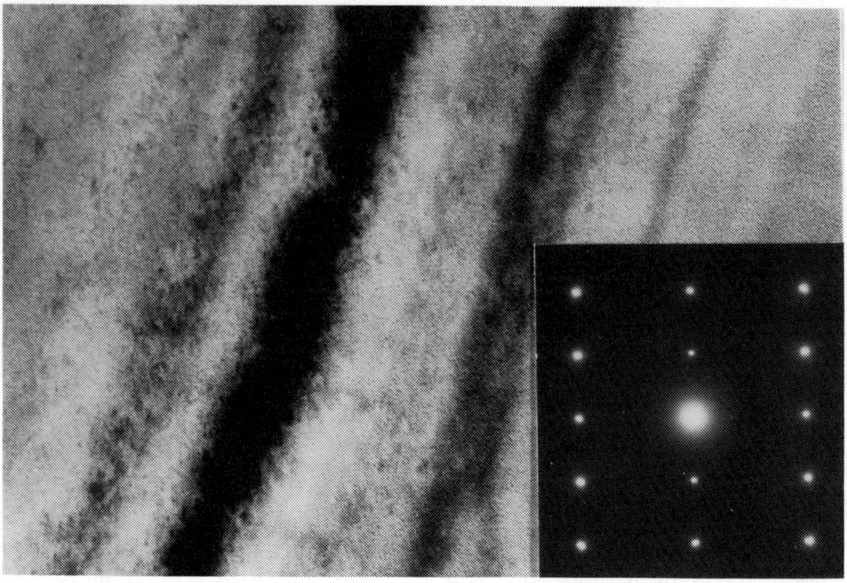

Fig. 4 TEM micrograph of single crystal sapphire doped with MgO shows no indication of any precipitation reaction; [10$\bar{1}$0] **zone axis.**

The result shown in Fig. 1, where marked improvements are observed only with MgO doping, is in substantial agreement with Crane`s work[15]. Crane considered both size and modulus differences in solute/dislocation interactions and concluded enhanced strengthening can be achieved for dopants having large ionic radii i.e., $Mg^{2+}$. Furthermore, Radford and Pratt`s work demonstrated that Mg^{2+} doping has a more pronounced effect on solute/dislocation interactions than Ti^{4+} doping.[14] Radford and Pratt found the greatest hardening occurred for $Mg^{2+} > Ti^{4+} > Cr^{3+} > Fe^{3+} > Ni^{2+}$; Ni^{2+} being the weakest hardener.

If the tensile strength of sapphire at elevated temperatures is determined by dislocations and the interaction of the dislocations with initial flaws, then the effect of dopants on high temperature strength may be to increase the K_{TH} by inhibiting dislocation activity. The fracture strength in solution hardened fibers will then be higher since dislocation assisted crack growth takes place at a higher stress.

CONCLUSIONS

The high temperature strength of single crystal alumina fibers can be significantly improved by MgO doping. However, no complete explanation for the strengthening mechanism in alumina can be offered at the present time. Transmission electron microscopy of MgO fibers demonstrates an absence of precipitation. Strengthening by solution hardening remains plausible. Experimental results indirectly indicate that MgO may decrease susceptibility to slow crack growth by raising K_{TH}. Additional experiments are clearly needed to understand the effect of MgO and other dopants on the long term mechanical properties of sapphire.

REFERENCES

1. R. E. Tressler and R. L. Crane, "Sapphire Filament Mechanical Property Considerations of Importance to Alumina Reinforced Matrices," Proc. 1st Int. Conf. Comp. Mat. [ICM], Geneva, Switzerland, (1976) 85-102.
2. M. L. Kronberg, "Plastic Deformation of Single Crystal of Sapphire: Basal Slip and Twinning," Acta Met., 5 [507] (1957) 507-524.
3. B. J. Pletka, T. E. Mitchell and A. H. Heuer, Strengthening Mechanisms in Sapphire," in Fracture Mechanics of Ceramics, Vol.2, pp.181-194. Ed. R. C. Bradt, D. P. Hasselman and F. F. Lange, Plenum, New York, 1974.
4. D. M. Kotchick and R. E. Tressler, Deformation Behavior of Sapphire via the Prismatic Slip System," J. Am. Ceram. Soc., 63 [7-8] (1980) 429-434.
5. D. J. Gooch and G. W. Groves, "The Creep of Sapphire Filament with Orientations Close to the C-Axis," J. Mat. Sci., 8 (1973) 1238-1246.
6. J. S. Haggerty, "Production of Fibers by Floating Zone Fiber Drawing Technique," NASA CR-120948, NASA Lewis Research Center, Cleveland, OH, 1972.
7. J. A. DiCarlo, J. B. Hurst, G. N. Morscher, A. Sayir and G. Selover, Current Issues for SiC and Oxide Fiber Development, pp. 50.1-50.6 in HiTEMP Review 1990. Advanced High Temperature Engine Materials Technology Program, NASA CP-10051.

8. E. A. Jackman and J. P. Roberts,"The Strength of Single Crystal and Polycrystalline Corundum", *Phil. Mag.* , **7**[46] 809-811 (1955).

9. J. B. Wachtman, Jr., and L. H. Maxwell, "Strength of Synthetic Single Crystal Sapphire and Ruby as a Function of Temperature and Orientation", *J. Am Ceram Soc.*, **42** 432-433 (1959).

10. P. Shahinian, "High Temperature Strength of Sapphire Filament," J. Am. Ceram. Soc., 54 [1], (1971) 67-68.

11. H. Sayir, A. Sayir, and P. Lagerlof, Temperature - Dependent Brittle Fracture of Undoped and Impurity Doped Sapphire Fibers," Ceram. Eng. & Sci. Proc., 14 (1993) 581-590.

12. A. H. Heuer and J. P. Roberts, "The Influence of Annealing on The Strength of Corundum Crystals", *Proc. Brit. Ceram. Soc.*, **6** 17-27 (1966).

13. H. Sayir, "The Effect of Mg^{2+} and Ti^{4+} Doping and Mg^{2+}/Ti^{4+} Co-Doping on the Temperature Dependent Bend Strength of C-Axis Sapphire," M. S. Thesis, 1993. Case Western Reserve University. University Circle, Cleveland, OH 44106.

14. K. C. Radford and P. L. Pratt,"The Mechanical Properties of Impurity -doped Alumina Single Crystals", Proc. Brit. Ceram. Soc., 15 (1970) 185-202.

15. R. L. Crane, "An investigation of the Mechanical Properties of Silicon Carbide and Sapphire Filaments," Technical Report AFML-TR-72-180, Air Force Materials Laboratory, WPAFB, OH, 1972.

16. B. J. Pletka, T.E. Mitchell, and A.H. Heuer,"Precipitation in Titanium Doped Sapphire", Proc. Mech. Behav. Mater. Vol IV, pp (413-421) 1972

17. A. Sayir, "Time Dependent Strength of Sapphire Fibers at High Temperatures," in Advances in Ceramic Matrix Composites (Ed. N. Bansal), Am. Ceram. Soc., Westerville, OH, 1993.

18. S. A. Newcomb and R. E. Tressler, "Slow Crack Growth in Sapphire Fibers at 800 to 1500 °C," J. Am Ceram. Soc., 76 [10] (1993) 2505-2512.

19. S. A. Newcomb and R. E. Tressler, "High Temperature Fracture Toughness of Sapphire," To be submitted to the J. Am. Cearm. Soc.

20. M. Iwasa and R. C. Bradt, "Fracture Toughness of Single Crystal Alumina," pp. 767-769 in Structure and Properties of MgO and Al_2O_3 Ceramics. Ed. W. D. Kingery. Am Ceram Soc., 1984.

21. S. G. Roberts, H. S. Kim and P. H. Hirsch, "The Brittle Ductile Transition and Dislocation Mobility in Silicon and Sapphire," in 9th Int. Conf. of Strengths of Metals and Alloys, Haifa, Israel, July 1991.

22 M. V. Klassen Neklyuvada, V. G. Govorkov, A. A. Urusovskaya, N. N. Voinova, and E. P. Kozlovskaya,"Plastic Deformation of Corundum Single Crystals," Phys. Status Solidi, 39 [2] (1970) 679-688.

PROCESSING OF FINE-DIAMETER CONTINUOUS MULLITE FIBERS

S. Al-Assafi, T. Cruse, J.H. Simmons, A.B. Brennan, and M.D. Sacks, Department of Materials Science and Engineering, University of Florida, Gainesville, FL 32611.

ABSTRACT

Fine-diameter, continuous mullite fibers were fabricated by sol-gel processing using two different chelating agents. The effects of the chelating agents on weight loss behavior, mullite formation, and sintered fiber microstructure were examined. The molar ratio, R, of chelating agent to alumina precursor in the sol was varied in order to optimize continuous spinning conditions. The importance of aging conditions on the sintered fiber density was demonstrated.

INTRODUCTION

Mullite has excellent creep resistance and strength retention at high temperatures, as well as good chemical stability [1]. Thus, fine-diameter, continuous mullite fibers are of interest for the development of ceramic fiber-reinforced composites with improved high temperature properties in oxidizing atmospheres.

Mullite fibers are usually fabricated by sol-gel methods [2-6]. These processing routes allow for good control over sol rheological properties. Hence, sols with good spinnability and fibers with fine diameter are readily produced. Furthermore, the molecular or near-molecular scale of mixing of alumina and silica precursors results in mullite formation and fiber densification at relatively low temperatures [2,6]. The mullite transformation temperature is dependent on the type of sol-gel precursor. Single-phase precursors, which exhibit a nucleation-controlled transformation mechanism, form mullite at temperatures below 1000°C [7]. In contrast, diphasic precursors transform to mullite at higher temperatures (typically at or above 1200°C) by a diffusion-controlled transformation mechanism [8].

Glass-free mullite fibers have been prepared from diphasic sols by Venkatachari et al.[2]. Aluminum formoacetate and aluminum hydroxychloride hydrate were

To the extent authorized under the laws of the United States of America, all copyright interests in this publication are the property of The American Ceramic Society. Any duplication, reproduction, or republication of this publication or any part thereof, without the express written consent of The American Ceramic Society or fee paid to the Copyright Clearance Center, is prohibited.

used as alumina precursors and LUDOX® colloidal silica sols were used as silica precursors. Fibers were produced with fine-diameter, low porosity, and fine-grain size after heat treatment at temperatures up to 1500°C. However, fibers were limited to short lengths, as continuous spinning was not carried out. Tucker et al. used single-phase sols to fabricate continuous mullite fibers [3]. The silica and alumina precursors were tetraethyl orthosilicate and aluminum diisopropoxide acetoacetic ester chelate, respectively. The sintered fibers were extremely brittle and microstructure observations showed extensive porosity and microcracks. This was attributed to difficulties in removing volatiles during heat treatment of the large-diameter fibers (~110 μm after sintering).

In this study, sols prepared with two different chelating agents were used to produce continuous single-phase mullite fibers with diameters of ~10-25 μm. The effects of chelating agent and aging conditions on the processing and properties of fibers were investigated.

EXPERIMENTAL PROCEDURE

Silica sols were prepared by mixing tetraethyl orthosilicate (TEOS) with ethanol and subsequently carrying out partial hydrolysis with water additions (i.e., 1:2 molar ratio of TEOS:water). In method I, the silica sol was added slowly to an unhydrolyzed solution of aluminum-sec-butoxide (ASB) and triethanolamine, TEA, (i.e., the chelating agent). The TEA/ASB molar ratio was 0.3. The TEOS/ASB ratio was selected to yield a composition of ~72 wt% alumina/28 wt% silica after heat treatment. In method II, the same procedure was followed except that the TEA was replaced by acetylacetone (AcAc), another chelating agent. The purpose of using chelated aluminum alkoxides is to reduce the hydrolysis/condensation rate and thereby prevent local precipitation of aluminum hydroxide in the mixed sol [9].

After mixing of components, the sol was concentrated until the rheological conditions for continuous spinning were achieved. The sol was then extruded through a spinneret with 40 μm hole diameter and continuous fibers were collected by winding on a rotating drum. The fibers formed with TEA were aged at room temperature for 24 hours inside a chamber at relative humidity of 100%. In contrast, more rapid aging was carried out for fibers prepared with AcAc. The as-spun fibers were heated in the humidity chamber at ~0.7°C/min to temperatures up to 100°C and then cooled immediately to room temperature. In each method, aged fibers were subsequently sintered at 1500°C for 10 min.

Rheological measurements, thermal gravimetric analysis/differential thermal analysis (TGA/DTA), sink-float apparent density measurements [10], and scanning electron microscopy (SEM) were used to study the effects of the chelating agent and the aging conditions on fiber processing and structure.

RESULTS AND DISCUSSION

The effect of molar ratio, R, of AcAc/ASB on the sol rheological behavior was investigated (Fig. 1). Low values of R (<0.3) resulted in thixotropic sols which gelled readily when exposed to air. In contrast, high values of R (>0.5) resulted in approximately Newtonian flow behavior. In addition, precipitate formation was observed in these sols. The optimum R value for good spinning behavior and good stability of the sol was ≈0.3. As shown in Fig. 1, sols with this R value have slightly shear thinning flow behavior. As reported previously [6], the optimum spinning behavior for method I (TEA) fibers is also obtained when sols are slightly shear thinning.

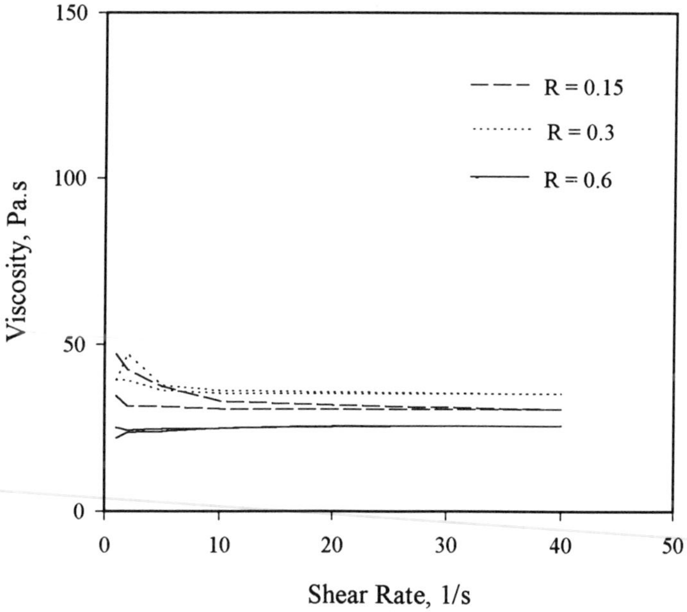

Figure 1 Plots of viscosity vs. shear rate for sols with different molar ratios, R, of AcAc/ASB.

Although good spinnability could be obtained for both method I and II sols, subsequent processing of green fibers was carried out much more easily using method II (AcAc). First, it was very difficult to avoid inter-filament sticking of green and sintered fibers prepared by method I. This was not a problem for method II fibers. Second, qualitative observations based on routine handling of the as-spun and aged fibers, indicated that the method II fibers had much better strength and flexibility. Third, it was easier to obtain fibers with circular cross-sections using method II.

The effect of the type of the chelating agent on the removal of volatile components was determined by TGA. Figure 2 shows plots of weight loss vs. temperature (heating rate 10°C/min) for fibers made with AcAc and TEA. Both samples were aged for 24 hours at room temperature in a humidity chamber prior to analysis. The data show that volatiles are removed at a lower temperature for fibers prepared with AcAc. This may be helpful in avoiding gas entrapment in the fibers when the pores close during sintering.

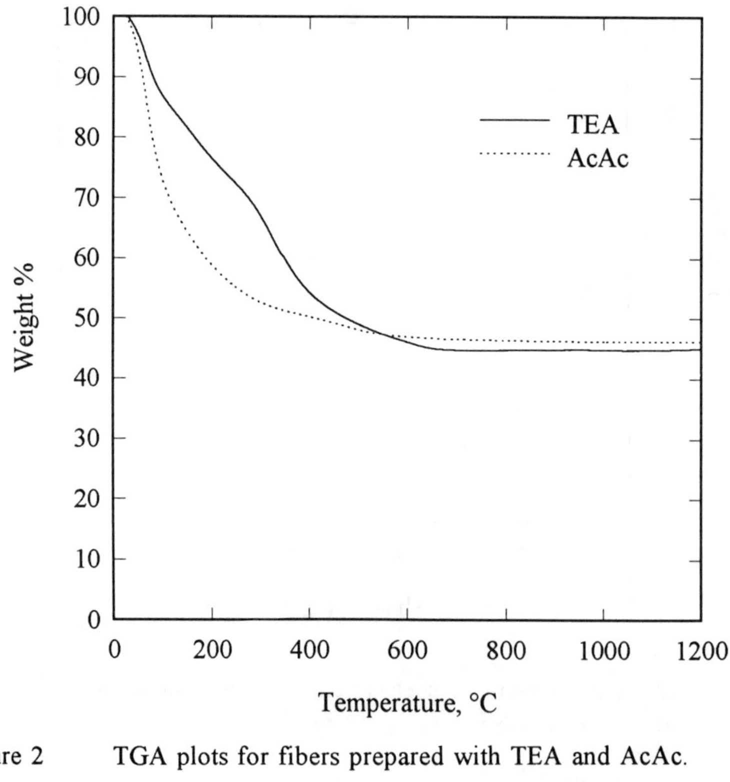

Figure 2 TGA plots for fibers prepared with TEA and AcAc.

Figure 3 shows the results of differential thermal analysis for fibers prepared with AcAc and TEA. Both fibers show an exotherm at ~980°C, although the peak is much stronger for the sample prepared with AcAc. These results indicate a significant enhancement in mullite formation in the method II (AcAc) fibers and suggests that more uniform (i.e., molecular-scale) mixing is achieved during sol-gel processing.

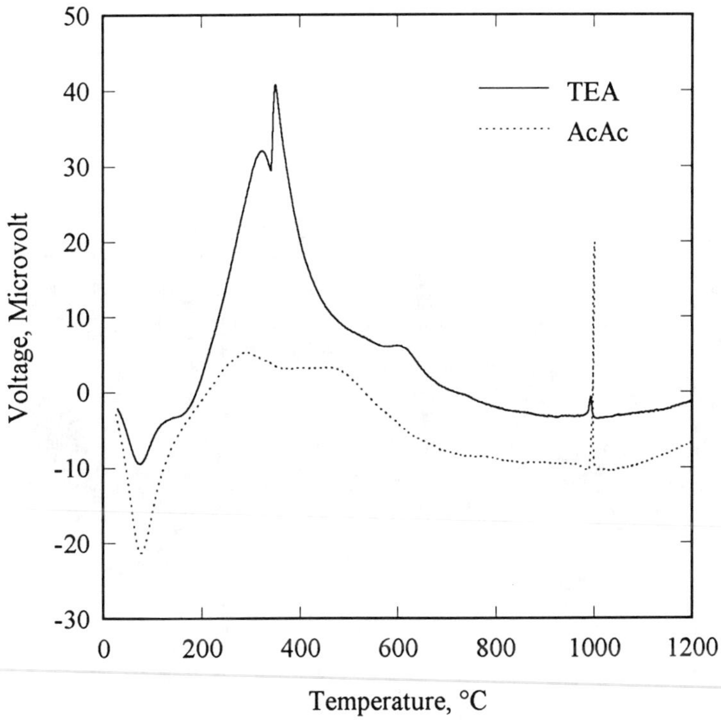

Figure 3. DTA plots for fibers prepared with TEA and AcAc.

Sintered fibers with high relative density (low porosity) were prepared by method II (AcAc) by utilizing a high temperature aging step. The final apparent density (determined by the sink-float method) depended upon the aging temperature. Fibers aged at 80°C and 100°C had apparent densities of ~2.9 and ~3.1 g/cm^3, respectively, after sintering at 1500°C for 10 min. In contrast, method I (TEA) fibers had considerably lower density (~2.7 g/cm^3) if aging was carried out rapidly (at 100°C).

The type of chelating agent used had a significant effect on the fiber microstructure after sintering at 1500°C. Fibers prepared with TEA (method I) had grain sizes >1 μm, as shown in Figure 4. In contrast, fibers prepared with AcAc (method II) had substantially smaller grain sizes, as shown in Figure 5.

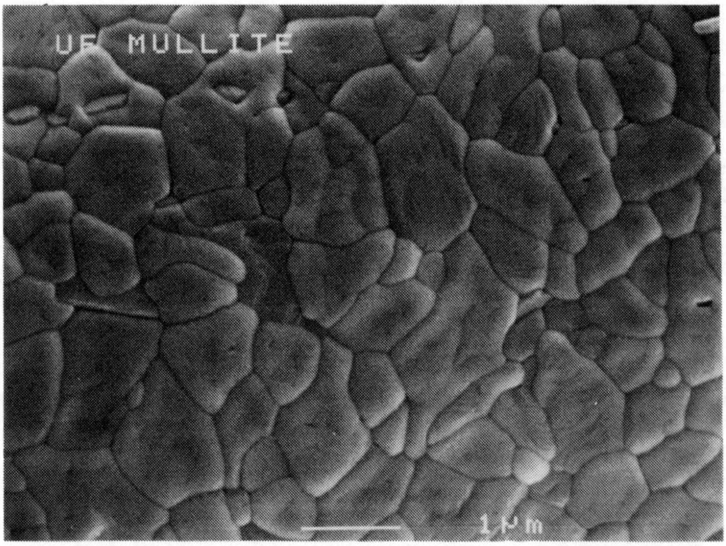

Figure 4. SEM micrograph for a method I (TEA) fiber sintered at 1500°C.

Figure 5. SEM micrograph for a method II (AcAc) fiber sintered at 1500°C.

SUMMARY

Continuous mullite fibers were prepared using acetylacetone (AcAc) and triethanolamine (TEA) as sol-gel processing additives. The use of AcAc in optimum concentration provided the following advantages: (i) enhanced sol stability, (ii) excellent spinnability, (iii) improved green fiber handling capability, (iv) reduced temperature for removal of volatile components, (v) enhanced mullite formation at low temperature (~980°C), and (vi) the development of sintered fibers with high relative density (≥98% of the theoretical value for mullite) and the retention of submicrometer grain size after heat treatment at 1500°C.

ACKNOWLEDGMENTS

This work was supported by DARPA under Grant No. N00014-91-J-4075. The authors gratefully acknowledge the assistance of R. Crockett and G. W. Scheiffele.

REFERENCES

[1] I. A. Aksay, D. M. Dabbs, and M. Sarikaya, "Mullite for Structural, Electronic, and Optical Applications," J. Am. Ceram. Soc., 74 [10] 2343-2358 (1991).

[2] K. R. Venkatachari, L. T. Moeti, M. D. Sacks, and J. H. Simmons, "Preparation of Mullite-Based Fibers by Sol-Gel Processing," Ceram. Eng. Sci. Proc., 11 [9-10] 1512 (1990).

[3] D. S. Tucker, J. S. Sparks, and D. C. Esker,"Production of Continuous Mullite Fiber via Sol-Gel Processing," Am. Ceram. Soc. Bull., 69 [12] 1971-1974 (1990).

[4] G. Kim, D. Lee, H. Lee, and S. Yoon,"A Study on the Development of Mullite Fibers Using the Sol-Gel Process," Materials Science and Engineering A:Structural Materials: Properties, Microstructure and Processing, 167 [1-2] 171-178 (1993).

[5] E. A. Richards, C. J. Goodbrake, and H. G. Sowman,"Reactions and Development in Mullite Fibers," J. Am. Ceram. Soc., 74 [10] 2404-2409 (1991).

[6] S. Al-Assafi, T. Cruse, J. H. Simmons, A. B. Brennan, and M. D. Sacks,"Processing of Sol-Gel Derived Continuous Mullite Fibers via Sol-Gel," Proceedings of the 18thAnnual Conference on Composites and Advanced Ceramics, Cocoa Beach 1994.

[7] D. X. Li, and W. J. Thompson, "Mullite Formation Kinetics of a Single-Phase Gel," J. Am. Ceram. Soc., 73 [4] 964 (1990).

[8] D. X. Li, and W. J. Thompson, "Kinetic Mechanisms for Mullite Formation from Sol-Gel Precursors," J. Mat. Res., 5 [9] 1963 (1990).

[9] T. Heinrich, F. Raether, H. Marsmann,"Growth and Structure of Single Phase Mullite Gels from Chelated Aluminum Alkoxides and Alkoxysilanes," J. N. Cryst. Sol., 168 14-22 (1994).

[10] ASTM Test Method D 3800, Procedure B,"Sink-Float Technique," American Society for Testing and Materials, Philadelphia, PA.

TERNARY OXIDE WHISKER GROWTH IN A REACTIVE ATMOSPHERE[a]

Ning-Wei Chen, Dennis W. Readey, and John J. Moore

Colorado Center for Advanced Ceramics
Colorado School of Mines
Golden, CO 80401

ABSTRACT

Whiskers are of interest for toughening polycrystalline ceramics. Nonoxide whiskers are available commercially but can oxidize and lose their effectiveness at high temperatures in oxidizing atmospheres. The availability of oxide whiskers would offer the potential of high toughness and oxidation resistant ceramic composites. Data are presented for the growth of ternary oxide whiskers of $3Al_2O_3{\cdot}2SiO_2$, $9Al_2O_3{\cdot}2B_2O_3$, and $2Al_2O_3{\cdot}B_2O_3$ in an HCl atmosphere. The possible mechanisms controlling the growth of these whiskers are examined.

INTRODUCTION

Whisker-reinforced ceramic composites are promising structural ceramics because of their high strengths and high fracture toughness values.[1] As a result, there has been a great deal of interest in the synthesis of mainly nonoxide whiskers.[2,3,4,5] and in particular, silicon carbide whiskers.[6,7,8,9,10] However, in oxidizing atmospheres, composites containing nonoxide whiskers oxidize and their properties degrade.[11] Therefore, there is an interest in synthesizing oxide whiskers.[12,13]

Whiskers that appeared to be mullite formed during the combustion synthesis of Al_2O_3 plus TiB_2 composites in an HCl atmosphere in a mullite tube.[14] These results prompted further investigation into the mechanism of the whisker growth and to the breadth of applicability of the mechanism to other oxides.

a Sponsored by the National Science Foundation, Grant No. CTS-9202948

To the extent authorized under the laws of the United States of America, all copyright interests in this publication are the property of The American Ceramic Society. Any duplication, reproduction, or republication of this publication or any part thereof, without the express written consent of The American Ceramic Society or fee paid to the Copyright Clearance Center, is prohibited.

MECHANISMS OF WHISKER GROWTH

Whisker growth frequently implies that the solid species are being transported through the vapor phase. The two primary mechanisms of whisker growth that have been described are the vapor-solid or VS technique[15] and the vapor-liquid-solid[9,16] or VLS process shown in Figures 1 and 2 respectively.

One mechanism of VS whisker growth is simply an anisotropic rate of growth that is very fast in one crystallographic direction and leads to an elongated crystal or whisker. This mechanism requires: 1.) a nucleation source; and 2.) a whisker growth rate limited by a gas-solid interface reaction rather than by transport through the vapor. A possible nucleation site is simply the presence of a screw dislocation at the substrate surface. Vapor diffusion can also lead to whisker or anisotropic growth since the rate of growth at the whisker tip can be very high because the diffusion flux to the surface, J, is inversely proportional to the whisker tip radius.[17]

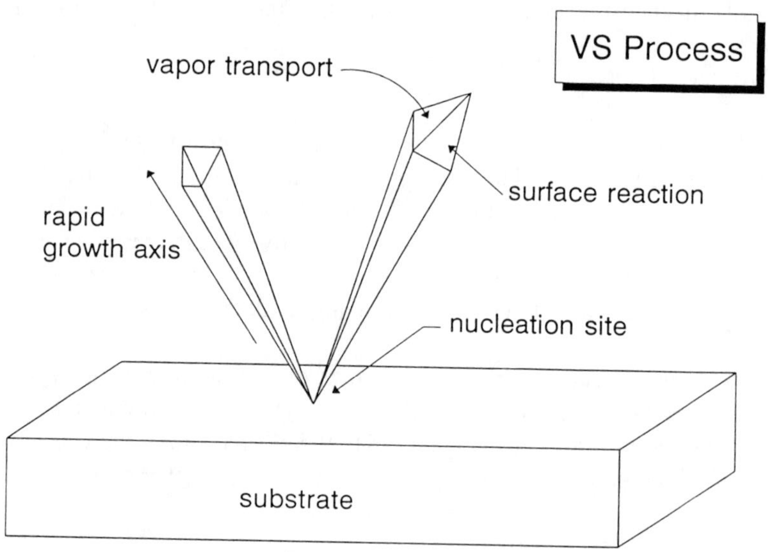

Figure 1. Schematic depiction of the vapor-solid (VS) mechanism of whisker growth.

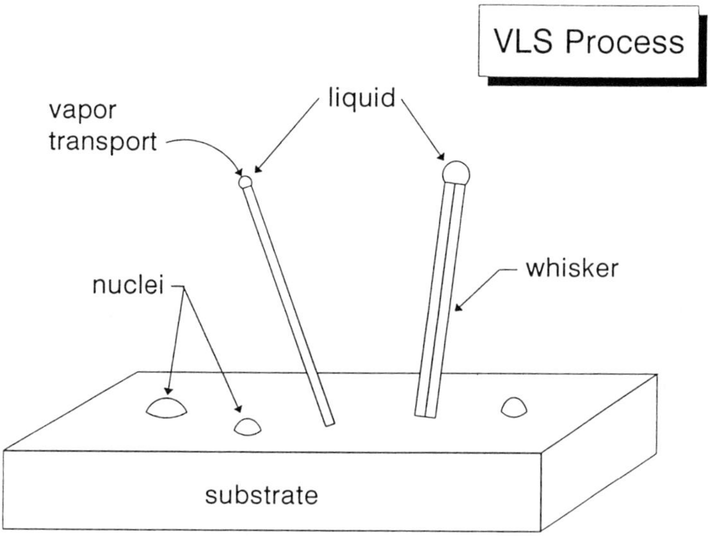

Figure 2. Schematic depiction of the vapor-liquid-solid mechanism of whisker growth.

In whisker growth by the VLS mechanism, the liquid drop becomes supersaturated with the solid constituents and precipitation of the solid occurs. The liquid drop is sufficiently small so that no other nucleation occurs and growth continues on the single existing nucleus, the whisker. In virtually all known cases of whisker growth by the VLS mechanism, some additive is present that forms a liquid phase at the whisker growth temperature. A frequent characteristic feature of the VLS process is the presence of a residual "liquid' ball on the whisker tip at the end of the process.

EXPERIMENTAL

In this research, ternary oxide whiskers were made by reacting end member powders at elevated temperatures in a reactive carrier gas that produces volatile species of the solid constituents. For the systems studied here, these reactions are mainly:

$$Al_2O_{3(s)} + 6\ HCl_{(g)} \rightleftarrows 2\ AlCl_{3(g)} + 3\ H_2O_{(g)}$$

$$B_2O_{3(s)} + 6\ HCl_{(g)} \rightleftarrows 2\ BCl_{3(g)} + 3\ H_2O_{(g)}$$

$$SiO_{2(s)} + 4\ HCl_{(g)} \rightleftarrows SiCl_{4(g)} + 2\ H_2O_{(g)}.$$

Thermodynamic calculations[18] were carried out for each solid and show that these are the primary reactions. The calculated gas partial pressures for alumina are shown in Figure 3. These vapor species then react at the growing whisker tip to form the ternary oxide. For example, mullite can be formed by:

$$6\ AlCl_{3(g)} + 13\ H_2O_{(g)} + 2\ SiCl_{4(g)} \rightleftarrows 3Al_2O_3{\cdot}2SiO_{2(s)} + 26\ HCl_{(g)}.$$

The oxide powders[b] were mixed and pressed into a pellet. Pellet samples were then placed into a small alumina ampule with a loose fitting plug. The ampule was placed into an alumina tube, evacuated and back-filled with pure HCl gas. The gas pressure was adjusted so that it would be approximately one atmosphere at the reaction temperature. Reactions were carried out for about one hour at the following temperatures:

$3Al_2O_3{\cdot}2SiO_2$	T = 1100°C-1500°C
$9Al_2O_3{\cdot}2B_2O_3$	T ≥ 1100°C
$2Al_2O_3{\cdot}B_2O_3$	T ≈ 900°C.

The resulting whiskers were analyzed by EDAX in the SEM and their crystal structure and orientation by TEM. In addition some experiments were carried out in Ar and some other experiments were carried out with a small single crystal of Al_2O_3 present in the reaction chamber.

RESULTS AND DISCUSSION

Whiskers of all three ternary compounds were obtained. Whiskers were found to grow on the surface of the pellets as shown in Figure 4. Figure 5 shows mullite and the aluminum borate whiskers. Figure 6 is a typical transmission electron micrograph and diffraction pattern confirming the presence of mullite as the whiskers formed in the alumina-silica system. . EDAX analysis confirmed the

---

b SiO_2 and B_2O_3, 99.9% pure, Cerac, Inc., Milwaukee, WI
Al_2O_3, 99.9% pure, Ceralox, Inc., Tucson, AZ

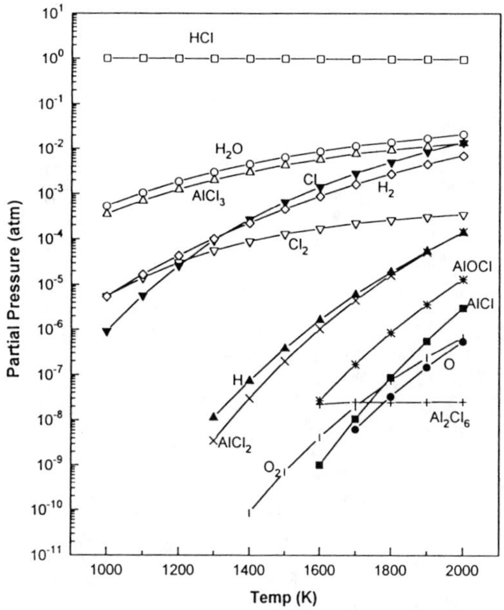

Figure 3. Partial pressures in the Al_2O_3 - HCl system as a function of temperature at 0.1 MPa total pressure.

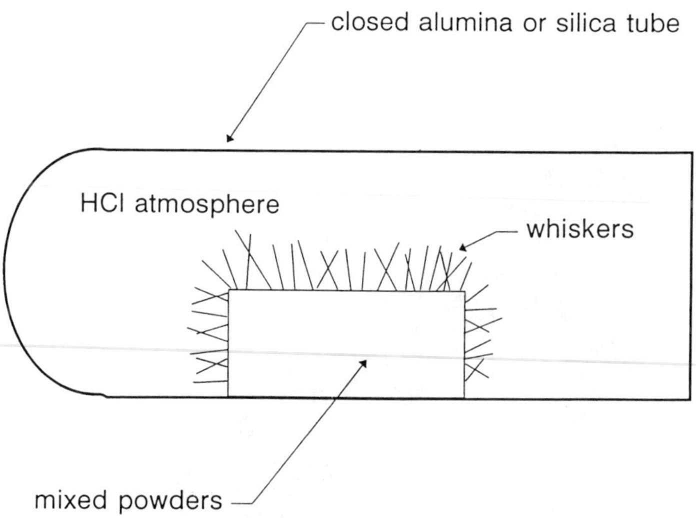

Figure 4. Schematic illustrating the experimental arrangement and the location of whisker growth.

Figure 5. Whiskers of $3Al_2O_3{\cdot}2SiO_2$, $9Al_2O_3{\cdot}2B_2O_3$, and $2Al_2O_3{\cdot}B_2O_3$ from left to right.

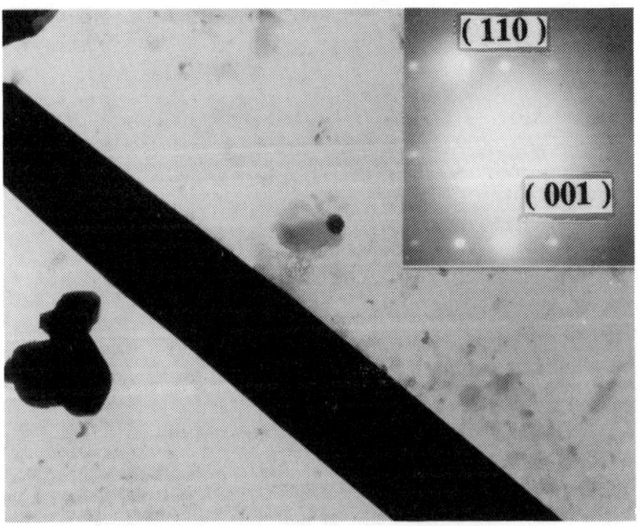

Figure 6. Transmission electron micrograph and diffraction pattern of mullite whisker.

presence of both Al and Si. Similar results were obtained for both species of the aluminum-borate whiskers. However, only the presence of Al could be confirmed by EDAX.

Notice that in all three systems, no residual "liquid" balls are found on the whisker tips implying that the VLS mechanism is not determining whisker growth. However, two other observations suggest that a second phase, perhaps or in some cases a liquid, may play an important role in whisker growth even if it does not operate in the classical VLS mode. The first observation is that $9Al_2O_3{\cdot}2B_2O_3$ whiskers are grown at temperatures above 1100°C while $2Al_2O_3{\cdot}B_2O_3$ is the phase grown at 900°C. The exact temperature, if one exists, that determines which phase grows has yet to be determined. However, inspection of the phase diagram, Figure 7[19] shows that above 1035 °C, $9Al_2O_3{\cdot}2B_2O_3$ is in equilibrium with a B_2O_3-rich liquid. Below this temperature, $2Al_2O_3{\cdot}B_2O_3$ is the solid phase in equilibrium with the liquid. These results suggest that whisker composition may be determined by the presence of an un-detected liquid phase: at least one that does not form a ball on the whisker tip.

To ensure that the HCl reactive atmosphere was necessary for whisker growth, mixed powders of Al_2O_3 with SiO_2 and Al_2O_3 with B_2O_3 in the correct ratios for the formation of mullite and $9Al_2O_3{\cdot}2B_2O_3$ were fired at 1200°C and 1100°C in argon respectively. No whisker formation was observed. However, elongated particles were produced in the Al_2O_3-B_2O_3 system indicating the anisotropic growth rates that exist. These results are shown in Figure 8.

In contrast, whiskers of both mullite, at 1200°C, and 9:2 Al_2O_3-B_2O_3, at 1100°C, were readily grown in an HCl atmosphere when SiO_2 and B_2O_3 powders, respectively, were placed on polycrystalline alumina, Figure 9. In addition, whiskers could be grown on single crystal alumina plates completely by vapor transport of SiO_2 or B_2O_3 in HCl. Figure 10 shows a single crystal of Al_2O_3 that has mullite whiskers growing on its surface. There are two features worth noting: first, the whiskers seem to grow with no regard for the orientation of the single crystal; second, there are obviously some rather featureless regions on the surface from which the whiskers seem to emanate. EDAX analysis suggests that these regions are rich in SiO_2. These mullite whiskers were grown below the temperature at which an equilibrium liquid phase exists in the Al_2O_3-SiO_2 system, 1585°C.[20]

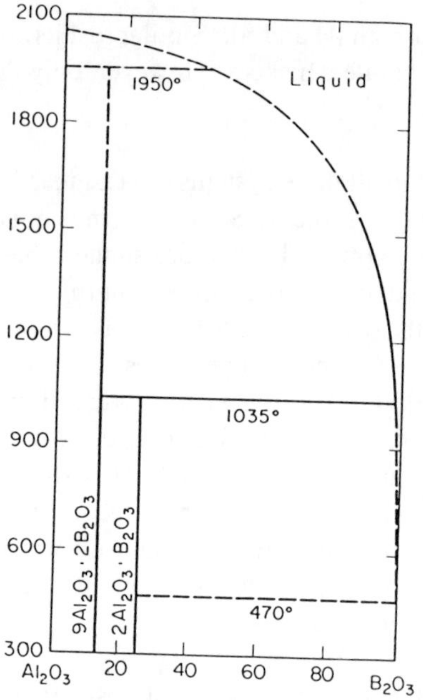

Figure 7. Al_2O_3-B_2O_3 phase diagram.[19]

Figure 8. Left: mixture of SiO_2 and Al_2O_3 powders fired in argon at 1200°C showing no whisker growth. Right: mixture of B_2O_3 and Al_2O_3 powder fired in argon at 1100°C showing elongated particles.

Figure 9. Left: mullite whiskers grown on polycrystalline alumina from SiO_2 powder in HCl at 1200°C. Right: aluminum-borate whiskers under the same conditions at 1100°C.

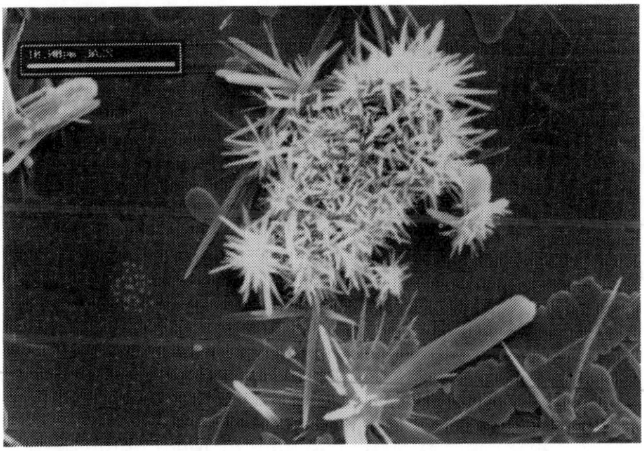

Figure 10. Mullite whisker growth on alumina single crystal in HCl in the presence of a second phase.

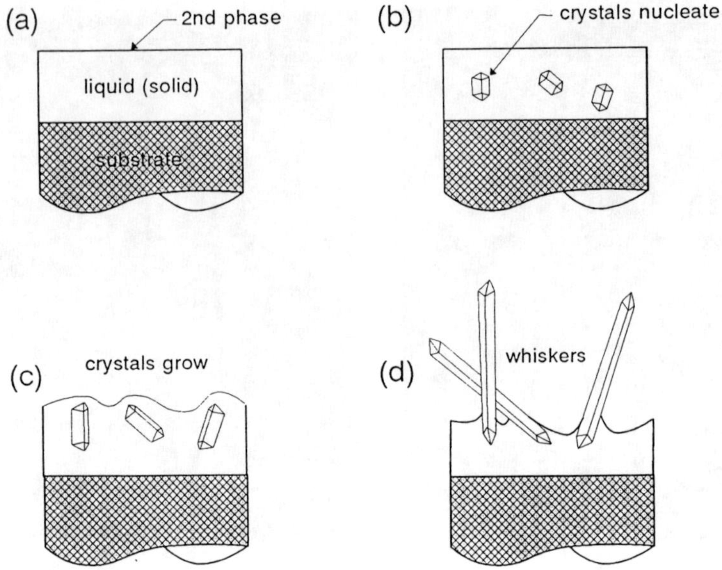

Figure 11. Schematic showing the role of the second phase in whisker nucleation and growth.

Either a solid or a liquid second phase appears to be necessary for whisker growth in these systems that does not seem to play the role of the liquid phase in the VLS mechanism of whisker growth. A plausible explanation is that the second phase acts as the nucleating agent for the whiskers which then grow by a VS process in a rapid crystallographic growth direction. Figure 11 sketches this proposed whisker nucleation mechanism.

CONCLUSIONS

Whiskers of $3Al_2O_3 \cdot 2SiO_2$, $9Al_2O_3 \cdot 2B_2O_3$, and $2Al_2O_3 \cdot B_2O_3$ have been successfully grown from mixed binary powders in an HCl atmosphere without the addition of "catalysts" or additives which could form a liquid phase and produce whiskers by a VLS process. No evidence for a residual liquid at the whisker tip, and, hence, the VLS process, was found. However, other observations suggest that the presence of a second phase, possibly either liquid or solid, may be important in nucleating whisker growth. The whiskers probably growth by a VS mechanism in the most rapid growth direction. Additional experiments are being performed to confirm this or another mechanism in these and other oxide systems. It is interesting to note that the similarity in whisker growth behavior in the alumina-silica and the alumina-boron oxide systems is probably more than accidental since there is a significant amount of solid solution reported between $9Al_2O_3{\cdot}2B_2O_3$ and $3Al_2O_3{\cdot}2SiO_2$.[21]

REFERENCES

1. G. C. Wei and P. F. Becher, "Development of SiC Whisker-Reinforced Ceramics," Am. Ceram. Soc. Bull., 64 [2] 298-304 (1985).

2. H. Wang and G. S. Fischman, "Whisker Growth and Composite Fabrication in the Si_3N_4-C System,

3. M-J. Wang and H. Wada, "Synthesis and characterization of silicon nitride whiskers," J. Mater. Sci. 25 1690-98 (1990).

4. R. Mathur and B. M. Gallois, "Morphological Evolution of Titanium Carbide Whiskers,"Mat. Res. Soc. Symp. Proc., 168 267-72 (Materials Research Soc., Pittsburgh), 1990.

5. W. J. Lackey, J. A. Hanigofsky, and G. B. Freeman, "Experimental Whisker Growth and Thermodynamic Study of the Hafnium-Carbon System for Chemical Vapor Deposition Applications," J. Am. Ceram. Soc. 73 [6] 1593-98 (1990).

6. J-G. Lee and I. B. Cutler, "Formation of Silicon Carbide from Rice Hulls," Ceramic Bull. 54 [2] 195-8 (1975).

7. S. R. Nutt, "Microstructure and Growth Model for Rice-Hull-Derived SiC Whiskers," J. Am. Ceram. Soc. 71 [3] 149-56 (1988).

8. J. V. Milewski, F. D. Gac, J. J. Petrovic, and S. R. Skaggs, "Growth of beta-silicon carbide whiskers by the VLS process," J. Mater. Sci. 20 1160-66 (1985).

9. W. E. Hollar, Jr. and J. J. Kim, "Review of VLS SiC Whisker Growth Technology," Ceram. Eng. Sci. Proc. 12 [7-8] 979-991 (1991).

10. A. Chrysanthou, P. Grieveson, and A. Jha, "Formation of silicon carbide whiskers and their microstructure," J. Mater. Sci. 26 3463-76 (1991).

11. D. W. Readey, "High Temperature Gas Corrosion of Ceramic Composites," Cer. Eng. and Sci. Proc. 13 [7-8] 301-318 (1992).

12. K. Okada and N. Otuska, "Synthesis of Mullite Whiskers and Their Application in Composites," J. Am. Ceram. Soc. 74 [10] 2414-18 (1991).

13. M. J. Readey, "Formation and Sintering Characteristics of Aluminum Borate Whiskers," J. Am. Ceram. Soc. 75 [12] 3452-56 (1992).

14. N. R. Perkins, "Synergistic Effects of Combustion Synthesis and Vapor Transport," M. S. Thesis, Colorado School of Mines, 1992.

15. W. B. Campbell, "Growth of Whiskers by Vapor-Phase Reactions," Whisker Technology, A. P. Levitt, ed., (Wiley, NY), pp. 16-46, 1970.

16. R. S. Wagner, "VLS Mechanism of Crystal Growth, " Whisker Technology, A. P. Levitt, ed., (Wiley, NY), pp. 47-133, 1970.

17. J. D. Jackson, Classical Electrodynamics, 2nd ed., (Wiley, N.Y.), p. 96, (1975).

18. T. M. Besmann, "SOLGASMIX-PV, A Computer Program to Calculate Equilibrium Relationships in Complex Chemical Systems," ORNL Manual TM-5775, Oak Ridge National Laboratory, Oak Ridge, TN, 1977.

19. E. M. Levin, C. R. Robbins, and H. F. McMurdie, Fig. 2339 in Phase Diagrams for Ceramists, Vol. 2. Edited by M. K. Reser. American Ceramic Society, Columbus, OH, 1969.

20. R. S. Roth, J. R. Dennis, and H. F. McMurdie, eds., Figs. 6442-50 in Phase Diagrams for Ceramists, Vol. 6. American Ceramic Society, Westerville, OH, 1987.

21. E. M. Levin, C. R. Robbins, and H. F. McMurdie, Fig. 763 in Phase Diagrams for Ceramists. Edited by M. K. Reser. American Ceramic Society, Columbus, OH, 1964.

Continuous-Fiber Ceramic Composites

PROCESSING AND MECHANICAL RESPONSE OF FIBER REINFORCED CERAMIC COMPOSITES FOR HIGH TEMPERATURE APPLICATIONS

Raj N. Singh*
Department of Materials Science and Engineering
University of Cincinnati
P.O. Box 210012
Cincinnati, OH 45221-0012

ABSTRACT

Dense fiber-reinforced ceramic composites (CMCs) capable of operation at very high temperatures are required in many structural applications. This requires reinforcement fibers and ceramic matrix materials with adequate mechanical properties and stability at high temperatures. An approach to the development of tough high temperature CMCs is illustrated. The important aspects of materials selection, processing, and fiber-matrix interfacial properties in the development of tough CMCs useful at high temperatures is described.

INTRODUCTION

Ceramic matrix composites (CMCs) are attractive for applications at elevated temperatures because of their low density, high strength, refractoriness, and toughness. Mechanical properties of these fiber-reinforced CMCs are dependent on properties of the reinforcing fibers, matrix, and fiber-matrix interface. In addition, issues related to thermomechanical and thermochemical compatibility are important contributors to the overall mechanical response of CMCs. Furthermore, processing conditions can have a significant influence on the mechanical behavior of composites because most CMCs are processed at relatively high temperatures where fiber properties may degrade because of the thermal or thermochemical effects. Therefore, a CMC useful as high-temperature structures requires selection of candidate reinforcing fibers, matrix materials, and reinforcement coatings with thermomechanical and thermochemical stability at application temperatures for extended time periods.

* Fellow, American Ceramic Society

To the extent authorized under the laws of the United States of America, all copyright interests in this publication are the property of The American Ceramic Society. Any duplication, reproduction, or republication of this publication or any part thereof, without the express written consent of The American Ceramic Society or fee paid to the Copyright Clearance Center, is prohibited.

An approach to the development of tough high temperature CMC's is schematically described in Fig. 1. The overall approach consists of an exploratory research phase and a developmental phase. In the research phase, composite constituents (fiber, matrix, and interface/coating) are selected based on the micromechanics and thermomechanical compatibility considerations, and appropriate processing methods to fabricate fully dense composites are developed. This is followed by the evaluation of thermal stability including environmental stability at elevated temperatures, and characterization of the mechanical properties at elevated temperatures. Once a promising composite system is demonstrated in the research phase, one can enter into the development phase in which further developments are made in fabricating prototype components and in characterizing engineering properties. We have used this approach in identifying a CMC materials system consisting of the zircon matrix uniaxially reinforced with continuous SiC fibers which is then used for illustrative purposes to demonstrate the applicability of this approach. Important aspects of materials selection, processing, interfacial properties, and mechanical behavior for the development of tough CMC's useful at high temperatures are described in this paper.

AN APPROACH FOR DEVELOPING TOUGH CMC's FOR HIGH TEMPERATURES

A tough and uniaxially fiber-reinforced ceramic matrix composite, when subjected to a tensile loading in the direction of fiber reinforcement, is expected to display stress vs strain or load vs deflection behavior as shown schematically in Fig. 2. An elastic behavior is displayed in the initial stage followed by an extended regime showing an inelastic behavior. There is a sudden load drop at the transition point from the elastic to inelastic regime that is characterized by the first evidence of matrix cracking. Beyond this point, in the inelastic region, the slope of the load-deflection curve progressively decreases as the sample elongates because of the fiber failure and matrix microcracking. In spite of fiber failures, the composite sample displays increasing load-bearing capacity reaching an ultimate in load-carrying capacity followed by a gradual load drop as an increasing number of fibers begin to break and pullout from the surrounding matrix phase. The first-matrix cracking stress (σ_{cr}) and strain (ε_{cr}), composite modulus (E), ultimate load-bearing stress (σ_u) and strain (ε_u), and work-of-fracture (WOF), i.e., the area under the load-deflection curve are important material parameters that can be used to design with composite materials. In addition, the fiber-matrix interfacial properties have important influence on the overall mechanical response of CMCs. It is desirable to create a composite having stress strain behavior shown in Fig. 2 not only at room temperature but also at elevated temperatures. This requires strong and creep resistant reinforcing fibers, optimum interfacial properties, stability of the composite constituents in oxidizing atmospheres, and thermomechanical/thermochemical compatibility of the composite constituents.

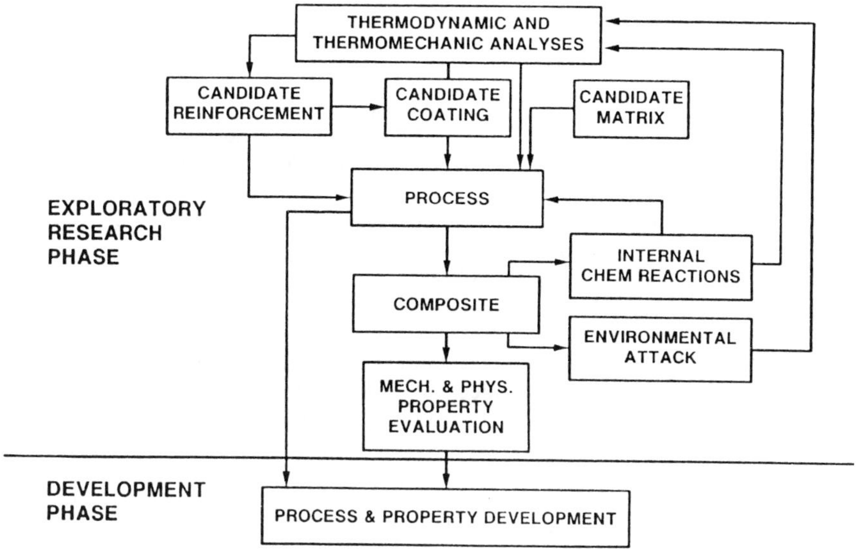

Fig. 1. A schematic showing an approach for developing high-temperature ceramic matrix composites.

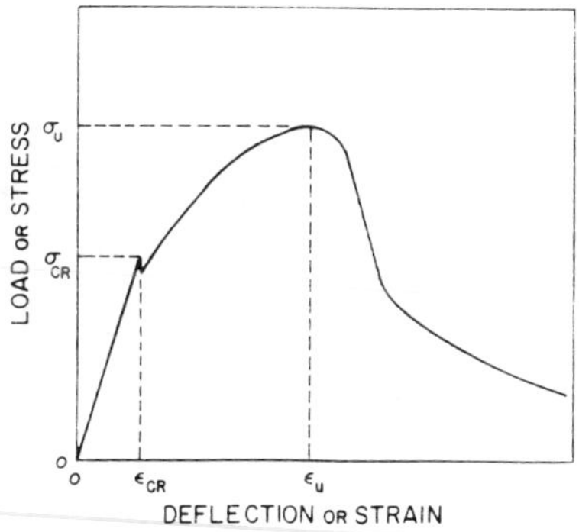

Fig. 2. A schematic drawing showing load-deflection behavior of a ceramic composite.

Selection of Materials System

Among the available reinforcing fibers, only silicon carbide (SCS-6) fibers have the necessary mechanical properties and adequate oxidative stability above 1200° C. This fiber was therefore chosen as a reinforcement material. An oxide ceramic matrix, zircon ($ZrSiO_4$), was chosen as a matrix material because of its inherrent stability in oxidizing atmosphere at high-temperatures. The zircon matrix is expected to provide protection for the SiC fibers, and its good thermal expansion match with the SCS-6 SiC fiber is expected to avoid problems of matrix cracking upon cooldown of composites from high processing temperatures because of the tensile residual stresses[1]. The zircon matrix was also chosen as a matrix material because of its thermochemical compatibility with the SCS-6 SiC fiber which is expected to inhibit fiber-matrix interaction and produce lower values of the fiber-matrix interfacial shear stress necessary for toughening by the fiber pullout mechanism. In addition, composites were also fabricated with a BN reinforcement coating to further avoid any possibility of the fiber-matrix reaction. As a consequence, we have chosen a composite system comprizing of the zircon matrix reinforced with SCS-6 SiC fibers because of the above considerations. A reinforcement coating of BN was also chosen based on these considerations although its oxidation resistance is not adequate.

Processing of Composites

A number of processing techniques are available for fabricating CMCs. These include filament winding followed by hot pressing, tape lamination followed by densification[2], infiltration into a fiber preform by sol-gel[3], or polymer precursor[4], melt infiltration[5], chemical vapor infiltration[6], and reactive oxidation of a metal[7]. Each of these techniques has certain advantages as well as disadvantages. An important consideration for cost-effective processing is an ability to fabricate fully dense composites with net-shape capability but without the application of external pressure. These requirements are only satisfied by two of the processing techniques, i.e., melt infiltration and reactive oxidation of metal (Lanxide™ process). Both of these processes have attracted considerable attention in recent years[5,7], but much more needs to be accomplished in order to apply these approaches on a commercial scale. Hybrid processing approaches are also promising in which one can take advantage of unique features of each of the approaches but at the same time avoiding their limitations.

Composites for this study were fabricated by the tape lamination approach in which SiC fibers were uniaxially aligned and then the zircon matrix was incorporated around each of the filaments by tape lamination[2,8]. This procedure ensured a uniform spacing between the fibers within and between the different layers. The final consolidation was done by hot-pressing between 1550° and 1600° C in a nitrogen atmosphere. These steps produced fully dense composites with little (<1%) porosity and uniformally distributed fibers as shown in Fig. 3. The composite samples with a fiber loading of 25% by volume had densities between 4.12 and 4.29 g/cm^3. The zircon matrix was fine-grained with an average grain size of 2.5 μm. The X-ray diffraction analysis of the fully consolidated composites showed zircon as the only matrix phase. Free zirconia phase was not detected by the X-ray diffraction analysis of the as-fabricated composites.

Fiber-Matrix Interfaces

Fiber-matrix interfacial properties have been shown to influence room temperature mechanical behavior of ceramic composites[9]. To study their influence on high-temperature mechanical behavior, two types of fiber-matrix interfaces were created in the fully consolidated zircon-SiC composites. In one case as-supplied (refered to as uncoated) fibers with the carbon-silicon outer layer were used, and in another case the BN-coated fibers were used. The BN coating of about 1-μm thickness was deposited by a low-pressure chemical vapor deposition technique[10] to alter the fiber-matrix interfacial properties from those composites reinforced with as-supplied fibers. The interface between an as-supplied (uncoated) fiber and surrounding zircon matrix was designated as interface A, and that between a BN-coated fiber and surrounding zircon matrix was designated as interface B. Interfacial properties were measured by the fiber pushout test[11,12].

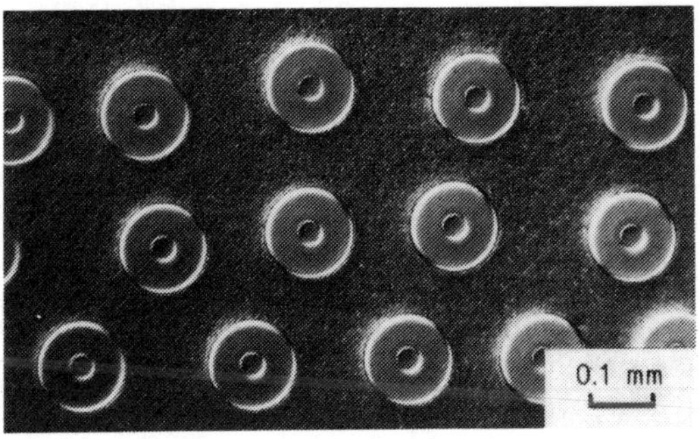

Fig. 3. Optical micrograph showing cross-section of a fully consolidated zircon-SiC composite.

Thermal Stability of Composites

Thermal stability of the composite constituents at elevated temperatures was determined by annealing samples at temperatures between 1200° and 1430° C for times up to 100 h[13]. The annealing was done either in a purified argon gas or in Ar-7% H_2 gas. The changes in composite characteristics caused by the annealing treatments were assesses by measuring phase composition using X-ray diffraction, mechanical properties, and fiber-matrix interfacial shear stress.

Mechanical Properties

Mechanical properties of composite samples were determined between 25° and 1500° C from load-deflection data obtained in a 3-point flexure mode. The uniaxially reinforced composite specimens, having typical dimensions of 3.2-cm length, 0.8-cm width, and 0.15-cm thickness, were individually hot-pressed and ground to a finish of 60 μm. The 3-point flexure tests were performed using a test fixture fabricated from thoriated-tungsten material. The lower support pins were 2.54 cm apart which resulted in a span-to-thickness ratio of about 17. All the tests were performed in a high-temperature furnace which was equipped with a tungsten-mesh heating element. The furnace temperature during testing was read by a W-Re thermocouple placed in the vicinity of the specimen, and the furnace temperature was controlled by a programmable digital controller utilizing the output from another W-Re thermocouple. All the tests were performed in a flowing Ar-7% H_2 atmosphere. A universal testing machine was used for mechanical tests which were performed at a crosshead rate of 0.0127 cm/min (0.005 in./min). Generally, four samples were tested at room temperature in the as-fabricated condition to establish the baseline properties, which were then used to compare with similar properties measured at higher temperatures.

Cross sections of failed monolithic zircon and zircon-SiC composites tested at progressively higher temperatures were examined by a SEM (scanning electron microscope) to determine the fracture morphology and the extent of fiber pullout. Particular attention was given to detecting changes in fracture morphology of the matrix material, the fiber, and the fiber-matrix interface as a result of the high-temperature testing.

RESULTS AND DISCUSSION

Mechanical and Interfacial Properties of As-Fabricated Composites

Mechanical properties of as-fabricated zircon matrix composites uniaxially reinforced with uncoated and BN-coated SiC fibers have been studied at room temperature[9]. The results for composite samples indicated a significant enhancement in strength and toughness values over those observed for a monolithic zircon as shown in Fig. 4 . In addition, composites reinforced with

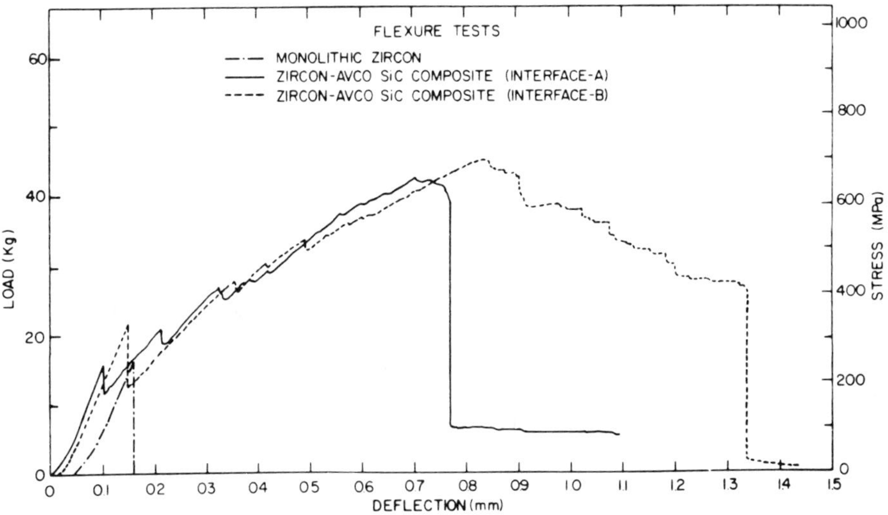

Fig. 4. Load-deflection behavior for a monolithic zircon and zircon-SiC composites.

Table 1. A summary of mechanical properties of the monolithic zircon and zircon-SiC composites.

Material	Monolithic (M) or composite (C)	Filament coating	Mechanical properties		
			σ_{cr} (MPa)	σ_u (MPa)	WOF (kJ m^{-2})
Zircon	M	-	281 ± 88	281 ± 88	1.2 ± 0.1
Zircon–SiC	C	Uncoated	287 ± 28	700 ± 56	18 ± 4.2
Zircon–SiC	C	BN-Coated	357 ± 43	681 ± 36	25 ± 5.6

Table 2. A summary of interfacial properties in as-fabricated and annealed composites.

Reinforcement/ coating/matrix	Annealing treatment		Interfacial shear stress	
	Temp. (°C)	Time (h)	Debond (MPa)	Frictional (MPa)
AVCO SiC/none/zircon	As-fabricated		39.0 ± 4.0	16.0 ± 4.0
	1350	100	15.0 ± 3.0	10.0 ± 2.0
AVCO SiC/BN/zircon	As-fabricated		18.0 ± 2.0	15.0 ± 2.0
	1350	100	8.0 ± 3.0	3.5 ± 2.0

BN-coated fibers were tougher than the composites containing uncoated fibers. Table 1 gives a summary of the average mechanical properties such as critical stress for first matrix cracking (σ_{cr}), ultimate composite strength (σ_u), and work of fracture (WOF). The results indicate that there is an insignificant influence of the fiber-matrix interface on the critical stress for first matrix cracking and ultimate strength, but the WOF was higher for composites reinforced with BN-coated fibers in comparison to composites reinforced with as-supplied SiC fibers. Mechanical properties of the monolithic zircon were also measured as given in Table 1, but in all cases composites were significantly stronger and tougher than the monolithic zircon. A significant amount of fiber pullout was displayed by composite samples (see Fig. 5) which is consistent with the enhanced WOF.

A summary of average interfacial shear strengths for each of the two composites is given in Table 2. An average shear strength of 39 MPa was measured for the first push in a zircon composite reinforced with uncoated SiC fibers, and an average value of 16 MPa was measured for the second pushout case. In contrast, composites reinforced with BN-coated SiC fibers (interface B) produced lower values of the interfacial shear strength. An average shear strength of 18 MPa for the first push and 15 MPa for the second push were obtained. The lower interfacial strengths for BN-coated fibers in comparison to uncoated fibers are because of the different locations of interfacial sliding. The interfacial sliding in composites reinforced with uncoated SiC fibers was between the two carbon coatings on as-supplied SCS-6 silicon carbide fibers, whereas it was between the BN coating and the zircon matrix in composites reinforced with BN-coated fibers[13]. The interfacial shear strength measured for the first push was significantly higher than the second push value in composites containing uncoated fibers. In contrast, composites fabricated with BN-coated fibers not only produced lower values of the interfacial shear strength but the difference between the first push value and the second push value was insignificant. This would indicate that there is a small contribution of bonding or interlocking to the first push value of the interfacial shear strength in samples containing uncoated SiC fibers, whereas samples containing BN-coated fibers essentially show frictional coupling. A careful analysis of the load-deflection data obtained during pushout tests in terms of a pushout model indicated that the fiber-matrix interface in both of these composites is frictionally coupled[14], i.e., the contribution of bonding to interfacial strength is insignificant in zircon composites reinforced with either uncoated or BN-coated SiC fibers. The higher interfacial shear strength in composites with uncoated fibers can be attributed to interlocking because of surface roughness between the two as-deposited carbon layers on SiC fibers, which required higher pushout loads for initial interfacial sliding than in the subsequent pushout tests[13,14].

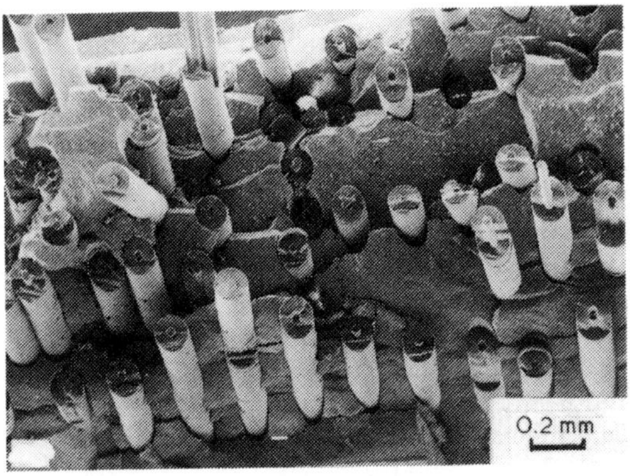

Fig. 5. Fractured cross-section of an as-fabricated zircon-SiC composite showing fiber pullout.

Table 3. Influence of annealing treatments on mechanical properties of zircon-SiC composites.

Reinforcement/ coating/matrix	Annealing treatment		Mechanical properties		
	Temp. (°C)	Time (h)	σ_{cr} (MPa)	σ_u (MPa)	WOF (kJ m^{-2})
None/none/zircon	As-fabricated		267	267	1.2
	1300	100	240	240	1.1
	1430	20	304	304	0.9
AVCO SiC/none/zircon	As-fabricated		273	635	24
	1200	100	218	675	31
	1300	100	285	740	49
	1350	100	240	901	50
	1430	20	207	328	12
AVCO SiC/BN/zircon	As-fabricated		395	645	26
	1200	100	240	647	40
	1300	100	210	478	22
	1350	100	203	436	29
	1430	20	218	305	10

Mechanical and Interfacial Properties of Annealed Composites

Internal thermochemical stability of the composite constituents at elevated temperatures was determined by annealing composites between 1200° and 1430° C in a flowing inert atmosphere for times up to 100 h and then measuring mechanical and interfacial properties at room temperature[13]. This study demonstrated excellent internal thermal stability of zircon-SiC composite constituents up to the annealing temperature of 1350° C for times up to 100 h as shown by Figs. 6 and 7 and Table 3. Composites fabricated with zircon matrix and SiC fibers were internally stable to 1350° C for an extended time. But, these composites displayed a significant amount of strength degradation when annealed at 1430° C for 20 h because of degradation in fiber properties caused by annealing at 1430° C although a toughened composite-like behavior was still maintained. The degradation in fiber properties caused by annealing at 1430° C is not unexpected because of the possible presence of free silicon in the as-fabricated SCS-6 SiC fibers[15]. To verify this hypothesis, uncoated SiC fibers were annealed between 1000° and 1430° C for 20 h in an inert atmosphere, and their strength was measured at room temperature. The dependence of fiber strength on annealing temperature is shown in Fig. 8. The as-fabricated fiber strength of about 3.9 GPa was maintained up to the annealing temperature of 1300° C. However, fibers annealed at 1430° C for 20 h produced a significant reduction in strength to a value of 2.5 GPa. These results, along with data on composites annealed up to 1350° C for 100 h, suggest that composites reinforced with SCS-6 SiC fibers may have an ultimate useful temperature limit of about 1350° C for long term applications.

Fiber-matrix interfacial properties, as characterized by interfacial shear strength measurements and determination of the sliding interfaces, are also indicative of internal thermal stability in zircon-SiC composites. Composites containing uncoated fibers in the zircon matrix showed a decreased interfacial shear strength after annealing at 1350° C for 100 h. But, this decrease was still not significant enough to influence mechanical properties. There was no change in the location of interfacial sliding because of the annealing treatment. Composites reinforced with BN-coated fibers produced lower values of the interfacial shear stress and displayed stable interfacial properties up to 1200° C, but samples annealed at 1350° C for 100 h produced much lower interfacial shear strength (see Table 2). This change was responsible for the lower strengths although composite toughness was maintained. These results indicated that although BN-coated fibers led to a more gradual drop in load beyond the point of ultimate load because of the lower interfacial shear strength, there was no other advantage in using BN coating to enhance thermal stability of zircon-SiC composites. Interface stability in this class of composites is expected to be good even beyond 1350° C because composites annealed at 1430° C for 20 h displayed toughened composite-like behavior although their strengths were reduced because of the degradation in fiber strength as discussed previously.

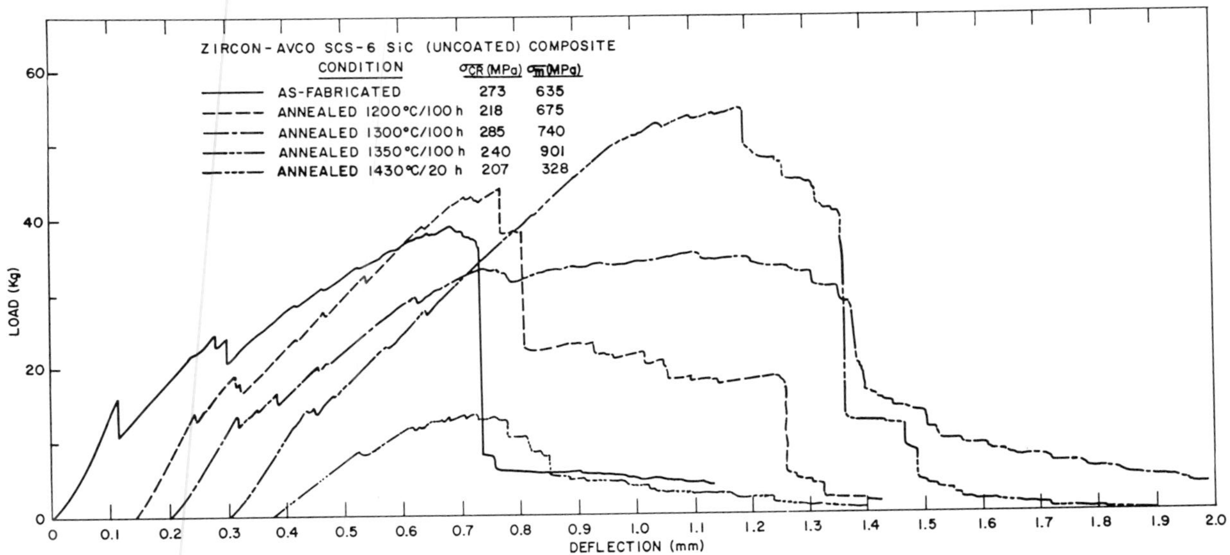

Fig. 6. Load-deflection behavior of an as-fabricated and annealed composites reinforced with uncoated SiC fibers.

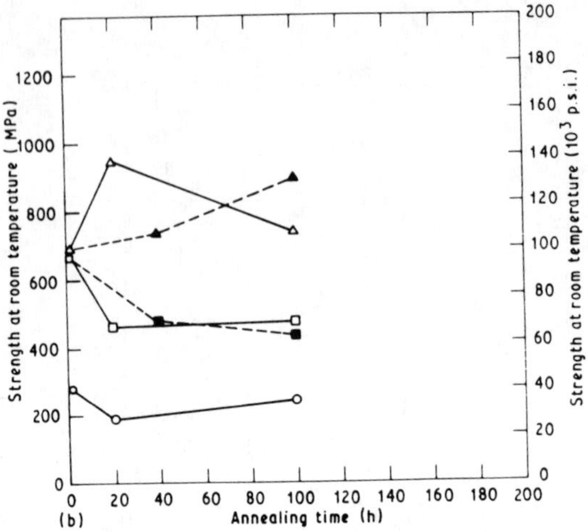

Fig. 7. Influence of annealing time on ultimate strength of (o) monolithic zircon and zircon-SiC composites with (△▲) uncoated and (□■) BN-coated fibers. Composites (▲■) annealed at 1350° C and (o△□) annealed at 1300° C.

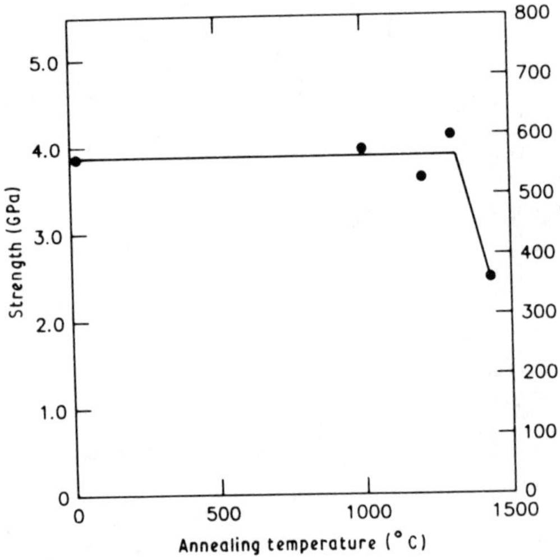

Fig. 8. Influence of annealing temperature on strength of SCS-6 SiC fibers.

Mechanical Properties at Elevated Temperatures

An attempt has been made to characterize high-temperature mechanical properties of zircon-SiC composites[16]. In order to accomplish this, short-term mechanical tests were performed on composite and monolithic samples between 25° and 1500° C in an inert atmosphere. Zircon composites reinforced with uncoated and BN-coated SiC fibers were also used in this investigation to study the influence of different fiber-matrix interfaces on high-temperature mechanical properties[16].

The load-deflection data were obtained for the monolithic zircon and zircon SiC composites at 25°, 1160°, 1315°, 1395°, and 1477° C. The results for the monolithic zircon samples are shown in Fig. 9. These results for the monolithic zircon indicate that a significant amount of plasticity is displayed by samples beyond about 1400° C, and essentially brittle behavior is observed below about 1300° C. In contrast to the brittle behavior of the monolithic zircon, the composite samples reinforced with as-supplied (uncoated) SiC fibers displayed a toughened composite-like behavior throughout the test-temperature range of 25° to 1477° C, as shown in Fig. 10. The load-displacement curve at 25° C shows an initial elastic portion, followed by a sudden load drop at the point of first-matrix cracking. Beyond this, there is an extended regime of increasing load-bearing capacity because most of the reinforcing fibers are still intact and can carry additional loads. A maximum in load-carrying capacity is displayed as the composite is further deformed because of the imposed motion of the crosshead. After this, there is a gradual decrease in load as more of the intact fibers begin to break and pullout from the matrix. The area under the load-displacement curve is a measure of the work-of-fracture (WOF), i.e., a measure of composite toughness, which is much higher for composite samples than for samples of the monolithic zircon. A similar behavior was displayed for composites reinforced with BN-coated fibers[16].

A number of composite characteristics, such as first-matrix cracking strength (σ_{cr}), composite strength and strain at the point of maximum load, Young's modulus (elastic modulus), and WOF, were calculated from the load-deflection data. The results for composites and monolithic zircon matrix are summarized in Table 4. The monolithic zircon matrix failed in a brittle manner up to the test temperature of 1315° C and its strength remained the same between 276 and 369 MPa for tests performed at 25°, 1160°, and 1315° C. There is a decrease in strength for tests performed at 1395° and 1477° C. The strain (based on the original dimensions) to failure monotonically increased from a value of 0.1% at the test temperature of 25° C to a value of 0.57% for testing at 1477° C. Similarly, the elastic modulus decreased from a value of 255 GPa at 25° C to a value of 48 GPa at 1395° C. These results for the monolithic zircon indicate that there is an insignificant change in strength up to the test temperature of about 1300° C. However, beyond the test temperature of about 1300° C the strength of

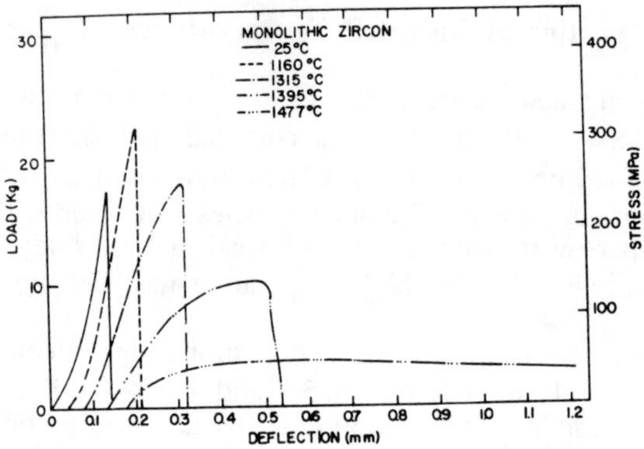

Fig. 9. Load-deflection behaviors for a monolithic zircon ceramic between 25° and 1477° C.

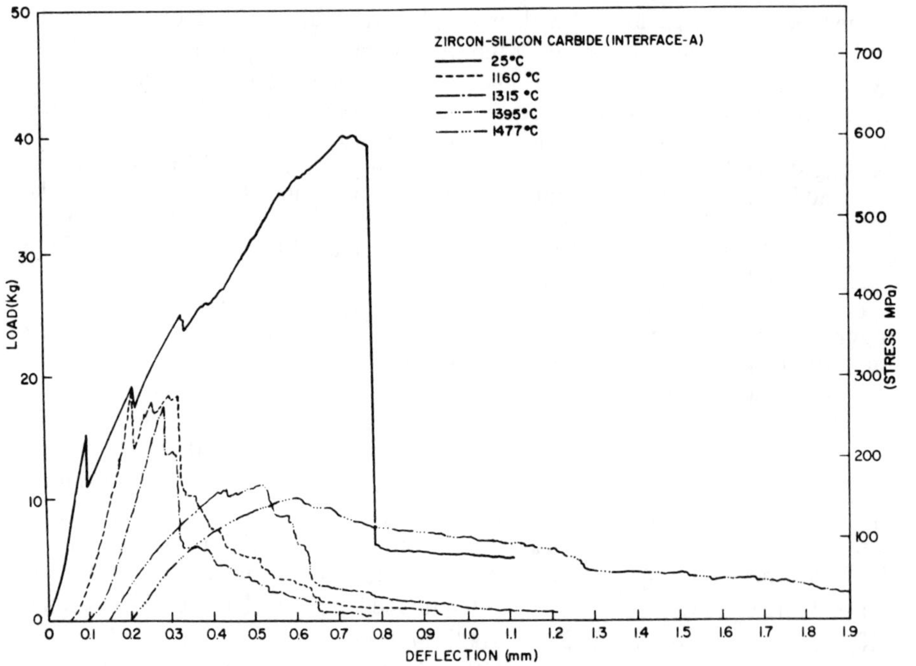

Fig. 10. Load-deflection behaviors for uniaxially reinforced zircon-matrix composites reinforced with uncoated SiC fibers between 25° and 1477° C.

Table 4. A summary of mechanical properties of the monolithic zircon and zircon-SiC composites at elevated temperatures.

Sample	Monolithic or composite (M or C)	Interface type (A or B)	Test temperature (°C)	Mechanical properties				
				σ_{cr} (MPa)	σ_u (MPa)	ε_u (%)	Elastic modulus (GPa)	WOF ($kJ\ m^{-2}$)
1-A[a]	M	None	25	—	281 ± 88	0.15 ± 0.02	195 ± 23	1.2 ± 0.1
2-A	M	None	1160	—	369	0.19	188	1.2
3-A	M	None	1315	—	276	0.35	111	1.7
4-A	M	None	1395	—	152	0.58	48	2.0
5-A	M	None	1477	—	65	0.57	—	—
1-B[a]	C	A	25	287 ± 28	700 ± 56	1.08 ± 0.13	244 ± 6	18 ± 4.2
2-B	C	A	1160	348	348	0.26	281	3.8
3-B	C	A	1315	262	262	0.24	108	2.3
4-B	C	A	1395	—	202	0.55	68	3.4
5-B	C	A	1477	—	188	0.59	57	—
1-C[a]	C	B	25	357 ± 43	681 ± 36	1.06 ± 0.03	235 ± 26	25 ± 5.6
2-C	C	B	1160	267	368	0.23	228	2.7
3-C	C	B	1315	—	315	0.31	147	2.9
4-C	C	B	1395	—	297	0.63	70	3.8
5-C	C	B	1477	—	154	0.95	43	—

[a]An average of four samples was tested at 25 °C.

the monolithic zircon is significantly reduced because of the plastic deformation. Similarly, the WOF varied between 1.2 and 2.0 kJ/m^2 over a temperature range of 25° to 1395° C.

Composites reinforced with uncoated SiC fibers produced first-matrix cracking strength between 287 and 348 MPa for tests performed between 25° and 1315° C. Beyond 1315° C, the first-matrix cracking strength was not calculated because of the plastic deformation and a lack of clear evidence for matrix cracking. The ultimate composite strength decreased significantly from a value of 700 to 348 MPa for tests performed at 25° and 1160° C, but then it gradually decreased from 262 to 188 MPa for tests performed between 1315° and 1477° C. The composite modulus remained high up to 1160° C and then monotonically decreased to 57 GPa at 1477° C. These composite strength and modulus values are higher than the corresponding values for the monolithic zircon at all testing temperatures between 25° and 1477° C, except that some of the composite strengths at 1160° and 1315° C are similar to those of the monolithic zircon. A similar behavior was displayed by composite samples reinforced with BN-coated SiC fibers as also summarized in Table 4.

The dependence of ultimate strength on temperature for the monolithic zircon and zircon-SiC composites is shown in Fig. 11. The composite at 25° C is much stronger than the monolithic zircon. At a test temperature of 1160° C, all of the samples, whether monolithic or composite, approach a similar strength value whereas at still higher test temperatures composite samples are stronger than the monolithic zircon. These results indicate that the maximum benefit of fiber reinfocement on mechanical properties of zircon matrix may be for enhancing strength at lower temperatures (< 1100° C) and for enhancing strength and possibly creep resistance at very high temperatures (> 1300° C). The extent of improvement in high-temperature mechanical properties may depend on high-temperature strength and creep resistance of the SiC reinforcing fibers.

An attempt was made to calculate composite strength as a function of temperature based on an approach in which composite strength was taken to be equal to the fiber strength times the volume fraction of fibers. In this approach, data for the temperature dependence of fiber strength were taken from the literature[17]. The calculation was done for a zircon composite reinforced with 25%(vol.) silicon carbide fibers (Fig.11). All of the calculated strengths are slightly higher than the measured strengths at lower temperatures. However, the results at higher temperatures are very close to the predicted values. Such an agreement can be considered good, keeping in mind that the actual fiber strength in composites may be lower than the values used in the calculations because of the processing-induced damage. A number of other factors such as statistical distribution of fiber strength and influence of plastic deformation at elevated temperature may also have influenced the strength values, which were ignored in this simple-minded calculation.

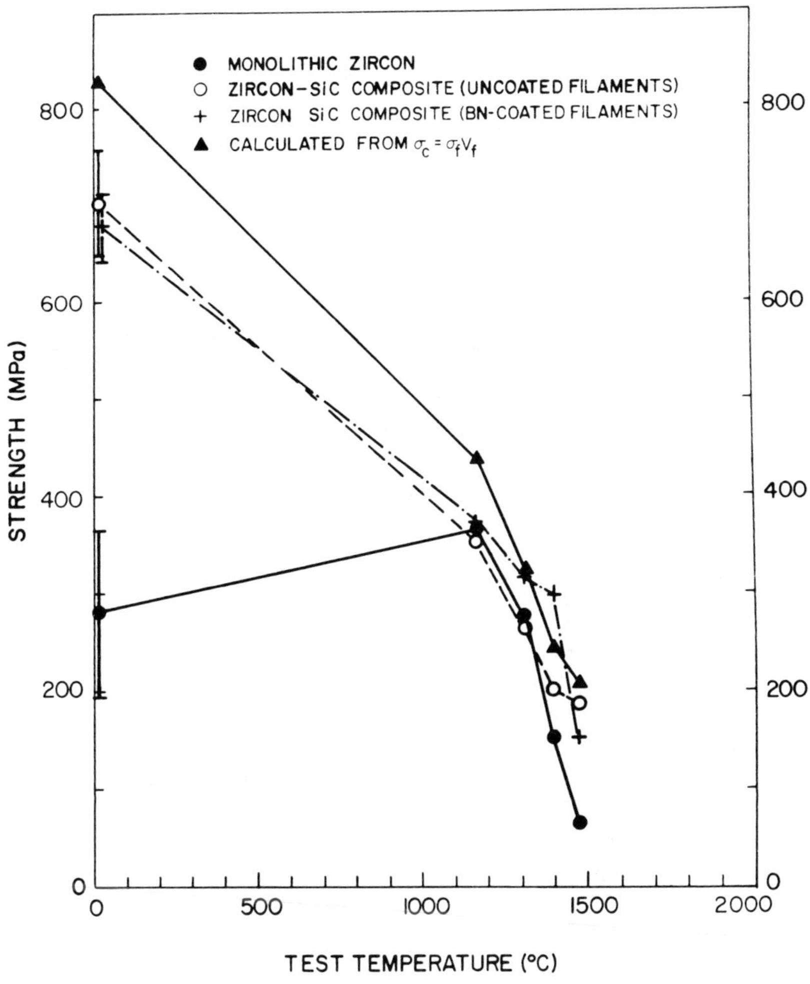

Fig. 11. The dependence of strength on temperature for a monolithic zircon and zircon-SiC composites. Also shown is the calculated dependence of composite strength on temperature.

Fracture Morphology

The failed cross sections of the monolithic zircon and composite specimens tested between 25° and 1477° C were examined by SEM to determine the fracture morphology and the extent of fiber pullout. The results for the monolithic zircon are shown in Fig. 12. The fracture at 25° and 1160° C is primarily transgranular, as shown in Figs. 12a. An increasing amount of the intergranular fracture is evident in a sample fractured at 1315° C, but the transgranular fracture is still present. A predominantly intergranular fracture is observed for the sample tested at 1395° C, as shown in Fig. 12b. These observations are consistent with the load-displacement data for the monolithic zircon as a function of temperature; i.e., essentially elastic behavior was observed for samples tested up to 1160° C, and an increasing evidence of plasticity was seen in specimens tested above about 1200° C. It appears that plasticity in zircon samples results from intergranular motion or grain boundary sliding. The grain boundary sliding also appears to be a major contributor to extensive plasticity of the zircon sample tested at 1477° C although this sample was not examined by SEM because the load-deflection test was stopped before the sample fractured.

The WOF, which is a measure of composite toughness and fiber pullout, progressively decreased as the samples were tested at increasingly higher temperatures (Table 4, Fig. 13). In contrast, the WOF for monolithic zircon was fairly constant between 25° and 1395° C. The extent of fiber pullout was qualitatively assessed by examining the fractured cross section of composites by SEM. The results for zircon composites reinforced with uncoated SiC fibers, which were fractured between 25° and 1395° C, are shown in Fig. 14. A significant amount of fiber pullout is evident for the sample tested at 25° C (Fig. 14a). Somewhat less pullout is seen for a sample tested at 1160° C. Composites tested at 1315° and 1395° C show even less fiber pullout (Fig. 4b). A similar behavior was displayed by composites reinforced with BN-coated SiC fibers; i.e., the extent of fiber pullout decreased with increasing test temperature. These results on fiber pullout are in agreement with the lower WOF values for composites tested at the higher temperatures. In addition, a significant amount of matrix cracking was observed between fibers for tests performed at the lower temperatures (<1160° C). This type of interfiber matrix cracking pregressively decreased in tests performed at the higher temperatures (> 1200° C). This change in the fracture morphology appears to be because of the stress/strain accommodation by plastic deformation at the higher test temperature.

(A)

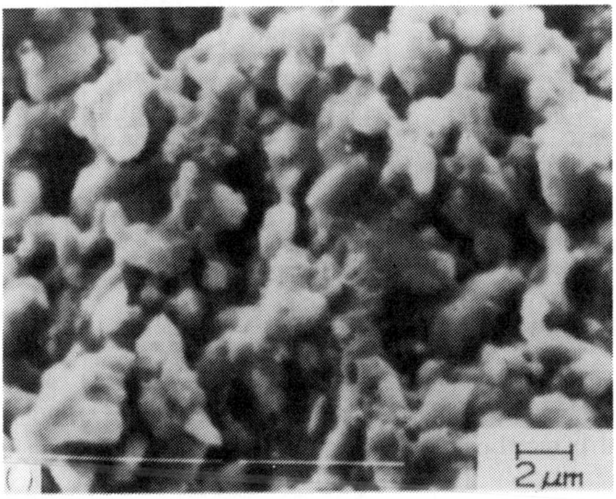

(B)

Fig. 12. SEM fractographs showing fracture morphology for a monolithic zircon ceramic tested at (a) 25° C, and (b) 1395° C.

Morphological stability of the fiber-matrix interface as a function of the test temperature was studied by examining interfacial regions of fractured samples by SEM. Such an examination can also provide information about changes in matrix characteristics in the vicinity of the reinforcing fiber because of the high-temperature testing. The fiber-matrix interface, comprising of two layers of the carbon-silicon coatings in contact with the zircon matrix, appears intact at all of the test temperatures. The outer of these two coatings is strongly bonded to the zircon matrix which results in interfacial sliding between the two carbon layers[13]. The zircon matrix in the vicinity of the fiber-matrix interface shows different fracture morphologies as a function of the test temperature. A transgranular fracture is seen in the zircon matrix for sample tested at 25° and 1160° C, respectively. In contrast, the samples fractured at 1315° and 1395° C show a significant amount of intergranular fracture. The monolithic zircon samples also displayed a similar fracture morphology as a function of the test temperature, which suggests that the presence of reinforcing fibers did not influence matrix fracture morphology in composite samples. The composite samples containing BN-coated SiC fibers showed a similar matrix morphological behavior as a function of the test temperature, i.e., essentially transgranular fracture between 25° and 1160° C, and then increasing amount of the intergranular fracture at 1315° and 1395° C. The fiber-matrix interface remained stable up to the testing temperature of 1395° C. However, the interfacial sliding in composites reinforced with BN-coated fibers was between the zircon matrix and the BN coating[13].

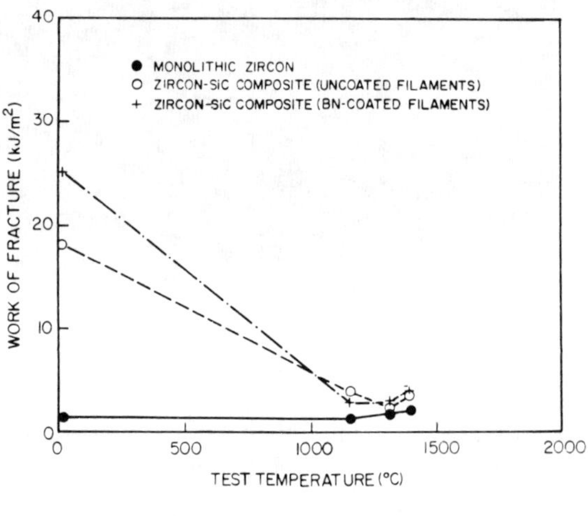

Fig. 13. The dependence of work-of-fracture (WOF) on temperature for a monolithic zircon and zircon SiC composites.

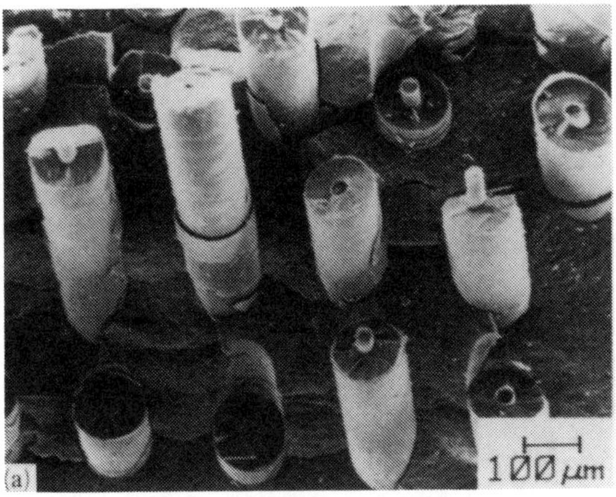

(a)

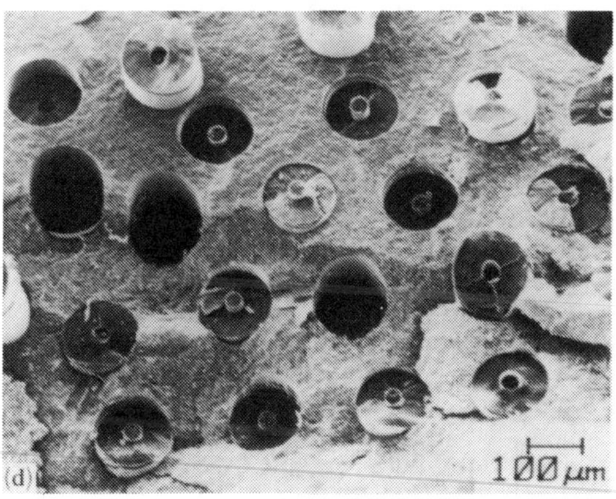

(b)

Fig. 14. SEM fractographs showing fracture morphology for zircon composites reinforced with uncoated SiC fibers and tested at (a) 25° C, and (b) 1395° C.

SUMMARY

Mechanical properties of the monolithic zircon ceramics and zircon-matrix composites uniaxially reinforced with either uncoated or BN-coated silicon carbide fibers were measured in a flexure mode at temperatures between 25° and 1477° C. Monolithic zircon ceramics were weak and exhibited a brittle failure up to the test temperature of about 1300° C. An increasing amount of the plastic deformation was observed before failure above about 1300° C. In contrast, composites reinforced with either uncoated or BN-coated SiC fibers were stronger and tougher than the monolithic zircon at all test temperatures between 25° and 1477° C, except that some of the composite strengths at 1160° and 1315° C were similar to those of the monolithic zircon. The ultimate strength, elastic modulus, and WOF of composites decreased with increasing temperature. A transgranular matrix fracture was shown by the monolithic and composite samples tested up to about 1200° C, whereas an increasing amount of the intergranular matrix fracture was displayed by specimens tested beyond the 1200° C. Some of these results were consistent with the theoretical considerations, but in most cases a sufficient amount of experimental data were not available to make quantitative predictions. Thermal stability of zircon composites uniaxially reinforced with SCS-6 SiC fibers was demonstrated up to a temperature of 1350° C for 100 h. These results clearly demonstrated the usefulness of our approach in designing composites for applications to elevated temperatures.

Acknowledgment: This work was supported by a grant from the Department of Energy-Basic Energy Science (#DE-FG02-91ER45459).

References

1. R. N. Singh and A. R. Gaddipati, " Mechanical Properties of a Uniaxially Reinforced Mullite-Silicon Carbide Composite," J. Am. Ceram. Soc., 71, C-100 (1988).
2. R.N. Singh, " Interfacial Properties and High-Temperature Mechanical Behavior of Fiber-Reinforced Ceramic Composites," Mater. Sci. and Eng., A 166, 185-198 (1993).
3. E. Fitzer and R. Gadow, " Fiber-Reinforced Ceramic Composites Fabricated via the Sol/Gel Route," Conference on Tailoring Multiphase and Composite Ceramics, Penn. State University, July (1985).
4. R. Naslain, in tailoring Multiphase and Composite Ceramics, Materials Science Research, Vol. 20, Plenum Press, New York, 145-164 (1986); Ed: R.E. Tressler, G.L. Messing, and G.C. Pantano.
5. K.L. Luthra, R.N. Singh, and M.K. Brun, " Toughened Silcomp Composites-Process and Preliminary Properties," Am. Ceram. Soc. Bull., 72, 79-85 (1993).
6. D.P. Stinton, A.J. Caputo, and R.A. Lowden, " Synthesis of Fiber-Reinforced SiC Composites by Chemical Vapor Deposition, " Am. Ceram. Soc. Bull., 65[2], 347-50 (1986).

7. M.S. Newkirk, H.D. Lesher, D.R. White, C.R. Kennedy, A.W. Urquhart, and T.D. Claar, " Preparation of Lanxide Ceramic Matrix Composites: Matrix Formation by the Directed Oxidation of Molten Metals," Ceram. Eng. Sci. Proc., 8[7-8], 879-85 (1987).
8. R. N. Singh, High-Temperature Mechanical Properties of a Uniaxially Reinforced Zircon-Silicon Carbide Composite," J. Am. Ceram. Soc., 73[8], 2399-2406 (1990).
9. R. N. Singh and A. R. Gaddipati, " A Uniaxially Reinforced Zircon-SiC Composite," J. Mater. Sci., 26, 957-962(1991).
10. R. N. Singh, " LPCVD of Boron Nitride from β-Trichloroborazine", in Tenth International Conference on Chemical Vapor Deposition (CVD-X), V. 87-7, p. 543 (1987), The Electrochem. Soc., Ed. G.W. Cullen.
11. M. Brun and R. N. Singh, " Effect of Thermal Expansion Mismatch and Fiber Coatings on the Fiber/Matrix Interfacial Shear Stress in CMCs," Advanced Ceramic Materials, 3[5], 506 (1988).
12. P. D. Jero, R. J. Kerans, and T. A. Parthasarathy, " Effect of Interfacial Roughness on the Frictional Stress Measured Using Pushout Tests," J. Am. Ceram. Soc., 74[11], 2793-801(1991).
13. R. N. Singh, " Influence of High-Temperature Exposure on Mechanical Properties of Zircon-Silicon Carbide Composites,"J. Mater. Sci., 26, 117-126 (1991).
14. R. N. Singh and M. Sutcu, " Determination of Fiber-Matrix Interfacial Properties in Ceramic Matrix Composites by a Fiber Pushout Technique," J. Mater. Sci., 26, 2547-2556 (1991).
15 M.K. Brun and M. Borom, " Thermomechanical Properties of Chemically Vapor Deposited Silicon Carbide Filaments," J. Am. Ceram. Soc., 72[10], 1993-96 (1989).
16. R. N. Singh, " Influence of Interfacial Shear Stress on First Matrix Cracking Stress in Ceramic Matrix Composites," J. Amer. Ceram. Soc., 73[10], 2930-37 (1990).
17. J. A. DiCarlo, " High Temperature Properties Of CVD Silicon Carbide Fibers", Paper Presented at Int. Conf. on Whisker- and Fiber-Toughened Ceramics, Oak Ridge, TN, June 7-9, 1988.

MICROSTRUCTURE AND MECHANICAL PROPERTIES OF POLYMER-DERIVED Si_3N_4/SiC FABRIC COMPOSITE

Y.-M. Pan, D.L. Davidson, S.T. Schwab, and R.C. Graef
Materials Engineering and Technology Division
Southwest Research Institute
San Antonio, Texas 78228

ABSTRACT

SiC fiber-reinforced Si_3N_4 matrix composites were fabricated through the utilization of preceramic polysilazanes and polymer infiltration/pyrolysis processing techniques. Microstructures and mechanical properties (modulus of rupture, compressive strength, and fracture toughness) were characterized. The K_{IC} and work of fracture, measured by the chevron notch method, were 15.6 MPa $m^{1/2}$ and 11.8 KJ/m^2, which compare favorably to those for hot-pressed, SCS-6 reinforced Si_3N_4. The toughening of these polymer-derived Si_3N_4 composites is associated with the substantial nonlinear behavior illustrated in the dimensionless load vs. load-point displacement plot. This is consistent with the rising R-curve behavior. The compressive strength of approximately 500 MPa was found to be independent of temperature up to 800°C; however, the linearity of the σ-ε curves does reflect the test temperature. Micromechanics experiments suggest that the high level of fracture toughness may be attributed to pseudo-plastic strain response of the composite material near the crack tip.

INTRODUCTION

The utilization of ceramics in structural applications has been seriously hampered by their brittle fracture behavior, high-flaw sensitivity, and resulting low-tensile strength.[1] Incorporating continuous fibers into a ceramic matrix provides a toughening mechanism that changes the failure mode of ceramics and lessens the propensity for catastrophic failure.[2] Considerable effort has been devoted to the development of effective methods for manufacturing ceramic matrix composites (CMCs), and several approaches with the potential for industrial utilization have been identified.[3] While chemical vapor infiltration (CVI) technology is currently viewed as the industry leader, it is a slow, complex process with many inherent difficulties, including corrosive gas environment, high cost of process tooling and substantial (15- 25%) residual porosity.[4] Polymer infiltration/pyrolysis (PIP) processing is a relatively new approach that is essentially a modification of those techniques currently used to manufacture organic-matrix and carbon-carbon composites.[3]

To the extent authorized under the laws of the United States of America, all copyright interests in this publication are the property of The American Ceramic Society. Any duplication, reproduction, or republication of this publication or any part thereof, without the express written consent of The American Ceramic Society or fee paid to the Copyright Clearance Center, is prohibited.

Because of its excellent strength at elevated temperature, resistance to thermal shock, and resistance to oxidation, silicon nitride is being considered for a wide variety of high temperature applications.[5] A particularly useful family of polymeric precursors to Si_3N_4 has been developed at the Southwest Research Institute (SwRI).[6,7] These polysilazanes have been shown to be effective matrix precursors for the PIP processing of fiber-reinforced Si_3N_4 composites.[8,9] This paper reports on the mechanical properties and fracture behavior of SiC fiber-reinforced Si_3N_4 composites, and describes the PIP method and the resultant composite microstructure as well.

COMPOSITE FABRICATION

Polymer infiltration/pyrolysis processing was employed to fabricate SiC fiber-reinforced Si_3N_4 matrix composites using the preceramic polysilazanes developed at SwRI as matrix precursors.[6,7] These polymers, known as perhydropolysilazanes (PHPS), contain only the elements silicon, nitrogen, and hydrogen, and so convert to carbon-free silicon nitride when fired. These polymers are isolated as waxes or low-viscosity, thermosetting liquids that exhibit ceramic yields of 80% by weight or greater. All manipulations of uncured polysilazanes were carried out under anhydrous and anaerobic conditions in an inert atmosphere/vacuum manifold system or an argon-filled drybox.[10] Graphite-coated Nicalon fabric (Ceramic Grade, 8-harness satin weave, carbon coated with PVA sizing) was purchased from Dow Corning. PVA sizing was removed by soaking the fabric in hot, deionized water, rinsing with acetone, and drying at 180°C for 24 hr.

Nicalon fabric was saturated with PHPS by simply pouring the neat polymer (no solvent) over pre-cut plies. The prepregged plys were stacked, placed in a double-sided vacuum bag, and cured in an autoclave. The curing process conditions for optimum fabric infiltration were determined from rheological studies.[8] Pyrolyses were carried out in a Lindberg 1700°C horizontal tube furnace equipped with a 3-in O.D. alumina process tube. The composite was removed from the bag and transferred to the tube furnace, where it was heated to 1000°C under flowing anhydrous nitrogen. The maximum firing temperature was limited to 1000°C because the Nicalon material is known to degrade at high temperatures. The resultant polymer-derived matrix determined by x-ray diffraction analysis is essentially amorphous Si_3N_4. Reinfiltration was accomplished in a vacuum bag with a layer of polymer-saturated bleeder cloth surrounding the composite. The curing and firing procedures were the same as that described for the initial lay-up. A total of eight reinfiltration cycles were carried out to yield a composite with a density of roughly 2.4 g/cm^3 and an apparent porosity of about 5.6 %, measuring approximately 150 mm x 60 mm x 3 mm. Details of the composite fabrication procedure and the crystallization behavior of polymer-derived Si_3N_4 have been described previously.[8,9]

MICROSTRUCTURE CHARACTERIZATION

The microstructure of the PIP composite was characterized using scanning electron microscopy. The microstructure of the polished cross-section of the composite is presented in Figure 1. As seen in Figure 1a, the infiltration procedure

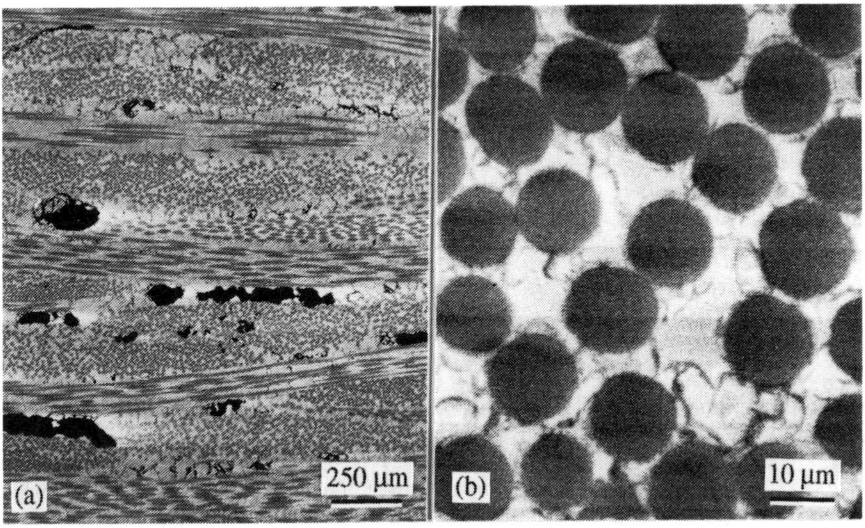

Figure 1. Cross-sectional view (a) and an enlarged area (b) of fiber bundles of polymer-derived Si_3N_4/SiC fabric composite.

appears to penetrate the fiber bundles effectively, although sizable voids are observed outside the fiber bundles. Higher magnification examination reveals the dense nature of the matrix within the fiber bundles and the presence of microcracks and small scars in the matrix around the fibers (Figure 1b). The microcracks are likely the result of the difference in thermal expansion between the SiC fiber and the Si_3N_4 matrix, while the small scars appear to be the result of differential polishing. Because dense matrix is observed inside the fiber bundles and void spaces are observed outside the fiber bundles, it seems likely that the polysilazane wets the individual fibers in each tow quite effectively, and that the polymer tends to shrink around the fiber (rather than away from the fiber) during pyrolysis. The preference for wetting and adhering to the fibers appears to continue through each stage of the densification process. In contrast to CVI method in which the fiber bundles are often sealed off before substantial matrix can be deposited within them, the matrix appears to grow outward from within each fiber bundle in the PIP technique.

The fiber/matrix interface is known to play a key role in the determination of composite properties. Auger analysis was performed on the fracture surface of the composite by centering the electron beam on either a Nicalon fiber or a matrix trough from which a fiber has been pulled out. The Auger concentration vs. sputtering time profile for the fiber of the composite indicates the presence of a carbon-rich layer of approximately 580 Å on the fiber surface.[8] The depth profile for the matrix trough also indicates a significant thickness of the carbon surface layer on the matrix trough. The carbon layer on the matrix trough is clearly derived from the fiber. It is apparent that this carbon-rich interfacial zone affects the mechanical properties of the polymer-derived composite, and perhaps promotes the observed fiber pull-out.

MECHANICAL PROPERTY EVALUATION

A variety of different test specimens were extracted from the fabricated composite for mechanical property evaluation at ambient and elevated temperatures. The fracture properties of the composite, as well as the strength and modulus, were determined through tests of notched and unnotched specimens. While fracture toughness was evaluated using both three-point chevron notch and compact tension specimens, strength and modulus measurements were made through flexural and compressive measurements. Fracture processes of this composite were also characterized using a dimensionless load-displacement method complemented by micromechanics analysis.

Flexural Strength

The flexural strength (modulus of rupture, MOR) measurements were made of the composite for the flatwise orientation where the crack plane across the fabric layers with the crack growth direction parallel to the molding direction. The MOR specimens were cut into bars of approximately dimensions 2.8x10x56 mm and tested in 4-point bending at room temperature with a cross-head speed of 0.02 mm/sec. The inner and outer loading points in these tests were separated by 22 mm and 44 mm. Flexural strength and elastic modulus values were determined from the load vs. displacement curves.

The average flexural strength and elastic modulus values were measured to be 207 MPa and 65 GPa, respectively. The measured flexural strength for the polymer-derived Si_3N_4 composite is much lower than that reported for hot-pressed, SCS-6 reinforced Si_3N_4 (500-600 MPa).[11] This low level of flexural strength appears to be due to the amorphous Si_3N_4 matrix and 5.6 % porosity. The stress vs. displacement curves of the PIP composite presented in Figure 2 shows a general feature that an initial linear elastic region is followed by a nonlinear load increase region to a maximum, and then a broad continuous load decrease region. This substantial nonlinear behavior that results from a non-brittle failure mode suggests that the PIP composite has a considerably high toughness. This agrees with the extensive fiber pull-out observed in the fracture surfaces.

Compressive Strength

Compression testing was performed with the compression axis parallel to the fabric layers. Samples were machined from the fabricated composite in the form of right circular cylinders (7.4 mm long x 2.9 mm diameter). To confine failure to the center section, small ring caps made of high-strength steel were honed to just fit over the ends of each sample. Compressive strength was measured in air at a strain rate of 10^{-4}/s and at room temperature and at 800°C. The average ultimate compressive strength was found to be approximately 496 MPa at room temperature and approximately 477 MPa at 800°C. It is apparent that the compressive strength of the polymer-derived ceramic matrix composite is independent of temperature up to 800°C; however, the linearity of the compressive stress (σ) - strain (ε) curves for the PIP composite reflects the effect of test temperature. As shown in Figure 3, while

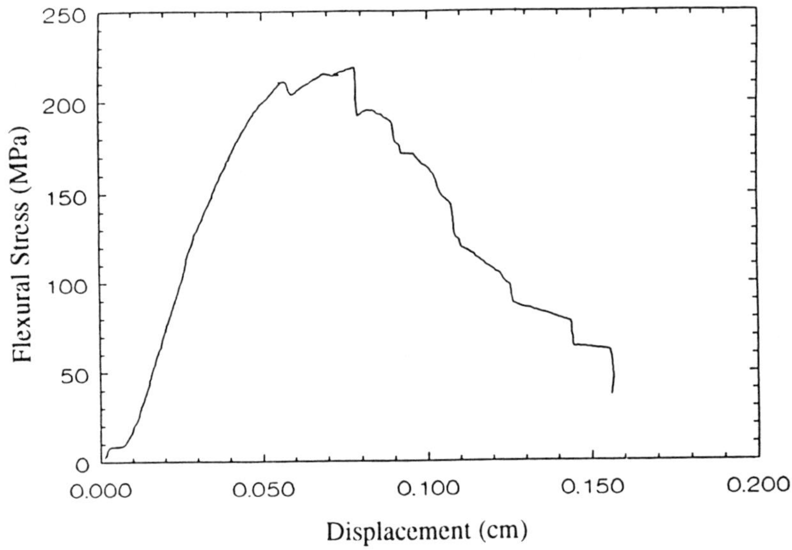

Figure 2. Flexural stress vs. displacement curve.

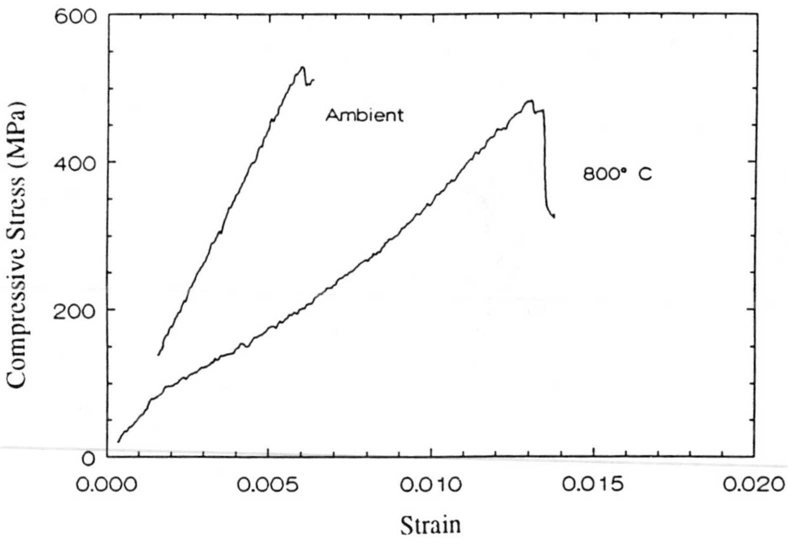

Figure 3. Compressive stress vs. strain curves tested at ambient temperature and 800°C.

the σ-ε curves of the room temperature tests were linear to failure, the 800°C tests yielded (i.e., the σ-ε curve deviated from its elastic loading line) well below the ultimate strength, possible due to softening of the amorphous Si_3N_4 matrix.

The damage introduced by compression loading in the polymer-derived ceramic matrix composite at room temperature is presented in Figure 4, and at 800°C temperature in Figure 5. It is apparent, as seen in Figure 4a & 5a, that the basic failure mechanisms are similar for both temperature tests as a result of the fabric confinement; however, the 800°C test results in much larger internal expansion and cracking. Inspection of the damage region in the cross-sections at a higher magnification shows that, while the damage mode of the room temperature specimens is associated with matrix microfracture and fiber microkinking (Figure 4b), the 800°C specimens have undergone extensive matrix cracking and have developed pulverized damage zones (Figure 5b). The high temperature failure is obviously more catastrophic, which may be associated with microstructure changes that occurs at the test temperature. Lankford[12] investigated compressive failure in fiber-reinforced ceramic matrix composites over a wide strain rates and showed that the compressive failure mechanism of matrix microfracture followed by gross failure through localized fiber kinking has been observed in laminated SiC-LAS-II composites at low strain rates.

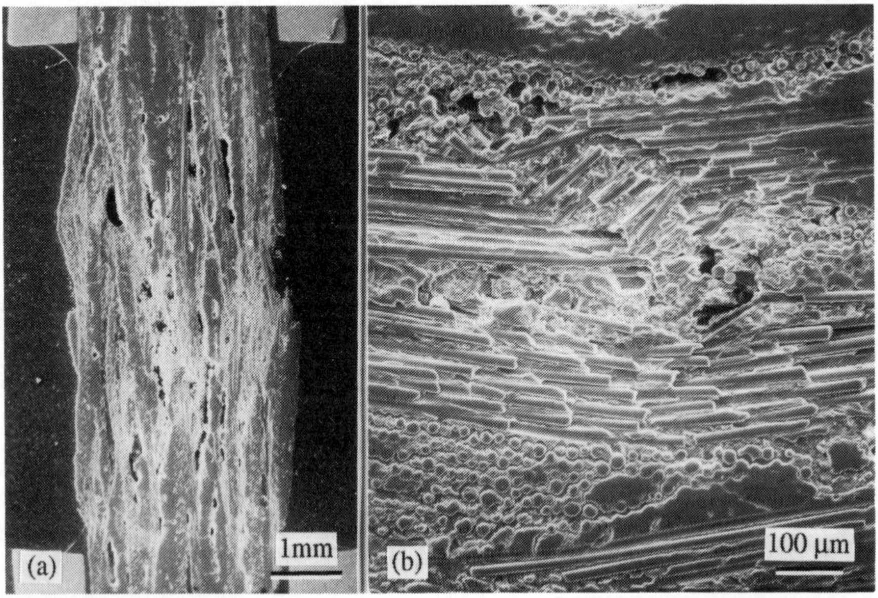

Figure 4. Cross-sectional view of failed room temperature compressive specimen (a) and an enlarged area (b) showing matrix microfracture and kinked fibers.

Figure 5. Cross-sectional view of failed 800°C compressive specimen (a) and an enlarged area (b) showing extensive matrix cracking and pulverized damage zones.

Fracture Toughness

Fracture toughness measurements were made in the edgewise orientation where the crack propagates across the fabric layers with the crack growth direction orthogonal to the fabric layers, approaching the fabric layers on edge. The dimensions of the fabricated composite and the dimensions of the 3-point bend specimens with chevron notch (CHN), as well as the compact tension specimens, required this orientation, which, fortunately, is also the most likely orientation for the high stress applications of this material. For the chevron notch specimens, the notches were cut using a 0.3-mm thick diamond blade. The dimensionless crack length ($\alpha_0=a_0/W$) was around 0.3 for accuracy of the stress intensity calculations.[13] Fracture toughness of the chevron-notched specimens was calculated from the expression, $K_{IC}(CHN) = PmaxYmin/(BW^{1/2})$, where Pmax is the maximum fracture load of the composite specimen during steady state crack propagation and Ymin is the minimum value of the shape factor.[14] Work-of-fracture, an energy term of toughness, for the chevron notch specimens was determined by integrating the areas under the load-displacement curves and dividing by the total projected fracture surface areas.

The K_{IC} and Work of Fracture values obtained for the PIP composite are 15.6 MPa $m^{1/2}$ and 11.8 KJ/m^2, which are well above that reported for monolithic Si_3N_4 (1.5-8.2 MPa $m^{1/2}$). Kodama, et al.[11] and Yang, et al.[15] have reported room temperature fracture toughness data for hot-pressed, SCS-6 reinforced Si_3N_4. The K_{IC} values obtained for the polymer-derived Si_3N_4 composite are substantially higher than those values reported by Kodama, et al. (7-13 MPa $m^{1/2}$), but below those values reported by Yang, et al. (18.5-38.2 MPa $m^{1/2}$). Nonetheless, the Work of Fracture values obtained for the PIP composite fall into the middle of the range reported by Yang, et al. (2.7-27.3 KJ/m^2). This fracture toughness data, which was measured by the chevron notch method, is not expected to be a material constant because of the possible existence of large-scale frontal process zones and the existence of complicated fiber pull-out and bridging fracture processes in the developing crack wake.

To complement the fracture studies made with the chevron notch bend specimen, a compact tension specimen (outer dimensions 19x19.8 mm by 2.8 mm thick) was tested at ambient temperature. The specimen was precracked using compression-compression cyclic loading, and tested using a SEM loading stage that enables observation and photography of the crack region while loads are applied. Fracture occurred at approximately 15 MPa $m^{1/2}$. This value agrees with toughness values obtained from the chevron notch specimens.

Nonlinear Fracture Behavior

Because of the complicated fracture processes of these polymer-derived materials, the usual toughness, K_{IC}, is not adequate to characterize the nonlinear fracture process. Crack propagation in ceramic composites has been studied previously on the basis of energy principles for which graphical methods have been outlined to determine the nonlinear fracture resistance parameters.[16] The dimensionless load-displacement method has been applied in the present study for the determination of the crack-growth resistance (Rc). Details of this method are presented elsewhere.[16]

The representative dimensionless load (Pr) vs. displacement (Ur) plot is presented for the PIP composite in Figure 6. As seen in Figure 6, the experimental (Pr-Ur) curve is in excellent agreement with the theoretical linear elastic curve during the initial stages of testing; however, substantial non-linear behavior dominates the later stages of crack growth. This is consistent with the rising R-curve behavior expressed by the crack growth resistance vs crack length, which was determined from the deviation of the experimentally observed curve from the theoretical curve using the graphical technique. The rapid increase of Rc values (from 6 KJ/m^2 at α=0.45 to 20 KJ/m^2 at α=0.9) during crack extension into the latter part of the specimen ligament is expected for the composites because the specimens never actually broke into two distinct pieces, but were always held together by the partially pulled out fibers bridging the wake region. The high toughness values and the rapidly increasing Rc values are, therefore, directly related to the fiber strengths, their partial pull out and the integrity retained by the composite during the fracture process.

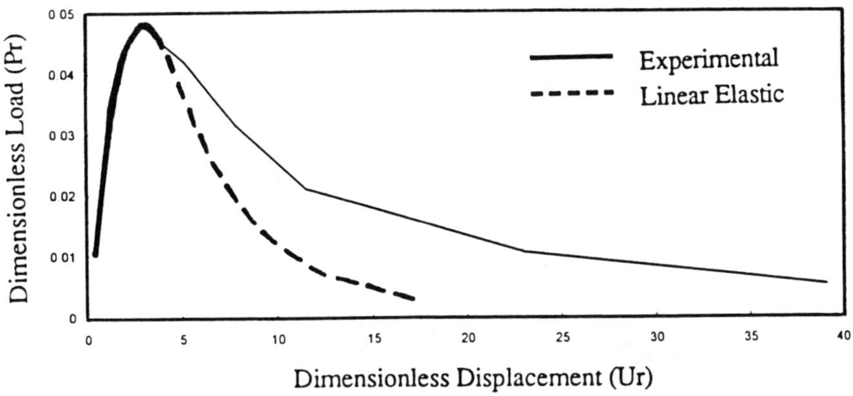

Figure 6 The dimensionless load vs. displacement plot showing nonlinear fracture behavior.

Micromechanics Analysis

The precracked compact tension specimen was loaded to failure using the SEM loading stage. Photographs were made as load increased. Displacements were measured using a machine vision implementation of the stereoimaging technique called DISMAP.[17] As increasing load was applied, the crack opened in the same manner as has been observed for metallic specimens, except that the crack opening displacement did not vary with distance from the crack tip in the same way. Above a stress intensity factor of approximately 6 MPa $m^{1/2}$, the crack grew in a stable manner until final fracture occurred at approximately 15 MPa $m^{1/2}$. Fiber matrix bonds were unbroken until the crack began to grow, when debonding occurred and some fibers broke as the crack passed, leaving other fibers unbroken. Ahead of the torn crack tip, fibers were still well bonded to the matrix.

As the compact tension specimen loaded to K=5.9 MPa $m^{1/2}$, an analysis of the near crack tip region is shown in Figure 7 where the crack and fibers ahead and behind the tip are seen. Measured displacements are shown overlaid on the photograph (Figure 7a). Measured crack opening displacements (COD) (Figure 7b) indicate that the crack tip was blunt; for metallic specimens, COD vs. square root of the distance from the crack tip usually gives a straight line relationship except when large scale plasticity occurs in the crack tip deformation zone. For this composite, the functional dependence shown indicates that some sort of pseudo plasticity is occurring ahead of the crack tip. The distribution of maximum shear strain surrounding the crack tip also indicates that strain is dispersed near the crack tip. For metallic materials, a large value of strain would have occurred immediately at the crack tip. Thus, the response of the near-crack tip material in this composite is indicative of a plasticity-like response even though debonding was not observed between fibers and matrix. A material response similar to that observed for stable crack growth also existed near the point of rapid crack growth (fracture).[9] With

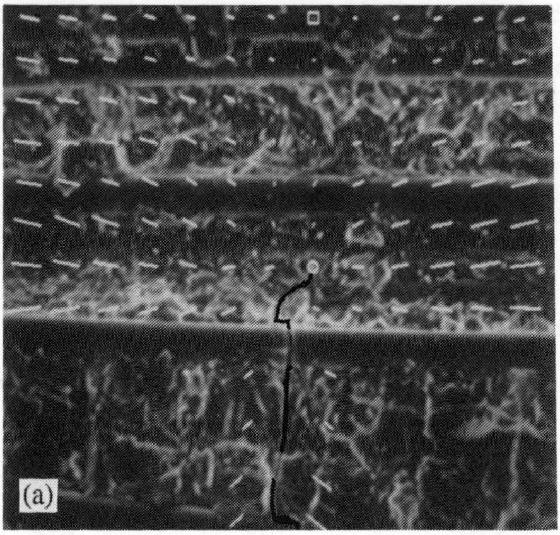

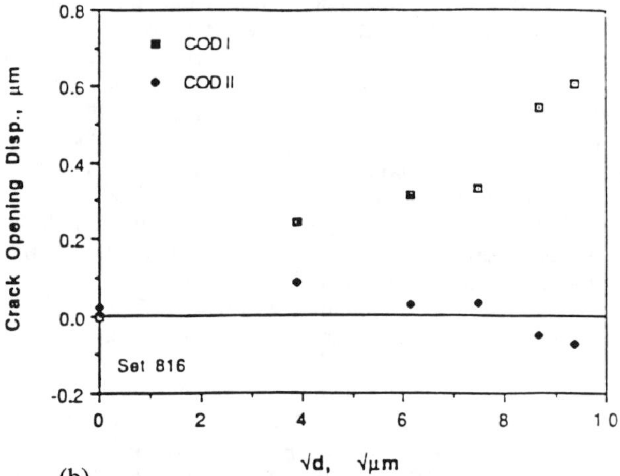

(b)

Figure 7. Micromechanics analysis of polymer-derived Si_3N_4/SiC composite compact tension specimen loaded to K=5.9 MPa $m^{1/2}$ showing (a) micrograph of the crack tip region and (b) COD for the crack in (a). Displacements shown on micrograph were measured each 15 µm, and the displacement magnitude are 20 times actual value.

increasing load, the crack grew about 350 μm and blunted even more. Also, several unbroken fibers are left in the wake of the crack. The strain distribution remained disbursed, although the level of strain increased.

Strains were measured in several fibers for load levels approaching fracture. Strains were then converted to stress using the Young's modulus of Nicalon fibers. At lower loads, these fibers were ahead of the crack tip, but at K=13.8 MPa $m^{1/2}$ the fibers were 260 to 270 μm behind the crack tip. At low stress intensity factors, the distances between the crack tip and fibers were 60, 75 and 90 μm, respectively for fibers 1, 2 and 3. Fiber stresses for fibers 1, 2 and 3 were measured to be 890, 620 and 510 MPa, respectively, which is a mild function of distance from the crack tip at low K levels. After the crack has passed, however, the stress in fiber 1 is approaching its fracture strength (~3,660 MPa) while the stress level in the other two fibers remains relatively low (~1,000 MPa). The reasons for this behavior are not currently known.

From this micromechanics analysis, we may conclude that: (1) strain is disbursed at a crack tip in this composite when compared to similar behavior in metallic materials; (2) matrix/fiber debonding occurs during crack passage, and crack bridging occurs behind the crack tip during stable crack growth; and (3) the high level of fracture toughness occurs mainly because of the pseudo-plastic strain response of material near the crack tip, and, perhaps, somewhat due to crack bridging.

CONCLUSION

Polymer-derived Si_3N_4 matrix composite reinforced with SiC fabric was fabricated using PIP method and characterized for microstructure and mechanical properties. The use of preceramic polymers as matrix precursors in the PIP processing allows the production of fiber-reinforced Si_3N_4 composites with useful mechanical properties. The fracture toughness of the PIP composites was measured to be higher than those for Si_3N_4 matrix composites reported in the literature. The high level of fracture toughness is attributed to their non-brittle fracture behavior. Micromechanics analysis that provides the information on the strain and crack opening displacement around the crack tip has been utilized to identify the fracture process of the polymer-derived composites.

ACKNOWLEDGMENT

This work was supported by the Southwest Research Institute Internal Research Program and the Materials Engineering and Technology Division. We acknowledge with gratitude the technical assistance of Mr. Stephen Salazar, Mr. Arthur E. Nicholas, Mr. John B. Campbell, and Mr. James F. Spencer.

REFERENCES

1. M.H. Lewis, "Ceramics: Applications and Limitations," in Materials at Their Limits, The Institute of Metals, London, 1-20, 1986.
2. R.L. Lehman, "Ceramic Matrix Fiber Composites," in Structural Ceramics, J.B. Wachtman, Jr. Ed., Academic Press, San Diego, 229-291, 1989.

3. J.R. Strife, J.J. Brennan, and K.M. Prevo, "Status of Continuous Fiber-Reinforced Ceramic Matrix Composite Processing Technology," Ceram. Eng. Sci. Proc., 11, 871-919, 1990.
4. R.D. Veltri and F.S. Galasso, "Chemical-Vapor-Infiltrated Silicon Nitride, Boron Nitride, and Silicon Carbide Matrix Composites," J. Am. Cer. Soc., 73, 2137-2140, 1990.
5. G. Ziegler, J. Heinrich, and G. Wötting, "Relationships Between Processing, Microstructure and Properties of Dense and Reaction-Bonded Silicon Nitride," J. Mater. Sci., 22, 3041-3086, 1987.
6. S.T. Schwab, "Polysilazane Precursors for Silicon Nitride and Resultant Products," U.S. Patent 5,294,425 (15 March 1994).
7. S.T. Schwab, R.C. Graef, D.L. Davidson, Y.-M. Pan, "Applications of Polysilazanes as Precursors to Silicon Nitride," Polymer Preprints, 32, 556-558, 1991.
8. S.T. Schwab, R.C. Graef, Y.-M. Pan, and D.L. Davidson, "Infiltration/Pyrolysis Processing of SiC Fiber-Reinforced Si_3N_4 Composites," NASA-CP 3175, Part 2, 721-738, 1992.
9. S.T. Schwab, D.L. Davidson, Y.-M. Pan, C.R. Blanchard, and R.C. Graef, "Processing and Properties of Polymer-Derived Si_3N_4 Composites," NASA-CP 3235 (in press).
10. D.F. Shriver, M.A. Drezdzon, The Manipulation of Air-Sensitive Compounds, 2nd Ed., John Wiley, New York, 1986.
11. H. Kodama, H. Sakamoto, and T. Miyoshi, "Silicon Carbide Monofilament-Reinforced Silicon Nitride or Silicon Carbide Matrix Composites," J. Am. Cer. Soc., 72, 551-558, 1989.
12. J. Lankford, "Compressive strength and Damage Mechanisms in a SiC-Fiber Reinforced Galss-Ceramic Composite," in Proceeding of the Fifth International Conference on Composite Materials (ICCMV), W. C. Harrigan, et al, Eds., TMS-AIME, Warrendale, 587-602, 1985.
13. Chevron-Notched Specimens: Testing and Stress Analysis, J.H. Underwood, et al, Eds. ASTM, STP 855, 1984.
14. M. Sakai and K. Yamasaki, "Numerical Fracture Analysis of Chevron-Notched Specimens: I, Shear Correction Factor, k," J. Am. Cer. Soc., 66, 371-375, 1983.
15. J.-M. Yang, S.T.J. Chen, S.M. Jeng, R.B. Thayer, and J.-F. LeCoustaouec, "Processing and Mechanical Behavior of SiC Fiber-Reinforced Si_3N_4 Composites," J. Mater. Res., 6, 1926-1936, 1991.
16. Y.M. Pan, M. Sakai, J.W. Warren and R.C. Bradt, "Toughness Anisotropy of a SiC/SiC Laminar Composite," in Tailoring Multiphase and Composite Ceramics, R.E. Tressler, et al, Eds., Plenum, New York, 631-638, 1986.
17. D.L. Williams, D.L. Davidson and J. Lankford, "Fatigue Crack Tip Plastic Strains by the Stereoimaging Technique," Experimental Mechanics, 20, 134-139, 1980.

STRENGTHENING AND TOUGHENING IN 1D-SIC/RBSN COMPOSITES

B. Kanka, J. Göring

German Aerospace Research Establishment (DLR), Materials Research Institute, D-51147 Cologne, Germany.

Abstract

A RBSN-based ceramic matrix composite was produced with fiber contents (Textron SCS-6) between 4 and 26 vol.%. A threefold increase of the critical stress level for first matrix cracking was achieved compared to the strength of the monolithic RBSN reference material. This strengthening effect is due to elastic reinforcing mechanisms, such as load transfer and residual compressive stresses in the matrix. At higher loads the composites exhibit a non-brittle deformation behaviour. This toughening effect is due to matrix cracks bridging by intact fibers. Because of their elastic character, the strengthening and toughening mechanisms exhibit no sensitivity with respect to cyclic fatigue.

I. Introduction

The inherent brittleness of monolithic ceramic materials is the most limiting factor for applications as structural components under high temperature/high stress conditions, especially in air crafts and space vehicles. Presently, the addition of continuous fibers is the most promising reinforcing concept in order to produce tough ceramic composites with an adequate damage tolerant fracture behavior.

Today carbon and silicon carbide composites, such as C/C, SiC/SiC and C/SiC, are discussed for air craft and space applications. This materials exhibit a relatively low strength level, due to micro- and macroscopic cracks caused by the anisotropic shrinkage during processing. For composites with a higher strength level a dense and possibly failure free matrix is necessary.

One system, which appears to be a promising candidate, is SiC fiber reinforced silicon nitride [1,2]. Extensive degradation and damage of the fibers is avoided due to moderate processing temperatures and the neglecting shrinkage during fabrication of reaction bonded silicon nitride (RBSN). Therefore RBSN is an potential candidate for a matrix material in a ceramic composite.

II. RESULTS AND DISCUSSION

II.1 PROCESSING AND MICROSTRUCTURE

1D-SiC/RBSN composite materials were produced using SCS-6 SiC fibers (Textron Specially Materials, Lowell, MA.). A detailed microstructural study of the fiber is given by Ning and Pirouz [3]. The SCS-6 fiber consists of multilayers deposited by chemical vapour deposition on a carbon core monofilament. The individual sublayers exhibit local fluctuations in the C/SiC ratio and consists of

To the extent authorized under the laws of the United States of America, all copyright interests in this publication are the property of The American Ceramic Society. Any duplication, reproduction, or republication of this publication or any part thereof, without the express written consent of The American Ceramic Society or fee paid to the Copyright Clearance Center, is prohibited.

faulted ß-SiC and or turbostratic graphite. A fine-grained silicon powder of high purity (Kema Nord F5) was used to fabricate the SiC/RBSN composites.
Figure 1 describes the processing route of two different types of SiC/RBSN composites employed in this study. The powders were first milled 15 hours in order optimize the particle size and distribution to obtain maximum green density and an enhanced conversion from silicon to silicon nitride during nitridation. Milling was performed in a polymer vessel, filled with silicon powder, thin silicon nitride balls, isopropanol and 6 wt.% of oil acid with the mass ratio $m_{Si}:m_{SN}:m_{ISO} = 1:1:2$. The milled powder was dried in a rota vapour and finally sieved. The silicon powder and the fibers were filled into a steel die. The volume fraction of the fibers in the green compacts was varied between 4 and 26 vol.%.

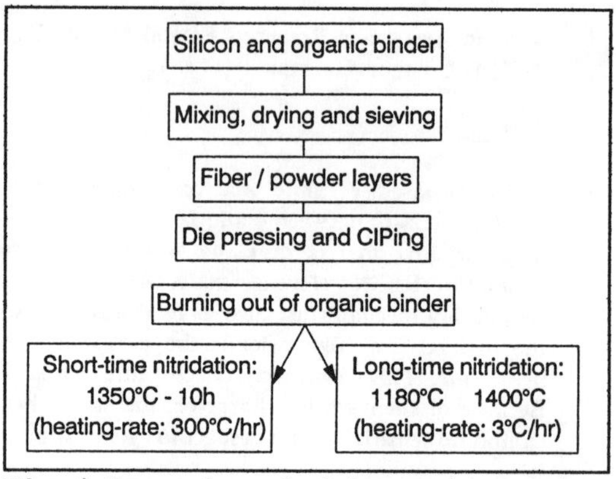

Fig. 1 Processing schedules used to produce two different SCS-6 SiC/RBSN composite materials

The fibre/powder layers in the steel die were uniaxially pressed at 15 bar in order to obtain 50mm x 5mm x h (h 2-5mm) shaped bars. The die-pressed bars were vacuum sealed in latex tubings and cold isostatically pressed (CIPed) at 250 MPa. After CIPing the organic binder was burned out at temperatures up to 400°C.
Two different nitridation schedules were performed in a gas mixture consisting of 90 vol.% N_2 and 10 vol.% H_2. The so-called "long-time nitridation" schedule was similar to the one normally employed in our institute for monolithic RBSN: Fast heating (300°C/hr) up to 1180°C followed by a further slower heating rate (3°C/hr) up to 1400°C (about 75 hr) and finally holding for 20 hours at a constant temperature of 1400°C. The modified "short-time nitridation" was carried out under a fast heating rate (300°C/hr) up to 1350°C holding for 10 hours.

The SiC/RBSN composites, fabricated under the two different nitridation schedules as well as monolithic reference material processed by a short-time nitridation, achieved densities of 75-80 % theoretical density. Typically for the microstructure after short time nitridation is the high residual silicon content (reaction rate: 55 to 60%). The $\alpha/(\alpha+\beta)$ - silicon nitride ratio is quite low after long time nitridation due to the low heating rate. It is significantly higher after short time nitridation, as gas-phase reactions benefit from the high heating rate. Beside this obvious difference in the matrix microstructure, the thermal stability of the fibers was the limiting factor for the mechanical behaviour of the composite materials. Figure 2 shows the fibers embedded in the RBSN matrix

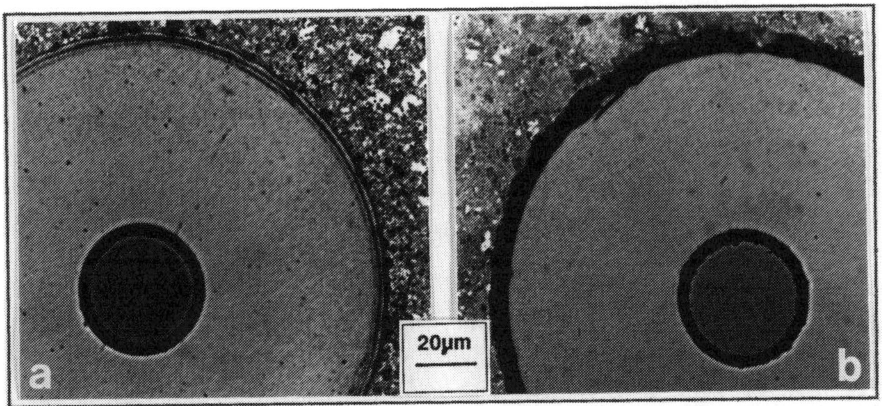

Fig. 2 SCS-6 Fibers embedded in the RBSN matrix after short time (a) and long time (b) nitridation shedules respectively

after the different nitridation schedules. After short time nitridation (see Fig. 2a) no morphological change of the fiber in the Fiber/matrix interface could be observed. After long time nitridation however, the outer, carbon rich layers of the SCS-6 filaments are totally degradated, resulting in a gap between the residual fiber periphery and the matrix (see Fig. 2b).

II.2 MECHANICAL TESTING

The room temperature flexural tests were carried out in a UTS testing machine using four point bending (40/20 mm span) and crosshead speeds of 0.5 mm/min for the monotonic and 1 mm/min for cyclic loading, respectively. The deflection was measured by three inductive extensiometers, compensating contact- and fixture deformations. During monotonic and cyclic testing load and deflection was stored by a microcomputer.

During flexural tests of SIC/RBSN produced by long-time nitridation, the composite failed at strength values of about 160 MPa. In this material a single crack propagates through the whole matrix material and under a further loading, the fibers were pulled out at a very low stress levels. This failure mechanism is due to the total debonding of the fiber/matrix interface, preventing any significant load transfer from the matrix into the fiber.

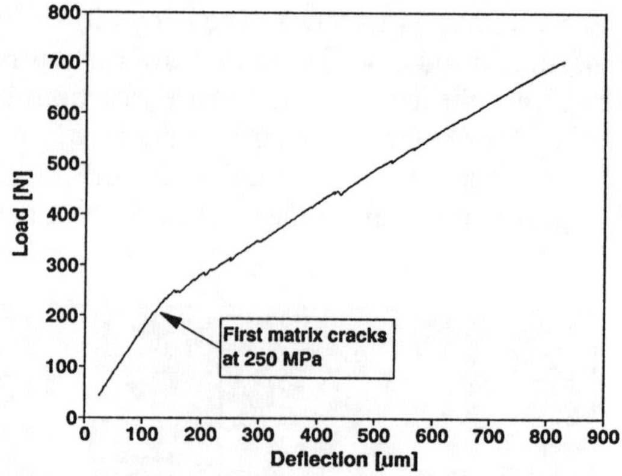

Fig. 3 Load/deflection behaviour of a SiC/RBSN composite containing 11 vol.% SCS-6 fibers

Figure 3 shows the load/deflection behaviour of a SiC/RBSN composite fabricated by a short-time nitridation process. Initially the deformation is linear-elastic up to the first matrix cracking. Further loading initializes a multiple matrix cracking without any fiber damage, reaching loads which are much higher than the matrix cracking load (see Fig.4). Using bend bars the stress values are only available in the linear-elastic region of the load/deflection curve. Nevertheless, the deformation behaviour in Figure 3 demonstrates the extensive failure tolerant fracture behaviour of the material.

Fig. 4 Multiple matrix cracking after extensive deformation. Note that no fiber damage has occured

Beside load and strength transfer, a main function of the matrix in a composite material suitable for high-temperature applications is to protect the fibers against corrosive attack from the environment. Once extensive matrix cracking has occurred (see Fig 4), protection of the fiber is no longer existing. Therefore the matrix cracking strength is of primary importance for technical applications of the composite materials.

Figure 5 shows the matrix cracking strength for various fiber contents. Compared to the strength of the monolithic reference material (about 140 MPa), a strong increase up to about 400 MPa for a SiC/RBSN composite containing 26 vol.% fibers was detected.

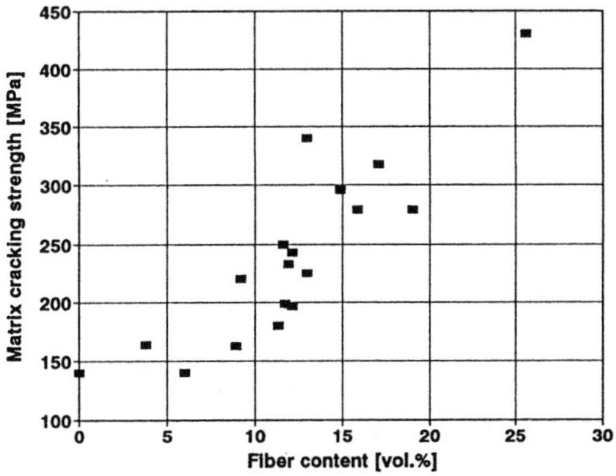

Fig. 5 First matrix cracking strength of SiC/RBSN composites as a function of SCS-6 fiber content

The strengthening effects in SiC/RBSN are discussed in [4]. An important strengthening mechanism is contributed by the higher elastic modulus of the fibers compared to the RBSN matrix (rule of mixings). This leads to an increasing Youngs modulus with higher fiber contents in the composite material (see Fig.6). Furthermore, the matrix cracking strain is increased significantly with the fiber content (Fig.7). This effect is probably due to the different thermal expansion coefficients between fibers and RBSN matrix, producing a residual compressive stress in the matrix which results in a higher matrix cracking strain with increasing fiber content. For this reinforcing mechanism an interface is mandatory strong enough to enable load transfer from the matrix into the fibers.

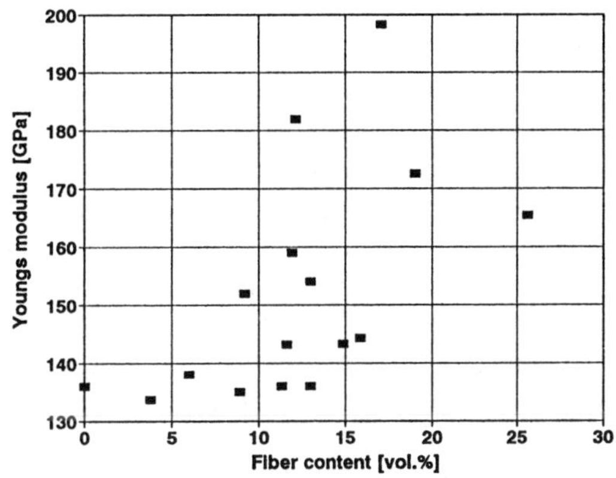

Fig. 6 Joungs modulus of SiC/RBSN composites (linear-elastic region) as a function of SCS-6 fiber content

The cyclic fatigue behaviour of a composite material is an interesting parameter for technical applications. Figure 8 shows selected segments of the load/deflection curve displaying cycles after first matrix cracking. Between 100 and 1078 cycles no significant change in the load deflection curve could be observed and even after 1078 cycles the shape of the deformation behaviour did not change. This data indicate, that the investigated SiC/RBSN material reacts insensitive against cyclic loading [5]. This is caused by the intact fibers after first matrix cracking forming elastic bridges to avoid further matrix cracking at the same stress level, while cyclic sensitive frictional effects are of minor importance for the reinforcing mechanisms.

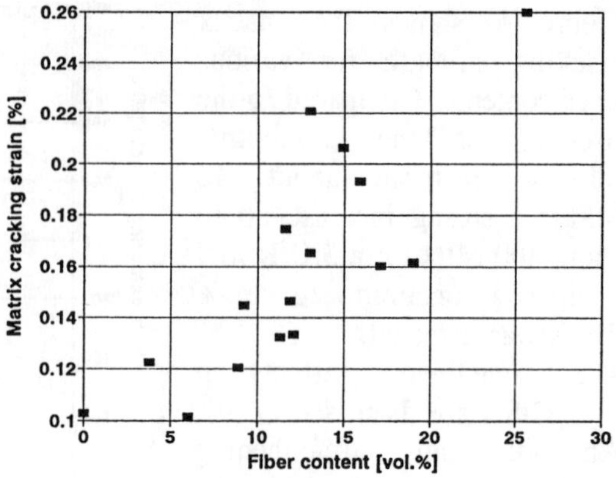

Fig. 7 Matrix cracking strain of SiC/RBSN composites as a function of SCS-6 fiber content

III. CONCLUSIONS

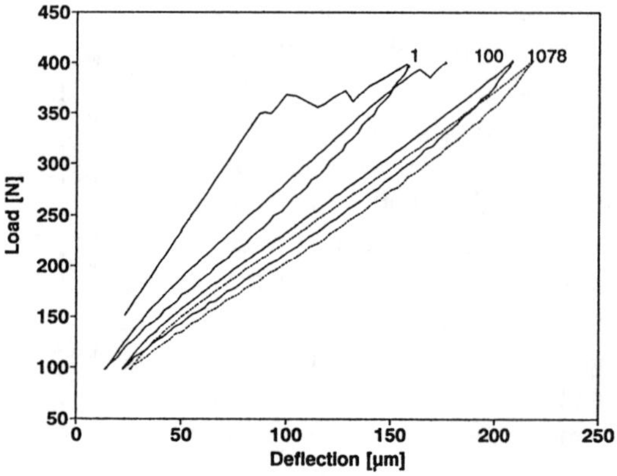

Fig. 8 Selected segments of the load deflection behaviuor of a SiC/RBSN composite containing 12 vol.% SCS-6 fibers under cyclic loading

A 1D SCS-6 SiC/RBSN composite was fabricated using a short time nitridation schedule to avoid the degradation of fiber properties. The interfacial fiber/matrix bonding is strong enough to transfer loads from the matrix into the fiber, but bond strength is still below the critical level where fiber damage would occur. This leads to an significantly increase in matrix cracking strength for composites containing a relatively low amount of fibers only. Furthermore, the matrix cracks were bridged by undamaged fibers leading to a failure-tolerant damage behaviour. Caused by the elastic nature of the reinforcing mechanism, the response of the SiC/RBSN composite is rather insensitive to cyclic fatigue.

In this stage of research the minor conversion rate to silicon nitride and the low theoretical density are still limiting factors for this material, and a further optimization is necessary. Nevertheless, the data obtained have demonstrated the advantages of a ceramic composite material employing a relatively strong interfacial fiber/matrix bonding.

IV. LITERATURE

[1] R.T.Bhatt, R.E.Phillips, *Laminate Behavior for SiC Fiber Reinforced Reaction Bonded Silicon Nitride Matrix Composites,* Journal of Composites Technology & Research, JCTRER, Vol.12, No.1, Spring 1990, 13-23

[2] H.Kodama, H.Sakamodo, T.Miyoshi, *Silicon Carbide Monofilament-Reinforced Silicon Nitride or Silicon Carbide Matrix Composites,* J. Am. Ceram. Soc. 72 [4] 551-58 1989

[3] X.J.Ning, P.Pirouz, S.C.Farmer, *Microstructural Analysis of the SCS-6 Silicon Carbide Fiber,* J. Am. Ceram. Soc. 76 [8] 2033-41 1993

[4] A.Chulya, J.P.Gyekenyesi, R.t.Bhatt, *Mechanical Behavior of Fiber Reinforced SiC/RBSN Ceramic Matrix Composites: Theory and Experiment,* NASA TM-103688, AVSCOM TR 91-C-004, 1991

[5] T.Fend, J.Göring, *Strength and Fatigue Behaviour of 2 D-Carbon/Carbon Under Shear Conditions,* will be published in the proceedings of ACerS 96th Annual Meeting & Exposition, Indianapolis 1994

EFFECTS OF IN-SITU FIBER STRENGTH CHARACTERISTICS ON MECHANICAL PROPERTIES OF SiC(f)/SiC COMPOSITES*

Dileep Singh, Jitendra P. Singh, and Michael J. Wheeler
Energy Technology Division, Argonne National Laboratory
Argonne, Illinois, 60439

Abstract

Nicalon-fiber-reinforced silicon carbide (SiC) matrix composites were tested in flexure at room and elevated temperatures. Measured strength of the composites decreased slightly from a room-temperature value of 400 MPa to 380 MPa at 1200°C. However, at 1300°C, strength decreased significantly to 290 MPa. This rapid decrease in strength at 1300°C is believed to be due to degradation in strength of the reinforcing fibers. In-situ fiber strength and fiber pullout distribution in fractured composites were estimated by fractographic techniques. Correlations were made between the measured strengths of composites and in-situ fiber strength characteristics to explain the mechanical properties of composites at room and elevated temperatures.

Introduction

The strength of reinforcing fibers is an important parameter in controlling the fracture behavior of continuous fiber ceramic composites (CFCCs). High fiber strength is required to facilitate successful transfer of load from matrix to fibers once a matrix crack is initiated and extended.[1] In addition, the amount of fiber pullout (which contributes to the toughening of the composite) is strongly influenced by the mean strength and the variability in strength of the reinforcing fibers.[2] Moreover, the ultimate load–bearing capacity of the composite is determined by fiber strength characteristics.[3] Therefore, fiber strength at room temperature and its retention at elevated service temperatures are important in the design and development of CFCCs with superior mechanical properties.

In the present study, mechanical characterization of Nicalon-fiber-reinforced SiC-matrix composites were conducted in flexure at room and elevated temperatures.

To the extent authorized under the laws of the United States of America, all copyright interests in this publication are the property of The American Ceramic Society. Any duplication, reproduction, or republication of this publication or any part thereof, without the express written consent of The American Ceramic Society or fee paid to the Copyright Clearance Center, is prohibited.

Single–fiber strength distribution of as–fabricated Nicalon fibers was obtained from bundle tests. Strength distributions of in-situ fractured fibers in Nicalon/SiC composites tested at room and elevated temperatures were assessed by measuring fracture mirror radii. Variations in the mechanical properties of composites evaluated as a function of test temperatures were correlated to the in-situ fiber strength characteristics.

Experimental Procedure

Material

Carbon–coated ceramic–grade Nicalon fibers were used to study the effect of fiber strength distribution on room- and elevated-temperature properties of Nicalon fiber-reinforced SiC composites. Bundle tests of Nicalon fiber tows were used to determine the single–fiber strength distribution. Typically, each tow consisted of 500 individual fibers with diameters ranging from 10–15 μm. The polymer–derived Nicalon fibers consist primarily of β–SiC crystallites, along with excess carbon and SiO, with an average length of 1.7 nm.[4] The reported density and elastic modulus (E_f) of Nicalon fibers are 2.55 g/cm^3 and 210 GPa, respectively.[5]

Nicalon–fiber–reinforced SiC-matrix composites were used to evaluate the strength distribution of fibers in the composites, which were fabricated at Oak Ridge National Laboratory by densifying multiple layers of Nicalon mats stacked in a graphite die. Chemical vapor infiltration (CVI), under forced conditions of thermal and pressure gradients, was used to densify the preforms with SiC.[6] The fiber content in the composites was about 40 vol.% and the resulting composites were ≈90% dense. For tests at elevated temperatures, the composites were overcoated with a 35 μm layer of SiC to prevent oxidation of exposed carbon coating on the fiber surface caused by cutting the composite specimens.

Bundle and Flexure Tests

The fiber bundle test[7] was used to determine single–fiber strength distribution of as–fabricated Nicalon fibers. Weibull parameters, needed to describe the strength distribution of fibers, were estimated from load–vs.–strain plots of a fiber bundle loaded in uniaxial tension. Bundle tests were conducted on a universal testing system¶ with the experimental setup shown schematically in Fig. 1. Tests were conducted on fiber tows with gage lengths of 27 to 100 mm under ambient conditions and at a crosshead speed of 0.5 mm/min.

¶Model 4505, Instron Corp., Canton, MA.

Flexural tests on bars (2.9 x 4.2 x 28.0 mm) of continuous Nicalon–fiber–reinforced SiC composites were conducted in the four–point–bend mode at room and elevated temperatures on the universal testing system. For room–temperature tests, loading and support spans were 9.53 and 19.05 mm, respectively, and tests were conducted at a crosshead speed of 1.27 mm/min under ambient conditions. For tests at elevated temperatures, SiC fixtures with loading and support spans of 12.7 mm and 25.4 mm, respectively, were used. Crosshead speed for these tests was 1 mm/min. The composite bars were held at the test temperature for 15 min in air prior to their fracture in vacuum. All specimens were loaded perpendicular to the layers of mats. Fractured composite specimens were examined by scanning electron microscopy (SEM)† to locate the origin of failure and establish the associated characteristic fracture surface morphology of the fibers.

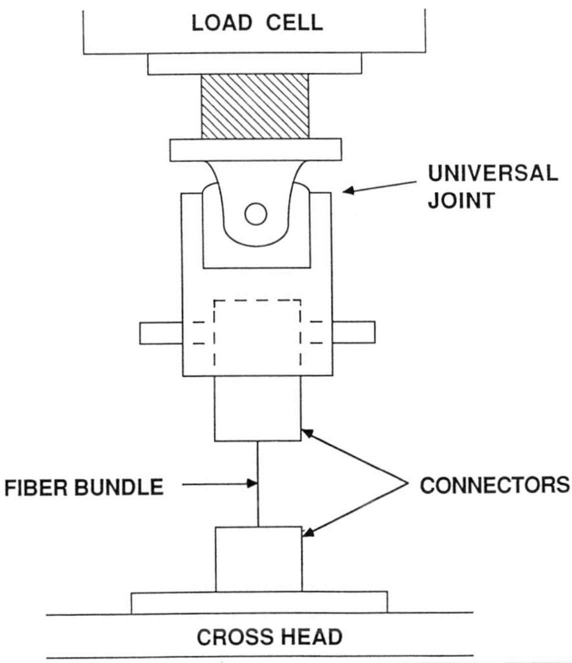

Figure 1. Schematic Representation of Experimental Setup Used for Bundle Tests.

---

†Model JXA–840A, JEOL Co., Ltd., Tokyo, Japan.

Results and Discussion

Strength Distribution of As–Fabricated Nicalon Fibers

The distribution function used to describe Nicalon fiber strength is given in Eq. 1. This formulation was used because strength distributions of fibers were obtained from fibers with various gage lengths, and it was appropriate to account for the size effect and represent the Weibull strength distribution at a standard gage length, L, taken to be 10 mm.

$$F(\sigma) = 1 - \exp\left[-\frac{L}{L_o}\left(\frac{\sigma}{\sigma_o}\right)^m \right] \tag{1}$$

where $F(\sigma)$ is the cumulative failure probability at an applied stress σ, σ_o is the scale parameter signifying a characteristic strength of the distribution, m is the Weibull modulus that characterizes flaw distribution in the material, and L_o is the fiber gage length at which the Weibull parameters are determined.

Figure 2 shows a typical load–strain plot obtained from a bundle test. Strain or displacement in the fiber bundle at a particular load was determined by subtracting the system displacement (grips, connectors, etc.) from the absolute displacement of the crosshead of the testing machine. Displacement due to system accessories was obtained by estimating system compliance. Weibull parameters (σ_o and m) of the as–fabricated Nicalon fibers were obtained from the maximum load P_{max}, slope of the linear portion of the load–strain behavior, S_o, slope of the line joining the maximum load to the origin, S_A, (as indicated on Fig. 2) and the specimen gage length, L_o, by solving the following equations:[7]

$$m = \frac{1}{\ln\left[\frac{S_o}{S_A}\right]} \tag{2}$$

$$P_{max} = S_o \varepsilon_o \left(\frac{1}{2.7 L_o m}\right)^{\frac{1}{m}} \tag{3}$$

$$\sigma_o = E_f \varepsilon_o \tag{4}$$

Results of seven tests conducted on fiber bundles with various gage lengths gave an average value of 7.1 for the Weibull modulus. The average value for the scale parameter, after correcting for a gage length of 10 mm with Eq. 1, was 3.45 GPa.

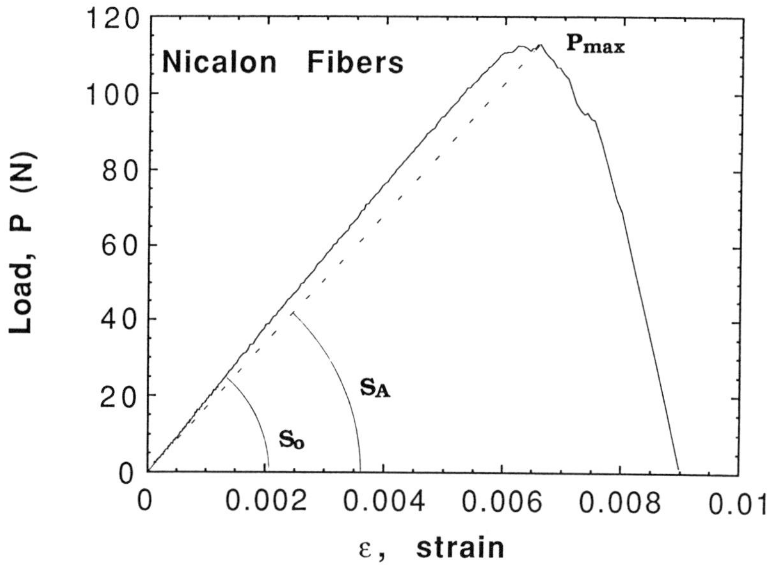

Figure 2. Typical Load–Strain Variation Obtained from a Bundle-Test of Nicalon Fibers.

These results are in accordance with reported values in the literature for Nicalon fiber strength distribution.[8]

Bulk Mechanical Property Characterization

Figure 3 shows the typical load-displacement behavior obtained from the flexure tests conducted at room temperature. More extensive fiber pullout and larger work-of-fracture was seen relative to the elevated temperature tests. The ultimate strength is obtained from the peak load. The results of mechanical property measurements are shown in Fig. 4. Measured strength shows a small decrease from a room-temperature value of 400 MPa to 380 MPa at 1200°C. However, at 1300°C, strength decreases significantly to 290 MPa. This rapid decrease in strength at 1300°C is believed to be due to degradation in strength of the reinforcing fibers.

In-Situ Fiber Strength Evaluations

SEM was used for fractographic evaluation to estimate in-situ fiber strengths in composites tested at room temperature and 1300°C. Specifically, critical flaws and the associated characteristic fracture markings were identified. In-situ fiber strength in the composites was evaluated from measurements of these characteristic

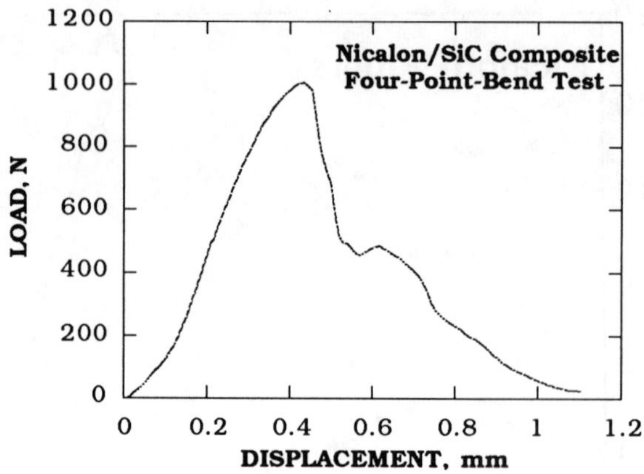

Figure 3. Typical Load-Displacement Behavior for Nicalon Fiber-Reinforced SiC Composites Tested at Room-Temperature in Flexure.

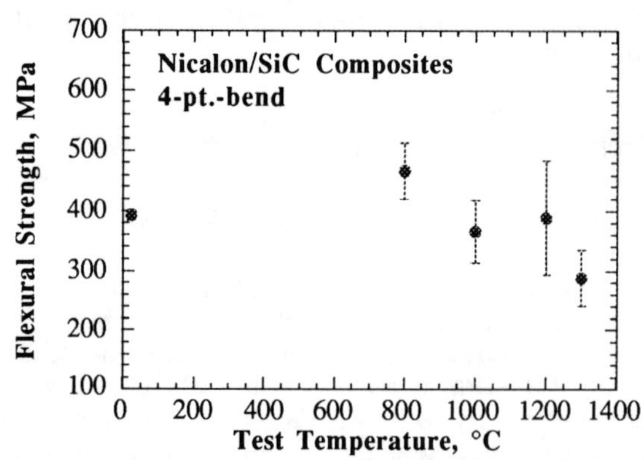

Figure 4. Variation of Ultimate Strength with Test Temperature for Nicalon Fiber-Reinforced SiC Composites.

markings. Typical fracture surface of a fiber in a Nicalon-fiber–reinforced SiC composite, tested in the four–point–bend mode, is shown in Fig. 5. Characteristic features associated with brittle failure, such as mirror (a smooth region around the fracture origin) and hackle (a region of multiple fracture planes), are clearly observable on the surface of fractured fibers. SEM investigation showed that most fibers failed from defects or flaws at the fiber surface.

For brittle materials, fracture surface features such as mirror radii can be correlated to tensile strength through the empirical relationship:[9]

$$\sigma_f \sqrt{r_m} = A_m \tag{5}$$

where r_m represents the mirror radii, σ_f is the tensile strength, and A_m is the mirror constant, which is related to the fracture toughness of the material. In the present study, A_m is taken as 3.5 $MPam^{1/2}$.[2] Strengths of more than 30 Nicalon fibers for each set of composite specimens were determined by measuring their fracture mirror radii and using Eq. 5. However, the Weibull scale parameter requires correction to account for effective fiber gage length of the in-situ fibers over which the stresses are acting. Therefore, it was assumed here that the in-situ fiber gage lengths are of the order of the fiber pullout lengths. To this end, fiber pullout lengths were measured on the room-temperature and 1300°C tested specimens (Fig. 6). The average values for pullout lengths for composites tested at room and elevated temperatures were ≈320 and 200 µm, respectively. Using the values of average pullout lengths, scale parameters at the standard gage length of 10 mm were evaluated. Table 1 lists the various Weibull strength distribution parameters. Values for scale parameters reported in Table 1 are for a gage length of 10 mm.

Figure 7 shows Weibull strength distribution plots for Nicalon fibers in three different conditions, based on the data in Table 1 and the use of Eq. 1. It is clear from this figure that the strength of Nicalon fibers decreases significantly (≈60%) from the as-fabricated state to their incorporation in composites. There is an additional 25% drop in strength when the composites are exposed to 1300°C. These drops in strength are believed to be due to mechanical and/or thermal damage to the fibers during processing. Strength degradation of Nicalon fibers due to high temperatures is well documented in the literature, and losses of more than 30 and 70% in tensile strength have been reported for ceramic–grade Nicalon fibers by exposure at 1000°C and 1200°C, respectively, for 12 h in a wet–air atmosphere.[10] This is attributed to microstructural and stoichiometric changes that occur in the fibers at elevated temperatures. Thermochemical calculations made by Luthra[11] have shown that SiC-based fibers are inherently unstable both in terms of its chemical composition and structure. Similar changes, although not as severe, are expected for the carbon–coated Nicalon fibers investigated in this study. Higher processing temperatures (≈1200°C) and longer exposure times (>24 h) used in the fabrication of the composites make it difficult to prevent fiber degradation. Moreover, Okamura et al.[12] showed that formation of an SiO_2 film can also contribute to the reduction of both tensile strength and Young's modulus of the fibers.

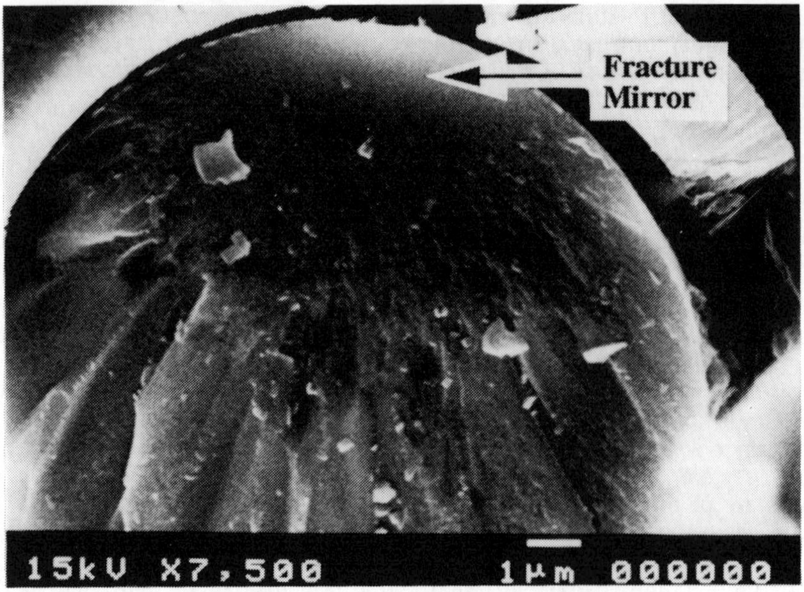

Figure 5. Characteristic Features on a Fractured In-Situ Nicalon Fiber.

Table 1. Weibull Strength Parameters of As-Fabricated and In-Situ Nicalon Fibers in Composites Tested at Room-Temperature and 1300°C.

Fiber Condition	Test Temperature (°C)	Scale Parameter, σ_o (GPa)	Weibull Modulus, m
As-Fabricated	24	3.45	7.1
In-situ	24	1.28	6.0
In-situ	1300	0.96	5.2

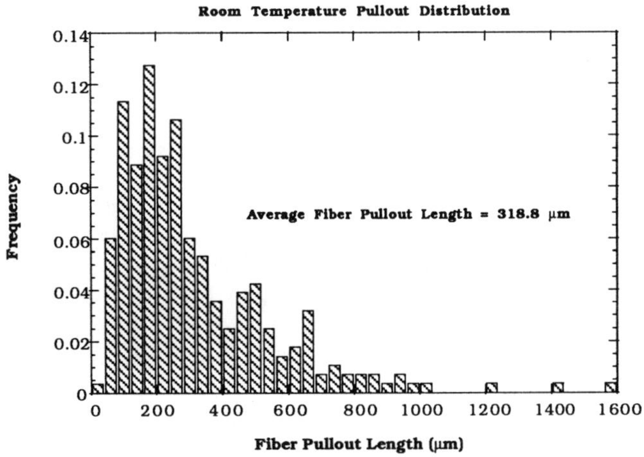

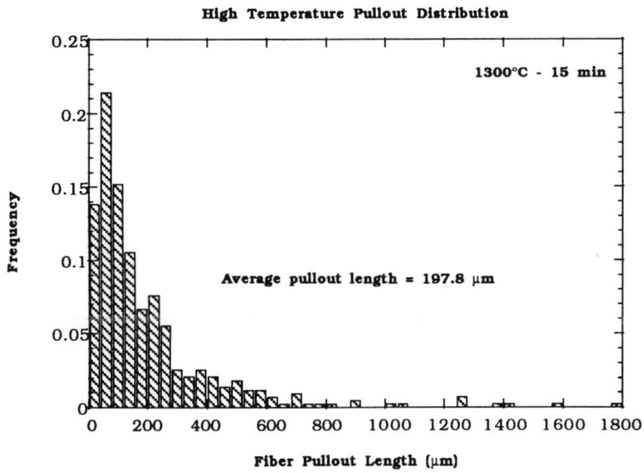

Figure 6. Distribution of Fiber Pullout Length in Composites Tested at Room-Temperature and 1300°C.

Conclusions

1. Strength distribution of as-fabricated Nicalon fibers was obtained from bundle tests, whereas the strength of in-situ fibers was evaluated from fracture mirror sizes.
2. In-situ strength of Nicalon fibers in composites is lower than the strength of as-fabricated Nicalon fibers.
3. Decrease in strength of Nicalon fibers after incorporation in composites is attributed to thermal and mechanical degradation of the fibers during fabrication.
4. High-temperature exposure (1300°C) further decreases in-situ fiber strength of Nicalon fibers, which is reflected in a drop in ultimate strength of the composite at those temperatures.

Acknowledgments

This work was supported by the U.S. Department of Energy, Advanced Research and Technology Development, Fossil Energy Materials Program, under Contract W–31–109–Eng–38. The authors thank D. P. Stinton of Oak Ridge National Laboratory for providing the composite specimens. The authors are grateful to Dr. Jules Routbort for permitting us to use the high-temperature testing facility, and to Martha Boling–Risser for providing assistance in using the high-temperature test facility.

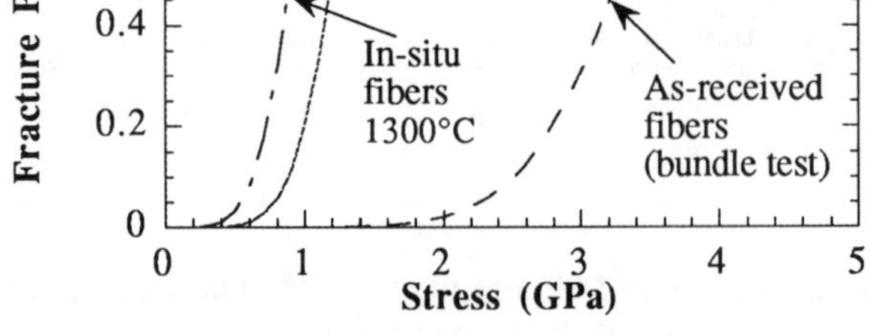

Figure 7. Weibull Strength Distribution Plots of Nicalon Fibers in As-Fabricated State, after Incorporation in Composites, and after Exposure of Composites to 1300°C.

References

1. A. G. Evans and D. B. Marshall, "The Mechanical Behavior of Ceramic Matrix Composites," Overview No. 85, Acta Metall., 37 [10] 2567–2583 (1989).
2. M. D. Thouless, O. Sbaizero, L. S. Sigl, and A. G. Evans, "Effect of Interface Mechanical Properties on Pullout in a SiC–Fiber–Reinforced Lithium Aluminate Silicate Glass Ceramic," J. Am. Ceram. Soc., **72** [4] 525–32 (1989).
3. T. Mah, M. G. Mendiratta, A. P. Katz, R. Ruh, and K. S. Mazdiyasni, "Room–Temperature Mechanical Behavior of Fiber–Reinforced Ceramic–Matrix Composites," J. Am. Ceram. Soc., **68** [1] C–27–C–30 (1985).
4. G. Simon and A. R. Bunsell, "Mechanical and Structural Study of High Performance Silicon Carbide Fibers," Sci. Ceram. **12** - Ceramurgia s.r.l, Faenza, 647-654 (1984).
5. D. J. Pysher and R. E. Tressler, "Ceramic Fiber for Composites," Center for Advanced Materials Newsletter, College of Earth and Mineral Sciences at Pennsylvania State University, **5** [1] 3 (1991).
6. D. P. Stinton, R. A. Lowden, and R. H. Krabill, "Mechanical Property Characterization of Fiber–Reinforced SiC Matrix Composites," Proc. 4th Annual Conf. on Fossil Energy Materials, Fossil Energy AR&TD Materials Program, ORNL/FMP–90/1, 3–13 (1990).
7. Z. Chi, T-Wei Chou, and G. Shen, "Determination of Single Fiber Strength Distribution From Fiber Bundle Testings," J. Mater. SiC., **19** [10] 3319-3324 (1984).
8. K. Goda and H. Fukunaga, "The Evaluation of the Strength Distribution of Silicon Carbide and Alumina Fibers by a Multi-Modal Weibull Distribution," J. Mater. Sci. **21** [12] 4475-4480 (1986).
9. H. P. Kirchner and R. M. Gruver, "Fracture Mirror in Alumina Ceramics," Phil. Mag. **27** 1433-1446 (1973).
10. T. J. Clark, R. M. Arons, and J. B. Stamatoff, "Thermal Degradation of Nicalon SiC Fibers," Ceram. Eng. Sci. and Proc., **6** [7-8] 576-588 (1985).
11. K. L. Luthra, "Thermochemical Analysis of the Stability of Continuous SiC Fibers," J. Am. Ceram. Soc., **69** [10] C-231-C-233 (1986).
12. K. Okamura, M. Sato, T. Matsuzawa, and Y. Hasgawa, "Effect of Oxygen on Tensile Strength of SiC Fibers," Polymer Preprints, **25** [1] 6–7 (1984).

OPTIMIZATION OF MICROWAVE–ASSISTED CHEMICAL VAPOR INFILTRATION

Daniel J. Skamser, Paul S. Day, Hamlin M. Jennings, D. Lynn Johnson
Department of Materials Science and Engineering, Northwestern University, Evanston, IL 60208-3108

ABSTRACT

Microwave heating was used to heat alumina fiber specimens as part of a forced–flow chemical vapor infiltration process. A hybrid microwave heating system was configured with the aid of a mathematical model so that a temperature gradient of 50°C was predicted to exist inside a specimen at a centerline temperature of 1000°C. Experiments were conducted to determine the optimum flow rate and gas mixture for the deposition of alumina matrix inside preforms of alumina fibers. Results show that high flow rates and lean gas mixtures lead to high density fiber composites with bulk densities as high as 85%. Different types of porosity were found in the dense specimens caused by initial packing non–uniformities between the fibers.

INTRODUCTION

High density composites have been made using a chemical vapor infiltration (CVI) process which includes both temperature and gas pressure gradients across the specimen. Stinton, et. al.[1] were the first to use this process by forcing reactant gases to flow through a planar preform from the cool to the hot side causing preferential deposition near the hot face and then filling of the preform from the hot to the cool side. They achieved both faster deposition rates and higher density composites than possible with isothermal/isobaric CVI.

Recently microwave heating has been used to create temperature gradients inside fiber specimens.[2-5] Microwaves penetrate a body and heat it volumetrically, while heat is lost only at the surface, resulting in temperature gradients. Deposition is fastest at the center because the temperature is the highest at the center. This technique has been used to produce both alumina composites[2,3] and SiC composites[4,5], but difficulties persist. For instance, alumina is difficult to heat with microwave energy because it is relatively transparent to microwaves at low temperatures and tends to overheat at high temperatures due to thermal runaway.[6]

Hybrid microwave heating, where both microwave and radiation heating are used simultaneously, has been used to solve the problems associated with microwave heating low loss dielectric materials such as alumina.[6-9] Models of the temperature profile produced by microwave heating have shown that increasing the ambient temperature around a low loss dielectric material heated with microwave energy leads

To the extent authorized under the laws of the United States of America, all copyright interests in this publication are the property of The American Ceramic Society. Any duplication, reproduction, or republication of this publication or any part thereof, without the express written consent of The American Ceramic Society or fee paid to the Copyright Clearance Center, is prohibited.

to higher critical temperatures inside the material before thermal runaway occurs.[6,7] Thus, a high loss material can be placed around a sample and act as a microwave susceptor which raises the ambient temperature around the sample. This technique has been used by De',et al.[8] to heat alumina compacts in a thermal insulation chamber coated with SiC powder. The microwaves heated the SiC, which caused heating of the alumina compacts until they reached temperatures where they absorbed microwave energy and heated themselves. Janney, et. al.[9] successfully used a SiC "picket fence" arrangement to microwave heat and sinter zirconia.

Skamser, et al.[3] have used hybrid microwave heating to heat alumina fiber samples as part of a forced–flow CVI process (MCVI). An external susceptor consisting of a mixture of Sr-doped lanthanum chromite (LSC) and alumina powder was placed around an alumina preform and heated with microwaves to 1100°C. Statistical experiments were designed using a Taguchi method to produce partially infiltrated specimens, and ANOVA analysis of the results showed that low utilization of reactants, low temperatures, high pressures, and high flow rates led to high density alumina fiber/alumina matrix composites.

The present work uses a similar setup as Skamser, et al.[3] to determine the optimum conditions for producing high density fiber composites. Experiments were conducted to determine the total flow rate of the reactant gases, the amount of reactant dilution, and the composition of the reactant gases which lead to high density composites.

EXPERIMENTAL PROCEDURE

The MCVI system (Figure 1) consists of a source of microwave energy and cavity, gas delivery system, quartz reactor tube, pressure control, and scrubber. Microwave power is supplied at 2.45 GHz by a magnetron with variable power control up to 3 kW.[a] A tunable single mode cavity[b] is maintained in a perturbed TM_{012} mode. A gas delivery system[c] is used to control flows of Cl_2 and a 1:1 ratio of CO_2/H_2 to a quartz reactor tube containing the sample. Chlorine gas is flown through a bed of aluminum particles heated to 290°C, forming $AlCl_3$ vapor. Nitrogen is added as both a carrier gas and a diluent, and is passed along with the Cl_2 through the chlorinator. The pressure is monitored both upstream and downstream of the sample with pressure transducers[d] and is controlled by a ballast valve located upstream of the vacuum pump[e] and connected to the scrubber system.

Figure 2 shows the hybrid heating setup found inside the microwave cavity. The susceptor consists of two concentric quartz tubes closed at the bottom and filled with two zones of powder. The upper zone is a mixture of LSC and alumina powder and is placed above a zone of alumina powder. The 15 mol% Sr-doped lanthanum chromite was made using a sol-gel technique based on the Pechini process.[10] LSC was chosen because it has a high electrical conductivity compared to alumina, and it does not degrade at high temperatures in air.[11,12] The LSC/alumina powder absorbs some of the microwave power and increases the ambient temperature surrounding the specimen. The bottom end of the sample is positioned level with the bottom of the LSC/alumina zone promoting both axial and radial temperature gradients. This

[a] GL119 Low Ripple Power Source and GL814 Power Control Unit, Gerling Lab., Modesto, CA.
[b] Model CMPR 250, Wavemat, Inc., Plymouth, MI.
[c] Model 1259C, MKS Instruments, Inc., Andover, MA.
[d] Baratron 122A, MKS Instruments, Inc., Andover, MA.
[e] LRMB 10603 Close Couple Pump, SIHI Co., Grand Island, NY.

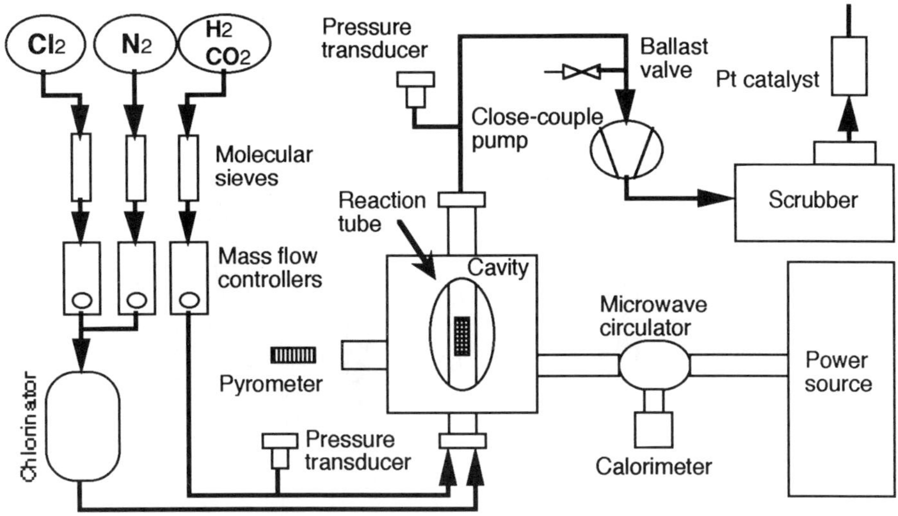

Figure 1 MCVI setup used to infiltrate alumina fiber specimens.

allows the reactant gases to pass through the sample and reach the hottest fibers near the outlet of the specimen before reacting prematurely at the inlet of the specimen.

Each sample consisted of a bundle of uniaxially aligned alumina fibers[f] which were placed in a quartz reactor tube so that the reactant gases had to pass through the specimen. Typical dimensions were 55–65 mm long and 25 mm in diameter with a fiber volume fraction ranging from 33 to 35%.

Temperature was measured using two optical pyrometers[g] at a wavelength of 950 nm. A Pt-coated blackbody lightpipe sensor was used to measure the internal temperature of the specimen. It was placed 20 mm below the tip of a closed end alumina tube located along the axis of the specimen. Also, an optical pyrometer measured the surface temperature of the susceptor in the LSC/alumina zone.

During the infiltration experiments the internal temperature inside the specimen was maintained at 1000±10°C and the downstream pressure was held at 100±1 torr. The total flow rate of the gases was varied between 200 and 400 sccm, with 400 sccm being the maximum flow rate without disrupting the specimen position. Nitrogen dilution was quantified as the ratio of the moles of nitrogen to the total moles of gases, (Ω), and was varied between 0.15 to 0.67. The composition of the reactant gas mixture was quantified as the ratio of the moles of $AlCl_3$ to the total moles of reactant gases ($AlCl_3 + H_2 + CO_2$), (ß), and was varied between 0.04 to 0.25.

Each infiltration experiment was run intermittently since unreacted $AlCl_3$ deposited on the downstream plumbing and constricted the flow of gases. The experiments were periodically and temporarily stopped, including a shutdown of the microwave power, so that the tubing could be cleared. Experiments were allowed to run until the upstream pressure reached 500 torr. Some experiments did not reach 500 torr because the dense specimens thermally expanded upon reheating and caused the reactor tube to fracture.

---

[f] Nextel 610™ Alumina Fibers, sized, 3M Co., Maplewood, MN.
[g] Model 100c with HF-2 Pyrometer and Pt Blackbody Lightpipe, Luxtron, Inc., Beaverton, OR.

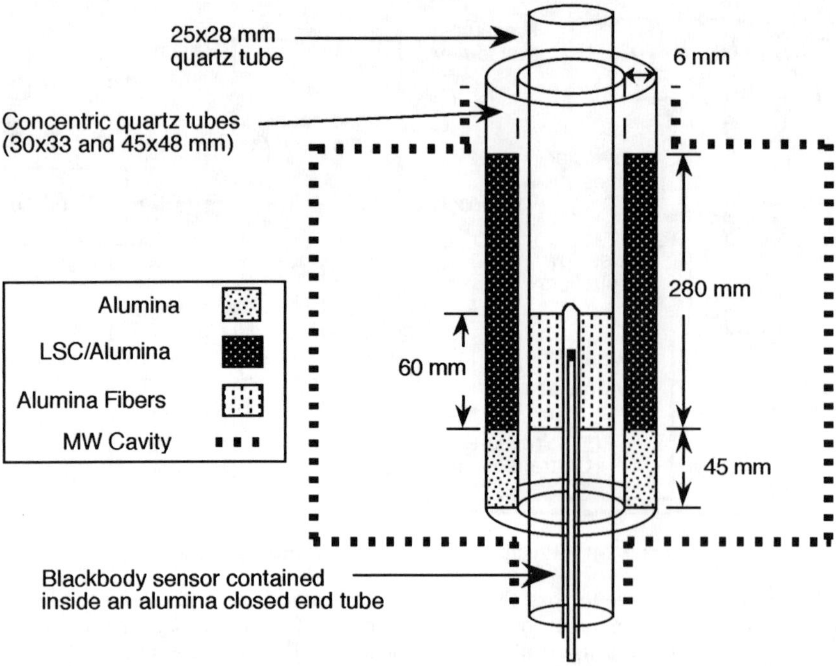

Figure 2 Hybrid heating setup used inside the microwave cavity.

Dense composites were sectioned into 9–10 disks, with the ends trimmed, and the bulk density was measured for each disk. The sectioned disks were polished and coated with 20 nm of gold so that the microstructure could be analyzed using SEM. The crystal structure of the deposits were determined by X-ray diffraction.

RESULTS AND DISCUSSION

Unstable heating occurred when the internal temperature of the specimen increased uncontrollably at a constant power level. Tests of heating stability were conducted to determine the optimum mixture of LSC and alumina powder to be used in the susceptor so that the threshold for unstable heating occurred at 1000°C. At the threshold of unstable heating the temperature gradients inside the specimen were at a maximum using the hybrid microwave heating approach. Figure 3 shows both the threshold temperature of the specimen and the corresponding temperature on the surface of the susceptor when unstable heating occurred. For a mixture of 38% LSC and alumina powder the internal temperature of the specimen could be held at 1000°C while having a temperature at the surface of the susceptor of 860°C. A model[13] predicts the temperature gradient across the radius of the specimen to be about 50°C at this temperature, which is an adequate temperature gradient for a CVI process.

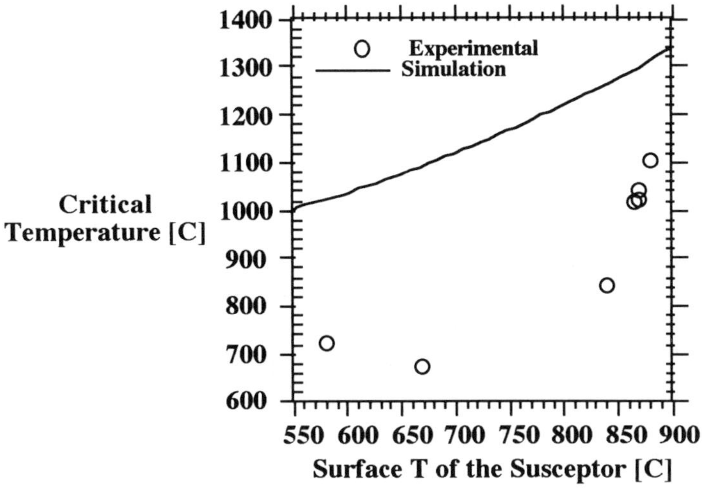

Figure 3 Experimental temperatures at the point of unstable microwave heating are compared to simulated critical temperatures[13] for various ratios of LSC to alumina (increasing to the right).

This model was developed by Skamser and Johnson[13] and simulates the temperature profiles that are produced in a hybrid microwave configuration similar to the experimental setup. Predicted critical temperatures from this model are much higher than those determined experimentally. The predicted critical temperature cannot be achieved experimentally because small changes in power cause large changes in the internal temperature in the vicinity of the critical temperature. Perturbations of power or electric field at any point in the specimen can lead to local runaway, which spreads throughout the entire specimen.

Table 1 shows the conditions and results of the experiments conducted to study the effects of nitrogen dilution, flow rate, and gas composition on the CVI process. The deposition rate of matrix inside the specimen is calculated as the total weight of matrix divided by the span of time of the experiment. The efficiency of the reaction is the ratio of the actual deposition rate divided by the calculated deposition rate for 100% conversion of the $AlCl_3$ into alumina. Even though some of the experiments did not reach an upstream pressure of 500 torr, the density profiles of the specimens were adequately compared to one another since all the specimens had a section with a peak density of at least 77%.

Figures 4–6 show the axial density profiles for each experiment and the effect of the separate process parameters on the degree of densification inside the specimens. Plot (a) of each set shows the extent of densification along the axis of the specimen. The specimen inlet is at 0 and the outlet is at 100% of the normalized length. The extent of deposition rises to a maximum near the inlet, remains constant between 20 to 50% of the length, and decreases steadily for the latter half of the specimen. Plot (b) of each set shows the extent of densification as a function of a given variable at intervals of 25, 50, 75% along the normalized length of a specimen using data from the first graph. The error bars are due to inaccuracies for the bulk density of each disk.

Table 1. Trial Conditions and Results

Trial	Ω[¥]	ß[§]	Flow Rate (sccm)	Time (hr)	Dep. Rate (g/hr)	Efficiency %	P final[¶] (torr)
1	0.15	0.15	200	25.1	1.25	36.0	380
2	0.33	0.15	200	40.9	0.91	33.3	313
3	0.67	0.15	200	63.3	0.48	35.9	341
4	0.15	0.15	300	22.1	1.45	27.7	510
5	0.15	0.15	400	18.7	1.74	24.9	510
6	0.15	0.04	200	39.5	0.80	85.7	461
7	0.15	0.09	200	43.1	1.00	48.1	485
8	0.15	0.20	200	26.2	1.40	30.2	350
9	0.15	0.25	200	37.1	1.11	19.2	532

¥Ratio of the moles of N_2 to the total moles of gas entering the specimen.
§Ratio of the moles of $AlCl_3$ to the total moles of reactants ($AlCl_3 + H_2 + CO_2$).
¶Final pressure upstream of the specimen before the experiment was terminated.
Internal temperature of the specimens were held at $1000\pm10^{\circ}C$.
Downstream pressure was held constant at 100 ± 1 torr.
Samples varied in length from 55 to 65 mm.
Initial solids fraction of the specimens varied from 0.33 to 0.35.

As Ω is increased the gas mixture becomes dilute with nitrogen, leading to longer processing times for densification of the specimen. Table 1 clearly shows a trend of decreasing deposition rates as Ω increases. At the same time, the efficiency of the reaction remains constant for all levels of Ω. This indicates that the nitrogen is simply acting as an inert carrier gas and does not change the reaction kinetics significantly. Figure 4 shows that the densification profile along the axis of the specimen remains fairly constant with Ω. It is concluded that the optimum amount of nitrogen dilution should be low because this results in faster densification without changing the density profiles of dense specimens.

Table 1 shows that increasing the total flow rate leads to faster deposition rates but with decreasing reaction efficiencies. The gas supply will be less depleted of reactants and contain proportionally less reaction byproducts that inhibit the reaction at smaller flow rates. Thus, the concentration of reactants inside a specimen will be higher downstream at higher flow rates and result in more uniform density profiles. At the same time, a high amount of the reactants will pass through the specimen without reacting and cause lower reaction efficiencies. Figure 5 shows that increasing the flow rate leads to more uniform and higher densities throughout the specimen. Overall, the optimum level of flow rate should be high because it leads to both uniform and high densities, and shorter processing times.

Increasing ß causes the efficiency of the reaction to decrease (Table 1).. Thermodynamic equilibrium calculations show that ß should be less than 0.20 for complete hydrolysis of $AlCl_3$ to alumina.[14] Thus, increasing ß decreases the amount of excess water available for reaction. This suggests that the efficiency of the reaction is limited by the concentration of the H_2/CO_2 gas entering the specimen. This is expected since it is well known that the rate limiting step for the hydrolysis of $AlCl_3$ is the formation of water on an alumina substrate.[15]

A maximum deposition rate occurs for ß equal to 0.20, similar to the maximum ß for complete $AlCl_3$ conversion to alumina predicted by thermodynamic calculations[14]. Both the amount of excess water and the efficiency of the reaction are at levels resulting in a maximum deposition rate. Lowering ß decreases the deposition

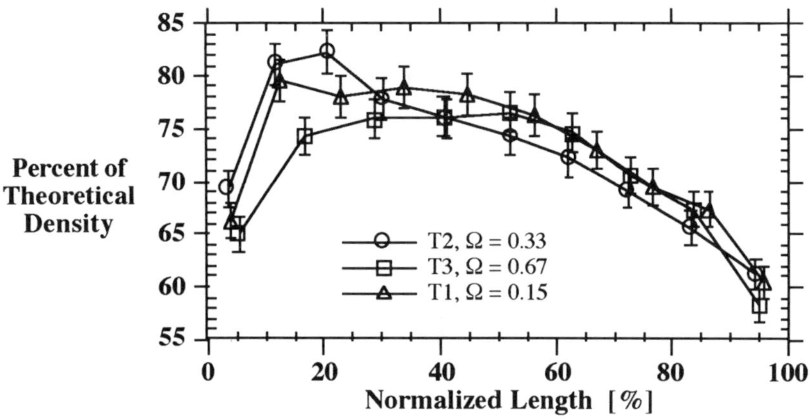

(a)

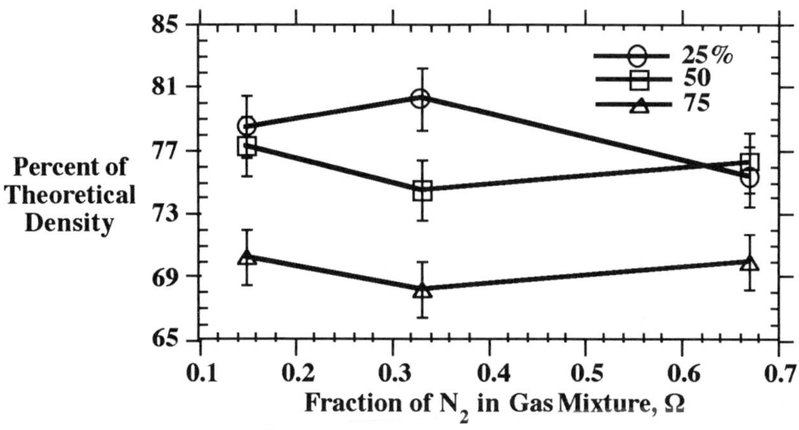

(b)

Figure 4 Plots showing the effects of nitrogen dilution on the extent of densification. Plot (a) shows the extent of densification along the axis of the specimen where the specimen inlet is at 0. Plot (b) shows the extent of densification at various locations along the normalized length of the specimen as a function of Ω.

rate because the concentration of $AlCl_3$ becomes lean. As ß is decreased, even though the reaction efficiency is high, the concentration of $AlCl_3$ is so low that the deposition rate is low. Also, at high levels of ß the deposition rate decreases since the efficiency is very low. In this case, most of the $AlCl_3$ passes through the specimen unreacted because the concentration of H_2/CO_2 is lean, resulting in a deficiency of water.

Figure 6 shows a minimum extent of densification for ß equal to 0.20. Since the deposition rate is highest at this ß, the reactant gases become depleted closer to the inlet of the specimen and result in a lower density near the outlet. For both ß>0.20 and ß<0.20, the deposition rate decreases leading to deeper penetration depths of the

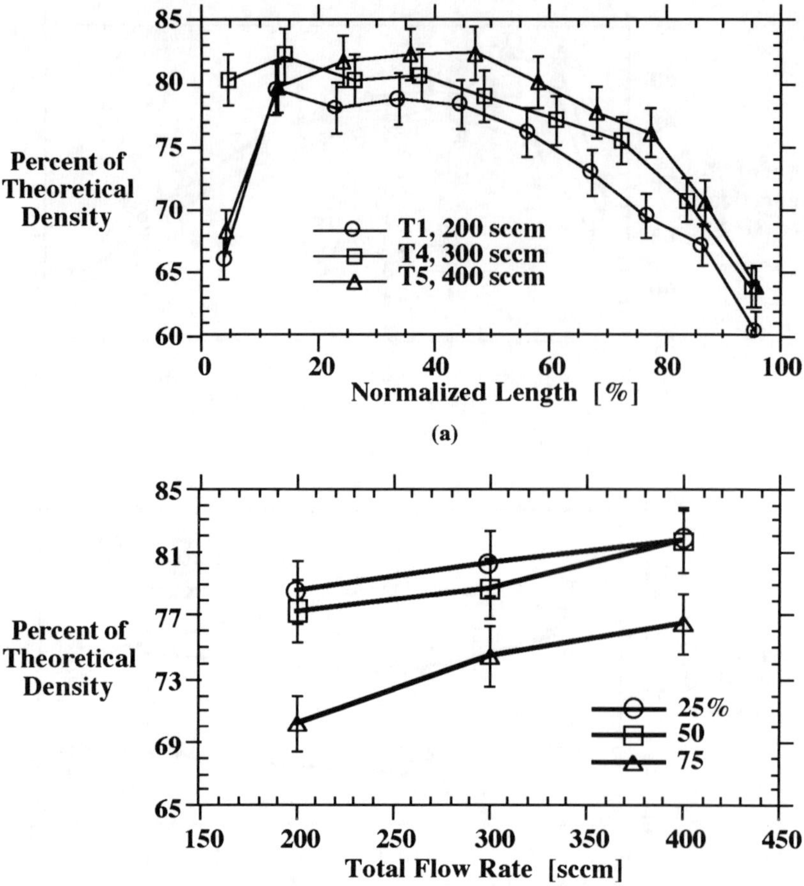

(b)

Figure 5 Plots showing the effect of total flow rate on the extent of densification, (refer to the caption of Fig. 4).

reactant gas into the specimen before reaction. This results in increased densities farther downstream into the specimen before closure of the porosity.

As ß decreases, the density profiles become more symmetric about the middle of the specimen. Since the concentration of H_2/CO_2 is increased, the water conversion reaction occurs more readily upstream of the middle. The hydrolysis of $AlCl_3$ is fast and efficient and depletes the gas mixture of $AlCl_3$ resulting in very little deposition occurring farther downstream. At ß equal to 0.04, the efficiency of the reaction is so high at 86% that most of the deposition occurs in the middle of the specimen, resulting in very low densities near the outlet. Thus an optimum ß is observed at about 0.10 which achieves the highest density and most uniform dense specimens.

The highest density achieved was about 85% of the theoretical density for dense alumina. This may be the highest achievable density since the porosity

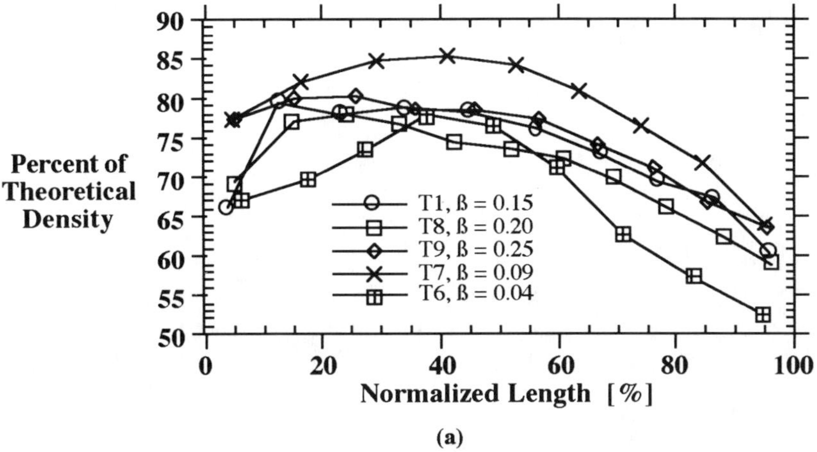

(a)

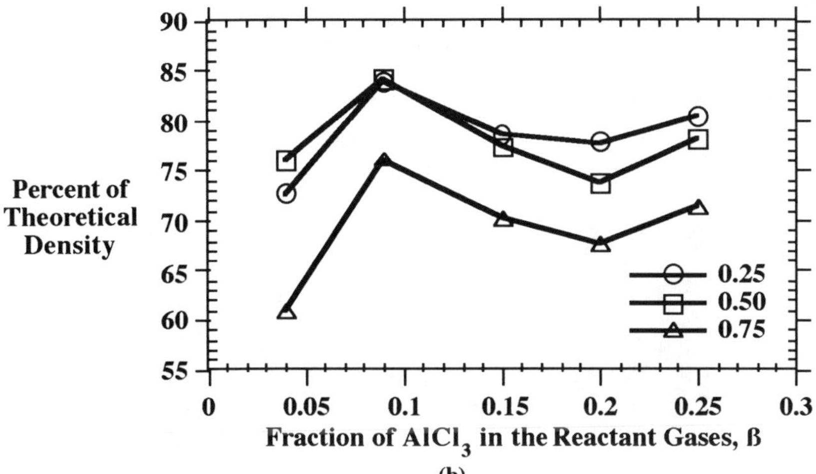

(b)

Figure 6 Plots showing the effect of reactant gas composition on the extent of densification, (refer to the caption of Fig. 4).

perpendicular to the axis of the fibers is depercolated and the gas permeability parallel to the axis of the fibers is very low. This causes the pressure drop across the specimen to increase greatly, (i.e., P_{up} > 500 torr), so that the process has to be aborted to avoid a gas phase reaction which forms alumina powder at the inlet of the specimen.

The deposits are α–Al_2O_3 and have a columnar crystal structure as shown in Figure 7a. The crystals are about 3 to 6 microns in diameter and are highly faceted. When two matrix deposits impinge upon one another they tend to trap small pores, (Figure 7b). These types of pores will always occur in a CVI process but are not considered to be detrimental to overall strength of the composite material due to their small size and the occurrence of much larger pores.

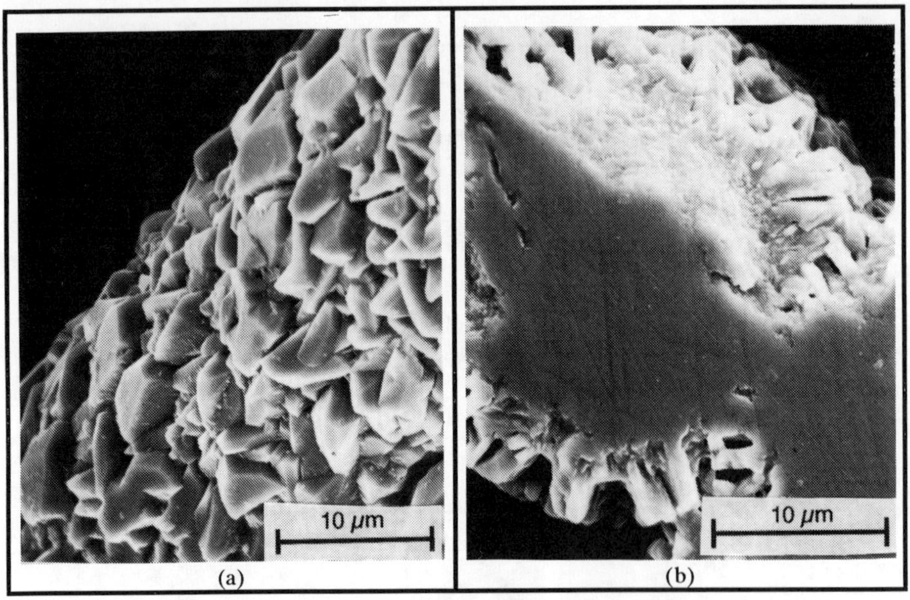

Figure 7 SEM micrographs of (a) the crystalline matrix deposits and (b) the matrix deposits of two fibers impinging upon one another.

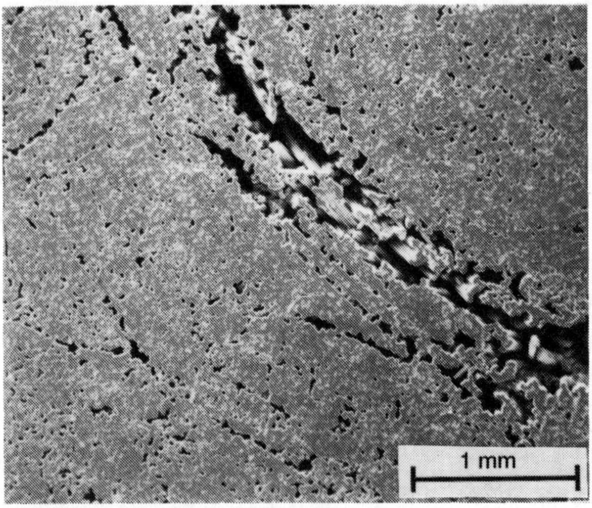

Figure 8 SEM micrograph of an 82% dense section of a specimen. Note the large defect caused by entangled fibers. Also note the larger inter–tow pores and the smaller intra–tow pores between fibers.

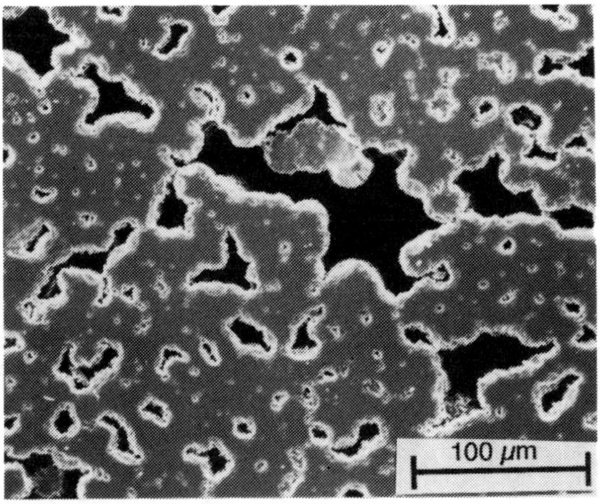

Figure 9 SEM micrograph of a large inter–tow pore and smaller intra–tow pores found in the lower left corner of Fig. 8.

There are three types of defects found in the 82% dense specimen that could effect both the strength and toughness, (Figure 8). The largest defect is about 0.3 to 0.6 mm wide and spans tens of millimeters through the specimen. This defect is caused by entanglement of fibers at the boundary between groups of tows. It is the most harmful defect, but can be avoided by proper design and arrangement of the fibers inside a non–deposited specimen.

Figure 9 shows a typical inter–tow pore found widespread throughout the dense composite. These pores are typically 50 to 100 microns wide and are found at the boundaries between tows. The final size and distribution of these pores will ultimately determine the strength and toughness of the dense fiber composites produced through CVI. They are difficult to eliminate because they result from the random layup of uniaxial fibers in a specimen. The final type of pore, also shown in Figure 9, is about 10 to 15 microns wide and is found everywhere in the dense composites. These intra–tow pores are formed between impinging fibers and will always occur in CVI processes.

CONCLUSIONS

Microwave–assisted CVI has been used successfully to produce an 85% dense alumina fiber/alumina matrix composite. A hybrid heating arrangement was used to heat the specimens, resulting in predicted temperature gradients across the specimens of 50°C at a process temperature of 1000°C. Nitrogen dilution of the reactant gases was shown to have a negligible effect on the extent of deposition inside the specimens. Optimum conditions for the CVI process should include high flow rates and gas mixtures with a ratio of moles of $AlCl_3$ to the total moles of reactants equal to about 0.10 to produce uniform dense composites. The strength and toughness will be limited by the size of the largest inter–tow pores, which will always be found in a CVI process.

ACKNOWLEDGMENT

This work was supported by the Advanced Research Projects Agency through the Office of Naval Research under Contract No. N00014-90-J-4020.

REFERENCES

[1]D.P. Stinton, A.J. Caputo, and R.A. Lowden, "Synthesis of Fiber-Reinforced-SiC Composites by Chemical Vapor Infiltration," *Am. Ceram. Soc. Bull.,* **65** [2] 347-50 (1986).

[2]M.S. Spotz, D.J. Skamser, P.S. Day, H.M. Jennings, and D.L. Johnson, "Microwave-Assisted Chemical Vapor Infiltration," *Ceram. Eng. and Sci. Proc.,* **14** [9-10] 753-760 (1993).

[3]D.J. Skamser, P.S. Day, M.S. Spotz, H.M. Jennings, and D.L. Johnson, "Hybrid Microwave-Assisted CVI of Alumina Fiber Composites"; in *18th Annual Conference on Composites and Advanced Ceramics,* (Cocoa Beach, FL January, 1994). The American Ceramic Society, Westerville, OH, 1994.

[4]P.S. Day, D.J. Skamser, M.S. Spotz, H.M. Jennings, and D.L. Johnson, "Microwave-Assisted Chemical Vapor Infiltration"; in *Silicon-Based Materials,* PAC Rim Meetings. (Honolulu, HI November, 1993). The American Ceramic Society, Westerville, OH, 1994.

[5]D.J. Devlin, R.P. Currier, R.S. Barbero, B.F. Espinoza, and N. Elliott, "Microwave Assisted Chemical Vapor Infiltration"; pp. 245-50 in Chemical Vapor Deposition of Refractory Metals and Ceramics II, Vol. 250, *Mater. Res. Soc. Symp. Proc.* Edited by T.M. Besmann, B.M. Gallois and J.W. Warren. Materials Research Society, Boston, MA, 1992.

[6]D.L. Johnson, D.J. Skamser, and M.S. Spotz, "Temperature Gradients in Microwave Processing: Boon and Bane"; pp. 133-46 in Ceramic Transactions, Vol. 36, *Microwaves: Theory and Application in Materials Processing II.* Edited by D.E. Clark, W.R. Tinga and J.R. Laia, Jr. The American Ceramic Society, Westerville, OH, 1993.

[7]M.S. Spotz, D.J. Skamser, and D.L. Johnson, "Thermal Stability of Ceramic Materials in Microwave Heating," *J. Amer. Ceram. Soc.,* accepted for publication (1994).

[8]A. De', I. Ahmad, E.D. Whitney, and D.E. Clark, "Effect of Green Microstructure and Processing Variables On the Microwave Sintering of Alumina"; pp. 283-88 in Microwave Processing of Materials II, Vol. 189, *Mat. Res. Soc. Symp. Proc.* Edited by W.B. Snyder, Jr., W.H. Sutton, M.F. Iskander and D.L. Johnson. Materials Research Society, Pittsburgh, PA, 1991.

[9]M.A. Janney, C.L. Calhoun, and H.D. Kimrey, "Microwave Sintering of Solid Oxide Fuel Cell Materials: I, Zirconia-8 mol% Yttria," *J. Am. Ceram. Soc.,* **75** [2] 341-46 (1992).

[10]L.-W. Tai and P.A. Lessing, "Modified Resin-Intermediate Processing of Perovskite Powders: Part II" *J. Mater. Res.,* **7** [2] 511-19 (1992).

[11]N.Q. Minh, "Ceramic Fuel Cells," *J. Am. Ceram. Soc.,* **76** [3] 563-88 (1993).

[12]S.T. Song, H.Y. Pan, Z. Wang, and B. Yang, "Synthesis, Properties and Application of High Conductive $LaCrO_3$-Based Ceramic Materials," *Ceram. Int.,* **10** [4] 143-46 (1984).

[13]D.J. Skamser and D.L. Johnson, "Simulation of Hybrid Heating"; *Microwave Processing of Materials IV,* Materials Research Society Symposium (San Francisco, CA April, 1994). Materials Research Society, Pittsburgh, PA, 1994.

[14]R. Colmet, R. Naslain, and P. Hagenmuller, "Thermodynamic and Experimental Analysis of Chemical Vapor Deposition of Alumina From $AlCl_3$-H_2-CO_2 Gas Phase Mixtures," *J. Electrochem. Soc.: SOLID-STATE SCIENCE AND TECHNOLOGY,* **129** [6] 1367-372 (1982).

[15]N.-H. Tai and T.-W. Chou, "On the Deposition Mechanism of Al_2O_3 in the CVI Process for Forming Ceramic Composites," *J. Mater. Res.,* **5** [10] 2255-56 (1990).

PROCESSING PARAMETERS OF PULSE CVI IN C/C COMPOSITES

Hee Dong Park, Hyeok Je Jeong, Young Min Choi and Jae Do Lee

Engineering Materials Lab.
Korea Research Institute of Chemical Technology
P.O. Box 107 Yousoung, Taejeon 305-606 KOREA

Abstract

Carbon/carbon composites were fabricated by polymer impregnation and carbonization, and followed by pulse CVI by using a propane gas.

The processing parameters of pulse CVI, such as temperature, gas concentration, flow rate, introducing time, evacuation time and the number of pulse, were studied using Robust Design method. The most effective parameters of pulse CVI to manufacture carbon/carbon composites are temperature, number of pulse and introducing time of reactant gas.

Introduction

Among the processes of fiber-reinforced composites, chemical vapor infiltration (CVI) has received considerable attention over the last 15 years.[1-4] In this process, a source gas is flowed through a fiber preform and the specific material is deposited as a matrix between the fibers. The largest advantage of CVI are near net shaping process and excellent adhesion of matrix to fiber. However, the process requires a lengthy operation time due to block of the flow path and generation the by-product gas from the deposition site.

Bryant[5] has proposed an unique process of "pulse CVD" in which cycles of evacuation of the vessel and introduction of the gas are repeated sequentially. Sugiyama et al.[6-8] applied this pulse CVD process to CVI expecting prominent reduction of operation time.

To the extent authorized under the laws of the United States of America, all copyright interests in this publication are the property of The American Ceramic Society. Any duplication, reproduction, or republication of this publication or any part thereof, without the express written consent of The American Ceramic Society or fee paid to the Copyright Clearance Center, is prohibited.

In the present paper, the processing parameters of pulse CVI using Robust experimental design method were studied for the preparation of carbon/carbon composites from a gas of propane.

Experiments

The preform samples were prepared by liquid impregnation of phenolic resin to 40-50 sheets of carbon cloth and followed by carbonization at 1000°C for 2hr. The only once impregnated and carbonized carbon/carbon preforms were used as the sample of this pulse CVI. The phenol based carbon/carbon samples were infiltrated with carbon from a gas system of propane and nitrogen using an apparatus of pulse CVI shown in Fig.1. 5 samples of 25mmx20mmx15mm were placed in the reaction chamber. The sequential steps of gas filling and evacuation were controlled by programmed controller via electromagnetic valve. The volume of reaction chamber was minimized by insertion of a bottom-closed inner tube.

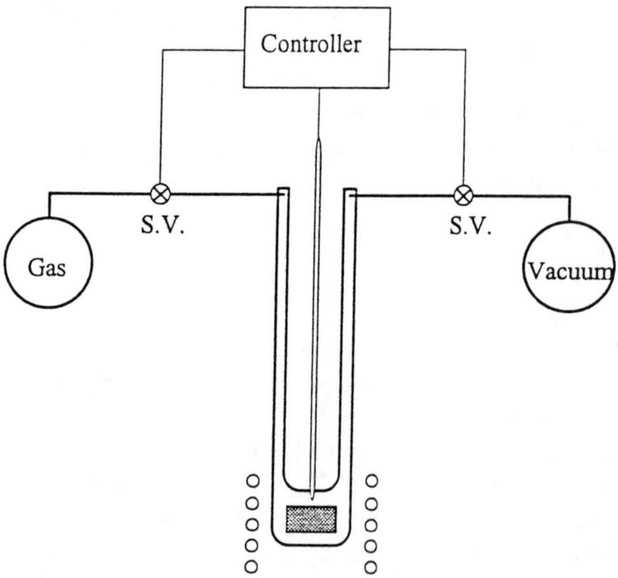

Fig.1. Schematic diagram of pulse CVI process system

In order to determine the important parameters in this pulse CVI process, Robust experimental design method was adapted. The processing parameters, such as temperature, gas concentration, flow rate, introducing time of reactant gas, evacuation time of product gas, and the number of pulse, were considered as effective parameters for the densification of carbon/carbon composites in a pulse CVI process. The factors and their chosen levels are listed in Table 1. Two setting are chosen for each parameter to cover the range of interest. The matrix experiment selected for this study is given in Table 2. It consists of eight individual experiments corresponding to the eight rows. The seven columns of the matrix represent the four factors and three interaction effect as indicated in the table. The entries in the matrix represent the levels of the factors. The matrix experiment of Table 2 is the standard orthogonal array L_8 of Taguchi and Wu[(9)].

The density and porosity were measured by water immersion method (ASTM-C20-87). Hitachi S2150 and Autopore II 9220 were used for SEM observation and measurement of pore size distribution, respectively.

Table 1. Factors and their levels of the experimental design

Factor		level 1	level 2
T	Temperature(°C)	900	1100
C	Gas concentration(%)	20	100
F	Flow rate(SCCM)	600	1800
TG	Introducing time(sec)	2	10
TV	Evacuating time(sec)	10	30
N	Number of pulse	1000	5000

Table 2. Matrix experiment of orthogonal array L_8

Exp.	T	C	F	TG	TV	TGx TV	N
#1	1	1	1	1	1		1
#2	1	1	1	2	2		2
#3	1	2	2	1	1		2
#4	1	2	2	2	2		1
#5	2	1	2	1	2		2
#6	2	1	2	2	1		1
#7	2	2	1	1	2		1
#8	2	2	1	2	1		2

Results and Discussion

Fig.2 shows the apparent density and porosity of pulse CVIed samples in L_8 experimental design. The phenol-based carbon/carbon samples untreated by pulse CVI has a apparent density of 1.34 g/cm^3 and a porosity of 21.0%. After conducting L_8 experiment of pulse CVI, the density increased and the porosity

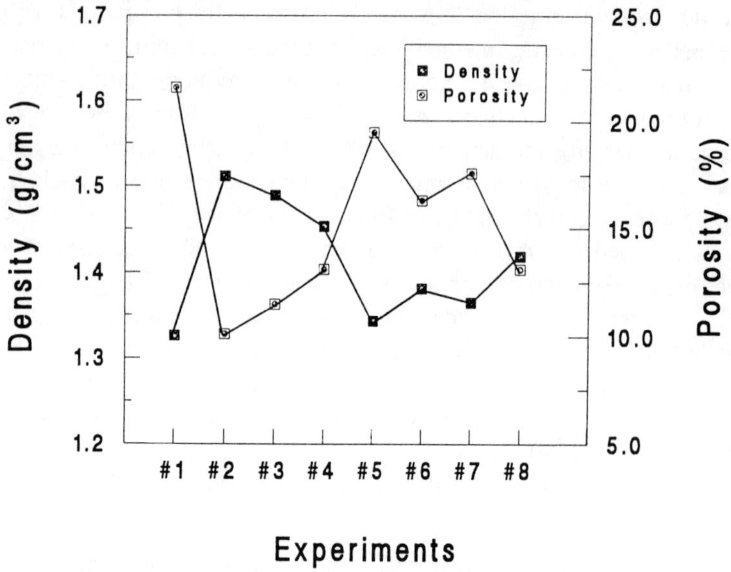

Fig.2. The apparent density and porosity of the carbon/carbon composites obtained from L_8 experimentation.

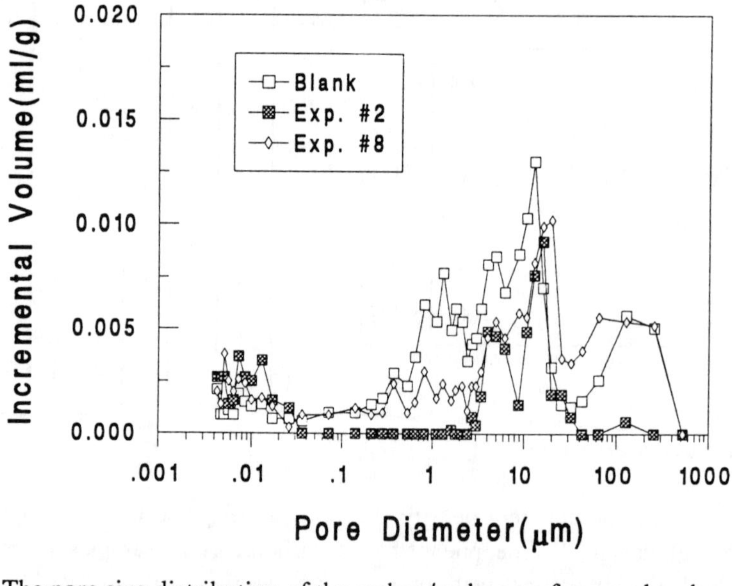

Fig.3. The pore size distribution of the carbon/carbon preform and carbon/carbon composites conducted by pulsed CVI experimentats.

decreased at most experimental conditions except for the experiment #1. In experiment #2, the density increased to 1.51 g/cm^3 and the porosity decreased to 10.1%, which shows the best results. The density increment is larger for experiments conducted at 900°C than those of at 1100°C. The reason may be attributed to the slow deposition rate or some desorption at higher temperature.

Fig.3 shows the pore size distribution of some samples measured by mercury porosimeter. The blank sample untreated to pulse CVI has a porosity of 21% including three ranges of pore size ; 30~400 μm, 0.3~20 μm, and 0.005~0.02 μm. After experiment #2, the pore size range of 30~400 μm has been disappeared and the ranges of pore sizes with 0.3~20 μm has been narrowed to the pore size range of 3~20 μm. On the other hand, the volume of very fine pore sizes with 0.005~0.02 μm increased a little bit. The other experiments #3 and #4 which were conducted at 900°C show a similar pattern. After experiment #8 conducted at 1100°C, the volumes of large pore sizes 30~400 μm and very fine pore sizes with 0.005~0.02 μm are increased somewhat, and the volume of pore sizes of 0.3~20 μm is decreased a little. There are also a similar pattern at the other experiments conducted at 1100°C.

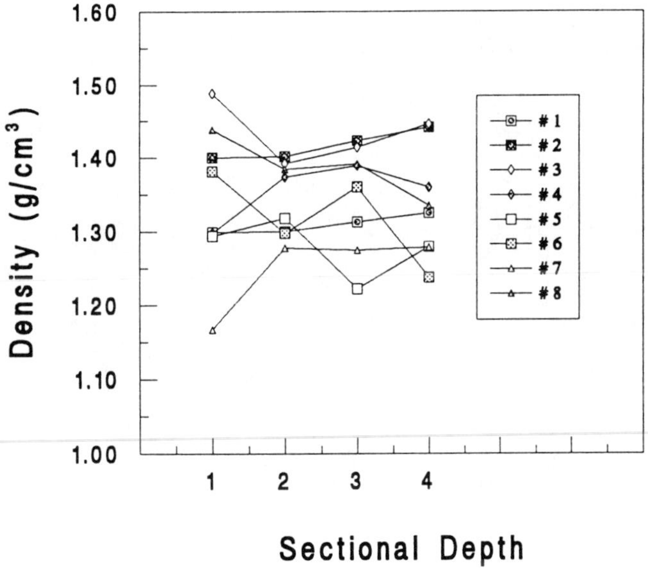

Fig.5. The density profile as a function of sectional depth for the carbon/carbon composites obtained from L_8 experimentation.
Sectional depth 1represents the surface and 4 irepresents the center

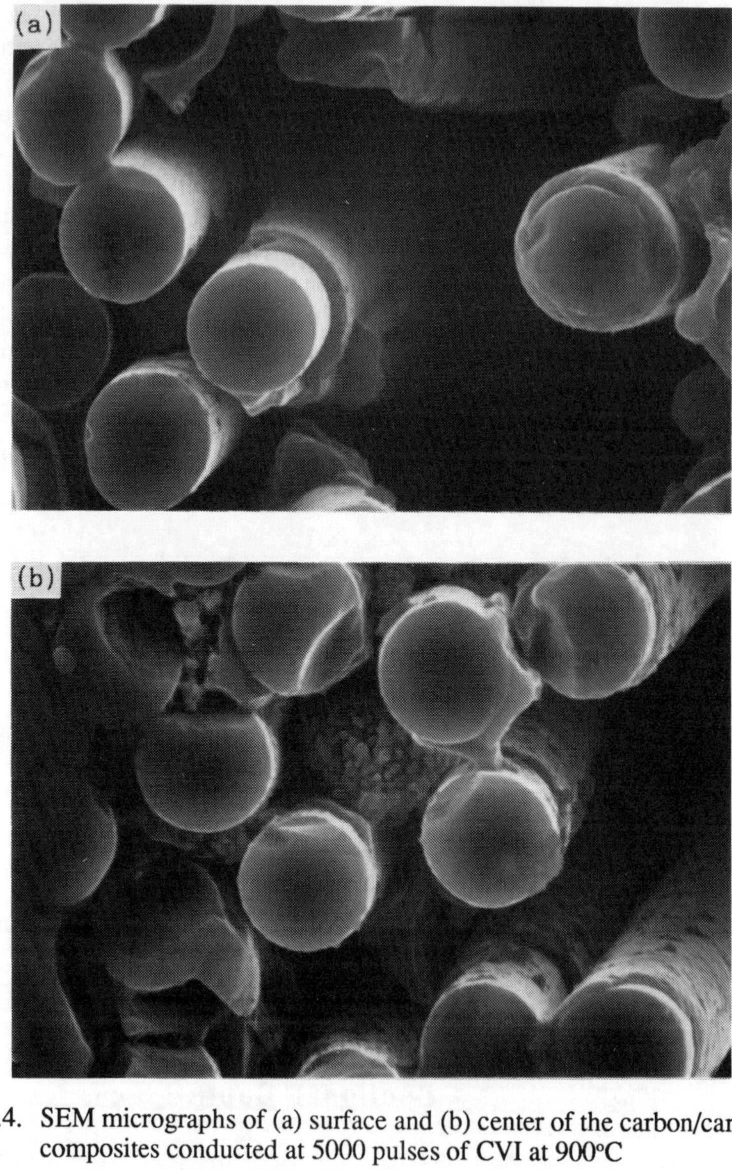

Fig.4. SEM micrographs of (a) surface and (b) center of the carbon/carbon composites conducted at 5000 pulses of CVI at 900°C

Fig.4 shows the cross-sections of surface and center after 5000 pulses of CVI at 900°C. About 1 μm thin coatings on the fiber can be seen at the surface regions, but a little less coatings at the center. From the SEM photographs, no marked difference of infiltration can be seen between the surface and the center of the same carbon/carbon composites.

From the SEM photographs the infiltration uniformity is confirmed a little bit. In order to measure the infiltration uniformity more detailed, the density of seven sliced samples of each one is measured after pulsed CVI and the density profile is shown in Fig.5. As mentioned earlier at results of density and porosity, the density profile of experiment #2 shows very uniform infiltration especially.

Fig.6 shows factor levels and 2-factors level interaction effects on the density of carbon/carbon composites obtained from L_8 pulse CVI experiments. The proper experimental conditions for pulse CVI are at low temperature, higher concentration of gas, higher introduction time, higher interaction of introduction and evacuation time, and higher number of pulses. However, parameters of flow rate and evacuation time is not much affected on the density.

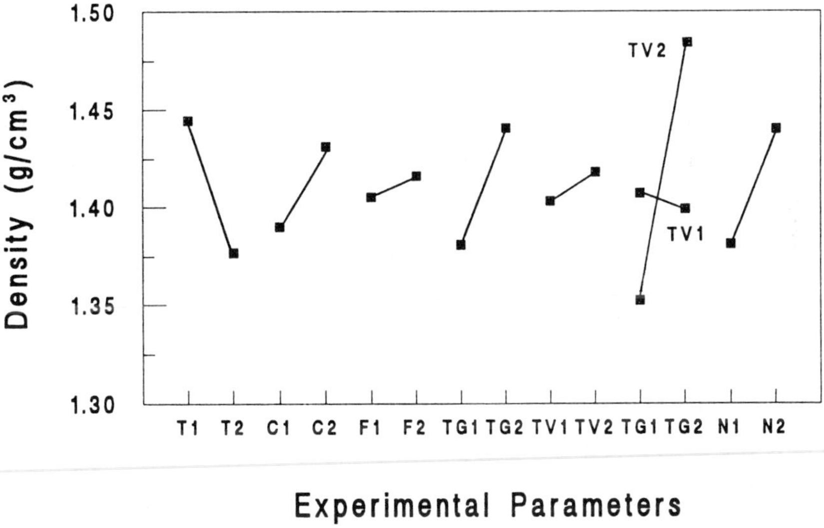

Fig.6. Factor level and 2-factor level interaction effects on the bulk density of carbon/carbon composites obtained from L_8 experimentation. (T_1=900°C, T_2=1100°C, C_1=20% C_3H_8, C_2=100% C_3H_8, F_1=600 SCCM, F_2=1800 SCCM, TG_1=2 sec, TG_2=10 sec, TV_1=10 sec, TV_2=30 sec, N_1=1000, N_2=5000)

Table 3 shows a result of analysis of variance of density in carbon/carbon composites obtained from L_8 pulse CVI experiments. The last column represent percent contribution of source to the density. The most important processing parameters in pulse CVI to density of the carbon/carbon composites are temperature, number of pulse, and introduction time of gas. The interaction of introduction and evacuation time and concentration of gas are also considered as somewhat important processing parameters for pulse CVI process.

Table 3. Analysis of variance of bulk density in carbon/carbon composites obtained from L_8 experimentation.

Source	Pool	Df	S	V	F	S'	rho%
T	[N]	1	0.0183	0.0183	241.6531	0.0182	27.69
C	[N]	1	0.0068	0.0068	89.4473	0.0067	10.17
F	[N]	1	0.0005	0.0005	6.2502	0.0004	0.60
TG	[N]	1	0.0142	0.0142	187.8790	0.0141	21.50
TV	[N]	1	0.0009	0.0009	11.6938	0.0008	1.23
TGxTV	[N]	1	0.0107	0.0107	140.8432	0.0106	16.09
N	[N]	1	0.0139	0.0139	183.2547	0.0138	20.97
e_1	[N]	0	0.0000	0.00008			
e_2	[Y]	8	0.0006	0.00008		0.0011	1.73
(e)		8	0.0006				
Total (Raw)	[-]	15	0.0658	0.0044			

Source : Factors and errors, Pool: Pooling toggle for source, Df: Degree of freedom, S: Sum of squares of source, V: Variance of source, F: F value of source, S': Pure sum of squares of source, rho%: Percent contribution of source, e_1: Interaction error, e_2: Repetition error, (e): Pooled error

Summary

The effects of processing parameters on the densification of carbon/carbon preforms in the pulse CVI process were examined using Robust experiment design. The results are summarized as follows.

1) The density increment of 12.5% was obtained after pulsed CVI at 900°C for 20hr.
2) The most effective processing parameters of pulse CVI on densification of carbon/carbon composites are temperature, number of pulse, and introducing time of reactant gas.

Acknowledgements

The authors are grateful for partial support of this work by Daewoo Heavy Industries LTD Aerospace Product Division.

References

1. A.J. Capto, W.J. Lackey and D.P. Stinton, Ceram. Eng. Sci. Proc. 6 (1989) 694.
2. H. Hannache, J.M. Quenisset, R. Naslain and H. Heraud, J. Mater. Sci., 19 (1984) 202.
3. R. Naslain and F. Langlais, Mater. Sci. Res., 20 (1986) 145.
4. H. Tauil, D. Bentseen, S. Baskaran and D.P.H. Hasselman, J. Mater. Sci., 20 (1985) 3201.
5. W.A. Bryant, J. Crystal Growth, 35 (1976) 257.
6. K. Sugiyama and T. Nakamura, J. Mater. Sci. Lett., 6 (1987) 331.
7. K. Sugiyama and Y. Ohzawa, J. Mater. Sci. Lett., 7 (1988) 1221.
8. K. Sugiyama and E. Yamamoto, J. Mater. Sci., 24 (1989) 3756.
9. G. Taguchi and Y. Wu, Introduction to off-line quality control, Meieki Nakamura-Ku Magaya, Japan, 1979.

STRENGTH AND FATIGUE BEHAVIOUR OF 2D-CARBON/CARBON COMPOSITES UNDER SHEAR CONDITIONS,

T.Fend, J.Göring,

German Aerospace Research Establishment (DLR), Materials Research Institute, D-51147 Cologne, Germany.

ABSTRACT

In this study flexural tests under cyclic shear loads (R›0) were performed with different two dimensional reinforced carbon/carbon materials, produced under different processing conditions. During fatigue testing a continuous increase of damage density (characterized via stiffness degradation) was observed, caused by the "wear out" of graphitic matrix carbon (characterized via energy absorption in load-deflection hysteresis loops) which eventually leads to a time-dependent failure. This study includes SEM fractography and a microstructural accessment by TEM.

I. INTRODUCTION

Today's C/C-Composites are used for different technical applications, such as aircraft brakes, rocket motors or heatshields for re-entry vehicles [1]. Besides their good thermal shock behaviour, remarkable damage tolerance and low specific weight, they offer excellent specific strength properties up to very high temperatures above 2000°C [2]. With oxidation resistant ceramic coatings C/C is a promising structural ceramic matrix composite material for all high temperature/low weight applications especially in the field of aircraft and spacecraft industry.

Besides the improvement of oxidation resistance and the development of time and cost saving production techniques, present research programs focus on the improvement of the weak interlaminar properties.

Because of the weak interlaminar bonding in 2D C/C, shear is one of the most significant failure mechanisms of this composite material (see Table 1).

To the extent authorized under the laws of the United States of America, all copyright interests in this publication are the property of The American Ceramic Society. Any duplication, reproduction, or republication of this publication or any part thereof, without the express written consent of The American Ceramic Society or fee paid to the Copyright Clearance Center, is prohibited.

Table 1: Ratio of Tensile and Interlaminar Shear Strength for different materials: GFRP/CFRP: glassfiber/carbonfiber reinforced polymer

Material	SiC/Sialon	GFRP	CFRP	C/C	Metal
σ_{TS}/τ_{ILS}	2.5	15-25	10-30	80-100	1.25

The main object of this paper is to study the damage and failure mechanisms of C/C composites under monotonic and cyclic shear loads at room-temperature. Besides short-term exposal at high service temperatures a range of 200°C‹T‹200°C is of primary importance for long-term orbit missions due to solar heating and cooling.

II. MATERIALS

The material investigated, CF222, is produced commercially by Schunk Kohlenstofftechnik, Gießen, Germany. It consists of woven carbon-fibre plies bound with a carbon matrix based on a thermosetting resin matrix precursor.

For this study the processing route of this material was slightly modificated in two ways thus introducing partly different microstructural properties. Including the non-modificated standard-material CF222, three types of C/C were employed. Abbreviations used throughout this paper are i) C/C-18, ii) C/C-21 (equivalent to CF222) and iii) C/C-24, corresponding to graphitization temperatures of 1800°C, 2100°C and 2400°C respectively.

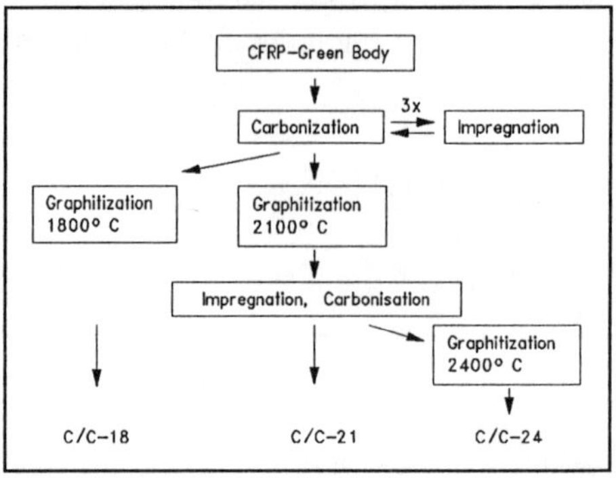

Figure 1: Processing route of the three C/C materials tested

The fibre bundles consisting of approximately 3K PAN based HT carbon fibres form a woven two-dimensional fabric (8H-satin weave) and can thus either be called 0°-bundles (warp direction) or 90°-bundles (fill-direction). They are impregnated with a phenolic resin getting a common CFRP-material. The next processing steps are illustrated in Figure 1. The various heat and densification treatments lead to a pure carbon matrix. A final impregnation step after a graphitization at 2100° (C/C-21) is used to reduce the porosity to approximately 5%. Additional data concerning density and porosity can be taken from Table 2.

Table 2: Density and Porosity of the C/C materials tested

	CFRP	C/C-18	C/C-21	C/C-24
Density (g/cm^3)	1.48	1.46	1.55	1.57
Porosity (%)	4.0	6.3	5.2	7.1

The porosity is caused by a 50% shrinkage of the phenolic resin during the carbonization and graphitisation process. The volume loss turns out to be highly anisotropic, because there is negligible shrinking of the bundles in the direction of the fibre axis. Thus, every cross-section of a 3K fibre bundle is separated by large *intra-bundle cracks* into several segments (see 2 in Figure 2) . Another important "weak link" of the material is defined by the *inter-bundle porosity* (see 1 in Figure 2), which typically develops between parallel fibre bundles. Eventually relative movements of adjacent 0° and 90°-bundles during carbonization lead to *local delaminations*.

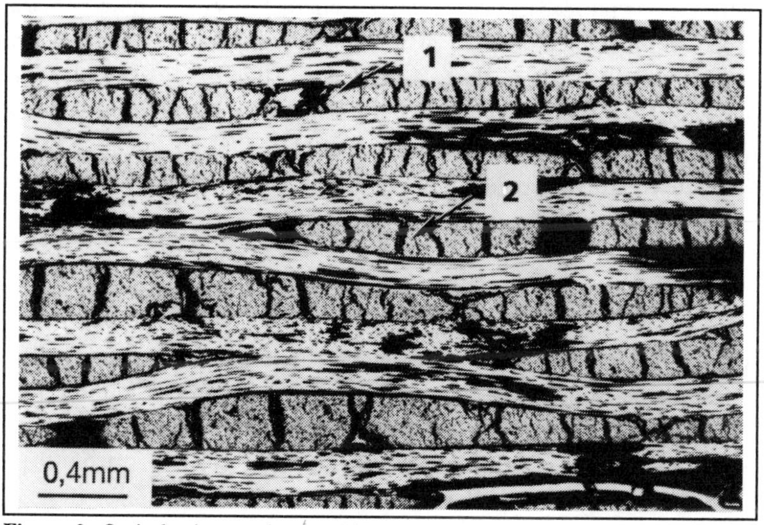

Figure 2: Optical micrograph of a C/C-21 sample (as received) displaying inter-bundle porosity (1) and bundle segmentation (2) both introduced during processing

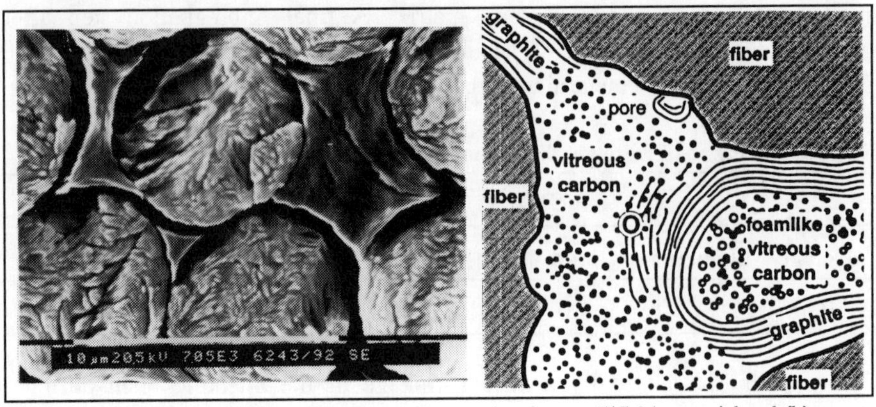

Figure 3: (a) SEM-micrograph of the intrabundle section from a C/C-21-material and (b) structural heterogenity of intra-bundle matrix

The heterogenity of carbon variants in the intra-bundle matrix is revealed through microstructural investigations by TEM [3]. In absence of stresses during carbonization materials involving phenolic resins are known to form a pure vitreous carbon matrix. Due to biaxial stress distributions in parts of the intra-bundle matrix *stress-oriented graphite* is formed especially in the vicinity of the fibre matrix interface (Figure 3). The specimen characterized in the SEM was heat-treated for 20 minutes at 600°C in air. Graphitic carbons are less oxidation resistant than vitreous carbon, thus, regions consisting of stress-oriented-graphite have nearly totally disappeared due to corrosive attack by environmental oxygen. Due to the higher graphitization temperature, the C/C-24 material exhibits the highest amount of stress oriented graphite of the materials tested and consequently C/C-18 the lowest.

III. EXPERIMENTAL METHODS

The 3-point bending-tests were performed with a standard digitally controlled *Wolpert*-testing machine attached to a data storing microcomputer. Short-beam dimensions (35*12*5mm) with a span-to-depth ratio of 5 were used to induce shear loads (ASTM D-2344). Shear-stresses τ can be calculated from the applied load F, the beam height h and the thickness b by

$$\tau = \frac{3}{4}\frac{F}{b \cdot h} \quad . \tag{1}$$

Note that the stress distributions of a short-beam sample do not agree well with classical beam theory used in ASTM D-2344 [4]. The maximum load is taken to calculate the Interlaminar Shear Strength (ILSS), though the validity of equation (1) is restricted to the linear elastic range. Nevertheless, the failure stress calculated by this equation is a useful parameter to compare the different C/C grades and is an appropriate tool to discuss the microstructural failure mechanisms, but it has its limits as a material parameter for design considerations.

Beam deflection was measured by three inductive extensiometers (see Figure 4) to compensate for contact effects and fixture deformations.

Monotonic bending experiments were performed with a crosshead-speed of 0.1 mm/min, cyclic experiments with 1-2 mm/min leading to a comparably low frequency of 0.3 Hz. Approximately 100 load-deflection values per second were taken and stored in order to calculate beam-stiffnesses and energy absorption from the hysteresis-loop of each load-deflection cycle.

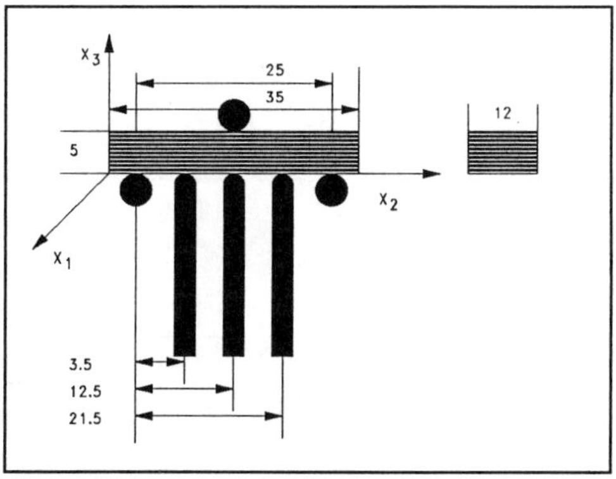

Figure 4: Short-beam bending device with deflection measurement by three inductive extensiometers (distances are given in [mm])

IV. RESULTS AND DISCUSSION

a.) Results of monotonic bending tests:

A total number of 30 samples for each type of material was tested to investigate the ILSS and the load-deflection behaviour. Both, ILSS and the shear-modulus turned out to be significantly dependent on the heat treatment temperature (see Table 2). The best strength performance was established for the reimpregnated C/C-21 material. Figure 5 shows a typical load-deflection plot of a C/C-21 short-beam sample.

Table 2: Elastic Constants and interlaminar shear strength (ILSS) of three different C/C materials in comparison with the corresponding CFRP material (mean values with standard derivations)

	CFRP	C/C-18	C/C-21	C/C-24
ILSS (Mpa)	22.7 ± 1	4.5 ± 0.9	8.3 ± 0.7	5.7 ± 0.6
Shear Modulus G_{13} (Gpa)	-	1.5	2.7	1.6
Young's Modulus, fibre direction E_{11} (Gpa)	-	79	100	82

The deformation behaviour can roughly be divided into 3 regions. The *linear elastic* range (1) is comparably small, because plies and bundles tend to slide even in case the shear load is relatively low. Sliding mechanisms dominate the *inelastic range* (2) of the diagram. Fractographic investigations of the samples in this stage of the experiment do not furnish informations about the deformation mechanism or a damage in the

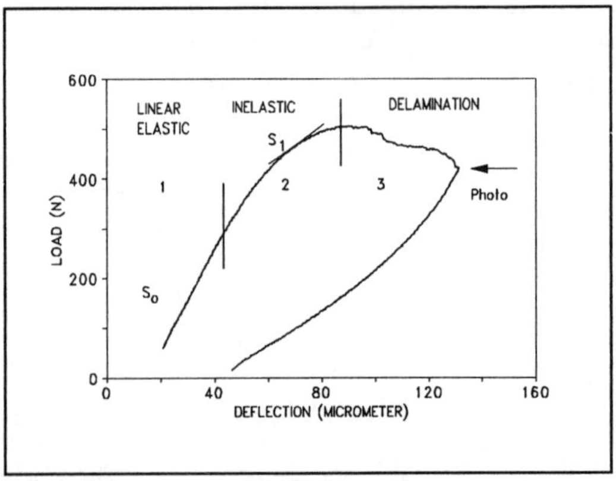

Figure 5: Load-deflection diagram of a C/C-21 short-beam sample

microstructure. In range (3) the specimen shows the typical crack pattern as displayed in Figure 6. Typically the crack propagation takes place in the interbundle interfaces (0°/0° or 0°/90°-bundles) or in the intra-bundle cracks induced during processing.

Cracks may be initiated at macroscopic flaws denoted as *inter-bundle porosity* (Figure 6).

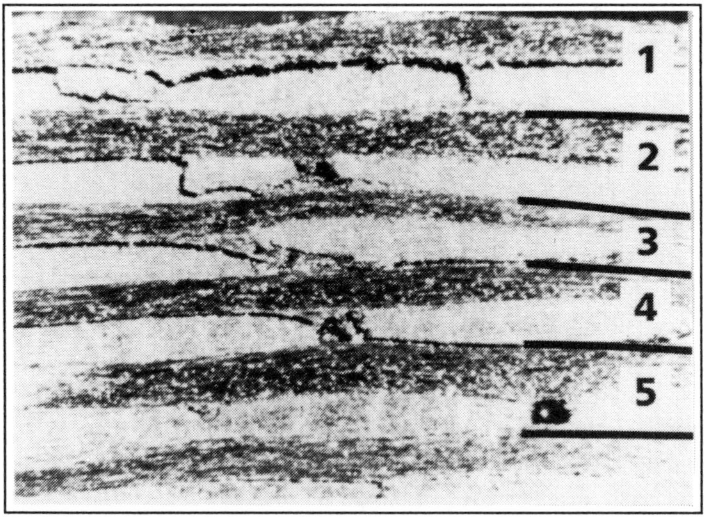

Figure 6: Optical micrograph of a fractured C/C-21 sample (in situ) after monotonic loading (plynumbers marked)

Within each diagram several tangent lines were drawn at different loads F_i in the inelastic range, comparing the slopes S_i with the initial slope S_0 (see Figure 5), and giving an amount of how much the material deviates from linear elastic behaviour. The following power function was used to describe the dependency of the ratios S_i/S_0 on the load level F_i in relation to the maximum load F_{max}.:

$$\frac{F_i}{F_{\max}} = 1 - c \cdot \left(\frac{S_i}{S_0}\right)^k$$

The following parameter values for c and k could then be calculated:
c=0.4 ± 0.1, k=1.41 ± 0.05 (C/C-18)
c=0.3 ± 0.1, k=1.83 ± 0.05 (C/C-21)
c=0.7 ± 0.1, k=1.70 ± 0.05 (C/C-24).

This allows an estimate for the maximum static load F_{max} for an actual sample from the initial slope S_0, any given load F_i and the corresponding slope S_i in region (2), before maximum load is reached. The validity of this procedure could be approved by applying it to the static load-deflection diagrams and it turns out to give the maximum load values within an accuracy of 3%. It was then used to calculate the true τ/τ_{max} values for a sample in the cyclic fatigue tests. Using this procedure the scatter in lifetime-diagrams turned out to be much lower than using the maximum stress value from the monotonic tests.

b.) Results of cyclic bending tests:

Cyclic shear loads were induced to investigate microstructural constraints for the mechanical behaviour of the material in the inelastic regime of the static load-deflection curve.

In Figure 7 the true normalized shear strength is plotted against the number of failure cycles. No significant difference in cyclic fatigue behaviour could be observed between the two material grades C/C-18 and C/C-21. For clarity only the values for C/C-21 are shown. The fatigue behavior of the C/C-24 samples was significantly worse.

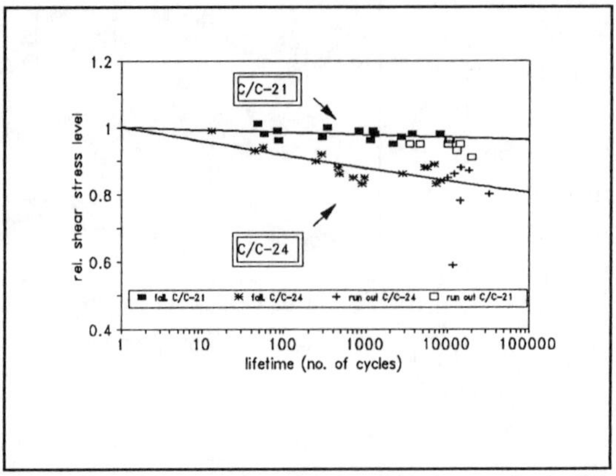

Figure 7: Lifetime diagram comparing the standard C/C-21 material with the C/C-24 version.

Figure 8 presents selected segments of the load-deflection curve of a C/C-24 sample resisting 963 cycles before failure.

The upper load level was 77% of the average maximum load from monotonic tests. In addition to a slight continuous decrease of the specimen stiffness a shape change of a single hysteresis can be seen. The integral value

$$E(n) = \oint_{loop\ n} F\,dx$$

describes the energy absorption of the material in each load cycle (deflection is denoted by x) and can be calculated numerically from the load-deflection data. The stiffness-reduction was calculated approximately from the average slope of each hysteresis loop. Figure 9 shows the development of these two values during the fatigue test of a C/C-24 sample that failed after 6856 load cycles. Along with the decrease in specimen stiffness, the energy absorption continuously increases, thus allowing to predict catastrophic damage several hundred cycles before it eventually occurs.

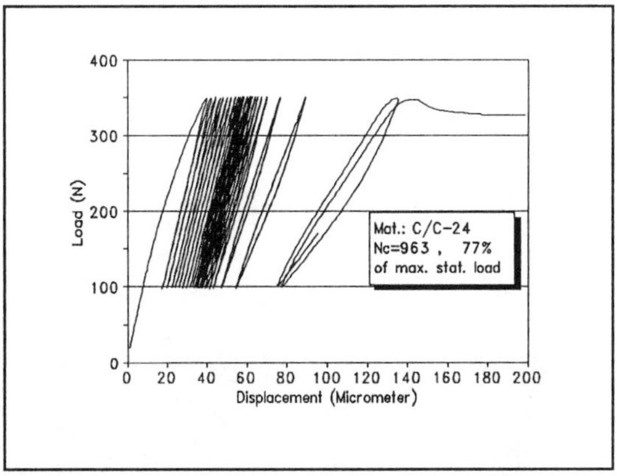

Figure 8: Load-displacement curve of a C/C-24 short beam sample during cyclic loading

The stiffness reduction may be interpreted as an increase of crack density as well as a propagation of cracks, which are initiated due to processing conditions. Thus, the continuous reduction in stiffness corresponds with a continuous increase in damage density.

The measured energy absorbtion is a sensitive indicator for frictional effects occurring during cycling. Repetitive loading and deloading of cracks under mode II conditions (shear load) leads to a sliding of crack surfaces. This sliding is dominated by friction and may consequently be interpreted as an energy dissipative process. The special shape of the stiffness and energy absorbtion curve in Figure 9 demonstrates the dynamics of damage accumulation during cycling. The strong decrease of the absorbed energy during the first 10-15 cycles can be explained by a "wear out" effect of crack- and microcrack surfaces, initiated during processing. The loss in frictional forces leads to an increasing stress intensity at crack tips resulting in an increase in damage density. During further cycles, wear-out occurs in existing and newly formed

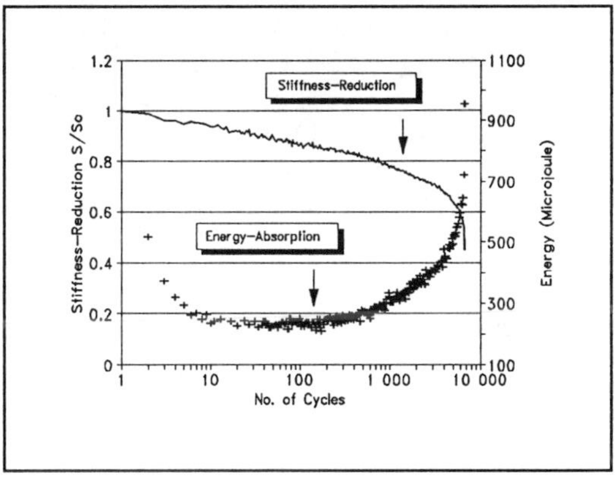

Figure 9: Stiffness reduction and energy absorption of a C/C-24 sample failing after 6856 cycles

surfaces, so that the total amount of "active" surfaces is continuously increasing thus resulting in an increase of the energy curve. In the last stage the energy curve increases exponentially and the stiffness drops down. This region may be explained by reaching a "critical damage state", in which crack coupling occurs, leading eventually to catastrophic interlaminar failure.

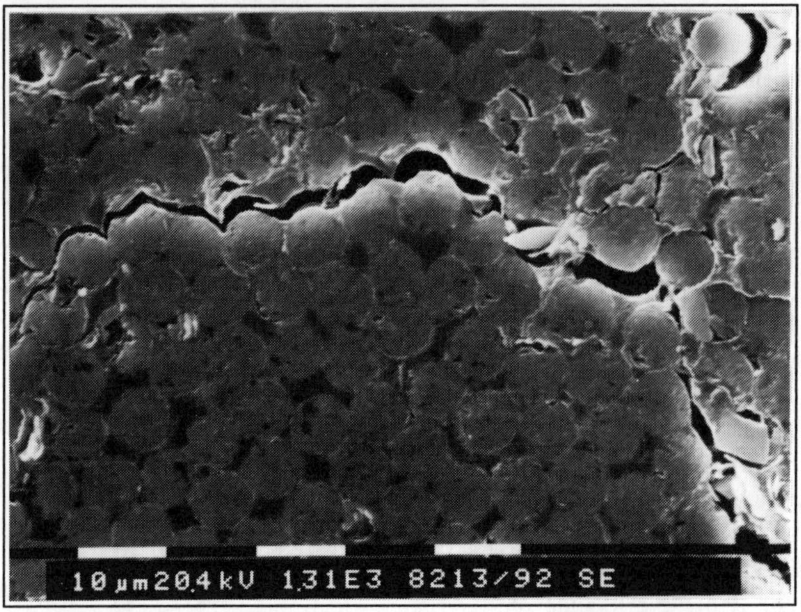

Figure 10: SEM-micrograph from the intrabundle section of C/C-21 sample after cyclic loading (failure after 8008 cycles)

After deloading, a small residual deformation could be observed, but the cracks were mainly closed and could not be identified in the SEM. So a special frame was used to store the deformation and crack opening during the tests in different defined strain stages. Figure 10 and 11 show intra-bundle cracks which were typically occurring in intact or failed specimens after cycling in the inelastic range of the load-deflection diagram (Figure 5). The cracks are propagating along the regions of stress-oriented graphite (compare Figure 2). TEM investigations have documented why graphitic carbon actually promotes crack propagation in intra-bundle regions [3]. Cracks can be deflected around fibres and vitreous carbon (Figure 12), splitting the basal planes of stress oriented graphite. These microstructural observations are in good agreement with the results shown in

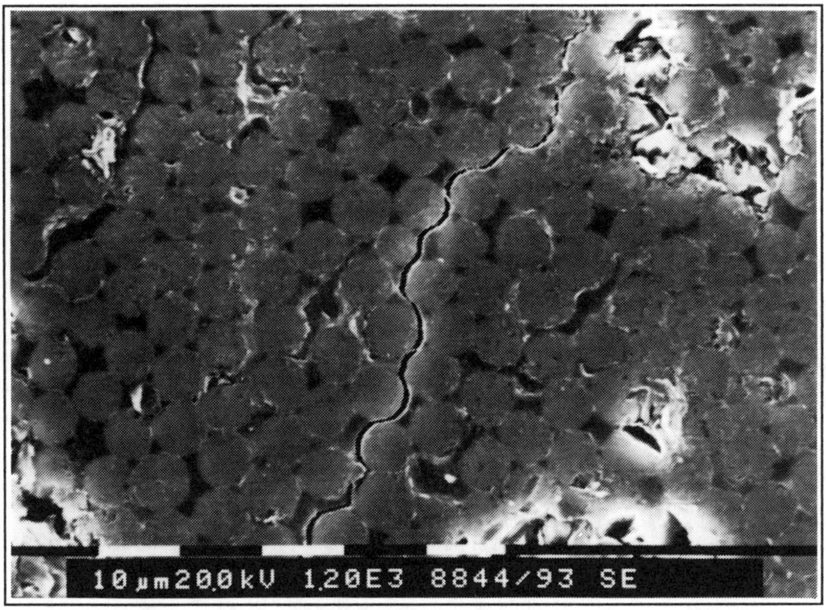

Figure 11: SEM-micrograph from the intrabundle section of C/C-21 sample after cyclic loading (no failure after 20000 cycles)

Figure 7, where the C/C-24 material containing the highest amount of stress oriented graphite, exhibits the worse fatigue behaviour.

V. CONCLUSIONS

A significant dependence of both, the monotonic and cyclic fatigue behaviour on the processing conditions and the microstructural failure mechanisms was established for three different 2D-C/C composites.
The shear strength was primarily controlled by large pores and cracks initiated during processing. The material grade exposed to an impregnation/carbonization step after graphitization shows the highest density and the best strength performance.

The cyclic fatigue behaviour was affected by the amount of stress-oriented graphite in intrabundle regions, which is introduced by the higher graphitization temperatures during fabrication. After cycling, cracks propagating along the basal planes of the stress oriented graphite were observed, which are deflected around

fibres and vitreous carbon.
The "wear out" of frictional effects at the crack surfaces causes an increase of the stress intensity at macro- and microscopic cracks, resulting in a continuous increase of damage density during cycling.

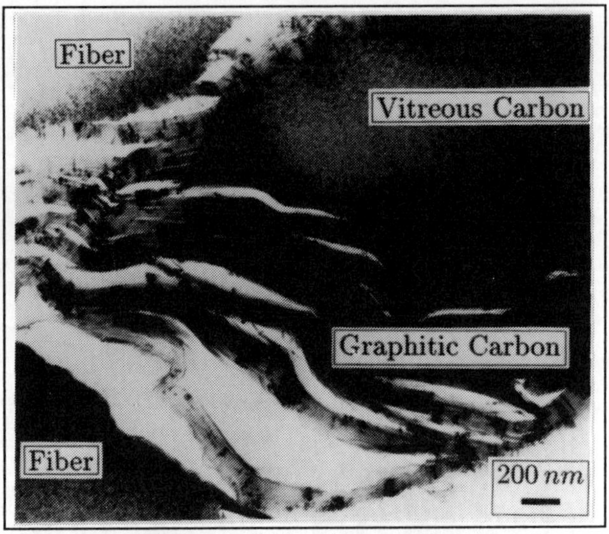

Figure 12: TEM-micrograph from intra-bundle section of a C/C-21 sample emphasizing the typical fracture behaviour of graphitic matrix carbon (Pleger, 1992)

The most interesting result of this investigations was the fact, that the shear strength was limited by macroscopic defects on a macroscopic [mm] scale, however the fatigue behaviour was controlled by nanoscale features, that is the graphite structure in defect free matrix regions of the composite materials.

Acknowledgement: The TEM-investigations cited in this paper are part of a study carried out by *Dr. H.R. Pleger*. His help is thankfully appreciated.

REFERENCES

[1] G. Savage: *Carbon-Carbon Composites,* Chapman & Hall, London (1993)

[2] J. D. Buckley: Carbon-Carbon, An Overview, *American Ceramic Society Bulletin,* **67,** 2 364-368 (1988)

[3] H.R. Pleger, W. Braue: *On the Structural Variety of Carbon in C/C-Composites by Means of TEM and Digital Image-Processing*, 5th International Carbon Conference, Essen/Germany (1992)

[4] J.M Whitney: *Experimental Characterization of Delamination Fracture,* in: N.J. Pagano, Interlaminar Response of Composite Materials Elsevier Science Publishers B.V., Amsterdam (1989)

MULLITE MATRIX CONTINUOUS FIBRE REINFORCED COMPOSITES BY SOL-GEL PROCESSING

J. Wu, F.R. Jones and P.F. James
Department of Engineering Materials
University of Sheffield
Sheffield S1 4DU, UK

ABSTRACT

A fibre reinforced crystalline mullite matrix composite was prepared by sol-gel processing. The unidirectional fibre composite was synthesised by single stage infiltration of carbon fibres with an alumina-silica sol ($3Al_2O_3 \cdot 2SiO_2$), followed by hot-pressing at 1550 °C under 15 MPa to 95% of full density. The composite exhibited non-brittle failure with a mean flexural strength of 703 MPa.

INTRODUCTION

Mullite has been extensively investigated [1-5] as a monolithic or composite matrix material in the recent years due to its excellent strength and creep resistance at elevated temperatures, good chemical stability and relatively low thermal expansion coefficient. The fibre reinforcement for ceramics is an effective method of overcoming their inherent brittleness. Fabrication of mullite ceramics and mullite matrix composites by conventional methods normally requires a high processing temperature in excess of 1700 °C with a long sintering time. Crystalline mullite can be produced by sol-gel methods using alkoxide precursors at a temperature as low as 980 °C [4,5]. Unfortunately, sol-gel processing of composite matrices using alkoxides alone is impractical due to the low solid content of the starting solutions, the excessively high shrinkage in drying and occurrence of crystallisation before densification is complete. If these disadvantages could be overcome, sol-gel processing would be very promising for ceramic composite preparation especially using continuous fibres as a reinforcement [6]. The sol-gel route also has the advantages that the consolidation of matrix and fibre can be realised by a single stage infiltration, so damage to the fibres is less and a uniform distribution of fibres

To the extent authorized under the laws of the United States of America, all copyright interests in this publication are the property of The American Ceramic Society. Any duplication, reproduction, or republication of this publication or any part thereof, without the express written consent of The American Ceramic Society or fee paid to the Copyright Clearance Center, is prohibited.

in the matrix can be easily achieved. Furthermore, no organic binders are needed so there is no burning-off problem and the green strength is high. More importantly, a lower processing temperature can be achieved. In previous work [7,8], a high solid yield sol (up to 40% volume of solid) of the stoichiometric mullite composition of $3Al_2O_3 \cdot 2SiO_2$ was developed to produce monolithic samples of densified mullite (density 3.13×10^3 Kg/m^3, nearly 99% of the theoretical density), and carbon fibre reinforced composites with a non mullitised matrix of alumina particles and silica glass were produced by single stage infiltration followed by hot-pressing.

In this paper, the fabrication of a hot-pressed crystalline mullite matrix carbon fibre reinforced composite is described and the composite is characterised by scanning electron microscopy (SEM) and transmission electron microscopy (TEM). The flexural strength, elastic modulus and fracture toughness of the composite are measured, and the fracture behaviour of the samples studied by SEM.

EXPERIMENTAL PROCEDURE

Grafil LXA unsized high strength carbon fibre tows (theoretical strength 4020 MPa, elastic modulus 224 GPa, density 1.81×10^3 Kg/m^3, diameter 8 μm) were employed as the reinforcement and α-alumina powder (Alcoa A1000 SG 0.5μm) and a silica sol (DuPont Ludox AS40 5-50 nm in size) were used as precursors to form the stoichiometric mullite composition sol of $3Al_2O_3 \cdot 2SiO_2$, for the composite matrix. Preparation of the sol was discussed in detail in earlier studies [7,8]. In brief, however, the pH value of Ludox silica sol was reduced to 3.0 by introducing dilute acid (HCl, 1M) and then a dispersion of α-alumina powder was added using magnetic stirring. The resultant sol had a high solid yield of 40 volume% (alumina-silica mixture), which could be converted to a bulk gel by further addition of dilute acid.

Unidirectional carbon fibre reinforced mullite matrix composite prepregs were fabricated using a single stage infiltration technique. Carbon fibre tows were impregnated by passing them through the sol. They were then wound onto a hexagonal drum to produce a green prepreg of composite of approximately 10 mm in thickness. Composites with fibre volume fractions in the range 0.35 to 0.55 were prepared by adjusting the solid content of the sol. The fibre volume fraction could be finally determined by completely burning off the carbon fibre in the composite prepreg at 900 °C in air. In this study a composite with fibre volume fraction of

0.53 was prepared. The green prepreg was normally dried in air or in an oven at 90 °C until the gel matrix of the composite formed, and then released from the drum. The composite samples were then fired at 1 °C/min. to 500 °C with a hold of 2 h for further drying. The dried samples were cut into circular discs 45 mm in diameter and laid in a graphite die ready for hot-pressing.

The hot-pressing of the carbon fibre composites were carried out in an induction furnace (Inductoheat Banyard Ltd). The graphite die containing the samples was completely sealed by graphite powder during hot-pressing to protect the carbon fibre from degradation. The composite samples were consolidated at a temperature of 1550 °C for 0.5 h with a ramp of 20 °C/min, at a pressure of 15 MPa, and then were allowed to naturally cool to room temperature in the furnace overnight.
The density measurement of the composite was carried out using a mercury immersion technique and the phases present were identified by XRD using a Philips 1710 X-ray Diffractometer with a scanning speed of 2° 2θ/min. SEM and TEM for microstructural analysis were performed on a JEOL JSM6400 and a Philips EM400T respectively. Thin sections for TEM were prepared by ion beam milling of samples mechanically dimpled to 30-40 μm thickness using fine grinding paste. The flexural strength, elastic modulus and toughness of the composites were determined at room temperature using three point bending on a Mayes Universal Tester with a strain rate of 0.2 mm/min. Rectangular test specimens cut from the hot-pressed cylinder samples were ground and polished to a span of 40 mm with a span to thickness ratio greater than 20. The reported mechanical properties are the mean values for eight specimens.

RESULTS

Fig. 1 shows the XRD results of the composite sample hot-pressed at 1550 °C for 0.5 h with a pressure of 15 MPa. The predominant phase present in the matrix of the composite is mullite with some unreacted alumina. The reinforcement phase of carbon fibre in the amorphous state does not show any XRD peak. Preliminary studies [7,8] indicated that the alumina particles started to react with silica to form mullite at a sintering temperature of 1450 °C in this α-alumina-silica system and above 1500 °C a substantial amount of mullite phase started to form at the expense of α-alumina and cristobalite. In the range of 1550-1600 °C the mullite phase became dominant with traces of alumina remaining embedded in the middle of the mullite grains, and small pockets of residual glass at mullite grain triple junctions. The high solid yield sol employed enabled composite samples to be produced with a green density of $1.50 \times 10^3\ \mathrm{Kg/m^3}$, nearly 60% of that calculated

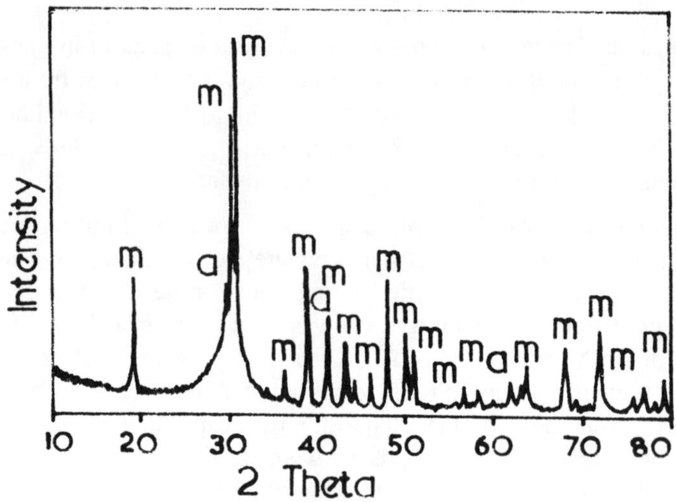

Fig. 1 XRD of composite hot-pressed at 1550 °C for 0.5 h under 15 MPa; Mullite (m) and alumina (a) peaks are shown.

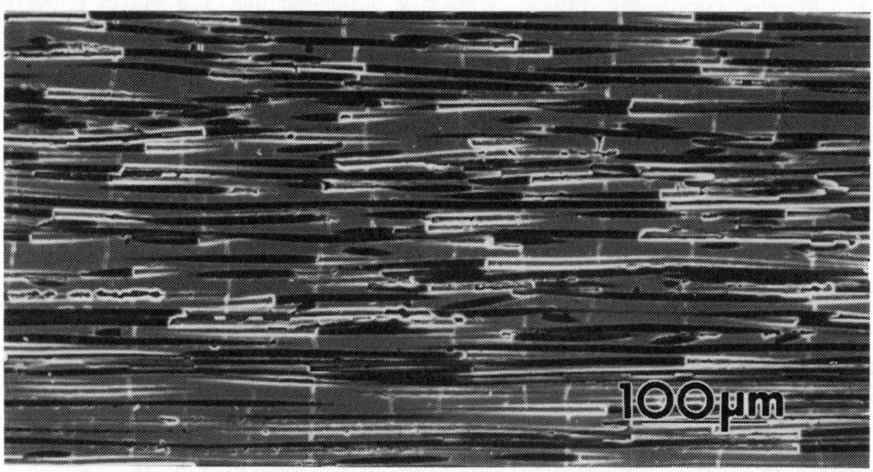

Fig. 2 SEM of polished surface (parallel to the fibre direction) of composite hot-pressed at 1550 °C for 0.5 h under 15 MPa.

(2.48×10^3 Kg / m^3) assuming that pure α-alumina has a theoretical density of 3.98 $\times 10^3$ Kg / m^3 and silica glass 2.22×10^3 Kg / m^3 [7,9]. Mercury density measurements indicated that the hot-pressed composite sample had a density of 95% of the calculated value (2.45×10^3 Kg / m^3) assuming that pure mullite has a theoretical density of 3.17×10^3 Kg / m^3 [1] and carbon fibre 1.81×10^3 Kg / m^3. The high level of densification of the composite was confirmed by SEM observations. Polished cross sections through the composite, parallel and perpendicular to the fibre direction are shown in Figs. 2 and 3 respectively. Figs. 2 and 3 reveal an uniform distribution of fibres in the mullite matrix and, apart from some voids due to polishing damage, show a highly densified microstructure. Some regularly spaced microcracks (average spacing 80 μm) were observed running perpendicular to the fibre direction. These are believed to result from the difference in thermal expansion coefficients between mullite ($\alpha = 4.8 \times 10^{-6} /°C$ average measured value for polycrystalline samples) and carbon fibre (α approximately zero) [10]. The large volume fraction of fibre consolidated in the mullite matrix did not appear to affect the phase development. The microstructure of the composite consisted of carbon fibres embedded in a densified mullite matrix with some extremely fine spherical unreacted alumina particles in the centre of mullite grains and fine pockets of residual silica-rich glass at the triple grain boundaries. TEM revealed a clear interface between fibres and mullite matrix as shown in Fig. 4 and no obvious signs of a chemical reaction or strong bonding having taken place. There was some evidence for an effect of the fibres on the morphology of the mullite grains; grains next to the carbon fibre surface tended to be elongated along the fibre direction, as shown in Fig. 4, indicating some preferred orientation, although this needs to be confirmed by further work.

A typical load-deflection curve of the hot-pressed composite samples is shown in Fig. 5. All samples exhibited non-brittle failure with extensive fibre pull-out on the tensile fracture surface (Fig. 6), suggesting a weak fibre/matrix interface. The flexural strength (σ) and elastic modulus (E) of the hot-pressed composite are shown in Table 1. The mean values for eight specimens determined by three point bending test are given; σ_O and E_O are the calculated theoretical strength and elastic modulus of the composite assuming that there is no contribution from the matrix. The fracture surface energy value is calculated from the areas under the stress-strain curves. The hot-pressing conditions yielded composite samples with flexural strengths, elastic moduli and fracture surface energy values in the range of 560-840 MPa, 90-170 GPa and 370-600 KJ / m^2 respectively.

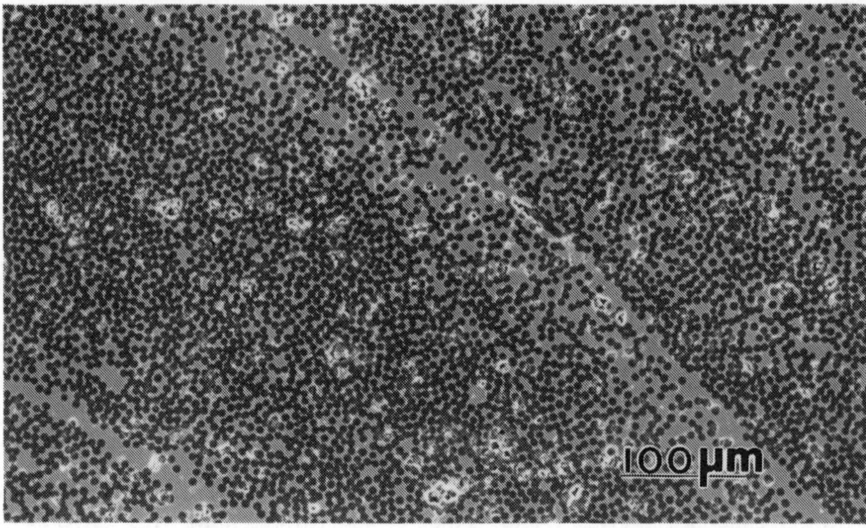

Fig. 3 SEM of polished cross section (perpendicular to the fibre direction) of composite hot-pressed at 1550 °C for 0.5 h under 15 MPa.

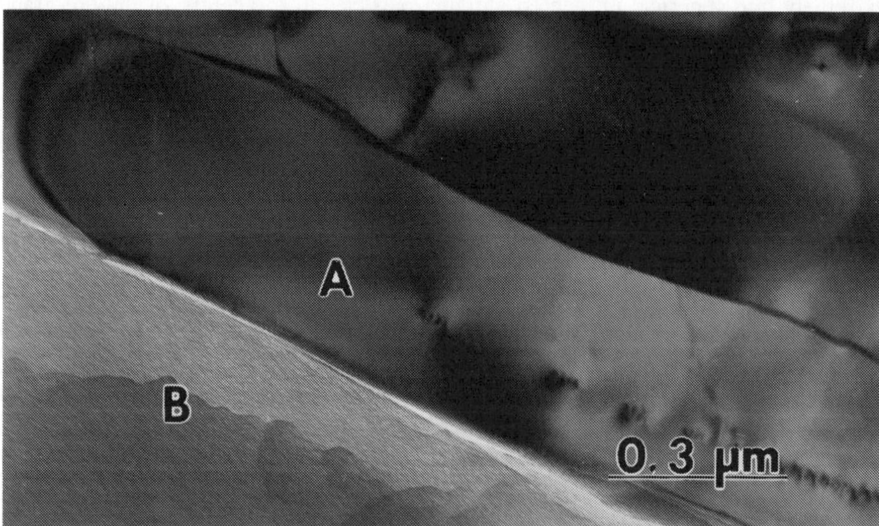

Fig. 4 TEM showing interface between carbon fibre and mullite grain in the composite hot-pressed at 1550 °C for 0.5 h under 15 MPa. A denotes mullite, B denotes fibre

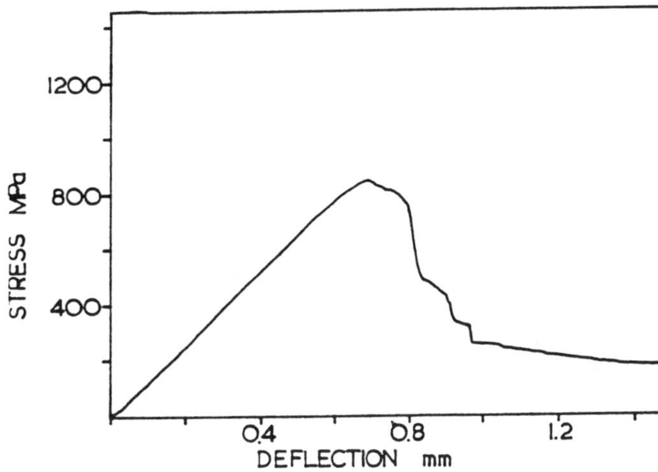

Fig. 5 Typical load vs. deflection curve (three point bending test) for the unidirectional carbon fibre reinforced mullite matrix composite using a bar specimen with a span of 30 mm and a span to thickness ratio greater than 20.

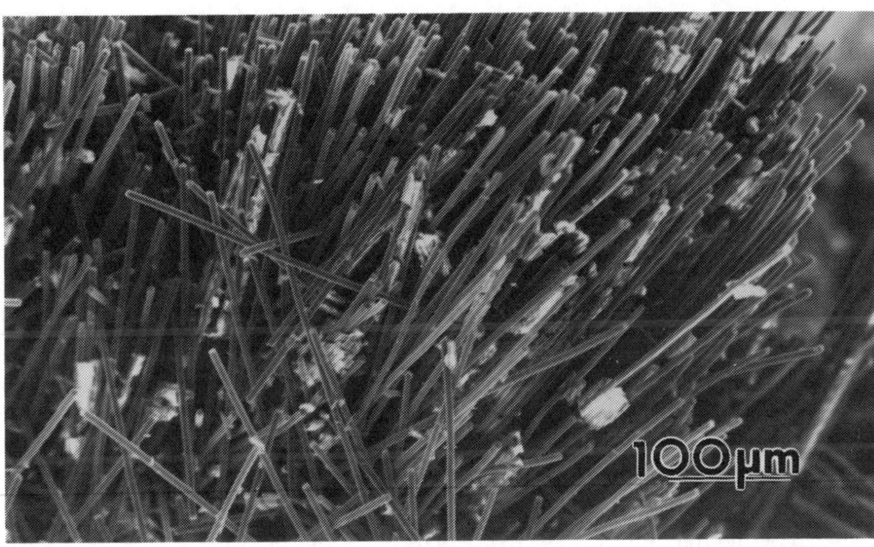

Fig. 6 SEM of the fracture surface of composite after three point bending test showing extensive fibre pull-out from mullite matrix .

Table 1 Properties of composites hot-pressed at 1550 °C for 0.5 h under 15 MPa (95% of theoretical density), mean values with standard deviations.

σ (MPa)	703±126
E (GPa)	119±28
Fracture Surface Energy (KJ/m^2)	512±104
σ_O (MPa)	2131
E_O (GPa)	119

DISCUSSION

The high sinterability of the matrix material has been discussed in previous work [8]. The matrix densification results from the viscous sintering of porous amorphous silica in the presence of rigid inclusions of crystalline alumina. The sintering rate of this process depends on the proportions of the glassy and the crystalline phases, the viscosity of the glass and the state of dispersion of alumina particles. Although the presence of alumina particles will inhibit the viscous flow and reduce the sintering rate, no sintering voids have been detected, probably because the alumina particles are well separated and do not form a scaffolding (network) in the glassy matrix. It is believed that at the sintering temperature of 1300°C the viscosity of the vitreous silica is low enough that collapse of pores in the glassy phase can occur and the glassy matrix can flow around the alumina inclusions, resulting in a highly densified microstructure [8]. Mullitisation started at above 1400°C, achieved by using submicron alumina and silica as precursors, following complete densification at the lower sintering temperature. The presence of occasional fine unreacted alumina regions embedded in the centre of mullite grains and residual silica fine pockets at mullite grain triple junctions is thought to be caused by the slow diffusion rate across the mullite matrix. In fact, separate experiments [10] suggest that the glass pockets could be reduced or eliminated altogether by adjusting the overall matrix composition to be slightly alumina rich. The flexural strength (σ) and Young's modulus (E) of the composite theoretically predicted by the rule of mixtures are given by equations (1) and (2),

$$\sigma = \sigma_m V_m + \sigma_f V_f \qquad (1)$$

$$E = E_m V_m + E_f V_f \qquad (2)$$

Where σ_m, E_m, V_m and σ_f, E_f, V_f refer to the strength, modulus and volume fraction of matrix and fibres respectively.

The average observed strength (703±126 MPa) of the composite are lower than the predicted values of 2131 MPa, using the rule of mixtures and neglecting the effect of the matrix. The presence of the thermally induced transverse cracks in the matrix suggest that the strength of the matrix is low and that it is reasonable to neglect the first term in equation (1). The lower strength of the composite than that predicted may be explained by slight degradation of the carbon fibres in the matrix during hot-pressing. However, the experimental value of the elastic modulus for the composite is the same as that predicted, neglecting the effect of the matrix. The non-brittle mode of failure of the composite was clear and confirmed by the fibre pull-out, indicating a comparatively weak interfacial bond. From TEM observations at the temperatures of hot-pressing studied, the interface between the carbon fibre and mullite matrix was sharply defined with no evidence of any strong interaction. This supports the observations of non brittle failure and of extensive fibre pull-out. The weak interfacial bonding is further confirmed by an independent measurement of the interfacial bonding strength of 9.8 MPa, determined assuming a strength value of the matrix of 110 MPa, a volume fraction of fibre of 0.53 and an average microcrack spacing of 80 μm from the thermal expansion mismatch mentioned above [10,11].

CONCLUSIONS

A modified sol-gel route was developed for processing fibre reinforced mullite composites ($3Al_2O_3 \cdot 2SiO_2$), using submicron alumina powder and colloidal silica sol as precursors. The method offered the following advantages: a low viscosity and a high solid yield sol for fibre infiltration, ensuring reduced processing shrinkage; viscous sintering by hot-pressing at a relatively low temperature with nearly full densification prior to mullite crystallisation; and a uniform distribution of fibres and adjustable volume fraction of fibres in the composites. The route successfully enabled high strength unidirectional continuous carbon fibre composites to be fabricated by a single stage infiltration followed by hot-pressing. The composites were highly densified and the toughening effect of the fibre reinforcement to the crystalline mullite matrix was significant. In addition, the infiltration is a simple single stage process avoiding the multiple impregnation often required in sol-gel processing [12].

A densification of 95% of the composite was achieved after hot-pressed at 1550°C under 15 MPa and the composite exhibited non-brittle failure with a mean strength of 703 MPa. The extensive fibre pull-out at the fracture surface and the sharply defined interface between fibres and mullite matrix in the composite as revealed by TEM, indicated relatively weak bonding between fibre and matrix, and was in agreement with the low measured interfacial bonding strength value of 9.8 MPa.

REFERENCES

(1) M.D. Sacks, H.W. Lee and J.A. Pask, "A Review of Powder Preparation Methods and Densification Procedures For Fabricating High Density Mullite, " Ceram. Trans. Vol.6. Mullite and Mullite Matrix Composites, Edited by S. Somiya, R.F. Davis, J.A. Pask 167-207 (1987).

(2) R.N. Singh and A.R. Gaddipati, "Mechanical Properties of a Uniaxially Reinforced Mullite-Silicon Carbide Composite, " J. Am. Ceram. Soc., 71 (2) C-100-3 (1988).

(3) R.N. Singh, "Role of Fiber-Matrix Interfacial Shear Stress on the Toughness of Reinforced Oxide Matrix Composites, "Mat. Res. Soc. Symp. Proc. Vol. 120 High Temperature and High Performance Composites (Edited by F.D. Lemkey, S.G. Fishman, A.G. Evans and J.R. Strife) 259-264 (1988).

(4) B.E. Yoldas, "Mullite Formation From Aluminium and Silicon Alkoxides, " Ceram. Trans. Vol. 6 Mullite and Mullite Matrix Composites (Edited by S. Somiya, R.F. Davis, and J.A. Pask) 255-262 (1987).

(5) A. Taylor and D. Holland, "The Chemical Synthesis and Crystallisation Sequence of Mullite, " J. Non-Cryst. Solids Vol. 152 No.1 1-17 (1993).

(6) M. Chen, F.R. Jones, P.F. James and J. E. Bailey, "The Role of the Interface on the Densification of Sol-Gel Processed Alumina and Mullite Fibre Composites, " Inst. Phys. Conf. Ser No. 111, 227-237 (1990).

(7) P.F. James, J. Wu, M. Chen and F.R. Jones, "Preparation of Mullite Using Transient Viscous Sintering of Alumina with Sol-Gel Derived Silica, " Proceedings of Int. Conf. on Advances in Amorphous State Chemistry, (Society of Glass Technology & Royal Society of Chemistry) Topical Issues in Glasses Vol. 1, 77-78 (1993).

(8) J. Wu, M. Chen, F.R. Jones and P.F. James, "Mullite and Alumina-Silica Matrices for Composites by Modified Sol-Gel Processing, " J. Non-Cryst. Solids Vol. 162 Nos.1-2 197-200 (1993).

(9) R.B. Sosman, "The Properties of Silica, " A. C. S. Monograph Series No. 37, New York: Chemical Catalog Co. , 289 (1927)

(10) J. Wu, unpublished results, Department of Engineering Materials, University of Sheffield, (1993).

(11) J. Aveston, G.A. Cooper and A. Kelly, "The Properties of Fibre Composites, " Conference Proc. NPL, IPC Science and Technology Press Ltd. 15-26 (1971).
(12) L.M. Sheppard, "Enhancing Performance of Ceramic Composites, " Am. Ceram. Soc. Bull. 71(4) 617-631 (1992).

ACKNOWLEDGEMENTS
Thanks are due to the Chinese Government and to the British Council for financial support (J. Wu)

FORMATION OF AN ORDERED BORON NITRIDE MATRIX FOR FIBER REINFORCED COMPOSITES

Cameron G. Cofer,* Dong-Pyo Kim, and James Economy
University of Illinois at Urbana-Champaign, Urbana, IL 61801

In recent work, we have fabricated carbon fiber/boron nitride matrix composites which exhibited excellent mechanical properties (900MPa in flexure). Using a borazine precursor, highly ordered boron nitride was formed at temperatures as low as 1200°C. Work currently being published suggests the formation of a liquid crystalline phase in the borazine oligomer which may account for the tendency for orientation in the resultant boron nitride. In this paper, new data will be presented including mass spectroscopy and x-ray diffraction which further characterize the ordering process. Its relationship to the final composite properties will then be discussed.

INTRODUCTION

Carbon/carbon composites (C/C) have found widespread commercial use and applicability. However, their very high fabrication costs and their inherent susceptibility to oxidation at 450°C place limits on their potential applications. We have recently demonstrated for the first time that we can use oligomers derived from borazine in order to form a boron nitride (BN) matrix in ceramic composites.[1,2] The BN matrix offers significant advantages over the carbon matrix including oxidation resistance, low dielectric constant, and very high ceramic yield.

Borazine oligomers are formed by reaction of borazine with itself at only 70°C. This reaction leads to a very stable BN on further heating at 1200°C. We observe a near theoretical d(002) spacing of 3.36Å, which is indicative of its high degree of order.[2]

A number of fiber reinforced BN matrix composites have been produced using a single impregnation of borazine oligomers (Table 1).[1,2] The carbon fiber/boron nitride matrix composites (C/BN) exhibit mechanical properties which are far higher than those of C/C made under similar conditions with only one impregnation. The yield of BN from the oligomer is approximately 90%, permitting fabrication of relatively dense composites with only one impregnation (Figure 1). This significantly reduces the processing costs compared to their C/C counterparts which usually require multiple reimpregnations followed by chemical vapor infiltration of carbon.

In addition to their mechanical attributes, these composites represent a significant advance over the current C/C materials because of their increased

To the extent authorized under the laws of the United States of America, all copyright interests in this publication are the property of The American Ceramic Society. Any duplication, reproduction, or republication of this publication or any part thereof, without the express written consent of The American Ceramic Society or fee paid to the Copyright Clearance Center, is prohibited.

Table I. Summary of Properties of Composites in this Study with a BN Matrix Annealed in Nitrogen 3 hours at 1200°C.

Fiber	Flexural Strength (MPa)	Flexural Modulus (GPa)	Strain to Failure (%)	Fiber Composition (wt%)
PAN T300*	610	80	0.90	93% C
Pitch P25*	500	60	0.95	99% C
Nicalon†	420	80	0.50	61% SiC 10% C 28% SiO_2
C-Nicalon†	550	60	1.00	Nicalon with a 200Å C coating
Sumica‡	380	80	0.50	85% Al_2O_3 15% SiO_2

* Amoco Performance Products, Atlanta, GA.
† Dow Corning, Midland, MI.
‡ Sumitomo Corporation, Tokyo, Japan.

oxidation resistance from 450 to 850°C.[1] Composites fabricated with insulating fibers, such as the Sumica/BN, also possess a low dielectric constant of 7.6 at 1 MHz.[2]

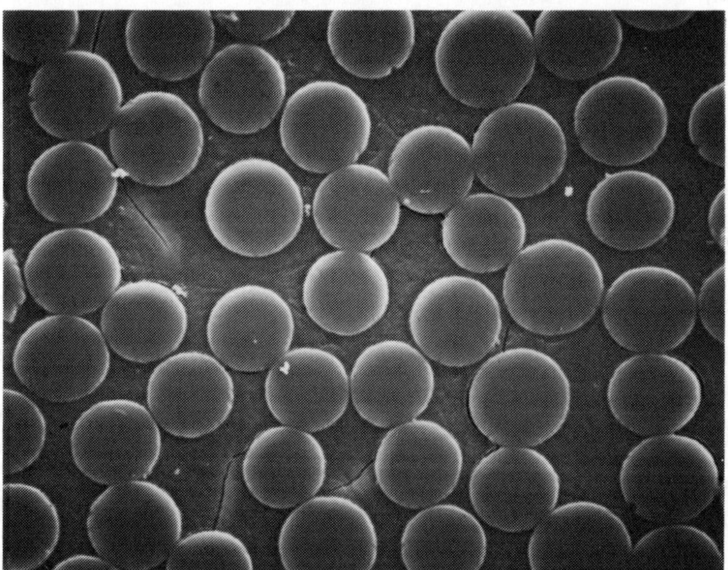

Figure 1. Polished cross-section of pitch P25/BN composite made with one oligomer impregnation.

We have observed that the flexural strength of pitch/BN increases from 200 to 710MPa upon heating from 1200 to 1500°C. This is surprising because of the anisotropy of pitch fiber versus the anticipated isotropic character of the BN matrix. Typically, the thermal mismatch between the anisotropic carbon fiber in an isotropic carbon matrix results in extreme interfacial stresses which cause a decrease in properties at increasing processing temperatures.[3] This has led us to consider the possibility that the borazine oligomer may form an ordered mesophase which could produce an aligned boron nitride at the fiber/matrix interface. This would tend to reduce problems of mismatch in thermal expansion coefficient at the fiber/matrix interface.

A discotic mesophase is known to form in certain carbons[4] and is essential in the formation of high modulus pitch carbon fibers. The carbon mesophase can be comprised of a wide range of polynuclear species including naphthalene (**I**), phenanthrene (**II**) and pyrene (**III**):

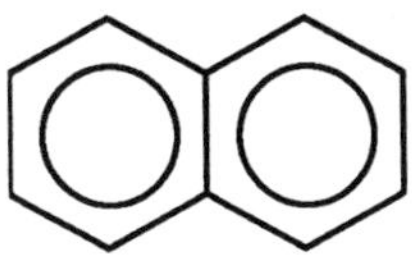

I

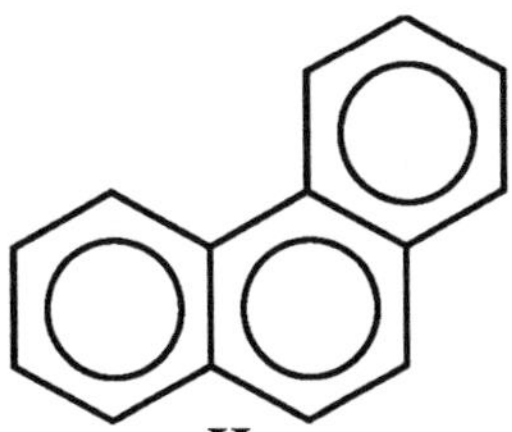

II

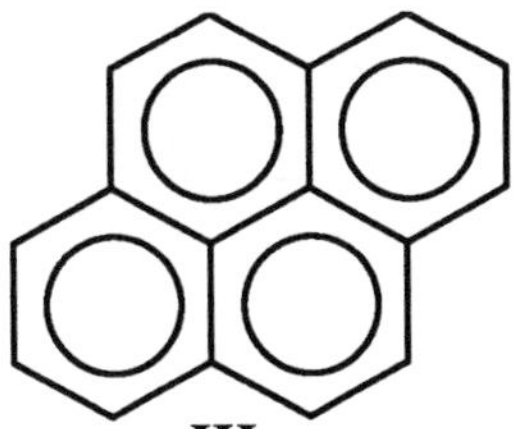

III

An analogous discotic phase in the boron-nitrogen system has not previously been reported.

In this paper, we present data supporting the formation of a discotic phase in the borazine oligomers. We then describe how this phase influences ordering in the formation of BN. Finally, the use of these oligomers in forming thin dielectric films of BN is described.

EXPERIMENTAL

Borazine monomer which had been prepared in our laboratory was heated at 70°C in a 3 ounce pressure reactor vessel under dry nitrogen for up to 48 hours. Under these conditions the borazine underwent self-condensation reactions with accompanying hydrogen evolution to produce a low molecular weight polymer. This oligomer was used as a precursor to form coatings, films and composites (described in more detail elsewhere[1]).

Mass spectroscopy was performed using a VG 70-VSE mass spectrometer with HP 590 gas chromatography (MS/GC), controlled with a VGOPUS data system. X-ray diffraction (XRD) data was collected with a Rigaku DmaxII Geiger Flex using Cu K_α radiation at 5°/min with a 0.01/sec sampling rate. Thermogravimetric Analysis (TGA) was performed using a TA instruments model 951 with a horizontal quartz tube and balance sensitivity of 0.01μg.

RESULTS AND DISCUSSION

Evidence for Liquid Crystallinity

Based on the earlier composites work,[1] the potential to form a liquid crystalline phase was suspected. It was first observed that a melt of borazine oligomers treated at 5°C for two weeks exhibited ordering based on both birefringence with polarizing microscopy and XRD.[5] Hot-stage XRD of the 5°C oligomer demonstrated the presence of a partially reversible transition near 70°C where the isotropic structure is formed on heating.

In order to understand the chemical structure which might account for the liquid crystalline character, mass spectroscopy has been applied to the oligomers as-produced at 70°C (Figure 2). It should be noted that oligomers obtained by heating benzene do not form a liquid crystalline phase, whereas oligomers from a 50/50 mixture of benzene and naphthalene do form a mesophase.[6] The first major mass/charge (m/z) peak at 131 identifies the naphthalene analog suggesting a very different process than that observed in benzene. The second peak at 158 is indicative of a biphenyl analog. The combination of B- and N-centered trimers, observed at 172, in combination with the naphthalene analog are suggestive of the

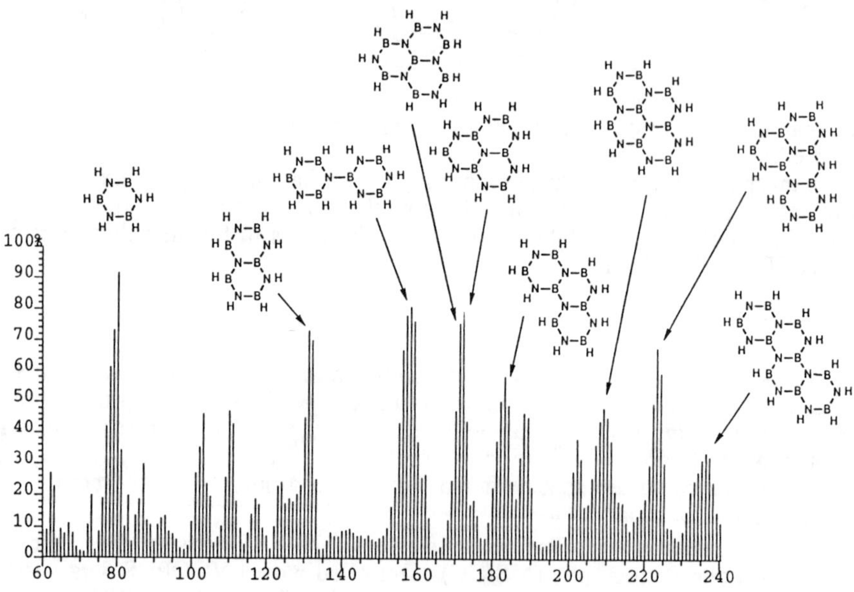

Figure 2. Mass spectra of borazine oligomers after 20 hours of polymerization at 70°C.

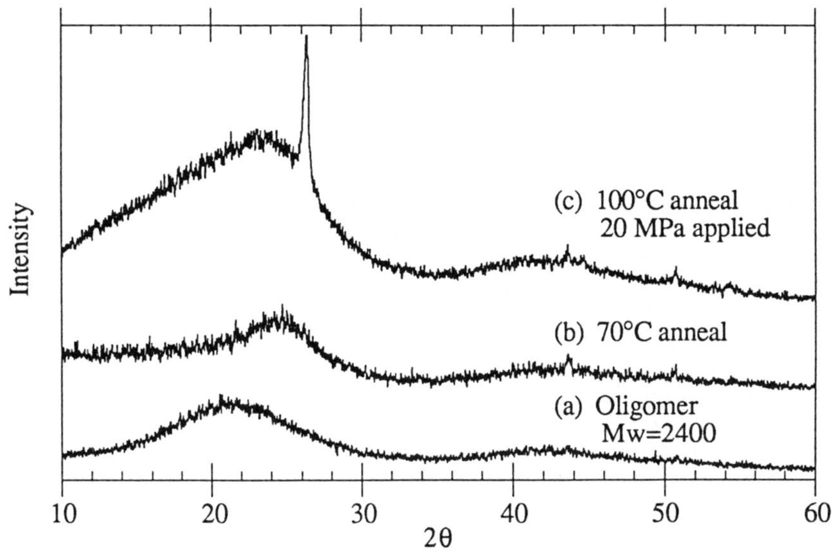

Figure 3. XRD of borazine oligomers after low temperature treatments.

potential for orientation and stacking like that observed in mesophase pitch.[4] A number of additional structures analogous to those observed in the carbon mesophase are also observed in the borazine oligomers.

Ordering in the isotropic melt appears to be promoted by the application of an external pressure. Figure 3 displays XRD scans for a series of oligomers that have undergone various low temperature processing conditions. The first curve, 3a, is that of the oligomer with an average molecular weight of 2400. Here, a broad amorphous peak centered at 4.10Å is observed. Upon further heat treatment at 70°C in the absence of pressure, the specimen becomes insoluble, and a diffuse diffraction pattern with a peak centered at 3.56Å is observed suggesting modest ordering (Figure 3b).

At a temperature of 100°C, 20 MPa of pressure was applied and a very sharp peak at 3.37Å emerged (Figure 3c). Hence, a second mechanism appears to exist for forming crystalline boron nitride, even at very low temperatures of 100°C. Presumably, this process may be related to the formation of a small fraction of liquid crystalline BN oligomer which converts to the stable BN structure.

Conversion to BN

The treatment of the borazine oligomers at low temperatures (50-250°C) appears to have a profound impact on the ordering in the boron nitride which is formed. Specimens shown in 3b, in which pressure had not been applied, were heated to 800, 1000, and 1200°C under nitrogen and held for three hours at each temperature. Re-examination using XRD is shown in Figure 4. While some ordering is observed in boron nitride at 1200°C, a relatively large d(002) spacing of 3.41Å is observed. Furthermore, the peak is quite broad with a FWHM of 1.61°(2θ).

Specimens from 3c, in which ordering had been induced by the application of pressure, behaved very differently. In these specimens, pressure was only applied up to 400°C and subsequently heated to 800, 1000, and 1200°C as above. Shown in Figure 5, two distinct peaks are observed in the 800°C specimen. The major peak is at 3.42Å while the second smaller peak is at 3.33Å, corresponding to the crystal structure which had formed at low temperatures under pressure. Ordering is further accelerated at 1000°C where a turbostratic BN is formed with d(002) of 3.35Å. Upon further heating to 1200°C a stable BN is formed with a more well-defined crystal structure.

These results suggest that the history of the oligomeric phase has a major effect on the ability to produce a stable boron nitride. The application of pressure appears to induce alignment of the discotic-like borazine oligomer. Ordered crystallographic

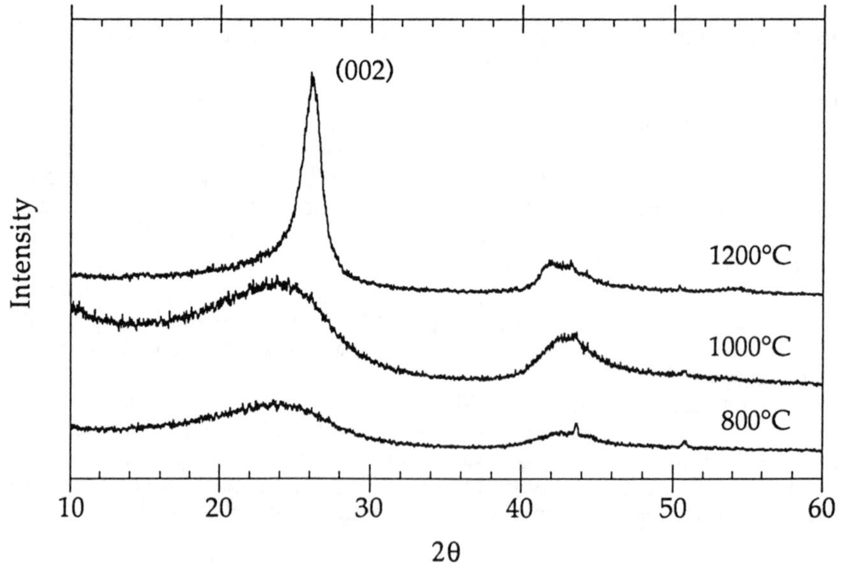

Figure 4. XRD of specimens with no applied pressure heated to 800, 1000, and 1200°C.

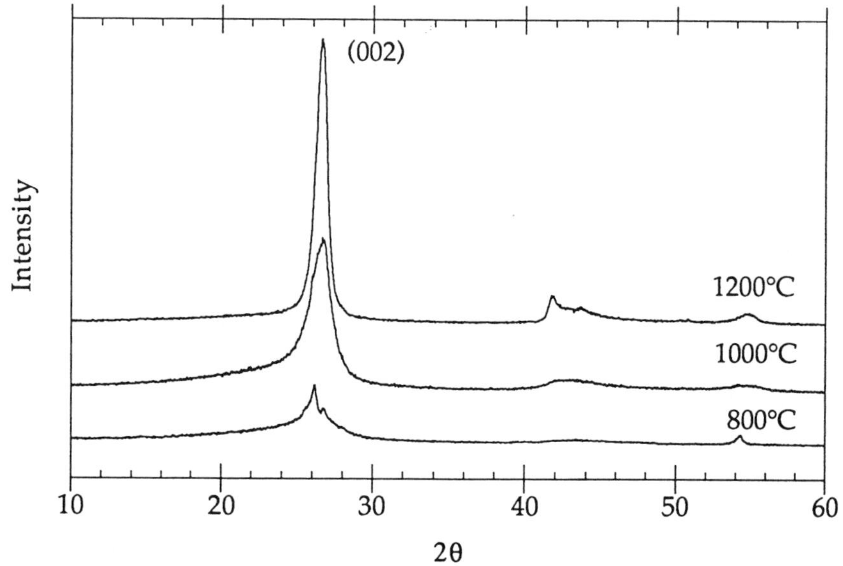

Figure 5. XRD of specimens with 20MPA pressure applied from 70 to 400°C and subsequently heated to 800, 1000, and 1200°C.

domains which are generated at the low temperatures may serve as nucleation sites during additonal stabilization and conversion to boron nitride.

The stability of the boron nitride product to moisture is dependent on the d(002). Shown in Figure 6 is a TGA of several boron nitride products exposed to nitrogen with 100% relative humidity at 700°C. Here, it is observed that BN with interlayer spacings greater than 3.40Å are aggressively attacked by moisture. In contrast, the BN with near theoretical (002) is quite resistant even in this very severe environment.

Formation of Thin Films of BN

The ability to form thin dielectric films of BN would be an important advantage in the field of microelectronic devices. We have found that solutions of borazine oligomers can be used to dip coat substrates such as silicon. Thus, a very uniform 13μm thick BN film was produced by dip coating onto a silicon wafer and then heating to 1200°C (see Figure 7). X-ray diffraction of the thin film indicates the formation of a very well defined (002) peak after processing to 1200°C. The availability of this kind of process affords the microelectronics technologist with access to a highly refractory coating with a dielectric constant of only 4.5 and a thermal conductivity near that of graphite.

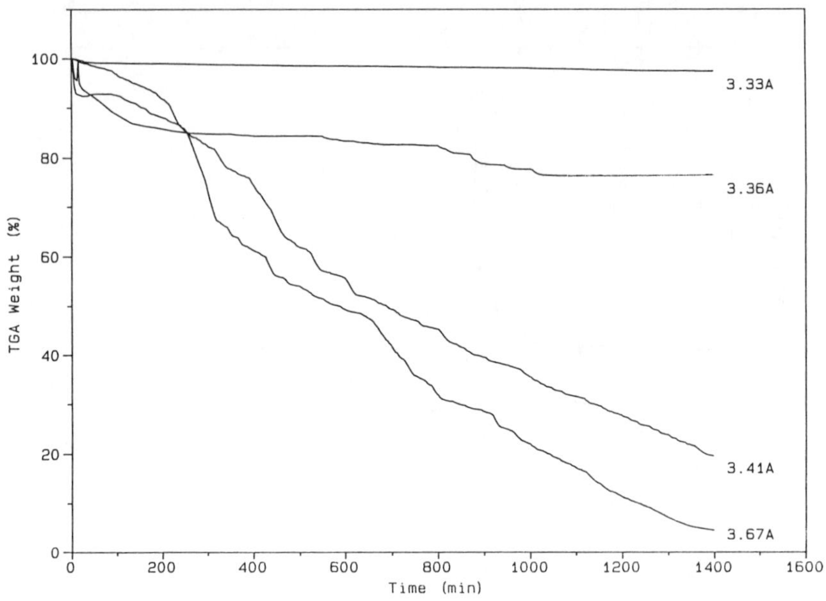

Figure 6. TGA of BN specimens with various d(002) in Steam/Nitrogen at 700°C

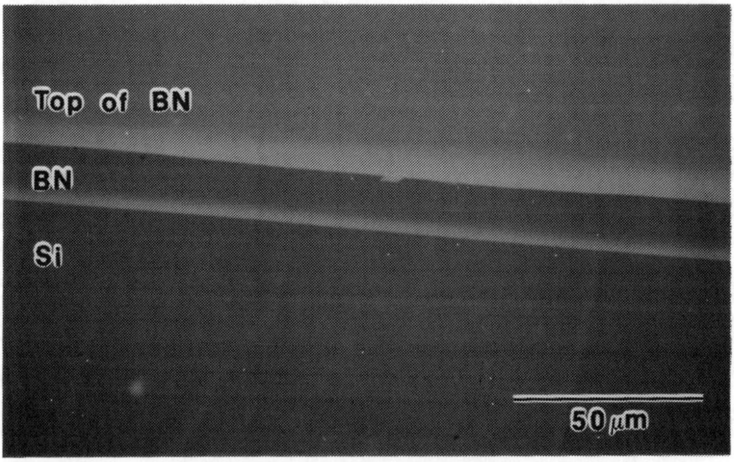

Figure 7. SEM micrograph of cleaved BN film and silicon substrate.

CONCLUSIONS

New evidence is presented supporting the existence of the first all inorganic liquid crystalline phase. One can now provide a rational interpretation of the unexpected high mechanical properties of the C/BN composites. The mesophase observed in the borazine oligomer may result in alignment of boron nitride at the fiber/matrix interface. Aligned boron nitride would exhibit a thermal expansion which would more closely match that of the anisotropic carbon fiber.

Potential applicability of the borazine oligomers appears broad including formation of very stable BN films, coatings, and matrices for composites. The presence of a liquid crystalline phase may provide a route to achieve a stable BN by processing at much lower temperatures than 1200°C.

ACKNOWLEDGEMENTS

The authors wish to thank Aaron Saak for assistance with specimen characterization and Emmanuel Dimotakis for the preparation of borazine. Research was performed with financial support provided by ONR/DARPA.

REFERENCES

1. D. Kim and J. Economy, "Fabrication of Oxidation-Resistant Carbon Fiber/Boron Nitride Matrix Composites," *Chem. Mater.* **5** [9] 1216-1220 (1993).
2. D. Kim, C. G. Cofer, and J. Economy, "Fabrication and Properties of Ceramic Composites with a Boron Nitride Matrix," accepted, *J. Am. Cer. Soc.*
3. G. Savage, *Carbon-Carbon Composites,* Chapman and Hall, London (1993).
4. H. Marsh, F. Dachille, J. Melvin, and P. L. Walker, Jr., "The Carbonization of Anthracene and Biphenyl under Pressures of 300MNm^{-2}," *Carbon* **9**, 159-177 (1971).
5. D. Kim and J. Economy, "Occurrence of Liquid Crystallinity in a Borazine Polymer," accepted, *Chem. Mater.*
6. E. Fitzer, K. Mueller, and W. Schaefer, "The Chemistry of the Pyrolytic Conversion of Organic Compounds to Carbon," *Chemistry and Physics of Carbon, Vol. 7*, Marcel Dekker, 237 (1971).

MICROSTRUCTURAL CHARACTERIZATION OF SILICON CARBIDE FIBER-REINFORCED ALUMINUM NITRIDE COMPOSITES

K. Park, *S. Hsu, *T. Vasilos, and C. Sung
Center for Advanced Materials and *Department of Chemical and Nuclear Engineering, University of Massachusetts, Lowell, MA 01854

ABSTRACT

SiC fiber-reinforced AlN composites were fabricated by hot-pressing at 1650°C and 17.6 MPa in vacuum. The microstructure and chemistry of matrix, fiber, matrix/fiber interface in as-fabricated and crept composites were characterized using analytical transmission electron microscopy, in order to find out the relationship between the microstructural features and the mechanical properties measured at room- and high-temperatures and to understand the reinforcing effects of the SiC fiber in the matrix at room- and high-temperatures. Room-temperature mechanical properties and high-temperature creep properties of the composites were investigated by four-point bending tests.

INTRODUCTION

In addition to SiC and Si_3N_4 ceramics [1-3] for high-temperature structural applications such as heat engines and gas turbines, AlN is also considered a potential candidate. It is difficult to sinter pure AlN, since it is a covalent material with a wurtzite crystalline structure (a = 0.311 nm and c = 0.498 nm). Therefore, several sintering additives, such as CaC_2, CaO, MgO, Y_2O_3, and $Ca(NO_3)_2$, have previously been used to fabricate fully dense AlN [4, 5]. In addition, the AlN has high thermal conductivity, high electrical resistivity, and a thermal expansion coefficient close to Si to make it a prime candidate for semiconductor substrates [6].

Fiber-reinforcement may be one of the promising strategies to strengthen AlN. In this program, an attempt was made to improve the mechanical properties of AlN by the incorporation of SiC (SCS-6) fibers in an unidirectional array. The SiC fibers are one of the most important reinforcements for ceramic- and metal-matrix composites due to high tensile strength (3450 MPa), high tensile modulus (400 GPa), and low density (3.0 g/cc). The SiC fiber (15 vol%)-reinforced AlN composites were fabricated by hot-pressing in vacuum. The microstructure and chemistry of as-fabricated and crept composites, especially matrix/fiber interface,

To the extent authorized under the laws of the United States of America, all copyright interests in this publication are the property of The American Ceramic Society. Any duplication, reproduction, or republication of this publication or any part thereof, without the express written consent of The American Ceramic Society or fee paid to the Copyright Clearance Center, is prohibited.

were characterized using analytical transmission electron microscopy, in order to understand the relationship between the microstructural features and the mechanical properties measured at room- and high- temperatures as well as the reinforcing effects of the SiC fiber in the matrix at room- and high-temperatures. Interfacial characteristics of the composites are the most important factors influencing the mechanical properties of the composites.

EXPERIMENTAL

The materials used in this study were (1) commercial AlN powders with an average particle size of 1.8 µm and specific surface area of 2.5 m^2/g (Advanced Refractory Technologies, Buffalo, NY), (2) SiC (SCS-6) fibers (TEXTRON Specialty Materials, Lowell, MA), and (3) $Ca(OH)_2$ and Al_2O_3 as sintering additives. These sintering additives were selected so that a low densification-temperature could be used in order to minimize the fiber degradation.

AlN powders combined with 6 wt% $Ca(OH)_2 \bullet Al_2O_3$ sintering additives were mixed with methanol to form a slurry. The resulting slurry was milled using Si_3N_4 balls and then dried in a drying oven. The SiC fibers were wound onto plastic foils unidirectionally with a density of 20 fibers/cm and then fixed using a sprayed Krylon type adhesive in order to avoid contact between the fibers. The fiber foils were dried and then cut into small sections (54 × 54 mm). The dried mixture powders (AlN powders and sintering additives) were spread between the fiber foils in the press die cavity (54 × 54 mm). The die cavity and punch were coated by a thin layer of BN in order to avoid reaction with a graphite die. Applied pressure was 5.9 MPa when furnace temperature reached at 1200°C. Applied pressure increased up to 17.6 MPa at 1600°C. It was maintained at 17.6 MPa and 1650°C for 2 hours and then cooled. The fabricated composites were machined into rectangular bars (3 × 4 × 45 mm) with fibers parallel to the long axis.

Cross-sectional TEM specimens were sectioned from the composites with a low-speed diamond saw and mechanically ground to ~300 µm. Circular discs of 3 mm in diameter were core drilled from the ground sections, mechanically ground to ~120 µm, and then dimpled to ~30 µm. The specimens were ion milled with 5 kV Ar^+ ions at an incident angle of 12° until perforation was achieved. A light carbon film was evaporated on the specimens in order to prevent charging in the electron microscope. The microstructure and chemistry of the matrix, fiber, and matrix/fiber interface of the composites were investigated by a bright field image, selected area diffraction (SAD), convergent beam electron diffraction (CBED), and energy dispersive X-ray spectroscopy (EDXS) using a Philips EM400T TEM and a Noran ultrathin window Micro-Z detector.

The mechanical properties of the composites were measured at room temperature by four-point bending tests. High-temperature creep properties were also measured by four-point bending tests at 1100°C and bending stress of 46.3 MPa for 48 hours under oxidizing condition.

RESULTS AND DISCUSSION

1. As-fabricated Composite

The densification of pure AlN is extremely difficult because of covalent bonding of AlN. The sintering additives of $4Ca(OH)_2•Al_2O_3$ were therefore used to fabricate fully dense AlN. The composites hot-pressed with 6 wt% $4Ca(OH)_2•Al_2O_3$ additives reached densification up to 99.3 % of theoretical density. Pure AlN hot-pressed at 1700°C and 17.6 MPa without sintering additives reached 83.3% theoretical density. The AlN matrix in as-fabricated composite contained a liquid phase at grain boundaries and a high density of dislocations within the grains, as shown in Fig. 1. Most of the additives used were present in the liquid phase at the grain boundaries. CBED and EDXS were used to investigate the microstructure and chemistry of the liquid phase, respectively. The liquid phase was identified as a $Ca_{12}Al_{14}O_{33}$ with a cubic structure (a = 1.198 nm).

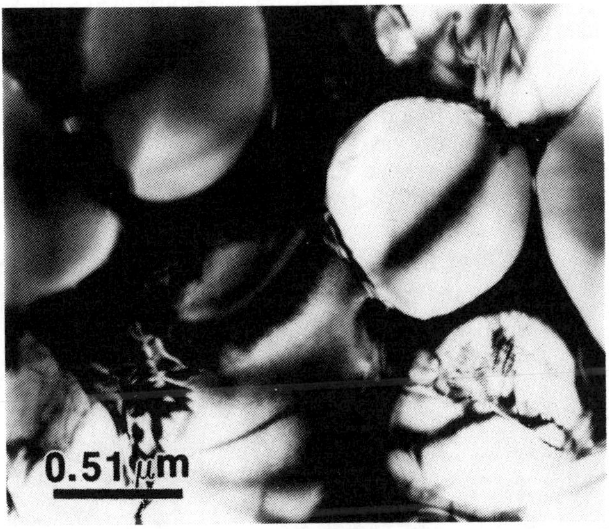

Fig. 1. Bright field image showing a $Ca_{12}Al_{14}O_{33}$ liquid phase at matrix grain boundaries and a high density of dislocations within the matrix grains.

It was found that the SiC fiber in as-fabricated composite consists of several different layers, i.e., core carbon layer, inner carbon layer, SiC layer, and outer carbon layer. Figs. 2(a), (b), and (c) show a bright field image, CBED pattern, and EDXS spectrum of SiC layer in the SiC fiber, respectively. The SiC layer revealed columnar grains and a high density of twins. The SiC grains in the fiber are clearer than those in as-fabricated (unprocessed) SiC (SCS-6) fiber, because C in the SiC layer was diffused into grain boundaries and segregated at the grain boundaries [7]. Since Al in AlN matrix diffused into the fiber during composite fabrication, the SiC grains contained a small amount of Al. Although the layer contained Al, the SiC layer has the same crystalline structure, ß-SiC (a = 0.435 nm), as the SiC layer without Al in as-fabricated SiC (SCS-6) fiber.

Fig. 3(a) shows a bright field image of interface (marked by arrows) between the SiC layer and outer carbon layer in the fiber. No interacted layer was observed at the interface. The SAD pattern obtained from the outer carbon layer is shown in Fig. 3(b). This SAD pattern consists of distinctive arcs and randomly distributed spots, indicating the preferred alignment of turbostratic carbon blocks and the existence of SiC particles, respectively.

In an attempt to determine the nature of the reaction between the fiber and matrix during composite fabrication and to understand the reinforcing effects of SiC fiber in the AlN matrix, interfacial regions were investigated. The AlN matrix reacted with outermost carbon layer in the fiber, Fig. 4, showing a bright field image of SiC layer, outer carbon layer, and AlN matrix. It was found that outer carbon layer consists of three sublayers. The unreacted sublayers A, B, and C and the reacted layer D formed by the reaction of outermost carbon layer in the fiber and AlN matrix are indicated in Fig. 4. The measured thicknesses of the unreacted sublayers A, B, and C and the reacted layer D are ~1.5, ~0.28, ~0.3, and ~1.3 µm, respectively. In a previous study [8], the outer carbon layer in as-fabricated SiC (SCS-6) fiber was found to have a thickness of ~3 µm and to consist of three sublayers whose thicknesses are ~1.6, ~0.2, and ~1.3 µm, respectively, from the interface between the SiC layer and outer carbon layer

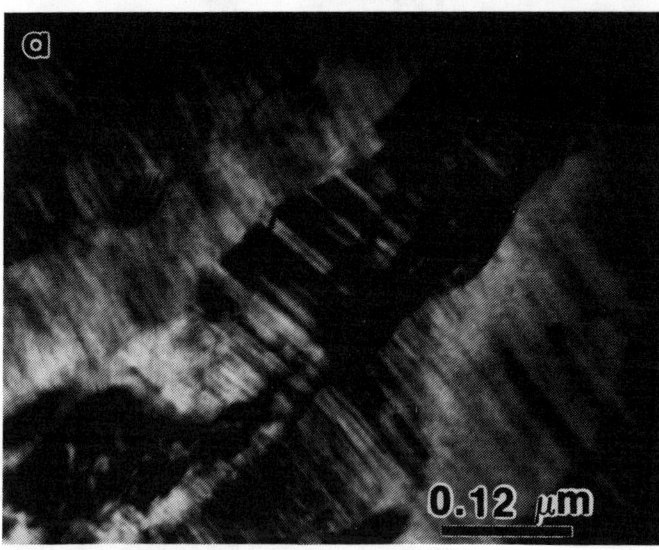

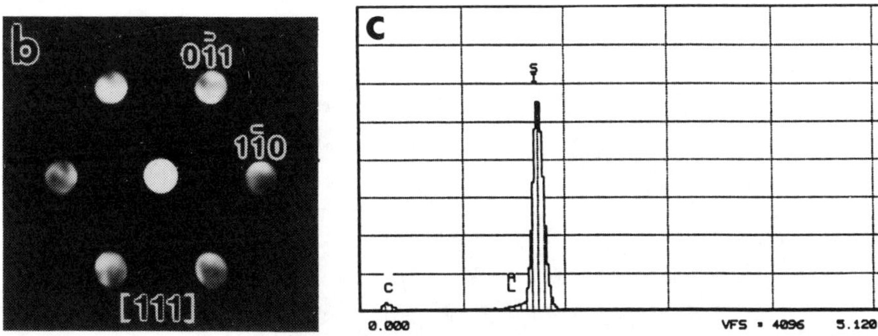

Fig. 2. (a) Bright field image, (b) CBED pattern, and (c) EDXS spectrum obtained from the SiC layer in SiC fiber.

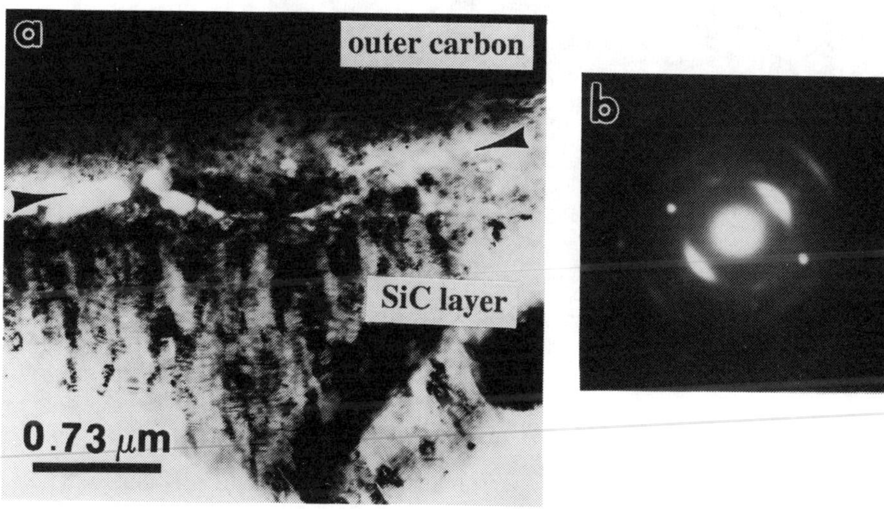

Fig. 3. (a) Bright field image showing an interface (marked by arrows) between the SiC layer and outer carbon layer of the fiber. (b) SAD pattern obtained from the outer carbon layer.

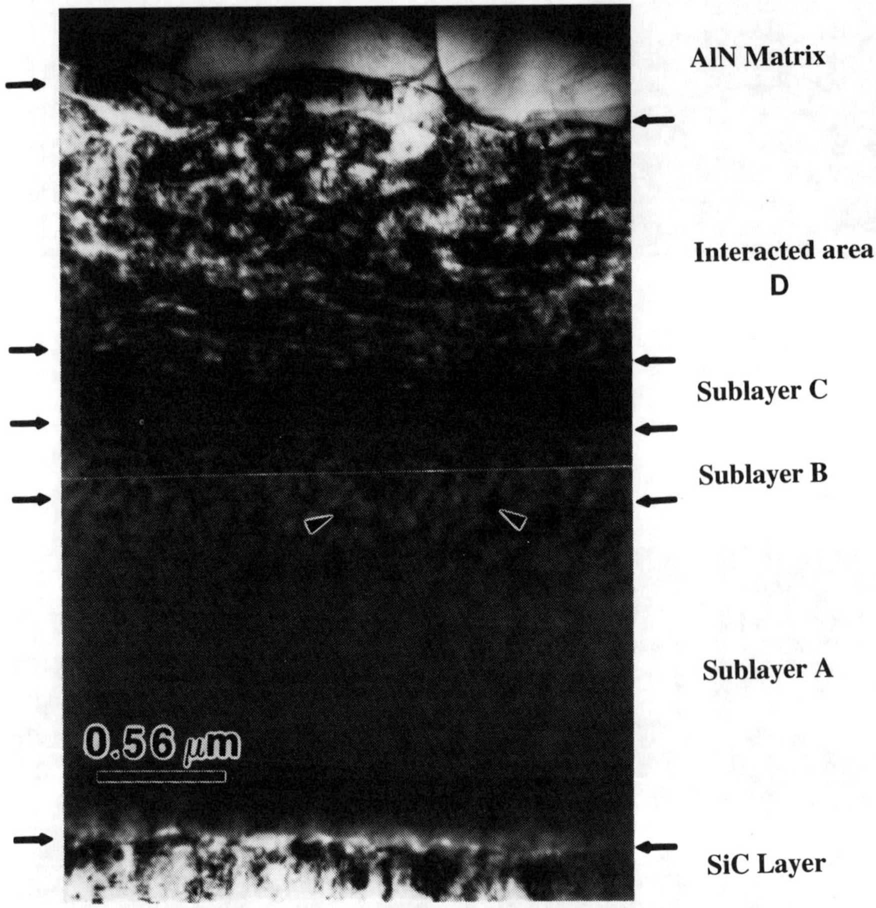

Fig. 4. Bright field image of SiC layer, outer carbon layer, and AlN matrix. The outermost carbon layer in the fiber reacted with the matrix. The unreacted sublayers A, B, and C and the reacted layer D are indicated.

towards the matrix radially. These results indicate that the thicknesses of the outermost carbon layer and AlN matrix interacted during composite fabrication are ~1.0 and ~0.3 μm, respectively.

Most of the reacted layer D consist of C-Al-Ca-O (bright area) and Si-Al-C-Ca-O (dark area) phases. The bright areas with C-Al-Ca-O phase were formed by the reaction of turbostratic carbon in the outermost carbon layer and Al, Ca, and O in the matrix, whereas the dark areas with Si-Al-C-Ca-O phase were formed by the reaction of SiC particles in the outermost carbon layer and Al, Ca, and O in the matrix. Although the bright areas contained Al, Ca, and O along with C, they have the turbostratic carbon structure, which is a structure of the outer carbon layer in as-fabricated SiC (SCS-6) fiber. At present, structure of the Si-Al-C-Ca-O phase is unidentified. SiC particles (marked by arrowheads) are embedded in the turbostratic carbon at the unreacted sublayers. The unreacted sublayer B seems to be the weakest and to contain the smallest amount of SiC particles among the unreacted sublayers.

The load-deflection curves for the composite and monolithic AlN measured at room temperature by four-point bending test are shown in Figs. 5(a) and (b), respectively. The fracture strength and its corresponding fracture strain of the composite measured at room temperature are 430 MPa and 0.87%, respectively, whereas the fracture strength and its corresponding fracture strain of the monolithic AlN measured at room temperature are 371 MPa and 0.46%, respectively. The incorporation of SiC fibers into an AlN matrix improved the mechanical properties significantly. Especially, the incorporation of the fibers caused a twofold increase in the fracture strain. It is believed that this improvement would result from crack deflections around the SiC fibers. The elastic modulus of the composite did not follow a rule of mixtures probably because of the reaction of AlN matrix and outermost carbon layer in the SiC fiber discussed previously.

2. Crept Composite

The microstructure of the SiC/AlN composites showed significant changes after creep tests, which were performed by four-point bending at 1100°C and bending stress of 46.3 MPa for 48 hours under oxidizing condition. Fig. 6(a) shows a bright field image of $Ca_{12}Al_{14}O_{33}$ (dark area) and Al_2O_3 (bright area) phases at matrix grain boundaries without a crack in crept composite. The microstructure and chemistry of the Al_2O_3 phase were identified by CBED and EDXS, respectively, as shown in Figs. 6(b) and (c). Pores and second phases such as $Ca_{12}Al_{14}O_{33}$ and Al_2O_3 were also observed at and around matrix grain boundaries with cracks in the crept composite. The pores, cracks, and Al_2O_3 phase were formed by the penetration of oxygen during creep test. The cracks were generally associated with pores and propagated along the matrix grain boundaries.

Fig. 7 shows a bright field image of core carbon, SiC layer, and reacted layer formed by the reaction of inner carbon layer and SiC layer in the fiber during creep test. The reacted layer consists of two sublayers, i.e., A and B with thicknesses of ~0.7 and ~0.8 μm, respectively. In a previous study [8], the

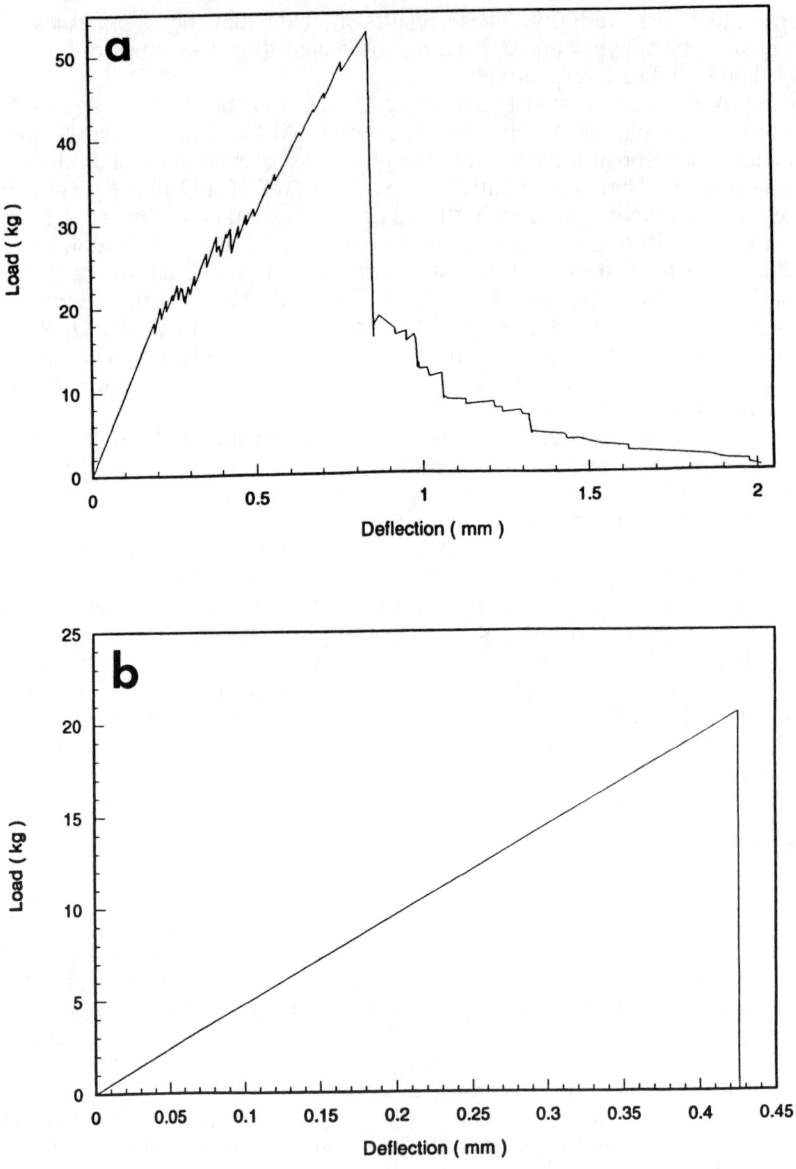

Fig. 5. Load-deflection curves for (a) the composite and (b) monolithic AlN measured at room temperature by four-point bending test.

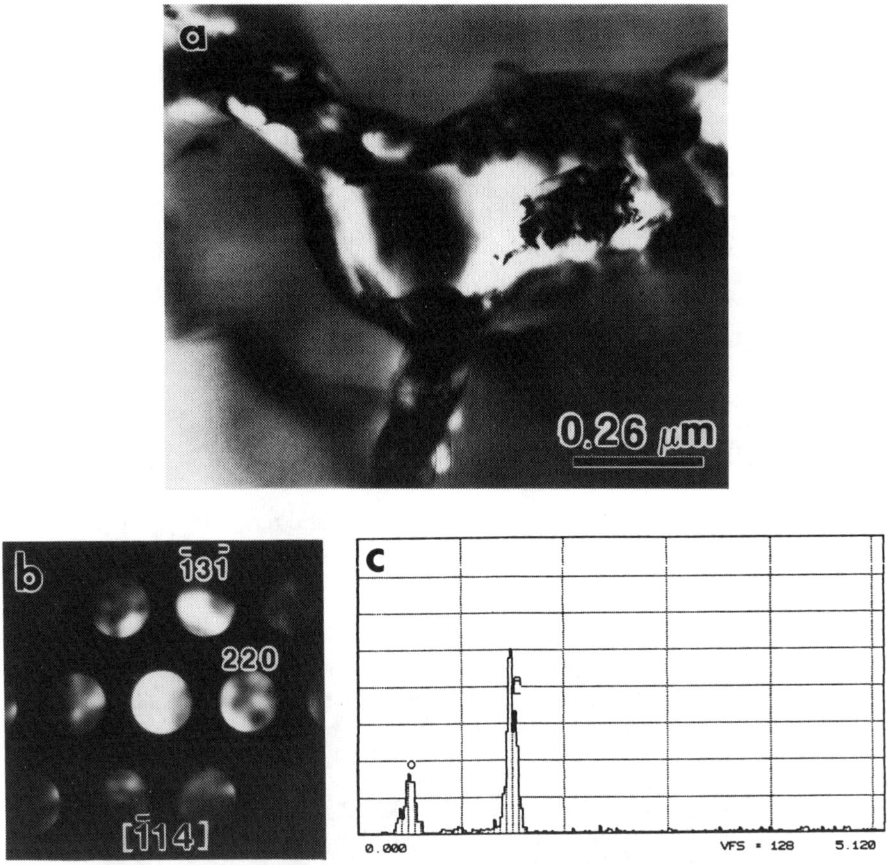

Fig. 6. (a) Bright field image showing $Ca_{12}Al_{14}O_{33}$ (dark area) and Al_2O_3 (bright area) phases at matrix grain boundaries without a crack in crept composite. (b) CBED pattern and (c) EDXS spectrum obtained from the Al_2O_3 at the matrix grain boundaries.

thickness of inner carbon layer in as-fabricated (SCS-6) SiC fiber was ~1.5 μm. The SiC layer shown in Fig. 7 contained a small amount of Al because of the diffusion of Al in the AlN matrix during composite fabrication and creep test. The reacted sublayers A and B contained minor elements such as Si, O, and Al along with C. The concentration of oxygen in the sublayer A is significantly lower than that in the sublayer B. It was determined that the sublayers A and B have cubic and amorphous structures, respectively. Also, basal planes of the core carbon were found to be oriented randomly.

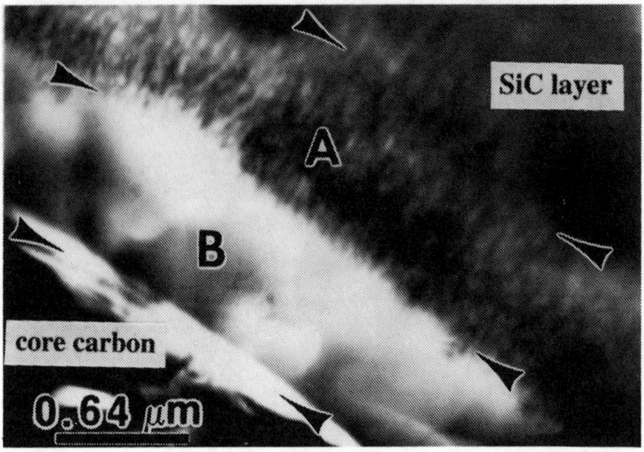

Fig. 7. Bright field image of core carbon, SiC layer, and reacted layer formed by the reaction of SiC layer and inner carbon layer in the fiber during creep test.

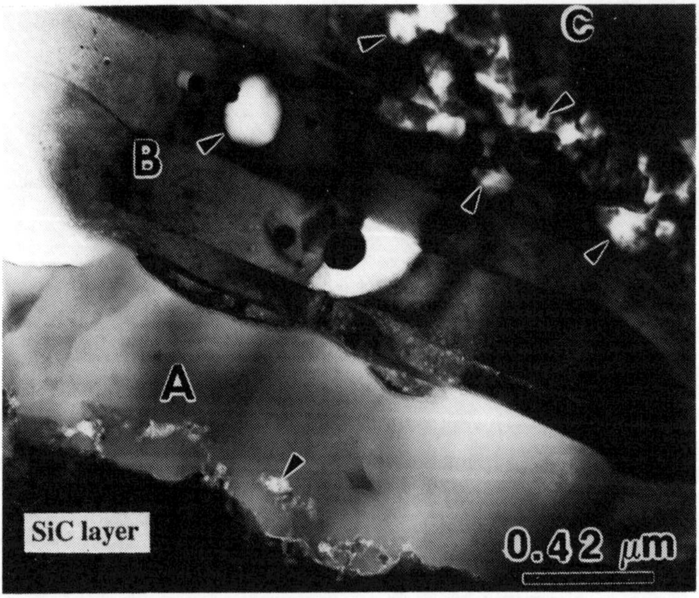

Fig. 8. Bright field image of SiC layer and reacted layer formed by the reaction of outer carbon layer in the fiber and matrix.

In order to determine the nature of the reaction between the fiber and matrix during creep tests, interfacial regions were investigated. The SEM micrographs of crept composite showed that most of cracks were initiated at the interfacial regions between the fiber and matrix and then propagated into the matrix. Attempts to make a TEM sample with cracks in the reacted regions were not successful. Therefore, the reacted regions without a crack were investigated. It was observed that the AlN matrix reacted with outer carbon layer in the fiber during creep tests. Fig. 8 shows a bright field image of SiC layer and reacted layer formed by the reaction of outer carbon layer in the fiber and matrix. The reacted layer consists of three sublayers, i.e., A, B, and C. The sublayers A, B, and C were identified as SiO_2 phase with an amorphous structure, Al-Si-Ca-O phase, and Al-N-O phase with a wurtzite structure, respectively. At present, structure of the sublayer B is not determined. The Al_2O_3 phase (marked by arrowheads) formed during creep test was observed at the reacted layer.

The deflection-time curves for the composite and monolithic AlN measured at 1100°C and bending stress of 46.3 MPa for 48 hours under oxidizing condition are shown in Fig. 9, indicating that the incorporation of SiC fibers into an AlN matrix degraded severely the high-temperature creep properties under oxidizing condition. This would be due to the bloating of the composite, the development of pores and Al_2O_3 at the matrix grain boundaries, and the development of pores and oxide phases such as SiO_2, Al-Si-Ca-O, and Al_2O_3 at the interfacial regions

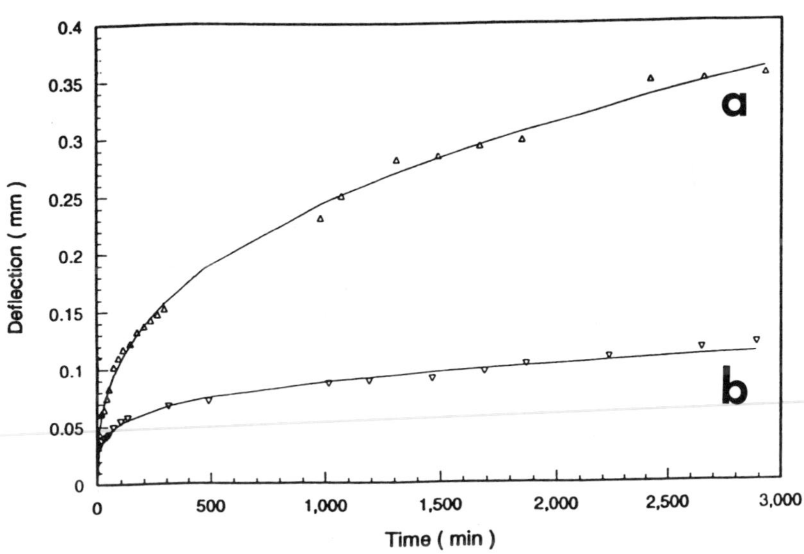

Fig. 9. Deflection-time curves for (a) the composite and (b) monolithic AlN measured at 1100°C and bending stress of 46.3 MPa for 48 hours under oxidizing condition.

between the outer carbon layer in the fiber and matrix. The bloating and cracks were observed at and around the reacted regions by visual observations and SEM micrographs.

CONCLUSIONS

The microstructure of matrix, fiber, and matrix/fiber interface in SiC fiber-reinforced AlN composites appeared to be changed significantly after creep tests. For as-fabricated composite, the matrix contained a $Ca_{12}Al_{14}O_{33}$ phase at grain boundaries and a high density of dislocations within the grains. The outermost carbon layer in the fiber reacted with the matrix during composite fabrication. The mechanical properties of the composite measured at room temperature were superior to those of monolithic AlN because of crack deflections around the SiC fibers.

In contrast, for crept composite, pores, cracks, and second phases such as $Ca_{12}Al_{14}O_{33}$ and Al_2O_3 were observed at and around matrix grain boundaries. The outer carbon layer in the fiber reacted with the matrix during creep tests to produce three sublayers, i.e., SiO_2 phase with an amorphous structure, Al-Si-Ca-O phase, and Al-N-O phase with a wurtzite structure. The incorporation of SiC fibers into an AlN matrix degraded high-temperature creep properties severely because of the bloating of the composite, the development of pores and Al_2O_3 at the matrix grain boundaries, and the development of pores and oxide phases such as SiO_2, Al-Si-Ca-O, and Al_2O_3 at the reacted layer formed by the reaction of outer carbon layer in the fiber and matrix.

REFERENCES

[1]. H. Kodama, H. Sakamoto, and T. Miyoshi, "Silicon Carbide Monofilament-Reinforced Silicon Nitride or Silicon Carbide Matrix Composites", J. Am. Ceram. Soc. **72**, 551 (1989).

[2]. Y. Ukyo and S. Wada, "High Strength Si_3N_4 Ceramics", J. Ceram. Soc. Jpn. Inter. Ed. **97**, 858 (1989).

[3]. I. Tanaka and G. Pezzotti, "Delayed-Failure Resistance of High-Purity Si_3N_4 at 1400°C", J. Am. Ceram. Soc. **75**, 1023 (1992).

[4]. Y. Kurokawa, K. Utsumi, and H. Takamizawa, "Development and Microstructural Characterization of High-Thermal-Conductivity Aluminum Nitride Ceramics", J. Am. Ceram. Soc. **71**, 588 (1988).

[5]. N. Kuramoto and H. Taniguchi, J. Mater. Sci. Lett. **3**, 471 (1984).

[6]. W. Werdecker and F. Aldinger, "High Performance Aluminum Nitride Substrate by Tape Casting Technology", in Proceedings of the 35th Electronic Components Conference. IEEE Components, Hybrids, and Manufacturing Technology Society, Piscataway, NJ, 1984.

[7]. X. J. Ning, P. Pirouz, and R. T. Bhatt, "Effect of High Temperature Annealing on the Microstructure of SCS-6 SiC Fibers", Mat. Res. Soc. Symp. Proc. **250**, 187 (1992).

[8]. K. Park and C. Sung, unpublished work.

PSEUDO 3-D REINFORCEMENT WITH STRETCH-BROKEN CARBON FIBERS IN A BOROSILICATE GLASS MATRIX

G. M. Crosbie, J. M. Nicholson, and L. A. Deering
Research Laboratory, Ford Motor Company, Dearborn, MI 48121

ABSTRACT

To compare delamination of "stretch-broken" vs. "unbroken" carbon-fiber reinforced glass matrix composites, six-ply Corning 7740 Pyrex® slurry-coated lay-ups were hot-pressed. Both uni-axial and cross-ply samples were fractured in short-beam three-point bending with both longitudinal and transverse orientations of outermost plies in tension. The stretch-broken fibers reduce the delamination viewed in SEM micrographs, and only these fibers prevented brittle failure in the uni-axial transverse [90°] case. A three-factor analysis of load-deflection curve values shows that the stretch-broken fibers reduce the proportional limit and ultimate loads. Delamination resistance results from the interlocking of broken filaments (distributed at angles to the fiber axis) to bridge across adjacent layers of the 2-D laminates.

INTRODUCTION

Among the light-weight (<2.2 g/cm^3) composite materials, carbon fiber reinforced glass matrix composites provide alternatives that have higher specific stiffness and are more dimensionally stable and temperature resistant than metals or polymers, without brittle fracture.[1,2,3] Such composites[4,5] could be employed more widely if a means were proven to obtain 3-D strengthening of the composites while achieving low overall cost and high manufacturing rates. In a review of fiber architectures for CMC's,[6] the potential was described for through-thickness strength and shear resistance with stretch-broken fibers. This would afford a linear fiber assembly alternative to costly and slow 3-D braided fiber reinforcement.

We present a fabrication method and fracture results for the stretch-broken tow reinforcement in carbon fiber reinforced glass matrix composites made from uni-axial and bi-directional lay-ups of bristly (or hispidulous[6]) fiber tows and compare them to ones with unbroken (or smooth tow) filaments.

To the extent authorized under the laws of the United States of America, all copyright interests in this publication are the property of The American Ceramic Society. Any duplication, reproduction, or republication of this publication or any part thereof, without the express written consent of The American Ceramic Society or fee paid to the Copyright Clearance Center, is prohibited.

EXPERIMENTAL

Three factors (denoted by Roman numerals) were tested at each of two levels, respectively: **(I)** "smooth tow" <u>vs.</u> "stretch-broken tow" fiber type; **(II)** uni-axial <u>vs.</u> cross-ply samples; and **(III)** longitudinal [0°] <u>vs.</u> transverse [90°] fiber orientation of the outermost ply put in tension by bending.

We prepared the glass matrix composites using two carbon fiber types which differed in bristly nature. These are identified as non-hispidulous and hispidulous. Although hispidulous means "to be covered with stiff short hairs," in this work, it denotes fibers that have filaments that stick out as shown in Fig. 1. This hispidulous fiber is a T-300 type staple spun yarn.* The comparison "smooth tow" fiber is an HMU-type† used earlier by others.[1]

The composites were made following the general procedures outlined by others[1,4,7] with additional detail provided here. After immersion (~18 cm) in an agitated slurry, the wet fiber tows covered with glass frit were wound onto a hexagonal drum with full-length slots between the faces. The slurry was comprised of 500 g glass frit§ and 3.34 g dispersant‡ added to 250 mL d. water. A custom lay-up apparatus, using a calibrated winder with controlled speed (~30 cm/sec), brake-drag fiber tension (~0.6 N), and forced-air drum drying was employed in mat fabrication. After die cutting and stacking (uni-axial or cross-ply) of 6 plies, 6.35 cm diameter stacks were held under vacuum for 24 h and then vacuum hot-pressed in graphite dies (with Mo lining) at 855°C for 20 min under 6.8-7.9 MPa (980-1150 psi) axial pressure. During cool-down, the pressure was held until 692°C was reached (~10 min). As pressed, the fraction of fibers was 30 wt.% for the smooth fiber composites and 22.5 wt.% for the hispidulous fibers.

---

* Stretch-broken carbon fiber, Heltra GRAFITEX™ 1680 denier T-300 staple spun yarn. Courtaulds Advanced Materials, Quakertown, PA [Now closed, per Courtaulds Aerospace, Coventry, UK. Similar fibers are believed to be available from Zoltek, St. Louis, MO and from s.a. Schappe, Meximieux, France.]

† Carbon fiber, HMU-PVR-3K (Lot 396-IV). Hercules, Inc., Advanced Materials and Systems Company, Composite Products Group, Magna, UT.

§ Borosilicate glass frit, -325 mesh. No. 7740 Pyrex®. Corning, Inc., Materials Business, Main Plant 21-3, Corning, NY. [Nominally, 80.5 SiO_2, 12.9 B_2O_3, 3.8 Na_2O, 0.4 K_2O, 2.2 Al_2O_3; ρ=2.23 g/cm^3; CTE=3.25×10^{-6}/°C, E=63 GPa, ann.pt.= 560°C]

‡ Low molecular weight carboxylic acid ammonium-salt dispersant. Dispex® A40. Allied Colloids, Inc., Suffolk, VA.

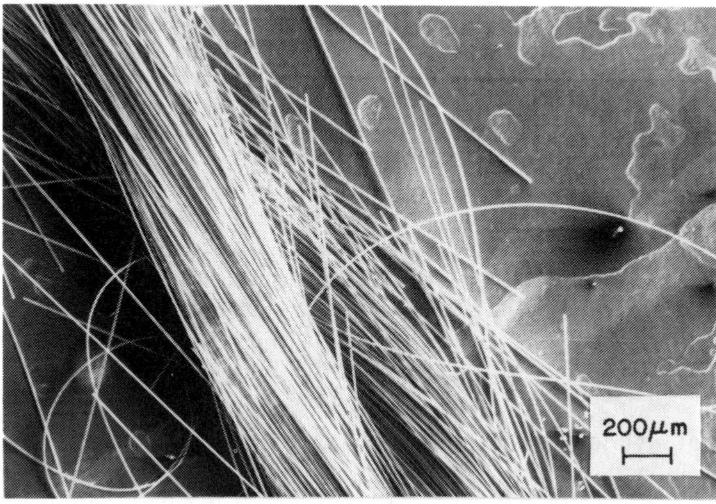

Fig. 1. Stretch-broken carbon fiber showing broken filaments distributed at angles around the main tow, giving a bristly or hispidulous character which can interlock with similar adjacent fibers in a composite.

Mechanical testing was carried out in short-beam 3-point bending to maximize delamination shear stresses. Bars were cut with the outermost fiber ply (of both uni-axial and cross-ply) oriented to the tensile stress both longitudinally and transverse. Diamond-sliced bars were 3.175 cm long × 0.635 cm wide × 0.39-0.44 cm "as-pressed" thickness. A 3-point SiC fixture (with a span of 1.905 cm and a nose radius of 0.15 cm) was used in an Instron test machine with a crosshead speed of 0.05 cm/min.

RESULTS AND DISCUSSION

As seen in previous longitudinal-axis flexure and tensile testing of carbon-fiber reinforced borosilicate glass composites, a pseudo-ductile fracture behavior was observed for both fiber reinforcement types tested. That this general behavior occurs has been previously discussed[3,8] in terms of weak bonding of continuous fibers to the glass matrix allowing pull-out toughening.

Fabrication Evaluation

While the design of this experiment allows us to make self-standing comparisons based on close control of processing within this work, the fabrication we used probably differs in several details from others' fabrication treatments. In particular, the densification was achieved at a lower temperature (855°C) than that used by others (>1200°C,[1] 1200°C,[2] or 1000-1400°C[9]), presumably allowing less degradation of the carbon fiber.[1,3] The glass did appear to permeate the tows.

Further evidence of the reasonableness of our fabrication is from mechanical properties: For the baseline case (uni-axial, longitudinal, smooth fiber), the apparent modulus of rupture (at the proportional limit) was between 200 and 260 MPa (30 and 40 ksi). Thus, our process interpretation was adequate to produce baseline[1,4] material of structural value.

Fracture Observations — Smooth Tow vs. Stretch-Broken (Factor **I**)

Uni-axial stacking, fibers oriented transverse to tensile $[90°]_6$*:* Transverse strengths of high toughness composites are generally very low.[8] The weak interfaces allow debonding. Transverse cracks propagate along the interfaces and through the adjacent matrix. In the Fig. 2 pair comparing two levels of Factor **I** (fiber type), the "smooth tow" sample in Fig. 2a broke in a brittle manner, exposing the individual filaments. In contrast, the stretch-broken fibers of Fig. 2b, bridged the transverse crack as it propagated, keeping the sample from brittle failure.

In Fig. 2b (with backscattered electron imaging), we see that we did achieve some of the overlap splaying that we wanted, with a halo of filaments around the twisted tows. With through-focusing on an optical microscope, one can trace lines back to see that the splayed filaments in overlapping regions originate from different parent tows.

Uni-axial stacking, fibers oriented longitudinal to tensile $[0°]_6$*:* Longitudinal strengths of weakly bonded fiber composites are high, because the fibers help carry the load and the weak interfaces are parallel to the tensile direction. Failure in bending, therefore, occurs by delamination. The delamination of another Factor **I** (fiber type) comparison pair is shown in the Fig. 3 pair. As the "smooth fiber tows" were not twisted, the individual filaments separated in hot-pressing, allowing many flat planes of delamination as seen in Fig. 3a. In contrast, in Fig. 3b, the stretch-broken fibers reduce the apparent delamination.

The effect of Factor **III** ([0°] vs. [90°] orientation of outermost layer in tension from bending) is seen in comparison of the Fig. 2 and 3 pairs.

Cross-ply stacking, outermost fibers longitudinal to tensile $[0°/90°]_3$*:* The outermost layer of cross-ply stacking behaved as the in-tension surfaces of the uni-axial ones did, in both orientations. In each image of the Fig. 4 pair, the

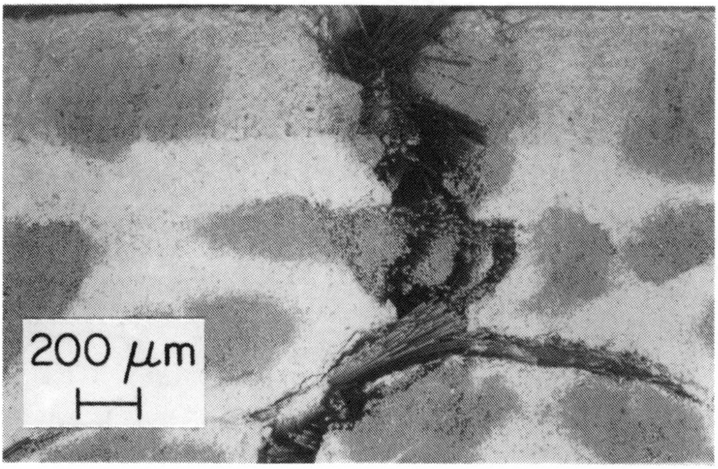

Fig. 2a. Brittle fracture of uni-axial, transverse oriented $[90°]_6$ smooth fiber carbon fiber reinforced borosilicate glass matrix composite.

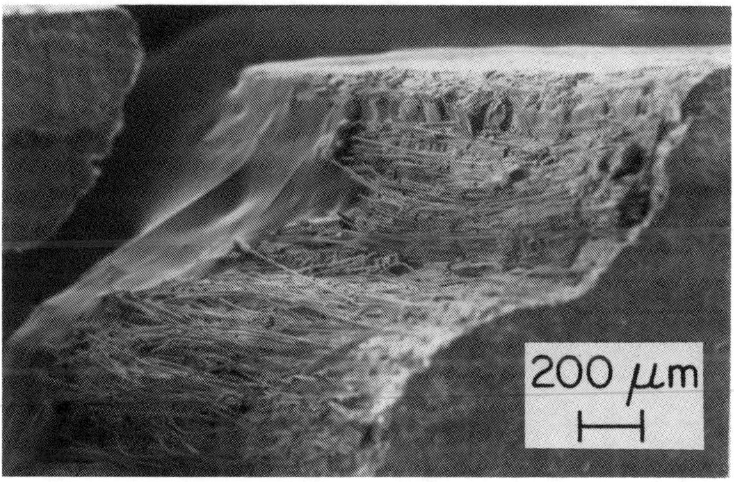

Fig. 2b. Pseudo-ductile fracture of uni-axial, transverse oriented $[90°]_6$ hispidulous (bristly) fiber tow composite, showing splayed filament halos around each tow and filaments bridging across the fracture.

Fig. 3a. Delamination fracture of uni-axial, longitudinal oriented $[0°]_6$ smooth fiber carbon fiber reinforced borosilicate glass matrix composite.

Fig. 3b. Fracture of uni-axial, longitudinal oriented $[0°]_6$ hispidulous (bristly) fiber tow composite, showing reduced delamination.

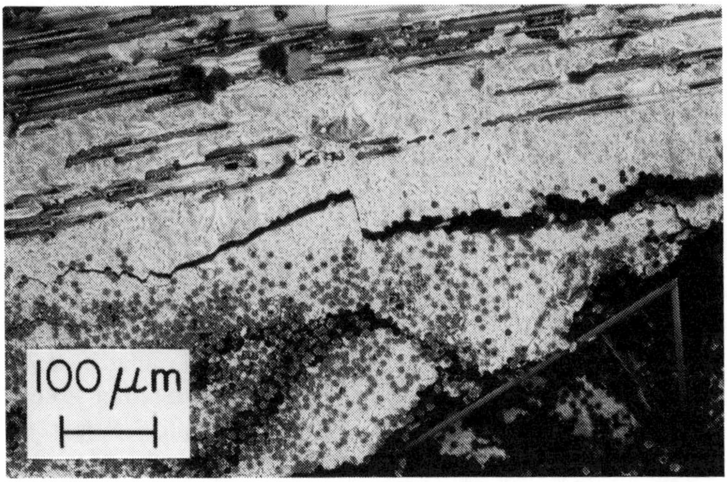

Fig. 4a. Fracture of cross-ply, longitudinal top-ply oriented $[0°/90°]_3$ smooth fiber composite, with cracks between top (0°) and 2nd (90°) plies.

Fig. 4b. Fracture of cross-ply, longitudinal top-ply oriented $[0°/90°]_3$ hispidulous (bristly) fiber tow composite, with splayed, bridging filaments.

top half of the image corresponds to the longitudinal plies of Fig. 3 and the lower half to the transverse plies of Fig. 2. In Fig. 4a, the "smooth tow fibers" show inter-ply cracking (between these first two plies) and transverse cracking in the second ply (bottom right of image). In Fig. 4b, filaments bridge the plies and transverse separation is seen in the twisted fiber bundles of the second ply.

The hybridizing effect of Factor **II** (uni-axial vs. cross-ply) is seen in comparison of Fig. 3 and 4 pairs.

Mechanical Properties

That the combination of two brittle materials can make a pseudo-ductile one is the remarkable qualitative incentive for use of continuous fiber (CF) CMC's over whisker or *in situ* reinforced ceramic matrix composites. Automotive applications require significant multiple-direction strength at the proportional limit. It is in this regard that CF mechanical properties fall off abruptly with some fiber architectures and orientations. In previous[1] tension tests to compare cross-ply [0°/90°] samples to uni-axial [0°] composites (Factor **II** in the present work), the ultimate strength was reduced by more than 2× and the proportional limit strength by 10×.

In this work, five mechanical properties were read from the load-deflection curves for factorial analysis. Those properties are: 1) ultimate load, 2) crosshead displacement at ultimate load, 3) proportional limit load, 4) crosshead displacement at the proportional limit load, and 5) an approximate load (at large deflection) for the pseudo-ductile fractures.

Because of the small number of samples tested (8) and as-pressed surface irregularities, there is need for caution in analysis of this exploratory set. Because the failure loads differ by large amounts (90× across the set), we report these values (aware of their statistical limitations) to help point to trends. Combinations in factorial tests reduce principal effect errors.

Proportional Limit and Ultimate Load Properties: Each of the factors (**I, II,** and **III**) had a negative effect on the proportional limit load (Table I and the associated bar graph in Fig. 5 factorial analysis) and the ultimate load (Table II and Fig. 6). The values for proportional limit load range from a high of 954 N to a low of 11.8 N; for ultimate load, from 1034 N to 12 N. The deflections at the respective load limits are given in Tables III and IV and related Figs. 7 and 8. The proportional limit deflections (Table III) are in roughly same proportions as the proportional limit loads (Table I), with a higher HMU filament stiffness (vs. the stretch-broken T-300) but lower vol.%.

Table I.

		Proportional Limit Load (N)	
		Longitudinal	Transverse
Non-Hispidulous	Non-Cross-ply	954.	60.
	Cross-ply	280.	360.
Hispidulous	Non-Cross-ply	490.	11.8
	Cross-ply	240.	122.

Figure 5.

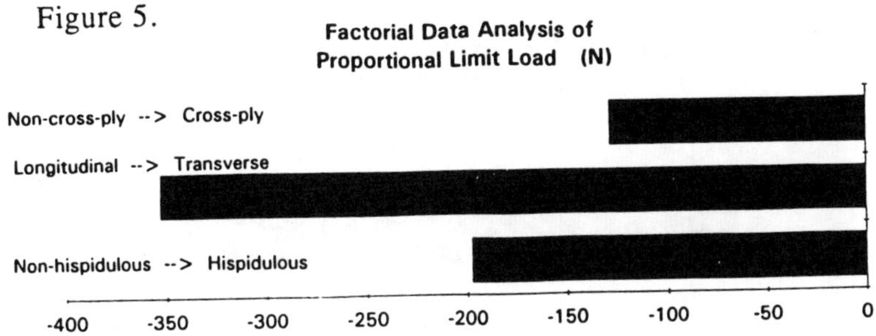

Table II.

		Ultimate Load (N)	
		Longitudinal	Transverse
Non-Hispidulous	Non-Cross-ply	1034.	78.
	Cross-ply	478.	448.
Hispidulous	Non-Cross-ply	504.	12.
	Cross-ply	276.	220.

Figure 6.

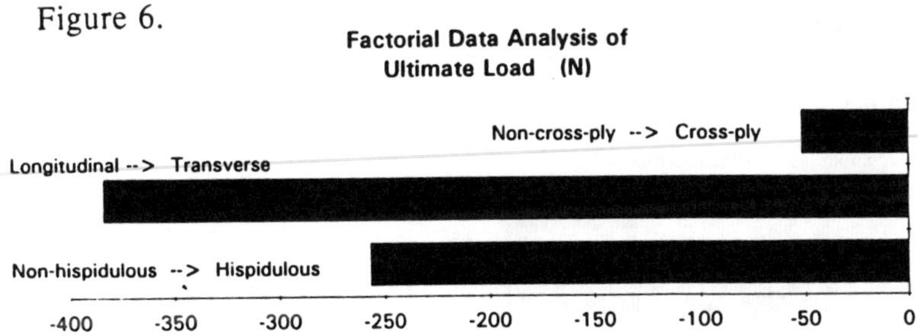

Table III.

		Crosshead Displacement at Proportional Limit Load (mm)	
		Longitudinal	Transverse
Non-Hispidulous	Non-Cross-ply	0.268	0.055
	Cross-ply	0.110	0.130
Hispidulous	Non-Cross-ply	0.185	0.005
	Cross-ply	0.150	0.110

Figure 7. Factorial Data Analysis of Crosshead Displacement at Proportional Limit Load (mm)

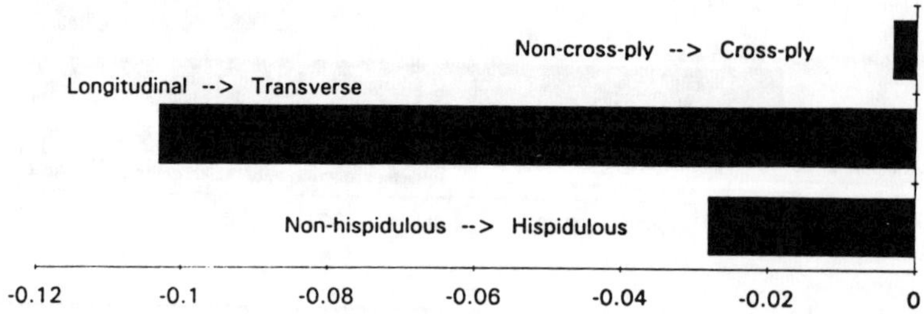

Table IV.

		Crosshead Displacement at Ultimate Load (mm)	
		Longitudinal	Transverse
Non-Hispidulous	Non-Cross-ply	0.325	0.070
	Cross-ply	0.230	0.265
Hispidulous	Non-Cross-ply	0.205	0.045
	Cross-ply	0.215	0.315

Figure 8. Factorial Data Analysis of Crosshead Displacement at Ultimate Load (mm)

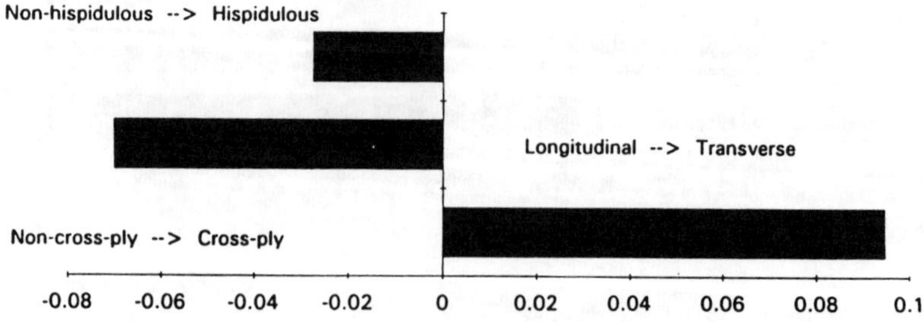

The strongest factor for reductions in both proportional limit load (Fig. 5) and ultimate load capability (Fig. 6) is for Factor **III**, the orientation of the outermost ply, shown by the middle bar length in each graph.

Although the bristly fibers change the quality of the [90°] fracture from brittle to pseudo-ductile (Fig. 2 comparison), they contribute little to transverse load capability. There are relatively few filaments which overlap and interlock regions and these are at shallow angles to the long fiber tow axis (Fig. 1). They do "bridge" across, but carry little load.

As expected from broken filaments,[6] the longitudinal loads (Tables I and II, "Longitudinal" column) are lower for the bristly (hispidulous) fibers than for the smooth (non-hispidulous) ones. However, these decrements (by about half in each [0°] case) are larger than the fraction of filaments that are stretch-broken. It is possible that the cooling rate we used may have worsened residual stresses due to multi-directional splayed fibers in the bristly fiber cases (Factor **I** = "bristly") similar to the previously attributed [0°/90°] expansion mismatch due to ply anisotropy (Factor **II** = "cross-ply") cases which increased matrix cracking in angle-ply and cross-ply composites.[4] Alternatively, the glass may not have permeated everywhere into the stretch-broken tows.

It is in the cross-ply cases, that the mechanical properties are most similar between the bristly and non-bristly reinforcement. In Table I for proportional limit, these values are 280 N and 240 N, which correspond to apparent (proportional limit) modulus of rupture values of 76 MPa (11 ksi) and 65 MPa (9.4 ksi). It appears that the long fibers are carrying the load and the irregularity of the inter-ply surfaces of the smooth fiber tows contribute more than the few overlapped splayed filaments. The large deflection load of the hispidulous fiber cross-ply transverse $[90°/0°]_3$ case appears to exceed that of the smooth fiber (120 N vs 106 N).

CONCLUSIONS

In the uni-axial cases, the bristly fibers are responsible for a transition from brittle to pseudo-ductile behavior (Fig. 2b *cf.* Fig. 2a) in the transverse [90°] orientation, due to overlap of stretch-broken and splayed filaments. Because there are relatively few overlaps and those are at low angle to the tow axis, the contribution to transverse mechanical properties is small in this realization. For the longitudinal [0°] case, the bristly composites made here were only half the proportional limit (apparent MOR of 132 MPa or 19.3 ksi) and ultimate load capability as the smooth fiber ones (MOR of 260 MPa or 37.5 ksi). The extent of delamination was reduced (Fig. 3b *cf.* Fig.3a).

In the cross-ply versions, the bristly fibers reduced the inter-ply delamination (Fig. 4b *cf.* Fig. 4a) and the mechanical properties were similar to smooth fiber ones. For the longitudinal orientation of fibers in the outermost ply in tension $[0°/90°]_3$ case, apparent (proportional limit) modulus of rupture values are 76 MPa (11 ksi) vs. 65 MPa (9.4 ksi), for the bristly ones.

The cross-ply cases appear best suited for further development with the bristly fibers. With less tightly twisted yarns and shorter broken filaments lengths, the bristly-fiber $[0°/90°]_x$ architectures may compete in properties with the more expensive needled and braided 3-D composites.

ACKNOWLEDGMENTS

The authors are grateful for the assistance of Ort Tool and Die (Erie, Michigan), the fiber vendors, and B. N. Juterbock, P. M. Reeber, Y. T. Liu, W. T. Donlon, and C. F. Johnson of the Ford Research Laboratory.

REFERENCES

1. K. M. Prewo, "Carbon Fiber Reinforced Glass Matrix Composite Tension and Flexure Properties," *J. Mater. Sci.*, 23 [8] 2745-52 (1988).

2. K. M. Prewo, "A Compliant, High Failure Strain, Fibre-Reinforced Glass Matrix Composite," *J. Mater. Sci.* 17 [12] 3549-63 (1982).

3. S. M. Bleay and V. D. Scott, "The Role of the Interface in the Mechanical Behaviour and Oxidation of Carbon Fibre Reinforced Pyrex Glass," *Carbon* 29 [7] 871-76 (1991).

4. V. C. Nardone and K. M. Prewo, "Tensile Performance of Carbon-Fibre-Reinforced Glass," *J. Mater. Sci.*, 23 [1] 168-80 (1988).

5. R. Morrell and L. N. McCartney, "Measurement of Properties of Brittle Matrix Composites," *Br. Ceram. Soc. Trans.* 92 [1] 1-7 (1993).

6. F. K. Ko, "Preform Fiber Architecture for Ceramic-Matrix Composites," *Am. Ceram. Soc. Bull.* 68 [2] 401-14, 1989.

7. K. M. Prewo, J. J. Brennan, and G. K. Layden, "Fiber Reinforced Glasses and Glass-Ceramics for High Performance Applications," *Am. Ceram. Soc. Bull.* 65 [2] 305-13+22 (1986).

8. A. G. Evans and D. B. Marshall, "The Mechanical Behavior of Ceramic Matrix Composites," *Acta Metall.* 37 [10] 2567-83 (1989).

9. J. R. Strife, J. J. Brennan, and K. M. Prewo, "Status of Continuous Fiber-Reinforced Ceramic Matrix Composite Processing Technology," *Ceram. Eng. Sci. Proc.* 11 [7-8] 871-919 (1990).

FREEZE GELATION: A MODIFIED SOL-GEL METHOD FOR THE FABRICATION OF CERAMIC-MATRIX COMPOSITES

J.M. Chant, R.G. Cooke and B. Harris,
School of Materials Science,
University of Bath,
Bath BA2 7AY
UK

*A novel process involving the rapid freezing of silica-based sols ("**freeze gelation**") permits the fabrication of ceramic-matrix composite components cheaply, to near net shape, at low sintering temperatures and with low shrinkage levels. This paper explores the influence of matrix composition on the microstructures and mechanical properties of unidirectional, carbon-fibre/silica composites obtained by filament winding.*
Transmission and scanning electron microscopy and X-ray diffraction techniques were used to characterise the composites. Tensile and flexural modes of testing were complemented by acoustic emission studies and fractographic analysis.
Measured mechanical property levels and acoustic emission responses are related to mechanisms of fracture in the different testing modes and to microstructural features.

INTRODUCTION

The aerospace industry's search for materials which offer good mechanical and environmental performance at elevated temperatures has been the primary driving force behind the recent advances in the fabrication of ceramic matrix composites (CMCs). There remain, though, many other application areas where the engineering demands are more modest and where ceramics or fibre-reinforced ceramics could profitably replace metals or polymers. Candidate CMCs for such applications must offer moderate but reliable stiffnesses and strengths at modest temperatures; in this context, reliability implies a practical level of fracture toughness and resistance to environmental degradation as well as consistency in properties. Cost is also an important factor. The work described here is part of a research project conceived to fulfil these needs via sol-gel processing, with particular emphasis being placed on the flexibility of shaping methods and on the economics of component manufacture [1].

To the extent authorized under the laws of the United States of America, all copyright interests in this publication are the property of The American Ceramic Society. Any duplication, reproduction, or republication of this publication or any part thereof, without the express written consent of The American Ceramic Society or fee paid to the Copyright Clearance Center, is prohibited.

Fabrication methods applicable to CMCs have been reviewed by many authors [2,3] and most successful techniques involve processing at high temperatures and pressures to achieve an adequate degree of matrix densification. Unfortunately, hot pressing inevitably causes physical and chemical interactions between fibres and matrices, the effects of which are rarely predictable or desirable. The possible degradation of the reinforcing fibres must also be considered.

Sol-gel processing involves physico-chemical reactions which occur at much lower temperatures than those of the conventional hot-pressing routes [4]. The controlled destabilisation of a colloidal sol to produce a three dimensional network of colloidal particles (*gel*) is of practical interest because the high surface area presented by the solid particles permits sintering of the gel at a relatively low temperature. However, the principal disadvantage of the sol-gel method is the inherently large shrinkage of the gel which occurs during drying. Although the incorporation of high volume fractions of inert fillers may reduce the volume shrinkage to as low as 7% from typically 20 to 25% [3], this feature of sol-gel processing has largely restricted application of the approach to the fabrication of chemically pure powders, fibres and coatings and small monoliths for optical applications.

A sol-gel processing route developed at Bath University [5] involves rapidly freezing a silica sol to produce a porous, easily-handled green body which may then be infiltrated and sintered. The principal advantages of the *freeze gelation* process are its low cost, low shrinkage levels (typically less than 1%) and low sintering temperatures. It is applicable to both continuous- and short-fibre composites and is extremely flexible in the range of shapes it is capable of producing. The shaping technologies available are essentially those used in the fabrication of polymer-matrix composites; the principal shaping routes used to date have been casting (for short-fibre-reinforced systems) and filament winding for continuous-fibre-reinforced composites. Examples of components manufactured by freeze gelation include gas-burner vanes, wound tubes for pressure containment, porous tubes for air bearings in polymer film manufacture and burst discs for protecting chemical processing equipment.

The freeze gelation process is illustrated schematically in figure 1. The primary silica sol is rapidly frozen to well below 0°C. The freezing process causes the formation of ice crystals (from the original aqueous solvent), with a consequential small increase in the volume occupied by the water. A gel is formed in the regions between the ice crystals. On warming back to room temperature and drying off the water, the solid exhibits minimal shrinkage and may be readily handled. Residual porosity, in the regions previously occupied by the ice crystals, is large (e.g. 10μm) and open, presenting the possibility of infiltration followed by sintering to densify the bodies. Practically, different types of filler are added to the initial sol both as a processing aid and as a means of exerting some influence over the properties of the matrix.

The potential of the freeze gelation technique and the influences of various processing parameters on processability and composite mechanical properties have been discussed elsewhere [1]. The choice of the silica sol, chemical and physical compatibility of reinforcing fibres and fillers, freezing (*i.e.* gelation) conditions, and liquid phase infiltration and sintering schedules have been optimised for particular composite systems. In particular, the rôles of different matrix fillers have been explored. In short-fibre reinforced composites, it has been demonstrated that the addition of a filler to the silica sol improves the handleability of the green component and decreases the initial gel porosity. The drawbacks are that the volume fraction of fibres which can be incorporated is reduced and the increase in viscosity of the sol mixture hinders processing to a small degree. It is worth noting that the reinforcing fibres in short-fibre systems themselves act as fillers in that they enhance green strength *etc.*. Moreover, the incorporation of even small volume fractions of fibres improves processability and reduces waste rates to such a degree that their use will be cost-effective even in applications where their presence is not necessary to achieve specified levels of mechanical performance.

In this paper, the influence of glass-ceramic filler particles on the microstructures and mechanical properties of a unidirectional composite system are explored.

EXPERIMENTAL

MATERIALS FABRICATION

Unidirectional composites consisting of carbon fibres in a sol-gel-derived filled silica matrix were manufactured by filament winding. The matrix components are described in table 1. The carbon fibres (Torayca T300) were passed through a desizing furnace at approximately 750°C and then through a bath containing a mixture of colloidal silica (HT50) and glass-ceramic filler particles. The volume fraction of filler particles in the sol bath was varied from 0 to ~0.30, the latter figure being the highest fraction practically achievable. The primary crystalline phase of the particles is β-spodumene, and the particles were selected on the basis of their having a nominally low thermal expansivity. The impregnated fibre tows were wound onto a hexagonal mandrel of width 100mm until the desired specimen thickness had been achieved, cut from the mandrel, stacked between PTFE-coated stainless steel plates and then placed in a press designed to keep the wet windings flat and the fibre tows in their correct orientation, whilst minimising potential damage to the fragile wet windings through over-application of pressure. The press was then briefly immersed in liquid nitrogen and subsequently maintained below 0°C for at least two hours.

Table 1 Details of matrix components

Designation	Supplier	Description and use
HT50	Monsanto	PRIMARY MATRIX PHASE Colloidal silica of mean particle size 125nm
D30	Monsanto	MATRIX INFILTRANT Colloidal silica of mean particle size 7nm
NLAS	CDML	MATRIX FILLER Low expansion glass ceramic particles of mean size 100nm. Primary crystal phase is β-spodumene ($LiAlSi_2O_6$).

On removal from the press, the plates were allowed to thaw to room temperature. These green composites, which were readily handled, were then subjected to a repeated liquid-phase infiltration and sintering cycle. Details of the infiltrant (D30) are given in table 1. Infiltration with D30 was performed under pressure of ~1MPa, and all specimens were sintered at a maximum temperature of 750°C in argon.

MICROSTRUCTURAL ANALYSIS

Samples were mounted in epoxy resin and polished to a finish of 0.5µm for optical microscopy. They were then coated with a thin layer of carbon in an evaporation unit for examination in the scanning electron microscope (SEM). Argon ion bombardment was used to thin samples for analysis in the transmission electron microscope (TEM). Samples of the composites were ground into a fine powder for X-ray diffraction (XRD) analysis using a Cu Kα radiation source.

MECHANICAL TESTING

Plates were cut into strips of nominal dimensions 100mm x 10mm for mechanical testing. The time taken for a 150kHz pulse to travel the length of a specimen was measured by means of a digital monitor. Values of dynamic modulus (E_{dyn}) could then be calculated from the approximate relation $E_{dyn} = c^2\rho$ where c is wave speed and ρ is specimen density. Flexural (three point bend) testing was carried out at span-to-depth ratios of 25 to 30, and interlaminar shear testing was carried out at a span-to-depth ratio of ~5. In both cases, tests were performed at a crosshead speed of 0.5mm min^{-1}. Tensile data were obtained from unwaisted specimens to which aluminium plate end tabs had been bonded with an epoxy adhesive. Strain gauges were glued to the opposite (widest) sides of tensile specimens. Transverse strain gauges permitted determination of Poisson's ratios. Tensile specimens were loaded to failure at a strain rate of ~0.05% min^{-1}.
Acoustic emission (AE) activity during flexural and tensile testing was monitored with a 150kHz PZT transducer, coupled to the specimens with petroleum jelly. A Marandy MR1004 digital data acquisition system was used to store and subsequently manipulate AE, load and strain gauge data.
Fracture surfaces from the flexural and tensile tests were coated with carbon for SEM examination.

RESULTS AND DISCUSSION

MICROSTRUCTURAL ANALYSIS

The most striking feature of the microstructures when observed at low magnifications is the network of open, columnar porosity (figure 2). The pores are up to 40µm wide and highly directional. Volume fractions of porosity determined by Archimedes measurements and by image analysis of polished sections lie in the approximate range 0.15 to 0.35, although localised volumes of the composite samples appear to be almost completely dense. An essential feature of the freeze gelation process is that the ice crystals which form during the initial freezing subsequently melt and leave pores in the volumes they previously occupied. The strong directionality of the observed porosity therefore arises from the direction of the local thermal gradient in the sample during freezing, *i.e.* through the thickness of the specimens. The presence of reinforcing fibres does not affect the pore structures; figure 3 is an SEM micrograph of a transverse section through a composite showing fibres 'bridging' the columnar pores. The apparent fibre damage in this micrograph is attributed to the metallographic preparation route, and does not necessarily indicate that fibre degradation occurred during fabrication.

The fact that the porosity is so pronounced indicates that liquid phase infiltration of the porous, green composites is insufficient to fill the porosity completely. Discrete boundaries between uninfiltrated regions and infiltrated regions are visible in the optical microscope using differential interference contrast imaging. In the SEM, infiltrated pores are readily discerned. Figure 4 is a transverse section micrograph taken from an apparently well densified region, and reveals significant cracking of the silica infiltrant (dark grey phase). At higher magnifications (figure 5), it is seen that cracking of the infiltrant in this way can lead to a lack of contact between fibre and matrix. This is undesirable since it reduces the scope for composite toughening by fibre/matrix interfacial friction during fibre pullout.

Porosity volume fraction, pore shapes and size distributions, efficacy of liquid phase infiltration and degree of infiltrant cracking were all unaffected by the amount of filler incorporated with the original colloidal silica sol.

XRD revealed the presence of two crystalline phases, α-quartz and β-spodumene, and some amorphous material (figure 6). The β-spodumene derives from the filler particles which must crystallise during sintering, and the relative intensity of its peaks varies with the filler content as expected. Figure 7 is a TEM micrograph which shows these crystalline filler particles within a completely densified region of silica. Selected area electron diffraction confirmed the identity of the β-spodumene, and that the HT50 sol-gel silica has crystallised to α-quartz (figure 8). Fine-scale porosity was seen in regions where the primary silica particles had not fused completely (figure 9) and these regions were only partly crystalline. The D30 sol-gel silica used to infiltrate the coarse pores in the green composites was not very consolidated (figure 10) and was frequently cracked. Electron diffraction showed it to be amorphous.

Although the distribution of the reinforcing fibres appeared to be uniform and independent of matrix composition, measured fibre volume fractions, V_f, varied approximately inversely with the volume fraction of filler in the matrix. Thus, V_f ranged from 0.50 in composites containing no filler to 0.10 in composites containing 0.28 volume fraction of filler in the matrix. This is a practical consequence of the winding process used; higher concentrations of filler particles in the sol-bath increased the viscosity of the sol mixture, and more of the mixture adhered to each fibre tow and was laid down on the winding with each pass.

MECHANICAL PROPERTIES

Two different types of response to flexural loading were observed. Samples containing between ~10vol% and 20vol% of filler in the matrix behaved in a brittle, linear manner and generated only small amounts of acoustic emission (AE). Outside this matrix composition range, the primary specimen failure mode was shear delamination even though the highest practicable span to depth ratio was employed during the tests. Figure 11 shows a typical vurve of flexural stress versus mid-point deflection with cumulative acoustic emission event counts superposed. This example is from a sample which had 1vol% of filler in its matrix. The deviation from linearity in the curve just before the peak stress is reached is accompanied by a sudden increase in AE activity, indicative of permanent damage occurring in the specimen. Appreciable post-peak load-bearing ability is shown; in this region of the curve, each load drop and subsequent reloading is associated with a delamination event. These delaminations were frequently visible even to the naked eye during testing. The frictional sliding over each other of large areas of separated laminae evidently consumes energy and contributes to the toughness of the material, and is reflected in the increase in the number of acoustic emission events detected. Figure 12 is an SEM micrograph from a fractured specimen which shows two separated laminae which have slid over each other in this way. In contrast, figure 13 shows the fracture surface of a specimen which failed in a brittle manner; note the absence of any fibre pullout or sliding between laminae.

All samples exhibited a similar response to tensile loading. Figure 14 is a typical tensile stress/strain curve, again with cumulative AE counts superposed. There is an initial linear region followed by a gradual decrease in elastic modulus. The proportional limit stresses in the composites studied ranged from 25% to 90% of peak stresses (but typically around 30MPa) and no relationship was found between matrix composition and proportional limit stress. A rapid increase in the number of AE counts accompanied the deviation from linearity in the manner described above for flexural tests. Strain to failure varied between 0.20 and 0.50%. Poisson's ratio values were in the range 0.10 to 0.30.

All the tensile specimens failed within their gauge lengths and the overall appearance of the fracture surfaces was fibrous. Fibre pullout lengths were typically ~100s μm. In relatively dense matrix regions, transverse matrix cracks were seen with fibre bridging over open cracks and pullout of fractured fibres over these regions (figure 15). In more porous regions, transverse matrix cracking in the conventional sense did not occur. However, localised fracture of the matrix ligaments which separated columnar pores formed plate-like regions of matrix perpendicular to the fibres (figure 16). In some cases fibre bridging was seen between these matrix plates, followed by fibre fracture and pullout processes similar to the fibre bridging/fracture/pullout across transverse matrix cracks which occur in 'conventional' ceramic-matrix composites [5]. In other cases the matrix 'platelets' appeared to adhere strongly to fibres and did not contribute to composite toughening in this way.

Tensile, flexural and dynamic measurements of elastic modulus agreed well (figure 17), and the moduli are seen to decrease as the amount of filler in the matrix is increased. This modulus reduction corresponds to the variation of V_f with matrix filler content described earlier; indeed, measured moduli accord well with rule of mixtures predictions. Tensile strength showed a similar dependence on matrix composition (figure 18), and this is also attributed to fibre volume fraction effects. In this case, however, the measured strengths are less than half those predicted by the rule of mixtures even when the matrix contributions to strength are ignored, and we believe that some degradation of the fibres occurs during fabrication.

The variation of flexural strength with matrix composition does not show the same trend as tensile strength (figure 18). As the volume fraction of filler in the matrix is increased, flexural strength increases to a maximum (reaching ~400MPa where the matrix contains ~15vol% filler) and then subsequently decreases. Interlaminar shear strength (ILSS) shows a similar dependence on matrix composition (figure 19). The relationship between flexural strength and ILSS is confirmed in figure 20 which shows a strong linear dependence between the two. One can use this relationship to explain qualitatively the variation in flexural damage mode with matrix composition described in the first paragraph of this section. In composites which have a matrix filler content either below ~10vol% or above ~25vol%, the ILSS of the composite is sufficiently low that shear failure is unavoidable during flexural testing and the result is the type of flexural stress / deflection curve shown in figure 11. By contrast, if the matrix contains between ~10vol% and ~25vol% of filler, then the composite ILSS is high enough for shear failure to be avoided during flexural testing and the composite behaves in a brittle manner.

CONCLUSIONS

- Freeze gelation is an economic, practical method of fabricating unidirectional, silica-matrix composites at low temperatures.
- Matrix composition has been optimised for a system comprising carbon fibres in a silica matrix with glass-ceramic (β-spodumene) filler particles.
- The distinctive patterns of directional porosity are artefacts of the freeze gelation process. Although liquid-phase infiltration is an inefficient method of filling this porosity, more sophisticated infiltration methods (*e.g.* CVI) would inevitably breach the 'low-cost' concept behind freeze-gelled composites.
- The observed fibre bridging, fracture and pullout across columnar, transverse pores indicates the potential for appreciable composite toughening. Transverse matrix cracking occurs in dense matrix regions.
- Degradation of the reinforcing fibres during processing needs to be minimised if the potential of these composites is to be realised.

Future Work will consider:

- The interplay between fabrication conditions, distributions of pore shapes and sizes and their effects on mechanical properties and composite fracture behaviour.
- The applicability of 'conventional' CMC mechanical property - microstructure models and, where appropriate, modifications to accommodate the porosity in these materials.

REFERENCES

1 R.S. Russell-Floyd, B. Harris, R.G. Cooke, J. Laurie, F.W. Hammett, R.W. Jones, and T.W. Wang, *The application of sol-gel processing techniques for the manufacture of fibre-reinforced ceramics*, J. Amer. Ceram. Soc. **76** (10) (1993) 2635.

2 J.A. Cornie, Y.M. Chiang, D.R. Uhlmann and A. Mortensen, *Processing of metal- and ceramic-matrix composites*, Ceram. Bull. (Amer. Ceram. Soc.) **65** (1986) 293.

3 J.R. Strife, J.J. Brennan and K.M. Prewo, *Status of continuous fibre reinforced ceramic matrix composite processing technology*, Ceram. Eng. Sci. Proc. **11** (1990) 871.

4 L.L.Hench and J.K.West, *The Sol-Gel process*, Chem. Rev. **90** (1990) 33-72.

5 A.G. Evans and D.B Marshall, *The mechanical behaviour of ceramic matrix composites*, Acta. Metall. **37**(10) (1989) 2567-2583.

ACKNOWLEDGEMENTS

We are indebted to the late Dr Richard Russell-Floyd and to Mr Frank Hammett for their contributions to this program. The financial assistance of the Science and Engineering Research Council and its ACME directorate is acknowledged.

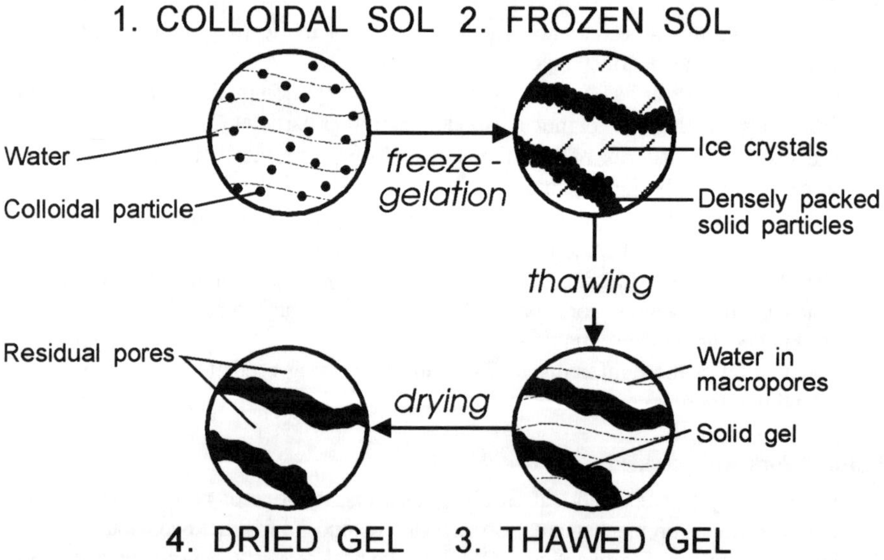

Figure 1 Schematic representation of the freeze gelation process.

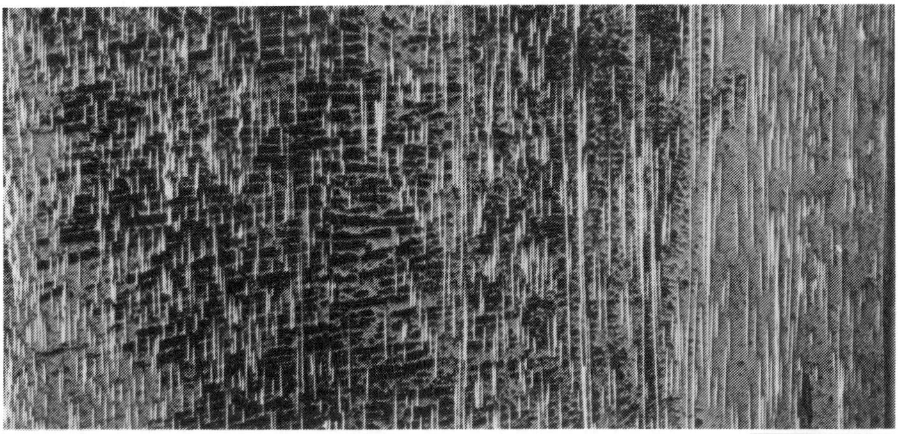

Figure 2 500 μm

Optical micrograph of transverse section revealing gross columnar porosity

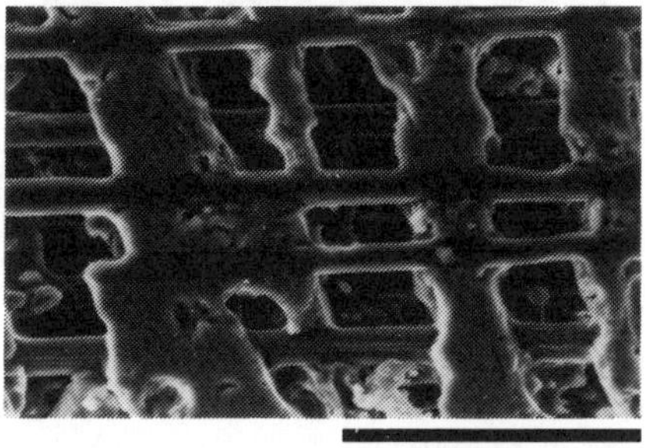

Figure 3 50 μm

SEM micrograph showing fibres 'bridging' columnar pores

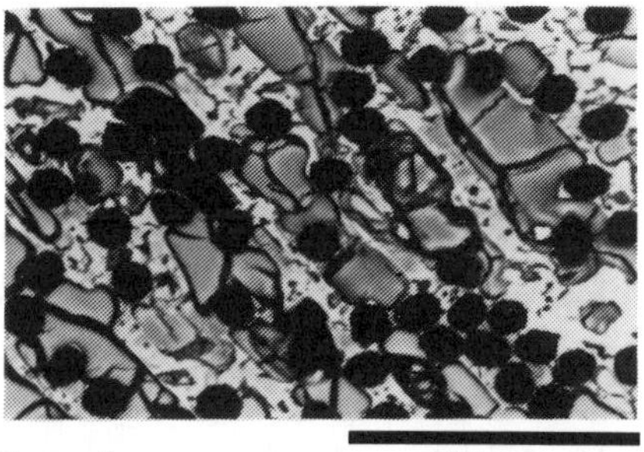

Figure 4 50 μm

Internal cracking of the silica infiltrant (dark grey phase)

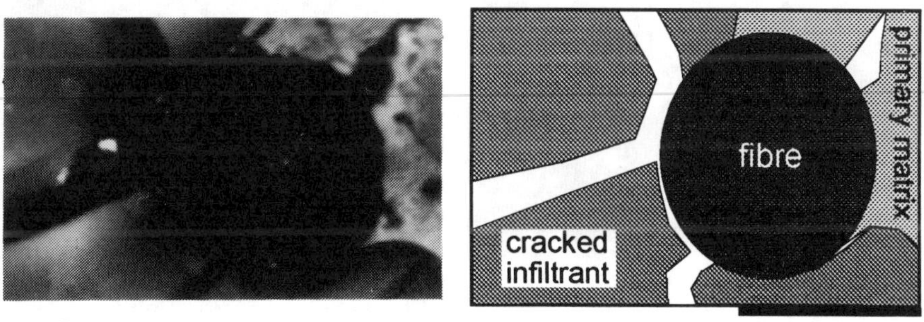

Figure 5 5 μm

Infiltrant cracking leading to poor fibre / matrix contact. RIGHT: key.

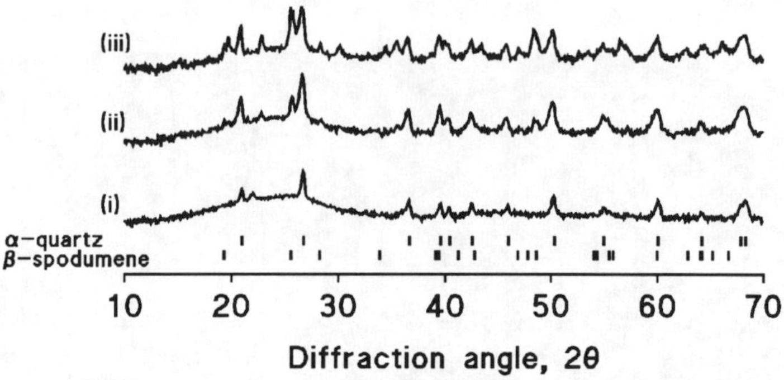

Figure 6 XRD traces of ground composite samples containing different volume fractions of NLAS filler in the primary matrix:
(i) 1%, (ii) 12% and (iii) 28%.
Reference peak positions are given for α-quartz and β-spodumene.

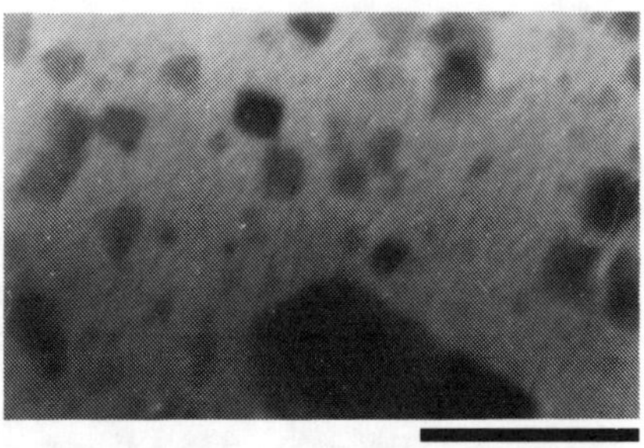

Figure 7 200 nm
TEM micrograph showing β-spodumene filler particles in α-quartz matrix.

Figure 8
Selected area diffraction image from region shown in figure 8.

Figure 9 200 nm
Fine scale pores between unconsolidated HT50 (primary) sol-gel silica particles.

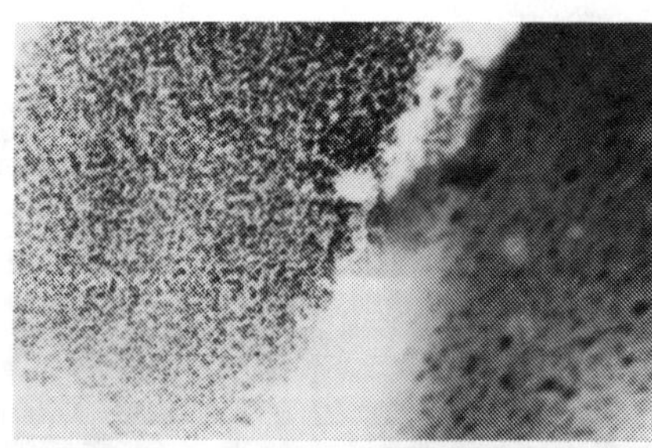

Figure 10 500 nm
Unconsolidated D30 (infiltrant) sol-gel silica particles, showing infiltrant cracking.

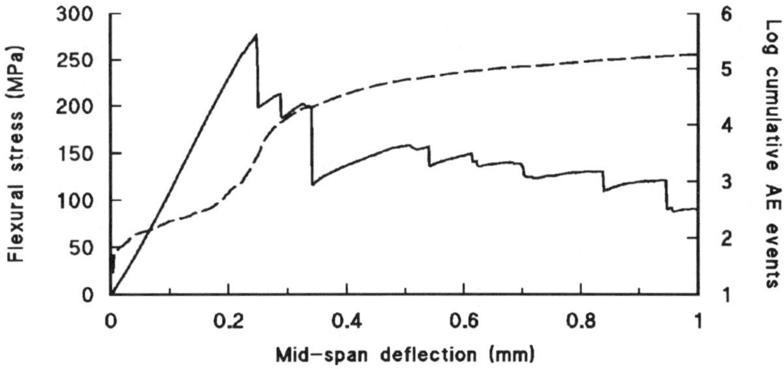

Figure 11 Typical flexural stress / mid-span deflection curve (solid line). Acoustic emission (AE) counts are represented by the dashed line.

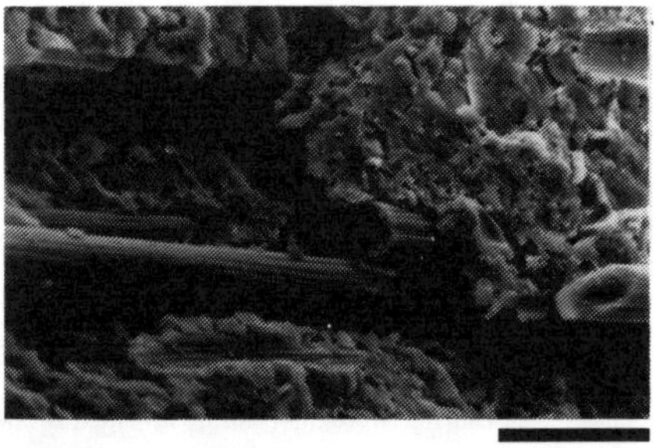

Figure 12 20 μm

Flexural fracture surface from a 'tough' specimen showing interlaminar failure.
(Matrix filler content = 1vol%).

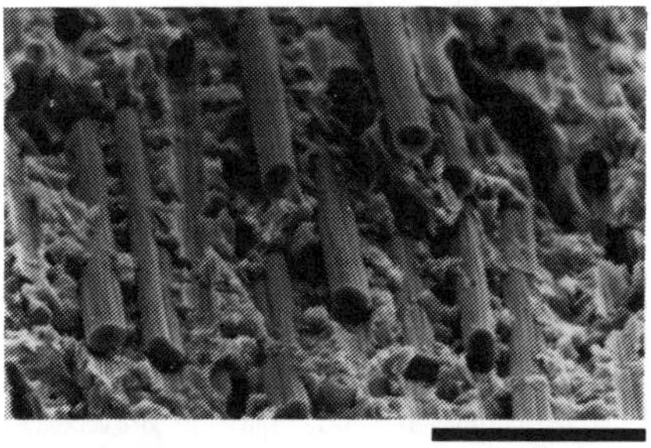

Figure 13 40 μm

Brittle fracture surface from a flexural specimen (tensile side).
(Matrix filler content = 14vol%).

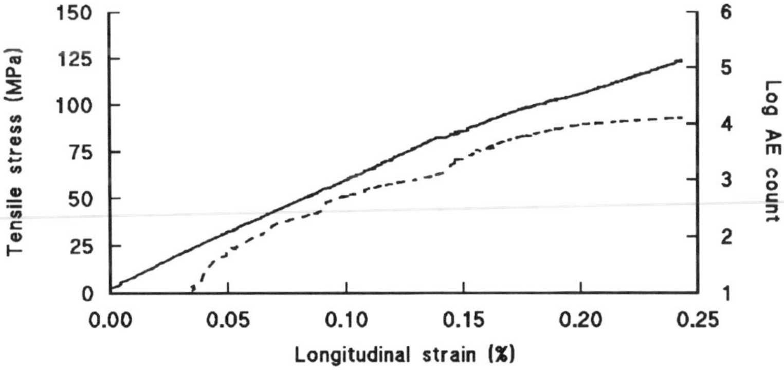

Figure 14 Typical tensile stress / strain response (solid line) and corresponding acoustic emission (AE) counts (dashed line).

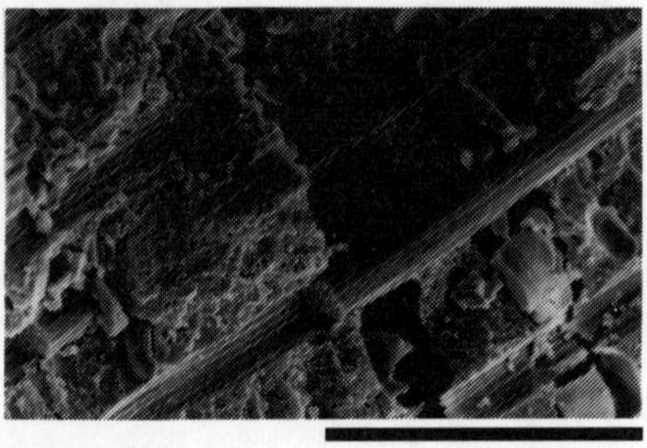

Figure 15 40 μm
Fibres bridging transverse matrix cracks and columnar pores in a tensile specimen.

Figure 16 100 μm
Transverse 'plates' of matrix adhering to fibres in porous regions (tensile).

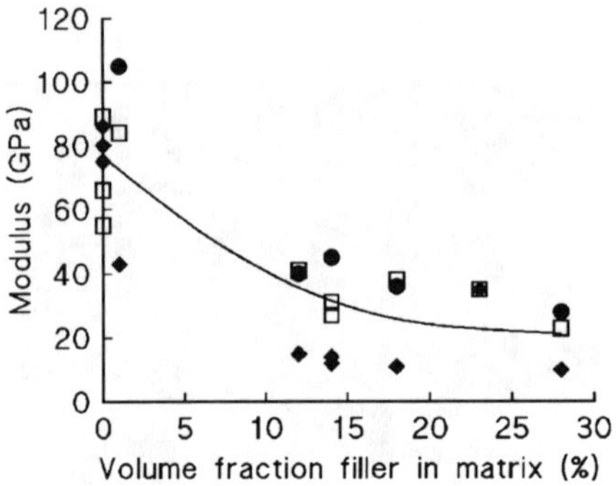

Fig 17 Effect of matrix filler content on composite elastic modulus.
● - tensile; □ - flexural; ♦ - dynamic.

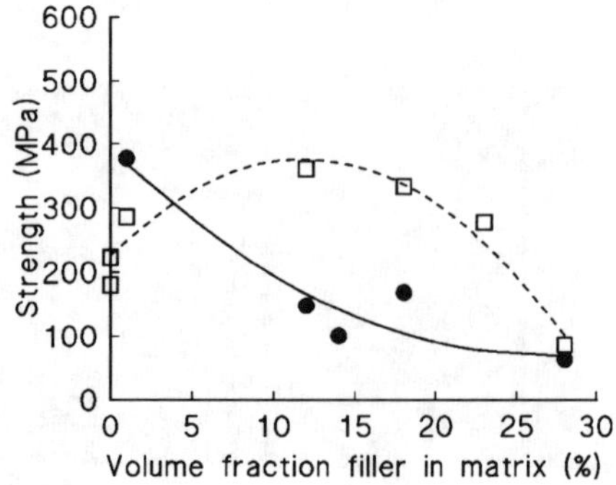

Fig 18 Effect of matrix filler content on composite strength.
● - tensile (solid); □ - flexural (dashed).

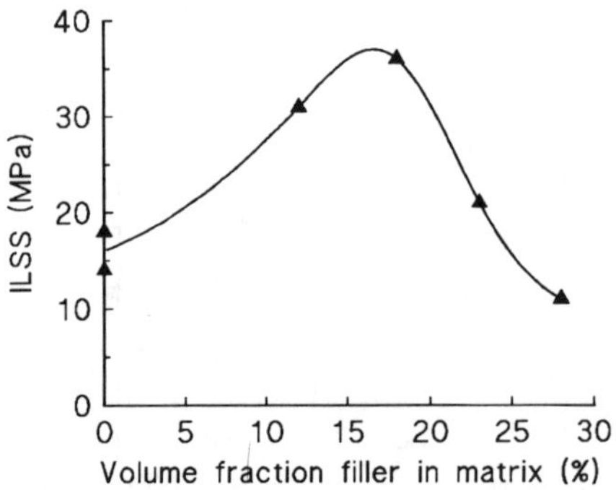

Fig 19 Variation of ILSS with matrix filler content.

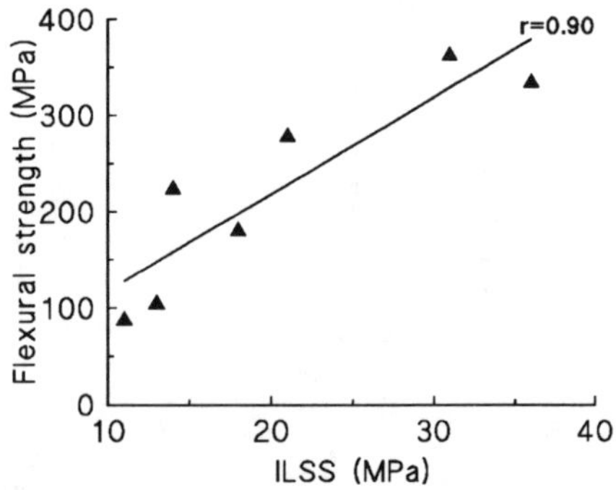

Fig 20 Variation of flexural strength with ILSS.

Particulate, Platelet, and Whisker Composites

PROPERTIES OF WHISKER-REINFORCED CERAMICS

P. F. Becher, C. H. Hsueh, H. T. Lin, K. B. Alexander,
T. N. Tiegs, W. H. Warwick, and S. B. Waters
Metals and Ceramics Division
Oak Ridge National Laboratory
Oak Ridge, Tennessee

Abstract

The paper summarizes the findings of a series of studies of SiC whisker-reinforced alumina composites emphasizing the properties of these composites and the influence of composition and microstructure. First, both the steady-state toughness and the R-curve response increase with whisker content and diameter. The room temperature flexure strength of aluminas with different grain sizes also increases with whisker content. Furthermore, the resistance to slow crack growth under both monotonic and cyclic loading is substantially improved by whisker reinforcement in comparison to that of alumina and other ceramics. The thermal conductivity and electrical resistivity exhibit a substantial increase and decrease, respectively, with whisker loadings above 10 vol.%. The thermal expansion coefficients decrease in a systematic fashion with increase in whisker content. However, the expansion coefficients are greater in the direction parallel to the hot pressing axis as compared to the perpendicular direction due to internal stresses and the preferred orientation of whiskers. The changes in mechanical and thermal properties due to SiC whisker additions result in substantial improvement in thermal shock resistance for samples of various sizes. At elevated temperatures, tests in air show that improvements in strength and resistance to strength degradation are associated with SiC whisker reinforcement. The limiting factors in elevated temperature mechanical response of the composite at 1200°C and above are related to surface oxidation-reaction processes and creep. On the other hand, the creep resistance of the composite is far greater than alumina with a similar fine grain size at temperatures ≥ 1200°C. Furthermore by increasing the grain size of the alumina matrix, the creep

To the extent authorized under the laws of the United States of America, all copyright interests in this publication are the property of The American Ceramic Society. Any duplication, reproduction, or republication of this publication or any part thereof, without the express written consent of The American Ceramic Society or fee paid to the Copyright Clearance Center, is prohibited.

resistance of the composite can be substantially increased. In essence, SiC whisker-reinforced alumina composites provide an example of a system whose properties can be tailored through control of the microstructure and composition.

Introduction

The brittle nature and low fracture toughness found with most ceramics have led the ceramics community to seek different approaches to improving overall mechanical performance. The introduction of continuous fibers, whiskers, platelets or elongated grains offers one approach to increase the toughness of various ceramics. The toughening mechanisms involve formation of a zone behind the crack tip in which the reinforcing phase bridges the crack and is subsequently pulled out of the matrix. The manner (i.e., magnitude and rate) in which toughening effects develop as a small crack grows can influence the degree to which the resistances to fracture, thermal shock, impact damage, and slow crack growth are improved. In order to assess this, it is useful to look not only at the fracture toughness and R-curve (rising fracture resistance with crack extension) behavior, but also to determine a variety of mechanical properties for the same composite system.

Earlier studies showed that the incorporation of strong microscopic whiskers into a ceramic matrix can result in a three- to four-fold increase in steady-state (long, $\geq 1000\ \mu m$, cracks) fracture toughness.[1-3] Certain whisker criteria can be used to optimize the toughening contribution. For instance, increases in the whisker diameter are beneficial in promoting frictional bridging and pullout processes. Similar effects are seen in silicon nitride ceramics where elongated β-phase grains can also bridge cracks[4,5] and increases in diameter of these larger elongated grains enhance the toughening effect.[6,7] Reinforcement morphology also plays a role. Smooth, straight whiskers are a more effective reinforcement. In alumina containing 20 vol% SiC whiskers a steady-state toughness of ~ 8.5 MPa√m was achieved with smooth ~ 0.7 μm diameter whiskers versus ~ 6.5 MPa√m for ~ 1.5 μm diameter whiskers with a very rough, corrugated surface morphology.[8] This can be a result of increased whisker strength by minimizing stress concentrations that reduce whiskers' strength. A smooth interface associated with such whiskers also circumvents mechanical interlocking of the interface. As a result, both interfacial debonding and pullout should be enhanced.

Toughening Response: Whisker Reinforcement

As we shall see, crack bridging processes evolve during crack growth. This leads to an increase in toughness with increase in crack length.[9,10] Immediately behind the crack tip, a bridging stress associated with elastic stretching of the whisker develops. Both the bridging stress and the corresponding interfacial shear stress increase rapidly with increasing distance from the crack tip. When the interfacial shear stress becomes sufficiently high, interfacial debonding occurs, and frictional bridging commences when the debonded whisker remains in contact with the matrix. Contact between the debonded whisker and matrix will be promoted when the matrix has the larger thermal expansion coefficient or when the whisker is inclined to the crack plane. The frictional bridging stress in the partially debonded whisker now increases at a somewhat slower rate (i.e., than that for an elastically bridging whisker) due to load transfer to the matrix. Thus, a frictional bridging whisker fractures further behind the crack tip and is accompanied by pullout when the point of fracture is not in the plane of the main crack.

The stress intensity factor associated with bridging processes along a line crack, K, is:[11,12]

$$K = K_0 + \left(\frac{2}{\pi}\right)^{0.5} \int_0^{D_{bz}} \frac{P\,dx}{x^{0.5}} \tag{1}$$

where K_o is the intrinsic toughness of the matrix, D_{bz} is the bridging zone length, x is the distance behind the crack tip, and P is the bridging stress profile. At this point, a simple, yet accurate, relationship between the crack opening at position x behind the crack tip:[13]

$$u = \frac{2\left(1-\nu_c^2\right) K_o \left[D_{bz}^2 - \left(D_{bz} - x\right)^2\right]^{0.5}}{\left(\pi D_{bz}\right)^{0.5} E_c} \qquad (2)$$

where E_c is the Young's modulus and ν_c is the Poisson's ratio of the composite will be used in establishing the definitions for the bridging stress profiles and zone lengths for each bridging mechanisms. Here, the results for frictional and pullout processes are only summarized; a detailed description is given elsewhere.[10]

The frictional bridging stress as a function of the crack opening displacement is:

$$P_{fb} = 2f\left[\frac{2\tau u E_w E_c}{d_w(1-f)E_m}\right]^{1/2} \qquad (3)$$

and is dependent upon the whisker diameter (d_w), Young's modulus (E_w) and content (f), the frictional shear resistance of the debonded interface (τ), and the Young's modulus of the composite, E_c and matrix, E_m. The interfacial shear stress is a product of the interfacial coefficient of friction and radial stress imposed on the interface due to mismatch in thermal expansion between matrix and whisker (compressive for SiC whiskers in alumina) and will decrease with increase in whisker content.[14] With increase in the crack opening displacement, the pullout bridging stress for a pullout length of l_{po} decreases as:

$$P_{po} = 4 f \tau \frac{\left(l_{po} - 2u\right)}{d_w} \qquad (4)$$

the embedded whisker end is extracted. The total bridging stresses for each process obviously will increase with whisker content.

The overall toughening effect due to whisker bridging is contingent upon the determinations of the sizes of both frictional and pullout bridging zones. One can start by considering the mechanics for uniaxially-aligned whiskers normal to the crack surface and then account for the fact that whiskers are biaxially-aligned. When whiskers are not normal to the crack surface, their resistance to fracture is reduced (i.e., their "effective" strength is reduced) due to bending stresses at the crack surface. For biaxially-aligned whiskers, the average fracture strength is ~ 60% of the uniaxial strength, σ_s.[15] The frictional bridging zone length, D_{fb}, is:

$$D_{fb} = 0.002\pi\left[\frac{d_w(1-\lambda f)E_m\sigma_s^2}{2\left(1-v_c^2\right)K_o\,\tau\,E_w}\right]^2 \qquad (5)$$

where λ represents the fraction of whiskers that form effective bridges. Then pullout length and crack opening displacement, 2u, can be equated to determine the size of the pullout bridging zone, D_{po}:

$$D_{po} = 0.0056\,\pi\left[\frac{d_w\,\beta(1-\lambda f)E_m\,\sigma_s}{2\left(1-v_c^2\right)\tau\,K_o}\right]^2 \qquad (6)$$

where β is the fraction of the debonded whisker length that is actually pulled out. It is immediately apparent that whisker diameter and strength, as well as the thermal expansion mismatch with the matrix that controls τ, can be used to vary the size of the pullout zone. Furthermore, whisker lengths (and distribution of lengths) will be a factor; this condition is modified as whisker fracture can occur at any point along the debonded length.

Based on the above discussions, it should be possible to tailor the R-curve response of a specific matrix reinforced with whiskers of a given compound, e.g., SiC, by manipulating whisker properties, characteristics, and whisker content. First, frictional bridging will lead to a rapid increase in fracture resistance, K_R, for small crack extensions. The corresponding rate of increase in K_R should also rise substantially with whisker content. Secondly, for large crack extensions where pullout dominates, the R-curve rises more slowly with crack extension, but will be shifted upwards with increase in whisker content. Finally, increases in whisker diameter should shift the R-curve upward while extending the frictional bridging zone size. As a result of frictional bridging, the composite with the larger diameter whisker will exhibit a stronger R-curve effect when the crack extends up to 200 μm in length. However, increases in whisker diameter have negligible effects on the rate of increase in fracture toughness where pullout dominates. In fact, the measured R-curves for SiC whisker-reinforced fine grained aluminas are consistent with these predictions, Figure 1. These observations suggest a number of factors (e.g., whisker content, strength, diameter and length, thermal expansion mismatch stress, or interfacial debonding stress) that will modify both how quickly the fracture toughness will increase as a crack extends and the magnitude of increased toughness. One should not, then, expect that all combinations of whiskers and matrix materials will provide the same effects. At the same time, there is some latitude in selecting compositions that might prove useful in tailoring other properties (e.g., thermal properties, hardness).

Finally, it was noted earlier that one can use additional approaches in combination with whisker-reinforcement to alter the toughness of the alumina-based composites. For instance, the alumina-matrix grain size can be a factor in the level of toughness achieved.[16-18] This grain size effect is well documented in alumina and other noncubic ceramics, where anisotropy introduces local residual stresses and microcracking,

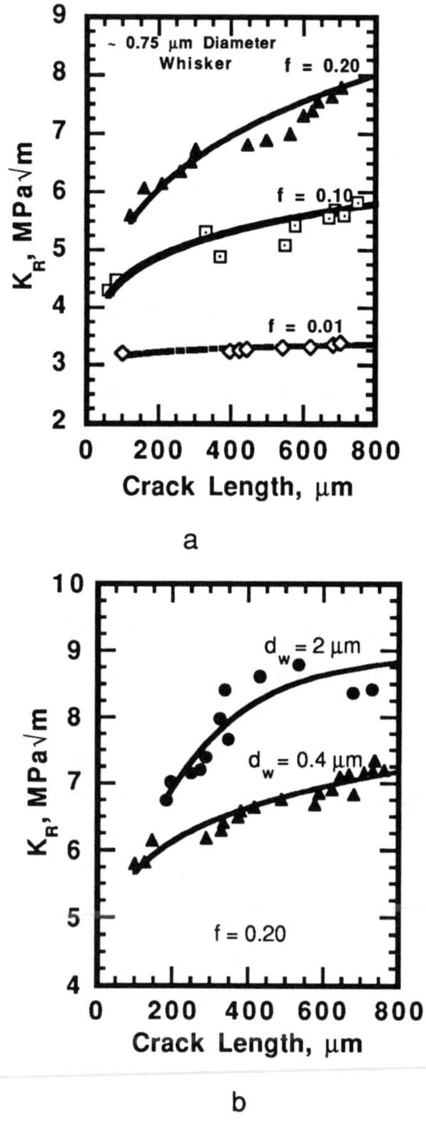

Figure 1. The fracture toughness/resistance, K_R, of SiC whisker-reinforced fine grained alumina increases with crack extension due to the evolution of crack bridging processes. This R-curve response increases with increase in whisker content (a) and with whisker diameter (b).

and is attributed to crack bridging by matrix grains. Subsequently, it was shown that toughness of SiC whisker-reinforced aluminas can be raised an additional 20 to 60% by increasing alumina grain size.[1c]

Mechanical and Other Properties of Whisker-Reinforced Aluminas

Whisker-reinforced aluminas can also exhibit substantial improvements in fracture strength[19], strength distributions (higher Weibull modulus),[20] and resistance to thermal shock[21] and slow crack growth under static,[19,22] and cyclic[23,24] loading conditions as compared to unreinforced aluminas. Some of these improvements in mechanical properties are shown in Table I.

Room Temperature Fracture Strength

The room temperature fracture strengths of hot pressed whisker-reinforced aluminas with fairly coarse surface finishes (i.e., 180 grit diamond resinoid bonded grinding wheel) are quite impressive. In fine grained alumina composites, strengths of ~ 700 to nearly 900 MPa are obtained for whisker loadings of ≥ 20 vol. %, Figure 2. A very interesting feature is the strengthening effects achieved in large grained alumina composites where strengths of up to 600 MPa have been attained. Recall that in unreinforced aluminas large grain sizes are used to gain increases in steady-state toughness, but, as shown here, this results in greatly reduced strengths. As noted earlier, when whisker-reinforced, these large grained aluminas exhibit greater toughness than the composites with finer grained alumina.

Slow Crack Growth Under Static and Cyclic Loading

The retention of strength as a result of damage/crack growth is another important consideration. The slow growth of cracks required for fast fracture (i.e., see times to failure for cyclic loading of alumina Table 1); often applied stresses of only > 50% of the fast fracture strength

Table I. Mechanical Performance of Reinforced and Unreinforced Alumina

Hot Pressed Alumina (G=1.5 μm) -20 Vol % SiC Whiskers

Temperature	22°C	1000°C	1200°C
Flexure Strength, MPa[a]	800	775	550
Fracture Toughness, MPa√m	8.3	8.1	10

Hot Pressed Alumina (G = 1.5 μm)

Temperature	22°C
Flexure Strength, MPa[a]	550
Fracture Toughness, MPa√m	2.8

Cyclic Fatigue Response (Tension-Tension with $\sigma_{min} = 0.1\sigma_{max}$)[b]

Hot Pressed Alumina (Grain size =1.5 μm) -20 Vol% SiC Whiskers

Maximum Stress, MPa	435	330	+367[c]	+396	+414
% of Fracture Strength	100	75	84	91	95
Cycles	1	1.1×10^5	$+4 \times 10^4$	$+3 \times 10^4$	$+6 \times 10^2$
	Failed	No Failure	NF	NF	Failed

Sintered Alumina (Grain size = 5 μm)

Maximum Stress, MPa	318	249	221	207	211
% of Fracture Strength	100	78	69	65	66
Cycles to failure	1	500	4×10^3	3×10^4	7×10^5

are sufficient to result in time-dependent failures. The chemical activity of reactive species (e.g., water) and temperature at the crack tip are controlling factors. Therefore increases in relative humidity at room temperature would promote crack growth at even lower stress levels. When the starting flaw size is the same, this is equivalent to stating that slow crack growth will be initiated at applied stress intensities more than

---

[a] Surfaces prepared with 180 grit diamond resinoid bonded surface grinding wheel with the grinding direction parallel to the length of the four point flexure bar.
[b] K. C. Liu, ref. 25; tension-tension testing at room temperature of alumina-25 wt. % SiC whisker composite fabricated by Cercom, Vista, CA.
[c] Stress cycled under these conditions for number of cycles shown; no failure. Testing continued after changing the maximum stress.

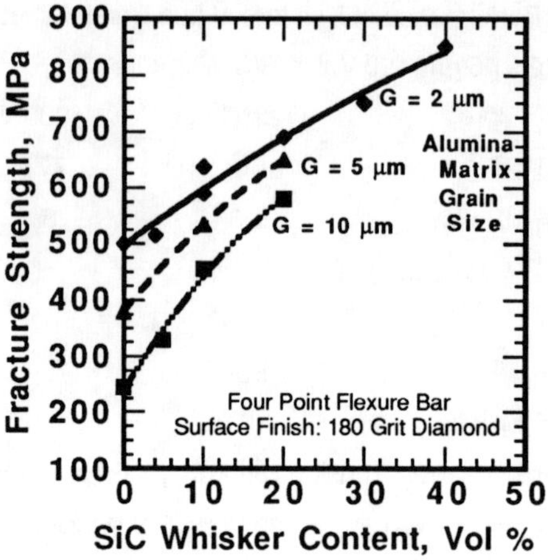

Figure 2. Flexure strengths increase with whisker content even in larger grain sized aluminas.

one-half the fracture toughness value. For an alumina ceramic with a fracture toughness of 4.5 MPa√m, crack growth should occur at applied stress intensities ≥ 2.5 MPa√m, Figure 3. One of the beneficial effects of SiC whisker reinforcement of alumina is the fact that stress intensity to initiate slow crack growth (~ 7 MPa√m) is a factor of two to three times greater than that required in unreinforced alumina, Figure 3.[19,23] Not only will whisker-reinforced aluminas withstand higher stresses and avoid fast fracture, but time-dependent failures will be avoided even when exposed to to quite high stress levels. This is reflected in Table 1 by the absence of failures in the composite even when subjected to peak cycle stress < 90% of the fast fracture strength. This is quite impressive for any class of materials. However, alternating opening and closing of the crack under cyclic loading can degrade the whisker and reduce the effectiveness of the crack bridging processes. The results for tension-tension cyclic fatigue do reveal that cyclic loading promotes slow crack growth to a degree as compared to static loading conditions, Figure 3.

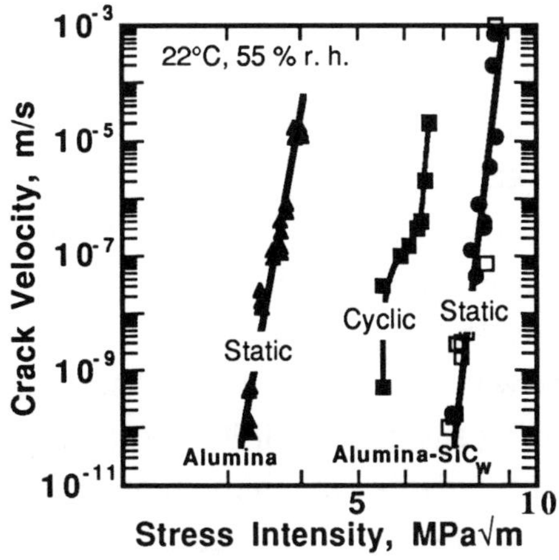

Figure 3. Whisker-reinforcement raises resistance to slow crack growth, under both static and cyclic loading. For composites, closed symbols-28 vol% whiskers; open symbols-20 vol% whiskers.

However, even under cyclic loading conditions, the reinforced alumina is far less susceptible to slow crack growth than the unreinforced alumina subjected only to static loading conditions.

Thermal and Electrical Properties

The thermal expansion behavior of composites can be a critical issue in design of components where one seeks to minimize expansion-contraction mismatch or maintain specific clearances between components subjected to temperature variations. In addition, tailoring the thermal expansion coefficients may be an important consideration in optimizing resistance to stresses generated during heating-cooling cycles or imposed by other thermal gradients. In composites, the expansion coefficients will be a product of the expansion coefficient of each phase and the volume fraction of each phase present (e.g., rule-of-mixtures).[26] The response is compounded when longitudinal axes of the reinforcing fibers or whiskers are oriented within the body.

Models of linear thermal expansion coefficients (α) of composites containing uni-axially aligned fibers indicate that α values will depend upon orientation with respect to fiber axes.[26-28] This is due to internal stresses brought on by mismatch in linear thermal expansion coefficients and elastic properties between fiber and matrix. These internal stresses are a function of (1) reinforcement content[26,28-30], (2) aspect ratio of reinforcements[27,31], and (3) degree of preferred orientation of the reinforcement axis.[31] The elastic strains, associated with internal stresses couple with the expansion/contractions generated during heating and cooling of the composite. These combined strains dictate the expansion coefficients of the composites. Furthermore, the axial internal stress in the whisker (and as a result, in the matrix) is greater than that normal to the whisker axis as a result of the axi-symmetric geometry of the whisker. This leads to an anisotropy in the expansion coefficients of composites with aligned fibers.[28]

The thermal expansion behavior suggested by analyses based on the above discussion is, in fact, observed in the SiC whisker-reinforced aluminas. First, the lower thermal expansion coefficient of SiC, as compared to alumina, leads to a decrease in the linear thermal expansion coefficients of the composites with increase in SiC whisker content, Figure 4. The internal stresses, due to the low thermal expansion coefficient of SiC versus alumina, introduced in the post densification cooling cycle place the whiskers in compression and the matrix in tension. As a result of this and the preferred orientation of the whiskers, the expansion coefficient of the composite in the direction parallel to the hot pressing axis (HPA) should be greater than that perpendicular to HPA for each composition due to the fact that the longitudinal axis of the whiskers lies, more or less, in a plane normal to

HPA. This is, indeed, observed in the SiC whisker-reinforced alumina, Figure 4.

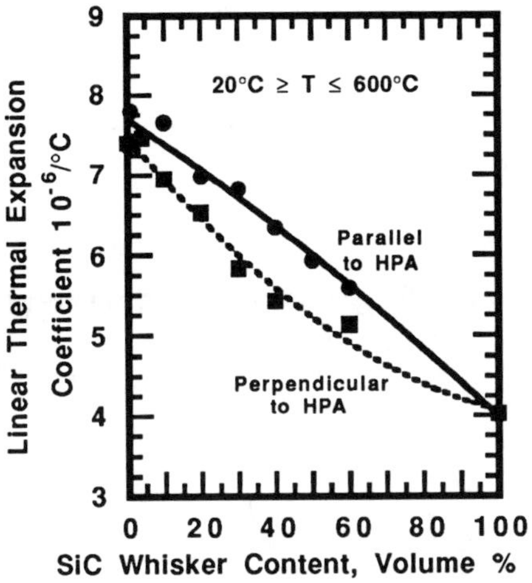

Figure 4. The thermal expansion coefficients (α) decrease with whisker content. Alignment of the whiskers during hot pressing results in an orientation dependence of α.

As a result of its much greater thermal and electrical conductivity as compared to alumina, the addition of SiC whiskers will also raise thermal conductivity and lower electrical resistivity of alumina-based composites. However, neither property will exhibit a simple rule-of-mixtures law dependent on whisker content. When the conducting phase (e.g., SiC) content exceeds the percolation limit (i.e., in the range of 10 to 20 vol% of this phase), much more substantial changes in these properties can be expected.[32] Above this whisker loading, interconnected highly conductive whisker paths are established. Such behavior is observed in the measured electrical resistivity[20] and thermal conductivities[33] of the SiC whisker-reinforced aluminas where rapid changes in electrical resistivity and thermal conductivity are observed as the whisker content is decreased below 20 vol.%, Figure 5. Enhancing the degree of preferred orientation of whiskers could result in establishing a difference in the whisker content to exceed the percolation limit for measurements parallel versus perpendicular to the pressing axis. This might allow one to select a composition that leads to a much higher electrical and thermal conductivity in the plane normal to the hot pressing axis versus in the direction of the HPA.

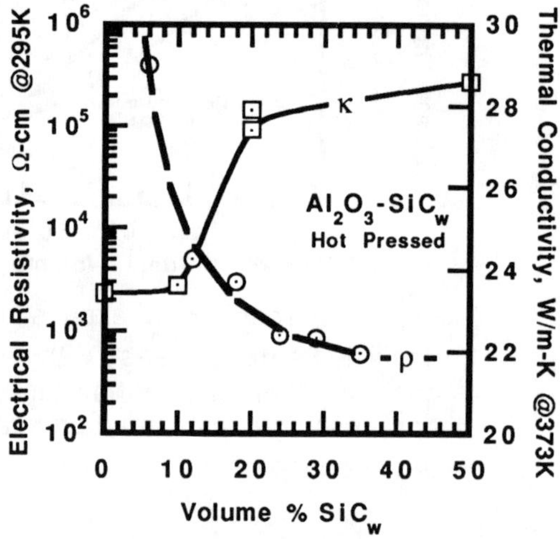

Figure 5. Both the electrical (ρ) and thermal (κ) conduction change significantly with increase in whisker contents above 10 vol.% related to the formation of interconnected SiC whisker paths.

Thermal Shock Resistance

Crack growth due to stresses generated by either fixed thermal gradients or rapid heating or cooling can lead to either a loss in strength or time dependent failures just as with mechanical stresses.[34] One of the more popular test techniques to evaluate thermal stress/shock resistance is the down-quench test involving measurement of strength retained after the sample is quenched from an elevated temperature, T, into a bath held at a fixed, but lower, temperature, T_B. Loss of strength after quenching is assumed to be a result of crack growth initiated when the thermal tensile stress, σ_T, equals the material's unshocked fracture strength σ_o.[34c-e] The magnitude of the thermal stress is dependent not only on the magnitude of the temperature drop (i.e., $\Delta T = T - T_B$) but also on various material properties (e.g., α, the linear thermal expansion coefficient; E, the Young's modulus; *k*, the thermal conductivity; ν, the Poisson's ratio), the characteristic (or shortest) heat transfer length *a* of the specimen (i.e., herein equal to the thickness *t* of the flexure bar), as well as the surface heat transfer coefficient *h* of the specific quenching media.

The critical temperature change, ΔT_c, that produces a tensile stress resulting in a loss in fracture strength after quenching is defined as:[34c-e]

$$\Delta T_C = [\,\sigma_o\,(1-\nu)\,/\,\alpha\,E\,]\,f\,(ah/k) \tag{7}$$

where the last term, $f\,(ah/k)$, reflects the role of heat transfer in establishing the temperature gradient in the sample. For a circular rod, this term can be approximated as:[34b,35]

$$f\,(k/th) \sim 1.5 + 4.67(k/th) - 0.5\exp(-51k/th) \tag{8}$$

where t is the diameter of the rod. Note that when the k/th term is small, ΔT_C becomes an independent sample size (e.g., Eq'n 8 reduces to a constant (~ 1.5). However, for large k/th values, ΔT_C exhibits an inverse dependence on the minimum sample dimension (e.g., the thickness of the test sample here). For a given material, this implies that the critical ΔT value should first decrease and then transition to become more or less independent of size as the thickness of the test specimen increases. Both the ΔT_C values and the critical sample thickness above which ΔT_C is independent of thickness will be functions of the thermal and mechanical properties of the material.

Based on increases observed in thermal conductivity, fracture strength, and toughness and decrease in thermal expansion, the SiC whisker-reinforced aluminas should exhibit greater thermal shock resistance than the unreinforced alumina. Measurements of ΔT_c values based on strength losses after quenching samples into boiling water baths confirm the greater thermal shock resistance of the composite as compared to unreinforced alumina.[36] Also as predicted, the ΔT_c values are dependent upon thickness t of the flexure bar in both the composite and the alumina ceramic, Figure 6. At small sample thicknesses, there is a

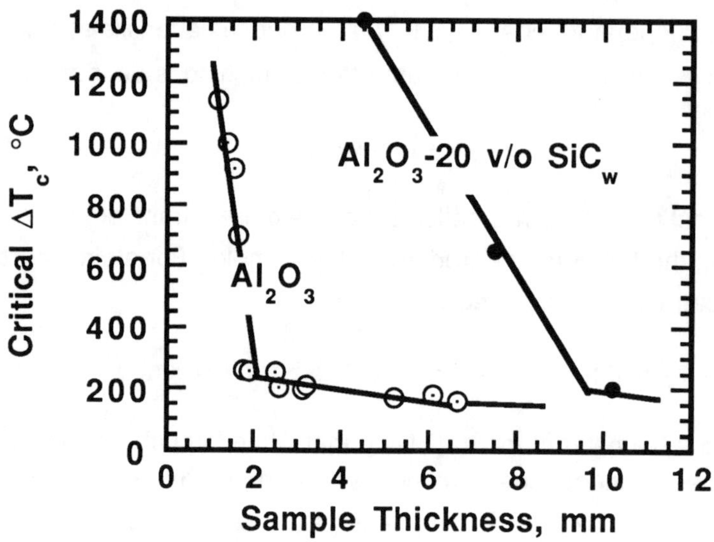

Figure 6. The thermal shock resistance is susbstantially increased by whisker-reinforcement. Even sections that are multiple millimeters thick do not undergo strength losses after quenching from several hundred degrees centigrade into a boiling water.

rapid decrease in ΔT_c values with increase in sample thickness followed by a transition to a region where ΔT_c becomes essentially independent of sample size. As expected, the composite exhibits greater resistance to thermal shock at much larger sample sizes. Finally, the composite also exhibits excellent resistance to repeated thermal shock cycles;[36a] this may not be surprising in view of its excellent resistance to mechanical fatigue. The ability to tailor thermal properties (e.g., Figures 4 and 5) combined with the excellent mechanical properties of whisker reinforced ceramics obviously provide an opportunity for developing thermal shock resistant structural ceramics.

Elevated Temperature Mechanical Properties

Previous studies show that the high flexure strengths (650 to 800 MPa) at room temperature are retained to quite high temperatures (e.g., 1000 to 1100°C in air) in SiC whisker-reinforced fine grained (< 5 μm) alumina.[37] Above 1100°C, the strengths decrease with increase in test temperature. Similarly, the fracture toughness of this same composite remains constant for temperatures up to ~1100°C in air; above 1100°C, the measured toughness increases. Each of these changes above 1100°C is due to the contributions of creep deformation primarily associated with grain boundary sliding as a result of the fine grain size of the alumina matrix.

Tests involving slow crack growth and surface reactions in air at elevated temperature revealed that alumina-20 vol% SiC whisker composites have remarkable resistance to strength degradation. At test temperatures of 800, 1000, and 1100°C, the fracture strengths were virtually unaffected after exposing the composites to applied flexure stresses equivalent to two-thirds the fast fracture strengths for times up to 1000 h.[37] Delayed failure tests at 1000 and 1100°C resulted in no failures in samples exposed to stresses ≤ 75% of the fast fracture strengths for times up to 14 weeks.[22] Only when the applied stress was > 75% of the fast fracture strength at 1000 and 1100°C, was slow crack growth induced failure observed. At temperatures of 1200°C and above, the fine grained alumina composites exhibited time dependent failure over an applied stress range of 40 to 100% of the fast fracture strength. At applied stress levels < 75% of the fast fracture strength, failed samples exhibit permanent deformation indicative of creep-related failure. These observations are indicative of the beneficial effects of whisker reinforcement on the elevated temperature of mechanical properties. As seen above, degradation of mechancial properties at elevated temperatures is associated with the onset of creep. Thus studies of the role of microstructure and composition on creep processes would prove useful in improving the elevated temperature mechanical performance of these composites. Studies do show that creep rates of fine grained alumina are greatly reduced by incorporating SiC whiskers; at 1200 (Figure 7) and 1300°C the creep rates are two orders of magnitude less

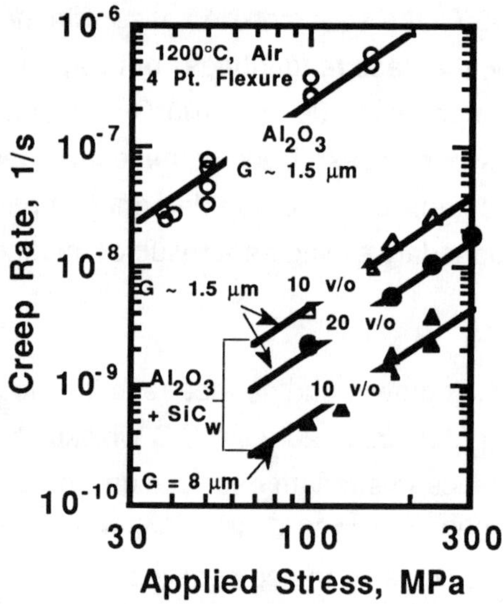

Figure 7. Creep rates at elevated temperatures are greatly reduced with the addition of whiskers. Matrix grain size can also be increased to obtain further creep resistance.

than those of the unreinforced aluminas.[38] Microscopy studies indicate that grain boundary sliding in the fine grained matrix is the primary mode of creep in these materials. A major contribution to reduced creep in reinforced alumina is a result of grain boundary pinning of the grain boundaries that results both from whiskers penetrating across grain boundaries and those that lie in the grain boundary planes.

However, creep resistance of SiC whisker reinforced alumina composites in air can also be degraded when whisker content is raised to 30 vol% or greater.[39] This is a result of both increased cavitation associated with whiskers and enhanced surface oxidation and glass formation. At the same time, the use of certain densification aids (e.g., yttria) that interact with the oxidation of SiC whiskers and form amorphous phases are also found to increase creep rates.[39] On the other hand, increasing the matrix grain size can result in greater creep resistance in the composite.[40] Attention to the formulation and design of the composite provide an approach to increasing the resistance to creep and extend the life of composites at elevated temperatures.

Conclusions

The successful application of whisker-reinforcement to improve the mechanical reliability of ceramics demands that attention be paid to the influences of whisker characteristics, matrix microstructure, and chemical composition. Progress in the area of reinforced ceramics is providing a wealth of new insights into the toughening of ceramic systems. The theoretical descriptions of the toughening response in whisker-reinforced ceramics provide details on how to design tougher materials and directions where more insight is needed, e.g., interfacial property-structure relationships. In addition, models for whisker reinforcement point to a need to develop techniques to synthesize whiskers where size and strength can be altered in a controlled manner.

As described herein, significant improvements in the fracture strength and its statistical distribution can be achieved in whisker-reinforced aluminas. These reinforced materials also exhibit fracture strengths that are much more resistant to subsequent degradation by crack growth due to thermal shock, and subcritical and cyclic fatigue as compared to unreinforced ceramics. At elevated temperatures, these systems offer considerable potential as they can exhibit excellent strength retention and creep resistance. Attention must also be given to changes in phase content resulting from exposure to various environments at elevated temperatures. For example in whisker-reinforced alumina, increasing the whisker content may be attractive for enhancing the fracture toughness but can result in degradation of creep performance due to the promotion of oxidation and creep rates. Clearly, tailoring compositions and microstructures of these composites offers an avenue to developing advanced high temperature ceramics and ceramics for various structural applications.

Acknowledgements

The authors note the assistance of V. M. Gibson in the preparation of the manuscript. The research was sponsored by the U.S. Department of Energy, Assistant Secretary for Energy Efficiency and Renewable Energy, Office of Transportation Technologies, as part of the Ceramic Technology Project of the Materials Development Program, under contract DE-AC05-84OR21400 with Martin Marietta Energy Systems, Inc. and by the Division of Materials Sciences, Office of Basic Energy Sciences, U.S. Department of Energy under Contract No. DE-AC05-84OR21400 with Martin Marietta Energy Systems, Inc.

References

1. a. P. F. Becher, C. H. Hsueh, P. Angelini, and T. N. Tiegs, "Toughening Behavior in Whisker Reinforced Ceramic Matrix Composites," *J. Am. Ceram. Soc.*, 71 [12] 1050-61 (1988).
   b. P. F. Becher, "Microstructural Design of Toughened Ceramics," J.*Am. Ceram. Soc.* 74(2) 255-69 (1991).

c. P. F. Becher, E. R. Fuller, Jr., and P. Angelini, "Matrix Grain Bridging in Whisker-Reinforced Ceramics," *J. Am. Ceram. Soc.* 74(9) 2131-35 (1991).

2. P. Sajgalik and J. Dusza, "Reinforcement of Silicon Nitride Ceramics by β-Si_3N_4 Whiskers," *J. Eur. Ceram. Soc.* 5: 321-6 (1989).

3. a. J. Homeny, W. L. Vaughn, and M. K. Ferber, "Processing and Mechanical Properties of SiC-Whisker-Al_2O_3-Matrix Composites," *Am. Ceram. Soc.Bull.* 65(2) 333-38 (1986).

b. G. H. Campbell, M. Rühle, B. J. Dalgleish, and A. G. Evans, "Whisker Toughening: A Comparison Between Al_2O_3 and Si_3N_4 Toughened with SiC," *J. Am. Ceram. Soc.* 73(3) 521-30 (1990).

4. F. F. Lange, "Fracture Toughness of Si_3N_4 as a Function of the Initial α-Phase Content," *J. Am. Ceram. Soc.* 62 (7-8) 428-30 (1979).

5. C. W. Li and J. Yamanis, "Super-Tough Silicon Nitride with R-Curve Behavior," *Ceram. Sci. and Engr Proc.*, 10 (7-8) 632-45 (1989).

6. T. Kawashima, H. Okamoto, H. Yamamoto, and A. Kitamura, "Grain Size Dependence of the Fracture Toughness of Silicon Nitride Ceramics," *J.Ceram. Soc. Japan*, 99: 1-4, (1991).

7. M. Mitomo, "Toughening of Silicon Nitride Ceramics by Microstructural Control," pp. 101-07 in Proc. Science of Engineering Ceramics, S. Kimura and K. Niihara, eds., Ceramic Society of Japan, Tokyo, 1991.

8. P. F. Becher, K. B. Alexander, C. H. Hsueh, H. T. Lin, T. N. Tiegs, and W. H. Warwick, "The Design and Properties of Ceramics Reinforced by Whiskers and Similar Microstructural Features," Proc. UNIDO Workshop on Advanced Ceramics, M. Miriani ed., in press.

9. C. H. Hsueh and P. F. Becher, "Some Considerations of Bridging Stresses for Fiber-Reinforced Ceramics," *Compos. Engng.*, 1 [3] 129-143 (1991)

10. P. F. Becher, C. H. Hsueh, "The Influence of Reinforcement Content and Diameter on the R-Curve Response in SiC Whisker-Reinforced Alumina," submitted for publication.

11. B. R. Lawn and T. R. Wilshaw, in Fracture of Brittle Solids, p. 85, Cambridge University Press, London (1975).

12. I. N. Sneddon and M. Lowengrub, in Crack Problems in the Classical Theory of Elasticity, p. 27, John Wiley, New York (1969).

13. D. B. Marshall, B. N. Cox, and A. G. Evans, "Mechanics of Matrix Cracking in Brittle-Matrix Fiber Composites," *Acta. Metall.* 33(11) 2013-21, 1985.

14. C. H. Hsueh and P. F. Becher, "Thermal Expansion Coefficients of Unidirectional Fiber-Reinforced Ceramics," *J. Am. Ceram. Soc.*, 71 [10] C438-41 (1988).

15. M. R. Piggott, in Load Bearing Fiber Composites, p. 126, Pergamon Press, Elmsford, New York (1980).

16. M. V. Swain, "R-Curve Behavior in a Polycrystalline Alumina Material," *J. Mater. Sci. Lttr.* 5:1313-15 (1986).

17. S. T. Bennison and B. R. Lawn, "Role of Interfacial Grain-Bridging Friction in the Crack-Resistance and Strength Properties of Nontransforming Ceramics," *Acta Metall.* 37(10) 2659-71 (1989).

18. G. Vekinis, M. F. Ashby, and P. W. R. Beaumont, "R-Curve Behavior of Al_2O_3 Ceramics," *Acta Metall. Mater.* 38(6) 1151-62 (1990).

19. P. F. Becher, T. N. Tiegs, J. C. Ogle, and W. H. Warwick, "Toughening of Ceramics by Whisker Reinforcement," pp 61-73 in Fracture Mechanics of Ceramics, Vol. 7, R. C. Bradt, A. G. Evans, D. P. H. Hasselman, and F. F. Lange, eds., Plenum Publishing Corp., New York, 1986.

20. J. F. Rhodes, "Alumina and Alumina Composites," Gorham Int'l Conference, Monterey, Calif., April, 1985.

21. T. N. Tiegs and P. F. Becher, "Thermal Shock Behavior of an Alumina-SiC Whisker Composite," *J. Am. Ceram. Soc.* 70(5) C-109-111 (1987).

22. P. F. Becher, P. Angelini, W. H. Warwick, and T. N. Tiegs, "Elevated Temperature Delayed Failure of Alumina Reinforced with 20 Vol % SiC Whiskers," *J. Am. Ceram. Soc.* 73(1) 91-96 (1990).

23. R. H. Dauskardt, D. Yao, B. J. Dalgleish, P. F. Becher, and R. O. Ritchie, "Cyclic Fatigue-Crack Growth in a Silicon Carbide Whisker-Reinforced Alumina Composite: Role of Load Ratio," *J. Mater. Sci.* . 28: 3258-3266 (1993).
24. L. X. Han and S. Suresh, "High-Temperature Failure of an Alumina-Silicon Carbide Composite under Cyclic Loads: Mechanisms of Fatigue Crack Tip Damage," *J. Am. Ceram. Soc.* 72(7) 1233-38 (1989).
25. K. C. Liu and C. R. Brinkman, pp. 359-72 in Semi-Annual Report of the Ceramics Technology for Advanced Heat Engines Project, TM-11586, Oak Ridge National Laboratory, Oak Ridge, TN, September, 1990.
26. D. K. Hale, "The Physical Properties of Composite Materials," *J. Mater. Sci.* 11: 2105-41 (1976).
27. Y. Takao and M. Taya, "The Effect of Variable Fiber Aspect Ratio on the Stiffness and Thermal Expansion Coefficients of a Short Fiber Composite," *J. Compos. Mater.* 21: 140-56 (1987).
28. C. H. Hsueh and P. F. Becher, "Thermal Expansion Coefficients of Unidirectional Fiber-Reinforced Ceramics," *J. Am. Ceram. Soc.* 71(10) C-438-C-441 (1988).
29. Z. Li and R. C. Bradt, "Thermoelastic Anisotropy in SiC, Residual Stresses in Monolithic SiC Ceramics and SiC Whisker Reinforced Composites," *Mater. Sci. Form.* 34-36: 511-23 (1988).
30. C. H. Hsueh, "Effects of Aspect Ratios of Ellipsoidal Inclusions on Elastic Stress Transfer of Ceramic Composites," *J. Am. Ceram. Soc.* 72(2) 344-47 (1989).
31. Y. Takao, T. W. Chou, and M. Taya, "Effective Longitudinal Young's Modulus of Misoriented Short Fiber Composites," *Trans. ASME*, 49 (3) 536-40 (1982).
32. S. De Bondt, L. Froyen, and A. Deruyttere, "Electrical Conductivity of Composites: A Percolation Approach," *J. Mater. Sci.* 27: 1983-88 (1992).

33. a. P. H. McCluskey, R. K. Williams, R. S. Graves, and T. N. Tiegs, "Physical Properties of Structural Ceramics," pp. 274-83 in Ceramic Technology for Advanced Heat Engines Project Semiannual Progress Report, ORNL/TM-11116, Oak Ridge National Laboratory, March 1989.

b. R. E. Taylor and H. Groot, "Thermophysical Properties of SiC-Reinforced Al_2O_3," PRL 692, Properties Research Laboratory, West Lafayette, IN, July 1983.

34. a. S. S. Mason, "Thermal Stress: I," *Mach. Des.* 30 114-20 (1958).

b. J. C. Jaeger, "On Thermal Stresses in Circular Cylinders," *Phil Mag.*, 36(257) 418 (1945).

c. W. R. Buessem, "Thermal Shock," pp. 460-83 in High Temperature Technology, L. E. Campbell (ed.), Chapman and Hall, New York, 1956.

d. W. D. Kingery, "Factors Affecting Thermal Stress Resistance of Ceramic Materials," *J. Am. Ceram. Soc.* 38(1) 3-15 (1955).

e. D. P. H. Hasselman, "Unified Theory of Thermal Shock Fracture Initiation and Crack Propagation in Brittle Ceramics," *J. Am. Ceram. Soc.* 52 (11) 600-604 (1969).

35. P. F. Becher, D. Lewis III, K. R. Carman, and A. C. Gonzalez, "Thermal Shock Resistance of Ceramics: Size and Geometry Effects in Quench Tests," *Am. Ceram. Soc. Bull.*, 59(5) 542-8 (1980).

36. a. T. N. Tiegs and P. F. Becher, "Thermal Shock Behavior of an Alumina-SiC Whisker Composite," *J. Am. Ceram. Soc.* 70(5) C-109-111 (1987).

b. P. F. Becher and W. H. Warwick, "Factors Influencing The Thermal Shock Behavior of Ceramics," pp. 37-48 in Thermal Shock and Thermal Fatigue Behavior of Advanced Ceramics, G. A. Schneider and G. Petzow eds., Kluwer Academic Publishers, Dorddrecht, The Netherlands 1993.

37. P. F. Becher and T. N. Tiegs, "Temperature Dependence of Strengthening by Whisker Reinforcement: SiC Whisker Reinforced Alumina in Air," *Adv. Ceram. Mater.* 3(2) 148-54 (1988).
38. H. T. Lin and P. F. Becher,"Creep Behavior of SiC Whisker Reinforced Alumina," *J. Am. Ceram. Soc.* 73(5) 1378-81 (1990).
39. H. T. Lin and P. F. Becher, "High Temperature Creep Deformation of Alumina-SiC Whisker Composites," *J. Am. Ceram. Soc.*, 74(8) 1886-93 (1991).
40. H. T. Lin and P. F. Becher, "Grain Size Dependence of the Creep Response in SiC Whisker Reinforced Aluminas," submitted to *J. Am. Ceram. Soc.*

ELASTIC ANISOTROPY IN HOT-PRESSED Si_3N_4 COMPOSITES: ROLE OF IN-SITU β-Si_3N_4 GRAINS AND ADDED β-Si_3N_4 WHISKERS

Murli H. Manghnani, Edward S. Fisher, Feng-ying Li, and Ru-ju Wang
Mineral Physics Group, School of Ocean and Earth Science and Technology, University of Hawaii, Honolulu, HI 96822

Jitendra P. Singh and Jules L. Routbort
Argonne National Laboratory, Argonne, IL 60439

ABSTRACT

The elastic anisotropy in hot-pressed samples of β-Si_3N_4 monolithic polycrystalline composites and in composites with 5 to 35% β-Si_3N_4 whiskers have been investigated at room temperature by high precision acoustic interferometery. All of the samples, including the monolithic Si_3N_4, show anisotropy characterized as transverse isotropic about the hot-pressing (HP) direction and are described using the five elastic stiffness moduli for hexagonal symmetry. The principal anisotropy is revealed in the shear modulus difference, $(C_{66} - C_{44})/C_{44}$, which increases from 7% in the monolithic sample to 15% in samples with 10% or more whiskers. All the stiffness moduli except for C_{44} increase with the addition of whiskers. Young's modulus in the HP direction is 5% smaller than in the HP plane in all samples. The whiskers enhance the shear anisotropy produced by the in-situ elongated β-grains.

INTRODUCTION

About two decades ago Lange[1] reported the beneficial effects of elongated β-Si_3N_4 grains (henceforth, designated as β-grains) on the mechanical properties of the monolithic ceramic matrix. The elongated β-grains were grown in-situ by hot pressing 90% α-Si_3N_4 powder at 1750°C for 2 hours, whereas the exact same preparation schedule for 90% β powder produced equi-axed grains. Although both final products were over 99% of theoretical density and 100% β phase Si_3N_4 crystal structure, the high α powder also produced the highest strength and fracture toughness. Lange also noted that the axial directions of the elongated grains were

To the extent authorized under the laws of the United States of America, all copyright interests in this publication are the property of The American Ceramic Society. Any duplication, reproduction, or republication of this publication or any part thereof, without the express written consent of The American Ceramic Society or fee paid to the Copyright Clearance Center, is prohibited.

parallel to the *c* axes of the hexagonal β crystals and were also predominately perpendicular to the uniaxial hot-pressing (HP) direction. There have been a large number of publications since Lange's report that describe the growth of the elongated β grains in Si_3N_4 containing either MgO or Y_2O_3 sintering aid materials[2-5]. Many of these studies describe aspects of the anisotropy of physical properties. The results for fracture toughness show that for a crack propagating parallel to HP direction, the ceramic material is generally significantly tougher than for a crack moving perpendicular to HP direction. This difference is attributed to greater deflection of a crack propagating along grain boundaries transverse to the elongated grains.

The effect of the elongated grains on the elastic properties of hot-pressed Si_3N_4 is, however, not at all clear. For example, Yeheskal and Gefen[6] report that a hot-isostatically pressed sample, that had apparently been preformed by unidirectional die pressing, had a Young's modulus, E, of 320 GPa in a direction perpendicular to HP direction and 306 GPa parallel to HP direction. This sample contained 6.5% of α phase and an undefined wt. % of Y_2O_3 sintering aid. Lee and Bowman[5], however, report the opposite anisotropy for a sample hot pressed with 10% MgO, and give an E value of 332 GPa parallel to HP direction and 315 GPa in the perpendicular direction. (The higher values for E of the latter samples[5] indicate that they contained a greater α phase concentration; this conclusion is based on the data[6] that indicate an E value of 362 GPa for α phase versus 312 GPa for 100% β phase). Buljan et al.[7] report no elastic anisotropy in their samples that were hot pressed with 7.5% Y_2O_3 + Al_2O_3 sintering aids. The reasons for these differences may be related to differences in sample preparation, or there may be errors or uncertainty in relating the HP direction to the directions of acoustic wave propagation. The major factor may, however, be in the assumptions involved in calculating an E value from acoustic velocity data taken in two different directions. The raw data generally used are, respectively, the compressional wave velocity V_P in each of the two orthogonal directions and a shear wave velocity V_S again in each of two directions. The values of E in each direction are then generally calculated from the equation derived for an isotropic solid, where

$$\rho V_P^2 = \lambda + 2G \tag{1}$$

$$\rho V_S^2 = G, \quad \text{and } \lambda = \text{Lamé constant} \tag{2}$$

$$E_{ISO} = 2G\left(1 + \frac{\lambda}{2\lambda + 2G}\right) = G\left(\frac{3\rho V_P^2 - 4\rho V_S^2}{\rho V_P^2 - \rho V_S^2}\right) \tag{3}$$

Since V_P and V_S are generally slightly or substantially different in each different direction, the use of the above equations to evaluate $E_{\parallel}$ or $E_{\perp}$ is immediately not applicable, but they are often used without reference to the fact that the results are merely approximations. There is, however, an even greater oversight in the use of these equations (the use of a single value for V_S), whereas in any anisotropic solid there are two normal transverse modes along any single direction of wave propagation. In most of the investigations of polycrystalline aggregates, the measurements usually involve only one measured V_S for each direction; the second, V_{S2}, for shear polarization normal to the first V_S, is mostly disregarded for unknown reasons. Consequently, the values given for $E_{\parallel}$ and $E_{\perp}$ in hot pressed samples will depend not only on the accuracy of measuring V_P but also on which of the two generally different V_S values are possible when a texture or preferred grain orientation is present in the sample. This can result in incorrect characterization of the elastic anisotropy in the composite, particularly when its microstructure contains elongated β-Si_3N_4 grains or in which short fibers are preferentially oriented perpendicular to the HP direction. The present investigation was performed to obtain a better understanding of the extent of elastic anisotropy in hot pressed monolithic Si_3N_4 and in samples of Si_3N_4 whisker reinforced composites. Whereas preliminary study[8] of similar samples did not indicate significant effects of whisker content on elastic properties, the use of more precise ultrasonic velocity measurement techniques and measurements of the two different shear wave velocities in the present study reveal significant increases in moduli and changes in anisotropy.

EXPERIMENTAL PROCEDURES

Sample Preparation

The procedures for preparing and characterizing the monolithic Si_3N_4 and composites containing 5%, 10%, 15%, 20% and 35% Si_3N_4 whiskers are described[8]. All samples contained 2.5% MgO as sintering aid, the whiskers after final processing were about 15 microns long x 0.6 microns diam., and the samples were hot pressed for 2 hours at a temperature sufficiently close to 1750°C to yield 100% β phase as monitored by x-ray diffraction[9].

The discs produced by hot-pressing were cut into rectangular bars, each with 3 sets of orthogonal faces, with 3 to 4 mm between faces. These faces were then ground and polished (finally, with 1 micron diamond paste), so as to be flat and have uniform thickness between parallel faces for measuring velocities of the acoustic waves to be propagated in each of the three orthogonal directions. The measured densities of each of the six samples are listed in Table I.

Table I. Acoustic wave velocities (in units of km/s) measured in HP direction and 2 orthogonal directions, as function of β-Si_3N_4 whisker content. The propagation and polarization directions are with respect to the three axes (*a,b,c*) of the specimen. The *a* and *b* directions are ⊥ to HP direction and *c* direction is ∥ to HP direction.

Velocity	prop. dir.	polar. dir.	vol. % β-Si_3N_4 whiskers					
			0	5	10	15	20	35
V_1	*a*	*a*	11.373	11.490	11.608	11.521	11.524	11.500
V_2	*b*	*b*	11.380	-	-	-	-	11.500
V_3	*c*	*c*	11.131	11.220	11.337	11.226	11.249	11.234
V_4	*c*	*a*	6.159	6.122	6.113	6.123	6.118	6.074
V_5	*c*	*b*	6.158	6.122	6.112	6.124	6.107	6.085
V_6	*a*	*c*	6.160	6.128	6.119	6.135	6.114	6.075
V_7	*a*	*b*	6.366	6.484	6.564	6.502	6.512	6.525
V_8	*b*	*c*	6.158	-	-	-	-	6.071
V_9	*b*	*a*	6.391	-	-	-	-	6.529
			% difference in velocities					
V_3			0.01	0.00	0.01	0.01	0.18	0.18
$V_4 - V_6$			0.01	0.10	0.10	0.20	0.08	0.00
$V_4 - V_8$			0.01					0.05
$V_7 - V_9$			0.39					0.08
$V_7 - V_4$			3.40	5.90	7.40	6.20	6.40	7.40
$V_7 - V_3$			2.20	2.30	2.30	2.60	2.40	2.30
Density, ρ (g/cm^3)			3.154	3.172	3.179	3.183	3.179	3.178

Wave Velocity Measurements

The acoustic wave velocities were measured by the phase comparison method[10]. The sample face normal to the direction of wave propagation is placed on the top end face of a round fused silica buffer rod, the bottom end of which is permanently bonded to a thin wafer quartz crystal piezoelectric transducer. The sample is acoustically bonded to the buffer rod with either a stop-cock grease for longitudinal waves or a viscous clear compound for transverse waves. The wave velocity, V, is calculated from

$$V = 2tf / n,$$

where t is the thickness between parallel faces, f is the carrier wave frequency at which an integral number of wave lengths, n, within the sample are in phase. Because of dispersion of velocity with wave frequency, the true free space velocity, V, is established at carrier frequencies above 30 MHz.

Computing the Elastic Moduli

The wave velocities measured in each of the six samples are listed in Table I, arranged among the three orthogonal Cartesian coordinates, *a*, *b* and *c*. The *a* and *b* coordinates are perpendicular to the HP direction. The *c* coordinate is parallel to the HP direction. Among the longitudinal wave velocities, V_1, V_2 and V_3, those in the HP deviation, parallel to the *c* coordinate, V_3, are clearly smaller than V_1, measured parallel to the a coordinate in all six samples, and smaller than V_2 in the *b* direction of the monolithic Si_3N_4 (0% whiskers) or in the sample containing 35% whiskers. Secondly, it is to be noted that in each sample, V_1 and V_2 are the same within 0.1% in the 0% and 35% whisker samples, indicating that the elastic anisotropy is similar to that for hexagonal single crystals or for transverse-isotropic symmetry in polycrystalline aggregates with fiber textures. In conformance with the elastic stiffness moduli for hexagonal symmetry, the V_P values parallel to HP direction are designated as defining the C_{33} modulus via

$$\rho V_3^2 = C_{33} \tag{4}$$

where ρ is the sample density. The two velocities perpendicular to HP direction, V_1 and V_2, are converted to the C_{11} modulus

$$\rho V_1^2 = \rho V_2^2 = C_{11} \tag{5}$$

If we are correct in designating HP direction as the axis of 6-fold symmetry, analogous to the hexagonal *c* axis, the two transverse (shear mode) velocities propagated parallel to HP direction, V_4 and V_5, should be equal to each other within the error of measurement and designated as defining the C_{44} modulus in the hexagonal system.

$$\rho V_4^2 = \rho V_5^2 = C_{44} \tag{6}$$

whereas the transverse mode velocities propagated in the *a* and *b* perpendicular directions, V_6, V_7, V_8 and V_9 should differ, depending on the directions of particle displacement, *i.e.*, polarization direction. The velocities of the transverse modes propagated perpendicular to HP direction but polarized parallel to HP direction should conform to

$$\rho V_6^2 = \rho V_8^2 = \rho V_4^2 = C_{44} \tag{7}$$

When the polarization and propagation directions are both perpendicular to the HP direction

$$\rho V_7^2 = \rho V_9^2 = C_{66} = (C_{11} - C_{12})/2 \tag{8}$$

From measurements of V_1 through V_9 in each sample we should then determine whether the assumed transverse-isotropic symmetry is, indeed, the correct elastic symmetry for hot pressed aggregates of Si_3N_4 and calculate values of C_{11}, C_{12}, C_{33} and C_{44}. These four moduli are not sufficient to establish values for the anisotropic Young's moduli, $E_{\|}$ or $E_{\perp}$, with respect to the HP direction, nor to calculate other elastic parameters of engineering interest. These parameters can, however, be semi-quantitatively predicted from judicious estimates of the fifth stiffness modulus, C_{13}.

RESULTS

Conformation with Transverse Isotropy

The major test of the transverse isotropic assumption is in the measurements of the shear moduli for the two directions of polarization. The six velocities involved in these cross-checks are given in equations (6), (7) and (8) and the measured values and differences, *i.e.*, disagreements, are tabulated in Table I. These cross-checks have been completed only for the monolithic Si_3N_4 sample and the 35% whisker composite. For the C_{44} modes the velocities agree to within 0.01% for the monolithic sample and to within 0.20% for the composite samples. For the C_{66} modes, where both wave propagation and wave polarization are perpendicular to HP direction, the disagreement in the monolithic sample is somewhat larger but still within 0.40%, whereas the disagreement is less than 0.10% for the 35% composite. These disagreements are considered small deviations, since we are dealing here with textured polycrystals rather than single crystals and they are certainly small relative to the magnitude of $(C_{44} - C_{66})$. This suggests that the whiskers enhance the preferred orientation of the elongated β grain axes. The other factor that confirms that the anisotropic treatment of elastic properties is appropriate is the differences between the velocities of the C_{44} and C_{66} modes, as represented by the differences $V_7 - V_4$. In the monolithic Si_3N_4 this difference is 3.4%, or about 10 times as large as the cross-check difference $V_7 - V_9$. As noted in Table I, the $V_7 - V_4$ difference increases by a factor of two with 10% whiskers and remains so up to 35% whiskers.

These results support the conclusion that the whiskers do in fact enhance the preferred orientation of the elongated β grains over that of the in-situ microstructure.

Evaluation of Anisotropic Elastic Moduli

In view of the above results, the elastic moduli of hot pressed monolithic Si_3N_4 containing near 100% β phase and composites containing β-Si_3N_4 whiskers should be reported with reference to a hexagonal Cartesian coordinate system in which the HP direction becomes the *c* axis. The four directly measured moduli, C_{11}, C_{22}, C_{33}, and C_{44} and the C_{12} value derived from Equation 5, are listed in Table II as a function of whisker content up to 35%, and are plotted in Figs. 1 and 2. Both of the longitudinal moduli increase significantly with the addition of 10 vol. % of whiskers and drop off slightly from 10% to 35%. The difference between C_{11} and C_{33} remains in the range of 5.5 ± 1.0% as whisker content increases, with C_{11} being larger in magnitude. This evidently reflects a larger stiffness modulus along the elongated in-situ or whisker grain axis, since these axes lie primarily perpendicular to the HP direction.

Table II. Elastic stiffness moduli (in units of GPa) with reference to *a, b, c* directions; HP direction is along *c* axis.

Modulus	vol. % Si_3N_4 whiskers					
	0	5	10	15	20	35
C_{11}	407.9	424.3	428.4	422.5	422.2	420.3
C_{33}	390.7	399.2	408.5	401.1	402.3	401.1
C_{44}	119.6	118.9	118.8	119.4	118.8	117.8
C_{66}^*	127.8	133.3	137.0	134.6	134.8	135.3
C_{12}	152.3	157.7	154.4	153.3	153.0	149.7
$(C_{11} - C_{33})/C_{33}$	4.4%	6.3%	4.9%	5.3%	4.9%	4.8%
$(C_{66} - C_{44})/C_{44}$	6.9%	12.1%	15.3%	12.7%	13.5%	15.1%

* $C_{66} = (C_{11} - C_{12})/2$

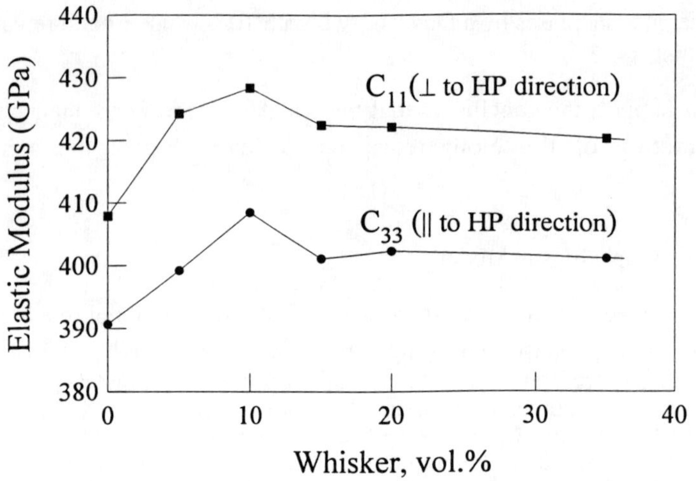

Figure 1. Longitudinal moduli, C_{11} and C_{33}, perpendicular and parallel, respectively, to HP versus volume % of Si_3N_4 whiskers in *B*-Si_3N_4.

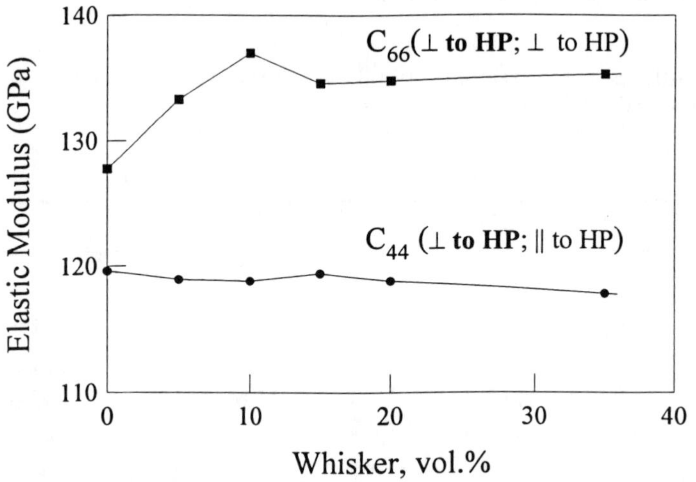

Figure 2. Shear moduli, C_{44} and C_{66}, versus volume % of Si_3N_4 whiskers.

The difference between the C_{66} and C_{44} shear moduli is, however, strongly affected by the addition of β-Si_3N_4 whiskers. As noted in Table II, this difference increases by over a factor of 2, *i.e.*, 6.9% to 15.3%, with the addition of the first 10% of whiskers. The C_{44} shear modulus, which represents a shearing of the hot pressed planes relative to each other, is essentially unaffected by the whisker addition, whereas the C_{66} modulus, representing a distortion of the HP planes, increases in strength with whisker additions.

DISCUSSION OF RESULTS

Estimates of Compliance Moduli and Engineering Parameters

Because the data that we have obtained from the three orthogonal directions of wave propagation permit only the evaluation of four of the five stiffness moduli, the evaluation of the compliance moduli, S_{ij}, the Young's moduli, and Poisson's ratio requires that we first make a reasonable estimate of the magnitude of the C_{13} modulus, which cannot be extracted from the present data. In making such an estimate, we first note that the C_{12} moduli, given in Table II, remains nearly constant with the addition of whiskers. Secondly, because the difference $(C_{11} - C_{33})$ is relatively small (i.e., $C_{11} \sim C_{33}$), we may assume $C_{44} = (C_{33} - C_{13})/2$. From this, we estimate C_{13} to be 150 ± 10 GPa. This estimate can, of course, be guided to some extent by the values of the Young's moduli, E, and bulk moduli, K, that result from the subsequent calculations and by comparison with value of E and K that have been quoted in the literature for isotropic β-Si_3N_4. For evaluating the Young's moduli parallel ($E_{\parallel}$) and perpendicular ($E_{\perp}$) to HP direction, we have

$$E_{\parallel} = 1/S_{33} = \left[C_{33}(C_{11} + C_{12}) - 2C_{13}^2\right]/(C_{11} + C_{12}) \tag{9}$$

$$E_{\perp} = 1/S_{11} = 2C_{66}\left[C_{33}(C_{11} + C_{12}) - 2C_{13}^2\right]/(C_{11}C_{33} + C_{13}^2) \tag{10}$$

As an approximation, we use

$$E_{ISO} = (E_{\parallel} + 2E_{\perp})/3 \tag{11}$$

as the value of E that would be measured in an isotropic polycrystal.

For evaluating the bulk moduli, K, we first calculate the compressibilities, parallel ($\beta_{\parallel}$) and perpendicular ($\beta_{\perp}$) to HP direction:

$$\beta_{\parallel} = (S_{33} + 2S_{13}) = (C_{11} + C_{12} - 2C_{13})/\left[C_{33}(C_{11} + C_{12}) - 2C_{13}^2\right] \tag{12}$$

$$\beta_{\perp} = (S_{11} + S_{12} + S_{13}) = (C_{33} - C_{13})/\left[C_{33}(C_{11} + C_{12}) - 2C_{13}^2\right] \tag{13}$$

Thus,

$$K = \frac{1}{\beta_{vol}} = \frac{1}{2\beta_{\perp} + \beta_{\parallel}} \tag{14}$$

or $$K = \frac{1}{S_{33} + 2S_{11} + 2S_{12} + 4S_{13}}$$

The results of computing Equations 9-14, assuming C_{13} = 150 GPa, are given in Table III and plotted in Figs. 3 and 4. Young's modulus, perpendicular to HP direction, $E_{\perp}$, is significantly larger and the difference between $E_{\perp}$ and $E_{\parallel}$ remains at about 5% as whiskers are added to the composites. There is, however, a significant increase in both $E_{\perp}$ and $E_{\parallel}$ values with addition of the first 10% of whiskers. Although there is no experimental evidence for explaining the maxima in the elastic modulus values near 10%, whisker composition, a correlation of the measured ultrasonic and density data, suggests that the maxima may, in part, be attributed to densification (packing) during hot-pressing.

Table III. Calculated Young's moduli, compressibilities and bulk moduli for Si_3N_4 composites containing β-Si_3N_4 whiskers, assuming $C_{13} = 150$ GPa .

Parameter	vol. % Si_3N_4 whiskers					
	0	5	10	15	20	35
$E_{\parallel}$ (GPa)	310	322	331	323	324	322
$E_{\perp}$ (GPa)	325	340	347	341	341	340
$(E_{\perp} - E_{\parallel})/E_{\parallel}$ (%)	4.8	5.6	4.8	5.6	5.2	5.6
$\beta_{\parallel}$ $(10^3/\text{GPa})$	1.50	1.46	1.46	1.48	1.48	1.47
$\beta_{\perp}$ $(10^3/\text{GPa})$	1.38	1.33	1.34	1.35	1.35	1.37
K (GPa)	235	240	242	239	239	238

The compressibility is largest in the HP direction and is relatively unaffected by whisker additions. Overall, bulk modulus is only slightly affected by whisker additions.

Comparison with Previous Results

As noted in Table IV, the present measurements and calculated anisotropic parameters for the monolithic sample are in good agreement with the results of Yeheskal[6] and in contradiction to other results[5]. From our estimated results, the

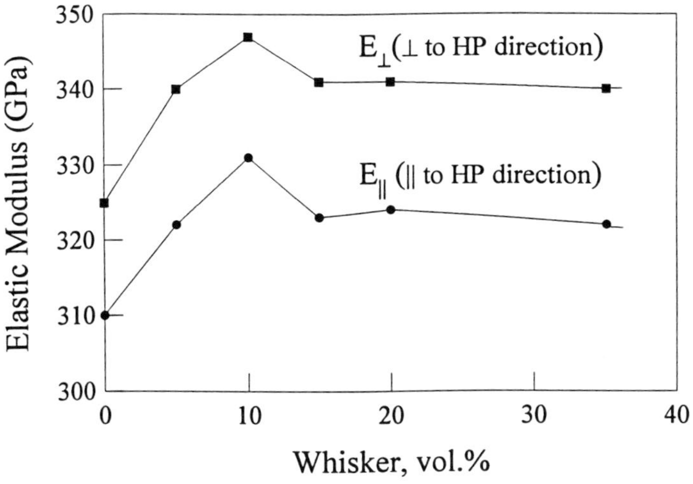

Figure 3. Anisotropic Young's moduli, $E_{\parallel}$ and $E_{\perp}$ to HP as a function of volume % of Si_3N_4 Whiskers, assuming C_{13} = 150 GPa.

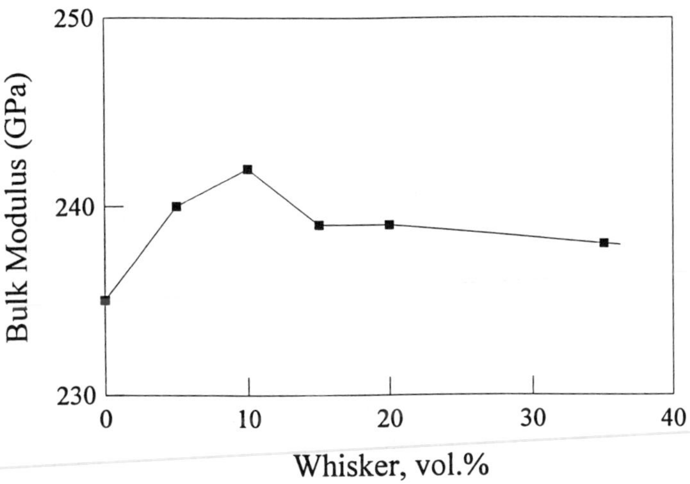

Figure 4. Bulk modulus, K, as a function of volume % Si_3N_4 whiskers, assuming C_{13} = 150 GPa.

average value of E, E_{ISO}, in a randomly oriented polycrystalline-monolithic β-Si_3N_4 would be about 320 GPa and the average shear modulus, $G_{ISO} = (C_{44} + 2C_{66})/3$, would be 125 GPa. The calculated value of bulk modulus, K, is 235 GPa, which is in excellent agreement with that estimated from the average values of E and G for isotropic aggregates.

Table IV. Comparison of $E_{\parallel}$, $E_{\perp}$, E_{ISO}, K and G (shear modulus) values (in GPa) for monolithic Si_3N_4.

Reference	$E_{\parallel}$	$E_{\perp}$	E^{*}_{ISO}	K	G^{**}_{ISO}
Present study	310	322	320	235	125
6	307	320		234	122
6			310	220	125
5	322	315			
11			317	215	126

* $E_{ISO} = (E_{\parallel} + 2E_{\perp})/3$

** $G_{ISO} = (C_{44} + 2C_{66})/3$

CONCLUSIONS

The elastic moduli of hot-pressed monolithic β-Si_3N_4 composites measured at ambient temperature exhibit anisotropic behavior with transverse isotropy in the plane perpendicular to the hot-pressing (HP) direction. Young's modulus parallel to the HP direction, $E_{\parallel}$, is about 5% smaller than $E_{\perp}$. These values increase with addition of β-Si_3N_4 whiskers up to 10 vol. % and decrease slightly to 35 vol. % whiskers. The largest effect of the texture in the elongated β-grains in the monolithic material and the whiskers in the composite is in the shear modulus anisotropy, where the shear stiffness in the HP plane, C_{66}, is about 7% higher than C_{44} (shear stiffness between HP planes) in the monolithic sample and increases to 15% with addition of 10% whiskers. Measurements of the single-crystal elastic constants, C_{ij}, of Si_3N_4 and the variations in the crystallographic texture (preferred orientation) in monolithic Si_3N_4 and whisker-reinfored Si_3N_4 composites would be necessary in order to explain the observed anisotropy in the elastic properties.

ACKNOWLEDGMENTS

We thank John Balogh for technical help in experimental work. We are grateful to Dr. Zhuang Li for reviewing the manuscript. Support for this research was provided by the University of Hawaii, Materials and Applied Science Program. School of Ocean and Earth Science and Technology contribution No. 3573, and by the U.S. Department of Energy, Office of Fossil Energy, Advanced Research and Technology Development Materials Program, under Contract W-31-109-Eng-38.

REFERENCES

1. F.F. Lange, Relation Between Strength, Fracture Energy and Microstructure in Hot Pressed Si_3N_4, J. Am. Ceram. Soc. 56[10], 518-522 (1973).

2. L.J. Bowen, R.J. Weston, T.G. Carruthers and R.J. Brook, Hot Pressing and the α–β phase transformation in silicon nitride, J. Mat. Science 13, 341-350 (1978).

3. K-R Lai and T-Y Tien, Kinetics of β Si_3N_4 Grain Growth in Si_3N_4 Ceramics Sintered under High Nitrogen Pressure, J. Am. Ceram Soc. 76[1], 91-96 (1993).

4. D. Suttor and G.S. Fishman, Densification and Sintering Kinetics in Sintered Silicon Nitride, J. Am. Ceram. Soc. 75[5], 1063-67 (1992).

5. F. Lee and K.J. Bowman, Texture and Anisotropy in Silicon Nitride, J. Am. Ceram. Soc. 75[7], 1748-55 (1992).

6. O. Yeheskel and Y. Gefen, The Effect of the α Phase on the Elastic Properties of Si_3N_4, Mat. Sci. and Eng. 71, 95-99 (1985).

7. S.T. Buljan, J. Baldoni and M.L. Huchabee, Si_3N_4-SiC Composites, Am. Ceram. Soc. Bull. 66, 347-52 (1987).

8. C.Y. Chu and J.P. Singh, Mechanical Properties and Microstructure of Si_3N_4 Whisker Reinforced Si_3N_4 Matrix Composites, Ceram. Eng. Sci. Proc. 11[7-8], 709-720 (1990).

9. C.P. Gazzara and D.R. Messier, Determination of Phase Content of Si_3N_4 by X-Ray Diffraction Analysis, Am. Ceram. Soc. Bull. 56[9], 777-780, (1977).

10. E.S. Fisher and C.J. Renken, Single Crystal Elastic Moduli and h.c.p.-b.c.c. Transformation in Ti, Zr and Hf. Phys. Rev. A135, 482-94 (1964).

11. E.S. Fisher, M.H. Manghnani, J.F. Wang and J.L. Routbort, Elastic Properties of Al_2O_3 and Si_3N_4 Matrix Composites with SiC whiskerReinforcement J. Am. Ceram. Soc. 75[4], 908-914 (1992).

PROCESSING AND CHARACTERIZATION OF SiC PLATELET / SiC COMPOSITES

J. Cao, W. J. MoberlyChan, L. C. De Jonghe, B. Dalgleish, and M. Y. Niu

Materials Science Division, Lawrence Berkeley Laboratory, and Dept. of Materials Science and Mineral Engineering, University of California, Berkeley, CA 94720

Hot pressed β-SiC and SiC matrix composites containing encapsulated α-SiC platelets were prepared and investigated. The microstructures were characterized using electron microscopy, Auger electron spectroscopy, and x-ray diffraction. Prior to hot pressing, the platelets were either encapsulated with hydrated aluminum sulfate or yttrium hydroxycarbonate (later calcined to form alumina or yttria) from aqueous solutions, or oxidized to form a silica layer. The effect of these interfacial layers on toughness was described.

1. INTRODUCTION

SiC is an attractive material for high temperature structural applications due to its high strength, low creep rate, good thermal conductivity, and high oxidation resistance. A major drawback of SiC as a structural material resides in its modest fracture toughness. The incorporation of reinforcement phases to enhance toughness may be effective, but must be evaluated in conjunction with the role of sintering additives to ease potential densification difficulties.

Pressureless sintering of SiC to high density at temperatures upwards of 2000°C, using small amounts of boron and carbon, was first reported in 1975 [1]. Later, it was found that the addition of aluminum and carbon could also assist in densification [2]. The combination of Al, B, and C has been extremely effective as sintering additives [3, 4], providing enhanced densification particularly at lower sintering temperatures. Lin et al. [5] hot pressed silicon carbide to 96% theoretical density at 1650°C by adding 6wt% Al metal, 1wt% B metal, and 1wt% C-black. The possible role of these additives was not fully understood, and was only briefly discussed based on limited optical and x-ray analysis. The low sintering temperature, however, allows consideration of a range of reinforcement phases that would not be compatible at higher sintering temperatures either for chemical or

To the extent authorized under the laws of the United States of America, all copyright interests in this publication are the property of The American Ceramic Society. Any duplication, reproduction, or republication of this publication or any part thereof, without the express written consent of The American Ceramic Society or fee paid to the Copyright Clearance Center, is prohibited.

microstructural reasons. The low sintering temperature also significantly reduces manufacturing cost.

The dispersion of second-phase particles can improve the fracture toughness of ceramics. However, the choice of toughening phase(s) in a SiC matrix composite is restricted. High temperature stability and chemical compatibility of the second phases are necessary requirements during densification. It has been demonstrated that TiB_2 [6, 7] and TiC [8, 9] particles can be effective in raising the fracture toughness of SiC. Both composites exploit the thermal expansion coefficient mismatch between the particulate and the matrix phases to generate a compressive hoop stress and a tensile radial stress around the particles, causing crack deflection and an increment in fracture toughness. However, at elevated temperatures, the internal stresses are relieved by creep, with a corresponding degradation of the mechanical properties [10].

SiC platelets have been shown to be effective reinforcements in alumina [11, 12], silicon nitride [13], and mullite [14]. The apparent chemical compatibility of α-SiC platelets make them an ideal reinforcing phase in a β-SiC matrix. A high aspect ratio of the platelets could potentially provide a significant toughening [15]. However, without a protective barrier, the reinforcement of a β-SiC matrix with α-SiC platelets is likely to be negated by the $\beta \rightarrow \alpha$ phase transformation and strong bonding between the matrix and the platelets. In such a case, the resulting mechanical properties would be expected to be similar to those of monolithic silicon carbide. Therefore, fabrication of useful SiC platelet/SiC composites requires a means of protecting the platelets. Production of an interfacial phase can protect the integrity of the platelets during sintering and hopefully also achieve a weak bonding between the platelets and the matrix, thereby promoting crack deflection and/or debonding and pullout.

The isolation of the platelets from the matrix can be accomplished by the encapsulation of the platelets with an appropriate oxide prior to incorporation in the (matrix) β-SiC powder. A layer of silica, for example, is readily introduced by oxidation of the platelets in air. Yttria, alumina, or titania coatings from aqueous solutions onto a number of ceramic substrates have been demonstrated by De Jonghe and co-workers [11, 16-19]. A preliminary use of alumina-coated SiC platelets in a SiC matrix has shown promising mechanical properties, yielding in some instances a fracture toughness of over 7 $MPam^{1/2}$ [20].

The present research studies hot pressed SiC composites reinforced with encapsulated SiC platelets using sintering additives similar to Lin's [5]. Transmission electron microscopy, scanning electron microscopy, auger electron spectroscopy, and x-ray diffraction are used to study the microstructures. The relative effects of the processing and the nature of interfacial layers of alumina, yttria, and silica in α-SiC platelet / β-SiC composites are presented.

2. EXPERIMENTAL

2.1 Encapsulation of SiC Platelets in Oxides

The SiC platelets used in this study were provided by Third Millennium Technologies, Inc. The crystal structure was determined to be predominantly 4H α-phase. The platelets were a -400 mesh grade, with diameters ranging from 3 to 30 microns (average 10 microns), and thickness ranging from 0.5 to 3 microns.

2.1.1 Production of a silica layer on SiC platelets

SiC platelets were oxidized in air at 1300°C for 3 hours to form a layer of silica. The oxidation kinetics are known to follow parabolic kinetics [21-23], from which the thickness of the silica layer was estimated to be about 100 nm.

2.1.2 Encapsulation with alumina

Kapolnek and De Jonghe[17] and Mitchell and De Jonghe[11] obtained heterogeneous precipitation of hydrated basic aluminum sulfate on powder surfaces using urea decomposition in aqueous solutions of salts. The SiC platelets used in this study were coated with this alumina precursor extending the procedures described in [11, 20] as follows.

The concentrations of the aqueous suspension were: 0.1 M/L hydrated aluminum sulfate (crystals, Fisher Scientific), 0.2M/L urea (Mallinckrodt), 10g/L platelets, and 0.75g/L PVP K30 (a polymeric dispersant, GAF Chemical Corporation). The suspension was vigorously stirred and maintained at 88°C for 24 hours. The coated platelets were annealed in inert gas at 880°C for 6 hours to convert the coating to crystalline Al_2O_3.

2.1.3 Encapsulation with yttria

The hydrolysis of yttrium salts by urea decomposition was first used to synthesize monodispersed particles of yttrium hydroxycarbonate [24]. Garg and De Jonghe[16] extended this method to encapsulate powders and whiskers of silicon nitride with a yttria precursor. This method had also been used to coat alumina fibers [18]. In the present work, SiC platelets were coated in the same manner with the following modifications of the processing parameters.

A small amount of polymeric dispersant (PVP K30) was dissolved at room temperature in distilled water prior to the addition of SiC platelets. This suspension was thoroughly dispersed with magnetic stirring. An aqueous solution of yttrium nitrate (NOAH Teck. Co., TA) and urea was mixed with the suspended platelets in a flat bottom distilling flask. The final suspension contained 0.1 M/L yttrium nitrate, 0.75 M/L urea, 10 g/L SiC platelets, and 0.75 g/L PVP K30. The mixture was slowly heated on a heating mantel, capped with a water-cooled reflux condenser, to a final temperature of 65°C and held for 20 hours. Magnetic stirring was maintained to prevent the platelets from settling or agglomerating. After the reaction run, the suspension was allowed to settle, and the supernatant was discarded. The coated platelets were re-sedimented in water to remove any free precipitates, rinsed twice in acetone, and air dried at room temperature. The coated platelets were then calcined at 650°C for 12 hours to convert the coating to crystalline yttria.

2.2 Processing of Composites

Submicron size β-SiC powder (BSC-21, Ferro) was mixed in toluene with 6wt% metallic aluminum powder (H-10, Valimet, CA), 0.6wt% boron (high purity, sub-micron particle size, Callery Chemical Co., PA), and 4wt% apiezon (which was found to yield 50% carbon upon pyrolysis). The slurry was ultrasonically agitated for 5 minutes, followed by wet screening through a 325 mesh sieve. After drying, the powder mixture was screened through a 200 mesh sieve. Appropriate quantities of α-SiC platelets with the various types of encapsulations (see Table 1) were then dry-mixed with the β-SiC powder, and cold compacted into 38 mm diameter discs at a uniaxial pressure of 35 MPa. The composites were hot pressed in graphite dies lined with graphite foil, at 1690°C, for 45 minutes, at 50 MPa, and under flowing argon gas.

Table 1. Processing, densities, and Fracture Toughness of silicon carbide and SiC Platelet/SiC Composites

Sample designation	SiC platelet (vol%)	Platelet surface treatment	Density (g/cm^3)	Fracture toughness (relative value)
#1	0	/	3.17	1
#N	20	none	3.11	1.2
#5	10	oxidized	3.18	1.1
#Y	14	yttria coated	3.18	1.2
#2	8	alumina coated	3.20	1.2
#A	20	alumina coated	3.23	1.6
#HFA	20	etched and alumina coated	3.23	1.5
#A2	28	alumina coated	3.27	not measured

2.3 Mechanical Testing

A four-point bend testing was used to measure fracture toughness. Beams of 2.5 by 2.8 by 28 mm were cut from the hot pressed billets. The tensile surfaces were polished to a 3 μ finish, and the tensile edges were beveled to eliminate edge flaws. An elliptical flaw was then introduced on the tensile surface with a Knoop indentation. The flaw was made parallel to the hot pressing direction and perpendicular to the long axis of the beam. The beams were tested at an outer span of 25.4 mm (1 in.) and an inner span of 9.5 mm (3/8 in.). The crosshead speed was of 0.05 mm/min. The fracture toughness was determined from the relation [25]:

$$K_{1c} = \frac{2}{\sqrt{\Pi}} \sigma_f \sqrt{a}$$

where σ_f is the fracture stress and **a** is the initial flaw radius of the indentation measured on the fracture surface using a scanning electron microscope (SEM).

3. RESULTS AND DISCUSSION

3.1 Encapsulation of SiC platelets

The microstructural development of the coating of alumina precursor has been described elsewhere [11]. Figure 1 shows an SEM micrograph of alumina coated platelets after calcination. The coating presents a good coverage on the platelets and has a thickness of 1 to 2 μm. Experiments also show that etching of the platelets with HF, prior to the application of alumina coating, does not significantly affect the morphology nor the thickness of the alumina coating under identical coating conditions.

Figure 2 exhibits an SEM image of the yttria coating on SiC platelets. This coating also completely covers the surface. X-ray diffraction determines that the coating has been converted to crystalline yttria after calcination. The heating rate and the soaking temperature of the calcination process have to be carefully controlled to avoid extensive cracking of the deposit. Annealing at excessive temperatures (e.g., 900°C) induces film cracking. The insert in Figure 2 shows a crack, the edge of which indicates the yttria film is approximately 0.25 micron thick.

Urea decomposes upon heating in aqueous solutions, thereby causing the pH of the system to progressively increase [26]. Yttrium hydroxycarbonate (YBC) forms by the hydrolysis of yttrium nitrate in a urea-containing aqueous solution [24]. The pH decreases upon formation of yttrium hydroxycarbonate. Measurements indicate that the pH of the suspension drops initially due to the ionization of yttrium nitrate. Subsequently, the pH rises as a result of continued urea decomposition. After two hours, the pH stabilizes at approximately 5.5, due to the rate balance between urea decomposition and precipitation of the yttria precursor, until the coating run is terminated at 20 hours. The constant pH should correspond to a yttrium hydroxycarbonate concentration between the homogeneous nucleation and the heterogeneous nucleation limits, as is traditionally shown in a LaMer [27] diagram. Therefore, nuclei could form heterogeneously either on the surface of the platelets or on existing precipitates. Simultaneously, the already existing precipitates will grow. Since the concentration never reaches the homogeneous nucleation limit, free precipitation in the solution is minimized. The morphology of the obtained coating is consistent with such a mechanism. Experimental observation shows that the coating initially appears as discrete particles on the surface of the SiC platelets. These precipitates grow, meanwhile the population of the precipitates increases with time. Eventually the coating covers the entire surface.

3.2 Sintering and Characterization

SiC matrix composites with up to 35wt% of alumina-coated SiC platelets (approximately 28vol% of SiC platelets) are hot pressed to nearly full density at 1690°C and 50 MPa. Densities are measured using the Archimedes principle in distilled water, and the results have been listed in Table 1. As the exact quantities of the encapsulation phases are difficult to measure and the quantities of sintering additives are relatively large, the theoretical densities of the composites are not accurately known. However, the relative densities of the SiC composites containing encapsulated platelets are estimated to be 98% and up. The sintered densities are also examined by optical and scanning electron microscopy of polished surfaces. Figure 3 shows a polished surface of the composite with 35wt% alumina coated platelets. The bright areas are the alumina coatings. Very few, small isolated pores are observed in the matrix.

Figure 4 shows the typical grain sizes of silicon carbide, ranging from submicron to 2 microns. Most of the grains are equiaxed. Less than 10% of the grains are elongated along the planes of most stacking defects. Electron diffraction indicates that all these grains have a cubic crystal structure (β-phase). The low processing temperature apparently suppresses the $\beta \rightarrow \alpha$ phase transformation. Figure 5 depicts a typical grain boundary of β-SiC grains. Lattice fringes from both grains extend into the interface because this grain boundary is tilted. However, it appears that the grain boundaries are free of amorphous phase.

The sintering additives, especially Al, are present in concentrations well above their solubilities in SiC. Secondary phases are therefore expected to form. Figure 6 is a fractograph of a "monolithic" silicon carbide (without the addition of SiC platelets). Secondary phases with dimensions of about 10 μm are apparent. It can be also noted that the secondary phases are preferentially oriented with the long axis perpendicular to the hot pressing direction. Moreover, TEM determines that these "pockets" are comprised of many grains approximately 1-2 μm in size. This suggests that these regions exist as liquid during hot pressing. With the presence of aluminum, carbon, and boron, (and probable oxygen from the surfaces of Al and SiC powders), two liquid phases are plausible at the sintering temperature. One is associated with the eutectic reaction between Al_2O_3 and Al_4O_4C [28]; the other one is $Al_8B_4C_7$ [29].

Minor peaks in bulk X-ray diffraction data confirms the presence of Al_4O_4C and $Al_8B_4C_7$. Electron diffraction determines both ternary phases form within the "pockets" of secondary phases. XRD also indicates the existence of Al_2O_3 and Al_4C_3. EDS analysis shows that the secondary phase regions contain Al, but are absent of Si. (The EDS spectrometer used is not sensitive to B, C, and O.) The average atomic ratio of Al vs Si over a large area is found to be 8:100, which is close to the atomic ratio of the raw materials (10:100). Figure 7 shows Auger electron spectra acquired of various regions of the fracture surface. Figure 7a is acquired of the matrix. Figures 7b and 7c analyze two neighboring grains inside a single secondary phase region (Figure 6). Figure 7b indicates mainly Al and O, with minor amounts of C and B; Figure 7c indicates mainly Al and C, with minor quantities of B and O. Therefore, the microstructural observations and the spectra results conclude that the secondary phase regions are composed of not only multiple grains, but also multiple phases.

Sintering of a composite with a large volume fraction of reinforcement phase, such as platelets, can be hindered by the percolation of the second phase [30]. Making composites with encapsulated reinforcement phases have been shown to ease densification[11, 17]. Typically, the coated phase would have a large volume fraction and becomes the matrix of the composite. However, a similar principle applies in the densification of coated SiC platelet / SiC composites. The oxide coating prevents the platelets from physically contacting each other. In addition, creep of the oxide coating during hot pressing reduces stress shielding and eases sliding of the platelets relative to each other. All these factors facilitate densification. A SiC composite containing 35wt% alumina-coated platelets (28vol% SiC platelet) was successfully hot pressed at 50 MPa to nearly full density at a temperature below 1700°C (Figure 3).

The coatings are also essential to protect the integrity of the platelets (i.e. to prevent the $\beta \rightarrow \alpha$ phase transformation, or the sintering together of the platelets and the matrix). Experimental observation shows that a composite made with the uncoated platelets suffers from sintering reaction between the platelets and the matrix grains. In this case, the composite acts similar to monolithic SiC because of lack of a phase barrier, as will be shown in mechanical testing. The coating of alumina separates the α-SiC platelets from the β-phase matrix (Figure 3). However, the coated yttria film (Figure 2) has formed clumps after sintering, and the yttria coating is no longer continuous (Figure 8). A reaction may also have occured at the platelet/yttria interface. The silica layer formed by oxidation of the platelets was not resolved by SEM and may have been reduced during processing. (TEM analysis on these concerns are currently underway.)

3.3 Mechanical Testing

Fracture toughness is measured by a four-point bend test, with a well defined elliptical flaw made by Knoop indentation on the tensile surface (Figure 9). The measured fracture toughness results have been summarized in Table 1. Fracture toughness of the alumina-coated platelets reinforced SiC composites are significantly increased over that of monolithic SiC. Addition of 25wt% alumina-coated SiC platelets (approximately 20vol% platelets) achieves a 60% increase in the fracture toughness. Etching SiC platelets prior to the coating process, shows no effect in mechanical behavior of the composite. Figure 10 displays the fracture surface of the composite. Crack path deflection is present at the platelet surface. This crack deflection alters the crack tip stress intensity, resulting in an improvement in toughness [15].

Platelet encapsulation proves to be essential in this composite system. As has been shown in the previous section, the addition of uncoated SiC platelets leads to sintering together of the platelet and the matrix. It is observed that the crack breaks through most of the uncoated platelets, resulting in a minimal increase in toughness over that of monolithic SiC (Table 1). A thin layer of silica, introduced by oxidation, also does not produce effective separation between the matrix and platelets, with only a limited increase in toughness. Two possible reasons can be cited. First, the oxidation product has been considered as a mechanically strong protective layer [21-23], thereby suggesting the bonding between silica and silicon carbide is very strong. Strong interfacial bonding is well known to be disadvantageous in ceramic composites. Second, there is a possibility that the silica layer is completely reduced by sintering additives during hot pressing.

The composite with yttria-coated platelets shows very limited increase in toughness over that of monolithic SiC (see Table 1). The bonding strength between SiC and yttria is not well known. Moreover, the morphology of the yttria coating has changed during sintering, and it does not fully separate the platelets from contacting the matrix. Therefore, the platelets may have been partially sintered together with the matrix.

SUMMARY

Silicon carbide and SiC matrix composites with up to 35% of coated SiC platelets were hot pressed to near full density at 1690°C and 50 MPa. Liquid phase sintering was believed to play a role in densification. Encapsulation of SiC platelets with oxide coatings prevented direct contact of the platelets (platelet network), and eased densification of the composites. An alumina coating proved to be most effective to protect the integrity of SiC platelets, and to promote crack deflection, thereby resulting in increased fracture toughness.

Acknowledgment

The authors gratefully acknowledge the help of J. P. Daniel and T. D. Mitchell, Jr. with developing the coating processes. This work is supported by the Division of Materials Science, Office of Basic Energy Sciences, United States Department of Energy, under Contract No. DE-AC03-76SF00098.

Figure 1

Figure 2

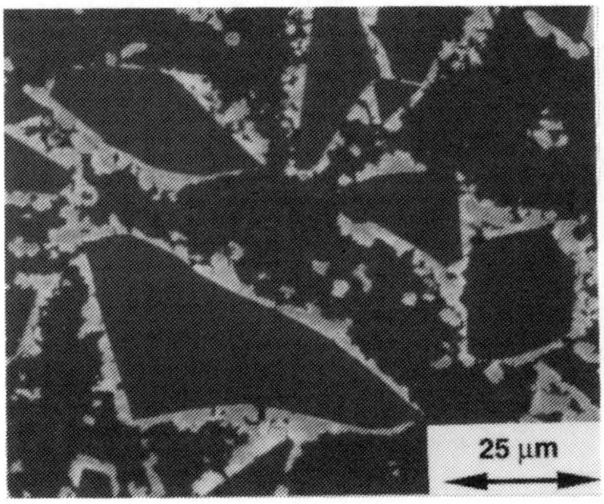

Figure 3

Figure 4

Figure 1 Scanning electron micrograph (SEM) of alumina coated platelets.

Figure 2 SEM of yttria-coated SiC platelets. The yttria film completely covers the platelets. The inserted picture shows a crack (induced by calcination at 900°C) through the coating. From the side view of the crack, the coated film is determined to be approximately 1/4 micron thick.

Figure 3 SEM micrograph of a polished surface of the SiC composite with 35wt% of alumina-coated SiC platelets in a direction perpendicular to the hot pressing direction. The composite is nearly full dense. The coating acts as a barrier to protect the integrity of the platelets.

Figure 4 Transmission electron micrograph (TEM) of SiC matrix grains. The grain size ranges from sub-micron to two microns.

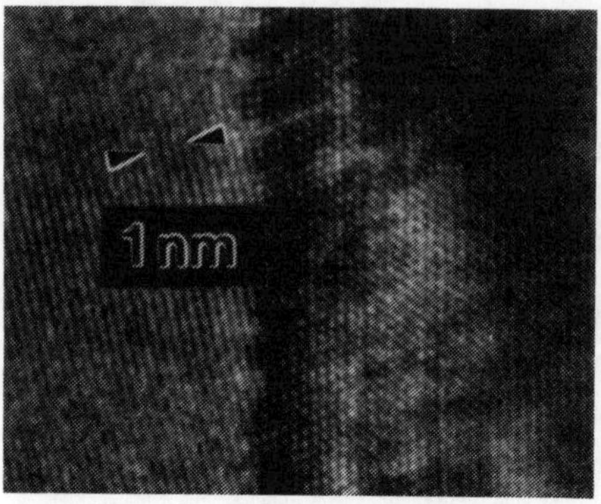

Figure 5

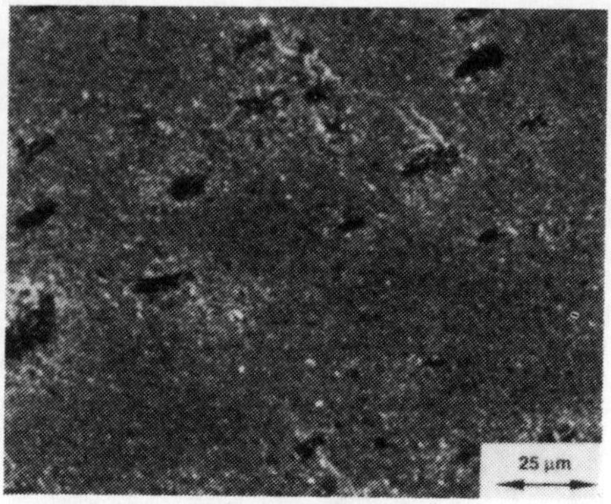

Figure 6

Figure 5 High resolution electron micrograph (HREM) of a typical SiC-SiC grain boundary. No amorphous interphase was observed.

Figure 6 SEM micrograph of the fracture surface of "monolithic" silicon carbide. Regions of secondary phase (approximately 10 microns long) are evident, which are determined to be composed of many 1-2 microns size grains.

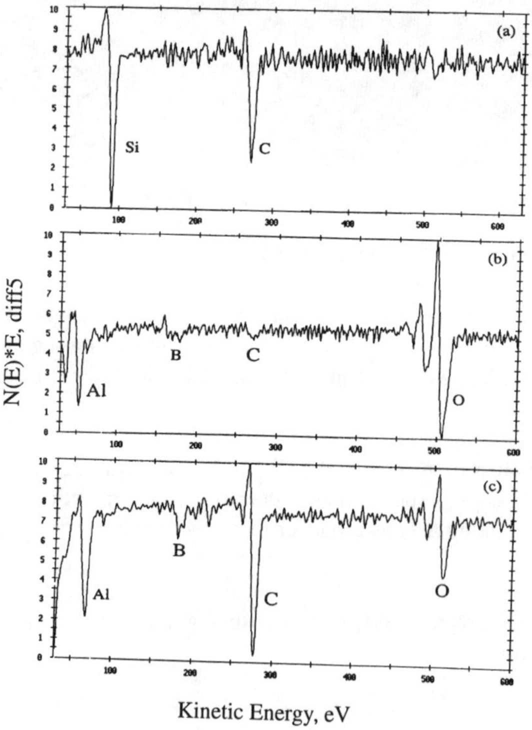

Figure 7 Auger electron spectra of the fracture surfaces of the "monolithic SiC" specimen are acquired (a) in the matrix (away from the secondary phases), and of two grains, (b) and (c), inside a single region of the secondary phase.

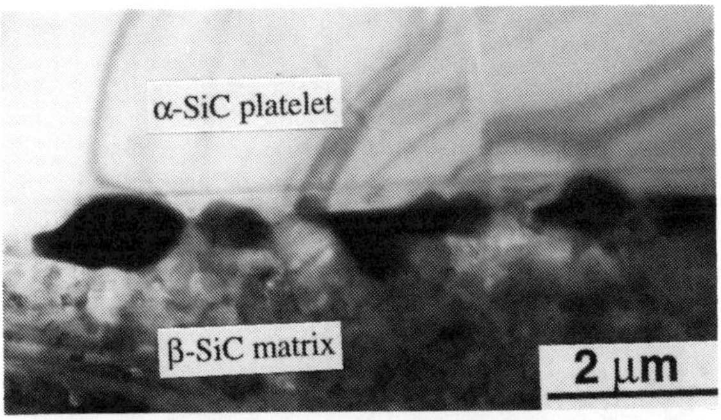

Figure 8

Figure 9

Figure 10

Figure 8 TEM micrograph of SiC composite with yttria-coated SiC platelets.

Figure 9 SEM micrograph of an elliptical flaw induced by Knoop indentation.

Figure 10 Fractograph of alumina-coated SiC platelet / SiC composite. Some crack deflection has occurred at the platelets.

References

[1] S. Prochazka, "The Role of Boron and Carbon in the sintering of Silicon Carbide", Special Ceramics, Vol. 6, British Ceramic Research Association, pp.171-81 (1975).

[2] W. Bocker, H. Landfermann, and H. Hausner, "Sintering of alpha silicon carbide with additions of aluminum". Powder Met. Int., 11 (2) 83-85 (1979).

[3] H. Tanaka, "Sintering of Silicon Carbide", in Silicon Carbide Ceramics-1, edited by S. Somiya and Y. Inomata (Elsevier Science Publishers LTD, 1992).

[4] H. Tanaka and Y. Inomata, "Normal Sintering of Al-doped Beta SiC", J. Mater. Sci. Letters, 4, (1985) 315-17.

[5] B. Lin, M. Imai, T. Yano, and T. Iseki, "Hot-Pressing of Beta-SiC Powder with Al-B-C Additives", J. Am. Ceram. Soc., 69 [4] C-67-C-68 (1986).

[6] Y. Ohya, M. J. Hoffmann, and G. Petzon, "Sintering of in-situ Synthesized SiC-TiB_2 Composites with Improved Fracture Toughness", J. Am. Ceram. Soc., 75 (9) 2479-83 (1992).

[7] Y. Murata, and G. Weber, "Sintered Silicon Carbide-Titanium Diboride Mixtures and Articles Thereof", US Patent 3,340,020 (1982).

[8] G. Wei and P. Becher, "Improvements in Mechanical Properties in SiC by the Addition of TiC Particles", J. Am. Ceram. Soc., 67 (8) 571-74 (1984).

[9] M. Janney, "Microstructural Development and Mechanical Properties of SiC and of SiC-TiC Composites", Am. Ceram. Soc. Bull., 65 (2) 357-62 (1986).

[10] B. Lin, T. Yano and T. Iseki, "High - Temperature Toughening Mechanism in SiC / TiC Composites", J. Ceram. Soc. Japan, 100 (4) 509-513 (1992).

[11] T. D. Mitchell, Jr., L. C. De Jonghe, "Alumina-SiC Composites from Coated Powders", J. Am. Ceram. Soc., in press (1994).

[12] Y. Chou and D. J. Green, "Silicon Carbide Platelet/Alumina Composites: 2, Mechanical Properties", J. Am. Ceram. Soc., 76 (6) 1452-58 (1993).

[13] N. Claussen, "Ceramic Platelet Composites", in Proceedings of the 11th RISO International Symposium on Metallurgy and Materials Science 1990, structural Ceramics Processing, Microstructure and Properties. Edited by J. J. Bentzen, J. B. Bilde-Sorensen, N. Christiansen, A. Horsewell, and B. Ralph. RISO National Laboratory, Roskidle, Denmark, 1990.

[14] C. Nischik, M. M. Seibold, N. A. Travitzky, and N. Claussen, "Effect of Processing on Mechanical Properties of Platelet-Reinforced Mullite Composites", J. Am. Ceram. Soc., 74 (10) 2464-68 (1991).

[15] K. T. Faber and A. G. Evans, "Crack Deflection Processes - 1. Theory", Acta Metall., 31 (4) 565-76 (1983).

[16] A. K. Garg and L. C. De Jonghe, "Microencapsulation of silicon nitride particles with yttria and yttria-alumina precursors", J. Mater. Res., 5 (1) 136-42 (1990).

[17] D. Kapolnek and L. C. Dejonghe, J. Europ. Ceram. Soc., 7 (1991) 1-7.

[18] J. P. Daniel and L. C. De Jonghe, "Coating of Ceramic Fibers with Yttria", A. Ceram. Soc. Ann. Meet., Indianapolis, April 1994, Paper No. SIII-105-94.

[19] M. gopal, W. J. MoberlyChan, and L. C. De Jonghe, "Synthesis of Rutile and Anatase at Low Temperatures", ibid, Paper No. SVIIa-8-94.

[20] T. Mitchell, Jr., L. C. De Jonghe, W. J. MoberlyChan, and R. O. Ritchie, "Silicon Carbide Platelet/Silicon Carbide Composites", Submitted to J. Am. Ceram. Soc.

[21] Z. Zheng, R. E. Tressler, and K. E. Spear, J. Electrochem. Soc., 137 (3) 845-58 (1990).

[22] Z. Zheng, R. E. Tressler, and K. E. Spear, J. Electrochem. Soc., 137 (9) 2812-16 (1990).

[23] J. Li, P. Eveno, and A. M. Huntz, "Oxidation of SiC", Werkstoffe and Korrosion. 41, 716-25 (1990).

[24] M. Akinc and D. Sordelet, Adv. Ceram. Mater., 2 (3A) 232 (1987).

[25] P. C. Paris, and G. C. Sih, ASTM STP, 381, 1965, 30-81.

[26] J. E. Blendell, H. K. Bowen, and R. L. Coble, Am. Ceram. Soc. Bull., 63 (6) 797-802 (1984).

[26] V. K. LaMer and R. H. Dinegar, J. Am. Chem. Soc., 72 (11), 4847 (1950).

[27] L. M. Foster, G. Long, and M. S. Hunter, J. Am. Ceram. Soc., 39 (1) 1-11 (1956).

[29] Y. Inomata, H. Tanaka, Z. Inoue, and H. Kawabata, J. Ceram. Soc. Japan, 88 (6), 353-55 (1980).

[30] F. Lange, L. Atteraas, F. Zok and J. R. Porter, Acta Metall., 39 (2) 209-19 (1991).

NEAR-SURFACE STRUCTURE OF THE IN-SITU NITRIDED SILICON CARBIDE

Q.Liu, S.L.Wen, D.L.Jiang and J.H.She
Shanghai Institute of Ceramics, Shanghai 200050, P.R.China

ABSTRACT

Near-surface structures of the in-situ nitrided SiC were investigated by TEM. It was found that a dense nitrided layer was formed on the SiC surface. The nitrided layer has a composition gradient consisting the outmost β-Si_3N_4 layer, the medium SiC/Si_3N_4 layer, and the matrix SiC layer. Furthermore, fiber-like carbon grains were induced in the nitrided layer during the nitridation. It is the surface layer that contributes to the increase in bending strength of the materials at room temperature which in turn attributes to the formation of a dense layer.

INTRODUCTION

SiC ceramics is an important candidate for engineering materials, because of its good oxidation resistance, high hardness, low thermal expansion, and excellent high temperature resistance, etc. But on the other hand, the intrinsic brittleness of this materials prohibit its applications. The main purpose of the research on SiC ceramics is to enhance basic strength and fracture toughness. One of the approaches is to fabricate multiphase composed ceramics using dispersed particles or fibers/whiskers, for example, SiC-TiC[1-2], SiC-Al_2O_3[3], and SiC-Si_3N_4[4-5], to improve the mechanical properties of the SiC matrix. Another one is to use surface treatment techniques to modify the structural, chemical and mechanical properties of the near-surface region of the SiC matrix. It has been reported that surface hardening for the sintered and reaction-bonded SiC materials was available by nitrogen ion implantation or by annealing in an NH_3 atmosphere, as a reaction gas, instead of N_2[6-7]. Recent research indicated that SiC ceramics fabricated by using high temperature isostatic pressing (HIP) with in-situ nitridation has the strength as high as 1000MPa[8]. This is due to the fact that the nitrided layer is formed near the

To the extent authorized under the laws of the United States of America, all copyright interests in this publication are the property of The American Ceramic Society. Any duplication, reproduction, or republication of this publication or any part thereof, without the express written consent of The American Ceramic Society or fee paid to the Copyright Clearance Center, is prohibited.

Table I. Effects of HIP with in-situ Nitridation on Relative Density, Open Porosity, Strength, and Toughness of SiC.

Properties	Before HIP with in-situ Nitridation	After HIP with in-situ Nitridation
Relative Density(%)	90.6	98.5
Open Porosity(%)	8.3	0.3
Bending Strength(MPa)	582 ± 78	907 ± 110
Toughness ($MPam^{1/2}$)	5.7	8.4

* With 3wt% Al_2O_3 additive in the specimens.

surface of the SiC. Therefore, the present paper mainly gives the results of the investigation on the near-surface microstructure of the in-situ nitrided SiC by TEM.

EXPERIMENT

Preparation of Specimens

The nitrided SiC ceramics specimens were prepared as follows:
(1) The green compacts (6x8x45 mm) were encapsulated with a glass seal and then hot-isostatically pressed (HIP) at 1850°C and 200MPa in argon.
(2) The HIPed-specimens were removed from the glass sealing, machined into rectangular bars (3x4x36mm) and then subjected to the second HIP at 200MPa in nitrogen and 1850°C for 1h.

Table I illustrates the effects of HIP with the in-situ nitridation on relative density, open porosity, strength, and toughness of the SiC. Bending strength was measured by three-point bending tests. Fracture toughness was determined using the single-edge notch beam technique (SENB). Otherwise, the detail of preparation of the materials, measurements of the mechanical properties, and the roles of Al_2O_3 additive have been published elsewhere[9-11].

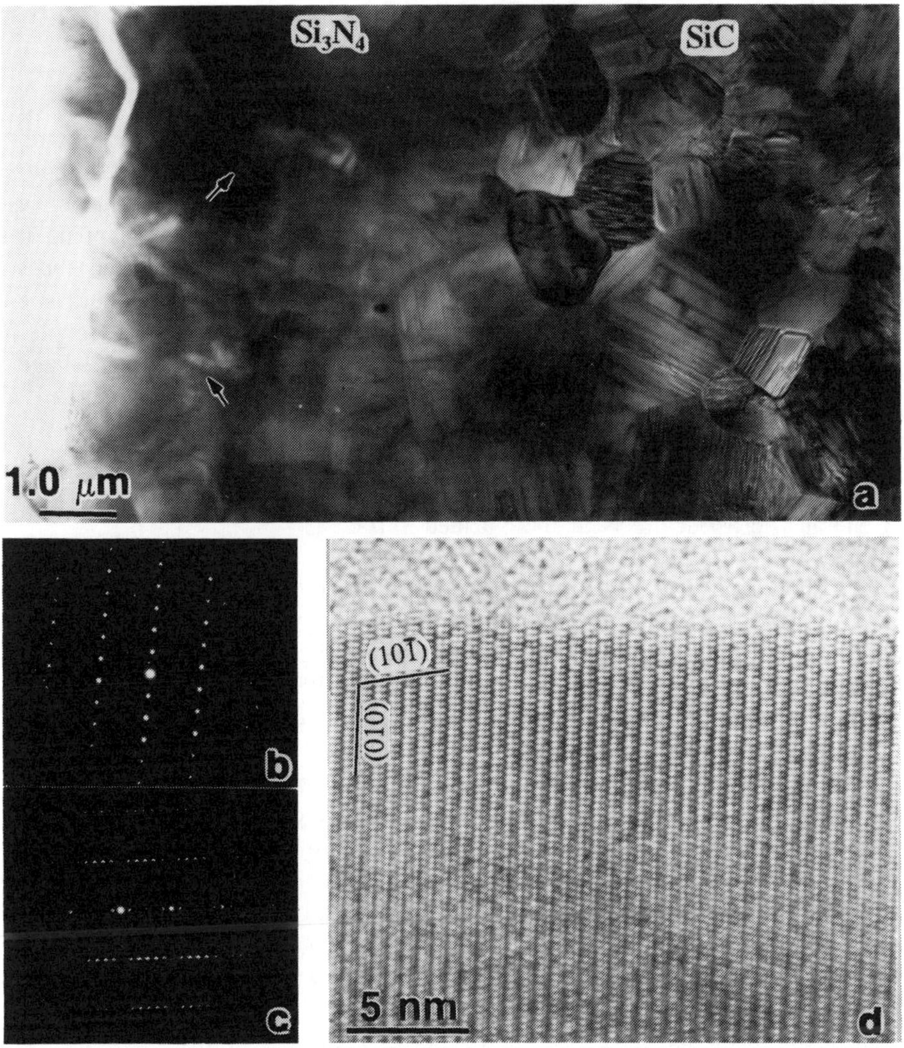

Fig. 1. Microstructure Near the Nitrided Surface of SiC.

(a) Si_3N_4 layer is on the left with fiber-like grains (indicated by arrows), but SiC Matrix is on the Right;

(b) Electron Diffraction Corresponding to the [101] Zone Axis in β-Si_3N_4;

(c) Electron Diffraction Corresponding to the [010] Zone Axis in α-SiC (6H).

(e) Lattice Image of the Top Surface on the Nitrided-Layer.

TEM Examination

A cross-sectional specimen for TEM examination was prepared as described here. Two pieces of the nitrided-SiC (3mm thick) that were glued together with the nitride surfaces facing each other were cut (with a diamond saw) perpendicularly into 500μm thick slices. These slices were mechanically polished to a thickness of about 40μm, and then sustained a concave lapping at a thickness of about 10μm in the thinner region. Finally, the polished slices were ion-milled using the Gatan ion beam thinner (Model-600) until the specimen was perforated. The prepared specimen, which was set on a double-tilt holder, was observed using a JEM-200CX TEM with an operating voltage of 200KeV.

RESULTS AND DISCUSSION

General Microstructure Near the Nitrided Surface

Fig.1 shows a typical structure near surface of the nitrided-SiC, where a dense nitrided-layer on the left is easily identified by the contrast different from that of SiC on the right. The nitrided-layer was distinguished to be of β-Si_3N_4 phase according to the selected area electron diffraction patterns. Furthermore, the top surface on the nitrided-layer usually appears to be the $(10\bar{1})$ planes in β-Si_3N_4 crystals which means the $(10\bar{1})$ planes preferentially grow at the beginning of nitridation reaction. In the matrix α-SiC (6H), the grain boundaries are lucid, but in the Si_3N_4 layer, the grain boundaries become confused, that implies that the nitridation reaction occur along the grain boundaries by N atoms diffusion. In the micrographs of TEM, some "fiber-like" grains can be seen in the nitrided-layer(indicated by arrows in Fig.1) and proved to be of carbon. The crystal structure of the carbon is of hexagon (12H) with a=2.456Å, c=20.088Å based on the electron-microscopy analysis.

Formation of the Nitrided Layer and Fiber-Like Carbon Grains

From the thermodynamic analysis for the phase stabilities in the Si-C-N-O system[12], SiC is unstable at both the high temperature and N_2 pressures. The equation for the nitridation can be written as

$$3\ SiC(s) + 2\ N_2 \dashrightarrow Si_3N_4(s) + 3\ C \qquad (1)$$

The change of the Gibbs free energy for the reaction is

$$\Delta G^\circ_{T,P} = -125500 + 72.34T - 2RT\ln P_{N2}$$

Under the present experimental conditions (1850°C, 200MPa of N_2 pressure), SiC should have been nitrided into Si_3N_4 with some carbon as by-product.

As mentioned above, there are many fiber-like carbon grains in the nitrided layer. These carbon grains always occur at the grain boundaries with a little glassy phases nearby (seen in Fig.2). Furthermore, Al_2O_3 as sintering additive can react with the silica from the SiC surface to from the melted aluminosilicate or a silica-rich liquid phase under given conditions of temperature and N_2 pressure[13]

$$SiO_2(s) + Al_2O_3(s) \longrightarrow \text{aluminosilicate(l)} \quad (2)$$

These aluminosilicate were considered to play a role of the grain boundary phase having positive effects on the densification of HIPed-SiC via a solution-reprecipitation mechanism. Also the C from equation (1) can react with silica to produce SiO(s), CO(g), and Si_3N_4(s) as shown here[14]:

$$SiO_2(s) + C(s) \longrightarrow SiO(s) + CO(g) \quad (3)$$

$$3\ SiO(s) + 3\ C(s) + 2\ N_2(g) \longrightarrow Si_3N_4(s) + 3\ CO(g) \quad (4)$$

$$SiO_2(s) + 2\ SiC(s) + 2\ N_2(g) \longrightarrow Si_3N_4(s) + 2\ CO(g) \quad (5)$$

The changes in the Gibbs free energy for the above reactions are showed in the Fig.3. These reactions are actually occurred under the experimental conditions. As a result, the CO gas phase may escape via grain boundaries, but carbon grains did not remain at the grain boundaries until a dense layer of Si_3N_4 had formed and the channels for gas diffusion were gradually blocked by it. An investigation on the effect of carbon grains on the mechanical properties is proceeding.

Surface Strengthening

In comparison with the unnitrided-SiC, the nitrided one has a high bending strength which might result from the formation of the dense Si_3N_4 layer on the surface of the specimen. As well-known, the basic strength and toughness of SiC ceramics are more or less lower than that of Si_3N_4. However, the formation

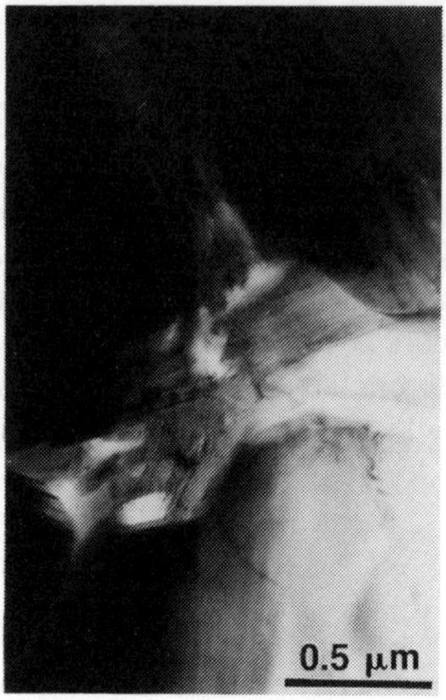

Fig.2. Existence of Carbon Grains at the Grain Boundary.

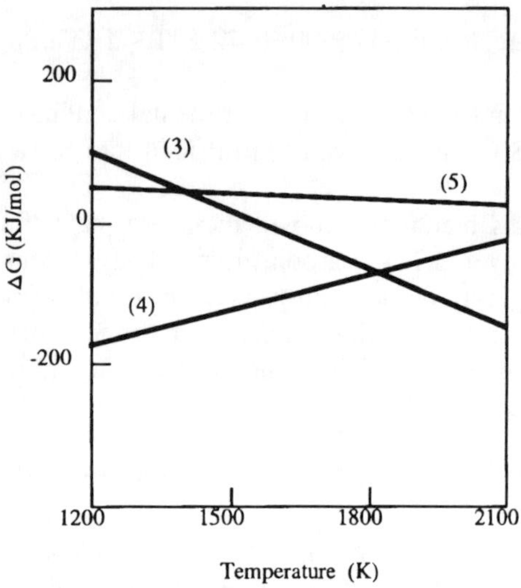

Fig.3. Changes in the Gibbs Free Energy of Equations from (3) to (5).

of Si_3N_4 layer is beneficial to the surface strengthening.

TEM examination on the SiC matrix showed that grains are about 1μm and oriented randomly. But for the Si_3N_4 layer, the grain size is increased and grain boundaries become confused that suggests that the surface nitridation occurs along the grain boundaries by atomic diffusion and the grains are connected closely. Finally, a dense Si_3N_4 layer forms with the strong bonds between grains. On the other hand, the existence of compressive surface stresses, owing to the differences of thermal expansion and elastic modulus between Si_3N_4 and SiC, may have some influence on the surface strengthening.

CONCLUSION

Summarily, based on the examination and analysis of near-surface microstructure for the in-situ nitrided SiC ceramics, some conclusions can be drawn:

1. A dense β-Si_3N_4 layer is formed on the surface of the in-situ nitrided SiC ceramics, and $(10\bar{1})$ planes of β-Si_3N_4 crystal grow preferentially on the top surface.

2. Many fiber-liker carbon grains are generated as the by-products of nitridation and other related chemical reactions.
3. The increases in the bending strength and toughness of the nitrided-SiC are mainly caused by the formation of the dense Si_3N_4 layer.

ACKNOWLEDGEMENT

This work was supported by The State Key Lab. of High Performance Ceramics and Superfine Microstructure, Shanghai Institute of Ceramics, Chinese Academy of Sciences, P.R.China.

REFERENCES

[1] G.C.Wei and P.F.Becher, "Improvements in Mechanical Properties in SiC by the Addition of TiC Particles", J. Am. Ceram. Soc., 67, 57(1984).
[2] D.L.Jiang, J.H.Wang, Y.L.Li and L.T.Ma, "Studies on the Strengthening of Silicon Carbide-Based Multiphase Ceramics:I, The SiC-TiC System", Mater. Sci. Eng., A109, 401(1989).
[3] Y.Katsumura and M.Kobayashi, "Some Properties of Al_2O_3-SiC Sintered Compact", J. Jpn. Soc. Powder Metall., 35, 137(1988).
[4] S.T.Baljan, J.G.Baldoni, and M.L.Huckabee, "Si_3N_4-SiC Sintered Composites", Am. Ceram. Soc. Bull., 66, 347(1987).
[5] H.Nakamura, S.Umebayashi, and K.Kishi, "Mechanical Properties of Hot-Pressed Si_3N_4-SiC Composites", J. Ceram. Soc. Jpn., 97. 1517(1989).
[6] S.G.Roberts and T.F.Page, "The effects of N^+_2 and B^+ Ion Implantation on the Hardness Bahavior and Near-Surface Structure of SiC", J. Mater. Sci., 21,457(1986).
[7] T.Arahori and N.Iwamoto, "Surface Nitridation of Silicon Carbide Ceramics", J. Ceram. Soc. Jpn., 97, 1347(1989).
[8] D.L.Jiang, J.H.She and S.H.Tan, "Strengthening of Silicon Carbide Ceramics by Surface Nitridation during Hot Isostatic Pressing", J. Am. Ceram. Soc., 75, 2586(1992).
[9] J.H.She, D.L.Jiang, S.H.Tan, and J.K.Guo, "Improvement of Presintered Silicon Carbide Ceramics by Hot Isostatic Pressing", Mater. Res. Bull., 26, 1277(1991).
[10] J.H.She, D.L.Jiang, S.H.Tan, and J.K.Guo, "Microstructure and Properties of Hot Isostatically Pressed SiC Ceramics with An Al_2O_3 Addition", Mater. Lett., 14, 240(1992).
[11] J.H.She and D.L.Jiang,"Hot Isostatic Pressing of Presintered Silicon Carbide Ceramics", J. Europ. Ceram. Soc., 7,243(1991).

[12] H.Wada, M.J.Wang and T.Y.Tien, "Stability of Phases in the Si-C-N-O System", J. Am. Ceram. Soc., 71(10), 837(1988).
[13] I.A.Aksay and J.A.Pask, "Stable and Metastable Equilibria in the System SiO_2-Al_2O_3 ", J. Am. Ceram. Soc.,58,507(1975).
[14] S.Bandyopadhyay, A.S.Sanyal and J.Mukerji, "Parameters Controlling the Synthesis of β-Silicon Nitrde from Silica", pp.55-66, Key Engineering Materials, Vol.89-91, Ed. M.J.Hoffmann et al., Trans. Tech. Publications, Switzerland, 1994.

Microstructure & Properties of SiC-AlN Multiphase Ceramics

Y.B. Pan S.H. Tan D.L. Jiang J.H. Wang

Shanghai Institute of Ceramics,
Chinese Academy Science
1295 Ding Xi Road, Shanghai 200050, China

Abstract

It is that AlN and SiC mixture could form solid solution at the temperature from 1800°C to 2100°C, its result will be conducive to important benefits for the improving to study and develop on the silicon carbide ceramics. The effect of AlN as a mainly additive phase on silicon carbide ceramic were investigated in this paper.

For the optimum hot press(HP) process, SiC and AlN mixture formed solid solution at the 1950°C--2050°C in Ar environment. The properties of SiC-AlN composition were that bending strength more than 600 MPa and fracture toughness more than 7 $MPa.m^{1/2}$ at the room temperature(R.T) could be received, at the same time the strength hold ascertain value from R.T. to 1400°C in air.

The dense samples were examined by metallograph, X-ray diffraction (XRD), scanning electron microscope (SEM) & transmission electron microscope (TEM) to determine the fracture structure, interface phase, crack spread etc. It is found that the aluminum and yttrium had been well-distributed to silicon carbide matrix and formed solid solution and thin film around SiC grains by SEM & TEM observation, and crack deflection & bowing could be a main contribution to fracture toughness improved.

To the extent authorized under the laws of the United States of America, all copyright interests in this publication are the property of The American Ceramic Society. Any duplication, reproduction, or republication of this publication or any part thereof, without the express written consent of The American Ceramic Society or fee paid to the Copyright Clearance Center, is prohibited.

1. Introduction

Companion with the rapidly development of the science and technology, it is request to the advanced basic materials. The multiphase ceramics have been widely studied to the structure ceramics as improving method for strength and toughness.

Silicon carbide ceramics have recently received much attention, for example sealing ring; heat exchanger and engine etc; specially in resistance to abrade, corrode & high temperature environment. However, the application was limited on the original interface more brittle, so it is key that improving the crystal interface property between SiC grains with the strength and toughness. Since Cutler & Miller[1] show that the 2H polymorphous of SiC and AlN form an extensive solid solution at the temperature range from 1800°C to 2100°C, and exhibit analogous lattice parameters. The people have gone on a lot of researching works around SiC-AlN mix system. Those papers were published that including the formation SiC-AlN mixture powder by organometallic precursors[2,3], carbonthermal reduction[4,5], as well as phase diagram[6,7,8].

The main purpose is to prepare dense SiC-AlN multiphase ceramic material by the commonly hot press sintering method, then to study microstructure and mechanical properties of their composition. At the same time, AlN additive phase effect & mechanism analysis on increasing properties were studied through X-ray diffraction, SEM & TEM analysis technology.

2. Experimental Procedure

The raw powder material of SiC & AlN grain average size are respectively with 0.6, 3μ by conventional powder weight distribution measurement, which powder size & shape shown as Fig.1. Then the mixture of α-SiC, AlN & Y_2O_3 powder were mixed and sintered in the graphite dies. Through hot pressure(HP) sintering process that 40 MPa pressure, it is successful that obtain dense multiphase ceramic material.

Both the flexural strength [room temperature(R.T) up to 1400°C] and fracture toughness measurements were made from bars that were cut from the specimens of the HP SiC-AlN multiphase ceramic, then ground dimension to 3*4*36 mm^3 measuring strength bars & 5*2.5*26 mm^3 fracture toughness testing bars which notch depth 2.5 mm, and polished by using various grades of diamonds paste as well as corrosion by alkali solution so that further observation microstructure & crack change phenomenon by the single edge notch beam (SENB) method.

(a) SiC (b) AlN

Fig.1 The shape & dimension of the original SiC and AlN powder by SEM

The microstructure; crack spread and interface for the SiC-AlN multiphase ceramics were examined by the XRD; SEM & TEM.

3. Results & Discussion

3.1 Mechanical Properties

The composition had one massratio SiC:AlN:Y_2O_3=85:14.5:0.5 (wt%) which HP at the three difference holding temperature, that respectively 1950°C, 2000°C, 2050°C. The multiphase ceramic could be obtained densification sample at each sintering temperature. Their physical and mechanical properties showing as Table 1 to 3.

Table 1 The properties of SiC-AlN composite ceramic

	Density (g/cm^3)	Porosity (%)
HP 1950°C, (A)	3.24	$5.6*10^{-3}$
HP 2000°C, (B)	3.26	$10.5*10^{-3}$
HP 2050°C, (C)	3.27	$4.5*10^{-3}$

Table 2 The bending strength & fracture toughness of SiC-AlN at the room temperature

A		B		C	
σ* (MPa)	K_{1C}** ($MPa.m^{1/2}$)	σ (MPa)	K_{1C} ($MPa.m^{1/2}$)	σ (MPa)	K_{1C} ($MPa.m^{1/2}$)
640	7.6	620	7.1	610	8.1

* σ: Bending strength (surface not polished)

**K_{1C}: Fracture toughness by indentation strength bending (ISB) method testing.

Table 3 The properties for HP 2050°C (C) at the high temperature

Temp. (°C)	R.T.	1100	1400
σ (MPa)	610	680	640
K_{1C} ($MPa.m^{1/2}$)	8.1	10	12

According to Table 1, we can say that this SiC-AlN multiphase ceramic had formed dense bodies under each sintering conditions. According to fabrication technology we can say that owning to rather than widely temperature range else. There are similar to fracture surface with A, B, C samples. Their section have a mount of rising and falling peak structure, and form unruly fracture surface shape. At the same time the microstructure A, B, C relation to SiC grains dimension is not alteration comparing with temperature change of 100°C range through SEM observation.

Many researchers have indicated that could improve fracture toughness by crack deflection, crack bowing and crack branching. In order to clear observation microstructure and crack extend, the surface of SiC-AlN multiphase ceramic were polished surface by each grade diamond and corroded by the alkali solution after the bars had been tested. It is found that the SiC-AlN composite ceramic existed crack deflection &

bowing, the AlN exist around SiC grain which form better interface layer (that discussion as following 3.2). The way of crack expand is along to interface between SiC grains, tortuous and very shorten. There are some photo for crack spread and SENB trace by SEM, that shown as Fig.3;4. Therefore, this composition system could be achieved satisfaction properties, the bending strength steady more than 600 MPa from R.T. to 1400°C in air, and fracture toughness more than 7 $MPa.m^{1/2}$ that maximum value 8.1 for sample C.

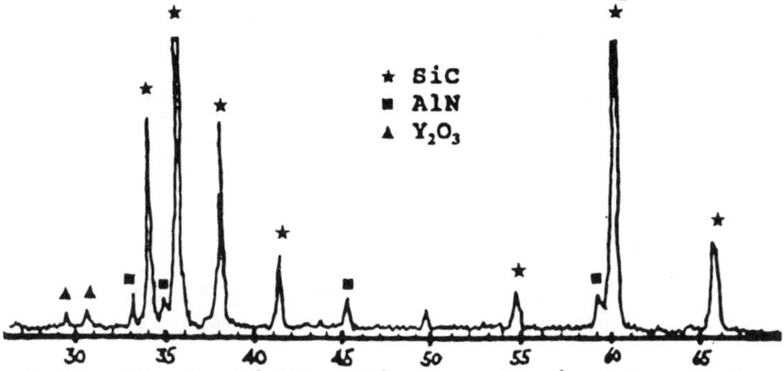

Fig.2 The X-ray diffraction analysis of sample (A)

(a) A (b) C

Fig.3 The fracture surface of sample (A) (C) by the SEM micrograph

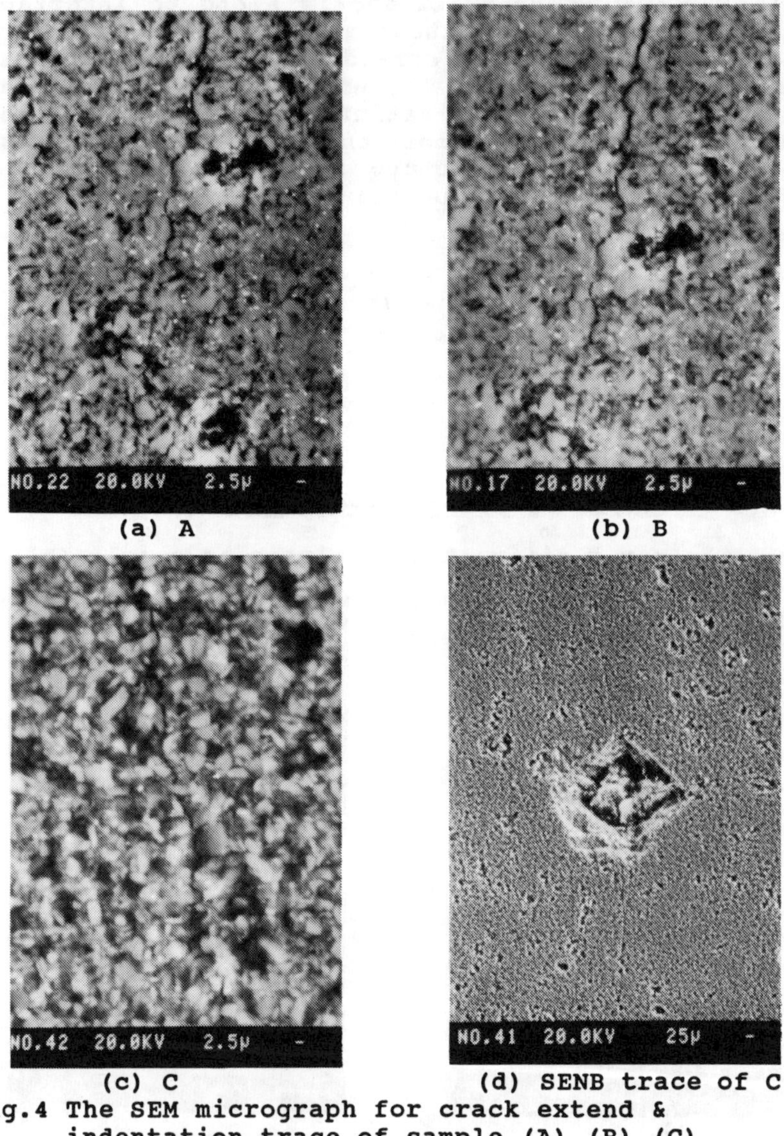

(a) A (b) B

(c) C (d) SENB trace of C

Fig.4 The SEM micrograph for crack extend & indentation trace of sample (A) (B) (C) after SENB measurement

3.2 The effect of AlN in the multiphase

The first that AlN composition in this SiC-AlN multiphase ceramics is the main purpose to form solid solution resulting in increasing properties of SiC crystal phase, and contribution to its strength and toughness. We could find that the aim had been successful achieved from results that AlN is not only abundant and well-distributed surround SiC grains but also better interface. The photo show as Fig.5;6 by SEM & TEM analysis.

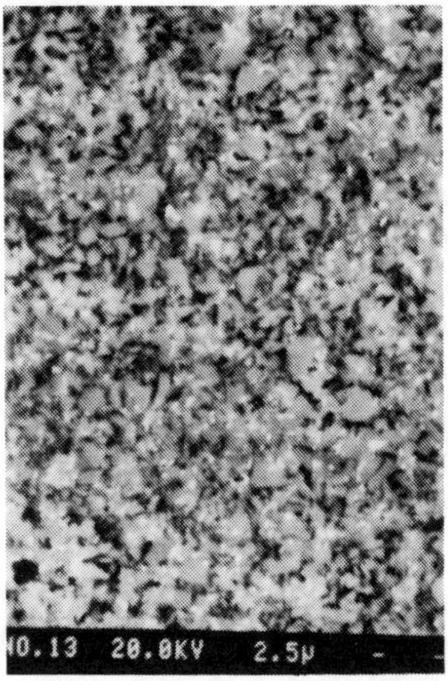

Fig.5 The SEM micrograph of sample A

On the other hand, the average dimension is about 1μ for sintered dense SiC-AlN multiphase ceramic in which not to be more than 2μ. But the original dimension of AlN powder is 3μ, what happen?

The solid vaporization pressure of AlN and SiC have achieved 14 & $1.23*10^{-3}$ mmHg[9] at the 1900°C, of course the vaporization rate of AlN solid is far exceed than it of SiC in limits from 1950 to 2050°C. Therefore, imagination that AlN pass through procedure: the first vaporization from large AlN powder surface; the second deposition and diffusion on the surface of SiC grains;

and at the least formation SiC-AlN solid solution in the sintering process. Because of this factor, AlN-SiC composition form obstruction layer on the SiC surface, its protection thin film restrain excessive growth and keep on SiC crystal size in the sintering multiphase ceramic is the same as original size of SiC powder.

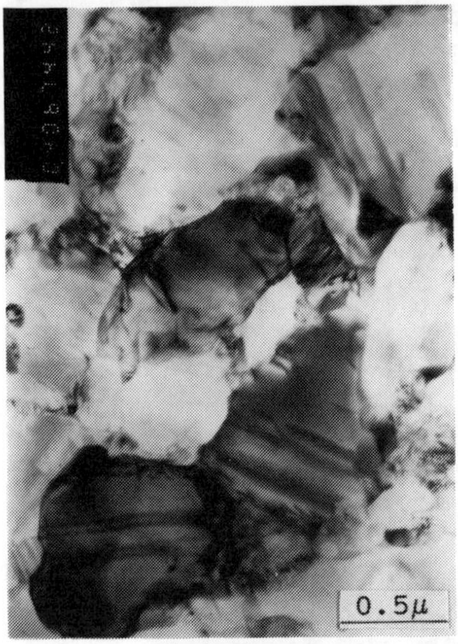

(a) A

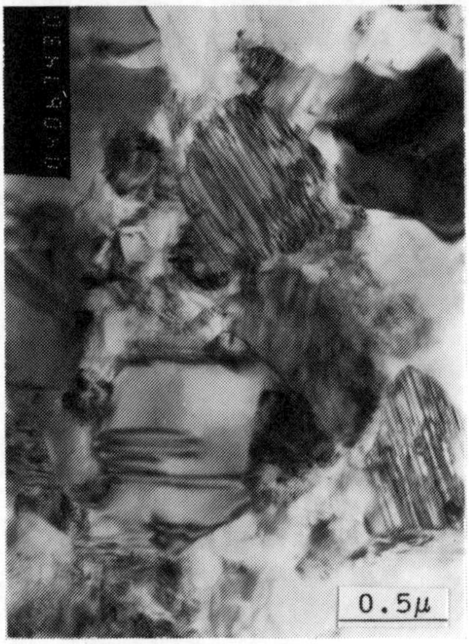

(b) B

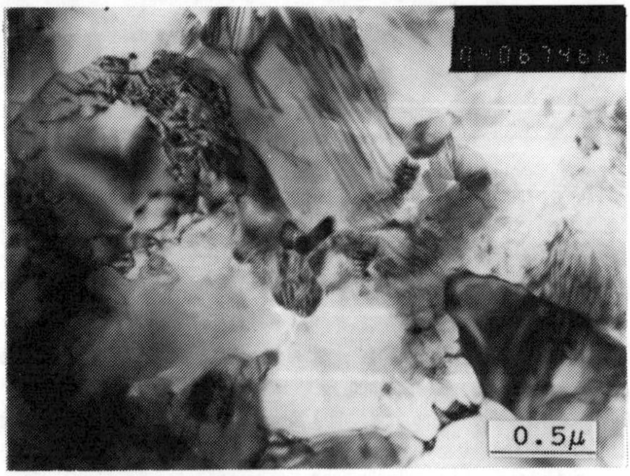

(c) C

Fig.6 The TEM photographs of sample (A) (B) (C) microstructure

According to above list results, the AlN phase contain many-side effects in the SiC-AlN multiphase ceramic. It can restrain growth of the SiC crystal, and form solid solution with SiC, the further improving crystal interface properties between SiC grains. So we can say that AlN is owing to improve properties in the SiC-AlN multiphase ceramics.

4. Conclusion

It is clear shown that SiC-AlN multiphase ceramic owning better physical and mechanical properties, their bending strength more than 600 MPa from room temperature to 1400°C in air, and fracture toughness more than 7 $MPa.m^{1/2}$ which maximum 8.1 $MPa.m^{1/2}$ at the room temperature. This composition system have widely HP sintering range, in which Y_2O_3 as sintering aid. The AlN is rather than important that going through vaporization, deposition and diffusion, and final formation solid solution & protection layer to the surround SiC grains. The phenomenon results in AlN well-distributed and restraint SiC crystal growth during the HP sintering process. At the same time, it is found that occurring crack deflection and bowing in the course for the crack extend, its tortuous crack is rather shorten along SiC crystal interface so that improving its material mechanical properties relation to composite SiC-AlN system.

REFERENCE:

1. B.Cutler; P.D.Miller US Pat. 4141740 1979
2. L.V.Interrante US AD-A206998 1985
3. Terence J.Clark; Robert E.Johnson US Pat. 4687657 1987
4. M.Mitomo; M.Tsutsumi; Y.Kishi J. Mater. Sci. Let. 7(1988) 1151--1153
5. Yoshiynji Sugahara; Ken-Ichi Sugimoto; Hiroyuki Takagi etc J. Mater. Sci. Let. 7(1988)795--797
6. Laurance L. Oden; R.A.McCune J. Am. Ceram. Soc. 73[6](1990)1529--1533
7. Shih-Yee Kuo; Anil V.Virkar J. Am. Ceram. Soc. 73[9](1990)2640--2646
8. R.J.Oscroft; Derek P.Thompson J. Am. Ceram. Soc. 74[8](1991)2327--2328
9. <Handbook of High-melting Point Compound> China Industry Publish Society 1965

ANELASTIC CREEP RECOVERY OF ALUMINA REINFORCED WITH SiC WHISKERS OR PARTICULATES

Weizhong Gu,[¶] John R. Porter[§] and Terence G. Langdon[¶]
[¶]Departments of Materials Science and Mechanical Engineering, University of Southern California, Los Angeles, CA 90089-1453
[§]Rockwell International Science Center, Thousand Oaks, CA 91360

ABSTRACT

Experiments were conducted to determine the anelastic response of samples of monolithic Al_2O_3 and Al_2O_3 composites reinforced with SiC whiskers or particulates. For each material, the load was removed during high temperature creep and the subsequent load-displacement response was recorded. The results show that the monolithic material and a composite with 5 vol % of SiC particulates exhibit no anelastic response on load removal, but there is significant anelastic recovery in composites reinforced with 15 vol % of SiC whiskers or with 15 vol % of SiC particulates. The results are interpreted in terms of the hypothesis that there is an anelastic creep recovery whenever there is some form of interconnecting network within the material.

INTRODUCTION

The incorporation of a discontinuous reinforcement into a ceramic matrix, in the form of whiskers or particulates, is an important procedure for introducing toughening and strengthening [1]. It is also well established that ceramic composites usually exhibit a significant improvement in their creep properties at elevated temperatures by comparison with the monolithic materials.

The present investigation was conducted to examine the anelastic creep response in alumina composites reinforced with either SiC whiskers or SiC particulates. The creep behavior of alumina composites has been examined in several investigations [2-10] and the results show that the stress exponents and activation energies for creep are often significantly higher in the composites than in the

To the extent authorized under the laws of the United States of America, all copyright interests in this publication are the property of The American Ceramic Society. Any duplication, reproduction, or republication of this publication or any part thereof, without the express written consent of The American Ceramic Society or fee paid to the Copyright Clearance Center, is prohibited.

unreinforced alumina. An earlier report described evidence for anelastic creep recovery in an alumina composite containing 15 vol % of SiC whiskers [11] and the present investigation extends these results to cover alumina reinforced with SiC particulates.

EXPERIMENTAL MATERIALS AND PROCEDURE

The experiments were conducted on specimens of monolithic Al_2O_3 and on a series of Al_2O_3 composites. The composites were fabricated using a colloidal pressure filtration method [12] and subsequent hot pressing into disks. High purity alumina powder was used with either SiC whiskers or SiC particulates, The whiskers had an average diameter of ~ 0.6 μm and an initial average length of ~ 16.6 μm, thus giving an average aspect ratio (A.R.) close to ~ 30. Some whiskers were used after ball milling for 24 hours to reduce the average length to ~ 6.3 μm and the aspect ratio to ~ 10. The whisker configurations in the unmilled condition and after ball milling for 24 hours are shown in Figs 1(a) and (b), respectively. The particulate composites were prepared using SiC particles with an average diameter of ~ 10 μm. Inspection after hot pressing showed that the composites with SiC whiskers contained a reasonably homogeneous distribution of whiskers but with a majority of the whiskers tending to lie in a plane essentially perpendicular to the hot pressing direction. The average grain size of the composites after hot pressing was ~ 2 μm, and this was slightly smaller than the grain size of ~ 5 μm for the monolithic material.

Test specimens were cut for all materials with lengths of 26 mm and cross-sections of 2.5×3.0 mm. Creep tests were conducted in air under conditions of constant load using a four-point bending rig with the test sample held between four loading pivots made of optically-polished sapphire. For each specimen, the load was applied in the hot pressing direction using an alumina loading ram, and the entire assembly was contained within a vertical split furnace with the temperature controlled at the selected value to within ± 2 K. Each specimen was held at the testing temperature for >30 minutes prior to application of the load. The vertical displacement of the loading ram was recorded using a linear variable differential transducer and a computer-controlled data acquisition system. The outer fiber stresses and strains were calculated using the procedure of Hollenberg *et al.* [13].

In order to investigate the anelastic response, the load was applied for a selected period of time and then removed so that only a small residual load (typically, $\sim 8\%$ of the total load) remained on the specimen in the unloaded condition: this small residual load was required to maintain the integrity of the loading assembly

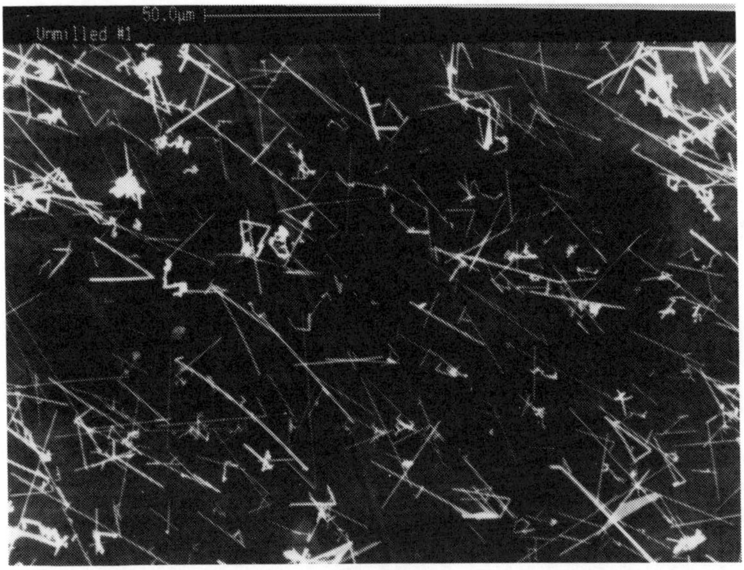

(a)

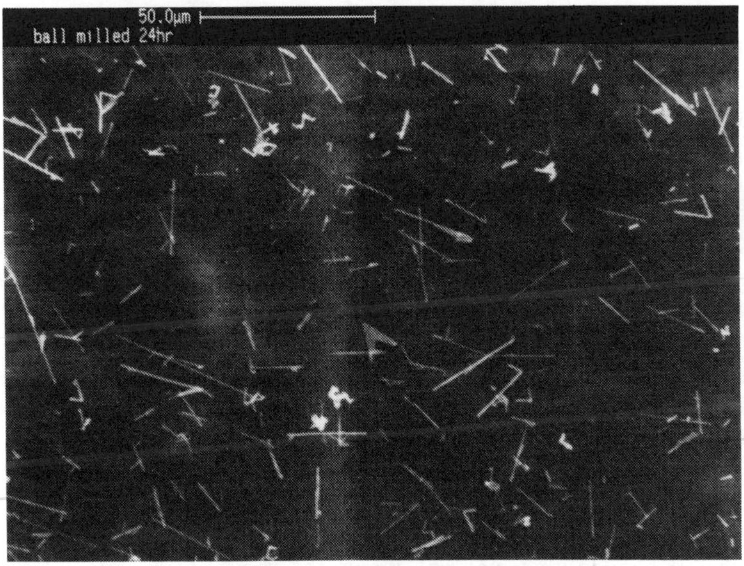

(b)

Fig. 1 Whisker configuration (a) unmilled and (b) after ball milling for 24 hours.

and to permit rapid reloading. After holding for a period of time in the unloaded condition, each specimen was reloaded to the original level. For several specimens, this procedure was repeated three or four times to provide information on the magnitude of the recovered strain as a function of the total creep strain within the specimen.

EXPERIMENTAL RESULTS

Several tests were conducted to determine the nature of the anelastic response in the monolithic Al_2O_3. In each experiment, conducted over a range of stresses and at 1723 K, there was no evidence for any anelastic recovery.

A typical example for the monolithic alumina is shown in Fig. 2 plotted as the creep strain, ε, versus the total time, t, for a specimen tested at an absolute temperature, T, of 1723 K and under an applied stress, σ, of 40 MPa. For this specimen, the load was removed three times, and on each removal there was a very small elastic recovery which is just visible in Fig. 2 and thereafter the creep strain remained essentially constant with no evidence for anelastic recovery. Close inspection of Fig. 2 shows a slight increase in strain with increasing time following the third stress removal and this is attributed to the small residual stress which remained on the specimen even in the unloaded condition.

The result shown for the monolithic alumina in Fig. 2 is in marked contrast to the very significant anelastic recovery which was recorded in all experiments on an alumina composite containing 15 vol % of SiC whiskers. Figure 3 shows an example of strain versus time for a composite tested at the same temperature of 1723 K and under an applied stress of 45 MPa: in this example, the composite was fabricated using unmilled whiskers so that the aspect ratio was ~30 and the stress was removed a total of four times during the creep test.

Two conclusions may be reached by comparing Figs 2 and 3. First, although both tests were conducted at the same temperature and the stresses were similar but with a slightly higher stress on the composite material, the rate of accumulation of creep strain is very much slower in the composite. For example, it appears by extrapolation in Fig. 3 that the composite will attain a creep strain of $\varepsilon = 0.01$ after a testing time of $\sim 3 \times 10^5$ s (~83 hours) whereas Fig. 2 shows that the monolithic material attains the same creep strain at the same temperature but under a slightly lower applied stress after a time of only $\sim 1 \times 10^3$ s (~16 minutes). Thus, the introduction of SiC whiskers leads to a very significant reduction in the overall creep rate. Second, there is a large anelastic recovery following each load removal in the composite material and this is in

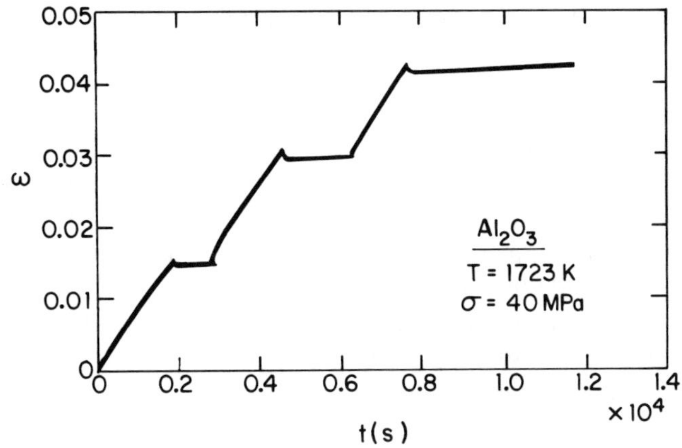

Fig. 2 Strain versus time for monolithic alumina showing the effect of load removal.

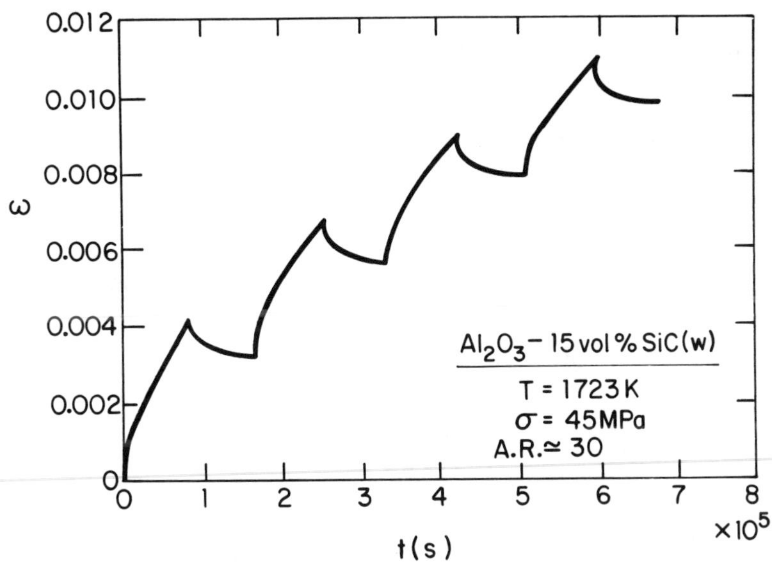

Fig. 3 Strain versus time for Al_2O_3 composite with 15 vol % of SiC whiskers having a large aspect ratio showing the effect of load removal.

direct contrast to monolithic alumina where there is no evidence for an anelastic response.

Although the anelastic creep recovery shown in Fig. 3 is for a composite with 15 vol % of SiC whiskers having an average aspect ratio of ~30, there was also a very similar anelastic recovery in composites having 15 vol % of SiC whiskers where the whiskers were ball milled prior to fabrication to reduce the aspect ratio to ~10. This is illustrated by the partial plot of strain versus time shown in Fig. 4 for a composite with the smaller aspect ratio tested under the same experimental conditions of stress and temperature as for the specimen shown in Fig. 3. For the single reduction in stress depicted in Fig. 4, the anelastic response is clearly defined and extends through a time period of almost 20 hours.

The situation with alumina composites containing particulates was more complex. Figure 5 shows the plot of strain versus time recorded by the data acquisition system for a specimen with 15 vol % of SiC particulates tested at a temperature of 1723 K under an applied stress of 30 MPa. Following a stress reduction, there is a clear anelastic creep recovery in this material. However, there was no anelastic response in composite specimens having only 5 vol % of SiC particulates, as demonstrated in Figs 6 - 8 for specimens tested at 1723 K under

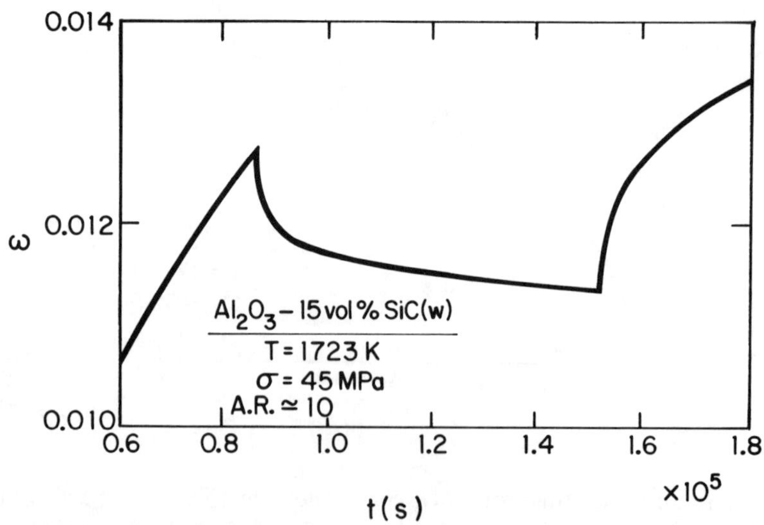

Fig. 4 Strain versus time for Al_2O_3 composite with 15 vol % of SiC whiskers having a small aspect ratio, showing the effect of load removal.

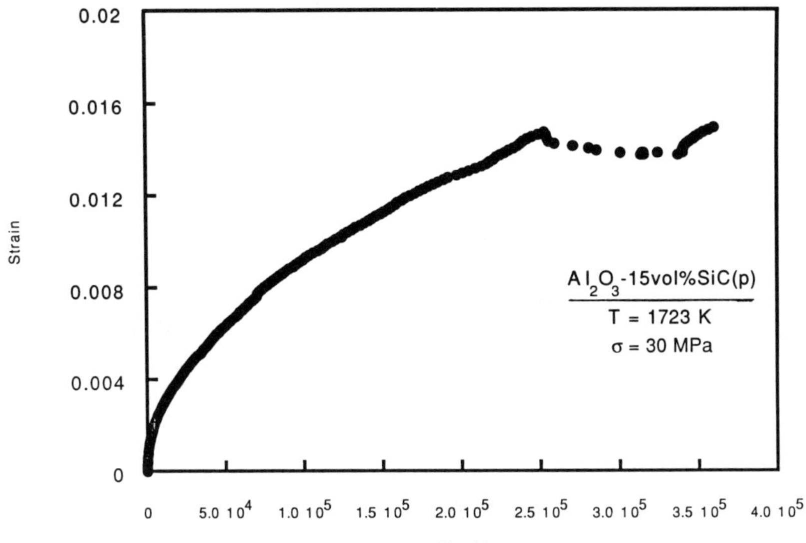

Fig. 5 Strain versus time for Al_2O_3 composite with 15 vol % of SiC particulates showing the effect of load removal.

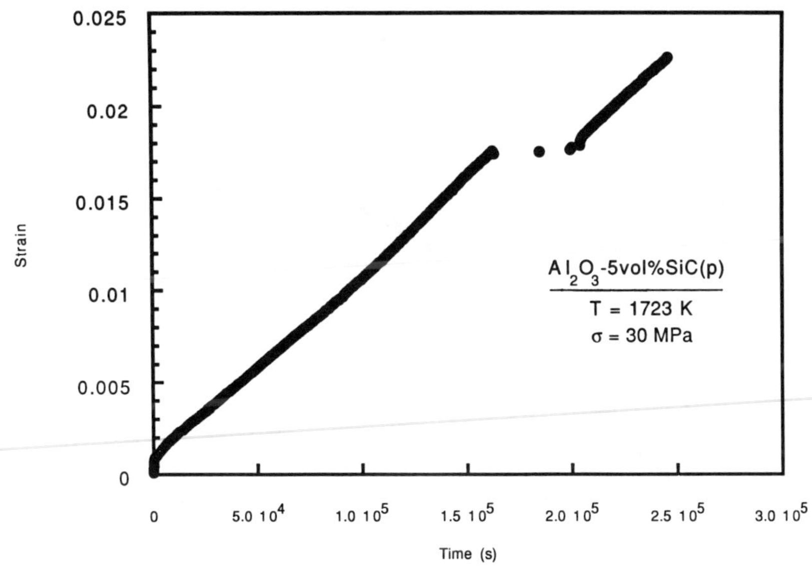

Fig. 6 Strain versus time for Al_2O_3 composite with 5 vol % of SiC particulates tested at an applied stress of 30 MPa, showing the effect of load removal.

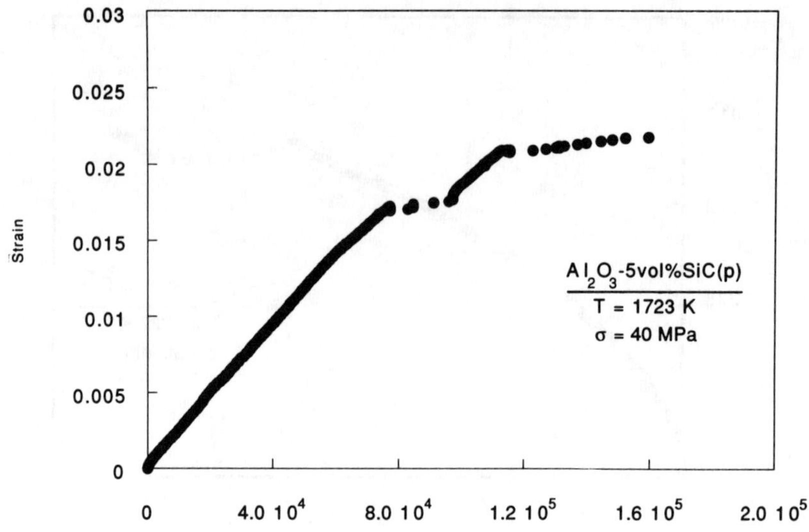

Fig. 7 Strain versus time for Al_2O_3 composite with 5 vol % of SiC particulates tested at an applied stress of 40 MPa, showing the effect of load removal.

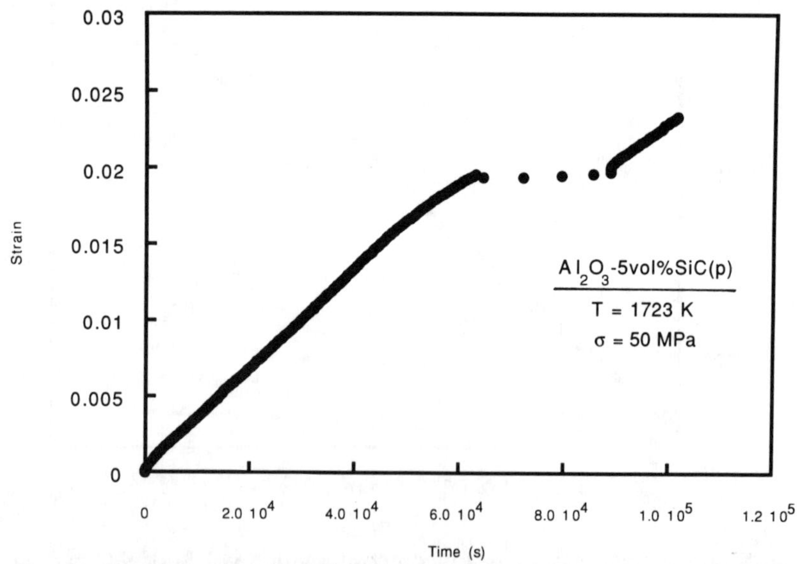

Fig. 8 Strain versus time for Al_2O_3 composite with 5 vol % of SiC particulates tested at an applied stress of 50 MPa, showing the effect of load removal.

applied stresses of 30, 40 and 50 MPa, respectively: Figs 6 and 8 show a single stress reduction and Fig. 7 shows two reductions.

DISCUSSION

The presence or absence of an anelastic creep recovery is clearly dependent upon microstructural features in the material. No anelastic recovery is observed in the monolithic alumina or in an alumina composite reinforced with 5 vol % of SiC particulates. However, when testing under similar conditions of stress and temperature, there is a large anelastic recovery in composites containing either 15 vol % of SiC whiskers or 15 vol % of SiC particulates. In addition, the nature of the anelastic response is similar in the alumina composites with 15 vol % of whiskers or particulates, as shown by the plots of recovered strain versus time after load removal in Figs 9-11 for the composite with 15 vol % of SiC particulates and for the composites with 15 vol % of SiC whiskers with either a large or a small aspect ratio, respectively. Furthermore, inspection of Figs 10 and 11 shows that the total anelastic response is essentially independent of the creep strain at which the load is removed, since the datum points for the four interruptions in loading for each composite tend to superimpose.

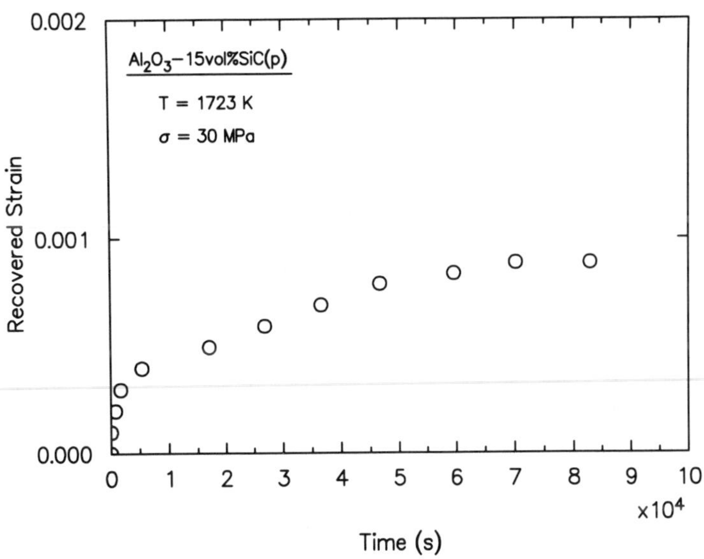

Fig. 9 Recovered strain versus time after load removal for Al_2O_3 composite with 15 vol % of SiC particulates.

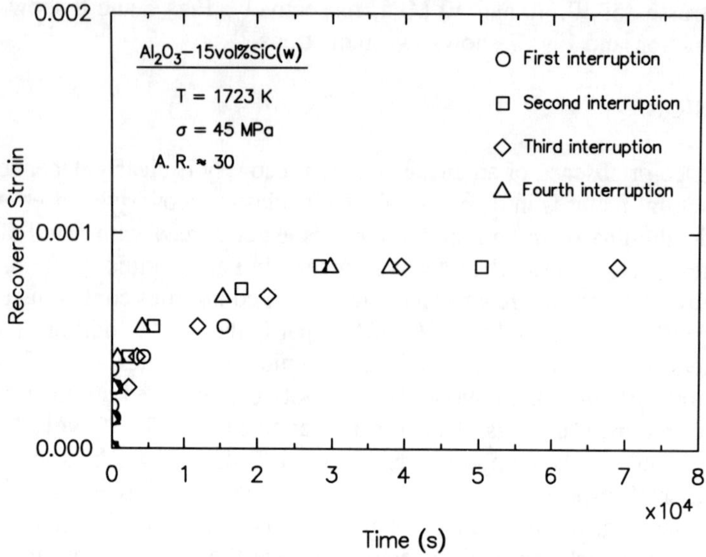

Fig. 10 Recovered strain versus time after load removal for Al_2O_3 composite with 15 vol % of SiC whiskers having a large aspect ratio.

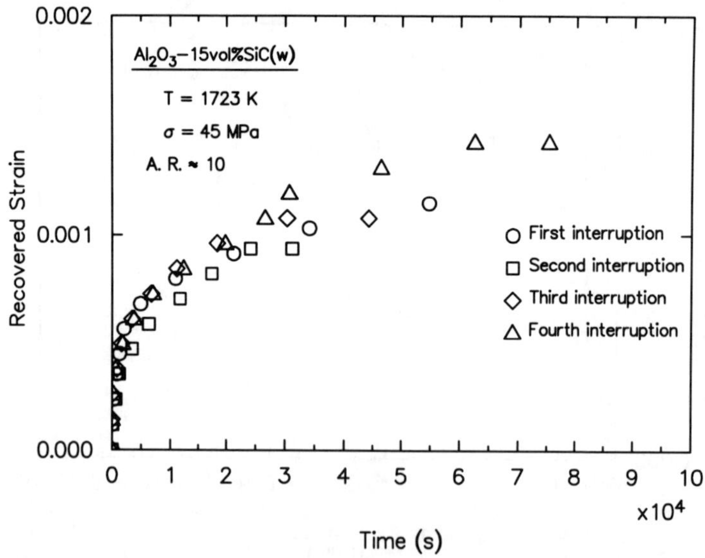

Fig. 11 Recovered strain versus time after load removal for Al_2O_3 composite with 15 vol % of SiC whiskers having a small aspect ratio.

The experimental results can be explained in terms of the hypothesis that an interconnecting network may be formed within the composites during processing. This hypothesis was initially developed to explain, at least in part, the additional creep resistance occurring in whisker-reinforced ceramic composites [14], but the presence of an interconnected network will clearly introduce an anelastic effect which will be absent in the unreinforced material. This explanation is supported also by experimental results on a metal matrix composite consisting of an Al-5 wt % Mg alloy reinforced with 26 vol % of randomly oriented Al_2O_3 fibers, where the fibers were known to form an interconnected network and the anelastic recovery on load removal was very similar to that observed in the present experiments [15]. The same argument can be used also to explain the difference in anelastic creep response between the composites with 5 and 15 vol % of SiC particulates. Whereas the composite with 5 vol % of particulates will contain a distribution of essentially discrete particles, the composite with 15 vol % of particulates is very close to the percolation threshold of ~16 vol % which, according to site percolation theory, leads to >90% of the particles forming part of a continuous network through the material [16].

CONCLUSIONS

1. Experiments show there is no anelastic creep recovery on load removal in monolithic alumina or in an Al_2O_3 composite containing 5 vol % of SiC particulates, but there is a significant anelastic recovery in Al_2O_3 composites containing either 15 vol % of SiC whiskers or 15 vol % of SiC particulates.

2. The results are interpreted in terms of the hypothesis that anelastic creep recovery occurs whenever there is some form of interconnecting network within the material.

ACKNOWLEDGEMENT

This work was supported by the U.S. Army Research Office under Grant No. DAAL03-91-G-0230.

REFERENCES

1. K. Xia and T.G. Langdon, "The Toughening and Strengthening of Ceramic Materials Through Discontinuous Reinforcement," J. Mater. Sci. (in press).
2. A.H. Chokshi and J.R. Porter, "Creep Deformation of Alumina Matrix Composites Reinforced with SiC-Whiskers," J. Am. Ceram. Soc., 68 [6] C144-45 (1985).

3. P. Lipetzky, S.R. Nutt and P.F. Becher, "Creep Behavior of an Al_2O_3-SiC Composite," Mater. Res. Soc. Symp. Proc., 120, 271-277 (1988).
4. S.R. Nutt, P. Lipetzky and P.F. Becher, "Creep Deformation of Alumina-SiC Composites," Mater. Sci. Eng., A126, 165-72 (1990).
5. A.R. de Arellano-López, F.L. Cumbrera, A. Domínguez-Rodríguez, K.C. Goretta and J.L. Routbort, "Compressive Creep of SiC-Whisker-Reinforced Al_2O_3," J. Am. Ceram. Soc., 73 [5] 1297-300 (1990).
6. H.-T. Lin and P.F. Becher, "Creep Behavior of a SiC-Whisker-Reinforced Alumina," J. Am. Ceram. Soc., 73 [5] 1378-81 (1990).
7. P. Lipetzky, S.R. Nutt, D.A. Koester and R.F. Davis, "Atmospheric Effects on Compressive Creep of SiC-Whisker-Reinforced Alumina," J. Am. Ceram. Soc., 74 [6] 1240-47 (1991).
8. H.-T. Lin and P.F. Becher, "High-Temperature Creep Deformation of Alumina-SiC-Whisker Composites," J. Am. Ceram. Soc., 74 [8] 1886-93 (1991).
9. A.R. de Arellano-López, A. Domínguez-Rodríguez, K.C. Goretta and J.L. Routbort, "Plastic Deformation Mechanisms in SiC-Whisker-Reinforced Alumina," J. Am. Ceram. Soc., 76 [6] 1425-32 (1993).
10. K. Xia, J.R. Porter and T.G. Langdon, "The Influence of Processing on the High Temperature Mechanical Properties of a Whisker-Reinforced Alumina Composite," pp. 253-64 in *Processing, Fabrication, and Manufacturing of Composite Materials*. Edited by T.S. Srivatsan and E.J. Lavernia, The American Society of Mechanical Engineers, New York, NY, 1992.
11. W. Gu, J.R. Porter and T.G. Langdon, "Evidence for Anelastic Creep Recovery in Silicon Carbide-Whisker-Reinforced Alumina," J. Am. Ceram. Soc. (in press).
12. J.R. Porter, F.F. Lange and A.H. Chokshi, "Processing and Creep Performance of SiC-Whisker-Reinforced Al_2O_3," Am. Ceram. Soc. Bull., 66 [2] 343-47 (1987).
13. G.W. Hollenberg, G.R. Terwilliger and R.S. Gordon, "Calculation of Stress and Strain in Four-Point Bending Creep Test," J. Am. Ceram. Soc., 54 [4] 196-99 (1971).
14. J.R. Porter, "Dispersion Processing of Creep Resistant Whisker-Reinforced Ceramic-Matrix Composites," Mater. Sci. Eng., A107 127-32 (1989).
15. T.L. Dragone and W.D. Nix, "Steady State and Transient Creep Properties of an Aluminum Alloy Reinforced with Alumina Fibers," Acta Metall. Mater., 40 [10] 2781-91 (1992).
16. F.F. lange, L. Atteraas, F. Zok and J.R. Porter, "Deformation Consolidation of Metal Powders Containing Steel Inclusions," Acta Metall. Mater., 39 [2] 209-219 (1991).

PROCESSING AND PROPERTIES OF Al_2O_3/SiC WHISKER COMPOSITES

C.M. Mroz, D.J. Bray, T.M. Mroz, A.B. Jedlicka, A. S. Barnes, D.M. Sperber
Advanced Refractory Technologies, Inc., Buffalo, NY 14207

Abstract

The effect of dispersant type and amount on the viscosity of SiCw, Al_2O_3, and Al_2O_3/SiCw slurries was examined. Hot pressed samples produced from well dispersed slurries exhibited higher final densities than those samples containing SiCw agglomerates. The mechanical properties of samples possessing higher final densities and homogeneous microstructure were higher than those measured on lower density samples with inhomogeneous regions. This paper investigates the preparation of Al_2O_3/30 vol. % SiCw slurries and hot pressed Al_2O_3/SiCw composites as well as the mechanical properties exhibited by these composites.

Introduction

To increase the mechanical properties of single phase ceramic materials, the addition of a reinforcing phase, such as silicon carbide whiskers (SiCw) can dramatically increase the mechanical properties of the resultant composite over those exhibited by the single phase monolithic.[1,2,3] However, to assure the fabrication of a homogeneous composite, a uniform dispersion of the SiCw in the matrix material (i.e. alumina, zirconia, silicon nitride, molybdenum disilicide, etc.) is required. Inadequate dispersion of the SiCw in the matrix will result in the production of a composite with varying and unpredictable mechanical properties.

Several studies have been conducted in which commercially available SiCw have been incorporated into an alumina matrix, by aqueous and non-aqueous processing, to a form ceramic matrix composite.[4,5,6] SiCw, however, have been synthesized by a variety of techniques and possess an array of physical and chemical properties. Therefore, SiCw from one supplier can not be directly substituted for another into a fabrication process.

To the extent authorized under the laws of the United States of America, all copyright interests in this publication are the property of The American Ceramic Society. Any duplication, reproduction, or republication of this publication or any part thereof, without the express written consent of The American Ceramic Society or fee paid to the Copyright Clearance Center, is prohibited.

This paper examines the dispersion of M grade SiCw* in water and in an aqueous alumina slurry to produce a dispersed feed material for the preparation of Al_2O_3/SiCw composites. The microstructure and mechanical properties of these composites are also included.

Raw Material Preparation

Two different alumina powders were used as matrix materials during the course of this program: Ceralox HPA 0.5** and Alcoa A-16SG***. Both powders are α-Al_2O_3 and have an average particle size of about 0.45 μm and a surface area greater than 8 m^2/g. The Ceralox powder also contains 0.5 wt.% MgO as sintering aid. Three types of SiCw were included in this study: as-received (AR) M grade, acid-washed (AW) M grade, and American Matrix Incorporated (AMI). Composites fabricated with AMI SiCw were included to provide comparisons between data from this study and previous literature results. M grade SiCw are produced by the same synthesis method that was previously used for the production of AMI SiCw. The physical and chemical characteristics of the M grade SiCw used in this study are shown in Table I.

Table I:
Physical and Chemical Characteristics of M Grade SiCw

Stoichiometry	β-SiC
Ave. Length (μm)	11.8
Ave. Diameter (μm)	1.2
Ave. Aspect Ratio	9.9
Free Carbon (%)	ND
Free SiO_2 (%)	2.1

After acid washing with hydrofluoric acid, the AW M grade SiCw contained no measurable silica. The AMI SiCw used in this study were slightly finer (1.0

* Advanced Refractory Technologies, Inc. Buffalo, NY, 14207

** Ceralox Corporation, Tuscon, AR 85706

*** Alcoa Industrial Chemicals Division, Bauxite, AR 72011

μm) in diameter and longer (18 μm) in length than the M grade SiCw.

SiC Whisker Dispersion

To effectively disperse the SiCw in an aqueous slurry, an initial study was conducted to investigate the settling characteristics of a 10 wt. % SiCw slurry in distilled water.[7,8] The packing density of the SiCw was first measured as a function of pH. The pH of the SiCw slurry was adjusted by the addition of ammonium hydroxide (NH_4OH). All samples were allowed to settle for 24 hours. As the pH increased, the packing density of the SiCw also increased. With the addition of 1 wt. % Emphos CS-1361* to the slurries, the packing density of each slurry tripled. After the Emphos addition, the packing density decreased as the pH increased. These results are shown in Figure 1 and indicate that whisker dispersion can be increased by adjusting the slurry pH and the addition of Emphos.

The effect of Darvan 821-A** dispersant additions on the viscosity of a 70 wt. % solids Al_2O_3 slurry was also examined. The slurry pH was first adjusted to 10 and the slurry viscosity was measured as a function of 1, 1.5, and 2 wt. % Darvan additions. It was determined that the viscosity of the Al_2O_3 slurry was minimized by additions of Darvan 821A below 1.5 wt. %. These results are shown in Figure 2.

Slurry Preparation

Aqueous slurries of Al_2O_3 and SiCw were prepared by first adding the alumina powder with distilled water in a 500 ml Nalgene bottle. The slurry pH was adjusted to 10, and 1.5, 2.0 or 2.7 wt. % Darvan 821-A was added. The slurries were ball milled for 24 hours with 1/2" alumina media. SiCw were then added to produce an Al_2O_3/30 vol. % SiCw slurry at 70 wt. % solids. The slurries were milled for an additional 60 minutes and slurry viscosity was measured (Figure 3). Each slurry became dilatant as the shear rate increased. The slurry containing 1.5 wt. % Darvan exhibited a lower viscosity than the systems containing higher Darvan additions.

To increase the dispersion of the SiCw in the slurry, an additional dispersant

* Witco Corporation, Houston, TX 77045

** R.T. Vanderbilt company, Inc., Norwalk, CT 06856

was added after the SiCw were milled with the Al_2O_3. Emphos was added to the slurry as 1.0 wt.% of the SiCw and the slurry was milled for an additional 30 minutes. Due to the nature of the Emphos, an anti-foaming agent (Foamaster VF*) was also added to reduce the amount of foam produced during milling. The viscosity was then remeasured. As is shown in Figure 4, the viscosity of the slurry decreased with the addition of the Emphos. The Al_2O_3/SiCw slurry was then poured into a container and 2-3 drops of Daxad CP-2** were added to flocculate the slurry. The slurry was then dried for 12 hours at 125°C and the dried powder was screened to - 60 mesh.

Composite Fabrication and Hot Pressing Results

Consolidated parts were produced by hot pressing the prepared powder in argon at a pressure of 17 MPa and a temperature of 1800°C for 30 to 60 minutes. After consolidation, the density of each hot pressed part was determined by the Archimedes method. These results are shown in Table II. A portion of the hot pressed sample was mounted and polished to a mirror finish with 0.5 μm diamond paste. These samples were used for optical examination of the composite microstructure as well as mechanical properties testing (hardness, fracture toughness). A separate sample was used for SEM examination of the fracture surface.

Hot pressing results indicate slightly higher final densities were obtained for samples processed with as-received (AR) and acid-washed (AW) M grade SiCw than AMI SiCw. The average densities of the M grade SiCw samples fabricated at a higher temperature (1900°C) and pressure (19 MPa) did not significantly change as compared to those processed at the standard conditions. No significant differences in density were observed between samples fabricated with as-received (AR) and acid-washed (AW) M grade SiCw. While signs of whisker toughening (i.e. whisker pullout and crack bridging) are evident in the fracture surface of the samples, areas of porosity are also present. Optical and scanning electron micrographs of a representative sample are shown in Figure 5.

Samples produced with the addition of Emphos and additional milling time obtained final densities of greater than 99% of theoretical and exhibited a more

* Henkel Corp., Ambler, PA 19002

** Organic Chemicals Division, W.R. Grace & Co., Lexington, MA 02173

uniform microstructure than samples using Darvan only. It appears that the combination of Emphos and additional milling time disperses the SiCw agglomerates more effectively than Darvan alone. It is these SiCw agglomerates which contribute to the porosity in the final hot pressed parts. It should also be noted that the additional 30 minute milling time did not appear to significantly decrease the length of the SiCw in the final composite. Artifacts suggesting SiCw toughening are also apparent in these samples. Optical and SEM micrographs are shown in Figure 6.

Table II
Hot Pressing Mixes, Conditions, and Results for Al_2O_3/SiCw Composites

SiCw Type	Al_2O_3 Type	Temp. (°C)	Pressure (MPa)	Ave. % Theo. Density
AMI	A-16SG	1800	17	96.5
AMI	HPA	1800	17	96.8
M Grade	A-16SG	1800	17	97.1
M Grade	HPA	1800	17	97.3
M Grade	HPA	1900	19	96.0
AW M Grade	HPA	1900	19	97.2
AW M Grade	HPA	1800	17	97.3
AW M Grade + Emphos	HPA	1800	17	99.0

Mechanical Properties Testing

Fracture toughness and hardness of the composites were measured using a Shimadzu HMV-2000 with a diamond Vickers indenter. All samples were measured parallel to the hot pressing direction, which is perpendicular to the whisker orientation. For hardness testing, a 200 gram load was applied with a dwell time of 15 seconds. Ten indents per sample were obtained and used to calculate the average hardness and standard deviation. The equation used for determination of the sample hardness is:

$$H_V = (1.8544\ P)/d^2$$

where:
- H_V = Vickers Hardness (GPa)
- P = load (N)
- d = diagonal length of the indent (m)

An elastic modulus of 404 GPa was determined by the ultrasonic vibration technique* on an Al_2O_3/30 vol.% SiCw hot pressed composite. This value was used for determining the sample fracture toughness.

The fracture toughness of the samples was also measured using the Shimadzu HMV-2000 with a diamond Vickers indenter. The standard load used for fracture toughness measurements was 2000 grams. Five fracture toughness values per sample were used to calculate the average fracture toughness. Fracture toughness was calculated using the following equation:[9]

$$K_C = 0.0141 \cdot (P/a^{3/2}) \cdot (E/H_V)^{2/5} \log (8.4a/c)$$

where:
- E = elastic modulus (GPa)
- H_V = Vickers hardness (GPa)
- P = load or force (N)
- $2a$ = indentation diagonal length (m)
- $2c$ = radial crack length (m)
- K_C = fracture toughness ($MPa{\cdot}m^{1/2}$)

Mechanical properties testing was also conducted on a commercially available Al_2O_3/30 vol.% SiCw reinforced cutting tool insert**. The mechanical properties results of these samples as well as those for the M grade and AMI SiCw containing samples are shown in Table III.

Hardness for the Emphos-dispersed M grade sample was measured at a load of 2000 grams because the indentations produced with the 200 gram load were difficult to measure.

* J.W. Lemmens, Inc., St. Louis, MO 63074-6051

** WG-300 cutting tool insert, Greenleaf Corp., Saegertown, PA 16433

Table III:
Mechanical Properties of ART Fabricated Al_2O_3/30 vol. % SiCw Composites
(Standard Deviation in Parentheses)

SiCw Type	Al_2O_3 Type	Ave. % Theo. Density	Hardness (GPa)	Fr.Toughness ($MPa{\cdot}m^{1/2}$)
AMI	A-16SG	96.5	19.4 (0.8)	5.0 (0.3)
AMI	HPA	96.8	18.8 (0.7)	4.9 (0.3)
M Grade	A-16SG	97.1	20.8 (1.3)	5.2 (0.2)
M Grade	HPA	97.3	21.5 (1.6)	5.4 (0.3)
AW M Grade	HPA	97.3	20.1 (1.5)	5.4 (0.2)
AW M Grade + Emphos	HPA	99.2	19.0 (1.5)$^{\alpha}$	6.0 (0.6)
WG-300	NA	99.0	23.2 (1.2)	6.1 (0.1)

$^{\alpha}$ measured at 2000 gram load

Mechanical Properties Results and Load Dependance

A test was conducted using the WG 300 sample in which the hardness and fracture toughness were measured at loads of 200, 2000, and 10,000 grams. The 10 kg. tests were conducted using a Leco V-100-A2 hardness tester with a diamond Vickers indentor. These results, shown in Table IV, indicate that there is a load dependance associated with hardness and fracture toughness of this sample. While the hardness values decreased, fracture toughness increased significantly with increasing load. This effects of load on the hardness of CMC composites has previously been reported for SiCw-reinforced Si_3N_4 composites[10] and are consistent with the results shown in this study. It was also concluded that the proportionally shorter cracks at the higher indentation loads were responsible for the increase in K_C as load increased. While the absolute lengths of the cracks are proportionally shorter at the larger loads, the cracks themselves are quite tortuous as a result of the SiCw reinforcement.

Literature values for hardness and fracture toughness determination by indentation include a range of testing loads, from 100 grams to 20 kgs. It appears that no standard sample testing load exists for this property measurement. In addition to testing differences, there are a variety of published formulas which can be used for the calculation of fracture toughness by indentation.[9] Therefore, it is difficult to compare the results obtained in one study with results from another.

Table IV:
Effect of Indentation Load on Sample Hardness and Fracture Toughness for Greenleaf WG 300 Cutting Tool Insert

	200 gr. load	2000 gr. load	10 kg. load
Hardness (GPa)	23.2 $\pm$ 1.2	22.0 $\pm$ 0.7	20.7 $\pm$ 0.4
Fr. Toughness ($MPa \cdot m^{1/2}$)	5.8 $\pm$ 0.2	6.1 $\pm$ 0.1	7.7 $\pm$ 0.3

Conclusions

The density of hot pressed Al_2O_3/30 vol. % SiCw composites is dependant on the dispersion of the SiCw in the alumina. As the SiCw agglomerates are reduced, the hot pressed density of the final composite increases. The dispersion of the SiCw in an Al_2O_3 slurry can be increased by the addition of a second dispersant after the SiCw have been added.

The mechanical properties of the M grade SiCw composites are higher than those produced with AMI SiCw. As the final density of the hot pressed composites increased, the mechanical properties improved due to the improved composite homogeneity. The fracture toughness of the high density (99%) composites were comparable to that exhibited by the WG-300 cutting tool insert.

A load dependance for hardness and fracture toughness was observed with the WG-300 cutting tool insert. As the applied load increased, the fracture toughness appeared to increase while the sample hardness decreased.

Acknowledgements

The authors wish to acknowledge the contributions of Dr. James Varner of the New York State College of Ceramics at Alfred University for composite mechanical properties testing and data interpretation.

Cooperative sponsorship of portions of this work from the New York State Energy Research and Development Authority is gratefully acknowledged.

References

1. J. Homeny, W.L. Vaughn, "Whisker-Reinforced Ceramic Matrix Composites," *MRS Bulletin*, 66-71, Oct. 1 - Nov. 15, 1987.

2. E.D. Whitney, P.N. Vaidyanathan, "Microstructural Engineering of Ceramic Cutting Tools," *Am. Ceram. Soc. Bull.*, **67** [6] 1010-1014, 1988.

3. E. Yasuda, T. Akatsu, Y. Tanabe, "Influence of Whiskers' Shape and Size on Mechanical Properties of SiC Whisker-Reinforced Al_2O_3," *J. Ceram. Soc. Japan,* **99** [1] 51-57 (1991).

4. G.C. Wei, P.F. Becher, "Development of SiC-Whisker-Reinforced Composites," *Am Ceram. Soc. Bull.*, **64** [2] 298-304 (1985).

5. J.R. Porter, F.F. Lange, and A.H. Chokshi, "Processing and Creep Performance of SiC-Whisker-Reinforced Al_2O_3," *Am. Ceram. Soc. Bull.*, **66** [2] 343-47 (1987).

6. T.N. Tiegs, P.F. Becher, "Sintered Al_2O_3-SiC-Whisker Composites," *Am. Ceram. Soc. Bull.*, **66** [2] 339-342 (1987).

7. D.M. Sperber, "Measurement of SiC Whisker Dispersion using Packing Density," internal report, Advanced Refractory technologies, Inc., October, 1993.

8. M.D. Sacks, H-W. Lee, O.E. Rojas, "Suspension Processing of Al_2O_3/SiC Whisker Composites," *J. Am. Ceram. Soc.*, **71** [5] 370-379 (1988).

9. C.B. Ponton, R.D. Rawlings, "Vickers Indentation Fracture Toughness Tests: Part 1 Review of Literature and Formulation of Standardised Indentation Toughness Equations," *Materials Science and Technology*, **5**, September 1989, 865-872.

10. A.G. Solomah, L. Esposito, "Indentation Fracture of Silicon Carbide Whisker-Reinforced Silicon Nitride Ceramic Matrix Composites (CMC's)" *Ceramic Engineering and Science Proceedings; 16 the Annual Conference on Composites and Advanced Ceramic Material*, January 7-10, Cocoa Beach, FL, 712-721 (1992).

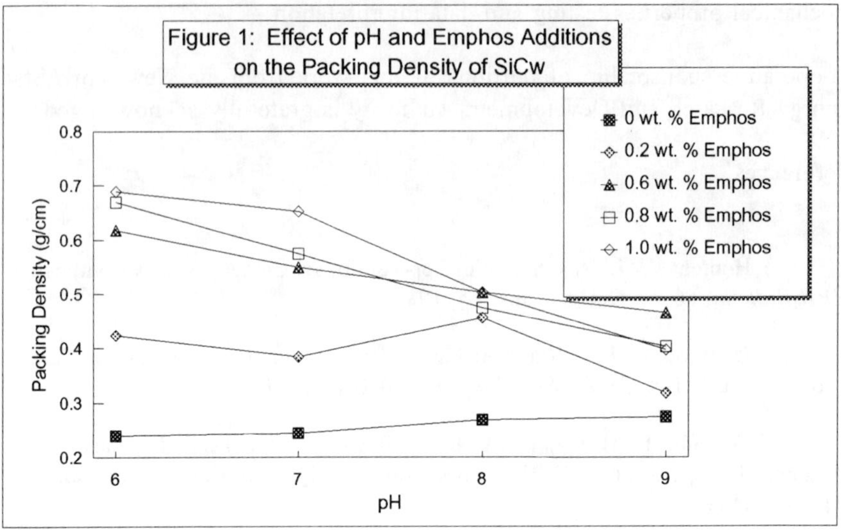

Figure 1: Effect of pH and Emphos Additions on the Packing Density of SiCw

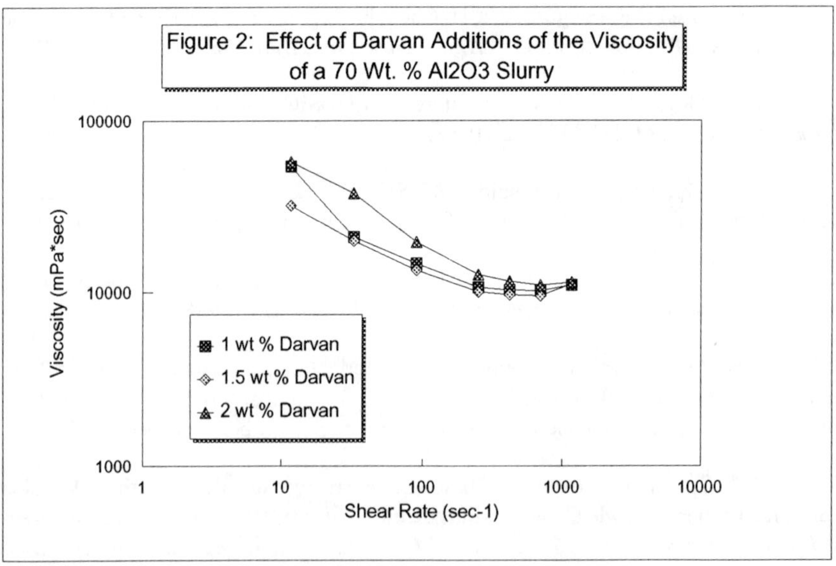

Figure 2: Effect of Darvan Additions of the Viscosity of a 70 Wt. % Al2O3 Slurry

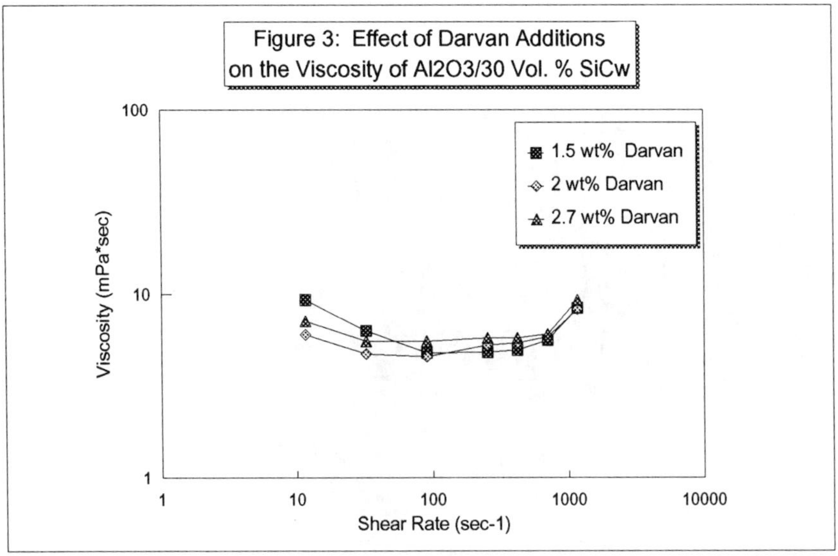

Figure 3: Effect of Darvan Additions on the Viscosity of Al2O3/30 Vol. % SiCw

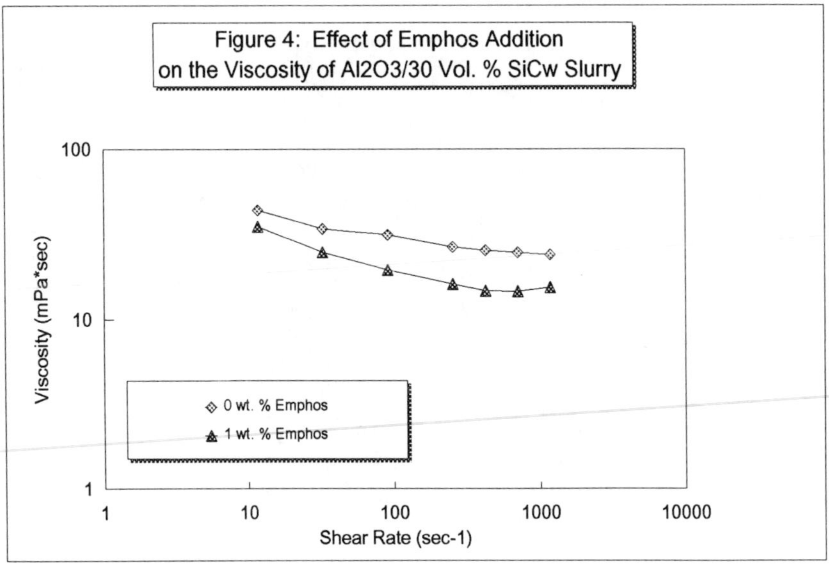

Figure 4: Effect of Emphos Addition on the Viscosity of Al2O3/30 Vol. % SiCw Slurry

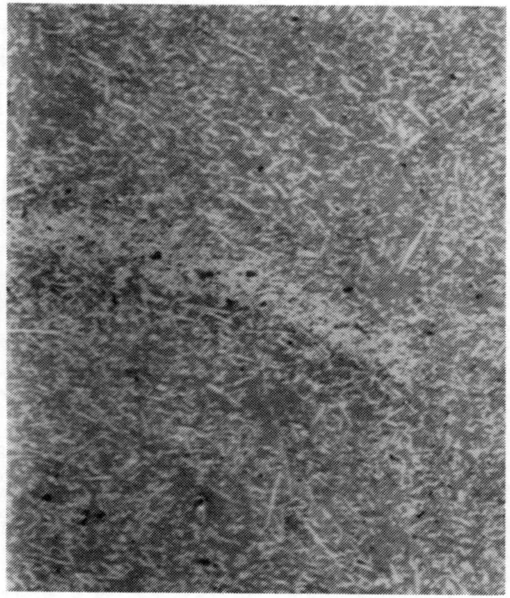

Optical Micrograph
Magnification = 500 X

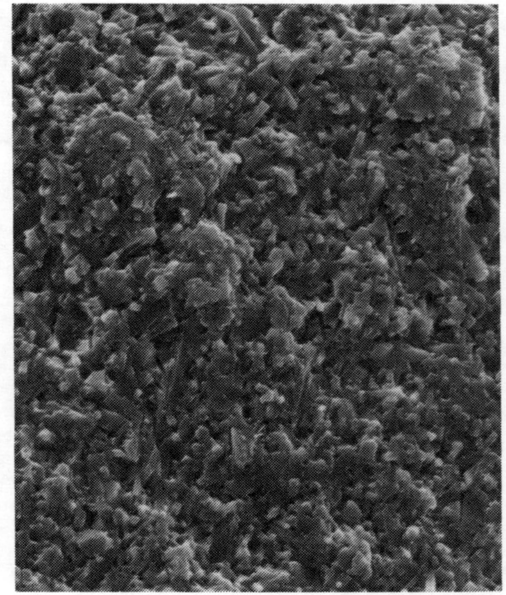

Scanning Electron Micrograph
Magnification = 1000 X

Figure 5: Microstructure of Hot Pressed Al_2O_3/30 vol. % SiCw Composite - 97 % Density

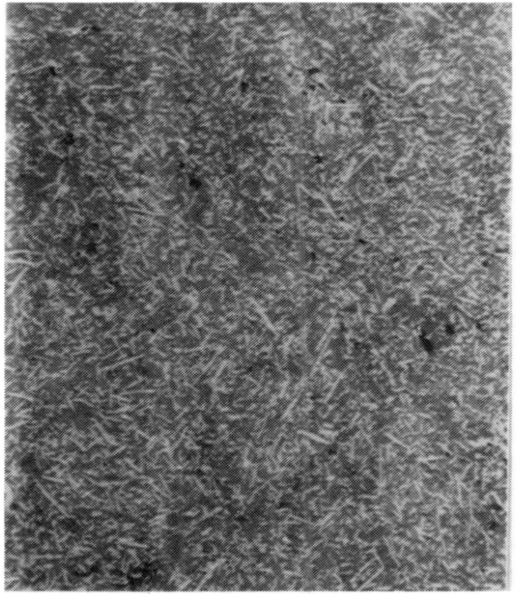

Optical Micrograph
Magnification = 500 X

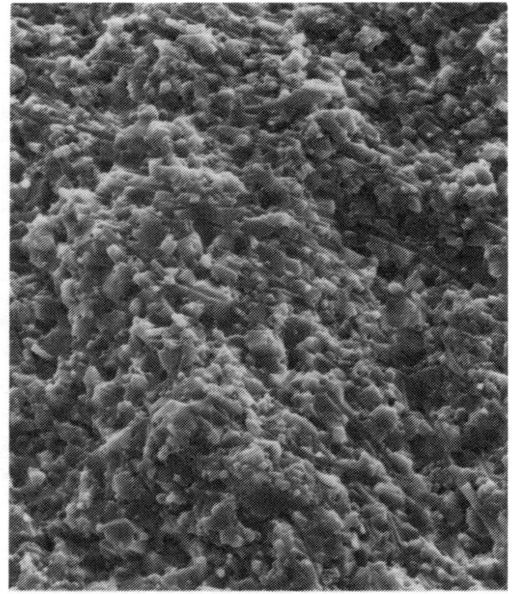

Scanning Electron Micrograph
Magnification = 1000 X

Figure 6: Microstructure of Hot Pressed Al_2O_3/30 vol. % SiCw Composite - 99 % Density

IMPROVEMENTS IN SiC_w/Al_2O_3 COMPOSITES THROUGH COLLOIDALLY STABILIZED SUSPENSIONS

Melissa J. Crimp
Department of Material Science and Mechanics
Michigan State University
East Lansing, MI 48824-1226

Dean A. Oppermann
Department of Material Science and Mechanics
Michigan State University
East Lansing, MI 48824-1226

Mingli Zhang
Department of Material Science and Mechanics
Michigan State University
East Lansing, MI 48824-1226

ABSTRACT

Through manipulation of colloidal parameters, suspensions of SiC(whisker)/ Al_2O_3 were prepared, at 5, 10 and 20 vol% SiC whisker, using processing conditions established in *Stable Suspension*©. Utilizing Hogg, Healy and Furstenau's modifications to DLVO theory, this program predicts stability conditions for composite suspensions. Variations in the suspension pH induce changes in the attractive/repulsive interactions between components. This type of interaction in turn influences the packing and green density. Composite suspensions were prepared, freeze dried, then cold consolidated. The distribution of the SiC whiskers within the Al_2O_3 matrix was determined from SEM micrographs and the composite green density correlated to the extent of homo- versus heterostability within the composite suspension. The green density of the pure Al_2O_3 and the 5

To the extent authorized under the laws of the United States of America, all copyright interests in this publication are the property of The American Ceramic Society. Any duplication, reproduction, or republication of this publication or any part thereof, without the express written consent of The American Ceramic Society or fee paid to the Copyright Clearance Center, is prohibited.

vol% SiC whisker additions was the highest at the pH of maximum stability for each interaction. In contrast, at whisker additions of 10 and 20 vol%, the green density is the highest at a pH of low heterostability.

INTRODUCTION

Design or tailoring of composite microstructures to achieve improved or new properties presents processing challenges. As with many other high-tech ceramics, limited or near zero porosity is commonly desired, along with a substantial degree of homogeneity. However, homogenous dispersions of the components and low porosity are difficult to achieve. Researchers at Oak Ridge National Laboratory have developed SiC whiskers in an Al_2O_3 matrix, made via hot pressing at 1850-1900°C. While achievement of homogeneous mixing was not fully realized at significant whisker volume fractions (e.g. 20%) [1], promising results and progress have been achieved using a relatively unagglomerated, 0.2μm diameter Al_2O_3 matrix.

Particle distribution and packing are considered critical in achieving quality ceramic products. Varying the pH, which changes the surface potentials of the components, may alter the distribution of materials. By using colloidal techniques, agglomerate-free, sintered compacts can be produced with increased strength and Weibull modulus [2]. More uniform packing improves sintering conditions, promoting sintering at lower temperatures and shorter times [3]. Studies have demonstrated these improvements by achieving >99% theoretical density at lower than normal sintering temperatures without the use of sintering aids which often degrade the mechanical properties of sintered ceramics [4]. Sintering of non-uniformly packed green bodies, however, results in fluctuation of localized densification rates which can lead to warpage, formation of microcracks, and a lack of dimensional control in sintered compacts [5]. Through increased control of the powder dispersion during processing, this study attempts to maximize the green density of cold consolidated Al_2O_3/SiC_w composites.

The behavior of particles in an aqueous suspension is controlled by electrical surface charges at the liquid/powder interface. The strength of these charges can be altered by changing the pH of the suspension which will, in turn, change the ion double layer distance. Charge differences can be evaluated by measuring the zeta potential of the suspension. Large ion double layer distances correspond to higher zeta potentials (> +/-25mV). These potentials create a significant repulsive force which leads to a stable suspension according to DLVO theory [6,7]. While this phenomena is well known for single component systems, the effects of a second component on system dynamics have not been widely evaluated. In multi-component systems, not only does the homostability of SiC whisker to SiC whisker and Al_2O_3 to Al_2O_3 need to be accounted for, but heterostability between SiC whisker and Al_2O_3 also becomes important.

For example, in a hetero-unstable state, a SiC whisker would be attracted to an Al_2O_3 particle and similarly, Al_2O_3 particles would be attracted to SiC whiskers each forming agglomerates. This is one example of two particle types, particulate and whisker, having a uniform dispersion with respect to each other. However, this hetero-unstable state would form green compacts having lower green densities than green compacts of components from completely stable, or unagglomerated, suspensions.

This research aims to determine ideal processing conditions for two-component whisker-powder suspensions. Conditions such as packing density, uniformity and particle interactions can be optimized by utilizing *Stable Suspension*© which employs modifications of the Hogg, Healy, and Fuerstenau refinement of DLVO theory [8] to predict the state of stability in the composite system. It is expected that optimizing the processing will improve the green density, sintered density and strength of a sintered ceramic composite.

EXPERIMENTAL PROCEDURE

Materials

The α-Al_2O_3 supplied by Sumitomo Industries, New York, type AKP30, has an average particle diameter, reported by the manufacturer, of 0.3μm. The β-SiC whisker supplied by Tokai Carbon Co. of Tokyo, Japan, type TWS100, has an average aspect ratio of 22.2 with a range of 8.3 to 50, having a whisker diameter of .3-.6μm and length .5-1.5 μm, as reported by the manufacturer.

Procedures

The zeta potential was collected for each material as a function of pH and used to calculate the stability ratios for homo- and heterocoagulation, $W_{SiC\text{-}SiC}$ (and $W_{Al2O3\text{-}Al2O3}$) and $W_{SiC\text{-}Al2O3}$, respectively [9,10]. Once the optimum pH for coating of unflocculated SiC whisker by unflocculated Al_2O_3 was established, two non-optimum pHs were chosen for contrast purposes. Suspensions were prepared, freeze dried, cold pressed at 15 MPa, and cold isostatically pressed at 275 MPa.

Freeze Drying

The freeze drying process begins by first shell freezing at -40°C followed by sublimation at 223 K and 0.5 Pa. The resulting particulate product was shaken for 5 minutes in a Nalgene bottle to disrupt the shell structure of the composite material. The approximate volume of total solids to liquid in freeze drying was 1:10 [11].

Zeta Potential Measurements

The zeta potential data for both the Al_2O_3 and the SiC whisker were collected in 0.001 N KNO_3 electrolyte and the pH adjusted by titrating either acid (HNO_3) or base (KOH).

Agglomerate Analysis

For each volume fraction and pH, green samples were broken, gold coated and the fracture surfaces examined in the SEM. From these micrographs, the individual SiC whiskers and Al_2O_3 particles were redrawn and plots of the number of Al_2O_3 particles per length of whisker were generated to quantitatively show any component distribution characteristics as a function of volume fraction SiC whisker and pH.

RESULTS AND DISCUSSION

Zeta potential versus pH curves at 0.001N KNO_3 for both Al_2O_3 and SiC whisker are shown in Figure 1. The iso-electric point (i.e.p.) of Al_2O_3 is approximately pH 8.5 and for SiC whisker, pH 6.0. Zeta potential data is input into *Suspension Stability*© as an approximation of the surface potential which is an integral assumption in predicting the composite stability. A graph of the stability calculations as a function of pH for the SiC_w/Al_2O_3 composite are shown in Figure 2. In order to show several different suspension conditions, three separate pHs were tested. Results are shown in Table 1 where stability is indicated by $\log(W) > 10$ [12].

A pH of 4 was chosen representing high stability for all interactions. A pH of 7 was chosen for low homostability of the Al_2O_3 to Al_2O_3, low homostability of the SiC whisker to SiC whisker, and high heterostability of Al_2O_3 to SiC whisker. A pH of 9 represents the case where Al_2O_3 agglomerates with itself and the SiC_w/Al_2O_3 interaction borders between stability and instability, while the SiC whisker remains unagglomerated. This situation indicates that Al_2O_3 particles coat

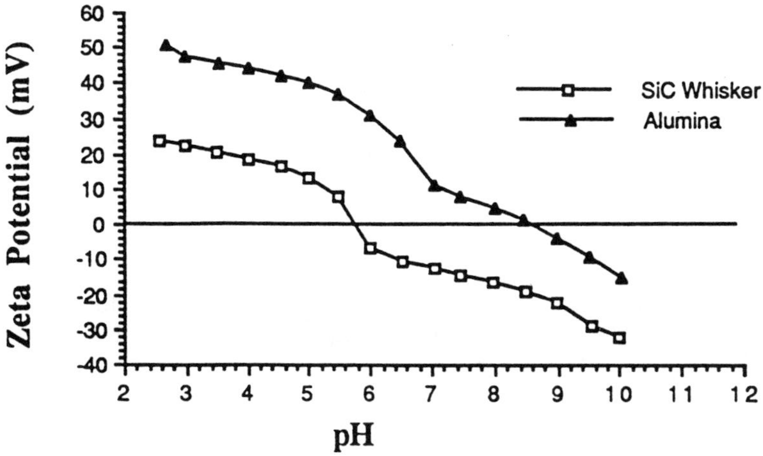

Figure 1: Zeta potential versus pH for SiC_w and Al_2O_3 in 10^{-3}N KNO_3.

individual SiC whiskers. Ching-li Hu et al made Al_2O_3/SiC whisker composites by free sintering of coated powders [13]. SiC whiskers were coated with a thick

Table 1. Dispersion state for SiC_w/Al_2O_3 composites as a function of pH.

Material \ pH	4	7	9
SiC_w/SiC_w	Stable	Unstable	Stable
Al_2O_3/Al_2O_3	Stable	Unstable	Unstable
SiC_w/Al_2O_3	Stable	Stable	Stable/Unstable

cladding of fine-grained Al_2O_3 powder by controlled heterogeneous precipitation in a concentrated suspension of whiskers. After calcination, a loose powder consisting of coated SiC whiskers, with a decreased aspect ratio (length to diameter ratio) was obtained. This lower aspect ratio increased the critical percolation volume fraction of whisker reinforcement and the efficiency of composite packing, producing higher green densities in the as-pressed composite [14,15]. The sinterability was thus significantly higher than that for similar composites formed from mechanically mixed powders. Therefore, in the current study, by causing controlled heterocoagulation, the aspect ratio was decreased and the composite properties followed the results of 14 & 15.

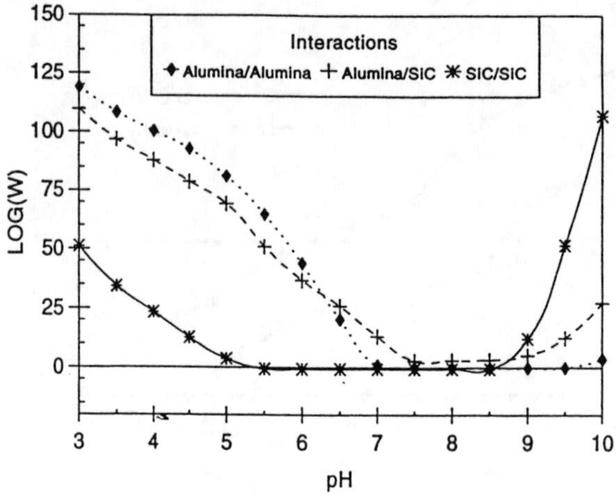

Figure 2: Stability calculations for SiC_w/Al_2O_3 composite suspensions as a function of pH.

The CIPed green density is plotted versus suspension processing pH for the pure Al_2O_3 as well as the 5, 10, and 20 vol% SiC_w/Al_2O_3 composites in Figure 3. For the unreinforced Al_2O_3, the maximum green density was found at pH 4 which is consistent with powder packing theory. For 5 vol% reinforced Al_2O_3, results also showed pH 4 to maximize the green density. In contrast, the 10 and 20 vol% SiC whisker reinforced Al_2O_3 achieved a higher green density at pH 9. The stability calculations at this pH indicate that SiC whiskers were coated by Al_2O_3 flocs.

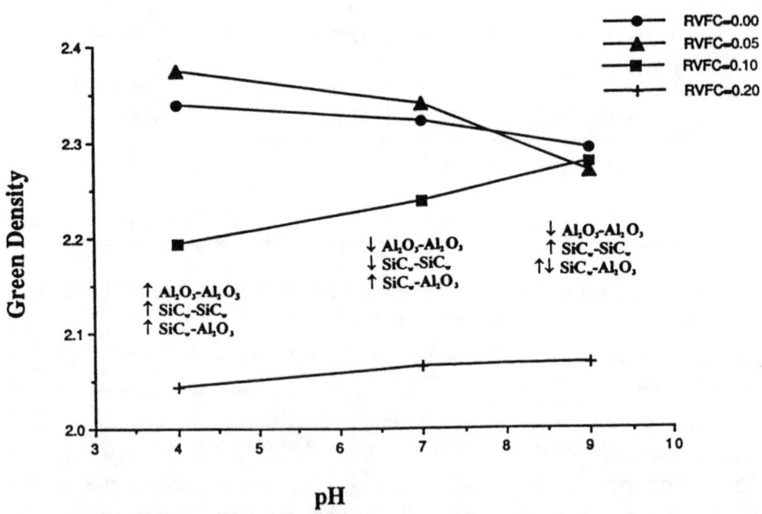

Figure 3: Freeze dried and cold pressed green density measurements as a function of pH for 0, 5, 10, 20 vol% SiC_w/Al_2O_3 composites.

These coated SiC whiskers have a decreased aspect ratio which serves to increase the packing and therefore increase the green density in agreement with the results of Ching-li Hu et al [13]. Since the pH 9 condition has the maximum density for the 10 and 20 vol% reinforcement cases and the pH 4 has the maximum density for the 0 and 5 vol% reinforcement cases, there must be a critical vol% where the optimum pH condition shifts. Below this critical vol% there is not enough whisker present to hinder the stable Al_2O_3 packing and above the critical vol% the highest density is where the whisker packing is maximized.

SEM photomicrographs of the 20 vol% SiC whisker reinforced Al_2O_3 composite at pH 4, 7, and 9 are shown in Figures 4-6, respectively. The SiC whiskers remained unaffected by the processing although there are no obvious differences in the distribution of the whisker versus pH. Therefore, analysis was done by tracing the outline of the SiC whiskers, counting the number of Al_2O_3 particles per μm length of whisker, and plotting these results versus pH as shown in Figure 7. From this data, the highest incidence of particles in contact with SiC whiskers occurred in the composite processed at pH 9. This is in agreement with the stability calculation predictions.

While the highest density in whisker particle systems was achieved at intermediate

Figure 4: SEM micrographs of 20 vol% SiC_w/Al_2O_3 processed as pH 4.

Figure 5: SEM micrographs of 20 vol% SiC_w/Al_2O_3 processed at pH 7.

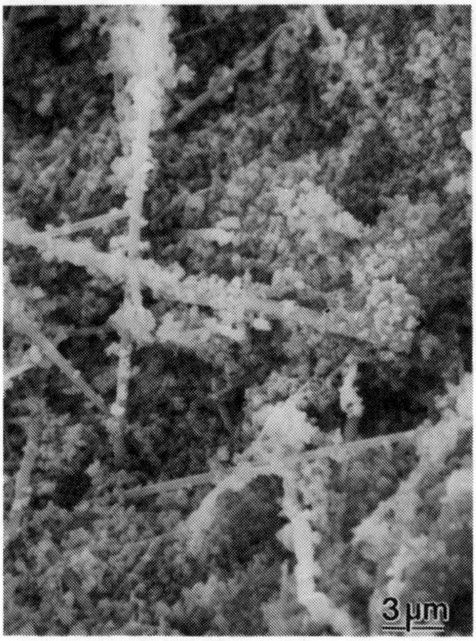

Figure 6: SEM micrograph of 20 vol% SiC_w/Al_2O_3 processed at pH 9.

volume fractions under conditions where powder flocs coated the reinforcement whiskers, particle-particle systems reached maximum density when stability was maximized for all interaction cases [16]. In contrast, when processing continuous fiber reinforced composites, improvements in both the fiber distribution and the matrix coating are obtained where stability between the matrix and fiber is minimized [17].

CONCLUSIONS

For 10 and 20 vol% reinforced SiC_w/Al_2O_3 composites, maximum green densities occurred when agglomerated Al_2O_3 coated the individual SiC whiskers. This maximum density occurred at pH 9, where the stability calculations indicated

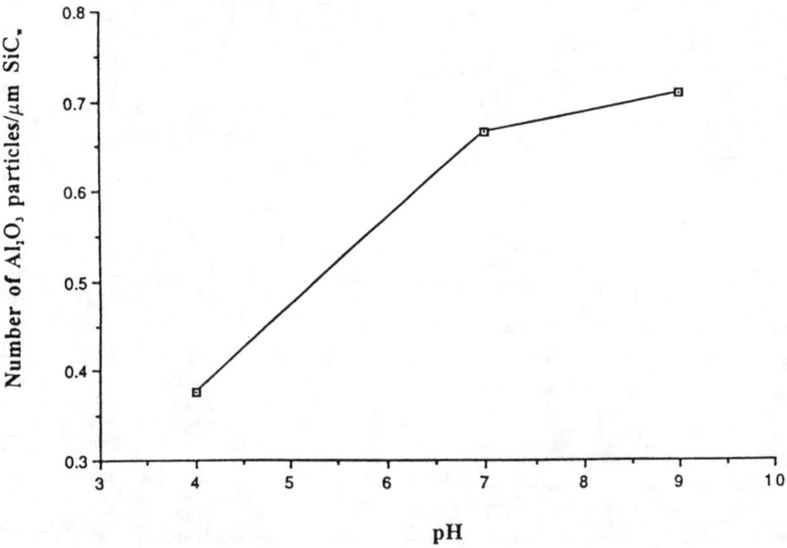

Figure 7: Coverage of SiC whisker by Al_2O_3 particles as a function of pH.

maximum stability for the SiC whisker and SiC_w/Al_2O_3 with low stability for Al_2O_3. The unreinforced and the 5 vol% reinforced Al_2O_3, although exhibiting identical stability values, reached maximum density at pH 4. These differences may be explained by the combined effects of particle packing and particle agglomeration.

ACKNOWLEDGEMENTS

This work was sponsored by the State of Michigan through the Michigan State University Research Excellence Funds and the 3M Corporation.

REFERENCES

1. P. R. Becher and G. C. Wei, Comm. J. Am. Ceram. Soc., "Toughening Behavior in SiC-Whisker-Reinforced Alumina", **67**, C267-C269 (1984).

2. Kulig, M. and Greil, P., "Surface Chemistry and Suspension Stability of Oxide-Nitride Powder Mixtures", *J. Mater. Sci.*, **26,** 216 (1991).

3. A. Roosen and H. K. Bowen, "Influence of Various Consolidation Techniques on the Green Microstructure and Sintering Behavior of Alumina Powders", *J. Am. Ceram. Soc.*, **71**, 970 (1988).

4. E. A. Barringer and H. K. Bowen, "Formation, Packing and Sintering of Monodisperse TiO_2 Powders", *Comm. Am. Ceram. Soc.*, **65**, C-199, (1982).

5. A. Pechenik and G. J. Piermarini, "Fabrication of Transparent Silicon Nitride from Nanosize Particles", *J. Am. Ceram Soc.,* **75**, 3283 (1992).

6. B. V. Deraguin and L. D. Landau, "Theory of the Stability of Strongly Charged Lyophobic Sols and the Adhesion of Strongly Charged Particles in Solutions of Electrolytes", *Acta Physichim., URSS*, **14**, 633 (1941).

7. E. J. W. Verwey and J. Th. G. Overbeek, Theory of the Stability of the Lyophobic Colloids, Elsevier, Amsterdam (1948).

8. R. Hogg, T. W. Healy and D. W. Fuerstenau, "Mutual Coagulation of Colloidal Dispersions", *Trans. Faraday Soc.*, **62**, 1638 (1966).

9. B. A. Wilson and M. J. Crimp, "Prediction of Composite Colloidal Suspension Stability Based Upon the Hogg, Healy, Fuerstennau Interpretation", *Langmuir*, **9**, 2836, (1993).

10. B. A. Wilson, Master of Science Thesis, "Prediction of Colloidal Suspension Stability for SiC/Si_3N_4 and $FeAl/Al_2O_3$ Fiber Systems Using Material and System Parameters", Michigan State University, Results, 65, (1992).

11. M. Real, *Proc. Brit. Ceram. Soc.*, **38**, 59 (1986).

12. R. J. Hunter, Foundations of Colloid Science: Volume 1, Clarendon Press, Oxford, (1991).

13. C-L. Hu and M. N. Rahaman, "SiC-Whisker-Reinforced Al_2O_3 Composites by Free Sintering of Coated Powders", *J. Am. Ceram. Soc.*, **76**, 2549 (1993).

14. E. A. Holm and M.J. Clima, "Two-Dimensional Whisker Percolation in Ceramic Matrix-Ceramic Whisker Composites", *J. Am. Ceram. Soc.*, **72**, 303-305 (1989).

15. T. N. Tiegs and P.F. Becher, "Sintered Al_2O_3-SiC-Whisker Composites", *Am. Ceram. Soc. Bull.*, **66**, 339-342 (1987).

16. D. A. Oppermann and M. J. Crimp, "Green Density Variations in Relation to Colloidal Packing in SiC/Si_3N_4 Composites", *Proc. of the Ceramic Matrix Composites Symposium: American Ceramic Society Annual Meeting* (1994).

17. B. A. Wilson, C. J. Suydam, M. J. Crimp and M. A. Crimp, High Temperature Intermetallic Alloys V, I. Baker, R. Darolia and D.J. Whittenberger eds., MRS Proceedings, **288**, 891-896 (1993).

THE EFFECT OF MICROSTRUCTURE ON MECHANICAL PROPERTIES OF ALUMINA-BASED COMPOSITES BY PARTICULATE-PRECIPITATED DISPERSION

Byung-Koog Jang, Manabu Enoki and Teruo Kishi
Research Center for Advanced Science and Technology, The University of Tokyo
4-6-1, Komaba, Meguro-ku, Tokyo 153, Japan

ABSTRACT

Alumina-based ceramic composites which second phases have a microstructure of equiaxed grains as YAG or elongated grains as $LaAl_{11}O_{18}$ by reaction of Al_2O_3 and additives during hot-pressing, were fabricated using particulate-precipitated dispersion. The Al_2O_3/SiC/YAG hybrid composites having the equaxied second phase were fabricated in the temperature range from 1000 to 1800°C using SiC and Y_2O_3 powders as additive. Then, YAG (yttrium aluminum garnet, $Y_3Al_5O_{12}$) phase was formed as the second phase from the reaction between Al_2O_3 and Y_2O_3 above 1400°C hot-pressing temperature. The grain size of monolithic Al_2O_3 increased with increasing hot-pressing temperature and hot-pressing time, while that of Al_2O_3/SiC/YAG hybrid composites was effectively restrained due to the incorporation of YAG and SiC particulate into Al_2O_3 matrix. Also, Al_2O_3/LBA(lanthanum-β-alumina, $LaAl_{11}O_{18}$) composites having an elongated second phase were successfully fabricated using Al_2O_3powder as additives. The properties of hot-pressed bodies, such as microstructural observation by SEM & TEM, phase analysis by XRD, strength, toughness, toughening mechanism etc., were investigated. The mechanical properties of flexural strength and fracture toughness of composites were higher than that of monolithic Al_2O_3. Also, the composites having elongated grains showed higher toughness due to grain bridging than the composites having equaxied second phase.

1. INTRODUCTION

Generally, Al_2O_3 ceramics have excellent properties such as easy sintering, high hardness, low electrical conductivity, oxidation resistance, good chemical stability, etc.(1). Many studies have been focused on particulate-dispersed Al_2O_3

To the extent authorized under the laws of the United States of America, all copyright interests in this publication are the property of The American Ceramic Society. Any duplication, reproduction, or republication of this publication or any part thereof, without the express written consent of The American Ceramic Society or fee paid to the Copyright Clearance Center, is prohibited.

composites in order to improve both room and high temperature mechanical properties[2],[3]. Among the fabricating Al_2O_3-based ceramic composites which were dispersed particulate, whisker and platelet, particulate reinforcement becomes more attractive for achieving fully dense sintered bodies as well as enhancing the mechanical properties by controlling microstructural development[4]. The attractiveness of these composites is not only the microstructural control of matrix, but also the possibility of enhancement of mechanical properties. For alumina-based composites, it can be expected that the morphology of Al_2O_3 grains influences the mechanical properties, thus, the microstructural control of composites is very important. In this present work, microstructural controlled composites composed of either equiaxed grains or elongated grains as the second phase in a ceramic-matrix was demonstrated. This work discusses the processing and fabrication of the composites by simultaneous consolidation of micro and nano particulates of equiaxed grains as well as elongated grains dispersed into the matrix. Evaluation of microstructural characteristics, mechanical properties and fracture behavior of composites of Al_2O_3-based composites are compared to monolithic Al_2O_3.

2. EXPERIMENTAL PROCEDURE

2-1. Fabrication of Al_2O_3/SiC/YAG Hybrid Composites

Starting powders were high-purity Al_2O_3 powder with average particle size of 0.2μm, ultrafine ß-SiC powder with average particle size of 0.3μm and a Y_2O_3 power with a purities of 99.9%. α-Al_2O_3 as the matrix, 5vol% ß-SiC, 20vol% YAG phase as dispersed phase was considered as a model composites composition. This composition was homogenized by ball milling in methanol solution using a polyethylene jar and Al_2O_3 balls for 24h. After ball milling, the slurry was dried. The mixed powders were uniaxially hot pressed at the temperature range of 1000 to 1800°C for 0-20h under 30 MPa pressure in Ar atmosphere.

2-2. Fabrication of Al_2O_3/$LaAl_{11}O_{18}$ Composites

Al_2O_3 powder used as the matrix and La_2O_3 powder as the additive. This composition was homogenized by ball milling for 24h in methanol solution using polyethylene mill and Al_2O_3 balls. After ball milling, the slurry was dried and then, sieved through 200 mesh. The mixed powders were uniaxially hot pressed in graphite dies coated with BN-spray in the temperature range of 1400 to 1800°C for 1h under 30 MPa pressure in an Ar atmosphere. The $LaAl_{11}O_{18}$ phase precipitated as the second phase from the reaction of Al_2O_3 and La_2O_3. The studied composition range for composites was 0 to 20 vol% of $LaAl_{11}O_{18}$ as a second phase.

2-3. Evaluation of Properties

The hot-pressed specimens were cut to nominally 3 mm by 3 mm by 40 mm with diamond blades and polished with 3 μm diamond pastes and finally with 1μm size diamond pastes to obtain a mirror surface. The experimental densities were determined by the Archimedes technique using room-temperature water. The phases in the hot-pressed bodies were identified from X-ray diffraction (XRD) patterns. For evaluating the grain structure, thermal etching was performed at 1500°C in an Ar atmosphere. For microstructural analysis, the polished and thermally etched surfaces of specimens were observed using scanning electron microscopy (SEM) and transmission electron microscopy (TEM). TEM samples were prepared using conventional a thinning procedure including dimple grinding and finally argon ion milling. Flexural strength was conducted by the three-point bending test. Fracture toughness was determined using the Vickers indentation microfracture (IM) method and the single edge precracked beam (SEPB) method. In order to observe the crack behavior, a Vickers diamond pyramid was used to introduce a controlled crack generation on the polished surfaces. The behavior of crack propagation was investigated by SEM, TEM observation.

3. RESULTS AND DISCUSSION

3-1. Al_2O_3/SiC/YAG Hybrid Composites with Equaxied Microstructure

3-1-1. Fabrication and Sintering Behavior

The relative density of hot pressed monolithic Al_2O_3 and Al_2O_3/SiC/Y_2O_3 materials as a function of hot-pressing temperature are given in Figure 1. Both monolithic Al_2O_3 and Al_2O_3/SiC/Y_2O_3 system show increased density in the range of 55% to ≥99% with increasing hot-pressing temperature. The relative density of the monolithic Al_2O_3 hot-pressed at 1000°C is about 65% and that of the Al_2O_3 hot-pressed at 1200°C is 95% with no existence of open pores. Additionally, highly dense Al_2O_3 of ≥99% could be obtained beyond 1400°C. On the other hand, for Al_2O_3/SiC/Y_2O_3 system, the relative densities of composites hot-pressed below 1400°C were relatively low compared to monolithic Al_2O_3. However, the remarkable densification was occurred between 1400 and 1600°C. It seemed to be due to the formation of YAG phase as explained in Fig.2. These hot-pressed bodies reached near full density above 1600°C. This results mean that sintering of this composites should be carried out at higher temperature to obtain full density in comparison with monolithic Al_2O_3. The reason can be attributed to the very poor sinterability of SiC particulates[5].

X-ray diffraction results are presented in Fig. 2. From XRD analysis, Al_2O_3, SiC and Y_2O_3 phases were observed in the specimen hot-pressed at 1200°C in Fig. 2(B), whereas the Y_2O_3 phase disappeared above 1400°C, and instead, YAG phase, resulting from the reaction between Y_2O_3 and Al_2O_3 was detected and no evidence of any other phase was found. The solid-state reaction for formation of YAG phase can be described by $5Al_2O_3 + 3Y_2O_3 \rightarrow 2Y_3Al_5O_{12}$.

It is obvious that the YAG phase was successfully formed from the reaction between Al_2O_3 and Y_2O_3 above 1400°C as shown in Fig. 2(A). The phase diagrams for the Y_2O_3-Al_2O_3 system indicate a good solubility of yttria in Al_2O_3[6]. Figure 3 shows the TEM observation of Al_2O_3/SiC/YAG composites

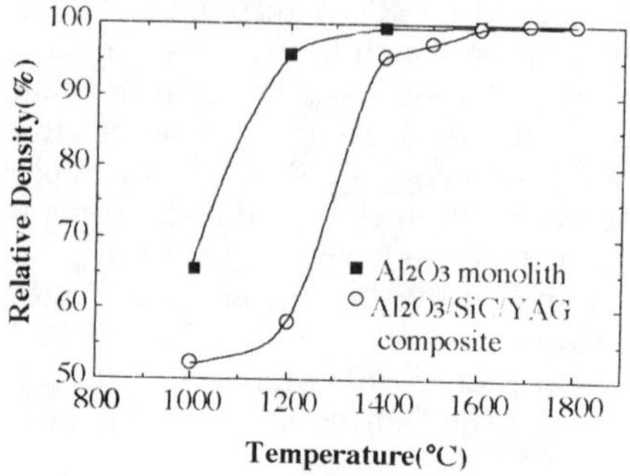

Fig. 1. Relative density of hot-pressed monolithic Al_2O_3 and Al_2O_3/SiC/YAG composites versus hot-pressing temperature

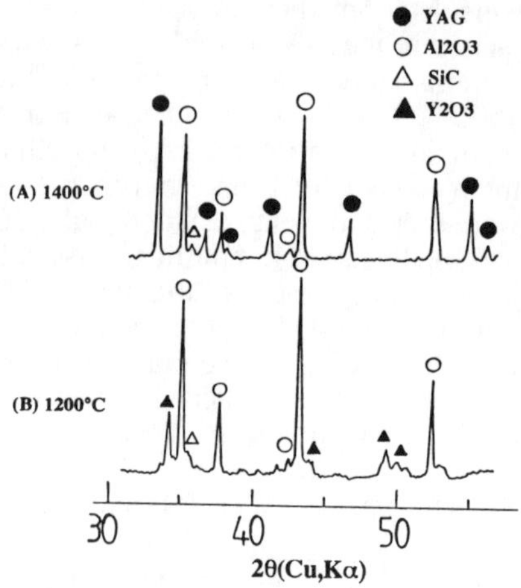

Fig.2. XRD analysis of hot-pressed composite of Al_2O_3/SiC/Y_2O_3 system

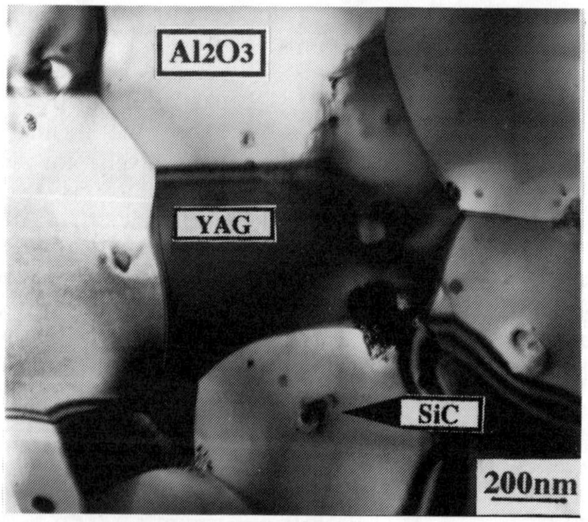

Fig.3. TEM micrograph of Al_2O_3/SiC/YAG composites

hot-pressed at 1600°C. As shown in TEM figure, the microstructural results were consistent with three phases, that is, the grains of bright contrast in center are Al_2O_3, while grains with dark contrast are YAG phase locating in Al_2O_3 grain boundaries, and nano-sized SiC particulates are homogeneously dispersed in intragranular or intergranular sites in both Al_2O_3 matrix and YAG grains or grain boundaries, respectively. Therefore, a hybridized-microstructural material is composed of micro-sized YAG and nano-sized SiC as a second phase is confirmed.

3-1-2. Characterization of Properties

Figure 4 shows the average grain size of the hot-pressed monolithic Al_2O_3 and Al_2O_3/SiC/YAG composites hot-pressed at 1600°C. The average grain size of the Al_2O_3 matrix of Al_2O_3/SiC/YAG composites was smaller than that of the monolithic Al_2O_3. The grain size of monolithic Al_2O_3 increased markedly with increasing hot-pressing time, whereas that of Al_2O_3 matrix in the Al_2O_3/SiC/Y_2O_3 system was comparatively restrained under the same hot-pressing condition. For monolithic Al_2O_3, the significant grain growth, resulting from abnormal grain growth, was observed above 10h of hot-pressing time. On the other hand, the significant grain growth of Al_2O_3 matrix for Al_2O_3/SiC/YAG composites does not rise even at 10h of hot-pressing time. In addition, the distribution of Al_2O_3 grain size in Al_2O_3/SiC/YAG composites is narrower, and the average grain size is smaller than those of monolithic Al_2O_3. Such microstructural properties are expectd to affect the mechanical properties. Figure 5 shows the relationship of flexural strength and matrix grain size of hot-pressed bodies. It appears that the flexural strength was proportional to the inverse of square root of matrix grain as described in the Hall Petch equation[7]. Flexural strength of the composites are higher

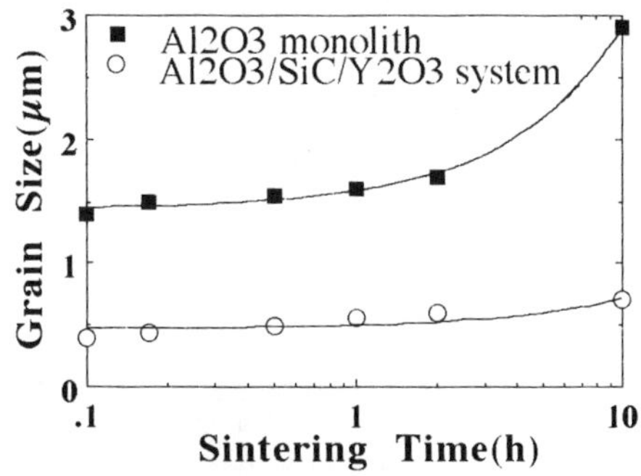

Fig. 4. Averaged grain size versus hot-pressing time of hot-pressed bodies at 1400°C

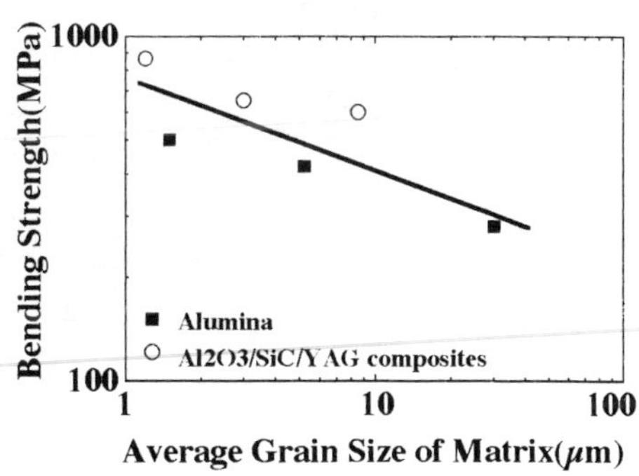

Fig. 5. Flexural strength dependence of grain size of monolithic Al_2O_3 and Al_2O_3/SiC/YAG composites

than that of monolithic Al_2O_3. The reason for improvement of flexural strength in composites is attributed to reductional the critical flaw size due to suppression of grain growth by the distribution of nano-sized SiC particulates and micro-sized YAG particulates. The significant decrease of critical flaw size in composites is probably due to disturbance of moving grain boundaries of Al_2O_3 grains. It has previously been reported that the strength of Al_2O_3 decreases with increased grain size[8].

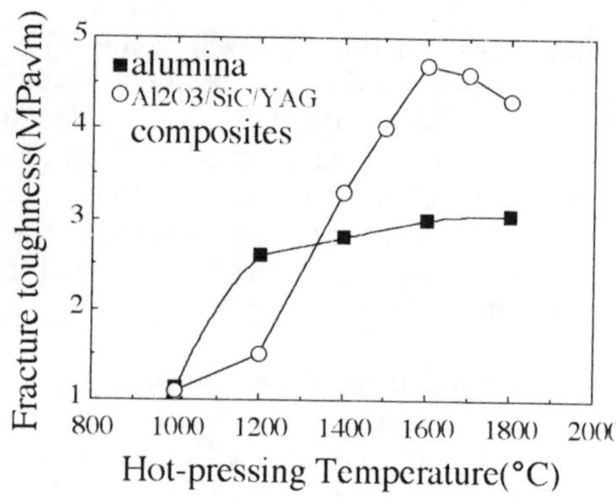

Fig.6. Fracture Toughness of hot-pressed monolithic Al_2O_3 and Al_2O_3/SiC/YAG composites versus hot-pressing temperature

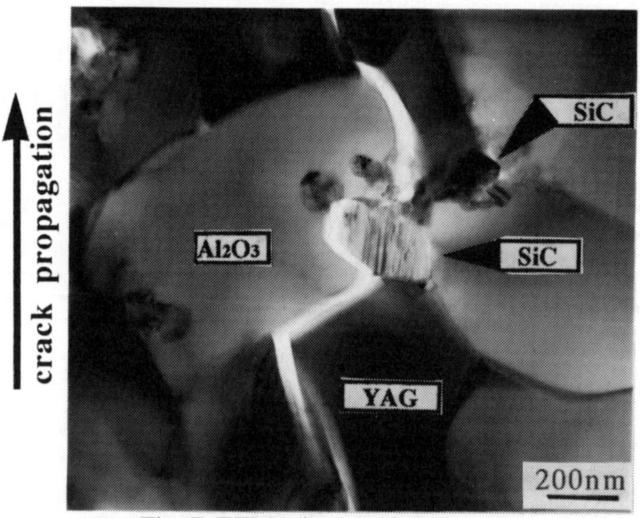

Fig. 7. TEM micrograph of crack deflection of Al_2O_3/SiC/YAG composites

Figure 6 shows the fracture toughness as a function of hot-pressing temperature. Fracture toughness of Al_2O_3/SiC/YAG composites increased with hot-pressing temperature because of increase of density, and reached approximately maximum value at 1600°C. It appears that lower fracture toughness below 1600°C is due to high porosity because of insufficient densification. Decrease of fracture toughness above 1600°C may be caused by internal residual stresses occurred during cooling. The overall fracture toughness of composites is higher than that of monolithic Al_2O_3. It is expected to exist any toughening mechanism, such as crack deflection by reinforcing SiC phases in Al_2O_3/SiC/YAG composites. Therefore, we try to observe crack propagation by TEM observation. Figure 7 shows the behavior of crack propagation which was introduced at the corners of a Vickers indentor. Propagating cracks were deflected at nano-sized SiC and micro-sized YAG particulates, which results in a complicated crack propagation and leads to a higher fracture energy. Thus, this crack deflection may be easily generated owing to disturbance of crack propagation at the dispersed second phases into matrix. It was reported that fracture toughness improves due to toughening by crack deflection at second phase in particulate-dispersed composites[9]. This behavior of crack deflection is expected to improve the fracture toughness.

3-2. Al_2O_3/$LaAl_{11}O_{18}$ Composites with Elongated Microstructure

3-2-1. Fabrication and XRD Analysis of Composites

Figure 8 shows the XRD analysis of composites hot-pressed at various temperature. In the case of samples hot-pressed at 1400°C, the $LaAlO_3$ phase were detected as a second phase, whereas in case of hot-pressed above 1500°C, $LaAl_{11}O_{18}$ phase, resulting from the reaction between Al_2O_3 and La_2O_3. The formation of $LaAl_{11}O_{18}$ phase can be described by $11Al_2O_3 + La_2O_3 \rightarrow 2LaAl_{11}O_{18}$. Figure 6 shows the grain structure of thermally etched surfaces of Al_2O_3/20vol%$LaAl_{11}O_{18}$ composites as compared with monolithic Al_2O_3. The Al_2O_3 grains are only equiaxed grains. However, the $LaAl_{11}O_{18}$ second phase of the composite show an elongated structure described as needle-like. This means $LaAl_{11}O_{18}$ phase have the hexagonal type occurred grain growth of special orientation. In addition, Al_2O_3 grains remained small in comparison with monolithic Al_2O_3. It can be explained that the grain growth of Al_2O_3 was restrained due to the existence of the $LaAl_{11}O_{18}$ second phase. Figure 10 shows the result of relative density as a function of hot-pressing temperature. The relative density is higher despite the 1400°C hot-pressing temperature, and increases with increasing hot-pressing temperature. Composites of elongated second phases are enerally difficult to obtain specimens having full density because of bad gsinterability[10],[11]. From density results, $LaAl_{11}O_{18}$ second phase seems to have good sinterability as well as elongated microstructure.

3-2-2. Mechanical Properties

Figure 11 shows the relationship of strength and precipitated LBA amounts for Al_2O_3/LaAl11O18 composites. The strength of the composite increases with increasing LBA amounts up to 10 vol%. Overall strength of the composite hot-

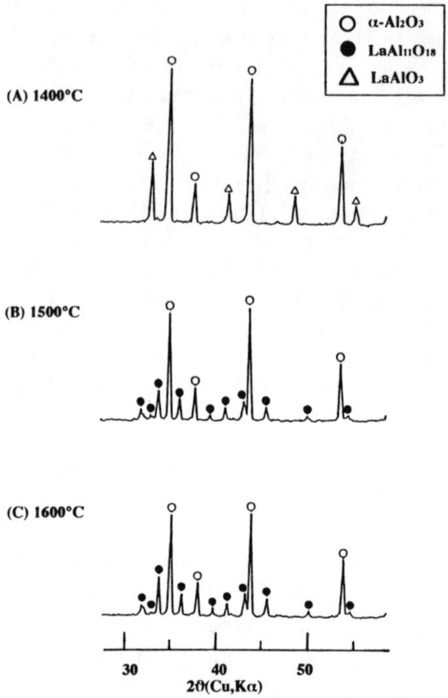

Fig.8. XRD analysis of hot-pressed bodies for Al_2O_3/La_2O_3 composition

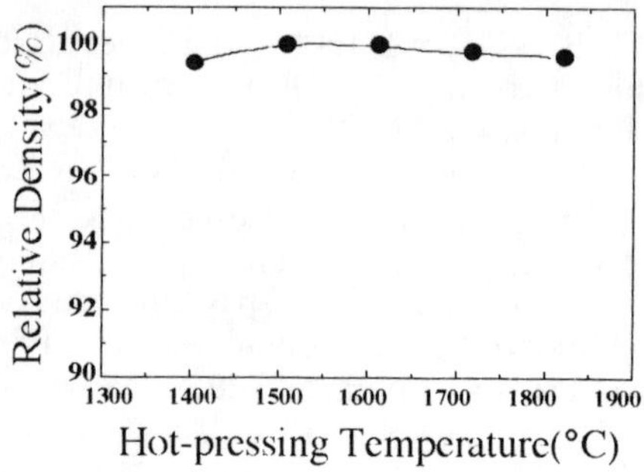

Fig. 10. Relative density of Al_2O_3/20vol%LBA composites

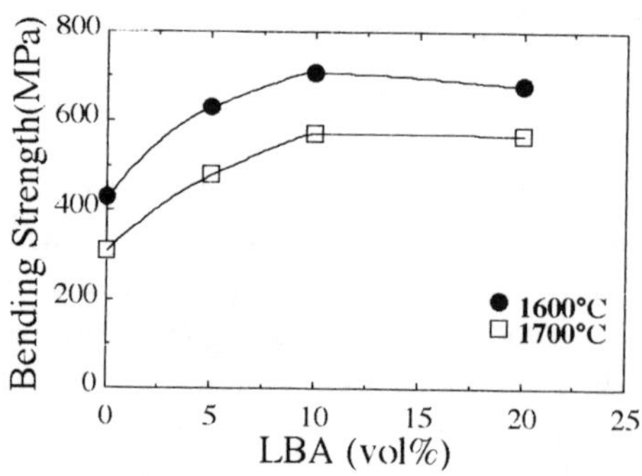

Fig. 11. Flexural strength versus LBA amounts of Al_2O_3/LBA composites

Fig. 9. SEM micrograph of Al_2O_3/LBA composites

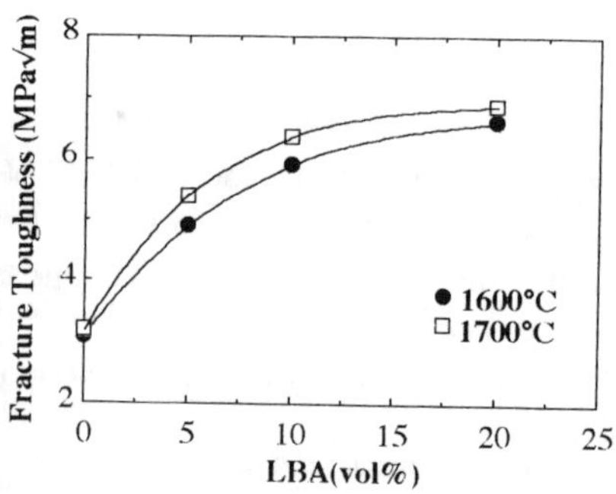

Fig. 12. Fracture toughness versus LBA amounts of Al_2O_3/LBA composites

pressed at 1700°C is lower than that at 1600°C. The reason contribute to the grain growth of Al_2O_3 matrix and/or second phase, that is, aspect ratio of LBA phase is

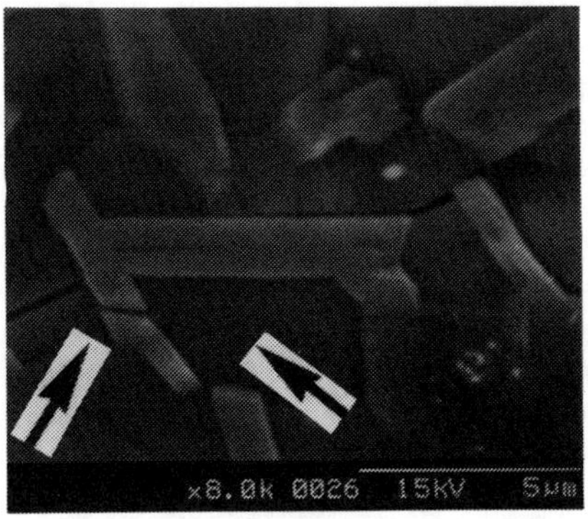

Fig. 13. SEM micrograph of crack propagation of Al_2O_3/20vol%LBA composites

larger with increased hot-pressing temperature. This is probability because the higher aspect ratio of the LBA phase may cause a larger critical flaw size. Generally, the strength of brittle ceramics is closely related to the critical defect size; higher strengths result from smaller critical flaw size[(12),(13)]. Figure 12 shows the relationship of fracture toughness and precipitated LBA amounts for $Al_2O_3/LaAl_{11}O_{18}$ composites. Fracture toughness increased with increasing temperature as well as LBA amounts. The improving tendency in fracture toughness corresponds with the increased aspect ratio of the LBA phase. It can be explained that an elongated structure having a larger aspect ratio is necessary to acquire a higher fracture toughness[(14)]. Thus, the formation of $LaAl_{11}O_{18}$ phase effectively enhanced the fracture toughness of the studied composites. It can be considered that such the enhanced toughness is relation to grain bridging at elongated LBA phase. In order to investigate the toughening mechanism, crack propagation was observed by SEM observation. Figure 13 shows the behavior of crack propagation which it was introduced at the corners of the Vickers indentor on thermally etched specimens. Propagating cracks show the deflected tendency and crack bridging at the elongated LBA second phase. At a higher volume fraction of LBA second phase, such crack bridging was frequently observed. The toughening effect of crack bridging of second phase is similar to that observed in Al_2O_3/SiC whisker composites[(15),(16)]. It can be considered that when a propagating crack undergoes crack deflection or crack bridging at second phases, improvement of toughness is achieved through the diminished stress intensity at crack tip, i.e., reduction of the crack driving force. Thus, improvement of toughness for ceramic composites can be expected from this toughening mechanism.

4. CONCLUSIONS

Fabrication of particulate-precipitated composites and influence of the morphology of the precipitated second phase on the mechanical properties and toughening mechanism have been investigated. Based on the present work, the following conclusions can be drawn.

(1) Al_2O_3/SiC/YAG hybrid composites having equiaxed structure, which consists of YAG with a micro-size particulate and SiC particulate with a nano-size were successfully obtained. Then, YAG (Yttrium aluminum garnet) phase as a second phase was formed by the reaction between Al_2O_3 and yttria in the Al_2O_3/SiC/Y_2O_3 composition. The distribution of alumina grain size of Al_2O_3/SiC/YAG composites is narrower, and for averaged grain size is smaller than those of monolithic Al_2O_3. Consequently, the grain growth of Al_2O_3 matrix grains in the Al_2O_3/SiC/YAG hybrid composites was effectively restrained. The mechanical properties, such as flexural strength and fracture toughness of Al_2O_3/SiC/YAG hybrid composites are higher than that of monolithic Al_2O_3. The improvements of mechanical properties are due to finely uniform microstructure and toughening mechanism providing crack deflection at nano-sized SiC and micro-sized YAG particulates.

(2) Al_2O_3/$LaAl_{11}O_{18}$ composites having elongated structure were successfully fabricated. Then, $LaAl_{11}O_{18}$ (lantanum-β-alumina) phase as a second phase was formed by the reaction between Al_2O_3 and La_2O_3 in the Al_2O_3/La2O3 composition above 1500°C hot-pressing temperature. Strength and toughness of composites increased with increasing the fraction of second phases. Fracture toughness increased with the increased aspect ratio of second phase. Improvements of fracture toughness attributes to toughening mechanism, that is, grain bridging and/or crack deflection at the elongated second phases.

REFERENCES

1. H.K. Bowen, “Ceramics as Engineering Materials: Structure-Property-Processing,” Mater.Res.Soc.Symp.Proc., **24** 1-11 (1984)

2. M. P. Harmer, H. M. Chan, and G. A. Miller, "Unique Opportunities for Microstructural Engineering with Duplex and Laminar Ceramic Composites," J. Am. Ceram. Soc., **75**[7] 1715-28 (1992)

3. A. G. Evans, "Perspective on the Development of High-Toughness Ceramics," J. Am. Ceram. Soc., **73**[2] 187-206 (1990)

4. D.J. Green, “Critical Microstructures for Microcracking in Al_2O_3-ZrO2 Composites,” J.Am.Ceram.Soc., **65**[12]610-14 (1982)

5. .S. Shinozaki, R.M. Williams, B.N. Juterbock, W.T.Donlon, “Microstructure Developements in Pressureless-Sintered ß-SiC Materials with Al, B and C additions,” Am.Ceram.Soc.Bull., **64**[10]1389-93 (1985)

6. E.M. Levin, C.R. Robbins, and H.F. McMurdie, Phase Diagrams for Ceramists, 1964. Edited by M.K. Reser. American Ceramic Society, Columnbus, O.H; Fig. 311.
7. J. H. Bucher, J. D. Grozier, and J. F. Enrietto, "Strength and Toughness of Hot-Rolled Ferrite-Pearlite" ; pp. 253-54 in Fracture, Vol. IV. Edited by H. Liebowitz. Academic Press, New York, 1969
8. P. Chantikul, S. J. Bennison, and B. R. Lawn, "Role of Grain Size in the Strength and R-Curve Properties of Alumina," J. Am. Ceram. Soc., **73**[8] 2419-27 (1990)
9. K T. Faber and A. G. Evans, "Crack Deflection Processes-II. Experiment," Acta Metall. **31**[4] 577-84 (1983)
10. Y. S. Chou, and D. J. Green, "Silicon Carbide Platelet/Alumina Composites: I. Effect of Forming Technique on Platelet Orientation," J. Am. Ceram. Soc., **75** [12] 3346-52 (1992)
11. C. Nischik, M. M. Siebold, N. A. Travitzky, and N. Claussen, "Effect of Processing on Mechanical Properties of Platelet-Reinforced Mullite Composites," J. Am. Ceram. Soc., **74**[10] 2464-68 (1991)
12. N. Padture and H. M. Chan, "Improved Flaw Tolerance in Alumina-1vol% Anorthite via Crystallization of the Intergranular Glass," J. Am. Ceram. Soc., **75** [7] 1870-75 (1992)
13. C. J. Russo, M. P. Harmer, H. M. Chan, and G. A. Miller, "Design of a Laminated Ceramic Composite for Improved Strength and Toughness"; presented at 93rd Annual Meeting of the American Ceramic Society, Cincinnati, OH, May 2, 1991
14. P. F. Becher, "Microstructural Design of Toughened Ceramics," J. Am. Ceram. Soc., **74** [2] 255-69 (1991)
15, J. Homeny, W. L. Vaugh, and M. K. Ferber, "Processing and Mechanical Properties of SiC-Whisker-Al2O3-Matrix Composites," Am. Ceram. Soc. Bull., **65** [2] 333-38 (1986)
16. P. F. Becher, C. H. Hsueh, P. Angelini, and T. N. Tiegs, "Toughening Behavior in Whisker-Reinforced Ceramic Matrix Composites," J. Am. Ceram. Soc., **71**[12] 1050-61 (1988)

FORMATION OF AL_2O_3-TIB_2 COMPOSITES USING A HIGH TEMPERATURE SYNTHESIS REACTION

Phillip Tefft
Dr. J.J. Moore
Dr. D.W. Readey
Colorado School of Mines
Golden, Colorado 80401

ABSTRACT

Porous Al_2O_3 + TiB_2 ceramics can be prepared by self-propagating high temperature synthesis (SHS). This process, also known as combustion synthesis, uses the inherent exothermicity of the reaction:

$$3TiO_2 + 3B_2O_3 + 10Al = 5Al_2O_3 + 3TiB_2$$

to form the products without additional heating after ignition.

The SHS reaction generates a combustion temperature in excess of 2000 °C which results in volatilization of the B_2O_3 and a high Al vapor pressure. Reaction kinetics are enhanced by the formation of liquid and gaseous phases.

Decreasing the combustion temperature below the melting point of Al_2O_3 by adding a diluent alters the morphology of the final product.

INTRODUCTION

Combustion synthesis is the process of allowing a highly exothermic reaction to continue to react after ignition due to its own heat generation. Figure 1 shows a typical time-temperature curve and a schematic drawing of a propagating reaction wave moving through the reactant pellet.

As can be seen, at the ignition point the reaction produces a large amount of heat, causing the material directly ahead of the wavefront to have its temperature raised to the ignition point (T_{ig}). This property allows substantial savings in processing and energy costs, and can be used to create a porous product directly, or a fully dense product when coupled with a consolidation process.[1]

To the extent authorized under the laws of the United States of America, all copyright interests in this publication are the property of The American Ceramic Society. Any duplication, reproduction, or republication of this publication or any part thereof, without the express written consent of The American Ceramic Society or fee paid to the Copyright Clearance Center, is prohibited.

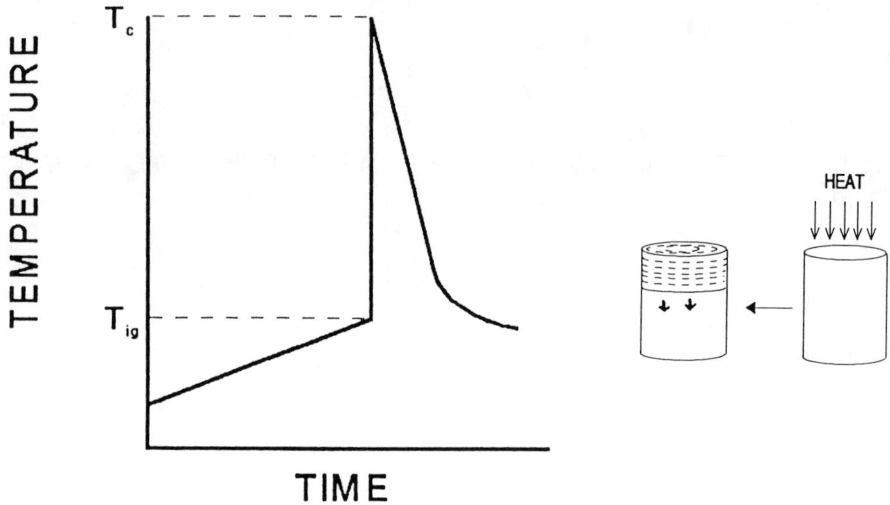

Figure 1. Time-temperature profile and schematic of a reactant SHS pellet in the propagating mode. Note each point in the pellet is subjected to basically the same time-temperature characteristics.

REACTION SYSTEM CHARACTERISTICS

The system under study:

$$3TiO_2 + 3B_2O_3 + 10Al = 3TiB_2 + 5Al_2O_3$$

has been investigated using the experimental parameters found in Table I. Examination of these parameters should help to determine the extent of vapor phase transport effects coupled with gravity induced convection effects on both microstructure and composition. To further the understanding of the reaction method, several tools have been used.

An enthalpy-temperature diagram (Figure 2) is a useful tool to evaluate a system's reaction potential. In such a figure, the enthalpy of reactants and products are plotted as a function of temperature in two separate lines, and the difference indicates the theoretical exothermicity of the reaction. Kinetics are not considered.

For this system Al and B_2O_3 melt well before the ignition temperature and B_2O_3 further vaporizes, producing substantial vapor pressure at the reaction front. Ignoring any heat losses, the adiabatic temperature can be determined from the H vs. T diagram as shown. Adding an inert diluent, i.e. Al_2O_3, to the system causes the reaction to produce less heat and therefore shows a markedly lower

Table I. Experimental Parameters
Standard = 1.38 MPa (2 Ksi) green density at
2 atmospheres chamber pressure

Pressure kPa/atm					
Argon	Vacuum	100/1	200/2	300/3	400/4
HCl	Vacuum	100/1	200/2	300/3	400/4
Green Density (MPa/Ksi)					
Argon	6.89/1	13.8/2	20.7/3	27.6/4	34.5/5
Dilutions Al_2O_3					
Argon	10%	20%			

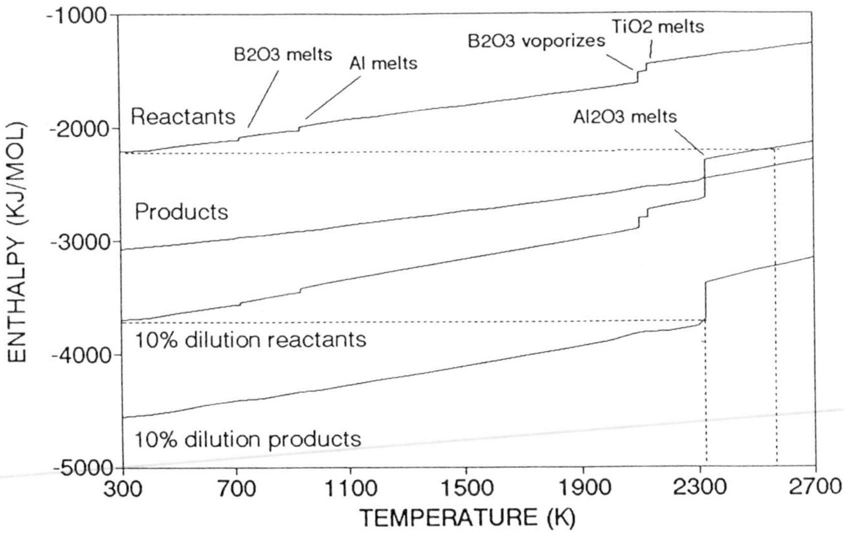

Figure 2. Enthalpy-temperature diagram for the TiB_2 + Al_2O_3 system for stoichiometric and 10% Al_2O_3 dilution. Note phase changes occur, and the adiabatic temperature denoted by the dotted lines is not the combustion temperature (T_c). Less heat is produced in the diluted reaction so the combustion temperature drops.

adiabatic temperature. Other combustion synthesis studies have concentrated on so called "gasless" reactions in which the components of the system have low vapor pressures even at the relatively high combustion temperatures produced.[2]

To evaluate our system more thoroughly, vapor species, as calculated by the SOLGAS software, have been plotted in Figure 3 to determine the extent of vapor phase transport for both the inert argon atmosphere and the reactive HCl atmospheres.[3] It is important to note that the software program evaluates the system on a free energy minimization basis, providing an approximation of vapor pressures only. Clearly, the inert atmosphere is just that, inert, since little to no vapor transport species are available at even high temperatures. Species that carry both titanium and aluminum are present at low temperatures in the HCl environment. This means that convection effects will tend to play a greater role in the reactive atmosphere as is expected.

Conducting these SHS reactions in different orientations with respect to gravity has also been used as a means of understanding the gravitational effects on this system. For each experiment samples were ignited using three different orientations: the normal mode has the combustion wave traveling in the direction of gravity, the inverse mode is opposite this, moving against gravity, and the horizontal mode moving perpendicular to gravity.

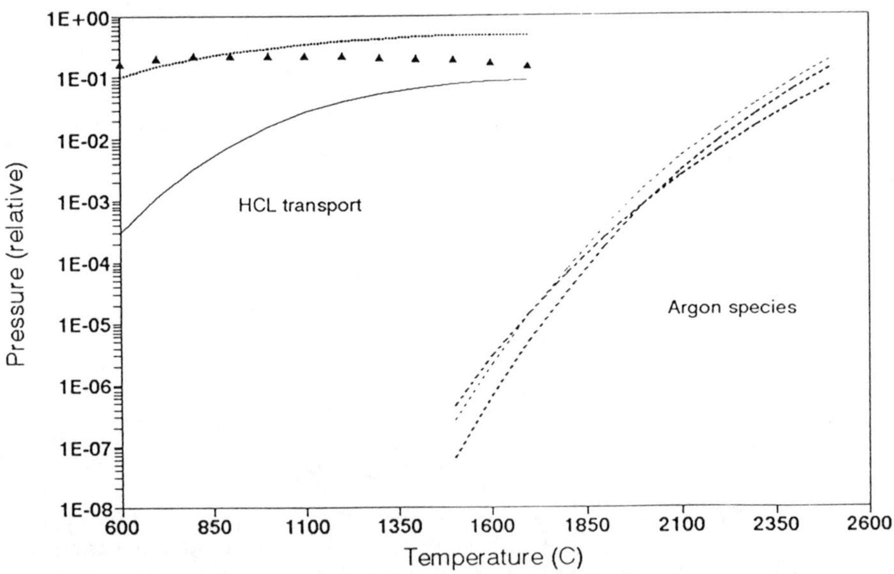

Figure 3. Plot of transport species developed from the reaction in inert and reactive atmospheres. Each curve represents several species, combined for readability.

EXPERIMENTAL PROCEDURE

Starting powders of TiO_2, B_2O_3 and Al (and the Al_2O_3 dilution) were mixed in a ball mill and pressed into 1.27 cm (0.5 in) diameter pellets at a variety of pressures as seen above. The pellets were dried for at least four hours in a furnace above 110 °C to drive off any developed moisture. Samples were then ignited under a variety of atmospheric conditions in a pyrex reaction chamber. All reactions were videotaped and monitored by thermocouple. Combustion temperatures were determined with a dual wavelength infrared pyrometer.

Reacted samples were fractured for viewing with a scanning electron microscope (SEM) and partially pulverized for x-ray diffraction measurements.

RESULTS

Green Density

The effects of green density as a function of pressing pressure on T_{ig} and T_c are shown in figure 4.

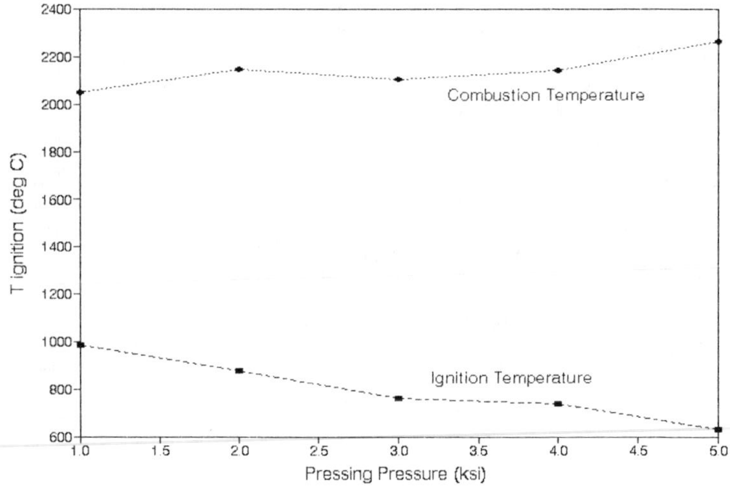

Figure 4. Effect of green density on the T_{ig} and T_c obtained for stoichiometric samples of 1.27 cm (0.5 in) diameter.

X-ray diffraction analysis of the products confirmed that green density did not affect the composition, and that reactions were at least 96% complete. The combustion temperature increased and the ignition temperature decreased with increasing green density.

The product microstructure was relatively unaffected as green density increased, although molten phases became more common. Typical product microstructures are shown in Figure 5.

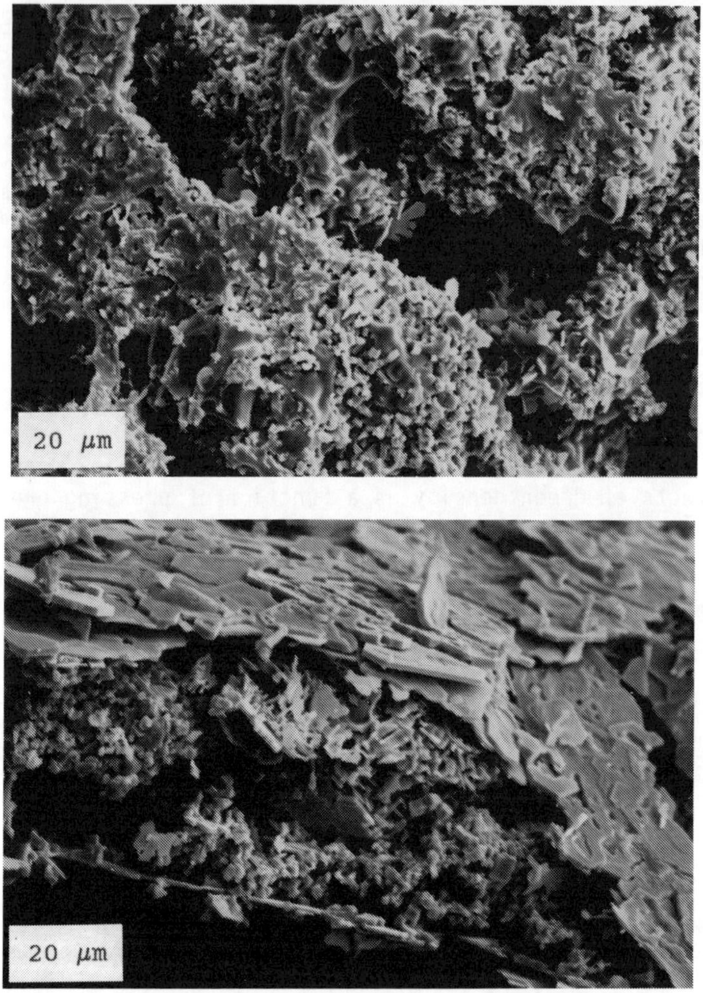

Figure 5. Typical SEM photomicrographs of green density samples. Green density (as a function of pressing pressure) increases top to bottom from 6.89 to 34.5 MPa respectively. Dark phases are alumina and the light phases and plates are TiB_2.

Effect of Gas Pressure in Argon and HCl

As gas pressure increased within the chamber, the combustion temperature increased and the ignition temperature decreased as seen in Figure 6, regardless of gas reactivity. Thus, there is little difference in temperature analysis between the argon and HCl atmospheres. Microstructural effects in argon are shown in Figure 7.

X-ray diffraction analysis indicates that gas pressure did not significantly affect product composition.

Reactive atmosphere

The effects of reactive atmosphere on the combustion temperature are shown in Figure 6, and the effects on product microstructure are shown in Figure 8 (compare with those in Figure 7). It should be noted that the sample reacted in vacuum correlates with vacuum for the argon atmosphere.

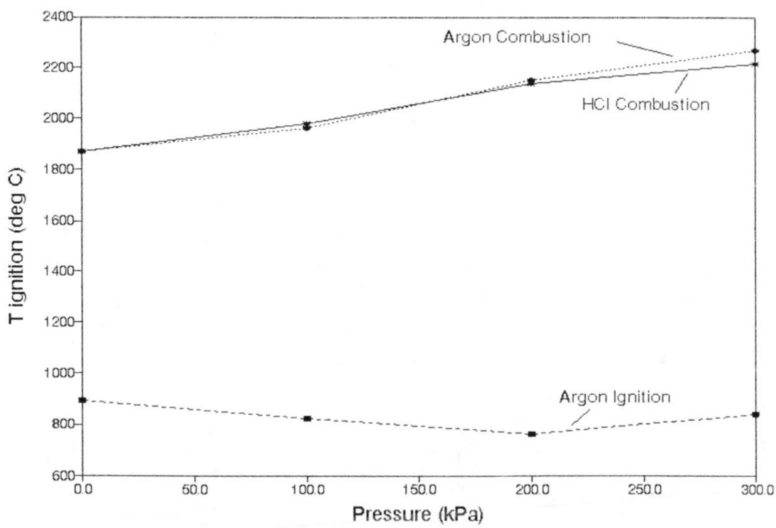

Figure 6. The effect of gas (Ar and HCl) on the temperature attained. Note that no ignition temperatures were determined for the HCl atmosphere.

Analysis of the x-ray data showed a decrease in alumina peak height, partial reduction of the titanium dioxide to Ti_3O_4, and the addition of a mystery phase that has yet to be identified that grows larger with higher HCl pressure. Furthermore, aluminum and boron containing ash-like deposits were found on the walls of the reaction chamber.

Effect of Adding a Diluent

The use of alumina as a diluent resulted in a small amount of partially reduced titanium dioxide in the product, identified in the diffraction patterns, as well as somewhat less than complete reaction in some cases, although still generally 90-95% complete.

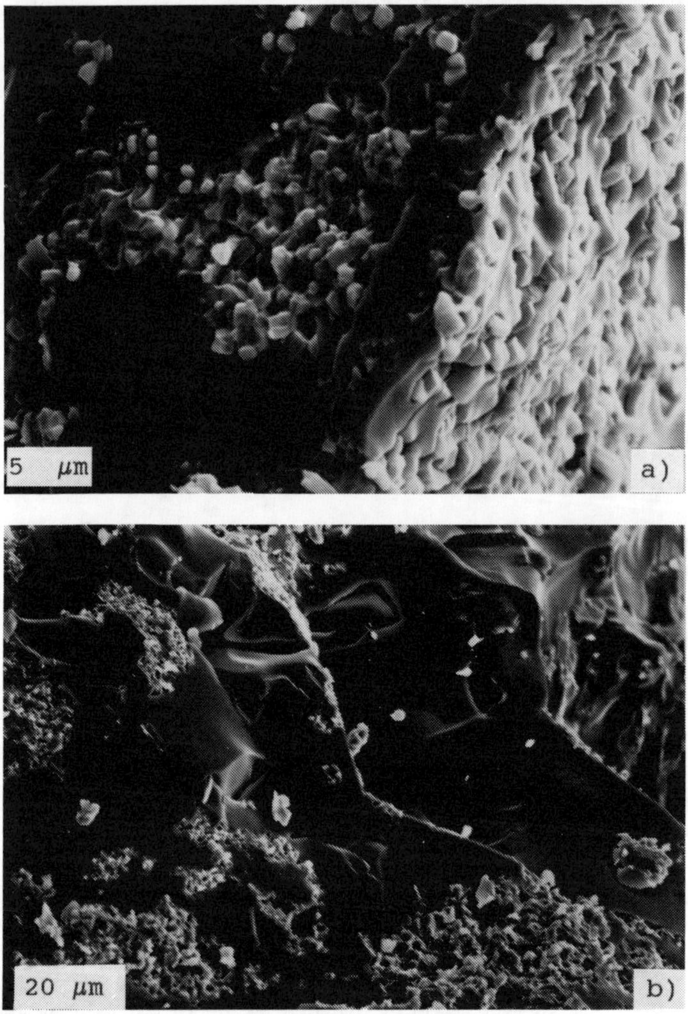

Figure 7. Typical microstructures of products synthesized in a) vacuum and b) 300 kPa (3 atm) of argon. Dark phases are alumina, light phases are TiB_2.

Videos of the propagating reaction front identified an unstable spin propagation mode in which the reaction front slows down and nearly quenches out before reigniting. Typical microstructures of the diluted species are shown in Figure 9.

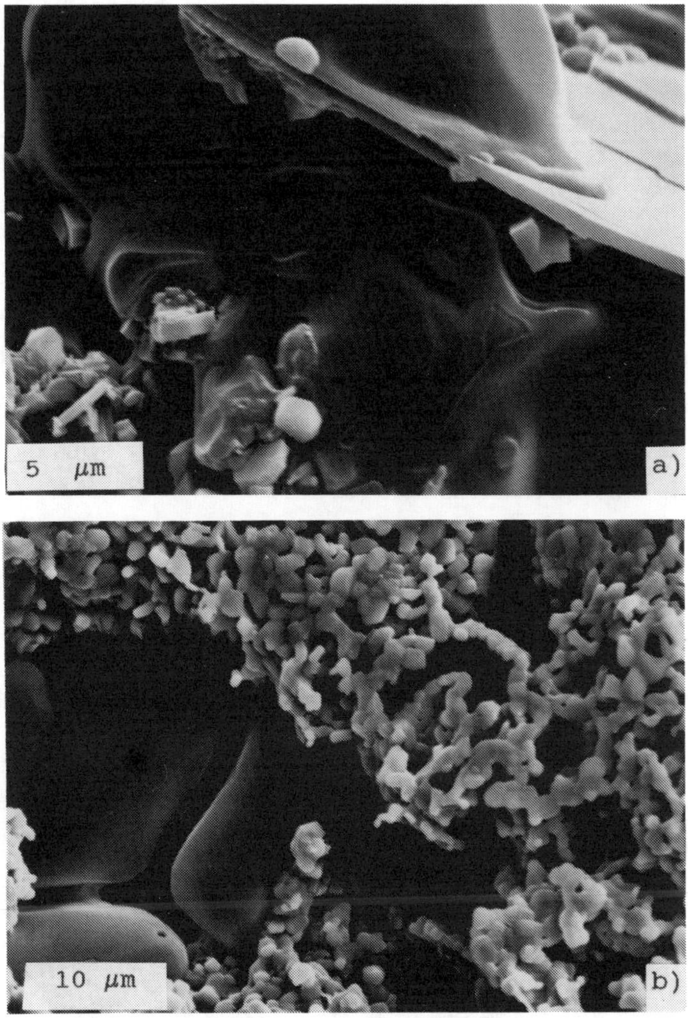

Figure 8. Typical microstructures of products in a) 200 kPa (2 atm) and b) 300 kPa (3 atm) of HCl gas. Dark phases are alumina and light phases are TiB_2.

Results From a Quenched Sample

A propagating front reaction which becomes unstable if too much diluent is added, may result in quenching of the reaction. This condition allows an opportunity to examine the combustion front and the possible reaction mechanisms. Typical macrophotographs of pore

Figure 9. Typical microstructures of products in which a) 10% alumina and b) 20% alumina dilutions were made. Dark phase is alumina, and light phase is TiB_2.

formation ahead of the reaction front are shown in Figure 10. In Figure 11, SEM photomicrographs of the same area indicate uncombusted particles undergoing sintering just ahead of the combustion front, and the infiltration of liquid into the voids.

Gravity has clearly caused molten phases to flow into pores ahead of the wavefront.

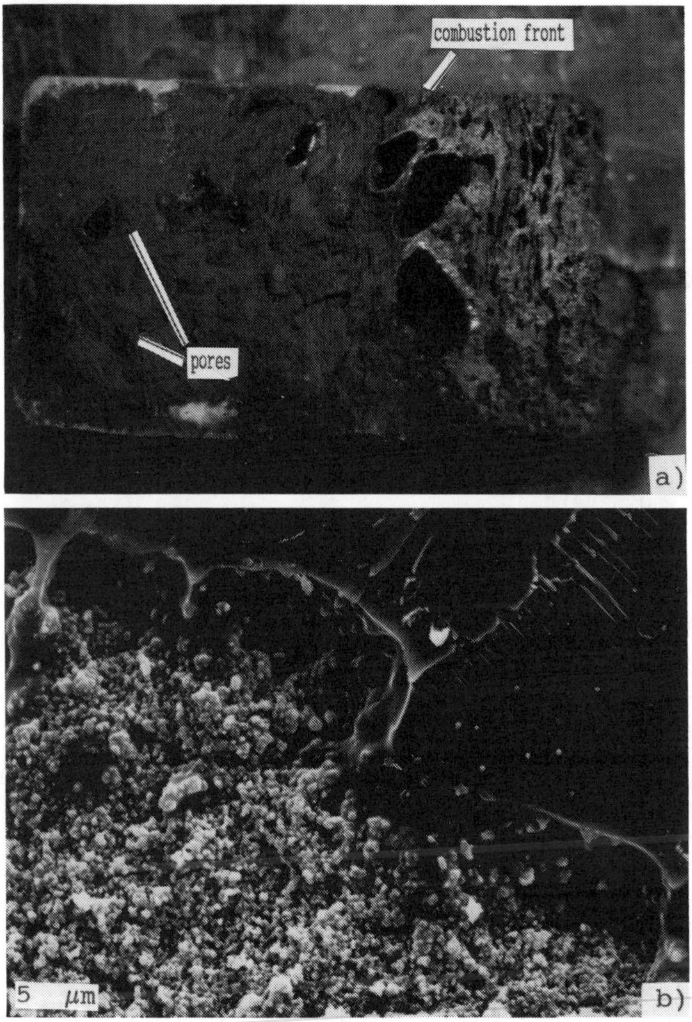

Figure 10. Structures of a quenched sample. Notice a) pores ahead of quenched front and b) liquid infiltration into partially sintered powders at the reaction front.

CONCLUSIONS

That green density effects are seen is not too surprising. Contact between reacting particles and the ability of the gas and liquid to escape affect how the reaction progresses, even if the effect is not a large one.

Much more interesting is the apparent effect of the gas pressure on the reaction. Increased chamber pressure must compensate for the pressure created within the pellet, perhaps keeping the gaseous components in place long enough to react more completely. Convection must also be affected by the change in pressure, and its effects can also be seen in the orientational differences between like samples. The powder deposited on the chamber walls in the HCl samples is further proof of the role of vapor transport coupled with convection. Clearly reactants and heat can be carried away from the pellet during reaction.

Pore formation ahead of the combustion front was an unexpected result, and clearly indicates that a high degree of heat transfer within reacting samples exists. Enough heat must reach the porous areas to liquify and vaporize the aluminum and the boron oxide respectively. The extent of liquid phases observed around these pores supports this conclusion. The role of this preheating and the extent of prereaction are unknown, as is the effect this had on the propagating wavefront.

FUTURE EXPERIMENTS

Conducting these reactions under microgravity will confirm whether or not the conclusions made above are valid, by eliminating the gravity-driven effects. Without convection and gravity-induced liquid flow, microstructures should be noticeably altered if these processes contribute to the synthesis of the reaction.

Further experiments will be designed that determine the role of pore formation ahead of the reaction front. In systems such as this where vapors, liquids and solids interact, it is difficult to construct a physical or mathematical model. If heat distribution is such that pores and liquids form well ahead of the reaction front, this must also be incorporated into any useful model(s).

REFERENCES

1. Munir, Z.A. and Anselmi-Tamburini, "Self-Propagating Exothermic Reactions," Material Science Reports V3 #7,8, May 1989.

2. Aldushin, A.P. and Matkowsky, B.J. "Combustion of Porous Samples with Melting and Flow of Reactants," Northwestern University.

3. Besmann, T.M., "Solgasmix-PV for the PC," Oak Ridge National Laboratory, 1989.

SYNTHESIS OF MULLITE COMPOSITES FROM POWDER MIXTURES OF ALUMINA, SILICA AND ZIRCONIA

Yung-Jen Lin, Kuan-Chieh Su
Department of Materials Engineering
Tatung Institute of Technology, Taipei, Taiwan

ABSTRACT

Mullite and its composites have potential as good structural ceramics. However, the difficulty in sintering mullite composites needs to be overcome. Considering the sintering problem, the fabrication of dense mullite composites via viscous flow of amorphous silica appears to be an excellent method. In this study, mullite/yttria-zirconia composites were viscous and reaction sintered from mixtures of alumina, amorphous silica and 3 mol% yttria-zirconia powders, which were processed by conventional procedures such as wet ball-milling and dry-pressing. The results showed that the densification and phase formation depended on compositions as well as sintering temperatures. Dense (~ 97% T.D.) mullite/yttria-zirconia composites could be achieved when samples contained 20 vol% of yttria-zirconia and sintered at 1500°C for 2 hours in air. The addition of yttria-zirconia promoted densification and retarded the crystallization of cristobalite. It also lowered the mullitization temperature from 1500°C to 1400°C. Microstructure of mullite/20 vol% yttria-zirconia sintered 1500°C showed that the mullite grains were equiaxed or elongated. The zirconia grains were mostly located at grain boundaries, indicating the inhibiting effects of zirconia on mullite grain growth.

INTRODUCTION

Mullite is a structural ceramic with excellent high temperature properties. It has good chemical stability, low density, low thermal expansion and high creep resistance (1-4). However, the low fracture toughness and poor sinterability of crystalline mullite impose difficulties on its applications in the industries (4,5). In order to increase the fracture toughness, mullite composites with particulate or whisker second phase reinforcements were attempted (4, 6-10). Among various composite systems, the mullite/zirconia composites have drawn much attentions. On one hand, there is no oxidation problems as in carbide-reinforcing composites. On the other hand, partially stabilized zirconia (especially yttria partially stabilized

To the extent authorized under the laws of the United States of America, all copyright interests in this publication are the property of The American Ceramic Society. Any duplication, reproduction, or republication of this publication or any part thereof, without the express written consent of The American Ceramic Society or fee paid to the Copyright Clearance Center, is prohibited.

zirconia) could offer toughening effect through its tetragonal to monoclinic phase transformation (transformation toughening) (11,12). Nevertheless, the fabrication of dense mullite/zirconia composites is a challenge.

The difficulty in obtaining dense mullite composites lies mostly in the poor sinterability of mullite. Conventional fabrication from mullite powders usually requires sintering at temperatures greater than 1600°C or even hot pressing in order to achieve dense pieces (13). Various alternative approaches have been extensively investigated to improve the fabrication processes (i.e., pressureless sintering at lower temperatures)(1, 14, 15). One promising route of fabricating dense mullite is to take the advantages of viscous flow of amorphous silica to densify the compacts first and then, reaction-sinter to form the desired structures (16-18). The green compacts usually consist of alumina and amorphous silica of sub-micron sizes or colloidal forms. At suitable temperature ranges the silica softens and flows viscously so that the alumina particles could rearrange and the whole compact densifies. After the densification is completed, the compacts are brought to higher temperatures to mullitize the alumina and silica in-situ.

The idea of viscous sintering was also applied in the fabrication of mullite-zirconia composites. Sacks et al. (16) started with composite powders of sub-micron alumina/silica and yttria-zirconia/silica. Dense mullite/zirconia composites were successfully obtained below 1500°C. Others prepared the mullite/zirconia composites from solution precursors (9,19) or reaction-sintered mixtures of alumina/zircon to form mullite/zirconia composites (20-23). Some of these processes (composite powders and solution precursor processes) required extensive efforts or special precursors, and others (such as reaction sintering of alumina/zircon) were lack of freedom in the composition of zirconia.

In this research, synthesis of mullite/yttria-zirconia composites from powder mixtures of alumina, yttria-stabilized zirconia and amorphous silica was attempted. Starting from the powders of these components, the process could be carried out in conventional powder processing procedures, such as mixing, ball-milling and dry-pressing. This would be a simpler and less expensive way of fabrication of mullite/zirconia composites if dense mullite/zirconia could be achieved with comparable properties.

EXPERIMENTAL PROCEDURES

The amorphous silica powder was prepared according to Stober process (24). The TEOS* (tetraethyl orthosilicate) was added into ethanol** (95%) stirred with a magnetic stirring bar. Ammonia hydroxide*(~28%) was then added into the solution as a catalyst. The solution was kept stirring for 1.5 to 2.0 hours for the reaction to complete. The volume ratio of ethanol, TEOS and ammonia hydroxide

---

* Fluka AG, Buchs, Switzerland.

** Taiwan Tobacco & Wine Monopoly Bureau, Taipei, Taiwan.

was 1000:61:77. After the reaction was completed, the silica powder was separated from the liquid by gravitational sedimentation, followed by drying under an infrared light. The dried cake was then ground with agate mortar & pestle to obtain free flowing powders. Initial organic burn-out was processed at 500°C/4hr. in a tube furnace with static air.

In addition to the silica powder, starting powders included commercially available alumina# and 3 mol% yttria-stabilized zirconia$. The weight ratio of alumina to silica was kept to 3:1in all the samples. The yttria-stabilized zirconia powder was added so that the final powder mixtures would contain 0 vol% (no zirconia addition for comparison) (denoted M), 10 vol% (denoted M10Z) and 20 vol% (denoted M20Z) of yttria-zirconia@. Table I is the characteristics of the powders.

Table I. Characteristics of Alumina and 3 mol% Yttria-Zirconia powders

	BET surface area (sq. m/g)	mean particle size (μm)	purity (wt%)
alumina (alpha)	15	~ 0.41	99.99
yttria-zirconia	30	~ 0.17	99.9

(BET and purity were supplied by the manufacturers, mean particle size were measured)

The powders were mixed by ball-milling of aqueous (de-ionized water) suspensions of 30 vol% solid contents. Milling was done with 3 mol% yttria-zirconia milling balls of 3 mm in diameter in polyethylene bottles. The pH of the suspensions was maintained at ~ 5 so that the surface zeta potentials of alumina and zirconia would be opposite to that of silica and hetero-flocculation could happen among these particles(25). Under this condition, better dispersion of silica among alumina and zirconia would be expected. Each composition was milled for 30 hrs. After milling was done, the slurries was dried and softly ground with agate mortar & pestle to obtain free flowing powders. The dry powder mixtures were then uniaxially dry-pressed in a WC mold, lubricated with stearic acid. Pressure applied was 285 MPa and pellet size was about 10 mm in diameter and 3 mm in thickness.

The pressed pellets were heat treated with heating rate of 3°C/min. to 800°C and held at 800°C for 1 hour to burn out the remaining organics. Sintering was performed by heating the pellets at temperatures ranging from 1300°C to 1500°C for 2 hours, followed by furnace cooling. The heating rate of sintering cycles was 10°C/min. All sintering was carried out in static air.

Baikalox CR15, Baikowski International Corp., Charlotte, NC, USA.

$ OZC-3YA, Osaka Cement Co., Ltd., Osaka, Japan.

@ Assuming the following densities: alumina : 3.92 g/cc; 3 mol% Y-ZrO_2: 6.06 g/cc; silica: 2.0 g/cc.

Silica powder was characterized with a laser diffraction particle size analyzer(Malvern Automeasure 4700, Worcs, UK) and a TGA(DuPont 2000). The densities of the sintered samples were measured by Archimede method using de-ionized water as the displacing medium. Phases were identified by XRD (Siemens D-5000) and microstructure was examined in SEM.

RESULTS

The silica powder synthesized with Stober process was spherical with narrow size range, Fig. 1. The mean particle size was 0.30 μm. All the particles were smaller than 0.51 μm. TGA analysis indicated that the weight of the powders did not change after 800°C. The total weight loss was about 10 wt%.

Figures 2 and 3 were the measuring results of bulk density and open porosity of the sintered pellets, respectively. The theoretical density of each sample was difficult to estimated because each sample usually consisted of several phases with undetermined proportions. Figures. 2 and 3 showed that the densification of the compacts was greatly enhanced by the addition of yttria-zirconia. The sample without yttria-zirconia (sample M) did not densify significantly until 1400°C while the sample with 20 vol% yttria-zirconia had reached nearly zero open porosity at 1400°C.

Figures 4(a-c) were XRD patterns of the samples sintered at different temperatures. Phase transformation depended on the temperatures as well as composition. In samples without yttria-zirconia (sample M), much cristobalite formed at 1300°C and 1400°C. There were no evidence of mullite below 1400°C. At 1500°C, significant amount of mullite formed with the expense of alumina and silica. There were still cristobalite present after 1500°C sintering although the peak became weaker.

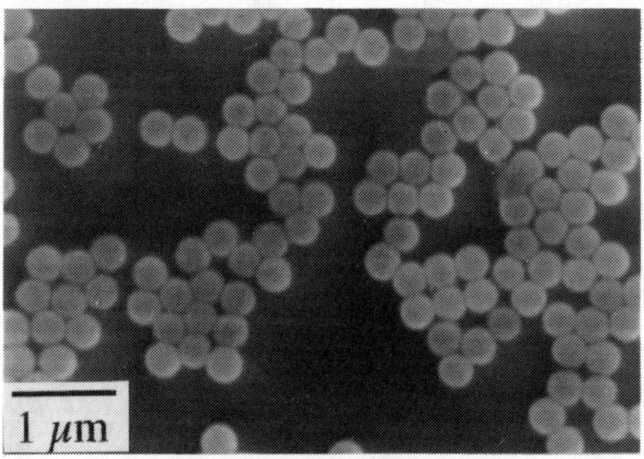

Fig. 1 SEM of Silica Spheres

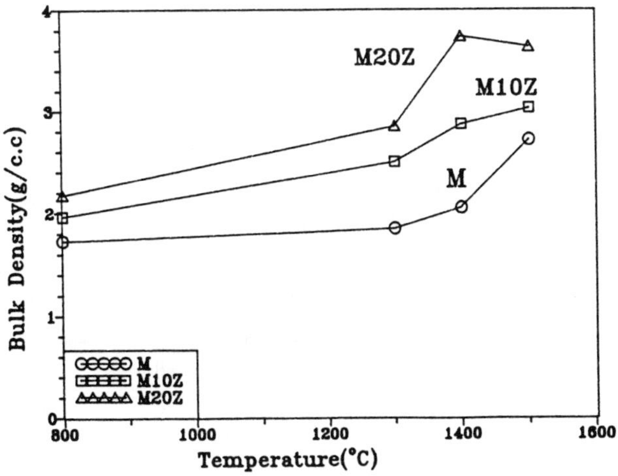

Fig. 2 Bulk Density of Samples (10°C/min heating rate, 2 hrs. sintering, in air). M: alumina/silica; M10Z: M + 10 vol% Y-ZrO_2; M20Z: M + 20 %Y-ZrO_2

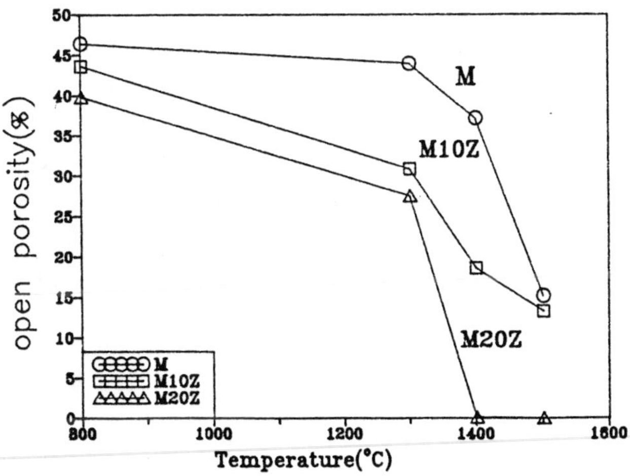

Fig. 3 Open Porosity of Samples (10°C/min heating rate, 2 hrs. sintering, in air). M: alumina/silica; M10Z: M + 10 vol% Y-ZrO_2; M20Z: M + 20 %Y-ZrO_2

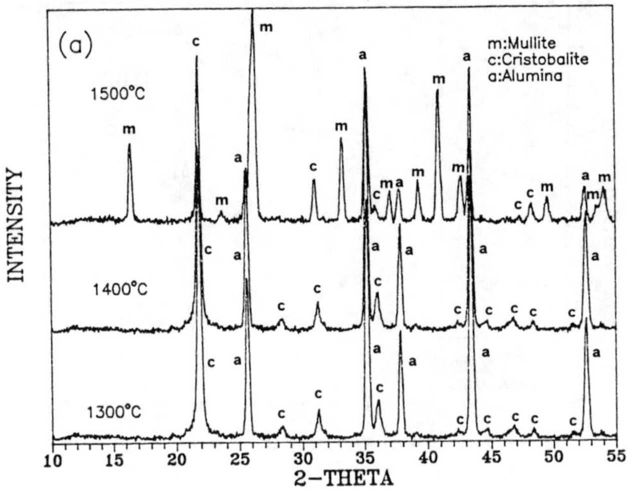

Fig. 4(a) XRD Pattern of M Samples (alumina + silica) (10°C/min heating rate, 2 hrs. sintering, in air).

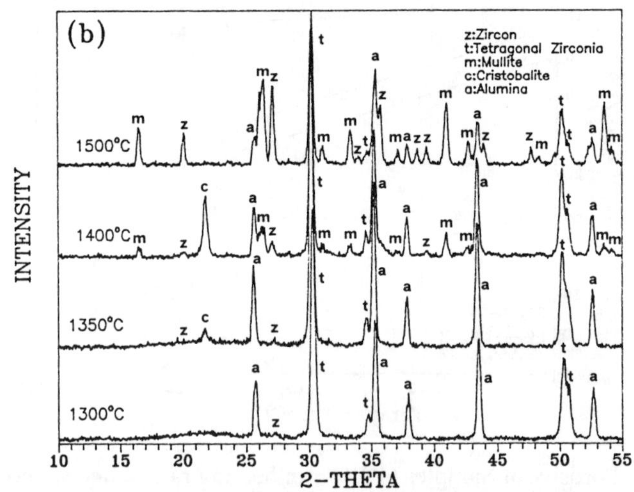

Fig. 4(b) XRD Pattern of M10Z Samples (M + 10vol%Y-ZrO_2) (10°C/min heating rate, 2 hrs. sintering, in air).

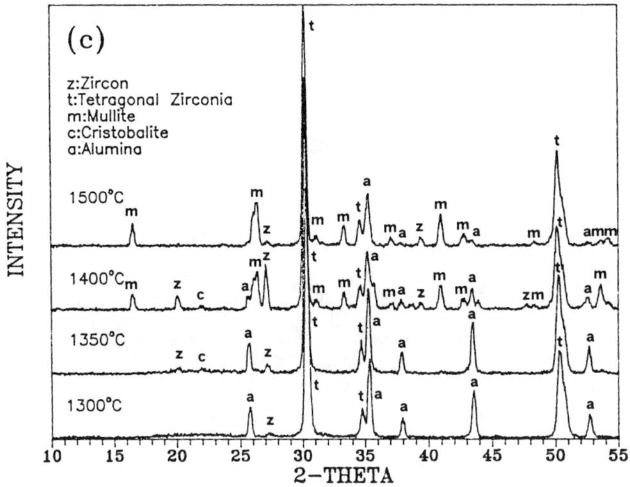

Fig. 4(c) XRD Pattern of M20Z Samples (M + 20vol%Y-ZrO_2) (10°C/min heating rate, 2 hrs. sintering, in air).

In samples with 10 vol% yttria-zirconia, trace of zircon in addition to alumina and zirconia was detected in the sample sintered at 1300°C. Cristobalite and zircon appeared after sintering at 1350°C. Mullite did not show until 1400°C. At 1500°C, mullite and zircon were both significant but the cristobalite vanished.

When samples containing 20 vol% yttria-zirconia, the zircon formation began at 1300°C. By 1350°C trace of cristobalite and zircon appeared. At 1400°C, mullite, zircon were both notable, and a little cristobalite also showed up. In the 1500°C sintered sample, there were mullite and a little zircon (in addition to zirconia and remaining alumina). No cristobalite was found. For samples with yttria-zirconia addition, the zirconia was tetragonal. Also, alumina was present in all the samples. The phases present in the samples were listed in Table II.

Table II. List of Phases in the Sintered Samples.

	M	M10Z	M20Z
1300°C/2hrs	C(+)[&], A(+)	A(+), Z(-), T(+)	A(+), Z(-), T(+)
1350°C/2hrs	------------	C(-), A, Z(-), T(+)	C(trace), A, Z(-), T(+)
1400°C/2hrs	C(+), A(+)	C, A, M, Z, T(+)	C(-), A, M, Z, T(+)
1500°C/2hrs	C, A, M(+)	A, M(+), Z(+), T(+)	A, M(+), Z(-), T(+)

C: cristobalite; A: alumina; M: mullite; Z: zircon; T: tetragonal zirconia.
[&]Symbol in the parenthesis indicated the peak intensity; +: strong, -: weak

Figure 5 is a SEM micrograph of the sample M20Z, showing fine grain structures. Equiaxed mullite grains were less than 1 μm and some grains were elongated. The zirconia grains (bright particles in Fig. 5) were smaller than 0.3 μm and existed mostly in the mullite grain boundaries.

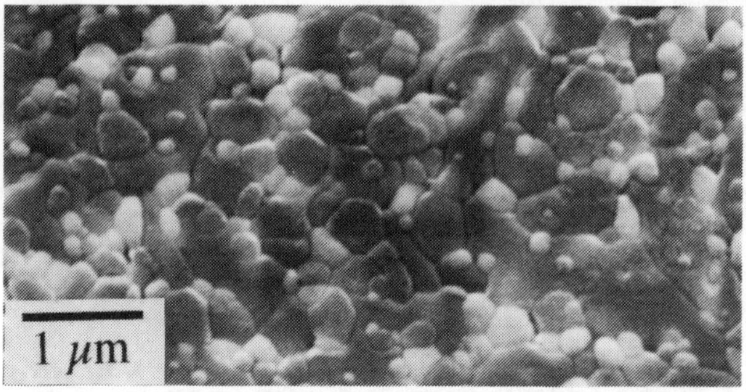

Fig. 5 SEM micrograph of M + 20vol%Y-ZrO_2 sintered 1500°C/2hrs.

DISCUSSIONS

After sintered at 1500°C for 2 hours, the sample with 20 vol. % yttria-zirconia (M20Z) reached a bulk density of 3.63 g/cc. This sample consisted of mullite and tetragonal zirconia with trace of zircon and remaining alumina. Since the zircon was very little and remaining alumina was about 1wt%(18), the theoretical density of this sample could be assessed by assuming 20 vol% of 3 mol% yttria-zirconia and 80% of mullite. Using rule of mixture, the theoretical density of this sample was calculated to be 3.74 g/cc. The density of this sample was therefore 97% of the theoretical density.

In comparison, dense mullite from reaction sintering of alumina and silica powder without the addition of yttria-zirconia seemed unattainable under current conditions. The alumina/silica samples (sample M) still contained 15 % open porosity after 1500°C/2 hrs sintering (see, Fig. 3). It was thus apparent that the addition of yttria-zirconia played a key role in the sintering process.

The major impact of the yttria-zirconia on the densification of the samples must have close relation with the interaction of yttria and/or zirconia in the alumina-silica system. From the phase diagrams of Al_2O_3-Y_2O_3-SiO_2 and Al_2O_3-SiO_2-ZrO_2 (26), the addition of either yttria or zirconia decreased the liquidus temperatures. Both would be expected to promote the densification of the samples by lowering the melting temperatures. In fact, yttria and zirconia had been used as additives in the sintering of mullite (27-29). The effect of yttria addition was even more effective. A eutectic point of less than 1400°C, which was below the most processing temperatures, existed in the Al_2O_3-Y_2O_3-SiO_2 system. Liquid phase was observed at relatively low temperatures when 1-2 wt% of yttria was added into alumina-silica system to improve sinterability (27, 28). It was therefore reasonable to assume that the densification improvement of alumina-silica-yttria-zirconia system resulted from the benefits of liquid phase sintering. The results shown in Figs. 2 and 3 were parallel with this assumption. Dramatic densification occurred at temperatures ≤ 1400°C. Although the amorphous silica in the samples was intended to improve the densification by a viscous flow at elevated temperatures, it showed that this was impractical if there were no elements such as yttrium and/or zirconium in the system. Without yttria and/or zirconia, the sample M sintered at 1500°C/2 hrs only reached 2.72 g/c.c. in the bulk density.

The role of yttria and zirconia in the sintering of these compacts was two folds. Firstly, the yttria and zirconia combined with alumina and silica and formed a viscous silicate glass phase at relatively low temperatures (< 1400°C), which enhanced the densification by viscous flow of the silicate glass phase. The significant solubility of yttria and zirconia in the aluminosilicate glass has been reported in the literature (30,31). Since the yttrium and zirconium were not strong glass formers, their dissolution into the aluminosilicate glass would lower the softening temperature and viscosity.

Secondly, the dissolution of yttria and zirconia in the silicate glass phase would hinder the early crystallization of the silicate phase and ensured liquid phase sintering to take place. In the viscous sintering of compacts containing amorphous silica, the most detrimental crystalline phase was cristobalite. When no yttria-zirconia was present, cristobalite formed at temperature (< 1300°C, Fig.4a) at which viscous sintering was unlikely to happen. As cristobalite formed, it consumed the silicate glass and made viscous sintering impossible.

The crystalline phases derived from this silicate glass often had limited solubility of either yttria or zirconia. The mullite and zircon contained essentially no yttria and mullite had little zirconia solubility(32,33). Although no information of the yttria and zirconia solubility in the cristobalite was found, it was expected to be very limited. Therefore, yttrium had to be rejected before any crystalline phase formed from the silicate glass phase whereas zirconia solubility would promote the zircon ($ZrSiO_4$) formation. Thus, in the crystallization of cristobalite, the crystallization process would be impeded by the requirement of yttria and zirconia rejection. The combination of enhancement of liquid phase sintering and prohibition of early cristobalite crystallization contributed the positive effects of the yttria-zirconia addition in the sintering of the alumina/silica system.

The hindrance of cristobalite crystallization by the addition of yttria-zirconia was obvious by comparing the XRD of these samples, Figs.4 (a-c). At 1300°C, much cristobalite has already formed if no yttria-zirconia was present. However, with more yttria-zirconia in the samples, the cristobalite became more difficult to form. The cristobalite formation was retarded till 1350°C in M20Z and M10Z, and the M20Z showed much less cristobalite than M10Z at this temperature. The formation of zircon was easier than cristobalite and started to show at 1300°C. Zircon became significant at 1400°C and above. However, it was also dependent on the amounts of yttria-zirconia. After sintered at 1500°C/2hrs, M20Z had less zircon than M10Z. Too much yttria seemed to inhibit zircon formation too. Addition of 20 vol% of yttria-zirconia could repress the crystallization of cristobalite and zircon.

On the contrary, the mullite phase was encouraged by the addition of yttria-zirconia. The mullitization temperature was lowered from 1500°C in sample M to 1400°C in samples M10Z and M20 Z. Mullite formed dominantly at 1500°C. As for the zirconia, it remained as tetragonal. Therefore, transformation toughening effect was probably still operating in these samples.

From the SEM micrographs of M20Z sample, the mullite grains were mostly small ($\leq 1\ \mu$m) and equiaxed. However, some larger mullite grains (a few microns) were elongated. The zirconia grains, which were mostly located at grain boundaries (Fig.5), were expected to retard the mullite grain growth.

CONCLUSIONS

The reaction sintering of mullite/zirconia composites from powder mixtures of alpha-alumina, amorphous silica and yttria-zirconia could be realized by conventional powder processing and pressureless sintering. In order to minimize the formation of zircon and cristobalite as well as to achieve high density, the samples had to be sintered at temperatures 1500°C (or above) and doped with suitable amounts of 3 mol% yttria stabilized zirconia. In this research, the sample containing 20 vol% yttria-zirconia could be sintered to high density (~97% theoretical density) and contained no other silicate phases except mullite at 1500°C for 2 hours in air.

Without yttria-zirconia addition, much cristobalite formed below 1300°C and existed up to 1500°C, where mullite appeared. When the samples contained 20 vol% yttria-zirconia, cristobalite became insignificant except sintering at temperatures around 1400°C. However, another silicate phase, zircon, started to form at 1300°C and attained notable amounts at 1400°C. The mullitization temperature was lowered by the addition of yttria-zirconia. The zirconia grains remained as tetragonal and were located at mullite grain boundaries. The mullite grains were less than 1 μm when equiaxed, but some were elongated.

ACKNOWLEDGMENT: This work had been supported by the National Science Council in Taiwan under grant number NSC82-0113-E-036-053-T

REFERENCES

1. I.A. Aksay, D.M. Dabbs, and M. Sarikaya, "Mullite for Structural, Electronic, and Optical Applications," J. Am. Ceram. Soc., 74 [10] 2343-58 (1991).

2. H. Schneider and E. Eberhard, "Thermal Expansion of Mullite," J. Am. Ceram. Soc.,73 [7] 2073-76 (1990)
3. P.C. Dokko, J.A. Pask and K.S. Mazdiyashi, "High-Temperature Mechanical Properties of Mullite Under Compression," J. Am. Ceram. Soc., 60 [3-4] 150-155 (1977).
4. S. Somiya, and Y. Hirata, "Mullite Powder Technology and Applications in Japan," Ceram. Bull, 70 [10] 1624-1632 (1991)
5. T.Il Mah and K.S. Mazdiyasni, "Mechanical Properites of Mullite," J. Am. Ceram. Soc., 66 [10] 699-703 (1983)
6. C. Nischik, M.M. Seibold, N.A. Travitzky, and N. Claussen, "Effect of Processing on Mechanical Properties of Platelet-Reinforced Mullite Composites," J. Am. Ceram. Soc., 74 [10] 2464-68 (1991)
7. K. Okada, N. Otsuka, R.J. Brook and A.J. Moulson, " Microstructure and Fracture Toulghness of Yttira-Doped Tetragonal Zirconia Polycrystal/Mullite Composites Prepared by an in Situ Method," J. Am. Ceram. Soc., 72 [12] 2369-72 (1989).
8. S. Blackburn and T.A. lawson, "Mullite-Alumina Composites by Extrusion," J. Am. Ceram. Soc., 75 [4] 953-57 (1992)
9. S. Rajendran and H.J. Rossell," Coprecipitation-Derived Mullite and Mullite-Zirconia Composites, " J. of Mater. Sci., 26 [21] 5815-22 (1991)
10. T. Tiegs, P. Becher, and P. Angelini, "Microstructures and Properties of SiC Whisker-reinforced Mullite Composites"; PP.463-472 in Mullite and Mullite Matrix Composites, Ceramic Transactions, Vol. 6. Edited by S. Somiya, R.F. Davis, and J.A. Pask. The American Ceramic Society, Inc., Westerville, OH, 1990.
11. C.A. Andersson and T.K. Gupta, "Phase Stability and Transformation Toughening in Zirconia"; pp. 184-201 in Advances in Ceramics, Vol. 3. Edited by A.H. Heuer and L.W. Hobbs. The American Ceramic Society, Westerville, OH, 1988.
12. A.G. Evans, "Toughening Mechanisms in Zirconia Alloys"; pp.193-212 in Advances in Ceramics, Vol. 12. Edited by N. Claussen, M. Ruhle and A.H. Heuer. The American Ceramic Society, Columbus, OH, 1984.
13. M. Mizuno, M. Shiraishi, and H. Saito, "Microstructure and Bending Strength of Highly Pure Mullite Ceramics"; pp.413-24 in Mullite and Mullite Matrix Composites, Ceramic Transactions, Vol. 6. Edited by S. Somiya, R.F. Davis, and J.A. Pask. The American Ceramic Society, Inc., Westerville, OH, 1990.
14. M.D. Sacks, H. Lee, and J.A. Pask, "A Review of Powder Preparation Methods and Densification Procedures for Fabricating High Density Mullite"; pp. 167-207, ibid.
15. K. Okada, N. Otsuka and S. Somiya, "Review of Mullite Synthesis Routes in Japan," Ceram. Bull. 70 [10] 1633-40 (1991)
16. M. Sacks, N. Bozkurt, and G.W. Scheiffele, "Fabrication of Mullite and Mullite-Matrix Composites by Transient Viscous Sintering of Composite Powders," J. Am. Ceram. Soc., 74 [10] 2428-37 (1991)

17. W.G. Fahrenholtz, D.M. Smith and J. Cesarano III, "Effect of Precursor Particle Size on the Densification and Crystallization Behavior of Mullite," J. Am. Ceram. Soc., 76 [2] 433-37 (1993).
18. P. Boch, T. Chartier, and P.D.D. Rodrigo, "High-Purity Mullite Ceramics by Reaction-Sintering"; pp.353-374 in Mullite and Mullite Matrix Composites, Ceramic Transactions, Vol. 6. Edited by S. Somiya, R.F. Davis, and J.A. Pask. The American Ceramic Society, Inc., Westerville, OH, 1990.
19. K. Rundgren, P. Elfving, H. Tabata, S. kanzaki, and R. Pompe, "Microstructures and Mechanical Properties of Mullite-Zirconia Composites Made from Inorganic Sols and Salts"; pp.553-566, ibid.
20. P. Boch T. Chartier, and J.P. Giry, "Zirconia-Toughened Mullite/The Role of Zircon Dissociation"; pp.473-494, ibid.
21. J.S. Moya, "Reaction Sintered Mullite-Zirconia and Mullite-Zirconia-SiC Ceramics"; pp.495-507, ibid.
22. A. Leriche, "Mechanical Properties and Microstructures of Mullite-Zirconia Composites"; pp.541-552, ibid.
23. N. Claussen, and J. Jahn, "Mechanical Properties of Sintered in Situ-Reacted Mullite Zirconia Composites," J. Amer. Ceram. Soc., 63 [3-4] 228-29 (1980)
24. W. Stober, A. Fink, and E. Bohn, "Controlled Growth of Monodispersed Silica Spheres in the Micron Size Range," J. Colloid and Interface Sci., 26, 62-69 (1968).
25. A. Bleier, and C.G. Westmoreland, "Effects of pH and Particle Size on the Processing of and the Development of Microstructure in Alumina-Zirconia Composites," J. Am. Ceram. Soc., 74 [12] 3100-11 (1991).
26. Figures 2586 and 772 in Phase Diagrams for Ceramist, complied by E.M. Levin, C.R. Robbins and H.F. McMurdie; ed. by M. K. Reser. The American Ceramic Society, Columbus, OH, 1969.
27. N. Ishibashi, K. Tahara, H. Kobayashi, T. Mitamura, and T. Akiba, "Effect of Metallic Salt Additives on the Properties of Mullite Sintered Body"; p.233 (2C48) in Extended Abstracts of Annual Meeting. The Ceramic Socety of Japan, 1989.
28. T. Kurihara, M, Horiuchi, A. Kuribayashi, Y. Takeuchi, S. Wakabayashi, "Effect of Additives on the Sinterability of Mullite Ceramics for Insulating Substrate," ibid, p.486 (3B42).
29. J. Asaumi, H. Yoshida, H. Kamiaka, and C. Yamagishi, "Fabrication of Mullite Composites"; pp.257-58 (3-1C09) in Extended Abstracts of 1st Autumn Symposium. The Ceramic Society of Japan, Tokyo, Japan, 1988.
30. Y. Chang and D.P. Thompson, "The Behaviour of ZrO_2 in Y_2O_3-Al_2O_3-SiO_2 Glasses," Br. Ceram. Trans. J. 87 [3] 107-110 (1988).
31. Y.-J. Lin, P. Angelini, and M.L. Mecartney, "Microstructural and Chemical Influences of Silicate Grain-Boundary Phases in Yttria-Stabilized Zirconia," J. Am. Ceram. Soc., 73 [9] 2728-35 (1990)
32. R. Torrecillas, S. De Aza, J.S. Moya, T. Epicier and G. Fantozzi, "Improved High-Temperature Mechanical Properties of Zirconia-Doped Mullite," J. Mater. Sci. Lett. 9, 1400-1402 (1990)
33. Yung-Jen Lin, Ph.D.Thesis, University of Minnesota, Minneapolis, MN(1990)

EVALUATION OF ZrO_2 TOUGHENED HIGH SODA DUPLEX CALCINED Al_2O_3 SINTERED BODIES

F.J.Arenas, M. Donat, I. Arenas and L. Centeno
Department of Materials Technology
University Institute of Technology, P.O. Box 40347
Caracas 1040-A. VENEZUELA

ABSTRACT

Composites consisting of a 0.5 wt% soda content alumina as the matrix and dispersed t-ZrO_2 particulates upto 20 wt% were processed by pressureless sintering. Results indicated that an improvement in fracture toughness (K_c= 6 ± 0.10 MPa.√m) can be achieved by incorporating 10 wt% of ZrO_2 for composites sintered at 1550-1600°C, while the hardness decreased with Zirconia addition. Microstructures showed an homogeneous dispersed particle distribution and transgranular fracture surface.

INTRODUCTION

The ceramics market encompasses a wide variety of applications for alumina, from glass and chinaware to spark plugs, bioceramics, and integrated circuits (1). Alumina is well suited for mechanical applications that require high wear resistance strength at low and high temperatures, and corrosion resistance. All varieties of calcined aluminas are used. When optimum hardness, density and wear resistance are required, the termed thermally reactive aluminas are used in compositions that are 95% alumina or more. When lower performance is adequate and when cost needs to be minimized, the lower priced normal-soda (0.1 to 0.6%) calcined aluminas are utilized in compositions as low as 85% alumina (2).
Recent developments include the use of zirconia (3-9) or silicon carbide fibers (10,11) to form composite alumina structures with improved mechanical strength, fracture toughness, and cutting behavior.

To the extent authorized under the laws of the United States of America, all copyright interests in this publication are the property of The American Ceramic Society. Any duplication, reproduction, or republication of this publication or any part thereof, without the express written consent of The American Ceramic Society or fee paid to the Copyright Clearance Center, is prohibited.

Most works have been done on the influence of zirconia on the mechanical properties of low-soda alumina. The densification and ZrO_2 toughening mechanism induced properties depend on factors such as purity, particle size distribution & morphology, structural nature and processing variables (3-9).
This paper examines the influence of zirconia particles addition on the densification and mechanical properties achieved on composites prepared from normal-soda calcined alumina using a conventional powder processing route. The objective is to assess the convinience of this raw material for certain applications with moderate mechanical performance requirements.

EXPERIMENTAL PROCEDURE

Commercially available metallurgical grade γ alumina (Interalumina C.A.) with 98.8 wt% typical minimum purity and a soda-content between 0.4-0.6 wt% was used. Its granulometry determined by Rotap process with Tyler sieves was: +100 mesh = 1% max., +200 mesh = 59% max., +325 mesh = 32 % max., -325 mesh = 12% max. Surface area determined by single-point BET method was 70 m^2/g.
Reactive grade t-ZrO_2 with a mean particle size below 2 μm was used as second phase particles addition in percentages up to 20 wt%.
Before mixing, γ-grade alumina was calcined at 1250°C for 1 h. and ball milled in alumina jars during 24 h. to obtain a powder with an average particle size of 2 μm as determined by Sedigraph. The x-ray diffraction pattern showed a transformation to α-alumina above 95% as determined semi-quantitatively and its morphology is shown in figure 1.

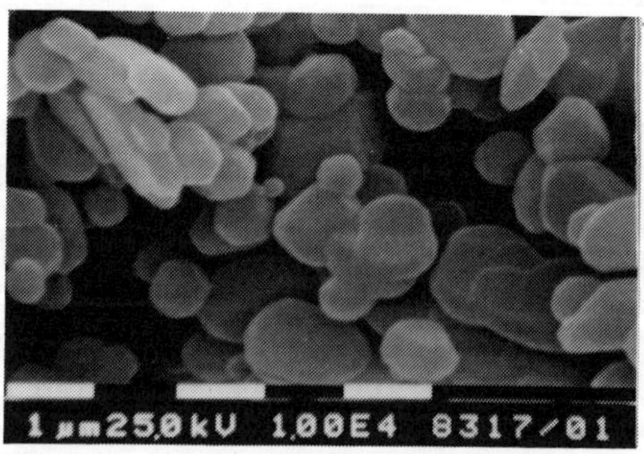

Figure 1. SEM micrograph of ball milled duplex calcined alumina

The compositions were prepared by wet-media mixing 1h. in plastic jars, dried, milled and uniaxially cold pressed in shapes of 25 mm. diameter at 350 MPa. The compacts were fired at temperatures between 1450-1600°C, 1 h. in ambient air in a $MoSi_2$ furnace. Heating rate (2°C/min.) was carefully controlled and the organic binder was eliminated at 350 °C.

Shrinkage and relative density of sintered samples were determined by dimensional change and Archimedes method using water as the immersion medium, respectively. The Vickers indentation technique with a 10 Kp-load was used for hardness and fracture toughness determination. For hardness and microstructural studies compact surfaces were polished and thermally etched by anneal at 1450°C-1 hour. The SEM evaluation was made on polished and fractured surfaces after Au-Pd sputtering.

RESULTS AND DISCUSSION

The effect of ZrO_2 particles dispersion on densification and mechanical properties of duplex calcined alumina compacts was studied as a function of its composition and sintering temperature. The results are illustrated in figures 2 - 4.

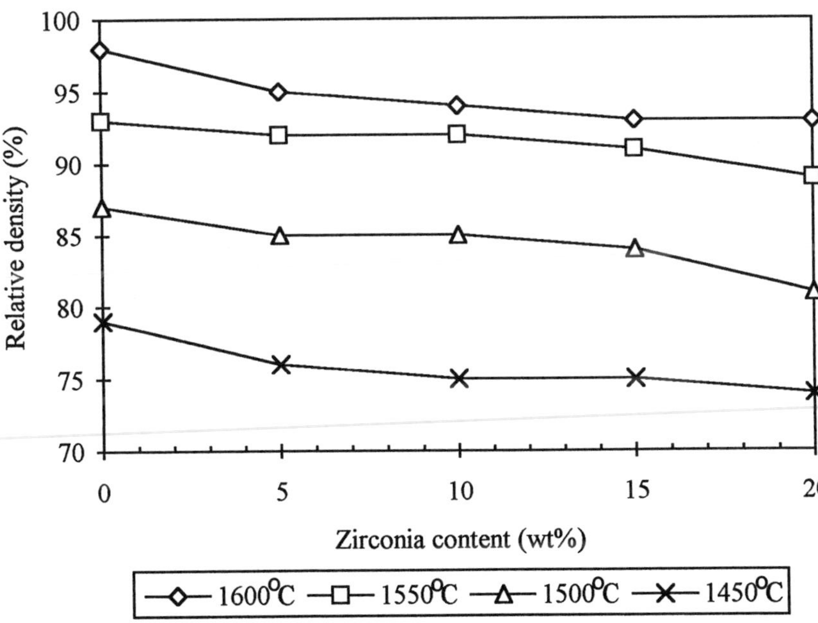

Figure 2. Densification behavior of pure alumina and Al_2O_3 - ZrO_2 compositions

For all sintering temperatures, it was found that composite density decreased as zirconia content in the alumina matrix increased. This can be explained by the t→m transformation on cooling which produces an increase in volume and development of porous zones (12) because of compact deformation on shrinkage. Compacts linear shrinkage follows a pattern similar to Osendi & Moya (13) although higher values were attained.

Hardness measured by Vickers indentation technique increased with the sintering temperature for all compositions (Figure 3), being this behaviour similar to densification. Therefore, it can be said that hardness depends on densification and the higher values corresponded to pure alumina and 5 wt% ZrO_2 compositions fired at 1600°C. Results obtained for this temperature (1000-1630 Kg/mm^2) are similar to Green (14) values and, decrease in hardness as ZrO_2 content increase can be ascribed to the increase in brittle nature m-ZrO_2 phase amount

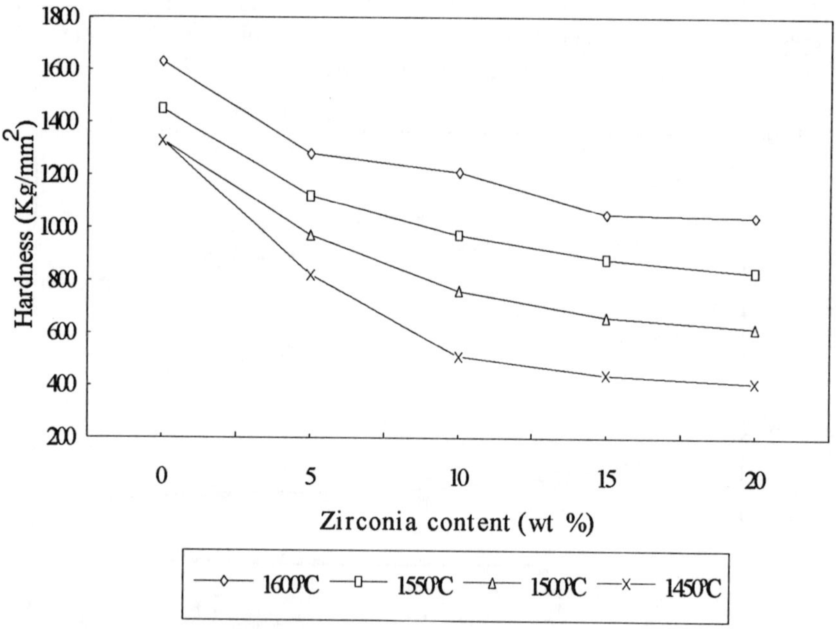

Figure 3. Dependence of hardness behavior on sintering temperature and ZrO_2 content

Figure 4 shows the dependence of toughness on sintering temperature and zirconia content. The highest values were attained by samples with 10-15 wt%, sintered at 1550 & 1600°C. Results are comparable to such obtained in composites prepared with low-soda content alumina as matrix material (4,8,13). The improvement of K_{ic} stems from the stress-induced phase transformation toughening.

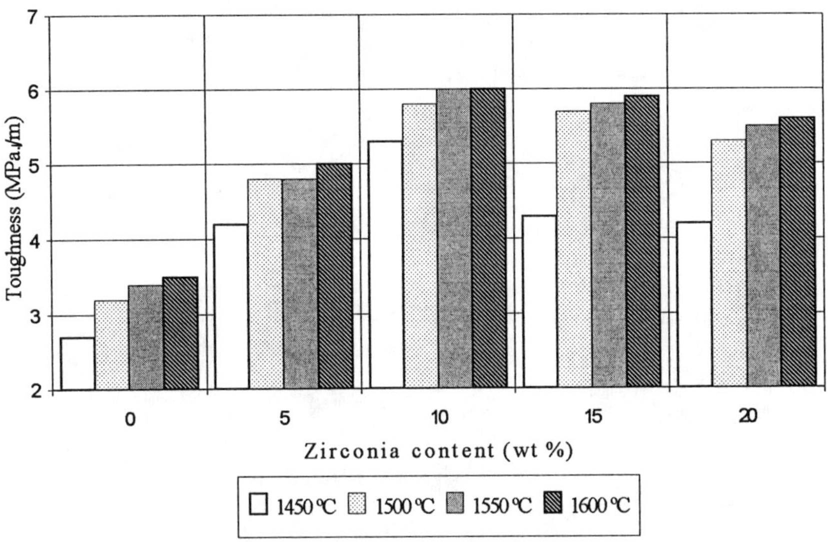

Figure 4. Dependence of fracture toughness on sintering temperature and ZrO_2 content

The microstructure of polished and fractured surface of Al_2O_3-10wt% ZrO_2 composition sintered at 1600°C-1h are shown in Fig. 5. Bright zirconia particles are homogeneously dispersed in the matrix (Fig. 5a) and were not necessarily located at grain boundaries but were also within alumina grains. This fact is better shown in figure 5b of the fracture surface of the same material where it can also be identified a transgranular type of fracture.

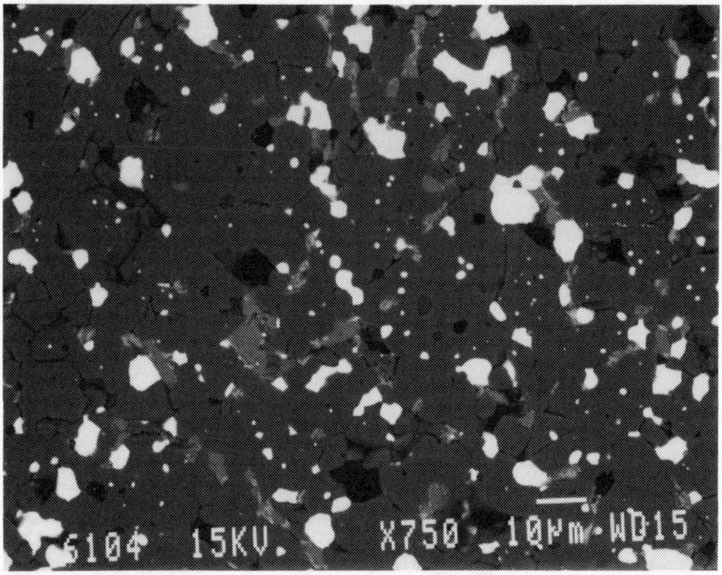

(a)

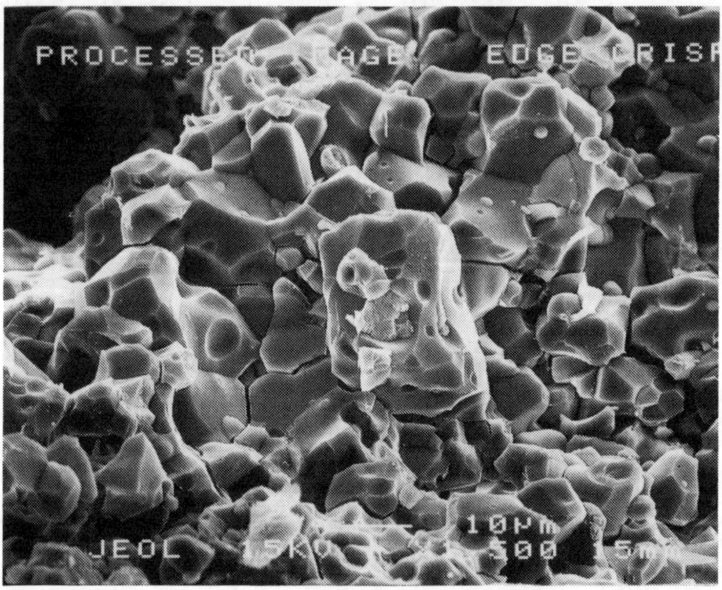

(b)

Figure 5. Scanning electron micrograph showing **(a)** ZrO_2 particle dispersion in matrix and **(b)** fracture surface of Al_2O_3 - 10 wt% ZrO_2 sintered compact

CONCLUSIONS

From this experimental study which is part of a research program to assess Venezuelan calcined normal-soda alumina as a ceramic grade raw material, it can be said that:

- Monolithic alumina compacts were sintered to greater than 98% of theoretical density at 1600°C for 1 h. Moderate toughness ($K_c \cong 3.5$ MPa.√m) and hardness values higher than 1600 Kg/mm^2 had been obtained.
- Zirconia particle reinforced alumina composites exhibiting a fracture toughness value of 6 MPa.√m and hardness higher than 1200 Kg/mm^2 have been fabricated by conventional powder processing route and pressureless sintering. The improvement in fracture toughness, for all sintering temperatures, compared to monolith alumina is possibly due to indent-induced transformation since a crack deflection mechanism can not be assured according to SEM results.
- At all sintering temperatures pure alumina exhibited higher densification and hardness.

Further work is being performed and will be published on the sinterability behavior of the Interalumina calcined alumina compared to other commercial aluminas and a treatment to obtain lower soda aluminas from Venezuelan raw materials (15,16).

ACKNOWLEDGEMENT

We wish to thank V.E. Robertson from JEOL Appl. Lab., Peabody, Mass., U.S.A. and Paulo Frías from Instituto de Ingeniería, Caracas, Venezuela for the SEM work carried out. This research work has been financed by I.U.T.-"Dr. Federico Rivero P." from the Venezuelan Ministry of Education.

REFERENCES

1. E. Dorre and H. Hubner, Alumina, Springer-Verlag, Berlin,1984.

2. T.J. Carbone, "Production processes, properties, and applications for calcined and high purity aluminas," Alumina Chemical, Science and Technology handbook, 107, 1990.

3. S.R. Witek and E.P. Butler, "Zirconia particle coarsening and the effects of Zirconia additions on the mechanical properties of certain commercial aluminas," *J. Am. Ceram. Soc.*, **69** [7] 523-29 (1986).

4. J.T. Lin and H.Y. Lu, "Grain Growth and mechanical property enhancement by adding ZrO_2 to Al_2O_3 matrix," *Ceramics International,* **14**, 251-58 (1988).

5. F.F. Lange, "Fabrication toughening, Part 4: Fabrication, fracture toughness and strength of Al_2O_3 - ZrO_2 composites," *J. Mater. Sci.*, **17** [1] 247-54 (1982).

6. K. Ranjbar, B.T. Rao, T.R. Rama Mohan and C.S. Harendranath, "Effect of chemically added Zirconia and Yttria on the mechanical properties of Zirconia-dispersed alumina," *Amer. Ceram. Soc. Bull.*, **73** [2] 63-66 (1994).
7. S. Bischoff, L. Michalowsky and G. Werner, "Structure formation and mechanical behaviour of Al_2O_3 / ZrO_2 (Y_2O_3) dispersion ceramics," *Ceramics International*, **16,** 121-27 (1990).
8. K. Ranjbar, B.T. Rao and T.R. Rama Mohan, "Effect of zirconia allotropes on fracture toughness of α-Alumina," *Ceramic Transactions,* **38**, 473-84 (1993).
9. A. Saha, D.C. Agrawal and A. Sharma, "Microstructure and mechanical properties of hybrid sol-gel processed Al_2O_3-ZrO_2 composites," *Ceramic Transactions,* **38**, 497-507 (1993).
10. P.F. Becher and C. Wei, "Toughening behavior in SiC-whisker reinforced alumina," *J. Am. Ceram. Soc.*, **67** [12] C-267-C-269 (1984).
11.C. Wei and P.F. Becher, "Development of SiC-Whisker-reinforced ceramics," *Am. Ceram. Soc. Bull.*, **64** [2] 298-304 (1985).
12. O. Sudre and F.F. Lange, "Effect of inclusions on densification: I, Microstructural development in an alumina matrix containing a high volume fraction of ZrO_2 inclusions," *J. Am. Ceram. Soc.*, **75** [3] 519-24 (1992).
13. M.I. Osendi and J.S. Moya, "Role of titania on the sintering, microstructure and fracture toughness of Al_2O_3-ZrO_2 composites," *J. Mat. Sci. letter,*7, 15-18 (1988).
14. D.J. Green, "Critical microstructures for microcracking in Al_2O_3 composites," *J. Am. Ceram. Soc.*, **65** [12] 610 -14 (1982).
15. I. Arenas and F. Arenas, "Sintering behavior of Al_2O_3-TiO_2-MnO doped with MgO or Cr_2O_3," *8th CIMTEC World Ceramic Congress Proceedings*, (1995).
16. M. Donat and A. Villani, "A treatment to obtain industrial aluminas from Venezuelan raw materials," *8th CIMTEC World Ceramic Congress Proceedings*, (1995).

FABRICATION BY COLLOIDAL FILTRATION OF ALUMINA PLATELET REINFORCED 3Y-TZP: MECHANICAL PROPERTIES

Isaac K. Cherian and W. M. Kriven
Department of Material Science and Engineering
University of Illinois at Urbana-Champaign
Urbana, IL 61801, U.S.A.

ABSTRACT

Composites of alumina platelet reinforced 3Y-TZP were fabricated using colloidal filtration for four different platelet sizes (3-5, 5-10, 10-15, and 15-25 μm). Data obtained by systematic investigations of the composite green body by mercury intrusion porosimetry (MIP) are presented along with room temperature mechanical properties, such as strength and fracture toughness of the sintered composite.

INTRODUCTION

In the development of ceramic matrix composites (CMC's), significant improvements in the mechanical properties of the composites have been achieved by the judicious use of reinforcements (1-3). The morphology of reinforcements such as fibers, whiskers, platelets and particulates provide for varying degrees of improvements with good processing. The best in properties are, however, obtained by fiber reinforcement, but at a higher cost due to the cost of the fiber and the requirement of high pressure densification.

Recent investigations of SiC and Al_2O_3 platelet based composites have produced properties between those of particulates and whiskers (4). Platelet based composites have the added advantage of low cost along with ease of processing (5). Alumina platelets offer the best in terms of reinforcing oxide matrix due to its oxidation resistance, unlike SiC.

Colloidal processing is a good technique for fabrication of two phase composites as compared to dry processing, since it has the advantage of uniform dispersion of second phase, narrow pore size distribution resulting in better densification characteristics and finally the reduction of strength degrading agglomerates (6). Sacks *et al* (7) had used this technique using SiC(w) and obtained good green state characteristics. Nischik *et al* (8) fabricated mullite composites reinforced with SiC and alumina platelets using colloidal filtration

To the extent authorized under the laws of the United States of America, all copyright interests in this publication are the property of The American Ceramic Society. Any duplication, reproduction, or republication of this publication or any part thereof, without the express written consent of The American Ceramic Society or fee paid to the Copyright Clearance Center, is prohibited.

and obtained improved mechanical properties, but they did not investigate the green state of the composite.

This paper is a systematic processing study of the effect of different platelet sizes on the green state, and mechanical properties of densified, colloidally processed alumina platelet reinforced 3Y TZP.

EXPERIMENTAL PROCEDURE

Processing

The composites were fabricated from sub micron 3Y-TZP powders (Tosoh Corporation, Tokyo, Japan) which had an equivalent spherical diameter of 0.3 micron and a specific surface area of $16 m^2/g$. Monocrystalline alumina platelets (Elf Atochem-France) used for this study consisted of four different size ranges of 3-5, 5-10, 10-15 and 15-25 microns. The thickness of these platelets were approximately one micron. Composite slips consisting of the above mentioned size ranges of platelets were prepared on a 30 vol % basis. The as-received platelets were ultrasonicated in deionized water to break up platelet agglomerates, and later sedimented to remove hard aggregates and particulates. The ultrasonicated platelets were then added into zirconia powders and ball milled in a Teflon bottle for 24 h using high purity zirconia milling media, and ammonium polymethacrylate dispersant. The slips were cast into disks, 4.5 cm in diameter and 1 cm thickness by casting a given volume of slip in a cylindrical tube containing a filter, across which a vacuum was held.

The samples were dried in air and bulk green densities were determined by the measurement of the disk dimensions. These samples were sintered at 1600°C in air for an hour and the sintered densities were determined by the Archimedes water displacement technique. Mercury intrusion porosimetry (MIP) was conducted on samples that were heat treated at 1100°C for one minute to determine the pore size distribution as a function of platelet size and volume fraction.

Mechanical Properties

The fired disks were machined into flexure bars of dimensions 25x2x1.5 mm, corresponding to configuration A of the MIL STD-1942. The machined bars had a surface finish of 37 microns and were annealed for 30 min at 1300°C to relieve stresses induced during machining. They were then fractured in four point bend configuration at a loading rate of 0.2 mm/min using a loading span of 10 mm and a support span of 20 mm. A minimum of seven bars were tested at each volume fraction of platelet loading.

The fracture toughness of the composite was determined by the chevron notch (CNB) method, utilizing the slice model of Munz et al (9). The CNB bend bars were of dimensions 25x3x4 mm, and had a slot width of 0.25 mm. The geometry of the notch was as follows; a_0 = 0.6 mm from the point to the bottom surface; a_1 = 3.5 mm which was the notch length from the slot starting point to the bottom surface of the bar. The bars were broken in four point configuration with a loading span of 10 mm and a support span of 20 mm at a loading rate of 0.05 mm/min. A minimum of four bars were tested at each volume fraction loading of platelets.

RESULTS

The green densities of all the samples were approximately 50% of the theoretical density and no significant change in the green densities was observed with platelet size or volume fraction. The fired densities of all samples were greater than 95% of theoretical density with the exception of samples containing 15 vol % or greater volume fraction of platelets in the size ranges 10-15 and 15-25 microns. These samples had large cracks resulting from densification.

Figures 1-3 show plots of the incremental volume against pore diameter for various volume fractions of platelets intruded by mercury into the composite specimens containing 3-5, 10-15, and 15-25 micron platelets. As can be seen from the plots, a very narrow distribution of pores was obtained with colloidal processing. However, as volume fractions of platelets were increased, the average pore diameter increased dramatically up to 15 vol% of platelets. Further loadings of platelets caused only a small increase in the average pore size. The average pore diameter for a given volume fraction of platelets remained at approximately the same value for all sizes of platelets studied. Figure 4 shows an SEM micrograph of the polished surface of the composite containing 5 vol % of 5-10 micron platelets that were uniformly dispersed as a result of colloidal filtration.

The variations of fracture strengths for composites containing 3-5, 5-10, 10-15 and 15-25 micron platelets are shown respectively in figures 5-8 . It can be seen that the strengths dropped with increase in platelet sizes and plateaued beyond 5 vol % loading, with the exception of composites containing 3-5 micron platelets.

Toughness values obtained as a function of platelet size and volume fraction loading in the composite are shown in figures 9-12. For composites containing 3-5 and 5-10 micron platelets, the average fracture toughness increased upto 15 vol % and dropped beyond it. For composites containing larger platelets of sizes 10-15 and 15-25 microns, the average value of toughness dropped with respect to unreinforced 3Y-TZP. The best values of toughness were had between 10 and 15 vol % loading for composites containing 5-10 micron sized platelets. The highest value of 8.3 MPa $m^{1/2}$ was obtained at 10 vol % for this composite which was a 16% increase over unreinforced 3Y-TZP.

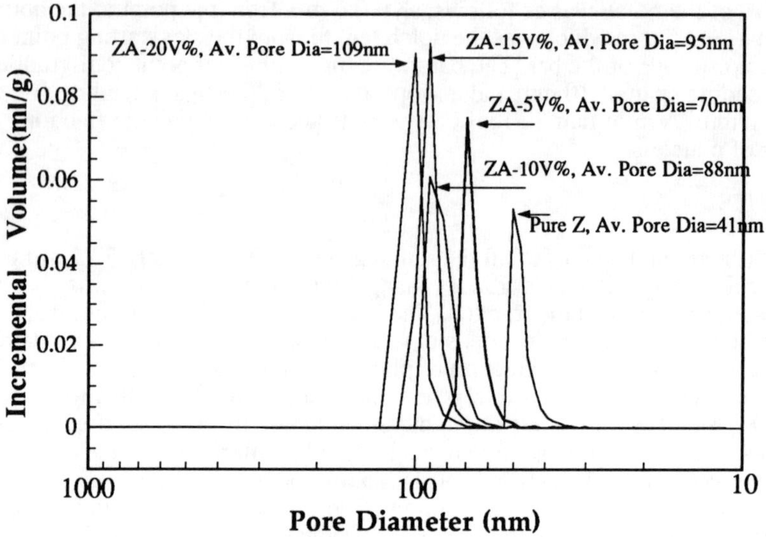

Fig. 1. Incremental Volume vs Pore Dia. of Composite containing 3-5 μm Platelets

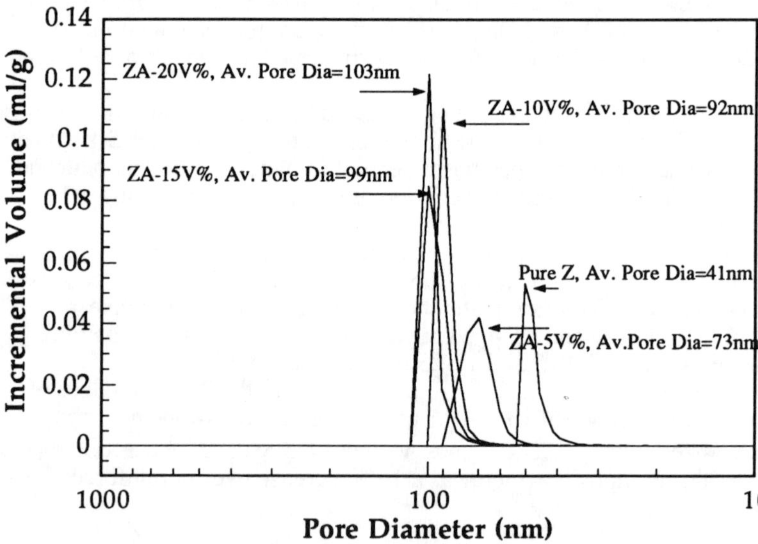

Fig. 2 Incremental Volume vs Pore Dia. of Composite containing 10-15 μm Platelets

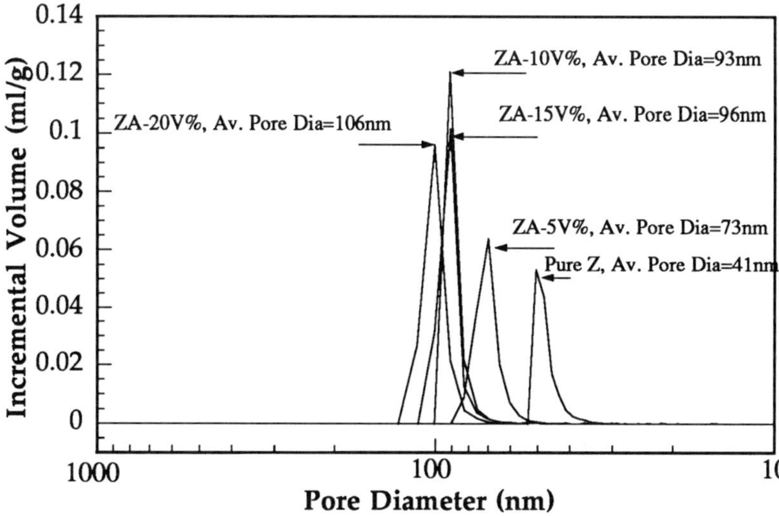

Fig. 3. Incremental Volume vs Pore Dia. of Composite containing 15-25 μm Platelets

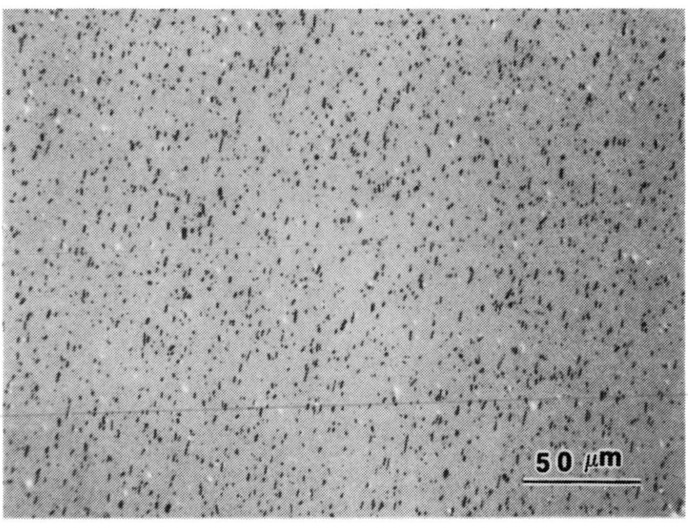

Fig. 4. SEM Micrograph of Polished Surface of 5-10 μm Platelet Composite Containing 5 vol % of Platelets

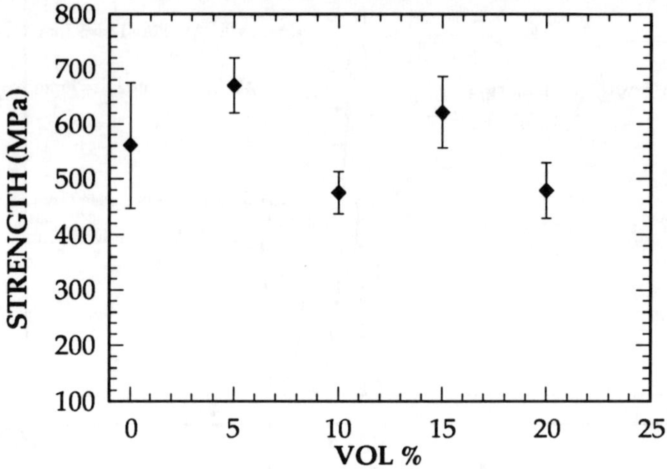

Fig. 5. Fracture Strengths vs Platelet Loading of 3-5 μm Platelet Composite

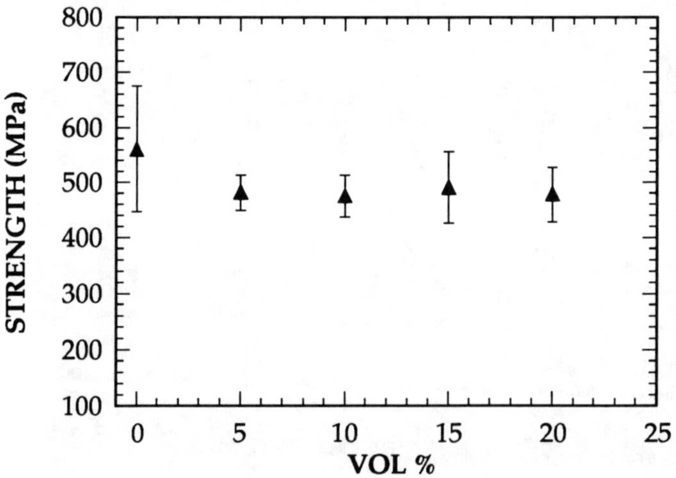

Fig. 6. Fracture Strengths vs Platelet Loading of 5-10 μm Platelet Composite

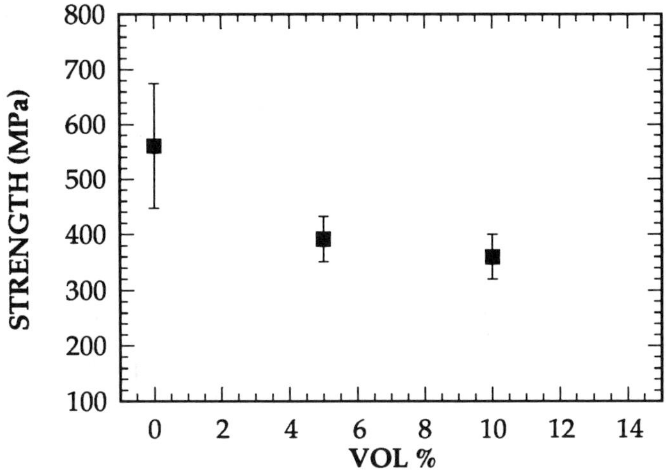

Fig.7. Fracture Strengths vs Platelet Loading of 10-15 μm Platelet Composite

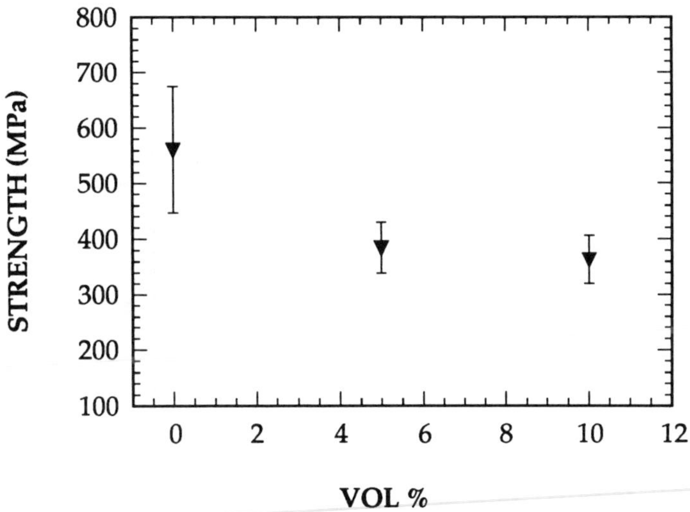

Fig.8. Fracture Strengths vs Platelet Loading of 15-25 μm Platelet Composite

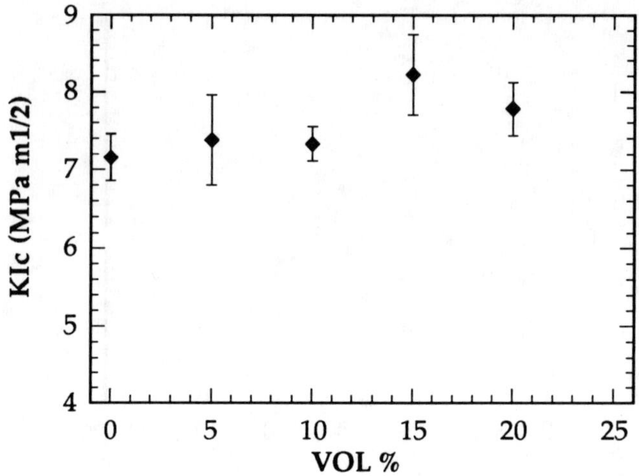

Fig.9. Fracture Toughness vs Platelet Loading of 3-5 μm Platelet Composite

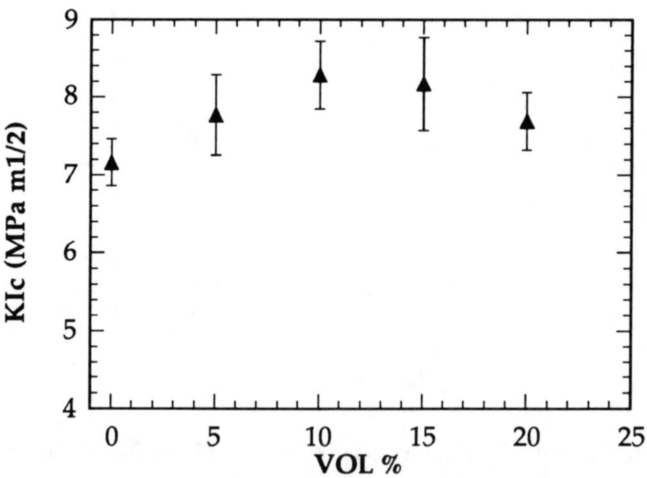

Fig.10. Fracture Toughness vs Platelet Loading of 5-10 μm Platelet Composite

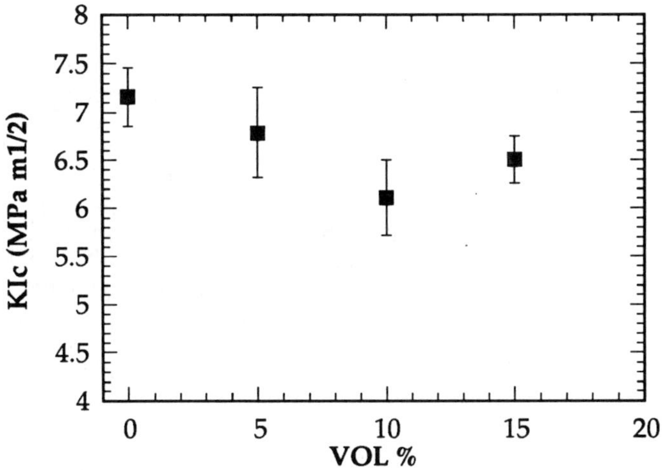

Fig.11 Fracture Toughness vs Platelet Loading of 10-15 μm Platelet Composite

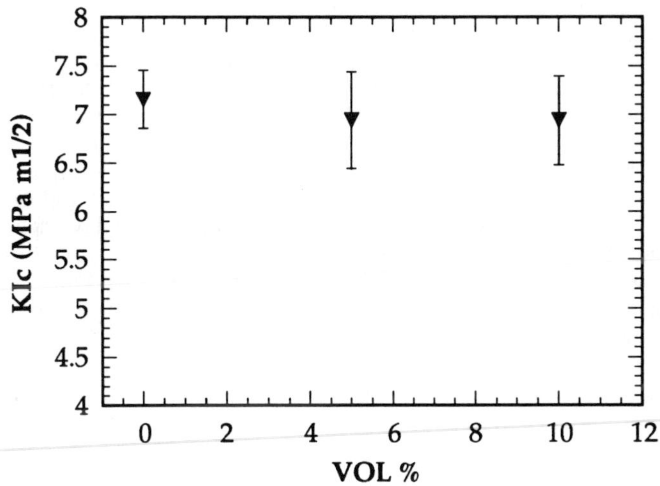

Fig. 12 Fracture Toughness vs Platelet Loading of 15-25 μm Platelet Composite

DISCUSSION

Processing

Composites containing up to 20 vol % platelets of sizes 3-5 and 5-10 microns sintered to greater than 95% of theoretical density, while composite samples containing greater than 15 vol % platelets of sizes 10-15 and 15-25 microns had large cracks in them. These cracked specimens were not tested for their mechanical properties. It is thought that the composites containing larger fraction and size cracked on account of large internal stresses arising due to larger aspect ratios of such platelets. Similar reasoning was put forth by Zhuang Li and Bradt (10) who calculated the internal stresses arising from SiC inclusion (platelet to whisker geometries) in alumina and found that the magnitude of the internal stresses scaled with the aspect ratios of the reinforcement. Constrained sintering (11-12) could be ruled out as a cause for cracking, since it would have affected composites containing smaller sized platelets as well, and higher densities would not have been possible.

MIP data was obtained on composite samples sintered at 1100°C for one minute. This was done in order to provide for some strength to the sample so that the intrusion of mercury would not crush the sample. The bulk densities of all specimens subjected to this heat treatment stayed at approximately 55% of theoretical, which was a gain of 5% compared to their bulk green densities. Obviously some very fine pores would have been lost, but the general characteristics of the pore structure were revealed. The pore sizes obtained by colloidal processing fell into a narrow range for each sample tested. With increasing volume fraction of platelets, the average pore diameter increased substantially upto 15 vol % of loading for all samples, while keeping the distribution of pores narrow. Beyond 15 vol % of platelet loading, the relative increase in the pore diameter was small and the pore size distribution began to overlap significantly. This would suggest that for the platelets, the onset of percolation would be around 15 vol %, where the platelets would have spanned the entire volume of specimen (13). Further increment in the platelet loading beyond the percolation limit would result in local distortions in packing, leading to flaws in the fired sample and consequently poorer fracture strengths in the composite. Thus one would expect relatively flaw free microstructures leading to higher sintered densities of composites below the percolation limit of the reinforcement. This aspect coupled with colloidal processing, which offers narrow pore sizes, renders composites with good sinterabilty and fired densities.

Mechanical Properties

The fracture strengths of the composites diminish with increasing platelet sizes. It is thought that platelets act as flaws causing lower strengths. This is because platelets are the largest heterogeneity in the microstructures when compared to the grain sizes of the matrix, which are less than one micron. Heussner and Claussen (14) provide a similar explanation for the drop in the fracture strengths of their composites containing platelets in the size ranges 2-40 microns, and consolidated by hot isostatic pressing. The trends in the strength data are somewhat different from those of the above authors in that a large plateau in strength was observed beyond 5 vol % loading for the composites containing 5-10 micron platelets and larger. The strengths were ~ 500 MPa for composites containing 5-10 micron platelets, while that of the 10-15 and 15-25 micron platelets plateaued at a value of 400 MPa.

The average toughness values of the composites for platelet sizes 3-5 and 5-10 micron platelets show a small gain in toughness when compared to the unreinforced 3Y-TZP (7.14 MPa $m^{1/2}$) over the entire range of platelet loadings. A maximum of 8.3 MPa $m^{1/2}$ in the toughness value was observed in the 5-10 micron platelet composite at a platelet loading of 10 vol %. This increase in toughness is believed to be due to alumina platelets providing for crack deflection mechanisms in the composite. Values of toughness obtained from the CNB method by Heussner and Claussen (14) for their composites were approximately 6 MPa $m^{1/2}$ for composites containing platelets in the range 5-20 vol %. Dey and Shetty (15) used a sintered composite containing 5-10 micron platelets in TZP and obtained a value of 6.64 MPa $m^{1/2}$ for their composite measured by the CNB method, at a platelet loading of 15 vol%. Composites obtained by colloidal filtration had better overall value of toughness, when compared to the above workers for platelets in the size ranges 3-5 and 5-10 microns. However in this study values of toughness decreased to less than unreinforced 3Y-TZP for composites containing 10-15, and 15-25 micron platelets. This is thought to be due to large scale formation of microcracks in the composite due to internal stresses arising due to larger geometry of the platelets. Further TEM investigations are needed to confirm this hypothesis

CONCLUSIONS

Fabrication of alumina platelet reinforced 3Y-TZP composites by colloidal filtration rendered narrow pore size distributions in the green state along with uniform distribution of platelets. The average pore sizes increased with addition of platelets and were around the same value for all platelets studied. Composites containing 3-5, 5-10 micron platelets sizes were sintered to greater than 95% of theoretical densities up to 20 vol % loading of platelets at 1600°C for 1h. Composites containing 10-15 and 15-25 micron platelets beyond 10 vol % loading cracked during sintering due to internal stresses that scaled with platelet size.

The fracture strengths of the composites decreased with increasing platelet sizes due to platelets acting as flaw sites. The value of the fracture strengths plateaued beyond 5 vol % loading for all sizes of platelets, with the exception of 3-5 micron platelets.

The toughness of the composites increased with addition of platelets of sizes 3-5 and 5-10 microns up to 20 vol % loading. The maximum in toughness of 8.3 MPa $m^{1/2}$ was obtained for 5-10 microns platelet sizes at 10 vol% loading. This by far is the best value obtained by the CNB method. Fracture toughness of composites containing 10-15 and 15-25 micron platelets decreased below that of the unreinforced 3Y-TZP possibly due to extensive microcracking in the composites, resulting from large internal stresses caused due to large platelets.

ACKNOWLEDGEMENT

The authors thank Drs. D. Zhu and C.H. Huang for valuable discussions. Elf Atochem of France is gratefully acknowledged for supply of platelets.

REFERENCES

1) P. F. Becher, "Microstructural Design of Toughened Ceramics," *J. Am. Ceram. Soc.*, **74** [2] 255-69 (1991).
2) D. B. Marshall and J. E. Ritter, "Reliability of Advanced Structural Ceramics and Ceramic Matrix Composites-A Review," *Am. Ceram. Soc. Bull.*, **66** [2] 309-317 (1987).
3) A. G. Evans and R. M. McMeeking, "On Toughening of Ceramics by Strong Reinforcements," *Acta Metall.*, **34** [12] 2435-41 (1986).
4) I. Wadsworth and R. Stevens, "Strengthening and Toughening of Cordierite by the Addition of Silicon Carbide Whiskers, Platelets and Particles," *J. Mater. Sci.*, **26,** 6800-6808 (1991).
5) N. Claussen. "Ceramic Platelet Composites"; in Proceedings of the 11th RISO International Symposium on Metallurgy and Material Science 1990, Structural Ceramics Processing, Microstructure and Properties. Edited by J. J. Bentzen , J. B. Bilde-Sorensen, N. Christiansen, A. Horsewell, and B. Ralph. RISO National Laboratory, Roskidle, Denmark, 1990.
6) F. F. Lange, "Powder Processing Science and Technology for Increased Reliability," *J. Am. Ceram. Soc.*, **72** [1] 3-15 (1989).
7) M. D. Sacks, H. W. Lee, and O. E. Rojas, "Suspension Processing of Alumina/Silicon Carbide Whisker Composites," *J. Am. Ceram. Soc.*, **71** [5] 370-79 (1988).
8) C. Nischik, M. M. Seibold, N. A. Travitzky, and N. Claussen, "Effect of Processing on Mechanical Properties of Platelet-Reinforced Mullite Composites," *J. Am. Ceram. Soc.*, **74** [10] 2464-68 (1991).
9) D. G. Munz, J. L. Shannon, Jr., and T. Bubsey, " Fracture toughness Calculations from Maximum Load in Four Point Bend Tests of Chevron Notch Specimens," *Int . J. Frac.*, **16**, R137-41 (1980) .
10) Zhuang Li , and Richard C . Bradt , "Micromechanical Stresses in SiC-Reinforced Al_2O_3 Composites", *J. Am. Ceram. Soc.*, **70** [1] 70-77 (1989).
11) F. F. Lange, "Constrained Network Model for Predicting Densification Behavior of Composite Powders," *J. Mater. Res.*, **2** [1] 59-65 (1987).
12) R. K. Bordia and G.W. Scheerer, "On Constrained Sintering-III Rigid Inclusions," *Acta Metall.*, **36** [9] 2411-2416 (1988).
13) R. Zallen, *The Physics of Amorphous Solids*; Ch. 4, pp. 135-204. Wiley, New York , 1983
14) K. H. Heussner and N. Claussen, "Yttria-and Ceria-Stabilized Tetragonal Zirconia Polycrystals Reinforced with Alumina Platelets," *J. Euro. Ceram. Soc.* **5**, 193-200 (1989).
15) S. Dey and D. K. Shetty, "Processing and Characterization of Alumina (Platelet) Reinforced Y-TZP (Matrix) Composites", p 461-71, Ceramic Transactions Vol. 38, Advances in Ceramic Matrix Composites.

MULLITE PHASE DEVELOPMENT IN PURE AND MULLITE MATRIX COMPOSITES

Young-Hyuk You and Avigdor Zangvil
Frederick Seitz Materials Research Laboratory and Dept. of Materials Science and Engineering, University of Illinois, Urbana, Illinois 61801

I. Introduction

The mullite phase development has been studied extensively for a long time using many different starting materials. Recently, the mullite formation kinetics and mechanisms were re-investigated using chemically prepared gels or powders because chemical compositions and mixing ratios of such starting materials were easy to control. However, numerous conflicting investigations have been reported about the mullite formation mechanisms.[1-8]

The mullite formation mechanisms in chemically derived starting materials can be classified into three general categories related to the "mixing scale" of starting materials: (1) Direct mullite formation from a single-phase gel or powder at around 980°C without intermediate phase[1,5,8] ; (2) The mullite formation through reaction between γ-Al_2O_3 (or θ-Al_2O_3) and amorphous silica at above 1250°C in a colloidal gel or powder.[3,4] ; (3) The mullite transformation through an intermediate Al-Si spinel phase.[2,7]

The main objective of the current study was to determine the microstructural/microchemical characteristics of mullite formation in mullite and mullite/zirconia composites with or without MgO, through direct microstructural observations especially for: (1) the relationship of mullite formation to the pre-mullite spinel phase, and (2) the effect of zirconia and MgO on the mullite matrix phase development.

II. Experimental Procedure

The mullite and mullite/zirconia composite powders were prepared through an alkoxide process. Mixtures of Al, Zr and Mg alkoxides and TEOS were refluxed for 24 hr and hydrolyzed in a strong acid water bath (PH~2) by dropping the sol. The details of the powder preparation methods and characterization of the powders were described elsewhere.[9,10] The chemical compositions after calcination for each powder are shown in Table 1.

The powders were calcinated and slightly crushed with alumina mortar and pestle, but the powders remained extremely agglomerated. The pellets, cold

To the extent authorized under the laws of the United States of America, all copyright interests in this publication are the property of The American Ceramic Society. Any duplication, reproduction, or republication of this publication or any part thereof, without the express written consent of The American Ceramic Society or fee paid to the Copyright Clearance Center, is prohibited.

pressed and CIPed under 22,500 psi, were sintered at up to 1600°C for 1 hr under vacuum in the range of $10^{-4} \sim 10^{-5}$ torr with a standard heating rate of 20°C/min.

Table 1. Chemical Compositions of the Powders

Powder	*Chemical composition (wt%)**			
	Al_2O_3	*SiO_2*	*ZrO_2*	*MgO*
M-RH	74.5	25.5		
MZ-RH	51.6	16.9	31.5	
MZMg-RH	48.7	18.1	32.3	0.9

* The chemical composition of the powders were analyzed by University of Illinois Micro. Analytical Laboratory.

The phase evolution temperatures were measured by DTA (Innovative Thermal Systems) and the phases were analyzed by XRD (Model D-Max Rigaku, Japan). For the SEM observation, the sintered samples were ground and polished down to 1 μm diamond and then thermally etched. The thermal etching conditions were 1350°C for 30 min under a vacuum of $10^{-4} \sim 10^{-5}$ torr. TEM specimens of the condensed sample were prepared by standard procedures. Two transmission electron microscopes were used: An analytical TEM/STEM and a conventional TEM (Models Philips EM 420 and CM 12, respectively). They were equipped with energy dispersive spectrometers (EDAX International), one being an ultra-thin window EDS. The details for the microcharacterization techniques were described elsewhere.[10]

III. Results and Discussion

(1) Pure Mullite System

The M-RH powder after calcination, including the composite powders, was characterized in our earlier study to have diphasic structure with several nm mixing scale of γ-Al_2O_3 particles and silica rich amorphous phase.[10] The DTA plot of M-RH conducted up to 1320°C, shows no exothermic peak corresponding to the mullite formation. XRD results, however, indicated that a significant amount of mullite phase had already formed in a powder taken from the DTA crucible after examination. As shown in Fig. 1, the crystal structure of the first forming mullite was orthorhombic in M-RH. As sintering temperature increased, the spinel phase gradually disappeared with the extensive mullite formation (Fig. 2). However, at 1100°C, the *maximum* amount of spinel phase was developed with mullite formation simultaneously; The spinel phase began to decompose at above 1200°C. From the above observations, it has become clear that the mullite development at the early stage of formation was independent of the spinel decomposition. This is consistent with Li and Thomson's[8] observation of simultaneous mullite and Al-Si spinel formation in a single phase gel. Another important result of Fig. 2 is the existence of an incubation time for the orthorhombic mullite crystallization at 1100°C after extended heating (e.g. 24 hr). The existence of an incubation time for the mullite crystallization indicates that the mullite formation is a time dependent diffusion controlled process.

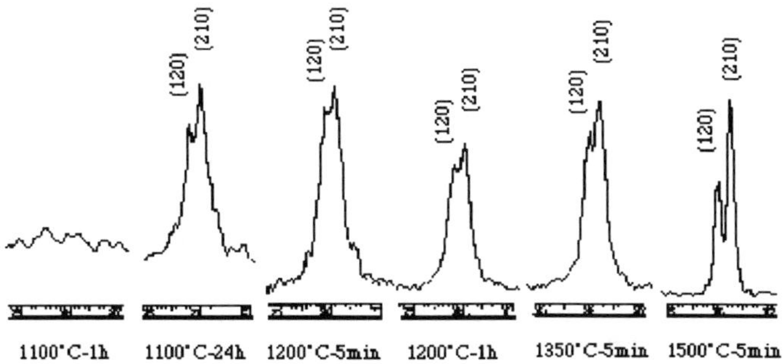

Fig. 1. Orthorhombic mullite formed directly in M-RH. Sintering conditions are indicated below the XRD patterns.

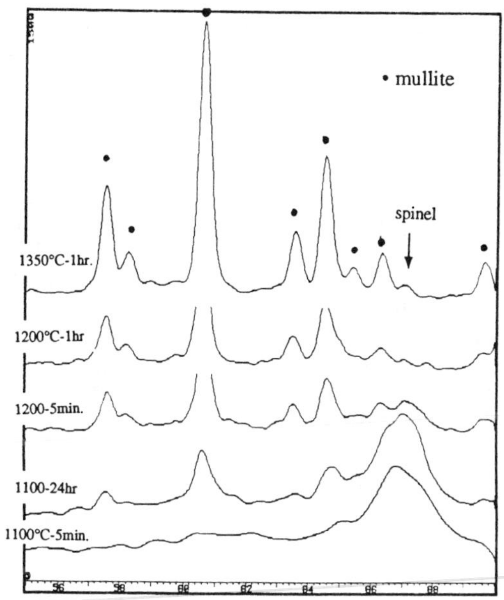

Fig. 2. Spinel phase begins to disappear at above 1200°C at which the majority of mullite crystallizes.

To investigate the structural and chemical composition changes of mullite during the transformation, the lattice parameters were measured from XRD patterns. Fig. 3 shows the changes of lattice parameters **a**, **b**, **c** and the unit cell volume of mullite with sintering temperature. The most dramatic change occurred in the **a** axis, which decreased rapidly right after mullite formation. On the basis of Cameron's relationship[11] between the **a**-axis cell dimension and the content of Al_2O_3 in mullite, the abnormally long **a**-axis of the first crystallized mullite was supposed to indicate 87 wt% (80 mol%) of Al_2O_3 in mullite. In this study, the chemical composition of the mullite at the early stage of formation was measured by analytical TEM, in addition to the XRD determination, as shown in Fig. 4. In contrast to the XRD-derived result, analytical TEM showed that the chemical composition of the mullite was close to 2:1. Therefore, it can be concluded that the abnormally long **a**-axis observed in XRD can not be used directly to evaluate the mullite chemical composition. Instead, the long **a**-axis is believed to be related to the pseudo-tetragonal crystal structure of mullite[12,13] or to an open crystal structure with ordered domains of oxygen vacancies which has been reported in alumina-rich mullite.[11,14,15]

The microstructures of the M-RH samples condensed at various sintering temperatures and times are shown in Fig. 5. Fig. 5 (a) shows a dark-field (DF) image of a poorly condensed M-RH sample (1100°C for 1h), obtained by using part of the {440} reflection ring. It shows the spinel particles with a size less than 10 nm. After isothermal annealing at 1100°C for 24 hr, the diffraction ring pattern of the spinel phase became more obvious and consisted of many diffraction spots (Fig. 5 (b)), compared to the diffuse ring pattern of the sample sintered at 1100°C for 1h. However, the growth of the spinel particles was limited to the several nanometer scale of the early-detected spinel particles. Fig. 5 (c) shows the extensive mullite formation with some amount of spinel particles after sintering at 1200°C/5min. The diffraction pattern shows numerous diffraction spots reflected from mullite grains and a very faint ring pattern of spinel particles.

Interestingly, the mullite grains at the beginning of formation are relatively large and show a broad size distribution from several nm to 300 nm (Table 2), with some nearly round and irregular shapes.

Table 2. Average Mullite Grain Size in M-RH, Measured by XRD and TEM

Temperature(°C)-Time	*XRD (nm)*	*TEM (nm)*
1100 - 24hr	24	30 (7-320)*
1200 - 5min	23	30 (5-250)
1200 - 1hr	29	-
1350 -5min	23	-
1350 - 1hr	27	40 (5-300)
1500 - 1hr	-	162
1600 - 1hr	-	647

* numbers in parenthesis indicate the observed mullite size range

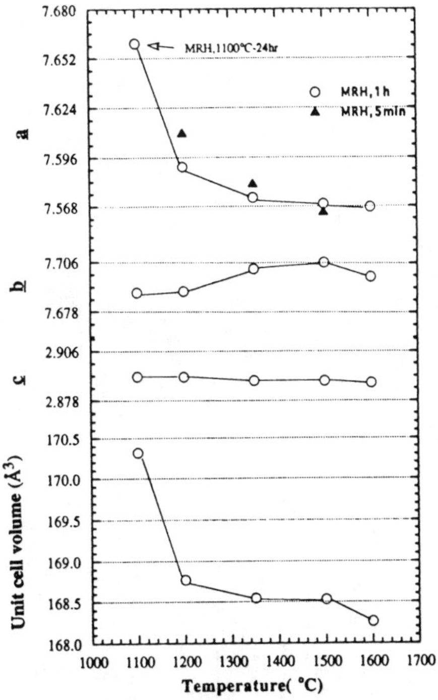

Fig. 3. Lattice parameters of the mullite developed in the M-RH samples as determined from XRD at different temperatures.

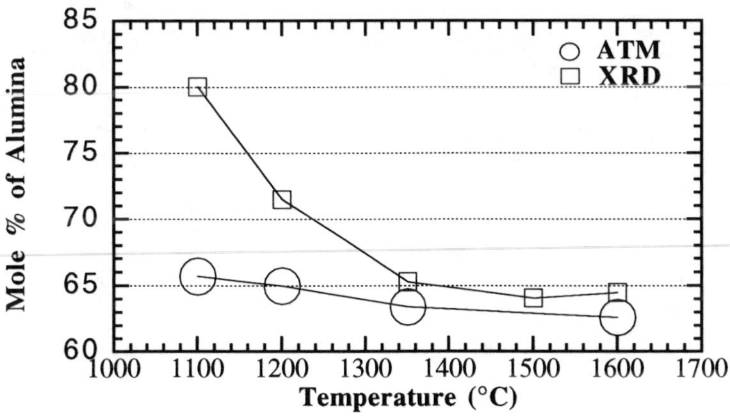

Fig. 4. Chemical composition changes of mullite with temperature, as determined by XRD and analytical TEM.

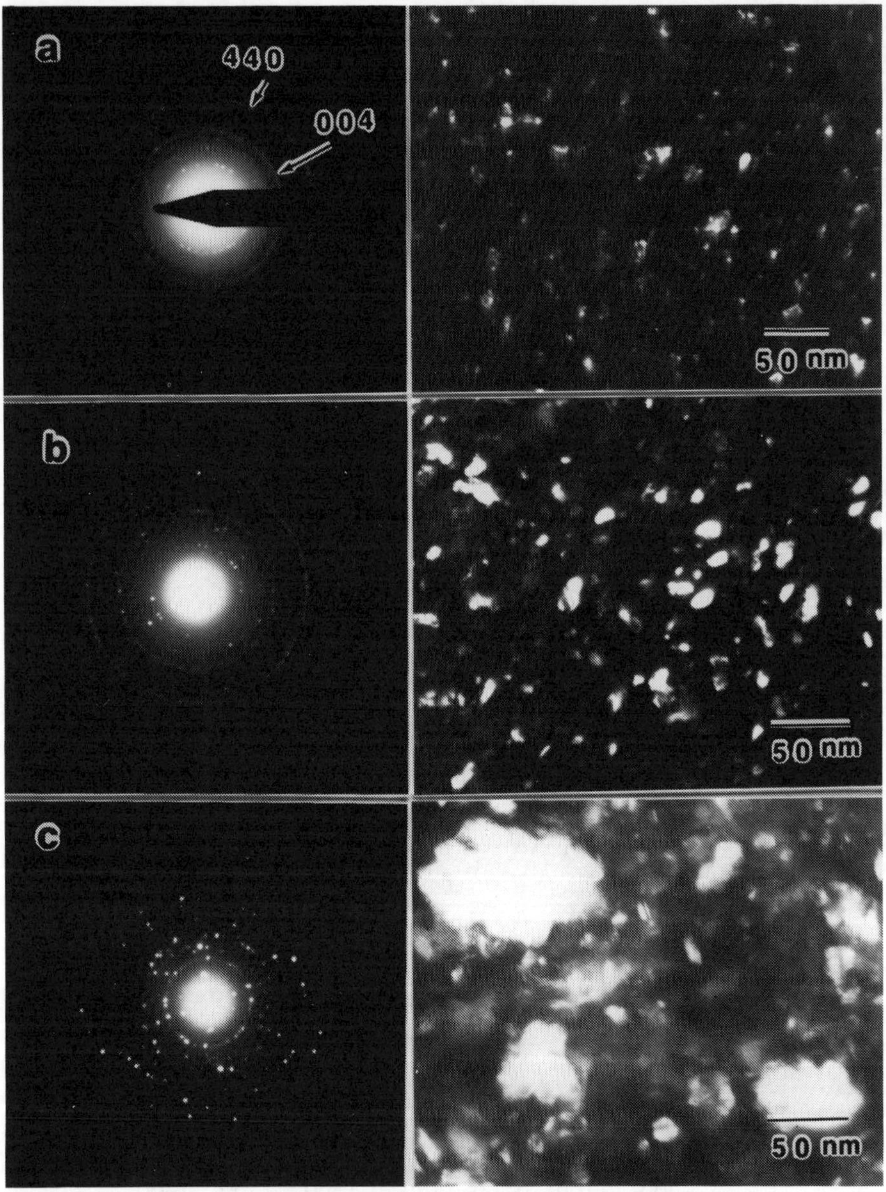

Fig. 5. SADP and DF micrographs of sample sintered at; (a) 1100°C for 1 hr, (b) 1100°C for 24 hr, and (c) 1200°C for 5 min.

The significant size difference between spinel particles and first crystallizing mullite grains suggested that the mullite was not directly transformed from the spinel particles. Although the TEM micrograph shows wide size distributions of the mullite grains, the average grain size estimated from the micrographs was close to that calculated from XRD broadening (Table 2).

The average mullite grain size was nearly constant at 23 to 29 nm, in samples sintered at up to 1350°C for 1hr where the mullite has nearly completely crystallized. Therefore, it is concluded that the grain size of the mullite developed at up to 1350°C is the same as that of the early nucleating mullite, and further grain growth is limited by poor particle contact, caused by the poor densification.

The large size variation of the mullite grains suggests that diffusion controlled (or interface diffusion controlled) grain growth was not a limiting factor for the mullite growth mechanism of M-RH. Instead, the difficulty of nucleation through short-range diffusion might be the limiting process while further growth of the mullite grains is blocked by poor contacts of particles.

As mentioned above, the mullite grains developed independently of the spinel phase and formed individually in the middle of the amorphous matrix and showed no particular characteristic crystal shapes as shown in Fig. 6. Thus the lack of particular grain shape of the mullite with irregular but more or less equiaxed grains indirectly proves again that the mullite was formed through a very short range (nanometers) solid state diffusion process.

As densification proceeds at above 1350°C, the mullite grains begin to come in contact with each other, resulting in polygonal shapes and the mullite grain size distribution became narrow at around 40 nm. At this temperature, no glassy phase was observed between the mullite grains as shown in Fig. 7. Occasionally, abnormal grain growth of the mullite was observed in a sample sintered at above 1500°C/1hr. It is believed that the abnormal grain growth in M-RH was caused by large size differences of the nucleated mullite grains at the beginning of crystallization, as described before.

(2) Composite Systems

The mullite phase development in the composite systems (mullite/zirconia and mullite/zirconia/MgO) was investigated and compared to that in the pure mullite system. The SEM micrographs of the composites, sintered at 1500°C for 1h, are shown in Fig. 8 along with that of pure mullite (a). While the zirconia addition had slight change on the mullite matrix microstructure, there was a significant difference in the microstructure of composites with and without MgO: equiaxed microstructure without liquid phase formation in composites without MgO; highly elongated and enlarged mullite grains with liquid phase formation in composites with MgO.

In this study, the zirconia effect on the mullite grain growth was investigated by measuring the grain size and solid solubility with sintering temperatures, in the composites without MgO. In mullite/zirconia composites, the exaggeratedly grown mullite grains which were frequently found in the pure system were not observed, indicating a refining effect of zirconia on the mullite grain size. The mullite grain sizes of the samples sintered at 1200°C and 1350°C for 5 min were estimated using the XRD peak broadening technique. However,

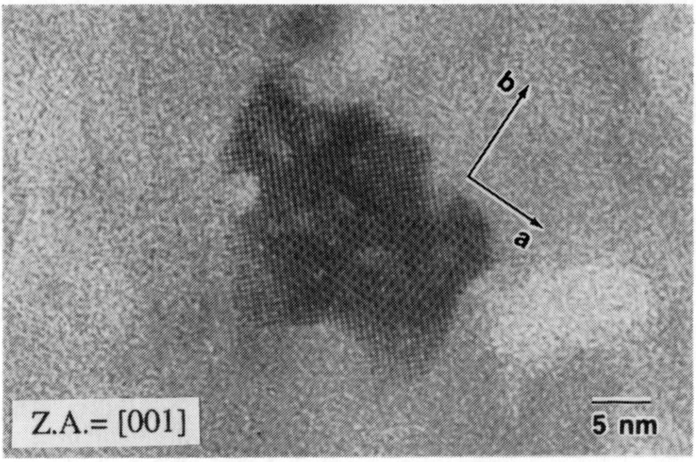

Fig. 6. A mullite grain nucleated in the middle of the amorphous phase. The mullite grain is parallel to the **c**-axis.

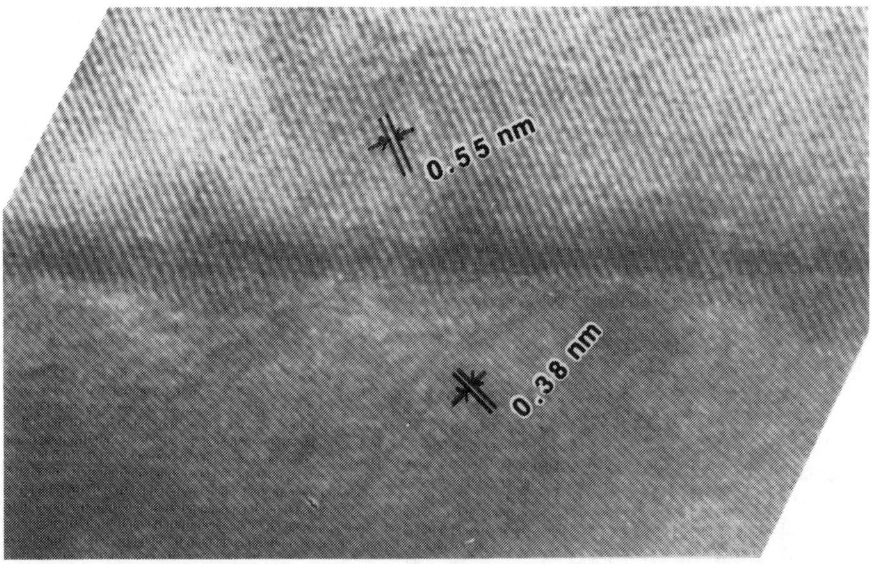

Fig. 7. HREM shows a liquid-phase-free boundary between mullite grains in MZ-RH sample sintered at 1500-1 hr. Line images of 5.50 Å and 3.82 Å correspond to (110) and (200) planes, repectively.

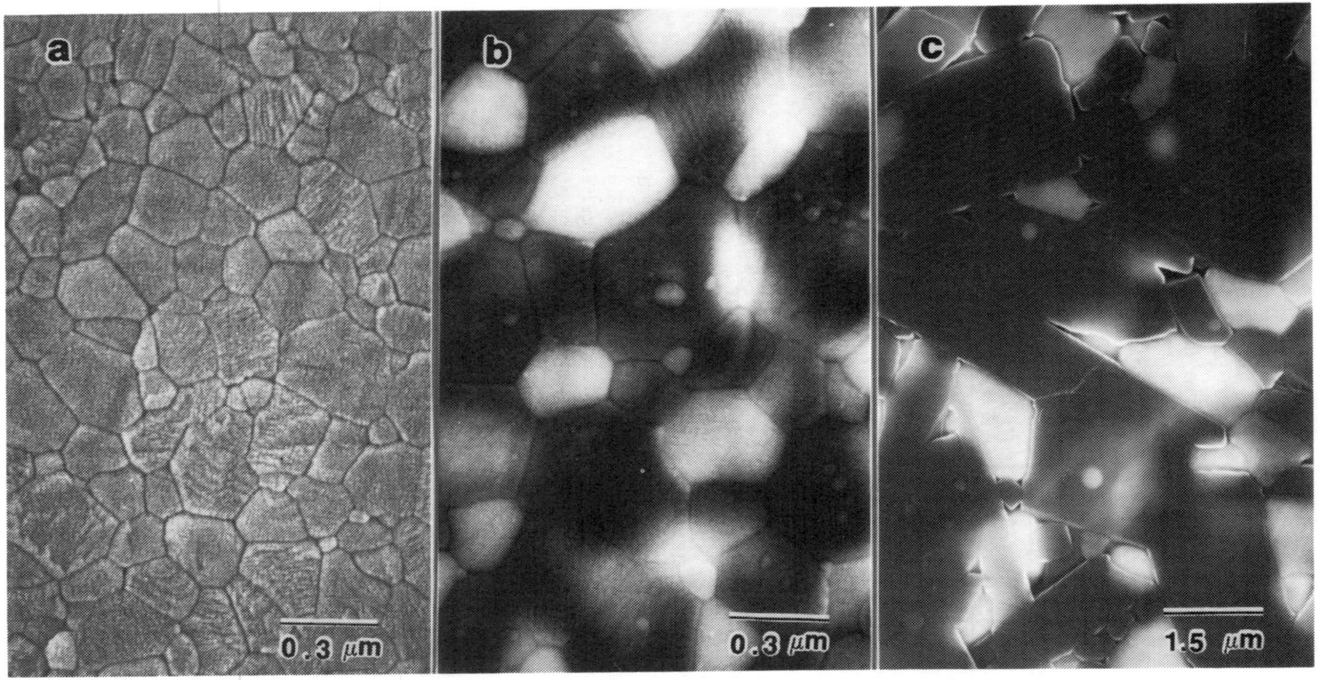

Fig. 8. Typical SEM micrographs of mullite and mullite/zirconia composites, sintered at 1500°C for 1 h. (a) M-RH (b) MZ-RH (c) MZMg-RH. The lighter grains are zirconia.

the average mullite grain sizes of the samples sintered at 1500°C and 1600°C were measured directly from SEM micrographs. The results are shown in Table 3.

Table 3. Average Mullite Grain Size (nm)* of the Composites with Sintering Conditions.

specimen	*1200°C -5min*	*1350°C -5min*	*1500°C-1h*	*1600°C-1h*
M-RH	23	23	160	620
MZ-RH	29	33	270	510
MZMg-RH	33	-	Elongated (AR*:3~5), L≈3 μm	Highly elongated (AR≥6)

*Aspect ratio

At the beginning of mullite formation, at 1200°C, the average mullite grain size was nearly identical in both pure and composite samples. However, the average mullite grain size of the composites at intermediate sintering temperatures (between 1350°C and 1500°C) was slightly larger than that of pure mullite. At above 1500°C, the average mullite grain size in the pure system (~0.6 μm) was larger than that of the MZ composites (~0.5 μm) in spite of smaller sizes of the mullite grains at the intermediate sintering temperatures.

The similar grain size distribution in both pure and composite samples at the mullite formation stage suggests that the nano size zirconia precipitation of less than 30 nm[10] had nearly no effect on the mullite grain growth after nucleation. In this stage, the mullite grain growth occurred through a nearly diffusionless process until it reached a size which was generally related to the agglomerate size and its growth was fast enough to entrap the precipitated zirconia or other phases such as amorphous pockets and spinel particles.

The solid solubility of the zirconia in mullite grains, investigated by EDS/TEM, was nearly constant at around 1wt% with temperature from the beginning of mullite formation at 1200°C to final sintering at 1500°C. Therefore the dependence of mullite grain growth on the change of zirconia solubility in mullite with temperature might be negligible.

At above 1500°C, where the amorphous phase has completely crystallized, grain growth should have occurred purely through solid state diffusion. At this stage, the large intergranular zirconia, located at the triple junctions of mullite grains, was very effective as grain growth inhibitor by pinning the mullite grains,[16] resulting in smaller grain size in the composites than that in the pure mullite, as shown in Table 3.

However MgO, added as stabilizer of zirconia, changed the microstructure of mullite and zirconia grains into elongated and enlarged ones, respectively, by forming a low-temperature-melting liquid phase. The liquid phase was determined by analytical TEM to be Mg-Si enriched (MgO: 13.1, Al_2O_3: 25.2, SiO_2: 57.4, and ZrO_2: 4.2 wt% of oxide). Therefore, in spite of the similar initial mullite grain sizes in composites with and without MgO, the mullite and zirconia grains later grow at a faster rate through liquid boundary phase diffusion in the composites with MgO.

Summary and Conclusion

(1) In the alkoxide-derived mullite and mullite/zirconia composites, with several nm mixing scale powder structure, the microstructural/microchemical characteristics of the mullite formation were investigated.

(2) In contrast to the nearly constant size of the spinel particles, the first appearing mullite grains showed a wide size distribution, from several nm to 300 nm, which was comparable to the primary particle size in the powder agglomerates. The average mullite grain size was constant for temperatures up to 1350°C and limited to the first appearing mullite crystal size. Therefore the mullite formation mechanism can be described as follows: The short range diffusion controlled nucleation is the rate limiting process for crystallization, and its growth after nucleation is so fast as to reach the primary particle size of the powder state. Further grain growth, however, is limited by poor contact of particles due to the poor densification.

(3) The mullite crystallized within the amorphous phase with irregular crystal shapes (but generally equiaxed) without the characteristic needle like morphology, indicating the short range of diffusion during transformation. Mullite crystallization at the early stage of formation was not related to the decomposition of nano size spinel.

(4) The addition of zirconia enhanced the grain growth of the mullite matrix only at the intermediate sintering temperatures (between 1200°C and 1500°C). At the final sintering stage, zirconia limited mullite grain growth by grain boundary pinning.

(5) Mullite/zirconia composites without MgO showed equiaxed grain growth and no liquid glassy phase was found at the grain boundaries. Composites with MgO, however, showed elongated mullite grains due to the formation of a Mg- and Si-enriched liquid phase at between 1350 and 1400°C.

Acknowledgment

This work is supported by DOE/Div. of Materials Science, under contract DEFG02-91ER45439

References

[1] J. A. Pask, X. W. Zhang, A. P. Tomsia and B. E. Yoldas, "Effect of Sol-Gel Mixing on Mullite Microstructure and Phase Equilibria in the α-Al_2O_3-SiO_2 system," *J. Am. Ceram. Soc.*, **70** [10] 704-7 (1987)

[2] K. Okada, N. Otsuka and J. Ossaka, "Characterization of Spinel Phase Formed in the Kaolin-Mullite Thermal Sequence," *J. Am. Ceram. Soc.*, **69** [10] c251~c253 (1986)

[3] W. C. Wei and J. W. Halloran, "Transformation Kinetics of Diphasic Aluminosilicate Gels," *J. Am. Ceram. Soc.*, **71** [7] 581-7 (1988)

[4] S. Sundarensan and I. A. Aksay, "Mullitization of Diphasic Alumino Silicate Gels," *J. Am. Ceram. Soc.*, **74** [10] 2388-92 (1991)

[5] D. X. Li and W. J. Thomson, "Mullite Formation Kinetics of a Single-Phase Gel," *J. Am. Ceram. Soc.,* **73** [4] 964-969 (1990)

[6] D. X. Li and W. J. Thomson, "Tetragonal to Orthorhombic Transformation During Mullite Formation," *J. Mat. Res.*, **6** [4] 819-824 (1991)

[7] J. C. Huling and G. L. Messing, "Epitactic Nucleation of Spinel in Aluminosilicate Gels and Its Effect on Mullite Crystallization," *J. Am. Ceram. Soc.*, **74** [10] 2374-81 (1991)

[8] D. X. Li and W. J. Thomson, "Effects of Hydrolysis on the Kinetics of High-Temperature Transformation in Aluminosilicate Gels," *J. Am. Ceram. Soc.*, **74** [3] 574-78 (1991)

[9] L. A. Paulick, Y. F. Yu and T-I. Mah, "Ceramic Powders from Metal Alkoxide Precursors," pp 121-9 in Advanced in Ceramics vol.21, Ceramic Powder Science, edited by Messing, Mazdiyasni, McCauley and Haber, Dayton, OH (1987)

[10] Y. H. You, "Microstructure Development in Alkoxide-Derived Mullite and Mullite/Zirconia Composites" Ph.D. Thesis, Univ. of Illinois at Urbana-Champaign (1994)

[11] W. E. Cameron, "Composition and Cell Dimensions of Mullite," *J. Am. Ceram. Soc.,* **56** [11] (1977)

[12] Ossaka, "Tetragonal Mullite-like Phase from Coprecipitated Gels," *Nature (London),* **19** [4792] 1000-1001 (1961)

[13] H. Schneider and T. Rymon-Lipinski, " Occurrence of Pseudo-tetragonal Mullite." *J. Am. Ceram. Soc.,* **71** [3] c162-c164 (1988)

[14] W. M. Kriven and J. A. Pask, "Solid Solution Range and Microstructure of Melt-Grown Mullite," *J. Am. Ceram. Soc.*, **66** [9] 649-54 (1983)

[15] Y.Nakajima and P. H. Ribbe."Twinning and Superstructures of Al-Rich Mullite." *Am. Mineral,* **66** 142-47 (1981)

[16] A. Zangvil, C. C. Lin and R. Ruh, "Microstructural Studies in Alkoxide-derived Mullite/Zirconia/Silicon Carbide-Whisker composites," *J. Am. Ceram. Soc.*, **75** [5] 1254-63 (1992)

EFFECTS OF ANNEALING ON THE MICROSTRUCTURE AND MECHANICAL PROPERTIES OF HOT PRESSED $BaAl_2Si_2O_8$ (BAS) AND $SrAl_2Si_2O_8$ (SAS) GLASS-CERAMICS

Jan J. Buzniak, Robert M. Dickerson, and K. Peter D. Lagerlof
Department of Materials Science and Engineering,
Case Western Reserve University, Cleveland, OH 44106

Narottam P. Bansal
NASA Lewis Research Center, Cleveland, OH 44135

ABSTRACT

The crystallization behavior, microstructural development during annealing, and the four point bend strength of hot-pressed $BaO{\cdot}Al_2O_3{\cdot}2SiO_2$ (BAS) and $SrO{\cdot}Al_2O_3{\cdot}2SiO_2$ (SAS) glass ceramics have been investigated. Both BAS and SAS show strength loss above the glass transition temperature (~900°C), suggesting the presence of residual glass along the grain boundaries in the hot pressed material. Annealing of BAS at temperatures above 1000°C, resulted in an increase of the bend strength when tested above its glass transition temperature. However, increasing porosity during annealing caused a decrease of the fracture strength at temperatures below the glass transition with respect to the as-hot-pressed material. The increased porosity is believed to be associated with the formation of gaseous reaction products during annealing.

To the extent authorized under the laws of the United States of America, all copyright interests in this publication are the property of The American Ceramic Society. Any duplication, reproduction, or republication of this publication or any part thereof, without the express written consent of The American Ceramic Society or fee paid to the Copyright Clearance Center, is prohibited.

INTRODUCTION

Barium aluminum silicate $BaO{\cdot}Al_2O_3{\cdot}2SiO_2$ (BAS) and strontium aluminum silicate $SrO{\cdot}Al_2O_3{\cdot}2SiO_2$ (SAS), both belonging to the feldspar family of ceramics, are potential structural materials for high temperature applications because of their good thermal shock behavior, oxidation resistance, and small coefficient of thermal expansion. There are two polymorphs common for both BAS and SAS; (i) a monoclinic phase (celsian in BAS), and (ii) a hexagonal phase (hexacelsian in BAS). Hexacelsian is metastable at all temperatures below the equilibrium transformation temperature (1590°C) and exhibits a reversible structural transformation into an orthorhombic phase at ~300°C. This transformation is accompanied by large volume decrease, which in turn strongly affects the fracture strength of the material. Thus, only celsian, which is stable throughout this temperature range, has a potential for use at temperatures up to 1500°C. Sintering studies at ambient atmosphere have shown that the hexagonal to monoclinic phase transformation is sluggish, which leads to the retention of a significant volume fraction of hexacelsian after sintering.[1-9] However, in SAS, the hexagonal to monoclinic transformation is fast and single phase monoclinic material can easily be formed.[14]

Previous investigators [1,10,11,12] suggested that the use of glass powder as a starting material and hot pressing as a compaction method may be advantageous because of viscous flow of the glass above its glass transition temperature. This technique is especially interesting for consolidation of glass matrix composites since the matrix easily can flow around the reinforcement during compaction.

The main objectives of this study were to determine the effects of annealing temperature, time, and environment on the microstructure and fracture strength of hot pressed BAS and SAS glass-ceramics.

EXPERIMENTAL PROCEDURE

Hot Pressing

Barium and strontium aluminum silicate glass powders (obtained from Ferro Corp.) with an average particle size of 2.5 μm [12,13] were hot pressed into 2 inch diameter discs in a graphite die using a GCA/Vacuum Industries (model 91-840) hot press. Three hot pressing runs each of BAS and SAS, respectively, were carried out in vacuum at 1250°C for 150 minutes under a pressure of 24 MPa. A preload pressure of 5 MPa was applied when the temperature was between 800°C and 850°C. After the hot pressing temperature was stabilized, the pressure was carefully increased until the final pressure was reached. Linear voltage displacement transducers (LVDT) were attached to the top and bottom rams of the hot press. These LVDT's facilitated continuous measurements of the ram displacement during hot pressing. The final density and porosity of each of the hot pressed samples were determined using Archimedes method (Mettler 33340 Density Kit). More details of the hot pressing procedure are reported in Ref.[12].

X-Ray Analysis

X-ray analysis of the hot pressed samples was carried out on a Philips XRD 3100 x-ray diffractometer with a Philips APD-3520 Processor using Cu Kα radiation. The majority of the diffractograms were obtained using a 2θ step scan of 0.05° increment and a speed of 1.2° per minute. Samples were prepared by crushing small cross-sectional pieces of the hot pressed discs into a fine powder, after which a small amount of the powdered sample was placed on a double-stick tape.

Mechanical Properties

Rectangular bars, 2.5 mm x 2.5 mm x 25 mm, were cut from the hot pressed discs. Opposing surfaces of the bend samples were ground parallel, followed by polishing using various grades of diamond paste down to 2μm. The fracture strengths of the test samples were determined in 4-point bending as a function of temperature using a SiC fixture with a 20 mm outer span and a 10 mm inner span at a cross-head constant displacement rate of 0.05 mm/min. Two or three samples were tested in air at each temperature between ambient and 1360°C. The high temperature fracture tests were performed using an MTS universal testing machine equipped with a high temperature furnace using Super Kanthal heating elements.

Annealing

Heat treatments of the polished bend bars were carried out; (i) in air using a CM Rapid Temp furnace equipped with Super Kanthal heating elements and a programmable temperature controller of accuracy better than +/-2°C of the set point value, and (ii) in a reducing atmosphere using GCA/Vacuum Industries hot press equipped with vacuum chamber, graphite heating elements and programmable temperature controller.

SEM/TEM Analysis

The microstructure of the fracture surfaces of samples heat treated at various temperatures were observed using a JEOL JSM-840A scanning electron microscope (SEM). A 10 nm thick layer of palladium was deposited onto the sample surface before viewing in the SEM. Foils for transmission electron microscopy (TEM) were prepared from each of the samples submitted using standard techniques. Disks with a diameter of 3 mm were cut using a Gatan ultrasonic disc cutter, which subsequently were ground and polished to a thickness of 80 microns with a 3 micron surface finish. The center of the polished discs were then dimple ground to ~30 μm thickness, and ion milling to electron transparency using Gatan ion mill. Each sample was carbon coated to prevent charging effects during TEM examination. All TEM was performed on a Philips 400T operating at 120 keV. A Kevex UTW detector with a Kevex Delta Class analyzer was used for qualitative XEDS compositional analysis.

RESULTS AND DISCUSSION

Hot Pressing

The hot pressing temperatures, pressures, times at temperature and final densities of three BAS and three SAS runs are shown in Table 1 [10]. Hot pressing of these glasses gives a relative density above 99% of the theoretical value (the theoretical density of monoclinic BAS and SAS are 3.39 g/cm^3 and 3.09 g/cm^3, respectively).

X-Ray Analysis

Results of X-ray diffraction analysis are shown in Fig.1. BAS samples before annealing and after annealing at 1000^oC for 24 hrs are shown. It can be seen from the graph that the residual hexacelsian phase initially present in BAS after hot pressing transformed to celsian during annealing.

Mechanical Testing

Samples from all BAS and SAS hot press runs were selected for mechanical testing in four-point bending. The decrease of the fracture strengths above ~900^oC (the glass transition temperature; is approximately 910^oC for BAS and 890^oC for SAS) of hot pressed BAS and SAS are shown in Fig's 2 and 3, respectively, where the fracture strength is shown as a function of temperature. This loss in fracture strength above the glass transition temperature is believed to be related to the onset of viscous creep of the residual glass at the grain boundaries giving rise to grain boundary sliding. The presence of glass at grain boundaries have been detected using transmission electron microscopy [12].

Annealing

The high temperature strength of BAS was improved through annealing at elevated temperatures (Fig. 4). Samples annealed at 1000^oC, 1200^oC and 1400^oC for 24 hrs were stronger than unannealed material when tested above 750^oC. The strength loss below 750°C is believed to be due to the formation of voids and local tensile stresses across grain boundaries after crystallization of the residual glassy phase; the specific volume of the glass is larger than that of the crystalline phase. SAS samples annealed at 1200°C and 1400°C for 24 hrs show no increase in the fracture strength when tested above 1000°C (Fig. 5). At 1000°C, we obtained mixed results. Samples annealed at 1400°C for 24 hrs were weaker than unannealed SAS, and those annealed at 1200°C for 24 hrs show a slight increase of fracture strength. The observation that there is no strengthening at high temperatures in SAS after annealing is believed to be due to grain coarsening and an associated increase in porosity, which is more pronounced than in BAS. The peak fracture performance at 500°C, which is observed in samples annealed at 1000°C for 240 hrs, is believed to be due to relaxation of residual stresses during annealing. The residual stresses are believed to be caused by the phase transformation of the small volume fraction of hexagonal SAS remaining in the material after hot pressing [12].

Microstructural Observations

SEM micro graphs of fracture surfaces of BAS and SAS bend bars before and after annealing at various temperatures and times in air are shown in Fig's 6 and 7, respectively. Both annealed and unannealed samples failed by intergranular fracture. Samples before annealing have smaller grains and almost no porosity. Samples annealed at 1200°C and 1400°C for 24 hrs still have small grains but the porosity increases rapidly, especially in SAS. Conversely, samples annealed at 1000°C for 240 hr's have little porosity but larger grains. The density and porosity results for this same annealing time (24 hrs), but different annealing temperatures are shown in Fig's 8 and 9 for BAS and SAS, respectively. The large porosity increase with temperature is believed to be due to the oxidation of residual carbon at the grain boundaries which generates a high gas pressure inside internal voids. Thus, CO/CO_2 gas formation, which in turn causes bloating, and degradation of the microstructure, is believed to lower the fracture strength at elevated temperatures. Normal grain growth with increasing annealing time is only observed at lower annealing temperatures due to grain boundary pinning due to the void formation at annealing temperatures above 1200°C. Annealing of BAS and SAS samples in reducing atmosphere (1400°C for 24 hrs) leads to significant corrosion of the exposed sample surfaces. X-ray mapping revealed that this degradation is associated with the evaporation of silica (Figure 10).

SEM analysis of BAS and SAS fracture surfaces also revealed several Al_2O_3 inclusions, probably introduced during attrition milling of the glass frit. It is believed that these inclusions are primarily responsible for lower than expected bend strengths of BAS and SAS, especially at lower temperatures, as well as the relatively large scatter of the data. TEM observations of BAS (Fig. 11 and 12), fully confirmed the SEM results, showing the presence of $\alpha-Al_2O_3$ and, in addition, revealed the presence of intergranular Fe containing particles. Since the intergranular Al_2O_3 particles are heavily deformed, both Al_2O_3 and Fe are believed to be impurities from the attrition milling of the glass powder. Also, most triple points in the electron transparent regions have some carbon-containing phase (or phases) in all samples (Fig. 12). The carbon rich areas are amorphous and/or crystalline. The source of carbon is believed to be from the surfactant used during attrition milling of the glass frit, which decomposes during hot pressing. BAS glass was also often observed in the samples annealed at 1200°C and 1400°C. Qualitative XEDS from the amorphous regions shows a comparative enrichment in barium.

CONCLUSIONS

BAS and SAS celsian glass powders can be hot pressed to approximately full density at 1250°C for 2.5 hrs. The glass powder undergoes rapid crystallization in conjunction with densification. X-ray diffraction shows that annealing at 1000°C for 24 hrs in air leads to a phase transformation to monoclinic from the hexagonal phase. The four-point bend strength of samples primarily containing the monoclinic phase is in the range 100-150 MPa at room temperature. After samples are annealed at temperatures above 1000°C, the four-point bend strength is reduced at temperatures below the glass transition, which is believed to be due to bloating caused by rapid oxidation of carbon at grain boundaries. The four-point bend strength of BAS at elevated temperatures can be increased through annealing at temperatures above 1000°C. However, the fracture strength of SAS at temperatures above 1000°C is not affected by annealing, primarily because of bloating. Several Al_2O_3 inclusions in hot pressed BAS and SAS samples cause lower than expected fracture strength, especially at room temperature. SEM and TEM analysis show that impurities are concentrated at grain boundaries and triple points. Al_2O_3 and Fe inclusions were probably acquired from the container and milling media during attrition milling, while the carbon impurities probably are due to surfactant used during the attrition milling.

ACKNOWLEDGMENTS

This research was in part supported by the NASA Lewis Research Center under the collaborative research grant NCC3-133 and NCC3-236.

REFERENCES

1. C.H. Drummond III and N.P. Bansal, "Crystallization Behavior and Properties of $BaO-Al_2O_3-2SiO_2$ Glass Matrices", NASA Report #185209, February 1990.
2. N.P. Bansal and M.J. Hyatt, "Crystallization Kinetics of $BaO-Al_2O_3-SiO_2$ Glasses", *J. Mater. Res.* 4 [5], 1257-65 (1989).
3. D. Bahat, "Heterogeneous Nucleation of Alkali Earth Feldspars in Glasses", *J. Mater. Sci.* 4, 855-60 (1969).
4. J.S. Moya Corral and A.G. Verduch, "The Solid Solution of Silica in Celsian", *Trans. J. Br. Ceram. Soc.* 77, 40-44 (1978).
5. W.F. Muller, "On Polymorphism of $BaAl_2Si_2O_8$", Electron Microscopy in Mineralogy, Ed. H.R. Wenk, pp. 354-60, Springer, New York, NY, 1976.
6. W.F. Muller, "Phase Transitions and Associated Domains in Hexacelsian ($BaAl_2Si_2O_8$)", *Phys. Chem. Minerals* 1, 71-82 (1977).

7. D. Bahat, "Kinetic Study on the Hexacelsian-Celsian Phase Transformation", *J. Mater. Sci.* 5, 805-10 (1970).

8. P. Gay, "A Note on Celsian", *Acta Cryst.* 9, 474 (1956).

9. E.J. Wuchina, "The Hexagonal-to-Monoclinic Phase Transformation in Barium Aluminum Silicate (Celsian)", M.S. Thesis, U. of Maryland, College Park, MD, 1991.

10. J.J. Buzniak, K.P.D. Lagerlof, and N.P. Bansal, "Hot Pressing and High Temperature Mechanical Properties of $BaAl_2Si_2O_8$ (BAS) and $SrAl_2Si_2O_8$ (SAS)", in "Advances in Ceramic-Matrix Composites", N.P. Bansal - Editor,Am. Cer. Soc., Westerville, OH, 789-801 (1993).

11. C.L. Walters, C.H. Drummond III, and N.P. Bansal, "Hot Pressing and Mechanical Strength of $SrO{\cdot}Al_2O_3{\cdot}2SiO_2$ Matrix", Proc. of 6th Annual HITEMP Review, Oct. 25-27, 1993, Cleveland, OH, NASA Conf. Publ. 19117

12. P. Beatrice, "Phase Evolution During Hot Pressing of Celsian Glass", M.S. Thesis, Case Western Reserve University, May 1992.

13. J. Buzniak, P. Beatrice, and K.P.D. Lagerlof, to be published

14. N.P. Bansal, and C.H. Drummond III, "Kinetics of Hexacelsian-to-Celsian Phase Transformation in $SrAl_2Si_2O_8$", *J. of the Am. Cer. Soc.* 76 (5), 1321-24 (1993)

Table 1. Hot Pressing of BAS and SAS Glasses

ID #	Temp. (°C)	Pressure (MPa)	Preload (MPa)	Time @ Temp. (min)	Density (g/cm^3)	Hexagonal Phase$ (Vol %)
				BAS Runs		
21	1250	24	11	150	3.38	>5*
22	1250	24	11	150	3.37	>5*
23	1250	24	11	150	3.36	>5*
				SAS Runs		
04	1250	24	11	150	3.08	> 2*
05	1250	24	11	150	3.07	> 2*
06	1250	24	11	150	3.09	> 2*

$ Note: The estimated error in the volume fraction of hexagonal phase is of the order of 2-5 vol. %.

*from ref.[13]

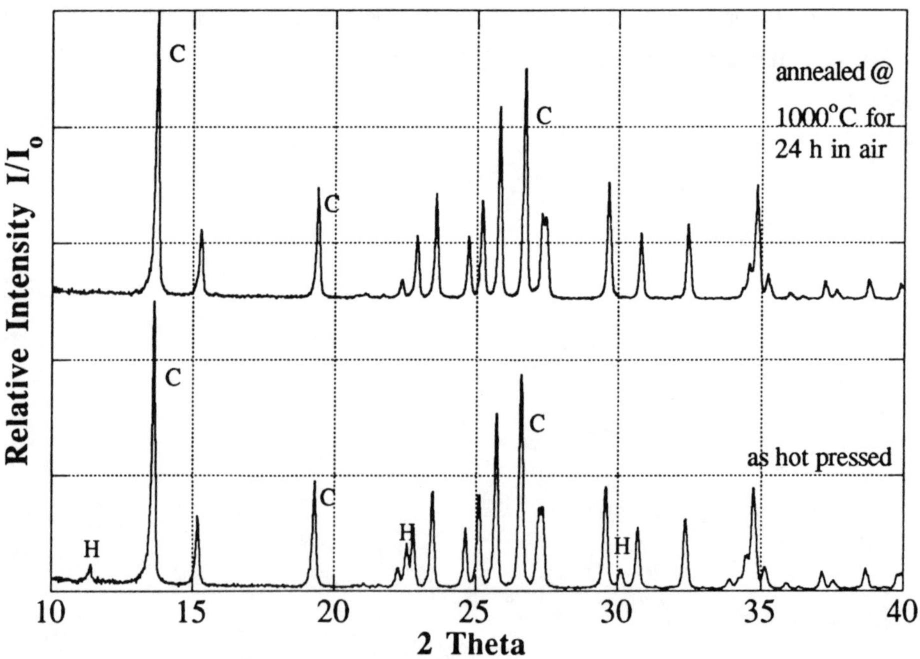

Figure 1. X-ray Diffraction Spectra Comparing Annealed vs Non-Annealed BAS

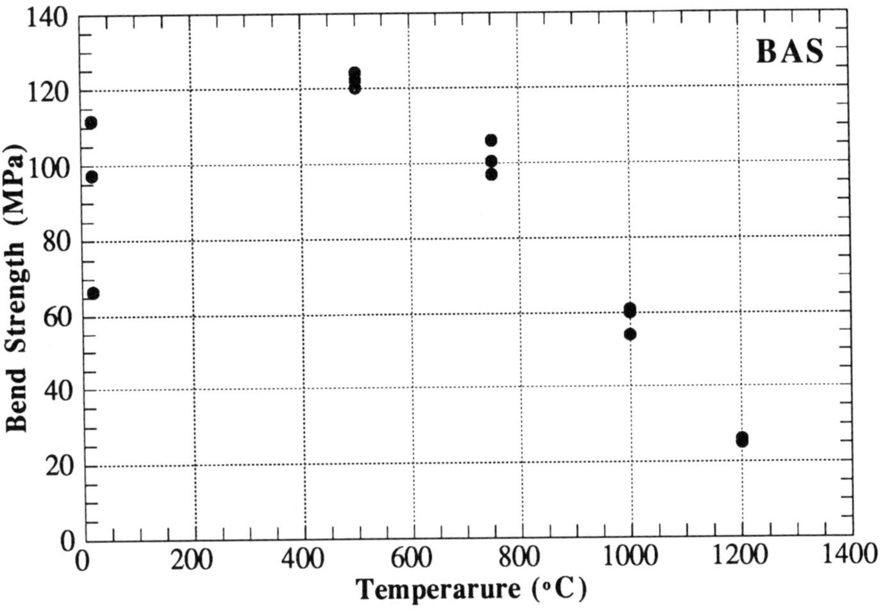

Figure 2. Effects of Temperature on Fracture Strength of BAS

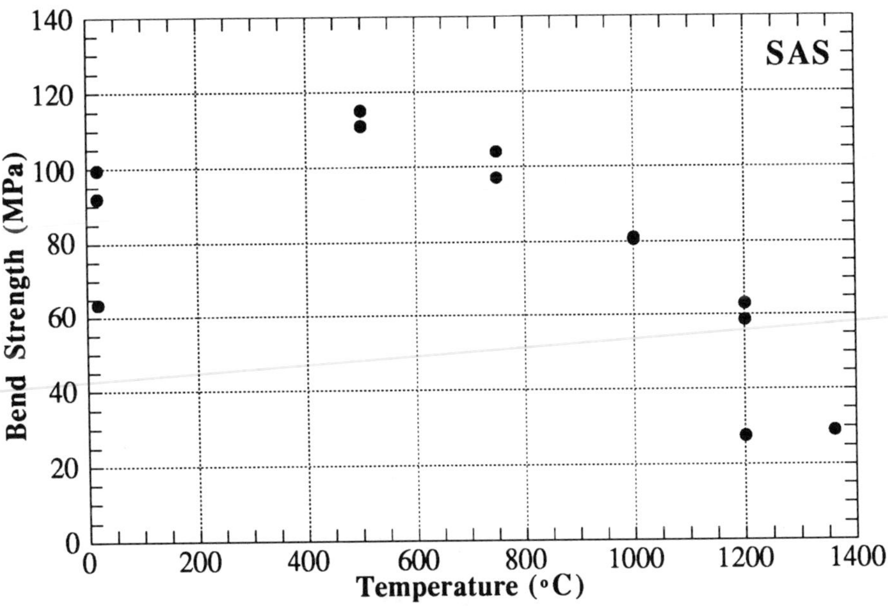

Figure 3. Effects of Temperature on Fracture Strength of SAS

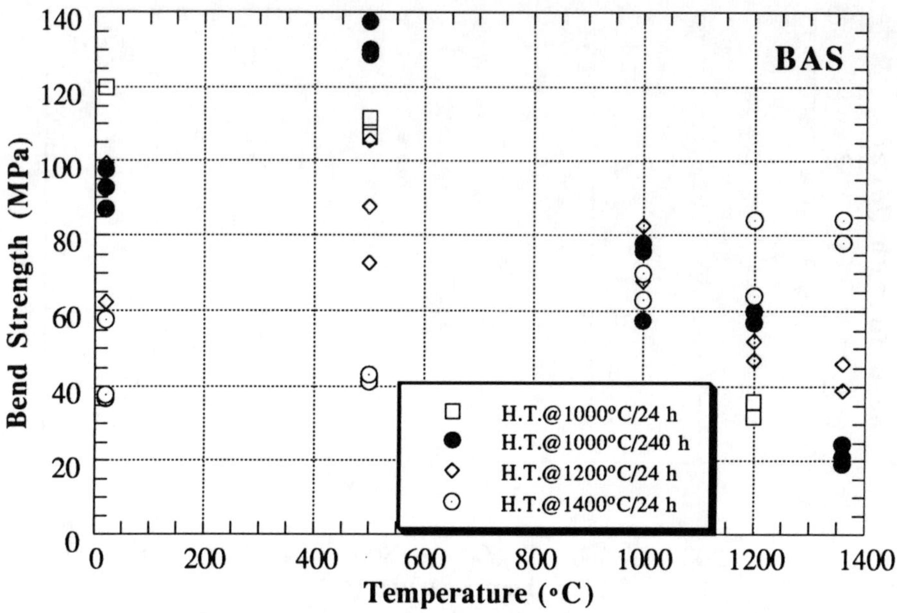

Figure 4. Effects of Annealing on the Fracture Strength of BAS

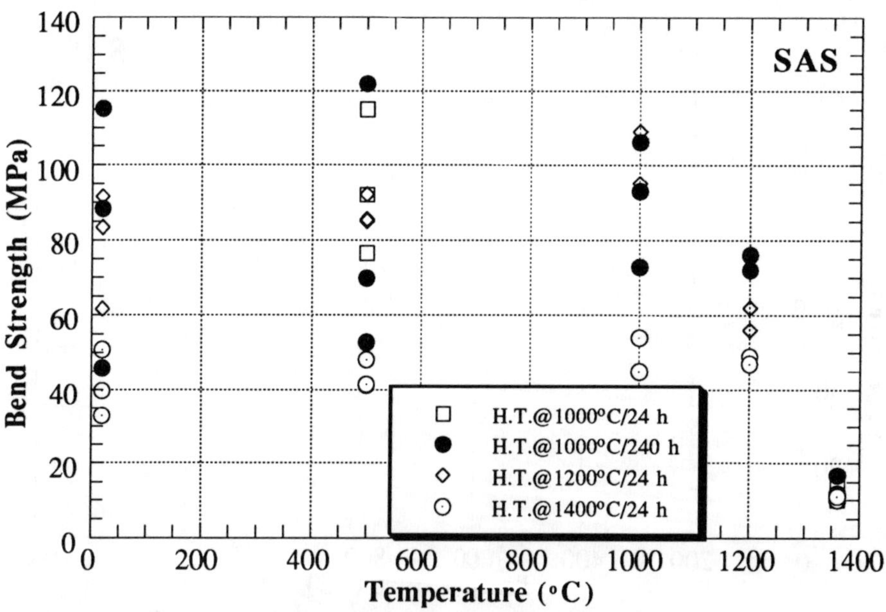

Figure 5. Effects of Annealing on the Fracture Strength of SAS

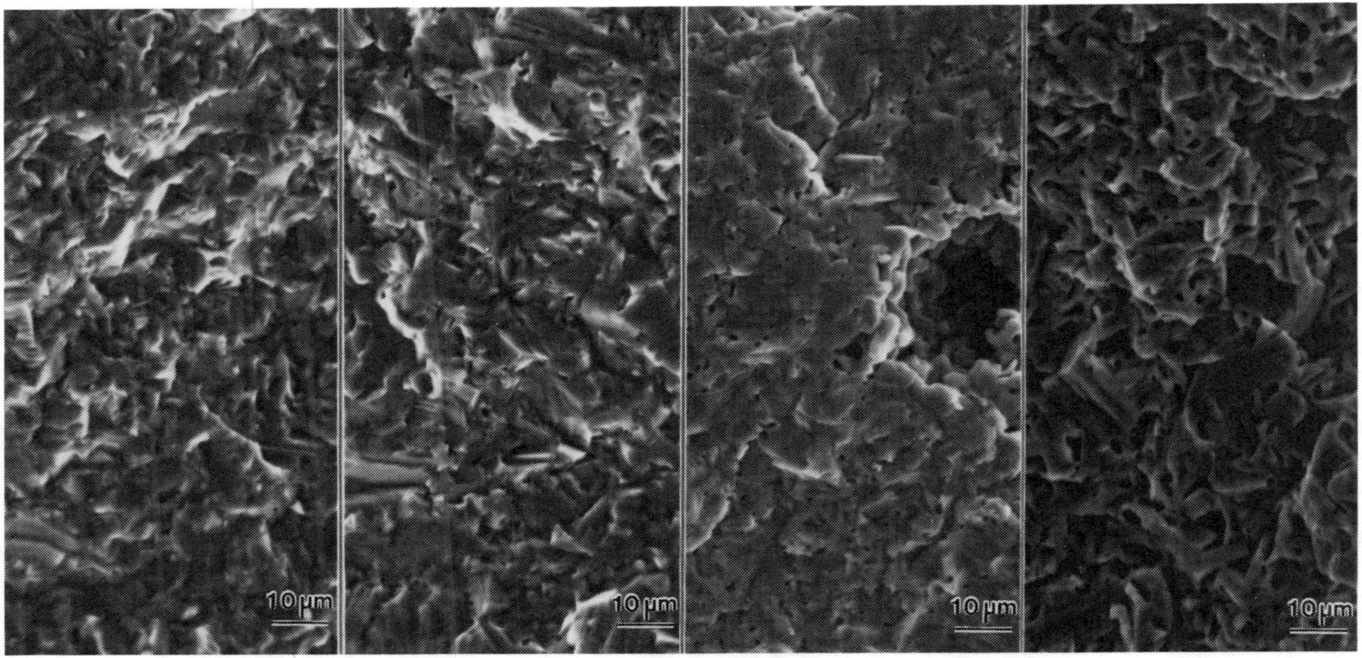

(a) (b) (c) (d)

Figure 6. SEM Micrographs of Fracture Surfaces of BAS Bars: (a) not annealed, (b) annealed @ 1000°C for 240h, (c) annealed @ 1200°C for 24h, (d) annealed @ 1400°C for 24h

(a) (b) (c) (d)

Figure 7. SEM Micrographs of Fracture Surfaces of SAS Bars: (a) not annealed, (b) annealed @ 1000ºC for 240h, (c) annealed @ 1200ºC for 24h, (d) annealed @ 1400ºC for 24h

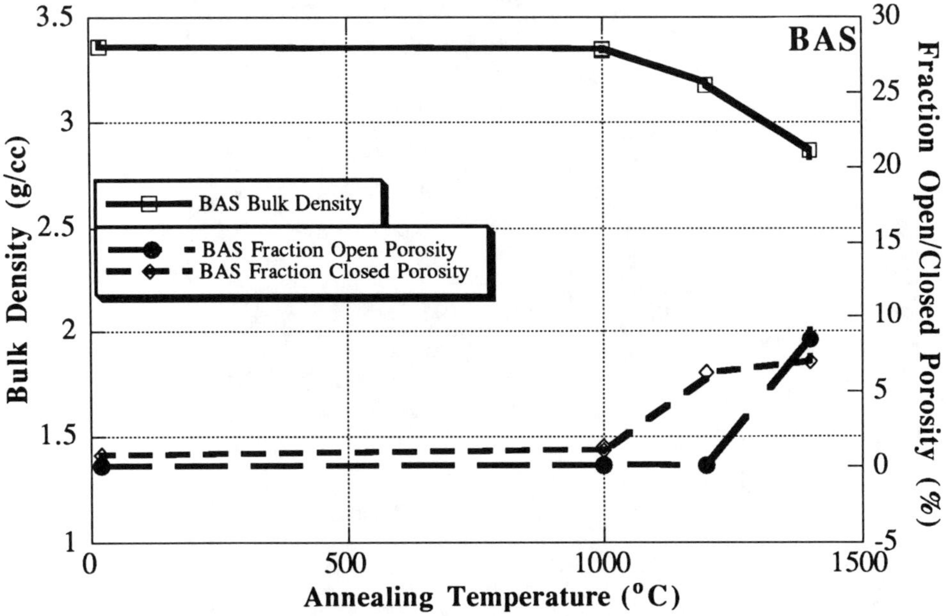

Figure 8. Effects of Annealing Temperature on the Density and Porosity of BAS

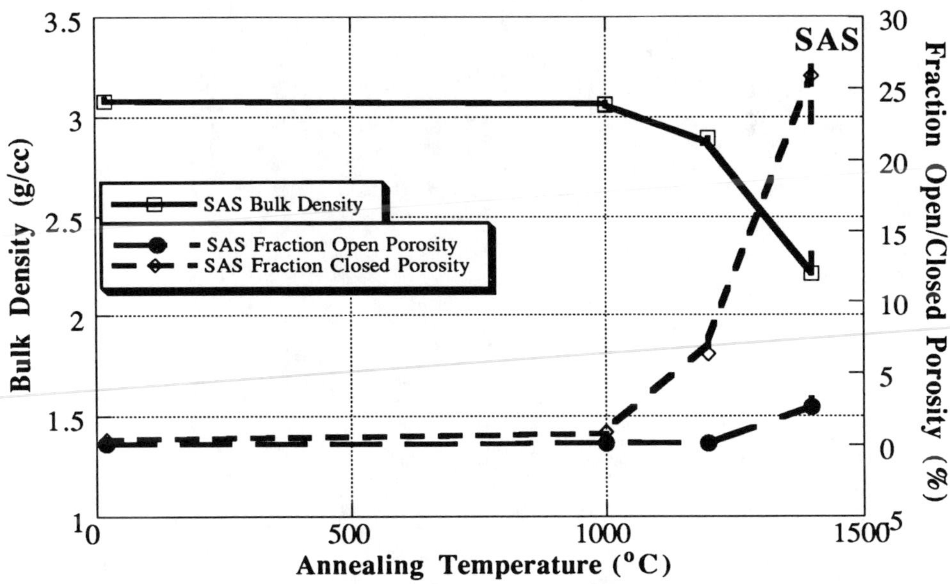

Figure 9. Effects of Annealing Temperature on the Density and Porosity of SAS

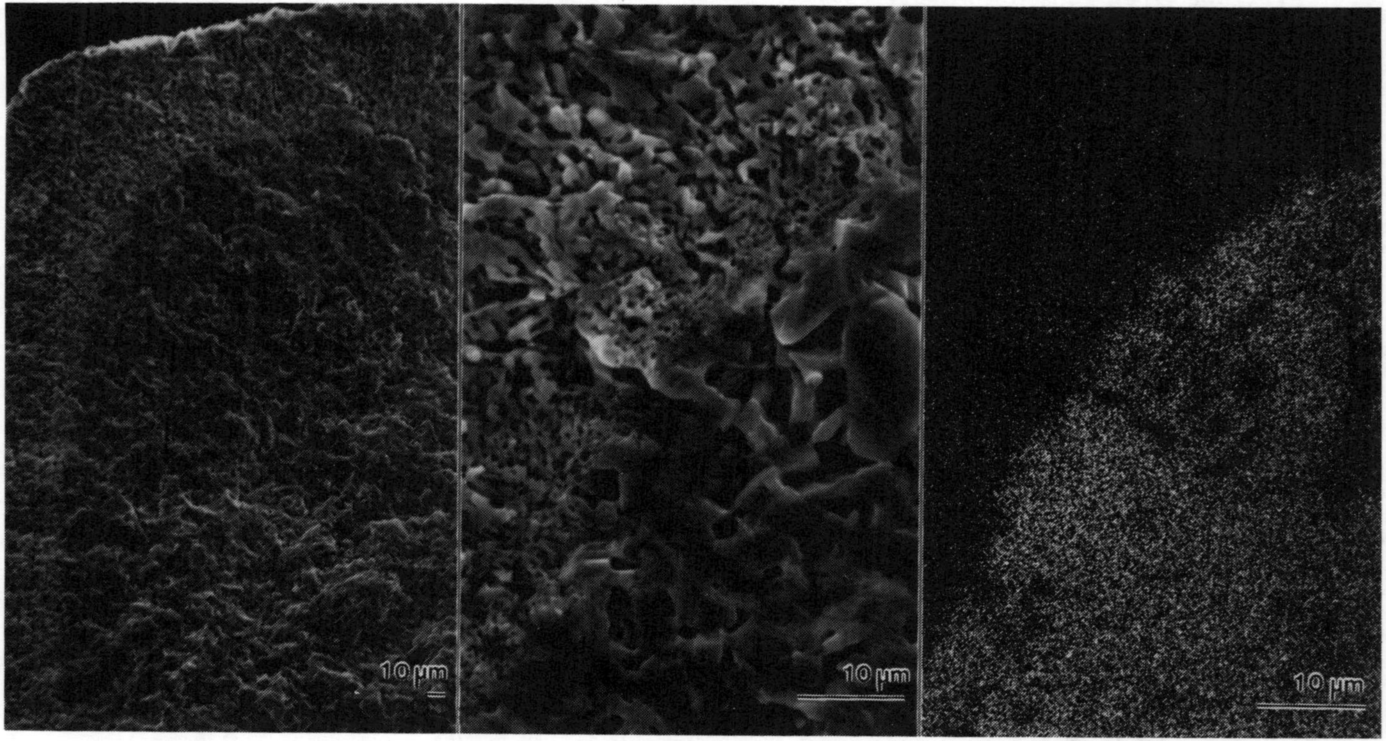

X-Ray Map of Si

Figure 10. SEM Fracture Surface of Sample Annealed in Reducing Atmosphere, Showing Region Severly Corroded and Depleted in Si

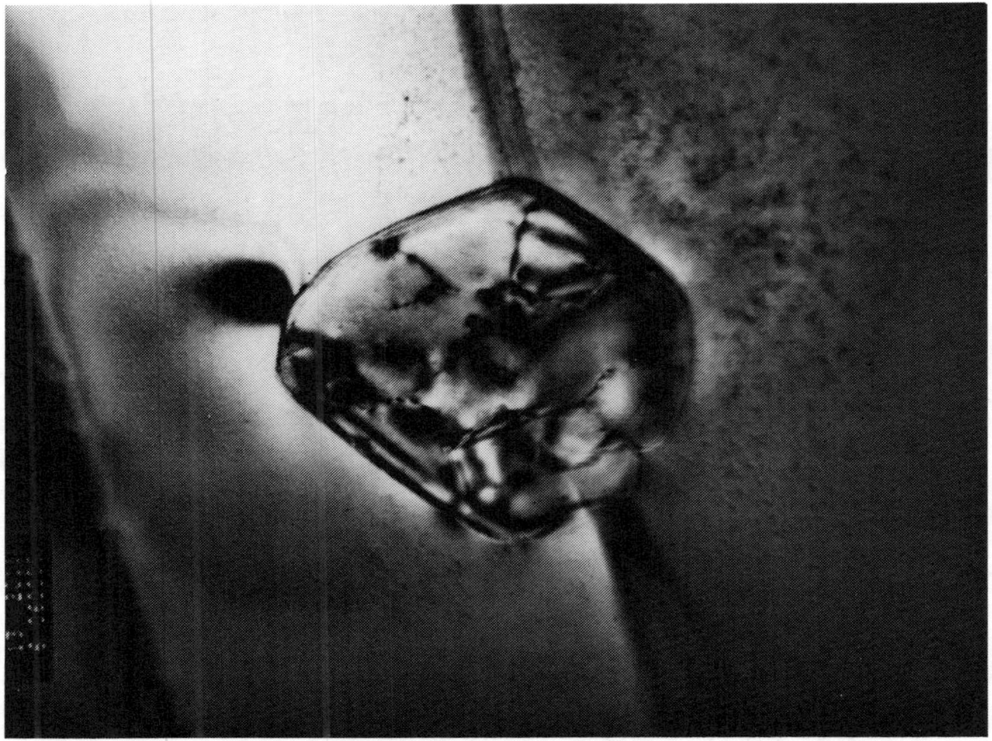

Figure 11. TEM Micrograph Showing Highly Deformed Intergranular Al_2O_3 Particle in Annealed BAS

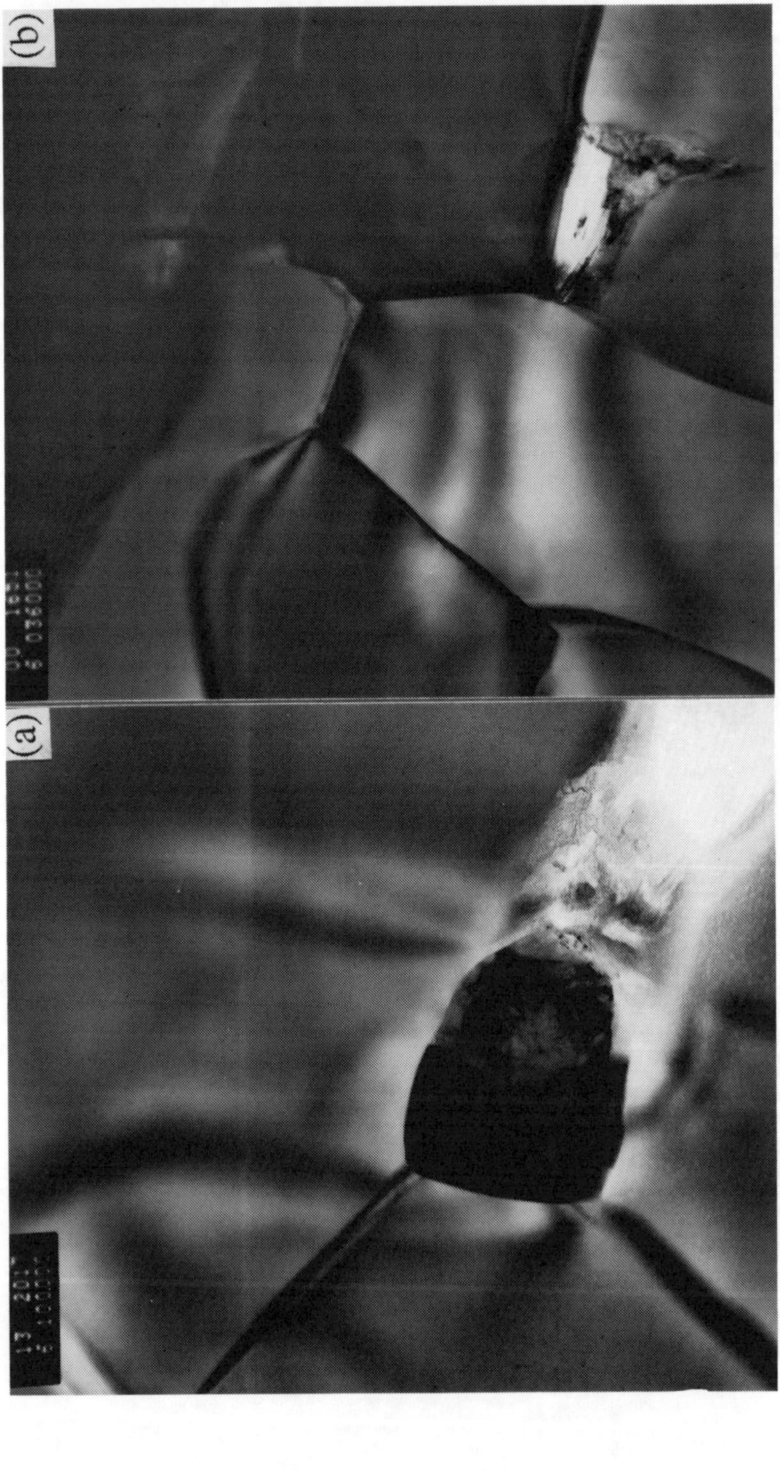

Figure 12. TEM Micrographs Showing Iron (a) and Carbon (b) Impurities on Grain Boundaries in Annealed BAS

POWDER PREPARATION AND SINTERING OF GRAIN GROWTH CONTROLLED Cr_2O_3-ZrO_2 COMPOSITES

Y.I Chen and J.J.Li, Opto-Electronics & Systems Lab., Industrial Technology Research Institute, Hsinchu, Taiwan, R.O.C.
M.S.Tsai, C.H.Yeh, and M.H.Hon, Department of Materials Science and Engineering, National Cheng Kung University, Tainan, Taiwan, R.O.C.

ABSTRACT

Zirconia is introduced in Cr_2O_3 to inhibit the abnormal grain growth of Cr_2O_3 during densification. The constituents of ceramic powders are mixed by ball milling and coprecipitating, respectively. Both normal sintering and hot press sintering are performed in non-oxiding atmosphere. An average grain size of 2-3 μm is obtained in normal sintering and an average grain size below 1.5μm with >99.5% theoretical density in hot press sintering. Mechanical properties of the composite are evaluated.

INTRODUCTION

Due to the volatilization of chromium oxide [1,2] as

$$Cr_2O_{3(s)} + 3/2\ O_{2(g)} = 2CrO_{3(g)} \quad (1)$$

high density Cr_2O_3 was difficult to produce under normal sintering in air. Reducing atmosphere [3] and dopants, such as TiO_2 [4] and MgO [3], were seemed to be the solutions to densify Cr_2O_3 during sintering. In our prior studies [5,6], TiO_2 was used as sintering additive to densify Cr_2O_3 under reducing atmosphere and 98% of theoretical density was achieved. Moreover, the abnormal grain growth of Cr_2O_3 was inhibited .by the incorporating of zirconium oxide grains as the second phase. The grain size of Cr_2O_3-ZrO_2 composites was reduced to 2-3μm with a dopant amount of 15 vol.% ZrO_2

To the extent authorized under the laws of the United States of America, all copyright interests in this publication are the property of The American Ceramic Society. Any duplication, reproduction, or republication of this publication or any part thereof, without the express written consent of The American Ceramic Society or fee paid to the Copyright Clearance Center, is prohibited.

The introducing of ZrO_2 grains also improved the mechanical properties, such as toughness and bending strength.

In this study, powders are prepared by ball milling and by coprecipitating. Normal sintering and hot pressing sintering are used for densification. Density and grain size of Cr_2O_3-ZrO_2 composites are measured. The mechanical testing focuses on the hardness.

POWDER PREPARATION

The composite powders were prepared by physical and by chemical methods.

In physical method, as shown in Fig.1, powders of Cr_2O_3, TiO_2, and ZrO_2 were mixed by ball milling with polyvinyl alcohol solution as the binder. The particles of Cr_2O_3 and ZrO_2 are sub-micron powders. The specific surface areas of the powders before mixing are $7.29 m^2/g$ and $11 m^2/g$, respectively. The ZrO_2 powders contained 3mol.% $YO_{1.5}$. The compositions of powders were

$$x \text{ vol.\% } ZrO_2 + (y \text{ wt.\% } TiO_2 + Cr_2O_3)$$
$$x = 7.5, 10, 15$$
$$y = 1, 3$$

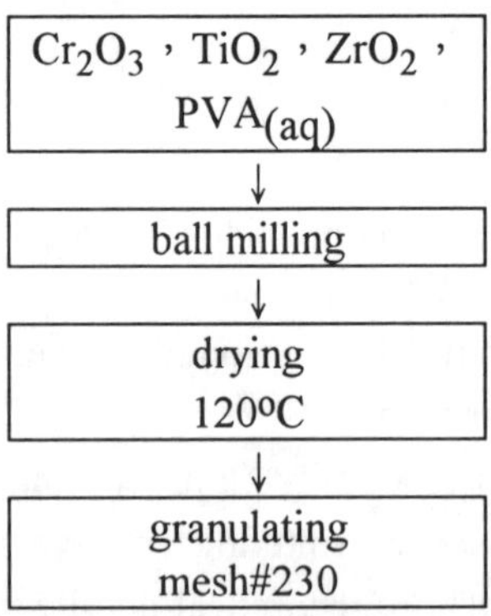

Fig.1 Physical Method of Powder Preparation

After mixing for 24 hrs., the powders were dried at 120°C, and sieved to 230 mesh. The composition of milling balls was the same as that of ZrO_2 used in this study.

In chemical method, as shown in Fig.2, $Cr(NO_3)\cdot 9H_2O$ and $ZrOCl_2\cdot 9H_2O$ solution with 2.6wt.% TiO_2 powders were coprecipitated by $NH_4OH_{(aq)}$. The $[Cr^{3+}]$ was fixed at 1.565M and the pH value of precipitating solution was controlled at 8.5 after coprecipitating. The coprecipitated Cr_2O_3 particle size is smaller than 0.5μm by TEM observation. The contents of coprecipitated zirconium oxide were 7.1, 14.2, and 21.3 vol.%. After coprecipitating, the powders were calcined at 650°C.

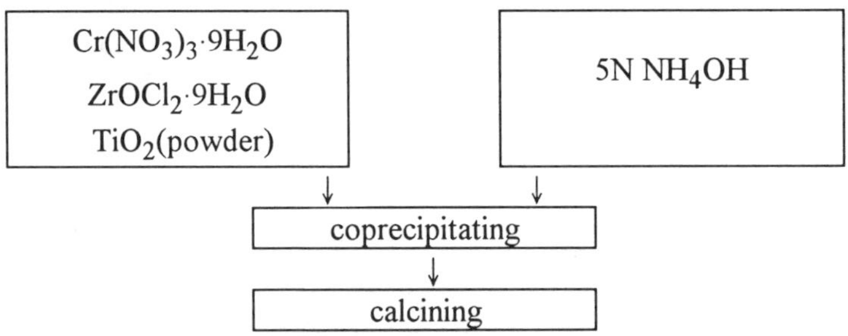

Fig.2 Chemical Method of Powder Preparation

SINTERING

The powders mixed by physical method were isostatically pressed at 196MPa to form green body and sintered at 1600°C for 1hr. The reducing atmosphere was formed by heating graphite powders in the aluminum oxide crucible.

Both type of powders were hot pressed at 1450°C for 1hr under a pressure of 30MPa in argon in a graphite mold.

The densities were measured by Archimedes' method. For the observation of microstructure, the composites were thermal etched at 1400°C for sintered samples and 1200°C for hot pressed samples. Grain size of Cr_2O_3 was estimated by lineal intercept method for two-phase polycrystalline ceramics [7].

RESULTS

Table I lists the measured densities, the relative densities and average grain sizes of Cr_2O_3-ZrO_2 composites produced by normal sintering. The powders are prepared by physical method.

Table I Density and Grain Size of Composites with Physical Method Powders after Normal Sintering

designation X-Y	measured density (g/cm^3)	theoretical density (g/cm^3)	relative density(%)	grain size of Cr_2O_3 (μm)
7.5-1	4.774	5.272	90.6	2.26
10-1	4.828	5.293	91.2	1.98
15-1	4.927	5.336	92.3	1.80
7.5-3	5.133	5.249	97.8	2.97
10-3	5.141	5.271	97.5	2.46
15-3	5.092	5.315	95.8	2.75

The designations "X-Y" in Tables I & II represent volume percent of ZrO_2 and weight percent of TiO_2. The theoretical densities of composites are calculated by the fraction of components. The relative density means the measured density divided by the calculated theoretical density.

Since TiO_2 plays the role of sintering additive, which promotes the densification of Cr_2O_3 so that the composites with 3wt.% TiO_2 dopants show denser and possess larger grain size than that with 1wt.% TiO_2 dopants. In a prior study [5] without the introducing of ZrO_2, the density with 3wt.% TiO_2 is higher than that with 1, 2, and 5 wt.% TiO_2 under normal sintering. Also shown in Ref.5 is the abnormal grain growth of Cr_2O_3 due to the effect of TiO_2. Large grain size up to 100μm is observed. However, the grain size of Cr_2O_3 can be controlled in the micrometer range as indicated in Table I and in Ref.6 by introducing ZrO_2 particles. Since the original Cr_2O_3 powders in this study is in sub-micron scale, which results the grain size can be reduced to 2-3μm with the ZrO_2 contents from 7.5 - 15 vol.%. For the composites with larger initial particle size 5.4μm before mixing as seen in Ref.6, the grain size is increased, e. g., 4μm for the sample with 7.5vol.%ZrO_2 and 10μm for the sample with 2vol.%ZrO_2.

Table II lists the density, grain size, and hardness of components produced by hot-press sintering. The powders are the same as that in normal sintering Except for the sample designated "7.5-3", all the other hot-pressing sintered samples show finer grain size than they do in the normal sintering. The reduction of grain size is contributed from the lower sintering temperature in hot-pressing than in normal sintering. The average grain sizes of hot-pressing sintered composites are in the range of 1.5-1.9μm. Moreover, the densities are higher than the calculated theoretical densities. The difference between 3wt.% and 1wt.% TiO_2 doped composites is not obvious. The hardness of hot-pressing composites is around 1600$HV_{0.3}$ with a indentation load of 300gf.

Table II Density, Grain Size, and Hardness of Composites with Physical Method Powders after Hot Press Sintering

designation X-Y	density (g/cm^3)	grain size (μm)	hardness (HV)
7.5-1	5.477	1.81	1615
10-1	5.399	1.86	1570
15-1	5.338	1.56	1682
7.5-3	5.168	4.5	1646
10-3	5.479	1.92	1595
15-3	5.351	1.54	1601

hardness testing load 300gf

For the sample with 7.5vol.%ZrO_2 and 3wt.%TiO_2, the density is 98.5% theoretical density only and the grain size is larger than that produced in normal sintering.

The grain sizes and hardnesses of hot-pressing composites with coprecipitated powders as listed in Table III are smaller than that with ball milled powders. However, the grain size is increased as the ZrO_2 content is increased from 7.1 to 21.3 vol.%, which reason is not realized.

Table III Density, Garin Size, Porosity and Hardness of Composites with Chemical Method Powders after Hot Pressing Sintering

No.	ZrO_2 (vol.%)	density (g/cm^3)	grain size (μm)	hardness (HV)
1	7.1	5.23	0.64	1509
2	14.2	5.36	1.09	1470
3	21.3	5.23	1.34	1445

TiO_2 : 2.6wt%

hardness testing load 300gf

CONCLUSION

1.Incorporating ZrO_2 particles can inhibit the abnormal grain growth of Cr_2O_3.

2.Both physical and chemical methods of powder preparation can be densified to >99.5% theoretical density by hot-press sintering.

3.Grain size of composites with physical method powder is 1.5-1.9μm after hot-press sintering, while using chemical method powders the grain size is 0.6-1.3μm.

4.Hardness of Cr_2O_3-ZrO_2 composites is 1400-1600$HV_{0.3}$.

REFERENCES

1.D.Caplan and M.Cohen, "The Volatilization of Chromium Oxide," J. Electrochem. Soc., 108 [5] 438 (1961).

2.R.T.Grimley, R.P.Burns, and M.G.Inghram, "Thermodynamics of the Vaporization of Cr2O3:Dissociation Energies of CrO, CrO2, and CrO3," J. Chem Phys., 34 [2] 664 (1961).

3.P.D.Ownby and G.E.Jungquist, "Final Sintering of Cr2O3," J. Am. Ceram. Soc., 55 [9] 433 (1972).

4.W.D.Callister, M.L.Johnson, I.B.Cutler, and R.W.Ure,Jr., "Sintering Chromium Oxide with the Aid of TiO2," J. Am. Ceram. Soc., 62 [3-4] 208 (1979).

5.Y.I Chen, J.J.Li, and S.H.Shieh, "Sintering Kinetics of Chromium Oxide," Proc. of the 1993 Annual Conf. of the Chinese Soc. for Mat. Sci., 2-131 (1993).

6.Y.I Chen and J.J.Li, "Microstructure and Mechanical Properties of Grain Growth Controlled Cr2O3-ZrO2 Composites," submitted to J. Mater. Sci.
7.J.C.Wurst and J.A.Nelson, J. Am. Ceram. Soc., 55 [2] 109 (1972).

In-Situ and Nanocomposites

IN-SITU FORMATION OF DUCTILE PHASE TOUGHENED CERAMICS BY PARTIAL REDUCTION REACTIONS IN THE Ni-Al-O SYSTEM

E. Üstündag, R. Subramanian, R. Dieckmann and S.L. Sass

Department of Materials Science and Engineering
Cornell University, Ithaca, NY 14853, U.S.A.

ABSTRACT

Metal-ceramic microstructures were obtained in the Ni-Al-O system by partial reduction of the spinel compound $NiAl_2O_4$. Electron microscopy studies were performed for microstructural characterization, phase identification and chemical analysis. Depending on the reduction temperature, two different morphologies of nearly pure Ni particles, equiaxed and rod-like, form within a ceramic matrix, which is either α-Al_2O_3 or a metastable "defect spinel" containing much less Ni than $NiAl_2O_4$. The fracture toughness of the Ni-Al_2O_3 mixture was significantly improved with respect to that of the original spinel phase. However, cracking at the original spinel grain boundaries, likely due to the large volume changes associated with the reduction reaction, makes it difficult to produce large specimens by this technique. The addition of ZrO_2 to spinel before reduction was found to prevent cracking. Initial results about the effect of ZrO_2 on the microstructures before and after reduction are presented.

INTRODUCTION

Engineering materials used in load bearing circumstances, particularly under extreme conditions (e.g. high temperatures or stresses), generally are formulated with complex microstructures. Such composite-like materials can possess properties unobtainable by either of the individual constituents. With metal-ceramic composites, for example, a material can be obtained with an optimum combination of the low density and high strength of the ceramic, and the toughness of the metal.

To the extent authorized under the laws of the United States of America, all copyright interests in this publication are the property of The American Ceramic Society. Any duplication, reproduction, or republication of this publication or any part thereof, without the express written consent of The American Ceramic Society or fee paid to the Copyright Clearance Center, is prohibited.

For high temperature service, where demanding conditions are often present, for example, inside jet turbine engines, a successful new material must have low density, high Young's modulus and good oxidation resistance, strength, fracture toughness, creep and fatigue resistance. The candidates include intermetallic compounds, and metal-ceramic, intermetallic-ceramic and ceramic-ceramic composites.

As will be explained later, partial reduction of $NiAl_2O_4$ results in two-phase mixtures of almost pure Ni and Al_2O_3-based oxide phases which potentially offer a solution to the brittleness of the ceramic component. Al_2O_3 is one of the most stable oxides providing excellent oxidation resistance and good high temperature mechanical properties. Ni-base systems in the form of Superalloys have already proven their usefulness in demanding environments. Hence, microstructures consisting of Ni and Al_2O_3 may be able to satisfy the good oxidation resistance and high temperature mechanical properties requirements of a successful high temperature structural material. Another important advantage of a material with such a microstructure is its high specific modulus. Al_2O_3 has a Young's modulus of 390 GPa [1] and a density of 3.98 g/cm^3 [2], which gives it a high specific stiffness. From phase diagram considerations, it can be shown that the two-phase mixture produced by partial reduction of $NiAl_2O_4$ is expected to contain ~ 20 vol.% Ni and ~ 80 vol.% Al_2O_3. Using values of the Young's modulus (200 GPa [1]) and the density (8.9 g/cm^3 [1]) of Ni, the Young's modulus and density of the Ni-Al_2O_3 mixture is calculated, from the rule of mixtures, to be ~ 350 GPa and ~ 4.96 g/cm^3, respectively. As a result, the specific modulus of the mixture is estimated to be ~ 70 GPa·cm^3/g. For comparison, the specific modulus of Superalloys is ~ 27 GPa·cm^3/g, which indicates that two-phase materials can be produced in the Ni-Al-O system with nearly three times larger specific modulus than those currently in use.

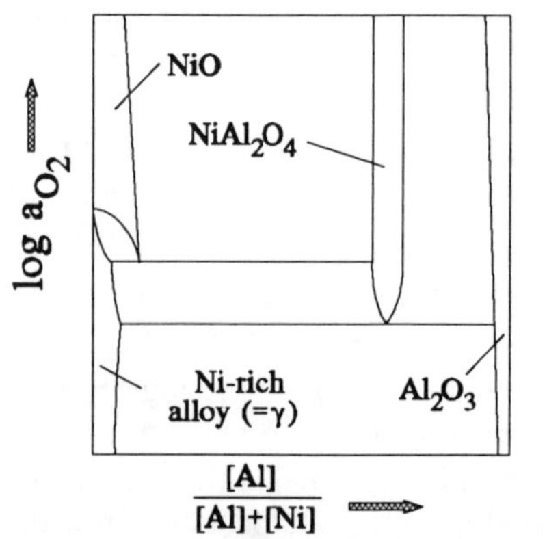

Fig. 1. Schematic log a_{O_2} vs composition phase diagram of the system Ni-Al-O at constant temperature.

THERMODYNAMIC BACKGROUND

The first requirement for a systematic study of reduction reactions to control metal-ceramic microstructures is knowledge of phase equilibria, in particular, the effect of the oxygen partial pressure. Detailed phase diagrams for the Ni-Al-O system at 940 and 1000 °C were obtained by Elrefaie and

Smeltzer [3]. Based on their work a schematic phase diagram for this system is shown in Fig. 1 for temperatures in the vicinity of 1200 °C. From Fig. 1 it is seen that when $NiAl_2O_4$ is reduced, a two-phase mixture of Ni-rich alloy (γ) and α-Al_2O_3 is expected to form.

EXPERIMENTAL PROCEDURE

The initial $NiAl_2O_4$ samples used in reduction were large-grained polycrystalline samples obtained in early attempts to grow single crystals by solidification of the $NiAl_2O_4$ melt (the details were reported previously [4,5,6]).

Subsequent samples were polycrystalline $NiAl_2O_4$ (with or without ZrO_2) prepared by hot pressing. The starting materials were powders of NiO (99 %, grain diameter < 20 μm) and α-Al_2O_3 (grain diameter < 60 μm). These were mixed together in a ball-mill to obtain a homogenous mixture, which was then reacted in air at 1600 °C for 4 days to yield $NiAl_2O_4$ which was subsequently crushed to powder and sieved. For all subsequent work $NiAl_2O_4$ powders of less than 6 μm grain diameter were used. For addition of ZrO_2, a water soluble compound, zirconyl nitrate ($ZrO(NO_3)_2{\cdot}2H_2O$), was added in appropriate amounts to $NiAl_2O_4$ powder and then wet ball-milled using teflon balls. This allowed a uniform distribution of the dopant within the polycrystalline spinel. The mixture was then dried and compacted in an alumina die in a uniaxial hot press for 2-4 hr, at 1600 °C, 23-30 MPa pressure, in air. Pure $NiAl_2O_4$ powders were hot pressed under the same conditions, except that the time was longer (6-8 hr), after which the sample was usually left at 1500 - 1600 °C without load being applied. The reason for the longer hot-pressing times for pure $NiAl_2O_4$ is the fact that it is more difficult to compact then the ZrO_2-doped material. The resulting density of all hot-pressed samples was better than 90 % of the theoretical density.

The hot-pressed samples were reduced at oxygen activities below the stability limit of the spinel (as specified in ref. [3]) and at temperatures between 1100 and 1350 °C [4,5,6]. The reduced samples were mechanically polished and then examined using backscattered electron imaging in a JEOL 733 electron microprobe. Transmission electron microscopy (JEOL 1200 EX) imaging and diffraction, and scanning transmission analytical microscopy (JEOL 200 CX) techniques were used to examine the microstructures, and to determine the phases present and their compositions.

EXPERIMENTAL RESULTS AND DISCUSSION

Pure $NiAl_2O_4$

Figs. 2 and 3 show two distinct morphologies of the Ni particles (white phase) in the two-phase mixtures produced at different reducing conditions. Fig. 2 for 1100 °C shows that the Ni is in the form of long rods. Fig. 3 for 1350 °C shows that the Ni is in the form of equiaxed particles.

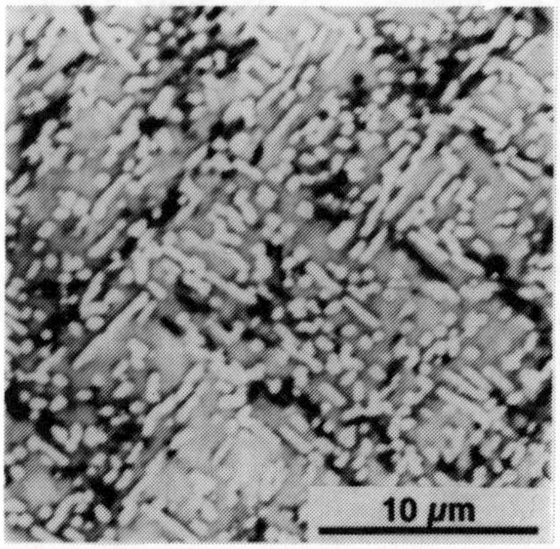

Fig. 2. $NiAl_2O_4$ reduced at 1100 °C for 2 hours at an oxygen activity of $10^{-15.6}$.

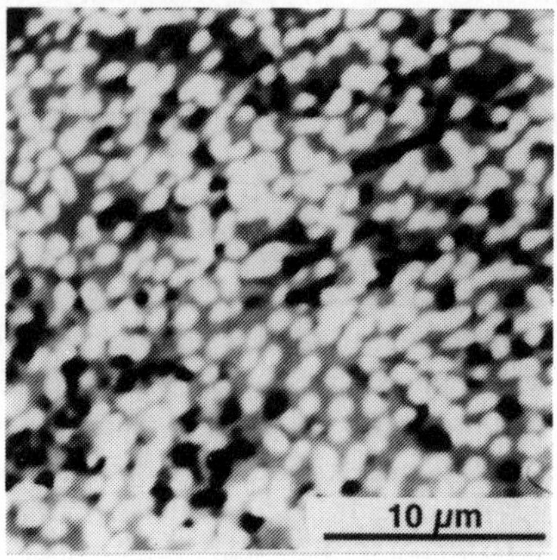

Fig. 3. $NiAl_2O_4$ reduced at 1350 °C for 0.5 hours at an oxygen activity of 10^{-10}.

Examination of the sample in Fig. 2, a part of which is shown with higher magnification in Fig. 4, shows several rod-like Ni particles embedded within a region identified as "defect spinel" which is adjacent to unreduced $NiAl_2O_4$. Such a microstructure was found just ahead of a totally reduced region which consists of similar rod-shaped Ni particles in a "defect spinel" matrix. The "defect spinel" is metastable and can be one of several phases, which are Ni-poor relative to $NiAl_2O_4$ [4,5,6]. It is known that metastable phases may form in the Ni-Al-O system [7,8]. The chemical com-po-si-tion of the "defect spinel" found in our work [4,5,6] falls within the range reported for the metastable phases mentioned in refs. [7] and [8]. The electron diffraction patterns [4,5,6] also have similarities to previous observations [7,8], from which it was concluded that the structures of the metastable phases are based on a pseudocubic array of oxygen ions. There are also extra reflections in the electron diffraction patterns from "defect spinel" [4,5,6] which may be evidence for ordering of point defects, in particular, cation vacancies [7,8].

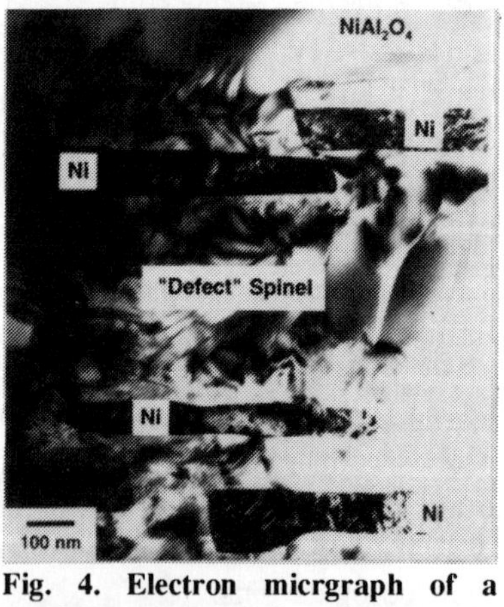

Fig. 4. Electron micrgraph of a partially reduced region of the sample shown in Fig.2.

The partially reduced microstructure shown in Fig. 4 as well as totally reduced regions were further examined by transmission electron microscopy to determine the orientation relationship between the crystal structures of the Ni rods and the "defect spinel matrix". It is observed [4,5] that the {111} planes of both materials are parallel (if the defect spinel is considered to be cubic) and that they are in a (111)[112] type twin orientation relationship (found in f.c.c. metals). The rod-like Ni particles are elongated with their long axis along the <112> directions of the regular spinel and "defect spinel".

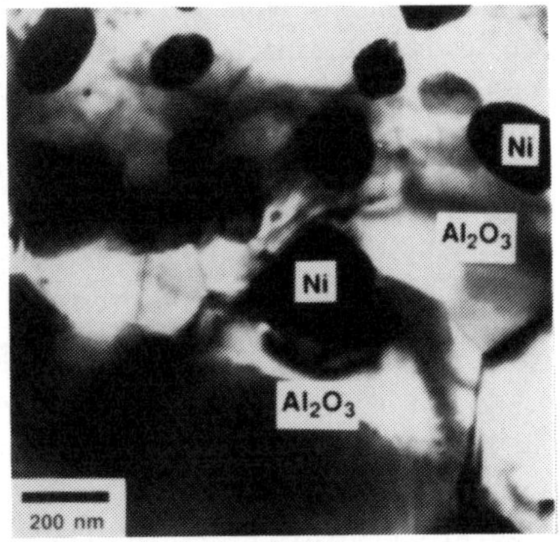

Fig. 5. Electron micrograph of the $NiAl_2O_4$ sample shown in Fig. 3.

Fig. 5 is a transmission electron micrograph of the sample shown in Fig. 3, from a region that is totally reduced. Imaging, electron diffraction and chemical analysis showed that equiaxed Ni particles with diameters of 0.02 - 0.5 μm are dispersed in a matrix consisting of equiaxed α-Al_2O_3 grains with diameters of 0.5 - 1 μm. The Ni particles are frequently located at the Al_2O_3 grain boundaries.

There is a large thermal expansion coefficient mismatch between Ni and Al_2O_3 which is calculated to produce a large hydrostatic tensile stress (about 2.7 GPa) in Ni particles upon cooling from reduction temperature [4,5,6]. To confirm the presence of this stress in the Ni particles, a sample reduced at 1300 °C with a microstructure similar to that shown in Fig. 3 was examined with X-ray diffraction. Fig. 6 shows part of the X-ray powder pattern obtained, with the expected positions of the diffraction peaks from bulk Ni and Al_2O_3 indicated. The observed Ni (200) peak is seen to be shifted towards higher interplanar spacings, indicating an increase in the average lattice parameter of Ni. The possibility of this increase resulting from Al and O dissolution in Ni was considered and it was found that the Ni peak shift due to residual thermal stresses

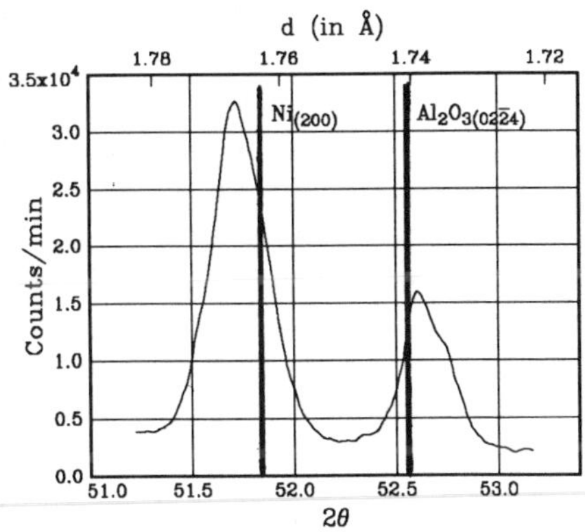

Fig. 6. Part of the X-ray diffraction spectrum of $NiAl_2O_4$ sample reduced at 1300 °C for 4 hours at an oxygen activity of $10^{-10.4}$. Ni (200) and Al_2O_3 (0224) peaks are shown along with their expected positions.

alone gives hydrostatic tensile stresses of 0.8 to 1.6 GPa (depending on the Al concentration in Ni) [4,5].

An interesting consequence of this large hydrostatic tensile stress within the Ni particles is that it places the surrounding ceramic phase in a state of compressive stress. The presence of such a stress can be detected from the slight shift in the position of the Al_2O_3 peak in Fig. 6 towards lower interplanar spacings, suggesting a smaller lattice parameter than in bulk Al_2O_3. The presence of a compressive stress in Al_2O_3 is potentially useful since such a stress would act to close up cracks that are propagating within the ceramic phase. Thus this state of stress should contribute to the improvement of the fracture toughness of the two-phase mixture compared to a monolithic ceramic.

In Fig. 5 it is clear that there are no cracks or voids along the interface between the Ni particles and the Al_2O_3. It is remarkable that the Ni-Al_2O_3 interface is able to survive the high hydrostatic tensile stresses developed upon cooling. According to Davidge and Green [9] there should be a critical size for the Ni particles above which cracks can open up at the Ni-Al_2O_3 interface. Our observations indicate that this interface survives intact up to a Ni particle diameter of at least 0.5 μm. However, when Ni particles become larger (e.g. above 1.5 μm as can be seen in Fig. 7), cracking occurs at this interface pointing to the necessity of keeping the Ni particle diameter below a certain size. Work is in progress to calculate this minimum Ni particle size as well as to model the cracking process due to the thermal expansion coefficient mismatch.

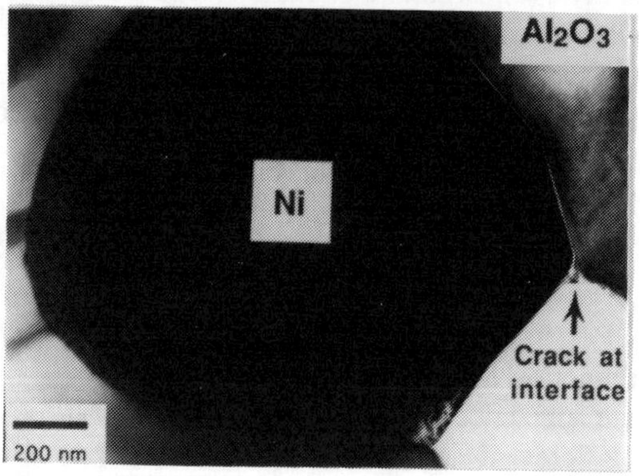

Fig. 7. Crack at the Ni/Al_2O_3 interface in a sample reduced at 1300 °C for 8 hours at an oxygen activity of 10^{-10}.

It has been observed that the interface between unreduced $NiAl_2O_4$ and the products of the reduction reaction, Ni and Al_2O_3, generally intact survives the large volume decrease (~ 19 %, calculated from the densities of the materials involved) associated with the reduction reaction [4,5,6]. However, as can be seen in Fig. 8, which shows a partially reduced sample with unreduced $NiAl_2O_4$ remaining in the middle of the grains (gray phase), cracking occurs at the original spinel grain boundaries, very likely due to this volume change. A similar phenomenon is observed in Fig. 9 in a sample completely reduced at 1300 °C. Another factor which might aid this process is the need to transport oxygen and/or

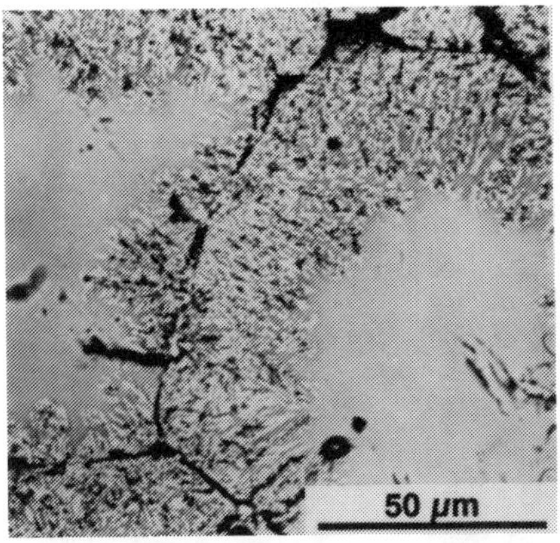

Fig. 8. Large-grained $NiAl_2O_4$ sample reduced at 1100 °C for 2 hours at an oxygen activity of $10^{-15.6}$.

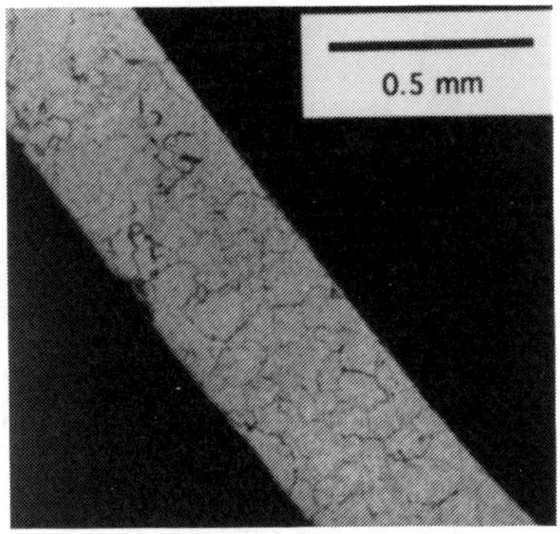

Fig. 9. Cracking due to reduction at 1300 °C for 2 hours at an oxygen activity of 10^{-10}.

metallic components between the location of the metal precipitates which are the products of the reduction reaction, and the external environment. If the specimen is macroscopic in dimension, mass transport over long distances is required. Considering the fact that diffusion is usually sluggish in ceramics with low defect concentrations, such as alumina, the lack of faster mass transport may contribute to the cracking process. Work is in progress to determine the rate limiting diffusion process and/or diffusing species as well as to model the stress state resulting from the conbined volume shrinkage due to reduction and diffusion of the related ions.

Since samples with extensive cracking at the grain boundaries will have quite poor mechanical properties, a means must be found to overcome this problem. The next sections deal with some initial results about the effect of ZrO_2 additions on the cracking and the mechanical properties of unreduced and reduced $NiAl_2O_4$ specimens.

$NiAl_2O_4$ with ZrO_2

The role of ZrO_2 in improving the mechanical properties of both dispersion-toughened ceramics, e.g. ZrO_2-toughened Al_2O_3 and precipitation-toughened ceramics, e.g. MgO-partially stabilized ZrO_2, has been extensively documented [10,11].

Our work on partial reduction reactions in the Fe-Mn-O system [5,6,12], also detailed in the preceding article in this book, has shown that ZrO_2 particles can have a considerable influence on the final microstructure and allow morphology control by serving as nucleation sites for the metal precipitation during the partial reduction of $(Fe_{1-x}Mn_x)O$.

Another important fact that led to the incorporation of ZrO_2 in spinel is that the diffusion of oxygen ions in stabilized ZrO_2 is very high. This has been utilized in the reduction of NiO [13]. If a continuous ZrO_2 phase along grain boundaries could be obtained in unreduced spinel, the enhanced diffusion of oxygen would help alleviate the cracking problem.

Since there was no data available on the solu-bil-ity of ZrO_2 in $NiAl_2O_4$, different amounts of ZrO_2 (2.5, 5 and 10 wt.%) were added to spinel powders and the mixtures were hot pressed. It was found that the addition of ZrO_2 greatly enhanced the sintering and samples with high densities were obtained. Fig. 10 shows a typical microstructure of the unreduced spinel with ZrO_2. ZrO_2 is found to be distributed in the form of distinct particles and there is no continuous network.

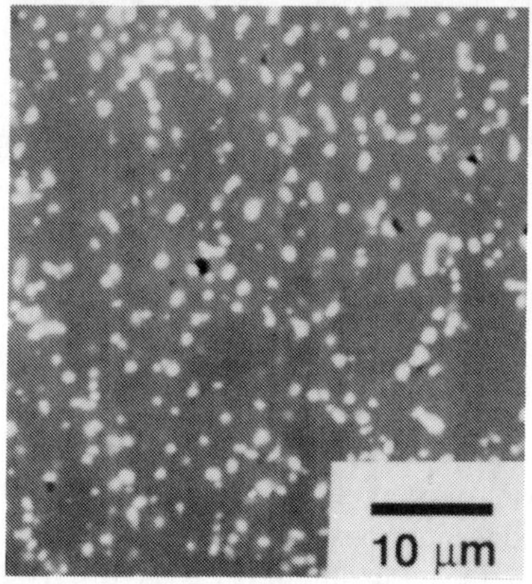

Fig. 10. Unreduced $NiAl_2O_4$(gray phase) with 5 wt.% ZrO_2(white particles).

When this sample (as well as those with 2.5 and 10 wt.% ZrO_2) was reduced at 1300 °C, however, the extensive cracking observed in Fig. 9 was completely eliminated. Fig. 11 shows such a sample with 10 wt.% ZrO_2.

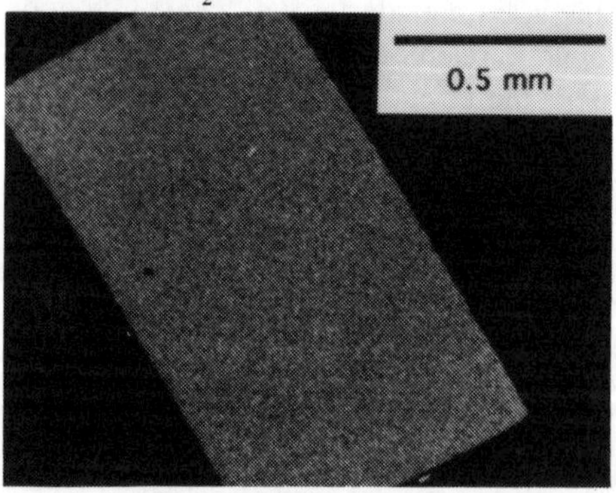

Fig. 11. $NiAl_2O_4$ with 10 wt.% ZrO_2 reduced at 1300 °C for 4 hrs. at an oxygen activity of 10^{-10}.

A partial explanation of how ZrO_2 helps in preventing the cracking was obtained when thermal etching was performed on pure spinel samples and those with 2.5 wt.% ZrO_2 to reveal the size of spinel grains in both samples (Figs. 12 and 13). It was found that the pure spinel sample (Fig. 12, same sample as shown in Fig. 9) has grain diameter of the order of 50-100 μm or larger, while the sample with ZrO_2 (Fig. 13) has much smaller grains (about 2 μm). Therefore, ZrO_2 appears to inhibit grain growth in spinel greatly enhancing transport along grain boundaries. Another effect of reduced grain size would be the increase in plastic deformation, i.e. creep, at the processing temperature. This in turn would accommodate the volume change due to reduction more easily . This effect of ZrO_2 in impeding grain growth in ceramics has been observed in other systems [14].

Fig. 12. $NiAl_2O_4$ sample thermally etched at 1200 °C for 2.5 hours in air.

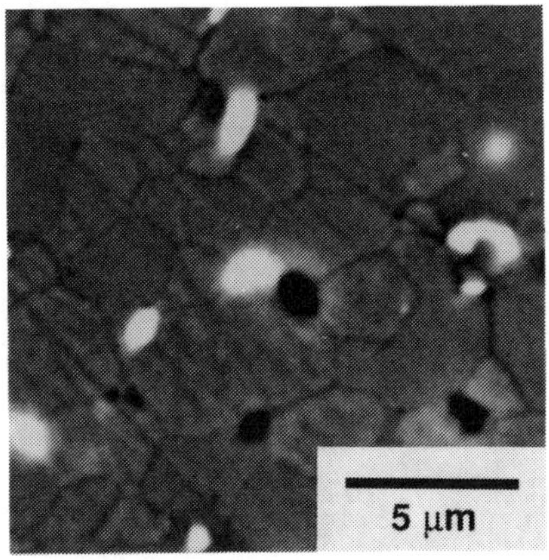

Fig. 13. $NiAl_2O_4$ with 2.5 wt % ZrO_2 thermally etched at 1200 °C for 1 hour in air.

Another observation in Fig. 13 is that ZrO_2 particles are mostly on grain boundaries. TEM studies have also confirmed this, as well as the spinel grain size shown in the same figure. The same studies have also indicated that the intergranular ZrO_2 particles are almost always monoclinic with extensive twinning. However, it is not clear at this point whether some of the tetragonal to monoclinic transformation in ZrO_2 particles is due to the TEM sample preparation which involves mechanical polishing as well as exposure of previously encapsulated particles during the final stages of ion thinning. Studies have shown that this phenomenon may occur in ceramics containing ZrO_2, such as Al_2O_3 [15]. Preliminary XRD analysis has indicated that there is indeed tetragonal as well as monoclinic ZrO_2 present. Systematic work is underway to confirm these findings and to determine the relative amounts of tetragonal and monoclinic ZrO_2 in unreduced spinel.

Fig. 14 is a TEM micrograph of a reduced sample with ZrO_2. Compared to Fig. 5, where the largest Ni particle was less than 0.5 μm in diameter, here there are much larger Ni particles (1-2 μm or more in diameter) as well as smaller ones similar to those in Fig. 5. There is apparently a bi-modal distribution in the Ni particle size in Fig. 14, with large Ni particles occuring along Al_2O_3 grain boundaries and small ones inside the grains. TEM also showed that a ZrO_2 particle was almost always associated with a large Ni particle, acting as a nucleation site for precipitation. This leads to the possibility of tailoring the reduced microstructure by controlling the distribution ZrO_2 particles, as was demonstrated in the model Fe-Mn-O system [5,12].

Preliminary Fracture Toughness Testing Results

The fracture toughnesses of the original spinel phase and the two-phase Ni-Al_2O_3 mixture were qualitatively compared [4,5,6] using the Vickers microindentation technique [16,17]. Pure $NiAl_2O_4$ sample showed extensive cracking, an indication of low fracture toughness. The reduced microstructure exhibited no cracking demonstrating a much higher fracture toughness. In fact it was "too tough" to be characterized by the indentation method.

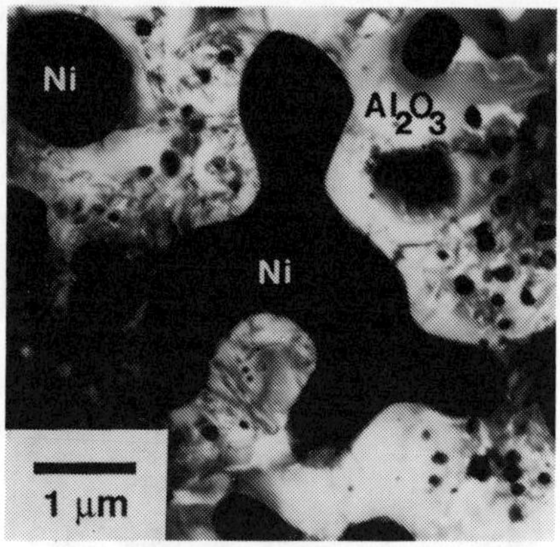

Fig. 14. $NiAl_2O_4$ sample with 2.5 wt % ZrO_2 reduced at 1300 °C for 4 hours at an oxygen activity of 10^{-12}.

Since the Vickers microindentation technique was inapplicable on reduced microstructures, a recent study was initiated using three-point bending to determine their fracture toughnesses. Fig. 15 is an SEM picture of the fracture surface of one of the tested samples. The white particles are Ni and the gray areas are Al_2O_3. The rough surface indicates extensive crack deflection due to the presence of the ductile phase, leading to high fracture toughness. However, there is not much evidence of pullout and/or plastic deformation by Ni particles, factors that could certainly increase the fracture toughness even more.

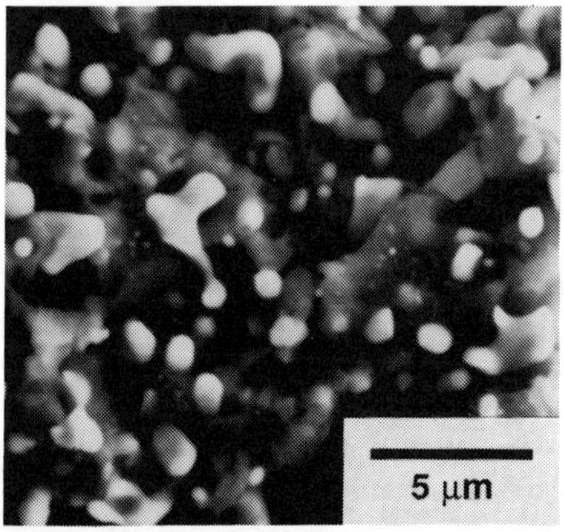

Fig. 15. The fracture surface of a sample with 2.5 wt % ZrO_2 reduced at 1300 °C for 4 hours at an oxygen activity of 10^{-12}.

Knoop indentation was used as the pre-cracking method and the pre-cracks obtained were not visible so that a precise determination of the fracture toughness of the material could not be made. A bridge-indentation setup [18] is currently being evaluated as a means for more reliable pre-cracks.

SUMMARY

Partial reduction of $NiAl_2O_4$ spinel yields a two-phase mixture of Ni-rich metal particles embedded in a ceramic oxide matrix. Two morphologies of Ni particles were observed: equiaxed particles embedded within an α-Al_2O_3 matrix and rod-like particles within an Al_2O_3-rich defect-form of the spinel phase. Electron microscopy studies indicate that the interface between the equiaxed Ni particles and Al_2O_3 can withstand large tensile stresses from a large coefficient of thermal expansion mismatch (up to a certain critical Ni particle size, above which debonding may initiate), and the interface between the unreduced $NiAl_2O_4$ and the two-phase mixture of Ni and Al_2O_3 can withstand large volume changes during the reduction reaction. Cracking was observed at the original spinel grain boundaries. Incorporation of ZrO_2 (as little as 2.5 wt %) in $NiAl_2O_4$ prevented cracking likely by inhibiting grain growth in spinel. Preliminary mechanical testing shows an improvement in the fracture toughness of the Ni-Al_2O_3 two-phase mixture compared to that of unreduced $NiAl_2O_4$.

ACKNOWLEDGEMENTS

This research was supported by the Office of Naval Research under Grant N00014-92-J-1526 and the Air Force Office of Scientific Research under Contract F49620-931-0235. The use of the Electron Microscopy and Materials Preparation Facilities of the Materials Science Center at Cornell University funded by the National Science Foundation under Grant No. DMR-91-21564 is gratefully acknowledged. The early work of J. Hillman, and the assistance of Y. Shapiro, S. Reeves, J.E. Tang and P. Ret is greatly appreciated.

REFERENCES

1. M. F. Ashby and D. R. H. Jones, Engineering Materials I, International Series on Materials Science and Technology, **34**, Pergamon Press, London (1986).
2. Engineering Materials Handbook, v. 4 - Ceramics and Glasses, ASM, Materials Park, OH (1991).
3. F. A. Elrefaie and W. W. Smeltzer, "Thermodynamics of Ni-Al-O System Between 900 and 1400 K," J. Electrochem. Soc., **128**, 2237-42 (1981).
4. E. Üstündag, R. Subramanian, R. Dieckmann and S.L. Sass, "In Situ Formation of Metal-Ceramic Microstructures in the Ni-Al-O System by Partial Reduction Reactions," Acta Metall. Mater. (submitted for publication).
5. E. Üstündag, R. Subramanian, R. Dieckmann and S.L. Sass, "In Situ Formation of Metal-Ceramic Composites and Ductile Phase Toughened

Ceramics Using Partial Reduction reactions" in: In-Situ Composites: Science and Technology, M. Singh and D. Lewis (eds.), The Minerals, Metals and Materials Society, Warrendale, PA, (1994) p. 97-113.

6. R. Subramanian, E. Üstündag, R. Dieckmann and S.L. Sass, "In Situ Formation of Metal-Ceramic Microstructures by Partial Reduction Reactions" in: Advances in Ceramic Matrix Composites, N.P.Bansal (ed.), Ceram. Trans., **38** (1994) p. 127-147.

7. P. Bassoul, A. Lefebvre and J. C. Gilles, "Periodic Antiphase Structures of Intermediate Phases Occurring During the Decomposition of Non-Stoichiometric Spinels in the Systems Ga_2O_3-MgO and Al_2O_3-NiO" (in French), J. Solid State Chem., **10**, 56-65 (1974).

8. P. Bassoul and J. C. Gilles, "Structure and Microstructure of the Metastable B Phase ($NiAl_{10}O_{16}$)," J. Solid State Chem., **58**, 383-397 (1985).

9. R. W. Davidge and T. J. Green, "The Strength of Two-Phase Ceramic/Glass Materials," J. Mater. Sci. **3**, 629-634 (1968).

10. N. Claussen, "Fracture Toughness of Al_2O_3 with an Unstabilized ZrO_2 Phase," J. Am. Ceram. Soc., **59** [1-2] 49-51 (1976).

11. D.L. Porter and A.H. Heuer, "Mechanisms of Toughening Partially Stabilized Zirconia (PSZ)," J. Am. Ceram. Soc., **60** [3-4] 183-184 (1977).

12. R. Subramanian, E. Üstündag, S.L. Sass and R. Dieckmann, "In Situ Formation of Metal-Ceramic Microstructures by Partial Reduction Reactions," Solid State Ionics (submitted for publication).

13. B. Bonvalot-Dubois, G. Dhalenne, J. Berthon, A. Revcolevschi, and R.A. Rapp, "Reduction of NiO Platelets in a NiO/ZrO_2(CaO) Directional Composite," J. Am. Ceram. Soc., **71** [4] 296-301 (1988).

14. J. Wang, "Effects of Transition Metal Oxide Addition on Sintering, Creep, and Fracture Properties in Alumina," PhD Thesis, Cornell University (1991).

15. A.H. Heuer, N. Claussen, W.M. Kriven, and M. Ruhle, "Stability of ZrO_2 Particles in Ceramic Matrices," J. Am. Ceram. Soc., **65** [12], 642-650 (1982).

16. B. R. Lawn, A. G. Evans and D. B. Marshall, "Elastic/Plastic Indentation Damage in Ceramics: The Median/Radial Crack System," J. Am. Ceram. Soc., **63** [9-10], 574-581 (1980).

17. G. R. Anstis, P. Chantikul, B. R. Lawn and D. B. Marshall, "A Critical Evaluation of Indentation Techniques for Measuring Fracture Toughness: I, Direct Crack Measurements," J. Am. Ceram. Soc., **64** [9], 533-538 (1981).

18. R. Warren and B. Johanneson, "Creation of Stable Cracks in Hard Metals Using 'Bridge' Indentation," Powder Metall., **27** [1], 25-29 (1984).

EFFECT OF DOPANTS ON THE IN-SITU FORMATION OF METAL-CERAMIC MICROSTRUCTURES BY PARTIAL REDUCTION REACTIONS

R. Subramanian, E. Üstündag, S.L. Sass and R. Dieckmann
Department of Materials Science and Engineering
Cornell University, Ithaca, NY 14853-1501

ABSTRACT

Metal-ceramic multi-phase microstructures can be obtained by partial reduction of ternary or higher ceramic oxides. This has been experimentally investigated in the Fe-Mn-O system by changing the oxygen partial pressure at constant temperature and total pressure. Dopants, such as Al_2O_3, CaO, Cr_2O_3 and ZrO_2, were added to the initial, polycrystalline oxide solid solution $(Fe_{1-x}Mn_x)O$. Electron microscopy studies were performed for microstructural characterization, phase identification and chemical analysis. It was observed that dopants strongly influence the location of the metal precipitation during reduction. The experimental observations are discussed based on solubility limits and the segregation of dopants.

INTRODUCTION

The performance of metal-ceramic composites, for a given chemical composition, is controlled by the size, morphology, and distribution of the different phases that are present. To produce multiphase microstructures economically, it is necessary to use relatively simple processing routes. For metal-ceramic composites, for example, *in situ* processing is desirable, where the ceramic and metal phases are formed in a single heat treatment step.

To the extent authorized under the laws of the United States of America, all copyright interests in this publication are the property of The American Ceramic Society. Any duplication, reproduction, or republication of this publication or any part thereof, without the express written consent of The American Ceramic Society or fee paid to the Copyright Clearance Center, is prohibited.

In this article, the influence of dopants on metal-ceramic microstructures formed by partial reduction reactions of the oxide solid solution $(Fe_{1-x}Mn_x)O$ will be discussed. The simplified notation $(Fe_{1-x}Mn_x)O$ is used instead of the more precise description - $(Fe_{1-x}Mn_x)_{1-\Delta}O$, where Δ is the deviation from stoichiometry. This oxide solid solution was chosen as a model system because relatively extensive information was available on phase stabilities [1-3], point defect structures [4,5], cation diffusion in $(Fe_{1-x}Mn_x)O$ [4,5] and on the kinetics of the internal reduction of $(Fe_{1-x}Mn_x)O$ [1,6]. By suitable choice of the initial oxide composition, oxygen partial pressure, reduction time, temperature, and type and concentration of dopants, it is possible to control the nucleation and growth of the metallic phase to produce different microstructures. Details on the effect of oxygen partial pressure, reduction time and temperature were already reported earlier [7,8,9].

The formation of metal-ceramic microstructures by partial reduction involves essentially the following two steps: (i) nucleation of metal and (ii) growth of metal nuclei. Homogeneous nucleation in a grain has a very high barrier. Nucleation at one- and higher-dimensional defects is energetically more favorable. Therefore, it is expected that grain boundaries and dislocations in grains are very important for controlling the morphology of the metal-ceramic microstructures produced by partial reduction from polycrystalline ceramic oxide solid solutions. Thus, the microstructure of the ceramic starting material must be very important for the nucleation of metal particles.

Addition of dopants which segregate or form compounds with the oxide matrix material has the potential to modify the kinetics of nucleation at different locations as well as the rate of the transport of cations and anions in the bulk, and along dislocations and boundaries. This may change the overall kinetics of metal formation and also its distribution in the metal-ceramic microstructures. If dopants segregate at boundaries and dislocations, the chemical compositions at such locations may be very different from those in an impurity-free material. As a consequence the interfacial energies associated with boundaries may change and consequently the barrier heights for the nucleation of metal. In addition, impurity segregation, especially of alio-valent ions, may lead to a significant modification of the transport properties at interfaces and dislocations. Furthermore, dopants dissolved in the bulk may alter the bulk transport properties. Dopants at boundaries and dislocations, in the form of precipitates or segregated, may provide heterogenous nucleation sites for the precipitation of metal. These factors may significantly influence the nucleation and growth of metal precipitates in partial reduction reactions of ceramic oxide solid solutions and compounds, and, hence, the metal-ceramic microstructure obtained.

THERMODYNAMIC BACKGROUND

For carrying out systematic studies on the use of partial reduction reactions to control metal-ceramic microstructures knowledge on phase equilibria, especially on the effect of the oxygen partial pressure on phase stabilities at constant temperature and pressure, is needed. Phase diagrams displaying log a_{O_2} versus cationic composition at different temperatures are very useful for selecting appropriate thermodynamic conditions for partial reduction reactions. A phase diagram of this type for the Fe-Mn-O system is shown in Fig. 1 [2]. In this system, metal-ceramic microstructures can be obtained by the partial reduction of oxide solid solutions with the rock-salt structure, i.e. of $(Fe_{1-x}Mn_x)O$. At oxygen activity values below the lower stability limit of the rock-salt structure phase, a metal alloy (almost pure iron) and a ceramic oxide with a higher manganese content than in the starting oxide, coexist. The oxygen activity is defined as $a_{O_2} = P_{O_2}/P°$ with $P° = 1$ bar. By control of the oxygen activity during the reduction, the composition of the starting oxide solid solution and the temperature of reduction, different metal-ceramic microstructures can be obtained.

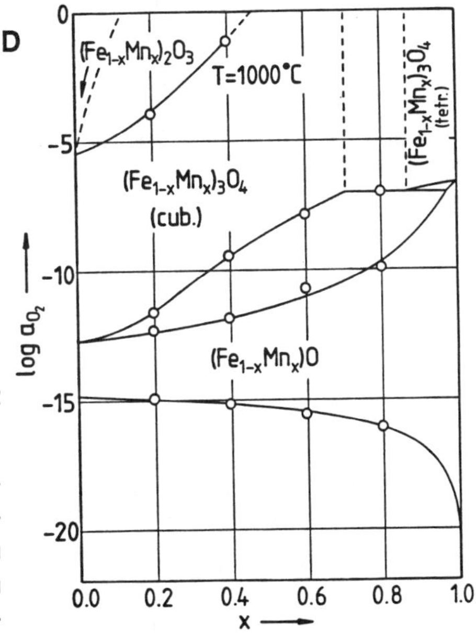

Fig. 1. log a_{O_2} vs composition phase diagram of the Fe-Mn-O system at 1000 °C.

EXPERIMENTAL PROCEDURE

The samples used for reduction were rectangular bars with a cross-section of 1 mm^2 and a length of 2-3 mm. An analysis of microstructural characteristics such as porosity was performed using an image analysis program of a JEOL 733 electron microprobe. Compositions of phases were determined by electron microprobe analysis. Transmission electron microscopy was used for microstructural evaluation and identification of phases. Details of the sample preparation are reported in Refs. [8], [9] and [10].

RESULTS

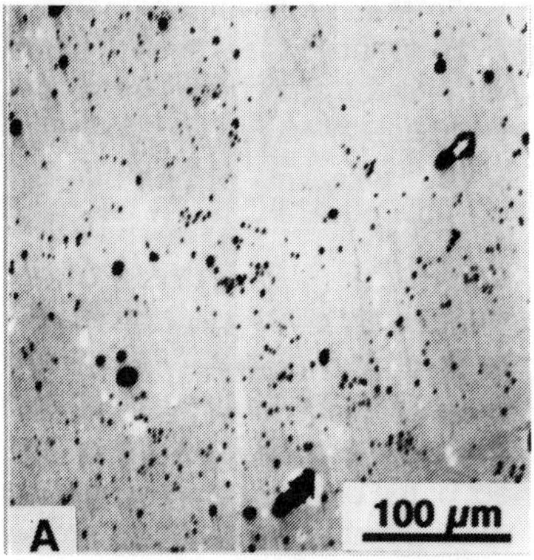

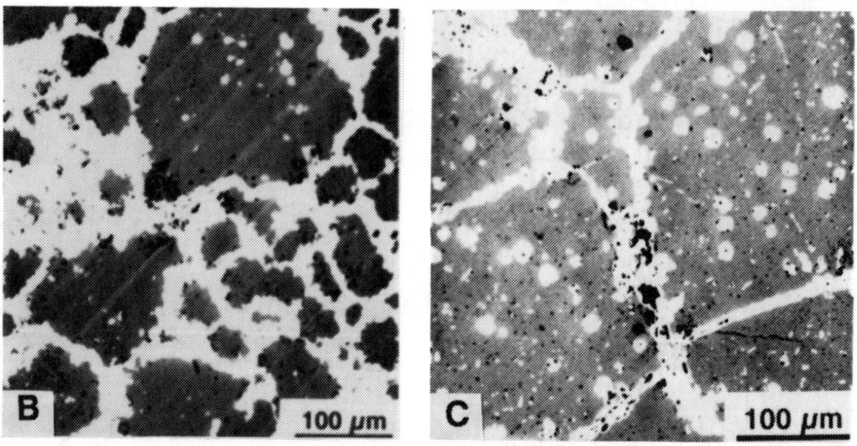

Fig. 2. Effect of ZrO_2 additions on metal-ceramic microstructures obtained by partial reduction of $(Fe_{0.6}Mn_{0.4})O$. The reduction conditions were T = 1000 °C, t = 4 hours, 5 % CO in "N_2": (a) undoped, (b) doped with 0.5 wt % ZrO_2, and (c) with 2 wt % ZrO_2.

Figs. 2a and b show the effect of the addition of ZrO_2 on the precipitation of metal during the partial reduction of $(Fe_{0.6}Mn_{0.4})O$. Typical features include a metallic phase (white), a ceramic oxide (gray) which has a higher manganese content (80 - 90 % Mn) than the initial oxide material, and approximately 2 to 4 % porosity (black). When an undoped ceramic oxide is reduced at 1000 °C for 4 hours in a gas atmosphere with 5 % CO in "N_2," ("N_2" = N_2 with ~ 200 ppm oxygen) significant metal precipitation was not observed (Fig. 2a). In samples doped with 0.5 wt % (Fig. 2b) and with 2 wt % ZrO_2 (Fig. 2c) metal precipitation at grain boundaries in thick layers is observed. An interesting feature in the reduced samples with 2 wt % ZrO_2 is the metal precipitation along grain boundaries as well as within grains. This behavior was observed for temperatures from 800 to 1200 °C and at different reduction times. For samples with no ZrO_2 additions, longer reduction times were needed for precipitation of metal

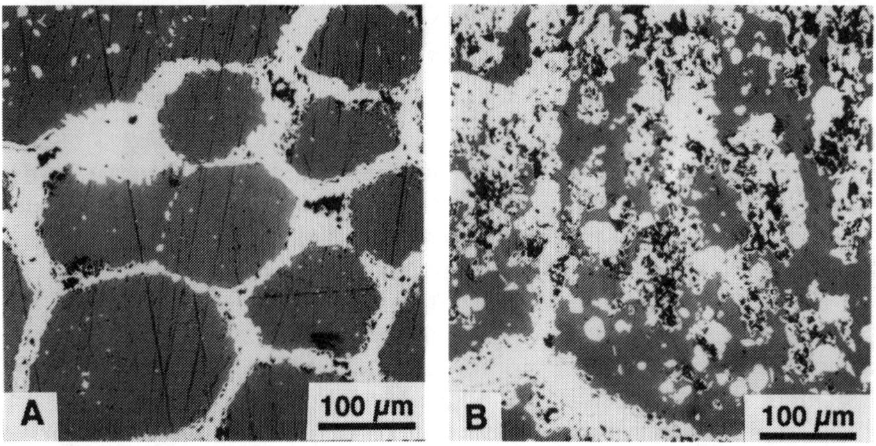

Fig. 3. Effect of CaO additions on metal-ceramic microstructures obtained by partial reduction of $(Fe_{0.6}Mn_{0.4})O$. The reduction conditions were T = 1100 °C, t = 4 hours, 5 % CO in "N_2": (a) doped with 0.1 wt % CaO and (b) doped with 1 wt % CaO.

along grain boundaries. Figs. 3a and b illustrate the effect of the addition of CaO and Figs. 4a and b show the effect of Al_2O_3 on the morphology of metal-ceramic microstructures obtained by partial reduction.

DISCUSSION

The basic processes involved in the precipitation of a metallic phase within a ceramic matrix have been discussed in the literature, e.g., [6,11-14]. As mentioned before, dopants present in polycrystalline $(Fe_{1-x}Mn_x)O$ may segregate at dislocations and interfaces and influence transport

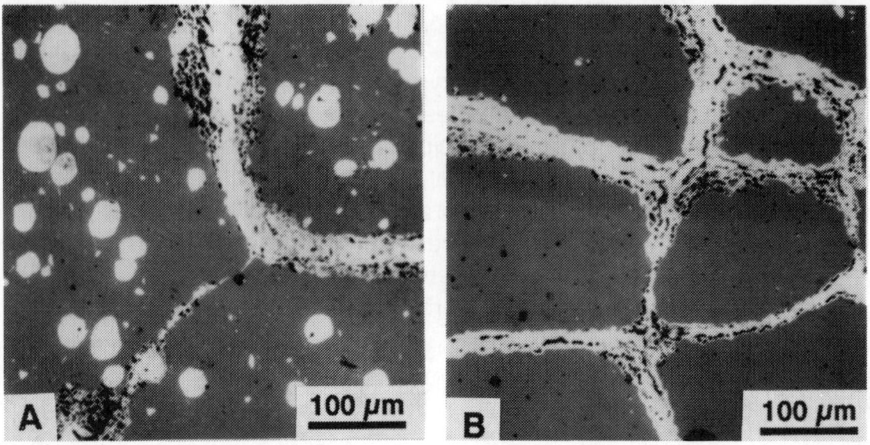

Fig. 4. Effect of Al_2O_3 additions on metal-ceramic microstructures obtained by partial reduction of $(Fe_{0.6}Mn_{0.4})O$. The reduction conditions were T = 1100 °C, t = 4 hours, 5 % CO in "N_2": (a) doped with 0.5 wt % Al_2O_3 and (b) doped with 4 wt % Al_2O_3.

properties. This leads to changes in the kinetics and location of the precipitation of metal. The photographs shown in Figs. 2-4 clearly demonstrate that the addition of dopants allows to control the location of precipitates. With increasing concentrations of dopants, the following two different trends were identified: (a) the precipitation of metal changes from predominant decoration of boundaries by metal precipitates to metal precipitation both at boundaries and within grains (e.g., in the cases of ZrO_2 or CaO additions), and (b) the precipitation of metal occurs at low dopant concentrations at boundaries and within grains and at high dopant concentrations predominantly at boundaries (e.g., in the cases of Al_2O_3 and Cr_2O_3 additions). In the following sections, it will be attempted to qualitatively explain the influences that different dopants have on the microstructure development during the partial reduction of polycrystalline $(Fe_{1-x}Mn_x)O$.

Effect of ZrO_2 Doping

As seen from Figs. 2a - c, in the case of the addition of a small amount of ZrO_2 (0.5 wt %) precipitation of metal during partial reduction occurs predominantly at grain boundaries. If the concentration of ZrO_2 is increased to 1 and 2 wt %, precipitation occurs both at boundaries and in grains. Assuming that ZrO_2 dopants act as nucleation sites for the precipitation of metal, one can explain the trend in the morphology change based on the limited, temperature dependent solubility of ZrO_2 in $(Fe_{1-x}Mn_x)O$. Upon cooling from the hot-pressing temperature to a lower reduction temperature, precipitation of ZrO_2 may occur either at boundaries and/or within grains (very likely at dislocations, especially at nodes) depending on the degree of supersaturation of ZrO_2 in the bulk. Since grain boundaries are the locations for easiest precipitation of dopants, the precipitation of dopants will occur predominantly at the boundaries in the sample with 0.5 wt % ZrO_2, where the supersaturation is relatively small. In the sample with 2 wt % ZrO_2, due to the larger supersaturation, precipitation of dopants may also take place at dislocations within grains. TEM observations of samples before reduction confirm this expectation. After reduction, TEM observations show metal precipitates within grains and at boundaries associated with ZrO_2 particles. A more detailed discussion is given in Ref. [10].

Effect of CaO Doping

Figs. 3a and b show the effect of CaO on the morphology development during the partial reduction of polycrystalline samples of $(Fe_{1-x}Mn_x)O$. A trend very similar to that for the addition of ZrO_2 dopants is observed. Metal precipitation during the partial reduction of polycrystalline $(Fe_{1-x}Mn_x)O$ occurs predominantly at grain boundaries for concentrations of CaO $<$ 0.25 wt %; much more metal is observed than in an impurity-free sample reduced under identical conditions. Metal is precipitated both at boundaries and within grains at higher concentrations of CaO. Although this trend is similar to the case of ZrO_2 additions, its explanation must be different because CaO must have an extended solubility in the isostructural $(Fe_{1-x}Mn_x)O$. The solubility limit for CaO in pure "FeO" is about 15 wt % at 1000 °C [15]; one may expect that this limit is similar in MnO and also in the solid solution $(Fe_{1-x}Mn_x)O$. Therefore, one may expect that the concentrations of CaO of up to 2 wt% used in this study lie well within the solubility limit for CaO in $(Fe_{1-x}Mn_x)O$ at 1100 °C. This is supported by the fact that we did not observe any CaO particles in our samples. Therefore, we conclude that an increased concentration of CaO at boundaries and at dislocations within grains must strongly influence the nucleation of metal and its subsequent growth. At lower concentrations of CaO the segregation of CaO is most effective at boundaries. Only at larger CaO contents, the segregation at dislocations within grains is significant to also allow metal precipitation within the grains. The segregation of CaO at boundaries and dislocations in $(Fe_{1-x}Mn_x)O$ is related to the limited solubility of CaO in $(Fe_{1-x}Mn_x)O$ which is caused by the fact that Ca^{2+} ions are significantly larger than Fe^{2+} and Mn^{2+} ions. This conclusion is supported by experimental observations of enhanced concentrations of CaO at interfaces in MgO [16]; Mg^{2+} ions are similar in size as Fe^{2+} and Mn^{2+} ions and significantly smaller than Ca^{2+} ions. Because interfaces are dislocation-rich and provide a more open space than dislocations within grains, segregation of CaO is likely to occur easiest at boundaries. Because CaO is beneficial for the nucleation and growth of metal particles during the partial reduction of $(Fe_{1-x}Mn_x)O$, at low CaO concentrations the metal precipitates are predominantly formed at boundaries and at higher CaO concentrations also (very likely at dislocation nodes) within grains, as experimentally observed.

Effect of Al_2O_3 and Cr_2O_3 Additions

Figs. 4a and b give examples of the effect of the addition of Al_2O_3 in different concentrations on the morphology of metal-ceramic microstructures formed by partial reduction of $(Fe_{1-x}Mn_x)O$. Before going into details it is worth noting that the effect of Cr_2O_3 on the location of the metal is practically identical to that of Al_2O_3. First, the results show that the addition of Al_2O_3 or Cr_2O_3 is beneficial for the formation of metal precipitates during partial reduction, i.e., less metal is formed in an impurity-free sample compared to Al_2O_3- or Cr_2O_3-doped samples under identical reduction conditions.

Partially reduced samples doped with 0.5 wt% Al_2O_3 or Cr_2O_3 show metal precipitates at boundaries as well as within grains (very likely formed at dislocation nodes). At this time it is still unknown whether secondary spinel phases were present in these samples. Further TEM work to answer this question is underway. In samples with Al_2O_3 or Cr_2O_3 concentrations of 2 wt % and above, spinel precipitates were observed at boundaries and within grains without any visible preference of location. The formation of spinel precipitates must depend on the solubilities of the compounds Al_2O_3 and Cr_2O_3 in $(Fe_{1-x}Mn_x)O$ at low oxygen activities ($\log a_{O_2} = -10$) at 1400 °C and at the lower reduction temperatures. Unfortunately, detailed information on this subject could not be found in the literature. It is worth noting that in samples with dopant concentrations of 2 wt % and above metal precipitates were formed by partial reduction almost entirely at boundaries. Therefore, it is unlikely that spinel inclusions significantly influence the formation of metal precipitates. As a consequence, one may conclude that the segregation of Al_2O_3 and Cr_2O_3 at boundaries and dislocations strongly influences the precipitation of metal during partial reduction in samples with small impurity additions.

The observation that at larger impurity concentrations metal forms largely at boundaries suggests that the nucleation capacity of boundaries can be affected by dopants much more than that of dislocations within grains. Currently, we do not have a conclusive explanation for this observation. Further TEM studies are needed to identify how boundary structures change with increasing dopant concentration. A more detailed discussion can be found in [10].

CONCLUSIONS

For the Fe-Mn-O system, it has been shown that dopants can be used to control the development of metal-ceramic microstructures during partial reduction. The development of such microstructures depends to a large extent on nucleation and transport phenomena which both can be influenced by the addition of dopants. Al_2O_3, CaO, Cr_2O_3, and ZrO_2 impurity additions enhance the kinetics of metal precipitation during partial reduction reactions. Precipitation of an iron-rich alloy is obtained only at grain boundaries if Al_2O_3 and Cr_2O_3 are present as dopants in high concentrations. Low concentrations of CaO and ZrO_2 lead to metal precipitation primarily at boundaries, while at higher concentrations, precipitation occurs at boundaries and also within grains.

ACKNOWLEDGEMENTS

The authors thank the Office of Naval Research for supporting this research under Grant N00014-92-J-1526. The use of the Electron Microscopy and Materials Preparation Facilities of the Materials Science Center at Cornell University, sponsored by the National Science Foundation under Award No. DMR-9121654 is gratefully acknowledged. Finally, the help of Protima Banerjee and Andrew Leonard in the preparation of samples is greatly appreciated.

REFERENCES

[1] "Internal Reduction of $(Fe,Mn)_{1-\Delta}O$ Mixed Oxides" (in German), H. Falke, Doctoral Dissertation, University of Hannover, Germany, [1987].

[2] "Thermodynamics of Iron Manganese Mixed Oxides at High Temperatures," P. Franke and R. Dieckmann, J. Phys. Chem. Solids, **51** (1) [1990] 49-57.

[3] "Nonstoichiometry and Thermodynamics of the Solid Solution $(Fe,Mn)_{1-\Delta}O$ at 1200 °C," R. Subramanian and R. Dieckmann, J. Phys. Chem. Solids, **54** (9) [1993] 991-1000.

[4] "Disorder and Transport in (Fe,Mn)O and in $(Fe,Mn)_3O_4$ Mixed Oxides" (in German), P. Franke, Doctoral Dissertation, University of Hannover, Germany, [1987].

[5] "Defect Structure and Transport Properties of Mixed Iron-Manganese Oxides," P. Franke and R. Dieckmann, Solid State Ionics, **32/33** (Part I) [1989] 817-823.

[6] "Internal Reduction of Iron-Manganese Mixed Oxides of the Type $(Fe,Mn)_{1-\Delta}O$" (in German), H. Falke, R. Dieckmann and H. Schmalzried, in: 'Prozeßkinetik und Prozeßtechnik im Hüttenwesen - Abschlußbericht eines Schwerpunktprogrammes der Deutschen Forschungsgemeinschaft,' Verlag Stahleisen GmbH, Düsseldorf, [1986] 1-13.

[7] "*In Situ* Formation of Metal-Ceramic Microstructures Including Metal-Ceramic Composites, Using Reduction Reactions," E. Üstündag, R. Subramanian, R. Vaia, R. Dieckmann and S. L. Sass, Acta Met. et Mater., **41** (7) [1993] 2153-2161.

[8] "*In-Situ* Formation of Metal-Ceramic Microstructures by Partial Reduction Reactions," R. Subramanian, E. Üstündag, R. Dieckmann and S.L. Sass, in: 'Advances in Ceramic Matrix Composites,' Ed. N.P. Bansal, Ceram. Trans., **38** [1994] 127-147.

[9] "*In-Situ* Formation of Metal-Ceramic Composites and Ductile Phase Toughened Ceramics Using Partial Reduction Reactions," E. Üstündag, R. Subramanian, R. Dieckmann and S.L. Sass, in: 'In-Situ Composites: Science and Technology,' M. Singh and D. Lewis, Eds., The Minerals, Metals and Materials Society, Warrendale, PA, [1994] p. 97-113.

[10] "*In-Situ* Formation of Metal-Ceramic Microstructures by Partial Reduction Reactions," R. Subramanian, E. Üstündag, S.L. Sass and R. Dieckmann, Solid State Ionics, (submitted for publication).

[11] "Internal Reduction of Wüstite Type Mixed Iron Manganese Oxides," R. Dieckmann and H. Falke, Metalurgia i Odlewnicto (Metallurgy and Foundry), **13** (1-2) [1987] 191-206.

[12] "Oxide Solid Solutions and its Internal Reduction Reactions," H. Schmalzried, Ber. Bunsenges. Phys. Chem., **88** [1984] 1186-1191.

[13] "Internal Solid State Reactions," M. Backhaus-Ricoult and H. Schmalzried, Progr. Solid State Chem., **22** [1993] 1-57.

[14] "On the Reduction of Oxide Solid Solutions" (in German), H.-J. Engell and H.K. Kohl, Z. Elektrochem., **66** (8/9) [1962] 684-689.

[15] "The Orthosilicate-Iron Oxide Portion of the System Cao-"FeO"-SiO_2," W.C. Allen and R.B. Snow, J. Am. Ceram. Soc, **38** (8) [1955] 259-280.

[16] "Chemical Aspects of Equilibrium Segregation to Ceramic Interfaces," P. Wynblatt and R.C. McCune, in: 'Surfaces and Interfaces in Ceramics and Ceramic-Metal Systems,' J.A. Pask and E.G. Evans, Eds., Materials Science Research, Vol. 14, Plenum Press, New York, [1980] 83-95.

IN-SITU MAGNETOPLUMBITE GROWTH IN SOL-GEL DERIVED ALUMINA MATRICES

D. D. Erickson and W. P. Wood, 3M Company, St. Paul, MN 55144-1000

ABSTRACT

The addition of a discrete secondary phase to ceramic bodies is well known to increase the fracture toughness of the composite body. In sol-gel processing, however, such additions generally result in increased porosity as the secondary phase opposes the shrinkage of the gel during drying and sintering. In the present work, in-situ growth of a hexagonal rare earth magnetoplumbite ($MgLnAl_{11}O_{19}$, where Ln = a lanthanide rare earth element) is accomplished within the sol-gel alumina matrix resulting in a body of near theoretical density. Mechanical properties will be presented and discussed with respect to microstructure.

INTRODUCTION

The addition of second phase particles into the matrices of polycrystalline ceramics has received considerable interest as a means to increase the fracture toughness of the ceramic[1,2,3,4]. However, the incorporation of particles using pressureless sintering processing has presented special problems; namely, these particles retard densification of the matrix resulting in increased porosity[5]. This is especially true for sol-gel processing where green densities are lower[6,7,8].

In this work, several sol-gel based alumina compositions doped with a lanthanide rare earth element such as La and a divalent spinel forming element such as Mg or Fe were identified to produce high density alumina-based ceramics containing a discrete magnetoplumbite phase of platelets or particles grown in-situ. Improvements in fracture toughness of the matrix have been observed in the presence of the in-situ magnetoplumbite platelet secondary phase without significant loss in microhardness. It has been found that changes in the α-alumina transformation temperature of the sol-gel matrix by the use of seeds or nucleating agents[9,10,11,12,13] will change the morphology of the magnetoplumbite secondary phase from platelets to equiaxed particles.

EXPERIMENTAL

For the purpose of this work, the microstructure and mechanical properties of compositions within the system of alumina with 3.5% La_2O_3, 1.5% MgO were studied. All samples were prepared using a 21% Al_2O_3-equivalent nitric-acid-stabilized boehmite sol (DISPERAL®, Condea Chemie GmbH, Hamburg, Germany). Two sols were prepared, the first sol, hereby designated as composition "A", contained 1194 gm of Disperal®, 71 gm of concentrated nitric acid and 3000 gm of deionized water. The second sol, designated composition "B", was identical to the first, except that 136 gm of deionized water were replaced with a water-based FeOOH (6.6 % as α-Fe_2O_3) slurry. Upon heating, these particles dehydrate and transform to hematite (α-Fe_2O_3). Thorough mixing of the sols was achieved using a Waring Industrial Blender set at high speed for one minute. The sols were poured into Pyrex® trays and allowed to dry overnight at 90°C to a friable solid.

To the extent authorized under the laws of the United States of America, all copyright interests in this publication are the property of The American Ceramic Society. Any duplication, reproduction, or republication of this publication or any part thereof, without the express written consent of The American Ceramic Society or fee paid to the Copyright Clearance Center, is prohibited.

The dried sols were crushed and screened to obtain particles in the -25 +40 fraction (U.S.A. Standard Testing Sieves). The screened materials were fed into a stainless steel rotary tube calciner in order to convert the components to anhydrous oxides (i.e. boehmite to a transitional alumina and FeOOH to α-Fe_2O_3). The rotary calciner consisted of a 15 cm diameter tube, 1.2 m long, having a 0.4 m hot zone at 700°C. The tube was inclined at 2.5° with respect to the horizontal to provide a residence time of approximately 5 minutes. The material was fed into the rotary kiln at approximately 30 gm per minute.

For both compositions, the dopants used to form the magnetoplumbite second phase (i.e. MgO and La_2O_3) were added by an impregnation process in which 66 grams of $La(NO_3)_3{*}6H_2O$ solution (28% La_2O_3 solids; Rhone-Poulenc, Inc., Shelton, CT) and 71.8 grams of $Mg(NO_3)_3{*}6H_2O$ solution (10.5% solids; Mineral Research and Development Corp., Harrisburg, NC) were combined with deionized water to make 325 cc of total solution. This solution was poured onto 500 grams of the freshly calcined alumina particles and continuously stirred to insure uniform distribution of the solution. The filling of the sub-micron diameter pores in the transitional alumina by capillary action is governed by the Washburn equation[14]. The amount of nitrate solution absorbed by the transitional alumina had been previously calculated based on water absorption measurements. The concentration of nitrates used in the impregnation was calculated so that the final composition of La_2O_3 and MgO in the fired bodies was as defined above. After impregnation, the particles were dried in a forced air oven at 90°C, and were then fed through the stainless steel rotary tube kiln previously described in order to convert the nitrate salts to oxides.

Both compositions were final fired using a silicon carbide rotary tube kiln. The silicon carbide tube was 8.9 cm in diameter and 1.32 m long and inclined 4.4° with respect to the horizontal. The tube had a 76 cm hot zone at 1430°C, and was rotated at 10.5 rpm which provided a residence time of 5 minutes. The calcined material was fed into the kiln at a rate of 30 gm per minute.

Density measurements were obtained using a Micromeritics AccuPyc 1330 helium pycnometer.

DTA analysis was obtained on -400 mesh calcined material using a Netzsch TG-DTA Apparatus 409 using alumina sample crucibles with an air reference. The samples were heated from room temperature to 1400°C at a rate of 5°C per minute.

X-ray diffraction (XRD) data was obtained using a Phillips vertical diffractometer with copper Kα radiation. Prior to the analysis, samples of the -400 mesh calcined material were suspended in a box kiln using platinum foil baskets and heated at a rate of 5°C/minute to the inset, peak, and outset temperatures calculated by DTA. Once the sample reached its designated temperature, it was removed from the box kiln and rapidly quenched to room temperature.

Microhardness and fracture toughness measurements were made on polished cross-sections of the fired pieces using a Leitz Miniload hardness tester fitted with a Vickers diamond indentor using a 500 gm load. Polished cross-sections of the ceramic pieces were obtained by mounting the samples in Epomet® (Buehler, Ltd., Lake Bluff, IL). The mounted samples were successively polished using Metadi® water-based diamond suspensions (Buehler, Ltd.) until no visible scratches were observed on the ceramic cross-sections at 100X under reflected polarized light.

Fracture toughness data was obtained by measuring the length of the cracks propagating from the apices of the Vickers diamond indents made with a 1000 gram load as observed under backscattered electron images obtained from a JEOL 840A scanning electron microscope (SEM). Fracture toughness values were calculated using the equation proposed by Evans and Charles[15]. Additional SEM characterization was undertaken using polished cross-sections and thermally etched fractured surfaces of the ceramic pieces. Thermal etching of freshly cleaved surfaces was carried out at 1225°C for 30 minutes. Immediately prior to the SEM characterization, the samples were coated with a gold conductive layer deposited using an Anatech Hummer VI sputtering system.

Samples for transmission electron microscopy (TEM) were prepared by mixing 3 to 4 acicular ceramic pieces with epoxy resin (Epo-Tek®, Epoxy Technologies, Inc., Billerica, MA) and allowing the mixture to cure in 3 mm diameter button-shaped Teflon® molds. One face on each of the buttons was polished until no visible scratching was observed at 100X under polarizing light. The opposite face was then polished to a 100 μm sample thickness, and dimpled to approximately 20 μm. The samples were then glued to a slotted grid (making sure that some of ceramic pieces were in contact with the glue) and top-side milled to perforation using a Gatan ion miller. The samples were examined using a Phillips CM30 TEM at 300 keV or JEOL 200CX TEM operation at 200 keV. X-ray microanalysis was performed using an EDAX energy dispersive x-ray spectrometer.

RESULTS & DISCUSSION

XRD characterization of both compositions indicated the presence of two phases: α-alumina and a magnetoplumbite of the form $MgLaAl_{11}O_{19}$. Because of the similarity in structure between beta-aluminas (i.e. $LaAl_{11}O_{18}$) and magnetoplumbites, positive identification of the magnetoplumbite phase was based on lattice parameter ratios reported by Versetgen and Stevels[16]. Magnetoplumbites are hexagonal unit cells consisting of spinel blocks with Al^{+3} and Mg^{+2} ions in tetrahedral and octahedral sites separated by mirror planes perpendicular to the c-axis and containing the lanthanide ion. The difference between the magnetoplumbites and β-aluminas is that the mirror planes containing the large cations are more closely packed for the magnetoplumbite than for the β-aluminas.[17]

SEM characterization of the fractured and thermally etched surfaces of both compositions are shown in Figures 1 and 2. In composition A, without α-Fe_2O_3 (Figure 1), a second phase of platelets (approximately 1 μm in diameter and 0.05 μm in thickness) is visible distributed within the matrix . These platelets are not evident in composition B containing α-Fe_2O_3. The matrix of B clearly shows signs of sub-micron α-alumina grains. Normally the exposed grains after fracturing are faceted. However, the thermal etching has rounded the facets. Examination of the polished cross-sections by SEM using backscattered electron imaging (BEI) revealed high atomic number platelets (which appear acicular in cross-section) in composition A as shown in Figure 3. Characterization of composition B from Figure 4 reveals a high atomic number phase that is more equiaxed in shape.

TEM characterization of composition A shows 3-5 μm α-alumina grains and 1 μm diameter magnetoplumbite platelets (which appear acicular in shape in the TEM foil) uniformly distributed within the matrix. The platelets appear to be trapped within the large alumina grains shown in Figure 5. Upon closer examination of the α-alumina grains, many of the platelets appear along sub-micron low angle grain boundaries within the α-alumina grains as shown in Figure 6. The electron diffraction pattern (SAD) in Figure 7 was taken encompassing a diffracting α-alumina grain. The SAD shows orientation between the magnetoplumbite and the α-alumina. Orientation relationships of rare earth magnetoplumbite structures grown on α-alumina substrates are known and have been reported in the literature[18]. Elemental analysis of the acicular inclusions using x-ray microanalysis indicated the presence of magnesium and lanthanum which was not detected in the surrounding matrix. TEM characterization of composition B (Figure 8) shows sub-micron faceted α-alumina grains. Trapped along triple points are irregularly shaped inclusions, which when analyzed by x-ray microanalysis indicated the presence of Mg, Fe, and La. The presence of these elements was not detected in the alumina grains. Because of the small and random orientation of the sub-micron α-alumina grains, we have not been successful in determining whether a crystallographic orientation exists between adjoining magnetoplumbite particles and α-alumina grains.

Differences in the growth habit of magnetoplumbite in these materials can be explained using a combination of DTA and XRD data. Figures 9 and 10 show an enlargement of the alpha-alumina transition for the two samples and the calculated onset, peak and outset temperatures. In either case, the α-alumina transition temperature has been pushed to higher temperatures than that reported for an non-seeded, non-modified transitional alumina sample of approximately 1200°C. However, as expected, the presence of the iron oxide in composition B has significantly reduced the α-alumina transformation over composition A. X-ray diffraction (Figures 11 and 12) indicates that the magnetoplumbite phase in composition A forms simultaneously with α-alumina. In this case, it is believed that the magnetoplumbite has the opportunity to form in a desired morphology before the surrounding α-alumina matrix has fully formed. This differs from composition B where α-alumina forms before any detectable quantity of the magnetoplumbite phase thereby restricting the growth of the magnetoplumbite to regions along the grain boundaries.

In Table 1A is shown a compilation of density, hardness and fracture toughness data (for both hardness and fracture toughness, the number of measurements = 8) obtained on the two experimental samples. Summarized in Table 1B, is an analysis of variance (ANOVA) of the toughness measurements made on samples A and B. The data collected indicates a statistically significant difference in toughness between the two samples. As reported by Faber and Evans[19], the morphology of the second phase appears to play a role in the resulting toughness of the ceramic body. They found that high aspect ratio rods are the most effective phase to increase toughening followed by disc-shaped particles and spheres, respectively. The data on the experimental samples supports the conclusions of Faber and Evans.

CONCLUSION

It has been shown that a secondary phase in the form of platelets or particles can be successfully grown in-situ in a sol-gel ceramic matrix without significant loss in density or microhardness. Furthermore, the morphology of the secondary phase appears to influence the fracture toughness of the ceramic body. The growth habit of the rare earth magnetoplumbite was found to be dependent on the α-alumina transformation temperature. For high transformation temperatures, the magnetoplumbite grows as platelets distributed along low angle α-alumina grain boundaries.

With low α-alumina transformation temperature, the magnetoplumbite forms more equiaxed particle shapes.

ACKNOWLEDGMENTS

The authors would like to thank Myles Brostrom for his x-ray diffraction work; Dr. Chris Goodbrake for his TEM characterization, interpretation, and sample preparation assistance; and Dr. Barry Carter and the Center for Interfacial Engineering at the University of Minnesota for access on the TEM and TEM sample preparation equipment.

---

[1]K.T. Faber and A.G. Evans, "Crack Deflection Processes - I. Theory", Acta. Metall. 31[4], 565-575 (1983).
[2]K.T. Faber and A.G. Evans, "Crack Deflection Processes - II. Experiment", Acta. Metall. 31[4], 577-584 (1983).
[3]K. Niihara, A. Nakahira, and M. Inoue, "Sintering Process and Mechanical Properties for Oxide-Based Nanocomposites", Mat. Res. Soc. Symp. Proc. Vol. 271, 589-593 (1992).
[4]T. Sato, A. Shiratori, and M. Shimada, "Sintering and Fracture Behavior of Composites Based on Alumina-Zirconia (Yttria) - Nonoxides", J. Phys. C1, 2 [47] pp. 733-737 (1986).
[5]L.C. Stearns, J. Zhao, and M.P. Harmer, "Processing and Microstructure Development in Al_2O_3-SiC Nanocomposites", J. Eur. Ceram. Soc., 10, 473-477 (1992).
[6]B.I. Lee and L.L. Hench, "Silicon Carbide/Silica Gel Matrix Composites", pp. 321-336 in Science of Ceramic Chemical Processing, Edited by L.L. Hench and D.R. Ulrich, Wiley, New York (1986).
[7]B.I. Lee and S.Y. Park, "Sol-Gel Processing of SiC-Whisker-Reinforced Silica-Based Ceramic Composites", J. Am. Ceram. Soc., 72[12] 2381-2385 (1989).
[8]D.E. Clark, "Sol-Gel Derived Ceramic Matrix Composites", pp. 237-246 in Science of Ceramic Chemical Processing, Edited by L.L. Hench and D.R. Ulrich, Wiley, New York (1986).
[9]M. Kumagai and G. L. Messing, "Enhanced Densification of Boehmite Sol-Gels by α-Alumina Seeding", J. Am. Ceram Soc., 67 [11] c230-c231 (1984).
[10]J. L. McArdle and G. L. Messing, "Transformation and Microstructure Control in Boehmite-Derived Alumina by Ferric Oxide Seeding", Adv. Ceram. Mater. 3 [4] 387-392 (1988).
[11]J. L. McArdle and G. L. Messing, "Transformation, Microstructure Development, and Densification in a α-Fe_2O_3-Seeded Boehmite-Derived Alumina", J. Am. Ceram. Soc., 76 [1] 214-222 (1993).
[12]M. Kumagai and G. L. Messing, "Controlled Transformation and Sintering of a Boehmite Sol-Gel by α-Alumina Seeding", J. Am. Ceram. Soc., 68 [9] 500-505 (1985).
[13]W. A. Yarbrough and R. Roy, "Microstructural Evolution in Sintering of AlOOH Gels", J. Mater. Res., 2 [4] 494-515 (1987).
[14]E. W. Washburn, "The Dynamics of Capillary Flow", Phys. Rev., 17 [3] 273-283 (1921).
[15]A. G. Evans and E. A. Charles, "Fracture Toughness Determinations by Indentation", J. Am. Ceram. Soc 59 [7-8] 371-372 (1976).
[16]J.M.P.J. Verstegen and A.L.N. Stevels, "The Relation Between Crystal Structure and Luminescence in β-Alumina and Magnetoplumbite Phases", J. Luminescence 9,(1974) 406-414.

[17]T. Gbehi, J. Thery, D. Vivien and R. Collongues, and G. Dhalenne and A. Revcolevschi, "Neodymium Sites in Calcium Lanthanide Hexaaluminates, Potential Laser Materials", J. Solid State Chem., 77, 211-222 (1988).
[18]K. J. Vaidya, C. Y. Yang, M. DeGraef, and F. F. Lange, "Heteroepitaxy of Rare-Earth Hexa-Aluminates on Sapphire", J. Mater. Res. 9 [2] 410-419 (1994).
[19]Faber and Evans, "Crack Deflection Process - II. Experiment", 577-584.

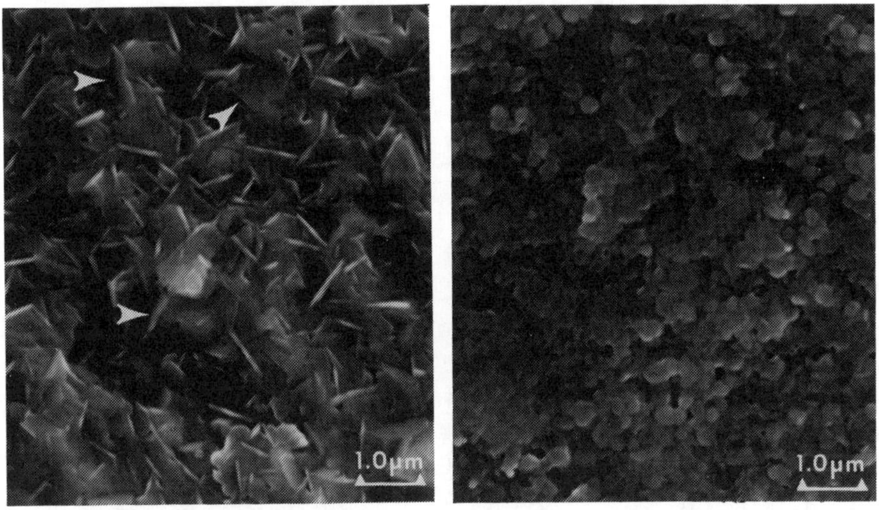

Composition A　　　　Composition B

FIGURES 1 & 2: The fractured and thermally etched cross-sections show evidence of platelet morphology (as indicated by the arrows) in A. The platelet morphology is absent in B.

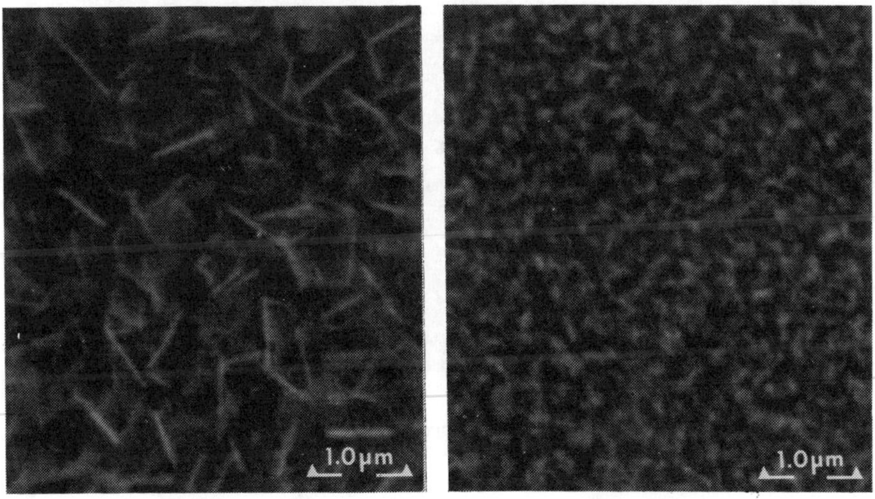

Composition A　　　　Composition B

FIGURES 3 & 4: Backscattered electron images (BEI) show high atomic number platelets (acicular in cross-section) in A and equiaxed particles in B as brighter than the surrounding matrix.

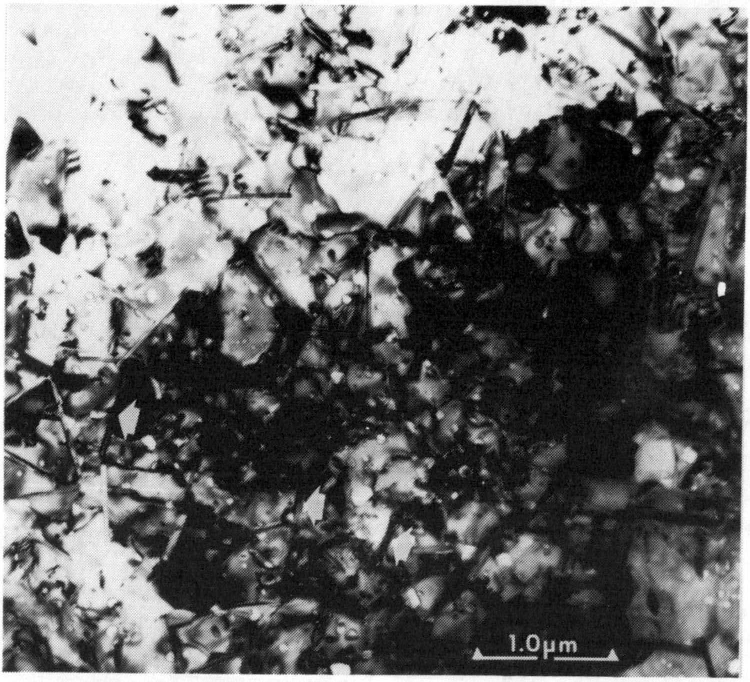

FIGURE 5: Shown by transmission electron microscopy (TEM) is a diffracting α-alumina grain from composition A. The arrows show examples of platelets (acicular in cross-section) within the matrix of the grain.

FIGURE 6: A higher magnification of the α-alumina grain shown in Figure 5 reveals platelets trapped along low angle grain boundaries.

FIGURE 7: The SAD pattern of Figure 5 shows an orientation between the α-alumina (intense spots) and the magnetoplumbite (weak, closely spaced spots).

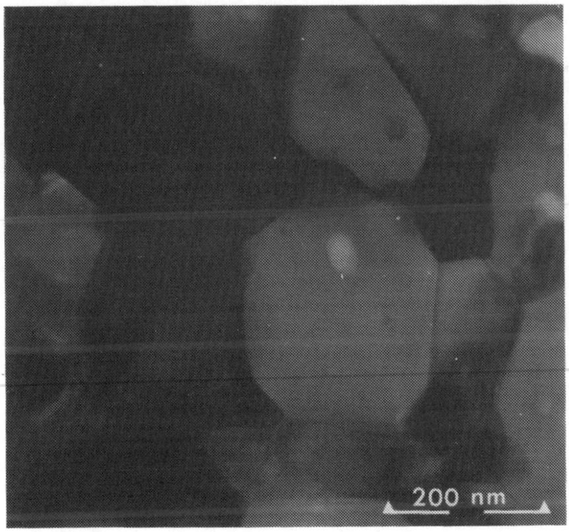

FIGURE 8: Shown by TEM is the microstructure of composition B. The image reveals the sub-micron faceted α-alumina grains and the magnetoplumbite trapped along triple points.

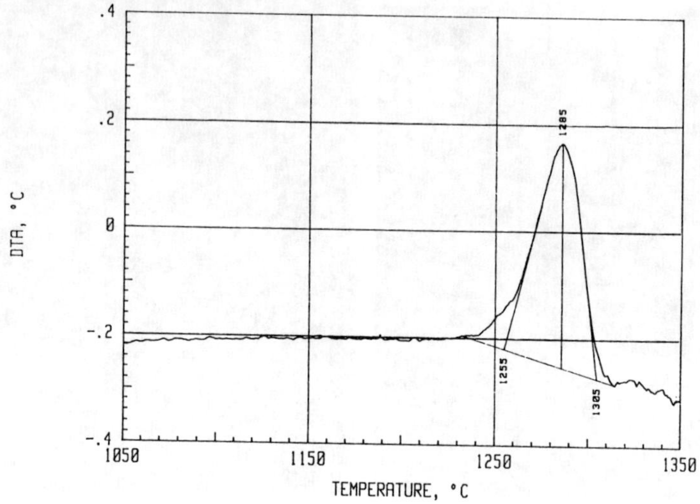

FIGURE 9: Shown is the differential thermal analysis (DTA) plot of composition A. The onset, peak and outset temperatures are indicated.

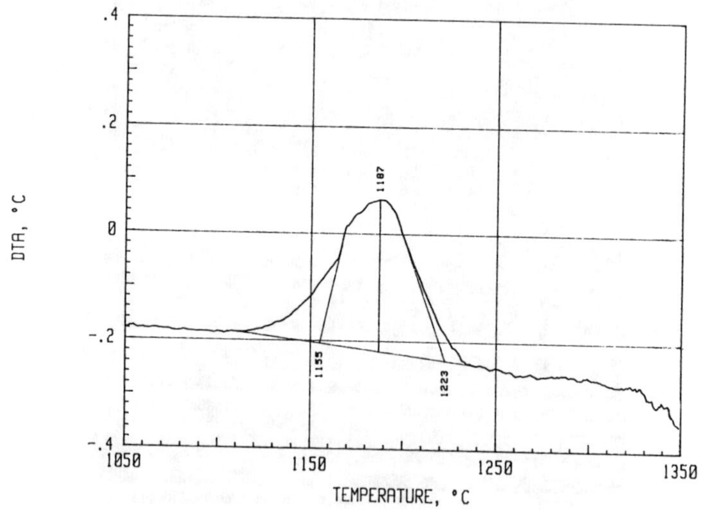

FIGURE 10: Shown is the differential thermal analysis (DTA) plot of composition B. The onset, peak and outset temperatures are indicated.

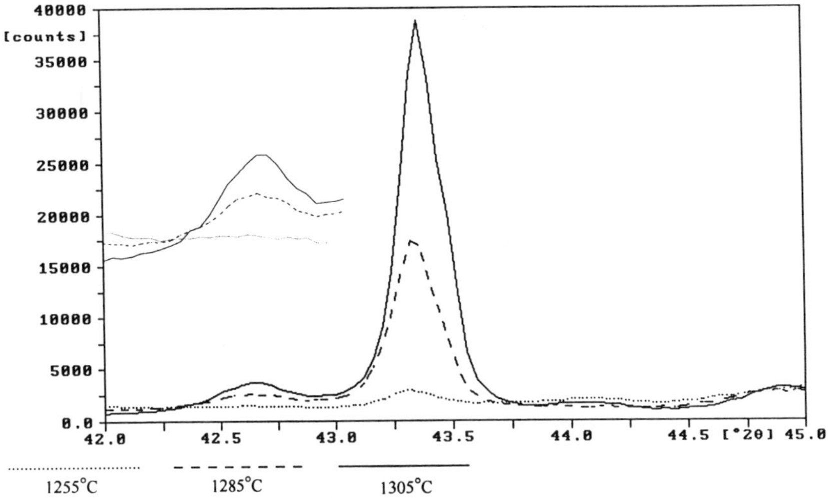

FIGURE 11: X-ray diffraction (XRD) of the 113 α-Al_2O_3 and 205 $MgLaAl_{11}O_{19}$ peaks of A as a function of the inset, peak and outset temperatures calculated by DTA (see Figure 9). An enlargement of the 205 $MgLaAl_{11}O_{19}$ peak is shown above the XRD plot. The data indicates that the α-Al_2O_3 is forming simultaneously with the $MgLaAl_{11}O_{19}$.

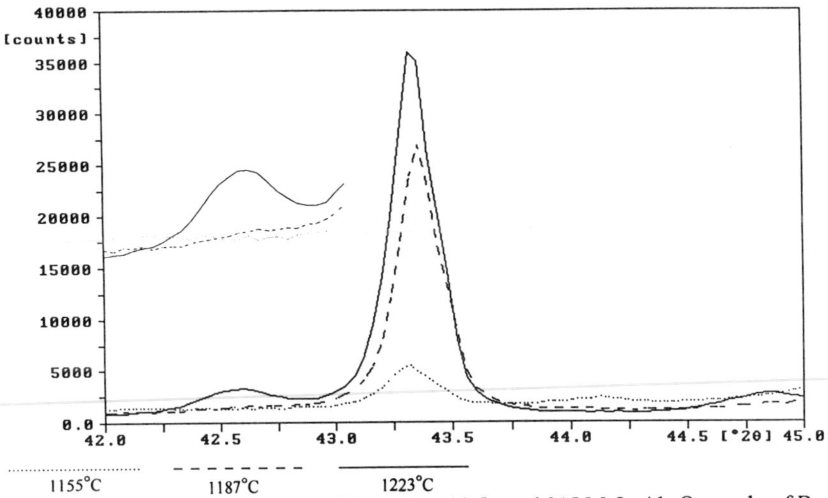

FIGURE 12: X-ray diffraction (XRD) of the 113 α-Al_2O_3 and 205 $MgLaAl_{11}O_{19}$ peaks of B as a function of the inset, peak and outset temperatures calculated by DTA (see Figure 10). An enlargement of the 205 $MgLaAl_{11}O_{19}$ peak is shown above the XRD plot. The data indicates that α-Al_2O_3 is forming before $MgLaAl_{11}O_{19}$.

TABLE 1A:
Density, microhardness, and fracture toughness for compositions A and B

SAMPLE	DENSITY (g/cc)	K_{1C} ($MNm^{-3/2}$)	Hv (GPa)
A	3.89	3.7 (s = 0.2)	19.6 (s = 0.7)
B	3.96	3.3 (s = 0.2)	22.0 (s = 1.5)

TABLE 1B:
ANOVA of fracture toughness for compositions A and B

```
MTB > print c1 c2

 ROW      A      B

   1    3.7    3.5   INDIVIDUAL 95 PCT CI'S FOR MEAN
   2    4.0    2.9   BASED ON POOLED STDEV
   3    3.4    3.0   ------+---------+---------+---------+
   4    3.6    3.3                  (-------*------)
   5    3.6    3.4    (------*------)
   6    3.5    3.3   ------+---------+---------+---------+
   7    4.1    3.6       3.25      3.50      3.75      4.00
   8    3.8    3.4

MTB > aovo c1 c2

ANALYSIS OF VARIANCE
SOURCE      DF        SS        MS        F        p
FACTOR       1    0.6806    0.6806    11.78    0.004
ERROR       14    0.8087    0.0578
TOTAL       15    1.4894


 LEVEL       N      MEAN     STDEV
A            8    3.7125    0.2416
B            8    3.3000    0.2390

POOLED STDEV =    0.2403
```

ENHANCED *IN SITU* TOUGHENING OF SILICON NITRIDE BY THE ADDITION OF β-Si_3N_4 WHISKERS

Veena Tikare*
Case Western Reserve University
Cleveland, Ohio 44106

William A. Sanders
Analex Corporation
Brookpark, Ohio 44142

and

Sung R. Choi*
Cleveland State University
Cleveland, Ohio 44115

ABSTRACT

Silicon nitride ceramics were processed with the addition of small quantities of β-Si_3N_4 whiskers to a commercially available α-Si_3N_4 powder. These whiskers grew preferentially during liquid phase sintering into very large, elongated β-grains. Silicon nitride samples of the same compositions, but without the whiskers were also processed. The fracture toughness and R-curve behavior of the whisker seeded samples were compared to those of unseeded samples. The fracture toughness of the seeded samples was higher because grain pullout and crack bridging were enhanced by the very large, elongated β-grains. R-curve behavior was present in both seeded and unseeded silicon nitrides.

INTRODUCTION

Silicon nitride ceramics are candidate materials for high temperature structural applications such as advanced automobile and aircraft engines and heat exchangers. However, like all ceramics, they are brittle and flaw intolerant. These limitations have lead to a great deal of interest in toughening silicon nitride ceramics. *In-situ* toughened silicon nitride ceramics, also called self-reinforced silicon nitrides, have received much attention in recent years as

*NASA Resident Research Associate at Lewis Research Center, Cleveland, Ohio 44135. Please address all correspondence to NASA.

To the extent authorized under the laws of the United States of America, all copyright interests in this publication are the property of The American Ceramic Society. Any duplication, reproduction, or republication of this publication or any part thereof, without the express written consent of The American Ceramic Society or fee paid to the Copyright Clearance Center, is prohibited.

they are easier to process than whisker or fiber reinforced silicon nitrides. The mechanical properties of these *in-situ* toughened silicon nitride ceramics are attractive. They are among the toughest of monolithic ceramic materials (Ref. 1), and some silicon nitrides retain good mechanical properties to temperatures as high as 1350 °C (Ref. 2) depending on the composition and amount of the grain boundary phase (Ref. 3).

The high toughness of silicon nitride ceramics is derived from their unique microstructure of interlocking, needle-like grains surrounded by a glassy grain boundary phase. Cracks propagate predominantly through the grain boundary phase around the elongated silicon nitride grains. This type of fracture yields high toughness by the grain pullout and crack bridging mechanisms (Ref. 4). Furthermore, toughness can be enhanced by promoting growth of large, elongated grains.

At sintering temperatures, the grain boundary phase is a molten silicate surrounding silicon nitride grains. Both densification and grain growth occur by solution-precipitation. Smaller silicon nitride grains dissolve in the liquid silicate phase and precipitate onto larger grains leading to grain growth. Initially, the α-Si_3N_4 grains or particles dissolve and precipitate onto the β-Si_3N_4 particles which are invariably present in the starting composition. Later, when all the Si_3N_4 has transformed to the β-phase, Ostwald ripening occurs (Ref. 5).

The disadvantage of *in-situ* toughening is that prolonged heat treatment at high temperatures is required to promote β-Si_3N_4 grain growth. These heat treatments are not only expensive, but also lead to decomposition of Si_3N_4 to its constitutive elements, Si and N, resulting in lower density and poor mechanical properties. In this investigation, small quantities of β–Si_3N_4 whiskers were used to seed the growth of the large β-Si_3N_4 grains rather than using prolonged heat treatments to obtain grain growth. Unseeded silicon nitride samples of the same compositions were also processed and sintered under the same conditions. The microstructural evolution, fracture toughness, and R-curve behavior of both seeded and unseeded silicon nitride ceramics will be presented and contrasted to determine if seeding yields silicon nitride ceramics with improved properties.

MATERIAL PROCESSING AND TESTING PROCEDURE

Material Processing

Silicon nitride ceramics of two different compositions with and without β-Si_3N_4 whiskers were prepared. One composition contained 6 wt% Y_2O_3 and 2 wt% SiO_2 as sintering aids, and the other 8.6 wt% Sc_2O_3. They were designated as 6Y and 8Sc, respectively. A commercially available silicon nitride powder (Kemanord Engineering Ceramics, Sweden) was milled in ethanol with the appropriate sintering aids and 1 wt% stearic acid in a silicon nitride mill with silicon nitride media for 300 h. The samples with whiskers were designated as 6Y-W and 8Sc-W and contained 4 and 3 wt% whiskers, respectively. The samples without whiskers were prepared by drying the slurry in an apparatus employing a hot water bath and vacuum to remove the ethanol, followed by drying in an oven at 110 °C. This dried powder contained some soft agglomerates which were crushed using hot-pressed silicon nitride hardware and sieved through a 150 μm screen. The samples with the whiskers were prepared by mixing commercially available β-Si_3N_4 whiskers (UBE-SNWB, UBE Industries Ltd., Japan) with the slurry by ultrasonic agitation after milling. This mixture was dried. Again, the dried powder was gently crushed, but not sieved to prevent segregation of the whiskers and powder. The powders were uniaxially pressed at 25 MPa to fabricate preform bars (57 x 7.5 x 5 mm). These bars were vacuum sealed in latex tubing and isostatically cold-pressed at 420 MPa. The pressed bars were sintered in groups of 15 in a tungsten resistance furnace for 4 hr at 2140 °C and 2.5 MPa nitrogen overpressure.

Interrupted sintering cycle experiments were conducted to examine microstructural evolution in the seeded and unseeded samples. The sintering cycle was interrupted at 1800, 1900 and 2150 °C. All samples used for microstructural characterization were polished using standard ceramographic techniques to <1 μm diamond paste. The polished samples were etched using a plasma. Plasma etching attacks β-Si_3N_4 phase much more than α-Si_3N_4 or the glassy grain boundary phase. A detailed description of the etching procedure is given elsewhere (Ref. 6). Test bars were machined by grinding along the longitudinal axis of sintered bars with a 400-grit diamond grinding wheel.

Testing Procedure

Fracture toughness was measured by the single-edge-precrack-beam (SEPB) method (Ref. 7). The prospective tensile surface of a flexural beam was indented with a Vicker's indenter under an indentation load of 294 N. The indented sample was loaded in a double anvil type geometry until the indentation crack popped open to form a sharp, single edge precrack. A detailed description of this technique is given elsewhere (Ref. 8). The precracked samples were loaded in a four-point bend geometry. The fracture toughness values were calculated as:

$$K_{IC} = \frac{3PY(l_o - l_i)\sqrt{\pi a}}{2WD^2} \tag{1}$$

where P was the fracture load, l_o and l_i were the outer and inner spans, respectively (l_o= 18mm and l_i=10mm), D was the sample depth, and W was the sample width, and Y was the geometric factor given by (Ref. 9):

$$Y = 1.122 - 1.121\left(\frac{a}{D}\right) + 3.74\left(\frac{a}{D}\right)^2 - 3.873\left(\frac{a}{D}\right)^3 + 19.05\left(\frac{a}{D}\right)^4 - 22.55\left(\frac{a}{D}\right)^5 \tag{2}$$

where a was the crack length. Nine specimens of each composition were tested.

The R-curve behavior was characterized by the strength-indentation method developed by Krause (Ref. 10). Flexural specimen of dimensions 3 by 4 by 25 mm were indented on the prospective tensile surface by a Vicker's indent. The indent was placed so that one diagonal was parallel to the long edge of the specimens. The indentation load was varied between 49 and 490 N in order to obtain flaws of differing sizes. The indented specimens were stressed to failure in a four-point geometry with inner and outer spans of 8 and 18 mm, respectively.

In the Krause method, a power law R-curve behavior of the form given below is predicted:

$$K_r = ka^m \tag{3}$$

where Kr is fracture resistance, a is the crack length, and k and m are empirically determined constants. The R-curve exponent, m, was obtained from the strength-indentation results. The fracture stress, sf, is proportional to the indentation load, P, as:

$$\sigma_f \propto P^{-\beta} \tag{4}$$

where b is obtained by linear least squares best fit to the logarithm sf versus logarithm P plot. The value for b obtained form such a plot is used to calculate the R-curve exponent, m, as:

$$m = \frac{1-3\beta}{2+2\beta} \tag{5}$$

Three samples were tested at each indentation load, P.

RESULTS AND DISCUSSION

Microstructure

Particle size analysis of the starting Si_3N_4 powder revealed that 90% of the powder was under 0.9 mm and 50% was under 0.45 mm. The β-Si_3N_4 whiskers ranged from 5 to 60 mm in length and 0.2 to 3 mm in diameter as shown in Fig.1. The whiskers were much larger than the starting powder size. For a complete description of the β-Si_3N_4 whiskers the reader is directed to Homeny et al. (Ref. 11)

Figure 1.—SEM micrograph of β-Si_3N_4 whiskers used to seed the starting powder.

The microstructures of the seeded 6Y silicon nitride, 6Y-W, at 1800, 1900 and 2140 °C are shown in Figs. 2(a) to (c), respectively. These are scanning electron microscope (SEM) photographs of polished and plasma etched samples. Plasma etching attacked β-Si_3N_4 much more than α-Si_3N_4 or the glassy phase. At 1800 °C, most of the sample was α-Si_3N_4 and plasma etching attacked only a few features. One of these, shown in Fig. 2(a), is clearly a β-Si_3N_4 whisker. At 1900 °C, in addition to the β-Si_3N_4 whiskers, few small β-Si_3N_4 grains were present, but again most of the sample was composed of fine α-Si_3N_4 particles. At 2140 °C, both the β-Si_3N_4 whiskers and the α-Si_3N_4 particles were growing. The microstructures of unseeded and seeded 8Sc silicon nitride, 8Sc and 8Sc-W, at 2140 °C are shown in Figs. 3(a) and (b), respectively. The 8Sc-W sample has a few grains which were much larger than others surrounding them and much larger than the ones in 8Sc sample. Examination of the large grains in Fig. 3(b) reveals that they had an inner region of a different contrast than the outer region. The inner region is the β-Si_3N_4 whisker used as seeds and the outer region is grain growth of the whisker during the sintering. The final sintered microstructure of 6Y, 6Y-W, 8Sc, and 8Sc-W are shown in Figs. 4(a) to (d), respectively. The seeded samples, 6Y-W and 8Sc-W, had a few very large grains surrounded by grains of the same size as those in the unseeded samples.

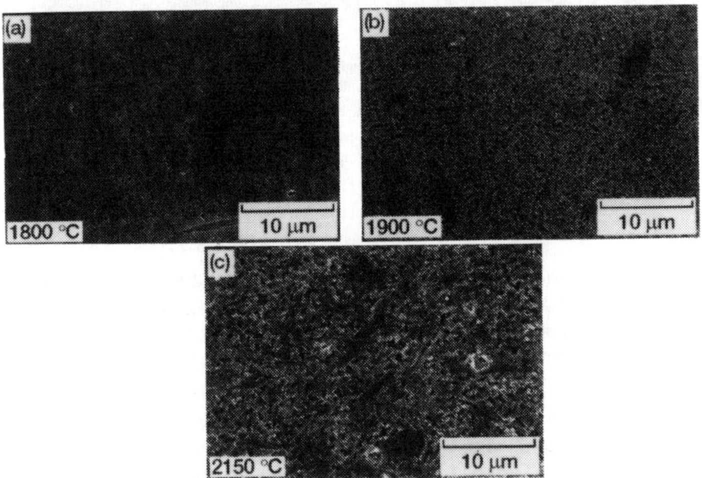

Figure 2.—SEM micrographs of 6Y-W at (a) 1800 °C, (b) 1900 °C and (c) 2150°C.

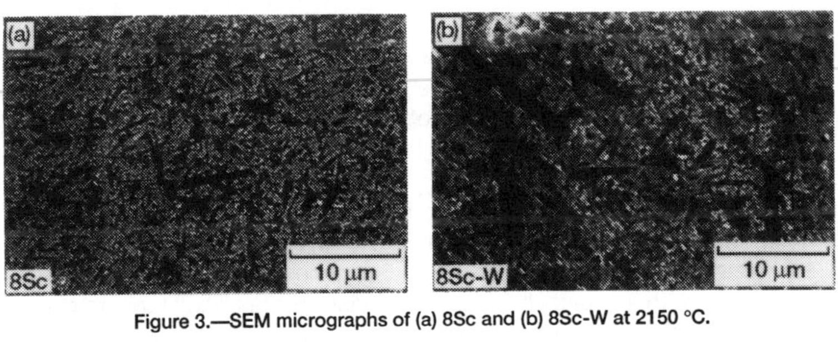

Figure 3.—SEM micrographs of (a) 8Sc and (b) 8Sc-W at 2150 °C.

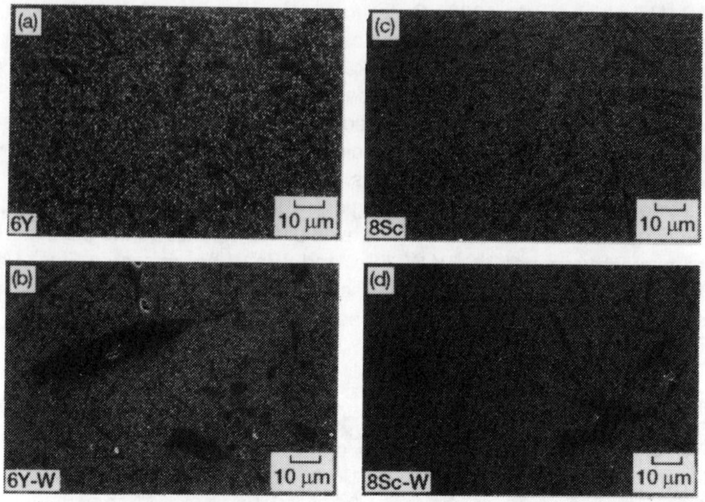

Figure 4.—SEM micrographs of final microstructures of (a) 6Y, (b) 6Y-W, (c) 8Sc, and (d) 8Sc-W silicon nitride ceramics.

Based on these results, we conclude that the β-Si_3N_4 whiskers introduced in the starting powder, survived the green processing. They were not broken down during mixing and pressing. These whiskers grew by the solution-precipitation mechanism. Initially the α-phase and later, the smaller β-grains dissolved and precipitated onto the β-whiskers preferentially to give exaggerated grain growth of the whiskers. Concurrently, the β-particles present in the starting powder were seeds for growth of the smaller grains. Initially the α-Si_3N_4 dissolved and precipitated on the β-particles. Once most of the α-phase has transformed to the β-phase, Ostwald ripening occurred with preferential growth of the whisker-seeded grains. The resulting microstructure consisted of a few very large, elongated grains surrounded by many smaller, elongated grains.

Mechanical Properties

The fracture toughnesses of the silicon nitride ceramics with and without whisker additions obtained by SEPB are given in Fig. 5. The fracture toughness of the 6Y-Si_3N_4 with whiskers (6Y-W), 8.7 MPa$\sqrt{m}$, was almost double that of the 6Y-Si_3N_4 without whiskers (6Y), 4.5 MPa$\sqrt{m}$. Similarly, the fracture toughness of the 8Sc-W was higher than that of 8Sc although the increase was not as dramatic as that between the 6Y and 6Y-W silicon nitride.

The results of the strength-indentation tests for 6Y, 6Y-W, 8Sc and 8Sc-W are plotted in Fig. 6(a) as fracture stress, sf, versus indentation load, P, on logarithmic scales. The fracture stress is the average fracture stress of 3 to 5 samples. The data for each silicon nitride ceramic has been fit with a straight line using the least squares fitting technique as shown in Fig. 6(a). The slope of this line is the exponent –β given in Eq. (4). A material with $\beta = 1/3$ has no R-curve. The fracture stress, σ_f, of a material with $\beta < 1/3$, is less sensitive to

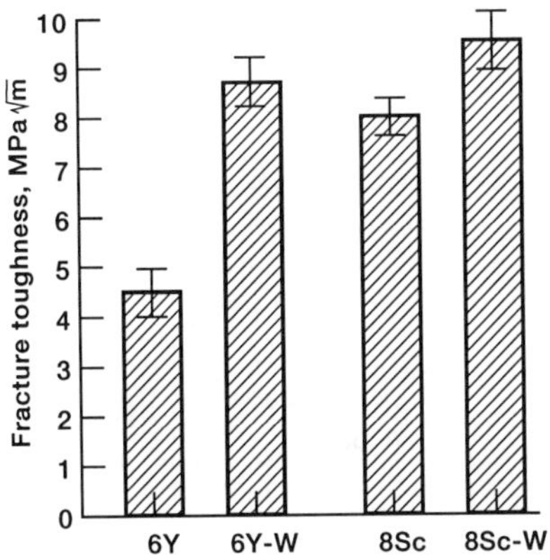

Figure 5.—Average fracture toughness of the silicon nitride ceramics tested by SEPB. The error bar represent one standard deviation.

indentation load, P, indicating flaw tolerance and R-curve behavior. The R-curve exponent, m, for each ceramic was calculated using Eq. (5). The constant k was calculated by substituting the SEPB fracture toughness, the corresponding crack length and m in Eq. (3). The calculated R-curves for 6Y, 6Y-W, 8Sc and 8Sc-W are plotted in Fig. 6(b). All four silicon nitride ceramics exhibited R-curve behavior. Seeding silicon nitride increased fracture toughness for small and large cracks, 50 to 2000 μm. In the case of 8Sc, seeding increased the fracture toughness of small cracks more than that at long cracks. The increase in fracture toughness at small crack sizes is of particularly interest as natural flaws are of these magnitudes. Higher fracture toughness in this range increases the fracture stress for these natural flaws yielding more reliable components.

SEM analysis of the fracture surfaces (shown in Fig. 7(a) to (d)) revealed that fracture was predominantly intergranular. While, pullout was seen in both seeded and unseeded samples; The seeded samples had a much more tortuous fracture surface with extensive pullout of the large, elongated β-grains. Crack trajectories of indentation cracks at the indent surface were examined to determine the toughening mechanisms. Figures 8(a) and d show that crack bridging and grain pullout are the toughening mechanisms in both seeded and unseeded samples. However, more effective bridging and pullout are achieved by the large, elongated grains in the seeded samples.

The bimodal microstructure achieved by seeding with whiskers increased the toughness of both 6Y and 8Sc silicon nitride by promoting grain pullout and crack bridging. The increase in toughness at the small crack sizes was particularly significant as natural and service induced flaws are in this size range. Failure of the component will occur either by fracture originating from these flaws or slow crack growth of these small flaws. For both types of failure, higher fracture toughness is desirable. In the first case, the fracture stress at a given flaw size is higher for a higher toughness material. For the second case, the stress intensity required to propagate a crack at a given velocity is higher for the for a higher toughness material.

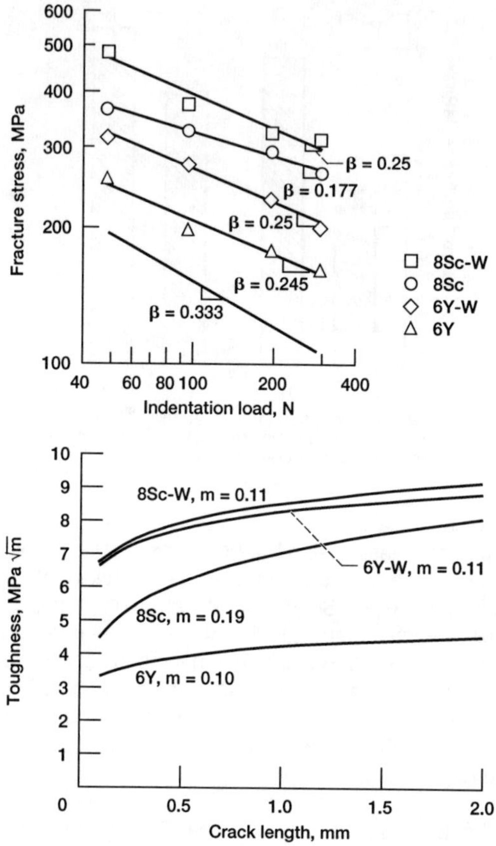

Figure 6.—(a) Indentation-strength results. (b) Calculated R-curves.

It is not clear why the presence of a few large grains increased the fracture toughness of the 6Y-W composition over that of the 6Y composition so dramatically; whereas, only a modest increase in toughness was realized for the 8Sc-W composition over that of the 8Sc composition. Crack deflection was greatly enhanced by the large, elongated β-Si_3N_4 grains in 6YW; whereas, the large β-grains in 8Sc-W did not increase crack deflection as dramatically as shown in Fig. 8. The 8Sc silicon nitride had a more transgranular fracture with more grain pullout than 6Y as shown in Fig. 7. The fracture toughness of 8Sc was already high. While, seeding 8Sc increased pullout and crack deflection; the increase was not as dramatic as that observed in 6Y.

Figure 7.—SEM micrographs of the fractured surfaces of (a) 6Y, (b) 6Y-W, (c) 8Sc and (d) 8Sc-W.

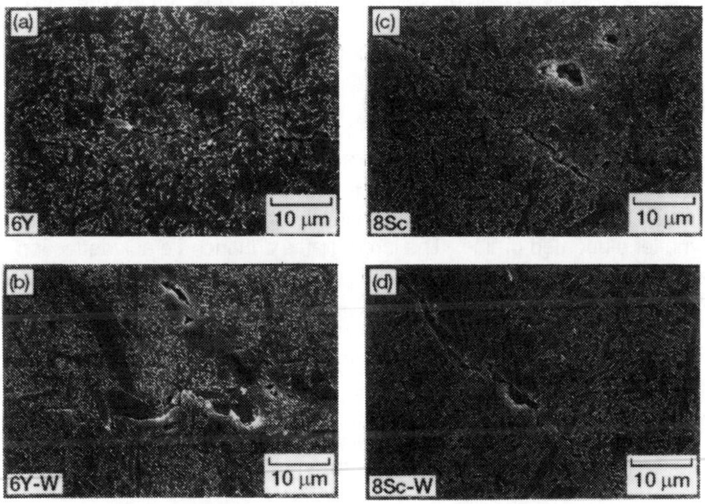

Figure 8.—SEM micrographs of indentation cracks showing crack trajectories of (a) 6Y, (b) 6Y-W, (c) 8Sc and (d) 8Sc-W.

A possible explanation for the difference in fracture toughness of the 6Y and 8Sc materials in spite of their similar grain size may be the differences in their grain boundary phases. X-ray diffraction analysis of 6Y and 8Sc revealed that the grain boundary phase in 8Sc had devitrified to $Sc_2Si_2O_7$. In contrast, no crystalline silicate phase was detectable in 6Y. Perhaps crystallization of the grain boundary phase in 8Sc created some residual stresses which enhanced crack deflection in the 8Sc silicon nitride ceramic.

In-situ toughening of silicon nitride ceramics has been achieved primarily through coarsening of all the grains. Prolonged heat treatments at temperatures exceeding 1850 °C in N_2 overpressure environments or prolonged heat treatment under HIP'ing conditions with glass-encapsulated samples are primarily the techniques used to achieve grain growth in silicon nitride ceramics. The data given in this paper suggests that seeding the starting powder with a small amounts of well dispersed β-Si_3N_4 whiskers and sintering with conventional techniques yields the desirable microstructure (a few large, highly elongated grains surrounded by smaller grains) and high toughness without the prolonged heat treatments.

We have only used this seeding technique with cold pressed compacts that were sintered in a N_2 overpressure furnace. However, it should also work with any silicon nitride which relies on the presence of a liquid phase for densification and grain growth. Silicon nitride ceramics made by colloidal techniques, injections molding, cold compaction, etc. can be easily modified to used seeded compositions. These can then be hot pressed, HIP'ed with glass-encapsulation or other encapsulation methods or sintered in N_2 overpressure atmospheres at temperatures where liquid phase sintering occurs.

CONCLUSIONS

High toughness silicon nitride ceramics can be processed by seeding the starting powder with small quantities of β-Si_3N_4 whiskers. The β-whiskers grow faster than the surrounding grains to give a microstructure consisting of a few large, highly elongated grains surrounded by much smaller elongated grains. The large grains enhanced crack deflection, grain pull-out and crack bridging leading to higher toughness. R-curve behavior was preserved in seeded silicon nitrides.

REFERENCES

1. C.W. Li and J. Yamanis, "Super-Tough Silicon Nitride with R-Curve Behavior," Ceram. Eng, Sci. Prod. 10[7–8] pp. 632–645 (1989).

2. C.J. Lin, M.G. Jenkins, M.K. Ferber, "Evaluation of Tensile Static, Dynamic and Cyclic Fatigue Behavior For a HIPed Silicon Nitride at Elevated Temperatures," Mat. Res. Soc. Symp Proc. Vol .287, Mat. Res. Soc. (1993)

3. W.A. Sanders and L.E. Groseclose, "Flexural Stress Rupture and Creep of Selected Commercial Silicon Nitrides," J. Am. Ceram. Soc., 76 [2] 553–556 (1993).

4. V. Tikare, S.R. Choi and J.A.Salem, "Crack Growth Resistance of *In Situ* Toughened Silicon Nitride Ceramics," In-Situ Composites: Science and Technology, ed. by M. Singh and D. Lewis, The Minerals, Metals & Materials Soc. 193–210 (1994)

5. K.R. Lai and T.Y. Tien, "Kinetics of βSi_3N_4 Grain Growth in Si_3N_4 Ceramics Sintered under High Nitrogen Pressure," J. Am. Ceram. Soc., 76 [1] 91–96 (1993).

6. D.R. Hull, T.A. Leonhardt, and W.A. Sanders. "Plasma Etching a Ceramic Composite," Microstructural Sci. 19 671–679 (1992).

7. T. Nose and T. Fujii, "Evaluation of Fracture Toughness for Ceramic Materials by a Single-Edge-Precracked-Beam Method," J. Am. Ceram. Soc., 71 [5] 328–333 (1988).

8. S.R. Choi, A. Chulya, J.A. Salem, "Analysis of Precracking Parameters for Ceramic Single-Edge-Precracked-Beam Specimens," pp 73–88, Fracture Mechanics of Ceramics, Vol. 10, edited by R.C. Bradt et al., Plemun Press, NY (1992).

9. Stress Intensity Factors Handbook, Vol. 1, pp 349–360, edited by Y. Murakami, Pergamon Press, N.Y. (1987).

10. R.F. Krause, "Rising Fracture Toughness from the Bending Strength of Indented Alumina Beams," J. Am. Ceram. Soc., 71 [5] 338–342 (1988).

11. J. Homeny, L.J. Neergaard, K.R. Karasek, J.T. Donner and S.A. Bradley, "Characterization of β-Silicon Nitride Whiskers", J. Am. Ceram. Soc., 73[1]102–105 (1990).

α'-β'-SIALON CERAMICS THROUGH MICROSTRUCTURE TAILORING

Hao Wang*, W-Y.Sun, H-R.Zhuang, P-L.Wang, and T.S.Yen
The State Key Lab of High Performance Ceramics, Shanghai Institute of Ceramics, Shanghai 200050, China

ABSTRACT

Basing on the phase relationship study of the Y-Si-Al-O-N system, α'-β'-sialon with YAG as the grain boundary phase is designed, and processed by two-step gas pressure sintering (GPS). The influence of the processing conditions on the microstructure and further on the mechanical properties are investigated. With increasing of the second step GPS temperature and holding time, the microstructure with elongated grains can be developed. The flexural strength increases from 612 to beyond 900 MPa, and the fracture toughness from 3.9 to 7.2 MPa.m$^{1/2}$. Crack bridging and crack deflection are identified to be the two main strengthening and toughening mechanisms for the materials.

INTRODUCTION

Silicon nitride and sialon ceramics have attracted increasing attentions in recent years because of their excellent properties, such as, high strength and fracture toughness, good wear resistance, good oxidation resistance, and are considered as the most hopeful structural ceramics for the application in the high temperature engineering[1-2]. It is well known that the microstructure with elongated grains in Si_3N_4-based ceramics will increase the strength and fracture toughness of the materials by crack bridging[3] and/or crack deflection[4]. However, the development of this kind of microstructure is related to the starting powders, the type and amount of additives and the processing parameters. The gas pressure sintering, which is based on liquid-phase sintering mechanism, is beneficial to such microstructure development[5-6]. The use of the high nitrogen pressure can reduce the thermal decomposition of Si_3N_4, which offers a more flexible choice

To the extent authorized under the laws of the United States of America, all copyright interests in this publication are the property of The American Ceramic Society. Any duplication, reproduction, or republication of this publication or any part thereof, without the express written consent of The American Ceramic Society or fee paid to the Copyright Clearance Center, is prohibited.

of processing parameters and a possibility to tailor the microstructure through processing. This work is to study the effect of the processing parameters on the development of microstructure in gas pressure sintering of α'-β'-sialon ceramics, and to find the relation between the mechanical properties of material and its microstructure.

EXPERIMENTAL PROCEDURE

The starting powders used were high α content Si_3N_4 (UBE-10, $\alpha > 95.5\%$ $O \approx 1.57wt\%$). AlN (laboratory made, $N > 32.5\%$, $O \approx 1.5wt\%$), Al_2O_3 ($>99.95\%$), Y_2O_3 ($>99.9\%$). The composition was designed to be 50wt% α'-sialon + 50wt% ß'-sialon + additional 9wt% YAG ($Y_3Al_5O_{15}$). The oxygen

Table I. Weight Percentages of the Starting Materials

Component	Si_3N_4	AlN	Al_2O_3	Y_2O_3
Composition (wt%)	81.1	8.0	4.4	6.5

contents of the silicon nitride and aluminum nitride powders were taken into account in computing the composition, and the weight percentages of the starting materials are shown in table I.

The mixtures of powders were milled in ethanol for 24h in an alumina jar using well sintered silicon nitride grinding media. After being dried, the powder mixtures were dry-pressed into bars under 20 MPa and then isostatically pressed under pressure of 200 MPa. The specimens were gas pressure sintered by a two-step process. In the first step, the temperature was raised to 1800°C and held 2h under a N_2 pressure of 0.10 MPa. In the second step, two sets of experiment were designed. One set of specimens was heated at 1900°C for 0.5h, 1.5h and 3.0h under 1.5MPa N_2 pressure; and another set was heated at the temperature of 1900°C, 1950°C and 2000°C respectively for 1.5h under the same N_2 pressure. All samples were then devitrified by heat treatment at 1350°C for 24h under N_2 atmosphere.

The sintered materials were ground and polished, and then etched for observation by scanning electron microscopy (SEM). Etching of the grain boundary glassy phase was done by immersing the specimens into molten NaOH for 3 min. An image analysis system Kontron IBAS KAT 386 was used to quantitatively characterize the microstructure. The average grain diameter (d), the real aspect ratio (R) and the volume of glassy phase (V_g) were chosen as the main parameters to characterize the microstructure of the samples. The grain diameter (d) was directly determined from the shortest grain diagonal in its 2-dimensional image. The aspect ratio R was calculated by L/d, where L was the grain length obtained from the longest diagonal. Because it was dependent on the grain orientation, the apparent ratio was smaller than the real value. For the statistical consideration there were about 10% of the grains cut parallel to their crystallographic c-axes, so the average aspect ratio of the 10% grain was used to represent the real ratio. A total of 800 to 1000 grains were analyzed for each specimen. The relative amount of the intergranular glassy phase was obtained by measuring the area of the etched region.

The bending strength was determined by three point method at room temperature with a polished specimens of 3×4×40mm in size , and the Vickers hardness (HV10) and the indentation fracture toughness (K_{IC}) were measured by using a Vickers diamond indenter under 98N[7].

RESULTS AND DISCUSSION

Microstructure Characteristics

After the first step sintering, the sample reaches about 97% theoretical density, and its microstructure is shown in Fig.1. The grains distribute fairly uniform in size and are very fine with the average diameter of 0.35μm. The real aspect ratio is low, about 2.6. These suggest that the homogeneous nucleation occurs during the process of solution and reprecipitation. The rate of grain growth is low and the elongated grains are not yet developed in this step, because the sintering temperature is not high enough.

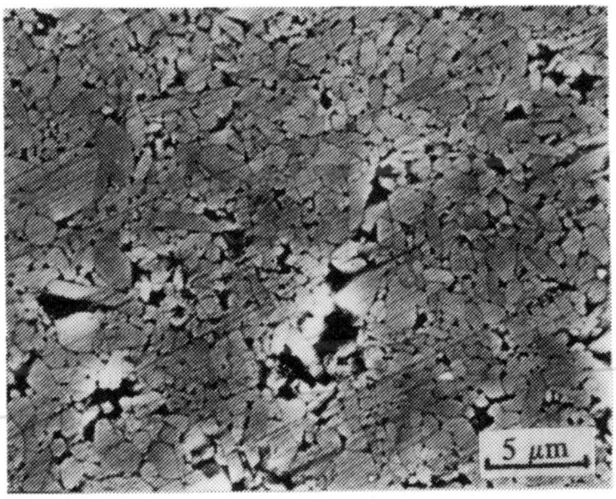

Fig.1 The microstructure of the sample sintered for 2h at 1800°C.

After the second step sintering at 1900°C for different soaking time, the polished and etched surfaces

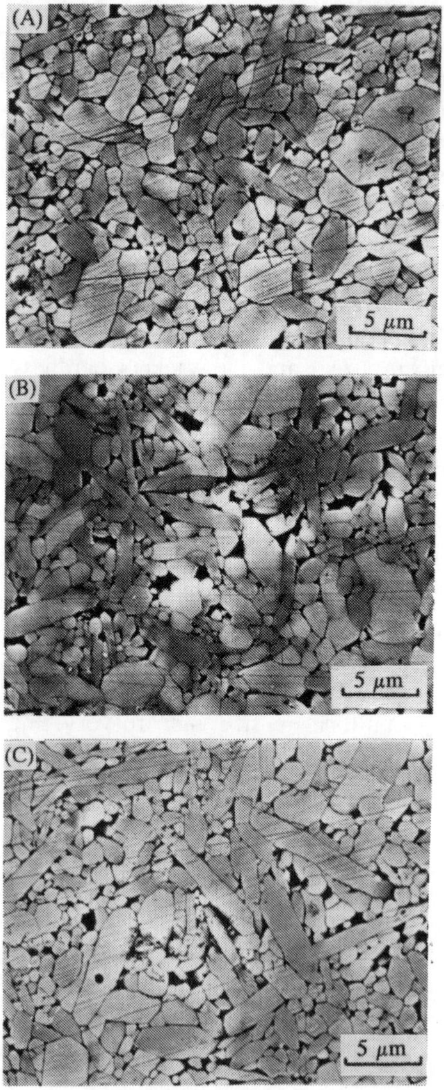

Fig.2 The microstructure of the samples after the second step sintering at 1900°C for (A) 0.5h, (B) 1.5h, and (C) 3.0h.

of the materials are shown in Fig.2. The growth of elongated grains in small equaxed matrix grains is observed. The microstructures are quantitatively characterized by image analyzer, and the average grain diameter (d), length (L), real aspect ratio (R) and volume of glassy phase are summarized in table II. The results show that the samples are close to be fully densified, attaining about 99% theoretical density, and the grain size and real aspect ratio reach a high level, which means the rapid grain growth occurs and the anisotropic microstructure forms in this sintering step. The changes of the average grain diameter and aspect ratio with the soaking time are shown in Fig.3. The microstructure characteristics, d and R, increase with increasing of the sintering time. This indicates that the grain growth will be accelerated by the long time sintering, however the rate in length direction is larger than that in width direction, which lead to the increasing of aspect ratio as well as the diameter.

The microstructures of the specimens after the second step sintering at three different temperatures for 1.5h are shown in Fig.4, and also compared with that in Fig.1. Some of the micro-structures characterization results

Table II. Density of Materials, Average Diameter (d), Average Length (L), Real Aspect Ratio (R) of Grains, and Volume of Glassy Phase with Soaking Time

Sintering Temp. (°C)	Soaking Time (h)	Density (g/cm^3)	d (μm)	L (μm)	R	Glass Volume (%)
1800	2.0	3.21	0.35	0.58	2.6	13.14
1900	0.5	3.24	0.44	0.80	4.0	11.65
1900	1.5	3.26	0.45	0.82	4.2	10.94
1900	3.0	3.27	0.48	0.88	5.4	9.73

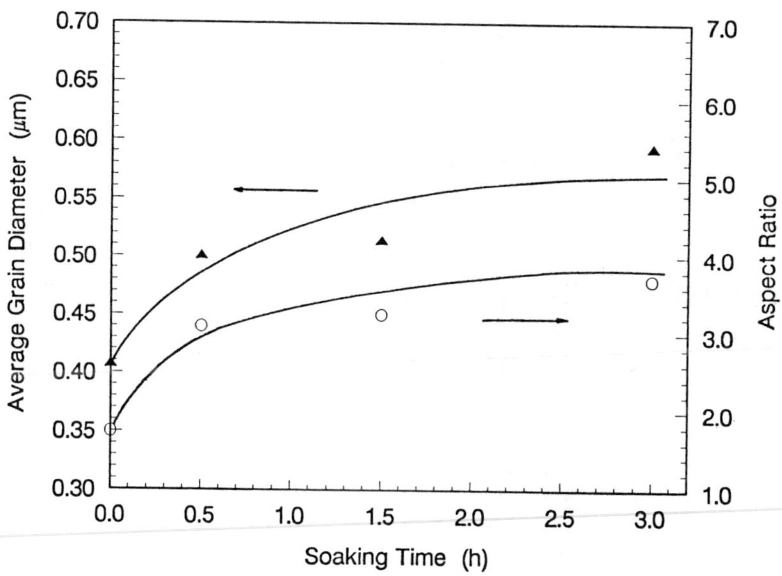

Fig.3. Average grain diameter and real aspect ratio as a function of the soaking time for 0.5h, 1.0h and 1.5h at 1900°C

Fig.4. The microstructure of the samples after the second step sintering for 1.5h at (A) 1900°C, (B) 1950°C, and (C) 2000°C.

are given in table III. The grains develop well and the samples get a high density. With the increasing of the sintering temperature in the second step, the average grain diameter and grain length have a gradual increase, while the aspect ratio has a slightly decrease, as seen in Fig.4. During the gas pressure sintering, the grain growth is due to the solution of the small grains and precipitation on the large grains. The liquid at the high sintering temperature has a low viscosity and makes the materials diffusion faster, so the change of grain diameter with increasing the sintering temperature is larger than that with the holding time. Meanwhile the globularization process occurs as well as the grain growth in order to reduce the interface energy. Wötting[8] reported the glodularization rate is primarily dependent on the viscosity of the liquid phase. Lower viscosity at high temperature will accelerate the glodularization processes, and leads to a decrease of aspect ratio.

Mechanical Properties

The bending strength, fracture toughness and indentation hardness of samples are given in table IV. the specimen which is sintered only by one step process gives very low mechanical prop-

Table III. Density of Materials, Average Diameter (d), Average Length (L), Real Aspect Ratio (R) of Grains, and Volume of Glassy Phase with Sintering Temperature

Sintering Temp. (°C)	Soaking Time (h)	Density (g/cm^3)	d (μm)	L (μm)	R	Glass Volume (%)
1800	2.0	3.21	0.35	0.58	2.6	13.14
1900	1.5	3.26	0.45	0.82	4.2	10.94
1950	1.5	3.27	0.47	0.88	4.0	9.89
2000	1.5	3.27	0.49	0.89	3.8	9.33

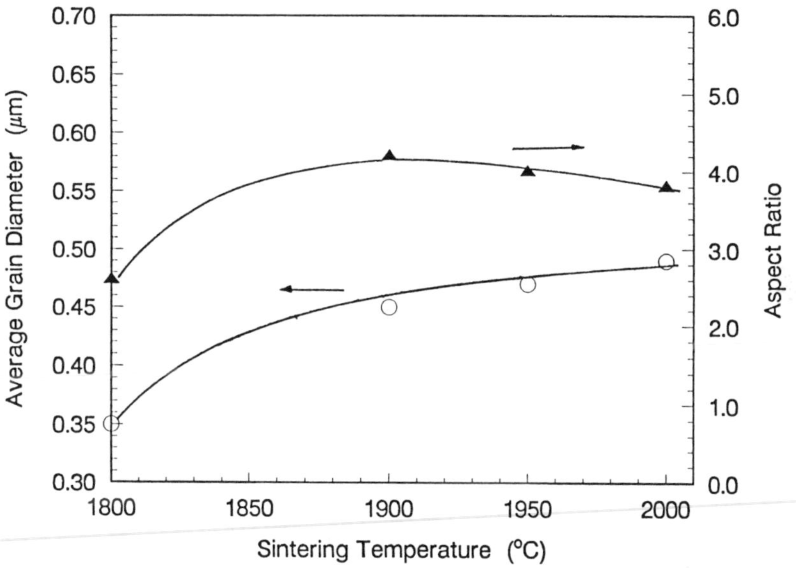

Fig.5. Average grain diameter and real aspect ratio as a function of the sintering temperature in the second step at 1900°C, 1950°C and 2000° for 1.5h

Table IV. Flexual Strength, Fracture Toughness and Indentation Hadness of Materials Sintered with Different Processing Conditions

Sintering Temp. (°C)	Soaking Time (h)	Strength (MPa)	Toughness (MPa $m^{1/2}$)	Hardness (HV10)
1800	2.0	612.1	3.9	1594
1900	0.5	797.0	5.8	1773
1900	1.5	844.2	6.1	1816
1900	3.0	896.4	7.0	1860
1900	1.5	844.2	6.1	1816
1950	1.5	869.2	6.5	1818
2000	1.5	925.1	7.2	1883

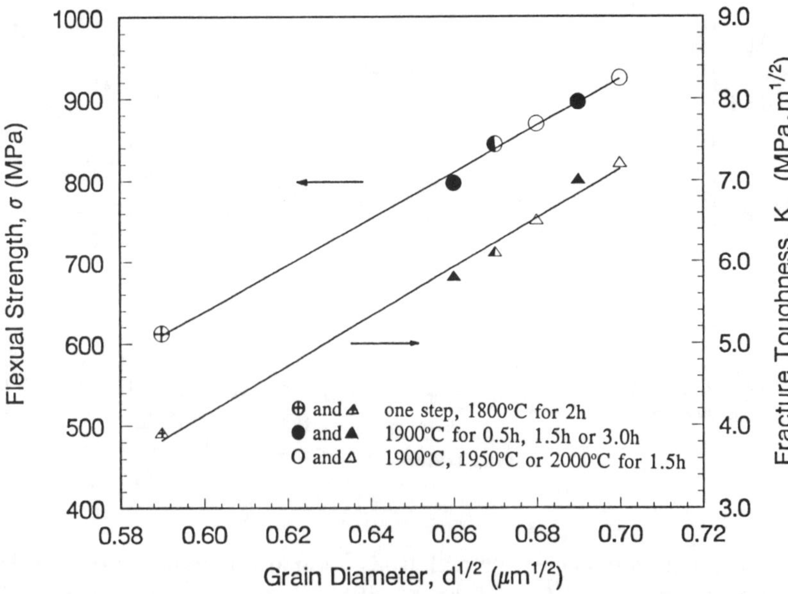

Fig.6. Relation between the mechanical properties and square root of the average grain diameter

erties. This is attributed to the fact that the microstructures are neither well developed nor fully dense. Those after the second step sintering, the strength and fracture toughness show a substantial increase, and the values have a gradually increase trend with the prolonging holding time and with the increasing of the sintering temperature. It has been reported that the crack deflection and crack bridging are the two main toughening mechanisms in the Si_3N_4-based ceramics. The development of microstructures with high aspect ratio is responsible for the high strength and high toughness because the interlocking of elongated grains provides crack deflection associated with fracture energy consumption. High aspect ratio is necessary for improving the toughness, however the mechanical properties obtained in the present work are not strictly proportional to the aspect ratio. Increasing the sintering tempera-ture, the samples have an increase of the flexural strength and fracture toughness, despite there is a slight decrease of the grain aspect ratio. This implies that the crack bridging plays an important role in strengthening and toughening of the materials. Fig.5 is the relation between the grain size and the bending strength or fracture toughness. It shows the linear relation exists between the square root of the grain diameter and the mechanical properties. Larger grains with high aspect ratio are more efficient in inhibiting the crack propagation by crack bridging mechanism.

The specimens show a very high hardness value because of the high volume of α'-sialon phase in the materials. The increase of the indentation hardness as the change of the sintering temperature and soaking time is related to the slight increase of α' and the decrease of the intergranular glassy phase.

SUMMARY

The microstructure of α'-β'-sialon can be tailored in gas-pressure sintering. The elongated grains are developed after the two-step sintering. The average grain diameter and aspect ratio will gradually increase with the prolonging the soaking time, and the aspect ratio will slightly decrease at very high sintering temperature. High mechanical properties can be obtained in the materials with well developed microstructure. The crack bridging and crack deflection might be the two main strengthening and toughening mechanisms for the materials.

REFERENCES

1. T.Ekström and M.Nygren, "SiAlON Ceramics," J.Am.Ceram.Soc., 75 (1992) 259.
2. H. Wang, W.Y. Sun, H.R. Zhuang, T.S. Yen "Preparation of R-α'-β'-Sialons (R=Sm, Gd, Dy, Y and Yb) by Pressureless Sintering," J. Europ. Ceram.

Soc., 13 (1994) 461.
3. K.T.Faber and A.G.Evans, "Crack Deflection Processes-II. Experiment," Acta Metall., 31[4] (1983) 577.
4. P.F.Becher,C.H.Hsueh,P.Angeline, and T.N.Tiegs,"Toughening Behaviour in Whisker-Reinforced Ceramic Matrix Composites," J.Am. Ceram. Soc., 72[12] (1988) 1050.
5. M.Mitomo and S. Uenosono,"Gas Pressure Sintering of β-Silicon Nitride", J.Mater.Sci., 26 (1991) 3940.
6. M. Mitomo, S. Uenosono "Microstructural Development During Gas-Pressure Sintering of α-Silicon Nitride" J. Am. Ceram. Soc. 75(1) (1992) 103.
7. G.R. Anstis, P. Chantikul, B. R. Lawn and D. B. Marshall, J. Am. Ceram. Soc., 64(1981) 533.
8. G.Wötting, B.Kanka and G.Ziegler, "Microstructure Development, Microstructural Characterization and Relation to Mechanical Properties of Dense Silicon Nitride", pp83 in Non-oxide Technical and Engineering Ceramics., Edited by S.Hampshire. Elsevier, London, 1986.

FABRICATION OF POLYMER-DERIVED MICRO-NANO-STRUCTURED Si_3N_4/SiC-COMPOSITES

Axel Greiner*, Peter Kroll and Ralf Riedel
Technische Hochschule Darmstadt
Fachbereich Materialwissenschaft
Fachgebiet Disperse Feststoffe
Hilpertstr. 31/D, D-64295 Darmstadt, Germany
* And Max-Planck-Institut, Stuttgart

Joachim Bill
Max-Planck-Institut für Metallforschung
Institut für Werkstoffwissenschaft
Pulvermetallurgisches Laboratorium
Heisenbergstr. 5, D-70569 Stuttgart, Germany

ABSTRACT

Polyhydridochlorosilazane and polymethylvinylsilane have been used as polymer precursors to produce micro-nano-structured Si_3N_4/SiC-composites. The polysilazane was pyrolysed to amorphous Si-C-N powders while polyvinylsilane was used to coat α-Si_3N_4 powder with a SiC-forming precursor. After pressureless liquid phase sintering and subsequent gas pressure sintering of the compacted as-synthesized powders in the presence of nitrogen atmosphere, the obtained microstructures were investigated by scanning and transmission electron microscopy (SEM, TEM). Nanosized SiC inclusions embedded in microcrystalline Si_3N_4 grains were analysed by elemental mapping with electron spectroscopic imaging (ESI).

INTRODUCTION

During the last decade, monolithic Si_3N_4 gained increasing interest as components for applications in heat engines and gas turbines. To improve the

To the extent authorized under the laws of the United States of America, all copyright interests in this publication are the property of The American Ceramic Society. Any duplication, reproduction, or republication of this publication or any part thereof, without the express written consent of The American Ceramic Society or fee paid to the Copyright Clearance Center, is prohibited.

outstanding properties of this material, SiC containing Si_3N_4 based materials have been developed and characterized. The obtained fine grained Si_3N_4/SiC composites exhibit advanced properties, e. g. superplastic behaviour[1] and high fracture strength[2]. Beyond this, SiC doping of Si_3N_4 matrices leads to important changes in thermal and electric conductivity, creep behaviour, thermal stability and oxidation resistance.

As Niihara[2] previously suggested, nano composites can be subdivided into four types: i) intergranular, ii) intragranular, iii) inter/intragranular and iv) nano/nano, as shown in Figure 1. According to the literature Si_3N_4/SiC-composites can be generated by the classical powdertechnological route - liquid phase sintering of additive containing Si_3N_4/SiC-powder mixtures[3,4].

Dispersing nanosized SiC-inclusions in the Si_3N_4 matrix leads to inter/intragranular micro-nano-materials, as prepared recently by hot-pressing of amorphous CVD Si-C-N powders[2].

This work reports on preparing bulk materials of mixed inter/intra-type Si_3N_4/SiC composites by polymer pyrolysis of appropriate starting materials.

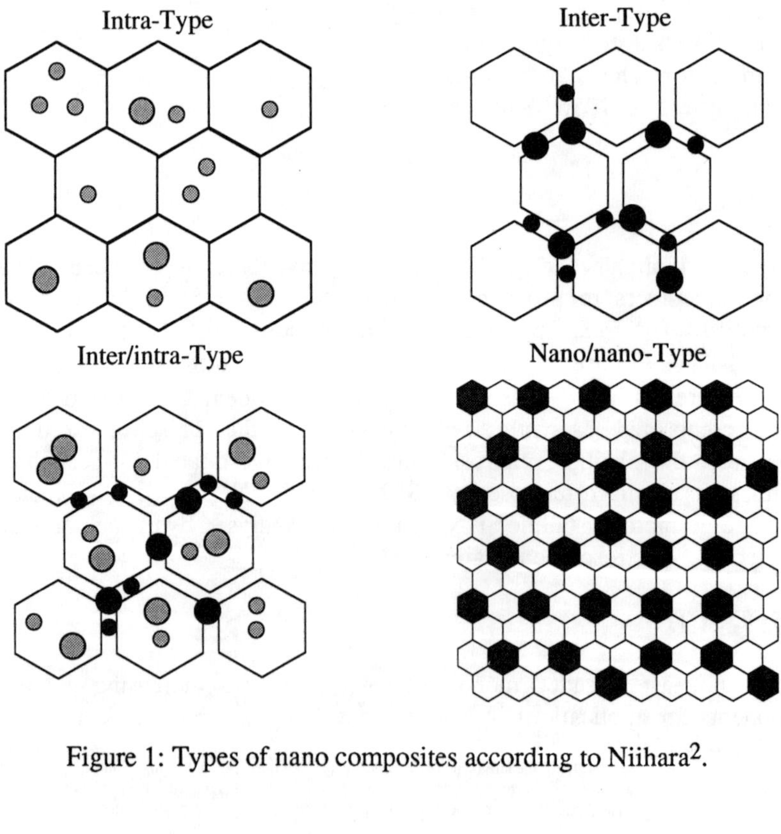

Figure 1: Types of nano composites according to Niihara[2].

EXPERIMENTAL PROCEDURE

Polymethylvinylsilane derived specimens

All work, as described below, was done under argon atmosphere. Polymethylvinylsilane was synthesized as reported previously[6]. To improve homogeneity of the SiC distribution in the Si_3N_4 matrix, the polymer precursor was precipitated out of a liquid phase onto a mixture of α-Si_3N_4 powder and sintering aids. The polymer content was adjusted to yield a 20 wt% coating after removal of the solvent. The residue was ball milled, sieved and cold-isostatically pressed at 630 MPa into green compacts which were subsequently pyrolysed at 1000°C under argon atmosphere. The amount of sintering aids was calculated to be 15 wt% after pyrolysis (Al_2O_3 : Y_2O_3 = 2.5 : 12.5 wt%).

Polysilazane derived specimens

Polyhydridochlorosilazane (VB33, Hoechst AG) was pyrolysed under argon atmosphere at 1000°C, as reported by M. Seher et al.[5]. The as-produced amorphous $Si_{3+x}N_4C_{x+y}$ (silicon carbonitride) was mixed with 3.5 wt% Y_2O_3 and 3.5 wt% Al_2O_3 and attrition milled to receive powder with an average particle diameter of 0.68 µm. Green bodies of 14 mm in length and 15 mm in diameter were prepared by cold-isostatical pressing at 630 MPa.

Sintering

Finally the compacted bodies were pressureless presintered at 1700°C under 0.1 MPa N_2 for 1h and subsequently gas pressure sintered at 1900°C under 10 MPa N_2 for 1h. Phase analyses of each product were carried out by X-ray diffraction with Cu K_α radiation (Model D5000, Siemens). The C, N and O contents were measured by high temperature combustion methods using C/S Analyser Model 244 and N/O Determinator Model TC 436 (Leco Corp.).
SEM-micrographs were taken on a Stereoscan 200 scanning electron microscope (Cambridge Instruments). For TEM-micrographs and investigations with electron spectroscopic imaging (ESI) a prototype of the Zeiss EM 912 OMEGA transmission electron microscope (Carl Zeiss) was employed.

RESULTS

Pyrolysis

The synthesis of polysilanes is based on a Wurtz-Fittig analogous dechlorination reaction as shown in equation 1 for the formation of polymethylvinylsilane.

$$n\ (CH_2=CH)CH_3SiCl_2 + m\ (CH_3)_2SiCl_2 \xrightarrow[-2(n+m)NaCl]{+2(n+m)Na} [-(CH_2=CH)CH_3Si-]_n[-(CH_3)_2Si-]_m \quad (1)$$

The coated powders were easy to densify by cold isostatic pressing, because the coating acted as lubricant and binder, resulting in ductile, machineable compacts with low abrasion. Since polymethylvinylsilane has a chemical reactive vinyl group, strengthening of the compacted bodies takes place at elevated temperatures, due to crosslinking reactions. Therefore, crack free pyrolysed bulk materials with a density of 2.3 g/cm^3 (68 % o. th.) were received.
Preparation and pyrolysis of polysilazanes is a main interest in materials science[7,8]. After crosslinking and decomposition pyrolysis results in X-ray amorphous residues. Depending on the pyrolysis conditions the reaction product is obtained as powder or hard foam. Decomposition and crosslinking of the polymers starts in the temperature range between 400 and 500°C, according to the simplified reaction equation 2:

$$[-CH_3SiH-NH-]_{4n-m-1}[-CH_3SiN-]_m[-CH_3SiCl-NH-]_l \xrightarrow{400-1000°C/Ar} Si_{3+x}N_4C_{x+y} + H_2 + CH_4 + \ldots \quad (2)$$

The pyrolysis residue was mixed with sintering additives and attrition-milled. The sieved Si-C-N powder was subsequently compacted to green bodies with a density of 1.5 g/cm^3 (44 % o. th.), showing remarkable abrasion under mechanical load.

Sintering

Both materials revealed, despite of their different preparation, the same sintering and crystallization behaviour when heated to temperatures above 1000°C. Crystallization of β-SiC and, in the case of amorphous Si-C-N materials, crystallization of α-Si_3N_4 occurs. At temperatures beyond 1440°C excessive carbon reacts with Si_3N_4 to form additional SiC and N_2 (Equation 3).

$$Si_3N_4 + 3C \xrightarrow{T \geq 1440°C} 3SiC + 2N_2 \tag{3}$$

At 1350°C liquid phase sintering starts by particle rearrangement. At elevated temperatures further densification occures by dissolving of α-Si_3N_4 in the secondary phase, accompanied by simultaneous reprecipitation of β-Si_3N_4. Owing to α/β-tranformation and subsequent grain growth during gas pressure sintering, nanosized SiC crystallites are included in microcrystalline β-Si_3N_4 grains. Figure 2 shows a SEM micrograph of polished and plasma etched micro/nano composite material. Microcrystalline SiC can be found in the intergranular areas (arrows), while nanocrystalline SiC inclusions are clearly detected within the Si_3N_4 matrix grains.

Figure 3A shows a TEM micrograph of Si_3N_4 with SiC inclusions (arrows). ESI investigation of the same area, given in Figure 3C-D, shows the elemental distribution of N (Figure 3B, Si_3N_4), C (Figure 3C, SiC) and O (Figure 3D, additives).

The phase composition of the gas pressure sintered composites listed in Table I was calculated from elemental analysis.

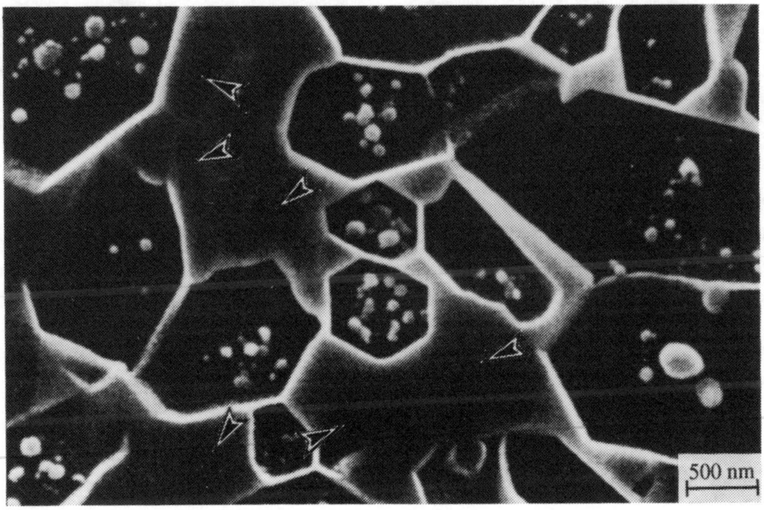

Figure 2: SEM micrograph of sintered (1700°C, 0.1 MPa N_2) and gas pressure sintered (1900°C, 10 MPa N_2) Si_3N_4/SiC-composite.

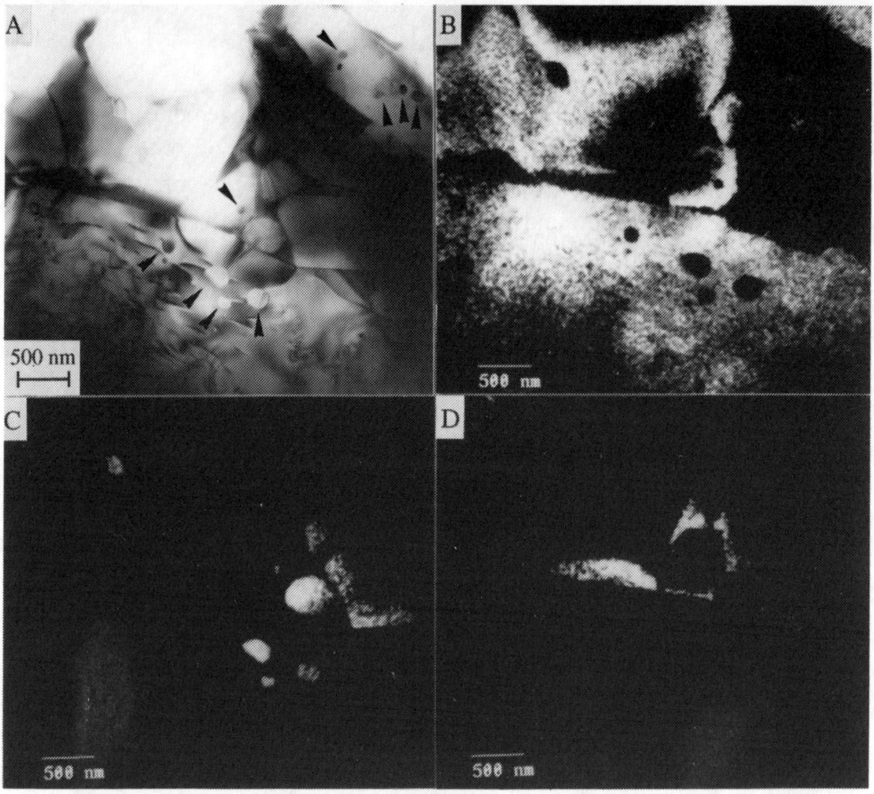

Figure 3A-D: TEM micrograph of Si_3N_4 grain with SiC inclusions and microcrystalline SiC grains at the grain boundary (A). ESI investigation of the same sample area showing elemental mapping of nitrogen (Si_3N_4) (B), carbon (SiC) (C) and oxygen (Additives) (D)

Table I: Phase composition of gas pressure sintered specimens

	Si_3N_4	SiO_2	SiC	C [wt%]	Y_2O_3	Al_2O_3
Polysilazane derived material	69.9	0.0	21.2	0.0	4.5	4.4
Polysilane derived material	67.5	3.6	11.9	0.0	14.2	2.8

Embedding of nanosized SiC crystallites in β-Si_3N_4 grains

The embedding of nanosized SiC crystallites in the Si_3N_4-matrix can be explained with two different models. As Niihara[2] already suggested, SiC particles can serve as crystallization nuclei for β-Si_3N_4 grains. Due to subsequent grain growth, a lot of nuclei are trapped in every Si_3N_4 grain, while SiC crystallites with diameters above 0.3 μm are located in the grain boundary phase to give inter/intra-type composite structures. The second explanation is, that β-Si_3N_4 and β-SiC crystallize independently of each other, and nanosized SiC particles are inclosed in growing β-Si_3N_4 grains (Figure 4).

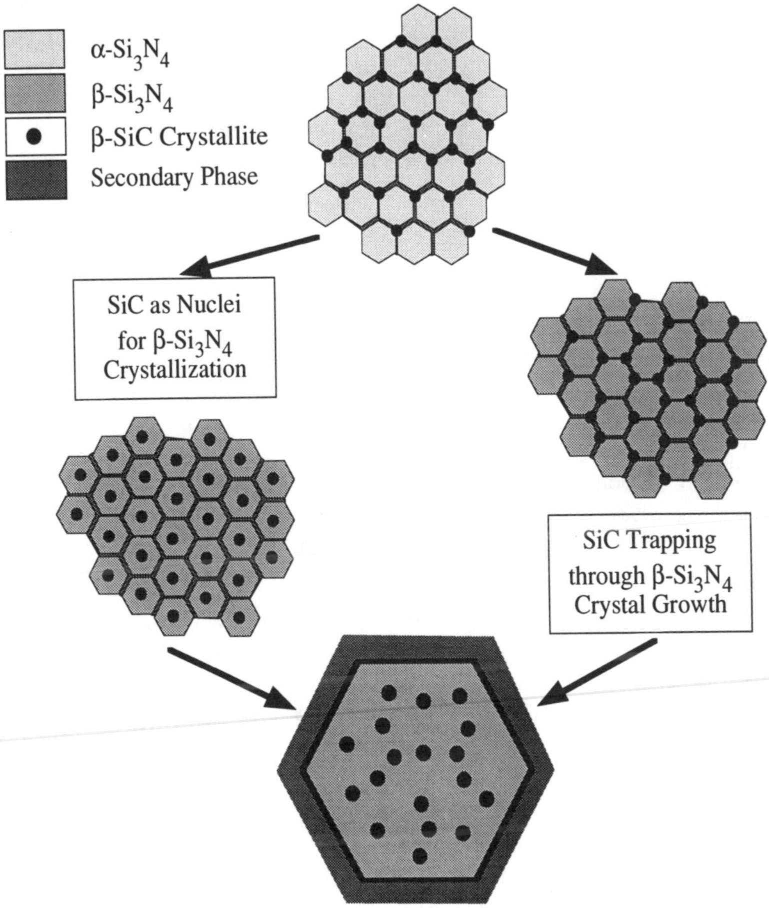

Figure 4: Embedding of SiC in a Si_3N_4 matrix

CONCLUSIONS

Dispersion of β-SiC in β-Si_3N_4 crystallites depends on the grain size of the β-SiC particles. Therefore, methods providing a high amount of nanosized SiC particles during the solution-diffusion-reprecipitation process of Si_3N_4 liquid phase sintering, result in a high density of SiC inclusions. In this work it is unequivocally shown, that materials prepared from polymer precursor pyrolysis provide a great number of nanosized SiC crystallites, small enough to get trapped in the matrix grains during grain growth of β-Si_3N_4. Polymer pyrolysis enables the production of Si_3N_4/SiC micro/nanocomposites in large quantities due to the fact that Si-based non-oxide polymers such as polysilanes and polysilazanes are now commercially available. The SiC content of Si_3N_4/SiC composite ceramics can be influenced either by changing pyrolysis conditions of polysilazanes or by varying the amount of polymer coating, leading to taylormade ceramic materials. Optimisation of polymer derived ceramics in terms of densification and control of microstructure will be the focal points in future investigations.

ACKNOWLEDGEMENT

The Authors wish to thank Dr. V. Szabó for preparing the TEM and ESI photographs.

REFERENCES

1. F. Wakai, Y. Kodama, S. Sakaguchi, N. Murayama, K Izaki, K. Niihara, *A Superplastic covalent crystal composite*, Nature, **344** [6265], 421-3, (1990).
2. K. Niihara, *New Design Concept of Structural Ceramics - Ceramic Nanocomposites* -, J. Ceram. Soc. Jp., **99** [10], 974-82, (1991).
3. F.F. Lange, *Effect of Microstructure on Strength of Si_3N_4-SiC Composite System*, J. Am. Ceram. Soc., **56** [9], 445-50, (1973).
4. Y. Akimune, T. Ogasawara, N. Hirosaki, *Influence of Starting Powder Characteristics on Mechanical Properties of SiC-particle/Si_3N_4 Composites*, J. Ceram. Soc. Jp., **100** [4], 463-7, (1992).
5. M. Seher, J. Bill, R. Riedel, F. Aldinger, *Processing and Properties of Carbon Containing Silicon Nitride Ceramics Derived From the Pyrolysis of Polyhydridochlorsilazanes*, Key Engineering Materials Vol. 89-91, Proceedings of the International Conference on Silicon Nitride-Based Ceramics Stuttgart, October 4-6, 1993, M.J. Hoffmann, P.F. Becher and G. Petzow [Eds.], Trans Tech Publications, Switzerland, (1994), p. 101-6.

6. R. Riedel, A. Kienzle, V. Szabó, J. Mayer, *Hydroboration of polymethylvinylsilane - a novel route to silicon boron carbide ceramics*, J. Mater. Sci., **28**, 3931-8, (1993).
7. D. Seyferth, G.H. Wiseman, *High-Yield Synthesis of Si_3N_4/SiC Ceramic Materials by Pyrolysis of a Novel Polyorganosilazane*, J. Am. Ceram. Soc., **67** [7], C-132-3, (1984).
8. R. Riedel, M. Seher, G. Becker, *Sintering of Amorphous Polymer-Derived Si, N and C Containing Composite Powders*, J. Europ. Ceram. Soc., **5**, 113-22, (1989).

TENSILE CREEP FAILURES OF Al_2O_3/SiC NANOCOMPOSITE

Tatsuki Ohji
National Industrial Research Institute of Nagoya
Nagoya 462, Japan

Atsushi Nakahira, Takeshi Hirano and Koichi Niihara
The Institute of Scientific and Industrial Research, Osaka University
Ibaraki, Osaka 567, Japan

ABSTRACT

The dispersion of ceramic particles with nanometer size into the ceramic matrix grains or grain boundaries gives rise to substantial improvements of the mechanical properties, particularly at high temperatures. However, creep performance, which is one of the most important mechanical properties for high temperature engineering materials has not been investigated for the ceramic nanocomposite. This paper reports tensile creep performance of alumina dispersed with 17 vol% silicon carbide nanoparticles at 1200°C. Compared to the monolithic alumina, the nanocomposite exhibited excellent creep resistance. The minimum creep rate of the nanocomposite was about three orders of magnitude lower and the creep life was 10 times longer than those of the monolithic. The improved creep resistance is caused by intergranular silicon carbide nanoparticles which rotate and plunge into alumina grains as grain boundary sliding proceeds.

INTRODUCTION

Niihara and his coworkers,[1-5] have revealed that dispersion of ceramic particles with nanometer size into the ceramic matrix grains or grain boundaries causes substantial improvements of the mechanical properties. For example, dispers–

To the extent authorized under the laws of the United States of America, all copyright interests in this publication are the property of The American Ceramic Society. Any duplication, reproduction, or republication of this publication or any part thereof, without the express written consent of The American Ceramic Society or fee paid to the Copyright Clearance Center, is prohibited.

ing 5 vol% silicon carbide nanoparticles into alumina increased the room temperature fracture strength more than three times. Similar strength improvements have been identified in alumina/silicon nitride, magnesia/silicon carbide, silicon nitride/silicon carbide nanocomposites. Some of these nanocomposites exhibited a remarkable increase in the fracture toughness as well. The benefits of ceramic nanocomposites can be identified not only at room temperature but also at elevated temperatures. Compared to the monolithic ceramic materials, the materials into which nano-size ceramic particles were dispersed have exhibited little strength degradation at high temperatures. The high strengths observed at room temperature for the alumina/silicon carbide and silicon nitride/silicon carbide nanocomposites were maintained up to 1200°C and 1400°C, respectively. The most interesting example is the case of magnesia/30 vol% silicon carbide nanocomposite. While the fracture strength of monolithic magnesia polycrystalline gradually decreases with raising temperature, the strength of the nanocomposite which two times higher than that of the monolith at room temperature, tends to increase rather than decrease.

The ceramic nanocomposites also possess good creep resistance. Nakahira and Niihara[6] investigated the flexure creep behaviors for alumina/17 vol% silicon carbide nanocomposite and monolithic alumina at 1100°C to 1400°C, and revealed that the creep rate of the nanocomposite was four orders of magnitude lower than that of the monolith. The magnesia/30 vol% silicon carbide nanocomposite demonstrated a creep rate 3 orders of magnitude lower than that of the monolithic material in compressive creep tests at 1450°C.[7] Silicon nitride/32 vol% silicon carbide nanocomposite showed very long creep life.[8] At 1400°C and 200 MPa, a specimen of the nanocomposite did not fail after 1000 hours in tension, while a monolithic Si_3N_4 exhibited only 0.56 hour life time.

The present paper deals with tensile creep and creep rupture behaviors of alumina/silicon carbide nanocomposite at temperatures of 1200°C to 1300°C.[9] The alumina/17 vol% silicon carbide nanocomposite and alumina polycrystalline whose grain size were equal each other were used to clearly understand the difference in the behaviors. Transmission Electron Microscopy (TEM) study was carried out to understand the boundary structure and the microstructural evolution during creep. Effects of the nanoparticle dispersion on the creep and creep rupture behaviors are investigated.

MATERIALS

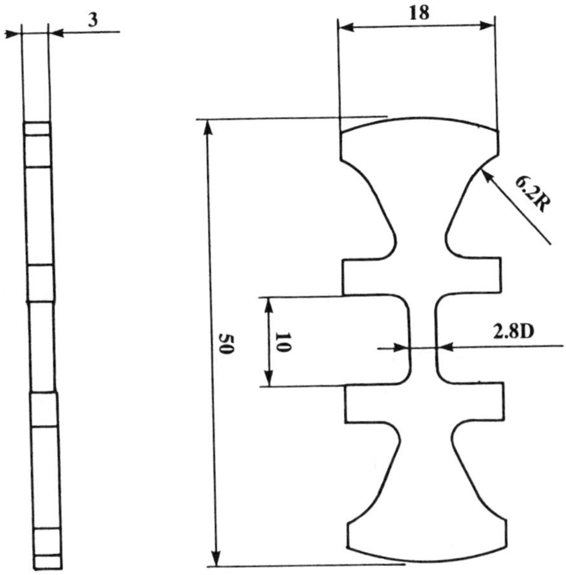

Figure 1. Geometry and dimensions of tensile creep specimen.

The alumina/17 vol% silicon carbide nanocomposite (hereafter nanocomposite) and the alumina polycrystalline (monolith) were fabricated by hot–pressing for this study. For the nanocomposite, the highly pure γ–alumina (Asahi Chemical Co. Ltd., Tokyo, Japan) and β–silicon carbide (Ibiden Co. Ltd., Ogaki, Japan) were employed as starting powders. The particle sizes of both powders were on the average within 100 nm. The powders were mixed by the ball–milling using alumina balls in a plastic pot, first in ethanol for 12 h and then for 24 h after drying them. The mixed powders were hot–pressed using a graphite die of a 50 mm diameter. In the nanocomposite, higher temperature is required for sintering, since the densification and grain growth are significantly suppressed by silicon carbide nanoparticles. Thus, the hot–pressing was carried out at 1800°C in nitrogen atmosphere with an applied pressure of 30 MPa for 1 h. The temperature was raised and lowered at 25°C/min. The hot–pressing continued during cooling down to 1000°C. Grain size of the obtained material was about 2 μm. The monolith was hot–pressed at 1500°C so that the grain sizes of the two materials are equivalent. The other procedures and starting alumina powder are the same as those of the nanocomposite. The obtained relative density was more than 98% in both of the materials. Four hot–pressed disks were prepared for the nanocomposite and two were for the monolith in total.

CREEP TESTING

The tensile creep testing procedures are the same as reported elsewhere,[10] except for the specimen. Since the total length available for preparing a tensile specimen was limited to the diameter of the hot–pressing die, 50 mm, a relatively small specimen was used. The geometry and dimensions of the specimen are shown in Fig. 1. One specimen was taken from one hot–pressed disk. The plates (50 mm by 18 mm by 3 mm) cut from these billets are roughly ground by a 200 grit diamond wheel into the shape of Fig. 1, and then finished by a 400 grit diamond wheel at a feed speed of 2 mm/min with a cutting depth of 20 μm in a numerically controlled cylindrical grinding. Finally, the gauge length area of each specimen is hand–polished along its length with a diamond paste; first 50 μm depth is removed by a 1000 grit paste and then another 30 μm is by 2000 grit. The purpose of this hand–polishing is to eliminate the effects of residual stress and surface flaws produced in the machining process. In this process, it was ensured that the surface has been removed uniformly over full gauge length area.

There are two target arms at the ends of the gauge section. The elongation of the distance between them was measured by an optical extensometer (Model 4100, Zimmer OHG, Darmstadt, FRG). The specimen was gripped in the hot zone of a furnace by retaining a shoulder of the specimen with two silicon carbide rollers. The ratio of the parasitic bending strain to the average tensile strain was suppressed to less than 4% at 50 MPa and less than 2% at 100 MPa. The creep tests were conducted in air atmosphere. The creep curves were found to fluctuate; the scatter of the measurement for the elongation was normally within 2 μm.

CREEP BEHAVIORS

A creep curve of the monolith at 1200°C and 50 MPa is shown in Fig. 2. The curve consists of primary, steady state and very small tertiary creep regimes. Similar creep curve behavior has been reported for tensile creep of polycrystalline alumina.[11,12] The specimen lifetime was some 150 hours, and some 4% of creep strain was obtained at fracture. A large number of microcracks, which

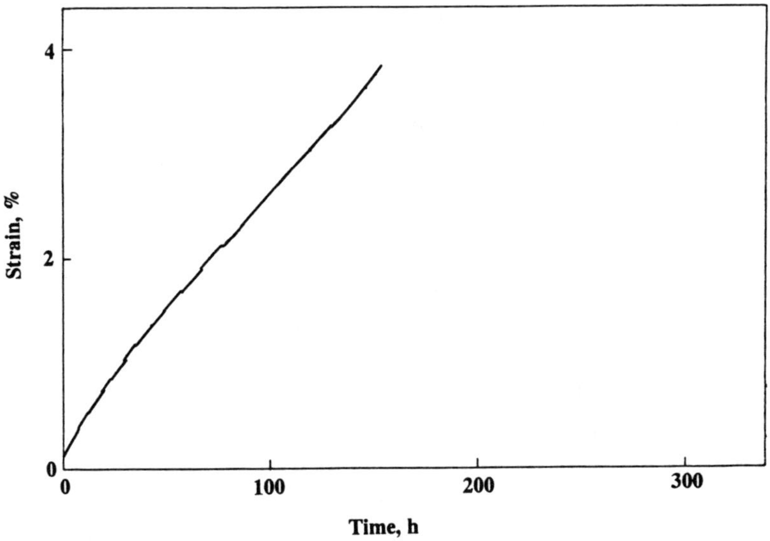

Figure 2. Tensile creep curve of the monolith at 1200°C and 50 MPa.

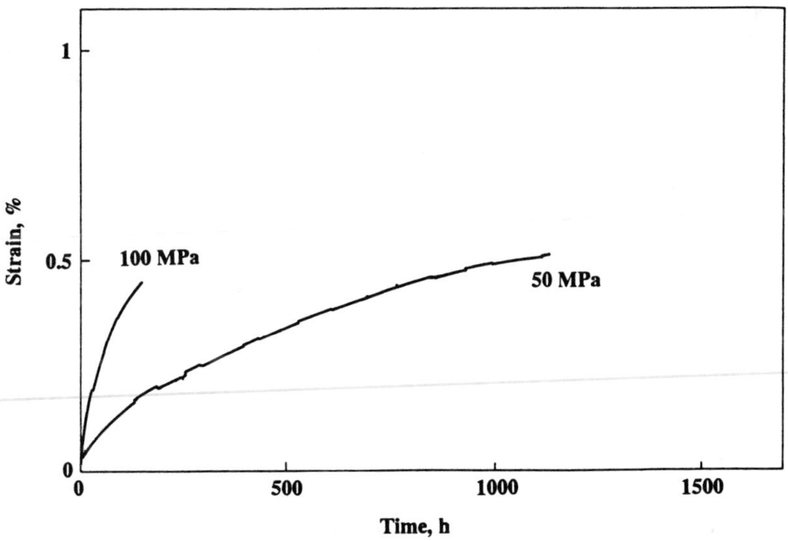

Figure 3 Tensile creep curves of the nanocomposite at 1200°C. The applied stress is 50 and 100 MPa.

have been very frequently reported in crept samples of polycrystalline alumina,[13–18] were also identified with optical microscopy in this case.

Creep curves of the nanocomposite are shown in Fig. 3. Compared to the monolith, the nanocomposite exhibited excellent creep resistance. At 1200°C and 50 MPa, the creep life is 1120 hours, which is 10 times longer than that of the monolith. The creep strain at fracture was 0.5 %, which is 8 times smaller. The most important difference is that transient creep was predominant in the nanocomposite.

Figure 4 shows stress dependencies of steady state creep rates for the monolith and the nanocomposite. One specimen of the monolith was used for a step–wise loading test (30, 50, 70, and 100 MPa) at this temperature, to obtain a creep exponent. This method can be used only in the case that the steady state creep appears quickly and is stable. Therefore, it was not applicable to the nanocomposite where little steady state creep was observed. For the same reason, the rates at fractures were plotted, and then the rigorous comparison is difficult. Crudely, however, the creep rate of the nanocomposite is about three orders of magnitude lower than that of the monolith. The stress exponent of the monolith obtained by the step–wise loading is 2.2. This value agrees pretty well with those reported for creep exponent of polycrystalline alumina by a number of researchers.[12,19–22] The deformation mechanism to yield this exponent, however, has not fully established. For the nanocomposite, only three data points were available, and the plots were widely scattered. In addition, all the fractures occurred in the transient creep regimes, which means premature termination of creep process. Then the calculated creep exponent significantly lacks dependability. A threshold stress, which has been confirmed to exist in creep for many of particle–dispersion strengthened materials,[23] could not be identified due to the small number of the available data.

The flexure creep test revealed that the creep rate of the nanocomposite was four orders of magnitude lower than that of the monolith at 1100°C through 1400°C,[6] as already stated in the introduction. In tension, however, the difference was some three orders of magnitude. This discrepancy was attributable to the difference in nature between flexure and tensile tests. Due to the thickness–wise stress gradient in flexure, the specimen lifetime is always longer than that in tension. While a creep rate continues to decrease with time (transient creep), a larger lifetime gives a lower creep rate. The nanocomposite showed transient

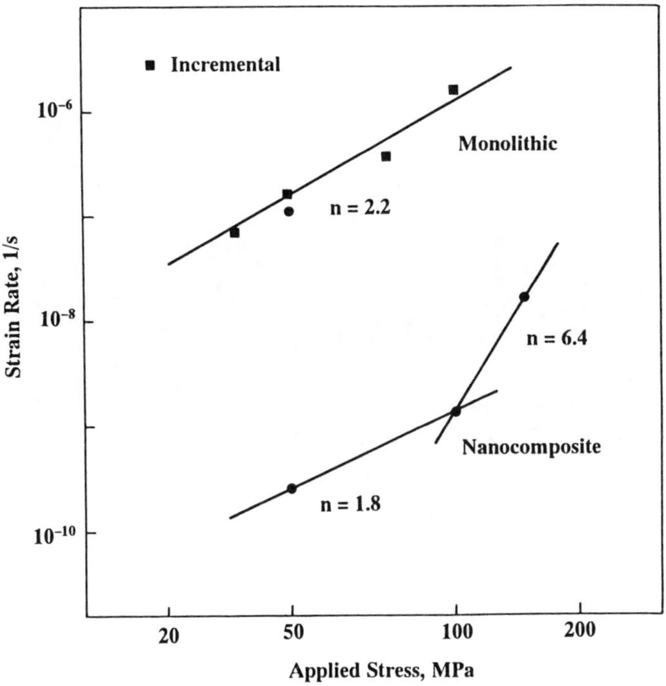

Figure 4. Stress dependencies of steady state or minimum creep rates in tensile tests for the monolith and the nanocomposite. The temperature is 1200°C. The n value is a stress exponent for creep rate.

creep until fracture while the monolith did steady state and accelerated creeps. Then the difference in creep rate between the nanocomposite and the monolith is larger in flexure than in tension.

FRATOGRAPHY

The fracture surfaces were examined with scanning electron microscopy. All the surfaces exhibited wavy area, so called creep damage zone, whether the nanocomposite or the monolith. As an example, the fracture surface of the nanocomposite at 1200°C and 100 MPa is shown in Fig. 5. The creep damage zone was expanded by lowering the applied stress or raising the temperature. Similar creep damage zones have been observed in tensile creep studies for

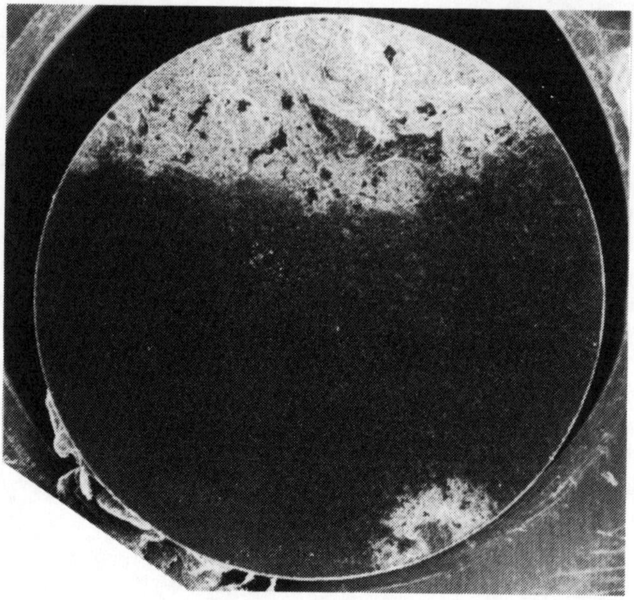

Figure 5. Fracture surface of the nanocomposite in a creep test at 1200°C and 100 MPa. The diameter is 2.8 mm.

alumina[18] and silicon nitride[24–27]. In the case of silicon nitride, the appearance of creep damage zone has been sometimes ascribed to oxidation of the specimen surface.[26,27] However, creep damage zone of alumina specimen which does not suffer from oxidation, can be considered to be a consequence of coalescence of microcracks.[18,25]

MICROSTRUCTURAL CHANGE DURING CREEP

TEM observations were carried out for the crept and fractured nanocomposite samples. One of the most characteristic changes of the microstructure during the creep is a rotation of intergranular silicon carbide particles accompanied with grain boundary sliding and small cavity formation around the particles (Fig. 6). These observations suggest the following mechanism for grain boundary sliding and its inhibition.[9] When creep occurred, the upper alumina grain slid downwards over the lower grain. In order to keep this sliding, the intergranular particle which supposedly had been on a straight alumina–alumina

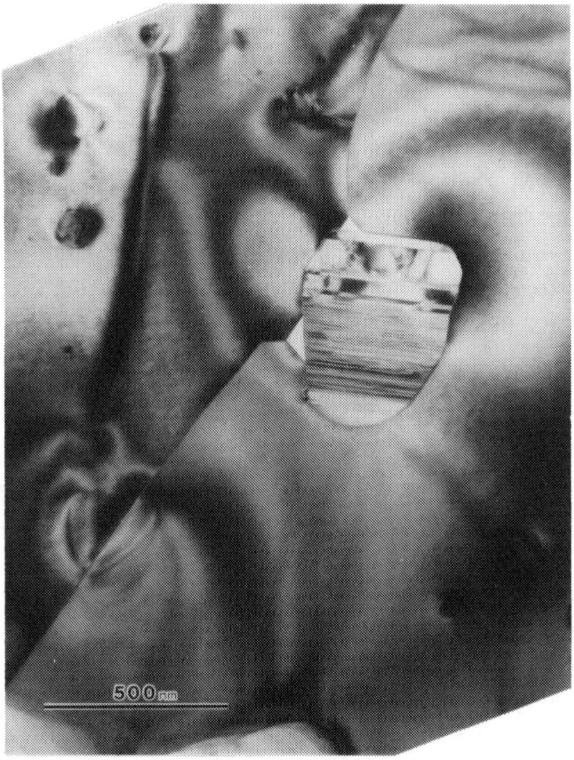

Figure 6. Transmission electron micrographs of microstructures of the nano-composite tested at 1300°C and 50 MPa (Ref. 9).

grain boundary, rotated counterclockwise, and plunged into the lower grain. As byproducts of this process, the strain contrast contours and the small cavities were generated, and the grain boundary was curved. It is obvious that this plunging increases the pinning effect of the particle and consequently the creep resistance, resulting in a transient creep.

The grain boundary diffusion can be considered the most predominant deformation mechanism of alumina polycrystalline at temperatures around 1200°C. Raj and Ashby[28] calculated sliding rate with diffusional accommodation in case that secondary particles are present on the grain boundaries and showed that creep is not slowed if the diffusivity (product of diffusion coefficient and boundary thickness) of the particle–matrix interface is equal to that of the matrix–matrix interface. This suggests that the diffusion ability in the silicon carbide–alumina interface of the nanocomposite is significantly lower than that in the alumina–alumina interface. This speculation is supported by the direct observation of the interfaces by TEM; the alumina–alumina interface contained markedly wider non–crystalline structure than the silicon carbide–alumina interface.[29] The strong particle–matrix interface is supposedly attributable to a great internal compression stress acting on the interface during cooling down due to the thermal expansion mismatch.[29]

SUMMARY

Tensile creep rupture behaviors of alumina–17 vol% SiC nanocomposite and monolithic alumina were investigated at 1200°C to 1300°C and at 50 to 150 MPa. Compared to the monolithic alumina, the nanocomposite exhibited excellent creep resistance. The minimum creep rate of the nanocomposite is about three orders of magnitude lower than that of the monolith. At 1200°C and 50 MPa, creep life of the nanocomposite is 10 times longer than that of the monolith. The important difference is that the nanocomposite demonstrated solely transient creep while slight some accelerated creep as well as steady state creep were observed in the monolith.

In the fractographic study, the creep damage zone was identified for both the materials, suggesting that the cavity nucleation and crack coalescence governed the failures. The TEM study provided some evidences that the nano–particles at the interfaces substantially suppressed the grain boundary sliding. With the sliding proceeding, the particles rotate and the suppression increases, leading to the transient creep behavior. The role of the intragranular nano–particles, which will be more important in creep at higher temperatures, is still unknown.

REFERENCES

1) K. Niihara and A. Nakahira, "Strengthening of Oxide Ceramics by SiC and Si_3N_4 Dispersions"; pp. 919–26 in Proceedings of the Third International Symposium on Ceramic Materials and Components for Engines. Edited by V.J. Tennery. American Ceramic Society, Westerville, OH, 1988.
2) K. Niihara, A. Nakahira, G. Sasaki and M. Hirabayashi, "Development of Strong Al_2O_3/SiC Composite"; pp. 129–34 in Proceedings of the First MRS International

Meeting on Advanced Materials, Vol. 4. Edited by M. Doyama, S. Somiya, and R.P.H. Chang. Materials Research Society, Pittsburgh, 1989.

3) K. Niihara, T. Hirano, A. Nakahira, K. Ojima, K. Izaki and T. Kawakami, "High–Temperature Performance of Si_3N_4–SiC Composites From Fine, Amorphous Si–C–N Powder"; pp. 107–12 in Proceedings of the First MRS International Meeting on Advanced Materials, Vol. 5. Edited by M. Doyama, S. Somiya, and R.P.H. Chang. Materials Research Society, Pittsburgh, 1989.

4) K. Niihara and A. Nakahira, "Particle–Strengthened Oxide Ceramics, Nanocom posite"; pp. 637–64 in Advanced Structural Inorganic Composites. Edited by P. Vincentini. Elesevior Sci. Pub. Co., Trieste, 1990.

5) K. Niihara, "New Design Concept of Structural Ceramics –Ceramic Nanocompos ite–," The Centennial Issue of the Ceramic Society of Japan, J. Ceram. Soc. Jpn., 99 [10] 974–82 (1991).

6) A. Nakahira and K. Niihara, "Microstructure and High–Temperature Mechanical Properties for Al_2O_3/SiC Nanocomposites"; unpublished work.

7) E. Yasuda, Q. Bao and K. Niihara, "The Effects of Fine Particles on the Creep of MgO at High Temperatures," J. Ceram. Soc. Jpn., 100 [4] 514–19 (1992).

8) K. Niihara, T. Hirano, K. Izaki and F. Wakai, "High Temperature Creep Deforma tion of Si_3N_4/SiC Nanocomposite"; this proceedings.

9) T. Ohji, A. Nakahira, T. Hirano and K. Niihara, "Tensile Creep Behavior of Alumina/Silicon Carbide Nanocomposites"; unpublished work (submitted to J. Am. Ceram. Soc.).

10) T. Ohji and Y. Yamauchi, "Long–Term Tensile Creep Testing for Advanced Ceramics," J. Am. Ceram. Soc., 75 [8] 2304–307 (1992).

11) M.K. Ferber, M.G. Jenkins, and V.J. Tennery, "Comparison of Tension, Com pression, and Flexure Creep for Alumina and Silicon Nitride Ceramics," Ceram. Eng. Sci. Proc., 11 [7–8] 1028–45 (1990).

12) A.G. Robertson, D.S. Wilkinson, and C.H. Caceres, "Creep and Creep Fracture in Hot–Pressed Alumina," J. Am. Ceram. Soc., 74 [5] 915–21 (1991).

13) W. Blumenthal and A.G. Evans, "High Temperature Failure of Polycrystalline Alumina: II, Creep Crack Growth and Blunting," J. Am. Ceram. Soc., 67 [11] 751–59 (1984).

14) B.J. Dalgleish, E.B. Slamovich, and A.G. Evans, "Duality in the Creep Rupture of Polycrystalline Alumina," J. Am. Ceram. Soc., 68 [11] 575–81 (1985).

15) S.M. Wiederhorn, B.J. Hockey, and R.F. Krause, Jr., and A.K. Jakus, "Creep and Fracture of Vitreous–Bonded Aluminium Oxide," J. Mater. Sci., 21, 810–24 (1986).

16) A.K. Jakus, S.M. Wiederhorn, and B.J. Hockey, "Nucleation and Growth of Cracks in Vitreous–Bonded Aluminium Oxide at Elevated Temperatures," J. Am. Ceram. Soc., 69 [10] 725–31 (1986).

17) P.F. Becher, P. Angelini, W.H. Warwick, and T.N. Tiegs, "Elevated Tempera ture–Delayed Fracture of Alumina Reinforced with 20 vol% Silicon Carbide Whiskers," J. Am. Ceram. Soc., 73 [1] 91–96 (1990).

18) D.S. Wilkinson, C.H. Caceres and A.G. Robertson, "Damage and Fracture Mechanisms During High Temperature Creep in Hot–Pressed Alumina," J. Am. Ceram. Soc., 74 [5] 922–33 (1991).
19) R.M. Cannon, W.H. Rhodes, and A.H. Heuer, "Plastic Deformation of Fine–Grained Alumina (Al_2O_3): I, Interface–Controlled Diffusional Creep," J. Am. Ceram. Soc., 63 [1–2] 46–53 (1980).
20) S.M. Johnson, B.J. Dalgleish, and A.G. Evans, "High Temperature Failure of Polycrystalline Alumina: III, Failure Times," J. Am. Ceram. Soc., 67 [11] 759–63 (1984).
21) A.H. Chokshi and J.R. Porter, "Creep Deformation of an Alumina Matrix Com posite Reinforced with Silicon Carbide Whiskers," J. Am. Ceram. Soc., 68 [6] C144–45 (1985).
22) H–T. Lin and P.F. Becher, "Creep Behavior of a SiC–Whisker–Reinforced Alumina," J. Am. Ceram. Soc., 73 [5] 1378–81 (1990).
23) E. Arzt, M.F. Ashby, and R.A. Verrall, "Interface Controlled Diffusional Creep," Acta Metall., 31, 1977–89 (1983).
24) T. Ohji, S. Sakai, M. Ito, Y. Yamauchi, W. Kanematsu, and S. Ito, "Yielding Phenomena of Hot–Pressed Si_3N_4," High Temp. Tech., 5 [3] 139–44 (1987).
25) T. Ohji and Y. Yamauchi, "Tensile Creep and Creep Rupture Behavior of Mono lithic and SiC–Whisker–Reinforced Silicon Nitride Ceramics," J. Am. Ceram. Soc., 76 [12] 3105–12 (1993).
26) M.K. Ferber and M.G. Jenkins, "Evaluation of the Strength and Creep Fatigue Behavior of Hot–Isostatically Pressed Silicon Nitride," J. Am. Ceram. Soc., 75 [9] 2453–62 (1992).
27) M.K. Ferber, M.G. Jenkins, T.A. Nolan, and R.L. Yeckley, "Comparison of the Creep and Creep Rupture Performance of Two HIPed Silicon Nitride," J. Am. Ceram. Soc., 77 [3] 637–65 (1994).
28) R. Raj and M.F. Ashby, "Grain Boundary Sliding and Diffusional Creep," Metall. Trans., 2 [4] 1113–27 (1971).
29) T. Ohji, A. Nakahira, T. Hirano and K. Niihara, "Inhibition Mechanism of Creep in Alumina/Silicon Carbide Nanocomposites"; unpublished work.

HIGH TEMPERATURE CREEP BEHAVIOR OF Si_3N_4/SiC NANOCOMPOSITES

Takeshi Hirano, Atsushi Nakahira and Koichi Niihara
ISIR, Osaka University, 8-1 Mihogaoka, Ibaraki, Osaka 567, Japan
Kansei Izaki
Mitsubishi Gas Chemical Co., Tsukuba, Ibaragi 300-42, Japan
Fumihiro Wakai and Tatsuki Ohji
NIRIN, Nagoya, Aichi 462, Japan

ABSTRACT

Creep behavior has been investigated for Si_3N_4/30vol%SiC nanocomposites prepared by hot-pressing amorphous Si-C-N powder precursors. The starting powder for the nanocomposites was prepared by vapor-phase reactions. Creep tests were conducted in tension at 1300 to 1400℃. Although Si_3N_4/30vol%SiC nanocomposites were characterized by small grain size, their creep resistance were higher than those of monolithic Si_3N_4. Transmission electron microscopic observations revealed that many cavities were formed for monolithic Si_3N_4, but fewer and smaller cavities for Si_3N_4/30vol%SiC nanocomposites. The reasons for the superior creep resistance of Si_3N_4/30vol%SiC will be associated with microstructural characteristics.

I. INTRODUCTION

Since Si_3N_4 ceramics show high strength, toughness and thermal shock fracture resistance, it is recognized as a prime candidate for certain structural applications at elevated temperatures. To date, Si_3N_4 ceramics have been used, for example, as turbocharger components in some automobiles. Wider application, however, requires higher toughness, strength and creep resistance.

Si_3N_4 could be sintered with the use of additives such as Y_2O_3 or Al_2O_3. These additives react with SiO_2 normally present on Si_3N_4 powders to form a glassy phase that enhances liquid-phase sintering. The retained intergranular glassy phase inherent in this process governs the high-temperature creep behavior[1,2]. Recently, SiC particles and whiskers have been added to improve the mechanical properties of Si_3N_4[3-11]. This approach will give the potential not only increasing

To the extent authorized under the laws of the United States of America, all copyright interests in this publication are the property of The American Ceramic Society. Any duplication, reproduction, or republication of this publication or any part thereof, without the express written consent of The American Ceramic Society or fee paid to the Copyright Clearance Center, is prohibited.

the fracture toughness at room temperature, but also improving creep resistance at elevated temperatures. However, a limited number of creep studies have been performed on particles or whisker reinforced Si_3N_4. In the case of SiC particle, early studies suggested that SiC additions result in improved creep resistance[12,13], but in contrast more recent works have indicated that the presence of SiC particle dispersions could detrimentally affect to the creep behavior of hot-pressed Si_3N_4[14,15]. In the case of SiC whisker, a considerable improvement of creep resistance was reported for SiC whisker reinforced Si_3N_4 by Baldoni and Buljan[15]. There are some reports, however, that the addition of SiC whiskers to the Si_3N_4 sometimes result in little or negative effects on the creep performance[16,17].

Niihara and colleagues [18-23] have investigated Si_3N_4/SiC nanocomposites, in which the nano-sized SiC particles were dispersed within the matrix grains and/or at grain boundaries. These materials were made by hot-pressing amorphous Si-C-N precursor powders prepared by vapor-phase reactions. Though either Al_2O_3 and Y_2O_3 or Y_2O_3 were used as sintering additives, these nanocomposites exhibited the improved fracture strength and hardness at elevated temperatures [18,21-23]. These improved mechanical properties at elevated temperatures for the nanocomposites were attributed to the intergranular fine SiC particles[18,20-23].

The main purpose of this study is to investigate the creep behavior of Si_3N_4/SiC nanocomposites and to make clear the microstructure-creep properties relationship.

II. EXPERIMENTAL PROCEDURE

(1) Materials

The starting amorphous Si-C-N precursor powders for the Si_3N_4/SiC nanocomposites were prepared by vapor-phase reaction of $[Si(CH_3)_3]_2NH$- NH_3-N_2 system at 1000℃, followed by 4h heat-treatment in N_2 at 1350℃[18,19]. The C/N ratio in the amorphous powder was adjusted so that SiC content was about 30vol%. The starting powder for the monolithic Si_3N_4 was a commercially available α-Si_3N_4 powder made by Toyo Soda Manufacturing Co. Ltd., Tokyo, Japan. Some properties of starting powders are summarized in Table I. The starting powders were mixed with either 8wt%Y_2O_3 or 2wt%Al_2O_3+6wt%Y_2O_3 as sintering aids in a plastic bottle using Si_3N_4 balls and ethanol. The dried mixtures were hot-pressed at 1700-1800℃ for 1 to 4h in N_2 under 34MPa applied pressure. The hot-pressed samples were 60mm in diameter and 5mm in thickness. Some room-temperature mechanical properties are summarized in Table II. The specimens are labeled by code showing the composite or monolith (SN: monolithic, SCN: composites) and sintering additives (AY: 2wt%Al_2O_3+6wt%Y_2O_3, 8Y: 8wt%Y_2O_3). For instance, a Si_3N_4/ 30vol%SiC nanocomposite containing 8wt%Y_2O_3 is labeled as SCN8Y.

(2) Creep Testing and Microstructural Observations

Creep specimens were machined from the hot-pressed samples by #800 diamond wheel and diamond tip. The configuration of the creep specimen is shown in Fig.1.

Table I. Properties of Starting Powders

		Si-C-N	α-Si_3N_4 TS-7
Chemical composition	N	32.2wt%	
	O	<2wt%	1wt%
	C	9.7wt%	<0.1wt%
	Al	< 20ppm	<10ppm
	Ca	< 50ppm	<10ppm
	Fe	< 50ppm	50ppm
Crystal structure		amorphous	α-Si_3N_4 97%
Specific surface area (m^2/g)		18	12

Table II. Mechanical Properties of Monolithic Si_3N_4 and Si_3N_4/SiC Nanocomposites Used in the Present Work.

Specimen Name	Composition (wt%) SiC	Al_2O_3	Y_2O_3	Density (g/cm^3)	Vickers hardness (GPa)	Fracture toughness ($MPa \cdot m^{1/2}$)	Fracture strength at R.T. (MPa)
SNAY	0	2	6	3.24	15.0	6.8	998
SN8Y	0	0	8	3.27	15.0	6.0	927
SCNAY	30	2	6	3.27	17.8	6.5	1314
SCN8Y	30	0	8	3.31	17.8	6.1	1240

The specimen has a cylindrical gauge section with a diameter of 3.0mm and a length of 14mm and two target arms at the end of gauge section. The elongation of the distance between them was measured by an optical extensometer. The creep experiments were conducted in air using the tensile creep testing machine equipped with $MoSi_2$ heating elements. To assure thermal stability, equipment and specimen were heated at the test temperature for 20h before testing. During the holding time, 5.1MPa was applied by the jigs[24].

Microstructure was observed by scanning electron microscopy (SEM) and transmission electron microscopy (TEM). The microchemical analysis of intergranular phases was attempted using a energy dispersive X-ray (EDX) analyzer. Sections for SEM and TEM were taken from the gauge sections of tensile

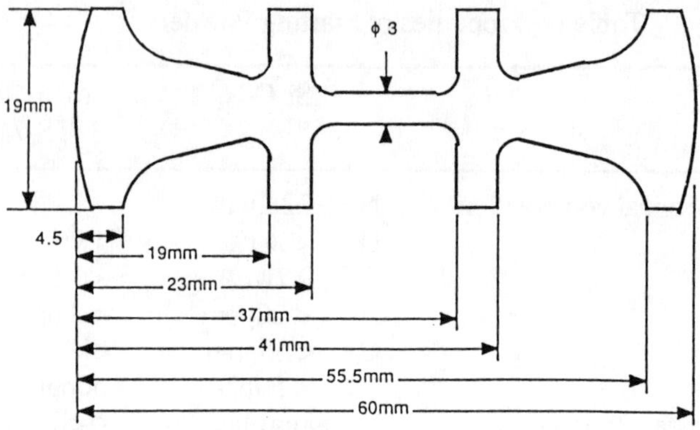

Fig. 1. Geometry and dimensions of creep specimen.

specimens after high-temperature creep tests and compared with those from as-hot-pressed samples. As intergranular phases strongly affect the creep behavior of Si_3N_4 and Si_3N_4 based composites, XRD analysis was also conducted for all specimens.

3. RESULTS AND DISCUSSION

(1) Microstructural Analyses of As-hot-pressed Samples

Fig.2 shows representative SEM pictures of as-hot-pressed specimens[24]. In the monolithic Si_3N_4, large elongated grains were frequently observed. The morphology of Si_3N_4 grains of SNAY and SN8Y was fibrous, with an aspect ratio of 3 - 8 and a diameter of 0.5 - 1.5μm, regardless of the sintering aids. In the Si_3N_4/SiC nanocomposites, the growth of elongated β-Si_3N_4 grains was suppressed by the presence of fine SiC particles and fine equiaxed grains and small rod-like grains were observed. There was little difference in microstructure between SCN8Y and SCNAY.

Fig. 3 shows the XRD profiles for the crushed as-hot-pressed specimens[24]. This XRD profiles show that 2wt%Al_2O_3+6wt%Y_2O_3 doped specimens (SNAY and SCNAY) have no clear crystalline grain boundary phases, and that 8wt%Y_2O_3 doped specimens (SN8Y and SCN8Y) contains crystalline phases such as $Y_5N(SiO_4)_3$, Y_2SiO_5 and $Y_2Si_2O_7$. Smith and Quackenbush [25] reported that the grain boundary phase of sintered Si_3N_4 with 2wt%Al_2O_3+6wt%Y_2O_3 was amorphous, while that of sintered Si_3N_4 with 6wt%Y_2O_3 included crystalline phases such as $Y_5N(SiO_4)_3$ and $Y_2Si_2O_7$. Their observation is in good agreement with the present analyses. As the complete crystallization of grain boundary phases are thermodynamically difficult[26], both SN8Y and SCN8Y may contain the glassy phases. However, due to crystallization of grain boundary phases, the amounts of glassy phases of SN8Y and SCN8Y are thought to be smaller than those of SNAY and SCNAY.

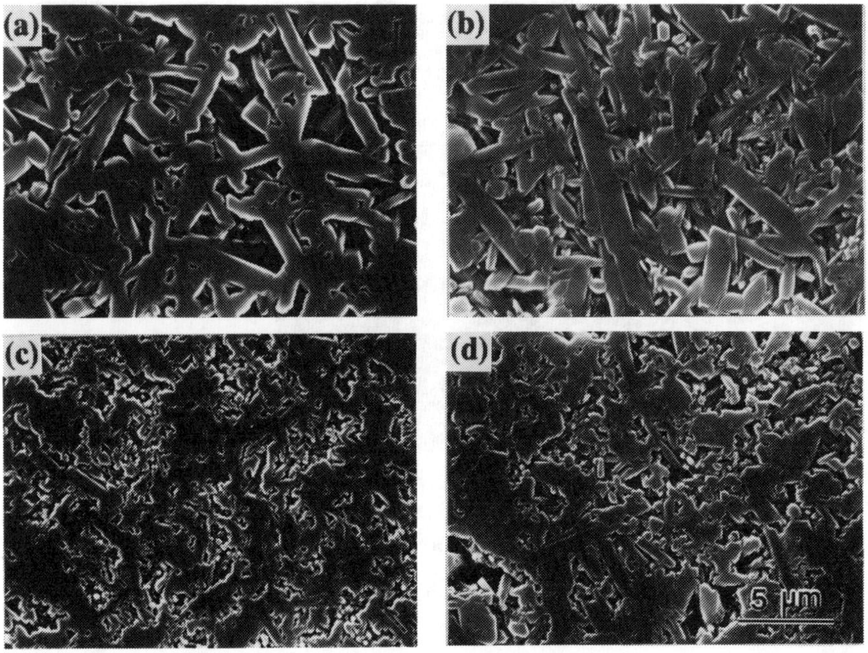

Fig. 2. Microstructure of monolithic Si_3N_4 and Si_3N_4/SiC nanocomposites used in the present study.
(a) SNAY, (b)SN8Y, (c) SCNAY and (d) SCN8Y

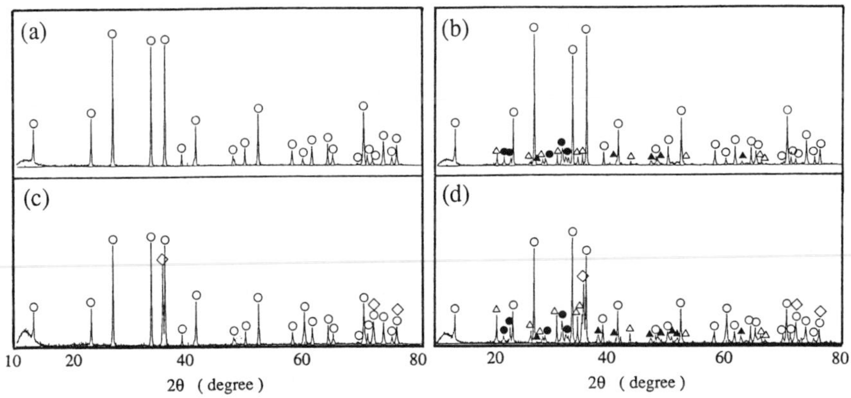

Fig. 3. X-ray diffraction profiles of as-hot-pressed samples used in the present study. (a) SNAY, (b) SN8Y, (c) SCNAY and (d) SCN8Y.
(○) β-Si_3N_4, (●) $Y_5N(SiO_4)_3$, (△) Y_2SiO_5, (▲) $Y_2Si_2O_7$, (◇) β-SiC.

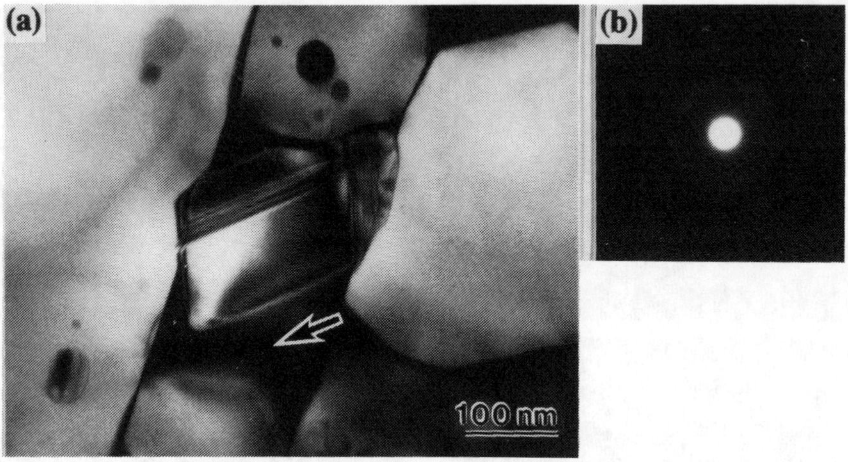

Fig.4. TEM micrograph of SCNAY
(a) Intergranular SiC particle plunging into β-Si_3N_4 grain .
(b) Microdiffraction pattern taken from the triple point (indicated by an arrow) of Fig.4 (a).

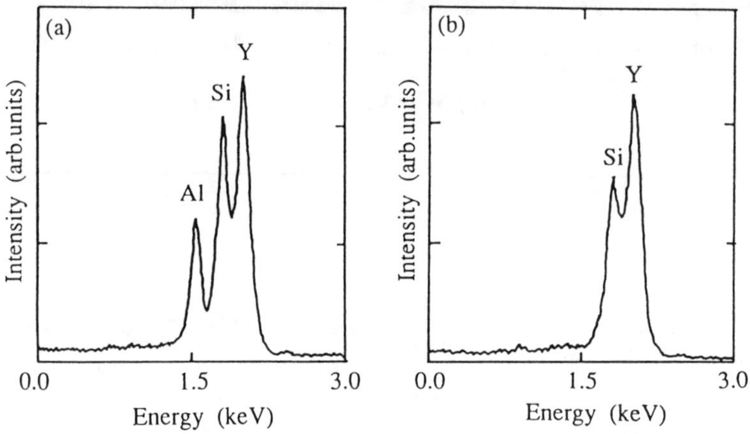

Fig. 5. Energy dispersive X-ray spectrums from the grain boundary phase at the triple points of SCNAY (a) and SCN8Y (b).

TEM observations for SCNAY and SCN8Y nanocomposites indicated that nano-sized SiC particles smaller than 100nm were mainly dispersed within Si_3N_4 grains and fine SiC particle, larger than approximately 100nm, were dispersed at the grain boundaries, and that some intergranular SiC particles are plunging into the β-Si_3N_4 grains as shown in Fig.4(a)[27]. Niihara et al. [20,22] reported that these intergranular SiC particles are directly connected with Si_3N_4 grains. The microdiffraction pattern taken from the grain boundary triple points of SCNAY indicated that the grain boundaries are amorphous as shown in Fig.4(b). The same results were also obtained from the SNAY. These results agree with those from XRD.

As the intergranular films were too narrow to be detected by EDX analysis, EDX spectra were taken only from the grain boundary phase at the triple points. The grain boundary triple point of SCNAY was found to contain Y, Si and Al (Fig.5(a))[27]. The similar result was also obtained from the SNAY. The EDX spectra from the pockets are thought to represent those from most intergranular films, since the pockets are linked with intergranular films without any barriers. Therefore, both triple points and intergranular amorphous films in SNAY and SCNAY are assigned to be the yttrium aluminum silicate glass. On the contrary, the EDX spectra from SCN8Y indicated that the grain boundary phase contain Y and Al (Figure 5(b))[27]. The same results were also obtained for the SN8Y. However, the chemical compositions of the triple points in SN8Y and SCN8Y may be different from those of intergranular films, since XRD profiles indicated that both SN8Y and SCN8Y contain crystalline grain boundary phases and such crystallization of grain boundary phases may affect the chemical composition of residual glass [28]. Kuroda et al. [29] reported that the amount of nitrogen in the intergranular films is much smaller than that of the grain boundary crystalline phase of $Y_5N(SiO_4)_3$ for Y_2O_3 doped Si_3N_4.

(2) Creep Behavior

A summary of creep tests is shown in Table III. The creep rate stated here is either the strain rate taken from an "apparently" straight region of creep curve or the minimum strain rate. During the holding time, the SNAY was deformed at low applied stress(5.1MPa from the weight of the jigs) and fractured when weights (equivalent to 56MPa) were added[27]. The SN8Y was fractured in 0.4h after 200MPa was applied, but did not show the deformation during the holding time. As both SNAY and SN8Y did not show the steady state creep, the creep resistance cannot be compared using the steady state creep rate. But as the SN8Y was not deformed at low applied stress during the holding time, the creep resistance of the SN8Y is thought to be superior to that of the SNAY[27]. Pasto et al. also reported the same results that the strain rate of Al_2O_3 and Y_2O_3 doped Si_3N_4 was faster than that of only Y_2O_3 doped Si_3N_4 by two orders of magnitude[30]. The difference in the creep resistance between SNAY and SN8Y may be caused by the different grain boundary structure and chemical composition as discussed in the previous section. Several researchers have shown that the dominant creep mechanism of Si_3N_4 is the grain boundary sliding due to the softening of amorphous phase[1,2]. It is reasonable that the SN8Y which includes the intergranular crystalline phases showed the higher creep resistance than the SNAY which has glassy grain boundaries.

In spite of the different applied stress, the tensile creep rate of SCN8Y is much slower than that of SCNAY as shown in Fig.6. This difference may come from the same reason discussed for SNAY and SN8Y, because the grain boundary

Table III. Summary of Tensile Creep

Specimen	Test Temperature (°C)	Applied Sresss (MPa)	Strain Rate (1/s)	Life Time (h)	Total Strain (%)	Note
SNAY	1400	56	-----	0	0.78	1
SN8Y	1400	200	-----	0.4	0.26	1
SCNAY	1400	100	6.7×10^{-8}	74	2.48	2
SCN8Y-a	1400	200	4.7×10^{-9}	500	1.21	3
SCN8Y-b	1400	200	4.7×10^{-9}	1000	1.70	3
SCN8Y-c	1300	200	5.1×10^{-10}	500	0.11	3,4
	1350	200	1.1×10^{-10}	500	0.25	

Notes: 1. Stage III creep only
2. Fast transition to stage III
3. Test interrupted before failure
4. Temperature cycled

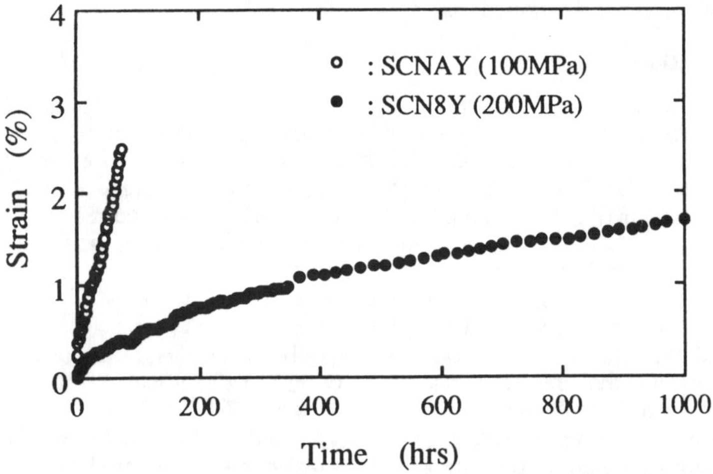

Fig.6. Time-strain curves of SCNAY and SCN8Y tested at 1400℃.

phases and chemical composition of SCNAY and SCN8Y are similar to those of SNAY and SN8Y, respectively, as described in the previous section. Furthermore, the the creep resistance of SCN8Y is higher than other 8wt%Y_2O_3 doped hot-pressed Si_3N_4 such as Ceradyne HPSN or NCX-34[24,31] which are thought to have similar grain boundary phases to SCN8Y. The creep rate of NCX-34 is, for

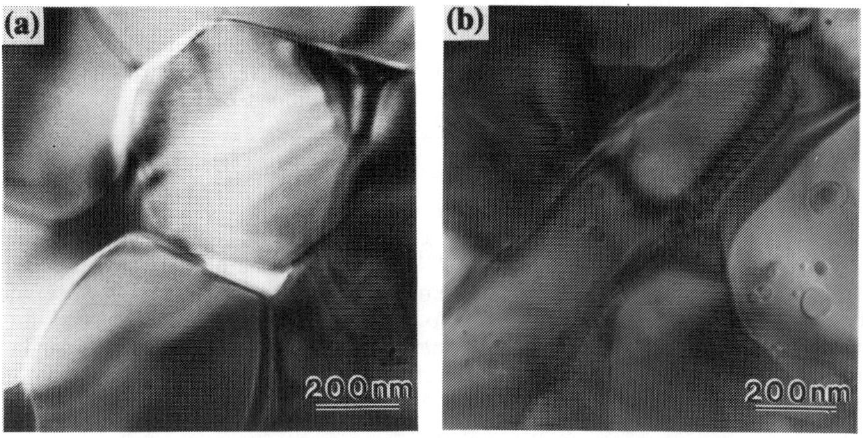

Fig.7. Microstructural observations of crept specimens.
(a) Example of facet-sized cavities in SNAY.
(b) Nucleation of grain boundary crack introduced by grain boundary sliding in SCNAY.

example, reported to be approximately 3 x 10^{-7}/s under 200MPa at 1350℃ [31], which is faster than that of SCN8Y (1.1 x 10^{-10}/s under 200MPa at 1350℃) by three orders of magnitude. Although an accurate comparison of the creep resistance is difficult because of the difference in size of the specimen and the method, the creep rate of SCN8Y in this study seems to be obviously lower than those of 8wt% Y_2O_3 doped hot-pressed Si_3N_4. As the creep rate is strongly affected by the grain size, it is not rationalized only by the intergranular crystalline phase that the SCN8Y showed higher creep resistance than other 8wt%Y_2O_3 doped hot-pressed Si_3N_4 which have larger matrix grains than SCN8Y. The grain boundary sliding is the most predominant deformation mechanism of Si_3N_4 at high temperature. Therefore, the cause of low creep rate of SCN8Y may be attributed to the inhibition of grain boundary sliding. One of the possible mechanism that contributes to inhibiting the grain boundary sliding may involve intergranular SiC particles which bond directly in the atomic scale with Si_3N_4 grains[24,27]. However, more detail microstructural observation is needed to clarify the effects of intergranular SiC particles.

TEM observations revealed that large cavities were observed in the SNAY in spite of lower applied stress and shorter creep duration. Representative cavities are shown in Fig.7(a). Similar cavities were also found in SN8Y and SCNAY, however, the size and the population of cavities are smaller than those in SNAY. In the SCNAY the nucleation of grain boundary crack introduced by grain boundary sliding was also found as shown in Fig.7(b). These microstructural observation indicated that in SCNAY cavity formation is slower than in SNAY. Theoretical consideration have indicated that the presence of large grains can induce stress concentrations, which enhances cavity nucleation at grain boundaries[17,32]. Therefore, the cause of smaller and fewer cavities in SCNAY may partly be attributed to the small grain size. In SCN8Y, however, the size and population of cavities were significantly smaller than SCNAY. This result also indicated that grain boundary sliding in the SCN8Y was strongly suppressed.

IV. SUMMARY

High temperature creep behavior of hot-pressed Si_3N_4/30vol%SiC nanocomposites were investigated. Although Si_3N_4/30vol%SiC containing 8wt%Y_2O_3 was characterized by small grain size, the remarkable creep resistance was observed. The cause of low creep rate may be due to the inhibition of grain boundary sliding. One of the possible mechanism that contributes to inhibiting the grain boundary sliding may involve intergranular SiC particles which bond directly in the atomic scale with Si_3N_4.

Acknowledgement

The authors gratefully acknowledge Mr.T.Sekino at Osaka University for his discussion and help. A part of this work was supported by the Japan Ministry of Education under a Grant-in-Aid for Science Research No. 03555157 and 04205099.

References

1. F.F.Lange, B.I.Davis, and D.R.Clarke, "Compressive Creep of Si_3N_4/MgO Alloys," J.Mater. Sci., 15,601-18 (1980).
2 T.Chartier, J.-L.Besson, and P.Goursat, "Microstructure, Oxidation and Creep Behavior of a β'-Sialon Ceramics," Int. J.High Tech.Ceram., 2 [1] 33-45 (1986).
3. F.F.Lange, "Effect of Microstructure on Strength of Si_3N_4-SiC Composite System," J.Am.Ceram.Soc.,56 [9] 445-450 (1973)
4. K.Ueno and S.Sodeoka, "Fracture Toughness of SiC Whisker Reinforced Si_3N_4 Ceramics," Yogyo-Kyokai-Shi, 94 [9] 981-985(1986)
5. P.D.Shalek, J.J.Petrovic, G.F.Hurley and F.D.Gac,"Hot-Pressed SiC Whisker Si_3N_4 Matrix Composites," Am.Ceram. Soc.Bull.,65 [2] 351-356 (1986)
6. P.Greil, P.Petzow and H.Tanaka, "Sintering and HIPing of Silicon Nitride-Silicon Carbide Composites," Ceramics International, 13 19-25 (1987)
7. S.T.Buljan, J.G.Baldoni and M.L.Huckabee, "Si_3N_4-SiC Composites," Am.Ceram.Soc.Bull.,66 [2] 347-352 (1987)
8. H.Kodama, T.Suzuki, H.Sakamoto and T.Miyoshi, "Toughening of Silicon Nitride Matrix Composites by the Addition of Both Silicon Carbide Whiskers and Silicon Carbide Particles," J.Am.Ceram.Soc,73 [3] 678-683 (1990)
9. G.Sasaki, H.Nakase, K.Suganuma, T.Fujita and K.Niihara, "Mechanical Properties and Microstructure of Si_3N_4 Matrix Composite with Nano-Meter Scale SiC Particles," J.Ceram.Soc. Japan, 100 [4] 536-540 (1992)
10. Y.Akimune, T.Ogasawara and N.Hirosaki, "Influence of Starting Powder Characterization on Mechanical Properties of SiC Particle/Si_3N_4 Composites,"

J.Ceram.Soc. Japan, 100 [4] 463-467 (1992)
11. Y.Akimune, N. Hirosaki and T.Ogasawara, "Mechanical Properties and Microstructure in Sintered and HIPed SiC Particle/Si_3N_4 Composites," J.Mater.Sci.,27, 6017-6021(1992)
12. N.L.Parr, G.F.Martin and E.R.W.May, "Preparation , Microstructure, and Mechanical Properties of Silicon Nitride," in Special Ceramics, vol. 1, p.102. Br. Ceram. Res. Assoc, Stoke-on-Trent (1960)
13. N.L.Parr, G.F.Martin and E.R.W.May, "A Simple Creep Machine for Operation at High Temperature," in Special Ceramics, vol. 1, p.76. Br. Ceram. Res. Assoc, Stoke-on-Trent (1960)
14. J.M.Birch and Wilshire, "The compression creep behavior of silicon nitride ceramics" J. Mater. Sci. 13, 2627 (1978)
15. J.G.Baldoni and S.T.Buljanl, "Creep and Clack Growth Resistance of Silicon nitride Composites, " ;pp.786-795 in Ceramic Materials and Components for Engines. Edited by V.J.Ternnery. American Ceramic Society, Columbus, OH,1989
16. B.J.Hockey, S.M.Widerhorn, W.Liu J.G.Baldoni and S.-T.Buljan, "Tensile Creep of Whisker-Reinforced Silicon Nitride," J.Mater. Sci.,26,3931-3939(1991).
17.T.Ohji and Y.Yamauchi, "Tensile Creep and Creep Rupture Behavior of Monolithic and SiC-Whisker-Reinforced Silicon Nitride Ceramics," J.Am. Ceram. Soc, 76 [12] 3105-3112 (1993).
18. K.Niihara, T.Hirano, A.Nakahira, K.Ojima, K.Izaki and T.Kawakami, "High-Temperature Performance of Si_3N_4-SiC Composites from Fine, Amorphous Si-C-N Powder," pp.107 in proceedings of MRS International Meeting on Advanced Materials, vol.5 (Materials Research Society, Pittsburgh), 1989.
19. K.Niihara, T.Hirano, A.Nakahira, K.Suganuma, K.Izaki and T.Kawakami, "Nanostructure and Thermomechanical Properties of Si_3N_4/SiC Composites Fabricated from Si-C-N Precursor Powders," J.Jpn.Soc.Pow.& Pow.Mett.,36 [2] 243-246 (1989).
20. K.Niihara, K.Suganuma, A.Nakahira and K.Izaki, "Interfaces in Si_3N_4-SiC Nano-Composite," J.Mater.Sci.Lett.,9,598-599 (1990).
21. K.Niihara, K.Izaki and T.Kawakami, "Hot-Pressed Si_3N_4-32% SiC Nanocomposite from Amorphous Si-C-N Powder with Improved Strength above 1200℃," J.Mater.Sci.Lett.,10,112-114 (1990).
22. K.Niihara, T.Hirano, A.Nakahira and K.Izaki, "The Correlation between Interface Structure and Mechanical Properties for Silicon Nitride Based Nanocomposites,"; pp.103-111 in Grain Boundary Controlled Properties of Fine Ceramics. Edited by K.Ishizaki, K.Niihara, M.Isotani and R.G.Ford. Elservier Applied Science, London and New York, 1992.
23. K.Niihara, A.Nakahira, T.Hirano, K.Izaki and F.Wakai "High Temperature Deformation of Si_3N_4 Matrix Nanocomposites," pp.42-46 in proceedings of 11th. Symposium on High-Temperature Materials,Hakone(Japan),1993.
24. K.Niihara, T.HIrano, K.Izaki and F.Wakai, "High Temperature Creep/ Deformation of Si_3N_4/SiC Nanocomposites," Proceeding of 1993 PAC RIM meeting of the American Ceramic Society, in print
25. J.T.Smith and C.L.Quackenbush, "Phase Effects in Si_3N_4 Containing Y_2O_3

or CeO_2: I Strength," Am.Ceram. Soc. Bull., 59 [5] 529-537 (1980).
26. R.Raj and F.F.Lange, "Crystallization of small quantities of glass (or a liquid) segregated in grain boundaries," Acta. Metall., 29, 1993-2000 (1981).
27. K.Niihara, T.HIrano, K.Izaki and F.Wakai, "Effects of Grain Boundary Structure on Creep Behavior for Si_3N_4/SiC Nanocomposites," Proceeding of International Workshop on Fine Ceramics'94 , in print
28. M.M.Chadwick and D.S.Wilkinson, "Microstructural Evolution in Annealed and Crept Silicon Nitride," J.Am.Ceram.Soc., 76[2], 376-384 (1993).
29. K.Kuroda, T.Kamino and T.Imura, "Microstructural and Microchemical Characterization of Intergranular Phase in Yttria-Doped Hot-Pressed Silicon Nitride," J.Electron Microsc.,vol.35[4], 378-383 (1986).
30. A.E.Pasto, W.C.VanSchalkwyk and F.M.Mahoney, "Creep Behavior of Yttria and Alumina-Doped Silicon Nitride," ;pp.776-785 in Ceramic Materials and Components for Engines. Edited by V.J.Ternnery. American Ceramic Society, Columbus, OH,1989.
31. D.C.Larsen, J.W.Adams, L.R.Johnson, A.P.S.Teotia, and L.G. Hill, Ceramic Materials for Advanced Heat Engines; Ch.8.Noyes Publications, Park Ridge, NJ.1985.
32. C.H.Hsueh and A.G.Evans, "Creep Fracture in Ceramic Polycrystals II. Effects of Inhomogeneity on Creep Rupture," Acta Metall., 29, 1907-1917 (1981)

Layered Composites and Joining Methods

MIMICKING NATURE'S CERAMIC/CERAMIC LAMINAR COMPOSITES BY ELECTROPHORETIC DEPOSITION

P.S. Nicholson, P. Sarkar, O. Prakash and G. Wang
Ceramic Research Group, Department of Materials Science and Engineering, McMaster University, Hamilton, Ontario, Canada.

ABSTRACT
3m/o Yttria-stabilised zirconia (TZ3Y + Al_2O_3 plates)/Al_2O_3 laminar microcomposites were fabricated by electrophoretic deposition (EPD). To enhance the mechanical properties, the individual layers were deposited with controlled thickness and sequence. The microstructural features introduced were alumina platelets in thick (>30μm) TZ3Y layers with multiple thin (5μm) alumina layers between. The fracture behaviour of and crack paths in, these materials, were evaluated using indentation at room temperature and four-point bend tests at 1300°C.

I. INTRODUCTION

The search for tough structural ceramics has led to the development of laminar ceramic composites. These composites are presently synthesised by tape casting [1,2], slip casting [3], centrifugal casting [4] and dough rolling [5]. Recently the present authors [6-8] have used an EPD technique to synthesise laminar composites. The room temperature strength [2] and fracture toughness [4,5] are significantly improved compared to ceramic monoliths. One important factor controlling the performance of a ceramic laminate is the laminate thickness because it controls the length of crack propagation before interception of an interface. In addition, the layer microstructures can be designed so as to make cracks deflect to further enhance the toughness of the composites.

The present paper demonstrates the synthesis of laminar microcomposites with designed layer thickness and designed layer microstructure.

II. EPD OF CERAMIC POWDERS

EPD is two processes. The first process (electrophoresis) is the motion of particles in a stable suspension under the influence of an electrical potential gradient. The second process is deposition of these particles in a dense layer. To achieve a dense, uniform and homogeneous microstructure, the EPD suspension must be colloidally

To the extent authorized under the laws of the United States of America, all copyright interests in this publication are the property of The American Ceramic Society. Any duplication, reproduction, or republication of this publication or any part thereof, without the express written consent of The American Ceramic Society or fee paid to the Copyright Clearance Center, is prohibited.

stable. The stability of a colloid suspension, according to Derjaguin-Landau-Verwey-Overbeak (DLVO) theory is determined by the balance between the repulsive (double layer repulsion) and attractive forces (van der Waals attraction) which the particles experience as they approach each other. Van der Waals attraction always exceeds the double layer repulsion at small distances since it is a power law interaction ($W \sim 1/D^n$, where W is the interaction potential energy, D is the distance between two particles, and n is a power-law exponent ≥ 1) whereas double layer interaction energy rises much more slowly as D approaches 0. Figure 1 shows schematically the various types of interaction potentials that can occur between the colloidal particles. The van der Waals interaction potential is insensitive to electrolyte concentration and pH of the suspension whereas double layer interaction is extremely sensitive to this factor. Therefore, the total interaction energy curve can be manipulated by controlling the pH/concentration of electrolyte in the suspension.

Highly charged surfaces in a dilute electrolyte have long range repulsion that peaks at some distance thus providing an energy barrier against coagulation. There is a secondary minimum which increases in depth with increasing electrolyte concentration. The potential energy minimum at particle contact is known as the primary minimum. For a colloid suspension, the thermodynamic equilibrium state is the situation when two particles are in contact (ie., deep primary minimum). The energy barrier ahead of that condition may be too high for particles to overcome during reasonable time therefore, a dispersed suspension can be considered "kinetically" stable as opposed to "thermodynamically" stable.

For the present systems, particles are positively charged and move towards the cathode under an applied DC voltage. Figure 2(a)-(c) schematically illustrate the sequence of deposition. Figure 2(a) shows a suspension of well-dispersed positively charged particles. When a DC field is applied between the two immersed electrodes (Figure 2(b)), particles migrate towards the cathode, the electric double layer around the moving particles polarising and "tailing" towards the anode. If one considers a monolayer of particles deposited on the cathode (Figure 2(b)) on which incoming particles can attach if the applied potential is high enough to overcome the particle-particle repulsive energy barrier then deposition is controlled by particle-particle interaction (Figure 2(c)). Therefore the electrode reaction is not necessary for the deposition process. In fact a dialysis membrane inserted to isolate the electrode from the suspension but not the solvent, collects the deposit (Figure 3) and ions pass through to complete the circuit.

III. EXPERIMENTAL PROCEDURE

The following materials were prepared using the EPD technique.

(i) Laminar composites with alternating layers of Al_2O_3 and TZ3Y (95 v/o Al_2O_3 and 5 v/o TZ3Y).

(ii) Thick (>30μm)layers of TZ3Y containing randomly dispersed alumina platelets with bundles of thin (<5μm) Al_2O_3 layers interspersed therein. The overall composition is 95 v/o TZ3Y and 5 v/o Al_2O_3.

To fabricate the laminar composites, the two suspensions (containing Al_2O_3 and TZ3Y/Al_2O_3 respectively) must have the same ζ-potential sign otherwise they cannot be deposited sequentially.

The ζ-potential of Al_2O_3 and TZ3Y in absolute alcohol as a function of pH was measured using the laser-Doppler technique and is shown in Figure 4. It shows that EPD can be performed in the pH range 3 to 5 where the ζ-potential is maximum. In the present study, deposition was performed at pH ~3.5.

The green density of the materials synthesised was 60% theoretical. The deposits were fired at 1550°C for 6hrs and their microstructure characterized by optical and electron microscopy. Mechanical properties were evaluated via Vickers hardness indentation and four-point bend testing.

IV. RESULTS AND DISCUSSION

Figures 5-6 show the microstructures of the two materials studied. Figure 5 is a >100 layer laminated composite of ZrO_2 (bright)/ Al_2O_3 (dark). TZ3Y layers are ~6μm thick and have a grain size ~0.5μm. The grain size in alumina layers is ~3μm. The interface uniformity is in the submicron range. Figure 6 shows a sample in which the Al_2O_3 platelets are dispersed randomly in TZ3Y layers and thin Al_2O_3 layers of a finer grain-size (~1.5μm compared to ~3μm in Figure 5) are interspersed.

The fracture behaviour of these materials was evaluated using indentation and four point bend tests. Indentations were made by a Vickers indenter under a load of 10 kgf.

Radial cracks formed at the indentation corners. The extent of propagation of the individual cracks depends on the position and orientation of the indents with respect to the layer interfaces. For example, a crack directed into a tough layer extends for a shorter distance than the cracks originating at the other corners of the indent. No delamination was observed at any interface, confirming their perfection.

Crack-interface interactions provide insight into the induced stress state in the individual layers. A crack approaching an interface orthogonally will travel unimpeded across the interface. However, at ≠90° approach-angles, the crack will tend to move towards or away from the interface as it approaches a compliant or stiff layer respectively (Figure 7(a)). This behaviour is explained as follows. The stress concentration at a crack tip approaching an interface induces a stress at the interface. To enable strain continuity across the interface, the stiffer layer has to strain more than it would if unconstrained and the compliant layer less so. This results in tension in the stiffer layer and compression in the compliant one (see Figure 7(b)). The preferred direction of growth of a crack approaching such a stress field can be predicted [9-12]. Figure 7(c) schematically shows that, in the compliant phase compressive field, the crack will propagate parallel to the compression direction. In contrast,it will propagate transverse to the tensile field in the stiff phase.

Accordingly, in the Al_2O_3/TZ3Y system (see Figure 7(a)), the crack (propagating from left to right direction) will turn towards the interface (i.e. becomes transverse to the induced tensile stress in the Al_2O_3 (dark) layer) as it approaches the TZ3Y layer, and becomes more horizontal as it approaches the stiffer alumina layer on the other side of the TZ3Y layer (i.e. the crack aligns with the compressive stress induced in the TZ3Y layer). This protocol is clearly illustrated in Figure 8 for an indent crack propagating through a ZrO_2/porous-La_2O_3.11Al_2O_3/dense-Al_2O_3-ZrO_2-La_2O_3.11Al_2O_3/porous La_2O_3.11Al_2O_3 composite. The crack ran from the top right hand corner in ZrO_2. On entering the porous La_2O_3.11Al_2O_3 (compliant) it experiences compression as it approaches the next interface (stiff,Al_2O_3/ZrO_2/La_2O_3.11Al_2O_3) and turns to run parallel to it. It is deflected into the stiff layer there to suffer deflection due to the inhomogeniety of the layer. The tensile stress before the next interface turns the crack towards the orthogonal. It arrests in the next porous La_2O_3.11Al_2O_3 and reinitiates at a new location in response to the tension in the final ZrO_2 layer. It finally arrests in this layer.

To guarantee off-normal crack approach to an interface, Al_2O_3 platelet inclusions were co-deposited in the TZ3Y layers (Figure 9). Crack bridging was observed near the platelets in the micrographs. Crack deflection is significant under the influence of multiple alumina layers dispersed in Al_2O_3-platelet-TZ3Y layers (Figure 10). Note, the crack advances in a meandering manner, fluctuating between the layer interior and its interfaces. This occurs because, for a thin alumina layer, the two interfacial regions with induced tension are in close proximity and are bordered by the induced compression in the two neighbouring ZrO_2 layers.

Four-point bend tests were conducted on the two sample types and monolithic TZ3Y at 1300°C. Typical load/deflection curves are shown in Figure 11. The elastic moduli of the two composite materials are different due to the different compositions. The maximum bending stresses and "work-to-fracture" are listed in Table 1. The laminated composites both have superior strength. The "work-to-fracture" (taken as the curve area) of the ZrO_2-Al_2O_3-platelet/Al_2O_3-thin-layers is 3-4.5 times higher than the Al_2O_3-ZrO_2 composite indicating superior toughness.

The superiority is attributed to the randomly-oriented, closely-spaced Al_2O_3 platelets in the ZrO_2, and the Al_2O_3 layers that offer an easy path for the crack (Figures 10,12), thereby improving the overall toughness. Graceful failure/delamination/chipping of these materials was observed in the bending failure. Delamination occurs along the thin Al_2O_3 layers (Figure 12).

V. CONCLUSIONS

It is clear from the above results that layered ceramic materials can have strengths and toughnesses superior to monolithic ceramics. These properties can be further enhanced by suitable design of the composite microstructure such as controlling layer thickness, exploiting induced elastic stresses and introducing platelets to cause crack deflection. Precise control of the scale of the microstructure and orientation of platelets is possible by EPD. The latter technique can be used to design specific ceramic laminate materials for specific applications.

REFERENCES:

1. P. Boch, T. Chartier and M. Huttepain, "Tape Casting of Al_2O_3/ZrO_2 Laminated composites", J. Am. Ceram. Soc., **69**, C-191 (1986).
2. T. Chartiful, J.L. Besson and P. Boch, "Mechanical Properties of ZrO_2-Al_2O_3 Laminates Composites", in Advances in Ceramics, Vol **24B**, 'Science and Technology of Zirconia III'. Edited by S. Somiya, N. Yamamoto, and H. Yanagida, American Ceramic Society Inc., 1131 (1988).
3. J. Requena R. Moreno and J.S. Moya, "Alumina and Zirconia/Alumina Multilayer Composites Obtained by Slip Casting", J. Am. Ceram. Soc., **72**, 1511 (1989).
4. D.B. Marshall, J.J. Ratto and F.F. Lange, "Enhanced Fracture Toughness in Layered Microcomposites of Ce-ZO_2 and Al_2O_3, J. Am. Ceram. Soc., **74**, 2979 (1991).
5. W.J. Clegg K. Kendall, N. McN. Alford, T.W. Botton and J.D. Brichall, "Simple Way to Make Tough Ceramics", Nature, **347**, 455 (1990).
6. P. Sarkar, X. Huang and P.S. Nicholson, "Stractural Ceramic Microlaminates by Electrophoretic Deposition", J. Am. Ceram. Soc. **75** 2907 (1972).
7. P. Sarkar,X. Huang and P.S. Nicholson, "Electrophoretic Deposition and Its Use To Synthesize YSZ/Al_2O_3 Micro-laminate Ceramic Composites", Ceramic Eng. and Sci. Proceeding **14** 707 (1993).
8. P.S. Nicholson, P. Sarkar and X. Huang, "Electrophoretic Deposition and Its Use To Synthesize ZrO_2/Al_2O_3 Micro-laminate Ceramic Composites", J. Mat. Sci. **28** 6274 (1993).
9. M-Y He and J.W. Hutchinson, "Crack Deflection At An Interface Between Dissimilar Elastic Materials", Int. J. Solid Structures **25** 1053 (1989).
10. C. Atkinson, "Stress Singularities and Fracture Mechanics", Appl. Mech. Rev. **32** 123 (1979).
11. P.D. Ewing, J.L. Swedlow and J.G. Williams, "Further Results on the Angle Crack Problem", Int. J. Fract. **12** 85 (1976).
12. J.L. Swedlow, Criteria for growth of the angle crack. In Cracks and Fractures, ASTM-STP 601, 1976.

TABLE 1 : 1300°C Mechanical Property Data for the Al_2O_3/TZ3Y Laminar Composites and Monolithic TZ3Y.

	STRENGTH (MPa)	AREA UNDER CURVE (kJ/m^2)
TEMPERATURE	1300°C	1300°C
Al_2O_3/TZ3Y	272	10
ZrO_2-Al_2O_3 platelet/ Al_2O_3-thin-layers	265	32
TZ3Y	120	7

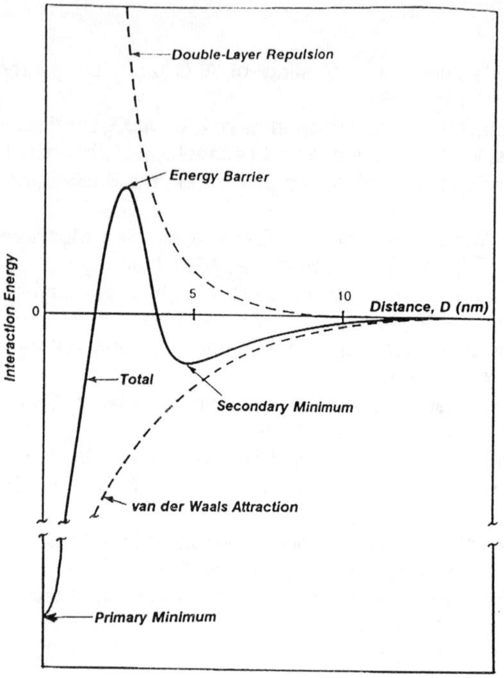

Figure 1. Schematic of energy versus distance profiles of DLVO interaction between colloid particles in suspension.

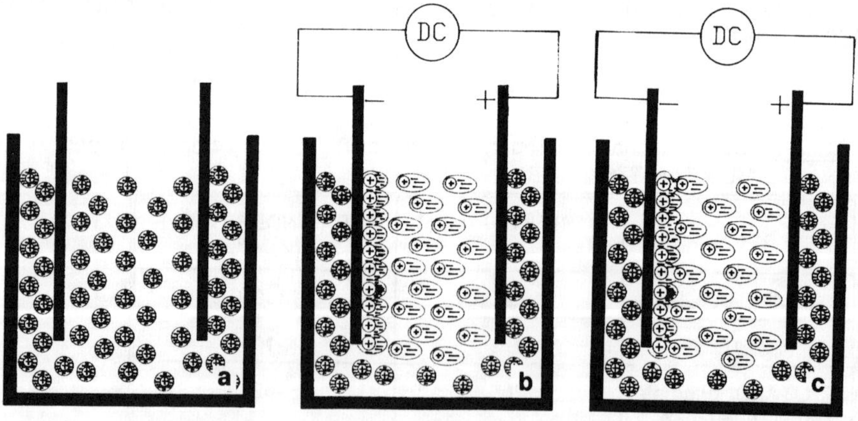

Figure 2. Schematic of electrophoretic deposition sequence: (a) well dispersed suspension of positively charged particles; (b) applied DC field deposits a monolayer of particles on the cathode surface and (c) incoming particles coagulate on the monolayer.

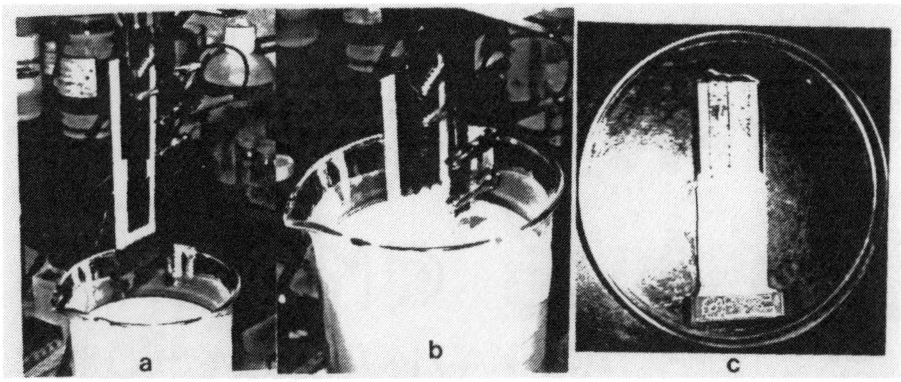

Figure 3. Electrophoretic deposition of Al_2O_3 on a dialysis membrane: (a) Graphite electrode surrounded by dialysis membrane; (b) EPD onto the dialysis membrane and (c) deposited Al_2O_3 on the dialysis membrane.

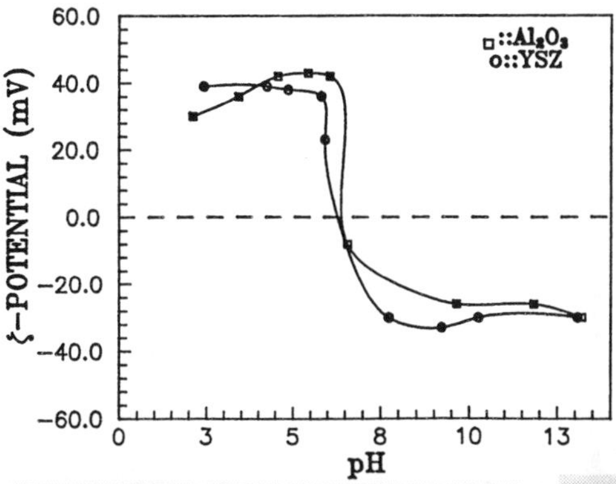

Figure 4. ζ-potential .vs. pH of Al_2O_3 and YSZ in absolute alcohol suspension.

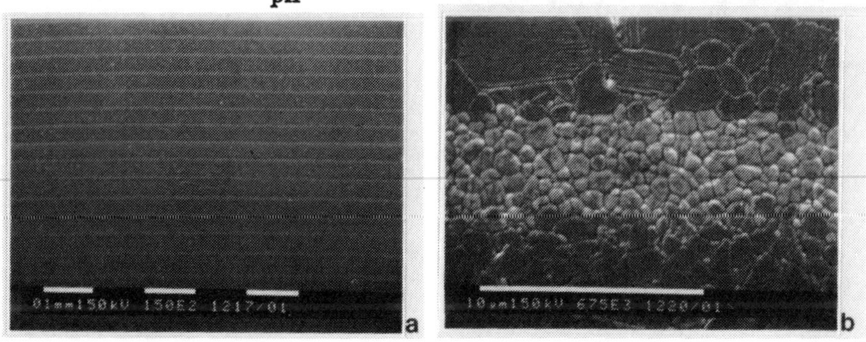

Figure 5. Micrograph of Al_2O_3 (grey)/YSZ (white) laminate (type A): (a) low magnification and (b) high magnification.

Figure 6. Micrograph of the $ZrO_2+Al_2O_3$ platelet/thin Al_2O_3 laminate showing the randomly dispersed Al_2O_3 platelets in the YSZ.

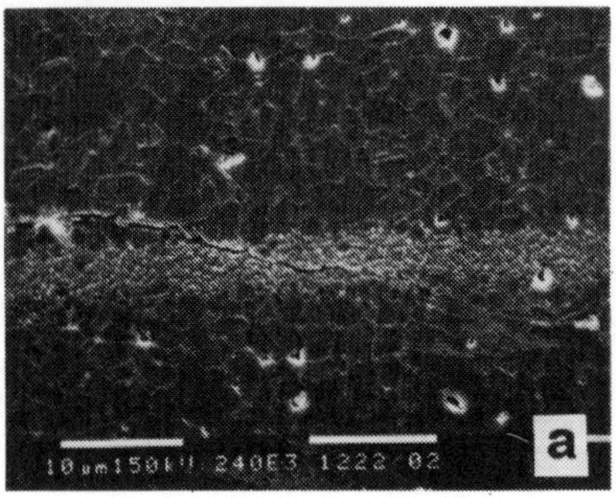

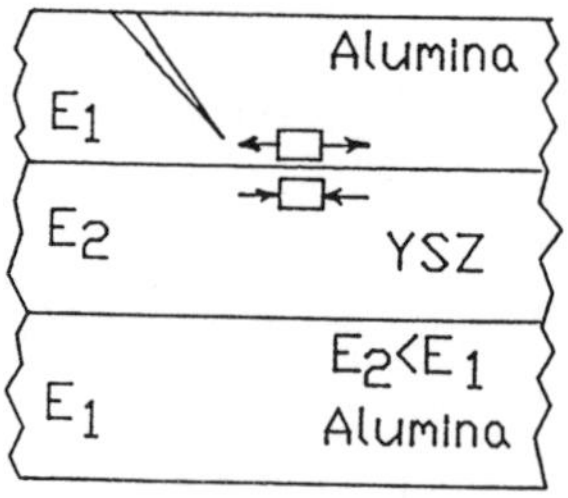

b

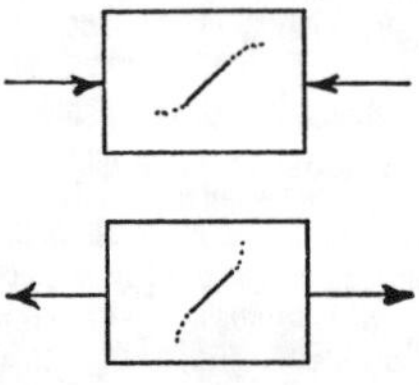

c

Figure 7. (a) Indentation crack in a Al_2O_3/YPSZ sample showing the change in crack growth direction as it approaches the interface; (b) residual stresses resulting from difference in elastic moduli and (c) growth direction of inclined cracks in compressive and tensile stress fields (E ≡ elastic modulus).

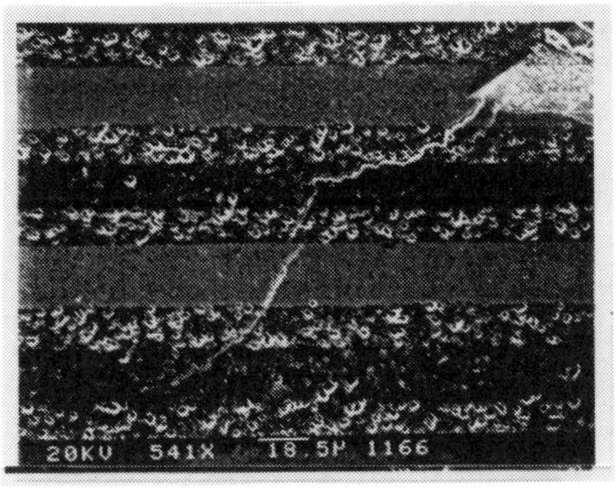

Figure 8. Indent crack propagation in a ZrO_2/porous-La_2O_3.11Al_2O_3/dense-Al_2O_3-ZrO_2-La_2O_3.11Al_2O_3 composite.

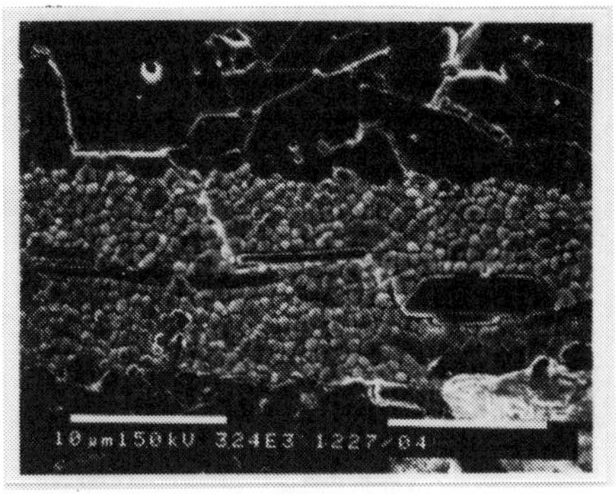

Figure 9. Indentation crack path showing crack deflection due to alumina platelets.

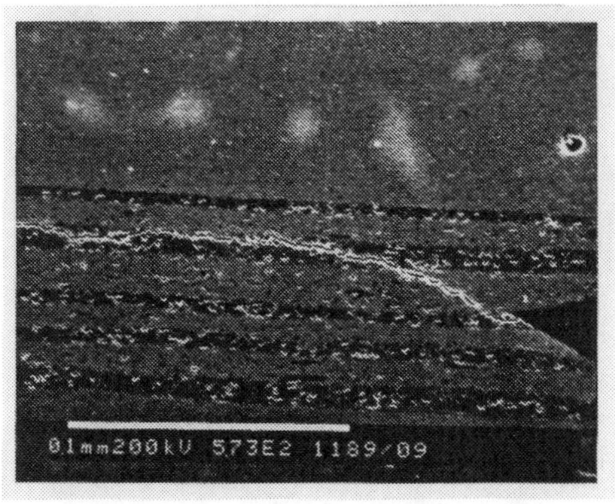

Figure 10. Indentation crack path in a ZrO_2-Al_2O_3 platelet/thin Al_2O_3 composite sample. The crack is attracted to an Al_2O_3 layer. Inside the Al_2O_3 layer, the crack switches between the two interfacial regions which are under residual tension.

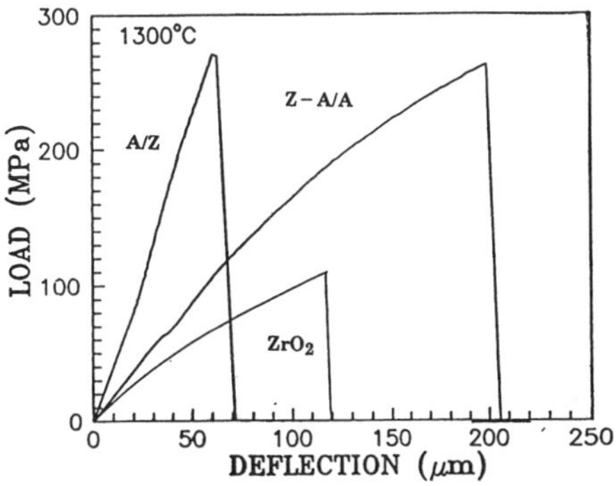

Figure 11. Load .vs. deflection for Al_2O_3/ZrO_2, ZrO_2-Al_2O_3 platelet/multiple-thin-Al_2O_3-layers and monolithic TZ3Y samples at 1300°C.

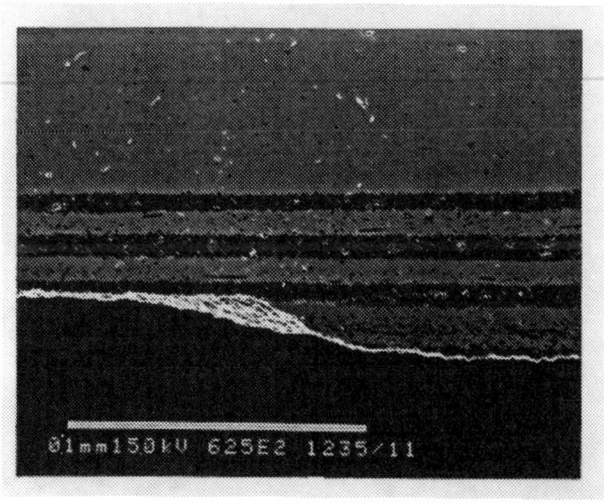

Figure 12. Fracture surface of ZrO_2-Al_2O_3 platelet/thin-Al_2O_3 composite sample tested in 4-point flexure, showing crack stepping and material chipping.

RESIDUAL STRESS AND CRACK PROPAGATION IN LAMINATED COMPOSITES

R-M.F. Yttergren, K. Zeng and D.J. Rowcliffe
Department of Materials Science and Engineering,
Royal Institute of Technology,
100 44 Stockholm, Sweden.

ABSTRACT

Residual stress distributions in several laminated ceramic composites were measured by an indentation technique. The materials included alumina-zirconia laminated composites, containing strong interfaces, and alumina-porcelain laminated composites with both weak and strong interfaces. The residual stress in these materials originates from the mismatch of the thermal properties, differences in elastic properties, and different shrinkage of the laminates during sintering. An experimental technique is presented which gives a direct view of the residual stress state in the materials. Values of residual tensile stress are presented as a function of position relative to the interface in each material.

INTRODUCTION

The increasing demand for highly reliable materials with a wide range of properties has resulted in an extensive development of composites. Hard materials with high wear resistance are often used as matrices, in which fibres, platelets and whiskers are introduced to increase toughening or strength. Fabrication of laminated composites of alternate ductile and brittle layers with various thicknesses is another way to control the properties [1]. Laminated composites may offer special advantages for ceramics since small dimensions (thin layers) should result in a small defect size. During fabrication of composites with two dissimilar materials residual stresses will arise due to a mismatch in thermal coefficient of expansion and differences in elastic properties [1-4]. In the case of composites of different ceramics, any

To the extent authorized under the laws of the United States of America, all copyright interests in this publication are the property of The American Ceramic Society. Any duplication, reproduction, or republication of this publication or any part thereof, without the express written consent of The American Ceramic Society or fee paid to the Copyright Clearance Center, is prohibited.

differences in shrinkage during sintering will also contribute to the residual stress field [5]. Proper selection of constituents can be used to create desirable patterns and levels of residual stress. The residual stress field strongly influences the mechanical behaviour of the composite. For example, the residual stress field has been shown to control crack propagation at the interface between two dissimilar materials [6-7]. There are three possibilities: the crack may arrest, it may penetrate the interface, or it may deflect into the interface. It is also possible that the stress field close to the interface results only in cracks that are parallel to the interface.

Residual stress fields and crack propagation can be studied by Vickers indentation [1,4,8-11]. On indenting unstressed and stressed materials, the cracks originating from the corners of an indent will have different lengths, depending on the magnitude and sign of the residual stress. Hsueh and Evans [4] evaluated an approach of measuring the residual stress in a ceramic matrix by comparing crack lengths growing from indentations made at various distances from a metal cylinder. The residual stress was then related to the radial distance from the axis of symmetry. Marshall and Lawn used an indentation crack size approach to evaluate the residual stress in tempered glass [8-9]. Lawn and Fuller [10] studied the changes of the indentation shape due to stresses in proton-irradiated glass. In recent work [11] a residual stress field was introduced in glass by indentation using high loads. Then using low loads, small indentations were made at various angles and distances from the cracks extending from the large indentation. It was then possible to map the residual stress field created on the material surface by the indentation.

The purpose of this work is to present an application of the indentation approach to study the residual stress and crack propagation in layered ceramic composites. The residual stress fields in layered ceramic composites have been mapped. The investigation focuses on alumina layers adjacent to fully stabilised zirconia or either of two types of porcelain.

RESIDUAL STRESS ANALYSIS

The derivation of residual stress values from indentation crack lengths [11] can be summarised as follows. For an indentation in an otherwise unstressed material the stress intensity factor at the tip of the radial crack is:

$$K_c = \chi_r P / c_0^{3/2} \quad (1)$$

where c_0 is the crack length obtained at the load P. χ_r is a dimensionless constant which is related to the ratio between elastic modulus and hardness. If an external residual stress field is applied to the material, the stress intensity factor becomes:

$$K_c = \chi_r P / c_1^{3/2} + \sigma_r \Phi c_1^{1/2} \quad \text{for tensile stress (2)}$$

$$K_c = \chi_r P / c_2^{3/2} - \sigma_c \Phi c_2^{1/2} \quad \text{for compressive stress (3)}$$

where Φ is a constant related to the crack geometry and loading condition. c_1 is the crack length subjected to tensile stress and c_2 is the crack length subjected to compressive stress. Only surface cracks are considered in this analysis reducing it to one of plane stress, i.e. $\Phi = \sqrt{\pi}$. By inserting Eq. (1) in Eq. (2) and Eq. (3) respectively the residual stress is obtained from:

$$\sigma_r = K_c \frac{1-(c_0 / c_1)^{3/2}}{\Phi c_1^{1/2}} \quad \text{tensile stress (4)}$$

$$\sigma_c = -K_c \frac{1-(c_0 / c_2)^{3/2}}{\Phi c_2^{1/2}} \quad \text{compressive stress (5)}$$

To perform this analysis, the crack lengths are measured for both stressed and unstressed material [11]. If the cracks in the stressed material are longer than the cracks in the unstressed material it implies that a tensile residual stress is acting perpendicular to the crack direction. If the cracks are shorter than in the reference case, a compressive stress is acting on the crack. However, it should be noticed that Eq. (4) and Eq. (5) are derived from the same relation as the indentation-bending test relation [12]. In that relation, if the crack lengths under stress exceed about 2.52 times the original crack length, the failure is spontaneous if the applied stress reaches the failure stress. Therefore, in the present analysis, Eq. (4), is also limited to the maximum crack length $c_1<2.52c_0$. If the crack length is longer than 2.52 c_0, the stress estimated from this method is not accurate. In the alumina-zirconia composite some cracks did exceed this limit without failure, and these data were not included. This method can still be used to estimate residual stresses which open or close cracks perpendicular to the interface even though the cracks extend along the stress gradient. However, the stress is assumed to be constant parallel to the whole interface at any distance from it. Thus stress fields normal to the interface can be measured accurately.

EXPERIMENTAL WORK

The multilayered alumina-zirconia composites used in this investigation were produced by warm pressing tape cast [5] alumina and zirconia sheets. In order to establish the residual stress behaviour in alumina-zirconia multilayered composites, different types of composites were fabricated, designated as 3A3Z-9, 3A3Z-15 and 5A5Z-15 (Table 1).

Table 1. Designation of Alumina-Zirconia Composites.

Composite	Total number of layers	Number of blocks	Number of layers in each block
3A3Z-9	9	3	3
3A3Z-15	15	5	3
5A5Z-15	15	3	5

To fabricate alumina-porcelain composites, pressed and sintered alumina discs were placed in a cylindrical mould and dental porcelain slurry was added by vibration. The whole composite was then fired in a dental furnace. Both Procera-All-Ceramic-Porcelain (PACP)* and Vitadur-N-Porcelain (VNP)** were used in this investigation.

All indentations were made in the alumina layers within the load range of 0.5-2N, in a microhardness machine. The indentations were made starting from the interface and gradually increasing the distance from the interface until the crack lengths were of the same size as the reference crack lengths or until the outer surface was reached. To minimise slow crack growth, the material was immediately moved to an optical microscope to measure indentation size, crack lengths and distance from the interface. The orientation of the indentation and cracks relative to the interface were noted.

---

* Procera Sandvik, Stockhom, Sweden
** Vita Zahnfabrik, Bad Sackingen, Germany

The same experimental procedure was followed for a three layer laminated alumina, both on the surface and on the cross-section, in order to establish reference crack sizes for the alumina in the multilayered alumina-zirconia composites. The residual stress states in the three composites were then estimated using the analysis described above.

During the experimental work it was found that the lowest load to obtain a measurable crack was 0.5N for the alumina-porcelain composites and 1N for the alumina-zirconia composites. Low loads are preferred, in order to prevent the indentations from becoming too large compared to the layer thickness.

RESULTS AND DISCUSSION

During the debinding and sintering cycle differences in shrinkage, thermal coefficient of expansion and elastic properties result in internal stresses. In previous work, it was found that the mean linear shrinkage of alumina and zirconia tapes was 13% and 23%, respectively, under the same sintering conditions [5]. The shrinkage effect can be seen in Fig. 1.

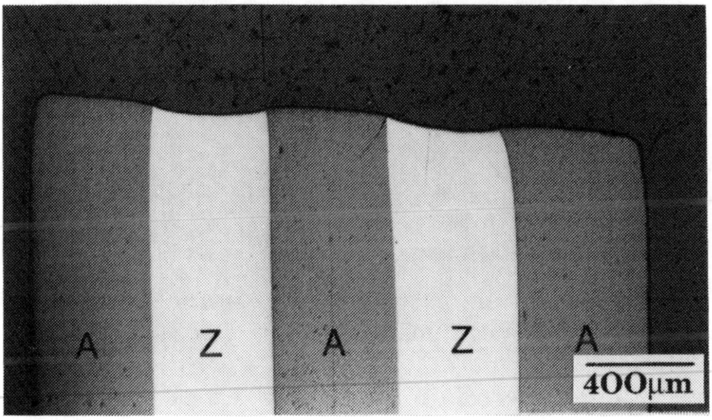

Fig. 1. The edge of an alumina-zirconia composite (3A3Z-15) showing the differential shrinkage.

Crack lengths measured in unstressed material were 23.5μm and 40μm for alumina in the alumina-porcelain composites and in the alumina-zirconia composites, respectively. The fracture toughness of alumina was 3.15MPa·m$^{1/2}$ in the alumina-porcelain composites [13], and 2.5MPa·m$^{1/2}$ in the alumina-zirconia composites. The latter value was obtained by indentation measurements using different loads in unstressed alumina.

The crack propagation was found to be significantly different in the alumina in the three different composites. Some typical crack patterns are shown in Figs. 2-4. In the case of alumina-zirconia composites (Fig. 2) no cracks extended from the indentation corners, normal to the interface until the distance from the interface was a large fraction of the layer thickness. Cracks perpendicular to the interface were typically only detected in the 5A5Z-15 composite where the alumina block is thick (about 850μm). This confirms that there are strong compressive stresses parallel to the interface in the alumina layers.

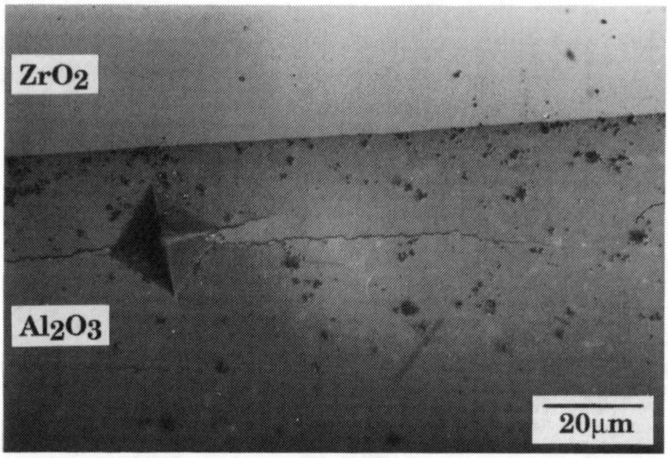

Fig. 2. Long cracks parallel to the interface in the alumina layer in alumina-zirconia composite.

For the two alumina-porcelain composites, cracks propagated from all four indentation corners both close to and away from the interface (Figs. 3 and 4). The lengths of the cracks growing away from the interface were of the same magnitude as the reference crack length. Cracks growing towards the interface were longer than the reference crack if the indent was close to the interface. At a larger distance these cracks became the same size as the reference crack. In the alumina-VNP composite (Fig. 3) cracks normal to the interface crossed it and continued to grow in the same direction through the porcelain layer. It is believed that the porcelain in this composite is weaker than the interface between alumina and the porcelain. The opposite was found in the alumina-PACP composite. A crack penetrating the interface stopped and delamination occurred. This interface is believed to be weaker than the PACP. In other work, it was found that PACP has a higher bending strength than VNP [14].

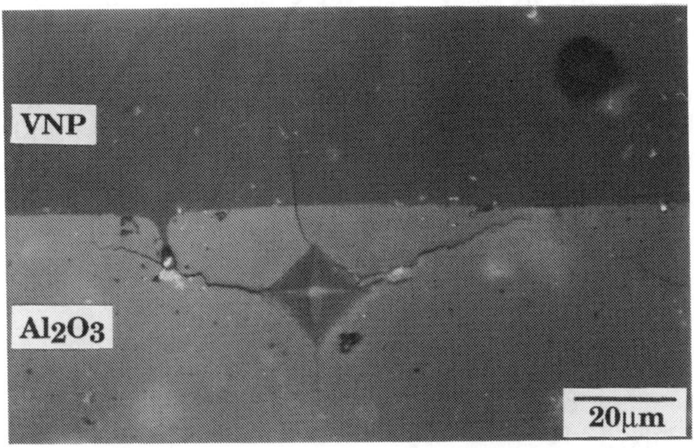

Fig. 3. Crack running into the porcelain in alumina-VNP composite.

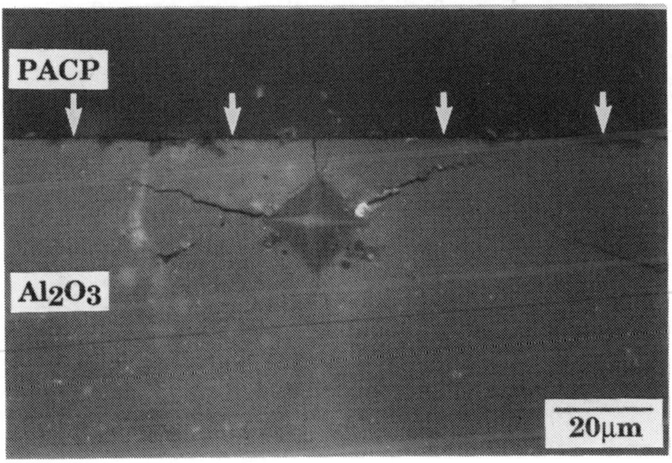

Fig. 4. Delamination (arrowed) at interface in alumina-PACP composite.

In figure 5 the length of cracks parallel to the interface is plotted as a function of the distance from the interface for two alumina-zirconia composites (only the outer alumina blocks have been considered in the 3A3Z-15 composite). Longer cracks indicate a higher residual stress. Some very long cracks (>100μm) were observed in the alumina-zirconia composites. However, such very long cracks were not detected in the alumina-porcelain composites. In the alumina-zirconia composites a stress peak (longer cracks) is detected at a distance of about 100μm from the interface. At present, there is no clear explanation of this behaviour. Note that it is not possible to calculate the residual stress close to the interface in the three block composites because the cracks exceed the crack length limit for the validity of Eq. (4). An electron microscopy investigation to explore the nature of the bonding at the interface is currently in progress.

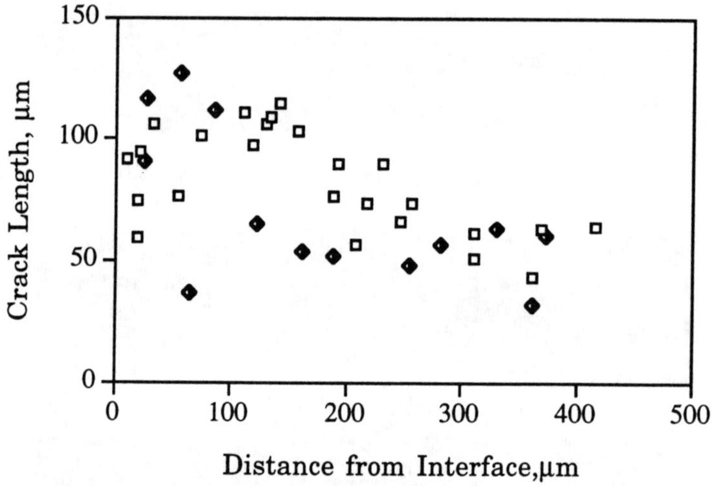

Fig. 5. Crack length versus distance from the interface in alumina-zirconia composites. The left vertical axis represents the interface and the right one represents the free surface (▫ 3A3Z-9, ◆ 3A3Z-15).

Figures 6 and 7 present the residual stress (calculated from Eq. (4)) versus the distance from the interface. In the alumina-porcelain composites the residual stress decreases rather quickly as the distance from the interface increases. It is also notable that the residual stress in the alumina adjacent to VNP falls more rapidly than in the alumina

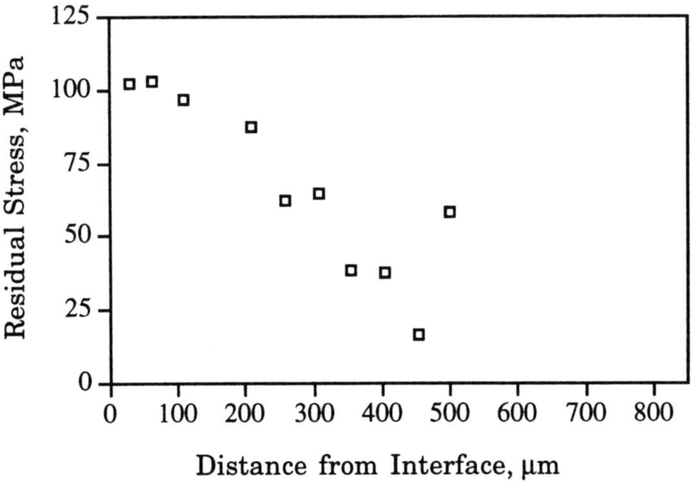

Fig. 6. Residual stress normal to the interface in the alumina layer, in a 5A5Z-15 alumina-zirconia composite.

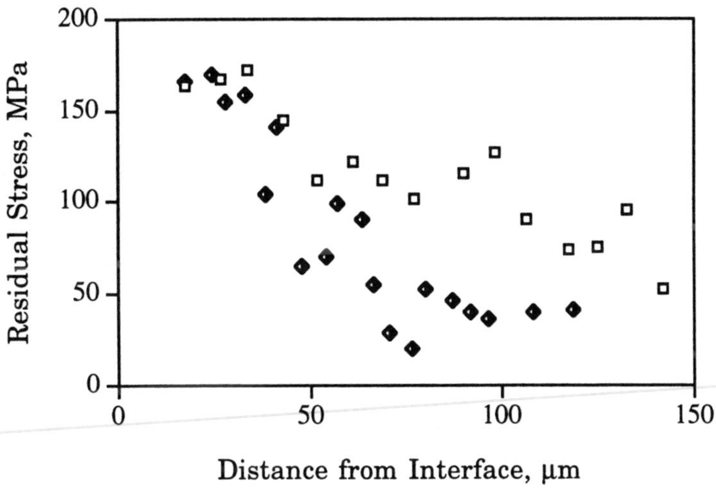

Fig. 7. Residual stress normal to the interface in the alumina layer, in alumina-porcelain composites (□ Al_2O_3-PACP, ◆ Al_2O_3-VNP).

adjacent to PACP. In the alumina-zirconia composites the residual stress extends over significantly larger distances. These differences could be a result of the fabrication methods. In the alumina-porcelain composites, the alumina disc is a dense body which is highly resistant to deformation. The alumina-zirconia composites are made by sintering tapes. In the former case the residual stresses arise only from differential contraction and modulus effects and this may lead to the steep stress gradients. In the alumina-zirconia composites, both expansion/modulus mismatch and differential shrinkage effects are present. However, the expansion/modulus effect would lead to stress fields of the opposite sign from those observed [5]. Thus the shrinkage effect is dominating in the alumina-zirconia composites.

CONCLUSIONS

Delamination occurs for an indentation crack running from alumina into porcelain in a composite made with high strength porcelain. In a similar composite made with a lower strength porcelain, the crack passes through the interface into the porcelain, with no delamination. For similar indentation loads, no cracks are generated normal to the interface in alumina-zirconia laminated composites. This indicates higher compressive stresses parallel to the interface. Residual tensile stresses generated normal to the interfaces open long cracks parallel to the interfaces in all three materials. Residual tensile stresses and distributions can be measured by an indentation technique. The stress levels are somewhat higher in the alumina-porcelain composites and fall off much more rapidly than in the alumina-zirconia composites. This may be related to differences in the methods of fabrication.

REFERENCES

1. Z.Chen and J.J. Mecholsky, Jr., "Toughening by Metallic Lamina in Nickel/Alumina Composites," *J. Am. Ceram. Soc.*, **76** [5] 1258-1264 (1993).
2. A.G. Evans, M.C. Lu, S. Schmauder and M. Rühle, "Some Aspects of the Mechanical Strength of Ceramic/Metal Bonded System," *Acta Metall.*, **34** [8] 1643-1655 (1986).
3. C.H. Hsueh and A.G. Evans, "Residual Stresses in Metal/Ceramic Bonded Strips," *J. Am. Ceram. Soc.*, **68** [5] 241-248 (1985).
4. C.H. Hsueh and A.G. Evans, "Residual Stresses and Cracking in Metal/Ceramic System for Microelectronics Packaging," *J. Am. Ceram. Soc.*, **68** [3] 120-127 (1985).
5. R-M.F. Yttergren et al., to be published.

6. M.Y. He and J.W. Hutchinson, "Crack Deflection at an Interface Between Dissimilar Elasic Materials," *Int. J. Solids Structures.* **25** [9] 1053-1067 (1989).
7. M.Y. He , F.E. Heredia, D.J. Wissuchek, M.C. Shaw and A.G. Evans, "The Mechanics of Crack Growth in Layered Materials," *Acta metall. mater.*, **41** [4] 1223-1228 (1993).
8. D.B. Marshall and B.R. Lawn, "An Indentation Technique for Measuring Stresses in Tempered Glass Surfaces," *J. Am. Ceram. Soc.*, **60** [1-2] 86-87 (1977).
9. D.B. Marshall and B.R. Lawn, "Measurement of Nonuniform Distribution of Residual Stresses in Tempered Glass Discs," *Glass Techn.* **19** [3] 57-58 (1978).
10. B.R. Lawn and E.R. Fuller, Jr., "Measurement of Thin-Layer Surface Stresses by Indentation Fracture," *J. Mat. Sci.*, **19** 4061-4067 (1984).
11. K. Zeng and D. Rowcliffe, "Experimental measurement of Residual Stress Field Around a Sharp Indentaion in Glass," *J. Am. Ceram. Soc.*, **77** [2] 524-530 (1994).
12. D.B. Marshall and B.R. Lawn, "Flaw Characteristics in Dynamic Fatigue: The Influence of Residual Contact Stresses" *J. Am. Ceram. Soc.*, **63** [9-10] 532-536 (1980).
13. K. Zeng, K. Breder and D. Rowcliffe, "Dynamic Fatigue of an Al_2O_3/SiC Whiskers Composite in Water," *Ceram. Eng. Sci. Proc.*, **12** [9-10] 2233-2250 (1991).
14. A.Odén et al., to be published.

Transient Liquid Phase Bonding of Silicon Ceramics Via Microdesigned Ni-Based Interlayers

B. J. Dalgleish, A. P. Tomsia and A. M. Glaeser
Center for Advanced Materials, Lawrence Berkeley Laboratory and
Department of Materials Science and Mineral Engineering
University of California, Berkeley, California 94720

Abstract

Ceramic-ceramic joining of silicon nitride and silicon carbide ceramics has been achieved by exploiting a multilayer metallic interlayer designed to form a thin layer of a transient liquid phase. Au-Cu/Ni/Au-Cu interlayers have been used to form Si_3N_4/Ni/Si_3N_4 and SiC/Ni/SiC bonds at 950°C. In both cases, the maximum four-point flexure strengths approach those of the parent ceramic, consistent with observations of exclusively ceramic failure. The average room temperature bend strength of joined Si_3N_4 assemblies was 770 MPa, while that for the joined SiC assemblies was 260 MPa. Results of fractographic and chemical analyses are used as a basis for discussing aspects of processing and mechanical properties.

Introduction

Efficient use of fossil energy requires materials that can function in high temperature, corrosive environments, such as those found in advanced combustion and engine systems. Ceramics and ceramic composite materials have the required heat and corrosion resistance for these extreme environments. However, utilization of ceramic materials in such environments requires cost-effective and reliable manufacturing of ceramic components and structures.

To utilize ceramics effectively in either exclusively ceramic structures or in multi-material assemblies such as advanced gas turbines, there is a need to develop reliable and effective means of joining ceramics to themselves and to advanced structural metals in a manner that is compatible with their prolonged use at elevated temperature. The need for new and more capable joining approaches has been recognized for some time. At present, no joining method can make reliable refractory joints between the ceramics that are candidates for high-temperature combustion and propulsion systems.

To the extent authorized under the laws of the United States of America, all copyright interests in this publication are the property of The American Ceramic Society. Any duplication, reproduction, or republication of this publication or any part thereof, without the express written consent of The American Ceramic Society or fee paid to the Copyright Clearance Center, is prohibited.

Ceramic joining for high temperature applications is difficult because of incompatibilities in the requirements for processing, joining, and ultimate performance. In order to avoid component degradation and excessive and deleterious reaction between components during joining it is desirable to process at as *low* a temperature as possible. Under these conditions, kinetics and in some cases thermodynamics are an ally, preventing rapid reaction or making reactions unfavorable. At the same time, it is necessary or desirable for the material forming the joint to retain a significant fraction of its room temperature strength to as *high* a temperature as possible during subsequent use. These contradictory constraints result in a narrow (perhaps vanishingly narrow) processing-condition window. The more disparate the properties of the components, the greater the challenge of identifying suitable joining methods. Conventional joining methods typically require joining temperatures that *exceed* the intended use temperature. Phase equilibria studies indicate that such approaches are likely to fail in providing refractory joints because at the requisite high joining temperatures, deleterious reactions are thermodynamically favorable, and the kinetic barriers that provide some protection at low temperatures are absent.

Our efforts have focussed on developing a new technique for fabricating ceramic joints at low temperatures that is compatible with subsequent use at high temperatures [1-8]. The method, partial transient liquid phase (PTLP) bonding, utilizes multilayer metal interlayers, *e.g.*, a thin cladding of a lower melting point metal such as Cu or Au on a thicker core of a higher melting point metal such as Ni, Pd, Nb, etc. The components within the interlayer are selected to produce an alloy or compound that is capable of supporting use at elevated temperature, potentially, a temperature approaching or exceeding the joining temperature. As is the case in conventional transient liquid phase (TLP) bonding [*e.g.*, 9], the interlayer material solidifies at the joining temperature. However, unlike conventional transient liquid phase bonding, 1) only a small (or partial) layer of the interlayer becomes liquid at the bonding temperature, 2) much larger differences between joining and use temperatures are possible, and 3) there is greater flexibility in interlayer design.

BACKGROUND

The potential of Si_3N_4 for high-temperature structural applications has prompted considerable effort in developing methods of producing strong Si_3N_4/Si_3N_4 and Si_3N_4/metal joints, and the difficulties that arise when attempting to develop such joints for high-temperature use are well documented. Si_3N_4 reacts with many high melting point metals to form silicides and nitrides [*e. g.*, 10]. Extensive phase equilibrium studies by Schuster and coworkers attest to the reactivity, and define temperatures for the onset of specific Si_3N_4/metal reactions [11, 12]. Although limited reaction may have a beneficial effect on strength, excessive reaction is detrimental [10, 13]. The roles of temperature and atmosphere in controlling reactions and inhibiting Si_3N_4 decomposition are known. Decreasing the joining temperature will limit the rate of reaction, and may, in some cases, thermodynamically inhibit reactions as well. In some instances, the bonding of a high performance metallic alloy to a ceramic such as Si_3N_4 is desired. If the joining temperature in such cases is excessive, the properties of the metallic component degrade due to "overaging" of the alloy. Thus, in such situations, the processing temperature constraints become even more severe, and this increases the difficulty of finding a suitable joining approach.

As discussed in more detail in the companion paper [14], our initial efforts focussed on joining alumina ceramics with several distinct multilayer interlayers. Multilayer Cu/Pt/Cu interlayers [1] were explored because of the promising high-temperature properties of diffusion-bonded Al_2O_3/Pt/Al_2O_3 assemblies [15, 16]. Multilayer Cu/Nb/Cu interlayers were used [5], and in general, produced higher average strength than comparably processed Cu/Pt/Cu interlayer bonded assemblies. This was thought to be due to the beneficial effect of Nb on the wetting of Cu on alumina. Interlayers of Cu/80Ni20Cr/Cu were also used to join Al_2O_3 [6, 14] in an effort to confirm the benefits of reactive metal incorporation in the TLP layer, and to allow a comparison of properties with those obtained in prior work on Cu/Ni/Cu interlayer bonded assemblies [4].

Our most recent work, summarized in the present paper, has tried to incorporate the lessons learned from our prior work on PTLP bonding of alumina ceramics, and the results of prior silicon nitride joining studies, in developing strategies for PTLP bonding of silicon ceramics. The initial results obtained using Ni core layers with a Cu-Au alloy based transient liquid layer have been encouraging [7].

INTERLAYER DESIGN

In designing an interlayer that might be appropriate for joining Si_3N_4 and SiC, several combinations of transient liquid formers and Ni core layers were considered. The candidates were selected because of their ability, or (data lacking) their expected ability to satisfy a number of criteria thought critical to success. These are:

1. The coating must melt at temperatures below 1100°C, and preferably at <1000°C to minimize reaction and dissociation of the ceramic;
2. The molten coating must fill the capillary gap between the ceramic and the interlayer; wetting will occur if the *sum* of the contact angles formed by the coating material on the metal core and on the ceramic is less than 180°;
3. To further inhibit the dissociation of Si_3N_4 and SiC, the solubility of Si in the coating material should be low, and the activity of dissolved Si ($\underline{Si}$) should be high. Si eutectics are not desirable because of their potentially detrimental effect on service temperature and their inhibiting effect on isothermal solidification.
4. The solubility of the coating in the Ni core layer, and the diffusivity of the coating material in Ni should be high to minimize the time required for isothermal solidification and homogenization.

Among the candidate TLP formers, one of the more attractive was a Cu-Au alloy. The phase diagram [17] for this system shows a melting point minimum at ≈910°C at a composition of ≈56 at % Au-44 at % Cu. A TLP of this composition thus allows joining to be performed at <1000°C. The Cu-Ni and Au-Ni phase diagrams show that there is extensive solubility of both Cu and Au in Ni, and thus, one can expect that a mildly solution hardened alloy dominated by the properties of Ni will emerge from homogenization. The solidus temperature of the resulting ternary alloy will obviously depend upon the relative thicknesses of the TLP and core layers, however, a liquidus temperature substantially higher than the eutectic temperature of the commonly used Ag-Cu eutectic based brazed alloys can be expected [18], and a correspondingly greater temperature capability for the joined assembly was anticipated.

EXPERIMENTAL PROCEDURES

Materials:

The Si_3N_4 (SN84 grade, NGK Insulators, Ltd., Nagoya Japan) used in this study is pressureless sintered and contains Y_2O_3, MgO and ZrO_2 as sintering aids. The quoted flexural strength of this material, determined by four-point bend test, is 970 MPa at both room temperature and at 1000°C. The as-received material was cut into blocks roughly 16 × 16 × 20 mm for bonding, and into beams of roughly 2.7 × 2.7 × 40 mm dimensions for bend tests. One block face was fine-ground and subsequently diamond polished on a copper lap, finishing with 1 μm diamond paste.

The SiC material (Hexalloy, Sohio Engineered Materials, Niagara Falls, NY) used is pressureless sintered α-SiC containing boron and excess carbon as sintering aids. The processing of this materials is described in [19]. Blocks and beams were prepared using procedures paralleling those used for Si_3N_4.

The Ni foil (Puratronic grade; Johnson Matthey Material Technology, Hertfordshire, UK) was ≈25 μm thick, and reportedly 99.994+ wt% pure with respect to metallic impurities. Foils were degreased in acetone, bright-dipped in dilute nitric acid, rinsed in distilled deionized water, and dried in a warm air stream just prior to assembly of the bond components. The Cu, Au, and Ti evaporation sources for the TLP layer were in the form of pellets of Puratronic grade material. No additional precleaning was performed, evolution of contaminants being expected during premelting of the pellets.

Coating

Just prior to deposition of the TLP layer, the substrates were ultrasonically cleaned by sequential immersion in tricloroethane, acetone, and isopropyl alcohol, and dried in a stream of pure nitrogen. The TLP layer was deposited directly onto the polished face of the Si_3N_4 and SiC blocks by sequential evaporation of Ti (≈200 nm), Au (≈1150 nm), Cu (≈1640 nm) and Au (≈1200 nm) as measured by a vibrating crystal monitor. The sequence was designed to protect the Ti and Cu layers from oxidation. The relative thicknesses of Cu and Au yield a 1:1 atomic ratio. The Ti concentration within the TLP layer is ≈4 at %. The total thickness of each TLP layer was ≈4.2 μm. The average composition of the TLP layers and core foil was thus ≈79.8 at % Ni, ≈9.7 at % Cu, ≈9.7 at % Au, ≈0.8 at % Ti. Assuming that the small amount of Ti has a negligible effect, the ternary phase diagram for the Ni-Cu-Au system suggests a liquidus temperature of ≈1325°C for this alloy. We note that this system exhibits a solid-state miscibility gap; the highest critical temperature is estimated to be 970°C.

PTLP Bonding

For both Si_3N_4 and SiC bonding, the coated-block/foil/coated-block assembly was assembled just prior to loading in the hot press. The temperature was raised at a rate of 5°C/min to 950°C, maintained for 2 h (vacuum $\approx 2 \times 10^{-6}$ torr), and then lowered to room temperature at 2°C/min. A pressure of ≈8.7 MPa was maintained on the assembly. (More recent work has shown that bonds can also be produced using pressures more like those in conventional brazing processes.)

Beam Preparation

Beams were prepared from both bonded blocks and from as-received material (to measure the inherent strength) by diamond machining. For the Si_3N_4, the beams prepared from bonded blocks were 2.5 × 2.5 × 40 mm, and the beams prepared from the as-received material were 2.7 × 2.7 × 40 mm. For SiC, similar preparation steps were used, but beam dimensions were 3.0 × 3.0 × 50 mm. The tensile surfaces of all beams were fine-ground and finished by polishing with diamond. The final abrasive used was 3 μm diamond. The tensile face edges were beveled to remove flaws. The side surfaces of bonded beams were also fine ground and finished with 1 μm diamond. These polished side surfaces allowed for observations of crack path, and EDS and probe analysis of bonded sections.

Flexure Testing

Beams were tested at room temperature using four-point bending with an inner span of 8 mm, outer span of 25 mm and a displacement rate of 0.05 mm/min. Strengths were calculated from the load at failure using standard relationships derived for monolithic elastic materials.

Microstructural and Microchemical Characterization

The microstructure of cross sections and fracture surfaces were examined using scanning electron microscopy (SEM), and the chemistry of the interlayer region and fracture surfaces was evaluated using energy dispersive spectroscopy (EDS) and microprobe analysis. Crack path characterization was performed using optical microscopy and SEM.

RESULTS AND DISCUSSION:

Microchemical Characterization

The results of a microprobe line scan perpendicular to the Si_3N_4/interlayer interface and a representative microstructure are provided in Figure 1. At the left and right extremes, Si, N, and O (from sintering additives) are the dominant components, and the composition is to a good approximation independent of position. There is a roughly 40 μm thick central region corresponding to the interlayer, and within which the composition varies. Some homogenization is evident, with Au and Cu concentrations reaching 5-10 at % at the core, however, there is also a region nearer the Si_3N_4 where the Au and Cu concentrations are higher. The overall profile does not fit a simple diffusion process. Near the Si_3N_4, the Cu and Au concentrations are higher. During cooling, the composition of this region will lie within the bounds of the miscibility gap in this system. We believe the structure observed reflects phase separation that occurred during cooling. Bonding at a somewhat higher temperature or post-bonding annealing could eliminate this problem and lead to a more homogeneous composition profile, and presumably, a higher temperature capability. The other feature of note is a slight enhancement of the Ti concentration near the Si_3N_4/interlayer interface; this is consistent with observations in other systems, where reaction layers containing Ti, *e.g.*, TiN, are formed. In contrast to more typical reactive metal braze formulations, the amount of Ti is quite limited, and thus, the amount of Si liberated by the reaction of Ti with the Si_3N_4 is also reduced. Similar homogenization trends are expected in SiC assemblies, although the reaction product is more likely to be TiC. Parallel experiments examining the compositional evolution of the interlayer in bonded SiC assemblies are in progress; results will be reported elsewhere.

Fracture Behavior of Bonded Silicon Nitride

The results of room temperature flexure tests are shown in Figure 2. Unbonded SN84 Si_3N_4 had an average flexure strength approaching 1 GPa, consistent with the information provided by the vendor. The average bend strength of bonded assemblies was 770 (±200) MPa. The use of a high strength silicon nitride, coupled with the design of an interlayer that combines a small amount of reactive metal to induce wetting and reaction, and a relatively ductile core layer has allowed the fabrication of joints of quite high strength; many of the strengths observed exceed those obtained in tests of diffusion bonded assemblies and brazed assemblies. In virtually all samples, at least some amount of ceramic failure occurs. The degree or extent of ceramic failure

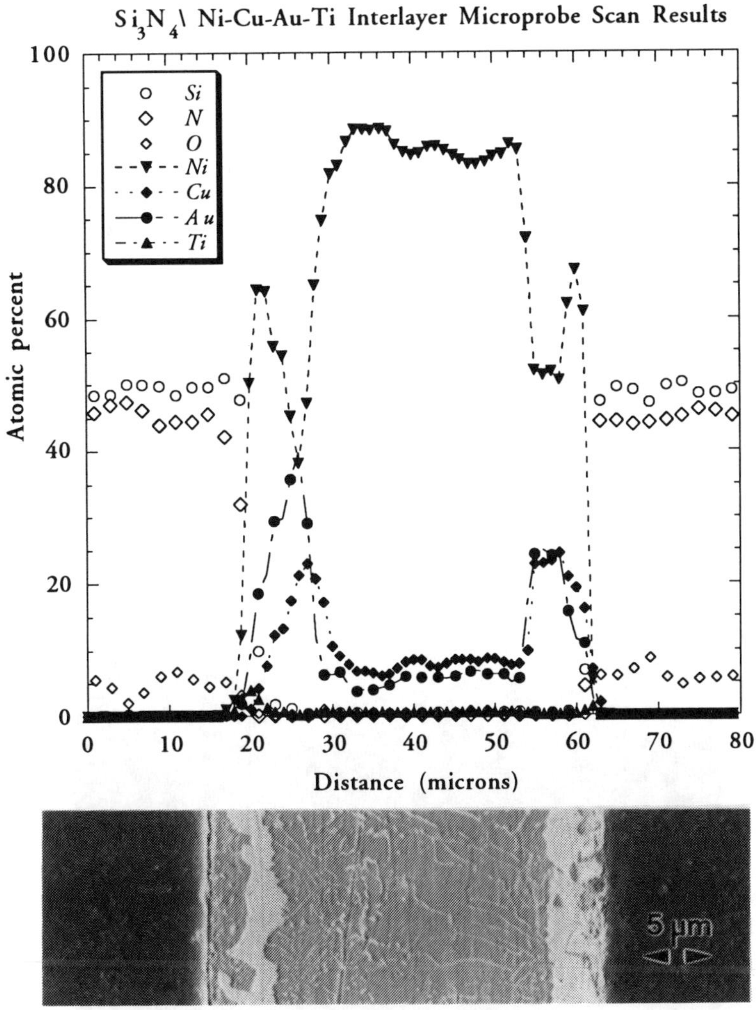

Figure 1 Microprobe scan results (above) and representative SEM micrograph (below) of the interlayer region of Si_3N_4 bonded with a Ni-Cu-Au-Ti interlayer after completion of the bonding cycle.

(as opposed to interfacial failure) tends to increase with the fracture stress. In several intermediate strength joints, failure involved extensive tearing of the metal foil, and there is evidence of the interfacial debonding switching from one ceramic-metal interface to the other, resulting in the formation of metal ligaments that bridged the crack. Such behavior is illustrated in the beam profile of Figure 3, a beam that

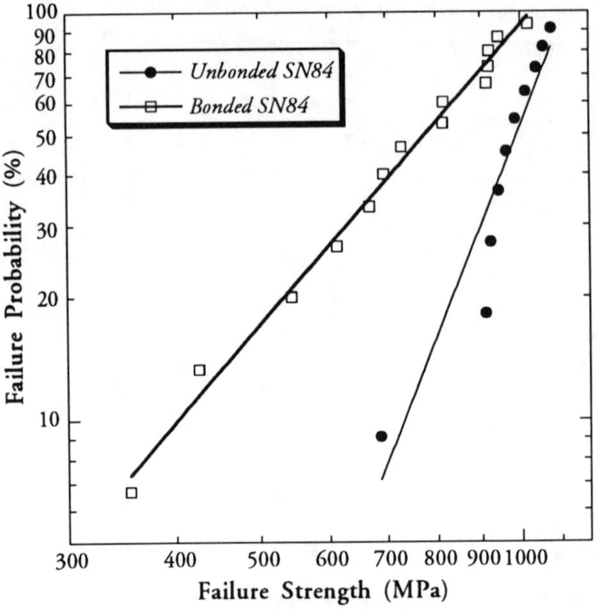

Figure 2

Plot of failure probability versus room temperature failure strength for unbonded and Cu-Au/Ni/Cu-Au interlayer bonded silicon nitride. The average strength of the bonded assemblies is high, however, steps must be taken to reduce the observed scatter. As in prior experiments using alumina [14], modification of the interlayer chemistry that further enhance wetting may be effective in increasing the average strength and decreasing the width of the strength distribution.

Figure 3 Reassembled bonded Si_3N_4 flexure specimen showing extensive tearing of the metallic interlayer during failure. The debonding switches between the two interfaces causing the development of bridging metallic ligaments. The tensile surface was at the bottom.

exhibited a failure strength of 614 MPa. For samples with bend strengths above ≈800 MPa, failure was primarily in the ceramic, and crack branching was extensive. This caused fragmentation of the ceramic, and secondary cracking was evident in many samples. The large amount of stored energy leads to a mode of failure in the highest strength samples that can best be described as "explosive".

Fracture Behavior of Bonded Silicon Carbide

The results of room temperature flexure tests on both unbonded and bonded SiC are shown in Figure 4. The unbonded SiC had an average flexure strength of 420 MPa with a standard deviation of ±30 MPa. For the bonded assemblies the average strength was 260 MPa, which may be useful for some applications. However, the standard deviation is relatively large (±130 MPa).

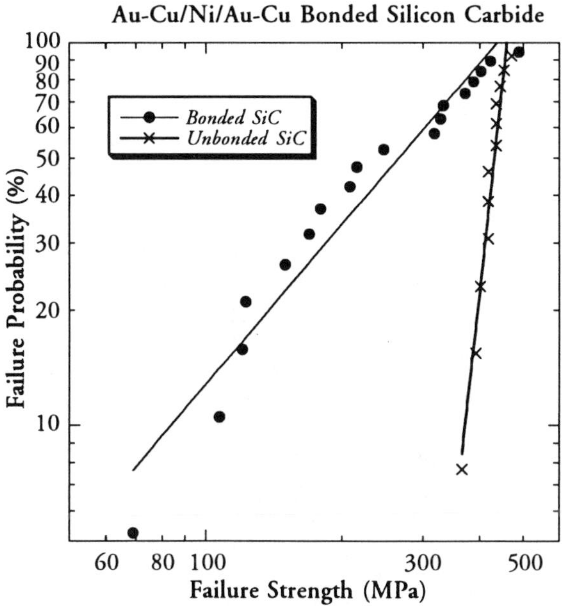

Figure 4

Plot of failure probability versus room temperature failure strength for unbonded and Cu-Au/Ni/Cu-Au interlayer bonded silicon carbide. As was the case for silicon nitride bonded with the same interlayer, a useful level of strength was obtained in many specimens, however, the scatter in strength is a concern for engineering applications, and steps to narrow the strength distribution are being explored.

Standard deviations that represent a substantial fraction of the mean strength have been observed in other PTLP bonded systems, *e.g.*, Cu/Ni/Cu [4]. In these systems, it was possible to link low strengths with the presence of regions along the ceramic/interlayer interface where ceramic-metal contact was poor, and failure was largely if not exclusively interfacial. As shown in the companion paper [14], the introduction of a sufficient amount of reactive metal into the TLP increased the average strength, decreased the standard deviation, and shifted the failure to the ceramic. For the Si_3N_4 joining, the Ti was incorporated only into the TLP layer. The resulting average strength was high, the scatter large, and the trend towards a progressively greater degree of ceramic failure with increasing strength suggests that in this case, interfacial defects caused by incomplete wetting may have played a direct or indirect role in initiating failure. Thus, modification of the interlayer chemistry may again be effective in improving the strength characteristics.

It is in this context, that the failure characteristics of SiC are surprising. In the lowest strength sample, SEM fractography revealed a large region of poor ceramic/interlayer contact. This sample failed primarily along the ceramic/interlayer interface, and the low failure strength (69 MPa) is not surprising. In some samples, metallic layers were exposed by fracture on exactly corresponding positions on the two fracture surfaces, indicating failure *within* the metallic interlayer. However, in most of the remaining samples, independent of the strength level achieved, failure occurred primarily in the ceramic, as illustrated in Figure 5. The observation of ceramic dominated failure suggest that the strength of the ceramic is degraded, however, since machining alone is not the cause, some combination of factors that is not yet understood may be responsible. One possibility is that when wetting is incomplete and interfacial flaws/gaps exist, these interact with and may amplify the residual stresses due to thermal expansion mismatch and those imposed during machining and cause the formation of a flaw population not present in the as-received (unbonded) material before or after machining. However, further work is required to identify failure origins and verify that failure actually initiates in the ceramic.

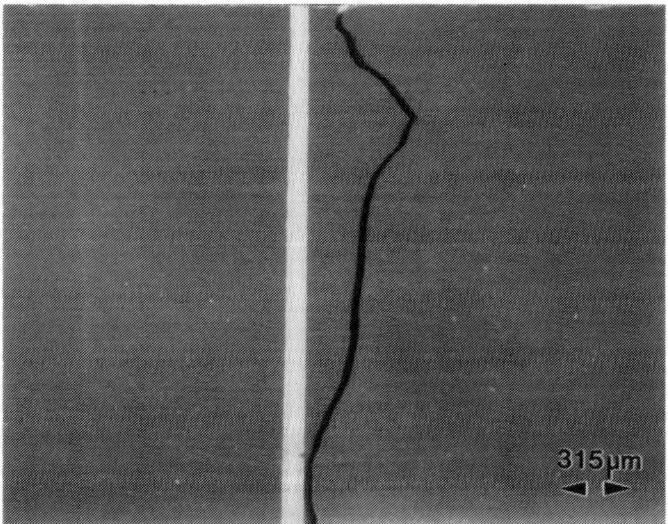

Figure 5 Reassembled bonded SiC flexure specimen with a bend strength of 333 MPa showing predominantly ceramic failure. Failure may have initiated at or near the interface; the tensile surface was at the bottom.

CONCLUDING REMARKS

Both Si_3N_4 and SiC ceramics have been successfully joined using the same Ni-based multilayer interlayers. The interlayers contained a thin layer of a Cu-Au based transient liquid phase to allow joining at temperatures below those associated with extensive Ni-Si_3N_4 reaction. The strengths observed in some joined assemblies were high, with many samples failing largely in the ceramic. The cause of the relatively larger scatter in bend strength for joined versus unbonded specimens is not known, and is currently under study. The homogenization of the interlayer was not complete under the conditions used, and rapid heating to elevated temperatures may cause local liquid formation. With adjustments in the processing conditions that promote greater homogenization, the retention of strength to temperatures higher than those currently accessible with Cu-Ag eutectic based alloys is anticipated. Opportunities to improve the strength distribution through modifications of the interlayer design and TLP chemistry to improve wetting are being explored, with the hope that the trends toward higher average strength and reduced scatter already achieved in alumina bonding with Ni interlayers can be duplicated. The temperature dependence of the joint properties must now be evaluated to determine their potential for use in high-temperature high-stress applications [20].

ACKNOWLEDGMENTS

This research was supported by the Director, the Office of Energy Research, Office of Basic Energy Sciences, Materials Sciences Division of the U.S. Department of Energy under Contract No. DE-AC03-76SF00098. We thank Dr. Ho Fang of Allied-Signal for providing the Si_3N_4 material. We are grateful for the contributions of M. L. Shalz to the success of the research program. Helpful discussions with R. M. Cannon Jr., Y. Iino, M. Koizumi, K. Nakashima, M. G. Nicholas, S. Peteves and R. O. Ritchie are acknowledged.

REFERENCES

1. M. L. Shalz, B. J. Dalgleish, A. P. Tomsia and A. M. Glaeser, "Ceramic Joining I. Partial Transient Liquid Phase Bonding of Alumina *via* Cu/Pt Interlayers," *J. Mater. Sci.*, **28**, [6], 1673-84 (1993).
2. M. L. Shalz, B. J. Dalgleish, A. P. Tomsia and A. M. Glaeser, "New Approaches to Joining Ceramics For High-Temperature Applications," *Ceramic Transactions,* **35**, 301-22, (1993)
3. A. M. Glaeser, M. L. Shalz, B. J. Dalgleish and A. P. Tomsia, "A Transient FGM Interlayer Based Approach to Joining Ceramics," *Ceramic Transactions,* **34**, 341-357 (1993).
4. M. L. Shalz, B. J. Dalgleish, A. P. Tomsia and A. M. Glaeser, "Ceramic Joining II. Partial Transient Liquid Phase Bonding of Alumina *via* Cu/Ni/Cu Multilayer Interlayers," *in press, Journal of Materials Science.*

5. M. L. Shalz, B. J. Dalgleish, A. P. Tomsia, R. M. Cannon and A. M. Glaeser, "Ceramic Joining III. Bonding of Alumina via Microdesigned Cu-Nb Interlayers," *in press, Journal of Materials Science.*

6. B. J. Dalgleish, A. P. Tomsia, K. Nakashima, M. Locatelli and A. M. Glaeser, "Low Temperature Routes to Joining Ceramics for High-Temperature Applications," *submitted to Scripta Metallurgica et Materialia.*

7. B. J. Dalgleish and A. M. Glaeser, "Ceramic Joining IV. Partial Transient Liquid Phase Bonding of Si_3N_4 *via* Cu-Au/Ni/Cu-Au Multilayer Interlayers," *in preparation for submission to Journal of Materials Science.*

8. M. L. Locatelli, K. Nakashima, B. J. Dalgleish, A. P. Tomsia and A. M. Glaeser, "Ceramic Joining V. Partial Transient Liquid Phase Bonding of Alumina *via* Cu/Ni and Cu/80Ni20Cr Interlayers," *in preparation for submission to Journal of Materials Science.*

9. D. S. Duvall, W. A. Owczarski and D. F. Paulonis, "TLP Bonding: a New Method for Joining Heat Resistant Alloys," *Welding Journal,* **53**, [4], 203-14 (1974).

10. M. Koizumi, M. Takagi, K. Suganuma, Y. Miyamoto and T. Okamoto, "Solid-State Bonding of Silicon Nitride to Metals Using HIP," pp. 1033-42 in HIGH TECH CERAMICS, P. Vincenzini, Ed., Elsevier Science Publishers B.V., Amsterdam (1987).

11. J. C. Schuster, F. Weitzer, J. Bauer and H. Nowotny, "Joining of silicon nitride ceramics to metals: the phase diagram base," (3rd International Conference on the Science of Hard Materials, Nassau, Bahamas, 9-13 Nov. 1987), *Mater. Sci. and Eng. A,* **A105-106**, 201-6 (1988).

12. F. Weitzer and J. C. Schuster, "Phase diagrams of the ternary systems Mn, Fe, Co, Ni-Si-N," *J. Solid State Chem.,* **70**, [2], 178-84 (1987).

13. M. G. Nicholas and D.A. Mortimer, "Ceramic/Metal Joining for Structural Applications," *Mater. Sci. and Tech.*, **1**, 657-665 (1985).

14. M. R. Locatelli, K. Nakashima, B. J. Dalgleish, A. P. Tomsia and A. M. Glaeser, "Transient Liquid Phase Bonding of Alumina Ceramics via Microdesigned Ni-Based Interlayers," this volume.

15. R. V. Allen and W. E. Borbidge, "Solid-state metal-ceramic bonding of platinum to alumina," *J. Mater. Sci.*, **18**, [9], 2835-43 (1983).

16. R. V. Allen, F. P. Bailey and W. E. Borbidge, "Solid-State Bonding of Ceramics with Platinum," *Plat. Metals Rev.*, **25** [4] 152-55 (1981).

17. BINARY ALLOY PHASE DIAGRAMS, Vol. 1, T. B. Massalski, ed., pp. 358-62, ASM International, Metals Park, OH (1990).

18. PHASE DIAGRAMS OF TERNARY GOLD ALLOYS, A. Prince, G. V. Raynor, and D. S. Evans, Eds., pp. 226-34, Institute of Metals, (1990).

19. J. S. Coppola and R. H. Smoak, "Method of Producing High Density SIlicon Carbide Product," U.S. Patent 4080415, March 21, 1978.

20. B. J. Dalgleish, A. P. Tomsia and A. M. Glaeser, unpublished research.

TRANSIENT LIQUID PHASE BONDING OF ALUMINA VIA MICRODESIGNED NI-BASED INTERLAYERS

M. R. Locatelli, K. Nakashima*, B. J. Dalgleish, A. P. Tomsia and A. M. Glaeser
Center for Advanced Materials, Lawrence Berkeley Laboratory and
Department of Materials Science and Mineral Engineering
University of California, Berkeley, California 94720
*work performed while on leave from Kyushu University, Fukuoka, Japan

ABSTRACT

Interlayers designed to form a thin layer of a Cu-rich transient liquid phase have been used to join alumina. Three multilayer interlayers were investigated: 1) Cu/Ni/Cu, 2) Cr/Cu/Ni/Cu/Cr, and 3) Cu/80Ni20Cr/Cu. In all three cases, bonding at 1150°C resulted in substantial chemical homogenization. However, marked differences in joint strength and failure mode were observed. High joint strengths were possible with all three interlayers. The highest average strength and smallest standard deviation were achieved using 80Ni20Cr core layers. The average four-point flexure strength of assemblies bonded using Cu/80Ni20Cr/Cu interlayers was ≈220 MPa, and failure proceeded primarily in the ceramic. Results of fractographic and chemical analyses, and wetting experiments are used to rationalize the observed bond strength characteristics of these systems.

INTRODUCTION

The ability to use and exploit the inherent property advantages of ceramics in a wide range of applications often hinges upon the ability to join ceramic components to other ceramic (or metal) components. The need for reliable, mass production oriented methods of joining that address the demanding requirements of high-temperature high-stress applications is particularly urgent.

Our efforts have focused on the exploration and development of new joining approaches that merge the processing advantages of liquid-state bonding methods (brazing) with the ability to use refractory metal interlayers normally associated with solid-state bonding methods (diffusion bonding). The approaches under investigation also seek to combine the manufacturing benefits of a low processing temperature

To the extent authorized under the laws of the United States of America, all copyright interests in this publication are the property of The American Ceramic Society. Any duplication, reproduction, or republication of this publication or any part thereof, without the express written consent of The American Ceramic Society or fee paid to the Copyright Clearance Center, is prohibited.

with the potential for subsequent use at temperatures that approach or exceed the joining temperature.

BACKGROUND

Numerous strategies for ceramic joining suitable for room temperature and relatively low temperature (≤500°C) applications are available, and details for specific ceramics can be found in several recent reviews [1-3]. The research effort responding to the need for strong joints for demanding high-stress, high-temperature applications has largely focused on solid-state diffusion bonding and liquid-based reactive metal brazing approaches. Many believe that these approaches offer the highest probability of success. Proponents of diffusion bonding have pointed to the flexibility of the approach, the potential for maintaining dimensional tolerances, and the ability to produce refractory joints as key advantages. Proponents of a brazing approach cite the greater flexibility in joint configuration, less stringent surface preparation requirements, freedom from the need to apply a load that may distort components, and the suitability of the method for mass production as key advantages. The ability to produce joints exhibiting high strengths at room temperature by both methods, has encouraged efforts to extend these approaches to the higher operating temperature regime.

For diffusion bonding and brazing, which typically use chemically homogeneous interlayers, efforts to increase the use temperatures for the joined assembly have respectively focused on the use of interlayers with higher solidus and liquidus temperatures. Solid-state diffusion bonding requires that void filling occur by plastic flow and diffusion. Studies in numerous systems indicate that an increase in bonding temperature is beneficial, and minimum processing temperatures of 0.5 T_m are suggested as a general guideline [1, 4]. However, "optimum" properties are often obtained when joining temperatures are in the 0.8-0.9 T_m range [5]. Thus, refractory metal interlayers require high processing temperatures, or high bonding pressures. An excessive bonding pressure may cause distortion of components. Conventional brazing requires the complete melting of the interlayer, and as a result, the brazing temperature of necessity exceeds the intended use temperature. Brazes with higher liquidus temperatures require higher brazing temperatures. At low processing temperatures, kinetic barriers can impede excessive and undesirable chemical interactions between the interlayer and ceramic. At high joining temperatures, such barriers are absent, and underlying thermochemical incompatibilities are more likely to drive the formation of excessively thick or defect-laden interfacial reaction layers. These reaction layers have the potential to degrade mechanical performance, or limit the temperature capabilities of the resulting joint, or possibly both.

Considerable research effort has been devoted to investigating the processing and properties of aluminum oxide-aluminum oxide joints formed by an intervening interlayer consisting of a relatively high melting temperature metal [2]. The

Al_2O_3/Ni system is one that has been the subject of several investigations, and both "brazing" oriented and diffusion bonded oriented studies have been conducted.

Brazing studies and wetting studies are closely linked since a sufficiently low (acute) contact angle for molten braze on the substrate of interest is required to drive capillarity-induced liquid redistribution during joining. Studies exploring the wetting characteristics of Ni on alumina substrates [*e.g.*, 5-9]. have shown that pure molten Ni generally forms obtuse contact angles on Al_2O_3. Thus, for successful brazing, reactive metal additions that promote wetting are necessary. Studies of the effects of Ti [7, 8] and Cr additions [7, 9] on the wetting behavior of Ni-rich alloys on Al_2O_3 have shown that Ti and Cr do reduce the contact angle, and at sufficiently high Ti and Cr levels, contact angles of ≤90° are obtained at 1500°C. However, the high processing temperatures have in part encouraged exploration of alternative solid-state joining routes.

Several studies of diffusion bonding of Al_2O_3 to Ni [11-13] have focused on the effects of time, temperature, and bonding pressure on the resulting failure strengths of joined assemblies. In general the results are consistent with trends observed in other ceramic/metal systems [*e.g.*, 3, 5]; the bonding temperatures required to form strong diffusion bonds are a significant fraction of the melting point of the interlayer, and increasing time and bonding pressure can be beneficial. Numerous factors, including the formation of interfacial reaction layers, can contribute to the development of a maximum bond strength at some intermediate temperature.

There have been a number of studies of diffusion bonding of Ni/sapphire couples. The bonding temperature, Ni purity, and atmosphere can play an important and interactive role in determining the microchemical and microstructural evolution of the Ni/Al_2O_3 interface [13, 14]. A TEM investigation [13] of Ni/Al_2O_3 diffusion bonds prepared at 1390°C under high vacuum conditions (10^{-8} atm) revealed an ≈1 μm thick reaction layer of nickel aluminate spinel, nominally $NiAl_2O_4$. When a specimen diffusion bonded under high vacuum conditions was subsequently annealed for 2 h at 1390°C under very high vacuum conditions (10^{-10} atm) the spinel layer disappeared [14]. Diffusion bonding experiments conducted under high vacuum conditions utilizing oxygen-free and oxygen-containing Ni foils have demonstrated the existence of a threshold oxygen activity for spinel formation at the interface [14], and that the source of the oxygen necessary to drive spinel formation can be the oxygen dissolved in the Ni. The threshold dissolved oxygen content for spinel formation increases with temperature, and is given by the expression

$$\left[\text{at. \% } \underline{\text{O}}_{\text{Ni(s)}}\right] = \exp\left(\frac{-7550 \pm 1200}{T} + 0.84\right)$$

which assumes values of ≈60 at. ppm and 115 at. ppm at 1000°C and 1150°C, respectively [14].

These findings, and others from work on more reactive systems, *e.g.*, Si_3N_4 and SiC, point to and reinforce the processing difficulties that are inherent to conventional joining approaches that rely on chemically homogeneous interlayers. These problems are amplified as the desired use temperatures for the joined assembly increases. As a result of these considerations, the research effort at Berkeley has focused on the development of ceramic-ceramic and ceramic-metal joining methods that allow low temperature processing and subsequent use at a relatively high temperature.

Recent work by Iino [15] and by the authors [16-19] has utilized microdesigned multilayer interlayers that develop a partial layer of a transient liquid phase (TLP); the method has been dubbed partial transient liquid phase (PTLP) bonding. A thin film (cladding) of a low melting point metal or alloy is deposited onto a much thicker foil (core layer) of a more refractory metal or alloy. The two metals or alloys are selected such that the more refractory core metal incorporates or consumes the less refractory cladding component, and produces a suitably refractory alloy. Ideally, such a method melds the best features of solid-state and liquid-based joining methods. The low melting point thin film mimics brazing, and relaxes constraints imposed on surface preparation. By design, only a small amount of liquid is formed, and it is formed only where it is needed. Diffusion of the less refractory metal into the refractory host, reaction of the less refractory metal with the refractory host, or a combination of these processes can lead to the disappearance of the liquid [19]. If necessary, pressureless annealing at higher temperatures can be used to fully homogenize the composition by interdiffusion.

There are numerous potential advantages to the approach. In conventional brazing, the melting point of the brazing alloy must exceed the ultimate use temperature; for high-temperature applications, this implies even higher temperature joining. Using a TLP, the joint is formed at low temperature, and processing is carried out at temperatures *less* than or equal to the ultimate use temperature. This minimizes degradation of the ceramic during bonding, and in cases where ceramics are bonded to temperature-sensitive structural metals, avoids overaging of the metal constituent. In PTLP bonding, the high melting point metal dominates the melting characteristics of the ultimately uniform interlayer, and can be selected to match the thermal expansion characteristics of the ceramic. The high melting point allows the component to be used at high temperature; reduced thermal expansion mismatch will reduce the likelihood of fatigue and failure if the assembly is thermally cycled.

INTERLAYER DESIGN

The simplest interlayer design entails selecting components for the refractory metal core and the cladding metal that form complete solid solutions. Work to date has focused on Cu as the TLP former due to its relatively low melting temperature, and ease of deposition. In addition, considerable prior research on $Cu-Al_2O_3$ diffusion bonding and brazing has been performed [6, 20-23], and several of these have included fracture studies of Cu/Al_2O_3 interfaces [6, 10, 20-23]. Cu is capable of forming strong bonds with Al_2O_3. In the present work, Ni and Ni-Cr alloys were selected as the interlayer core metals. Cu and Ni exhibit complete mutual solubility at elevated temperature, and thus, at temperatures above the melting point of Cu, one expects disappearance of the Cu-rich liquid due to diffusion of Cu into Ni. Ni was selected in part because of its attractive properties and relatively low cost, but also as a starting point for work involving Ni-based superalloys that have good oxidation and corrosion resistance.

EXPERIMENTAL PROCEDURES

Materials:

The polycrystalline Al_2O_3 (Coors, Golden CO) used was ≈98% dense and 99.5% pure and contains a small amount of silicate phases. The material was in the form of $19.5 \times 20.0 \times 22.5$ mm^3 blocks. The blocks were polished with successively finer grits of diamond in an oil slurry; the final polish used 6 μm diamond paste. Following polishing, samples were cleaned in organic solvents (*e.g.*, isopropyl alcohol), dried, and then air-annealed (≥2 h) at 1000°C to remove organic surface contaminants.

The Ni foil (Puratronic grade; Johnson Matthey Material Technology, Hertfordshire, UK) was ≈100 μm thick, and reportedly 99.994+ wt% pure with respect to metallic impurities. An oxygen analysis on the foil indicated levels were below the detection limit of the technique employed, nominally 10 wt ppm (≈37 at. ppm) dissolved oxygen. The 80Ni20Cr foil (Goodfellow, Malvern, PA) was ≈125 μm thick, and contains Si as its major impurity. Foils were cleaned in isopropyl alcohol, rinsed in nanopure distilled water (resistivity 18.3 MΩ cm), and dried with a heat gun prior to use.

The starting material for the Cu deposition was a 99.8% pure commercial Cu wire of unknown oxygen content. To prepare small Cu sources for deposition, cleaned pieces of wire were melted in W baskets under a vacuum ranging from 2 to 9×10^{-6} torr.

Coating

Thin layers of Cu were deposited directly onto the polished and cleaned Al_2O_3 blocks in an evaporator. After introducing the Al_2O_3 blocks, the chamber pressure was reduced to $\approx 3 \times 10^{-6}$ torr, and fluctuated between 7 and 10×10^{-6} torr during deposition. Weight gain and profilometry based methods were used to establish the thickness of the Cu layer. Typically the average Cu thickness was ≈3 μm ±≈0.3 μm. The resulting Cu/Ni and Cu/80Ni20Cr thickness ratio leads to average interlayer compositions of ≈94 at % Ni, 6 at % Cu, and 75 at % Ni, 21 at % Cr, and 4 at % Cu. Cr deposition was performed prior to Cu deposition using the same system; the conditions were designed to produce an ≈10 nm thick Cr layer.

PTLP Bonding

The block/foil/block assembly was prepared immediately after coating, and bonded in a graphite die in a hot press. The temperature was raised to 1150°C at 4°C/min, maintained at 1150°C for 6 h, and then lowered to room temperature at 2°C/min. During this cycle, the vacuum in the press was in the range of 8×10^{-6} to 2×10^{-5} torr, and a pressure of ≈5.1 MPa was maintained on the assembly.

Beam Preparation

Several plates were cut from the bonded block assembly using a diamond wafering saw. Subsequently, one surface of each plate was polished using progressively finer grit size (30, 15, 6 μm) diamond paste in an oil slurry. Beams of approximately square cross section (≈3.4 × 3.4 mm or 3.0 × 3.0 mm) were cut from the polished plates with the metal interlayer at the beam center. The tensile edges of flexure beams were beveled to remove machining flaws that could initiate failure.

Flexure Testing

Beams were tested at room temperature using four-point bending with an inner span of 8 mm, outer span of 25 mm and a displacement rate of 0.05 mm/min. Strengths were calculated from the load at failure using standard relationships derived for monolithic elastic materials. Prior work had determined that four-point bend strengths of (unbonded) Al_2O_3 beams prepared from the same source material and tested under the same conditions are of the order of 280 MPa.

Microstructural and Microchemical Characterization

The microstructure of cross sections and fracture surfaces were examined using scanning electron microscopy (SEM), and the chemistry of the interlayer region and fracture surfaces was evaluated using energy dispersive spectroscopy (EDS).

RESULTS AND DISCUSSION:

Microchemical Characterization

EDS line scans perpendicular to the ceramic/metal interface indicate that substantial homogenization occurs during the bonding cycle. For the Cu/Ni/Cu interlayer, homogenization is essentially complete, Figure 1a; for the thicker Cu/80Ni20Cr/Cu interlayer homogenization is substantial but not complete, Figure 1b. In both cases, the solidus temperature of the interlayer is well above the original processing temperature (1150°C); for the Cu/Ni/Cu interlayer the solidus temperature is ≈1430°C [24] and thus approaches 1453°C, the melting temperature of pure Ni.

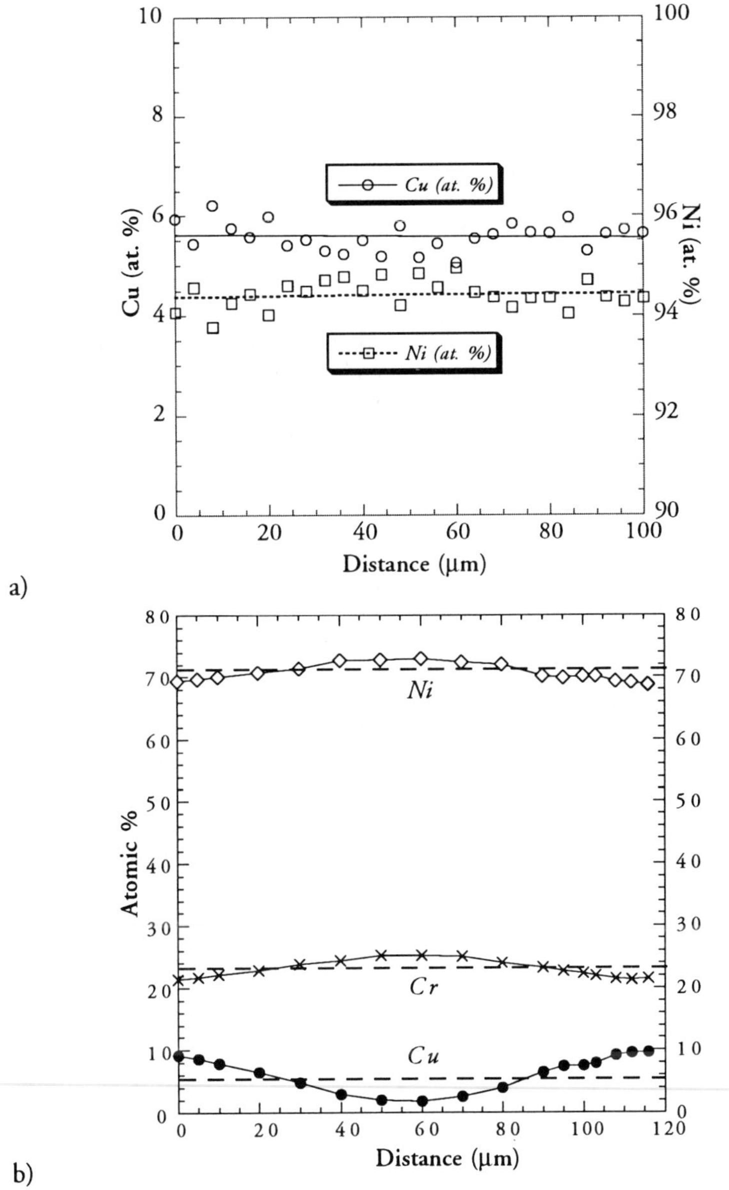

Figure 1 EDS scans of interlayer region of a) Cu/Ni/Cu and b) Cu/80Ni20Cr/Cu interlayers after bonding cycle. Substantial interdiffusion and homogenization has occurred.

Fracture Behavior of Cu/Ni/Cu Bonded Alumina

The results of flexure tests of as-bonded beams are shown in Figure 2. Failure strengths of as-processed beams ranged from ≈61 to 267 MPa with an average value of 160 MPa. The overall scatter in the failure strengths of as-processed beams, as measured by the standard deviation, is larger than observed in Cu/Pt/Cu [16] and Cu/Nb/Cu [18] bonded assemblies. The highest strength beams had strengths approaching the average strength of the unbonded ceramic, similar to behavior observed in other PTLP bonded systems. Failure in these high strength beams originated in the ceramic, and proceeded through the ceramic. Although the potential for forming high strength joints is demonstrated by such results, more generally, the bonded samples failed along one or both of the ceramic/metal interfaces. Fractography showed that although ceramic-metal contact was intimate over much of the interface, there were regions in which the contact was incomplete. Although these unbonded regions are not large individually, they occurred with sufficient frequency in some regions that collectively they acted as a large flaw. A bluish hue to the fracture surface provided visual indications of spinel formation at the alumina/interlayer interface during bonding, and spinel formation was verified by x-ray measurements. Post-bonding anneals (10 h at 1000°C) in air and in gettered argon appeared to intensify the blue hue and degrade the strength, with anneals in argon having the greater affect. A discussion of this issue is available in reference 17.

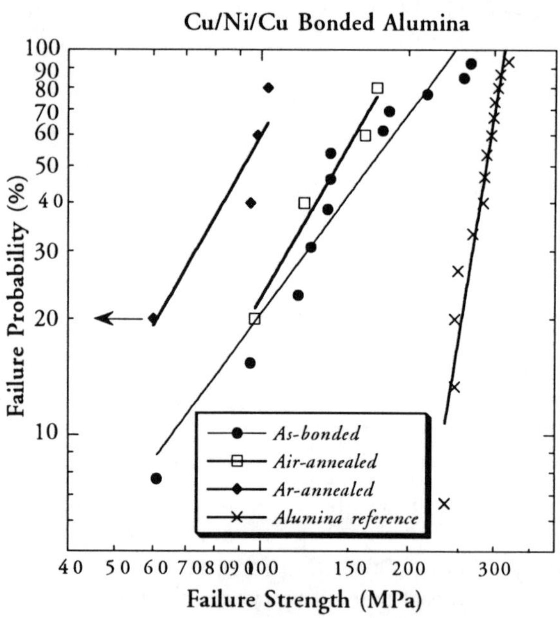

Figure 2

Plot of failure probability versus room temperature failure strength for Cu/Ni/Cu interlayer bonded alumina and unbonded alumina. In as-bonded specimens, strength variations were at least partially related to differences in the degree of ceramic/metal contact achieved during bonding. Anneals of 10 h duration at 1000°C in air and in gettered argon appeared to degrade strength. Spinel formation appears to weaken the interface and thereby contribute to the loss of strength [17].

The observations of regions indicating that ceramic/metal contact along the alumina/interlayer interface was not complete in samples with lower bend strengths suggested that reactive metal additions to the TLP layer might improve the strength distribution. Two methods of introducing a reactive metal were explored, one involving a Cr precoat and the other involving the use of an 80Ni20Cr core layer.

Sessile drop experiments were performed at 1150°C. The experiments examined the contact angle of pure copper on 99.5% alumina, copper with 10 wt % 80Ni20Cr added on 99.5% alumina, and of pure Cu on 80Ni20Cr. As shown in Figure 3, Cr additions have a beneficial effect on the wetting of a Cu-rich droplet, reducing the contact angle. The progressive decrease in contact angle may be a consequence of progressive dissolution of the 80Ni20Cr pieces embedded in the Cu, and longer time contact angles may more accurately reflect the full effect of the Cr addition.

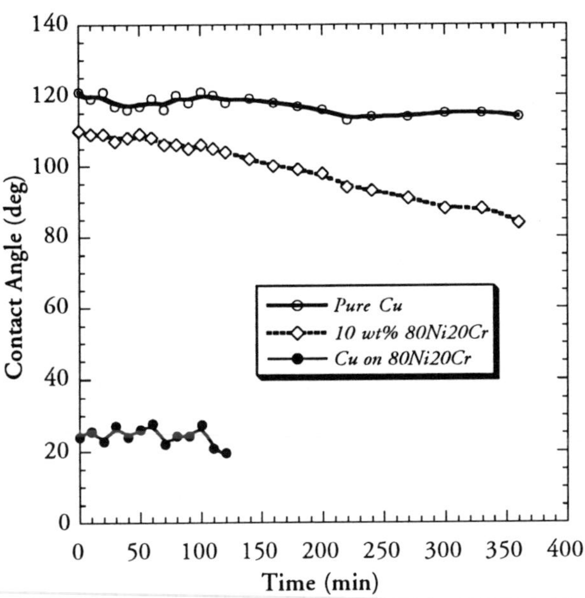

Figure 3

Wetting behavior at 1150°C of pure Cu and Cu with 10 wt % 80Ni20Cr alloy pieces added on the same 99.5% alumina used in the bonding study. Dissolution of the 80Ni20Cr gradually increases the Cr content, and is believed to contribute to the progressive decrease in contact angle. Pure Cu forms an acute contact angle on 80Ni20Cr, and the sum of the contact angles formed by the Cu-rich liquid on the alumina and 80Ni20Cr foil is well below the 180° value required for wetting.

When Cr was added as a thin precoat or bond layer on the Al_2O_3 prior to Cu deposition, the average strength of bonded assemblies (163 MPa) was virtually unchanged relative to the Cu/Ni/Cu bonded assemblies (160 MPa), Figure 4. However, a higher fraction of the total beams tested failed in the ceramic at stress levels approaching those required to fracture the unbonded ceramic beams, and the spread in strength, as measured by the standard deviation increased from 63 MPa to 84 MPa. Those samples with lower strengths failed along the interface. Possibly, the

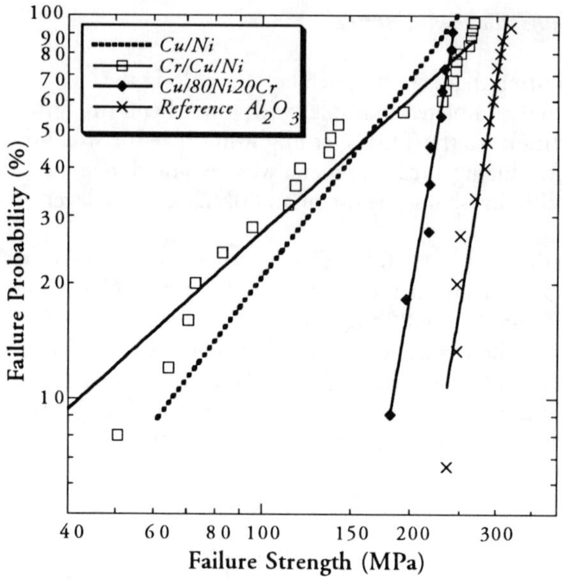

Figure 4

Comparison of room temperature fracture behavior of alumina bonded with Cu/Ni/Cu, Cr/Cu/Ni/Cu/Cr and Cu/80Ni20Cr/Cu multilayer interlayers. Note the cluster of high strength bonds achieved with the Cr coating, and the more consistently high strengths obtained using an 80Ni20Cr core layer.

coating was not sufficiently thick or sufficiently uniform to have the desired effect on the strength distribution.

Adding Cr as a precoat has the advantage of placing it where it will have maximum benefit. However, a thin Cr layer can be dissolved and incorporated in the Cu-rich liquid *and* in the Ni. The use of a Ni-Cr alloy for the core layer provides a more substantial Cr reservoir, but increases the yield stress. Ductile (low yield stress) interlayers are beneficial in reducing both the magnitude of the thermal expansion mismatch induced tensile stresses that develop in the ceramic during cooling from the bonding temperature, and the spatial extent of the zone of tensile stress. Despite an increased yield strength relative to pure Ni (some additional hardening due to Cu incorporation is expected for both core layers), the use of a 80Ni20Cr core layer resulted in a substantial improvement in the strength distribution. The average strength for 10 specimens approached 230 MPa, the highest average obtained to date in PTLP bonded assemblies. In addition, failure occurred either entirely within the ceramic, or primarily in the ceramic with limited crack propagation along the alumina/interlayer interface, suggesting that Cr "strengthens" the interface. The standard deviation in strength (±19 MPa) is comparable to that characterizing the unbonded alumina. The results and failure mode seem to suggest that the strength distribution reflects the flaw population of the ceramic, and use of a higher strength alumina might shift the strength distribution of bonded assemblies to correspondingly higher values. It is tempting to attribute the shift of the distribution relative to that of unbonded alumina to the effect of residual stress. Although the weakest samples failed within 0.1 mm of the interlayer, where the residual stresses would be most intense, others failed at distances several mm from the interlayer (where residual stress should be minimal), and at stresses lower than the lowest stresses observed in unbonded alumina. Further study of this issue is necessary.

SUMMARY AND CONCLUSIONS

Alumina ceramics have been joined using Ni-based multilayer interlayers. The interlayers are designed to form a thin layer of a transient liquid phase that allows joining at relatively low temperature while reducing only slightly the temperature capability of the interlayer relative to that of the pure core layer. Changes in interlayer design and TLP chemistry that improve wetting have been instituted, and the capability of forming strong joints has thereby been demonstrated. The temperature dependence of the joint properties must now be evaluated to determine their potential for use in high-temperature high-stress applications [25].

ACKNOWLEDGMENTS

This research was supported by the Director, the Office of Energy Research, Office of Basic Energy Sciences, Materials Sciences Division of the U.S. Department of Energy under Contract No. DE-AC03-76SF00098. We are grateful for the contributions of M. L. Shalz to the success of the research program. Helpful discussions with R. M. Cannon, Y. Iino, M. Koizumi, M. G. Nicholas, J. A. Pask, and R. O. Ritchie are acknowledged.

REFERENCES

1. K. Suganuma, Y. Miyamoto and M. Koizumi, "Joining of Ceramics and Metals," Ann. Rev. Mater. Sci., 18, 47-73, (1988).
2. G. Elssner and G. Petzow, "Metal/Ceramic Joining,"ISIJ *International*, **30** [12] 1011 (1990).
3. M. G. Nicholas and D. A. Mortimer, "Ceramic/metal joining for structural applications," *Mater. Sci. & Tech.*, **1**, [9], 657-65 (1985).
4. B. Derby, "Diffusion bonding," pp. 94-112 in JOINING OF CERAMICS, M. Nicholas, Ed., Chapman and Hall, London (1990).
5. R. V. Allen and W. E. Borbidge, "Solid-state metal-ceramic bonding of platinum to alumina," *J. Mater. Sci.,* **18**, [9], 2835-43 (1983).
6. M. G. Nicholas, R. R. D. Forgan and D. M. Poole, "The Adhesion of Metal/Alumina Interfaces," *J. Mater. Sci.,* **3**, [1], 9-14 (1968).
7. J. E. Ritter and M. S. Burton, "Adherence and Wettability of Nickel, Nickel-Titanium Alloys, and Nickel-Chromium Alloys to Sapphire," *Trans. A.I.M.E.*, **239**, [1], 21-26 (1967).
8. Yu. V. Naidich and V. S. Zhuravlev, "Adhesion, Wettability, and Reaction of Titanium-Containing Melts with High-Melting Oxides,' *Refractories* (USSR), **15**, [1-2] 55-60 (1974).

9. R. M. Crispin and M. G. Nicholas, "The wetting and bonding behavior of some nickel alloy-alumina systems," *J. Mater. Sci.,* **11**, [1], 17-21 (1976).

10. J. Klomp, "Solid-State Bonding of Metals to Ceramics," in SCIENCE OF CERAMICS, Vol. 5, edited by C. Brosset and E. Knopp, The Swedish Institute for Silicate Research, (1970); pp. 501-22.

11. M. C. Gee, "Surface Reaction Between Alumina and Nickel at High Temperatures," *Brit. Ceram. Proc.,* **34**, 261-71 (1984).

12. C. A. Calow and I. T. Porter, "The Solid-State Bonding of Nickel to Alumina," *J. Mater. Sci.,* **6**, [2], 156-63 (1971).

13. J. A. Wasynczuk and M. Rühle, "Microstructural Characterization of Ni/Al_2O_3 Diffusion Bonds," pp. 341-48 in CERAMIC MICROSTRUCTURES 86," J. A. Pask and A. G. Evans, Eds., Plenum Press, New York, (1987).

14. K. P. Trumble and M. Rühle, "The thermodynamics of spinel interphase formation at diffusion-bonded Ni/Al_2O_3 interfaces," *Acta Metallurgica et Materialia,* **39**, [8], 1915-24, 1991.

15. Y. Iino, "Partial transient liquid-phase metals layer technique of ceramic-metal bonding," *J. Mater. Sci. Lett.*, **10**, [2], 104-106 (1991).

16. M. L. Shalz, B. J. Dalgleish, A. P. Tomsia and A. M. Glaeser, "Ceramic Joining I. Partial Transient Liquid Phase Bonding of Alumina *via* Cu/Pt Interlayers," *J. Mater. Sci.*, **28**, [6], 1673-84 (1993).

17. M. L. Shalz, B. J. Dalgleish, A. P. Tomsia and A. M. Glaeser, "Ceramic Joining II. Partial Transient Liquid Phase Bonding of Alumina *via* Cu/Ni/Cu Multilayer Interlayers," *in press, Journal of Materials Science.*

18. M. L. Shalz, B. J. Dalgleish, A. P. Tomsia, R. M. Cannon and A. M. Glaeser, "Ceramic Joining III. Bonding of Alumina via Microdesigned Cu-Nb Interlayers," *in press, Journal of Materials Science.*

19. M. L. Shalz, B. J. Dalgleish, A. P. Tomsia and A. M. Glaeser, "New Approaches to Joining Ceramics For High-Temperature Applications," *Ceramic Transactions,* **35**, 301-22, (1993)

20. R. M. Crispin and M. G. Nicholas, "Alumina-Copper Diffusion Bonding," *Ceram. Eng. Sci. Proc.*, **10**, [11-12], 1575-81 (1989).

21. C. Beraud, M. Courbiere, C. Esnouf, D. Juve and D. Treheux, "Study of Copper-Alumina Bonding," *J. Mater. Sci.*, **24**, [12], 4545-54 (1989).

22. Y. Yoshino, "Role of Oxygen in Bonding Copper to Alumina," *J. Am. Ceram. Soc.*, 72, [8], 1322-27 (1989).

23. S. T. Kim and C. H. Kim, "Interfacial Reaction Product and Its Effect on the Strength of Copper to Alumina Eutectic Bonding," *J. Mater. Sci.,* 27, [8], 2061-66 (1992).

24. BINARY ALLOY PHASE DIAGRAMS, Vol. 2, T. B. Massalski, ed., pp. 1442-46, ASM International, Metals Park, OH (1990).

25. M. R. Locatelli, B. J. Dalgleish, A. P. Tomsia and A. M. Glaeser, unpublished research.

AN INNOVATIVE APPROACH TO JOINING METALS TO CERAMICS

Sumit Guha, Raouf Loutfy, James Withers and Panos Kiousis*
MER Corp., 7960 S. Kolb Rd., Tucson AZ 85706
*: Dept. of Civil Engg. & Mechanics, University of Arizona, Tucson,AZ 85721.

ABSTRACT

In contrast to most conventional approaches in modeling metal-ceramic couples, in the present investigation, the elastic behavior of the ceramic was modeled with a limit defined by the Drucker-Prager model. The resulting model affords excellent qualitative and quantitative ability to predict cracking in the ceramic component, if any, during the post-brazing cool-down stage. Several plane-stress geometries were investigated, and the numerical predictions compared with experimental results. In each and every case, the model successfully predicted cracking in the ceramic component, as well as predicted the crack initiation site which compared favorably with experimental observations. This model was successfully utilized in the design of a crack-free joint between Ceralloy 147-3 Si_3N_4 and 1040 Steel without using soft metals as interlayers.

INTRODUCTION

While researchers worldwide have focussed on functionally graded, low CTE or low yield strength interlayers and brazes as the important elements of designing low residual stress joints between metals and ceramics, the development of an accurate numerical model that could calculate interfacial residual stresses accurately enough to predict cracking, if any, and predict potential crack initiation sites in the ceramic component, has received less attention. The outline of such a numerical model is presented here and the model predictions validated with experimental observations.

In the past, several investigators have utilized finite element modeling (FEM) techniques to estimate the interfacial residual stresses when bonding ceramics to metals. Naka et al.[1] calculated the stress distribution for a Si_3N_4-stainless steel joint and indicated that the major

To the extent authorized under the laws of the United States of America, all copyright interests in this publication are the property of The American Ceramic Society. Any duplication, reproduction, or republication of this publication or any part thereof, without the express written consent of The American Ceramic Society or fee paid to the Copyright Clearance Center, is prohibited.

principal stresses were almost normal to the interface, and that the highest stresses were at the joint edge. Similar conclusions were drawn by Yamada et al.[2] and Suganuma et al.[3] from FEM calculations for Al_2O_3-steel couples joined using different interlayer materials. Unfortunately, for these calculations, an elastic behavior was assumed for both steel and the ceramic components which is inaccurate. Hamada et al.[4,5], although allowing for a plastic behavior for the metallic components of the joint, did not account for the temperature dependence of thermo-physical properties (coefficient of thermal expansion, Youngs modulus) and mechanical properties (yield strength etc.). Zhou et al.[6] accounted for the temperature dependence of thermo-physical and mechanical properties but did not account for any tension-compression strength asymmetry exhibited by most ceramics. Interestingly, none of the above numerical investigations attempted to predict whether a brazed joint will survive cracking during cool-down or correlate predicted crack initiation sites with experimental observations.

Numerical modeling investigations in the past have suffered from the following deficiencies:

- Unlimited load carrying ability for the ceramic - the assumption of a linear elastic behavior for the ceramic imposes no limit to the load carrying ability of ceramic elements in the mesh. Thus, ceramic elements continue supporting load well beyond their failure loads which is unrealistic.
- Tension-compression strength asymmetry of ceramic - most ceramics exhibit a few hundred MPa tensile strength but several GPa compressive strength. This failure strength asymmetry needs to be accounted for in order to obtain realistic quantitative estimates of interfacial residual stresses and predict whether a specific joint design will be resistant to cracking.
- Maximum Principal Stress as a measure of stress intensity - the maximum principal stress is a weak measure of the degree to which a ceramic is stressed, i.e. whether it will crack or not, and a poor indictor of potential crack initiation sites. Since a compressive confinement would lead to higher failure strength for the ceramic, a failure criterion that accounts for the effects of confinement should be used.

In the present model, the deficiencies alluded to above have been addressed as follows:

- a limit to the load carrying ability of ceramic elements was imposed,

- the tension-compression failure strength asymmetry was accounted for by modeling the ceramic with a Drucker-Prager failure criterion, and a non-associative flow rule,
- The Drucker-Prager criterion is expressed by Failure function intensity,

$$F = \sqrt{J_{2D}} - \alpha J1 - K \qquad \text{(eqn.1)}$$

where, for plane stress

$$J_{2D} = 1/6\{\sigma_x^2+\sigma_y^2+(\sigma_x-\sigma_y)^2\}+\tau_{xy}^2 \qquad \text{(eqn.2)}$$

$$J_1 = \sigma_x+\sigma_y \qquad \text{(eqn.3)}$$

$$K = \sigma_c(1/\sqrt{3} + \alpha) \text{ ; } \alpha = 1/\sqrt{3}*(\sigma_t-\sigma_c)/(\sigma_t+\sigma_c) \qquad \text{(eqn.4)}$$

σ_t and σ_c, respectively, represent the tensile and compressive strength of the ceramic. The temperature dependence of all material parameters were duly accounted for.

Previous FEM investigations on metal-ceramic joining have assumed von Mises elastoplastic behavior with isotropic hardening for metals. This assumption was retained but a closer representation of the real-life stress-strain behavior exhibited by steel, e.g. Luders yielding, non-linear hardening, finite material ultimate strength, was incorporated into the model. As in the case of the ceramic, the assumed material parameters for the metallic component were also temperature dependent. The new model was validated for a series of plane stress geometries involving brazing a plate of Si_3N_4 (grade Ceralloy 147-3) to a plate of 1040 mild steel.

EXPERIMENTAL TECHNIQUES

Since the accuracy of a quantitative model depends on the accuracy of material parameters used, material parameters were selected carefully, often in consultation with the manufacturer. The data utilized in these calculations for steel and Si_3N_4 are presented in Figure 1 and Table I. For Si_3N_4 grade Ceralloy 147-3, the manufacturer was able to provide only bend strength but no tensile strength data. Since both bend and tensile strength data was available in published literature for PY6 grade of ceramic, a temperature dependent tensile strength-to-bend strength ratio was computed for Si_3N_4 (see Table I). Tensile strength data of Ceralloy 147-3 at different temperatures were derived based on the manufacturer's bend

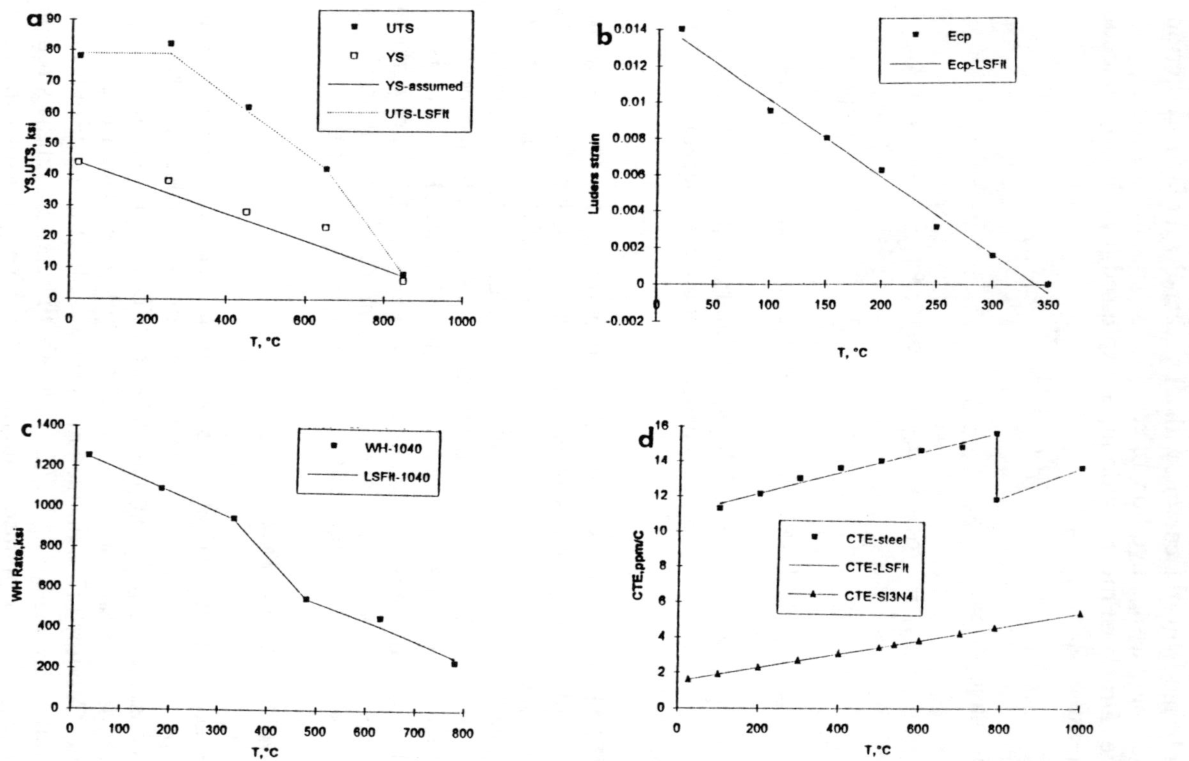

Figure 1. **FEM material parameter data: (a) yield/ultimate strength of steel (b) Luders strain in steel (c) work-hardening rate of steel (d) CTE for steel & Si_3N_4.**

strength data and the computed strength ratio. The compressive strength was assumed to be 400ksi (2.8GPa) at room temperature, and assumed to decrease at the same rate as the tensile strength. The α (= - 0.41) and K (see Table I) values were computed based on these strength values as per eqn.4.

The Si_3N_4 and steel plates were machined in-house as close as possible to the FEM geometries investigated and brazed using CuSil-ABA foils of 0.002" thickness. Brazing was carried out in a high vacuum furnace (10^{-5} Torr) at 900°C for 10 minutes. For numerical calculations, however, the stress-free "lock-up" temperature was assumed to be 780°C corresponding to the melting point of the braze filler metal. Brazed joints were furnace-cooled down to room temperature. The brazed pieces were square-shaped with dimensions of ≈1.5"x1.5" and a plate thickness of 0.3".

Table I. Strength (in ksi) Parameters for Si_3N_4.

T °C	σ_t PY6	σ_b PY6	ratio	σ_b 147-3	Adj. σ_t 147-3	Est.σ_c	K
25	89.8	125.2	0.72	95	68.1	400	67.2
1000	64.9	106.8	0.61	92	55.9	328	55.1
1200	45.0	99.7	0.46	67	31.2	183	30.8
1400	22.2	71.6	0.31	50	15.5	91	15.3

RESULTS & DISCUSSION

The results of FEM calculations for different joint configurations are shown in Figures 2-5 and experimental Si_3N_4-steel joints are shown in Figure 6. In each case, only one half of the structure was analyzed. Thus, the FEM mesh for the right half of each design is shown in Figures 2a-5a. The black triangle at the bottom of the mesh represents a hinge; rollers were placed at each node along the axis of symmetry. For the sake of numerical accuracy, a large number of nodes were utilized, i.e. 739 nodes for half-structures in Figures 2, 4 and 5, and 411 nodes for Figure 3. Further, due to uncertainity regarding the mechanical properties of the braze filler metal (the manufacturer, WESGO, was unable to supply us with high temperature data for CuSil-ABA), the effect of a thin (<100μm) braze

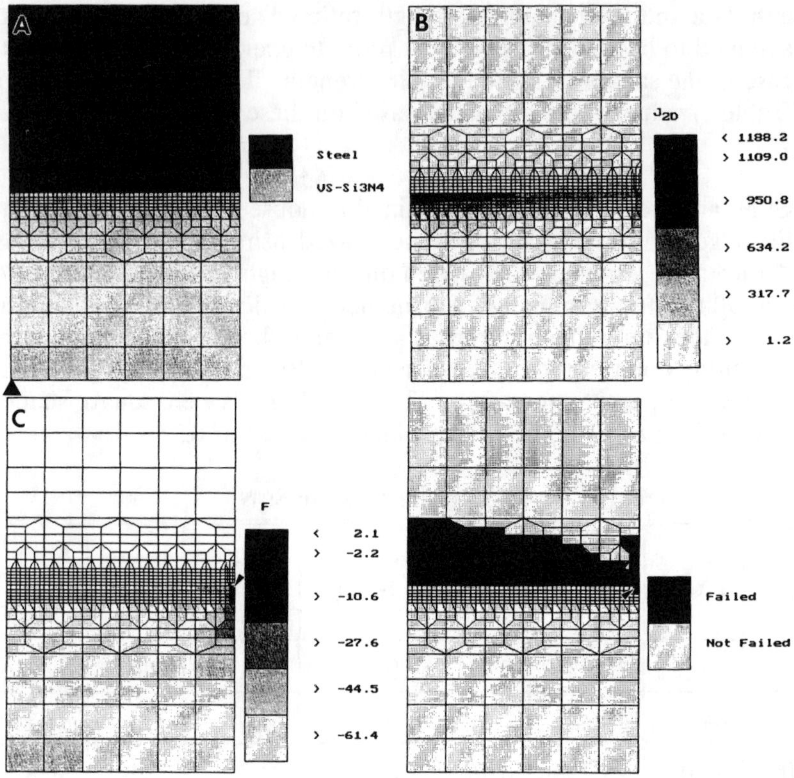

Figure 2. **Stress analysis for the case of steel bonded directly to Si_3N_4.**

layer on the overall stresses was ignored. Thus, the stresses reported here are a conservative estimate. Despite ignoring the effect of braze layer, the experimental results still correlated very well with model predictions.

Figure 2 represents the case when a rectangular plate of Si_3N_4 is brazed to a 1040 steel plate of similar dimensions. The resulting stresses (J_{2D}) and failure function intensity (F) are plotted in Figure 2b-2c respectively. J_{2D} expresses the amount of shear applied to an element, while J_1 expresses the confinement of the same element. Confinement-insensitive materials such as steel yield when shear (i.e. J_{2D}) reaches a fixed value. Confinement-sensitive materials such as ceramics fail at the worst combination of shear and confinement as is expressed by the eqn. 1. This rarely coincides with the maximum shear. Thus, although J_{2D} is highest at the center, the region that is at the maximum risk of failure is calculated to be at the edge, see arrow in Figure 2c. The "yield/no yield" diagram in Figure 2d indicates the cornerelements of the ceramic to have yielded (cracked), hence, cracks were expected to originate from the corner of the structure. This is borne out by practical experiments, see Figure 6a. The present analysis is designed to identify the temperature of the first failure in the ceramic. This is why the unconventional positive values of F for the failed elements are recorded. (F≥0 satisfies the ceramic failure criterion given above in equation 1). The cracking in the ceramic was predicted to occur at a temperature of 380°C.

Figure 3a shows a joint design where the angle in the steel between the interface and the free edge was 30° instead of 90°. The Failure function intensity values, F, are shown in Figure 3b which predicts cracking in ceramics (since $F > 0$) and a crack initiation site which is slightly inside the corner (arrowed). This prediction also correlates well with experimental observation, see Figure 6b.

The effect of two small grooves (position indicated with two black bars in Fig. 4b) near the edge of the joined piece is shown in Figure 4a-b where the spacing between the grooves is roughly twice the thickness of the grooves. It is clear that out of the five possible crack initiation sites (one at the corner and two at the root of each groove), the model predicts crack initiation from the inside edge of the second groove (arrowed). Experimentally, this prediction was also proven to be correct.

Finally, the stress invariant (J_{2D}), F values and "yield/no yield" plot for a design with multiple grooves are shown in Figure 5a-d where, the model

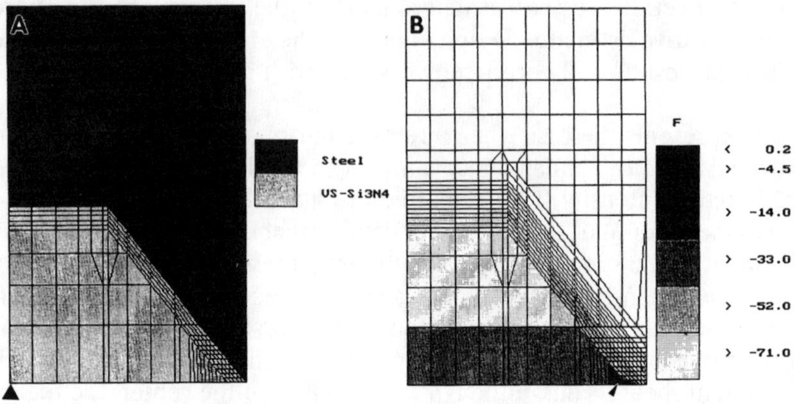

Figure 3. **Stress analysis for the case where steel is bonded to Si_3N_4 such that the angle between interface and free edge in steel is 30° instead of 90° (a) FEM mesh (b) F values.**

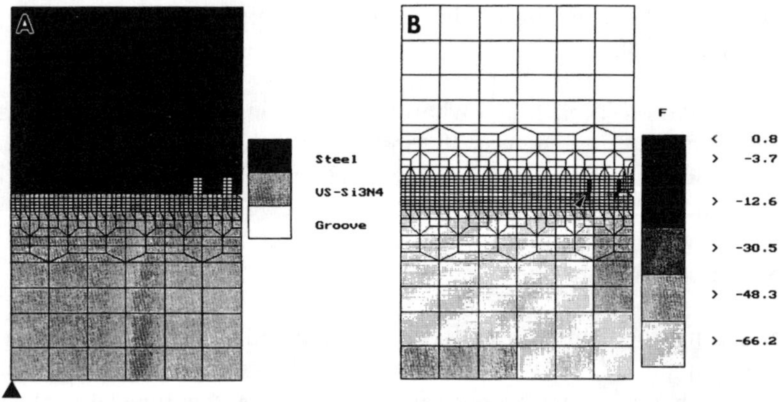

Figure 4. **Stress analysis results for the case where Si_3N_4 is bonded to a steel piece with two outer edge grooves (a) FEM mesh (b) F values.**

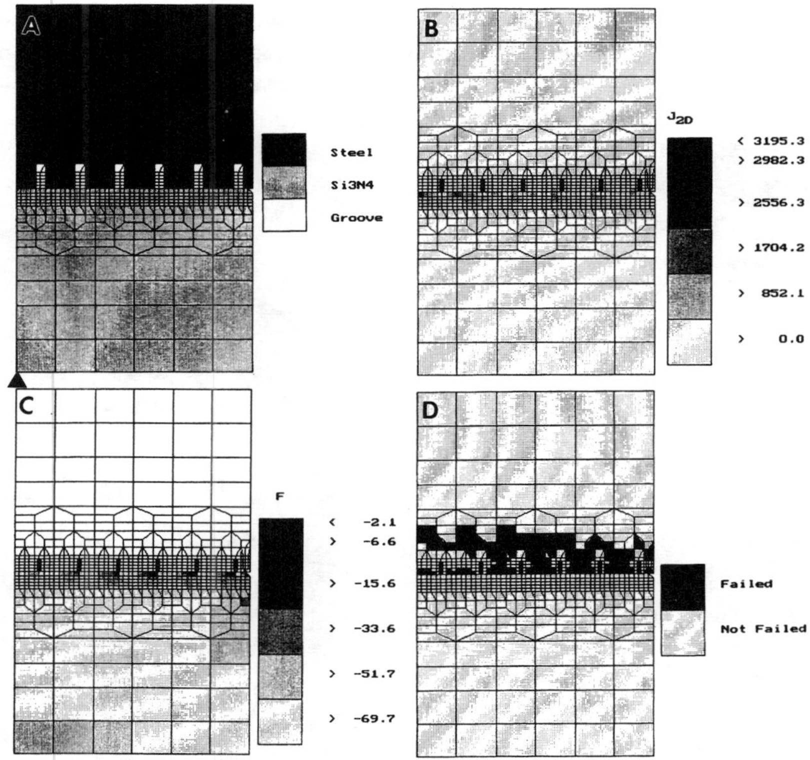

Figure 5. **Stress analysis results for the case where Si_3N_4 is bonded to a steel piece with regularly spaced grooves running throughout the steel piece (a) FEM mesh (b-c) J_{2D} & F values (d) yield diagram.**

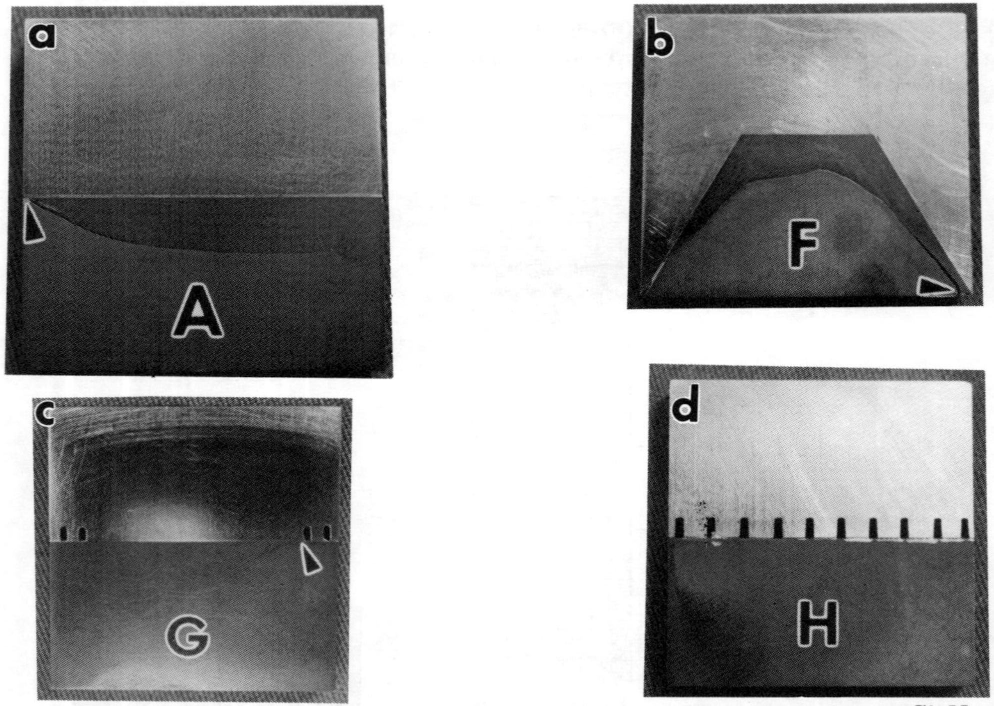

Figure 6. Practical brazed specimens of: (a) 1.5"x0.75" steel piece bonded to a Si_3N_4 piece of similar dimensions (b) steel brazed to ceramic with a 30° angle between interface and free edge in steel (c) Si_3N_4 brazed to steel with two outer edge grooves at each end (d) Si_3N_4 brazed to steel with regularly spaced grooves running throughout the steel.

predicts F < 0. The "yield/no yield" plot in Figure 5d indicates that, in contrast to previous joint configurations investigated, this design should be resistant to cracking during the post-brazing cool-down stage.

Experimentally, when Si_3N_4 was joined to a grooved steel piece, no cracking was observed. Thus, the experimental observations again validate the numerical calculations.The result was reconfirmed through a second set of specimens. Further, when the softer CuSil-ABA was substituted with a stronger braze, Cerametil, from Lucas Milhaupt Corp., the ceramic was still free of any cracks indicating that the joining technique was independent of the hardness of the braze. The significance of this interface design approach is that no soft metal interlayers are necessary to affect a crack-free bonding between the ceramic and the metal. This, in turn, would be beneficial by not limiting the high temperature performance of the joint; joints bonded with soft metal interlayers would be expected to be extremely weak at elevated temperatures.

The normal and shear stresses in the surface row of ceramic elements forjoint designs in Figure 2a and Figure 5a are shown in Figure 7a-b respectively. It is clear that when a ceramic is brazed to steel, the σ_y is essentially zero along the interface and reaches its highest tensile values only near the edge. By contrast, σ_x is compressive throughout with the highest values being attained at the center of the structure. The introduction of the grooves in steel dramatically alters the stress distribution. The resulting stresses are oscillatory with a net result of the reduction in Failure Intensity Function (F) values. Note that the stresses plotted in Figure 7 are average stresses within the element; the stress at a given gauss point within each element may be considerably higher than the average stress.

In summary, the good correlation established between model predictions and experimentals observations validate the proposed modeling approach. Further, the successful brazing of a grooved steel piece directly to a ceramic piece without the use of soft metal interlayers suggests an new approach to design of high strength (or low residual stress) ceramic-metal joints.

CONCLUSIONS

1. An improved approach towards accurate numerical estimation of the residual stresses at the interface of a ceramic-metal joint was

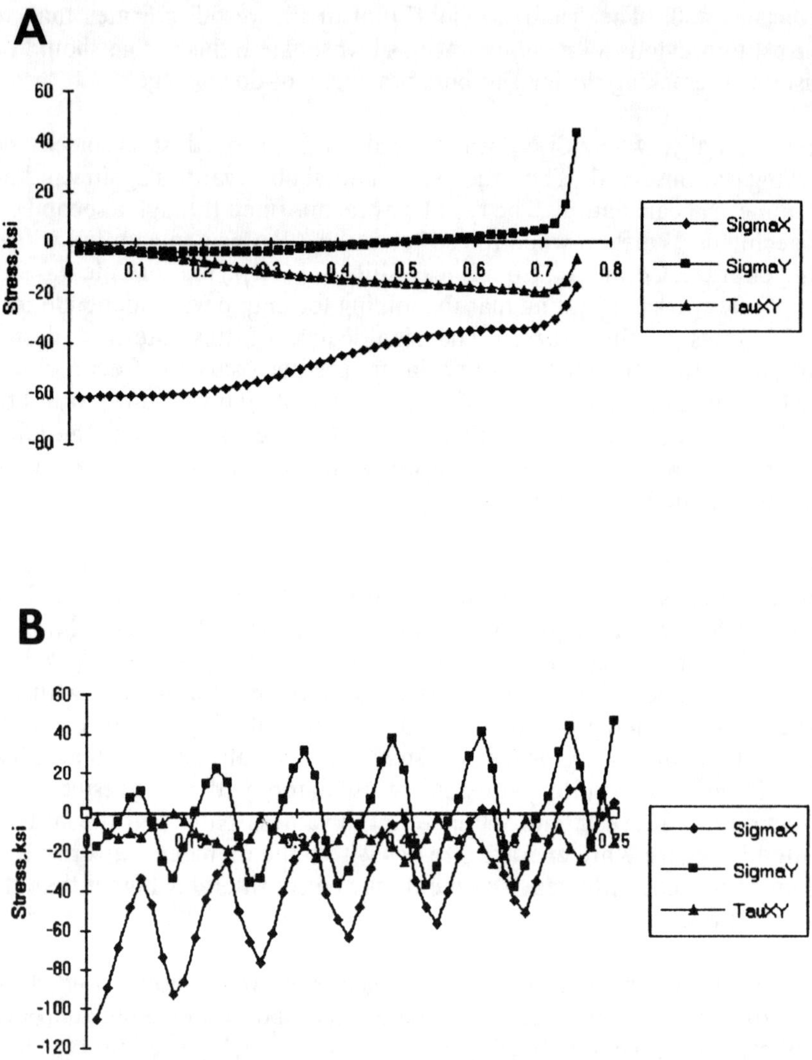

Figure 7. **Interfacial normal and shear stress distribution for: (a) steel brazed to ceramic (b) ceramic brazed to a steel piece with regularly spaced grooves running throughout the steel.**

presented which included:

- limiting the maximum stress tolerated by ceramic elements to be the failure strength of ceramic; the failure strength itself is determined by considering the tension-compression asymmetry,
- the ceramic failure criterion is based on a critical combination of confinement and shear rather than the maximum principal stresses.

2. The model was evaluated with a variety of interface designs between Ceralloy 147-3 Si_3N_4 and 1040 steel.
3. In each and every case, the model successfully predicted:
   - whether the ceramic will crack or not,
   - the crack initiation site.
4. The model was successfully utilized to joint ceramic to steel without using soft metal interlayers.

ACKNOWLEDGEMENTS

The financial support from Naval Sea Systems Command under contract no.N00024-91-C-4034 is acknowledged. Further, the contributions of program technical monitor Mr. G. Singh and MER personnel namely Mr. Tu Ngyuen and Mr. B. Howard are gratefully acknowledged.

REFERENCES

1. M. Naka, T. Tanaka, and I. Okamoto, Trans. Jpn. Weld. Res. Inst., 14(2), (1985) 85.
2. T. Yamada, K. Yokoi, and A. Kohno, J. Mater. Sci., 25 (1990) 2188.
3. K. Saganuma, T. Okamoto, Y. Miyamoto, M. Shimada, and M. Koizumi, Mater. Sci. Tech., 2(11), (1986) 1156.
4. K. Hamada, M. Kureishi, T. Enjo and K. Ikeuchi: IIW Doc.I-811-86/OE, 1986.
5. K. Hamada, M. Kureishi, T. Enjo and K. Ikeuchi: IIW Doc.I-812-86/OE, 1986.
6. Y. Zhou, F.H. bao, J.L. Ren, and T.H. North, Mater. Sci. Tech., 7 (1991) 863.

Ceramic-Metal Composites

CERAMIC-METAL COMPOSITES WITH INTERPENETRATING NETWORKS

N. Claussen, M. Knechtel, H. Prielipp and J. Rödel*
Technische Universität Hamburg-Harburg, Advanced Ceramics Group
21073 Hamburg, Germany
*Technische Hochschule Darmstadt, Nichtmetallische-Anorganische Werkstoffe,
61295 Darmstadt, Germany

ABSTRACT

The mechanical properties of metal reinforced ceramics with interpenetrating networks, especially Al/Al_2O_3 composites, are described. Tailoring these composites involves key parameters such as volume fraction of the ductile phase, mean metal ligament diameter or interfacial fracture toughness. The influence of these parameters on fracture strength and toughness is investigated. The current fracture mechanical understanding is presented and the influence of metal reinforcement on strength emphasized.

1. Introduction

In fabricating metal/ceramic interpenetrating networks with improved mechanical performance various methods can be applied, such as squeeze casting[1], directed metal oxidation[2], pressureless[3] and gas pressure assisted infiltration[4]. WC/Co for instance may exhibit fracture toughness as high as 15 $MPam^{1/2}$ due to the plastic yield of the ductile constituent[5]. Geometrically constrained metal deformation, as present in interpenetrating networks, can result in high uniaxial stresses[6]. This is the origin of toughnesses of up to 9.5 $MPam^{1/2}$ in Al/Al_2O_3 composites[7], inspite of a far lower yield strength of Al as compared to Co. The degree of ductile deformation depends on the interfacial properties between metal and ceramic as well as the geometry of the reinforcing phase. Futhermore the stiffness of the ceramic constituent predominantly accounts for high hardness, wear resistence and fracture strength, as evidenced in the Al/Al_2O_3 system, rendering a strength of 760 MPa[8].

Our current understanding, based on substantial amount of work both in the field of ceramics[9-13] and intermetallics[14-17], regards the stress displacement function p(u) of the crack bridging metal ligaments to uniquely determine the reinforcement characteristics. This fracture mechanical approach predicts our K_{Ic} data reasonable well, however, the strength data of our composites could not

To the extent authorized under the laws of the United States of America, all copyright interests in this publication are the property of The American Ceramic Society. Any duplication, reproduction, or republication of this publication or any part thereof, without the express written consent of The American Ceramic Society or fee paid to the Copyright Clearance Center, is prohibited.

satisfactorily be explained when assuming that initial flaw size and elastic properties of the matrix are the same as those of the composite.[18]

2. Processing and Characterization

Two alumina preforms with different microstructural scales (termed medium (m) and coarse (c)) were prepared using Alcoa CT 2000 (m) and CL 5000 (c) powders that were slip cast in to plates of dimensions 50 x 40 x 10 mm and sintered to densities between 60 and 90% TD. A third alumina was manufactured using the RBAO process[19], resulting in a small (s) microstructure that was sintered to 75% TD. The mean pore diameter of these preforms was measured by mercury porosimetry. Infiltration was carried out at temperatures well above the melting point of the metal constituent and applying an argon pressure of 15 MPa. Details are given elsewhere[20].

After cutting and grinding, bend bars with dimensions of 3 x 4 x 25 mm were obtained with the tensile side polished to a 3μm surface finish and the edges bevelled. 4 point bend tests were performed according to German DIN 51 110 standard with loading spans of 10 and 20 mm. With the same bar size, fracture toughness was determined using the SEPB method (DIN 51 109) with a precrack/bar width ratio ranging between 0.25 and 0.5. In order to avoid large scale bridging effects, the precracks were renotched, reducing the bridged portion of the crack to less than a fourth. Both strength and toughness values represent the mean of at least five measurements.

3. Results

A representative micrograph of an Al/Al_2O_3 c-composite with an aluminum volume fraction of 35% is provided in Fig. 1.

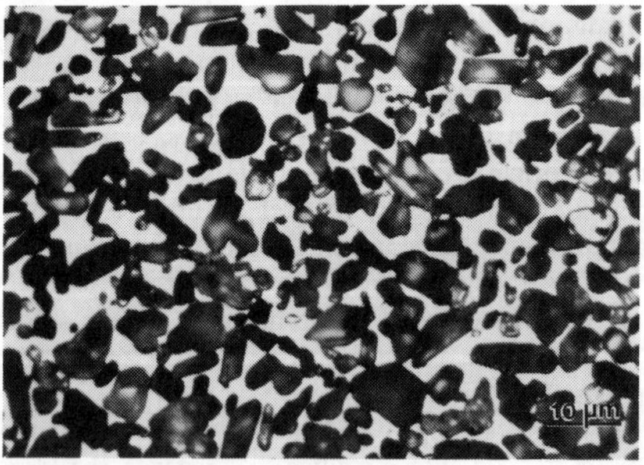

Figure 1: Representative optical micrograph depicting a coarse grained Al_2O_3 of 65% TD after Al infiltration.

Fracture toughness, K_{Ic}, and strength, σ_f, for the three preforms before and after infiltration are shown in Figs. 2 and 3. Besides the pronounced enhancement of σ_f and K_{Ic}, as compared to the matrix values, it should be emphasized that toughness increases with a factor of 2 to 4, whereas bend strength increases with roughly a factor of 4 to 5.

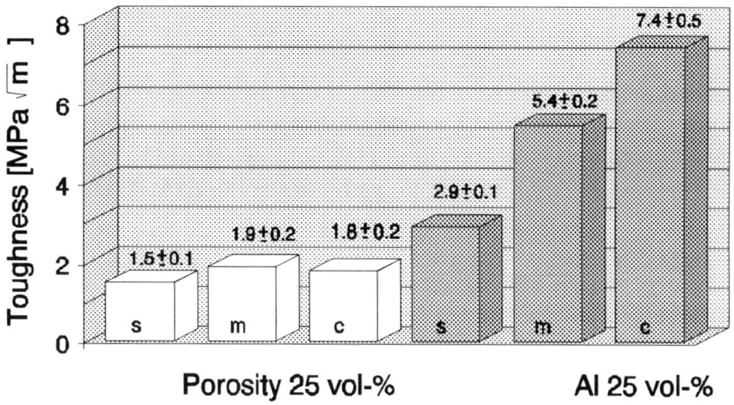

Figure 2: Fracture toughness of infiltrated and uninfiltrated Al_2O_3 of 25% porosity and different microstructures.

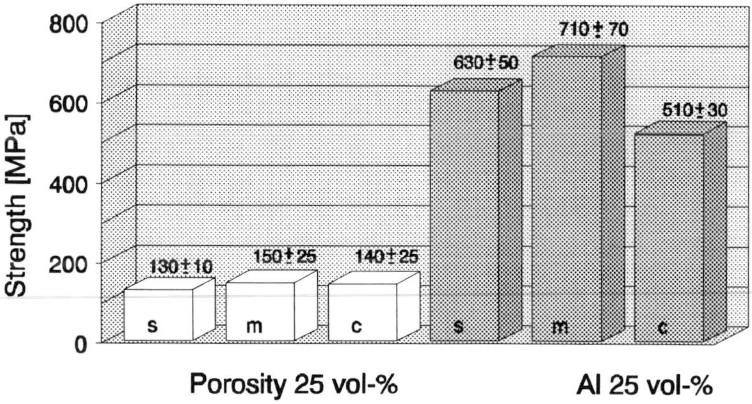

Figure 3: Bend strength of infiltrated and uninfiltrated Al_2O_3 of 25% porosity and different microstructures.

In addition to aluminum, metals with higher yield strength have also been infiltrated, such as Cu and Ni_3Al. These composites, however, exhibit a residual porosity of roughly 6%. Strength and toughness of c-composites with an initial density of 65% TD are shown in table 1.

	Al/Al_2O_3	Cu/Al_2O_3*)	Ni_3Al/Al_2O_3*)
K_{Ic} [$MPam^{1/2}$]	10.5	6.4	11.5
σ_f [MPa]	650	230	300

*) 6% residual porosity

Table 1: Toughness and strength of c-composites Al/Al_2O_3, Cu/Al_2O_3 and Ni_3Al/Al_2O_3 each with matrix density of 65% TD.

4. Discussion

The dependence of K_{Ic} on microstructural scale can be qualitatively described if only the different geometrical constraint on the metal ligaments is taken into account. As the bridging ligament diameter is decreased, plastic deformation is increasingly hindered. This leads to a diminished area enclosed by the p(u) function as well as a shift of its maximum to lower values of the openening u[9]. Hence the energy release rate, the square root of which is proportional to the plateau toughness, decreases with reduced microstructural scale. This corresponds with the K_{Ic} values of Fig. 2, since toughness determined by the SEPB method provides values near the plateau of the R-curve.

According to Griffith´s equation, σ_f increases at constant flaw size and loading geometry proportional to K_{Ic}. In R-curve materials, this equation gives an upper limit for σ_f when K_{Ic} is taken at a crack length where the first bridging ligament fails. Since the toughness at instability is certainly less than this plateau toughness, increase of σ_f can not exceed the increase in K_{Ic}, as long as the same flaw size is assumed to be present in both the preform and the composite. However flaws in the preform such as pores are filled with metal in the infiltrated composite. Once a crack is initiated from such a failure site, the crack is fully bridged. As the crack advances further, the relative volume fraction of the reinforcing phase decreases monotonically until it reaches a value equal to the metal content at long cracks. Therefore, the energy release rate becomes a function of crack length c and exhibits a maximum at small c. Accordingly, the R-curve shows a local maximum and the instability conditions lead to a larger bend strength.

In constrast to the high interfacial Al/Al_2O_3 fracture toughness, TEM investigations reveal microcracking at the interface in the Cu/Al_2O_3 composite. This accounts for a less effective flaw bridging as well as a lower K_{Ic} increase as in the Al/Al_2O_3 composite. In the case of the Ni_3Al/Al_2O_3 composite, the low strength presumably originates from the residual porosity, but until more detailed informations are available a satisfactory explanation can not be provided. Furthermore, the infiltration process needs to be optimized.

5. References

1. F. Lange, B.V. Velamakanni and E.A. Evans, J. Am. Ceram. Soc., **73** [2] 388-393 (1990).
2. M.S. Newkirk, A.W. Urquhart, H.R. Zwicker and E. Breval, J. Mater. Res. **1** [1] 81-89 (1986).
3. C. Toy and W.D. Scott, J. Am. Ceram. Soc., **73** [1] 97-101 (1990).
4. N. Travitzky and N. Claussen, J. Eur. Ceram. Soc., **9**, 61-65 (1992).
5. L.S. Sigl, Fortschrittberichte VDI, Reihe 5, Nr. 104.
6. A.G. Varias, Z. Suo and F. Shih, J.Mech. Phys. Solids, **39** [7] 963-86 (1991).
7. M.K. Aghajanian, N.H. Macmillan, C.R. Kennedy, S.J. Luszcz and R. Roy, J. Mat. Sci., **24** (1989) 658-670.
8. S. Wu, A.J. Gesing, N.A. Travitzky, and N. Claussen, J. Eur. Ceram. Soc., **7**, 277-281(1991).
9. L.S. Sigl, P.A. Mataga, B.J. Dalgleish, R.M. McMeeking, and A.G. Evans, Acta metall., **36** [4] 945-953 (1988).
10. B.D. Flinn, M. Rühle, and A.G. Evans, Acta metall., **37** [11] 3001-3006 (1989).
11. M.F. Ashby, F.J. Blunt, and M. Bannister, Acta metall. mater., **37** [7] 1847-1857 (1989).
12. P.A. Mataga, Acta metall., **37** [12] 3349-3359 (1989).
13. G. Bao and C. -Y. Hui, Int. J. Sol. Struct., **26** [5/6] 631-642 (1990).
14. H.C. Cao, B.J. Dalgleish, H.E. Deve, C. Elliott, A.G. Evans, R. Mehrabian, and G.R. Odette, Acta metall., **37** [11] 2969-2977 (1989).
15. H.E. Deve, A.G. Evans, G.R. Odette, R. Mehrabian, M.L. Emiliani, and R.J. Hecht, Acta metall., **38** [8] 1491-1502 (1990).
16. T.C. Lu, A.G. Evans, R.J. Hecht, and R. Mehrabian, Acta metall. mater., **39** 1853-1862 (1991).
17. L. Xiao and R. Abbaschian, Mat. Sci. and Eng., **A155**, 135-145 (1992).
18. G. Boa and F. Zok, Acta. metall. mater., **41** [12] 3515-3524 (1993).
19. S. Wu, D. Holz and N. Claussen, J. Am. Ceram. Soc., **76** [4] 970-980 (1993).
20. H. Prielipp, M. Knechtel, N. Claussen, S.K. Streiffer, H. Müllejans, M. Rühle and J. Rödel, submitted to J. Mat. Sci. and Eng.

FABRICATION OF FIBRE REINFORCED NICKEL ALUMINIDE MATRIX COMPOSITES BY REACTIVE PROCESSING

M. Downing and I. Horsfall

Materials Science and Engineering Group
Cranfield University, RMCS,
Shrivenham, Wiltshire, UK

Abstract

This paper describes the fabrication by reactive processing of short, and continuous, alumina fibre reinforced nickel aluminide matrix composites. The fibre is introduced into the aluminide system to increase toughness and high temperature strength.

The short fibre reinforced nickel aluminide is formed by squeeze casting a porous preform containing nickel powder and SAFFIL fibre with an aluminium or aluminium alloy melt. The continuous fibre reinforced nickel aluminide is formed by squeeze casting a jig containing nickel coated ALMAX fibre.

The short fibre reinforced composite (containing 10% and 20% volume fibre) reacted during infiltration with an aluminium melt to form a single phase intermetallic. Using an aluminium-copper melt the intermetallic formation was inhibited and a multi-phase composite was obtained. A preliminary study into reactive processing of this system by utilising a hot isostatic pressing (HIP) cycle is presented. HIP was required to prevent the formation of porosity due to an imbalance in the diffusive mobility of the various components. It was found that HIP was only effective on canned samples, the preferred encapsulation material being glass. The continuous fibre reinforced composite did not react to an intermetallic phase when infiltrated with an aluminium melt. Use of an aluminium-copper melt resulted in partial nickel-melt reaction producing various nickel-aluminium (-copper) phases. HIP was then used to form a two phase intermetallic matrix with no evidence of fibre damage.

To the extent authorized under the laws of the United States of America, all copyright interests in this publication are the property of The American Ceramic Society. Any duplication, reproduction, or republication of this publication or any part thereof, without the express written consent of The American Ceramic Society or fee paid to the Copyright Clearance Center, is prohibited.

Introduction

Intermetallic compounds are attracting considerable attention as engineering materials owing to their excellent high temperature mechanical and oxidation properties.[1] One particular area of concern limiting the increased usage of these materials is their poor ductility. By reinforcing them with ceramic fibres one could expect to increase the low temperature ductility whilst providing energy absorbing mechanisms during fracture.

Previous studies at Cranfield have shown the viability of using a squeeze casting process coupled with a fibre-containing metal powder preform to fabricate intermetallic matrix composites.[2-4] The technique of pressure casting for the production of reinforced metal matrix composites has been investigated by several other workers.[5-9] A comprehensive review of the solidification processing of metal matrix composites has been presented by Mortensen and Jin.[10] During the reactive infiltration or upon subsequent treatment, by a hot isostatic pressing route, the cast metal (namely aluminium or an aluminium-copper alloy) will react with the nickel powder or coating to produce an intermetallic matrix containing randomly oriented or continuous alumina fibres. The process route can be tailored such that fibre damage is minimised i.e. the aluminide does not form by reactive self-propagating synthesis.

This paper will describe the processing route by which a nickel aluminide intermetallic containing alumina fibre can be fabricated. Microstructural evaluations will be given throughout, together with the relevant processing conditions.

Preform Manufacture

The raw materials used for the fabrication of the preform were nickel powder, produced by Plasma-Technik Ltd and having a particle size of -325 mesh, SAFFIL alumina fibres, produced by ICI plc and having dimensions of 3 μm in diameter and 300-500 μm in length, and nickel coated ALMAX fibres, produced by Mitsui Mining Company Ltd and having a filament diameter of 11.7 μm with a nickel coat thickness of 1.2 μm.

The method of preform manufacture involves the dispersion of the powder and fibre in an aqueous media, containing an organic compound to aid dispersion and provide handleability of the subsequent preform. This suspension is then mechanically pressed against a filter, across which a pressure differential is set up, and the aqueous solution is removed.

The volume fraction of fibre throughout the work was either 10% or 20% by volume of the preform. This figure, whilst being low enough to cause no wettability problems in the casting stage, would offer some improvement of the mechanical properties of the aluminide. Preforms could also be manufactured with varying fractions of the metal powder and fibre.

The continuous fibre reinforced composite was produced by two methods that allowed for the fibre to experience the melt infiltration either perpendicular or parallel to the fibre orientation i.e. the fibre yarn was either wrapped around a bobbin such that the fibre would be exposed to the melt along its entire length or was cut into lengths of approximately 70 mm and bundled into a relatively thick walled steel tube such that the melt would infiltrate down the fibre length.

Melt Infiltration

The preform, measuring 100 mm in diameter and 15 mm thick, held in a die at 400 °C, was infiltrated with commercial purity aluminium (typically at a melt temperature of 800-1000 °C) in a squeeze casting rig. A pressure of 25 MPa was applied until melt solidification was achieved.

The nickel aluminide cast with the aluminium melt can be seen to possess a homogeneous dispersion of alumina fibres in a $NiAl_2$ matrix for the 10% volume fibre loading case with some evidence of porosity (see figure 1) and a Ni_2Al_3 matrix for the 20% volume fibre loading case (see figure 2). The phase determination is through energy dispersive X-ray spectrometry on a scanning electron microscope. The fibres show no damage resulting from the casting process.

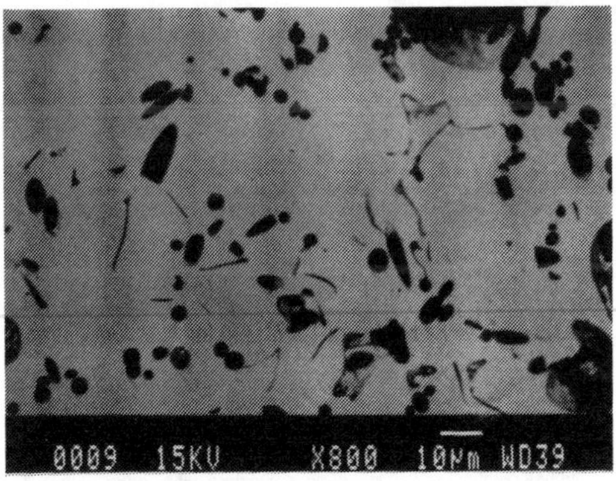

Figure 1 SEM micrograph of aluminium cast nickel preform containing 10% volume SAFFIL fibre.

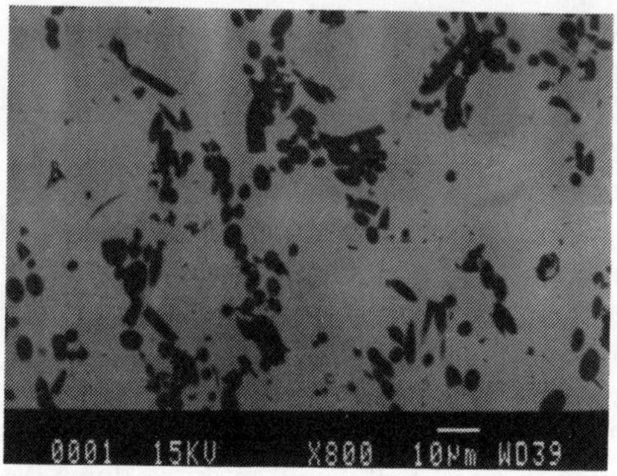

Figure 2 SEM micrograph of aluminium cast nickel preform containing 20% SAFFIL fibre.

The formation of an aluminide at the casting stage is not thought desirable as control of the microstructure is lost and there is a possibility of the excessive heat generated by the reaction adversely affecting the fibres. The use of a lower melting point alloy would reduce the level of melt/nickel reaction, and for this reason an aluminium-copper alloy, presently used at Cranfield in investigating the cubic structure of (reinforced) $Al_{67}Ti_{25}Cu_8$, was used. Copper alloying in NiAl alloys has been shown to enhance ductility.[11] With the use of an aluminium-copper alloy melt the resultant composite contains a matrix of nickel powder, aluminium-copper eutectic alloy and an alloy of aluminium-copper-nickel, with a homogeneous distribution of fibres (see figure 3). This composite contained 10% volume of fibre loading, shown as the dark phase. No fibre damage was evident. The nickel powder (light phase) can also be seen, with the eutectic aluminium-copper phase as the matrix.

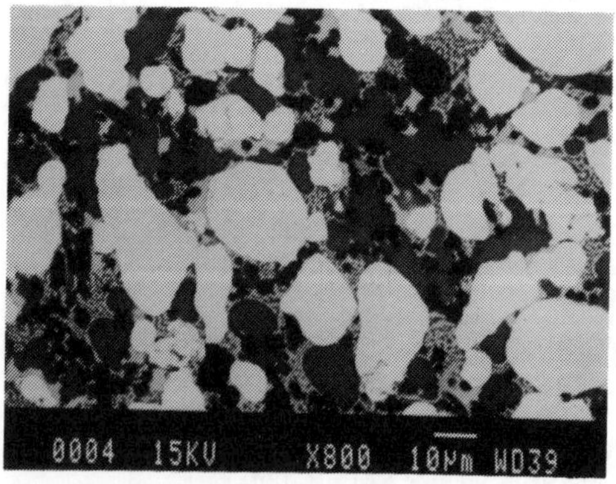

Figure 3 SEM micrograph of aluminium-copper cast nickel preform containing 10% volume SAFFIL fibre.

The nickel coated fibre was cast with an aluminium melt such that infiltration was perpendicular to the fibre orientation. This resulted in a compression of the fibre yarn and the resultant composite possessed a high fibre density showing little reaction between the nickel coat and the melt (see figure 4).

The aluminium-copper alloy has very good wetting properties and casting fluidity and the nickel coated fibres were cast with this alloy such that the infiltration was parallel to the fibre orientation. This resulted in a composite containing a distribution of fibres with a nickel coat in a matrix of copper containing $NiAl_2$ and $NiAl_3$ phases (see figure 5). A lower density of fibres results in the appearance of an aluminium phase.

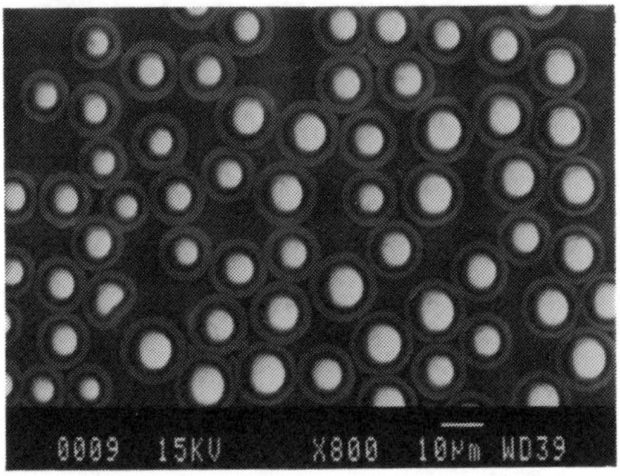

Figure 4 SEM micrograph of aluminium cast nickel coated ALMAX fibre.

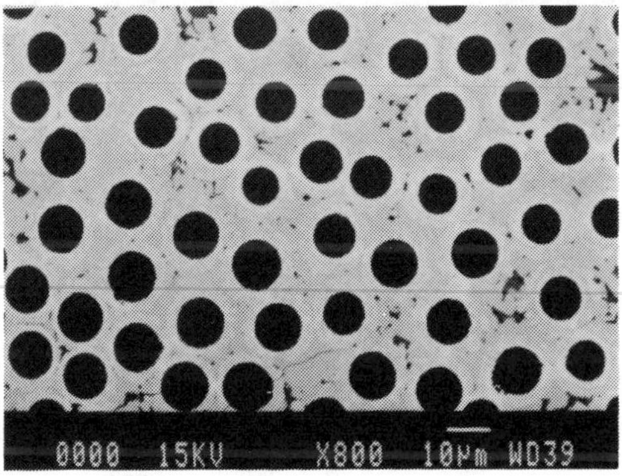

Figure 5 SEM micrograph of aluminium-copper cast nickel coated ALMAX fibre.

Hot Isostatic Pressing

A hot isostatic pressing (HIP) cycle was used to react the nickel and the melt whilst inhibiting the porosity developed in the cast material due to the relatively high mobility of the aluminium. (A pseudo-HIP process has been used previously to investigate the reaction synthesis of the NiAl intermetallic compound[12]). However, it was found necessary to encapsulate the cast material. The cast material could be vacuum encapsulated in soda glass which, whilst becoming plastic at the HIP temperature, would retain an effective pressure differential. This would also spall from the composite upon cooling.

The process cycle was of a type typical for glass encapsulated material, namely that the temperature would be applied prior to the application of pressure to soften the glass. At the full pressure of 200 MPa the temperature would then be increased to the working temperature and held for 4 hours before the work-piece is returned to standard temperature and pressure.

The short fibre reinforced nickel-aluminium-copper matrix composite was HIP'ed at a processing temperature of 625 °C. As can be seen in figure 6 the matrix is essentially a two phase matrix, with Ni at the original powder sites surrounded by a aluminium-rich NiAl matrix. There is some evidence of micro-porosity around the fibre sites and in the matrix. With a HIP processing temperature of 650 °C there is evidence of incipient melting and fibre pushing, with Ni, $NiAl_4$, Ni_2Al_3 and NiAl phases present (see figure 7).

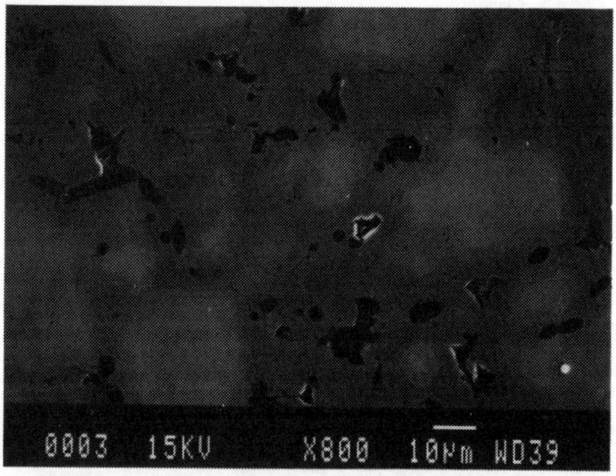

Figure 6 SEM micrograph of aluminium-copper cast nickel preform (10% volume SAFFIL fibre) HIP'ed at 625 °C.

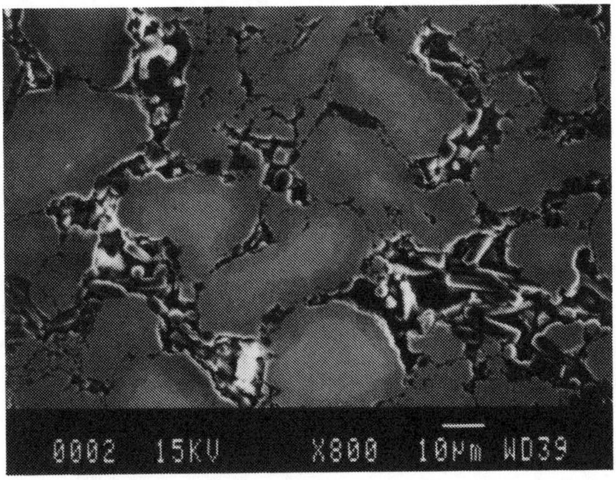

Figure 7 SEM micrograph of aluminium-copper cast nickel preform (10% volume SAFFIL fibre) HIP'ed at 650 °C.

The nickel coated continuous ALMAX fibre, cast with the aluminium-copper alloy, was HIP'ed at a processing temperature of 625 °C. As can be seen in figure 8 the nickel coating has fully reacted with the matrix, forming $NiAl_2$ and $NiAl_3$ phases. A third phase of aluminium (dark region) can also be seen but is distanced from the fibres. There is no evidence of fibre degradation.

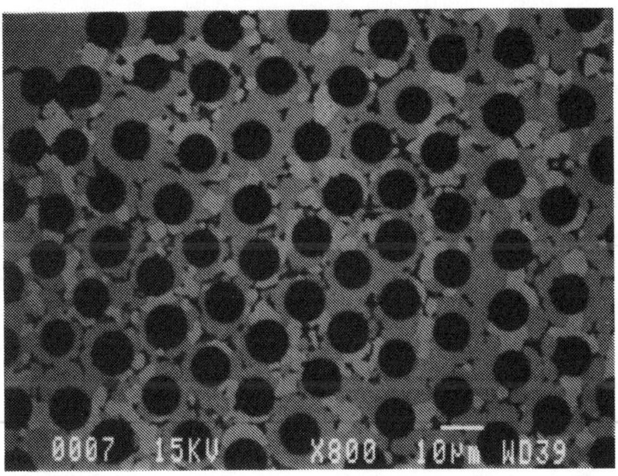

Figure 8 SEM micrograph of aluminium-copper cast nickel coated ALMAX fibre HIP'ed at 625 °C.

Discussion

The short exposure of the nickel powder and the alumina fibres to the aluminium melt results in excessive aluminide reaction but does not appear to degrade the fibres. For the aluminium-copper melt the composite shows a good dispersion of unreacted phases and the fibres appear undamaged.

Subsequent HIP treatment of the fibre containing nickel-aluminium-copper matrix composite results in widescale reaction between the matrix constituents. The processing parameters will need optimising to promote matrix reaction whilst avoiding any fibre degradation.

Analysis of the phases present when using the aluminium-copper melt suggest that the presence of copper and/or the lower amount of available aluminium limits the initial reaction in the casting stage. Upon subsequent HIP treatment the aluminium in the melt initially reacts with the nickel, with later diffusion of the copper.

It is important in the manufacture of the composites that no excess aluminium is present, as this will remain unreacted and may liquify at high temperatures. Apart from the deleterious effects on high temperature mechanical properties this also leads to fibre degradation. Earlier work[(3)] on titanium aluminides has shown that fibre degradation occurs at temperatures as low as 700 °C when excess aluminium is present.

The nickel coated continuous fibre composites showed little reaction with the aluminium melt when the density of fibre was high. The compression of the fibre yarn was believed to limit the amount of heat/aluminium available to the nickel and hence limited the aluminide reaction. With the aluminium-copper melt the nickel coat reacted with the melt to form a multi-phase matrix of copper containing Ni-Al phases. A subsequent HIP treatment resulted in complete diffusion of the nickel to form an aluminide matrix around the fibres. No fibre degradation was observed in either system during any of the processing.

Conclusion

With careful control of the preform manufacturing stage and the infiltration process to ensure homogeneity of the cast material it is possible to produce a short fibre reinforced composite material. This composite has the potential through HIP processing to form a single phase aluminide matrix containing randomly oriented short fibres.

The use of nickel coated continuous fibre in the infiltration process allows for the production of a single phase matrix around the coated fibre, which has the potential of full reactive processing via HIP, or the production of a partially reacted matrix which can be homogenised using HIP.

Acknowledgements

The authors wish to thank Dr D. Stephenson and Mr J. Hedge of Cranfield University, Bedford, for the HIP work, ICI Chemicals and Polymers for the SAFFIL fibre, Mitsui Mining Materials Company Ltd for the nickel coated ALMAX fibre, and to Mr R. Kimber for technical assistance. This work is generously supported by a Science and Engineering Research Council (SERC) grant, GR/H27779.

References

1. O. Izumi, **Intermetallics Compounds as Engineering Materials**, The First Pacific Rim International Conf. on Advanced Materials and Processing, eds. C. Shi, L. Hengde, and A. Scott, The Minerals, Metals and Materials Society (1992).

2. C.M. Friend, I. Horsfall, C.L. Burrows, **The Effect of Particulate:Fiber Ratio on the Properties of Short Fiber/Particulate Hybrid MMC Produced by Preform Infiltration**, J. Mat. Sci., vol. 26, pp 225-231 (1990).

3. I. Horsfall, S.J. Cundy, **A Process for the Fabrication of Ceramic Fiber Reinforced Titanium Aluminide**, Proc. 16th Annual Conf. on Composite and Advanced Ceramics, Cocoa Beach, FL, 7-10 Jan. (1992).

4. I. Horsfall, M. Downing, **Fabrication of Titanium and Nickel Aluminide Matrix Composites by Reactive Processing**, Proc. of In-Situ Composites: Science and Technology Symposium, TMS/ASM Materials Week '93, Pittsburgh, Pennsylvania, 19-20 Oct. (1993).

5. S. Nourbakhsh, F.L. Liang, H. Margolin, **An Apparatus for Pressure Casting of Fibre-Reinforced High-Temperature Metal Matrix Composites**, J. Phys. E: Sci. Instrum., vol. 21, pp 898-9902 (1988).

6. S. Nagata, M. Sakamoto, **Development and Applications of Metal Composites from Pressure Casting**, Materials and Design, vol. 10, no. 3, May/June (1989).

7. H. Fukunaga, X. Wang, **Preparation of Intermetallic Compound Matric Composite by Reaction Squeeze Casting**, J. Mat. Sci. Letters, vol. 9, pp 23-25 (1990).

8. H.C. Chen, M. Kaya, R.W. Smith, **Production of Intermetallic Matrix Composites by Reactive Infiltration Processes**, Proc. of Conf. on Advances in Production and Fabrication of Light Metals and Metal Matrix Composites, Edmonton, Canada, 23-27 Aug. (1992).

9. D.C. Dunand, J.L. Sommer, A. Mortensen, **Synthesis of Bulk and Reinforced Nickel Aluminides by Reactive Infiltration**, Met. Trans. A, vol. 24A, pp 2161-2170 (1993).

10. A. Mortensen, I. Jin, **Solidification Processing of Metal Matrix Composites**, Int. Mat. Reviews, vol. 37, no. 3, pp 101-128 (1992).

11. S.J. Jeon, H.-C. Lee, **Effect of Copper Alloying on the Deformation Behaviour of B2 NiAl Intermetallic Compounds**, Mat. Sci. and Eng., A153, pp 392-397 (1992).

12. P.H. Shingu, K.N. Ishihara, S. Umeda, **A Pseudo-HIP Process Applied to the Reaction Synthesis of Intermetallic Compounds**, Mechanical Behaviour of Materials - VI, Proc. of 6th Int. Conf., Kyoto, Japan, 29 July - 2 Aug. (1991).

THE INTERACTION OF ALUMINA FIBRES WITH TITANIUM ALUMINIDE MATRICES

I. Horsfall, M. Downing and K.D. Rogers
Cranfield University, RMCS,
Shrivenham, Wiltshire, UK

ABSTRACT

Three alumina fibre types have been incorporated into titanium trialuminide matrices. Alumina fibres represent a suitable reinforcement for such systems due to a good CTE match, limited fibre matrix interactions, and good thermal stability within the range of probable service temperatures. Previous work has assessed SAFFIL fibres in such systems, this paper presents work on two other reinforcement fibres, 3M Corporation's NEXTEL 610 and Mitsui Mining Materials' ALMAX. Both of these are alumina based continuous fibre although in this case they have been chopped and incorporated in the matrix in a pseudo random orientation.

INTRODUCTION

Titanium aluminides have been proposed as candidate materials for various high temperature components in gas turbine engines. [1-3] The major problem associated with the use of aluminides is their poor ductility, however this can be overcome by alloying.[4] In particular it has been shown that the addition of copper or chromium leads allows the formation of a $L1_2$ structure and with consequent improvement in compressive ductility. However this may also result in a loss of creep strength negating some of the materials initial advantage.[5]

The addition of fibres to a titanium aluminide material could be expected to improve stiffness, creep strength and, under the right conditions, toughness improvements might also be achieved. Fibres would also offset any reduction in high temperature creep strength caused by the ductilising effect of alloying additions.

To the extent authorized under the laws of the United States of America, all copyright interests in this publication are the property of The American Ceramic Society. Any duplication, reproduction, or republication of this publication or any part thereof, without the express written consent of The American Ceramic Society or fee paid to the Copyright Clearance Center, is prohibited.

Alumina fibres are a suitable reinforcement for titanium aluminides as they show a relatively small thermal expansion mismatch. Alumina is known to be compatible with titanium aluminide alloys having less than 50 at % titanium[6] and therefore reactions with a pure alumina fibre should be negligible. It has been shown that titanium aluminides with less than 50 at % titanium can react with certain reinforcement fibres if incipient melting occurs in the matrix due to the presence of pure aluminium and if the fibres contain significant silica content.[7]

In previous work it has been shown that titanium aluminide reinforced with alumina fibres can be produced by solid state reaction of a titanium in an aluminium matrix containing ICI SAFFIL fibres.[7] In this paper two other alumina fibres, Mitsui Mining Materials' ALMAX and 3M Corporation's NEXTEL 610, have been similarly incorporated into titanium aluminide matrices.

FABRICATION

The fabrication process for fibre reinforced titanium aluminide has been described in a previous publication.[7] A preform is made which consists of the reinforcement fibre and titanium metal powder. This is infiltrated with molten aluminium and subsequently reacted to form a titanium aluminide matrix around the fibres by a suitable heat treatment in a hot isostatic press. This process was used to produce a titanium trialuminide matrix with either SAFFIL, NEXTEL 610, or ALMAX fibres.

FIBRE PROPERTIES

Table I lists the properties as given by the respective manufacturers, of the 3 fibre types.[8,9,10] The NEXTEL and ALMAX fibres are essentially pure alumina and are consequently alpha phase, whilst SAFFIL contains 4% silica to stabilise the delta phase. It has been suggested that the use of a small crystal size delta alumina allows a smoother fibre surface and consequent higher strength, together with a reduction in the size of internal porosity. However the silica does provide a relatively reactive phase which can react with some matrix materials[7] and under some conditions may be removed leading to partial breakdown of the fibre structure.

Table I. Properties of the Fibre Types.

	SAFFIL	NEXTEL 610	ALMAX
Composition	96 Al_2O_3 4 SiO_2	>99 Al2O3 <0.3 SiO_2 <0.7 Fe_2O_3	>99.5 Al_2O_3
Crystal phase	Delta	Alpha	Alpha
Crystal size nm	50	500	500
Diameter μm	3	11	10
Fibre length mm	0.5	3	0.5
Elastic modulus GPa	300	373	324
Tensile strength MPa	2000	1900	1700

COMPOSITE MICROSTRUCTURE

Samples of composite containing each of the reinforcements were analyzed by X-ray diffraction. The materials were analyzed in three conditions: i) in the as cast condition; ii) cast and HIPed (650°C for 4 hours at 200MPa); and iii) cast, HIPed and then heat treated at 850°C for 6 hours. For the as cast condition large samples were used allowing the alumina peaks to be detected, but for the other two conditions smaller samples were used so that alumina peaks were not resolvable. Figures 1, 2, and 3 show the XRD spectra for SAFFIL, NEXTEL, and ALMAX reinforced titanium aluminide respectively. Table II lists the crystalline phases detected.

During casting some titanium trialuminide is produced probably around the edges of the titanium particles. The HIP treatment reacts most of the titanium and aluminium to titanium trialuminide. Subsequent heat treatment at 850°C produces TiAl and Ti_3Al in the ALMAX composite, probably as a result of a locally higher titanium content or large titanium particle in the analysis area.

The microstructures of the composites can be seen in figures 4-6 for the NEXTEL samples and figures 7-9 for the ALMAX samples. The phases seen in these microstructures were identified by energy dispersive X-ray analysis. In both cases the as cast samples show an even dispersion of fibres in an aluminium matrix with titanium particles which exhibit no visible reaction. The HIP heat

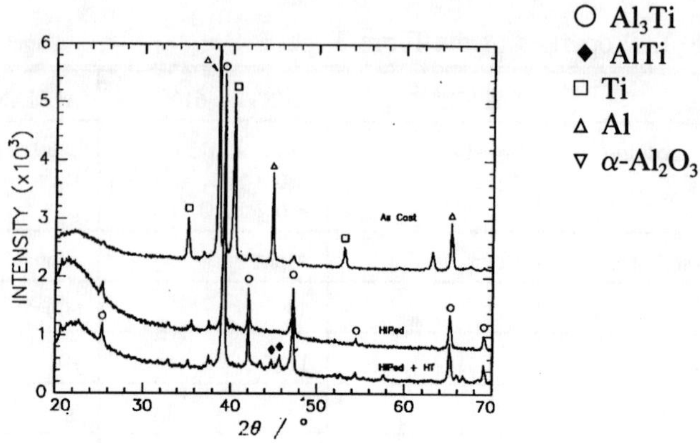

Figure 1 XRD spectra for SAFFIL fibre composite.

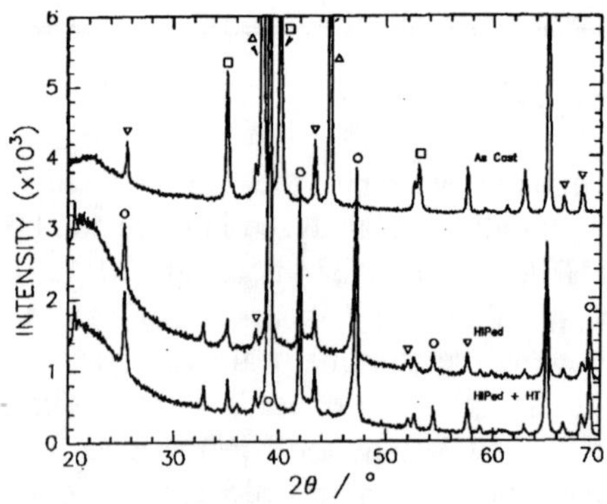

Figure 2 XRD spectra for NEXTEL 610 composite.

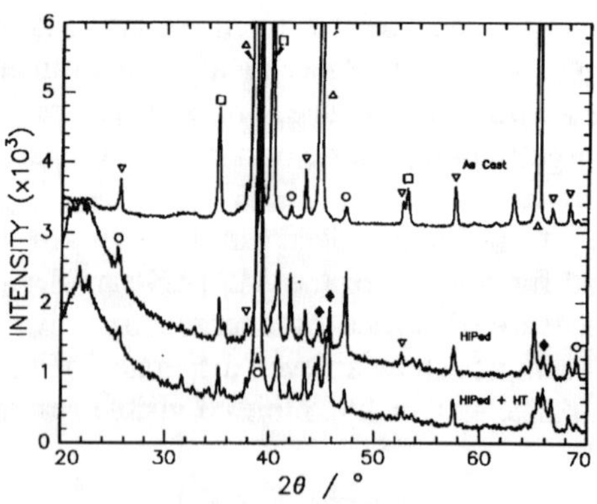

Figure 3 XRD spectra for ALMAX composite.

Table II. Main Crystalline Phases (X) and Minor Phases (x) Detected by XRD.

SAFFIL	As Cast	Cast and HIPed	Cast, HIPed and heated
Al	X		
Ti	X		
Al_3Ti	X	X	X
NEXTEL 610			
alpha-alumina	X		
Al	X		
Ti	X		
Al_3Ti	X	X	X
ALMAX			
alpha alumina	X		
Al	X		
Ti	X		
Al_3Ti	X	X	X
TiAl			x

treatment results in all the aluminium reacting to form titanium aluminides. Aluminium diffuses into the titanium particles producing a range of titanium aluminide phases with the centre of the titanium particles having typical compositions Ti-16Al with a few areas of pure titanium remaining. The main matrix phase has a composition of Ti-75Al, ($TiAl_3$). Heat treatment at 850°C for 6 hours produces more $TiAl_3$ at the expense of the titanium rich phases. No changes are observed in either of the fibre types.

A similar microstructure can be obtained by using a higher HIP processing temperature of 700°C for 4 hours at 200MPa. Figure 10 shows the structure of the NEXTEL 610 fibre composite following such a processing regime.

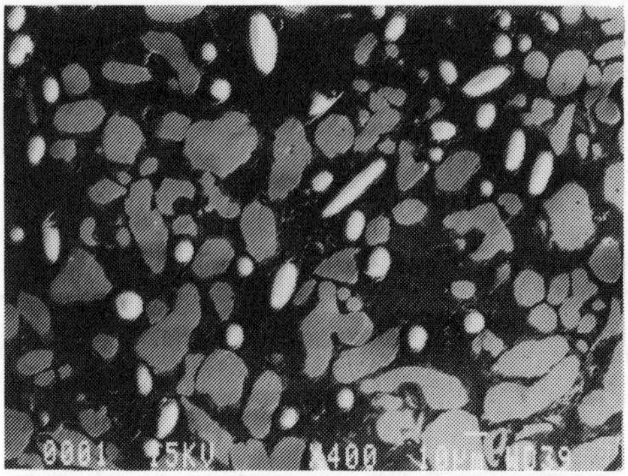

Figure 4 As cast composite containing NEXTEL 610.

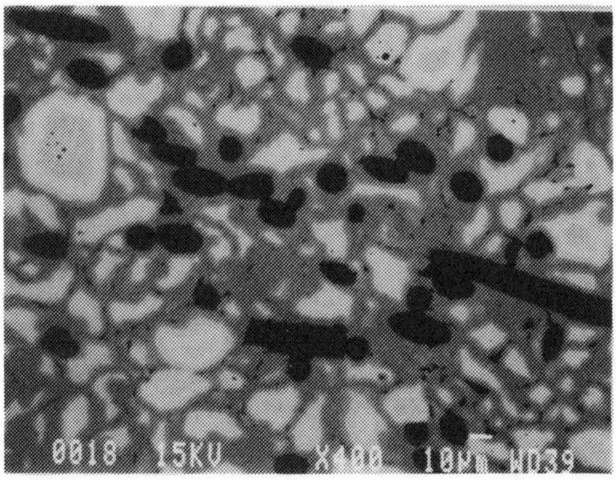

Figure 5 NEXTEL 610 composite after HIP treatment at 650°C for 4 hours at 200MPa.

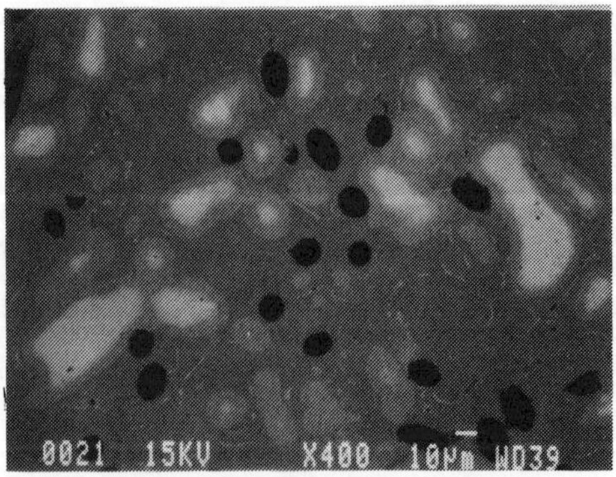

Figure 6 NEXTEL 610 composite after HIP, and heat treatment at 850°C for 6 hours.

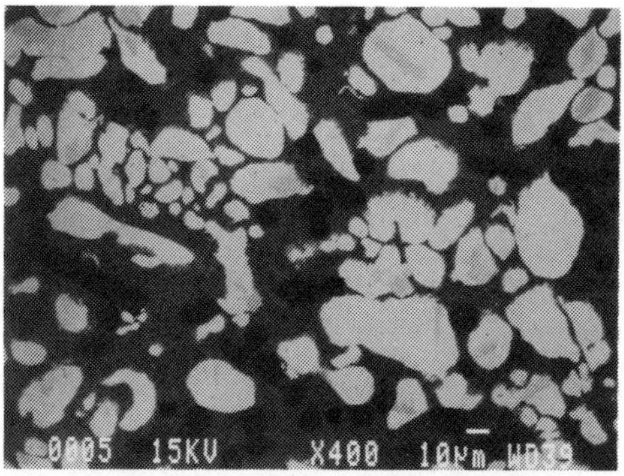

Figure 7 As cast composite containing ALMAX.

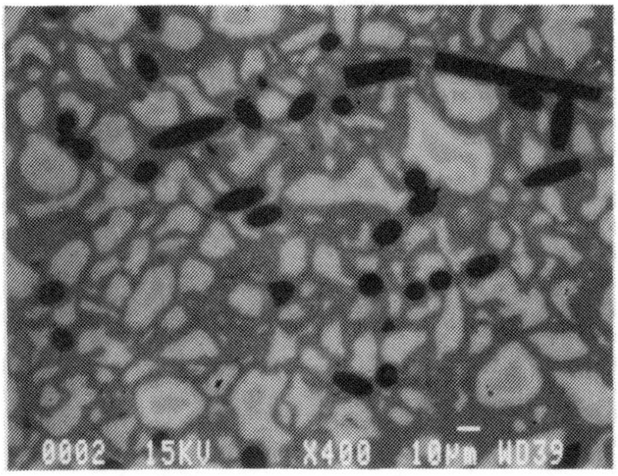

Figure 8 ALMAX composite after HIP treatment at 650°C for 4 hours at 200MPa.

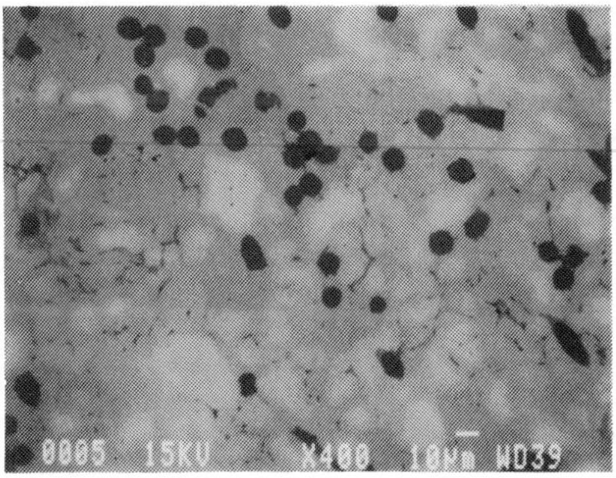

Figure 9 ALMAX composite after HIP, and heat treatment at 850°C for 6 hours.

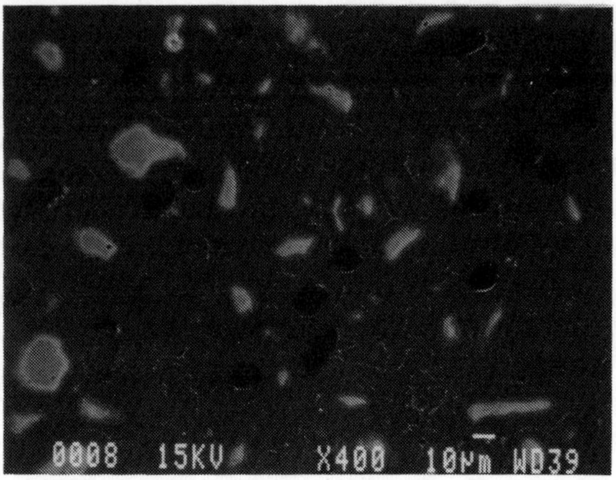

Figure 10 NEXTEL 610 composite after HIP treatment at 700°C for 4 hours at 200MPa.

DISCUSSION

NEXTEL 610 and ALMAX fibres have compositions which are essentially pure alumina. It has been shown that alumina is stable in titanium aluminides having less that 50 at % Ti and therefore the fibres should be chemically stable. Although there are titanium rich phases present in reaction processed titanium aluminide, these are confined to the centres of prior titanium particles and are therefore not in contact with the fibre surfaces.

It has been suggested that it is difficult to prevent grain growth in alumina fibres at elevated temperatures[8] and it has been shown[11] that grain growth and loss of strength may occur in NEXTEL 610 at temperatures in excess of 960°C in air. The maximum temperature achieved during the composite fabrication process was less than 1000°C, which is the melt temperature of the aluminium. This temperature was experienced by the fibres for only a few seconds as rapid cooling to the die temperature (400°C) would occur during infiltration. The HIP treatment was 650°C for 4 hours, which caused all the aluminium to take the form of titanium aluminide, predominantly Al^3Ti.

A subsequent heat treatment of 850°C for 6 hours was used to determine the effect of higher temperatures on the matrix and fibre. Some homogenisation of the matrix took place but no visible changes occur in the fibres which was as expected from previous studies.

CONCLUSIONS

NEXTEL 610 and ALMAX alumina fibre can be incorporated into titanium aluminide matrices by solid state reaction processing. The fibres are exposed to a chemical and thermal regime which is relatively benign.

Subsequent heat treatments at 850°C allows homogenisation of the matrix without causing gross fibre damage.

ACKNOWLEDGEMENTS

The authors wish to thank ICI PLC, Mitsui Mining Materials Company and 3M Corporation for their cooperation and to Mr R. Kimber and for his technical assistance. This work is generously supported by a Science and Engineering Research Council (SERC) grant, GR/H27779.

REFERENCES

1) H.A.Lipsitt, **Titanium Aluminides - Future Turbine Materials**, Proc. Advanced High Temperature Alloys Processing and Properties, Cambridge, Mass. 16-18 June (1985).

2) S.Ashley, **Titanium Aluminides: Tough Materials for Tomorrow's Engines**, Mechanical Engineering, Dec. pp 49-52 (1991).

3) F.H.Froes, C.Suryanarayana, D.Eliezer, **Review - Synthesis, Properties and Applications of Titanium Aluminides**, J. Mat. Sci., vol. 27, pp 5113-5140 (1992).

4) M.B.Winnicka, R.A.Ryan, **Microhardness and Compressive Mechanical Behaviour of $L1_2$ Titanium Trialuminides**, Met. Trans., vol 24A, April, (1993).

5) K.S.Kumar, M.S.Dipietro, J.D.Whittenberger, **Compression Response of Monolithic and Particulate-Reinforced Composites of $Al_{67}Ti_{25}Cr_8$ and $Al_{66}Ti_{25}Mn_9$,** Acta metall. mater., vol 41, no 5, (1993).

6) A.K.Misra, **Reaction of Ti and Ti-Al Alloys with Alumina**, Met. Trans. A, vol. 22A, March pp 715-721 (1991).

7) I.Horsfall, S.J.Cundy, **A Process for the Fabrication of Ceramic Fiber**

Reinforced Titanium Aluminide, Proc. 16th Annual Conf. on Composite and Advanced Ceramics, Cocoa Beach, FL, 7-10 Jan. (1992).

8) J.D.Birchall, J.A.A.Bradbury, J.Dinwoodie, **Alumina Fibres:Preparation Properties and Applications**, Handbook of Composites, vol 1-Strong Fibres, Elsevier, (1985).

9) Product information, **Nextel 610 Ceramic Fibres,** 3M Ceramics Materials Department, (1991).

10) Product information, **High Purity Alumina Fiber ALMAX - Technical Reference Manual and Handling Manual**, Mitsui Mining Company Limited, (1992).

11) Z.R.Xu, K.K.Chawla, X.Li, **Effect of High Temperature Exposure on the Tensile Strength of Alumina Fiber Nextel 610,** Fracture and Fatigue, ASM/TMS Materials Week, Pittsburgh, (1993).

SYNTHESIS OF Al_20_3-Al COMPOSITES WITH VARIED COMPOSITIONS VIA A SUBMICRON PRECURSOR

E. Sweet and S. N. Sinha
Department of Civil Engineering, Mechanics and Metallurgy
University of Illinois at Chicago,
842 W. Taylor Street, M/C 246
Chicago, IL 60607-7023

ABSTRACT

A submicron precursor of Al_20_3-5 wt% Al was synthesized by high energy milling, and the precursor was sintered at 1400°C in Argon-0.5% H_2 at a total pressure of 1.5 kPa. For the milling experiment, the particle sizes, which were measured by TEM directly as well as calculated by XRD patterns using the variance and integral breadths methods, ranged from 20-110 nm. The sintered samples exhibited 3 zones: (1) a dense, well sintered zone near the surface, (2) a porous composite in the center and (3) a transition zone with varying composition and properties. The microhardness of the zone near the surface (at 3 N) was 270 VHN in some cases, whereas that of the composite core ranged from 100-200 VHN.

INTRODUCTION

Fine grained ceramic metal composites with appropriate microstructures can offer many advantages over traditional ceramics. These materials will combine the potential advantages of ceramic composites such as increased toughness [1], with those parts made from fine grained precursor powder, such as easier forming and increased hardness [2,3]. Increased toughness in ceramic composites can probably be achieved if its microstructures incorporate architectures of Interpenetrating Phase Composites (IPC) [1].

Limited features of IPC materials [4-6] were observed in Al_20_35 wt% Al composites using a recent approach. In this approach, the precursor was synthesized by high energy milling and the resulting submicron composite precursor was sintered by Inductively Coupled Plasma. The results indicate that composite precursor had the features of IPC materials [6,7] but on a finer scale than observed in conventional processes.

Functionally gradient materials (FGM's) are materials that can combine the advantages of two or more types of materials in one part. Techniques used to produce FGM's are those in which a gradient of compositions or structures is formed in the part, either before, during or after final shaping. FGM's have been produced by a variety of methods; chemical vapor deposition, powder spray forming, and gas pressure combustive sintering are among the methods reported [8-10]. Reactive sintering promises to form gradients in a part after preliminary shaping [11].

To the extent authorized under the laws of the United States of America, all copyright interests in this publication are the property of The American Ceramic Society. Any duplication, reproduction, or republication of this publication or any part thereof, without the express written consent of The American Ceramic Society or fee paid to the Copyright Clearance Center, is prohibited.

In attempts to form FGM with fine grains in the Al_2O_3-5 wt% Al system, we have studied the sintering characteristics of Al_2O_3-5 wt% Al precursor in Argon-0.5% hydrogen at 1400°C and reduced pressure of 1.5 kPa (total). The sintering conditions, which were obtained by systematic trial experiments, were mildly oxidizing to Al_2O_3/Al system. The objective was to form a composite with varying concentrations of aluminum near the outer surface of the sample. The characteristics of the powder and the microstructure property relationship of the composites are also included.

EXPERIMENTAL

A. Synthesis of Submicron Powder

High purity (99.9% purity) α-alumina (AESAR) and aluminum powder (AESAR), 325 mesh, were premixed in the ratio of 95 and 5 wt%, ball milled for ten minutes with a small amount of alcohol, and then milled in an Attritor (Model 01-HD, Union Process, Ohio) for 6 hours at 550 rpm. The grinding media and milling condition, which were selected based on our earlier work [4], were as follows: 5 mm diameter yttria stabilized zirconia balls, ball to powder weight ratio of 7.9 to 1, 750 cc stainless tank and stirring rod, and 125 cc/min (at STP) of argon flow through the milling tank during milling. The milling tank was also cooled from outside by circulating cold water through outside jacket at 250 cc/min. The milling conditions were aimed to limit the temperature rise during milling to less than 25°C. The milled-precursor was always stored in an inert environment, and during milling special care was taken to minimize the exposure of powder to ambient conditions.

XRD measurements of powder and bulk were performed using a Siemens D5000 diffractometer operating at 40 kV with Cu Kα radiation. Powders were inspected using a JEOL JSM T-200 SEM at 25 kV. TEM observations were performed on powders at 100 kV using a JEOL 100-CX microscope. Samples for TEM were prepared by dipping a carbon or Parlodion® coated 400 mesh copper grid directly into milled powder.

B. Reactive Sintering

Milled powders were pressed into 6.3 mm by 7 mm cylindrical pellets using a Carver hydraulic laboratory press. No binders were used; pressures of 350-450 MPa were used for pressing. Pressed pellets were about 60% dense, and strong enough to machine; they could be stored in air indefinitely without deterioration.

Pellets were sintered in a vertical laboratory furnace under an atmosphere of argon and 0.5% hydrogen. The runs were performed at a total pressure of 1.5 kPa, with gas flowing at 150 ml (STP) per minute. After purging under vacuum for 10 h, temperatures were raised at 100°C/h to 1400°C, held at 1400°C for 6, 12 or 24 h, then lowered to room temperature at 100°C/h.

Sintered pellets were cut using a diamond saw, polished to 0.5 μm finish using water soluble diamond paste (Kay Industrial Diamond) and then tested for hardness using a Vickers microindenter [LECO] at 3 N loads. Samples were coated with gold/palladium, then observed using a JEOL JSM T-200 SEM at 25 kV. Microstructures were photographed, and grain sizes were measured using the ASTM standard circular intercept method [12].

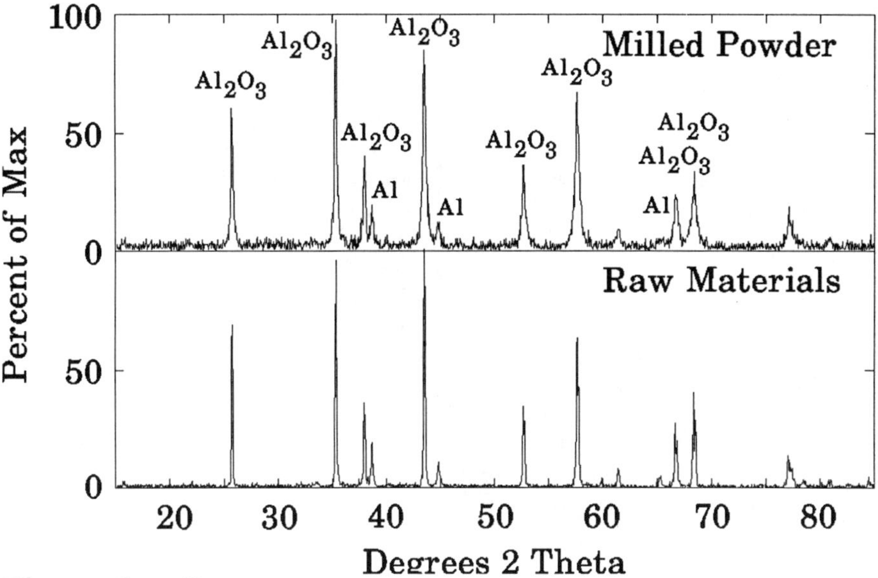

Figure 1: XRD of Al_2O_3 - 5 wt % Al composite powder and raw materials.

RESULTS AND DISCUSSION

A. Characterization of Powder

After the powder was milled, it was analyzed to determine composition, particle shape and size, and morphology. Composition of the powder was determined by powder XRD. Particle shapes and sizes were measured by direct TEM observation of powder mounted on copper grids; particle morphology was also determined using this method. Particle sizes were also determined using powder XRD. Two XRD data analysis techniques, the integral breadths and variance methods, were used to estimate powder size and strain [13].

1. Qualitative Analysis by XRD

Figure 1 shows Cu-Kα pattern of milled and unmilled powders. Patterns indicate both α-alumina and aluminum were present, in roughly the same amounts before and after milling. No additional phases or contamination peaks were seen [14]. Peaks of both phases were broadened presumably due to size effects.

The aluminum phase in the precursor was not mostly oxidized during the milling process presumably due to extensive cooling, presence of argon environment and choices of milling variables.

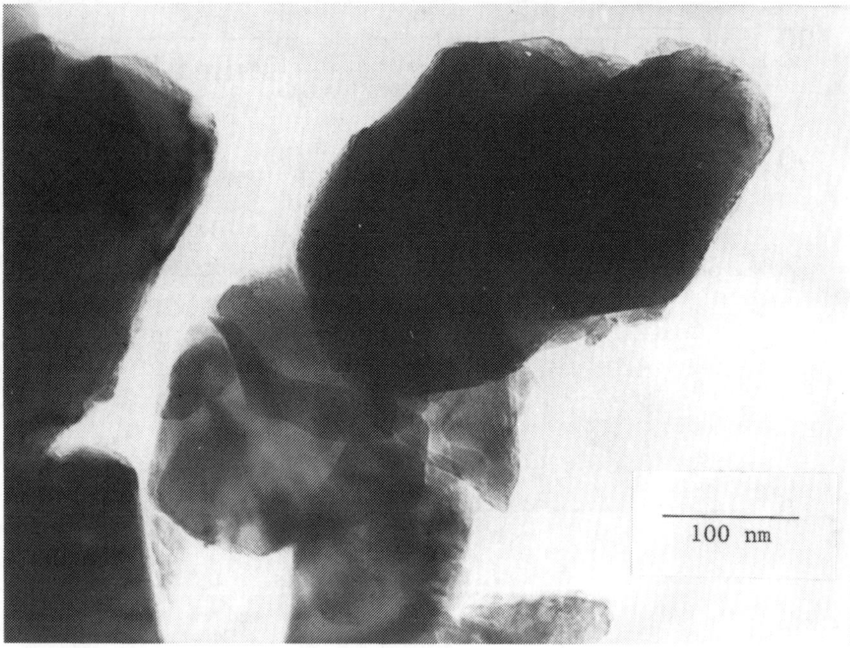

Figure 2: TEM micrograph of powder milled at 550 RPM for 6 hours.

2. SEM and TEM Observations of Composite Powder

SEM was the first method used to examine the structure of milled powders. Opaque agglomerates 100 to 400 nm across were observed. TEM at 100 kV penetrated these agglomerates, which were seen to be clusters of smaller particles. Although the high voltage may have affected details of the aluminum phase in the powder, it was an effective method for determining particle size and aspect ratio of the alumina particle which constituted 95 wt% of the sample.

Particle sizes and aspect ratios for the alumina phase were estimated by direct measurements of the particles selected from random areas of the sample. Since available instruments cannot measure the particle sizes in the nanoscale composite powder, we modified the known approaches based on TEM. Clusters and individual particles were examined by TEM individually at magnifications from 50,000 to 250,000; focus was varied slightly to detect particle edges. Large particles were examined closely, to determine whether they were agglomerates of smaller particles. A large number (300 or more) particles were observed to obtain a statistical estimate of size distributions. Aluminum phases could not be however measured by this approach because it deteriorated quickly on direct exposure of electron beams.

Figure 3 shows the observed sizes of about 300 alumina particles. The particles greater than 200 nm are likely to be agglomerates of smaller particles and we could not distinguish by our approach. Half of the particles measured were smaller than 75 nm; only 20% were larger than 200 nm. A

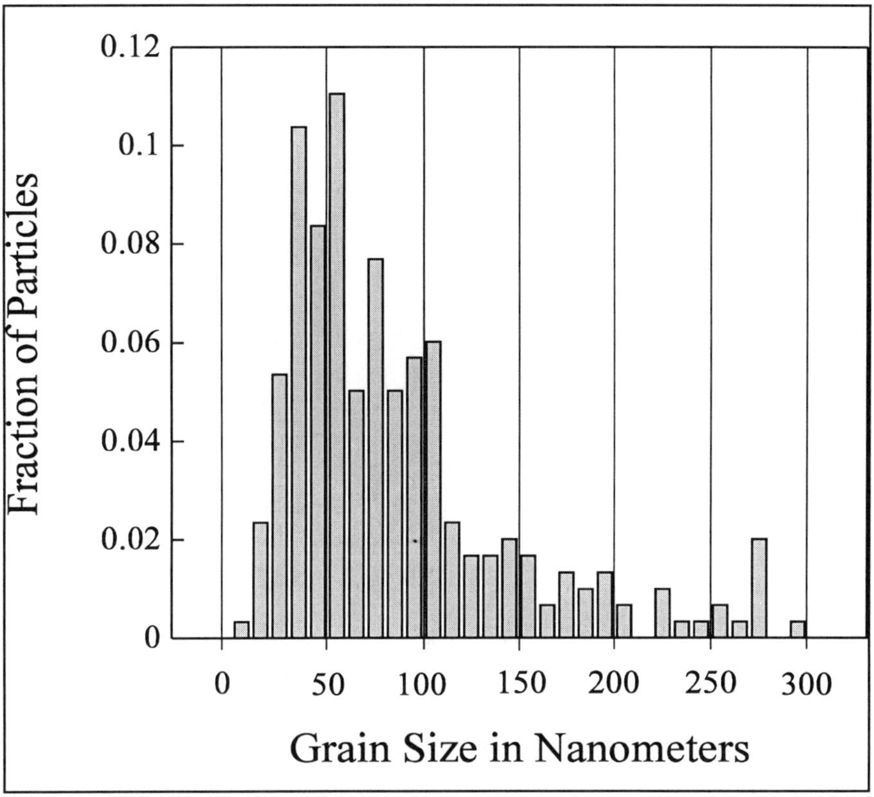

Figure 3: Particle Sizes of powder milled at 550 RPM for 6 hours, estimated from 300 TEM observations.

third of the particles were smaller than 50 mn, approaching nanocrystalline size ranges. Many of the smaller particles were the ones seen clinging to larger particles in Figure 2.

Only a few aluminum particles could be identified and data points were too few to obtain a reasonable statistical distributions. While too few aluminum particles were seen, their average sizes were very similar to those of the alumina particles, shown in Figure 2.

The aspect ratios, of the composite powder (alumina and aluminum particles), the ratios of long to short axes, were also estimated. Figure 4 summarizes the results from about 125 measurements. Particles were slightly elongated, the majority with shape factors of 1.25 to 1.75. While few particles were perfectly equiaxed, few were very elongated. There were a few particles with aspect ratios above 2.5; these may have been agglomerates.

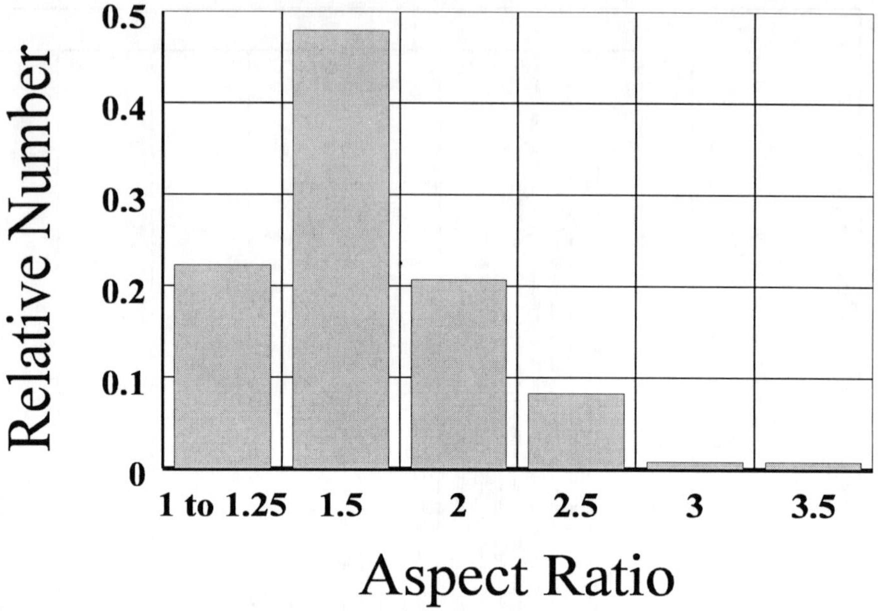

Figure 4: Ratios of maximum to minimum particle lengths for Al_2O_3 - 5 wt% Al composite powder, estimated from 125 observations.

3. Estimation of Alumina and Aluminum Particle Sizes by XRD

The particle sizes were also estimated by XRD, using two separate techniques. The first technique used the ratio of area of the peak and its height, referred to as the integral breadth [13]. Estimates were made for four separate alumina peaks and one aluminum peak. Also, the integral breadths of the four measured alumina peaks were plotted versus the sine of the diffraction angle in order to check if any of the peak broadening was due to lattice strain [13].

The second method used the square of the standard deviation of the peak, called the variance [13]. Which can separate peak broadening caused by small particle size from that contributed by lattice strain as well the instrument.

Table 1 shows that the results obtained by XRD using both methods agree reasonably. Also, both estimates overlap within the limits of TEM results where the median size was 75 nm. The particle size estimates for the α-alumina phase using XRD ranged from 20 to 109 nm, in good agreement with the TEM estimate of 75 nm. Both XRD techniques showed that lattice strain in the α-alumina phase was less than 0.5%.

The reasonable agreements between two XRD-based estimates and direct measurement by TEM show that the TEM based approach can be used reliably to measure particle sizes and aspect ratios of nano-scale composite powder.

Table 1: Estimated Sizes of Alumina and Aluminum Particles in the Precursor by Techniques using XRD Peaks.

Particles	XRD Peak 2θ Value	Sizes, nm	
		Integral Method	Variance Method
Alumina	25.6°	46-100	33-84
Alumina	35.2°	45-109	21-62
Alumina	57.6°	25-49	19-51
Alumina	68.2°	20-33	20-49
Aluminum	38.5°	63-268	36-260

For aluminum particles, the size estimates have a relatively large range. The probable reasons are difficulties in measuring Al peaks because of its low concentration (5 wt%). Also, XRD patterns show that the aluminum phase was textured in the present case. The microstructures of ICP sintered composites [5] indicate that the particle size of the aluminum phase was less than 100 nm.

These particle sizes are comparable to those of other composite precursors prepared using similar methods. Submicron precursor powders have been previously reported for attrition milled Al_2O_3 powders with 35-50 vol% Al [11]. Nanocrystalline Al_2O_3-Fe precursor powders with particle sizes of less than 20 nm have been produced by mechanical milling [15].

B. Reactive Sintering of Precursor in Argon-0.5% H_2 at Reduced Pressure

In attempts to establish processing conditions for the synthesis of functionally gradient materials based on alumina/aluminum composites, we conducted a series of trial sintering experiments. The objective of the experiments was to control the oxidation of the composite precursor during the consolidation process. In the preliminary sintering experiments in air at 800-1400°C, we observed that aluminum phase was completely oxidized. Sintering under argon gas also resulted in similar oxidation presumably because commercially pure argon gas have enough residual oxygen to oxidize the samples. Use of Argon-0.5% H_2 mixture at 1400-1500°C was quite successful [14] and the sintering experiments at 1400°C are included in the following sections.

After 1400°C heat treatment, the 6.3 mm diameter cylindrical sample showed three distinct zones: (1) 0.1-0.3 mm thick outermost layer, which was bright white in color, (2) an inside area which was dark grey and (3) a transition area between two areas.

Figure 5 shows effect of sintering period on the variation of grain size across the samples. Grain sizes in the center of the samples increased steadily over time, reaching 2.5 to 5 microns after 4 hours. The inside of the sample was quite porous dark gray, and contained both Al_20_3 and Al. Light microscopy revealed numerous bright flecks, 0.2 to 1 micron across.

The transition zone was not sharp in terms of porosity grain size, hardness. Porosity dropped sharply at the interface between the zones. The grain structure and grain size were in between the values for the other zones. The transition zone coarsened more slowly than the center, but

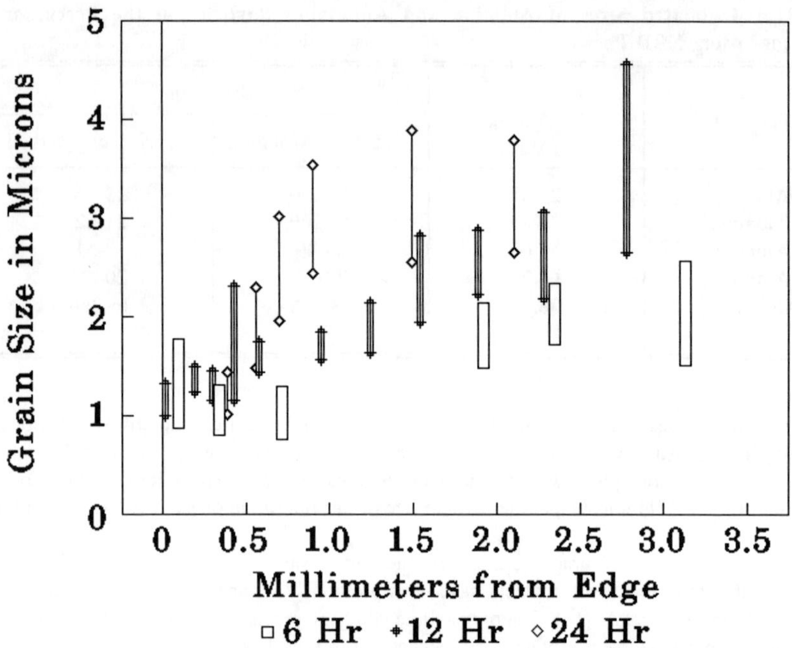

Figure 5: Grain sizes across sintered pellets varied over time. Note the convergence of grain sizes in the outer areas near the surface. The cylindrical samples were 6.3 mm in diameter.

increased steadily over time. The outer regions of the pellets, typically extending 0.1 to 0.3 millimeters inside of the outer edge, were bright white. XRD patterns of the outer layer showed that there was still some Aluminum present. This result is quite unexpected. The grain size of the outer parts of the samples was 0.7 to 2 microns; it remained stable from 6 to 24 hours of sintering time. Porosity was smaller than the ceramic metal regions, typically 15 to 40 percent. No bright flecks were seen using optical microscopy.

Figure 6 shows the microhardness of different parts of the samples over time. Central portions of samples had hardnesses of 100 to 200 MPa, with a large amount of scatter. These values shaded smoothly into the outer regions, with values of 180 to 270 MPa.

These values are comparable to similar work reported in literature. Attrition milled Al_2O_3/Al/ZrO_2 precursors have been used to form reaction bonded ceramics with hardnesses of 100 MPa at 60% of theoretical density, 250 MPa at 70% density and >1100 MPa at >99% density [16]. The same techniques were also used to form Al_2O_3 from Al_2O_3/Al precursors with grain sizes of 2-5 μm and hardnesses of 370 MPa [16].

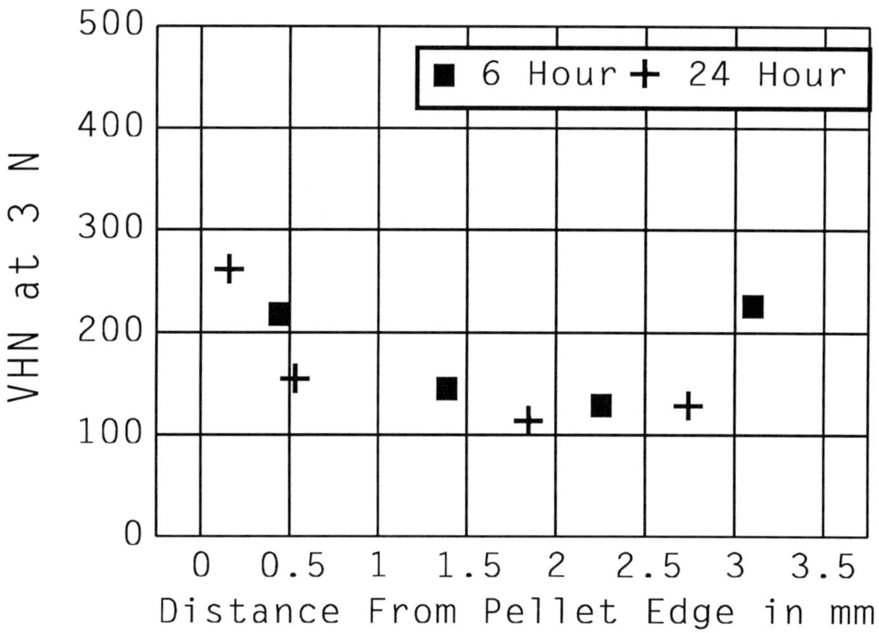

Figure 6: Vickers microhardness across sintered pellets.

CONCLUSIONS

A fine grained functionally gradient composite was prepared by sintering a submicron precursor powder in a reducing atmosphere. The submicron composite precursor powder was prepared by high energy milling of Al_2O_3 with 5 wt % Al. This precursor powder exhibited a particle size of 20-110 nm, as well as the presence of both of the starting phases. The size distribution and aspect ratios of the particles are summarized.

The composite powder was sintered in an atmosphere of Ar/0.5 % H_2 at 1400°C. The sintered composite exhibited three zones, ranging from a dense, well sintered outer region 0.1 to 0.3 mm thick through a transition zone to a porous composite interior.

Grain sizes of the sintered composite varied over time as well as across the composite. The outer regions exhibited grain sizes of 0.7 to 2 microns; these held steady over time. The inner regions exhibited grain sizes of 1.5 to 2.5 microns after 6 h, increasing to 2 to 4.5 microns after 24 h. Vickers microhardness of the sintered samples ranged from 100 to 200 VHN in the composite core, and approached 270 VHN in the outer regions.

REFERENCES

1. D. R. Clark, "Interpenetrating Phase Composite Materials," *J. Am. Ceram. Soc.*, 75 [4], 735-9 (1992).

2. M.S. Newkirk, H.D. Lesher, D.R. White, C.R. Kennedy, A.W. Urquhart and T.D. Elaar: "Preparation of LANXIDE^{TM} Ceramic Matrix Composites: Matrix Formation by the Directed Oxidation of Molten Metals" *Ceram. Eng. Sci. Proc.*, 8 [7-8], 879-885 (1987).

3. C. Suryanaranyana, D. Mukhopadhyay, S.N. Patanku and F.H. Troes: "Grain Size Effects in Nanocrystalline Materials" *J. Mater. Res.*, 7 [8], 2114-2117 (1992).

4. S.H. Kim and S. Sinha, "Microstructures and Interfaces in Al_20_3-5 wt% Al CMC Composites Synthesized by Plasma Sintering of Attrition-Milled Precursors," *Ceram. Eng. Sci. Pro.*, 13 [9-10]. 855-62 (1992).

5. S. Sinha and S.H. Kim, "Al_20_3-5 wt% Al Composites via Synthesized Precursors by High Energy Milling: ICP Sintering Characteristics," *Ceram. Eng. Sci. Pro.*, 14 [9-10] 768-75 (1993).

6. S. Sinha and S.H. Kim, "Al_20_3-5 wt% Al Composites by ICP Sintering Characteristics of Synthesized Precursor," accepted to *J. Mat. Sci.* (1993).

7. S.H. Kim, "Processing-Microstructure Relationships in Plasma Sintered Al_20_3 Composites via Attrition-Milled Powder," Ph.D. thesis, The University of Illinois at Chicago, Chicago, IL (1993).

8. T. Hirai, S. Toshio, "Preparation of SiC/C Functionally Gradient Nanocomposites," *Seramikkusu*, 26 [3] 216-17 (Japanese), (1991).

9. A. Kawasaki, H. Hirose, H. Hiroyuki, et al., "Fabrication of Sintered Functionally Gradient Material by Powder Spray Forming Process," *Funtai oyobi Funmatsu Yakin*, 37 [7], 922-8 (Japanese), (1990).

10. K. Tanihata, Y. Miyamoto, K. Matsushita, et al., "Fabrication of Cr_3C_2/Ni Functionally Gradient Materials by Gas-Pressure-Combustion Sintering," *Ceram. Trans.*, 34. 361-8 (1993).

11. N. Claussen, N. A. Travitzsky, S. Wu, "Tailoring of Reaction-Bonded Al_2O_3 (RBAO) Ceramics," *Ceram. Eng. Sci. Proc.*, 11 [7-8], 806-820, (1990).

12. ASTM E112-84 Standard Method for Determining Grain Size.

13. H.P. Klug and L.E. Alexander, X-Ray Diffraction Procedures for Polycrystalline and Amorphous Materials. John Wiley and Sons, New York, pp 661-694, 1974.

14. E.B. Sweet and S.N. Sinha, to be published.

15. P. Matteazzi, G. Le Caer, "Synthesis of Nanocrystalline Alumina-Metal Composites by Room-Temperature Ball-Milling of Metal Oxides and Aluminum." *J. Am. Ceram. Soc.*, 75 [10], 2749-55, (1992).

16. D. Holz, M. Röger, R. Janssen, N. Claussen, "Mechanical Properties of Reaction-Bonded Al_2O_3/ZrO_2 Composites," Advanced Ceramics Conference, Cocoa Beach, USA (1994).

PROPERTIES OF COMBUSTION SYNTHESIZED DENSE CERAMIC METAL COMPOSITES

H. J. Feng and J. J. Moore
Colorado Center for Advanced Ceramics
Department of Metallurgical and Materials Engineering
Colorado School of Mines
Golden, CO 80401

ABSTRACT

Combustion synthesis coupled with hot pressing has been used to produce dense, near net shape ceramic and ceramic-metal composites that could be used as high performance structural materials. The use of an in-situ liquid infiltration and hot pressing technique is discussed. The current research involves the combustion synthesis of $3TiC\text{-}2Al_2O_3\text{-}Al$ composites formed by the reaction of TiO_2-C-Al mixtures. Mechanical properties of the composites are presented and compared with similar materials produced by other methods. The control of the synthesis reaction, microstructural development, and mechanical properties are discussed together with the critical issues associated with this novel one step process to produce dense, near net shape, ceramic-metal composites.

INTRODUCTION

The application of a novel, one step, in-situ combustion synthesis technique for the economical production of ceramic-metal composites is described in this paper.

Combustion (self propagating high temperature) synthesis (SHS)[1] uses an exothermic reaction to provide the energy needed to synthesize ceramics, intermetallics and composites. The exothermic reactant mixture, normally in the form of powders, is pressed into a pellet of a certain green density and is subsequently ignited, either locally at one point (propagating mode) or by heating the whole pellet to the ignition temperature of the exothermic reaction (simultaneous combustion mode). A schematic representation of a typical temperature-time plot for a combustion reaction is given in Figure 1(a). The unreacted reactant mixture, at initial temperature, To, is heated to the ignition temperature,

To the extent authorized under the laws of the United States of America, all copyright interests in this publication are the property of The American Ceramic Society. Any duplication, reproduction, or republication of this publication or any part thereof, without the express written consent of The American Ceramic Society or fee paid to the Copyright Clearance Center, is prohibited.

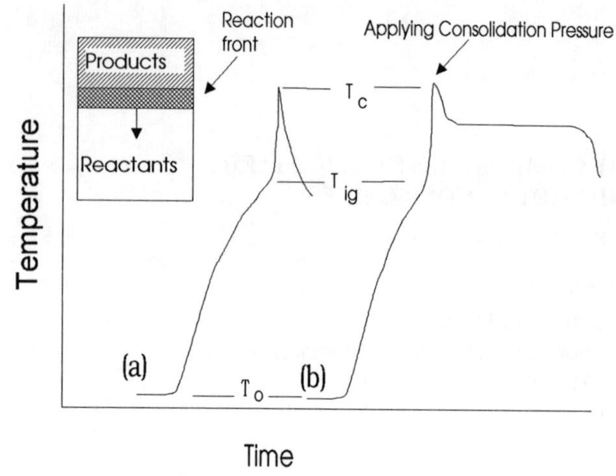

Figure 1. Schematic representation of the temperature-time curve during (a) SHS reaction (b) consolidating load applied to the sample immediately after the initiation of the SHS reaction and maintained at temperature for a certain time.

Tig, whereupon the reaction is initiated. Sufficient heat is released by the exothermic reaction, as indicated in Figure 1(a), in order to achieve a maximum or combustion temperature, Tc, and to sustain the reaction without the need for additional heat.

EXPERIMENTAL REACTION SYSTEM

One of the major limitations of combustion synthesis is that relatively high levels of porosity, e.g. 50%, remain in the product[2] (Figure 2). However, since this porosity is predominantly open porosity, i.e. >95%,[2] this can be used to advantage by simultaneously generating a liquid metal, e.g. Al, which rapidly infiltrates the open pores in the ceramic matrix immediately after/as the porous ceramic matrix is synthesized via the SHS reaction.

Several model composite systems have been investigated in which an excess amount of Al is used both as a reductant in the SHS reaction and as the liquid metal component which infiltrates the porous ceramic matrix to synthesize the ceramic-metal composite[2-6]. One example is given below:

$$3TiO_2 + 3C + (4 + x)Al = 3TiC + 2Al_2O_3 + xAl \quad [1]$$

where x was varied between 0 to 9[2]. The excess amount of liquid Al, i.e. xAl, generated at and ahead of the reaction front infiltrates the pores in the simultaneously synthesized

Figure 2. Porous nature of the ceramic composite produced from the combustion synthesis reaction $3TiO_2 + 3C + 4Al \rightarrow 3TiC + 2Al_2O_3$ conducted in the propagating mode.[2]

ceramic composite, and hence increase the relative density of the final composite.

This novel approach, i.e., SHS + in-situ liquid metal infiltration, has resulted in an increase in density of the composite from 45% of theoretical to approximately 75% of theoretical[2] in reaction (1) when conducting the SHS reaction in the propagating mode. Increasing the excess Al (i.e., x = 9) in the reaction stoichiometry lowers the combustion temperature and provides a means by which the product microstructure can be controlled. Incorporating a diluent, i.e. yAl_2O_3, into the reaction mix, as indicated in reaction (2), also lowers the combustion temperature and provides an additional control variable in the system, i.e.

$$3TiO_2 + 3C + 4Al + yAl_2O_3 = 3TiC + (2 + y)Al_2O_3 \qquad [2]$$

Therefore, a means of controlling Tc is provided when both an excess amount of Al and an Al_2O_3 diluent are used in the SHS reaction system, and the overall combustion synthesis reaction can be represented as:

$$3TiO_2 + 3C + (4 + x)Al + yAl_2O_3 = 3TiC + (2 + y)Al_2O_3 + xAl \qquad [3]$$

At the same time, igniting these SHS reactions in the simultaneous combustion mode and applying a consolidating, compressive load to the pellet as soon as the exotherm is recorded by a thermocouple, will provide a one-step process for producing ceramic-metal composites with densities in excess of 75% of theoretical.

EXPERIMENTAL PROCEDURE

Reactant powders (Table I) were thoroughly mixed using porcelain ball milling and pressed to various green densities, with varying values of x and y, as represented in the reaction system (3). The cylindrical pellets of 12.7mm (0.5 inches) in diameter and 25.4mm (1 inch) in length were dried for 1 hour at 120°C and ignited in the simultaneous combustion mode in a hot press schematically presented in Figure 3. The hot press was comprised of a uniaxial, two plunger graphite die system which was heated rapidly to the ignition temperature of the SHS reaction by means of a tungsten resistance coil. The graphite die and tungsten coil were packed with graphite felt and placed inside a steel reaction chamber. The Tig and Tc were determined using a W-W/5%Re thermocouple placed in the side wall of the graphite die, and interfaced to a chart recorder. As soon as the exotherm was observed on the chart recorder the consolidating load was applied via the two plunger graphite die system, and the temperature maintained at a certain level below Tc for up to 15 minutes as graphically presented in Figure 1(b).

SHS Table I -Physical Properties of Reactants and Products in Reaction (1)

Material	TiO_2	C	Al	TiC	Al_2O_3
Mpt°C	1830		660	3140	2050
Bpt°C	3000		2467	4820	2980
SG*	4.26	2.25	2.70	4.93	3.97
Size (μm)	-44	-44	-44		-44

*Specific Gravity

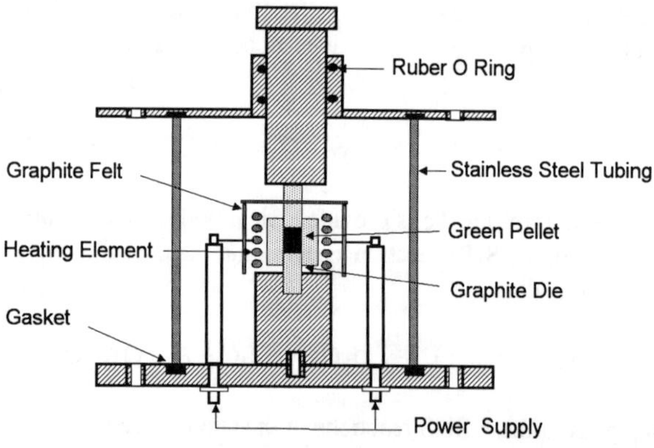

Figure 3. Schematic representation of the experimental system to provide simultaneous hot pressing and combustion synthesis in a graphite die. The reaction was conducted under an argon atmosphere inside the chamber.

RESULTS AND DISCUSSION

It was quickly established that, when the consolidating load was applied as soon as the exotherm was recorded on the thermocouple in the die, a minimum pressure of 24.8 MPa (3600 psi) applied for ten minutes at 1600°C was required to provide significantly improved densities compared with those produced in the propagating mode without pressing. Previous research[2] had also established the green densities needed to provide stable reactions in these SHS reactions conducted in the propagating mode. Therefore, a green density of approximately 65% of theoretical was used in each of these experiments. The relative densities achieved in the synthesized composite materials using this approach are recorded in Figure 4 as a function of xAl when $yAl_2O_3=0$, and as a function of yAl_2O_3 when x = 0. These data are also compared as a function of xAl with data obtained for the SHS reaction conducted in the propagating mode without pressure and for y = 0. The corresponding compressive strengths achieved in these composites are given in Figures 5 and 6 for reaction (3) as a function of x when y = 0 and as a function of y when x = 0 respectively.

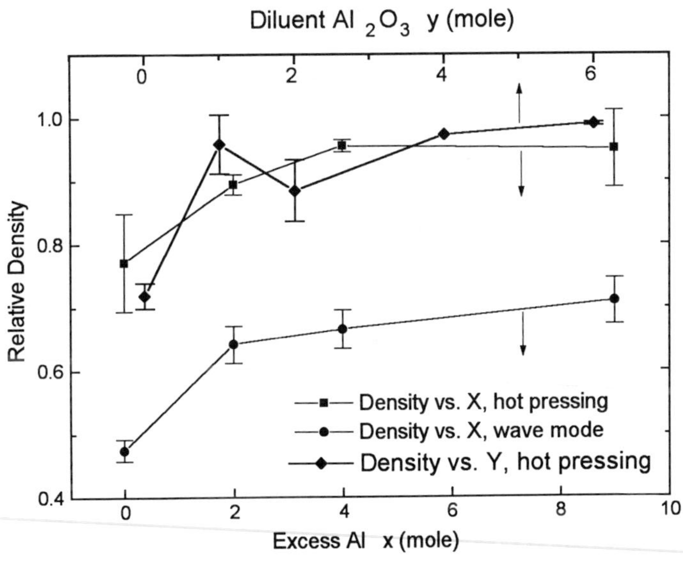

Figure 4. The effect of excess Al (xAl) when y = 0, (■), and Al_2O_3 diluent (yAl_2O_3) when x = 0, (♦), on the relative density of the $TiC-Al_2O_3-Al$ ceramic-metal composite using the simultaneous combustion mode and a consolidation pressure of 24.8 MPa (3600 psi) maintained at 1600°C for ten minutes. These data are compared with similar data (●) obtained for the SHS reaction (1) conducted in the propagating mode without pressure.

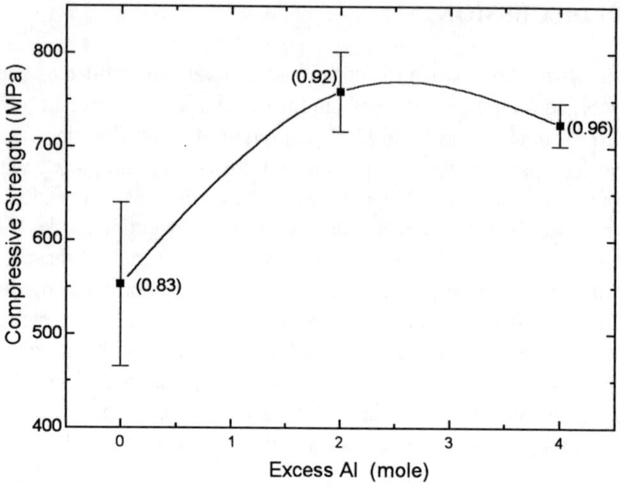

Figure 5. The effect of excess Al on compressive strength for TiC-Al_2O_3-Al ceramic-metal composites produced by simultaneous combustion synthesis and pressing in a graphite die, for reaction (3) for y = 0. The figures in parentheses indicate the relative density of the samples tested.

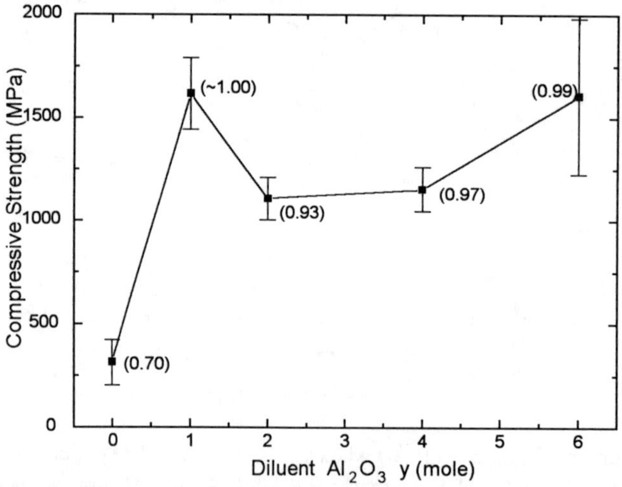

Figure 6. The effect of excess Al_2O_3 (diluent) in reaction system (3) when x = 0, (no excess Al) on the compressive strength for TiC-Al_2O_3 composites produced by simultaneous combustion synthesis and pressing in a graphite die. The figures in parentheses indicate the theoretical density of the samples tested.

The corresponding microstructures of these composites are given in Figure 7. The large plate of Al_2O_3 observed in Figure 7(a) indicates that the combustion temperature for the reaction stoichiometry when x = 0 exceeded the melting point of Al_2O_3 (2050°C). These large brittle plates of Al_2O_3 may provide channels for easy crack propagation. Therefore, some advantage is likely to be gained by controlling Tc to be less than 2050°C in order to produce fine Al_2O_3 particles. This can be achieved by increasing yAl_2O_3 and/or xAl in reaction system (3).

Applying a compressive load, simultaneously as the exothermic reaction is initiated resulted in fully dense ceramic-metal composites of $3TiC$-$2Al_2O_3$-$9Al$, (x=9, y =0) (Figure 4) and achieved compressive strengths of 758.4 MPa($1.1x10^5$ psi) (Figure 5). The excellent distribution of a fine Al network between TiC and Al_2O_3 particles is clearly evident in the $3TiC$-$2Al_2O_3$-$4Al$ (x = 4) ceramic-metal composite produced by this technique and shown in Figure 7(b). In this condition, increasing xAl has decreased Tc to below 2050°C, producing fine (1-2μm diameter) Al_2O_3 particles. Increasing the relative density of the composite from 83% to 96% produced a considerable increase in compressive strength and decrease in scatter of the data (Figure 5), even though the volume fraction of the soft ductile aluminum has increased. The terraced appearance of the fractured surface of the samples tested in the compression tests [Figure 7(c)] indicates a potential for increased toughness.

As indicated earlier, further refinement of the microstructure is possible by adding Al_2O_3 as a diluent to the reactant mix. This addition decreases Tc to avoid melting Al_2O_3 and, therefore, produces finer Al_2O_3 particles as indicated in Figure 7(b).

The low compressive strength recorded for the $3TiC$-$2Al_2O_3$ (x = 0, y = 0) composite was probably a result of low theoretical density, i.e. 83%, coupled with the large Al_2O_3 plates produced in this composite [Figure 7(a)]. Since the combustion temperature exceeded the melting point of Al_2O_3 in this reaction stoichiometry, i.e. x = 0, y = 2, the solidification shrinkage of Al_2O_3 possibly contributed to the higher porosity retained in this sample.

Again, considerable advantage can be gained by adding Al_2O_3 as a diluent in order to control the combustion temperature. The compressive strength data achieved for reaction stoichiometries in which x = 0 and y was varied between 0 and 6 (Figure 6) approach those predicted for fine grained composites of TiC and Al_2O_3, but also indicate a strong dependence on density.

However, considerable increases in compressive strength have been achieved in these samples compared with those composites in which no Al_2O_3 diluent was used. Typical microstructures of these composites [Figure 7(b)] reveal that Tc has been lowered to below 2050°C by incorporating excess Al_2O_3 as low as y = 1 in the reaction system (3).

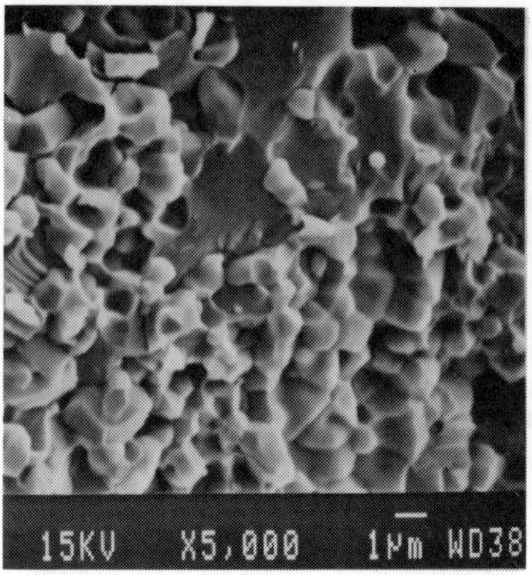

[a]

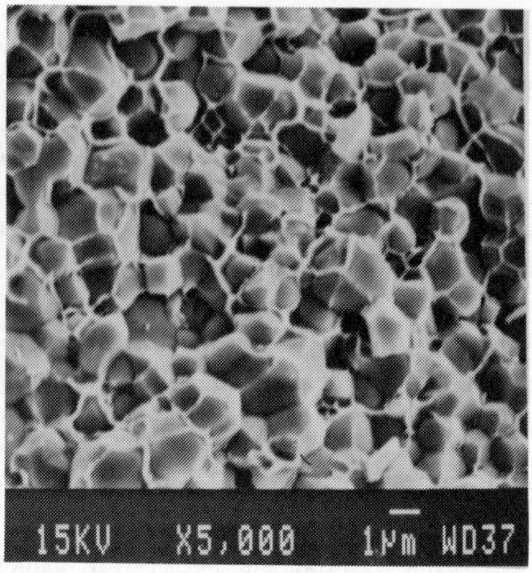

[b]

[c]

Figure 7. SEM photomicrograph of (a) $3TiC-2Al_2O_3$ (x = 0, y = 0), - note the large plate of Al_2O_3 at top center, (b) $3TiC-2Al_2O_3-4Al$ (x = 4, y = 0), (c) fractured surface of $3TiC-2Al_2O_3-4Al$ (x = 4, y = 0) samples tested in compression test, -note the terraced appearance of the fractured surface indicating crack deflection and a potential for increased toughness.

The elastic moduli of the composites are given in Figure 8 as a function of the amount of excess Al, i.e. xAl, for y = 0, and as a function of the amount of Al_2O_3 used as a diluent, i.e. yAl_2O_3, for x = 0. A significant increase in stiffness can be produced in these composites on increasing density and balancing the amounts of Al_2O_3 and Al. As expected, the trends follow those exhibited by the compressive strengths, as shown in Figures 5 and 6. Larger samples are currently being prepared in order to perform flexure strength and notched four point bend test fracture toughness evaluations on these composites.

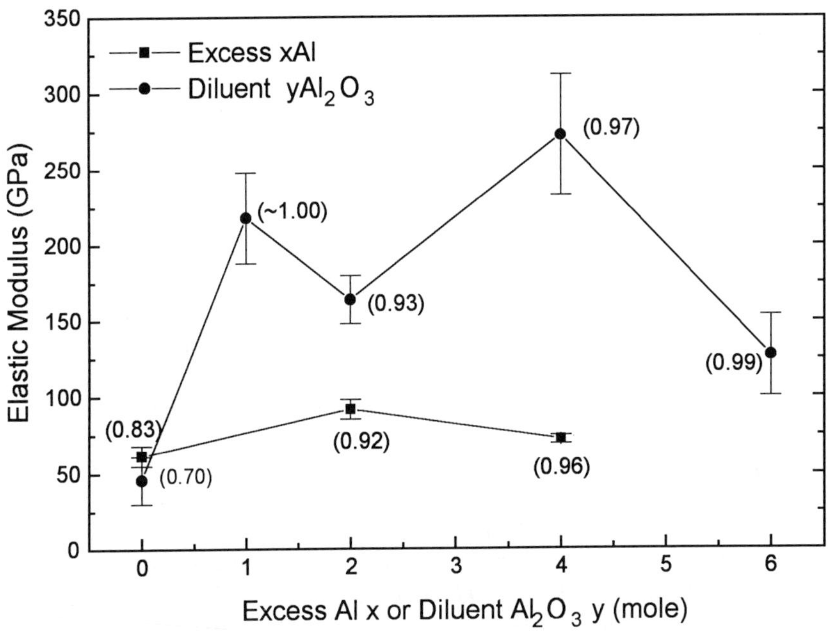

Figure 8. The effect of excess Al and relative density (figures in parentheses) when y = 0, and diluent Al_2O_3 and relative density when x=0, on the elastic moduli of composites produced from reaction (3) .

CONCLUSIONS

The application of simultaneous combustion synthesis in which a consolidating pressure is applied immediately after the SHS reaction is initiated has produced dense ceramic, i.e. $TiC-Al_2O_3$, and ceramic-metal, i.e. $TiC-Al_2O_3-Al$ composites. The application of a simultaneously applied load during the SHS reaction is beneficial in increasing the relative density of these ceramic and ceramic-metal composites. The addition of Al_2O_3 as a diluent lowers the combustion temperature and aids in control of both the particle size of the product phases and product density, facilitating an improvement in compressive strength and elastic modulus. The favorable distribution of a fine Al network in the ceramic-metal composite produced by this simultaneous SHS-in-situ liquid metal infiltration and pressing process should result in improved fracture toughness. This in-situ synthesis technique can be used in conjunction with Al_2O_3 as a diluent and an excess amount of Al to provide a ductile metal network to provide an affordable, low cost, process for the synthesis of novel three dimensional ceramic-metal composites.

ACKNOWLEDGMENTS

The authors are grateful for support of this work by NASA Microgravity Science Division - Code SN.

REFERENCES

1. Z. A. Munir and U. Anselmi-Tamburini, Mater. Sci. Report 3 (1989) 227-365.
2. H.J. Feng, J.J. Moore, D.G. Wirth, "The Combustion Synthesis of Ceramic-Metal Composites: The TiC-Al_2O_3-Al System", Met. Trans., 23A, 1992, pp 2373-2379.
3. H.J. Feng, J.J. Moore, D.G. Wirth, "Combustion Synthesis of TiB_2-Al_2O_3-Al Composite Materials" Symposium on Developments in Ceramic and Metal Matrix Composites, pub by TMS, 1992; , pp 219-239.
4. H.J. Feng, J.J. Moore, D.G. Wirth, "Combustion Synthesis of Ceramic Metal Composite Materials: The ZrB_2-Al_2O_5-Al System" International Journal Self-Propagating High Temperature Synthesis (SHS), 1, (2), 1992, pp 228-238.
5. H.J. Feng, J.J. Moore, D.G. Wirth, "The Combustion Synthesis of B_4C-Al_2O_3-Al Composite Materials" MRS on Synthesis and Processing of Ceramics: Scientific Issues Mat. Res. Soc. Symp. Proc., Vol. 249, 1992, pp529-532.
6. K.R. Hunter, J.J. Moore, D.G. Wirth, Titanium' 92, pub by TMS, 1993, pp2519-2527.

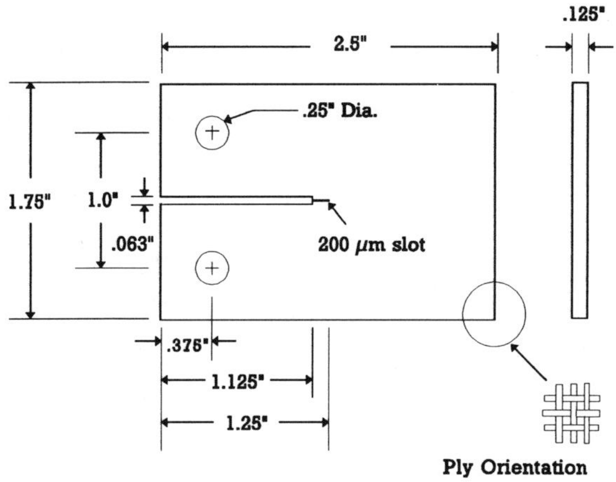

Fig.1 Compact Tension (CT) specimen design for 2-D SiC/SiC woven composite

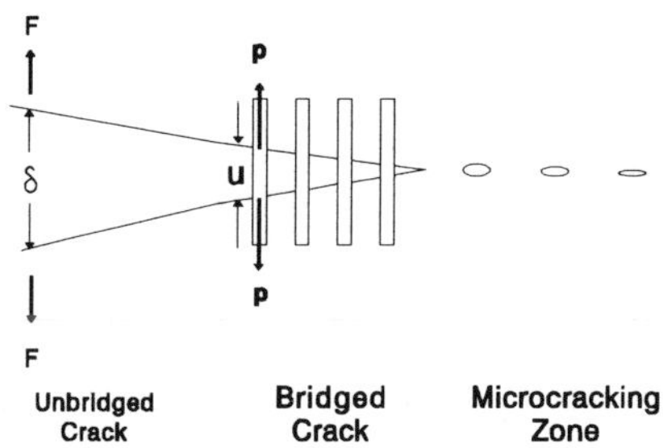

Fig.2 Relation between load-line displacement, δ, and crack tip opening, u.

infiltration of SiC matrix onto a woven uncoated Nicalon SiC fibrous preform was investigated. The mechanical properties and microstructure of this material have been described in previous papers [5,6]. A compact tension (CT) specimen geometry was used, see Fig.1 for dimensions. The crack mouth opening displacement (CMOD) was measured using an extensometer gauge attached to the specimen. The loading-line displacement(δ) was then obtained by assuming a linear crack profile. An applied CMOD rate of 25 μm/min was selected to enable in situ measurements of the CMOD in relation to the crack tip opening displacement, u, see Fig.2, which was accomplished by using a long-distance telescope (model Questar QM-100) fitted with a videotape recorder. The specimens were also polished to a 1 μm finish on a diamond wheel so that observations of crack growth and tip opening, u, could be performed without ambiguity.

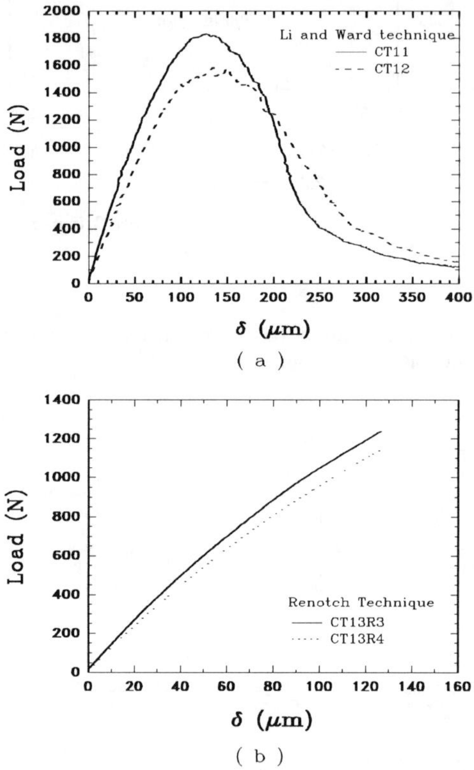

Figure 3 Load verse load-line displacement curves for (a) the Li and Ward technique and (b) the Renotch technique.

To determine the bridging stress-crack opening function, the specimen was first loaded to result in some crack plus bridging zone extension which was about 5 ~ 7 bundles or 7 ~ 11 mm in length. The specimen was next unloaded and then the initial notch was slowly extended by saw-cutting until the measured far-field compliance first began to show an increase. This is the point at which the renotched length just enters the bridging or cohesive zone. The F-δ curves, or load-load-line displacement curves, was then obtained for incrementally larger notch depths until the notch depth approached the end of the bridging zone, always ensuring at each step that there was no growth of either the crack or bridging zone during each loading. The *J*-integral, *J(δ)*, could be obtained from the area between two *F-δ* curves for two incrementally different crack (or notched) length. *J(u)* is then obtained from the measured relation between *u* and δ. Finally, *p(u)* is obtained using Eq.(2).

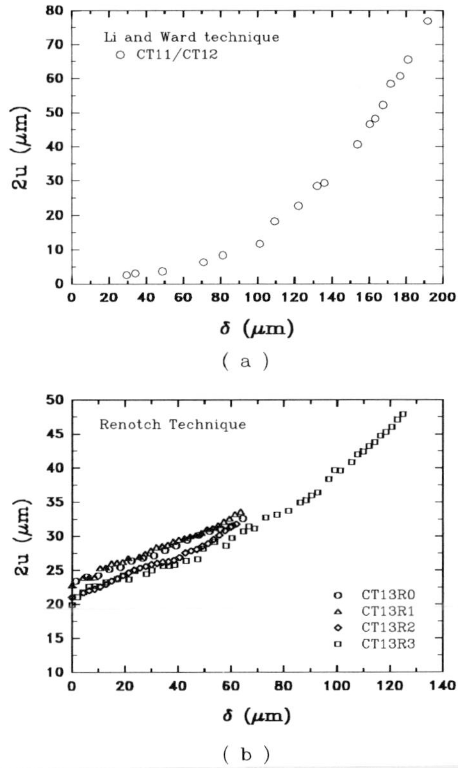

Figure 4 Relation between the load-line displacement, δ, and the crack tip opening, *u*, for (a) the Li and Ward technique and (b) the Renotch technique.

For comparison purposes, the test method of Li and Ward [1,2] was also conducted on two SiC/SiC specimens, with 22.111 mm (CT11) and 23.770 mm (CT12) of initial notch length respectively.

EXPERIMENTAL RESULTS

The load-displacement, F-δ, curves to fracture for the two specimens CT11 and CT12, tested to evaluate the $p(u)$ function using the Li and Ward technique [1,2], are shown in Fig.3(a). Firstly, note in Fig.3(a) that the two curves crossed each other, indicating some interaction of the crack with the far surface of the specimen. This cross-over point was taken as the upper limit of the validity of the $p(u)$ function using this technique. Secondly, this SiC/SiC composite suffers extensive microcracking in the crack tip process zone giving rise to a substantial process zone displacement, u, parallel to the applied load. To facilitate comparison with the result of the present technique the process zone displacement due to microcracking was included on the total measured value of u. Fig.6(a) shows the final result of $p(u)$ derived from Eq.(2) using the Li and Ward multiple-specimen approach. The closure pressure at first increases to a maximum value of about 350 MPa and then decreases to about 20 MPa after 55 μm of crack opening.

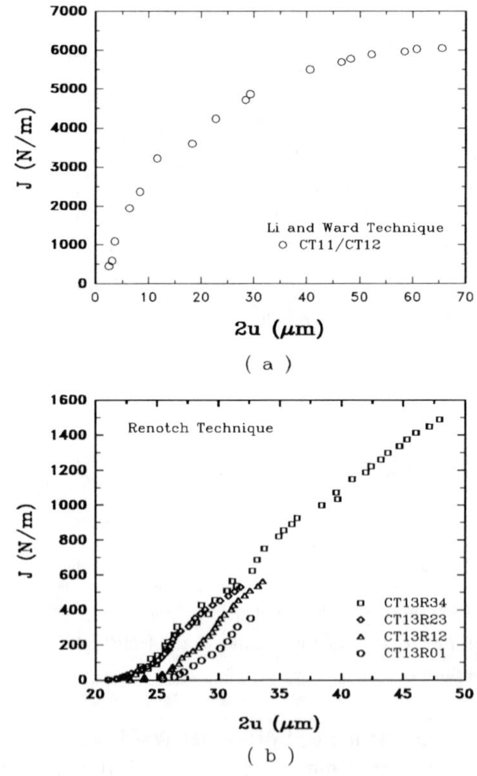

Figure 5 $J(u)$ curves for (a) the Li and Ward technique and (b) the Renotch technique.

In the renotch technique of this paper, the load-displacement, F-δ, curves corresponding to each stage of renotch are typically shown in Fig.3(b) with the fully developed bridging zone tip remaining at the same position. The area between the two curves was measured at each δ and applied to calculate $J(\delta)$. The relation between the displacement, δ, and u is shown in Fig.4(b) for each renotched crack length where u is measured directly at the major crack tip without ambiguity. Note the presence of a residual opening, u_o, at the crack tip when the specimen was first loaded to establish the bridging zone. Clearly the $p(u)$ function using this technique can only be obtained for $u > u_o$, see discussion below. The $J(u)$ can then be obtained as shown in Fig.5(b). Because one fiber bundle is renotched away at each step, the $p(u)$ functions corresponding to each bundle along the bridging zone is shown in Fig.6(b) by taking the slope of $J(u)$ at each value of u from Eq.(2). The residual crack opening is apparent on Fig.4(b) which limits the $p(u)$ function measurement to beyond about 20 μm of crack opening.

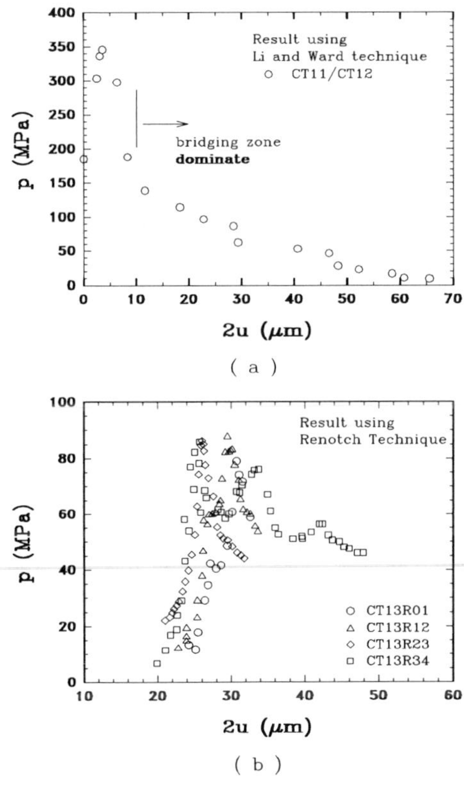

Figure 6 Final results of $p(u)$ curves for (a) the Li and Ward technique and (b) the Renotch technique.

DISCUSSION

An important question in the present analysis was the validity of the *J*-integral in the presence of repeated unloading steps. That is, the specimen was unloaded prior to each renotch to change the process zone size (by saw-cutting) and then reloaded. It was found, see Fig.7, that the load-displacement curves prior to and after unloading were identical at every renotch step. This established the validity of the *J*-integral approach proposed here.

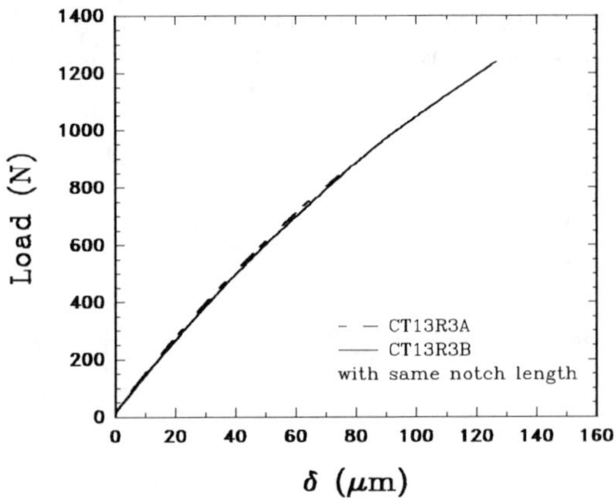

Fig.7 The load-load-line displacement curves with the same notch length in renotch technique.

The result of Fig.6(a) from the Li and Ward technique can be considered as the intrinsic *p(u)* function for SiC/SiC composite because of the absence of unloading and consequently no prior damage to the bridging fibers before testing. In contrast, the renotch technique must endure the residual opening because the material was unloaded after establishing the bridging zone before *p(u)* evaluation. Because of the general irreversibility of the bridging process, the *p(u)* function after unloading would exhibit the behavior shown schematically in Fig.8 where u^* is the crack tip opening prior to unloading and u_o is the residual opening. It is evident that only the part of the reloading *p(u)* curve for $u > u^*$ is characteristic of the intrinsic *p(u)* curve. This is phenomenologically similar to the behavior of the σ-ϵ curve of metals after unloading. Accordingly, in Fig.9, only the decreasing part of the *p(u)* function using the present method is plotted and compared to the *p(u)* curve obtained using the Li and Ward technique.

The result in Fig.9 indicates that the *p(u)* determined in the present technique agrees well with that obtained by the method of Li and Ward. This fundamental agreement between two methods further established the validity of the present approach. Although the results of the present technique were obtained over a narrower range of crack openings, *2u*, where compared with the Li and Ward technique, this does not represent a limitation of this proposed renotch technique. Data can be obtained in a single specimen over a wider range of u by limiting the value of u_o. This can be done by growing a smaller process zone in the first step of the renotch technique.

The area under the *p(u)* curve by definition is given by J_{SS}, where J_{SS} is the steady state or plateau value of the J_R curve. The area under the p(u) curve from the present result is about 5500 N/m. This is consistent with the previously measured value of J_{ss} for this material [5,6].

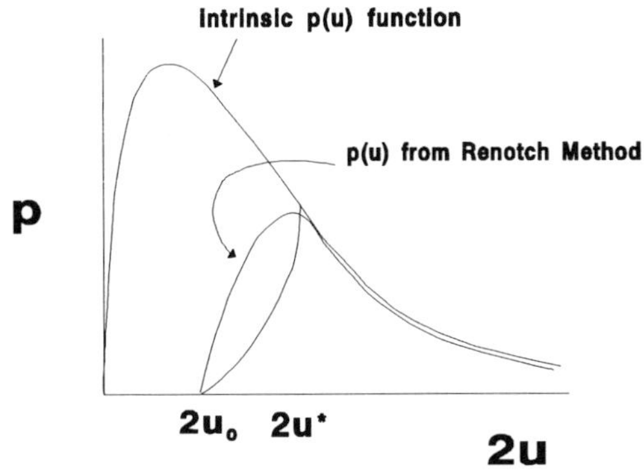

Fig.8 Schematic *p(u)* plot for the data obtained form renotch technique

Finally, the evaluation of *p(u)* function from both techniques are basically statistical averages of the behavior of a large number of fiber bundles. Further more each bundle contains about 150 fibers and consequently would exhibit statistical variation in the bridging stress function, *p(u)*, from bundle to bundle. It was recommended that at least four specimens be tested to evaluate the *p(u)* function in the Li and Ward technique, however, only two specimens were used in our test and consequently the result in Fig.6(a) does not reflect the scatter in the *p(u)* function. This scatter is more evident in Fig.6(b) from renotch method where each bundle exhibits its own *p(u)* function. There is potentially a substantial variation from bundle to bundle as would be expected. More work is needed to study the variability in the *p(u)* function and its consequent effect on the statistics of the strength distribution observed in this material.

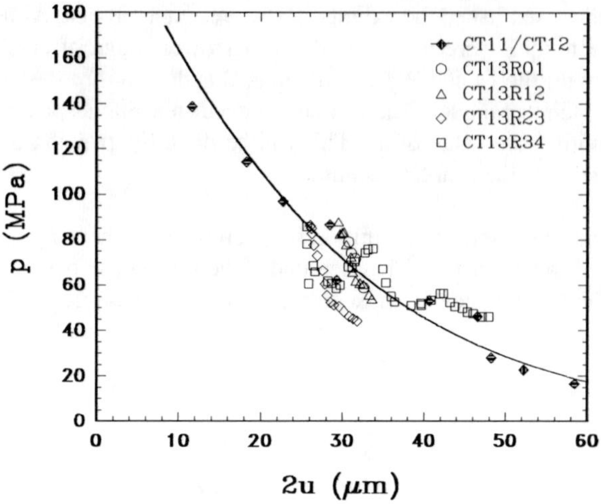

Fig.9 Comparison of *p(u)* curve results obtained from both Li and Ward technique (CT11/CT12) and renotch technique (CT13).

SUMMARY AND CONCLUSION

1 The *J*-integral technique from Li and Ward represents an intrinsic *p(u)* function. However, the multiple specimens are needed for just one *p(u)* result and no crack growth is allowed prior to the establishment of a steady-state process zone size.

2 The authors offer a *J*-integral technique with involving renotch procedure on a single specimen to investigate the *p(u)* function. The advantage of this technique is also that the *p(u)* function can be determined on a given crack whether in the as-fatigued, as-crept or statically fractured state.

3. The *p(u)* results from both technique *J*-integral techniques were essentially identical.

ACKNOWLEDGMENTS

The authors thank David Roach of E.I. duPont de Nemours & Co. Inc. for providing us with the machined compact tension SiC/SiC composite sample for this study. We also thank Dr. David Marshall for helpful discussions on the proposed new technique.

REFERENCES

1. V.C.Li and R.J.Ward, "A Novel Testing Technique for Post-Peak Tensile Behavior of Cementitious Materials," in Fracture Toughness and Fracture Energy - Test Methods for Concrete and Rock, 1989.

2. V.C.Li, C.-M. Chan and C.K.Y. Leung, "Experimental Determination of the Tension-Softening Relations for Cementitious Composites", Cement and Concrete Research, 17, 441-452 (1987).

3. B.N.Cox and D.B.Marshall, "The Determination of Crack Bridging Forces," Int.J.Frac. 49, 159-176 (1991).

4. X.Z.Hu, E.H.Lutz and M.V.Swain, "Crack-Tip-Bridging Stresses in Ceramic Materials," J.Am.Ceram.Soc. 74[8] 1828-32 (1991).

5. S.V.Nair and Y.-L.Wang, to be published.

6. S.V.Nair and Y.-L.Wang, "Failure Behavior of a 2-D Woven SiC Fiber/SiC Matrix Composite at Ambient and Elevated Temperatures," Ceram.Eng.Sci.Proc. 13[7-8] 433-41 (1992).

THERMAL SHOCK BEHAVIOR OF BORON NITRIDE/BORON CARBIDE COMPOSITE CERAMICS

W. Pompe and H.-A. Bahr, Max-Planck Society, Technical University, D-01069 Dresden, Germany

G. Kirchhoff and M. Holzherr, Fraunhofer Establishment for Materials Physics and Surface Technology, D-01069 Dresden, Germany

R. Ruh, Universal Technology Corporation, Beavercreek, OH 45431

A. Zangvil, University of Illinois, Urbana, IL 61801

The thermal shock behavior of hot-pressed B_4C-BN composites was investigated. A computer controlled laser beam was used to produce sharp, axisymmetric temperature profiles in thin circular disks, and the time-dependent temperature profile was measured by a high speed pyrometer. Experimentally, as the BN was added to B_4C, the figure of merit decreased to a minimum at ~20% BN, and then steadily increased with increasing amounts of BN. A similar curve was found when the figures of merit were calculated using strength, thermal expansion and Young's modulus data, except that the minimum was at ~25% BN.

INTRODUCTION

Thermal shock behavior is a characteristic of increasing importance. First, there is a need for new materials with improved performance under changing thermal conditions. Ceramic composites have the greatest potential to satisfy these requirements. It is the purpose of this investigation to show that materials with high thermal shock resistivity can be produced by developing a specific microstructure which provides an optimized relationship of strength, Young's modulus and thermal expansion.

Second, there is a problem of methodology. The thermal shock behavior is characterized by a coupled interaction of a temperature field and a stress field for the sample being investigated. Traditionally, heated samples are quenched in air, oil, water, liquid

To the extent authorized under the laws of the United States of America, all copyright interests in this publication are the property of The American Ceramic Society. Any duplication, reproduction, or republication of this publication or any part thereof, without the express written consent of The American Ceramic Society or fee paid to the Copyright Clearance Center, is prohibited.

or powder. An advantage of the quenching experiments is the well defined temperature difference between the sample and cooling media. However, it is difficult to determine the thermal stresses that lead to damage because the heat transfer between sample and cooling agent is usually subject to large uncertainties arising from the flow type, which depends on sample shape coolant. In view of this, it is necessary to set up model experiments with well defined thermal stresses, so that appropriate fracture mechanical tests can be performed. Such conditions are fulfilled by centrally heating a disk (Fig. 1) by means of an arc discharge [1], electron beam [2], conventional light [3,4] or laser light [5]. The temperature field can be calculated or measured by a pyrometer, which allows the calculation of the thermal stress field and its change with time. Thus, fracture mechanics analysis is feasible.

PREPARATION AND CHARACTERIZATION OF SAMPLES

Nine B_4C-BN compositions were prepared using a commercial 600 mesh B_4C powder of 99.9% purity (A.D. Mackay, New York, NY) and a commercial hexagonal (graphitic) BN powder of 99.8% purity and ≤1 μm particle size (No. 5717, Cerac, Inc., Milwaukee, WI). Impurity analyses of both powders have been given previously [6]. The desired compositions were prepared by dry ball milling in a plastic jar with Si_3N_4 balls for 7 h. Samples of 1.90 cm diameter by 1.5 cm thick were uniaxially hot pressed under vacuum (Model 3000, Vacuum Industries, TI Group, Nashua, NH) in graphite dies lined with graphite foil (Carbon Products Division, Union Carbide Corp., Chicago, IL) at 2100°C at a pressure of 35 MPa. Compaction during pressing was monitored by a micrometer dial gage attached to the ram. After specimens were removed from the die, the surfaces were ground and the density was determined by measuring dimensions and weight.

A listing of the samples, compositions and densities studied in this investigation is given in Table I. It is seen that B_4C is nearly theoretically dense and that as the BN content was increased, the density decreased, ultimately reaching a value of 77% for BN. X-Ray diffraction, SEM and TEM analyses revealed the hexagonal B_4C and hexagonal graphitic BN phases, with slight solubility of B_4C in BN and no intermediate phases [6]. BN particles are platey by nature and the faces, which are parallel to the basal planes, were found to be strongly oriented perpendicular to the hot-pressing direction. This orientation was apparent in thermal conductivity data as a function of composition [7]. Boron carbide had little or no anisotropy in thermal conductivity, but anisotropy increased steadily with increasing BN content, attaining a factor of 7 for pure BN. The porosity-corrected room temperature strength of B_4C-BN composites [6] is seen in Fig. 2. For a 2100°C hot-pressing temperature, B_4C

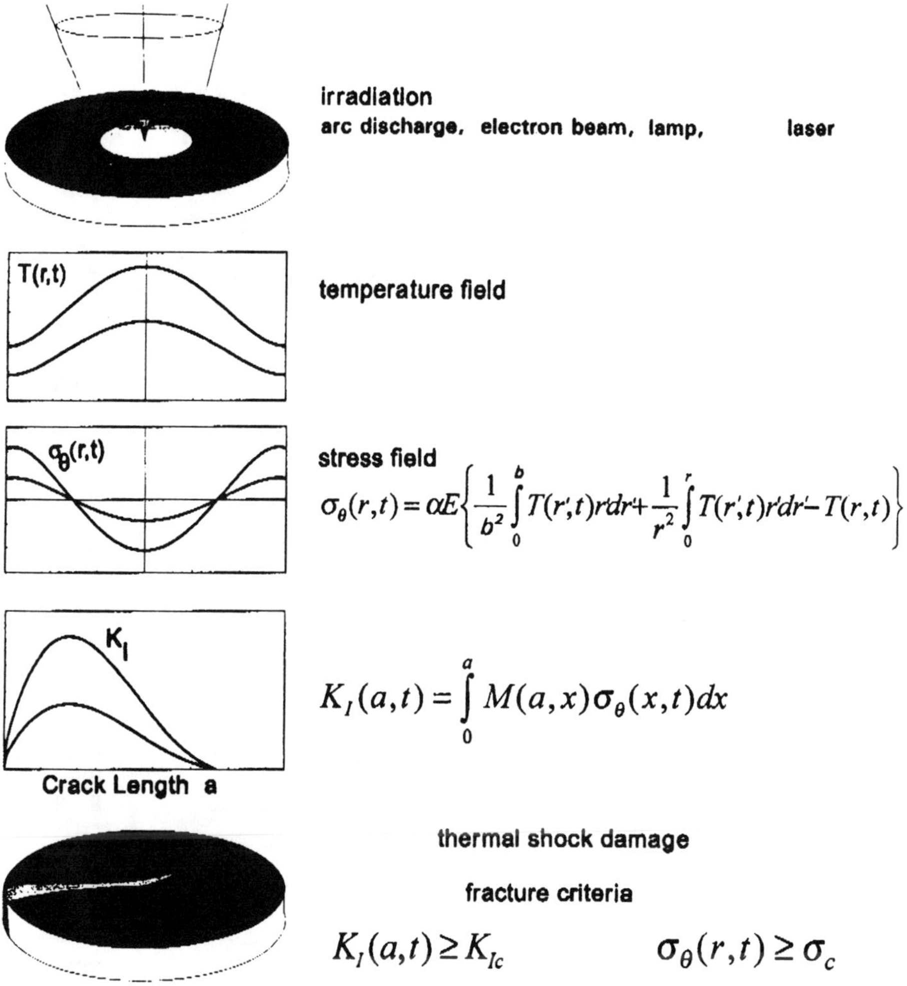

Fig.1. Diagram of thermal shock irradiation methods [1-5].

has the strength of 430 MPa. With BN additions the strength decreases sharply, reaching a value of ~150 MPa at 20% BN and then decreasing to a minimum of ~70 MPa at 80% BN.

Wafer shaped thermal shock and fracture toughness specimens were cut ~1 mm thick from billets using a diamond saw. They were further thinned for testing to ~0.5 mm by grinding.

Table I. B_4C-BN Samples, Compositions and Densities

Composition %B_4C	%BN	BN wt %	BN vol %	Theo ρ g/cm^3	Exp ρ g/cm^3	% Theo
100	0	0	0	2.52	2.51	99.6
					2.50	99.2
95	5	10.48	11.51	2.508	2.43	96.9
					2.46	98.1
90	10	19.83	21.54	2.495	2.37	95.0
					2.38	95.4
80	20	35.76	38.19	2.47	2.27	91.9
					2.25	91.1
					2.28	92.3
70	30	48.82	48.59	2.445	2.12	86.7
					2.17	88.8
					2.15	87.9
60	40	59.76	62.22	2.42	2.08	86.0
40	60	77.40	79.18	2.37	1.88	79.3
20	80	89.91	90.08	2.32	1.77	76.3
0	100	100.00	100.00	2.27	1.75	77.1

EXPERIMENTAL TECHNIQUE

The thermal shock behavior of B_4C-BN composites was investigated using a new test method based on [1-4]. The thermal shock testing equipment has been developed to generate high thermal stresses by defined heating. Thin circular samples (15 to 20 mm diameter by 0.3 to 0.5 mm thick) are irradiated on one side with a high power beam from a 1 kW Nd-YAG laser (wavelength of 1.06μm).

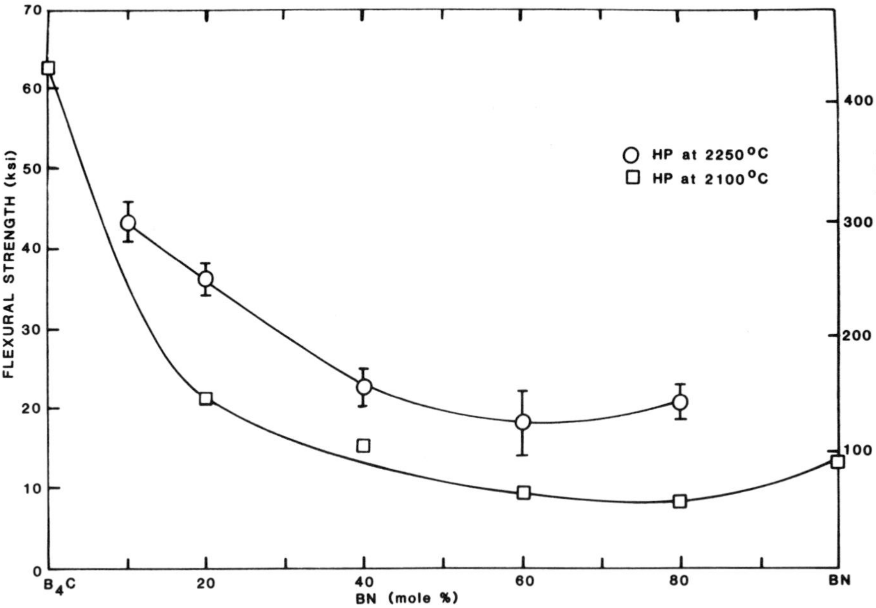

Fig.2. Flexural strength of hot-pressed B_4C-BN composites.

The experimental setup is illustrated in Fig. 3. The sample is fixed by two small ceramic tips to reduce heat losses. One of the tips serves as an acoustic wave guide, and an acoustic sensor is attached to its end. With this sensor and the coupled signal processing electronics, crack propagation or fracture can be detected.

A dynamic beam forming system between the laser optics and the sample provides high flexibility in producing various temperature distributions in the sample. This beam forming system consists of two electronically driven tilting mirrors that allow a rapid two-dimensional movement of the laser spot on the surface of the sample. The control signals for these two tilting mirrors are generated by a two-channel arbitrary waveform generator. By using special control functions (especially helical functions) for moving the laser spot over the sample, temperature distributions can be generated which produce very high stresses in the sample. A typical specimen of Si_3N_4 is heated from room temperature to ~1300°C in 0.3 to 0.5 s (Fig. 4).

Since the thermal stress field has to be derived from the transient temperature field, special effort has been devoted to temperature measurement. This was accomplished by a high speed pyrometer combined with an electronically controlled fast tilting mirror. By changing the tilting angle of the mirror, a series of pyrometer scans of the sample along the diameter is obtained during irradiation. Typical scan rates are 50 temperature profiles per second, with 50 values per profile.

A PC is used to control the whole thermal shock test, including programming the laser parameters and triggering the laser beam, programming and synchronizing the laser scanning system, starting the temperature measuring system, and acquiring the temperature data and acoustic emission data. The software has been written in language C, which allows the possibility of single thermal shock tests as well as thermal cycling experiments. Special software has been developed for calculating the time dependent stress distributions in the sample from the measured temperature data, computing the figure-of-merit **R** under thermal shock, and the stress intensity factor K_I for a given crack length. Further information about this thermal shock testing arrangement can be found in [5 and 8-10].

For a thin sample with axial symmetry, the stress distribution $\sigma_\theta(r)$ is related to the temperature distribution $T(r)$ by

$$\sigma_\theta(r)=E\left\{\frac{1}{b^2}\int_0^b \varepsilon(T(\bar{r}))\,\bar{r}\,d\bar{r}+\frac{1}{r^2}\int_0^r \varepsilon(T(\bar{r}))\,\bar{r}\,d\bar{r}-\varepsilon(T(r))\right\}, \quad (1)$$

from [11], where $\varepsilon(T)$ is the stress-free strain due to thermal expansion, **E** is Young's modulus, **b** is the radius of the sample, and **r** is the distance from the sample center.

A typical temperature distribution, obtained immediately before thermal shock failure of a silicon nitride disk, is shown on the left in Fig. 4. The $\sigma_\theta(r)$ as shown on the right has been obtained from equation (1) with dilatometric $\varepsilon(T)$ curves for silicon nitride. In the absence of complications such as macroscopic cracks or statistical size effects, the maximum stress at the instant of fracture is the strength σ_c of the material.

With the approximation $\varepsilon(T) = \alpha T$, equation (1) simplifies to yield the quantity $\sigma_\theta/E\alpha$ immediately from the temperature distribution. (α = thermal expansion coefficient.) Thus, at the instant of fracture, $\sigma_c/E\alpha$ is obtained, which is the thermal shock figure-of-merit **R**.

In this way, R can be measured without concern for σ_c, **E**, and α separately by using

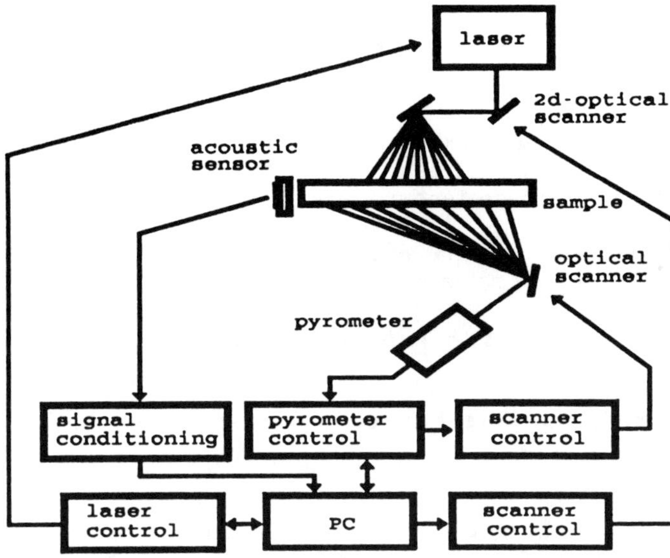

Fig.3. Schematic of thermal shock equipment with sample heating by high power laser.

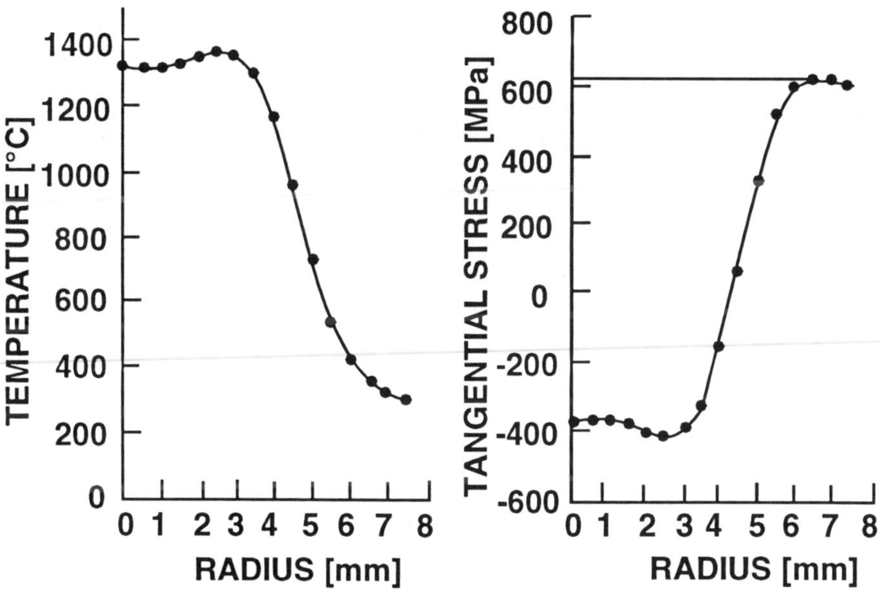

Fig.4. Temperature distribution, measured immediately before cracking, and calculated stress distribution.

$$R = \frac{2}{b^2} \int_0^b T(r)\, r\, dr - T(b) \quad . \tag{2}$$

For simplification, the maximum stress in the sample in equation (2) was replaced by the stress at the boundary, which is a good approximation for these test conditions (Fig. 4).

In the presence of a radial crack or notch of length **a**, the stress intensity due to thermal stress is given by

$$K_I = \int_0^a M(a,r)\, \sigma(r)\, dr \quad , \tag{3}$$

where **M(a,r)** is the weight function for this configuration [12].

RESULTS

The thermal shock behavior of boron nitride/boron carbide was investigated as a function of boron nitride content using small circular disks measuring 15 to 18 mm diameter and ~0.3mm thick. Boron nitride contents of 0 to 60% were tested in the laser-shock equipment. When the stress in the sample exceeded the strength of the material, fracture occurred and the samples broke into several pieces. From the temperature profile measured immediately before thermal shock fracture, the stress distribution in the sample and the figure-of-merit were calculated. In Fig. 5 the figures-of-merit thus determined are plotted versus boron nitride content. Despite the scatter in the figures-of-merit for the different compositions, significant dependence on boron nitride content is seen. Starting with ~150 **K** for pure boron carbide, the figure-of-merit decreases with increasing boron nitride content and reaches a minimum of 60 **K** at 20% boron nitride. Above 20%, the figure-of-merit increases with increasing content of boron nitride to a maximum value of 300 **K** for the 60% boron nitride composition.

The objective of further experiments was to obtain information on fracture toughness of the material. Therefore, some samples were notched with a diamond saw to a depth of 1.5 to 2.5 mm. The width of the notch was ~0.15 mm. During thermal shock testing, the samples were not destroyed. Only a crack from the boundary to the middle of the sample was observed on the video at the instant of high thermal loading (crack opening). The exact instant of crack propagation was determined by the attached acoustic emission system.

From the measured time dependent temperature distribution, the

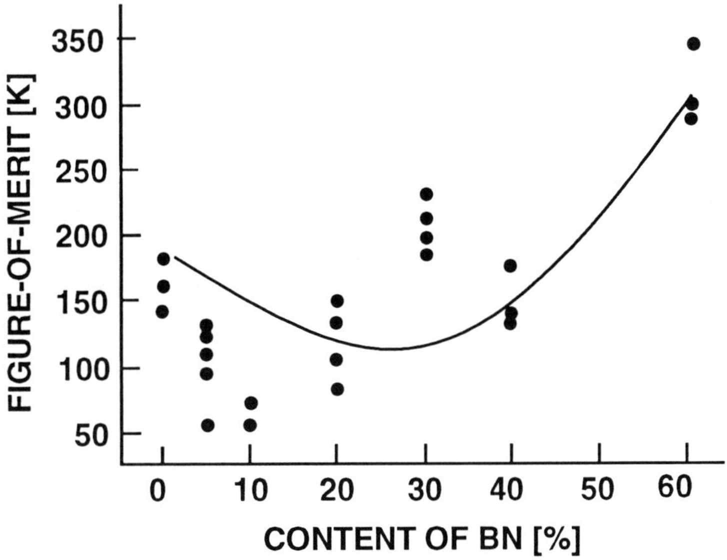

Fig.5. Figure-of-merit of B_4C-BN composites determined by laser-shock experiments and calculated.

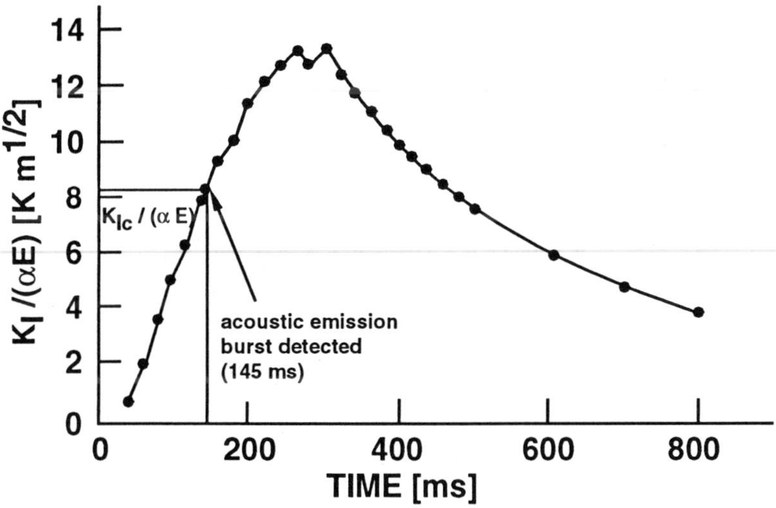

Fig.6. Normalized stress intensity factor versus time during thermal shock test of a notched B_4C-BN sample.

stress distribution in the sample and the stress intensity factor for the known notch length were calculated. In Fig. 6, for a composite with 30% boron nitride the stress intensity factor is plotted versus time. The value of the stress intensity factor at the instant of crack propagation gives the fracture toughness K_{Ic}.

From the temperature profile at the instant of cracking, the stress intensity factor can be calculated depending on the crack length (Fig. 7). Starting with a notch of 2.2 mm length, the crack propagation is discussed. Crack propagation begins as soon as the stress intensity factor K_I of a crack of length a_0 surpasses a critical value K_{Ic}. In the course of time, $K_I \geq K_{Ic}$, and continues at least as long as this condition is met. If this occurs on the upward slope of a curve, the energy release proportional to K_I^2 increases with crack propagation, which implies unstable propagation (black arrow). The surplus released energy gives rise to dynamic effects such that the crack can be driven to a length where its static K_I is below K_{Ic}. The crack stops for a moment until K_I reaches K_{Ic} again. Then the crack is dragged along the K_{Ic} line by the inflating K_I curve, which is indicated by the open arrow.

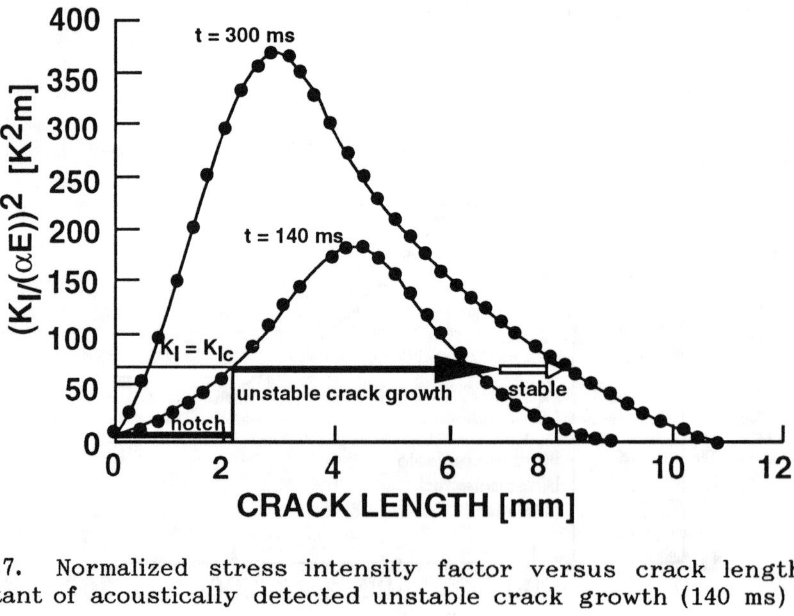

Fig.7. Normalized stress intensity factor versus crack length at instant of acoustically detected unstable crack growth (140 ms) and later.

DISCUSSION AND CONCLUSIONS

Boron carbide/boron nitride composites are materials with unusual thermal shock behavior. Whereas the strength of the composites decreases with increasing volume fraction of boron nitride, the thermal shock resistance increases for larger volume content of boron nitride. There is an additional peculiarity related to a minimum of figures-of-merit values for very small volume fraction of boron nitride.

A possible explanation for this behavior can be given by examining the microstructure with a high resolution scanning electron microscope. Figure 8 shows an SEM fractograph of a B_4C-10% BN sample. Predominantly intergranular fracture is seen, although some boron carbide grains may have transgranular fracture. Figure 9 shows a crack in a B_4C-30% BN sample and an enlarged view of part of the crack. This sample, which had a higher thermal shock resistance than the 10% BN sample, developed a main crack and secondary cracks, but did not fracture completely. Boron nitride plates are seen to bridge the crack and pull out, probably contributing to improved fracture toughness and thermal shock resistance of the composite.

As it follows from equation (2) and above, the change of the figures-of-merit values with the volume fraction of boron nitride can be understood as an interplay of strength, Young's modulus, thermal expansion and microstructure. The components differ considerably with respect to these quantities. Fully densified boron carbide can be characterized by a density of ρ_1 = 2.52 g/cm^3, Young's modulus E_1 = 450 GPa, and thermal expansion coefficient α_1 = 5 x 10^{-6} K^{-1}. For fully densified boron nitride the corresponding data are ρ_2 = 2.27 g/cm^3, E_2 = 50 GPa, and α_2 = 2 x 10^{-6} K^{-1} (parallel to the axis of compaction). As can be seen from the density data in Table I, the composite samples are not fully densified. Therefore, the porosity V_p of the composite influences the effective mechanical and thermal properties. For this two-phase system, the so-called self-consistent scheme (see e.g. [13]) was applied. This approach is based on the assumption that the heterogenous material surrounding the individual constituents of boron carbide, boron nitride and pores can be replaced by an "effective material" whose parameters are determined by putting them equal to the macroscopic values of E^* and α^* of the material.

In Fig. 10 the theoretical curves for the change of the effective values, as well as for the product of both, are plotted. Using the experimentally determined strength dependence of the composite on the boron nitride volume fraction [5], a theoretical curve for the figures-of-merit values can be calculated by applying equation (2).

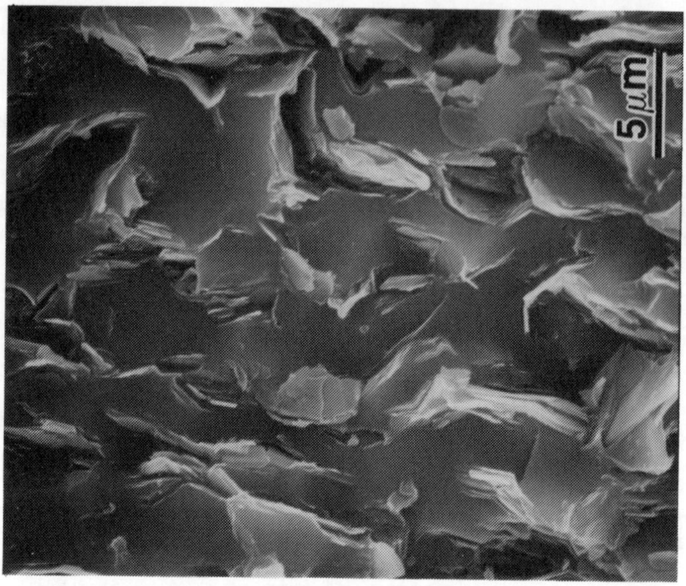

Fig.8. SEM fractograph of a B_4C-10% BN composite.

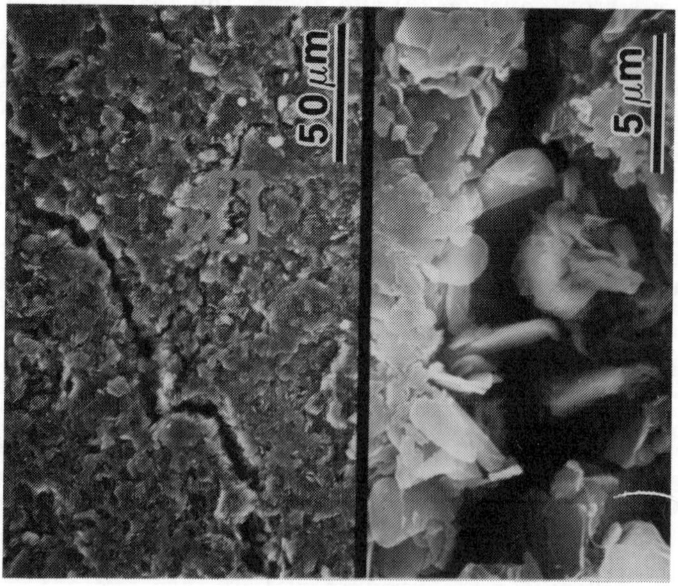

Fig.9. SEM fractograph of a B_4C-30% BN composite.

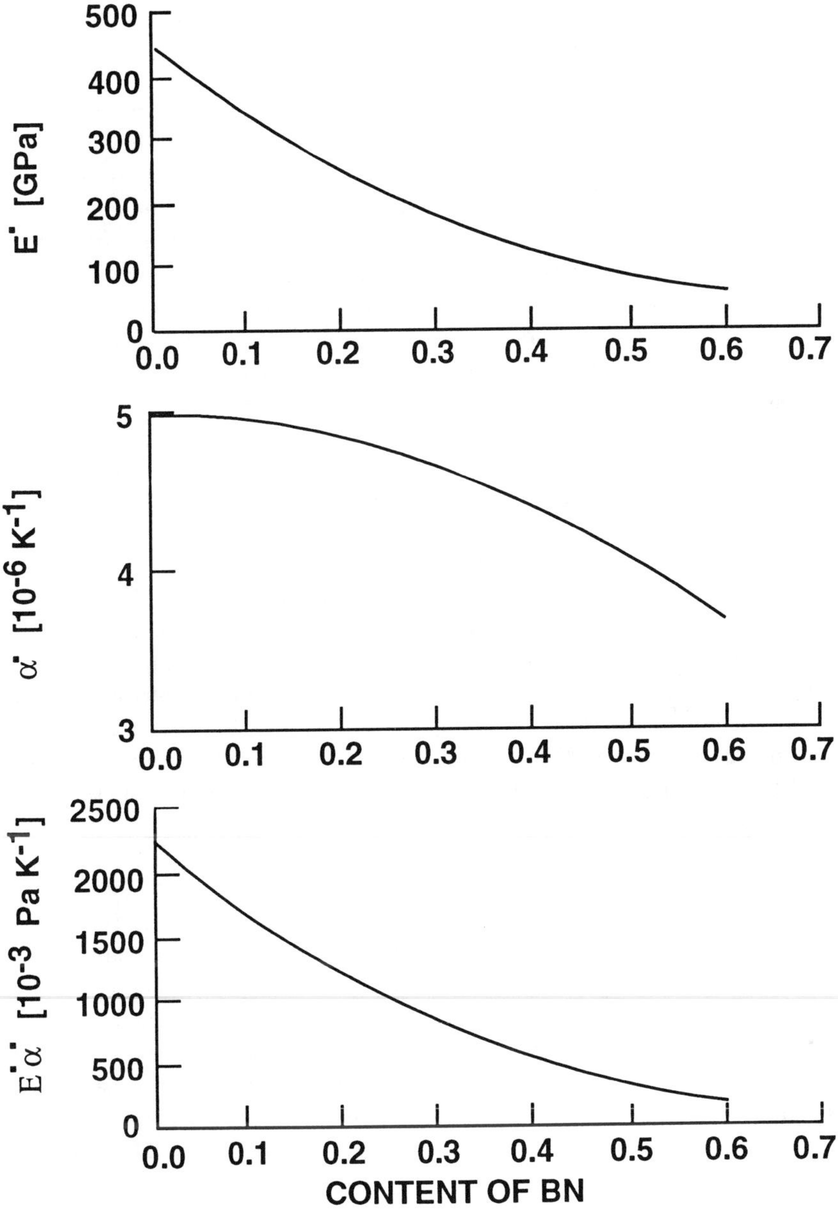

Fig.10. Effective Young's modulus and CTE calculated from the components' parameters including pores.

As can be seen from the comparison of the experimental data in Fig. 5, the model agrees surprisingly well with the observed behavior.

The composite is not only characterized by increasing figures-of-merit values with increasing boron nitride content; there also are indications of stable crack growth with increasing thermal shock load. Two stages of crack propagation are seen in Fig. 7. Whereas the first crack event is unstable because of the increasing stress intensity, with an increase of the crack length the following crack growth is stable. This could be the result of two effects. First, there is a decrease of the stress intensity near the center of the disk in Fig. 5, where K_{Ic} is assumed to be constant. Second, for a given microstructure, energy dissipation due to microcracking as well as R-curve behavior due to crack bridging cannot be excluded. These possibilities will be studied in further experiments.

In summary, it can be stated that boron carbide/boron nitride composites have very interesting thermal shock behavior. Applying the laser shock method, figures-of-merit values of the composite can be determined directly without any strength measurements. The interplay of strength, effective Young's modulus, effective thermal expansion and microstructure results in a distinct increase of the figures-of-merit values. This behavior can be explained in a microstructural model, taking into consideration the influence of the large differences of the physical properties of the components. Thus, the strength decrease with increasing boron nitride fraction is overcompensated by decrease of Young's modulus and the thermal expansion coefficient. Furthermore, disk shock tests offer the possibility to measure the thermal fracture toughness, as well as the change from unstable to stable crack growth during the thermal shock loading [14].

Acknowledgments

Experimental assistance and preparation of composites at the Materials Development Branch, Wright Laboratory are gratefully acknowledged. The authors thank W. Kreher and A. Vojta for calculation of the effective material parameters E* and a*; J. Paine for preparing the figures; and Y. Xu for assistance with the SEM studies, performed at the Center for Microanalysis of Materials, Materials Research Lab., University of Illinois. One author (A.Z.) acknowledges the support of the U.S. Dept. of Energy, Division of Materials Sciences, under Contract DE-FG02-91ER45439.

References

[1] S. Sato, K. Sato, Y. Imamura and J. Kon, "Determination of the Thermal Shock Resistance of Graphite by Arc Discharge Heating," *Carbon*, **13** [4] 309-25 (1975).

[2] C. Schubert, H.-A. Bahr and H.-J. Weiss, "Crack Propagation and Thermal Shock Damage in Graphite Disks Heated by Moving Electron Beam," *Carbon*, **24** [1] 21-28 (1986).

[3] G. A. Schneider and G. Petzow, "Thermal Shock Testing of Ceramics - A New Testing Method," *J. Am. Ceram. Soc.*, **74** [1] 98-102 (1991).

[4] G. A. Schneider, F. Magerl, I. Hahn and G. Petzow, "In Situ Observations of Unstable Crack Propagation and R-Curve Behavior in Thermally Loaded Disks"; pp. 229-42 in *Thermal Shock and Thermal Fatigue Behaviour of Advanced Ceramics*, NATO ASI Series, Series E: Applied Sciences, Vol. 241. Edited by G. A. Schneider and G. Petzow. Kluwer Academic Publishers, Boston, MA, 1993.

[5] G. Kirchhoff, "Thermal Shock Fracture by Laser Irradiation"; pp. 245-51 in *Thermal Shock and Thermal Fatigue Behaviour of Advanced Ceramics*, NATO ASI Series, Series E: Applied Sciences, Vol. 241. Edited by G. A. Schneider and G. Petzow. Kluwer Academic Publishers, Boston, MA, 1993.

[6] R. Ruh, M. Kearns, A. Zangvil and Y. Xu, "Phase and Property Studies of Boron Carbide-Boron Nitride Composites," *J. Am. Ceram. Soc.*, **75** [4] 864-72 (1992).

[7] R. Ruh, K. Y. Donaldson and D. P. H. Hasselman, "Thermal Conductivity of Boron Carbide-Boron Nitride Composites," *J. Am. Ceram. Soc.*, **75** [10] 2887-90 (1992).

[8] U. Rettig, U. Bast and G. Kirchhoff, "Eine Laserapparatur zur Messung der Temperaturwechselbeständigkeit von Keramik," (A Laser Apparatus for the Measurement of Temperature Fluctuations of Ceramics); pp. 135-40 in *Keramik in Wissenschaft und Praxis, Mechanische Eigenschaften keramischer Konstruktionswerkstoffe.* Edited by G. Grathwohl. DGM Informationsgesellschaft Verlag, 1993.

[9] G. Kirchhoff, M. Holzherr, U. Bast and U. Rettig, "Thermal Shock and Thermal Cycling Behaviour of Silicon Nitride Ceramics"; pp. 605-10 in *Proceedings of SILICON NITRIDE 93.* Edited by M. J. Hoffmann, P. F. Becher and G. Petzow. Trans Tech Publications, Ltd., Aedermannsdorf, Switzerland, 1993.

[10] G. Kirchhoff, M. Holzherr, I. Pflugbeil, H.-J. Weiss and H.-A. Bahr, "Fracture Mechanical Characterisation of Thermal Damage of Ceramics by Means of Laser Irradiation"; pp. 339-51 in *Proceedings of 2nd International Conference on Ceramics in Energy Applications*, London, England, Apr. 19-21, 1994.

[11] B. A. Boley and J. H. Weiner, *Theory of Thermal Stress.* John Wiley & Sons, Inc., New York, NY, 1960, 586 pp.

[12] T. Fett, "Direct Determination of Weight Functions from Reference Loading Cases and Geometrical Conditions," *Engin. Fracture Mech.*, **42** [3] 435-44 (1992).

[13] W. Kreher and W. Pompe, "Internal Stresses in Heterogeneous Solids"; pp. 94 and 122 in *Physical Research Vol. 9*, Akademie-Verlag, Berlin, Germany, 1989.

[14] W. Pompe, H.-A. Bahr, G. Schneider and H.-J. Weiss, "Modelling and Measuring of the Thermal Shock Behavior of Ceramics," *Ber. DKG*, **70** [1] 79-84 (1993). Corrections: see p. 356.

THERMAL SHOCK BEHAVIOR OF FIBER-REINFORCED COMPOSITES

Hongyu Wang* and Raj N. Singh*†
Department of Materials Science and Engineering
University of Cincinnati, Cincinnati, OH 45221-0012

Stephen C. Beecher* and Ralph B. Dinwiddie*
Metals and Ceramics Division
Oak Ridge National Laboratory, Oak Ridge, TN 37831-6063

ABSTRACT

The thermal shock behavior of three types of continuous fiber-reinforced ceramic composites (Nextel™ or Nicalon™ fiber-reinforced chemical vapor infiltrated or polymer-derived SiC matrix composites) was studied using the water quench technique. The thermal shock induced damage was characterized by both destructive and nondestructive techniques. As compared with monolithic ceramics, the continuous fiber-reinforced ceramic composites were capable of preventing catastrophic failure caused by thermal shock and were able to retain a significant portion of their original strength at $\Delta T = 1000°C$. The nondestructive techniques involved measuring the thermal diffusivity by the flash technique and determining the Young's modulus by the dynamic resonance method. It has been demonstrated that these nondestructive techniques can detect damage induced by thermal shock and are more sensitive in detecting damage in the early stage than the conventional destructive technique of measuring the retained strength.

INTRODUCTION

Continuous fiber-reinforced ceramic composites are being developed to meet the demands of high temperature applications and improve the reliability of ceramic materials. In many of their applications, fiber-reinforced ceramic composites are subjected to transient thermal conditions. Therefore, it is important to evaluate the thermal shock behavior of fiber-reinforced ceramic composites, understand the mechanisms of thermal shock failure, and improve their resistance to thermal shock damage.

---

* Member, American Ceramic Society
† Fellow, American Ceramic Society

To the extent authorized under the laws of the United States of America, all copyright interests in this publication are the property of The American Ceramic Society. Any duplication, reproduction, or republication of this publication or any part thereof, without the express written consent of The American Ceramic Society or fee paid to the Copyright Clearance Center, is prohibited.

Dense monolithic ceramics generally fail catastrophically once the temperature changes they are subjected to exceed a certain limit, the so-called critical temperature difference ΔT_c [1-3]. It has been observed that continuous fiber-reinforced ceramic composites posses superior resistance to thermal shock damage as compared with monolithic ceramics [3-6]. Their strengths decrease gradually over a range of temperature difference above ΔT_c and then become less affected by thermal shock at higher temperature differences up to ΔT = 1000°C. The fiber-reinforced composites were able to retain a significant portion of their original strength at ΔT = 1000°C [3].

Two criteria for evaluating the thermal shock behavior of continuous fiber-reinforced ceramic composites have been suggested: the ΔT for the onset of strength decrease (ΔT_c), and the ratio of the retained strength after quench at maximum ΔT to the original strength ($\sigma_{\Delta T=max}/\sigma_{\Delta T=0}$). It is expected that these two criteria are affected by many factors, including (i) the processing techniques used for forming matrices, which will affect the properties of the matrices, (ii) mismatch of properties, especially the thermal expansion coefficients, between fibers and matrices, (iii) interfacial properties, and (iv) fiber volume fraction and architecture. This paper reports our on-going effort in studying the thermal shock behavior of fiber-reinforced ceramic composites as a part of the Continuous Fiber Ceramic Composites (CFCCs) program with an emphasis on the nondestructive evaluation (NDE) techniques for assessing the damage to fiber-reinforced composites under thermal transient conditions.

EXPERIMENTAL

Three types of continuous fiber-reinforced composites were tested in the present study. These were 2-D woven Nextel™ or Nicalon™ fiber-reinforced SiC matrix composites. The matrices were formed by either chemical vapor infiltration (CVI) or polymer pyrolysis process, as summarized in Table 1.

Table 1. Fiber-Reinforced Composites Tested in the Present Study

Fiber	Matrix	Supplier
Nicalon™	CVI SiC	Oak Ridge National Lab.
Nextel™ 312	CVI SiC	Oak Ridge National Lab.
Nicalon™	polymer-derived SiC	Dow-Corning Corp.

The thermal shock damage to fiber-reinforced ceramic composites was induced by water quench. Details of the experimental procedures were reported elsewhere [3].

The destructive technique for evaluating the thermal shock behavior of fiber-reinforced composites involved measuring the retained strength after quenching by four-point bending. The critical quenching temperature difference (ΔT_c) was determined for each of the composites tested by progressively quenching the composite sample at different temperature differences. Specimens were cut from as-received billets. The specimen dimensions for the flexural test, listed in Table 2, were determined in part by the size and density distribution of the original samples.

Table 2. Specimen Dimensions for the Flexural Strength Measurement

Material	Specimen Dimension
Nicalon™ fiber/CVI SiC composite	3 x 6 x 50 mm
Nextel™ fiber/CVI SiC composite	4 x 6 x 50 mm
Nicalon™ fiber/polymer-derived SiC composite	2.6 x 4.6 x 45 mm

The flexural strength was measured using a mechanical testing machine with an inner span of 20 mm and outer span 40 mm. The crosshead speed was 0.5 mm/min. Due to the limitation on sample quantity, the original strengths were measured with three to five specimens, and the retained strengths after quench were obtained with one specimen for each data point (additional measurements will be taken when more samples are available).

The nondestructive techniques studied in the present research involved measuring the diffusivity (α) and Young's modulus (E) of the specimens before and after the water quench. The thermal diffusivity was measured using the Xenon Flash Thermal Diffusivity system at the High Temperature Materials Laboratory of Oak Ridge National Laboratory. The xenon flash lamp deposited a short pulse of energy on one surface of the sample and the temperature change on the opposite surface was measured by an infrared sensor. The thermal diffusivity was calculated by the following equation [7,8]:

$$\alpha = \frac{0.1388d^2}{t_{1/2}} \tag{1}$$

where d is the sample thickness and $t_{1/2}$ is the half rise time which is proportional to the square of the effective thermal thickness, L, through which heat is transferred:

$$t_{1/2} \propto L^2 \tag{2}$$

During this study, monolithic materials were evaluated for thermal shock behavior with this nondestructive method. They included Norton hot-pressed NC-203 SiC and Coors AD-998 Al_2O_3, as SiC and Al_2O_3 were candidate materials for the matrix of CFCCs. The critical quench temperature differences of the two materials, determined by retained flexural strength measurement, are 495°C for the NC-203 SiC and 200°C for the AD-998 Al_2O_3 [3]. Disc specimens of 12.7 mm diameter were core-drilled from plate samples. The thicknesses of the specimens were 7 and 4 mm for the SiC and Al_2O_3, respectively. The thermal diffusivity was measured as functions of both quench temperature difference and number of quench cycles. The cyclic quench was conducted at two temperature differences, one above and another below the ΔT_C, for both of the monolithic materials. Ten measurements were made for each data point to calculate the average and standard deviation.

The Young's modulus was measured using a dynamic mechanical resonance technique. Bar specimens used for the strength test were also used for the Young's modulus measurement. The fundamental resonant frequency f_r (Hz) of a specimen was detected using a GrindoSonic™ instrument (J. W. Lemmens Co., St. Louis, MO) and the Young's modulus E (Pa) was calculated according to the following equation [9,10]:

$$E = A\frac{\rho l^4 f_r^2}{t^2}T \tag{3}$$

where A = 0.94642, ρ is the density of the specimen (kg/m^3), l is the length (m), t is the thickness (m), and T is a correction factor given by:

$$T = 1 + 6.585(1 + 0.0752\mu^2)\left(\frac{t}{l}\right)^2 - 0.868\left(\frac{t}{l}\right)^4 - \frac{8.340(1 + 0.2023\mu + 2.173\mu^2)\left(\frac{t}{l}\right)^4}{1 + 6.338(1 + 0.14081\mu + 1.536\mu^2)\left(\frac{t}{l}\right)^2} \tag{4}$$

where μ is Poisson's ratio. This NDE technique of monitoring the change in Young's modulus has only been applied to the Nicalon™ fiber/CVI SiC composite in the present study.

RESULTS AND DISCUSSION

Destructive Test

The retained strengths of the three fiber-reinforced composites are plotted as a function of quench temperature difference in Fig. 1. The results are also summarized in Table 3.

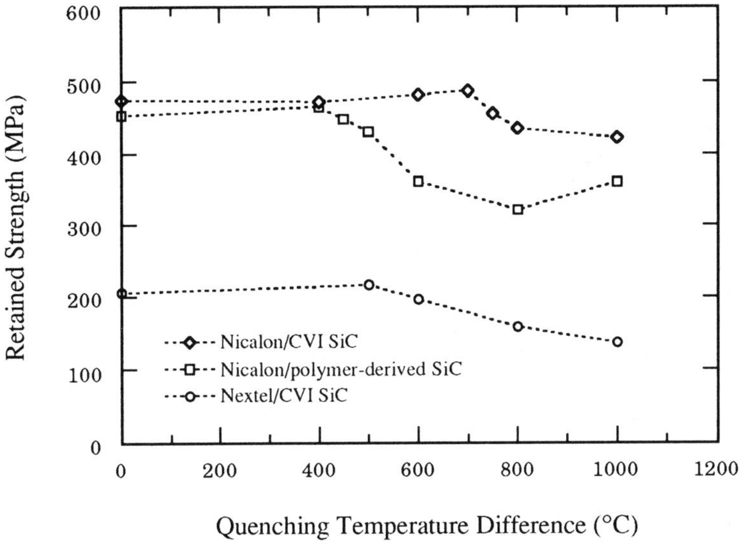

Figure 1. Thermal shock behavior of three continuous fiber ceramic composites.

Table 3. Thermal Shock Behavior of Three Fiber Reinforced Composites

Composite	Nicalon™ fiber/ CVI SiC	Nextel™ fiber/ CVI SiC	Nicalon™ fiber/ polymer-derived SiC
Original flexural strength (MPa)	474.5	206.8	453.4
Retained strength at ΔT = 1000°C (MPa)	422.2	136.5	359.9
$\sigma_{\Delta T=max}/\sigma_{\Delta T=0}$	89%	66%	79%
Onset ΔT (°C) for strength decrease	700	500	400

As compared with monolithic ceramics, composites exhibited a superior resistance to thermal shock. For all three composites, no abrupt strength decrease was observed. Instead, a gradual decrease in strength was seen at quench temperature difference above ΔT_C. After quenching at $\Delta T = 1000°C$, all three composites still retained a significant portion of their original strength.

The Nicalon™ fiber/CVI SiC composite showed an excellent resistance to thermal shock damage as judged by the two criteria proposed by the present authors [3]: the onset quench temperature difference for gradual strength decrease (ΔT_C) was high (700°C) and the ratio of the retained strength at maximum ΔT to the original strength (ΔT=0), $\sigma_{\Delta T=max}/\sigma_{\Delta T=0}$, was also high (89%). The two composites reinforced with Nicalon™ fibers showed similar original strengths which were higher than that of the Nextel™ fiber-reinforced CVI SiC composite, but the composite consisting of the matrix produced by CVI had a higher critical quench temperature difference than the composite with a polymer-derived SiC matrix. Further studies are being conducted to examine the influence of constituent properties and composite microstructure on the thermal shock behavior.

Nondestructive Tests

Thermal Diffusivity Measurement: The change in thermal diffusivity as a function of both quench temperature difference and number of quench cycle for the monolithic SiC and Al_2O_3 ceramics are shown in Figs. 2-5.

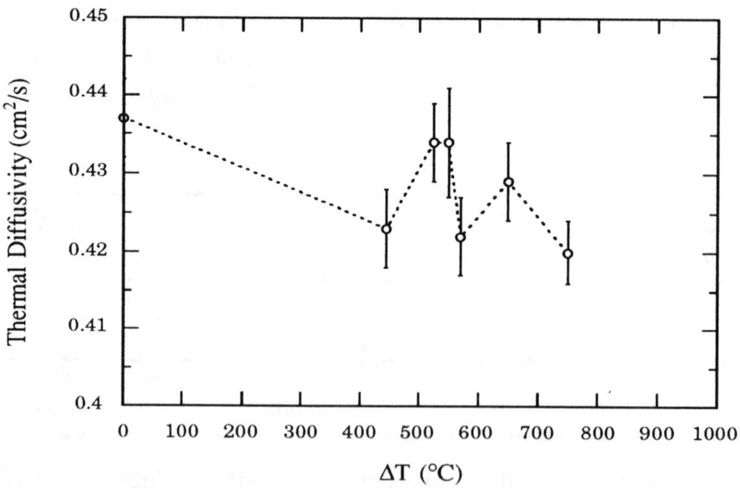

Fig. 2. Thermal diffusivity as a function of quench temperature difference for the NC-203 SiC.

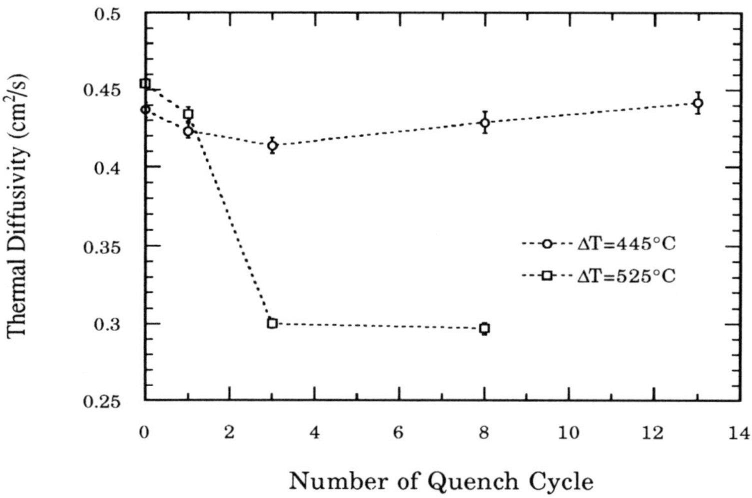

Fig. 3. Thermal diffusivity as a function of number of quench cycle for the NC-203 SiC.

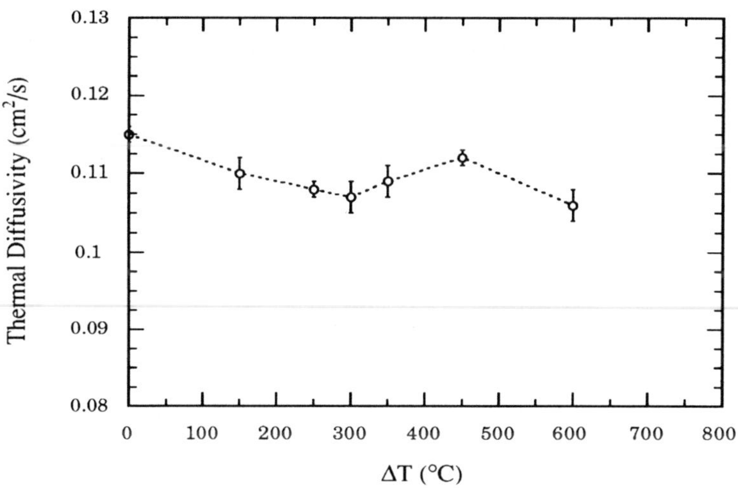

Fig. 4. Thermal diffusivity as a function of quench temperature difference for the AD-998 Al_2O_3.

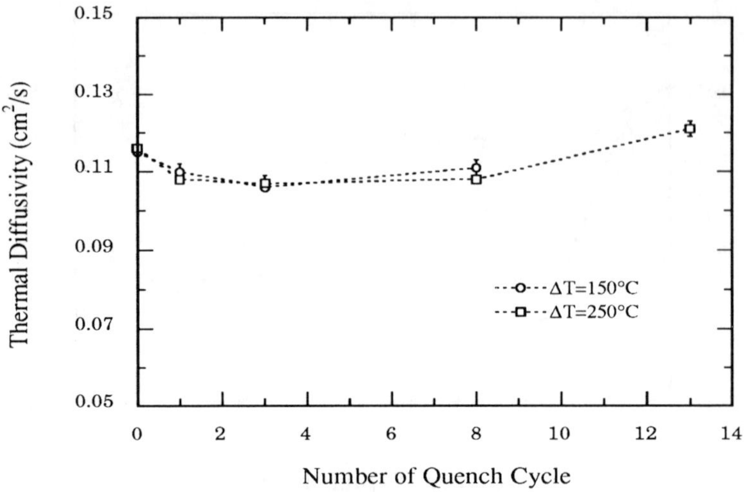

Fig. 5. Thermal diffusivity as a function of number of quench cycle for the AD-998 Al_2O_3.

The thermal diffusivity of all specimens decreased initially with quench temperature difference and up to three quench cycles. This might indicate that sub-surface microcracks were formed in the specimen or the surface cracks were still too small to affect the effective thermal thickness. After this initial decrease, the thermal diffusivity increased with increasing ΔT (up to 550°C for the SiC and 450°C for the Al_2O_3) and quench cycle (except for the SiC specimen under cyclic quenching at ΔT = 525°C). This may be attributed to the increase in the size of the surface cracks, which led to a decrease in the effective thermal thickness of the specimen because the flash light now could shine on the bottom of the cracks. The decreased effective thermal thickness would result in a decrease in the half rise time (Eq. (2)), and the thermal diffusivity would increase according to Eq. (1). The cause for the change in the thermal diffusivity of the SiC above ΔT = 550°C and the Al_2O_3 above ΔT = 450°C is still unknown.

The thermal diffusivity of the SiC specimen subjected to cyclic quenching at ΔT = 525°C (above its ΔT_c of 495°C) exhibited an over 30% decrease after the third quench and remained essentially unchanged up to eight cycles of quenching as shown in Fig. 3. A microstructural examination is being conducted to see if cracks perpendicular to the flash heat transfer direction were formed. During the water quench experiment, a disc specimen was found to break into two round halves through the center circular plane, hence it was possible that severe tensile thermal stresses developed in the thickness direction, which might cause the formation of cracks perpendicular to the flash heat transfer direction.

Therefore, the nondestructive method of measuring the thermal diffusivity for evaluating the damage due to thermal shock may have the advantage of providing information on the formation and development of cracks. Application of this NDE technique to continuous fiber-reinforced composites is under way and will be reported in the future.

Young's Modulus Measurement: The change in the Young's modulus of the Nicalon™ fiber/CVI SiC as a function of both quench temperature difference and number of quench cycle is shown in Figs. 6 and 7. Similar to the change in the retained strength, the Young's modulus of the composite decreased gradually with increasing quench temperature difference. However, the onset quench temperature difference for the decrease in Young's modulus seemed to be lower than that for the decrease in strength, which indicated that the damage might have occurred before it could be detected by the destructive method of strength measurement.

Fig. 7 clearly shows that a higher quench temperature difference results in a larger decrease in the Young's modulus. Although such an effect might not be very significant after the first quench, it becomes more profound with increasing number of quench cycles, which demonstrates the importance of studying the behavior of materials under cyclic thermal shock.

Further study is being conducted to explore other NDE techniques for assessing the damage to fiber-reinforced ceramic composites by thermal shock, correlate the findings with microstructural examination, and utilize finite element analysis (FEA) to predict the thermal shock behavior.

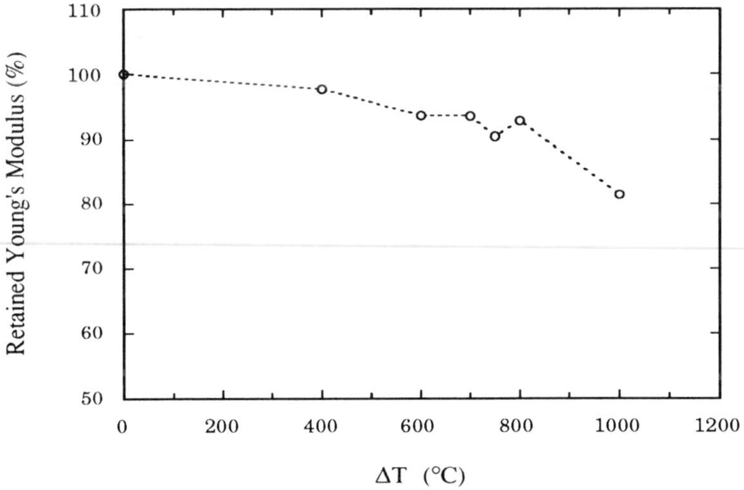

Fig. 6. The effect of quench temperature difference on the Young's modulus of Nicalon™ fiber/CVI SiC composite.

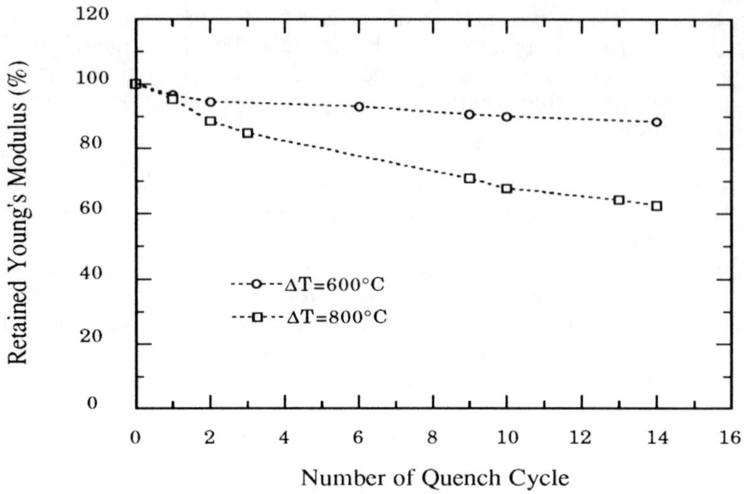

Fig. 7. The effect of cyclic quenching on the Young's modulus of Nicalon™ fiber/CVI SiC composite.

SUMMARY

The thermal shock behavior of three types of continuous fiber ceramic composites has been studied using both destructive and nondestructive techniques. As compared with monolithic ceramics, the fiber-reinforced composites exhibit superior resistance to thermal shock damage. Their strengths did not show a catastrophic decrease after the critical quench temperature difference. Among the three fiber-reinforced composites studied, the composite with Nicalon™ fiber reinforcement and CVI SiC matrix had the highest critical quench temperature difference and highest ratio of retained strength after quenching at maximum ΔT to the original strength. It has been demonstrated that NDE techniques of thermal diffusivity and Young's modulus measurement can detect the thermal shock damage. The NDE techniques were able to provide new insights not available from the destructive test. Further work is being carried out to correlate property measurement with microstructural examination and predict thermal shock behavior by finite element analysis.

Acknowledgment: The authors wish to thank Richard Lowden of Oak Ridge National Lab. and Andy Szweda of Dow Corning Corp. for providing the composite samples. The assistance of James Nagy in the thermal shock experiments is also gratefully acknowledged. This research is sponsored by the U.S. Department of Energy, Assistant Secretary for Energy Efficiency and Renewable Energy, Office of Industrial Technologies, Industrial Energy Efficiency Division and Continuous Fiber Ceramic Composites Program, and the Office of Transportation Technologies, as part of the High Temperature Materials Laboratory User Program, under contract DE-AC05-84OR21400 managed by Martin Marietta Energy Systems, Inc. and by an appointment to the Oak Ridge National Laboratory Postdoctoral Research Associates Program administrated jointly by the Oak Ridge National Laboratory and the Oak Ridge Institute for Science and Education.

References

1. W. D. Kingery, "Factors Affecting Thermal Shock Resistance of Ceramic Materials," *J. Am. Ceram. Soc.*, **38** [1] 3-15 (1955).
2. D. P. H. Hasselman, "Thermal Stress Resistance Parameters for Brittle Refractory Ceramics: A Compendium," *Am. Ceram. Soc. Bull.*, **49** [12] 1033-37 (1970).
3. H. Y. Wang, R. N. Singh, and R. A. Lowden, "Thermal Shock Behavior of Continuous Fiber Ceramic Composites (CFCCs)," *Ceram. Eng. Sci. Proc.*, to be published in 1994.
4. P. J. Lamicq, G. A. Bernhart, M. M. Dauchier, and J. G. Mace, "SiC/SiC Composite Ceramics," *Am. Ceram. Soc. Bull.*, **65** [2] 336-38 (1986).
5. R. T. Bhatt and R. E. Phillips, "Thermal Effects on the Mechanical Properties of SiC Fiber Reinforced Reaction-Bonded Silicon Nitride Matrix Composites," *J. Mater. Sci.*, **25**, 3401-07 (1990).
6. A. S. Fareed, B. Sonuparlak, C. T. Lee, A. J. Fortini, and G. H. Schiroky, "Mechanical Properties of 2-D Nicalon Fiber-Reinforced LANXIDE Aluminum Oxide and Aluminum Nitride Matrix Composites," *Ceram. Eng. Sci. Proc.*, **11** [7-8] 782-94 (1990).
7. W. J. Parker, C. P. Butler, and G. L. Abbott, "Flash Method of Determining Thermal Diffusivity, Heat Capacity and Thermal Conductivity, " *J. Appl. Phys.*, **32** [9] 1679-84 (1961).
8. ASTM Standard E 1461-92, "Standard Test Method for Thermal Diffusivity of Solids by the Flash Method." Annual Book of ASTM Standards, Vol 15.02, 1992.
9. G. Pickett, "Equations for Computing Elastic Constants from Flexural and Torsional Resonant Frequencies of Vibration of Prisms and Cylinders," *Proceedings, Am. Soc. Testing Mater.*, Vol. 45, pp. 846-65 (1945).
10. S. Spinner and W. E. Tefft, "A Method for Determining Mechanical Resonance Frequencies and for Calculating Elastic Moduli from These Frequencies." *Proceedings ASTM*, Vol 61, 1229-43 (1961).

EFFECTS OF YTTRIA STABILIZED ZIRCONIA ON TRANSFORMATION TOUGHENING OF MOLYBDENUM DISILICIDE COMPOSITES

Long-Ching Chen, Douglas Brooks, and Wolé Soboyejo
Department of Materials Science and Engineering,
The Ohio State University, Columbus, OH 43210-1179.

Richard Lederich
McDonnell Douglas Corporation, St. Louis, MO 63166-0516.

ABSTRACT

The effects of yttria stabilization (0-6 mol%) on the fracture toughness of molybdenum disilicide composites reinforced with 20 vol.% yttria partially stabilized zirconia (Y-PSZ) particles are presented. It is found that the fracture toughness increased from 3.8 MPa√m for the monolithic $MoSi_2$ to 4.8-5.7 MPa√m for the Y-PSZ reinforced composites. Examinations of the fractured specimens, using optical interference microscopy and laser Raman spectroscopy, showed that the stress-induced martensitic transformation of the Y-PSZ particles occurred only in the plastic wake of the composite stabilized with 2 mol.% yttria. However, the fracture toughness of the 2 mol.% Y-PSZ reinforced composite was in the same range as those of the other composites which did not exhibit evidence of transformation toughening. The results are rationalized using micromechanical models for the assessment of transformation toughening and microcrack shielding mechanisms.

INTRODUCTION

There has been significant interest in the development of high temperature structural material for the next generation of propulsion systems for aerospace vehicles. Amongst the candidate materials, molybdenum disilicide ($MoSi_2$) is a particularly promising candidate due to its combination of excellent high temperature oxidation/corrosion resistance, high melting point (2030°C), good thermal conductivity and moderate density (6.24 g/cm^3) [1,2]. However, the room temperature brittleness and limited ductility of $MoSi_2$ remain as major obstacles to practical applications.

Attempts to improve the fracture toughness of $MoSi_2$ via composite approaches have been carried out with the addition of either ductile [3-6] or brittle phases [7-9].

To the extent authorized under the laws of the United States of America, all copyright interests in this publication are the property of The American Ceramic Society. Any duplication, reproduction, or republication of this publication or any part thereof, without the express written consent of The American Ceramic Society or fee paid to the Copyright Clearance Center, is prohibited.

Ductile phase toughening of $MoSi_2$ is achieved by the addition of thermodynamically compatible refractory metals such as Mo, Nb and W. Brittle reinforcements such as ZrO_2, Al_2O_3, SiC, Si_3N_4, and WSi_2, exhibit attractive high temperature strength, chemical stability and oxidation resistance. They have therefore been suggested as potential reinforcements to improve the toughness of $MoSi_2$ without degrading its high temperature strength.

It is well known that partially stabilized zirconia (PSZ) undergoes 4% volume expansion during the transformation from tetragonal phase to monoclinic phase when it is subjected to stresses above the critical transformation stress [10]. Such a stress-induced martensitic transformation of the PSZ particles has been used to toughen brittle ceramics [11-15]. As a crack-containing PSZ-reinforced composite is loaded, the intensification of stress at the crack-tip can be sufficient to trigger the tetragonal-to-monoclinic phase transformation in the PSZ particles located in the vicinity of the crack. The resulting volume increase of the particles exerts a compressive shielding stress on the crack, thereby reducing the local stress intensity level at the crack tip. As a result, higher level of stress intensity is required to propagate the crack.

Pioneering efforts to improve the fracture toughness of $MoSi_2$ by the addition of Y-PSZ particles have been made by Petrovic and co-workers [16-18]. Based on indentation-type measurements on slurry-processed material, they reported 250-300% improvement of fracture toughness on hot pressed $MoSi_2$ composites . However, it is not clear whether the fracture toughness levels obtained from the indentation tests represent the actual mechanical properties of the $MoSi_2$ composites with 4-6 vol.% porosity. Moreover, the actual toughening mechanisms associated with the different levels of yttria stabilization have not been investigated using laser Raman spectroscopy techniques and optical interference techniques which have been shown to provide the transformation toughening parameters required for micromechanical modeling [15]. In the present work, dry-phase-blending and hot-isostatic-pressing (HIPing) techniques that can give a higher theoretical density (98-99 %) were used to fabricate $MoSi_2$ composites reinforced with yttria stabilized zirconia (Y-PSZ). The fracture toughness tests based on fracture mechanics experiments were conducted. The associated toughening mechanisms in $MoSi_2$+ Y-PSZ composites are elucidated using the parameters revealed by laser Raman spectroscopy and optical interference microscopy.

EXPERIMENTAL PROCEDURES

$MoSi_2$ composites reinforced with 20 vol.% Y-PSZ particles were fabricated using $MoSi_2$ powder procured from Cerac, Milwaukee, WI, and ZrO_2 powders supplied by Tosoh Corporation, Bound Brook, NJ. The $MoSi_2$ powders were screened to -325 mesh and had a particle size less than 45 μm. Four kinds of ZrO_2 powders containing 0, 2, 4, and 6 mol% yttria (designated respectively as TZ-0Y, TZ-2Y, TZ-4Y and TZ-6Y hereafter) were used in order to investigate the effect of Y_2O_3 on the stabilization of ZrO_2. These Y-PSZ powders had an average powder size about 0.7 μm. $MoSi_2$ and Y-PSZ powders were dry phase blended by ball milling for 24 hours, followed by HIPing at 1400 °C under 207 MPa pressure for 4 hours.

Duplicate Room temperature fracture toughness tests were conducted in air under three-point bend using single edge notched (SEN) bars with 6.35 mm x 6.35 mm square cross sections and 12.7 mm inner spans. Notches of 1.9 mm in depth were introduced by electro-discharge machining. Prior to the fracture test, the side surfaces of the SEN specimens were ground with emery paper and diamond polished to a surface finish of 1 μm. This was done to facilitate post-fracture analysis of the transformation zone using laser Raman microscopy and optical interference microscopy. The fracture toughness tests were carried out under load control in accordance with ASTM-E399 specifications. The specimens were loaded monotonically to failure at a loading rate that corresponded to a stress intensity increase rate of 0.92 MPa$\sqrt{m}$/sec at the notch-tip. The failure load was obtained from an automated data acquisition system, and used to calculate the fracture toughness.

Post-fracture laser Raman spectroscopy examination on the residual plastic wakes of the $MoSi_2$+ Y-PSZ composites was performed by SPEX Industries, Inc., Edison, NJ, using a Raman 500 microprobe under the following conditions: 488 nm laser excitation, 100 mW of power at the laser head (~ 10 mW at specimen), monochromator entrance width of 100 μm, and a probe diameter of ~ 1 μm. This Raman spectroscopy analysis was performed on the polished sides at various distances from the fracture surface (up to 500 μm) to reveal the variations of monoclinic zirconia content. From this examination, the presence of a transformation zone and its extent could be estimated.

X-ray diffraction showed that the TZ-0Y powder was purely monoclinic, while the TZ-6Y powder contained only tetragonal/cubic phase. Such a clear-cut difference of phase content made these two powders ideal materials to calibrate the relationship between the volume fraction of transformable zirconia in the composites and the Raman laser peak shift. The calibration was performed on five $MoSi_2$/TZ-0Y/TZ-6Y powder mixtures, consisting of 80-0-20, 80-5-15, 80-10-10, 80-15-5 and 80-20-0 volume percentage of the respective materials. Intensity ratios of the Raman peak shifts (192 cm^{-1} for the monoclinic phase and 264 cm^{-1} for the tetragonal/cubic phase) were used for the calibrations of Raman spectroscopy data. These were subsequently used to estimate phase volume fractions.

Optical interference microscopy examination of the plastic wakes around the cracks was carried out using a Zeiss interference microscope to determine the existence and extent of transformation zones near the fracture surfaces of fractured $MoSi_2$+ Y-PSZ specimens. Interference phenomena produced by a monochromatic green light from a thallium light source, resulted in fringes on the polished sides of the specimens when the fracture stresses were sufficient to induce the tetragonal-to-monoclinic phase transformation. The widths of the transformation zones were determined by identifying the points where the parallel fringe lines deviated from linearity. This deviation was caused by topographical changes associated with the volume expansion of the stress-induced martensitic transformations in the PSZ particles.

RESULTS

Typical micrographs of the as-fabricated $MoSi_2$+ Y-PSZ composites are presented in Figure 1. Due to the large difference in the powder size of $MoSi_2$ and PSZ, the PSZ particles were observed to agglomerate in the interstices between the larger $MoSi_2$ particles. This resulted in a dispersion of zirconia grains around the boundaries of the $MoSi_2$ grains after HIPing. Examination of Figure 1 shows that the distribution of Y-PSZ is more uniform in $MoSi_2$+20 vol.% TZ-0Y composite than the others. This resulted in more intergranular fracture along the $MoSi_2$ grain boundaries in the $MoSi_2$+20 vol.% TZ-0Y composite, as shown in Figure 2(a). In the other composites, cleavage fracture through $MoSi_2$ grains is the dominating fracture mode

(a)

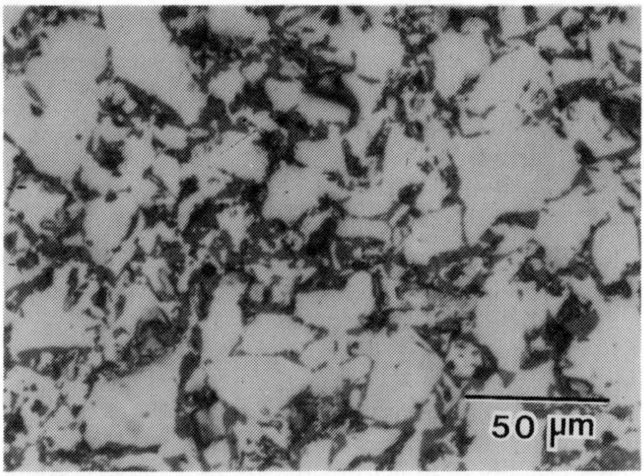

(b)

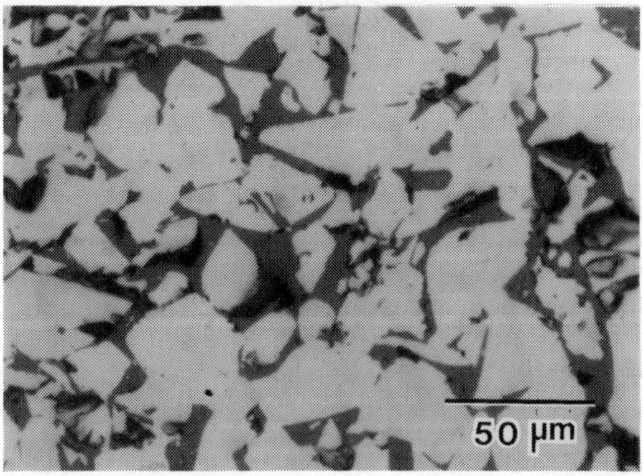

Figure 1. Microstructure of $MoSi_2$+ Y-PSZ composites. (a) $MoSi_2$+ 20 vol.% TZ-0Y; (b) representative micrograph for the other composites. Figure (b) was taken from $MoSi_2$+ 20 vol.% TZ-2Y.

(a)

(b)

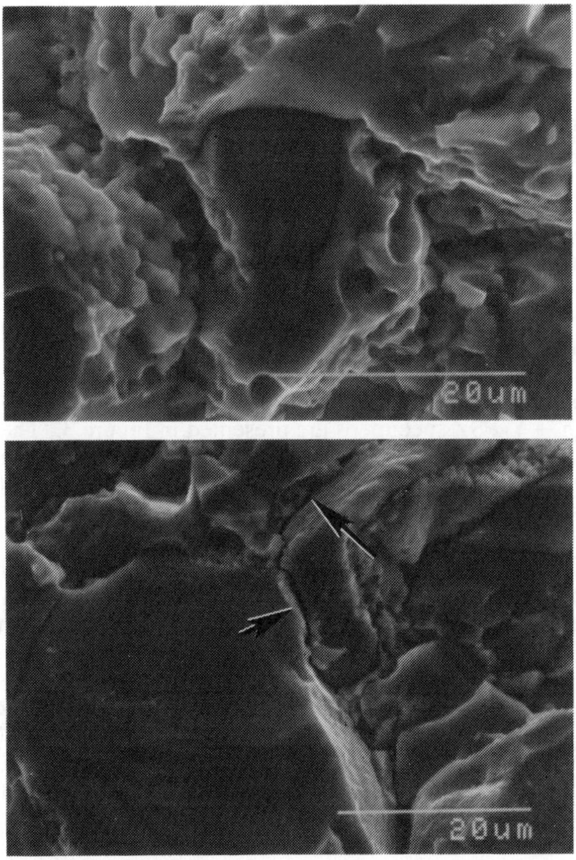

Figure 2. Fracture surface morphologies of $MoSi_2$+ Y-PSZ composites. (a) $MoSi_2$+ 20 vol.% TZ-0Y; (b) representative micrograph for the other composites. Figure (b) was taken from $MoSi_2$+ 20 vol.% TZ-2Y. The arrow points to the secondary cracking along $MoSi_2$ grain boundary.

and intergranular agglomeration of PSZ also promoted secondary cracking, as shown in Figure 2(b). The grain boundary secondary cracking is presumably as a result of the distribution of the zirconia particles around the grain boundaries, but may be partly due to residual stresses induced by the thermal shrinkage mismatch between the PSZ particles and $MoSi_2$ upon cooling from the HIPing temperature.

With the exception of the $MoSi_2$ + 20 vol.% TZ-0Y composite, the fracture surfaces were relatively flat, and there was limited evidence of toughening due to crack deflection. In contrast, the fracture surface of the $MoSi_2$ + 20 vol.% TZ-0Y was very rough, as shown in Figure 2 (a). This suggests that a greater amount of crack path tortuosity is associated with crack growth in this composite. The fractographic observations therefore indicate that toughening due to crack deflection and microcracking may occur in the $MoSi_2$+ PSZ composites, in addition to possible transformation toughening.

The fracture toughness values, K_Q, obtained from the three-point bend tests are presented in Table I. Since the amount of available material was very limited, the values of K_Q in Table I are average numbers from duplicate tests. The error ranges shown in Table I therefore correspond only to the deviation from the mean of two data points. Nevertheless, the data shows clearly that the fracture toughness of $MoSi_2$ is improved by the addition of PSZ particles. However, the degree of toughening is limited. Even in the best case of the TZ-0Y composite which has a fracture toughness of 5.7 MPa$\sqrt{m}$, the fracture toughness is only 50% higher than 3.8 MPa$\sqrt{m}$ obtained from the monolithic $MoSi_2$.

Typical optical interference micrographs obtained from the polished sides of the fractured $MoSi_2$+ Y-PSZ specimens are presented in Figure 3. The $MoSi_2$+ 20 vol.% TZ-2Y composite was the only material in which a transformation zone was detected. The zone can be distinguished by identifying the points of deviation from linearity in the optical interference fringes near the crack surface, as shown in Figure 3(a). The transformation zone size in the $MoSi_2$ +20 vol.% TZ-2Y composite was thus estimated to be approximately 120 µm. Other composites did not show the existence of transformation zone since their interference fringes remained straight throughout the polished surfaces, as shown in Figure 3(b). Since the stabilization of tetragonal ZrO_2 depends on the Y_2O_3 content, it is very likely that the tetragonal zirconia in the $MoSi_2$+ 20 vol.% TZ-4Y and $MoSi_2$+ 20 vol.% TZ-6Y composites was stabilized to a degree that the critical stresses required for the tetragonal-to-monoclinic transformation were actually higher than the fracture stresses experienced by these composites. Thus, no transformation toughening occurred in the $MoSi_2$+ 20 vol.% TZ-4Y and $MoSi_2$+20 vol.% TZ-6Y composites.

The variation of monoclinic zirconia content in the fractured $MoSi_2$+ 20 vol.% TZ-2Y composite, revealed by Raman spectroscopy, is shown in Figure 4. The volume percent of the monoclinic ZrO_2 drops from ~10 % near the crack surface to ~7% in the bulk. This indicates that only 3 vol.% of the entire material around the crack transformed during fracture. The percentage of monoclinic zirconia is observed to decrease significantly in the region between 100 µm and 150 µm from the crack. This agrees fairly with the transformation zone size, 120 µm, estimated by optical interference microscopy. In contrast, only monoclinic zirconia was detected in

Table I. Summary of fracture toughness values.

Material	K_Q (MPa$\sqrt{m}$)
Monolithic $MoSi_2$	3.8 ± 0.3
$MoSi_2$+ 20 vol.% TZ-0Y	5.7 ± 0.3
$MoSi_2$+ 20 vol.% TZ-2Y	5.0
$MoSi_2$+ 20 vol.% TZ-4Y	5.5 ± 0.2
$MoSi_2$+ 20 vol.% TZ-6Y	4.8 ± 0.2

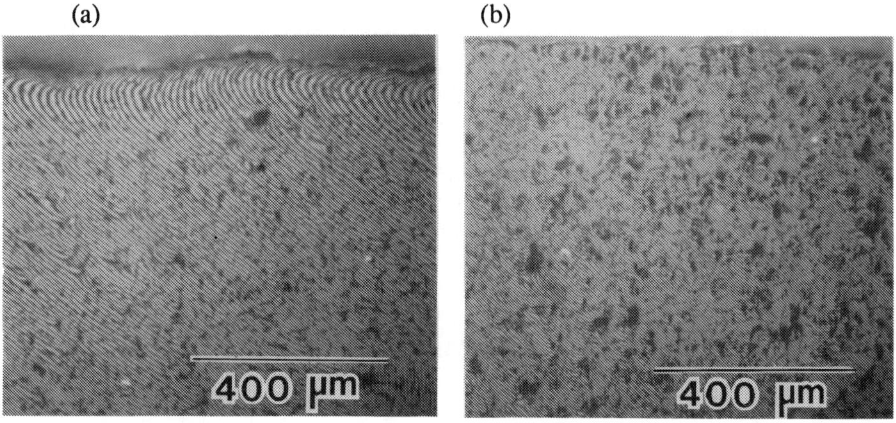

Figure 3. Optical interference micrographs near fracture surfaces of $MoSi_2$+ Y-PSZ composites: (a) $MoSi_2$+ 20 vol.% TZ-2Y; (b) $MoSi_2$+ 20 vol.% TZ-4Y.

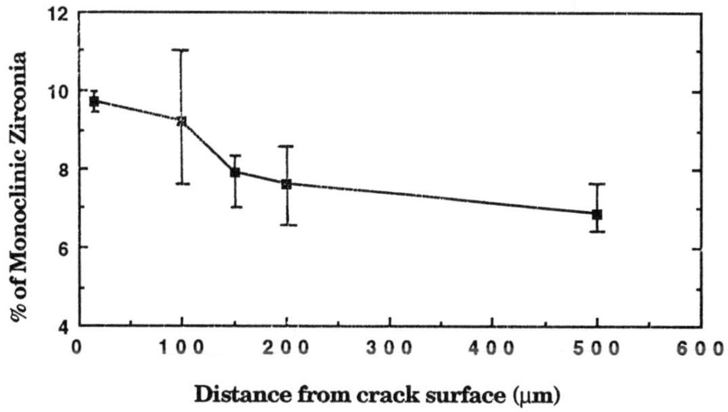

Figure 4. Volume fraction of monoclinic zirconia as a function of displacement from crack surface in $MoSi_2$+ 20 vol.% TZ-2Y composite.

$MoSi_2$+20 vol.% TZ-0Y composite by the Raman spectroscopy at locations near the crack and far from it. In the $MoSi_2$+TZ-4Y and $MoSi_2$+TZ-6Y composites, Raman laser spectroscopy revealed no evidence of monoclinic zirconia near the crack or far from it. These observations are consistent with the results of optical interference microscopy which indicate that no transformation zone exists in the $MoSi_2$ composites reinforced with TZ-0Y, TZ-4Y and TZ-6Y particles.

DISCUSSION

The fracture toughness data reported in this paper clearly demonstrate that the fracture toughness of monolithic $MoSi_2$ can be improved by the addition of ZrO_2 particles.

However, the highest fracture toughness that was achieved was only 5.7 MPa√m, a level that is well below that required for practical applications. This section will therefore focus on the discussion of toughening mechanisms and possible ways of improving the levels of transformation toughening.

Theoretical modeling of transformation toughening based on thermodynamic formulation has been conducted by Becher [13]. The critical stress, σ_t, to trigger the martensitic tetragonal-to-monoclinic phase transformation of zirconia, and the associated fracture toughness increase, ΔK_t, can be estimated using the following equations [13],

$$\sigma_t = \frac{\Delta G^{t\to m}}{\varepsilon^T} = \frac{\Delta S(T - M_S)}{\varepsilon^T} \tag{1}$$

$$\Delta K_t = 0.047\frac{1+\upsilon}{1-\upsilon}E_C f\left(\varepsilon^T\right)^2 \frac{K_I}{\Delta S(T - M_S)} \tag{2}$$

where K_I is the far field stress intensity factor and n is the Poisson's ratio, E_C is the elastic modulus of the composite, f is the volume fraction of transformed particles, e^T is the transformation volume strain, $\Delta G^{t\text{-}m}$ and ΔS are respectively the chemical free energy change and the entropy change associated with the tetragonal-to-monoclinic phase transformation of ZrO_2, T is the test temperature, and M_S is the temperature at which spontaneous transformation occurs.

According to Equations (1) and (2), the critical transformation stress, σ_t, and the degree of transformation toughening, ΔK_t , strongly depend on the spontaneous transformation temperature, M_S. As the Y_2O_3 content increases to enhance the stabilization of ZrO_2, M_S decreases rapidly. This results in an excessively high σ_t and a negligible ΔK_t . On the other hand, insufficiently stabilized ZrO_2 has a M_S higher than room temperature. The tetragonal-to-monoclinic transformation will occur during post-fabrication cooling of material, and no stress-induced transformation can be achieved when this occurs. There is therefore an optimal degree of stabilization at which the maximum fracture toughness improvement can be achieved.

The zirconia particles used in this study to reinforce $MoSi_2$ were stabilized with 0, 2, 4 and 6 mol.% of yttria. Both interference microscopy and Raman microscopy techniques showed that the stress-induced martensitic transformation occurred only in the $MoSi_2$+ 20 vol.% TZ-2Y composite. The other composites did not exhibit evidence of transformation, due to either the lack ($MoSi_2$+ 20 vol.% TZ-0Y) or the excess ($MoSi_2$+ 20 vol.% TZ-4Y and $MoSi_2$+ 20 vol.% TZ-6Y) of stabilization. Moreover, even in the case of $MoSi_2$+ 20 vol.% TZ-2Y where a transformation zone was observed, the size of the transformation zone (120 μm) and volume fraction of transformed phase induced by martensitic transformation (approx. 0.03) were small. The corresponding degree of transformation toughening was therefore very limited.

Systematic studies of the effect of Y_2O_3 contents on the martensitic tetragonal-to-monoclinic transformation of ZrO_2 has also been investigated by Becher [13] and Lange [19]. In Al_2O_3 + 20 vol.% ZrO_2 composites, they showed that the maximum room temperature toughness was obtained in the Al_2O_3+ Y-PSZ composite alloyed with 1 mol.% yttria. Since Al_2O_3 and $MoSi_2$ have a comparable strength level, their observations suggest that the optimal fracture toughness of $MoSi_2$+ PSZ composites is very likely to be accessed in the composite alloyed with 0 to 2 mol.% Y_2O_3. This is in agreement with the present observation.

In addition to manipulating the Y_2O_3 contents, an alternative approach of improving the degree of transformation toughening involves the use of larger zirconia particles. The size effect of PSZ particles has been investigated in other ceramic systems [20-22]. Generally, the thermal stability of the tetragonal phase increases as the particle size decreases. That is, a lower spontaneous transformation temperature, M_s, is observed when the PSZ particles have a smaller size. For example, using dilatometry technique Becher et al. [20] reported the M_s for fine-grained (0.4 μm) tetragonal ZrO_2 stabilized with 2- 2.5 mol.% Y_2O_3 could be as low as -269°C. Since the particle size (0.7 μm) used in the present work was comparable to that used in Becher's studies [20], it is therefore not surprising to note that the degree of toughening is low in the $MoSi_2$+ 20 vol.% TZ-2Y composite. Other approaches which involve the use of different stabilizers may also promote improved levels of transformation toughening.

The fracture toughness improvement in $MoSi_2$+ 20 vol.% TZ-2Y composite due to transformation toughening can be calculated by substituting the transformation zone size, 120 μm, measured by interference microscopy, and the transformed volume fraction, 0.03, indicated by Raman microscopy into the following equation, [13]

$$\Delta K_t = \frac{0.22\ E_C\ f\ \varepsilon^T \sqrt{h}}{1-v} \qquad (3)$$

where h is the transformation zone size, and the other parameters have been previously defined. The value of ΔK_t obtained from this substitution was 1.2 MPa√m. As the monolithic $MoSi_2$ has fracture toughness 3.8 MPa√m, the fracture toughness of $MoSi_2$+ 20 vol.% TZ-0Y based on model prediction is 5.0 MPa√m, which is exactly the same value measured in three-point bend test. However, it should be noted the value based on the model calculation represents upper toughness estimate, since Equation (3) assumed a constant average volume fraction, f, of martensitically transformed phase inside the transformation zone, i.e. f did not vary as a function of distance from the crack face. This is clearly not the case in the TZ-2Y composite, as shown in Figure 4 in which the volume fraction of transformed phase decreases gradually as the distance from the crack surface increases. This suggests that other toughening mechanisms may also contribute to give the overall fracture toughness level of 5.0 MPa√m. The errors inherent in the assumptions made in the simplified micromechanics analyses presented above may also contribute to the differences.

As shown in Table I, the fracture toughness data of the PSZ reinforced $MoSi_2$ composites are essentially similar regardless whether a transformation zone exists. This again provides further evidence of the role of other toughening mechanisms apart from transformation toughening. These mechanisms may include crack deflection and microcrack shielding . Fractographic evidence of crack deflection and microcrack shielding is presented in Figure 2. It shows clearly the presence of secondary cracks along $MoSi_2$ grain boundaries. These secondary cracks were generated during fracture toughness testing, and their creation required the consumption of additional energy. As a result, the fracture toughness of the PSZ reinforced composites should be higher than that obtained for monolithic $MoSi_2$.

A close examination on the fracture surface morphology (Figure 2) of the $MoSi_2$+ PSZ composites also reveals that the fracture surface of the TZ-0Y composite is very rough when compared with those of the other composites. This suggests that crack deflection may play a significant role in the toughening of this composite. However, since the zirconia in this composite undergoes a tetragonal-to-monoclinic phase transformation during post-fabrication cooling, no stress-induced martensitic transformations are expected in this composite, as observed. Nevertheless, at the transformation temperature of about 1200°C, where material strength is low, higher density of dislocations might be generated which could enhance room temperature 'plasticity.' Such plasticity may promote improved fracture toughness. This possibility is currently being evaluated in ongoing transmission electron microscopy studies.

SUMMARY

The results of an ongoing investigation on the transformation toughening of $MoSi_2$ using yttria stabilized zirconia particles are presented. Stress induced tetragonal-to-monoclinic phase transformation was observed only in the $MoSi_2$+ 20 vol.% TZ-2Y composite, but the degree of toughening in this composite was limited (1.2 $MPa\sqrt{m}$). The other composites ($MoSi_2$+ 20 vol.% TZ-0Y, $MoSi_2$+ 20 vol.% TZ-4Y, and $MoSi_2$+ 20 vol.% TZ-6Y) did not show evidence of transformation toughening, due to either the lack or the excess of stabilization. However, the fracture toughness of the $MoSi_2$+ 20 vol.% TZ-2Y composite was similar to that of the other composites that did not transform. Based on experimental evidence, toughening in the other composites that did not transform was attributed to crack deflection and microcrack shielding mechanisms.

ACKNOWLEDGMENTS

The authors would like to thank Dr. B. Jelus-Tyrer, Dr. F. Adar and Mr. H. Schaffer of SPEX Industries for their assistance with the Raman spectroscopy analysis. The research is supported by the Air Force Office of Scientific Research (Grant No. F49620-94-1-0107DEF), with Dr. C. Ward as Program Monitor.

REFERENCES

[1]A. K. Vasudévan and J. J. Petrovic, "A Comparative Overview of Molybdenum Disilicide Composites," *Mater. Sci. Eng.*, **A155**, pp. 1-17, (1992).

[2]E. Fitzer and W. Remmele, "Possibilities and Limits of Metal Reinforced Refractory Silicides, Especially Molybdenum Disilicide, " *Proc. 5th Int. Conf. on Composite Materials,* ICCM-V, AIME, Warrendale, PA, pp. 515-530, (1985).

[3]L.-C. Chen, N. Bahtishi, R. Lederich and W. O. Soboyejo, "Ductile Phase Toughening of $MoSi_2$: Effect of Reinforcement Morphology", paper presented at MRS meeting, April 1994.

[4]W. O Soboyejo, K. T. Venkateswara Rao, S. M. L. Sastry and R. O. Ritchie, "Strength, Fracture, and Fatigue Behavior of Advanced High-Temperature Intermetallics Reinforced with Ductile Phases," *Metall. Trans.*, **24A** [3], pp. 585-600, (1993).

[5]T. C. Lu, A. G. Evans, R. J. Hecht and R. Mehrabian, "Toughening of $MoSi_2$ with a Ductile (Niobium) Reinforcement," *Acta Metall. Mater.*, **39** [8], pp. 1853-1862, (1991).

[6]L. Shaw and R. Abbaschian, "Toughening $MoSi_2$ with Niobium Metal - Effects of Size and Orientation of Ductile Laminae," *Acta Metall. Mater.* , **42** [1], pp. 213-223, (1994).

[7]D. H. Carter and G. F. Hurley, "Crack Deflection as a Toughening Mechanism in SiC-Whisker-Reinforced $MoSi_2$," *J. Am. Ceram. Soc.*, **70** [4], pp. C79-81, (1987).

[8]D. H. Carter, J. J. Petrovic, R. E. Honnell and W. S. Gibbs, "SiC-$MoSi_2$ Composites," *Ceram. Eng. Sci. Proc.*, **10** [9-10], pp. 1121-1129, (1989).

[9]U. Ramamurty, A. S. Kim, S. Suresh and J. J. Petrovic, "Micromechanisms of Creep-Fatigue Crack Growth in a Silicide-Matrix Composite with SiC Particles," *J. Am. Ceram. Soc.*, **76** [8], pp. 1953-1964, (1993).

[10]R. Stevens, *Zirconia and Zirconia Ceramics* , Magnesium Elektron Ltd., Twickenham, UK, (1986).

[11]B. Budiansky, J. W. Hutchinson and J. C. Lambropoulos, "Continuum Theory of Dilatant Transformation Toughening in Ceramics", *Int. J. Solids Structures*, **19** [4], pp. 337-355, (1983).

[12]R. M. McMeeking and A. G. Evans, "Mechanics of Transformation-Toughening in Brittle Materials". *J. Am. Ceram. Soc.*, **65** [5], pp. 242-246, (1982.)

[13]P. F. Becher, "Toughening Behavior in Ceramics Associated with the Transformation of Tetragonal ZrO_2," *Acta Metall.*, **34** [10], pp. 1885-1891, (1986).

[14]A. G. Evans and R. M. Cannon, "Toughening of Brittle Solids by Martensitic Transformations," *Acta Metall.*, **34** [5], pp. 761-800, (1986).

[15]D. B. Marshall, M. C. Shaw, R. H. Dauskardt, R. O. Ritchie, M. J. Readey and A. H. Heuer, "Crack-Tip Transformation Zones in Toughened Zirconia," *J. Am. Ceram. Soc.*, **73** [9], pp. 2659-2666, (1990).

[16]J. J. Petrovic and R. E. Honnell, "Partially Stabilized ZrO_2 Particle-$MoSi_2$ Matrix Composites," *J. Mater. Sci.*, **25**, pp. 4453-4456, (1990).

[17]J. J. Petrovic, R. E. Honnell, T. E. Mitchell, R. K. Wade and K. J. McClellan, "ZrO_2-Reinforced $MoSi_2$ Matrix Composites," *Ceram. Eng. Sci. Proc.*, **12** [9-10], pp.1633-1642, (1991).

[18]J. J. Petrovic, A. K. Bhattacharya, R. E. Honnell, T. E. Mitchell, R. K. Wade and K. J. McClellan, "ZrO_2 and ZrO_2-SiC Particle Reinforced $MoSi_2$ Matrix Composites," *Mater. Sci. Eng.*, **A155**, pp. 259-266, (1992).

[19]F. F. Lange "Transformation Toughness: Part 4. Fabrication, fracture toughness and strength of Al_2O_3-ZrO_2 composites," J. Mater. Sci. **17**, pp. 247-254, (1982).

[20]P. F. Becher, G. Begun and E. F. Funkenbusch, "Transformation-Toughening Behavior in Ceramics," pp. 565-571 in *Advances in Ceramics* , Vol. 24, S. Sõmiya, N. Yamamoto, N. Yamamoto and H. Hanagida, eds., Am. Ceram. Soc., Westerville, Ohio, (1988).

[21]K. Tsukuma, Y. Kubota and T. Tsukidate "Thermal and Mechanical Properties of Y_2O_3Stabilized Tetragonal Zirconia Polycrystals," pp. 382-390 in *Advances in Ceramics.*,Vol. 12, N. Claussen, M. Rühle and A. Heuer, eds., Am. Ceram. Soc., Columbus, Ohio (1983).

[22]I. Mühle and W. Mühle "Size Effect on Transformation Temperature of Zirconia Powders and Inclusions," pp. 443-454, *ibid.*

MICROSTRUCTURAL CHARACTERIZATION OF BURNER RIG THERMAL FATIGUED SiC CONTINUOUS FIBER/Si_3N_4 CERAMIC COMPOSITES

T. Ertürk, K. Park* and C. Sung*
Department of Chemical and Nuclear Engineering and *Center for Advanced Materials
University of Massachusetts
Lowell, MA 01854

ABSTRACT

Hot pressed $[(0/90)_3/\bar{0}]_S$ SCS-6® or $[(0/90)_5/\bar{0}]_S$ SCS-9® SiC fiber reinforced crossply Si_3N_4 ceramic composite specimens having a fiber volume fraction of 0.30 were thermal cycled under impinged jet fuel flame, constant applied tensile stress and in the temperature range 500-1350°C. Nine SCS-9 and four SCS-6 SiC fiber reinforced specimens were thermal fatigued at the NASA Lewis Mach 0.3 atmospheric pressure burner rig test station. The SCS-6 SiC continuous fiber reinforced specimens failed outside the flame impinged zone by high thermal stresses due to steep temperature gradients in 20-40 cycles at constant applied stresses as low as 110 MPa. SCS-9 SiC fiber reinforced specimens, on the other hand, failed at the center of the flame impinged zone where high temperature gradients did not exist. The SCS-9 SiC fiber reinforced material showed runout conditions (1000 cycles without failure) up to 125 MPa constant applied stress. The number of cycles-to-failure declined abruptly at 140 MPa constant applied stress; at 168 MPa it was 5-42. The microstructure of a runout SCS-9 SiC fiber reinforced Si_3N_4 composite specimen was investigated using transmission electron microscopy. TEM results of the runout SCS-9 specimens showed no void formation in the matrix, fiber, or at the matrix/fiber interface. No crystallization of the matrix grain boundary glassy network was observed.

INTRODUCTION

Crystalline ceramics are ideal materials for high temperature applications because of their high melting point and slow diffusion kinetics. With advances in the understanding of their mechanical properties [1] and the role of interfaces to deflect matrix cracks [2], continuous fiber ceramic composites (CFCC's) have the potential of replacing nickel-based superalloys and other high temperature materials for use at elevated temperature corrosive environments. The scientific

To the extent authorized under the laws of the United States of America, all copyright interests in this publication are the property of The American Ceramic Society. Any duplication, reproduction, or republication of this publication or any part thereof, without the express written consent of The American Ceramic Society or fee paid to the Copyright Clearance Center, is prohibited.

community is poised to utilize these materials specifically in the area of advanced gas turbine engines [3].

The primary constraint precluding the application of ceramic matrix composites in advanced gas turbine engines is the lack of adequate material characterization and predictive modeling to substantiate sufficient life limiting mechanisms such as creep, fatigue, thermal shock induced rupture, thermal fatigue under constant applied stress, thermomechanical fatigue, corrosion and oxidation. Most potential CFCC applications involve concurrent thermal and mechanical cycling in hostile environments. In order to generate material data that can be used by the designer, experiments should be conducted under cyclic conditions simulating the component's service environment. However, research data simulating the cyclic service conditions of gas turbine engines such as burner rig thermal fatigue under jet fuel impingement and constant applied stress or better burner rig thermomechanical fatigue are not available.

The thermal fatigue study of CFCC's has been limited to a glass ceramic composite, photon heating and thermal cycling in the temperature range 500-1100°C [4]. Likewise, thermomechanical fatigue research of CFCC's has been limited to glass ceramic composites, photon [5] or induction [6] heating and thermal cycling between 500-1100°C. Such heating techniques lack the rapid heating and corrosion effects of jet fuel flame impingement. Thus, they are inadequate to simulate the complex interaction of the thermomechanical loading, erosive and corrosive environment of a gas turbine engine. Further, CFCC's in gas turbine applications are expected to operate in a larger cyclic temperature range (500-1350°C).

The burner rig thermal fatigue testing under constant applied stress is nearly the ideal test technique to simulate a jet engine thermal, mechanical and chemical environment, although in or out of phase concurrent cycling of the mechanical load would be desirable. The advantages of this test technique over furnace thermal cycling include the impingement of a jet fuel flame at high velocities (0.3 Mach), rapid heating of the specimen, and a constant applied stress. The high velocity impingement of jet fuel flame also probes material's resistance to erosion under service conditions. The specimen is subjected to thermal shock in both the flame impinged region and in regions outside the heated zone where high temperature gradients exist. Thus, failure may occur in either locations. High thermomechanical stresses outside the heated zone may bring about cycle by cycle damage accumulation in these regions.

EXPERIMENTAL PROCEDURE

The 3.175 mm thick 0.30 volume fraction $[(0/90)_3/\bar{0}]_S$ SCS-6® or $[(0/90)_5/\bar{0}]_S$ SCS-9® SiC fiber-reinforced Si_3N_4 crossply laminates were hot pressed at 1650°C under 17.2 MPa pressure for 1 hour by Textron Specialty Materials, Lowell, MA. The matrix composition was 92.5% Stark LC-12 Si_3N_4, 5.0% Y_2O_3, 1.5% MgO, and 1.0% Al_2O_3, all in weight percentage. The additives form a grain boundary glassy network to aid sintering. Rectangular specimens were machined to 152.4 mm x 12.7 mm x 3.175 mm dimension. The 142 μm diameter SCS-6 and 79 μm diameter SCS-9 SiC fibers, also fabricated by Textron by chemical vapor deposition, have a fine columnar β-SiC emanating radially from a 33 μm diameter carbon monofilament substrate. Only the SCS-6 SiC fiber have a carbon rich region outside the carbon core. The outer zone of the SCS-9 and SCS-6 fibers has stoichiometric SiC compositions. Both fibers have approximately 3 μm layered carbon rich coating at the surface.

Burner rig thermal fatigue tests under constant applied stress were conducted using the Mach 0.3 atmospheric pressure durability testing burner rig at the NASA Lewis Research Center, Cleveland, Ohio. The specimen temperature was maintained to ± 1.7°C and impinging gas velocity to Mach ± .02 [7]. The severe test conditions and specimen failures were recorded using a video recording camera.

The flame was produced using a simple "can-type" combustion chamber. Preheated air at 260°C and JP-4 fuel with an equivalence ratio of 0.5 was first pumped into the combustion chamber to initiate the combustion process. The flow of the compressed air and fuel were controlled using temperature and pressure sensors in the combustion chamber. Approximately 27 mm of the leading edge of the test specimen was impinged. The specimen was located 25.4 mm away from the exit nozzle. The flame yielded a symmetrical temperature profile above and below the centerline of the specimen.

The core region of the flame occupied 4 to 5 nozzle diameters immediately downstream from the nozzle where the velocity and temperature distribution of the exhaust gas remain constant [8,9]. The specimen is located 1 nozzle diameter downstream from the nozzle exit where the maximum exit gas velocity and temperature in both the horizontal and vertical directions are maintained. Beyond the core region there is a transition zone extending up to 10 nozzle diameters, followed by a fully developed region.

A load cell actuated by compressed air (Fairlane H4X4HC) was bolted to the loading frame. The pneumatic load cell accommodated the cyclic thermal expansion and contraction of the specimen while applying a constant stress up to 710 MPa (500 psig) for the specimen geometry used. The maximum axial displacement of the load cell was 57.2 mm. An end-limit switch activated by specimen failure electronically powered down the burner rig and periphery devices.

The specimens were gripped vertically by end loading using simple collet-type, self-aligning loading grips (ATS 713C). The grips had no active water- or air-cooling circuits. This is an advantage as cooled grips have caused thermal stress concentrations at the grips causing failure in the grip section [10]. No specimen failure in the grip section was observed. Copper bands were wrapped around the specimen in the grip area. There was no indication that slip occurred at the grips.

Specimen alignment was verified using ASTM Standard E1012-89. A 25.4 mm gage length extensometer was attached to one side of an aluminum specimen. Longitudinal strain measurements were recorded at approximately 300 MPa. This procedure was repeated for each side of the bar. Bending stresses were found to be less than 1% of the total stress indicating good load train alignment.

A feedback control system using a two-color optical pyrometer was used to control the specimen temperature. The pyrometer was focused on the leading edge of the flame impinged surface. The flame temperature near the leading edge during steady state heating was measured with a hand-held R-type ceramic coated thermocouple. No radiation shield was used around the tip of the thermocouple. The measured flame temperatures were within 3% of pyrometer readings of the specimen leading edge. Due to the severe temperature and high velocity flame environment, thermocouples or strain gages, or both, could not be attached to the specimen to collect experimental data.

A 66°C temperature difference along the specimen width in the flame impinged zone at 1350 °C was measured using the optical pyrometer. The most extreme temperature gradient occurred outside the flame impingement zone. The temperature gradient at the leading edge outside the heated section was estimated to be 529°C (985°F) within approximately 5 mm vertical distance. The trailing edge temperature gradient was 588°C (1091°F) within approximately 5 mm vertical distance. The temperature at the grips was approximately 28°C (83°F).

Although the pyrometer readings were in agreement with thermocouple flame temperature measurements, the location of the measurement along the length or width of the specimen had an uncertainty of approximately ±1 mm. Thus, a finite element analysis of temperature distributions and accompanying thermal stresses during both steady state and transient thermal loading was carried out, which will be the subject of a separate publication [11].

An experimental test matrix for 13 test specimens was developed incorporating the capabilities of the test system. A "set-up" test specimen was used to evaluate the transient temperature profile during heating and cooling. The time to run out was arbitrarily selected as 1000 cycles due to the cost of operating the test facility. The applied stress was varied from specimen to specimen to reflect the learning of the composite system behavior under the severe thermal cycling. Constant tensile stresses 16.8, 84, 110, 120, 125, 130, 140 and 168 MPa were applied to individual specimens during thermal cycling.

Thermal cycling of the specimens between 500-1350°C was accomplished by a computer-controlled hydraulic piston that rotated the burner rig nozzle into and out of the direction of the specimen. An electronic timing device controlled the duration of the thermal cycle, and a cycle timer recorded the number of heating and cooling half cycles to failure.

The length of the heating half cycle (60 seconds) was dictated by the test system's capabilities. The combustion chamber temperature and pressure control system required approximately 50 seconds to adjust process parameters and obtain feedback from the system's components. This required the heating half cycle to be at least 60 seconds. Incorporating a 60-second heating half cycle eliminated specimen temperature overshoot as well as minimized "under-temperature" conditions during start up. The test facility required between three to five heating half cycles to achieve desired temperature. The setup specimen required approximately 60 seconds to cool to 500°C by radiation and natural convection after a 60 second heating half cycle. A symmetrical time parameter of 60 seconds was selected for cooling.

The specimen leading edge cyclic temperature profile resulting from a 60-second 1350°C flame impingement, followed by a 60-second passive, ambient temperature cooling is shown in Figure 1. The temperature of the impinging gas was a function of the combustion fuel, feed flow of compressed air and burner rig pressure. The maintained exiting gas temperature of 1350°C is near the peak temperature capability of the facility.

Longitudinal TEM specimens (beam direction perpendicular to fiber direction) were sectioned from the SCS-9 runout specimen thermal cycled under 115 MPa constant applied stress using a low-speed diamond saw, and mechanically ground to ~300 μm. Circular discs of 3 mm diameter were core drilled from the ground sections, mechanically ground to ~120 μm, and then dimpled to ~30 μm. The specimens were ion milled with 5 kV Ar^+ ions at an incident angle of 12° until perforation occurred. A light carbon film was evaporated onto the specimens to prevent charging in the electron microscope. The microstructure and chemistry of the composite were investigated using a Philips EM400T TEM and a Noran ultrathin window Micro-Z detector.

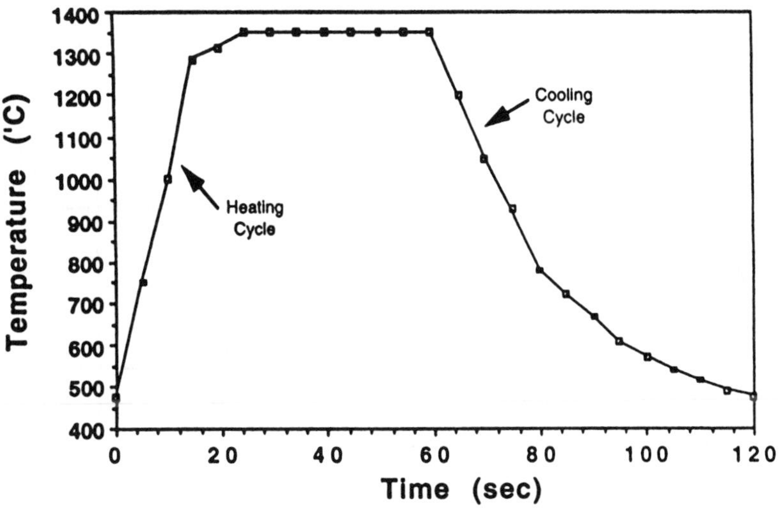

Figure 1. Transient specimen leading edge temperature profile during burner rig thermal cycling showing the heating and cooling half cycles.

RESULTS

The burner rig thermal fatigue lifetimes of nine $[(0/90)_5/\bar{0}]_S$ SCS-9 and four $[(0/90)_3/\bar{0}]_S$ SCS-6 SiC continuous fiber specimens under constant applied stress are given in Figure 2. The SCS-9 specimens showed superior burner rig thermal fatigue resistance at any constant applied stress, Figure 2. The SCS-6 specimens failed after 20-40 cycles at constant applied stresses as low as 110 MPa. The

SCS-9 specimens, on the other hand, showed run out conditions (1000 cycles without failure) up to 125 MPa constant applied stress. The number of cycles-to-failure in the SCS-9 continuous fiber specimens declined abruptly to 115 thermal cycles at 140 MPa constant applied stress. At 168 MPa applied stress fatigue life diminished to 5-42 cycles, Figure 2.

Two runout SCS-9 specimens thermal cycled under 84 and 105 MPa constant applied stress, were tension tested at room temperature. The tensile strength and moduli of the as received and thermal fatigue runout specimens was found to be 244 MPa, and 260 GPa (37.8 Msi), respectively. These indicate no strength or modulus degradation in the runout samples.

The video recording of each burner rig thermal fatigue test under constant applied stress provided information about the location and time of failure. SCS-9 specimens failed in the center of the hot zone during the heating half cycle. SCS-6 specimens, on the other hand, failed at the bottom edge of the hot zone where the temperature gradient is the greatest. Three SCS-6 specimens failed during cooling and the fourth during heating half cycle. The runout conditions exhibited by SCS-9 specimens up to 125 MPa constant applied stress indicate good burner rig thermal fatigue resistance in cycles between 500-1350°C.

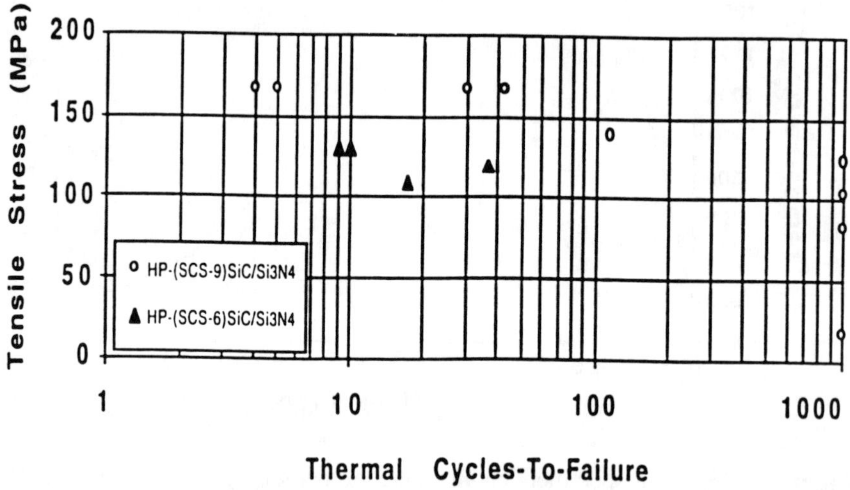

Figure 2. Burner rig thermal fatigue response of HP-(SCS-6) and HP-(SCS-9) SiC_f/Si_3N_4 crossply ceramic composites under constant applied stress.

Figure 3(a) shows a bright field image of β-Si_3N_4 matrix of the runout SCS-9 specimen thermal cycled under 115 MPa constant applied stress. The oriented matrix grains are elongated along the c-axis. This morphology could give enhanced fracture toughness at room temperature by providing a tortuous crack path. The grain boundaries contained an amorphous Si-Y-Mg-O phase, as shown in Figure 3(b). This indicates that flame impingement did not crystallize the grain boundary glassy network.

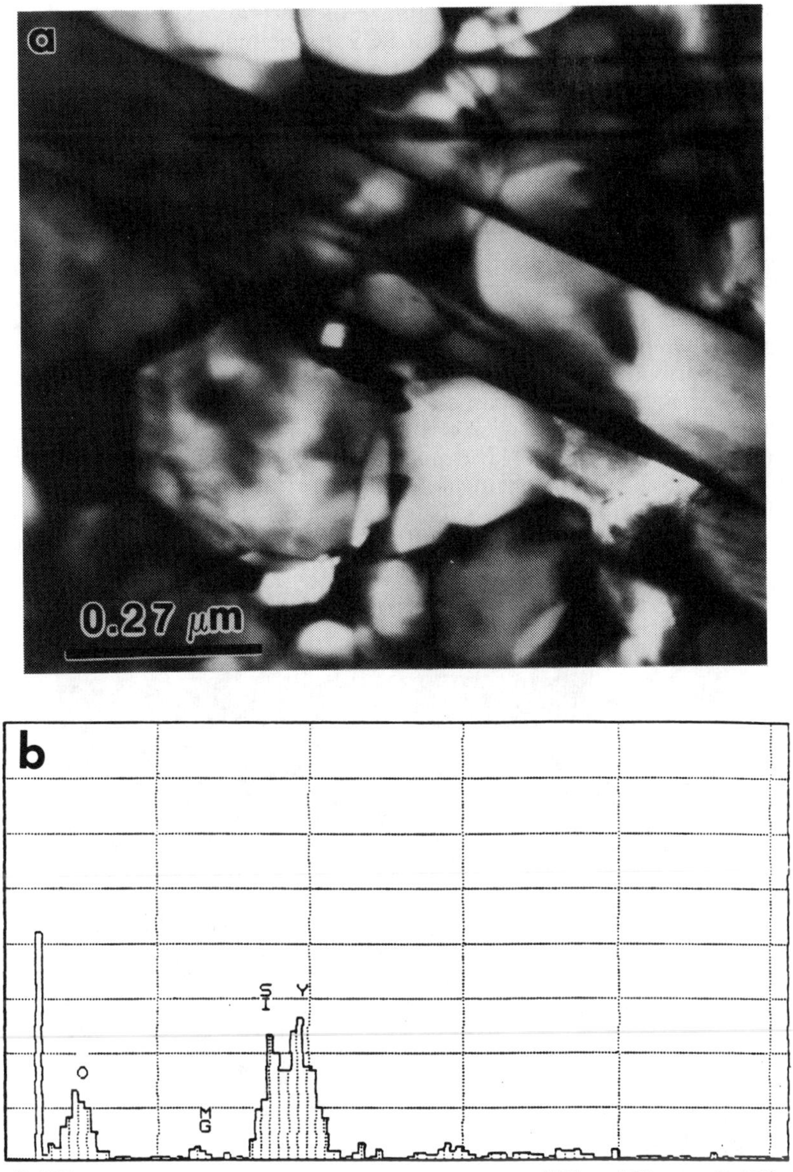

Figure 3. (a) Bright field image showing β-Si_3N_4 matrix in the runout SCS-9 specimen. (b) EDXS spectrum obtained from the amorphous Si-Y-Mg-O phase at the grain boundaries.

Figure 4(a) shows a bright field image of the interface between the Si_3N_4 matrix and the SiC fiber in the runout SCS-9 specimen. The thickness of the carbon coating in the SiC fiber is ~3.0 µm. No reaction zone exists at the outer carbon layer/Si_3N_4 matrix and outer carbon coating/SiC layer interfaces (see marked arrows). The interfaces seem to be relatively strong.

In a previous study [12], the ~3.0 µm thick outer carbon coating in as-fabricated (unprocessed) SCS-6 SiC fiber was found to consist of three sublayers having thicknesses ~1.6, ~0.2, and ~1.3 µm, respectively, from the outer carbon layer/SiC layer interface. In the present study sublayers within the outer carbon layer were not observed, Figure 4(a). This may be due to homogenization of sublayers during high-temperature thermal fatigue exposure.

The selected area diffraction (SAD) pattern obtained from the outer carbon coating is shown in Figure 4(b). The SAD pattern consists of distinctive arcs and randomly distributed spots. This indicates that turbostratic carbon blocks are aligned preferentially and SiC particles are embedded in the carbon layer.

Occasionally, weak areas were observed at outer carbon coating/Si_3N_4 matrix and outer carbon coating/SiC layer interfaces in the runout SCS-9 specimen, as shown in Figure 5(a). The chemistry and microstructure of the interfaces were examined by dispersive X-ray spectroscopy (EDXS) and convergent beam electron diffraction (CBED), respectively. Figure 5(b) shows an EDXS spectrum of the outer carbon coating/SiC layer interface, indicating the existence of a small amount of O along with Si and C. The interface retained its crystalline structure.

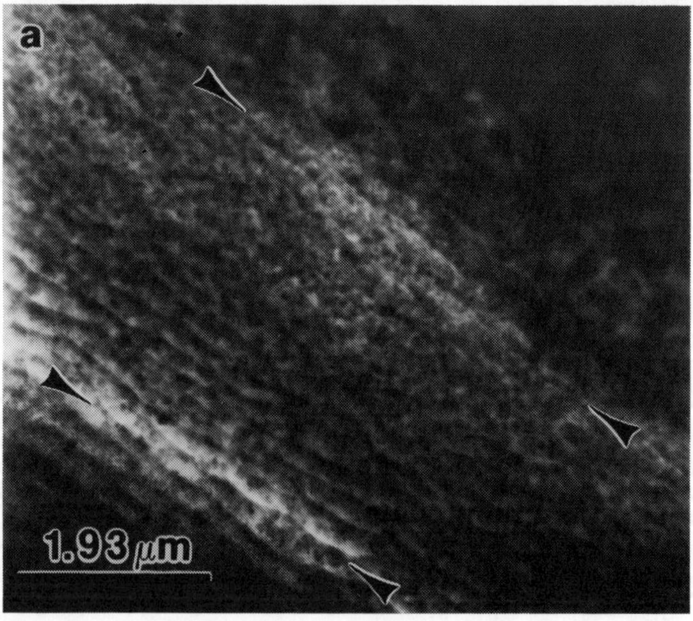

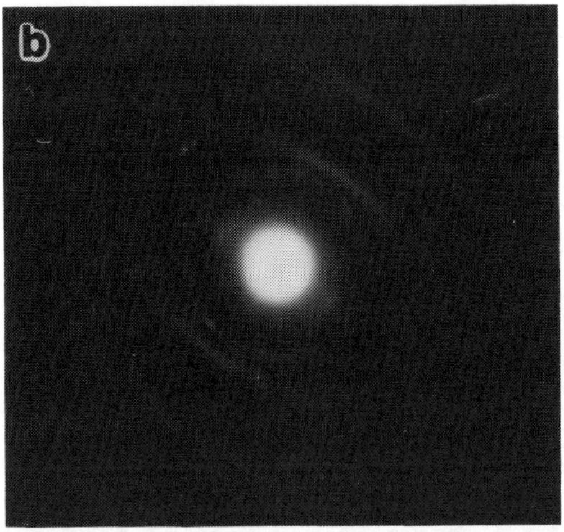

Figure 4. (a) Bright field image showing SiC layer, outer carbon layer, and matrix in the runout SCS-9 specimen. (b) SAD pattern obtained from the outer carbon layer.

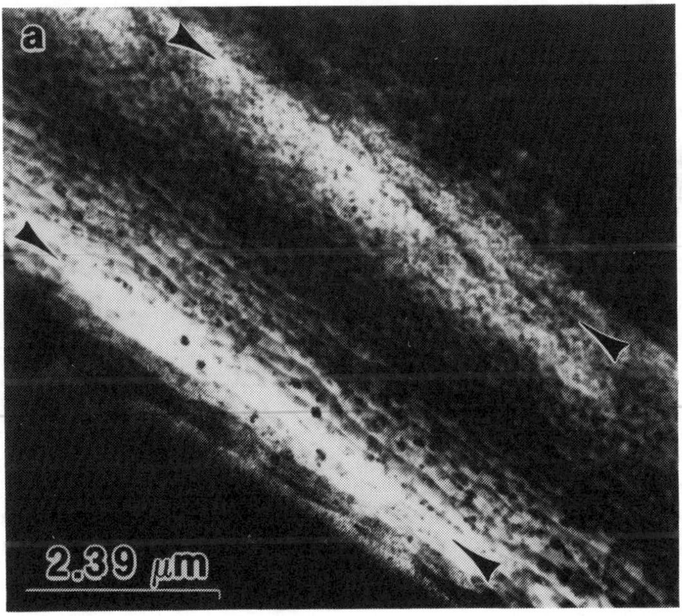

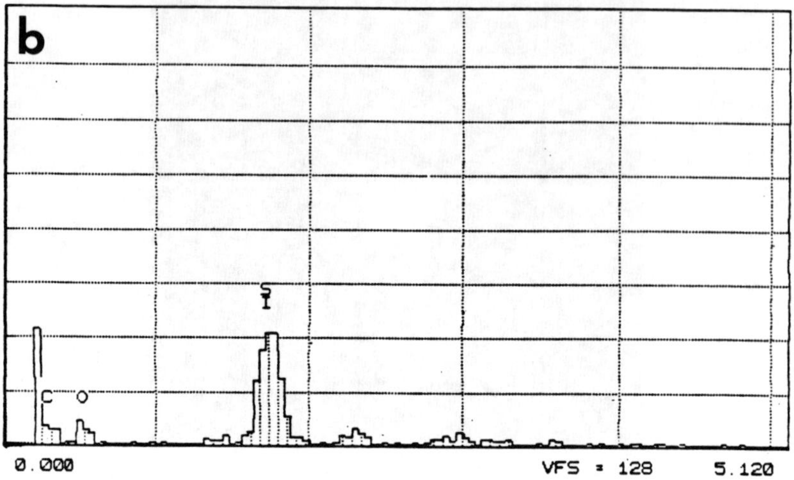

Figure 5. (a) Bright field image showing weak areas at outer carbon layer/matrix and outer carbon layer/SiC layer interfaces in the runout SCS-9 specimens. (b) EDXS spectrum obtained from the outer carbon layer/SiC layer interface.

DISCUSSION

The 13-ply SCS-6 and 21-ply SCS-9 SiC fiber 0/90 Si_3N_4 ceramic composites thermal cycled between 500-1350°C under impinged jet fuel flame and constant applied stress showed a different burner rig thermal fatigue behavior. SCS-9 SiC continuous fiber specimens showed superior burner rig thermal fatigue resistance than SCS-6 SiC fiber-reinforced Si_3N_4 at any constant applied stress. The SCS-6 SiC fiber reinforced specimens failed due to high thermal stresses induced by steep temperature gradients outside the flame impinged zone (thermal shock weakening). The SCS-9 SiC fiber-reinforced specimens, however, survived the thermal shock inside and outside the flame impinged zone, showing burner rig thermal fatigue runout conditions (1000 cycles without failure) at applied stresses greater than 50% of the room temperature strength.

If the failure occurs in the heated zone, burner rig thermal fatigue testing under constant applied stress crosses the boundary between elevated temperature tensile fatigue and creep testing. The room-temperature tension tests on the SCS-9 runout specimens showed no degradation of modulus or strength. Figures 3-5 show that no voids formed in the fiber, matrix or at the interface. Figure 3 indicates that no devitrification of the glassy grain boundary phase occurred. Figures 4 and 5 indicate that no interface reaction took place. The apparent homogenization of the sublayers in the outer carbon coating could not weaken the material under burner rig thermal fatigue conditions. The small amount of oxygen detected at the weak regions of fiber interfaces, is likely due to previous processing. Thus, the failure of the SCS-9 specimens can not be explained in terms of accumulation of cycle by cycle thermal fatigue or creep damage in the flame impinged zone.

Bhatt and Hull [13] studied the thermal instability and accompanying strength degradation in three chemically vapor deposited Textron SiC fibers after exposure to 1400°C under argon atmosphere for 1 hour. They found that strength degradation occurred due to recrystallization and grain growth of the Si-rich SiC grains near the surface of the fibers, where carbon diffused inward from the coating. In the present case, the SCS-9 fibers were exposed to 1350°C for more than 6 hours under constant applied stresses up to 125 MPa in the runout specimens. The stresses in the 0° plies would be higher than the applied stress. (Using the laminated plate theory, the stress in the 0° layers for an applied stress of 168 MPa is estimated to be 198 MPa.) Under these severe thermal cycle conditions incorporating constant applied stress the most likely mode of failure for the SCS-9 SiC fiber reinforced specimens would be the thermal instability (strength degradation by recrystallization and grain growth) of fibers. This is currently under investigation.

SUMMARY AND CONCLUSIONS

The SCS-9 continuous fiber runout specimens thermal cycled under constant applied stresses of 105 MPa showed no strength or modulus degradation. Likewise, no void formation, crystallization of the matrix grain boundary glassy phase, nor chemical reaction at the interface were observed. Furthermore, the amorphous grain boundary phase remained glassy after 1000 thermal cycles. Thus, the failure of SCS-9 SiC fiber reinforced specimens could not occur by cycle by cycle thermal fatigue damage accumulation. The abrupt decline in the burner rig fatigue life of the SCS-9 specimens in the constant applied stress range 140-168 MPa coupled with lack of microstructural damage and time of exposure at 1350°C indicate the likelihood of creep or thermal instability induced rupture of the fibers. On the other hand, the failure of the SCS-6 specimens is most likely due to cycle by cycle damage induced by high thermomechanical stresses outside the flame impinged zone.

ACKNOWLEDGMENT

The primary author is grateful to NASA Lewis Research Center, Cleveland, OH, for the use their burner rig test station.

REFERENCES

[1]. Evans, A., and Marshall, D., "The Mechanical Behavior of Ceramic Matrix Composites", *Acta Metall.*, **37** 2583 (1989).

[2]. Argon, A.S. and Gupta, V., "Control of Toughness of Composites Through Control of Strength of Fiber/Matrix Interfaces', *Damage and Oxidation of Protection in High Temperature Composites,* G.K. Haritos and O.O. Ochoa, (Eds.) ASME Book No. H0692B, 1991.

[3]. Kannmacher, K., and Groseclose, L., "Design and Analysis of Ceramic and CMC Components for Advanced Gas Turbines", presented at the International Gas Turbine and Aeroengine Congress and Exposition, Orlando, FL June 3-6, 1991, ASME 91-GT-156.
[4]. Zawada, L., and Wetherhold, R., "The effects of Thermal Fatigue on a SiC Fibre/Aluminosilicate Glass Composite," *J. Mater. Sci.* 26 654 (1991).
[5]. Butkus, L.M., Holmes, J.W., and Nicholas, T., "Thermomechanical Fatigue of a Silicon Carbide Fiber-Reinforced Aluminosilicate Composite", *J. Am. Ceram. Soc.*, **76** 2817 (1993).
[6]. Worthem, D.W. and Ellis, J.R., "Thermomechanical Fatigue of Nicalon/ CAS Under In-Phase and Out-Of-Phase Cyclic Loadings, *Ceram. Eng. Sci. Proc.*, **14**, [7-8] (1993) 292-300.
[7]. Deadmore, D.L., "Digital Temperature and Velocity Control of Mach 0.3 Atmospheric Pressure Durability Testing Burner Rigs in Long Time, Unattended Cyclic Testing", NASA Technical Memorandum 86959; March, 1985.
[8]. Beer, J.M., and Chigier, N.A., *Combustion Aerodynamics*, Robert E. Kreiger Publishing Company, (1972).
[9]. Lefebrve, A.H., *Gas Turbine Combustion*, McGraw-Hill (1983).
[10]. Holmes, J.W., "A Technique for Tensile Fatigue and Creep Testing of Fiber-Reinforced Ceramics", *J. Compos. Mater.*, **26** 932 (1992).
[11]. Ertürk, T., "Heat Transfer During Burner Rig Thermal Cycling of SiC Fiber Silicon Nitride Ceramic Composites: "A Finite Element Analysis," to be published.
[12]. Park, K. and Sung, C., unpublished work.
[13]. Bhatt, R.T. and Hull, D.R., "Microstructural and Strength Stability of CVD SiC Fibers in Argon Environments" *Ceram. Eng. Sci. Proc.*, **12** [9-10] (1991) 1832-1844.

INTERNAL DAMPING OF SILICON NITRIDE AND SILICON NITRIDE COMPOSITES WITH SILICON CARBIDE WHISKERS TO 1100°C

S. Yang and R. F. Gibson
Advanced Composites Research Laboratory
Department of Mechanical Engineering
Wayne State University
Detroit, MI 48202

G. M. Crosbie and R. L. Allor
Ford Research Laboratory
Ford Motor Company
Dearborn, MI 48121

ABSTRACT

In heat engines and in manufacturing, vibration damping is desirable for quietness and durability, but the usual polymer fixes fail at high temperatures. Ceramics and ceramic composites have potential for such applications, but little is known of their damping properties. With a special impulse-frequency response apparatus designed and built at Wayne State University, the temperature-dependent damping up to 1100°C was measured for hot-pressed Si_3N_4 with and without 5 volume percent tightly-bonded American Matrix SiC whiskers. With increasing temperature, the damping loss factor has an initial peak value of about 0.02 near 900° C. The thermal cycling environment was also simulated, and the damping peak was found to be slightly reduced with each thermal cycle.

INTRODUCTION

Internal damping is one of the most important properties of materials which are used in a vibratory environment. The inherent damping property is usually determined, from the materials science points of view, from the microstructure and micro-mechanics of a given material. Examples of this include the widely conducted research on the relationships between a material's damping behavior and such mechanisms as grain boundary sliding [1, 2], and dislocation movements [2, 3]. From the mechanics points of view, the damping property of a given material is closely related to its noise and vibration attenuation characteristics, which have recently become practically important in view of requirements for increasing quietness and durability for the structural applications of materials under vibratory loading. Two simple but instructive examples are the roles of the material damping in retarding amplitudes of the free and forced vibrations of a single degree of freedom system, as shown in Figures 1 and 2, respectively [4]. Even though the damping properties of various metal matrix and polymer matrix composite materials have been extensively studied [5], little is known about such behavior in ceramics and ceramic matrix composites. Because of their competitive performance (high strength, low density, good heat and environmental resistance), ceramics and ceramic matrix composites will

To the extent authorized under the laws of the United States of America, all copyright interests in this publication are the property of The American Ceramic Society. Any duplication, reproduction, or republication of this publication or any part thereof, without the express written consent of The American Ceramic Society or fee paid to the Copyright Clearance Center, is prohibited.

play an active role in substituting for metals and superalloys in advanced aircraft and automobile engines where high temperature and vibratory loading are the key operating conditions. So, the knowledge of the damping properties (especially the damping properties at high temperatures) of ceramics and ceramic matrix composites will be indispensable in their structural applications for increased performance and reduced noise and vibration levels in high temperature, vibratory environments.

MEASUREMENTS AND APPARATUS

Material damping is a manifestation of internal energy dissipation in the material. Damping is a property of viscoelastic and anelastic materials during cyclic deformation. It can be characterized by many different models such as logarithmic decrement, specific damping capacity, complex modulus notation, and hysteresis loop models [6]. The definition of damping used herein is the damping loss factor, and its expression by the notation of the half power band width is [7]

$$\eta = \frac{\Delta f}{f_n} \qquad (1)$$

where η is the so called damping loss factor, Δf is the half power band width of the nth mode resonant peak in the frequency response function, f_n is the resonance frequency of the nth vibration mode, and n is the mode number (Figure 3). The damping loss factor in this research is measured by the impulse-frequency response technique [8]. In order to realize the measurements at elevated temperatures, a specially designed experimental system was built. Figure 4 shows a cutaway drawing of the system [9]. In a typical experiment, the sample is heated up by the three zone quartz lamp heater; and the required temperature is obtained by adjusting the controls for the three zones independently. The vibration is excited by the electric solenoid hammer, and the response is measured by a remote, non-contacting electro-optical displacement follower. The analog signals are transformed to generate the frequency domain response spectrum by the Fast Fourier transform (FFT) analyzer, and finally, the computer interrogates the FFT analyzer to find the resonance frequency and the corresponding half power band width, then calculates the damping loss factor according to Equation (1). The system used here is suitable for testing cantilever beam samples under flexural vibrations, but its application can be easily extended to torsional vibration tests with slight modifications of the apparatus.

SAMPLES AND MATERIALS

Cantilever beam samples, with lengths of about 156 mm, width of 12.7 mm, but three different thicknesses of 1.524 mm, 3.048 mm, and 6.350 mm, respectively, of four different materials were prepared by Cercom, Inc. at the direction of Ford Motor Company. Table I lists the types of materials, structural compositions, and processing technology of these samples.

The purity of the matrix of the UBE matrix material is higher than that of the KemaNord [KN] matrix material. The materials reinforced with SiC whiskers are ceramic matrix composites, but those without whiskers can be considered in-situ composites. They are all classified as the silicon nitride based ceramics which are often considered to represent the toughest of the high - temperature (above 1000°C) ceramics intended for advanced heat engine use [10]. Thus, damping properties and the related research are of practical significance for the optimization of these materials in engine applications.

Table I. Identification, Composition, and Processing of Samples

Name*	Material	Whisker (vol. %)	Additive (wt. %)	Processing
KNw/SiC	Si_3N_4	SiC (5)	Y_2O_3/Al_2O_3 (6/3)	Hot-pressed
KNw/oSiC	Si_3N_4	----	Y_2O_3/Al_2O_3 (6/3)	Hot-pressed
UBEw/SiC	Si_3N_4	SiC (5)	Y_2O_3/Al_2O_3 (6/3)	Hot-pressed
UBEw/oSiC	Si_3N_4	----	Y_2O_3/Al_2O_3 (6/3)	Hot-pressed

* KNw/SiC = KemaNord low purity Si_3N_4 powder matrix with SiC whiskers,
KNw/oSiC= KemaNord low purity Si_3N_4 powder matrix without SiC whiskers:
UBEw/SiC= High puritySi_3N_4 powder matrix with SiC whiskers,
UBEw/oSiC= High purity Si_3N_4 powder matrix without SiC whiskers.

A schematic diagram of the microstructure of silicon nitride ceramics is shown in Figure 5. Such materials present the typically desired interlocking acicular β-silicon nitride grains and the grain boundary phases in relief and in cross-section. The powder purity is important because a low solubility in silicon nitride for many impurities leads to high concentrations of those impurities in grain boundary phases, and the carbon residues (which are found in some high purity powders) are considered detrimental to second phase oxynitride development, thermal stability, and mechanical properties. [10, 11]. The microstructures of silicon nitride composites reinforced with silicon carbide whiskers are shown in two optical micrographs in Figure 6 (a) and (b), where it is seen that the whiskers are rather uniformly distributed in the matrix for moderate purity KN powder and high purity UBE powder, respectively. For both purity levels, β-Si_3N_4 and a minor amount of β-SiC were determined present by x-ray diffraction (XRD). The initially high α-Si_3N_4 powder was converted to β-phase as is desired for needle-like grain growth. Since no other crystalline Si-N-Y-O phases were detected, the silicon nitride matrix grain boundary phase is assumed to have remained glassy [12].

RESULTS AND DISCUSSIONS

Experiments with 1.524 mm and 3.048 mm thick samples in first mode flexural vibration have been conducted in the following sequence: (1) damping measurements at room temperature, (2) damping measurements at elevated temperatures up to 1100°C, and (3) damping measurements under simulated thermal cycling to 1100°C.

Room Temperature Tests

The measured damping loss factors at room temperature are shown in Table II and Table III, respectively, for two different thickness samples. The measurement indicates that the room temperature damping is very small, with loss factors in the middle of the range 10^{-4} to 10^{-3}, for all the four materials with two different thicknesses. This small damping made the experiment difficult to conduct because it is almost in the same range as the air damping or other background damping [13]. It should be understood that damping, no matter large or small, is a property which is difficult to either experimentally measure or theoretically predict due to the fact that damping usually involves a variety of energy dissipation mechanisms, and these mechanisms are dependent on vibration parameters such as frequencies and amplitudes. These mechanisms are greatly influenced by material conditions such as impurities and cracks in specimens, as well as by environmental conditions such as temperature and humidity. Hence, even by experiment, exact damping is not easily obtained. Instead, the minimum measured damping is sought and considered to be closer to the real damping of the material than the maximum measured value.

Table II. Measured Damping Loss Factors of 1.524 mm Thick Samples at Room Temperature

Material	Mean Frequency (Hz)	Number of Tests	Loss Factor			
			Max (10^{-3})	Min (10^{-3})	Mean (10^{-3})	Std. Dev. (10^{-3})
KNw/SiC	119.693	32	0.667	0.344	0.491	0.109
KNw/oSiC	116.218	32	0.569	0.417	0.509	0.057
UBEw/SiC	118.214	32	0.582	0.358	0.492	0.060
UBEw/oSiC	117.557	32	0.632	0.397	0.519	0.075

Table III. Measured Damping Loss Factors of 3.048 mm Thick Samples at Room Temperature

Material	Mean Frequency (Hz)	Number of Tests	Loss Factor			
			Max (10^{-3})	Min (10^{-3})	Mean (10^{-3})	Std. Dev. (10^{-3})
KNw/SiC	235.612	32	0.600	0.387	0.518	0.066
KNw/oSiC	230.001	32	0.604	0.421	0.523	0.059
UBEw/SiC	232.999	32	0.604	0.349	0.486	0.067
UBEw/oSiC	231.012	32	0.595	0.422	0.506	0.053

Though near the measurement limit at room temperature, reproducible differences were measured among the materials tested. It is seen from Tables II and III that the reinforced composites, KNw/SiC and UBEw/SiC, have somewhat smaller damping than those of the corresponding ceramics, KNw/oSiC and UBEw/oSiC, but the T-test shows that it is not statistically significant. Also, the composites have higher natural frequencies than the ceramics because of the high stiffness of the SiC whiskers, and this is of statistical significance. The measured damping of a given material at two different frequencies (from two thickness samples) thus far shows no significant frequency dependence, although the frequency range covered is quite small.

High Temperature Tests

The curves corresponding to the mean measured damping loss factors as a function of temperature up to 1100°C for the four materials are shown in Figure 7, and the numerical results for the peak damping values are listed in Table IV. The maximum temperature of 1100°C is governed by the limitations of the heater system currently used. Only the results of the samples with 1.524 mm thickness under the first thermal cycle (which is because thermal cycling has been found to affect the measured damping), are presented, since the measured behavior at elevated temperatures of two different thickness samples still shows no difference.

From Figure 7, it can be seen that the damping properties are quite stable up to about 700°C, then the damping rises and reaches its peak value near 900°C. The peak loss factor values for the four materials are in the range 0.020-0.024, which is much greater than the room temperature damping and far beyond the possible air damping and background damping. It is believed that these peaks are due to the viscoelastic behavior of the amorphous grain boundary phase of the Si_3N_4 matrix material at these high temperatures. Contrary to room temperature results, the ceramic composites (KNw/SiC and UBEw/SiC), at the damping peak temperature, have slightly larger damping than that of the ceramics, which seems to indicate that the contaminants carried by the whiskers are affecting the viscosity of the glassy phase.

Table IV. Measured Damping Loss Factors of 1.524 mm Thick Samples at 900°C

Material	Mean Frequency (Hz)	Number of Tests	Loss Factor			
			Max (10^{-3})	Min (10^{-3})	Mean (10^{-3})	Std. Dev. (10^{-3})
KNw/SiC	108.023	16	21.95	20.38	21.00	0.545
KNw/oSiC	104.667	16	20.35	17.42	18.44	0.723
UBEw/SiC	107.785	16	24.01	22.72	23.51	0.448
UBEw/oSiC	104.038	16	21.97	20.33	20.93	0.500

Thermal Cycling Effects

As indicated earlier, repeated thermal cycling has been found to affect the damping properties of the four materials. Figure 8 shows such effects on the 1.524 mm thick UBEw/SiC samples for three cycles of thermal loading. This effect was observed by repeated heating of the sample from room temperature to 1100°C, then cooling it down to room temperature. Damping measurements were only conducted during the heating part of the thermal cycle, not during the cooling part. The materials seem to be stiffened, with higher natural frequencies and smaller damping loss factors, following each thermal cycle. One possible explanation for these results is that microcracks within the Si_3N_4 matrix are partially "healed" as a result of the viscoelastic flow at the peak temperature, and that the Coulomb friction damping due to crack interfaces rubbing together is thereby reduced. Changes in chemical composition and/or phase transformations such as crystallization of the amorphous phase are also possible explanations. Electron microscopy of samples before and after thermal cycling will be conducted in the future in order to better understand the responsible mechanisms.

Table V. Damping Loss Factors of 1.524 mm UBEw/SiC Samples vs. Thermal Cycles

Cycle Number	Mean Frequency (Hz)	Number of Tests	Loss Factor Max (10^{-3})	Loss Factor Min (10^{-3})	Loss Factor Mean (10^{-3})	Loss Factor Std. Dev. (10^{-3})
1	107.785	16	24.01	22.72	23.51	0.448
2	110.913	16	17.62	16.58	17.31	0.260
3	112.115	16	14.78	13.96	14.46	0.294

CONCLUSIONS

The damping behavior of hot-pressed silicon nitride and silicon nitride reinforced with silicon carbide whiskers is quite stable below 700°C. Even though the room temperature damping is quite small, the high temperature damping at and near its peak value around 900°C is large, and hence, these silicon nitride based ceramics should be able to play important roles as structural materials in advanced heat engines. This is not only because of their light weight, high strength, good heat and environment resistance, but also due to their good damping behavior which should lead to reduced noise and vibration levels at elevated temperatures.

This research also shows that the damping behavior of the tested materials is affected by repeated thermal cycling. The changes of damping properties under thermal cycling may imply changes of the microstructure, chemical composition, or phase transformation in the materials. The microstructures of the samples after testing should be further investigated to confirm this observation.

The state of knowledge regarding the damping properties of ceramics and ceramic composites is relatively sparse at this stage, and continued research is needed to resolve the issues raised in this work.

ACKNOWLEDGMENTS

The authors gratefully acknowledge the support of the Ford Motor Company through an unrestricted grant. Whisker dispersion and hot pressing were carried out at Cercom, Inc., Vista, California, under the direction of A. Ezis.

REFERENCES

[1] T. L. Tsai and R. Raj, "The Role of Grain-Boundary Sliding in Fracture of Hot-Pressed Si_3N_4 at High Temperatures," *the American Ceramics Society Bulletin,* 63[9-10], pp. 513-517, 1980.
[2] A. S. Nowick and B. S. Berry, *Anelastic Relaxation in Crystalline Solids*, Academic Press, New York, 1972.

[3] A. V. Granato, "Internal Friction Studies of Dislocations," *Internal Friction, Damping, and Cyclic Plasticity*, a Symposium Presented at the sixty-seventh meeting, ASTM, Chicago, IL, pp. 88-108, 1964.
[4] W. T. Thomson, *Theory of Vibration with Applications*, 3rd edition, Prentice-Hall International, Inc., 1988.
[5] V. K. Kinra and A. Wolfenden, Eds, *M³D: Mechanics and Mechanisms of Material Damping,* STP 1169, ASTM Publication, 1992.
[6] A. D. Nashif, D. I. G. Jones, and J. P. Henderson, *Vibration Damping*, John & Sons, Inc., 1985.
[7] R. F. Gibson, "Damping Characteristics of Composite Materials and Structures," *Journal of Materials Engineering and Performance*, 1(1) pp. 11-20, 1992.
[8] S. A. Suarez and R. F. Gibson, "Improved Impulse-Frequency Response Techniques for Measurement of Dynamic Mechanical Properties of Composite materials," *Journal of Testing and Evaluation*, 15(2), pp. 114-21, 1987.
[9] R. F. Gibson, R. Thirumalai, and R. Pant, "Development of An Apparatus to Measure Dynamic Modulus and Damping of Reinforcing Fibers at Elevated Temperature," *Proceedings of Society for Experimental Mechanics 1991 Spring Conference*, Milwaukee, Wisconsin, June 9-12, pp. 860-869, 1991.
[10] G. M. Crosbie, J. M. Nicholson, R. L. Predmesky, and E. D. Stiles, "Silicon Nitride Power Synthesis Program," *Proceedings of the 25th Automotive Technology Development Contractors' Coordination Meeting*, Society of Automotive Engineers, pp. 127 - 35, Warrendale, Pa., 1988.
[11] G. Woetting and G. Ziegler, "Power Characteristics and Sintering Behavior of Si_3N_4 Powders," *Powder Metall. Intl.*, 18 [1], pp. 25-32, 1986.
[12]R. L. Allor, G. M. Crosbie, E. L. Cartwright, and R. K. Govila, "Silicon Nitride Based Ceramic Composites," *Structural Composites Design and Processing Technologies, Proceedings of Advanced Composites Conference and Exposition,* ASM International, Materials Park, Ohio, pp. 555-641, 1990.
[13] G. G. Wren and V. K. Kinra, " On the Effect of an End-Mass on Beam Damping," *Experimental Mechanics*, 29, pp. 336-41, 1989.

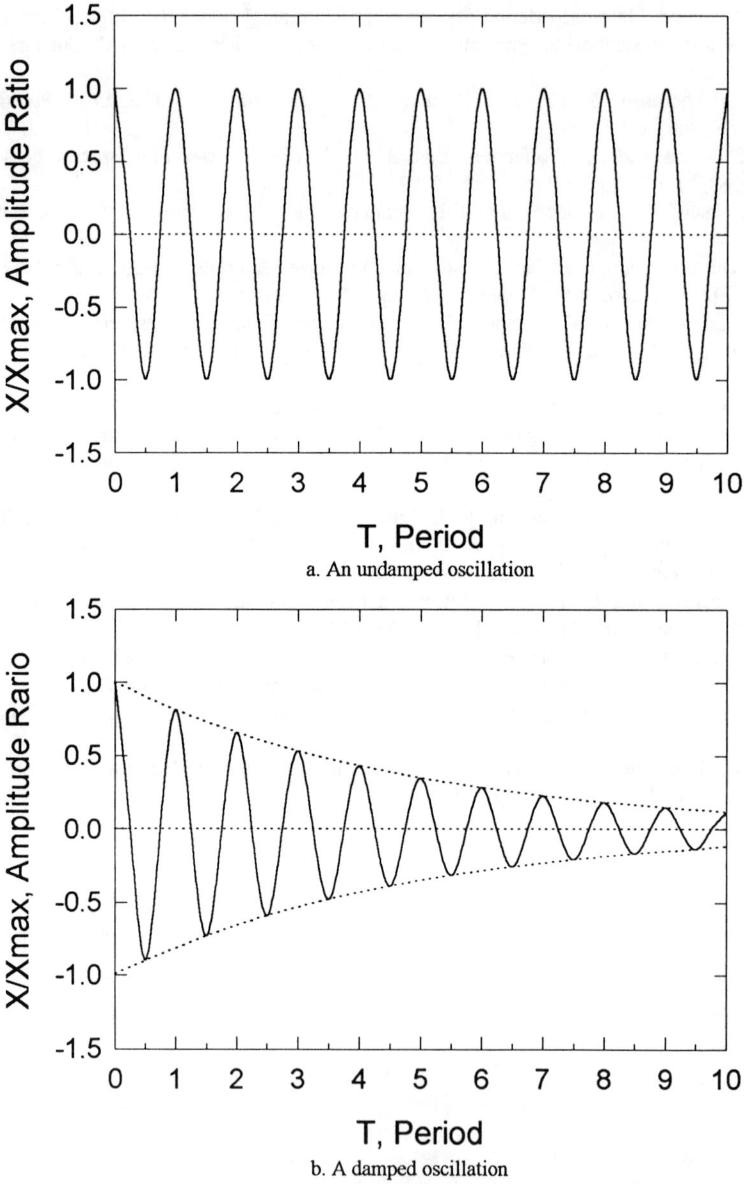

Figure 1. Free Vibrations without and with Damping

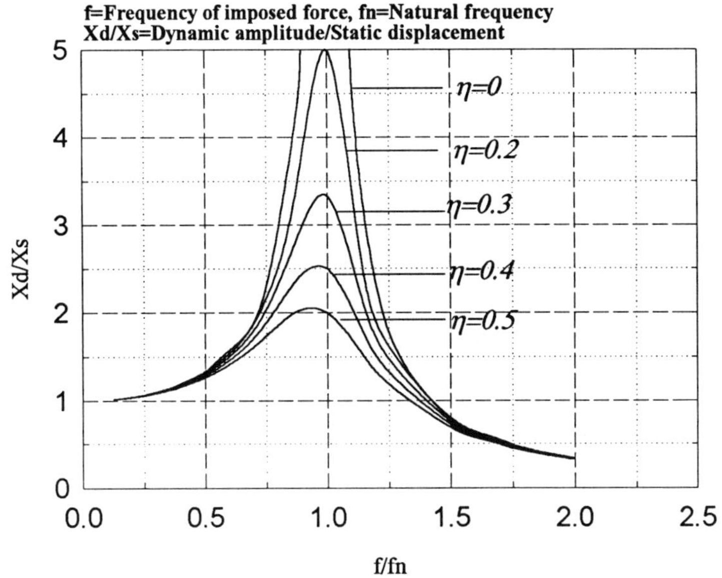

Figure 2. Foced Vibration Response for Different Damping Loss Factor, η

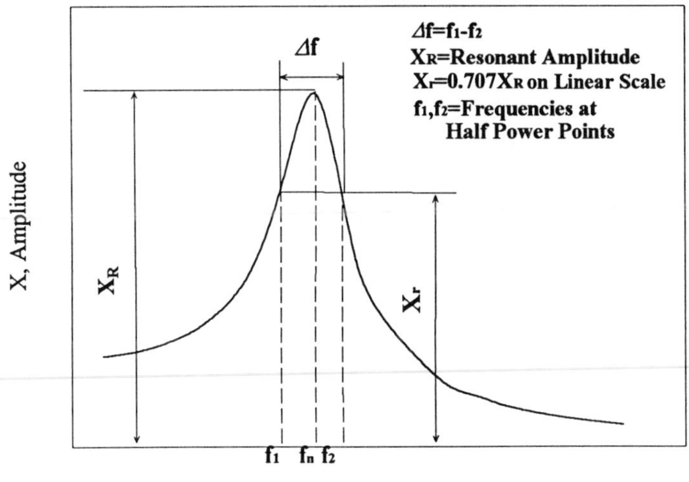

Figure 3. Definition of Parameters in Eq. (1) from Frequency Response Spectrum

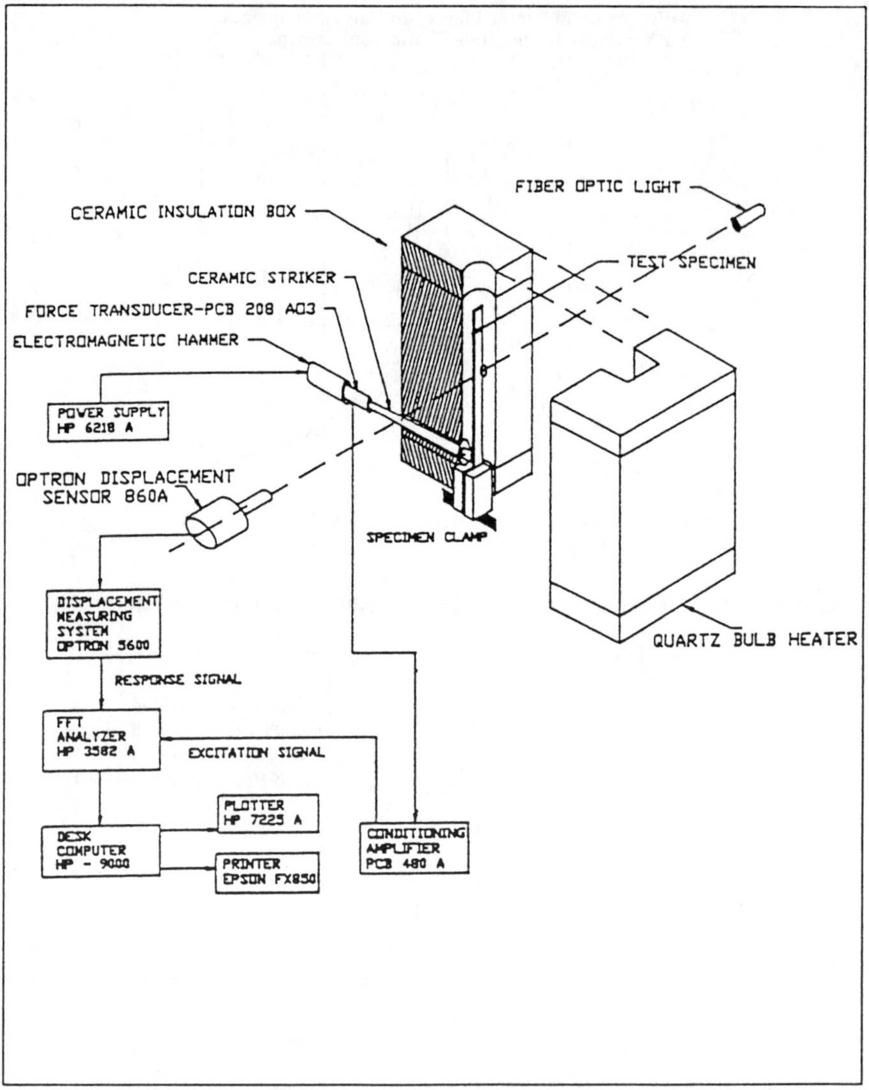

Figure 4. Cutaway Drawing of High Temperature Impulse-Frequency Response Test System [9]

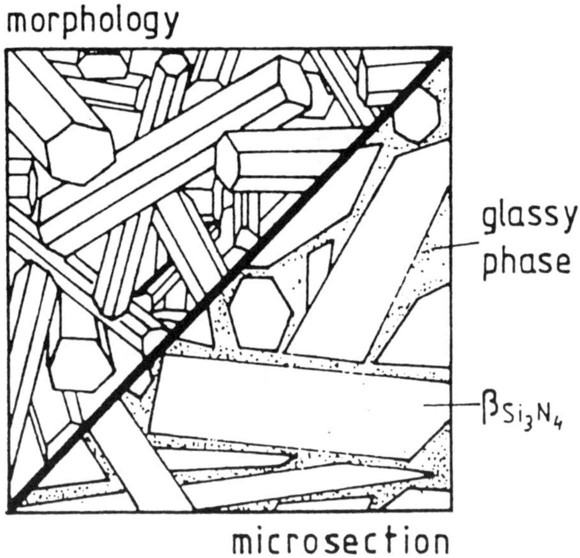

Figure 5. Microstructure of Dense Silicon Nitride from [11]

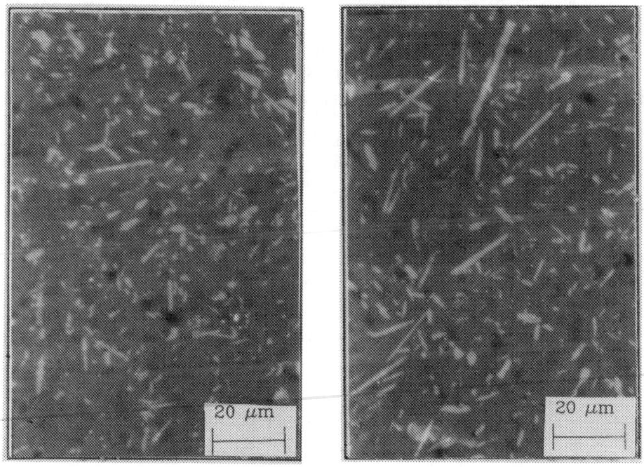

a. KNw/SiC

b. UBEw/SiC

Figure 6. Microstructures of Silicon Nitride with SiC Whiskers from [12]

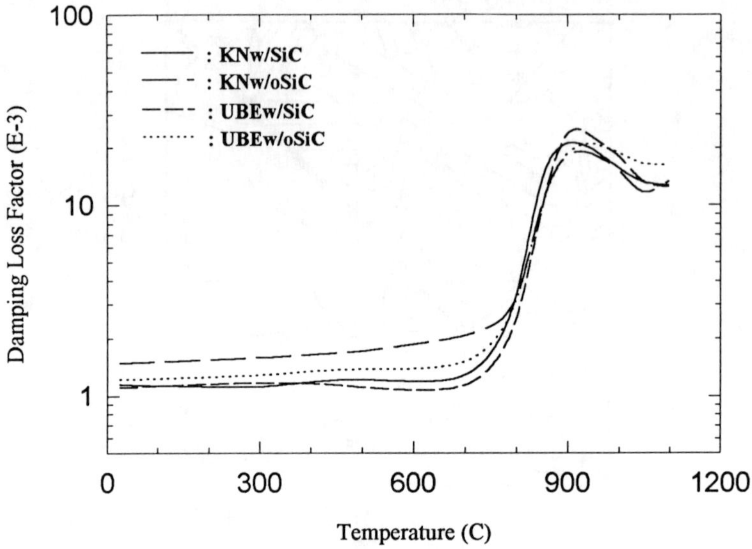

Figure 7. Damping Loss Factor of 1.524 mm Thick Samples vs. Temperature

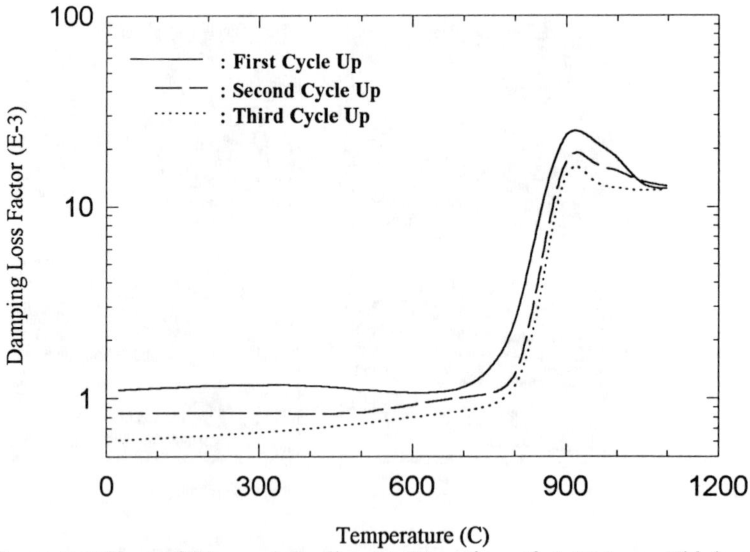

Figure 8. Effect of Thermal Cycling on Damping of 1.524 mm Thick UBEw/SiC Samples

TEM STUDIES OF ALUMINA CHOPPED FIBER/PARTIALLY-STABILIZED ZIRCONIA COMPOSITES

K. Park, *Y. Li, *T. Vasilos, and C. Sung
Center for Advanced Materials and *Department of Chemical and Nuclear Engineering, University of Massachusetts, Lowell, MA 01854

ABSTRACT

Alumina chopped fiber/partially-stabilized zirconia composites were fabricated by hot-pressing at 1550°C and 15.0 MPa in vacuum. The microstructure and chemistry of the composites were investigated using analytical transmission electron microscopy, in order to find out the relationship between the microstructural features and the mechanical properties measured at room- and high-temperatures and to understand the reinforcing effects of the alumina fiber in the matrix at room- and high-temperatures.

INTRODUCTION

Zirconia exists in four crystallographic forms, i.e., cubic, tetragonal, monoclinic, and orthorhombic structures [1,2]. Several additives, such as CaO, MgO, and Y_2O_3, were used to fabricate tough and strengthened zirconia [3-5]. Partially-stabilized zirconia has attracted considerable interest for structural applications because of excellent mechanical properties, such as high fracture strength and toughness. However, the mechanical properties of the zirconia are superior to those of most other ceramics mainly at ambient temperature. With increasing temperature, the mechanical properties of the zirconia are degraded severely because of reduced effectiveness of stress-induced transformation and softening of glassy grain boundary phases.

Fiber-reinforcement may be one of the promising strategies to strengthen zirconia-toughened ceramics at high temperature. In this study, we attempted to improve the high-temperature creep properties by the incorporation of alumina chopped fibers into a yttria doped partially-stabilized zirconia (Y-PSZ). The alumina chopped fiber (10 vol%)/Y-PSZ composites were fabricated by hot-pressing in vacuum. The microstructure and chemistry of the matrix, fiber, and matrix/fiber interface in the composites were characterized using analytical transmission electron microscopy, in order to correlate the microstructural features to the mechanical properties measured at room- and high-temperatures.

To the extent authorized under the laws of the United States of America, all copyright interests in this publication are the property of The American Ceramic Society. Any duplication, reproduction, or republication of this publication or any part thereof, without the express written consent of The American Ceramic Society or fee paid to the Copyright Clearance Center, is prohibited.

EXPERIMENTAL

The materials used in this study were (1) commercial 5.35 wt% yttria doped zirconia powders with an average particle size of 0.43 μm (HYS-3.0, Daichi Kigenso Kagaku Kogyo Co.) and (2) ALMAX alumina chopped fibers with a diameter of 10 μm (Mitsui Mining Co., New York, NY). The elastic modulus and tensile strength of the alumina fibers used are 324 GPa and 1765 MPa, respectively. Chopped fiber agglomerates were dispersed by blending with ethanol. Zirconia powders were then added into the blender and mixed with the fibers (10 vol%) to produce an uniform mixture. The resulting mixture was filtered, dried at room temperature, and then densified by hot-pressing. The applied pressure was ranged to 7.5 MPa at 1000°C and then 15.0 MPa at 1550°C, where it was held for 30 minutes before cooling. The fabricated composites were diamond-machined into rectangular bars ($3 \times 4 \times 55$ mm).

TEM specimens were sectioned from the composites with a low-speed diamond saw and mechanically ground to ~300 μm. Circular discs of 3 mm in diameter were core drilled from the ground sections, mechanically ground to ~120 μm, and then dimpled to ~30 μm. The specimens were ion milled with 5 kV Ar^+ ions at an incident angle of 12° until perforation was achieved. A light carbon film was evaporated on the specimens in order to prevent charging in the electron microscope. The microstructure and chemistry of matrix, fiber, and matrix/fiber interface in the composites were investigated by a bright field image, selected area diffraction (SAD), and energy dispersive X-ray spectroscopy (EDXS) using a Philips EM400T TEM and a Noran ultrathin window Micro-Z detector.

The mechanical properties of the composites were measured at room temperature by four-point bending tests. High-temperature creep properties were also measured by four-point bending tests at 1100°C and bending stress of 46.3 MPa for 24 hours in air.

RESULTS AND DISCUSSION

The composites were found to be densified up to 96.7% of theoretical density. The fibers were oriented slightly in the matrix. Fig. 1(a) shows a bright field image of the alumina fiber and Y-PSZ matrix with equiaxed fine grains. The grain sizes of the fiber and matrix are ~1.0 and ~0.5 μm, respectively. The alumina grains are more transparent than the Y-PSZ grains. A bright field image was obtained at higher magnification in order to find out the nature of the reaction at the interface between the fiber and matrix, as shown in Fig. 1(b), since the mechanical properties of the composites are critically dependent on the microstructure of the composites, especially interfacial regions between the fiber and matrix.

The interface appeared to be sharp, indicating no chemical reaction between the fiber and matrix. In addition, the bonding between two components seems to be relatively strong. Thus, the cracks initiated in the matrix can propagate through the fibers without deflection around the fibers, resulting in catastrophic failure of the composites. Fiber debonding and pullout were rarely observed in

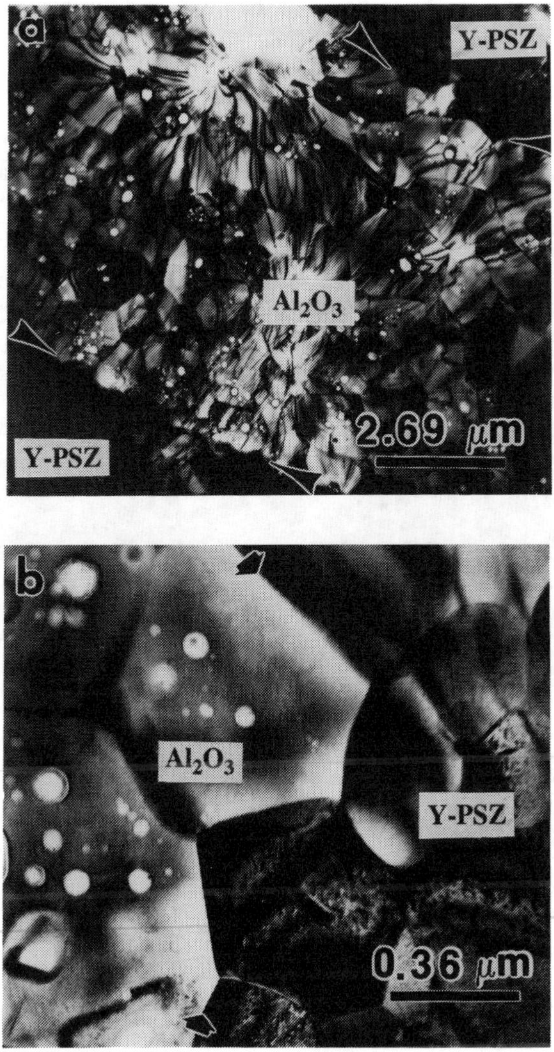

Fig. 1. (a) and (b) Bright field images showing interfacial regions between the alumina fiber and Y-PSZ matrix.

the specimens fractured by the bending tests at room temperature, primarily due to a strong interfacial bonding. Most of the residual pores present in the fibers were generally spherical but some of the pores were faceted. Such residual pores could act as stress concentrators to degrade mechanical properties. It is thus possible that cracks can be initiated at the pores in the fibers. Clustering of the fibers, which acts as a fracture-initiating flaw, was often observed.

In order to characterize the microstructure and chemistry of the Y-PSZ matrix, a bright field image, SAD pattern, and EDXS spectrum were obtained, as shown in Figs. 2(a), (b), and (c), respectively. The microstructure of the matrix was found to consist of tetragonal precipitates (a = 0.508 nm and c = 0.519 nm) dispersed in a cubic matrix (a = 0.513 nm), resulting in tough and strengthened

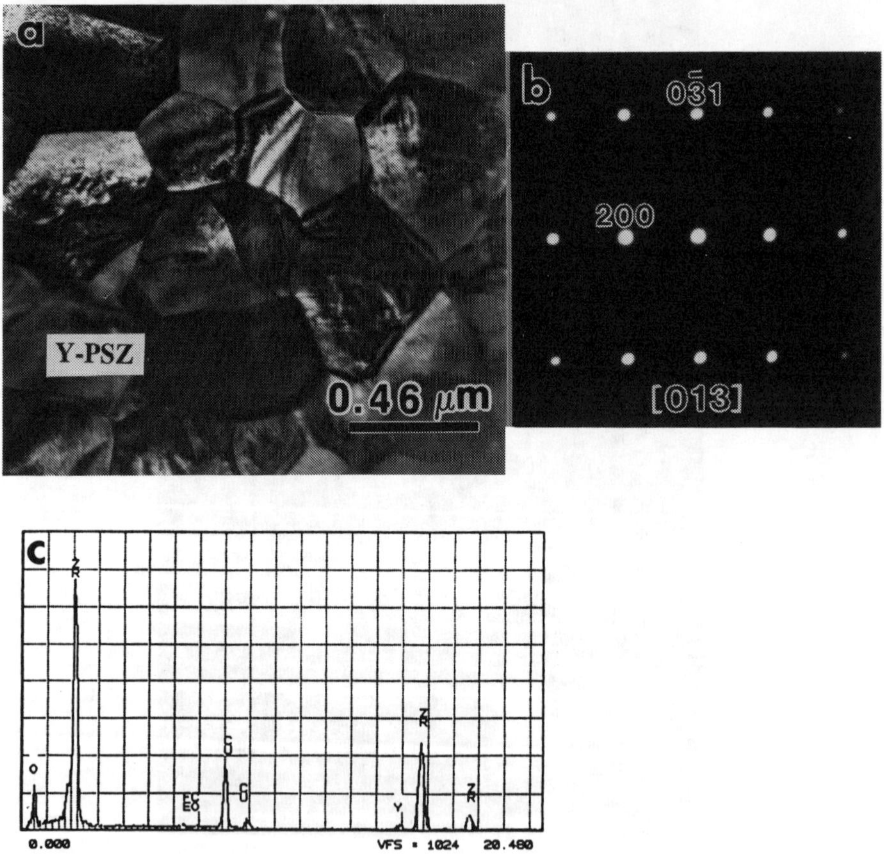

Fig. 2. (a) Bright field image, (b) SAD pattern, and (c) EDXS spectrum obtained from the Y-PSZ matrix.

zirconia. No liquid and second phases were observed at grain boundaries and within grains.

Figs. 3(a) and (b) show a bright field image and its corresponding SAD pattern of the alumina fiber, respectively. The alumina fiber was identified as α–Al_2O_3 with a rhombohedral structure. The segregation of Zr on the surface of pores in the fibers was observed, as shown in Fig. 3(c), showing an EDXS spectrum obtained from the surface of pores in the fibers.

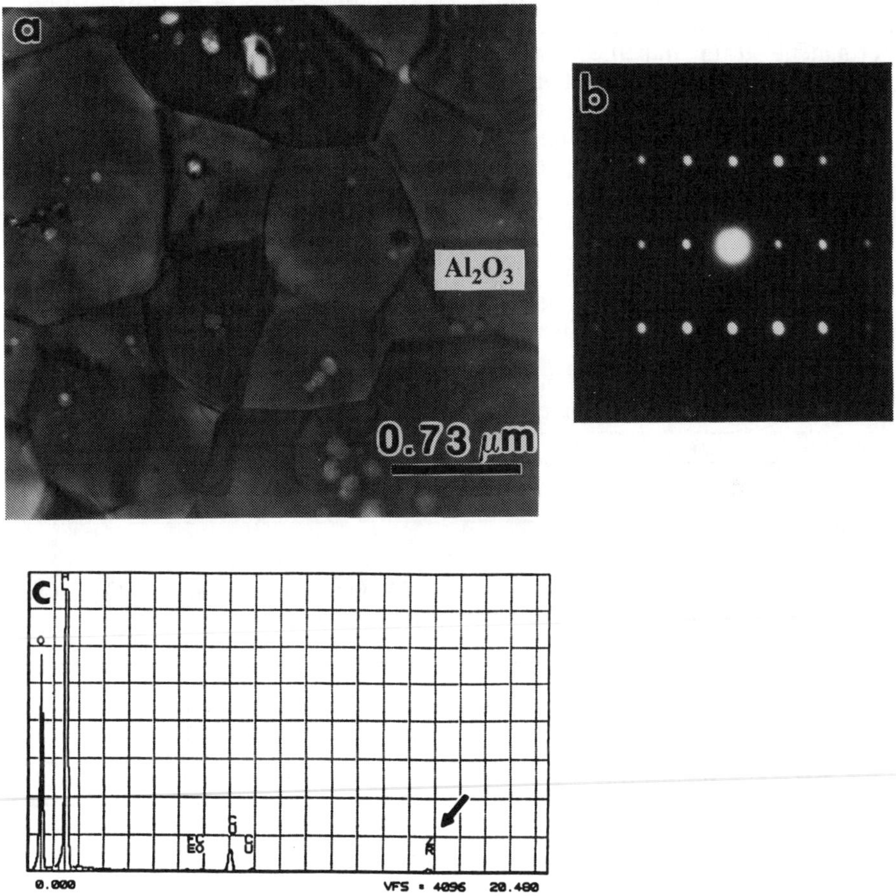

Fig. 3. (a) Bright field image and (b) its corresponding SAD pattern of the alumina fiber. (c) EDXS spectrum obtained from the surface of pores in the fiber.

The flexural strength and strain to failure of the composites and monolithic ZrO_2 measured at room temperature are shown in Table I. The flexural strength and strain to failure of the monolithic ZrO_2 measured at room temperature are nearly twice those of the composites, because of a strong bonding between the matrix and fiber, a high density of residual pores in the fibers, and clustering of the fibers. However, the mechanical properties of the composites and monolithic ZrO_2 measured at high temperature are markedly different from those at room temperature. The strain rate and permanent strain of the composites and monolithic ZrO_2 measured at 1100°C and bending stress of 46.3 MPa for 24 hours in air are shown in Table II. The composites have lower strain rate and permanent strain than the monolithic ZrO_2, indicating the benefit of fiber additions. The improvement in high-temperature creep properties probably results from the fact that the grain boundary sliding in the matrix is hindered by the fibers and the deformation in the matrix is restrained by the fibers. We believe that the alumina chopped fiber/Y-PSZ composites could be a potential candidate for high-temperature structural materials because of good creep resistance and interfacial stability at high temperature.

Table I. Flexural Strength and Strain to Failure of Monolithic ZrO_2 and Al_2O_3/ZrO_2 Composites Measured at Room Temperature.

Specimen	Flexural Strength (MPa)	Strain to Failure (%)	Elastic Modulus (GPa)
Monolithic ZrO_2	1112	0.67	183
Al_2O_3/ZrO_2 Composites	647	0.37	215

Table II. Strain Rate and Permanent Strain of Monolithic ZrO_2 and Al_2O_3/ZrO_2 Composites Measured at High-Temperature Creep Test.

Specimen	Strain Rate (mm/mm hr)	Permanent Strain (%)
Monolithic ZrO_2	8.1×10^{-5}	1.14
Al_2O_3/ZrO_2 Composites	1.9×10^{-5}	0.94

CONCLUSIONS

The incorporation of alumina chopped fibers into a partially-stabilized zirco matrix degraded room-temperature mechanical properties severely but improved hig temperature creep properties, compared to monolithic partially-stabilized zirconia. T interface between the alumina fiber and partially-stabilized zirconia matrix appeared be sharp, indicating no chemical reaction at the interface. The alumina chopp fiber/partially-stabilized zirconia composites could be a potential candidate for high-temperature structural materials.

REFERENCES

[1]. P. M. Kelly and C. J. Ball, "Crystallography of Stress-Induced Martensitic Transformations in Partially Stabilized Zirconia", J. Am. Ceram. Soc. **69**, 259 (1986).

[2]. R. Srinivasan, B. H. Davis, O. B. Cavin, and C. R. Hubbard, "Crystallization and Phase Transformation Process in Zirconia: An in situ High-Temperature X-ray Diffraction Study", J. Am. Ceram. Soc. **75**, 1217 (1992).

[3]. C. E. Curtis, "Development of Zirconia Resistant to Thermal Shock", J. Am. Ceram. Soc. **30**, 180 (1947).

[4]. R. C. Garvie and P. S. Nicholson, "Structure and Thermomechanical Properties of Partially Stabilized Zirconia in the CaO-ZrO_2 System", J. Am. Ceram. Soc. **55**, 152 (1972).

[5]. R. C. Garvie, R. H. Hannink, and R. T. Pascoe, "Ceramic Steel?", Nature **258**, 703 (1975).

Micromechanical Modeling and Design

EFFECT OF FLAW SIZE ON FIRST MATRIX CRACKING STRESS IN FIBER-REINFORCED CERAMIC COMPOSITES

S.Kumaria and R.N.Singh, Department of Materials Science and Engineering, University of Cincinnati, Cincinnati, OHIO 45221-0012 and R.T. Bhatt, NASA- Lewis Research Center, Cleveland, OHIO 44135.

ABSTRACT

Influence of flaws formed during processing or while in service on the first matrix cracking stress is important in assessing and predicting the performance and reliability of a ceramic composite. In the present investigation, the effect of change in the crack length on the first matrix cracking stress of ceramic matrix composites is studied. Silicon nitride and zircon matrix composites, uniaxially reinforced with SiC fibers were fabricated along with the corresponding monolithic silicon nitride and zircon. These composites and monolithic samples were microindented to create flaws of controlled size on the surface and were then tested in 3-pt. flexure mode. The experimental results clearly indicated an evidence of flaw-size dependent non-steady state and flaw size independent steady state first matrix cracking behaviors. These observations on the first matrix cracking behaviors were compared with the theoretical predictions.

1. INTRODUCTION

Ceramic matrix composites exhibit high load bearing capability and toughened behavior because of the reinforcing action of the strong fibers present in them. This behavior is in contrast to the mechanical response of the monolithic ceramic materials that undergo catastrophic failure upon application of stress. Consequently, they have low toughness and fracture stress in comparison to ceramic composites.

To the extent authorized under the laws of the United States of America, all copyright interests in this publication are the property of The American Ceramic Society. Any duplication, reproduction, or republication of this publication or any part thereof, without the express written consent of The American Ceramic Society or fee paid to the Copyright Clearance Center, is prohibited.

Even though ceramic matrix composites can resist high loads prior to total failure, the initiation of mechanical damage occurs at relatively low loads. This can be seen in the form of evenly spaced matrix cracks in uniaxially reinforced composites. The stress at which the matrix fractures for the first time is defined as the first matrix cracking stress of the composite. A knowledge of the first matrix cracking stress in composites is important because it represents the onset of composite failure and irreversible damage. Also, the fibers get exposed to processes like oxidation and corrosion which degrade mechanical properties[1].

Realizing the significance of first matrix cracking stress, theoretical work has been done to predict the effect of flaws (which may occur during processing or when in use) and other material properties on the matrix cracking stress [1,2]. The results from these investigations are summarized in Figure 1, which shows the dependence of first matrix cracking stress on the crack length [1]. It can be seen that in a monolith, the fracture stress decreases continuously as a function of the crack size. In contrast, for a composite, the first matrix cracking stress decreases up to a certain value of the crack length. This region is referred to as the non-steady state region. Beyond a certain value of the crack length, the first matrix cracking stress becomes independent of the crack length. This value of the matrix cracking stress is called the steady state matrix cracking stress and is defined by Marshall *et al.* [1] as,

$$\sigma_o = k\left[\left(1-\nu^2\right)K_c^M \tau E_f V_f^2 V_m (1+\eta)^2 / E_m R\right]^{1/3} \tag{1}$$

where k is a constant, ν is Poisson's ratio of the composite, K_c^M is the fracture toughness of the matrix, τ is the interfacial shear strength of the interface between the fiber and the matrix, E_f and E_m are the Young's moduli of the fiber and the matrix, V_f and V_m are the volume fractions of the fiber and the matrix, $\eta = E_f V_f / E_m V_m$ and R is the radius of the fiber.

The crack length corresponding to σ_o is approximately equal to one third of the normalized crack length, C_m, which is defined as,

$$C_m = k'\left[K_c^M E_m V_m^2 (1+\eta) R \Big/ \tau E_f V_f^2 \left(1-\nu^2\right)\right]^{2/3} \tag{2}$$

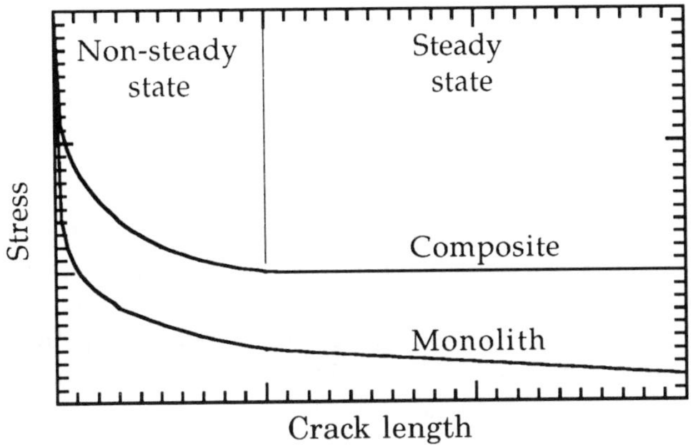

Figure 1. First matrix cracking stress as a function of the crack size.

This prediction of crack size independent first matrix cracking stress has important practical implications. It means that the composite properties become independent of the crack size and thus the presence of large flaws in a material has a limited effect on the load-bearing capabilitity of a composite.

Recognizing the importance of this prediction, the first objective of the present investigation was the fabrication and characterization of ceramic matrix composites to check the validity of the above predictions of the non-steady state and steady state behaviors in ceramic composites.

Also, the experimental values of first matrix cracking stress measured in an earlier investigation are different from the theoretical values [3]. One reason for this discrepancy could be that the size of pre-existing flaws in an as-processed composite is much smaller than the critical crack length needed to satisfy the steady state criterion. Thus, the experimental results correspond to the non-steady state region rather than the steady state region. To resolve this discrepancy between the experimental and theoretical results, the second objective was to measure the first matrix cracking stress of composites in the steady state region, by creating flaws of controlled size, and to compare with the corresponding theoretical values.

The micro-indentation technique was used to create flaws of controlled size in the composites and monolithic ceramic samples. These samples containing flaws of controlled size were tested by the 3-point bend method to obtain the first matrix cracking stress and fracture stress, respectively, for the composite and monolithic samples, and to demonstrate the influence of the flaw size on first matrix cracking strength.

2. EXPERIMENTAL

The two composite systems used were zircon and silicon nitride composites uniaxially reinforced with as-supplied silicon carbide fibers (Textron-SCS-6). The method of fabrication of the zircon matrix and silicon nitride matrix composites is described, respectively, in Singh and Gaddipati [4] and Chulya *et al.* [5]. The monolithic and composite samples were first ground to ensure that the two faces were parallel to each other. The samples were then cut to a width of 4mm and a length of 50mm. The thickness of the samples was 2.4mm. This was followed by polishing with diamond pastes of progressively increasing fineness, leading to a final polish with a 1μm size.

The polished samples were then indented with a Vickers diamond micro-indenter to create cracks at the center of the samples. A load of 3kg was chosen to create the cracks. This load was chosen because it resulted in well defined and symmetric crack patterns having consistent crack lengths. The size of the flaw on the sample surface was controlled by varying the number of indentations placed next to each other. A constant and optimum center-to-center distance was maintained between the indentations to ensure proper linking of cracks between adjacent indentations. The indentations were placed in such a way that the two diagonal cracks formed by the indentation were parallel and perpendicular to the fiber axis. A schematic of the cracks formed on a sample is shown in Figure 2. The total crack length as a result of the indentations is referred to as C_t in Figure 2. Flaws of the same size were created on the composite as well as the corresponding monolithic sample.

These samples containing cracks of different lengths were then tested in the 3-point bend flexure mode using a mechanical testing machine. A support span of 40mm and a cross-head speed of 0.2mm/min. was used for all the tests. Data acquisition was done at a rate of 10 pts./sec.

3. RESULTS AND DISCUSSION

A typical load-displacement curve for a silicon carbide fiber-reinforced silicon nitride composite sample is shown in Figure 3. The first matrix cracking occurs at the point where the load-displacement curve

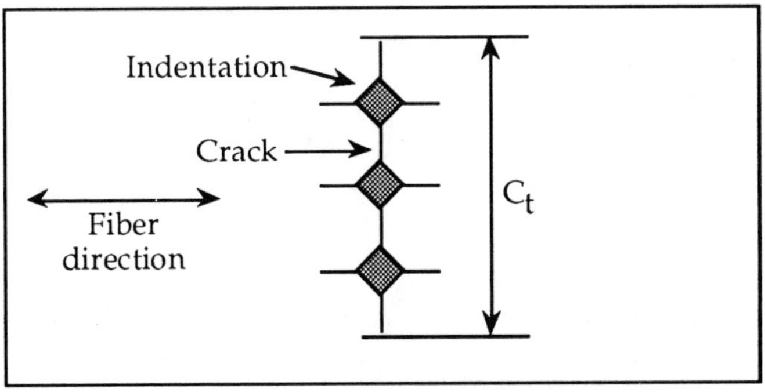

Figure 2. Schematic of equidistant indentations on a sample to create a flaw of total length C_t

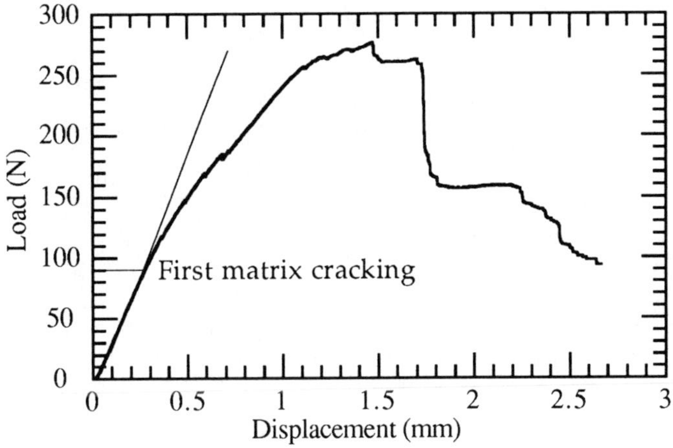

Figure 3. A typical load-displacement curve for a silicon nitride composite sample showing the point of first matrix cracking.

deviates from linearity. Figure 4a shows an SEM picture of a line of indentations that are joined by cracks. Figure 4b shows another set of indentations after the sample was tested by the 3-point bend method. It can be seen clearly that the matrix cracking occurs along the cracks created by the indentation method. The first matrix cracking stress was calculated from the load at which the deviation from linear behavior was observed.

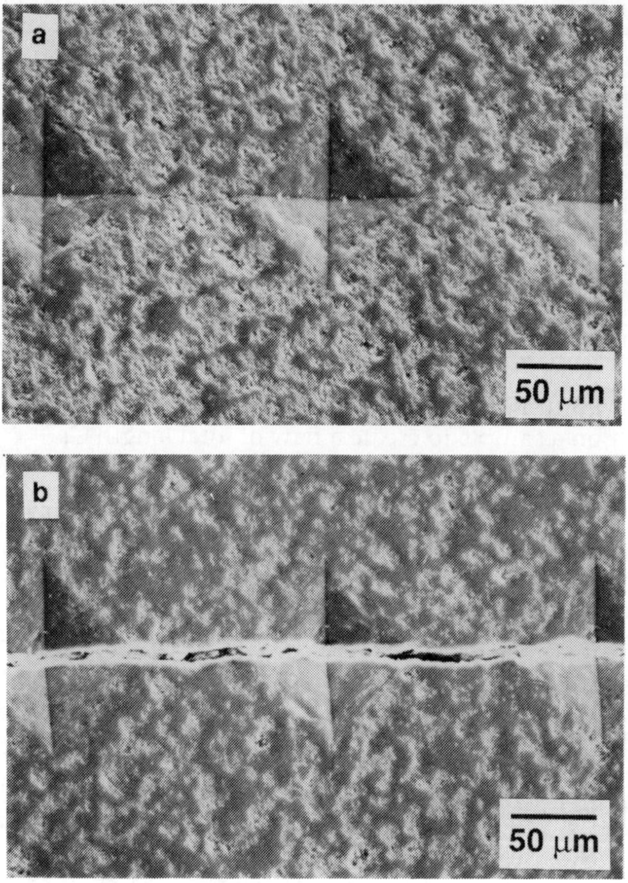

Figure 4. SEM pictures showing indentations on a silicon nitride composite sample before 3-point bent test. The indentations are joined together by well formed cracks (a). Same set of indentations after the 3-point bend test (b).

The values of the calculated first matrix cracking stress and fracture stress for the Si_3N_4/SiC composite and the monolithic silicon nitride, respectively, are plotted as a function of the crack length in Figure 5. It can be seen that the first matrix cracking stress decreases with increasing crack length for lower values of the crack size.

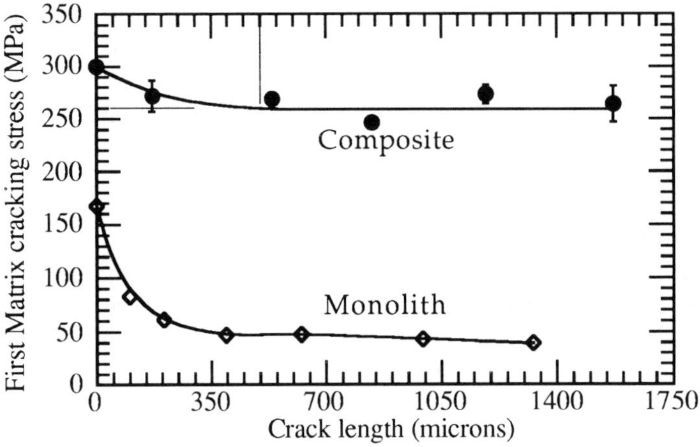

Figure 5. Dependence of first matrix cracking stress on crack length for the Si_3N_4/SiC composite.

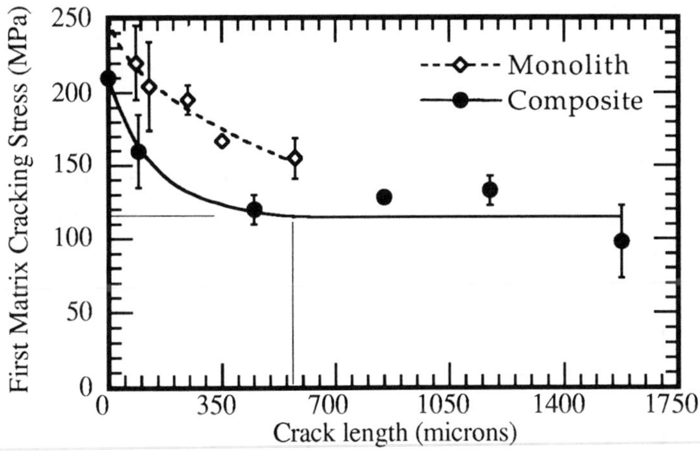

Figure 6. First matrix cracking stress as a function of crack length in the zircon/SiC composite.

Beyond a certain crack length, the matrix cracking stress becomes independent of the crack length. In contrast, the fracture stress of the monolithic silicon nitride shows a continued dependence on crack length and decreases with increasing crack length.

The results for the zircon matrix composite and the monolithic zircon are shown in Figure 6. Again, it can be seen that the first matrix cracking stress exhibits dependence on the crack size in the non-steady state region and becomes crack size independent in the steady state region. The fracture stress values for the monolithic zircon in the given crack size range are higher than the first matrix cracking stress values, even though they steadily decrease with increasing crack size. A possible reason for the higher values for the monolithic zircon could be the difference in processing conditions for the monolith and the composite, which may result in lower density of the composite in comparison to the monolith density.

The observed behavior of the first matrix cracking stress with varying crack size is in good agreement with the theoretical predictions of Marshall *et al* [1] and Danchaivijit & Shetty [2].

The experimentally obtained values of the steady state matrix cracking stress (σ_o) and the corresponding crack length ($C_m/3$) are compared with the theoretical values in Table I. The experimental steady state matrix cracking stress for the zircon composite is 112 MPa and this is in good agreement with the theoretical value of 118 MPa. The experimental value of 570 µm for $C_m/3$ is lower than the theoretical value of 1000 µm.

Table I. Comparison of the theoretical and experimental values of steady state matrix cracking stress (σ_o) and the critical crack length ($c_m/3$)

Composite	Steady state stress, σ_o (MPa)		Critical crack length, $C_m/3$ (µm)	
	Theoretical	Experimental	Theoretical	Experimental
Zircon/SiC	112	118	1000	570
Si_3N_4/SiC	148	260	470	500

In the case of silicon nitride composite, the difference between the experimental (260 MPa) and calculated (148 MPa) values of steady state stress is large, whereas the experimental (500 μm) and calculated (470 μm) values of $C_m/3$ are in good agreement.

The discrepancy between the experimental and the theoretical results could be due to a number of reasons. First, the theoretical calculations have been done under condition of tensile loading whereas the experimental work has been done in the 3-point flexure mode. Second, the effects of residual stresses and porosity that result from processing have not been considered. Further study is underway to characterize the influence of these parameters on the first matrix cracking behavior.

4. CONCLUSIONS

1. The technique of micro-indentation has been effectively used to create flaws of controlled size in ceramic matrix composites and monolithic ceramic materials.

2. It has been experimentally verified that the ceramic matrix composites exhibit non-steady state and steady state matrix cracking behaviors, which is in agreement with the theoretical predictions.

3. The first matrix cracking stress in ceramic matrix composites becomes flaw size independent above a critical flaw size.

REFERENCES

1. D.B. Marshall, B.N. Cox and A.G. Evans, "The Mechanics of Matrix Cracking in Brittle-Matrix Fiber Composites," *Acta Metall.*, **33**[11] 2013-2021 (1985).

2. S. Danchaivijit and D. K. Shetty, "Matrix cracking in Ceramic-Matrix Composites," *J. Am. Ceram. Soc.*, **76** [10] 2497-2504 (1993).

3. R. N. Singh, "Influence of Interfacial Shear Stress on First-Matrix Cracking Stress In Ceramic-Matrix Composites," *J. Am. Ceram. Soc.*, **73** [10] 2930-37 (1990).

4. R.N. Singh and A.R. Gaddipati, "A uniaxially reinforced zircon-silicon carbide composite," *J. Mater. Sci.*, **26** 957-962 (1991).

5. A. Chulya, J.P. Gyekenyesi, and R.T. Bhatt, "Mechanical behavior of fiber reinforced SiC/RBSN ceramic matrix composites: Theory and experiment," Am. Soc. Mech. Eng. (Paper). Publ. by ASME, New York, NY, USA. 17p.

THE PHYSICAL PROPERTIES OF MGO/SIC COMPOSITES: COMPARISON BETWEEN A MICROMECHANICS ANALYSIS AND EXPERIMENTAL RESULTS.

J. Luo, R. Taylor * and R. Stevens, School of Materials, University of Leeds, Leeds, LS2 9JT, U.K. *Dept. of Materials, University of Manchester.

MgO exhibits inferior thermal shock resistance as a consequence of poor mechanical properties combined with a large thermal expansion coefficient. Addition of SiC particles, having a high thermal conductivity, would be expected to alter the physical properties of the composites. Relevant physical properties, such as elastic modulus, thermal expansion coefficient and thermal conductivity, have been modelled. Experimental results have been compared with the computed evaluations. Good agreement has been obtained when factors such as remanent porosity and the interfacial structure between the SiC particles and the MgO are taken in consideration.

INTRODUCTION.

MgO is a good refractory material, but suffers from the poor thermal shock resistance due to its poor mechanical properties and high thermal expansion coefficient. SiC particles, widely used as a reinforcement phase in ceramic inclusion composites, has been chosen as the second phase when preparation of SiC particle reinforced MgO composites was undertaken, since SiC has a low thermal expansion coefficient and high thermal conductivity, both of which are of benefit to thermal shock resistance. Furthermore, addition of SiC particles is known to significantly improve the mechanical properties of MgO/SiC composites[1].

In this study, the physical properties of the MgO/SiC composites, such as Young modulus, thermal expansion coefficient and thermal conductivity are examined. The experimental results are compared with computed values using a micromechanics analysis based on a modified Eshelby model[2].

EXPERIMENTAL TECHNIQUES.

SiC particle reinforced MgO composites were prepared using hot pressing, at temperatures in the range 1500-1750°C and pressures of 10-25MPa. The hot pressing temperature and pressure used depends on the volume fraction of the SiC incorporated, which is in the range 0-40vol%. The fabrication procedure has been described in an earlier paper[1].

Young modulus of the composites was measured using a resonant frequency technique, and the thermal expansion coefficient obtained using dilatometry in the temperature range 100-850°C. The average value over this range was then calculated.The thermal diffusivity of the composites was measured in the temperature range of 200-1000°C using a laser flash technique.

To the extent authorized under the laws of the United States of America, all copyright interests in this publication are the property of The American Ceramic Society. Any duplication, reproduction, or republication of this publication or any part thereof, without the express written consent of The American Ceramic Society or fee paid to the Copyright Clearance Center, is prohibited.

RESULTS AND DISCUSSION.

Young Modulus.

The Young modulus values of the composites are listed in Table 1 and also shown in Figure 1, related to the volume fraction of the SiC. The experimental results indicated that the Young modulus of the composite slightly increased with addition of the SiC due to the higher Young modulus of the inclusions. The calculations, using micromechanics were made on the basis of fully dense composites, as shown in Figure 1. The predictions are higher than the experimental values, by 5-10%. This deviation is ascribed to the effect of the pores present in the fabricated composite. The relative density of the composites is in the range 97-99%, Table 1. The pores present will significantly decrease the elastic modulus of the composite material and this effect needs to be taken into account. Previousely, the effect of porosity on the elastic modulus of materials using the micromechanics has been calculated [2,3] and compared with the available data of some ceramics. The effect of the pores on the elastic moduli is significant, in that simply assuming the pores as being spherical and evaluating the effect using micromechanics causes a high deviation. Assuming the pores to be randomly oriented oblate sections, with a radius relationship: $a_1=a_2=10a_3$ we can obtain a good agreement between calculation and experiment, particularly when the volume fraction of the pores is less than 10% [2,3].

The effect of the pores in the present study was calculated using micromechanics, based on the materials having randomly oriented oblate ellipsoidal pores. The equation used for the calculation is as follows[2]:

$$E_p = \frac{(1-V_p)}{1+(1.95+4.34\nu)V_p} E \qquad (1)$$

where E_p and E are the Young moduli of materials with and without pores, respectively. V_p is the volume fraction of pores and ν is Poisson's ratio for the material. The total contribution of the SiC inclusion and the pores on the Young modulus of the composites was calculated and listed in table 1and also shown in Figure 1. As can be seen, after the effect of the porosity is taken into account, the calculated values are close to the experimental results, except for the 40 vol% SiC composite, in which the calculated value is 6% higher than the experimental one. This can be explained by the presence of microcracks in the composite [4]. Microcracks when present in materials have a significant effect on the decrease of the elastic modulus[2,5].

Thermal expansion coefficients

The experimental results for the thermal expansion coefficients (TEC's) of the composites are shown in Figure 2, related to the volume fraction of the SiC particles. Addition of the SiC steadily decreased the TEC of the composites. Micromechanics was used to calculate the thermal expansion coefficients values. According to the analysis made, the thermal expansion coefficient of the materials is unaffected by the presence of pores , but is influenced by the bulk modulus of the inclusions as follows[2]:

$$\alpha_C = (1-V_f)\alpha_M + V_f\alpha_I + F(V_f, \lambda) \tag{2}$$

where α_C, α_M and α_I are the TEC's of the composite, matrix and inclusion, respectively. V_f is the volume fraction of the inclusions. $F(V_f, \lambda)$ is a function of V_f and the ratio of the bulk modulus between the inclusions and the matrix, λ. When the bulk modulus of the inclusions is the same as that of the matrix, $F(V_f, \lambda)$ is zero, and the TEC of the composites satisfy the simple rule of the mixture. consequently, the TEC's of the composites were calculated by considering the effect of the difference of the bulk modulus between the matrix and the inclusions and ignoring the effect caused by the difference of the bulk modulus (or the rule of mixture). Both calculated results are shown in Figure 2. Good agreement is clearly shown between the experimental results and the predictions made by micromechanics. When the effect of the difference in the bulk modulus is not taken into account, a slight deviation is observed and this deviation increases with the increasing volume fraction of the SiC.

Thermal diffusivity and conductivity

Figure 3, shows the thermal diffusivity of the MgO/SiC particulate composites, measured over the temperature range 200-1000°C. The thermal diffusivities of all of the composites decrease with increasing temperature, since the thermal conduction mechanism of the two components, MgO and SiC, is dominated by phonon excitation. Addition of the SiC particle second phase has not had a significant effect on the thermal diffusivities of the composites.

The thermal conductivities (t^C) of the materials can be calculated by following equation[6]:

$$t^C = C_p \rho \upsilon \tag{3}$$

where C_p is the specific heat of materials; ρ is the density and υ the thermal diffusivity of materials. Since the specific heat of materials is not strongly dependent on the microstructure, C_p for the composite was calculated using the rule of the mixture based on the specific heats of MgO and SiC obtained from the literature[6,7]. The density of the composites at room temperature was used in the calculation, since the density decrease is less than 0.5% from room temperature to 1000°C. Finally, the thermal conductivity (TC) is calculated by Equation 3. The results are shown in Figure 4. The TC's of the composites decrease with increasing temperature by an inverse power relationship.

The thermal conductivity of the inclusion composites can also be calculated using the equivalent inclusion method. Considering a fully-dense composite with an isothermal interface, a prediction of the TC was made [8]:

$$t^C = \frac{3+(\lambda-1)(1+2V_f)}{3+(\lambda-1)(1-V_f)} t^M \quad (4)$$

where λ is the ratio of the TC's between the inclusions and the matrix, and t^M is the TC of the matrix. Since SiC has a higher TC than MgO, the predicted TC's of the composites will be higher than that of the MgO by Equation 4. The experimental results did not show this to be the case (see Figure 4). Porosity is an important factor which can decrease the TC of materials [9,10] and should be taken into account. Micromechanics can also be used to calculate the effect of pores present on the TC by the equivalent inclusion method and the relationship between pores and the TC of the material is given by [2]:

$$t^P = t^0 (1- V_p)/(1+2.13V_p) \quad (5)$$

where t^P and t^0 are the TCs of materials with and without pores, respectively. After the effect of the pores is considered, the calculated values of the TC in the composites are still higher than the experimental results. This further difference between the calculation and experiment is ascribed to the effect of the microstructure, since the calculation is based on the assumption of the isothermal interfacial condition.

Several factors contribute the effect of the microstructure on the TC of the inclusion composites. Firstly, there are interfacial thermal barriers present, which have been widely observed in ceramic composites[11-15].The interfacial debonding or microcracking was rarely observed in the MgO/SiC particulate composites, but other factors could well result in phonon scattering taking place on the interface. A large residual strain was generated in the MgO/SiC composites due to the mismatch of the TEC's. An investigation using high resolution electron microscopy demonstrated that lattice distortion and misfit occurred at the interface which will have increased the phonon scattering. In the presence of high shear strain levels the number of defects would be expected to increase in both the matrix and the inclusions, increasing the phonon scattering. A significant residual strain which was reflected in the lattice interspacing change could also result in phonon scattering.

Another factor which could decrease the thermal conductivity of the composites is the formation of precipitates and solid solutions[10,16]. TEM results indicate that lattice distortions and dislocations were present near the precipitates. In addition, the calculation was based on the thermal conductivity of pure SiC. In fact, the presence of impurities, present in all commercial grade material might significantly decrease the thermal conductivity of the SiC itself,[17,18].

An interfacial condition parameter was defined to express the effect of the microstructure on the thermal conductivity of particulate composites by the following equation[19,20]:

$$t^C = \frac{2\lambda(1-V_f)+\beta[3+(\lambda-1)(1+2V_f)]}{\lambda(2+V_f)+\beta[3+(\lambda-1)(1-V_f)]} t^M \quad (6)$$

where β is a constant which represents the interfacial conditions. When $\beta=0$, the interfacial condition is adiabatic, such as when interfacial debonding has taken place. $\beta=\infty$ represents the isothermal interfacial condition. Therefore β is in the range $0 - \infty$. An accurate value of β can only be obtained by experiment due to the complications arising out of the interfacial effects. Figure 5 shows the comparison between the experimental and calculated results. It can be seen that at $\beta=1$ the computed results are in good agreement with the experimental results.

The experimental results of two 20vol% SiC particle composites, MS3 and MS01, are shown in Figure 5c. The SiC particles in specimen MS01 have been thermally treated at 900°C for 1 hour to form a 10nm thickness oxide layer, before mixing with MgO powder, the thermal conductivity of MS01 being 10% lower than that of MS3. The decrease can be attributed to the formation of precipitates and the corresponding lattice distortion development at the interface. TEM results have demonstrated that many more precipitates of Mg_2SiO_4 are formed in the MS01 specimen[2]. Such precipitates also increase the lattice distortion and misfit strain in MgO grains. The interface between the SiC and the Mg_2SiO_4 also becomes distorted, eventually resulting in debonding which was observed at the interface between the SiC and the precipitates[2].

CONCLUSIONS

Addition of SiC particles increases the elastic modulus of the MgO-SiC composites. The SiC content and the presence of pores dominate the effect. The thermal expansion coefficients of the composites decrease significantly with increasing the content of the SiC particles due to their low value of TEC. The experimental results were in good agreement with the calculated values.

The thermal diffusivity and the thermal conductivity of MgO/SiC composites decrease with temperature in the range of 200-1000°C. Addition of SiC particles slightly decreases the thermal conductivity of the composites, even though pure SiC has a much higher thermal conductivity than the matrix. The decrease is attributed to the interfacial structure, defects, residual strain and precipitates in the composites. All of these factors can be expressed in terms of an interfacial condition parameter. When the effect of the interfacial condition and porosity is taken into account, the predictions made using micromechanics have been found to give good agreement with the experimental results.

REFERENCES.

[1] J. Luo and R. Stevens, Crack stopping mechanisms in MgO/SiC composites, in " The proceedings of TMS", 1993
[2] J. Luo, Ph.D. Thesis "Micromechanics of composites and MgO/SiC composites", University of Leeds, 1994
[3] J. Luo and R. Stevens, Thermal and mechanical stress fields in particulate composites,
[4] J. Luo and R. Stevens, Residual stress and microcracking in MgO/SiC particulate composite, J. Europ. Ceram. Soc., 12, 369-375, (1993)
[5] S. Nemat-nasser and M. Hori, ed. " Micromechanics: Overall properties of heterogeneous materials", Elsevier Science Publishers, 1993
[6] R. Morrell, ed.," Handbook of properties of technical and engineering ceramics, part I", Her Majesty's stationery Office, London, 1985
[7] Y. S. Touloukian, R. W. Powell, C.Y. Ho, and P. G. klemens ed." Thermal conductivity: nonmetallic solids", IFI, Plenum Press, New York, 1970
[8] H. Hatta and M. Taya, Equivalent inclusion method for steady state heat conduction in composites, Intl. J. Eng. Sci., 24, 1159-1172, (1986)
[9] L. M. Russell, L.F. Johnson, D.P.H. Hasselman, and R. Ruh, Thermal conductivity/diffusivity if silicon carbide-whisker-reinforced mullite, J. Am. Ceram. Soc., 70, C226-229, (1987)
[10] L. D. Bentsen, D. P. H. Hasselman and R. Ruh, Effect of hot pressing temperature on the thermal diffusivity/conductivity of SiC/AlN composites, J. Am. Ceram. Soc., 66, C40-41, (1983)
[11] H. Bhatt, K. Y. Donaldson, D. P. H. Hasselman and R. T. Bhatt, Role of the interfacial thermal barrier in the effective thermal diffusivity/ conductivity of SiC-fiber-reinforced reaction-bonded silicon nitride, J. Am. Ceram. Soc., 73, 312-16, (1990)
[12] R. Ruh, L. D. Bentsen, and D. P. H. Hasselman, Thermal diffusivity anisotropy of SiC/BN composites, J. Am. Ceram. Soc., 67, C83-c84, (1984)
[13] L. M. Russell, L. F. Johnson, D. P. H. Hasselman and R. Ruh, Thermal conductivity/diffusivity of silicon carbide whisker reinforced mullite, J. Am. Ceram. Soc., 70, c226-229, (1987)
[14] S. C. Beecher, R. E. giedd, D. G. Onn, R. M. Anderson and J. B. Wachtman, Thermal properties of mullite-cordierite composites, in "Thermal conductivity 20", ed. by, D. P. H. Hasselman and J. R. Thomas, Jr., Plenum Press, New York, 1989, PP119-127
[15] D. P. H. Hasselman, Thermal diffusivity and conductivity of composites with interfacial thermal contact resistance, in "Thermal conductivity 20", ed. by, D. P. H. Hasselman and J. R. Thomas, Jr., Plenum Press, New York, 1989, PP405-413
[16] H. Buhr, G. Müller, H. Wiggers, F. Aldinger, P. Foley and A. Roosen, Phase composition, oxygen content, and thermal conductivity of $AlN(Y_2O_3)$ ceramics J. Am. Ceram. Soc., 74, 718-23, (1991)
[17] L. M. Russell, K. Y. Donaldson, D. P. H. Hasselman, N. D. Corbin, J. J. Petrovic and J. F. Rhodes, Effect of vapor-liquid-solid and vapor-solid silicon carbide whiskers on the effective thermal diffusivity/conductivity of silicon nitride composites, J. Am. Ceram. Soc., 74, 874-77, (1991)

[18] P. H. McCluskey, R. K. Williams, R. S. Graves and T. N. Tiegs, Thermal diffusivity/conductivity of alumina-silicon carbide composites, J. Am. Ceram. Soc., 73, 461-64, (1990)
[19] Y. Benveniste and T. Miloh, The effective conductivity of composites with imperfect thermal contact at constituent interfaces, Int. J. Eng. Sci., 24, 1537-52, (1986)
[20] Y. Benvensite, Effective thermal conductivity of composites with a thermal contact resistance between the constituents: nondilute case, J. Appl. Phys., 61, 2840-2843, (1987)

Table 1.
Young modulus of SiC particle reinforced MgO composites as measured using the resonant frequency technique and calculated by micromechanics.

Sample No	SiC (vol%)	Relative density (%)	Experi-mental results (GPa)	Theoretical values (GPa
MgO	0	98	282	282
MS1	5	98	290	288
MS2	10	96.9	290	285
MS2	10	98.3	294	297
MS3	20	99.1	317	317
MS3	20	98.1	309	307
MS4	40	99.1	322	342

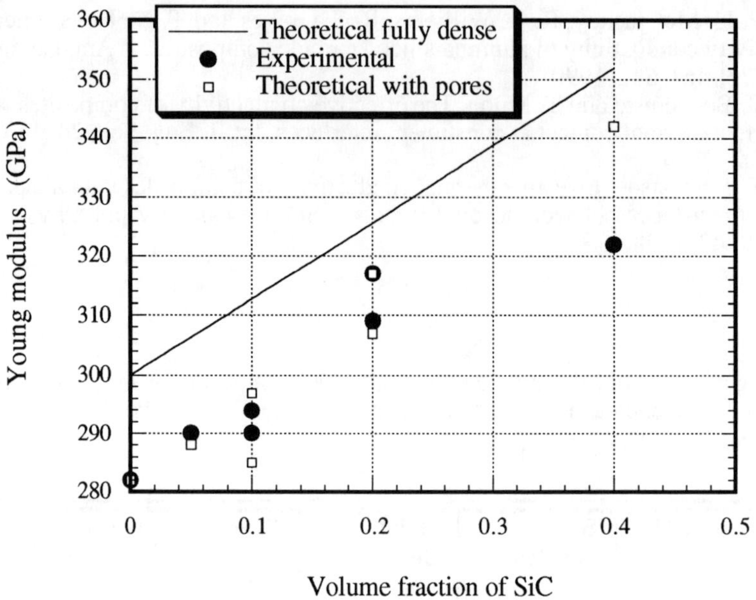

Figure 1

The experimental results of Young modulus related to volume fraction of SiC particles, compared with the evaluation by micromechanics

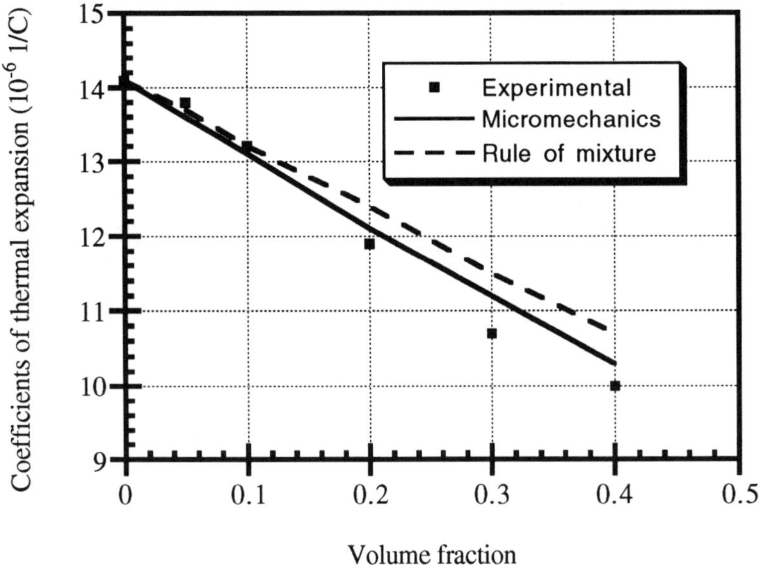

Figure 2

The Comparison between experimental and calculated results of the thermal expansion coefficients in MgO/SiC particulate composites

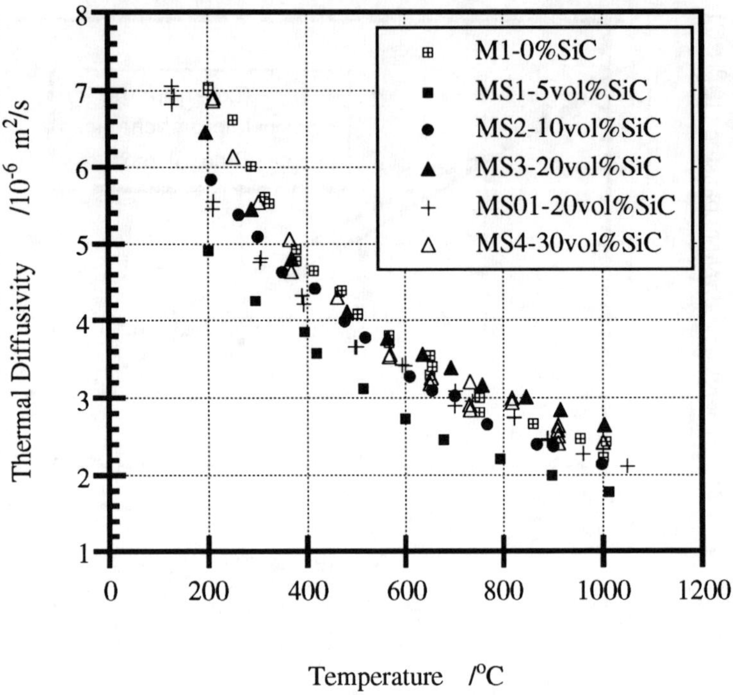

Figure 3

The thermal diffusitivity of MgO/SiC composites related to the temperature and the volume fraction of SiC particles.

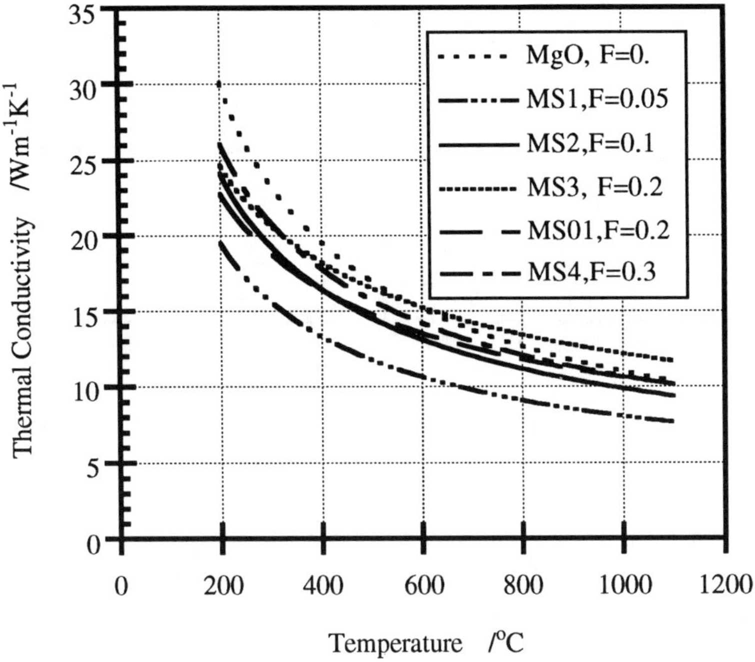

Figure 4

The thermal conductivity of MgO/ SiC composites related to the temperature and the volume fraction of SiC particles.

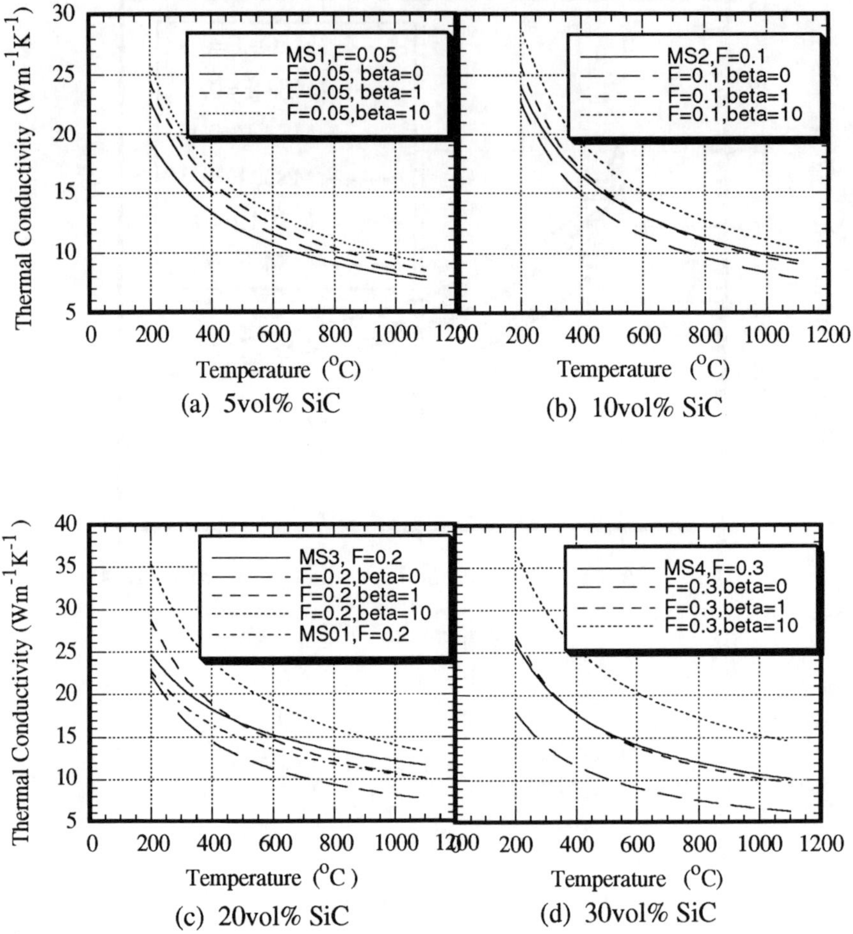

Figure 5

The comparison of thermal conductivities between the experimental and theoretical results in MgO/SiC particulate composites

INFLUENCE OF INTERFACIAL DEBONDING ON THE BEHAVIOR OF A LAMINATED SYSTEM

G. P. Tandon, AdTech Systems Research Inc., 1342 North Fairfield Rd, Dayton, OH 45432 & N. J. Pagano, WL/MLBM, Wright Patterson AFB, OH 45433

ABSTRACT

The influence of fiber-matrix interfacial debonding upon laminate behavior is examined through an analytical study for a glass matrix composite (Nicalon/1723 glass). Indirect evidence supports the notion of "micro-laminate effect" indicating that ply or layer elastic properties may not be treated as unique quantities, but are strong functions of interface behavior, which, in turn, is dependent on macroscopic state of stress. A recently developed variational model in conjunction with classical lamination theory is used to determine the micro- and macro-stresses in ±45 laminates subjected to residual curing stresses. Of particular concern are the onset of debonding and subsequent response. The results demonstrate the potential impact of micromechanical details on the response of brittle matrix composite laminate.

INTRODUCTION

In this work we shall consider the behavior of a laminated brittle matrix composite (BMC) material under the action of residual curing stresses and subsequent applied loading. Our emphasis will be to study the influence of micromechanical phenomenon, such as fiber-matrix debonding, upon the elastic and failure response of a laminated system. In such a body the (macroscopic) stresses acting on the layers are dependent upon the layer properties, orientations, and stacking sequence, as well as applied loading. Indirect evidence [1] supports the notion of "micro-laminate effect" indicating that ply or layer elastic properties may not be treated as unique quantities, but are strong functions of interface behavior, which, in turn, is dependent on macroscopic state of stress. Recent modeling [2] efforts and experimental [3] results have further shown that the composite transverse modulus may be extremely sensitive to fiber-matrix interface conditions. Thus, we cannot claim a 'priori' knowledge of the layer moduli as one does for continuous interface composites. Rather, an iterative procedure is required to execute layer-by-layer laminate stress field analysis. The final predicted stress state within a laminate must be consistent with the physical behavior of the interface and the details of the laminate loading and geometry. Our approach will be to employ a recently developed

To the extent authorized under the laws of the United States of America, all copyright interests in this publication are the property of The American Ceramic Society. Any duplication, reproduction, or republication of this publication or any part thereof, without the express written consent of The American Ceramic Society or fee paid to the Copyright Clearance Center, is prohibited.

variational model [4] in conjunction with classical lamination theory to determine the micro- and macro-stresses in composite laminates subjected to residual curing stresses followed by tensile or compressive loading.

METHODOLOGY

To demonstrate the micro-laminate effect in a real system and to illustrate the approach, we will consider $\pm 45^0$ laminates of Nicalon/1723 glass unidirectional layers with a 40 percent fiber volume fraction. The material properties used in the calculations, assuming both constituents to be isotropic, are listed as follows:

Material	E (GPa)	G (GPa)	α (μm/m °C)
Nicalon	200.0	77.0	3.2
1723 Glass	88.0	36.0	5.2

Let us denote the fiber direction by x_1 and let x_2 - x_3 be the transverse plane with angle θ measured from the x_3 axis. We also introduce a right-handed cylindrical coordinate system z, θ, r and employ contracted notation in the representation of the stress and strain components, i.e.,

$$\sigma_1 = \sigma_{zz},\ \sigma_2 = \sigma_{\theta\theta},\ \sigma_3 = \sigma_{rr},\ \sigma_4 = \sigma_{r\theta},\ \sigma_5 = \sigma_{rz},\ \sigma_6 = \sigma_{z\theta} \quad (1)$$

and the analogous relation for the engineering strain components ε_i (i =1, 2, .. 6).

We begin by computing the effective layer thermoelastic properties corresponding to a continuous fiber-matrix interface employing a concentric cylinder model [4][1]. Since we have assumed perfect bonding, the unidirectional layer is transversely isotropic. The effective coefficients of thermal expansion (CTE), α_i, and the stiffness coefficients, C_{ij}, for the undamaged layer are shown in Tables 1 and 2, respectively. By application of lamination theory then, we calculate the stress and strain components (away from the free edge region) within each layer for the condition consisting of a 600°C drop in temperature for the entire laminate to simulate the development of curing stresses. We then return to the micromechanics model to determine the stress fields within the constituent materials of the unidirectional layer.

Table 1: Effective CTE (μm/m/°C) for perfectly bonded interface conditions

α_1	α_2	α_3	α_4	α_5	α_6
3.95787	4.30981	4.30981	0	0	0

[1] Let r_f and r_m denote fiber and matrix radius, respectively, for the concentric cylinder model. For variational analysis, we divide the fiber region into a core and a concentric layer whereas the matrix was subdivided into three layers. The core and subsequent layer outer radii were given by $0.85\, r_f$, r_f, $r_f + (r_m - r_f)/10$, $r_f + (r_m - r_f)/2$ and r_m, respectively, for all the calculations reported in this study.

Table 2: Effective Stiffness Coefficient Matrix (C_{ij}, GPa) for perfectly bonded interface conditions

157.859	48.298	48.298	0	0	0
48.298	141.478	45.834	0	0	0
48.298	45.834	141.478	0	0	0
0	0	0	47.822	0	0
0	0	0	0	48.224	0
0	0	0	0	0	48.224

In Fig 1, we have plotted the variation of the non-zero stress components at the fiber matrix interface as a function of angle θ measured from x_3 axis. We will now assume that the initial failure will be governed by the maximum values of the pointwise stresses as a first approximation. Various failure scenarios may now develop depending on the magnitude and mix of the interfacial stresses, and the respective ultimate properties of the interface and constituents. As precise data reflecting strength and fracture characteristics and criteria of the constituents and interface tend to be lacking, we will assume reasonable bulk values where necessary to continue the analysis.

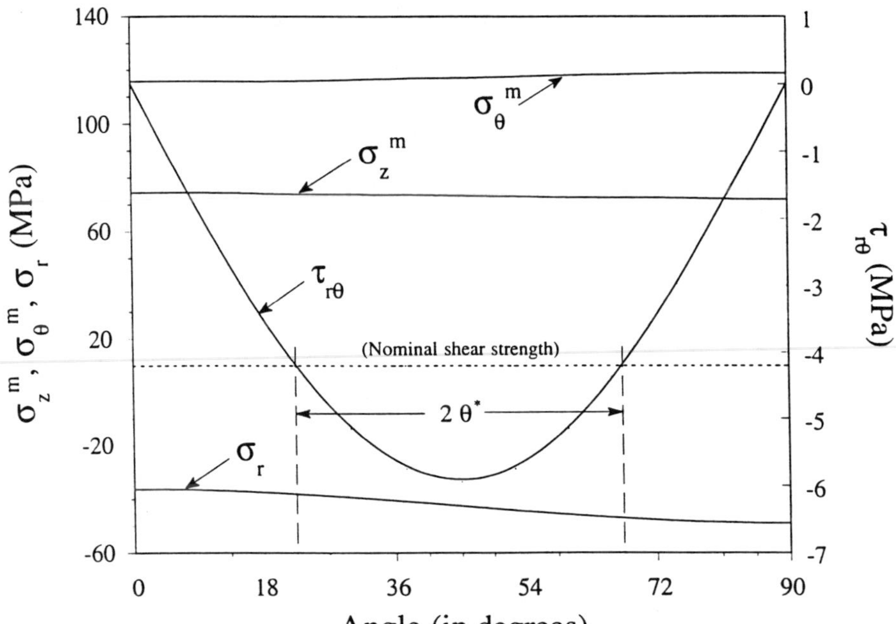

Fig 1: Variation of matrix stresses at the interface in 45^0 layer

As seen in Fig 1, the radial stress, σ_r, at the interface is compressive, whereas, the axial stress, σ_z, and hoop stress, σ_θ, in the matrix are tensile in nature. The failures caused by σ_z and σ_θ would tend to take the form of transverse (fiber bridging) and radial matrix cracks, respectively. Based on bulk matrix tensile strength of the order of 100 MPa, we could expect some radial cracks to initiate at the fiber-matrix interface. However, our previous work [5] has shown very little influence of radial cracking on the elastic response of the unidirectional layer. As such, for this study, we will only examine the influence of fiber-matrix debonding. The fiber-matrix interface will now be assumed to debond over the region where the interface shear stress exceeds the interfacial strength, although this is likely to depend on σ_r as well. We must now recalculate the effective layer thermoelastic properties corresponding to an interfacial debond of length $2\, r_f\, \theta^*$, where r_f is the fiber radius and $2\theta^*$ is the angle of debond (as suggested by Fig 1). Notice that the center of the debond is located at 45^0 from x_3 axis. A schematic of the composite cylinder model with partial fiber disbond is illustrated in Figure 2.

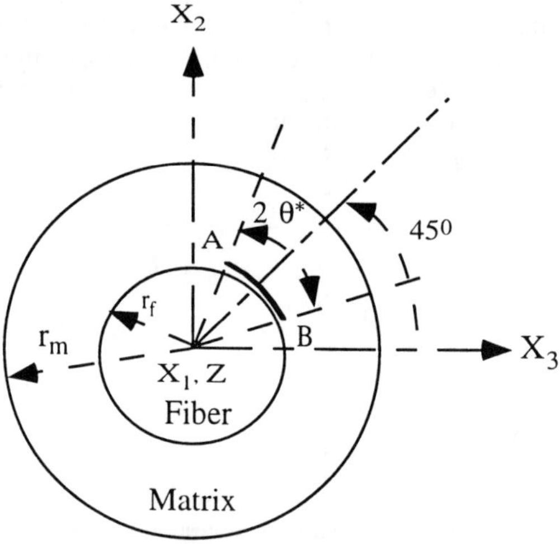

Fig 2: Schematic of partial fiber disbond ($2\theta^*$ is angle of debond)

As an example, the components of the "damaged" stiffness matrix are shown in Table 3 for a 45^0 debond angle. Notice the additional non-zero stiffness terms (namely, C_{14}, C_{24}, C_{34} and C_{56}) indicating the shear coupling phenomenon in the presence of damage. The composite engineering constants can now be defined in terms of the elastic stiffnesses. In Table 4, we have listed the effective coefficient of thermal expansion (CTE) for the debonded layer. Notice that we now have a non-zero, though small, value of α_4.

Table 3: Effective Stiffness Coefficient Matrix (C_{ij}, GPa) for 45 ^{0} debond angle

$$
\begin{bmatrix}
155.371 & 43.678 & 43.678 & -2.365 & 0 & 0 \\
43.678 & 130.161 & 39.985 & -4.393 & 0 & 0 \\
43.678 & 39.985 & 130.161 & -4.393 & 0 & 0 \\
-2.365 & -4.393 & -4.393 & 45.333 & 0 & 0 \\
0 & 0 & 0 & 0 & 46.311 & -1.831 \\
0 & 0 & 0 & 0 & -1.831 & 46.311
\end{bmatrix}
$$

Table 4: Effective CTE (μm/m/^{o}C) for 45° debond angle

α_1	α_2	α_3	α_4	α_5	α_6
3.95913	4.34897	4.34897	.079555	0	0

The effect of shear coupling, non-zero α_4, and (possible) anisotropy in the transverse plane must now be included in the lamination theory. Under the state of plane stress ($\bar{\sigma}_2 = \bar{\sigma}_4 = \bar{\sigma}_6 = 0$) the layer (or ply) stress-strain relation can be shown to reduce to

$$
\begin{aligned}
\bar{\sigma}_1 &= Q_{11}\,(\bar{\varepsilon}_1 - \bar{e}_1) + Q_{13}\,(\bar{\varepsilon}_3 - \bar{e}_3) \\
\bar{\sigma}_3 &= Q_{31}\,(\bar{\varepsilon}_1 - \bar{e}_1) + Q_{33}\,(\bar{\varepsilon}_3 - \bar{e}_3) \\
\bar{\sigma}_5 &= Q_{55}\,\bar{\varepsilon}_5
\end{aligned}
\tag{2}
$$

where

$$
\begin{aligned}
Q_{11} &= C_{11} + \frac{C_{12}\,(C_{21}\,C_{44} - C_{24}\,C_{41}) + C_{14}\,(C_{22}\,C_{41} - C_{21}\,C_{42})}{(C_{24}\,C_{42} - C_{22}\,C_{44})} \\
Q_{13} = Q_{31} &= C_{13} + \frac{C_{12}\,(C_{23}\,C_{44} - C_{24}\,C_{43}) + C_{14}\,(C_{22}\,C_{43} - C_{23}\,C_{42})}{(C_{24}\,C_{42} - C_{22}\,C_{44})} \\
Q_{33} &= C_{33} + \frac{C_{32}\,(C_{23}\,C_{44} - C_{24}\,C_{43}) + C_{34}\,(C_{22}\,C_{43} - C_{23}\,C_{42})}{(C_{24}\,C_{42} - C_{22}\,C_{44})} \\
Q_{55} &= C_{55} - \frac{C_{56}\,C_{65}}{C_{66}}
\end{aligned}
\tag{3}
$$

and C_{ij} are the components of the damaged (or degraded) stiffness matrix and $\bar{e}_i$ are the (damaged) effective expansional strains.

We then revisit the lamination theory to calculate the stress ($\bar{\sigma}_1$, $\bar{\sigma}_3$ and $\bar{\sigma}_5$) and strain ($\bar{\varepsilon}_1$, $\bar{\varepsilon}_3$ and $\bar{\varepsilon}_5$) components within each layer for condition consisting of a 600^0C drop in temperature. The remaining three non-zero strains are evaluated using plane stress conditions, namely, $\bar{\sigma}_2 = \bar{\sigma}_4 = \bar{\sigma}_6 = 0$, and are obtained as

$$(\bar{\varepsilon}_2 - \bar{e}_2) = \frac{(C_{21} C_{44} - C_{24} C_{41})(\bar{\varepsilon}_1 - \bar{e}_1) + (C_{23} C_{44} - C_{24} C_{43})(\bar{\varepsilon}_3 - \bar{e}_3)}{(C_{24} C_{42} - C_{22} C_{44})}$$

$$(\bar{\varepsilon}_4 - \bar{e}_4) = \frac{(C_{21} C_{42} - C_{22} C_{41})(\bar{\varepsilon}_1 - \bar{e}_1) + (C_{23} C_{42} - C_{22} C_{43})(\bar{\varepsilon}_3 - \bar{e}_3)}{(C_{22} C_{44} - C_{24} C_{42})} \qquad (4)$$

$$\bar{\varepsilon}_6 = -\frac{C_{65}}{C_{66}} \bar{\varepsilon}_5$$

With all six components of the layer strains known, we again go back to the micromechanics model to calculate the constituent stresses. The nature of the predicted stress field and energy release rate will then guide the next stage of analysis.

A word about failure criteria is in order here. For crack-tip stress fields, we shall employ a critical energy release rate criterion. In the event that a stress component, at a given location outside the crack tip zone, is finite, we will assume that failure will be governed by the maximum value of the stresses at the point as a first approximation. This, however, assumes that stresses can exceed the "strength' value in a crack-tip region. Hence, the two-pronged approach may not be entirely consistent. However, we must remark that both approaches can be employed under the stated conditions to successfully predict failure [6]. It would seem that these different methods of estimating failure conditions can be unified by use of an average stress or point stress criterion [7]. However, this would require a major speculation regarding an additional dimension, namely, the distance ahead of the crack tip over which the average stress or the point stress exceeds the critical value.

The cycle of operations will be completed if the calculated stresses do not cause any further damage initiation according to the failure criteria discussed above. Otherwise, we repeat the procedure, which is summarized in Fig 3, until the calculated stresses do not exceed their corresponding strength values and the energy release rate is below the critical value needed for crack advancement. In what follows, we have done a parametric study in which we have progressively debonded the interface and examined the effect of residual curing stresses on the elastic response of the laminate. The effect of subsequent external loading of the laminate can be examined in an entirely similar manner by superimposing the mechanical stresses due to the external loading on the residual stresses.

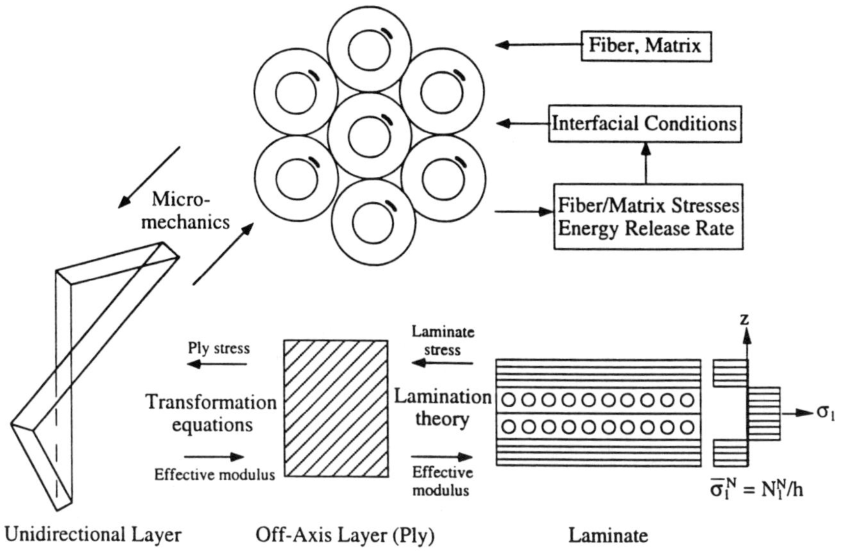

Fig 3: Micro-Laminate Effect in Brittle Matrix Composites

In Fig 4 we have plotted the variation of the effective Young's moduli of debonded unidirectional layer and of ± 45 symmetric laminate with the angle of debond ($2\,\theta^*$). As seen from Fig. 4, debonding at the interface is seen to have negligible effect on the longitudinal Young's modulus, E_{11}, of the unidirectional layer. On the other hand, the transverse Young's modulus, E_{33}, (= E_{22}, for the debond geometry considered) of the unidirectional layer is extremely sensitive to the angle of debond. Initially the decrease in transverse moduli is gradual but rapidly approaches the asymptotic unbonded value around $2\,\theta^* = 270$ degrees. The behavior of the laminate in-plane modulus ($E_x = E_y$) for this laminate is very similar to that of the transverse moduli of the layer as seen in Fig. 4. In Fig 5, we have plotted the variation of the effective coefficient of thermal expansion with interface debond angle. Similar to E_{11}, interface debonding has negligible influence on α_{11} (longitudinal CTE) of the layer. The transverse CTE ($\alpha_{33} = \alpha_{22}$) of the layer and the in-plane CTE ($\alpha_x = \alpha_y$) of the laminate, on the other hand, increase with angle of debond, as seen in Fig 5. The results in Figs 4 and 5 clearly demonstrate the strong influence of interface debonding on the effective thermoelastic properties. The ply or layer elastic properties should therefore not be treated as unique quantities, but are strong functions of interface behavior, which, in turn, is dependent on macroscopic state of stress.

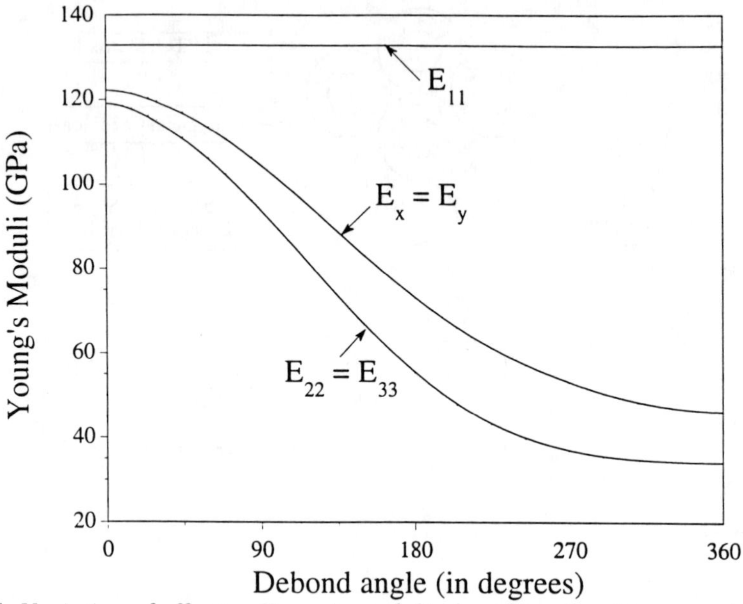

Fig 4: Variation of effective Young's moduli of unidirectional layer and $\pm 45^0$ *symmetric laminate with angle of debond*

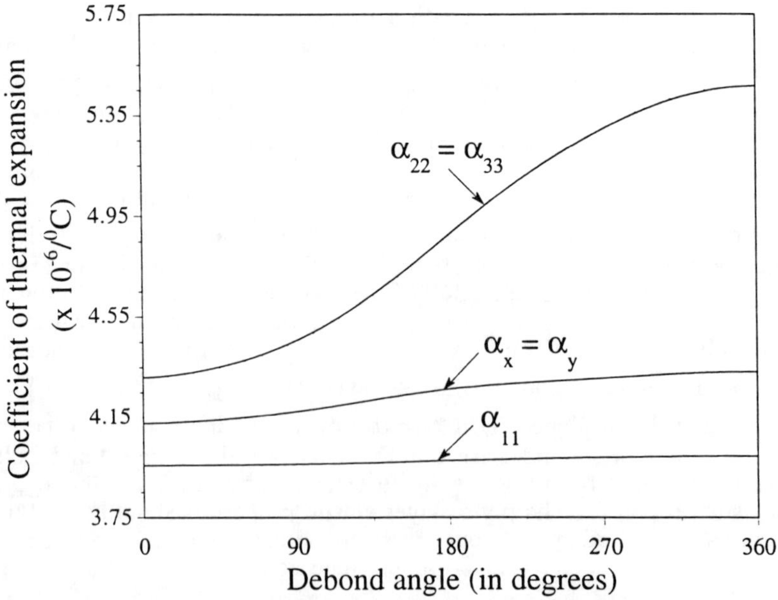

Fig 5: Variation of effective coefficients of thermal expansion of unidirectional layer and $\pm 45^0$ *symmetric laminate with angle of debond*

In Fig 6, we have plotted the variation in the normalized energy release rate at the crack tips labeled A and B (in Fig 3) as a function of angle of debond at the fiber-matrix interface for the condition consisting of a 600^0C drop in temperature for the entire laminate to simulate the development of curing stresses. The energy release rate is defined as

$$G = -\Delta P / \Delta A \qquad (5)$$

where P is the potential energy U - W, U is the strain energy, W is twice the work done by the prescribed tractions, and ΔA is the newly formed crack area. The calculation is actually performed as suggested by Eq (5), i.e., by computing ΔP corresponding to two neighboring values of crack length. If $\Delta\theta$ denotes the increment in debond angle, then ΔA is given as $r_f \Delta\theta\, l$, where l is the length of the composite. The reported values in Fig 6 have been normalized with respect to the fiber radius. As seen in Fig 6, the crack tip located at point A has a much higher value of the energy release rate compared to point B. The interface crack will therefore start to grow from the end A and will propagate when the strain energy release rate reaches the critical strain energy release rate of the interface, G_{iC}. The energy release rate passes through a maximum indicating that the crack, if it grows, may be arrested. However, based on reported measurements of approximately 1 N/m [8] for G_{iC}, (and assuming the fiber radius as 8×10^{-6} m) suggests immobility of the initial debond crack under the action of residual stresses alone.

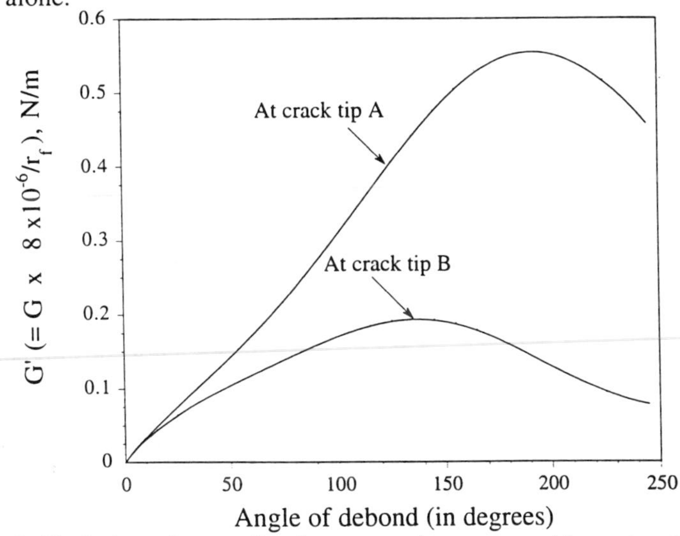

Fig 6: Variation of normalized energy release rate with angle of debond

Next we will examine the stress distribution along the bonded region of the interface. Both the normal stress, σ_z, and hoop stress, σ_θ, in the matrix at the interface are seen to decrease slightly with increase in the angle of debond.

Thus, the matrix stresses are being relieved as the debond length is increased. In Figs 7 - 9, we have plotted the distribution of the shear stress, $\tau_{r\theta}$, radial stress, σ_r, and the ratio of shear stress to radial stress, $\tau_{r\theta}/\sigma_r$, along the bonded portion of the interface for three different debond angles, namely, 5^0, 45^0 and 105^0, respectively. Both shear and radial stresses display sharp peaks near the vicinity of the crack tips as seen in Figs 7 - 9. The reported stress values near the crack tips represent an approximation of a singular stress field since singular values do not arise in the present model [4]. One can justify the use of point or average-stress failure criteria [7] or possibly rely on the ratios of the stress components, but not the stresses themselves to predict failure. For crack-tip stress fields, we are employing a critical energy release rate criterion for this study, as discussed earlier.

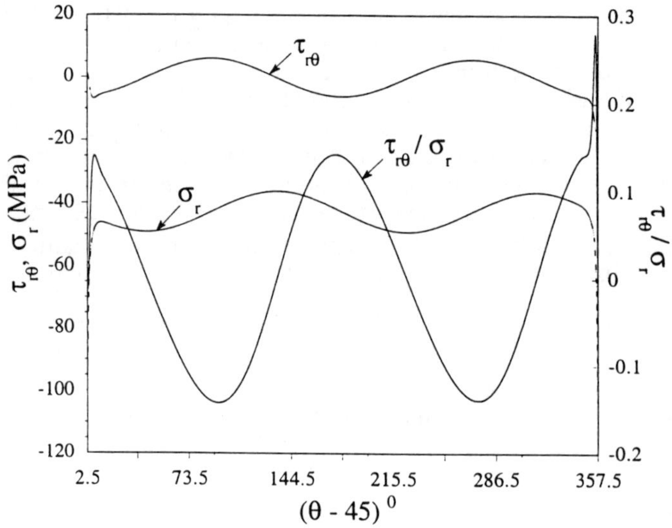

Fig 7: Variation of $\tau_{r\theta}$, σ_r and $\tau_{r\theta}/\sigma_r$ over the bonded portion of the interface for a debond angle of 5^0

Away from the crack-tip region, the magnitude of both the interfacial shear stress and radial stress is seen to increase with the angle of debond. Thus, debond predictions based on maximum shear stress values alone would not be meaningful. Since the radial stress at the interface is compressive and is also increasing in magnitude with increase in angle of debond, we have at hand a possible classic mechanism of stress induced strengthening of the interface. Similar observations of stress-induced strengthening has been observed by Kim [9] in studying the free-edge delamination behavior of symmetric laminates. What we should therefore examine is the ratio of the shear to the radial stress at the interface and when this ratio reaches a critical value, a new crack (or debond) is likely to initiate at that location. As expected, the maximum value of the ratio is seen to increase with angle of debond, as seen in Figs 7 - 9. Whether the calculated value of the ratio will incite a new failure depends upon the ultimate value which in turn may vary considerably within the present class of materials. For typical glass-ceramic composites, estimates of the interfacial shear strength range from 0 to 13.9 MPa [10], depending on the nature of the fiber surface treatment and test technique employed, such as indentation, pull-out, push-out or measurement of matrix crack spacing, amongst other material and processing variables. In view of this variability, we will not indulge in making any further prediction of subsidiary failure mechanisms. Suffice to say, it is hoped that the modeling can serve to determine the nature of the parameters that require measurement and to help establish the sensitivity of the response of these parameters, as well as to guide experimental efforts to attempt validation of the failure process hypothesized. The complexity of the process will demand an iterative approach between the analyst and experimentalist.

CONCLUDING REMARKS

We have outlined a methodology to study the influence of micromechanical phenomenon, such as interfacial debonding, upon the elastic and failure response of a laminated system. Our approach is to employ a recently developed variational model in conjunction with classical lamination theory to determine the micro- and macro-stresses in composite laminates subjected to residual curing stresses followed by tensile or compressive loading. The final predicted stress state within a laminate must be consistent with the physical behavior of the interface and the details of the laminate loading and geometry. Subsequent research to develop refined failure criteria will depend on the iterative correlation of modeling and experimental studies.

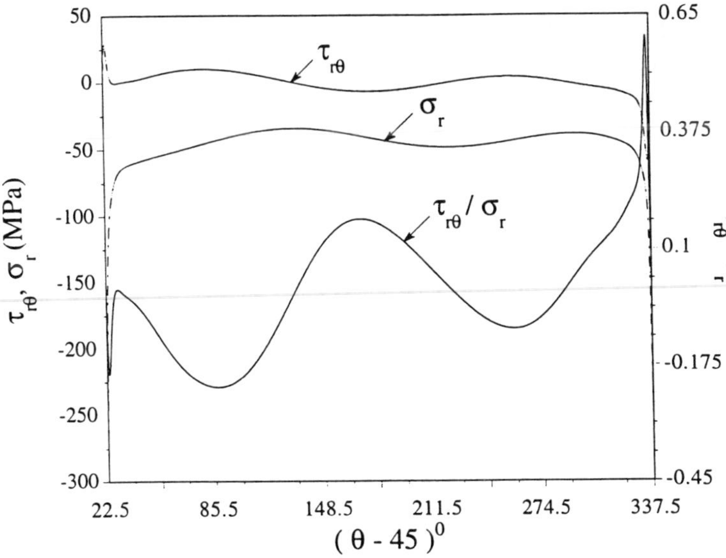

Fig 8: Variation of $\tau_{r\theta}$, σ_r and $\tau_{r\theta}/\sigma_r$ over the bonded portion of the interface for a debond angle of 45^0

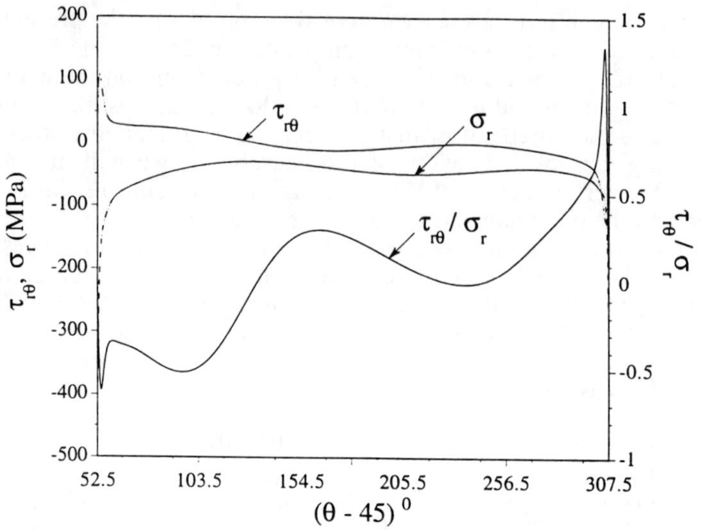

Fig 9: Variation of $\tau_{r\theta}$, σ_r and $\tau_{r\theta}/\sigma_r$ over the bonded portion of the interface for a debond angle of 105^0

REFERENCES

1) Pagano, N. J. , and Kim, R. Y. (1989) " A Micro-Laminate Effect in Brittle Matrix Composites ", Symposium on High Temperature Composites, Dayton, OH, pp 187-197
2) Tandon, G. P., and Pagano, N. J. (1993), " Effective Moduli of a Unidirectional Fiber Composite Containing Interfacial Arc Microcracks ", in Thermomechanical Behavior of Advanced Structural Materials, ASME, AD-Vol 34/AMD-Vol 173, pp 113-128
3). Kim, R. Y. (1992), " Experiment Observations of Progressive Damage in SiC/Glass Ceramic Composites ", *Cer. Sci. Eng. Proc.* , Vol 13, No. 7-8, pp 281-300.
4) Pagano, N. J., and Tandon, G. P. (1994), " 2 - D Damage Modes in Unidirectional Composites under Transverse Tension and/or Shear ", to appear in *Mechanics of Composite Materials & Structures*
5) Tandon, G. P. , and Pagano, N. J., (1993) " Modeling of Radial Cracking and Interfacial Debonding in Unidirectional Brittle Matrix Composites ", Mechanics of Composites Review, Dayton, OH, Dec 7 - 8, pp 98-106
6) Pagano, N. J. and Brown III, H. W. (1993), The Full-Cell Cracking Mode in Unidirectional Brittle Matrix Composites, *Composites 24,* p 69
7) Whitney, J. M. and Nuismer, R. J. (1974) " Stress Fracture Criteria for Laminated Composites Containing Stress Concentrations ", *J Compos Matl.*, Vol 8, pp 253-265
8) Kerans, R. J., Jero, P. D., Parthasarathy, T. A. and Chatterjee, A. (1990), " Determination of Fiber/Matrix Interface Mechanical Properties in Brittle Matrix Composites ", *Proc. MRS Symposium*, Vol 194, pp 263-270
9) Kim, R. Y. (1989), " Experimental Observations of Free-Edge Delamination ", in Interlaminar Response of Composite Materials, N. J. Pagano, Ed., Vol 5, Composite Material Series, Elsevier Science Publishers B. V., pp 111-160
10) Kerans, R. J., Hay, R. S., Pagano, N. J., and Parthasarathy, T. A. (1989), " The Role of The Fiber/Matrix Interface in Ceramic Composites ", *Am. Ceram. Soc. Bull.*, Vol 68, No. 2, pp 429-442

Numerical Simulation of Toughening of Alumina Particulate Glass Matrix Composite by Interfacial Precipitation

Kensuke Kageyama*, Manabu Enoki*, Teruo Kishi*,
Kazuhiro Ikuina** and Michiru Kimura**
*Research Center for Advanced Science and Technology, The University of Tokyo
, 4-6-1 Komaba, Meguro-ku, Tokyo 153, Japan
**NEC Corp.
, 1-1, Miyazaki 4-Chome, Miyamae-ku, Kawasaki, Kanagawa 216, Japan

ABSTRACT

Strengthening and toughening by interfacial precipitation are strongly connected with crack bowing and deflection. In the present study, three dimensional numerical simulation of these events was performed on ceramics particulate glass matrix composites with interfacial precipitation by calculating the equations for a crack bowing and deflection. This numerical simulation revealed that fracture toughness and strength increased with the addition of interfacial precipitation because a crack bowing emerged. These results are in agreement with experimental data for fracture toughness.

INTRODUCTION

Glasses are used to create high dense, multi layered substrates because they have low thermal expansion and low dielectric constant, and can be processed by low temperature sintering (~1273K).[1] The low fracture strength and toughness of glass, however, present a problem, because glass is intrinsically brittle and a crack growth resistance decreases from stress corrosion in air.[2] A ceramics particulate glass matrix composite thus was designed to enhance strengthening and toughening of glass.[3,4] The resultant strengthening and toughening of glass by ceramics particles is mainly attributable to load sharing caused by the increased macroscopic elastic modulus,[4] since fracture strength and toughness increase in proportion to that modulus. The fracture process, however, remained as unstable as glass, and the decreases in fracture

To the extent authorized under the laws of the United States of America, all copyright interests in this publication are the property of The American Ceramic Society. Any duplication, reproduction, or republication of this publication or any part thereof, without the express written consent of The American Ceramic Society or fee paid to the Copyright Clearance Center, is prohibited.

strength and resistance caused by stress corrosion became greater. To overcome these problems, a crack deflection and bowing[5,6] were suggested to improve the unstable fracture process. Although these studies did show that a larger particle enhanced fracture resistance by a crack bowing, the fine ceramics particulate glass composite therefore has too little effect on crack bowing and, the effect of the crack deflection on the fracture toughness was doubtful in the glass matirix composites[4,6].

A new method was used next to improve mechanical reliability: strengthening and toughening by interfacial precipitation. The idea behind interfacial precipitation is that an intermediate layer between the particle and the matrix stops crack extension at the interface and enhances fracture resistance by crack bowing. In only particulate composite is used, if a crack meeting a particle will barely extend into a particle and will deflect around the particle because a ceramics particle has much higher fracture toughness (about 4MPa/m) than that of glass (0.3~0.8MPa/m). In this case, the crack resistance is enhaced only by a crack deflection but a crack bowing. If interfacial precipitation create an intermediate fracture toughness between the particle and the matrix, it will more easily extend into the particle through the interfacial precipitation than without precipitation. In this case, the crack resistance is mainly enhaced by a crack bowing. We already investigated the microfracture of alumina particulate borosilicate glass composites[7,8] and this study revealed that interfacial precipitation improved both the average value and the scatter of bending strength because microcracks extended into anorthite at the interface diminished the unstable crack extension caused by stress corrosion around a particle. Strengthening and toughening by interfacial precipitation thus are intimately related to a crack bowing, which is supposed to be more effective to enhance the fracture toughness than a crack deflection. It is difficult, however, to quantitatively evaluate by experimental methods the effect of crack bowing and deflection on mechanical properties, although some researchers have observed the existence of crack bowing or deflection by SEM or TEM[3,4]. The purpose of the present study therefore was to construct a three dimensional model of crack bowing and deflection in ceramics particulate glass matrix composite and to clarify the effect of interfacial precipitation on fracture toughness and strength. The results of the present calculations also are compared with experimental data from the our prior works.

THEORY AND ALGORITHM

A circular and a plain crack in glass matrix were imaged with an applied mode I stress, σ_0. The stress intensity factor of that crack is expressed as[9].

$$K_I^0=2\sigma_0\sqrt{\frac{r}{\pi}} \tag{4}$$

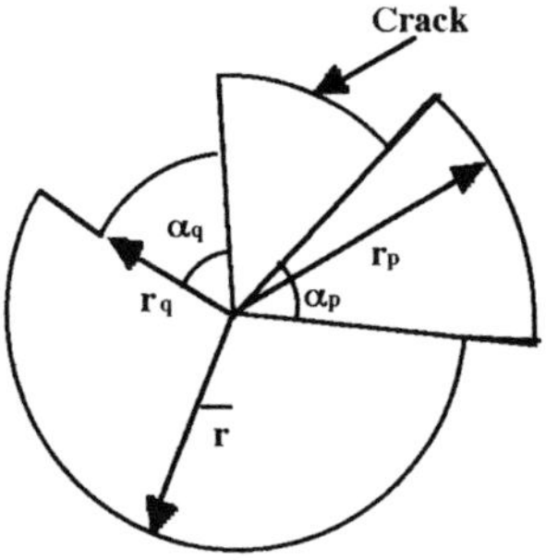

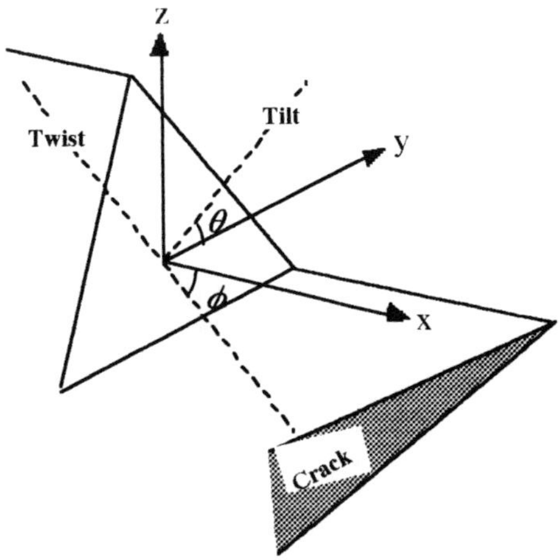

Figure 1. The geometries of bowing sections to calculate a stress intensity factor at the extrema of a bowing crack.

Figure 2. The geometry of a deflectiong section to calculate a stress intesity factor at a deflecting crack tip.

As a crack extends, the crack front demonstrates bowing and deflecting by the particles. For that process, Cox and Morris derived the following equations, which show the stress intensity factor of the protruding and retarded sections of a bowing plain crack[5].

$$K_I^p = K_I^0 \frac{2}{\pi} \tan^{-1}\{(3+\frac{\bar{r}}{r_p})/(1-\frac{\bar{r}}{r_p}) \tan\frac{\alpha_p}{2\alpha_0}\} \tag{5}$$

$$K_I^q = K_I^0 / [\frac{2}{\pi} \tan^{-1}\{(3+\frac{r_q}{\bar{r}})/(1-\frac{r_q}{\bar{r}}) \tan\frac{\alpha_q}{2\alpha_0}\}] \tag{6}$$

Here K_I^p, r_p and α_p represent fracture toughness, crack radius at the extrema and the angle of the protrusion section respectively, K_I^q, r_q and α_q are values for the retarded section (See Fig.1). r_0 is the average crack radius and K_0 is the applied stress intensity factor at the crack tip of the average radius. The value α_0 is a fitting parameter by which natural crack features are replaced with square-shouldered segments.

On the other hand, Faber et al.[6] derived equations that show the stress intensity factor of a deflecting crack. Those equations divided a crack deflection into two forms, the tilt (angle, θ) and the twist (angle, ϕ) (See Fig.2), and produce the following equation.

$$K_I^D = K_I^0 h_{11}(\theta,\phi) \tag{7}$$

$$K_{II}^D = K_I^0 h_{21}(\theta,\phi) \tag{8}$$

$$K_{III}^D = K_I^0 h_{31}(\theta,\phi) \tag{9}$$

where

$$h_{11}(\theta,\phi) = 2\nu\cos\frac{\theta}{2}\sin^2\phi + \cos^3\frac{\theta}{2}\cos^2\phi$$

$$h_{21}(\theta,\phi) = \sin\frac{\theta}{2}\cos^2\frac{\theta}{2}$$

and

$$h_{31}(\theta,\phi) = \cos\frac{\theta}{2}\left(\cos^2\frac{\theta}{2} - 2\nu\right)\sin\phi\cos\phi$$

Here, $K_I{}^d$, $K_{II}{}^d$ and $K_{III}{}^d$ are the stress intensity factors of the deflecting crack tip and ν is the Poison's ratio.

A crack front line was separated into nodes and the energy release rate of each node calculated by the above equations. A particulate glass composite with interfacial precipitation then was three-dimensionally simulated, with a crack extending statically and appearance of crack bowing and deflection by the particles.

First, The effect of bowing at the crack front of arbitrary profiles must be considered, but Eqs. (5) and (6) are applicable onlyfor a plain crack and not for an irregular crack. A crack thus was projected onto an x-y plain whose normal axis coincided with the direction of a tensile loading and the stress intensity factor of the projection was calculated. This approximation should be reasonable when the surface inequality of a crack are minimal. Although the inequality ranged from about 1/200 to 1/10 (crack radius = ~10 to ~100 μm and particle radius = ~0.5 to ~1 μm) and so may not be negligible, this approximation was presumed reasonable because most of fracture origins were observed as a nearly flat cracks in the particulate glass composites.

The second consideration is that the stress condition at the crack tip changes from mode I to a mixed mode with a crack deflection. The stress intensity factors at the deflecting crack tip thus were calculated using Eqs. (7)-(9) and substituting $K_I{}^p$ or $K_I{}^q$ from Eqs. (5) and (6) for $K_I{}^0$ after determining the effect of a crack bowing on the stress intensity factor. The stress condition at the bowing and deflecting crack tip could then be approximated.

Finally, the changes in the stress intensity factor when the crack tip extends to the next step in the direction of the tilt angle of (θ') have to be considered. The stress intensity factor at the next step was then expressed as follows[5].

$$K_{Itip}(\theta')=K_I^D h_{11}(\theta',0)+K_{II}^D h_{12}(\theta',0) \quad (10)$$

$$K_{IItip}(\theta')=K_I^D h_{21}(\theta',0)+K_{II}^D h_{22}(\theta',0) \quad (11)$$

$$K_{IIItip}(\theta')=K_{III}^D h_{33}(\theta',0) \quad (12)$$

where

$$h_{12}(\theta,\phi)=2\nu\sin\frac{\theta}{2}\sin^2\phi-3\sin\frac{\theta}{2}\cos^2\frac{\theta}{2}\cos^2\phi$$

$$h_{22}(\theta,\phi)=\cos\frac{\theta}{2}(1-3\sin^2\frac{\theta}{2})$$

and

$$h_{33}(\theta,\phi)=\cos\frac{\theta}{2}(\sin^2\phi-\cos^2\phi)$$

The energy release rate $G_{tip}(\theta')$ to a next step in the direction to the tilt angle θ' therefore can be expressed as

$$G_{tip}(\theta')=[\{K_{Itip}(\theta')\}^2+\{K_{IItip}(\theta')\}^2+\{K_{IIItip}(\theta')\}^2]\frac{1-\nu^2}{E} \quad (13)$$

The value of $G_{tip}(\theta')$ then was compared with G_C of the critical energy release rate ($G_C=K_{IC}{}^2(1-\nu^2)/E$) where K_{IC} represents the intrinsic fracture toughness of the material where crack tip was located (glass matrix, ceramics particle or interfacial precipitated phase). If $G_{tip}(\theta')$ is higher than G_C, the calculated section of a crack tip will extend; if $G_{tip}(\theta')$ is lower than G_C, the crack will stop and if all segments stop, the applied stress, $K_I{}^0$, will increase. Finally, if an unstable extension of the crack occurs, $K_I{}^0$, σ_0 and r are regarded as fracture toughness; $K_{IC}{}^c$ as fracture stress; σ_C and critical crack radius; r_C.
If the crack tip does not meet a particle, θ' is zero degree because the crack will extend in a straight path. If the crack tip meets a particle, θ' will be zero or a direction to tangential to the line of the interface and will be represented by the higher value of $G_{tip}(\theta')/G_C$. If the crack tip meets a intermediate layer (interfacial precipitation), the crack is presumed to extend straight , with no deflection (Fig.3), because the experimental data of AE measurements of the precipitated samples in our work showed that a microcrack extending into the interfacial precipitated anorthite, and an indistinct interface (amorphous) was observed by TEM between the glass matrix and the intermediate layer.[8]

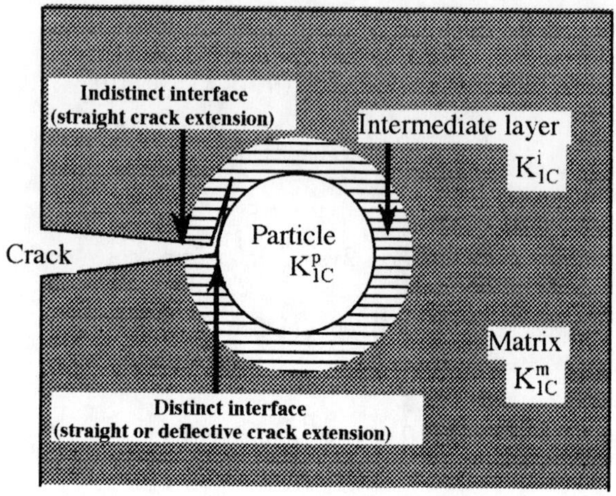

Figure 3. The micro structure of interfacial precipitated composites.

The initial parameters of fracture toughness of the glass matrix, $K_{IC}{}^{m}$, the ceramic particles, $K_{IC}{}^{p}$, and the intermediate layer, $K_{IC}{}^{i}$; volume fraction of particles, V_p and intermediate layer, Vi ; and the radius of the initial crack, r_0, and particles, r_p, were set. The value $K_{IC}{}^{i}$ was set at $0.6K_{IC}{}^{p}$, and $K_{IC}{}^{m}$ varied from 0.2 to $0.5K_{IC}{}^{p}$. The value for r_0/r_p varied from 5 to 50, and the nodal points of the crack front varied from 157 to 1570 as the crack size was adjusted.

Alumina particulate glass matrix composites with anorthite precipitation were studied in our prior work[7,8]. Anorthite crystal precipitates at the interface between alumina particle and glass matrix by following reaction.[1]

$$Al_2O_3+2SiO_2+CaO \rightarrow CaO\text{-}Al_2O_3\text{-}2SiO_2$$

Volume fraction of alumina particle, therefore, decreases as anorthite precipitates. Following equation then was presumed in the present simulation, because alumina particle decreases according to the change of atomic weight.

$$V_p = V_p^0 - \frac{102}{276} V_i \tag{14}$$

$V_p{}^0$ is initial volume fraction before fabrication.

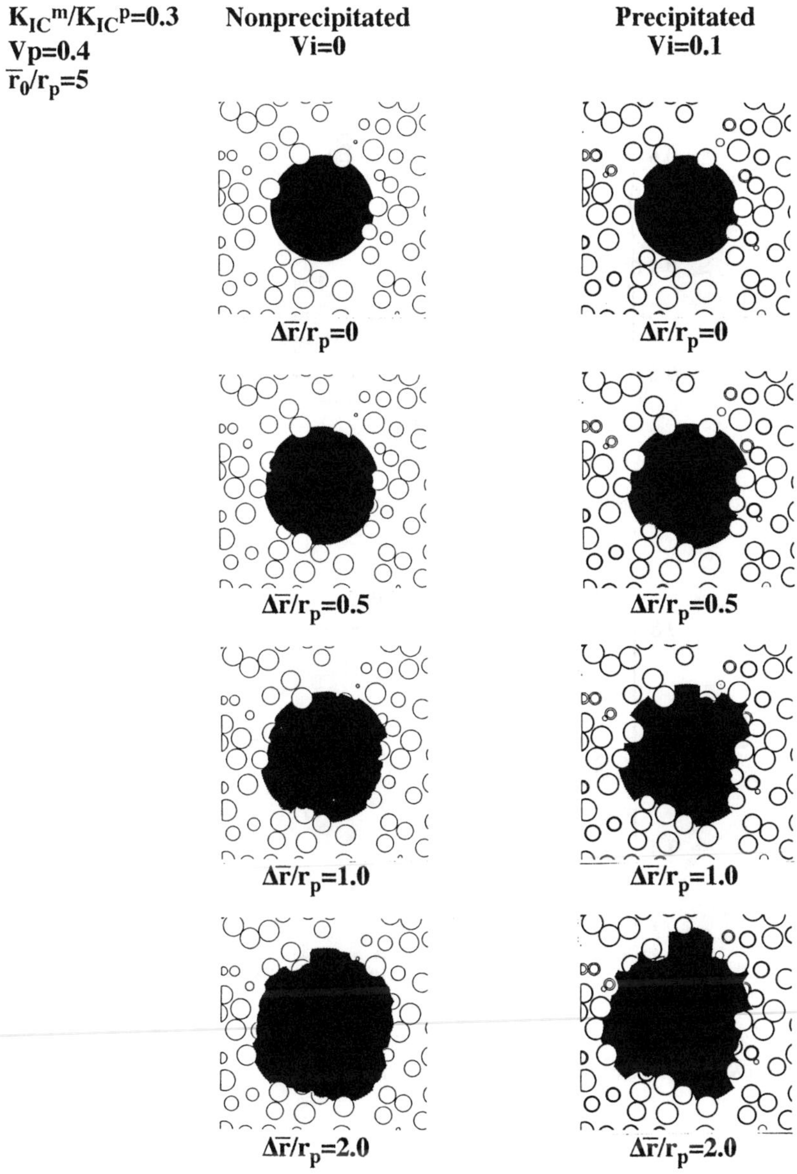

Figure 4. The difference of crack growth between nonprecipitated and precipitated. Black area means a crack which is observed in direction to vertical axis against crack plain and white circles means particles.

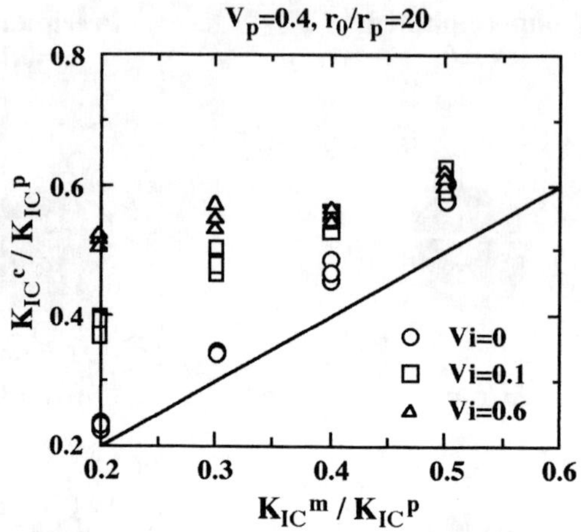

Figure 5. Relationship between fracture toughness of composites and matrix.

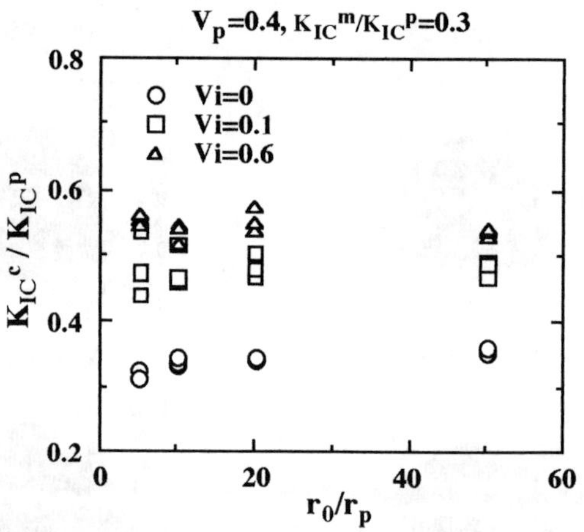

Figure 6. Dependence of fracture toughness of composites on the initial crack length.

RESULTS

Figure 4 shows examples of the fracture process in non precipitated and precipitated cases, as obtained by numerical simulation. In the case of the non precipitated sample, if a crack tip meets a particle, the crack will extend easily around the particle and so the particle apparently has no role as an obstacle. Though a crack deflection occurs at the interface, the effect of a crack deflection on a fracture resistance therefore is small. On the other hand, the behavior of the crack extension of the precipitated sample differs from the that of the non precipitated. A crack tip extends into the intermediate layer stopping the unstable extension. Notably, crack bowing appears by the intermediate layer (interfacial precipitation).

Figure 5 shows the effect of toughening caused by interfacial precipitation as the fracture toughness of the glass matrix varies. Even if the fracture toughness of the glass matrix is much lower than that of the particle (here, as in air), the non precipitated material shows little toughening caused by a crack deflection around the particles. The precipitated material, however, shows higher toughening caused by a crack bowing because of interfacial precipitation. When K_{IC}^{m} was set to $0.5K_{IC}^{p}$ (as in vacuum), however, little difference showed in the behavior of K_{I}^{0} between non precipitated and precipitated samples. In this case, Figure 5 shows fracture toughness increases from K_{IC}^{m} in both cases because some crack tips extend straight into a particle without interfacial precipitation instead of deflecting around the particle .

Figure 6 shows the dependence of K_{IC}^{c} on crack size. As crack size becomes grows, K_{IC}^{c} increases, similarly to the R-curve behavior in the non precipitated case since a crack deflection are effective because of the stable growth of the crack. The K_{IC}^{c} value of a short crack in the non precipitated case therefore is somewhat higher than the fracture toughness of the glass matrix. On the other hand, toughening caused by interfacial precipitation in the precipitated case is independent of crack size. The K_{IC}^{c} value of a short crack in the precipitated case is just as high as that of a long crack.

Figures 7 show the relationship between the volume fraction of interfacial precipitation and fracture toughness. The increase of fracture toughness is caused by interfacial precipitation since the volume fraction of precipitation is low.

DISCUSSION

Comparison with experimental data

Fracture toughness behaviors of numerically simulated Vi is in agreement with experimental data (Fig.7). If K_{IC}^{m} was much lower than K_{IC}^{p} (in air), fracture strength behavior of numerically simulated Vi was in agreement with the bending strength (Fig.8). Fracture strength was defined from $\sigma_c=K_{IC}^{c}/2\sqrt{(\pi/r_c)}$ and $\sigma_p=K_{IC}^{p}/2\sqrt{(\pi/r_0)}$. The fracture strength behavior in vacuum, however, differ from that in the experimental data. If K_{IC}^{m} was closer to K_{IC}^{p} (in vacuum), the bending strength decreased but numerical simulation showed that the fracture strength increased slightly. This inconsistency was caused by the porosity of the sintered body. In precipitated samples, the sintered densification decreased because of the decrease of

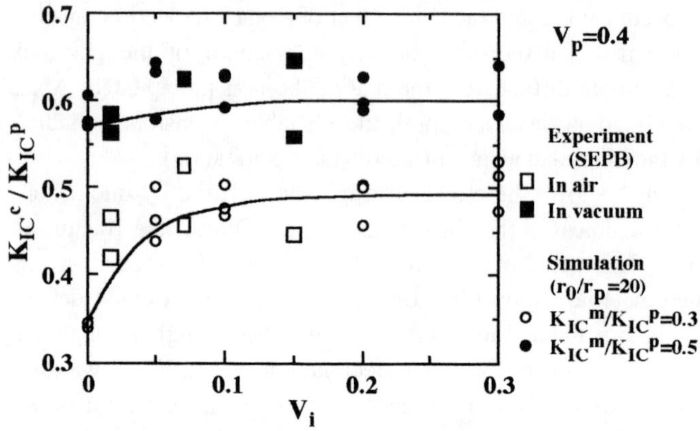

Figure 7. Comparison of fracture toughenss between results of numerical simulation and experimental data. White marks show results of numrical simulation and circles show where K_{IC}^m was set to $0.5K_{IC}^p$ and squares are set that K_{IC}^m was set to $0.3K_{IC}^p$ Black marks are experimental data in the prior work[7,8] and circles are data in vacuum and squares are data in air. Experimental data was normalized presuming that K_{IC}^p was 2.7 $MPam^{1/2}$. Vp is 0.4 and r_0/r_p is 20 in numerical simulation.

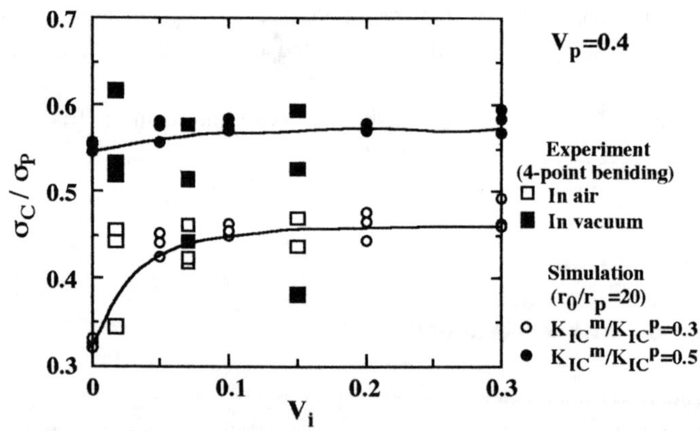

Figure 8. Comparison of fracture strength between results of numerical simulation and experimental data. White marks show results of numrical simulation and circles show set that K_{IC}^m was set to $0.5K_{IC}^p$ and squares are set that K_{IC}^m was set to $0.3K_{IC}^p$ Black marks are experimental data in the the prior work[7,8] and circles are data in vacuum and squares are data in air. Experimental data was normalized presuming that σ_C^p was 400 MPa. Vp is 0.4 and r_0/r_p is 20 in numerical simulation.

the glass matrix and the elastic modulus decreased, which reduced strengthening caused by load sharing. Higher-quality materials would have higher strength. If the process were to remain the same, a lower fraction of particles and precipitation should be attempted, since the effects of interfacial precipitation toughening can be expected even within that lower fraction, in which sintering would be better accomplished.

The experimental data can not, of course, show the pure effect of crack bowing and deflection on mechanical properties because these data contain the total effects of the three mechanisms already mentioned. More work on the interaction between these other mechanisms therefore must be undertaken, but the qualitative validity of the present simulation can be certified.

Role of interfacial precipitation on microfracture process

Only about crack bowing and deflection have been discussed in the present numerical simulation. Other factors, however, seem to influence on fracture toughness and strength. In particular, the inhomogenity of the elastic modulus and the thermal expansion causes internal stress and complicate stress distribution in materials. Strictly speaking, the influence of those change in stress distribution on a stress intensity factor must be calculated but calculations are very complicated. This present simulation therefore represent a special case, in which the crack tip can reach the interface between the glass matrix and the interfacial precipitation unless the crack tip is deflected form the interface by thermal stress.

The reason that interfacial precipitation improves the fracture toughness and strength can be proposed by the results of these numerical simulations. Apparently, interfacial precipitation causes toughening through crack bowing (pinning) by trapping crack tip in the particles rather than permitting it to extend along the particles. Interfacial precipitation toughening therefore is effective only in particulate composite with large inhomogenities. This phenomenon is ineffective in the materials with small inhomoginities such as ceramic particulate glass composite in vacuum and ceramic particulate ceramic matrix composite, because some microcracks extend into the particles in ceramic particulate glass composite in vacuum without interfacial precipitation and a crack deflection in ceramic particulate ceramic matrix composite is more effective than that in the present study. The process also may be ineffective for fiber-reinforced composites, because these composites are designed to enhance fracture toughness by fiber pull-out and interfacial precipitation might overstrengthen the bonding between the matrix and the fiber and prevents fiber pull-out.

CONCLUSIONS

Three dimensional numerical simulation of ceramics particulate glass matrix composites with interfacial precipitation was achieved in the present study by calculating equations for crack bowing and deflection.

(1) Fracture toughness and strength increased only slightly in particulate glass composites without interfacial precipitation because cracks extended along the particles and a crack deflection hardly enhanced the fracture toughness. On the other hand, fracture toughness and strength improved more with the addition of interfacial precipitation because cracks extended into the particles and a crack bowing enhanced the fracture toughness much more than a crack deflection without interfacial precipitation.

(2) These results were similar to the experimental data for fracture toughness, except in the case of bending strength, apparently because of the porosity of the sintered body. The numerical simulation also indicated that the effects of interfacial precipitation toughening are viable even within a lower fraction, where sintering is better accomplished even if the process remains the same.

REFERENCES

1. S. Nishigaki and J. Fukuta, Low-Temperature, Cofireable, Multilayered Ceramics Bearing Pure-Ag Conductors and Thier Sintering Behavior, *Advandes in Ceramics vol.26*, The American Ceramic Society, Inc., Westerville, Ohio, 199-215 (1989)

2. B.C.Bunker and T.A. Michalske, Effect of Surface Corrosion on Glass Fracrue, *Fracture Mechanics of Ceramics vol.8*, Prenum Press, NewYoik, 391-413

3. F. F. Lange, Fracture Energy and Strength Behavior of a Sodium Borosilicate Glass-Al_2O_3 Composites System, *J.Am.Ceram.Soc*, **5 4** (12) 614-20 (1971)

4. J. C. Swearengen, E. K. Beauchamp and R. J. Eagan, Fracture Toughness of Reinforced Glasses, *Fracture Mechanics of Ceramics vol.4*, 973-84, Plenum Press, New York, 973-84 (1978)

5. B. N. Cox and W. L. Morris, Monte Carlo Simlations of The Growth of Small Fatigue Cracks, *Engineering Fracture Mechanics*, **3 1** (4), 591-610 (1988)

6. K. T. Faber, A. G. Evans and M.D.Drory, A Statistical Analysis of Crack Deflection As A Toughening Mwchanism in Ceramic Matericals, *Fracture Mechanics of Ceramics vol.6*, Plenum Press, New York, 77-91 (1983)

7. K. Kageyama, M. Enoki, T. Kishi, K. Ikuina and M Kimura, Controlling Propagation of Microfractures by Anorthite Crystal Precipitation in Alumina Particulate Glass Composite, *J. Japan Inst. Metals*, **5 7** (7) 756-60 (1993)

8. K. Kageyama, M. Enoki, T. Kishi, K. Ikuina and M Kimura, Strengthning and Toughening of Ceramics Particulate Glass Matrix Composites by Interfacial Precipitation, *J.Am.Ceram.Soc*, submitted.

9. M. K. Kassir and G. C. Sih, Mechanics of Fracture 2, Noordhoff International Publishing, Leyden, Chapter 1 (1975)

Coating and Surface Modifications of Fibers and Composite Interfaces

INTERFACIAL CHARACTERISTICS OF MULLITE FIBER/BN COATING/MULLITE MATRIX COMPOSITES

K.K. Chawla*, Z.R. Xu*, J.-S. Ha*, E. Lara-Curzio**, M.K. Ferber**, and S. Russ**

*Department of Materials and Met. Eng., New Mexico Tech, Socorro, NM 87801, USA
**High Temperature Materials Laboratory, Oak Ridge National Laboratory, Oak Ridge, TN 37831, USA

ABSTRACT

The interface in mullite fiber (Nextel 480)/mullite matrix was engineered by using a thick BN (1 μm) coating on mullite fibers, such that deformation mechanisms conducive to toughness enhancement could be brought to play. Significant improvements in the mechanical properties of these mullite fiber/mullite matrix composites could be achieved by incorporation of BN interfacial coating and by using a colloidal processing route to make dense mullite matrix. An interfacial testing system with a flat-bottomed, diamond indenter was used to obtain the interface characteristics. Using a progressive debonding of the interface model, it was determined that the average interfacial frictional sliding shear stress in this composite was about 50 MPa. Fracture surfaces of these BN coated composites obtained in flexure test showed fiber pullout.

INTRODUCTION

Nonoxide fiber/nonoxide matrix composites, such as C/C, SiC/SiC, and SiC/Si_3N_4, generally show good low temperature strength, but oxidation resistance can be a major limitation. Nonoxide fiber/oxide matrix composites or oxide fiber/nonoxide matrix composites, e.g., C/glass, SiC/glass, SiC/alumina, SiC/mullite, and Al_2O_3/SiC, also

To the extent authorized under the laws of the United States of America, all copyright interests in this publication are the property of The American Ceramic Society. Any duplication, reproduction, or republication of this publication or any part thereof, without the express written consent of The American Ceramic Society or fee paid to the Copyright Clearance Center, is prohibited.

do not show high oxidation resistance because the permeability constant for the diffusion of oxygen is high, resulting in rapid oxygen permeation through the oxide matrix. Oxide fiber/oxide matrix composites are desirable for applications where stability in air at high temperature is a prime objective. In general, a strong fiber/matrix interface results in oxide matrices reinforced with uncoated oxide fibers, such as in the like/like systems (e.g., mullite/mullite) or in the mixed systems (e.g., Al_2O_3/SiO_2) where compound(s) forms at the interface, and toughness of these composites is not much improved. As a consequence, a barrier layer needs to be introduced to prevent chemical interaction between the fiber and matrix, and thus, prevent a strong interfacial chemical bonding.

OBJECTIVE

The objectives of this work were to use an interface engineering approach, involving the use of a suitable fiber coating, microstructural characterization of the interface, and determination of interfacial characteristics in a mullite fiber/mullite matrix system. The ultimate aim is to obtain a weak interfacial bond in order to achieve high work of fracture and a noncatastropic failure mode in mullite/mullite composites.

MATERIALS AND EXPERIMENTAL PROCEDURE

Nextel 480 fiber (3M Co.) has a composition (wt%) of 70% Al_2O_3, 28% SiO_2, and 2 % B_2O_3. The fibers were coated with a 1 μm thick BN by chemical vapor deposition. The mullite matrix was obtained via a powder synthesized in our laboratory by a diphasic gel route using a boehmite (ALOOH) powder (Dispal 11N7, Remet Chemical Corp., Chadwicks, NY) and an amorphous silica sol (SP-30, Remet Chemical Corp.) as precursors. The gel was prepared by dispersing the boehmite powder in distilled water to get a boehmite sol, which was then mixed with the silica sol to have the stoichiometric mullite composition (71.8wt.% Al_2O_3 and 28.2wt.% SiO_2), followed by gelling at room temperature. The gel obtained was dried at 70 ^{o}C and ground into powder form. The composites were fabricated by infiltrating uncoated and coated fibers, Nextel 480, with a slurry of the matrix powder prepared with isopropanol and an organic binder, and unidirectionally laying the infiltrated fibers on mylar tapes. After drying, the prepared tapes were cut, stacked, and heated at 700 °C for 2 h in air to remove the binder and the hydroxyl in the boehmite. Consolidation of composites was done by hot-pressing at 1300 °C and 35 MPa for 1 h in vacuum. The details of processing are reported in elsewhere [1-3].

Conventional mullite powders need temperatures above 1500 °C for considerable densification because of the low interdiffusion rates of silicon and aluminum ions in crystalline mullite [4, 5]. We selected BN as the coating material because it has been found to be efficient for providing a weak interface in mullite matrix composites [6, 7] and the mismatch between the radial thermal expansion coefficient of BN and that of mullite is minimized [8]. The use of a BN coating for ceramic matrix composite was first proposed by Rice [9].

Although there was some variability in the coating thickness because of the difficulty in controlling the CVD process, the thickness in the composite varied between 0.1 and 0.7 μm. X-ray diffraction analysis of the coatings showed that they were amorphous (i.e., turbostratic or nanocrystalline [10, 11]) in the as deposited state. The detailed chemistry of the coatings is proprietary and was not provided by the company. It is known that a BN coating from CVD usually contains some impurities such as carbon and oxygen [10,12,13].

The interface characteristics were determined by means of a dedicated interfacial testing system (ITS) at the Oak Ridge National Laboratory. The system has been described in ref. 14. The operation of this apparatus is very similar to that of a commercially available instrument, Nanoindenter, except that the ITS operates under displacement control conditions and loading is accomplished by pushing the sample upward into a fixed indenter. The testing procedure involves the following steps: 1) selection a fiber from the polished cross section of the composite with the aide of a microscope and a video system, 2) translation of the sample underneath a flat-bottomed indenter and 3) pushing the sample against a fixed indenter using a vertical translation stage. Different indenters (e.g., diamond or tungsten carbide) are available, and when testing composites containing fibers with diameters between 10 and 20 μm, -- including the tests reported here -- a flat-bottomed 10 μm in diameter diamond indenter is used.

With the current set-up, a fiber can be positioned underneath the indenter reproducibly within ±1 μm in the X-Y plane. During the test, both the load and fiber-end displacement are recorded continuously with 1mN and 0.01 μm resolution, respectively.

The operation of the ITS is fully computerized and automated for data acquisition and control, and user intervention is only required to select the fibers that will be tested. A schematic description of the ITS system is presented in Fig. 1. A schematic illustration of the sample geometry is presented in Fig. 2. This geometry allows for the determination of the interfacial parameters (e.g., debond and pushout stresses) as a function of the fiber embedded length using only one sample.

RESULTS AND DISCUSSION

The composites made with BN-coated fibers and the mullite powder synthesized in our laboratory showed a higher densification (e.g., 95% of theoretical density with fiber volume fraction of 0.16) than the composite made from the commercial mullite powder in spite of higher fiber content and lower hot-pressing temperature, 1300 °C [2, 3]. Such densification at this low temperature is not reported in previous studies on mullite matrix composites [6,7,15,16]. An optical micrograph of the polished surface of one of these composites is shown in Fig. 3a which demonstrates rather nonuniform fiber distribution in a dense mullite matrix. Figure 3b shows B^+ mapping of the coating by secondary ion mass spectrometry (SIMS) (17). Also, owing to the low processing temperature, the fibers maintained their original oval cross-section without deformation, see Fig. 3.

Typical raw load vs. fiber-end displacement curves from single fiber pushout tests are shown in Fig. 4 (a and b). Note that after fiber pushout, which coincides with load drop, the curve rises with upward concavity. This is because the 10 μm flat-bottomed indenter is very close in dimension to the minor axis of the fiber and soon after fiber pushout the indenter starts loading the matrix. Figure 5 shows the stress vs. fiber-end displacement curve after correcting for the contribution of the load train compliance to the total measured displacement. The load train compliance is determined from load vs.

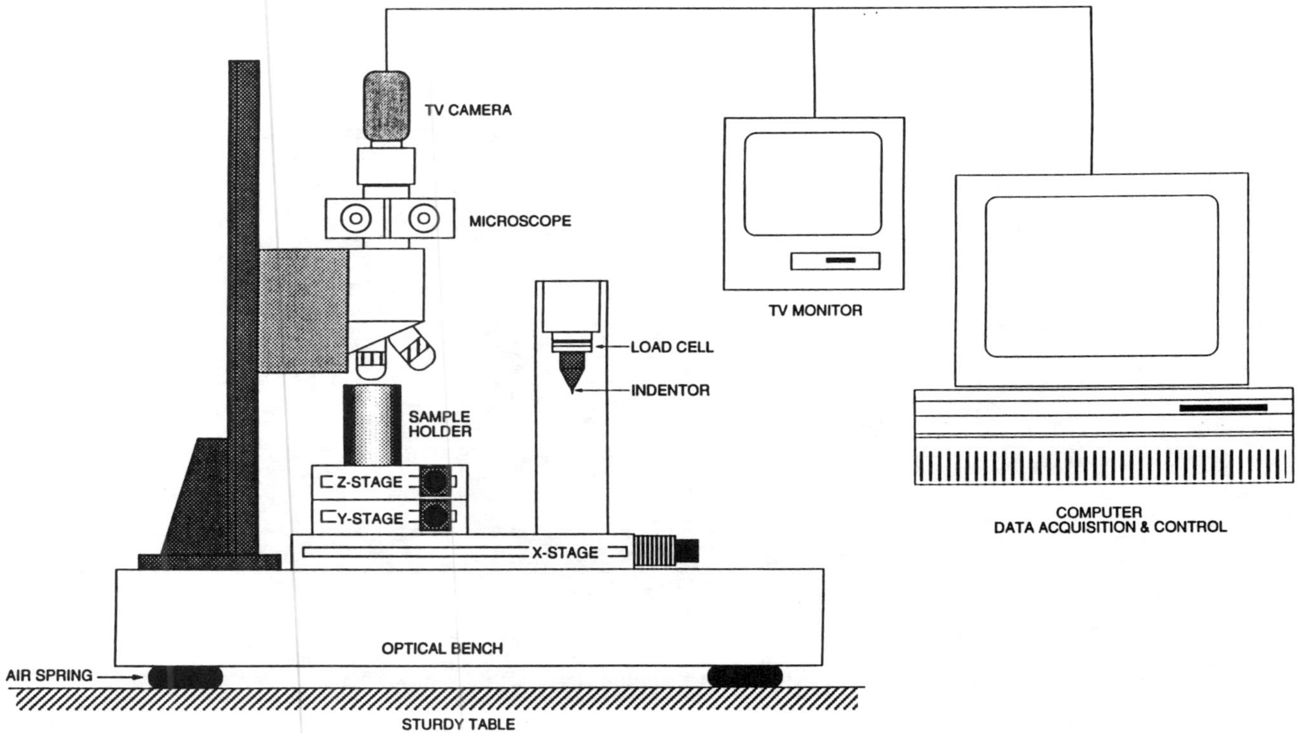

Fig. 1. A schematic of the interfacial testing system (ITS) system. ITS operates under displacement control conditions and loading is accomplished by pushing the sample upward into a fixed indentor.

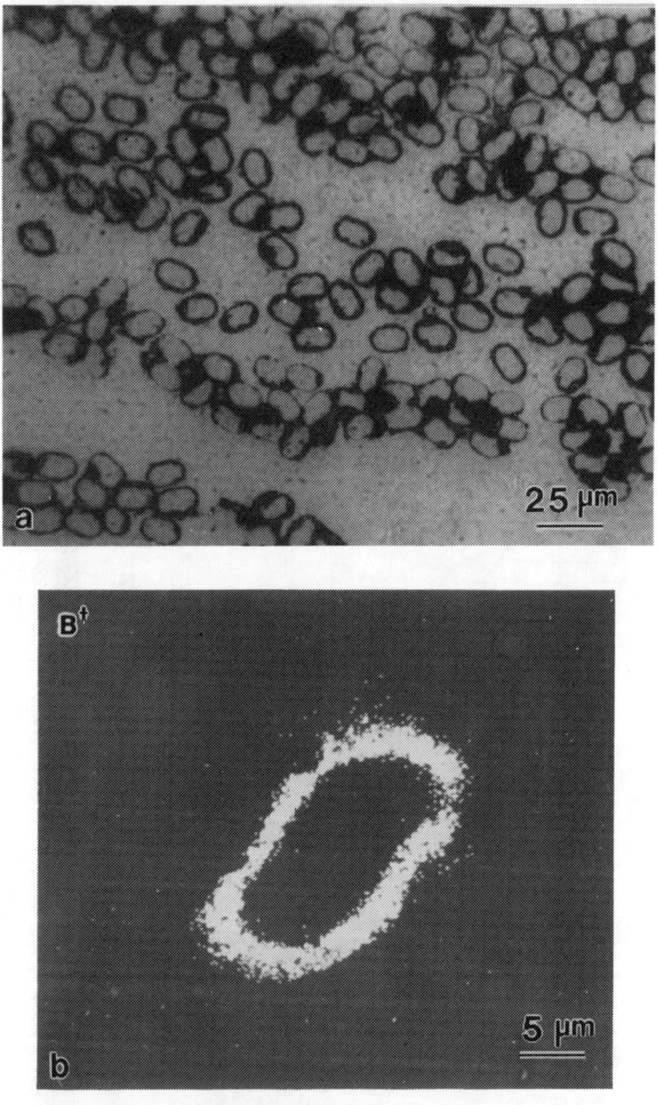

Fig. 3 (a) An optical micrograph of the polished surface of the composite. (b) B^+ mapping by secondary ion mass spectrometry (SIMS) showing the integrity of the BN coating after processing (17).

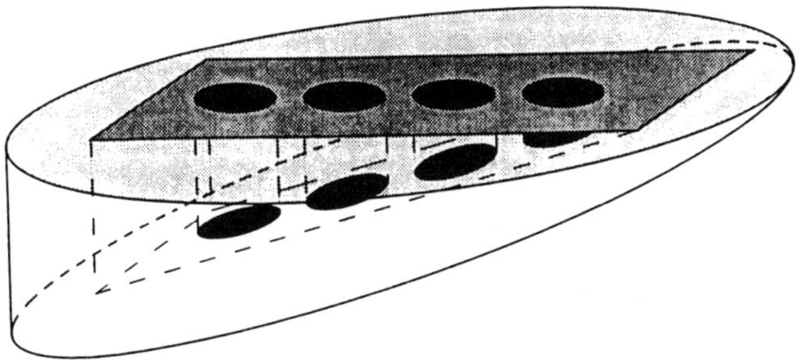

Fig. 2. Sample geometry for the interfacial testing system (ITS). The thickest part of the wedge is 700 μm and the diameter is 10 mm.

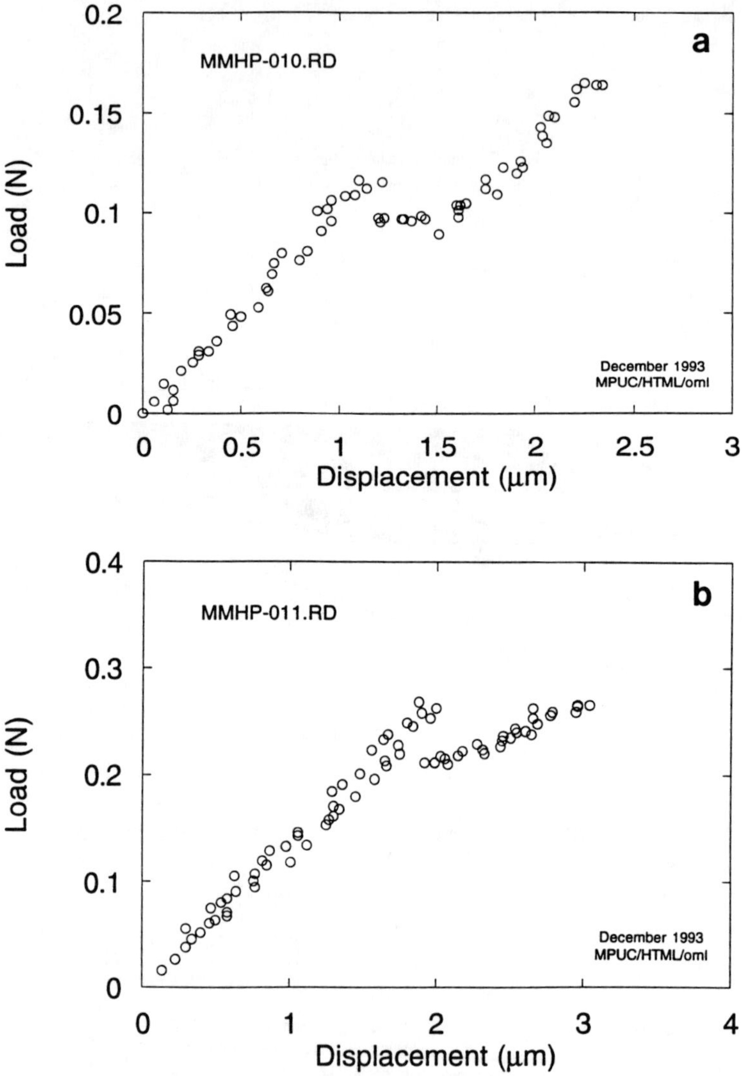

Fig. 4 (a and b) Typical raw load vs. fiber-end displacement curves from single fiber pushout tests. Note that after the fiber pushout, which coincides with load drop, the curve rises with upward concavity. This is because soon after debonding the indenter starts loading the matrix.

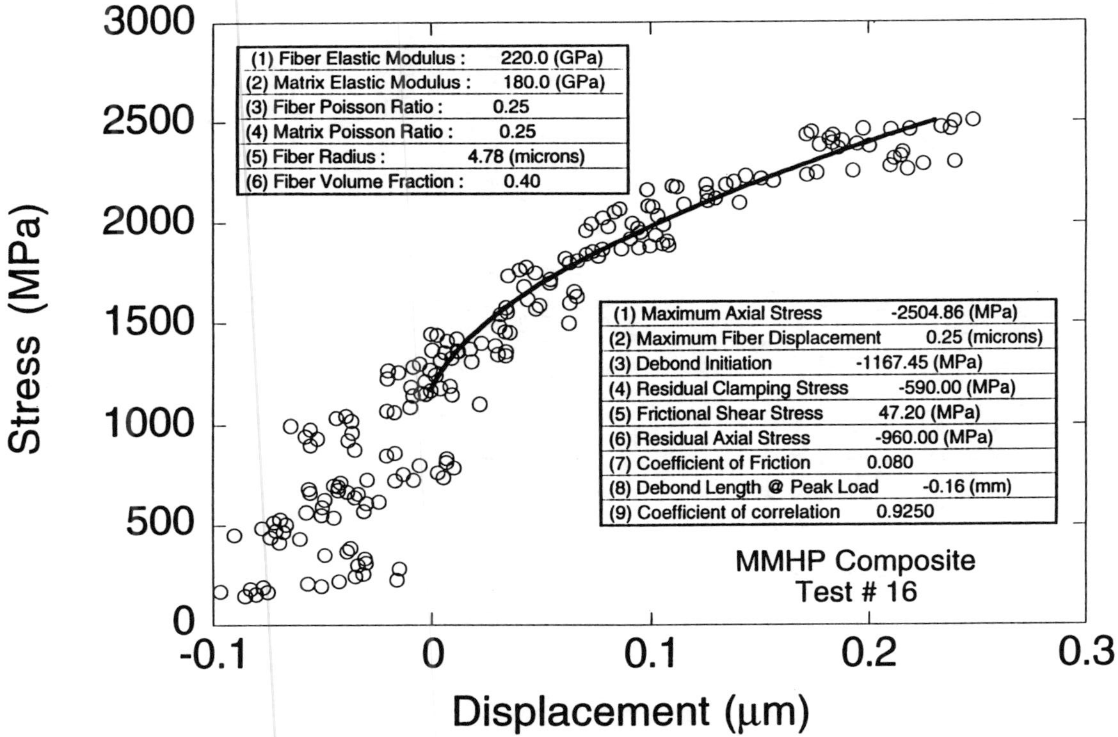

Fig. 5 Stress vs. fiber-end displacement curve after correcting for the contribution of the load train compliance to the total measured displacement. Also shown is the best fit of the data (from the initiation of debonding to the point of fiber pushout) using the progressive fiber debonding and sliding model of Hsueh (18).

displacement curves obtained by subjecting the sample to compression at various points along the wedge by means of a 500 μm diameter stainless steel rod. Typical values for the compliance of the ITS system and wedged continuous fiber reinforced ceramic composite sample are between 4 and 6 μm/N (14). Also shown is the best fit of the data (from the initiation of debonding to the point of fiber pushout) using the progressive fiber debonding and sliding model of Hsueh (18). It should be pointed out that the predictions of the strength-based model of Hsueh are equivalent to those of the energy-based model of Kerans and Parthasarathy (19,14). The frictional shear stress in the present system was about 50 Mpa.

Figure 6 shows the stress-displacement curves obtained in a three-point bend test while Fig. 7 shows the fracture surfaces of a BN coated fiber composite. Note the fiber pullout, which led to a higher work of fracture in the coated fiber composites than in the uncoated fiber composite.

CONCLUSIONS

The fiber-matrix interaction in mullite fiber-mullite matrix composites could be effectively controlled by the use of a low processing temperature and 1 μm thick BN coating. Composite processing temperature was decreased to 1300 ^{o}C by using a mullite powder synthesized via a diphasic gel route. This approach allowed densification with no fiber deformation and degradation. Thick BN coating survived processing and provided a weakly bonded interface which promoted fiber pullout during composite fracture. Fracture surfaces of BN coated composites obtained in flexure test showed fiber pullout.

A flat-bottomed, diamond indenter was used to determine the interfacial properties of the composite by means of single fiber pushout tests. Using a progressive debonding and sliding model, it was determined that the average interfacial frictional sliding shear stress in this composite was about 50 Mpa.

ACKNOWLEDGMENTS

This work was supported by the U.S. Office of Naval Research, contract No. N0014-89-J1459. Dr. S.G. Fishman was the Program Manager. Partial support was also obtained from the US Dept. of Energy, Assistant Secretary

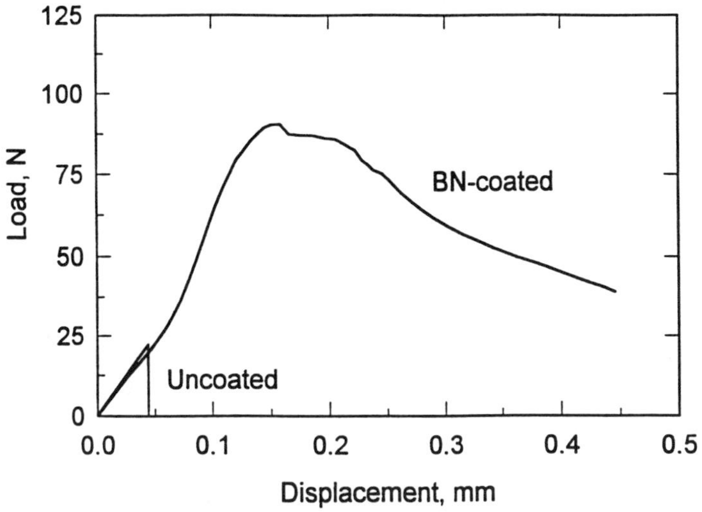

Fig. 6 Stress-displacement curves obtained in a three-point bend test. A higher work of fracture in the coated fiber composites than in the uncoated fiber composite can be seen.

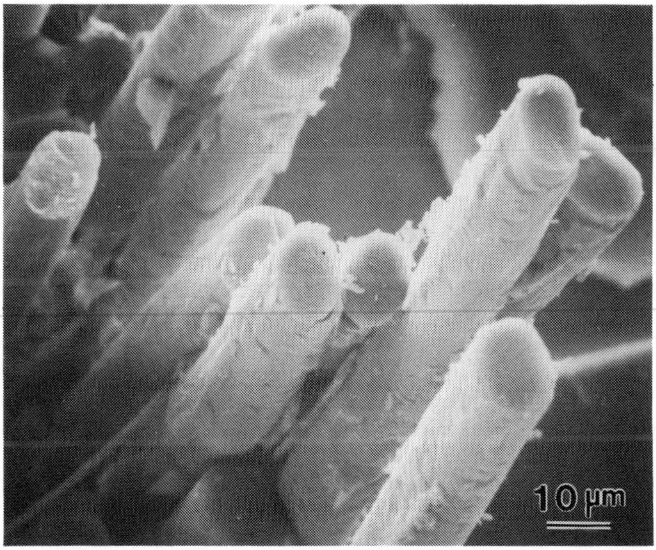

Fig. 7 Fracture surface of a BN coated fiber composite showing fiber pullout.

for Energy Efficiency and Renewable Energy, Office of Transportation Technologies, as part of the High Temperature Materials Laboratory User Program under contract DE-AC05-84OR21400 managed by Martin Marietta Energy Systems, Inc.

REFERENCES

1. J.-S. Ha and K.K. Chawla, *J. Mater. Sci. Lett.*, 12 (1993) 84.
2. J.-S. Ha and K.K. Chawla, and R. E. Engdahl, *Mater. Sci. & Eng.*, A161 (1993) 303.
3. J.-S. Ha and K.K. Chawla, *Ceramics International*, 19 (1993) 299.
4. M.D. Sacks, H.-W. Lee, and J.A. Pask, "A Review of Powder Preparation Methods and Densification Procedures for Fabricating High Density Mullite," in Ceramic Transactions, Vol. 6: **Mullite and Mullite Matrix Composites**, Edited by S. Somiya, R. F. Davis, and J. A. Pask, American Ceramic Society, Westerville, OH, 1990, p. 167.
5. I.A. Aksay, D.M. Dabbs, and M. Sarikaya, *J. Am. Ceram. Soc.*, 74 (1991) 2343.
6. R. N. Singh and A. R. Gaddipati, *J. Am. Ceram. Soc.*, 71 (1988) c-100.
7. J.R. Hellmann, D.J. Krause, R.E. Tressler, and D.P. Boldt, "Effect of High Temperature Exposure on Interfacial Shear Strength of Coated Silicon Carbide Fibers in a Mullite Matrix," in **High Performance Composites for the 1990's**, Edited by S.K. Das, C.P. Ballard, and F. Marikar, The Minerals, Metals & Materials Society, 1991,p.295.
8. D.M. Mattox, S.R. Gurkovich, J.A. Olenick, and K.M. Mason, *Ceram. Eng. Sci. Proc.*, 9 (1988) 1567.
9. R.W. Rice, *US Patent* 4,642,271, 1985.
10. W.Y. Lee, W.J. Lackey, and P.K. Agrawal, *J. Am. Ceram. Soc.*, 74 (1991) 2642.
11. T.T. Borek, X. Qiu, L.M. Rayfuse, A.K. Datye, R.T. Paine, and L.F. Allard, *J. Am. Ceram. Soc.*, 74(1991)2587.
12. R.N. Singh, in G.W. Cullen (ed.), *Proc. 10th Int. Conf. Chemical Vapor Deposition*, The Electrochemical Society, Penington, NJ, 1987, p. 543.
13. R. Naslain, O. Dugne, A. Guette, J. Sevely, C.R. Brosse, J.-P. Rocher, and J. Cotteret, *J. Am. Ceram. Soc.*, 74 (1991) 2482.
14. E. Lara-Curzio and M.K. Ferber, J. Mater. Sci., in press.
15. R.N. Singh and M.K. Brum, *Ceram. Eng. Sci. Proc.*, 8

(1987) 636.
16. O. Yeheskel, M. Balmer, and D.C. Cranmer, *Ceram. Eng. Sci. Proc.*, 9 (1988) 687.
17. K.K. Chawla, A. Choudhury, R. Venkatesh, and J.R. Hellmann, *Materials Characterization*, 31 (1993) 167.
18. C.-H. Hsueh, *J. Am. Ceram. Soc.*, 76 (1993) 3041.
19. R.J. Kerans and T.A. Parathasarathy, *J. Am. Ceram. Soc.*, 74 (1991) 1585.

INTERFACIAL PROPERTIES OF SiC (SCS-6) FIBER REINFORCED REACTION-FORMED SiC MATRIX COMPOSITES

P.J. Gaeta* and R.D. Sisson, Jr., Worcester Polytechnic Institute, Worcester, MA 01609; M. Singh and J.I. Eldridge, NASA Lewis Research Center, Cleveland, OH 44135

ABSTRACT

Fiber push-out tests were conducted on SiC (SCS-6) fiber reinforced SiC matrix composite samples to characterize the interfacial shear strength. Average debond shear and frictional sliding stresses were calculated. The effects of sample thickness and second phase addition in the matrix were evaluated. In addition to room temperature tests, a limited number of push-out tests were conducted at 800°C to ascertain any effect of residual stress on the debond shear strength of the interface. Tested samples were examined by SEM to determine the location of the interfacial failures.

INTRODUCTION

The performance of a composite is largely determined by the extent of any mechanical and/or chemical bonding at the fiber/matrix interface. A brittle matrix composite must have a relatively low interfacial shear strength for optimum properties. This requirement allows energy dissipation and crack bridging, a major toughening mechanism, to occur[1]. The fiber push-out test is a commonly used procedure for determining the interfacial shear strength and thus characterizing the interfacial properties of composites. The two main contributions to the interfacial shear strength of most composites are the debond shear stress and the frictional sliding stress. Both of these stresses may be calculated from the loads measured during the fiber push-out test. The objective of the present work is to examine the interfacial behavior of silicon carbide fiber reinforced silicon carbide matrix composites.

To the extent authorized under the laws of the United States of America, all copyright interests in this publication are the property of The American Ceramic Society. Any duplication, reproduction, or republication of this publication or any part thereof, without the express written consent of The American Ceramic Society or fee paid to the Copyright Clearance Center, is prohibited.

EXPERIMENTAL PROCEDURE

The reinforcing fibers used in this study were silicon carbide SCS-6 grade fibers, 142 µm in diameter with a carbon core surrounded by layers of β silicon carbide and a 3 µm outer carbon layer divided into three sublayers. The silicon carbide matrix was fabricated using a reaction-forming technique in which a porous carbon preform is infiltrated with liquid silicon. This technique has advantages such as reduced processing times and temperatures and near net shape forming over other processing methods[2]. There is some residual silicon present from this process which increases the modulus and thermal conductivity of the matrix but which limits the use to temperatures below 1415°C and may react to degrade the fibers. Alloy infiltrants can be added to reduce the activity of the silicon in the melt, thus protecting the fibers from reaction. The alloy also permits lower processing temperatures. In this work, silicon-1.7 at% molybdenum was chosen as the alloy infiltrant because it readily forms molybdenum disilicide, $MoSi_2$, with well known properties[3].

The processed composites were sliced perpendicular to the fiber axis into samples ranging in thickness from approximately 0.4 to 1.0 mm. The samples were then polished to a 1 µm finish and glued to a sample block containing 750 µm wide grooves that allow fibers to be pushed through unimpeded. For each sample, at least ten fibers were pushed in order to best represent the behavior of the system. The push-out test equipment is described in detail elsewhere[4]. In summary, a 100 µm diameter tungsten carbide, flat-bottomed cylindrical indenter was used to displace individual fibers in the composite slice. The sample block was mounted on a vertical translation stage which was aligned beneath the indenter. The stage then raised the sample block up at a constant speed, thereby applying a load to the tested fiber. The force required to displace the fiber and then to move the fiber against frictional resistance were continuously monitored via the signal from a load cell. In addition to the load signal, a magnified video recording and acoustic emission signal were recorded for each test in order to identify the features of the load profile. The limited number of high temperature push-out tests were performed at 800°C in vacuum (10^{-6} torr) in a separate test chamber described elsewhere[5].

RESULTS

Results were obtained for each test in the form of a load vs. time profile. A linear rise to the peak load was seen for all pure Si infiltrated samples, indicative of catastrophic debonding. Average debond and frictional shear stresses were calculated by dividing the measured load by the nominal contact area between the fiber and matrix. These stresses are given below in Table I. The debond stress seemed to remained constant and the frictional stress appeared to increase slightly with sample thickness; however, there was a large amount of scatter in the data.

Table I. Summary of Results for Pure Silicon Infiltrated Composite Samples

Thickness (mm)	Average Debond Shear Stress (MPa)	Average Frictional Shear Stress (MPa)
0.44	29.4 ± 9.7	15.7 ± 0.9
0.62	29.5 ± 12.6	21.6 ± 2.5

The profiles from the thinner sample showed large load drops upon fiber debonding followed by constant frictional loads. The profiles from the thicker sample showed smaller load drops and ramping of the frictional loads for many of the fibers; that is, the frictional load increased over time. It is believed that this is due to the increase in resistance as the interfacial layers crack or crumble. Typical load vs. time profiles for the 0.44 mm and 0.62 mm pure Si samples are shown in Figure 1.

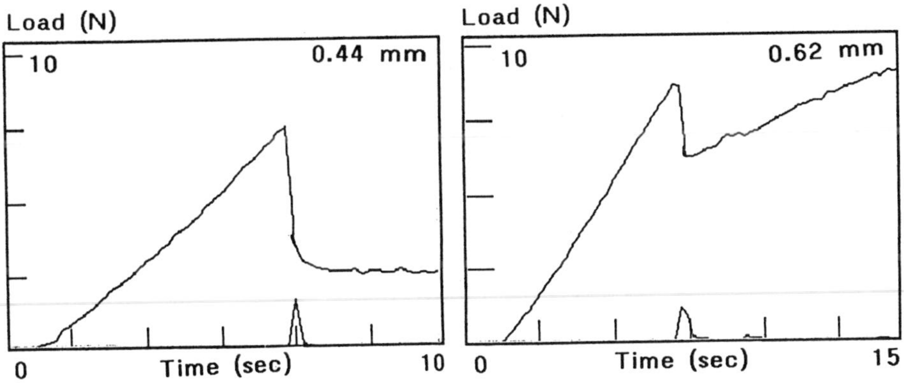

Figure 1. Typical Load vs. Time Profiles for 0.44 mm and 0.62 mm Pure Silicon Samples

Stress values for the Si-Mo infiltrated composite samples are given in Table II. The debond stress decreased with sample thickness; however, there was again a large variation in the data. In addition, the thicker sample showed progressive debonding. Despite this, the stress value reported was calculated from the final debond load using the same formulation as that for the catastrophic debonding, as a first approximation. A more complex analysis for this sample may be necessary. The frictional stress again increased with increasing sample thickness.

Table II. Summary of Results for Silicon-Molybdenum Alloy Infiltrated Composite Samples

Thickness (mm)	Average Debond Shear Stress (MPa)	Average Frictional Shear Stress (MPa)
0.43	85.2 ± 12.0	38.9 ± 8.5
0.95	56.9 ± 18.0	51.0 ± 18.2

Linear pre-debond behavior, pronounced load drops, and constant frictional loads were the features of the profiles for the thinner, 0.43 mm, silicon-molybdenum sample. However, non-linear behavior, indicative of progressive debonding, was evident for the thicker, 0.95 mm, Si-Mo sample. There was little or no load drop and an increase in the frictional load observed for nearly all fibers tested in the thicker sample. Typical load vs. time profiles are shown in Figure 2.

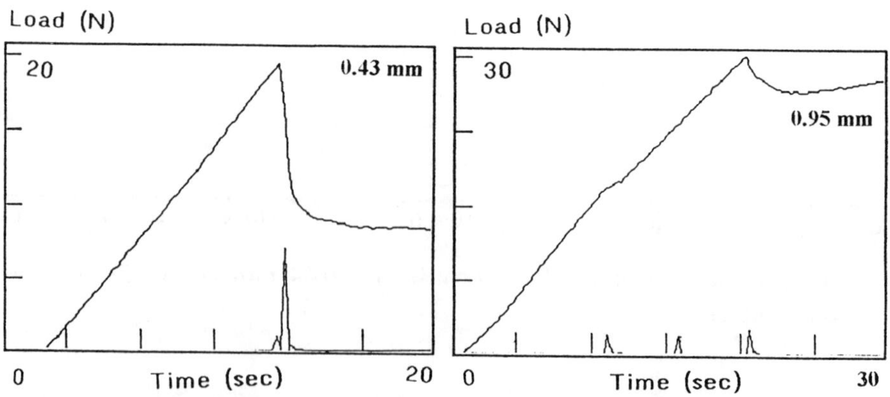

Figure 2. Typical Load vs. Time Profiles for Si-Mo Infiltrated Samples

High temperature push out tests were conducted on the 0.62 mm pure silicon and the 0.43 mm silicon-molybdenum infiltrated samples. For the pure Si sample, it was found that the debond and frictional shear stress increased as compared to the room temperature value. The load profiles showed a roughly linear rise to the peak load and a sharply ramping frictional load with little or no load drop. The high temperature debond stress decreased and frictional sliding stress increased as compared to the room temperature values for the Si-Mo sample. The high temperature profiles showed a roughly linear rise to the peak load, a small load drop, and ramping of the frictional load. Typical profiles are shown in Figure 3 and a summary of the high temperature results is given in Table III.

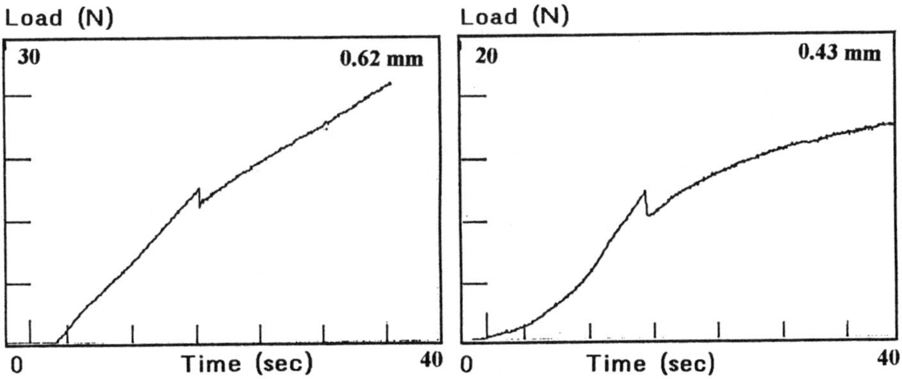

Figure 3. Typical Load vs. Time Profiles for Silicon and Si-Mo Samples Tested at 800 °C

Table III. Summary of Results From High Temperature Push-Out Tests

	Thickness (mm)	Average Debond Shear Stress (MPa)	Average Frictional Shear Stress (MPa)
Pure Si	0.62	39.4 ± 14.1	34.8 ± 14.3
Si-Mo	0.43	63.9 ± 9.1	47.6 ± 10.3

Samples were then examined by scanning electron microscopy. In all cases, it was found that failure occurred either between the fiber and the innermost carbon sublayer or between the carbon sublayers. It was also evident that the outermost carbon sublayer always remained bonded to the matrix. For the thinner, pure silicon infiltrated sample, a roughly equal number of fibers debonded at the two locations. When failure occurred between the carbon sublayers, the exposed sublayer was either cracked or broken, or remained bonded to the fiber. Figures 4 and 5 are examples of fibers debonding at the two locations. The fibers in the thicker pure Si sample debonded nearly every time at the fiber-innermost sublayer interface. Fiber surfaces were smooth and undamaged for both Si samples.

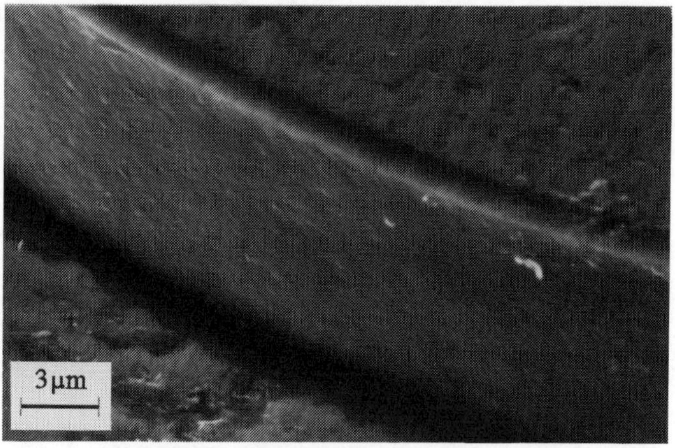

Figure 4. SEM Micrograph of Fiber Debonding Between Carbon Sublayers

Figure 5. SEM Micrograph of Fiber Debonding at Fiber/Innermost Carbon Layer

The fibers in the silicon-molybdenum infiltrated samples debonded in the same manner as those of the pure Si infiltrated samples; that is, for the thinner sample, failure occurred at either the interface between the fiber and innermost carbon sublayer or between the carbon sublayers and for the thicker sample failure occurred at the fiber/innermost carbon sublayer interface. In comparison to the silicon infiltrated samples, there was more matrix damage around the fibers and the fiber edges were slightly rougher. Figures 6 and 7 show fibers from these samples.

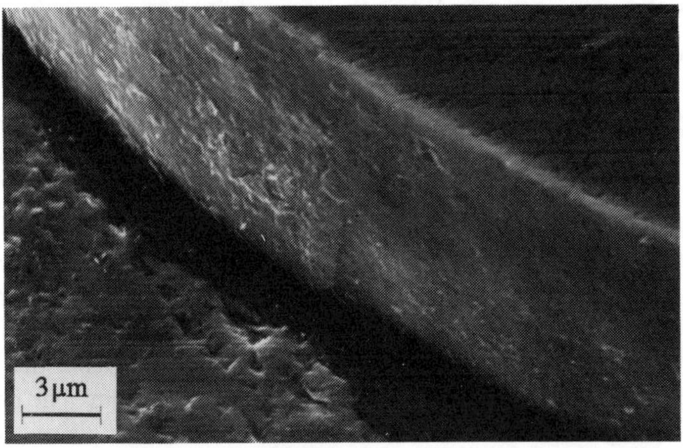

Figure 6. SEM Micrograph of Fiber Debonding at Fiber/Innermost Carbon Layer

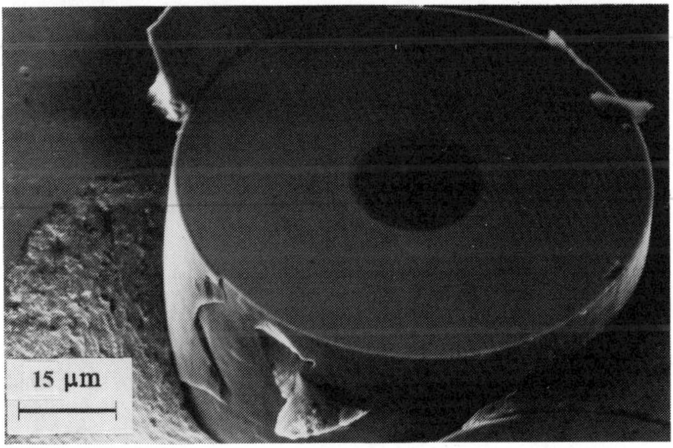

Figure 7. SEM Micrograph of Debonded Fiber Showing Matrix Damage

For the pure silicon infiltrated sample tested at 800°C, a roughly equal number of fibers debonded at the fiber-innermost sublayer interface and the interface between the carbon sublayers. The exposed sublayers were again cracked or broken and the fiber edges appeared relatively smooth.

For the high temperature Si-Mo infiltrated sample, failure occurred nearly always between the carbon sublayers. In addition, the exposed sublayers were badly damaged.

DISCUSSION

Although additional push-out tests are needed in order to complete the data sets, some trends were found. For example, the debond shear stress was greater for composite samples infiltrated with the silicon-molybdenum alloy as compared to the pure silicon. This could be explained by the presence of reactions taking place at the interface, such as the formation of carbides, which would increase the debond strength. However, no such reaction products have been detected thus far. An alternative explanation is that the formation of molybdenum disilicide in the matrix during processing may be increasing the thermal expansion coefficient of the matrix relative to the fiber. This would result in a compressive stress on the fibers upon cooling from the processing temperature, thus increasing the load needed to debond them. The graph below compares the thermal expansion coefficients for SiC, Si, C, and $MoSi_2$.

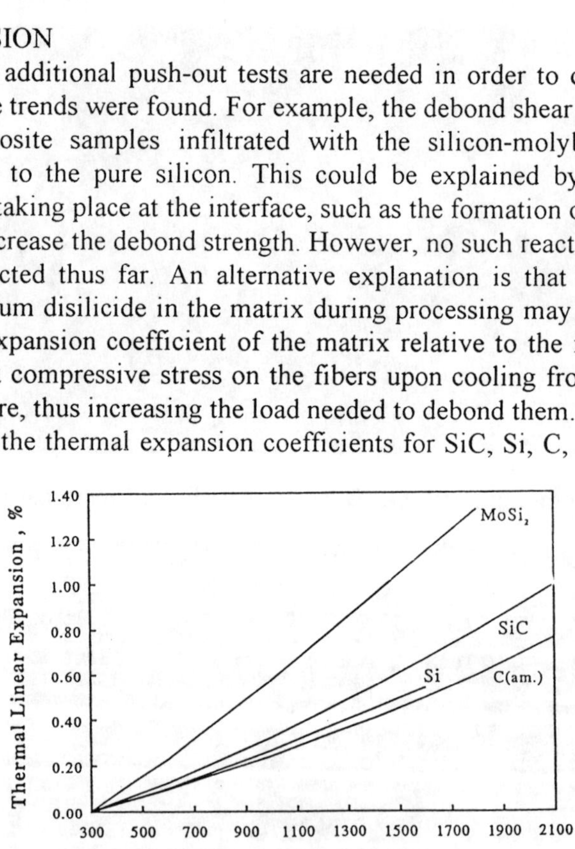

Figure 8. Thermal Expansion vs.Temperature for Composite Constituents[6]

The effect of thermal expansion mismatch was also seen in the push-out test results at high temperature. The debond shear stress for the Si-Mo infiltrated composite sample was found to decrease at high temperature in comparison to the room temperature value. It is believed that this was due to the compressive stress on the fibers from the CTE mismatch between the fibers and the matrix which was then relieved at the higher temperature, allowing the fibers to debond more easily.

The debond shear stress for the pure silicon infiltrated sample behaved in the opposite manner, i.e., the stress increased at high temperature relative to the room temperature value. This may again be explained by the CTE mismatch; that is, if the SiC fiber shrinks more than the SiC matrix containing free silicon, then a tensile stress at the fiber/coating interface may be created. This stress would then be relieved at elevated temperatures causing the debond stress to increase.

The locations of interfacial failure also appear to be influenced by the CTE mismatch. The percentage of fibers debonding between the carbon sublayers increased with increasing temperature. Therefore, if the residual stresses are relieved at the higher temperature, the carbon/carbon sublayer interface becomes the weakest link, as expected.

CONCLUSIONS

The interfacial behavior of SiC (SCS-6) fiber reinforced reaction-formed SiC matrix composites was examined. It was found that the use of an alloy (Si-1.7%Mo) infiltrant altered the debond and frictional shear stresses of the composite. This was likely due to the presence of molybdenum disilicide in the matrix and the resulting residual compressive stresses caused by the coefficient of thermal expansion mismatch. The results of push-out tests at high temperature supported this conclusion. The results of high temperature tests on the silicon infiltrated samples suggested the relief of residual tensile stresses. The shift in the location of interfacial failure for all samples at the elevated temperature was consistent with these trends thus emphasizing the importance of residual stresses in this composite system.

REFERENCES

1. A.G. Evans and D.B. Marshall, "The Mechanical Behavior of Ceramic Matrix Composites", Acta metall, 37 (10), 2567 (1989).

2. D.R. Behrendt and M. Singh, "Effect of Carbon Preform Pore Volume and Infiltrants on the Composition of Reaction-Formed Silicon Carbide Materials", J. Mat. Syn. and Proc., 2(2), 133 (1994).

3. R.P. Messner and Y. Chiang, "Liquid-Phase Reaction-Bonding of Silicon Carbide Using Alloyed Silicon-Molybdenum Melts", J. Am. Ceram. Soc., 73 (5), 1193 (1990).

4. J.I. Eldridge, "Desktop Fiber Push-Out Apparatus", NASA TM-105341, 1991.

5. J.I. Eldridge and B.T. Ebihara, "Fiber Push-Out Testing Apparatus for Elevated Temperatures, J. Mater. Res., 9 (4), 1035 (1994).

6. Y. Touloukian, R. Kirby, R. Taylor and T. Lee, Thermal Expansion: Non-Metallic Solids, Vol. 13, Plenum, New York, 1977.

PLASMA-ENHANCED CHEMICAL VAPOR DEPOSITION OF SILICON NITRIDE ON SILICON CARBIDE FIBERS

C. D. Stinespring, D. F. Collazos, and R. K. Gupta,
Department of Chemical Engineering, West Virginia University, Morgantown, WV 26506-6101;
D. N. Braski,
Oak Ridge National Laboratory, High Temperature Materials Laboratory , Oak Ridge, TN 37831-6064.

ABSTRACT

Near stoichiometric Si:N coatings were deposited by means of PECVD on SCS-6 SiC fibers which contained a carbon-rich coating. Weight loss associated with oxidation of the outer carbon-rich coating of the as-received SiC fibers was greatly reduced for the Si:N coated SiC fibers even after 10 h heat-treatment in oxygen at 800°C. Auger Electron Spectroscopy (AES) was used to obtain elemental compositions of the as-received and Si:N coated SiC fibers after heat-treatment. Negligible amounts of oxygen were found at the carbon-rich coating of the heat-treated Si:N coated SiC fiber. These results clearly prove the effectiveness of PECVD silicon nitride coating as an oxygen diffusion barrier.

INTRODUCTION

Fiber-reinforced ceramic matrix composites are ideal structural materials for heat-engine components because of their ability to withstand high temperatures, their low density, high toughness, and graceful failure[1]. It is known that fiber/matrix interfacial shear strength strongly influences load transfer from matrix to fiber and the overall fracture properties of the composite[2]. Generally, for a fiber-reinforced ceramic matrix composite, a relatively weak interface is desirable since it deflects the matrix crack around the fibers without breaking them and prevents catastrophic failure of the composite[3].

One system which shows promise as a structural component for high temperature applications is the SiC fiber-reinforced, reaction-bonded silicon nitride composite (SiC/RBSN). In the as-fabricated condition, the SiC/RBSN

To the extent authorized under the laws of the United States of America, all copyright interests in this publication are the property of The American Ceramic Society. Any duplication, reproduction, or republication of this publication or any part thereof, without the express written consent of The American Ceramic Society or fee paid to the Copyright Clearance Center, is prohibited.

composite has a desirable non-catastrophic stress-strain curve, load-bearing capability which greatly exceeds the first matrix cracking stress, and properties superior to that of unreinforced RBSN[4,5]. These excellent mechanical properties are attained by the use of high strength, high elastic modulus, SiC fibers having a carbon coating which provides the desired weak interfacial properties for assuring matrix crack deflection and load transfer. However, it has been shown severe degradation of the composite mechanical properties after exposure to an oxidizing environment in the temperature range of 400°C to 1000°C[6]. It was found that there was complete oxidation of the outer carbon-rich coating on the SiC fibers and this was responsible for severe degradation of the mechanical properties due to loss of bonding between the fiber and the matrix.

One method to protect the outer surface of the SiC fibers from oxygen attack is by means of hot isostatic pressing (HIPing)[7]. The purpose of this technique is to densify the reaction-bonded Si_3N_4 matrix to avoid oxygen diffusion through pore channels within the composite. However, this leads to a decrease in the strength of the SiC fibers due to the formation of coarsened SiC grains during HIPing at temperatures above 1400°C.

On the other hand, plasma-enhanced CVD can produce ceramics with better properties than those of ceramics fabricated by other processes. For example, the product of PECVD is stronger, harder, more corrosion, and oxidation resistant than the same compound consolidated by sintering or reaction-bonding. The improved properties of PECVD materials are attributable to fine grain size, high density and purity. Hence, a major advantage of PECVD is the ability to coat SiC fibers with a non-porous Si:N film, and this film has a similar coefficient of thermal expansion as the RBSN matrix (since the two are the same phase). Mismatches in CTE values have to be avoided since they cause cracks and catastrophic failure upon heating. Another advantage of plasma-enhanced CVD is the ability to deposit films at relatively low temperatures (300-500°C), so that fiber degradation is minimized.

The objectives of this study were to coat SiC fibers having an outer carbon-rich layer with a protective coating of silicon nitride and to evaluate the effectiveness of this coating as a diffusion barrier against oxygen attack. Scanning Electron Microscopy (SEM) and Scanning Auger Microprobe (SAM) were used to characterize and identify any oxygen contamination on the Si:N coated SiC fibers after heat-treatment in an oxidizing environment.

EXPERIMENTAL PROCEDURE

Materials Used

The filaments used in this study were SCS-6 SiC fibers (Textron Specialty Materials Division, Lowell, MA). These fibers were chemically vapor deposited from methyl-trichlorosilane on a resistively heated carbon substrate. The 142 μm diameter SiC fibers consist of an inner pyrolytic graphite-coated carbon core (34 μm) followed by a SiC sheath. The outer surface of the SiC fiber consists of a 4 μm carbon-rich coating with SiC particles embedded in it. The purpose of the carbon-rich coating is to heal the SiC crystalline surface for improved surface strength and to provide the desired fiber/matrix weak interfacial properties for assuring matrix crack deflection and graceful failure. Figure 1 shows a SEM micrograph of the as-received SCS-6 SiC fiber revealing the interior of the fiber and the carbon-rich coating. A more complete description of the fiber is given elsewhere[8].

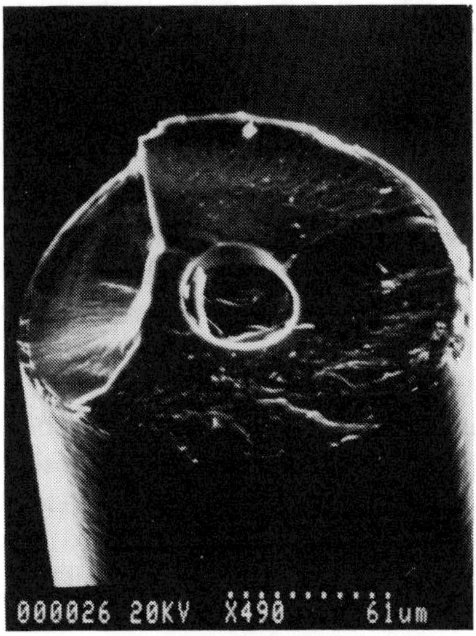

Figure 1. As-received SCS-6 SiC fiber

Silicon Nitride Deposition

Near stoichiometric Si:N coatings were deposited on SCS-6 SiC fibers in a pyrex tubular reactor (10.5 mm i.d.) by means of microwave-plasma CVD from a mixture of nitrogen and silicon tetrachloride. However, at this point, we are not sure whether the coatings are crystalline or amorphous. The deposition conditions were a $SiCl_4$:N_2 ratio of 0.003, a total reactor pressure of 280 Pa (2.1 torr), and 200-W microwave discharge power at 2450 MHz frequency. Preliminary experiments were done to determine the effect of deposition parameters on film Si:N ratio. Compared to ammonia, which is the most frequently used nitrogen precursor, the use of nitrogen has the advantage of producing silicon nitride films with a lower hydrogen content. Hydrogenated silicon nitride coatings causes severe film cracking due to the loss of Si-H groups

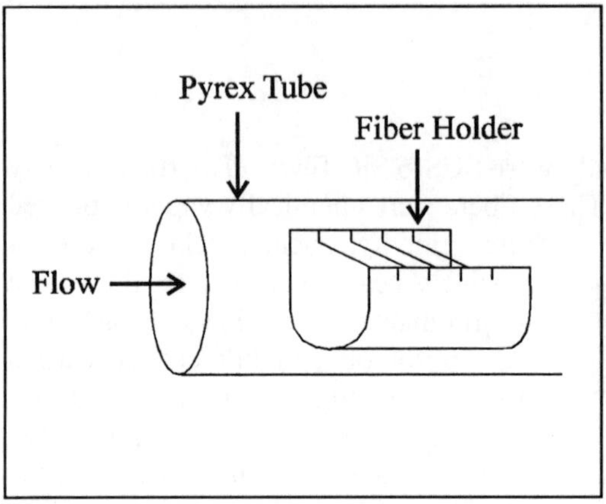

Figure 2. Fiber holder configuration

upon heating[9,10]. In this case, however, because of the less active nitrogen (high dissociation energy), the $SiCl_4$ to nitrogen ratio was kept low in order to get a near stoichiometric film. The SiC fibers were held by a slotted pyrex boat as shown in Figure 2. Axial and transverse coating thickness nonuniformities were found. However, this is actually an advantage since it allows fibers of different Si:N coating thicknesses to be prepared under the same conditions for further analysis. Figure 3 is a SEM micrograph of the silicon nitride coated SCS-6 SiC fiber. This picture also shows Si:N coating/fiber debonding after fracture. This behavior is typical of matrix crack deflection and graceful failure in a fiber-reinforced ceramic matrix composite. Hence, the desired weak interfacial properties of the outer carbon-rich coating remained intact after plasma deposition.

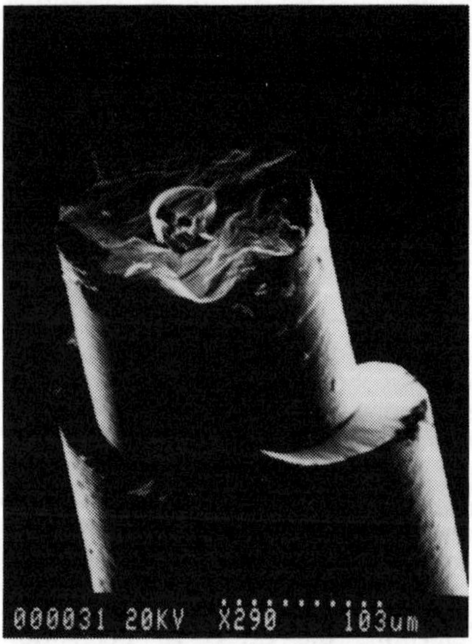

Figure 3. Si:N coated SiC fiber

Heat-Treatment

As-received and Si:N coated SCS-6 SiC fibers were heat-treated in a thermogravimetric analysis (TGA) furnace (Cahn Instruments, Cerritos, CA) at 800°C in oxygen for 10 hours. Medical grade oxygen (99.6%) was continuously fed through the TGA furnace at a rate of 50 cm^3/min. A platform of titanium foil suspended from a nichrome stir pan was used to secure the fibers in the TGA furnace. Heat-treatment effects were determined by measuring the weight of the fibers before and after heat-treatment.

Microstructure Characterization

Elemental compositions of the outer and fractured surfaces of the as-received and Si:N coated SCS-6 SiC fibers was obtained with the help of Auger electron spectroscopy (AES). All Auger electron spectroscopy (AES) analyses were conducted at Oak Ridge National Laboratory using a Perkin-Elmer PHI 660 scanning Auger microprobe. Generally, an accelerating voltage of 3 kV and beam current of 10 nA were used, but in a few instances, when specimen charging was observed, it was necessary to use 2 kV and 15 nA. All analyses were conducted with a chamber pressure < 2.7 x 10^{-7} Pa (< 2 X 10^{-9} torr). Outer surface layers were removed (since they have been exposed to air) so that underlying material could be analyzed by sputtering with argon ions for one minute at 3.5 kV. A sputtering rate of 20 nm/min was assumed based on past calibration for SiO_2 films. The sensitivity factors for 3 kV were those supplied with the PHI software, except for silicon which was determined by using a NIST-112b SiC standard. The sensitivity factor for silicon at 2 kV was also determined with the SiC standard while those for the other elements were estimated by comparing spectra from the sample at both 2 and 3 kV when specimen charging was not a problem. The accuracy of the AES results for this study was estimated to be ±3 at.%.

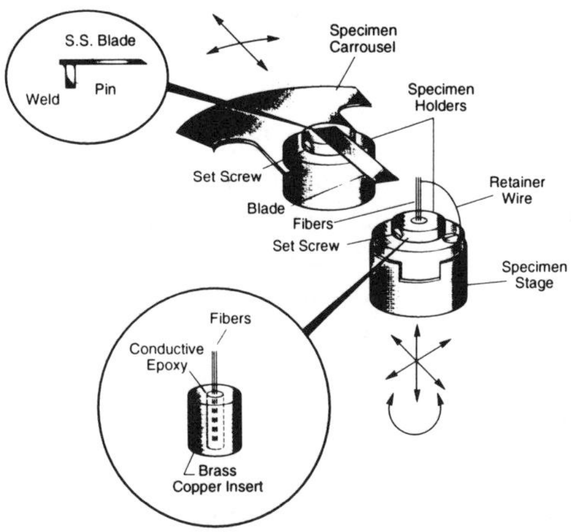

Figure 4. In-situ fracture apparatus

The as-received and Si:N coated SiC fibers analyzed in this study were fractured in situ by mounting them in small brass copper inserts using conductive epoxy as shown in Figure 4. The insert was then placed in a fracture holder and secured with a set screw. Fracture was initiated by moving a stainless steel knife across a filament to break it in bending. A thin copper wire was attached to the top of the filament with a bead of epoxy to prevent it from dropping into the specimen chamber after fracture. In-situ fracturing has the advantage of obtaining elemental composition through the thickness of the sample without the need for long sputtering times. The in-situ fracture and AES analysis are described in detail elsewhere[11].

RESULTS AND DISCUSSION

Heat-Treatment

The as-received SCS-6 SiC fibers had a 4.5 ± 1.0 % weight loss during oxygen heat-treatment at 800°C for 10 hours in the TGA. On the other hand, the Si:N coated SCS-6 SiC fibers had only a 0.5 ± 0.25 % weight loss. All coated fibers had different coating thicknesses varying from 80 μm to 8.0 μm. It was found that percent weight loss was independent of silicon nitride coating thickness. A previous study has shown that weight loss is associated with complete oxidation of the carbon-rich coating on the SiC fibers. Hence, these results clearly show the effectiveness of the PECVD silicon nitride coating as a protective barrier against oxygen attack.

Microstructure Characterization

Table I. Composition (atomic percent) of the as-received SiC fiber

Analysis Area		%Si	%C	%O	%N
Outer surface		5.2	80.1	14.7	-
Outer sfc (1 min sputter)		23.0	72.7	4.3	-
SiC profile	area 1	11.7	75.8	12.5	-
(after fracture)	area 2	12.8	77.4	9.2	-
	area 3	13.8	77.0	9.2	-
	area 4	10.7	82.7	6.6	-
	area 5	8.7	84.5	6.8	-
	area 6	4.1	92.0	3.9	-
Inner core		-	100	-	-

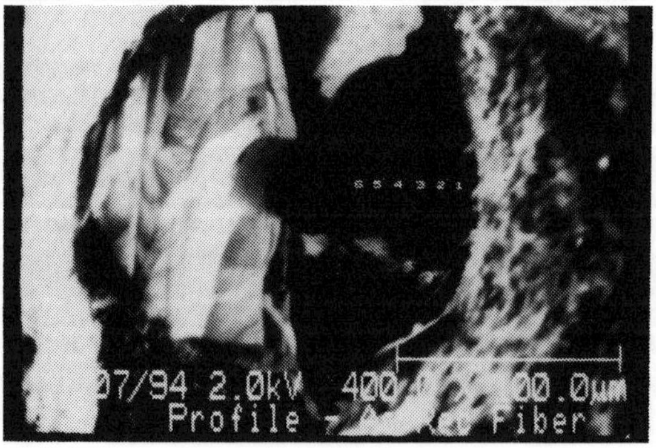

Figure 5. Fractured surface of the as-received SiC fiber

AES elemental characterization was done on the heat-treated and untreated as-received SCS-6 SiC fibers, and on a heat-treated Si:N coated SCS-6 SiC fiber of 60 μm coating thickness. The elemental compositions for the outer and fractured surfaces of the as-received SiC fiber are shown in Table I. Figure 5 is a SEM micrograph of the fractured surface of the as-received fiber showing six analysis points with their chemistry given in Table I. The outer surface shows the presence of a small amount of Si (5.2 at.%) along with C and O. Sputtering off about 20nm of material increased the Si and decreased the O. Thus, oxygen was incorporated at the fiber outer surface from exposure to air. The ratio of

Si/C in the SiC sheath was considerably less than 1, so it appears that the SiC fiber is non-stoichiometric. This observation is consistent with values obtained from Ning and Pirouz[8] for SiC regions near the carbon core (analysis points #4 - #6 in Fig. 5). The fiber inner core is 100% C.

Table II. Composition (atomic percent) of the heat-treated as-received SiC fiber

Analysis Area		%Si	%C	%O	%N
Outer surface		-	51.4	48.6	-
Outer sfc (1 min sputter)		-	53.8	46.2	-
SiC profile	area 1	10.3	84.3	5.4	-
(after fracture)	area 2	21.1	73.0	5.9	-
	area 3	14.3	81.8	3.9	-
	area 4	13.0	83.8	3.2	-
	area 5	7.9	81.1	11.0	-
	area 6	-	77.3	22.7	-
Inner core		-	100	-	-

Figure 6. Fractured surface of the heat-treated as-received SiC fiber

The elemental compositions of the outer and fractured surfaces (Fig. 6) of the 10 h heat-treated as-received SiC fiber are given in Table II. As expected, the outer surface has 3-4 times more oxygen than the as-received fiber. Also, a complete depletion of Si was detected at the outer surface. Si depletion was also found for SCS-6 SiC fibers heat-treated at 1200°C in N_2[12]. Elemental composition of the outer surface remained nearly the same after sputtering for 1 min. This may suggest that oxidation occurred through the thickness of the carbon-rich coating. The composition of the SiC sheath after heat-treatment is similar to the as-received except that O may be slightly lower. Analysis point #6 shows Si depletion and an increase in the amount of oxygen near the core. Again, only C is detected in the inner core.

The elemental compositions for the outer and fractured surfaces (Fig. 7) of the heat-treated Si:N coated SiC fiber are given in Table III. Note that this specimen is incorrectly labeled 80 μm coating thickness rather than 60 μm in the SEM micrographs. The outer surface is essentially C and O. Oxygen probably comes from exposure to oxygen gas during heat-treatment; however, it is not clear where all the carbon is coming from. The possible sources of carbon might be contaminants in the oxygen cylinder used for heat-treatment and/or backdiffusion

Table III. Composition (atomic percent) of the heat-treated Si:N coated SiC fiber

Analysis Area		%Si	%C	%O	%N
Outer surface		-	75.8	24.2	-
Outer sfc (1 min sputter)		32.9	16.2	7.82	43.1
Si:N profile	area 1	38.1	-	10.1	51.8
(after fracture)	area 2	36.4	-	11.1	52.5
	area 3	36.4	-	11.8	51.8
	area 4	33.5	-	11.9	54.6
C-rich coating	area 5	5.5	94.5	-	-
SiC profile	area 6	32.5	62.4	5.1	-
	area 7	28.1	66.9	5.0	-
	area 8	22.9	72.8	4.3	-
	area 9	15.8	80.8	3.5	-
Inner core	area 10	-	100	-	-
Torn C-rich coating (Fig.8)		-	100	-	-
Exposed surface (Fig.8)		4.3	94.7	1.0	-
Exp sfc (15 sec sputter)		9.4	89.2	1.4	-
Exp sfc (30 sec sputter)		10.0	88.4	1.6	-
Exp sfc (2.5 min sputter)		9.0	91.0	-	-

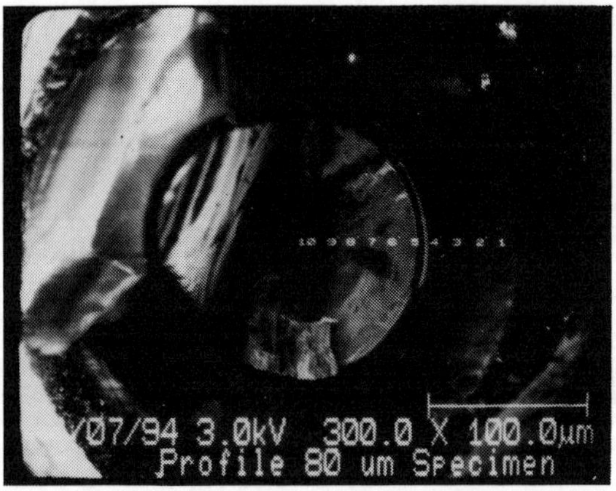

Figure 7. Fractured surface of the heat-treated Si:N coated SiC fiber

of pump oil during evacuation of the reactor after deposition. After sputtering for 1 min, the O and C decreased markedly, but not completely, and revealed Si and N at a close 3:4 ratio. The analyses through the Si:N coating (analysis points #1-#4)show reasonable ratios of Si/N and consistent O values from 10 to 12 at. %. Oxygen was probably incorporated in the Si:N coating during plasma deposition. The pyrex tube could also be a possible oxygen source. Also note that C was not present through the Si:N coating; this confirms that carbon was somehow incorporated at the outer surface only after deposition. The carbon-rich shows essentially C, some Si, and no oxygen . The SiC sheath again has non-1:1 ratios of Si:C with O values fairly consistent with those found in the heat-treated as-received filament and only slightly lower than those found in the as-received specimen. The inner core is 100% C. Additional analyses (Table III) were conducted on the carbon-rich coating which separated from the SiC fiber and the Si:N coating as shown in Figure 8. Only carbon and no oxygen were found in the carbon-rich coating. Immediately at the surface, exposed by the torn carbon-rich coating, the composition is C-rich with small amounts of Si and even less O. Sputtering increased the amounts of Si detected. It can be assumed that continued sputtering would have produced a composition similar to those found in the SiC

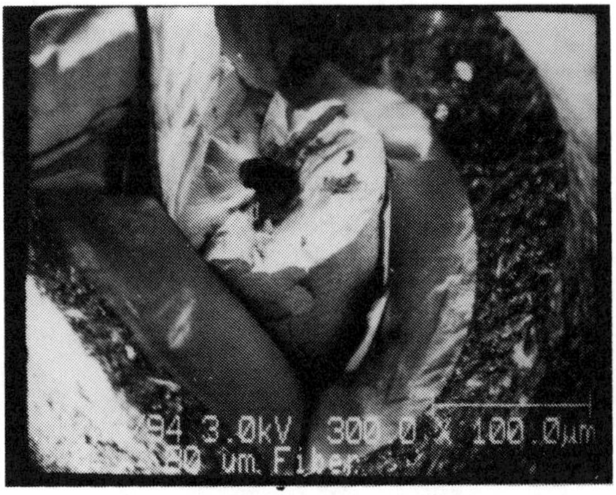

Figure 8. Carbon-rich coating that debonded from the heat-treated Si:N coated SCS-6 SiC fiber

areas.

These analyses confirm the data obtained from heat-treatment. The amount of oxygen at the carbon-rich coating of the heat-treated Si:N coated SiC fiber is negligible (0-1 at.%) in comparison to the amount of oxygen at the carbon-rich coating in the as-received fiber (48.6 at.%). Hence, this proves the effectiveness of the Si:N coating in protecting the carbon-rich coating from oxygen. The excellent properties of the Si:N coating as a diffusion barrier are attributable to its density (non-porous), its thermal stability (no hydrogen incorporation), and a good thermal expansion compatibility with the SiC fiber that resulted in a crack-free coating.

SUMMARY AND CONCLUSIONS

Near stoichiometric Si:N coatings were deposited on SCS-6 SiC fibers by means of PECVD. The microstructure characterization of both the as-received and the Si:N coated SiC fibers after 10 h exposure in oxygen at 800°C was characterized using AES. From the results of this study, the following conclusions can be drawn:

1. The weak interfacial properties of the outer carbon-rich coating on the SCS-6 SiC fibers remained intact after plasma deposition. This is probably because of the low processing temperature of PECVD.

2. The weight loss for the Si:N coated SiC fibers was much less than the weight loss for the as-received SiC fibers after heat-treatment in oxygen.

3. A considerable amount of oxygen (48.6 at. %) was present at the outer surface of the heat-treated as-received SiC fiber.

4. The elemental composition of the SiC sheath and inner carbon core remained nearly the same for all the specimens.

5. Elemental analyses on the heat-treated Si:N coated SiC fiber show consistent values of Si, N, and O through the thickness of the silicon nitride coating.

6. Negligible amounts of oxygen (0-1 at. %) were detected at the carbon-rich coating of the heat-treated Si:N coated SiC fiber. Therefore, silicon nitride coatings produced by PECVD were proven to be effective diffusion barriers against oxidation at 800°C.

7. Therefore, silicon nitride coatings produced by PECVD were proven to be effective diffusion barriers against oxidation at 800°C.

ACKNOWLEDGEMENTS

The authors would like to thank Mr. J. Harris and Ms. D. Berry for their assistance with the SEM, Mr. M. A. Mittnick from Textron Specialty Materials for the provision of SCS-6 SiC fibers, Dr. D. B. Dadyburjor for the use of the TGA furnace, and Dr. M. Madhukar for helpful discussions.

REFERENCES

1. A. J. Eckel, T. P. Herbell, "Ceramic Matrix Composites for Rocket Engine Turbine Applications," presented at the Gas Turbine and Aeroengine Congress, Cologne, Germany (1992).

2. D. B. Marshall, "Interfaces and Toughening in Ceramics," Journal de Physique, colloque C5, supplement 10, tome 49 (1988).

3. R. T. Bhatt, "Properties of Silicon Carbide Fiber-Reinforced Silicon Nitride Matrix Composites," Whisker-and Fiber-Toughened Ceramics, R. A. Bradley, et al., ed., ASM International, pp. 199-208 (1988).

4. R. T. Bhatt, and R. E. Phillips, "Laminate Behavior for SiC Fiber-Reinforced Reaction-Bonded Silicon Nitride Matrix Composites," J. Compos. Technol. Res., vol. 12, no. 1, pp. 13-23 (1990).

5. N. D. Corbin, G. A. Rossetti, Jr., and S. D. Hartline, "Microstructure/Property Relationships for SiC Filament-Reinforced RBSN," Ceram. Eng. Sci. Proc. 7, pp. 958-968 (1986).

6. R. T. Bhatt, "Oxidation Effects on the Mechanical Properties of a SiC-Fiber-

Reinforced Reaction-Bonded Si_3N_4 Matrix Composite," J. Am. Ceram. Soc., vol. 75, no. 2, pp. 406-412 (1992).

7. J. D. Kiser, and R. T. Bhatt, "Effect of Hot Isostatic Pressing (HIPing) on the Microstructure of SiC Fiber-reinforced RBSN Matrix Composites Containing an Oxide Additive," HITEMP Review 1990, NASA CP-10051, pp. 52.1 to 52-13 (1990).

8. X. J. Ning, and P. Pirouz, "The microstructure of SCS-6 SiC fiber," J. Mater. Res., vol. 6, no. 10, pp. 2234-2248 (1991).

9. W. A. P. Classen, H. A. J. Th. v. d. Pal, A. H. Goemaus, and A. E. T. Kuiper, "Characterization of Silicon-Oxynitride Films Deposited by Plasma-Enhanced CVD", J. Electrochem. Soc., vol. 133, no. 7, pp. 1458-1464 (1986).

10. L. J. Howard, W. R. Knolle, J. W. Osenbach, and J. L. Zell, "Electrical, physical, and chemical characteristics of plasma-assisted chemical-vapor deposited semi-insulating a-SiN:H and their use as a resistive field shield for high voltage integrated circuits," J. Appl. Phys., vol. 67, no. 11, pp. 6830-6843 (1990).

11. D. N. Braski, and M. C. Osborne, "The In-Situ Fracture and Auger Analysis of Nicalon Fibers," Proceedings of 1994 Amer. Cer. Soc. Meeting., Cocoa Beach, FL, Jan. (1994).

12. J. I. Eldridge, and F. S. Honecy, "Characterization of interfacial failure in SiC reinforced Si_3N_4 matrix composite material by both fiber push-out testing and Auger electron spectroscopy," J. Vac. Sci. Technol. A., vol. 8, no. 3, pp. 2101-2106 (1990).

INTERFACIAL BONDING AND FRICTION IN SiC FIBER/β' SiAlON COMPOSITES

Chao M. Huang, D. Zhu, Y. Xu, T. Mackin# and W. M. Kriven, Department of Materials Science and Engineering and #Department of Mechanical and Industrial Engineering, University of Illinois at Urbana-Champaign, Urbana, IL 61801

Interfacial mechanical properties of SiC fiber-reinforced, combustion synthesized β'-SiAlON composites were studied by a fiber push-out technique. Interfacial debonding and parameters were studied in terms of embedded fiber length. Stable, progressive interfacial debonding prior to fiber frictional sliding was observed in specimens with large embedded fiber lengths. Linear, shear-lag and progressive debonding models were used in the analysis of interfacial parameters. The coefficient of friction and the residual radial stress estimated from the progressive debonding model were 0.25 and 158 MPa, respectively, as compared to 0.26 and 102 MPa, respectively obtained from the shear-lag model. The radial residual stress extracted from either model was reasonablely close to that (125 MPa) calculated from the thermal expansion mismatch and cooling temperature range. An axial residual load (8.7 N) extracted from the progressive debonding model was compared well with that (6.7 N) obtained from a calculation based on thermal expansion mismatch. The interfacial fracture toughness was calculated to be 0.5 J/m^2. TEM interfacial characterization correlated with SEM observation of the interfacial debonding site, revealed that interfacial debonding was attributed to the weak physical bonding between the outermost carbon-rich layer of the SiC fiber and the matrix.

INTRODUCTION

The mechanical behavior of fiber-reinforced ceramic matrix composites (CMC's) is especially sensitive to the interfacial properties [1-3]. In recognition of this fact, increasing effort has gone into the micromechanical characterization of interfaces. To avoid the difficulties associated with the pull-out tests in fiber-reinforced ceramic matrix composites, a simple fiber push-out test was developed by Marshall et al. [4-5]. Several analytical models have been developed to assess interfacial properties from the single-fiber push-out experiment. A number of studies have used a simple force balance that assumes that the applied load is supported by a constant friction stress at the interface, along the sliding length of the fiber [4-9]. Therefore, the push-out tests are usually reported in terms of a peak load and a frictional load. These loads are averaged over the entire interface area

To the extent authorized under the laws of the United States of America, all copyright interests in this publication are the property of The American Ceramic Society. Any duplication, reproduction, or republication of this publication or any part thereof, without the express written consent of The American Ceramic Society or fee paid to the Copyright Clearance Center, is prohibited.

and reported as an interfacial bond strength, τ_d, or as a frictional strength/stress, τ_f. However, a source of some controversy is that, during fiber push-out, the frictional stresses increase locally due to a Poisson expansion of the fiber. The Poisson expansion of the fiber under compressive loads and the consequent increase of the normal stress across the interface lead to a nonlinear variation of the frictional shear stress along the embedded fiber length. Takaku and Arridge [10] proposed a shear-lag analysis that includes an approximate correction for the effect of Poisson contraction in fiber pull-out tests. A similar analysis has been proposed by Shetty [11] to determine the frictional stress, τ_f, for debonded interfaces in push-out tests, in which both Poisson effects and the radial residual stress due to the thermal expansion mismatch between the fiber and matrix have been taken into account. However, the shear-lag treatment neglects the residual axial strain and interfacial roughness, which have been shown to substantially affect the interfacial debonding and sliding. Recent theoretical work of Kerans and Parthasarathy [12] attempts to specifically address the contribution of interface roughness to the sliding friction, and the contribution of axial residual stress to the debonding, simultaneously. Kerans and Parthasarathy's model explicitly relates the interfacial parameters such as the coefficient of friction, residual stresses, and interfacial toughness to the experimental load-displacement curves.

The primary goal of this investigation was to extract the interfacial parameters in a newly developed SiC fiber-reinforced, combustion synthesized-β' SiAlON composite [13] using a fiber push-out technique. Previous studies [13] on the SiC/β' SiAlON composite revealed that the composite exhibited a significant increase in ultimate bending strength and work of fracture compared to monolithic β' SiAlON. However, the detailed interfacial mechanical properties remain unknown. In this paper, a comprehensive interfacial study was conducted, including comparisons between Shetty's shear-lag model and Kerans/Parthasarathy's progressive debonding model.

EXPERIMENTAL PROCEDURES

The material used in this study was a SiC fiber/combustion synthesized β' SiAlON composite. The typical properties of SiC fiber and β' SiAlON are given in Table 1. The widely quoted coefficient of thermal expansion (CTE) of SiC fibers in the radial direction is 2.6×10^{-6} [14], while the CTE of the β' SiAlON matrix is 3.2×10^{-6}. The β' SiAlON (Benchmark S-011, Benchmark Structural Ceramics Corporation, Buffalo, N.Y.) uniaxially reinforced with silicon carbide fibers (SCS-6, Textron Specialty Materials, Lowell, MA) was hot pressed in a N_2 environment at 1600 °C for 2 h under a pressure of 34 MPa. The detailed procedures for the composite fabrication are given in reference [13].

The specimens for push-out tests were prepared by cutting thin sections perpendicular to the fiber direction in the unidirectional composites. The as-cut sections were surface ground on both sides, and polished to a final finish of 1 μm. Six different thickness specimens were prepared, which ranged from 0.5 to 2.56 mm. The details of the fiber push-out technique for interfacial property characterization has been described elsewhere [13, 15, 16]. The tests were conducted in a universal testing machine (Model

4502, Instron Corp., Canton, MA) with an 100 N loading cell. A flat-end tungsten carbide probe, 100 μm in diameter was fixed into a cylinder which was threaded to the loading cell. The tests were conducted in ambient atmosphere with a constant crosshead displacement rate of 60 μm/min. The specimen for the test was mounted on a slotted alumina holder which was centered on a two-axis micropositioning stage. Typically, 7 to 8 fibers were pushed-out for each thickness of specimen. After push-out, the sample was then turned over for push-back tests. Following push-out/push-back tests, scanning electron microscopy (SEM, H-800, Hitachi Ltd., Japan) was used to examine the debonded interface and to locate the site of debonding. High resolution transmission electron microscopy (HREM, CM-12, Philips Electronic Instruments, Eindhoven, The Netherlands) examination was also conducted to determine the microstructure of the interface between the SiC fiber and the matrix. The specimen for HREM examination was prepared by the conventional manner of ion milling.

RESULTS AND DISCUSSION

Load-Displacement Behavior

Fig. 1. shows typical load-displacement curves obtained from push-out tests for different fiber embedded lengths. The curves exhibit many general features typical of a push-out test. The peak load ,P_p, corresponds to the stage at which the remaining bonded length (approximately equal to 1~2 fiber diameters) debonds unstably. For thin specimens, catastrophic debonding of remaining length was clearly indicated by a load drop, ΔP, following the peak load. However, such a load drop was not observed as the fiber embedded length increased. A similar phenomenon was reported by others during push-out tests in a SiC/glass composite [17, 18]. The explanation of this phenomenon is given in [19]. Following the load drop, for thin specimens, a small increase in load, up to P_{max}, is required to induced confined fiber sliding. As the fiber is displaced from its original position, subsequent load changes are associated with (i) sliding-induced changes in the sliding resistance (abrasion), (ii) asperity interactions during push-out, and (iii) a reduced fiber/matrix contact area.

Progressive Interfacial Debonding Behavior

In general, interfacial debonding can be either progressive or catastrophic, depending upon whether the load required for debonding initiation is smaller or greater than the load for debonding propagation [12]. In general, progressive or catastrophic debonding cannot be distinguished immediately from the apparent load-displacement curves, because of machine compliance. Effective load-displacement curves were obtained by substracting the system steady state compliance from the apparent load-displacement curves. The steady state compliance values were calculated using a numerical derivative over small increments of load and displacement. In all tests, the compliance values varied between 3.5 μm/N to 4.5 μm/N, and an average value of 4 μm/N was used. Fig. 2. gives the effective load-displacement curves for different embedded fiber lengths. Progressive debonding is clearly indicated when these curves depart from linearity with a decreasing slope. The load corresponding to such a slope change in the effective load-displacement curve, designated as P_i, was the critical load for the crack propagation or initiation in the presence of axial residual stress [12]. P_i should be independent of the embedded fiber length, as is clearly illustrated in Fig. 2. However, it is important to note that, P_i was not always detectable. This behavior was noted for the 0.5 mm thin sample, wherein progressive debonding was not observed. Apparently, thin specimens do not permit stable crack propagation.

SEM and TEM Observations

SEM examination (Fig. 3) of specimens containing pushed-out fibers showed that the site of debonding and sliding was located at the interface between the carbon surface layer of the fiber and the matrix. The pushed-out fiber retained its original nodular surface morphology (Fig. 4), but abrasion traces were visible (arrowed in Fig. 4). The diameter of the largest nodules was in the order of 10 μm. High resolution transmission electron microscopy (HREM) (Fig. 5) indicated that there was no chemical reaction layer between the fiber and matrix, which suggested that the interfacial bonding was likely to be physical adhesion caused during consolidation of the composite.

Interfacial Parameters Analysis

(1) Linear and shear-lag approaches: The experimental approach for measuring the interfacial properties consisted of measuring the loads (P_p and P_{max}) from push-out curves (Fig. 1) as functions of the embedded fiber lengths. The interfacial properties were then assessed by nonlinear regression best fits and linear fits of experimental data. The detail fitting routine of the shear-lag model was illustrated in reference [15].

Fig. 6 shows the peak debonding load, P_p, and the maximum frictional sliding load, P_{max}, as functions of the embedded fiber length, L, respectively. The best-fit curves from both the nonlinear and linear models are also drawn. With the exception of the thin specimens, both figures show that the regression curves fit the measured values quite well. The relatively low values of P_p measured for the thin specimen were likely to be caused by the pre-debonding induced during specimen preparation. Table 2. gives the interfacial parameters calculated from the linear model and shear-lag model, respectively. Both models predicted similar values of τ_d, while the shear-lag model gives a lower value of τ_f than that obtained from the linear approach. However, in both cases, the nonlinear model gives better fits to the data than does the linear model, as shown in Fig. 6. More importantly, the shear-lag approach also permits calculation of both the radial residual stress and the coefficient of friction. However, the most important interfacial property, viz., interfacial toughness, cannot be determined from the shear-lag model.

(2) Progressive debonding approach: Kerans and Parthasarathy's model [12] is arrived at assuming progressive debonding. The external load (P_a) is related to the external effective displacement (δ_{eff}) in terms of the interface toughness (G_i), the radial clamping stress induced by interfacial roughness (σ_r), the radial residual stress due to the thermal expansion mismatch (σ_N), the residual axial load (P_r), and the friction coefficient (μ). In addition, the interfacial critical debonding load (P_d) is directly related to the interfacial fracture toughness (G_i). The prediction of these parameters was done using a three-parameter non-linear least square (downhill simplex optimization) method on the individual effective load-displacement curves [20].

Fig. 7 summarizes the critical debonding loads (P_d), debonding peak load (P_p), and axial residual load (P_r) extracted from the fitting procedure as a function of the embedded fiber length. The predicted axial residual load (P_r), as shown in Fig. 8, was larger than the predicted critical load (P_d) by a factor of 10. The large values of P_r implied that the most of the crack initiation load (P_i, where $P_i = P_r + P_d$) is attributed to the unreleased residual axial compressive strain that was processed into the composite. The fitted P_r mean value (8.7 N) was reasonablely close to that (6.7 N) estimated from the thermal expansion mismatch between the fiber and matrix in the longitudinal direction. Both P_r and P_d showed very little variation with the embedded fiber length. However, the calculated P_i (10 N) from the predicted values of P_r and P_d was much lower than that (19 N) measured directly from the effective load-displacement curve.

The mean value of interfacial fracture toughness (G_i) (0.5 J/m^2) agreed fairly well with previous studies on SiC/glass (0.24 J/m^2 [21], 0.18 J/m^2 [22]) and SiC/zircon (0.8 J/m^2 [23]) composites. The low interfacial fracture toughness obtained for these composites was attributed to the carbon-rich outlayer of SCS-6 SiC fiber preventing significant bonding between the fiber and the matrix.

The mean values of the frictional coefficient (μ) obtained from the progressive debonding model (0.25) compared well with values extracted from the shear-lag model (0.26). The normal residual stress (σ_N, 158 MPa) measured from the fitting was higher than that obtained from the shear-lag model (102 MPa) and that estimated from the thermal expansion mismatch and cooling range (125 MPa). The frictional stress predicted from the fitting (40 MPa) was slightly higher than that obtained from the linear approach (34 MPa), and that computed using the shear lag model (26 MPa). The value of interfacial roughness amplitude (A) calculated from the magnitude of the "seating drop" [24] was 25 nm, which was consistent with values obtained by Jero and Kerans [24] for the same fibers in a glass matrix (15-20 nm). The clamping stress induced by the interfacial roughness (based on the calculation of interfacial roughness amplitude A) was 35 MPa.

CONCLUSIONS

The interfacial mechanical properties of a SiC fiber-reinforced, combustion synthesized β' SiAlON composite were characterized by the fiber push-out technique. The interfacial parameters were extracted through the analysis of push-out data using both shear-lag and progressive debonding models. The nonlinear shear-lag analysis permits evaluation of the coefficient of friction and the residual compressive stress for the composite, while the progressive debonding approach provided more comprehensive interfacial information in addition to the frictional parameters.

SEM examination of specimens containing pushed-out fibers indicated that debonding occurred at the interface between the outermost carbon layer of the fiber and the matrix. The low interfacial fracture toughness was consistent with TEM observation that there was no chemical bonding between the fiber and the matrix.

ACKNOWLEDGMENTS

Funding for this research was provided by the US Air Force of Scientific Research through a grant, number AFOSR-F49620-93-1-0227. The authors wish to thank Dr. Ronald J. Kerans from the Wright Laboratory , Wright-Patterson Air Force Base, OH, for providing the fitting routines and for his help in the progressive debonding analysis. Valuable discussions with Dr. Paul Jero from the same organization are appreciated. Use of the electron microscopy facilities at the Center for Microanalysis of Materials Research Laboratory at UIUC are acknowledged.

REFERENCES

1. R. J. Kerans, R. S. Hay, N. J. Pagano and T. A. Parthasarathy, "The Role of the Fiber-Matrix Interface in Ceramic Composites," *Ceram. Bull.* **68** (2) (1989) 429-442.

2. H. C. Cao, E. Bischoff, O. Sbaizero, M. Rühle, A. G. Evans, D. B. Marshall and J. J. Brennan, "Effect of Interface on the Properties of Fiber-Reinforced Ceramics," *J. Am. Ceram. Soc.,* **73** (1990) 1691.

3. A. G. Evans, F. W. Zok and J. Davis, "The Role of Interfaces in Fiber-Reinforced Brittle Matrix Composites, *Composites Science and Technology,"* **42** (1991) 3-24.

4. D. B. Marshall, "An Indentation Method for Measuring Matrix-Fiber Frictional Stresses in Ceramic Composites," *J. Am. Ceram. Soc.*, **67** (12) (1984) C-259-C-260.

5. D. B. Marshall and W. C. Oliver, "Measurement of Interfacial Mechanical Properties in Fiber-Reinforced Ceramic Composites," *J. Am. Ceram. Soc.,* **70** (8) (1987) 542-548.

6. M. K. Brun and R. N. Singh, "Effect of Thermal Expansion Mismatch and Fiber Coating on the Fiber/Matrix Interfacial Shear Stress in Ceramic Matrix Composites," *Advanced Ceramic Materials,* **3** (5) (1988) 506-509.

7. J. W. Laughner and R. T. Bhatt, "Measurement of Interfacial Shear Strength in SiC-Fiber/Si_3N_4 Composites," *J. Am. Ceram. Soc.,* **72** (10) (1989) 2017-2019.

8. R. N. Singh, "Fiber-Matrix Interfacial Characteristics in a Fiber-Reinforced Ceramic Matrix Composite," Cerm. Eng. Sci. Proc. **10** (7-8) (1989) 883-893.

9. G. Morscher, P. Pirouz, and A. H. Heuer, "Temperature Dependence of Interfacial Shear Strength in SiC-Fiber-Reinforced Reaction-Bonded Silicon Nitride," J. Am. Ceram. Soc., **73** (3) (1990) 713-720.

10. A. Takaku and R. G. C. Arridge, "The Effect of Interfacial Radial and Shear Stress on Fiber Pullout in Composite Materials," *J. Phys. D: Appl. Phys.,* **6** (1973) 2038-2047.

11. D. K. Shetty, "Shear-Lag Analysis of Fiber Push-out (Indentation) Tests for Estimating Interfacial Friction Stress in Ceramic-Matrix Composites," *J. Am. Ceram. Soc.*, **71** (2) (1988) C-107-C-109.
12. R. J. Kerans and T. A. Parthasarathy, "Theoretical Analysis of the Fiber Pull-out and Push-out Tests," *J. Am. Ceram. Soc.*, **74** (7) (1991) 1585-1596.
13. C. M. Huang, Y. Xu, D. Zhu, and W. M. Kriven, "Combustion-Synthesized β'-SiAlON Reinforced with SiC Monofilaments," J. Mater. Sci. Eng. in press.
14. R. W. Goettler and K. T. Faber, "Interfacial Shear Stresses in SiC and Al_2O_3 Fiber-Reinforced Glasses," *Ceram. Eng. Sci. Proc.*, **9** (7-8) (1988) 861-870.
15. J. D. Bright, D. K. Shetty, C. W. Griffin, and S. Y. Limaye, "Interfacial Bonding and Friction in SiC (Filament)-Reinforced Ceramic- and Glass-Matrix Composites," *J. Am. Ceram. Soc.*, **72** (10) (1989) 1891-1898.
16. P. D. Warren, T. J. Mackin and A. G. Evans, "Design, Analysis Application of an Improved Push-through Test for the Measurement of Interface Properties in Composites, " Acta Metall. Mater., **40** (6) (1992) 1243-1249.
17. J. D. Bright, S. Danchaiviji, and D. K. Shetty, "Interfacial Sliding Friction in Silicon Carbide-Borosilicate Glass Composites: A comparison of Pullout and Push-out Tests," *J. Am. Ceram. Soc.*, **74** (1) (1991) 115-122.
18. G. Rausch, B. Meier and G. Grathwohl, "A Push-out Technique for the Evaluation of Interfacial Properties of Fiber-reinforced Materials, " *J. Eurp. Ceram. Soc.*, **10** (1992) 229-235.
19. C. M. Huang, D. Zhu, T. Mackin, and W. M. Kriven, "Interfacial Properties of SiC Fiber Reinforced β'-SiAlON Composite," submitted to *J. Mater. Sci. Eng.*
20. R. J. Kerans, "Notes for Push-out Data Analysis," Unpublished results.
21. D. R. Mumm and K. T. Faber, "Fiber Debonding and Pullout Processes in Ceramic Composite," *Ceram. Eng. Sci. Proc.*, **13** (7-8) (1992) 70-77.
22. T. A. Parthasarathy, P. D. Jero and R. J. Kerans, "Extration of Interface Properties from a Fiber Push-out Test," *Scr. Metal. Mater.*, **25** (1991) 2457-2462.
23. S. K. Reddy and R. N. Singh, "Characterization of Interfacial Properties in SiC-Zircon Composites," *Ceram. Trans.* **38** (1993) 211-222.
24. P. D. Jero, R. J. Kerans, and T. A. Parthasarathy, "Effect of Interfacial Roughness on the Frictional Stress Measured Using Push-out Tests," *J. Am. Ceram. Soc.*, **74** (11) (1991) 2793-2801.

Table 1. Material data of SiC fiber and β'-SiAlON matrix.

	SiC - fiber (SCS - 6)	β' - SiAlON matrix
Diameter (μm)	140	N/A
Young's modules (GPa)	415	230
Poisson's ratio	0.15	0.25
Density (g/cm^3)	3.11	3.28
Coefficient of Thermal Expansion (10^{-6} /°C)	2.63	3.20

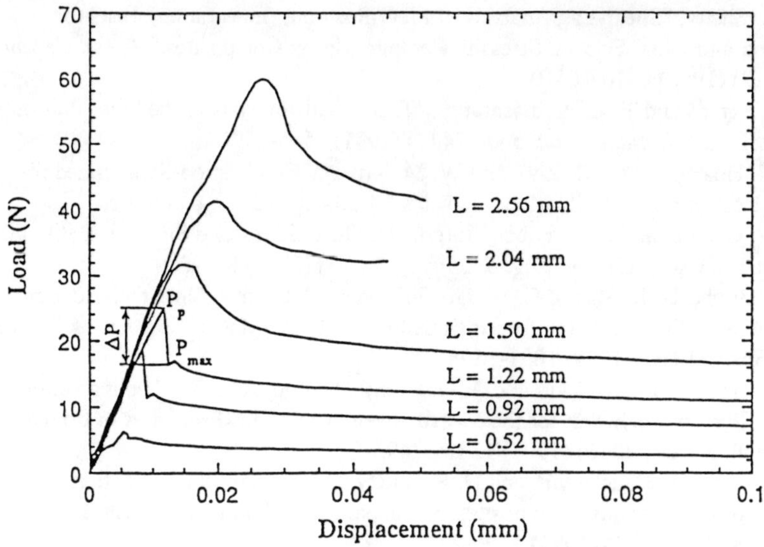

Fig. 1. Typical push-out load vs. displacement curves for different embedded fiber lengths.

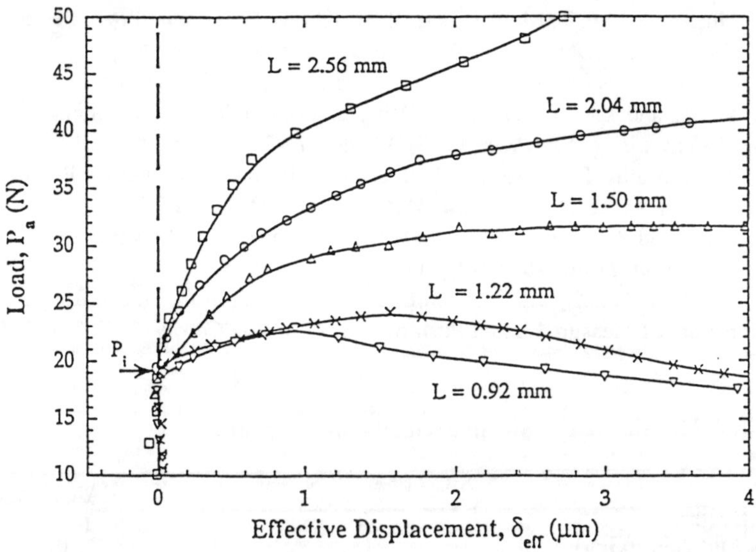

Fig. 2. Typical push-out load vs.effective displacement curves after subtraction of the load-train compliance.

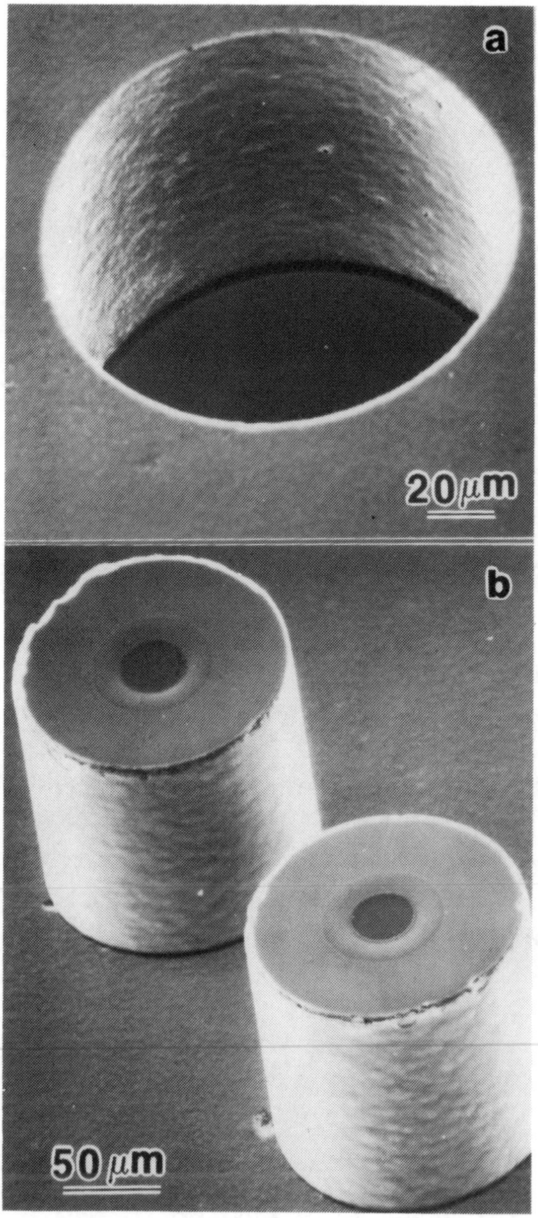

Fig. 3. SEM micrographs of SiC fiber after push-out showing: (a) top view, (b) bottom view.

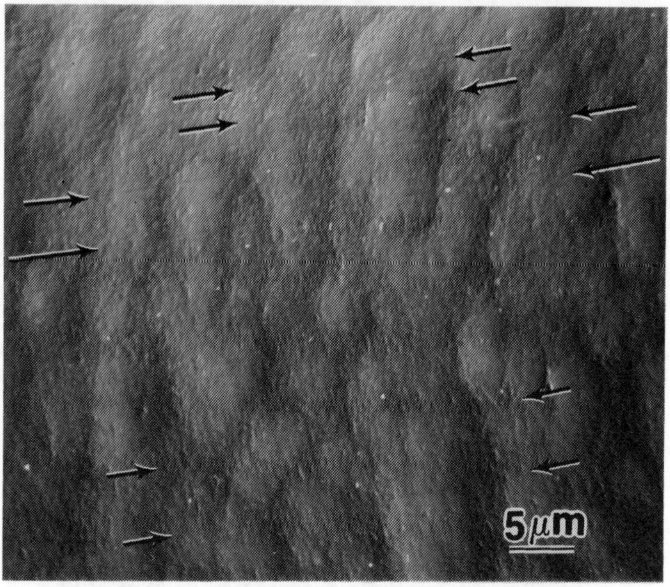

Fig. 4. SEM micrograph showing the surface morphology of a pushed-out SiC fiber and visible wear tracks (arrowed).

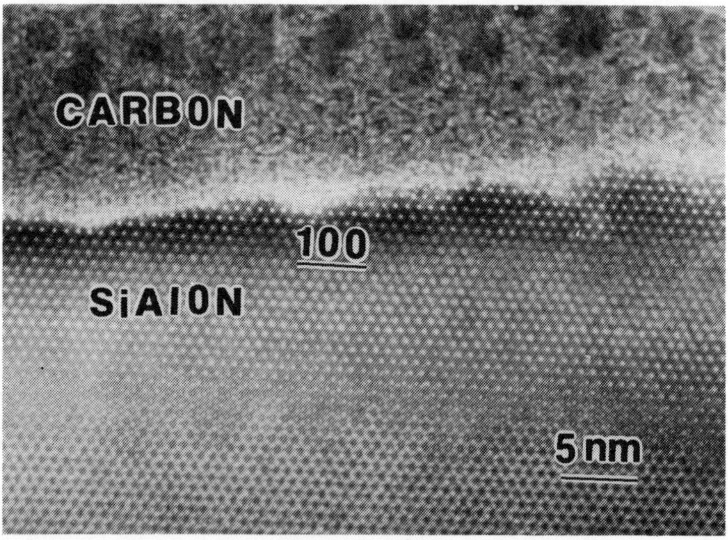

Fig. 5. High resolution transmission electron microscopy (HREM) images of the interfacial region in a SiC/β' SiAlON composite.

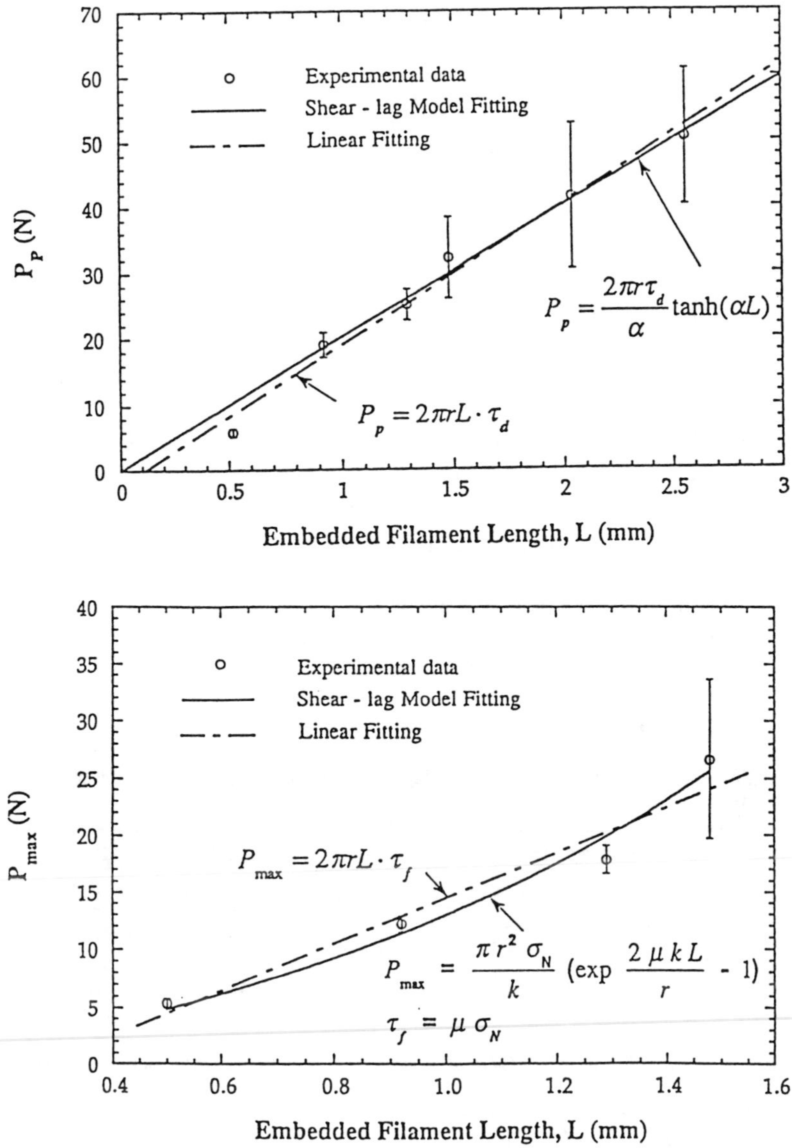

Fig. 6. Shear-lag model fitting of (a) the debonding peak load (P_p) and (b) the maximum load for frictional sliding (P_{max}) plotted as a function of embedded fiber length (t) for the SiC/β' SiAlON composite.

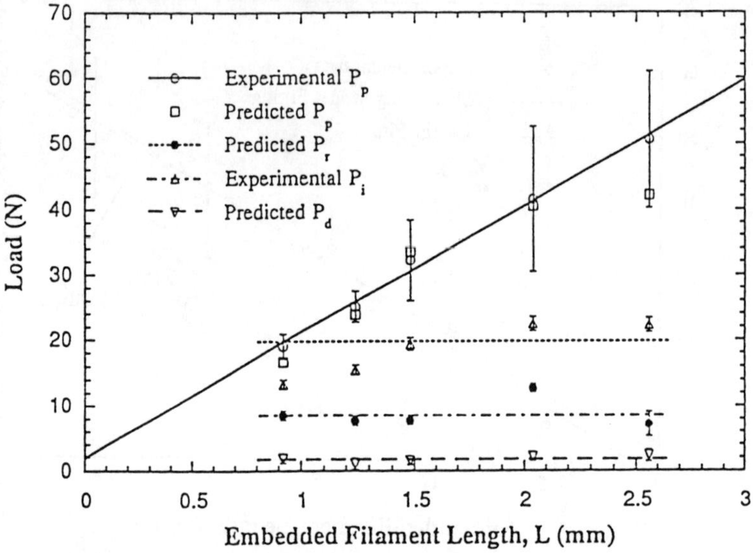

Fig. 7. The critical load for interfacial debonding propagation (P_d), axial residual load (P_r), debonding peak load (P_p) predicted by progressive debonding model are plotted as a function of embedded fiber length, as compared to experimental P_i ($P_i = P_d + P_r$).

Table 2. Interfacial parameters calculated by the linear model and shear-lag model.

Model	Debonding		Friction Sliding			
	τ_d (MPa)	SD (N^2)	τ_f (MPa)	μ	σ_N (MPa)	SD (N^2)
Linear	49	3.84	34	N/A	N/A	3.42
Shear-Lag	50	1.49	26	0.26	102	1.48

Table 3. Interfacial parameters calculated by the progressive debonding model.

P_r (N)	P_d (N)	G_i (J/m^2)	μ	A (nm)	σ_r (MPa)	σ_N (MPa)	τ_f (MPa)
8.5	1.7	0.5	0.25	25	35	158	40

Interfacial structure and precipitate formation in MgO/SiC composites

J. Luo and R. Stevens,
School of Materials, University of Leeds, Leeds, LS2 9JT UK

ABSTRACT

High resolution electron microscopy has been used to examine the interfacial structure of two MgO- 20vol% SiC composites, in one of which SiC particles were thermally treated to form a SiO_2 layer before adding into MgO. Both of the composites showed well bonded interfaces but in the first composite, the interface is not coherently bonded,while the coherent interface can be readily observed in the pre-oxidised SiC particle reinforced composite. The Mg_2SiO_4 phase is precipitated in the aggregated SiC particles in the both composites, which formed coherent and semi-coherent interfaces with the MgO grains and the SiC particles.

INTRODUCTION

The properties of ceramic composites are closely related to the interfacial structure. The interfacial bonding significantly influences the mechanical properties of composites by changing the toughening mechanisms such as crack deflection, crack bridging and pull-out. The earlier investigation indicated that the thermal conductivity of the MgO/SiC particulate composites did not increase by addition of SiC particles although SiC has a much higher thermal conductivity than MgO[1], which is also ascribed to the interfacial structure by forming thermal barriers. The mechanical properties of the composites significantly increased by addition of the SiC particles, and were also affected by changing the interfacial structure[2]. In addition, large residual stress is generated in the composites due to the high mismatch of the thermal expansion coefficients[3], which is expected to affect the interfacial structure.

In this paper, high resolution electron microscopy is used to examine the interfacial structure of the composites.

MATERIALS AND EXPERIMENTAL TECHNIQUE

Two 20vol% SiC particle reinforced MgO composites are fabricated using hot pressing. The fabrication details were described in an earlier paper[2]. The only

To the extent authorized under the laws of the United States of America, all copyright interests in this publication are the property of The American Ceramic Society. Any duplication, reproduction, or republication of this publication or any part thereof, without the express written consent of The American Ceramic Society or fee paid to the Copyright Clearance Center, is prohibited.

difference of the two specimens is on the thermal treatment of the SiC particle; in the first (called MS3), the SiC particles did not undergo any thermal treatment, but the SiC particles were thermally treated in 900°C in air for 1 hour before mixing with the MgO powder in the second one (called MS01). TGA results showed that a 10nm layer of SiO_2 formed on the thermally treated SiC particles. The two specimens were examined using Phillips 430 transmission electron microscopy (TEM). The TEM specimen were prepared by cutting a 100nm thick slides from the hot press disc, followed by grinding, polishing and ion beam thinning. The ion beam thinning is finished at 6^0 in order to have a large thin area for TEM examination.

RESULTS AND DISCUSSION

MS3 specimen

The interfaces between the matrix and the SiC particles are usually formed incoherent bonding, but no interfacial debonding occurs. In the composite, large SiC particles are normally located in the matrix grain boundaries, but a few SiC particles, which have particle size less than 100nm are intragrannularly located. The intragrannular SiC particles did not show coherent bonding with MgO. The residual stress which is generated in the SiC particles due to the mismatch of the thermal expansion coefficients between SiC and MgO , could not be relieved for the intragrannular particles. Hence, the strain contours in the intragrannullar particles generated by the mismatch is more likely to be observed in the TEM.

In certain areas with high concentration of SiC particles, the precipitates of Mg_2SiO_4 can be observed in the composite, as shown in Figure 1. However, X-ray diffraction results did not show any of the characteristic peaks of Mg_2SiO_4 and $MgSiO_3$ phases in the specimen, which indicated that the quantity of precipitates is too small. The precipitate shown in Figure 1 has a size of approximate 1μm. The interface between the precipitate and the MgO grain is coherent (or semi-coherent). The stand-off dislocations were observed on the interface due to the lattice misfit between $(111)_{MgO}$ and $(120)_{Mg2SiO4}$. The stand-off distance is related to the lattice misfit[4]. The lattice was distorted around the dislocation as a result of the residual strain generated by the dislocation[5].

The high magnification micrograph indicated the possible formation mechanisms of the precipitate. A possible explanation for the formation of Mg_2SiO_4 phase is that the precipitate is initiated by the oxidation of SiC and diffusion to the MgO, which was also observed in the MS01 specimen discussed later. The interface between the MgO and Mg_2SiO_4 is characteristic of the final stage of the formation of Mg_2SiO_4 phase by the diffusion of SiO_2 and its reaction with MgO. The Mg_2SiO_4 rather than $MgSiO_3$ is formed in the excess of MgO[6,7]. The SiO_2 could be formed by the oxygen impurity in SiC and/or by oxidation of SiC during hot pressing.

A coherent interface formed between the SiC particles and the Mg_2SiO_4 precipitate, as shown in Figure 1c. The Mg_2SiO_4 phase is nucleated on the SiC particles and grows epitaxially. This phenomenon was widely observed in metal/ oxide interface by metal oxide, oxide reduction or epitaxial deposition[8-13]. The misfit dislocation is generated on the coherent interface. As can be seen from the figure, the planar defects (or some dislocation concentration zones occurred in the precipitate. One reason to cause the defects is the residual strain generated by the misfit dislocations between the precipitate and MgO and /or SiC. The effect is not expected to be significant since the misfit dislocation is localised on the interface areas. Another important reason is the residual strain generated by the mismatch of the thermal expansion coefficient between the precipitate and the surrounded SiC particles. The large residual strain and the deformation would generate in the slightly soft precipitate (the Young modulus, 130GPa, Vickers hardness 80GPa) by the mismatch.

MS01 Specimen

As mentioned earlier, the MS01 specimen differs from MS3 on that in MS01 specimen, about 10nm thick SiO_2 layer formed on the SiC particles before mixing with MgO powder. Figure 2 shows a coherent interface between the MgO and the SiC in MS01 specimen. The SiO_2 layer did not appear on the interface but diffused into the matrix, reacting with MgO and forming the precipitates near the interfaces. An amorphous phase was not observed in the interface. The size of the precipitates ranges from several nm to 100nm. The small precipitates are observed in the MgO grains. The microanalysis results by EDX showed that the precipitates contain Si, Mg and O. The carbon element was not detected in the precipitates. X-ray diffraction examination on the specimen confirmed that there exists Mg_2SiO_4 phase but no $MgSiO_3$ in the composite. Microdiffractions also showed that the precipitates are likely Mg_2SiO_4 phase. The domination of Mg_2SiO_4 phase of the precipitates can be attributed to the excess of MgO phase[6,7]. Figure 3 showed an incoherent interface between the matrix and the SiC particle. Some of the original SiO_2 on the SiC diffused into the matrix and the resulting distortion of the lattice generates the strain contrast, while some SiO_2 still remains on the interface forming the amorphous layer.

The large precipitates (greater than 100nm in size) are also readily observed in the composite, forming coherent or semi-coherent interfaces with the SiC particles and the MgO grains, as shown in Figure 4. The incoherent interface can also be observed between the precipitate and the SiC, as shown in Figure 5. A micro-size Mg_2SiO_4 formed in the aggregated SiC particles, the precipitate being surrounded by several SiC particles. Interfacial debonding or cracking occurs between a SiC particle and the precipitate. The debonding (or crack) width is approximately 10Å, i.e. only several atoms wide. At the tip of the debonding, there are no dislocations. The debonded interface is not coherent; therefore, a possible explanation for the debonding lies in the high residual strain at the interface, generated by the mismatch of the thermal expansion coefficient between the Mg_2SiO_4 and SiC, which is relieved by the formation of a crack. A small inclusion is in the Mg_2SiO_4 grain. The high magnification of the small inclusion, as shown in Figure 5c, did not demonstrate any difference in lattice fringe with the Mg_2SiO_4. The lattice fringe is formed more likely by the Mg_2SiO_4 rather than by the inclusion.

Comparing the interfaces between the SiC and MgO in MS3 and MS01 specimens shows that no coherent interfaces were observed in MS3, but the coherent interface is readily observed in MS01. The requirement for the formation of the different interfaces is a matter for further investigation, but a possible explanation can be based on the formation of a fresh surface of the SiC in MS01, after SiO_2 layer diffused into the MgO to form the precipitates, enabling direct contact with MgO, subsequently the MgO grain recrystallised on the fresh SiC particle interface during sintering, growing epitaxially.

CONCLUSIONS

MgO/SiC particulate composites show well bonded interfaces between the matrix and the reinforcing phase. Mg_2SiO_4 phase is precipitated near the interfacial areas, forming coherent and semi-coherent interfaces with the MgO grains and the SiC particles. When the SiC particles were thermally treated to form a SiO_2 layer before mixing with MgO, more Mg_2SiO_4 precipitates are formed in range of several nanometre to 1μm in size in the composite.

ACKNOWLEDGEMENTS

Authors would like to thank Mr. A. Nichells for his help on TEM.

REFERENCE

[1] J. Luo, R. Taylor and R. Stevens, The physical properties of MgO/SiC composites: comparison between a micromechanics analysis and experimental results, In this proceeding

[2] J. Luo and R. Stevens, Crack stopping mechanisms in MgO/SiC composites, in " The proceedings of TMS", 1993

[3] J. Luo and R. Stevens, Residual stress and microcracking in MgO/SiC particulate composites, J. Europ.Ceram. Soc., 12,369-375, (1993)

[4] W. Mader and D. Knauss, Equilibrium position of misfit dislocations at planar interfaces, Acta Metall., 40, S207-215, (1992)

[5] M. Y. Gutkin, M. Militzer, A. E. Romanov and V. I. Vladimirov, Equilibrium position of misfit dislocations, Physica Status Solidi, 113(a), 337, (1989)

[6] G. W. Brindley and R. Hayami, Kinetics and mechanisms of formation of forsterite (Mg_2SiO_4) by solids state reaction of MgO and SiO_2 , Phil. Mag., 33,

505-514 (1965)
[7] W. A. Deer, R. A. Howie and J. Zussman, ed. "Rock-Forming Minerals: Vol.1a, Orthosilicates, 2ed.", Longman, London, 1982
[8] M. Backhaus-Ricoult and S. Hagège, Metal-ceramic interfaces in internally reduced (Mg,Cu)O, Acta Metall., 40, S267-274, (1992)
[9] J. Mayer, G. Gutekunst, G. Möbus, J. Dura, C. P. Flunn and M. Rühle, Structure and defects of MBE grown Nb-Al_2O_3 interfaces, Acta Metall., 40, S217-225, (1992)
[10] A. Trampert, F. Ernst, C. P. Flynn, H. F. Fischmeister and M. Rühle, High resolution transmission electron microscopy studies of the Ag/MgO interface, Acta Metall., 40, S227-236, (1992)
[11] K. L. Merkle, M. I. Buckett and Y. Gao, Atomic structure of heterophase interfaces, Acta Metall., 40, S249-257, (1992)
[12] D. X. Li, P. Pirouz and A. H. Heuer, S. Yadavalli and C. P. Flynn, The cahracterization of Nb-Al_2O_3 and Nb-MgO interfaces in MBE grown Nb-MgO-Nb-Al_2O_3 multilayers Acta Metall., 40, S237-247, (1992)
[13] P. Lu and F. Cosandey, Dislocation structures at Cu-MgO and Pd-MgO interfaces, Acta Metall., 40, S259-266, (1992)]

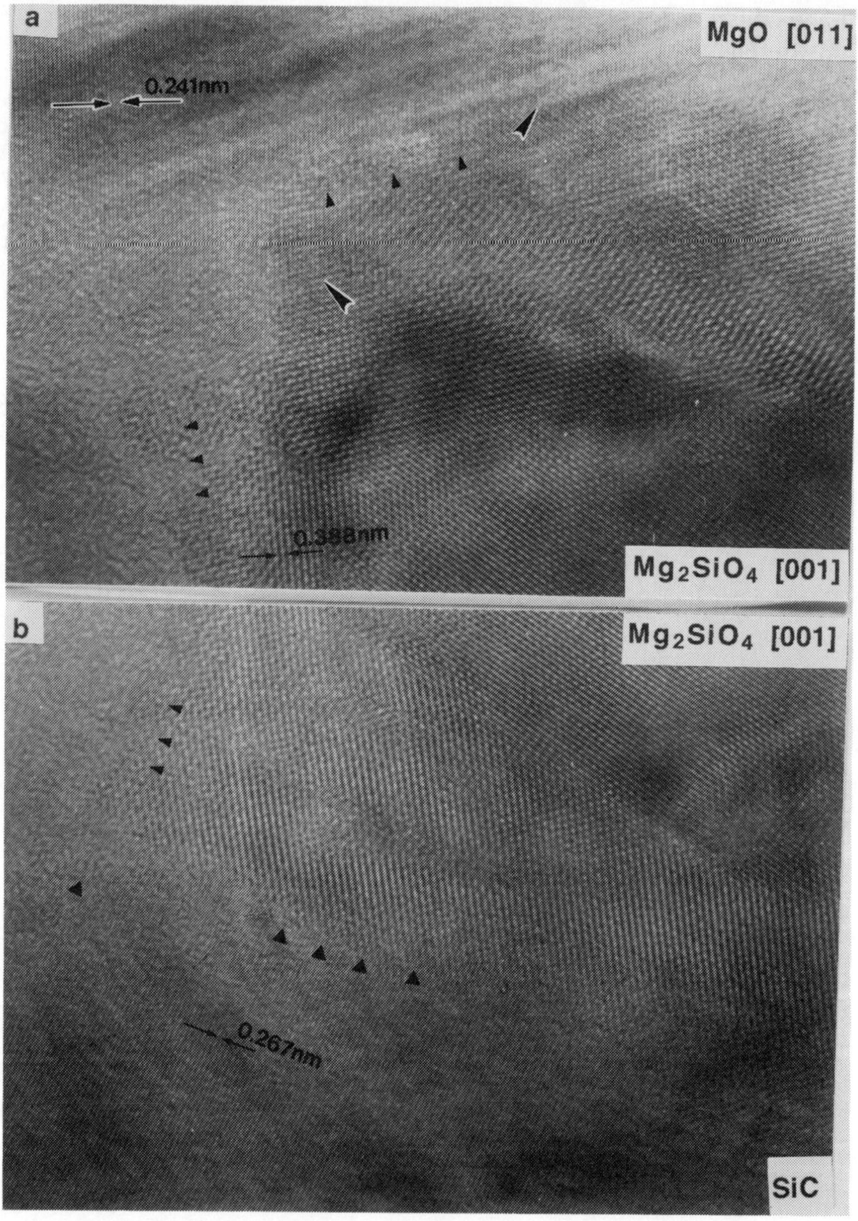
a
MgO [011]
0.241nm
0.388nm
Mg_2SiO_4 [001]
b
Mg_2SiO_4 [001]
0.267nm
SiC

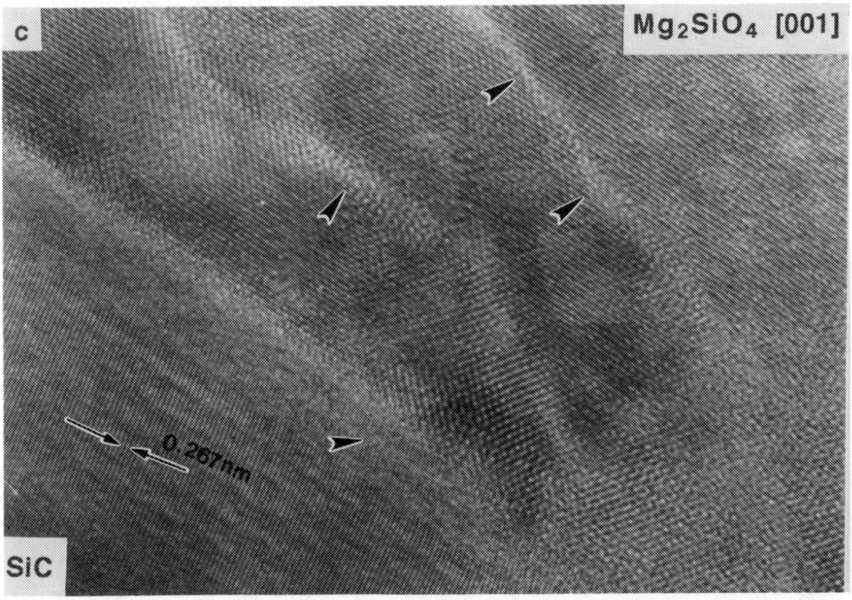

Figure 1 a,b , Mg_2SiO_4 precipitate was observed between the MgO grain and SiC particle in MgO/SiC composite. The micrographes illustrate the possible formation mechanism of Mg_2SiO_4 . Coherent bonding has taken place between MgO and Mg_2SiO_4 and between SiC and Mg_2SiO_4.
c, The interface between SiC and Mg_2SiO_4 . The SiC lattice fringe near the interface is distorted. Defect bands are similar to kink bands, which are able to accomodate the strains are observed at the interface of Mg_2SiO_4 . Bends of the defects can also be observed within Mg_2SiO_4.

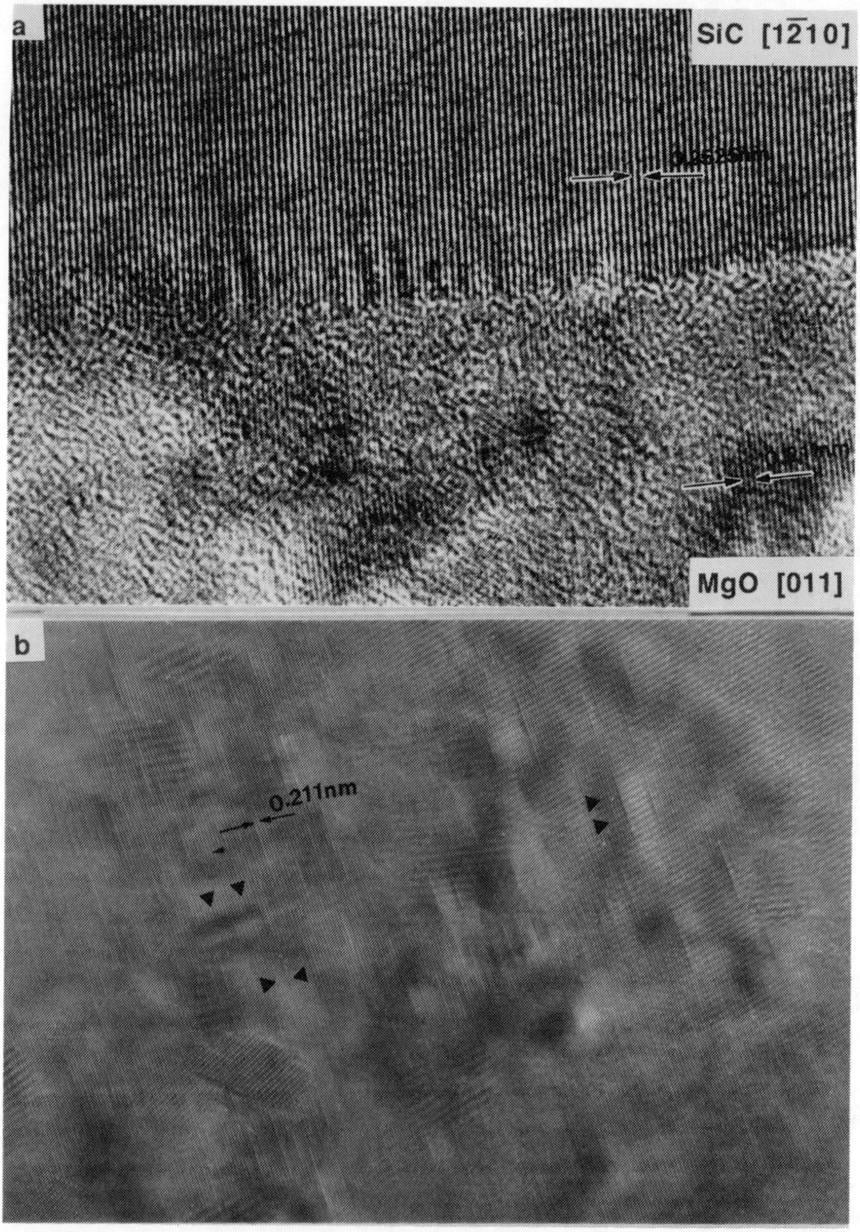

Figure 2 a, The micrograph shows coherent interfaces between a MgO grain and a pre-oxiised SiC particle in the composites. The precipitates, which have a size range of 1-10 nm, are formed in the MgO lattice.
b, The detail structure of the precipitates in the MgO lattice. The dislocations and the strain contrast are formed due to the inward diffusion of SiO_2 into MgO matrix and the formation of the precipitates.

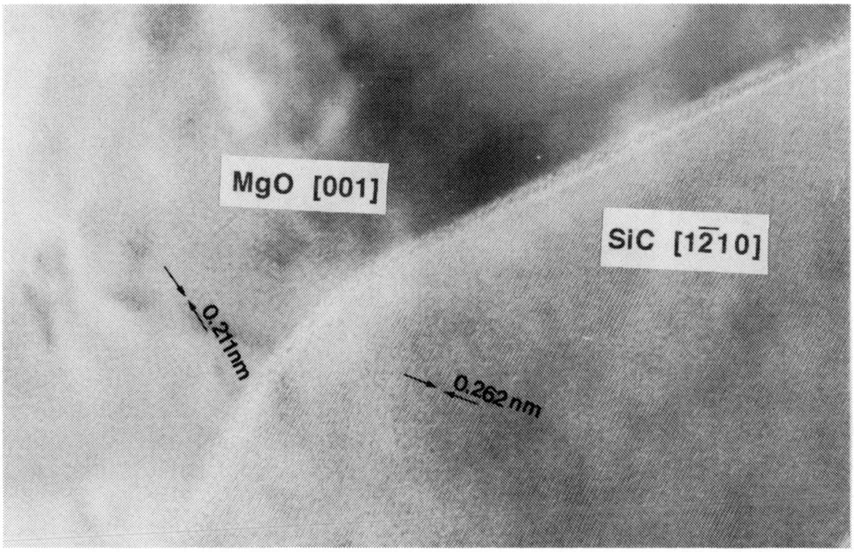

Figure 3, The incoherent interface between a MgO grain and a pre-oxidised SiC particle. A 10Å thick interface is seen to have formed. SiO_2 has diffused into the MgO and the resulting distortion of the lattice generates the strain contrast. Some SiO_2 still remains on the interface forming the amorphous layer.

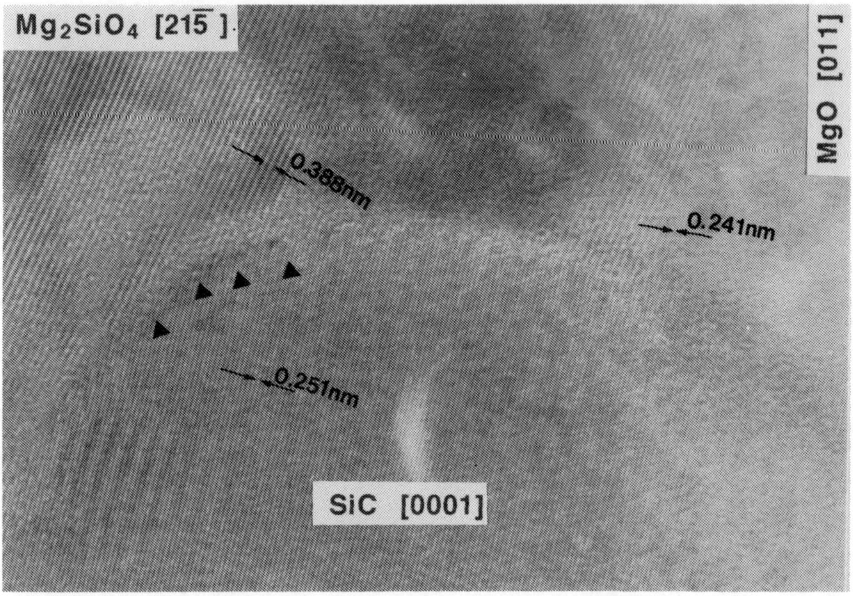

Figure 4, The precipitates of Mg_2SiO_4 can be readily observed in pre-oxidised SiC particle reinforced MgO composites. The micrograph shows the interfaces among MgO, SiC and the precipitate. Both coherent and incoherent interface are apparent.

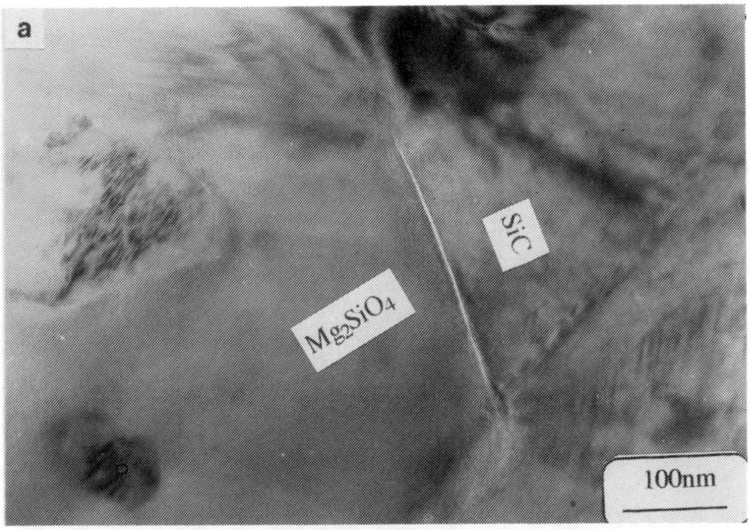
a
SiC
Mg2SiO4
100nm

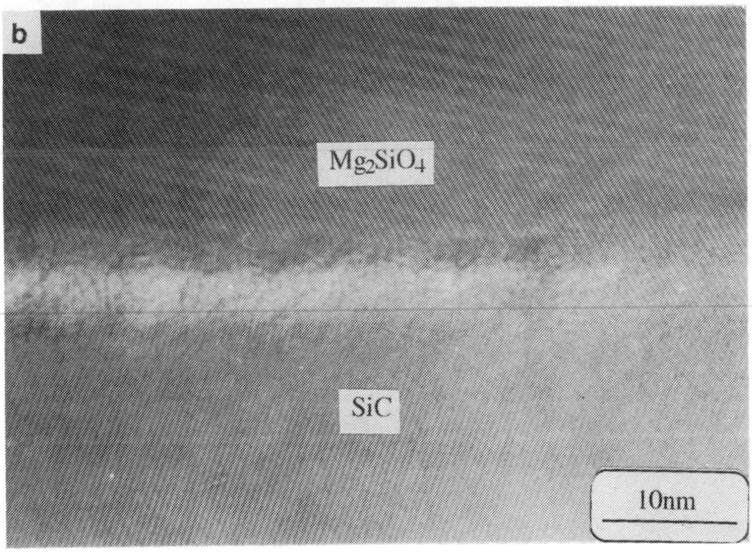
b
Mg2SiO4
SiC
10nm

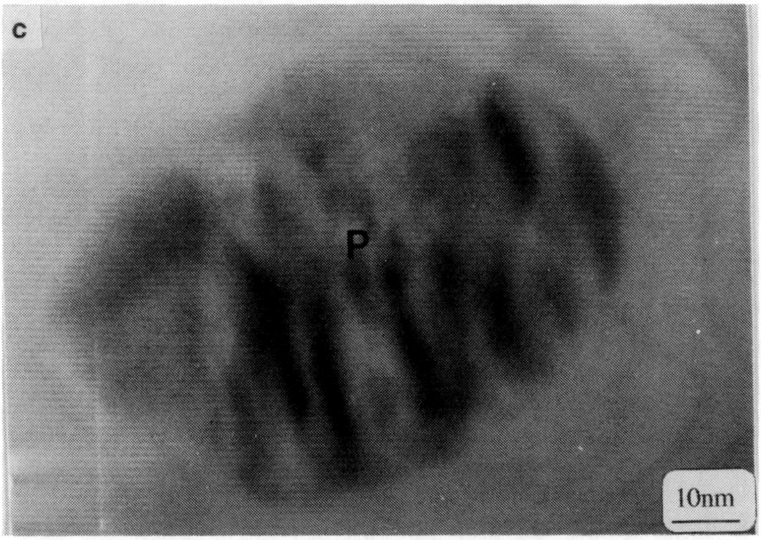

Figure 5 a, An interfacial debonding (or cracking) between a SiC particle and a Mg_2SiO_4 precipitate. b, The micrograph shows the detail structure of the debonded interface. c, The micrograph illustrates the small precipitate in the Mg_2SiO_4 , marked by p in figure 5 a.

INTERFACIAL DEGRADATION AND SURFACE MODIFICATION IN ZIRCONIA-COATED SAPPHIRE FIBERS *

M.A. Stough, J.R. Hellmann, and M.S. Angelone, The Pennsylvania State University, University Park, PA 16802

ABSTRACT

Previous work on zirconia (ZrO_2)-coated Saphikon fiber has demonstrated that interfacial degradation occurs in this system. Sapphire surface modification at ZrO_2 particle contact sites, termed pitting, may lead to significant fiber strength degradation. SEM (Scanning Electron Microscopy) and EPMA (Electron Probe Microanalysis) revealed that an impurity-based transient liquid phase forms in this system at 1400°C in air. Titanium, silicon and sodium present in trace quantities in the coating precursor sols segregate to the fiber/coating interface during processing. Local fiber surface dissolution from liquid formation is proposed as a mechanism contributing to fiber surface pitting.

INTRODUCTION

A wide interest in fiber-reinforced ceramic matrix composites (CMC) has developed over the past few decades. Improvements in fiber properties and processing routes have made available several technologically-advanced reinforcements for monolithic ceramic materials. For example, wide use of small diameter SiC fibers and whiskers have improved CMC mechanical properties by increasing strength and fracture toughness. However, recent developments in fiber processing have made single crystal sapphire more readily available as a strong, creep resistant reinforcement ideally suited for use in high oxygen ambients. Incorporation of sapphire fibers in oxidatively stable ceramic matrices (such as Al_2O_3) should result in a composite material with high strength, enhanced fracture toughness, and outstanding resistance to environmental degradation in high oxygen potentials.

However, interfacial coatings to inhibit fiber/matrix bonding and concomitant fiber degradation and low fracture toughness are required for further development of sapphire fiber reinforced alumina composites. Recent work [1-3] has con-

*Work sponsored by the National Aeronautics and Space Administration (NASA) HITEMP program under Grant Number NAGW-1381.

To the extent authorized under the laws of the United States of America, all copyright interests in this publication are the property of The American Ceramic Society. Any duplication, reproduction, or republication of this publication or any part thereof, without the express written consent of The American Ceramic Society or fee paid to the Copyright Clearance Center, is prohibited.

centrated on developing fiber coatings for ceramic matrix composites (CMCs), primarily alumina matrix composites. In particular, coatings have been developed to tailor fiber/matrix bonding, to serve as diffusion barriers, and to act as low-energy paths for crack propagation [1]; but, the role of coatings on fiber properties has been inadequately addressed to date.

In the present study, results are presented on the analysis of interface degradation in zirconia (ZrO_2)-coated single crystal sapphire/alumina composites. It appears that not only can the bare fiber and matrix interact adversely during high temperature processing but also the polycrystalline zirconia coating and sapphire fiber react to degrade the fiber as well. In collaboration with NASA-Lewis Research Center (LeRC), Cleveland, OH, we have examined the cause(s) of surface degradation (pitting) in ZrO_2-coated fibers before and after heat treating. A mechanism contributing to pitting has been identified and will be explained.

EXPERIMENTAL

Fractured fiber specimens from room temperature tensile tests were supplied to us from NASA-LeRC† (Table I). Additional coated fibers were processed at The Pennsylvania State University (PSU). The fibers processed at PSU (designated Sol #3) were coated with a ZrO_2 loaded-sol consisting of Tosoh ZrO_2 powder dispersed in a zirconium n-propoxide solution. Unsized, c-axis sapphire fiber

Table I. Fibers examined in this study and their designation

Designation	Description of sol vintage & content
Sol #1	circa 1990, used Strem§ 99.9% ZrO_2 powder
Sol #2	after sol #1, used Strem 99.9% ZrO_2 powder£
Sol #3	PSU 1993, used Tosoh¥ TZ-0 ZrO_2 powder

from Saphikon, Inc. was ultrasonically cleaned in acetone and methanol for 10 min. each prior to coating. Fiber samples approximately 8-10 cm long were dip-coated at a rate of 8.7±0.2 cm/s to produce a polycrystalline coating ≈ 2μm thick with grain size ≈ 1-2μm after conversion. Drying was performed in air at 110°C for 30 minutes. Conversion of the coating to monoclinic zirconia took place at 1000°C for 3-4h. At this point, the fibers would be ready for incorporation into a matrix (alumina). In this study, however, fibers were subjected to additional heat treatments that mimic temperature/time conditions that might be experienced during composite fabrication or in-service use. Fibers were annealed at 1400°C

† Fibers specimens were supplied by Martha Jaskowiak, NASA-LeRC
§ Strem Chemicals, Inc., Newburyport, MA 01950
£ same company, different vintage (hence lot number)
¥ Tosoh Ceramics Divition, Bridgewater, NJ 08807; mean particle size ≈ 0.4μm

for 8h and 20h in air in a high purity alumina tube furnace with internal diameter equal to 2.86 cm. Air was circulated over the fibers at a rate of ≈ 5 cc/sec to flush any possible contaminates and keep the atmosphere oxidative.

A chemical and structural characterization of the precursor sol and subsequent coating product was performed using inductively-coupled plasma emission spectroscopy (ICP-ES), powder X-ray diffraction (XRD), and Debye-Scherrer X-ray techniques. All sols (and the original ZrO_2 powders) were analyzed for minor and trace impurities using ICP-ES. Sol #3 was dried and characterized as monoclinic ZrO_2 using XRD.

Small sections of each fiber (≈ 5-10mm) were prepared for Scanning Electron Microscopy (SEM) examination and Electron Probe Microanalysis (EPMA) by carefully clamping one end of the fiber and, under a light microscope, flexing the other end of the fiber to spall some of the coating from the fiber. This technique can be assisted by lightly scraping the fiber end†. For SEM and EPMA work, complete 360 degree freedom of rotation about the c-axis of the fiber for ample viewing was attained by mounting the fibers on a wire spindle attached to a microscopy stub. Fiber specimens from NASA-LeRC (Sol #1 and #2) and PSU (Sol #3) were used for EPMA analysis; only those heat treated to 1400°C were examined, however, since those fibers tended to exhibit the worst surface modification. Specimen preparation for EPMA differed from SEM only by using carbon as a conductive surface coating rather than gold, respectively, to eliminate unwanted gold X-ray interference and shielding.

X-ray Photoelectron Spectroscopy (XPS) was used to probe the surface of as-received Saphikon fibers (see Table II) for possible impurity sources which might

Table II. Classification of XPS fiber samples

Sample	Sizing	Nominal Fiber Diameter (μm)	Saphikon Spool Number	Date Received
A	early-grade methylcellulose§	≈ 190	56448	4-28-89
B	Dow Methocel®	≈ 125	402	6-14-91
C	unsized	≈ 125	567	7-31-92

carry through the processing. Ten to twelve 3 cm long fibers of each sample in Table II were scanned across the outer 20 - 70Å layer of sized fiber. Sample C (unsized fiber) served as a reference.

† When necessary, a piece of cubic 9.5mol% yttria-stabilized zirconia disc was used to carefully scrape away the monoclinic zirconia coating to reveal the underlying pitted surface.

§ J. Fitzgibbon, Saphikon, Inc., personal communication, 3/94

RESULTS AND DISCUSSION

Interfacial degradation seen in the ZrO_2/Al_2O_3 system is not unprecedented. Others working with various coatings on sapphire have shown that reactions with, or modification of, the sapphire surface can occur during processing and/or heat treatment [1, 4]. In addition, modification of alumina was documented in Al_2O_3 platelet reinforced tetragonal zirconia polycrystal composites [5]. In the current study sapphire fibers coated with polycrystalline ZrO_2 exhibit pits on the fiber surface, surrounded by ridges, that become more pronounced as heat treatment temperature increases; see Figure 1. These features are similar to those in the aforementioned studies. Figure 1 also demonstrates that pitting occurs for each sol type (refer to Table I). Visually, 'pits' characterize the pock marks on the fiber surface and remnant ZrO_2 particles appear white. Pit size scales with ZrO_2 grain size and increases with increasing heat treatment temperature as the ZrO_2 particles coarsen. Hypotheses developed to explain pitting are outlined below. Special emphasis is given to the first hypothesis for which most results were gathered. Preliminary work on the second hypothesis will be highlighted.

Impurity-Driven Corrosion of the Fiber Surface

Based on the findings of Heussner and Claussen [5] and the work of Morgan and Koutsoutis [6], we postulate that an impurity-driven corrosion of the fiber surface occurs via liquid phase formation. Morgan and Koutsoutis showed that a liquid does form in the oxide ternary system of aluminum, titanium and sodium at 1350°C in the high alumina portion of the phase diagram. Heussner and Claussen attributed the pitting they witnessed to liquid phase evolution from silicon and magnesium impurities. Together, these studies suggest that common liquid-forming elements (Si, Na, Mg, Al) can cause transient liquids to form when present in small quantities. Therefore we chemically analyzed all precursor sols and ZrO_2 powders using Inductively-Coupled Plasma Emission Spectroscopy†. The results appear in Table III. Using Electron Probe Microanalysis (EPMA), pitted fiber surfaces were scanned for segregation of impurities to this region. At low accelerating voltages (4keV), we detected increased concentrations of impurity elements Si and Na relative to scans at higher voltages (12keV). At 4keV, characteristic X-rays of elements Si and Na originate generally no deeper than approximately 0.2μm from the surface, whereas at 12keV, signal predominantly arises from bulk sapphire. Since all EPMA work focused on pitted fibers heat treated at 1400°C, where estimated pit depth is of the same scale (tenths of microns), these results suggest that at 4keV we scan the pitted surface only. Therefore, the results corroborate that common liquid-forming elements do migrate to the coating/fiber interface where, as concentration increased, a transient liquid interphase formed during heat treatment.

A liquid phase rich in Si and Na (and possibly Ti) might dissolve alumina to form an aluminosilicate at the ZrO_2 particle/fiber surface contacts. This would

† work performed by Henry Gong, Materials Characterization Laboratory, PSU

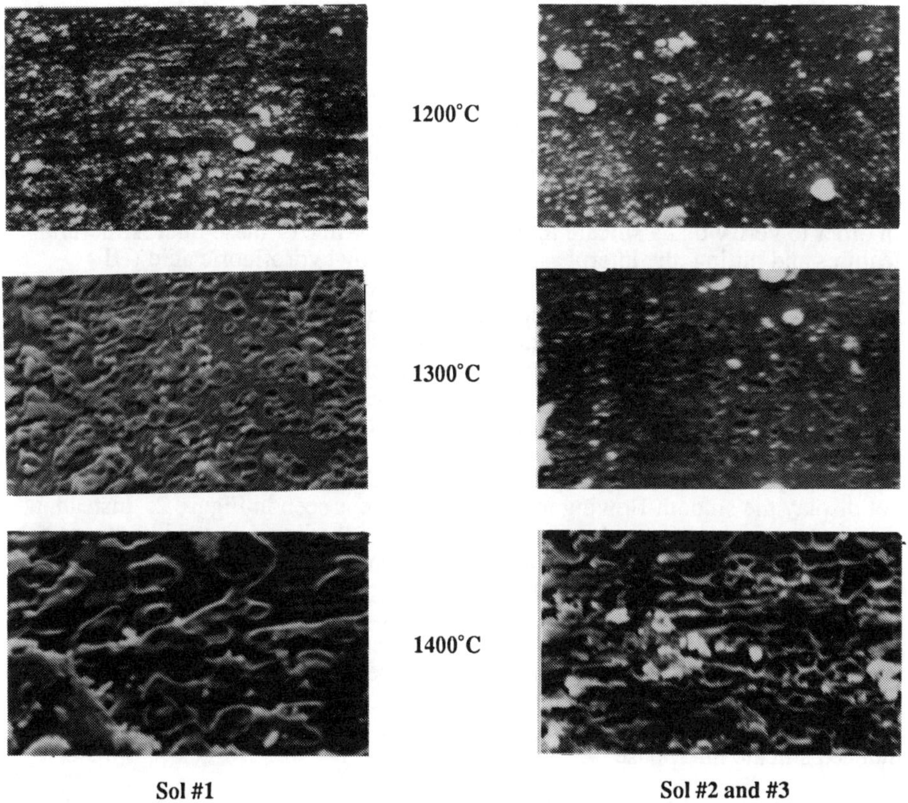

Figure 1 Scanning Electron Microscope (SEM) images of 'pits' which are characterized by the pock marks seen on the fiber surface. White particles are remnant zirconia grains. Pit size increases with heat treatment temperature and pitting generally occurs to the same degree for all sols studies. Sol #3 fibers were only processed at 1400°C and are visually similar to Sol #2 fibers shown.

Table III. ICP-ES identified impurities in precursor sols and powders (values are in weight percent)

Material	Si	Ti	Na	Hf
Strem ZrO_2	0.2	0.2	< 0.02	1.8
Sol #1	0.4	0.1	< 0.02	2.0
Sol #2	0.06	0.32	0.06	1.42
Tosoh ZrO_2	< 0.02	< 0.02	< 0.05	1.85
Sol #3	0.07	0.02	< 0.05	1.61

contribute to pit formation and cause liquid phase redistribution via capillarity within the polycrystalline coating/sapphire interface. Figure 2 illustrates the nature of liquid phase formation and reaction with the fiber surface of Sol #2 fibers heat treated at 1400°C for 20h†. The liquid is drawn to the ZrO_2 particle/fiber surface contact points where dissolution of the alumina then occurs. The interphase also wicks between ZrO_2 particles.

In order to verify that a silicate interphase is the source of these microstructural features and pitting, the interphase was removed by hydrofluoric acid (HF) etching. It is widely known that HF aggressively attacks silicate phases while remaining relatively benign to sapphire and zirconia. The fibers in Figure 2 were subjected to a 1min. etch in HF acid (48% HF/51% H2O) followed by soaking in distilled, deionized water for ≈ 5min., then examined again in the SEM. Figure 3 shows the effect of this etch on the interphase. Exact relocation was not possible (some ZrO_2 particles were removed during the etching and handling) but close examination of Figure 3 reveals that the liquid phase has been effectively removed by HF. The contact region between ZrO_2 particle and fiber surface does not display the smooth-flowing interphase meniscus seen in Figure 2. Instead, a sharper transition is seen. Also note that although the liquid-based interphase has been altered, the region surrounding each pit is still clearly delineated. We believe that sapphire surface reconstruction (ridging) also occurred, which may contribute to pit formation as well. A schematic of the HF etch process and results is shown in Figure 4. EPMA of HF etched fibers revealed a decrease in Si and Na impurity concentrations at the etched fiber surface by approximately 36 and 95 percent, respectively. This corroborates our microscopic interpretation of pitting being caused, at least in part, by sapphire dissolution in an impurity-induced silicate interphase.

During the investigation of Sol #1 and Sol #2 fibers, it was noticed that Sol #1 fibers exhibited significant surface modification beyond ordinary pitting. The presence of an additional crystalline phase at the interface between ZrO_2 coating and fiber surface has been found; see Figure 5. Previous work [7] had demonstrated that Sol #1 fibers exhibited significant strength degradation relative to Sol #2 fibers or uncoated fibers [2, 3]. This crystalline phase does not appear on later vintage fibers (Sol #2 and Sol #3). Compositional analysis performed using EPMA revealed that the phase contained 88.7 wt.% Al_2O_3, 8.5% CaO, 1.7% MgO, and 1.1% TiO_2. Although neither Ca nor Mg impurities are present in the precursor sols or powders (Table III), previous work by Norkitis [8] showed that the sizing on early vintage Saphikon fibers left residual Ca on the fiber surface after cleaning, as measured with Energy Dispersive Spectroscopy (EDS). We investigated the surface atomic states of the three fiber samples described in Table II using XPS. Without altering the as-received surface, XPS spectra show that fibers initially sized with the early-grade methylcellulose sizing contain detectable amounts of Ca in the outer most 20-70Å layer; see Figure 6. Correlation has been drawn between the vintage of sapphire fibers with the early-

† Fibers heat treated at 1400°C for 8h show similar behavior, but to a lesser degree.

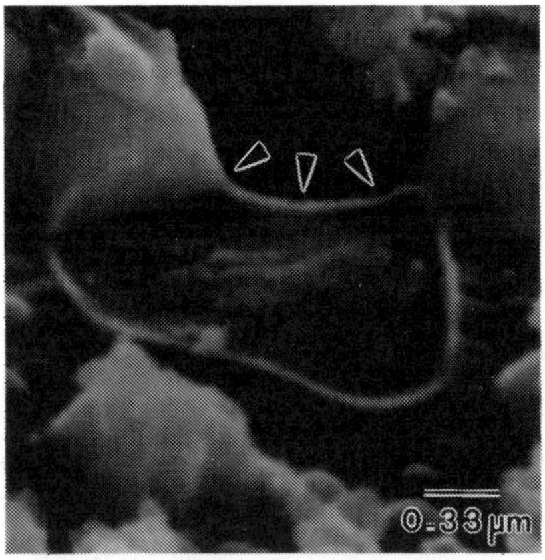

Figure 2 Liquid phase formation is evident on Sol #3 fibers heat treated at 1400°C for 20 h. Note capillary behavior of interphase at contact points (see arrows).

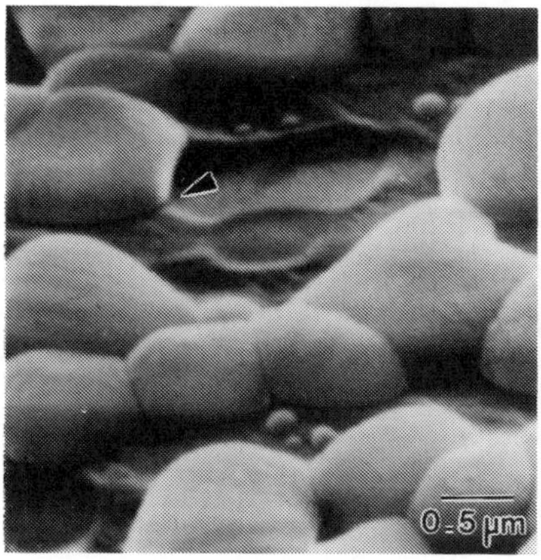

Figure 3 After etching in HF for 1 min., removal of interphase is noticed. Contact points (see arrow) appear less smooth and ridging becomes clearly evident.

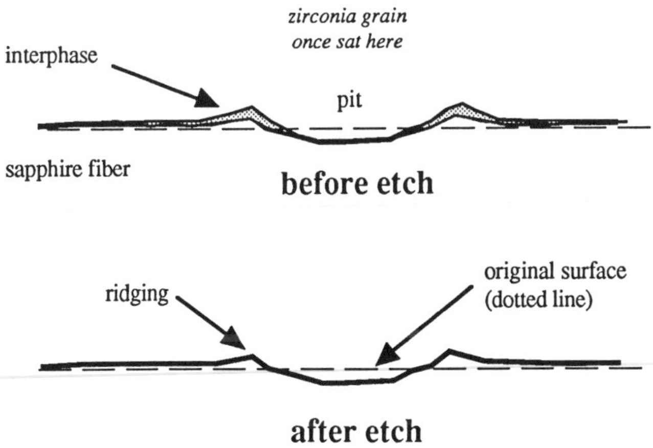

Figure 4 A schematic representation of the proposed HF etching effect. Before the etch, the pit consists of a portion of the sapphire surface dissolved in the interphase material which gathers at the junction points of zirconia and sapphire. After removal of the interphase, the dissolved region of the fiber surface is surrounded by ridging which is suspected due to surface reconstruction of the sapphire.

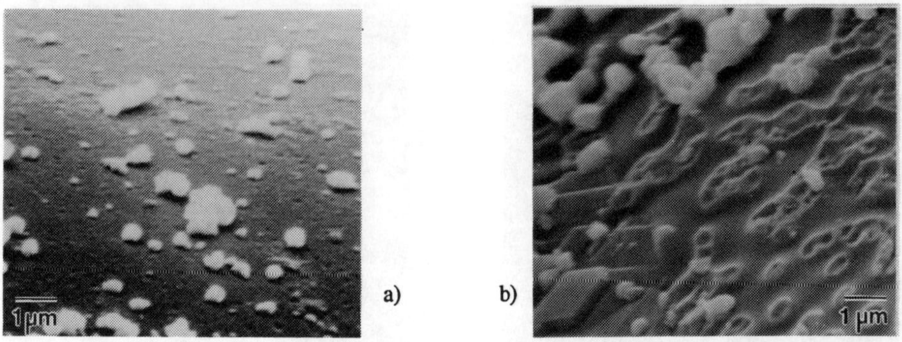

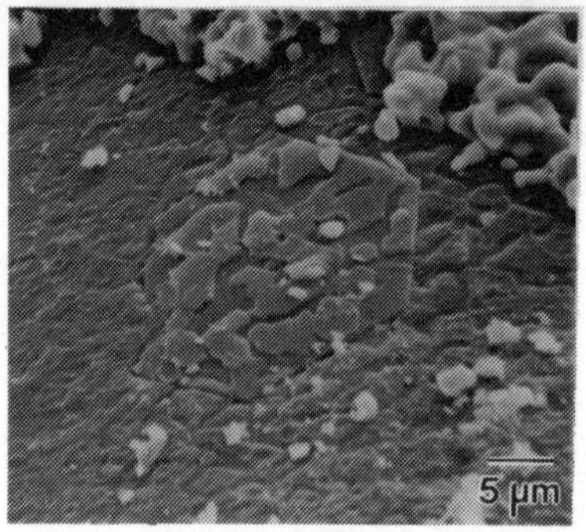

Figure 5 Second phase crystallite develops as a function of heat treatment temperature after 8h at (a) 1200°C, (b) 1300°C, at lower left, and (c) 1400°C, center. At 1400°C, the average size is ≈ 25µm on the fiber surface.

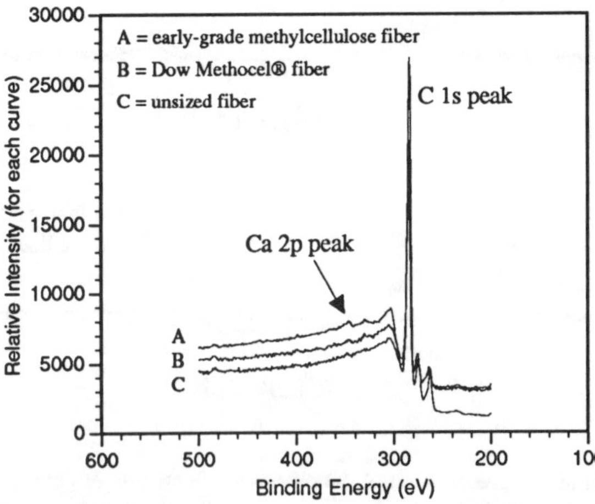

Figure 6 XPS spectra at the Ca 2p peak was used to identify the presence of calcium in the near surface (20 - 70Å) atomic states of fibers in Table II. UV/Ozone cleaning will be used to reduce the carbon content and assist quantification of Ca present.

grade methylcellulose sizing (1989) and the processing and strength testing of Sol #1 fibers (circa 1990; [7]). Therefore, we feel presence of the crystalline phase can be attributed to reaction of the sapphire with residual Ca which lingers on the fiber surface after sizing removal as well as other impurities in the sol.

High Mutual Solid Solubilities Exist

Pit evolution may also be a consequence of dissolution of sapphire into the zirconia coating itself. However, numerous inconsistencies exist in the literature concerning the solid solubility of Al_2O_3 in ZrO_2 and ZrO_2 in Al_2O_3. It has been stated that the solubility of zirconia in alumina ranges from "vanishingly" small [9] to several tenths of a weight percent [10] in the temperature range of interest here (to 1600°C). To date, no concerted study has addressed this issue; therefore, a Secondary Ion Mass Spectroscopy (SIMS) study was initiated on the diffusion behavior of a 9.5mol% Y_2O_3-stabilized ZrO_2/Al_2O_3 diffusion couple heat treated to 1000°C for 1000+ hours. Figure 7 shows that measurable penetration of Zr

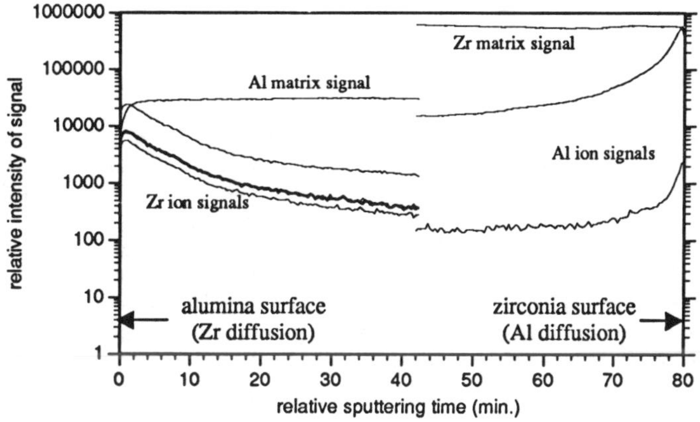

Figure 7 Depth profiling via SIMS shows that diffusion of Zr into alumina and Al into zirconia has occurred in this 9.5mol% Y_2O_3-stabilized ZrO_2/Al_2O_3 couple heat treated to 1000°C for 1000+ hours.

into alumina and Al into zirconia has occurred; however, a thorough quantification of this diffusion behavior requires standards with well known concentrations. Ion implantation was employed to create doped materials which could be used as standards. Various orientation sapphire single crystals were implanted† with Zr, and both monoclinic ZrO_2 and 9.5mol% Y_2O_3-stabilized

† implantation was done by Robert G. Wilson, Hughes Implantation Services, Malibu, CA 90265

ZrO_2 were implanted with Al. Implanted profiles, as determined by SIMS, resemble near-Gaussian distributions centered about a depth heavily dependent on the implant energy. Preliminary results on implanted samples annealed in air at temperatures up to 1600°C indicate that chemically-driven diffusion does occur in this system (i.e., in the case of implanted ion diffusion, this means inward diffusion as opposed to diffusion towards the free surface). Also, the SIMS data for diffusion of Zr in Al_2O_3 at 1600°C for long-times (> 25h) suggests a measurable equilibrium solubility limit is being approached; accordingly, this work continues.

Future work will involve determination of diffusion coefficients as a function of temperature for Zr in both A-plane and C-plane orientation alumina and Al in zirconia (both monoclinic and cubic). Also, solubility limits from long-time diffusion anneals will be determined using SIMS. Possible effects of higher solubility are 1) diffusion of Zr into the sapphire creates an unfavorable volumetric strain which is relieved to some extent by upheaval of the sapphire surface around zirconia particles, 2) alumina may dissolve in zirconia at the particle/fiber contact sites. The strain mismatch across the interface might facilitate this latter mechanism since zirconia exhibits a martensitic transformation at ≈ 1050°C.

CONCLUSIONS

Surface modification and interfacial degradation occur in polycrystalline ZrO_2-coated Saphikon fiber systems. Pit size generally increases as heat treatment temperature increases (or time at temperature increases). Chemical and microprobe analysis reveals that Si, Na, and some Ti impurities are present in the precursor sols and segregate to the fiber surface during processing and heat treatment. Migration of these impurities to the interface between ZrO_2 particles and sapphire causes a liquid to form, which can locally dissolve the alumina at particle contact points. In addition, certain fibers displayed second phase crystallization from impurities present in the sol as well as fiber sizing. Other mechanisms contributing to pitting are currently being evaluated.

REFERENCES

1. J.B. Davis, J.P.A. Löfvander, A.G. Evans, E. Bischoff and M.L. Emiliani, "Fiber Coating Concepts for Brittle Matrix Composites," *J. Am. Ceram. Soc.*, **76** [5] 1249-57 (1993).

2. M.H. Jaskowiak, W.H. Philipp, J.I. Eldridge, J.A. Setlock, M.E. Norkitis and J.R. Hellmann, "Progress in Sapphire/Al_2O_3 Composite Interfaces and Fiber/Fiber Coatings Interactions," 5th Annual HITEMP Review, NASA-CP 10104, pp. 59-1 to 59-14 (1992).

3. M.H. Jaskowiak, J.I. Eldridge, J.B. Hurst and J.A. Setlock, "Interfacial Coatings for Sapphire/Al2O3," 4th Annual HITEMP Review, NASA-CP 10082, pp. 84-1 to 84-11 (1991).

4. J.B. Davis, E. Bischoff and A.G. Evans, "Zirconia Coatings for Sapphire Fiber-Reinforced Composites," pp. 631-38 in Ceramic Transactions, Vol. 19, ed. M.D. Sacks, The American Ceramic Society, Westerville, OH (1991).

5. K.H. Heussner and N. Claussen, "Yttria- and Ceria-stabilized Tetragonal Zirconia Polycrystals (Y-TZP, Ce-TZP) Reinforced with Al2O3 Platelets," *J. Euro. Ceram. Soc.*, **5** 193-200 (1989).

6. P.E.D. Morgan and M.S. Koutsoutis, "Phase Studies Concerning Sintering in Alumina Doped with Ti4+," *J. Am. Ceram. Soc.*, **68** [6] C156-58 (1985).

7. M.H. Jaskowiak, M.J. Hyatt, W.H. Philipp, J.I. Eldridge and J.A. Setlock, "Effects of ZrO2 Interfacial Coatings in Al2O3/Al2O3 Composites," 3rd Annual HITEMP Review, NASA-CP 10051, pp. 60-1 to 60-12 (1990).

8. M.E. Norkitis, "Critical Evaluation of Tin Oxide as an Interfacial Modification Agent in the Alumina/Alumina Composite System," M.S. Thesis, The Pennsylvania State University (1994).

9. A.H. Heuer, Personal Communication, 5/93.

10. A.H. Heuer, N. Claussen, W.M. Kriven, and M. Rühle, "Stability of t-ZrO_2 Particles in Ceramic Matrices," *J. Am. Ceram. Soc.*, **65** [12] 648 (1982).

MULLITE FORMATION AT THE INTERFACE OF CaO-Al_2O_3 FIBER REINFORCED Al-Si ALLOY COMPOSITES

Yun-Mo Sung,[1] Stanley A. Dunn,[2] and James A. Koutsky[2]
[1]Department of Materials Science and Engineering
[2]Department of Chemical Engineering
University of Wisconsin-Madison, Madison, WI 53706

Inviscid melt spun (IMS) calcia-alumina (CA) fiber reinforced aluminum-silicon alloy (4032) composites were produced using melt infiltration technique. The composites processed at 927 °C showed formation of an interphase region of about 10 to 20 μm thickness, as well as excellent interfacial bonding. Energy dispersive x-ray spectroscopy (EDS) analysis gave averaged compositions of this interphase region at 74 wt% of Al and 26 wt% of Si, which corresponds to the composition of mullite ($3Al_2O_3 \cdot 2SiO_2$). The formation of mullite at the interface was confirmed by x-ray diffraction (XRD).

1. INTRODUCTION

Metal matrix composites (MMC's) have been developed to improve specific mechanical properties such as modulus, creep, and wear resistance of structural components.[1-10] The major applications of these MMC's are aerospace parts, automobile components, and sports equipment. The first high-strength, high-modulus fiber to be used in MMC application was boron fiber.[4] This fiber, however, shows limitations in the application to MMC's. For instance, boron

To the extent authorized under the laws of the United States of America, all copyright interests in this publication are the property of The American Ceramic Society. Any duplication, reproduction, or republication of this publication or any part thereof, without the express written consent of The American Ceramic Society or fee paid to the Copyright Clearance Center, is prohibited.

fibers react with molten aluminum, and the mechanical properties of boron / aluminum composites degrade over the long term at high temperatures (> 480 °C). These drawbacks have led to the development of SiC fibers.[5] Due to good bonding between SiC fibers and reactive metals and the small loss of strength degradation during high-temperature processing, SiC fiber reinforced metal matrix composites have become viable engineering materials.

Alumina fiber is another ceramic fiber which is widely used for reinforcing metal matrices such as Al and Mg.[6] The chemical and oxidation resistance of alumia can overcome fiber degradation during both composite fabrication and use. Since strength and stiffness remain high at elevated temperatures, the composites creep less and have significantly better fatigue resistance than the unreinforced metals. The hardness of alumina accounts for the exceptionally good wear properties of their MMC's. In spite of these excellent properties of ceramic fibers as reinforcements for metals such as aluminum, titanium, and magnesium, they have only limited use due to their high cost which is related to their production processes.

The Inviscid Melt Spinning (IMS) process has been successfully applied to fiber production using low-viscosity melts of inorganic materials such as ceramics and metals.[11-17] Furthermore the IMS process gives a very rapid production rate of fibers and thus considerably reduces fiber cost. The formation of a stabilizing sheath on the surface of inviscid molten stream by introduction of a reactive gas such as propane, can prevent the Rayleigh break-up and stabilize the molten stream until solidification occurs. Calcia-alumina fiber (CA) was the first ceramic fiber spun by IMS. Various compositions of CA were spun into fibers.[13-15] The fibers for this study have a eutectic composition of 46.5 wt% CaO and 53.5 wt% Al_2O_3. This fiber would be suitable for reinforcing aluminum matrices since it can retain about 80% of strength up to approximately 800 °C. X-ray diffraction (XRD) patterns show non-crystallinity in the as-spun IMS CA fiber.[18] The properties of this amorphous IMS CA fiber are listed in Table I.[19,20] Aluminum alloys containing high percentages of silicon (12 wt%) were selected as matrix materials for the IMS CA fiber in order to try to transform some portion of the amorphous CA fiber to mullite ($3Al_2O_3 \cdot 2SiO_2$) phase using a reaction between Al_2O_3 in the CA fiber and Si in the Al-Si matrix.

Table I. Physical Properties of As-Spun IMS CA Fibers

Tensile Strength (MPa)	Young's Modulus (GPa)	Density (g/cm^3)	Fiber Diameter (μm)	Glass Transition Temp. (°C)	Crystallization Peak Temp. (°C)
519	100	2.74	200 –300	~850	~980

Mullite has been well known as an excellent high temperature material which can maintain most of its strength up to ~1400 °C and shows the lowest creep rate up to ~1400°C.[21] The alumina fiber reinforced aluminum composites containing alloying elements such as Li, Mg, and Cu gave compounds such as $LiAlO_2$, $MgAl_2O_4$, and $CuAl_2O_4$ at the interface.[22,23]

The objective of this study is to fabricate IMS CA fiber reinforced Al-Si alloy composites using melt infiltration and investigate the interphase formation (mullite) at the interface of the composites.

2. EXPERIMENTAL PROCEDURES

A graphite mold consisting of three pieces: a top piston, a main cylinder, and a bottom plate, was designed for the CA / Al-Si composite fabrication. A high grade graphite with low-ash content (grade ECV, National Carbon, Cleveland, OH) was used to reduce the wetting of the Al alloy melt with the graphite mold. Al alloy (4032) was provided by Alcoa Technical Center (Alcoa Center, PA) and contained 12.0 wt% Si as the major alloying element, and 0.96 wt% Cu and 0.05 wt% Fe as minor alloying elements. CA fibers (~200 – 300

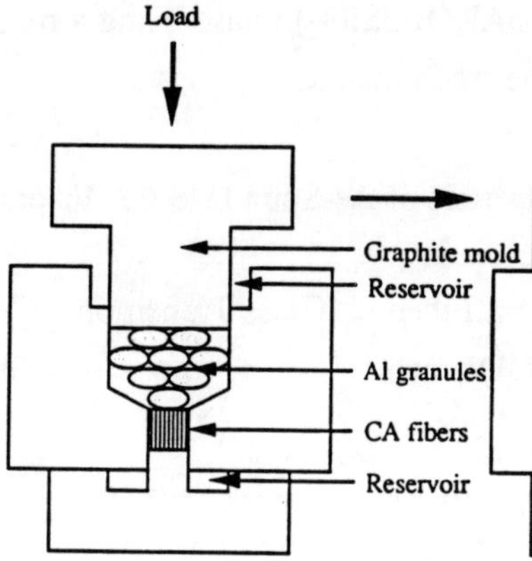

(a)

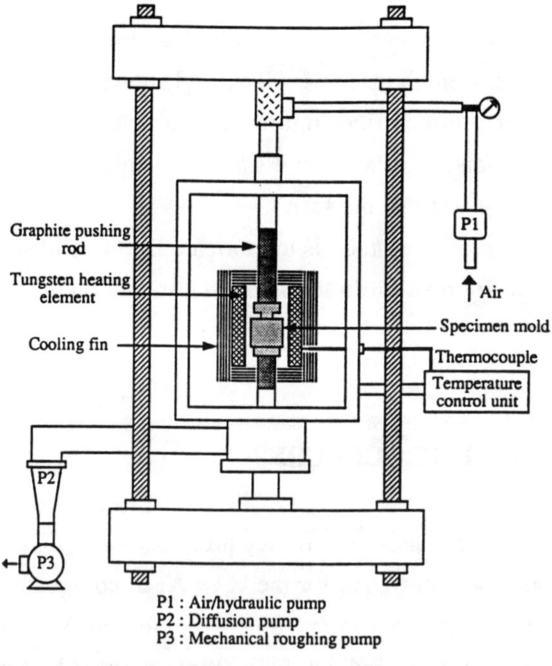

(b)

Fig. 1. Schematics of the graphite mold (a) and hot press (b) used for composite fabrications.

μm diameter) were preloaded parallel to the mold axis and Al-Si alloy chunks were loaded at the top of the CA fibers. This mold was placed in a hot press machine (Physical Science Lab at University of Wisconsin-Madison) which had a tungsten heating element and a diffusion pump for high vacuum (~10^{-6} torr). The weight of the top piston of the hot press was sufficient to break the Al_2O_3 film on the surface of the aluminum melt and obtain melt infiltration . Additional hydraulic pressure was not needed. Fig. 1 shows schematics of the graphite mold (a) and hot press (b) used for the composite fabrication.[24] Melt infiltration was done at 927 °C. The temperature of 927 °C, which is slightly lower than the crystallization temperature of this CA fiber (~980°C), was used to obtain interface / interphase chemical reactions without causing a crystallization and consequently, a degradation of mechanical properties of this fiber, as was shown in previous study.[19] The graphite mold containing CA fibers and Al-Si alloy chunks was heated to 927 °C for 2 hrs and maintained at that temperature for 2 hrs. Cooling this mold to room temperature took approximately 2.5 hrs. The nipple shape CA/Al-Si composite was easily released from the graphite mold. For scanning electron microscopy (SEM) studies a sample surface perpendicular to the fiber direction was polished using 400 and 600 grid SiC papers (CARBIMET, Buehler, Germany) and 1 μm Al_2O_3 powder (TEXMET, Buehler, Germany). A Jeol JSM 35-C SEM was used for secondary electron imaging (SEI), to investigate wetting behavior and morphology of interface / interphase and backscattered electron imaging (BEI), to obtain compositional contrast at the interface region. Energy dispersive x-ray spectroscopy (EDS: Noran Co. Middleton, WI) was also used with SEM analysis to provide the composition profiles across the interfacial zones of the composite. XRD (Nicolet Stoe Transmission / Bragg-Brentano Stoe Co., Germany) with a Cu-Kα source, a 20s time constant, a 15 to 65° scan range, and 0.02° step size, was performed for phase identification. The XRD analysis directly on the composites would not detect the new interphase formed due to the small volume percentages (< 5%). Some of the CA fibers could be separated from the composite by dipping the composite into a solution of sodium hydroxide (40 wt%) and distilled water (60 wt%) for about 10 min.. To identify the crystalline phase on the surface of these fibers, powdered CA fibers were analyzed. The resulting XRD patterns were identified using the Joint Committee on Powder Diffraction Standards (JCPDS) card to confirm the formation of mullite ($3Al_2O_3 \cdot 2SiO_2$). The peak positions in terms of two-theta for mullite, listed in decreasing order of intensity, are 26.27 (1), 25.95 (2), 40.79 (3), 35.28, 16.43 (4), 33.23 (5), 60.71 (6), 42.59 (7), 57.56, 30.96 (8), 64.57 (9), and 53.88, 36.99 ° (10).

3. RESULTS AND DISCUSSION

The interface of CA/Al-Si composites processed at 927 °C were primarily investigated using SEM. Fig. 2 shows the SEI (a) and BEI (b) of CA/Al-Si composite processed at 927 °C for 2 hrs. A new interphase of about 10 to 20μm thickness as well as an excellent wetting was observed at the fiber-matrix interfacial zone. Evidently, for 927 °C processing, the work of spreading is increased due to the decreased surface tension of liquid caused by Si accumulation at the interface, and the decreased interfacial tension between solid and liquid caused by compound formation at the interface. Therefore, the CA / Al-Si composite produced at 927 °C should show better wetting then the composite processed at lower temperatures. Furthermore, it was observed that the interphase was formed inside the CA fiber. This implies that matrix elements diffused into the original boundaries of the CA fibers to form the new phase. BEI of this composite showed that the interphase was depleted of the high atomic number element (Ca) since the interphase region is darker than the CA fibre area. Fig. 3 shows the blown-up image of the interphase (a) shown in Fig. 2 and a compositional scan for the corresponding interphase (b). The interphase is a polycrystalline compound with an approximately 1 μm grain size. The EDS for the composite shows a high Si content and Ca depletion in the interphase. The amount of Ca in the interphase region was negligible. The averaged amount of Al and Si in the interphase was 74 and 26 wt%, respectively, possibly indicating the formation of mullite ($3Al_2O_3{\cdot}2SiO_2$). Silicon accumulation (~35 wt%) was also observed at the interface.

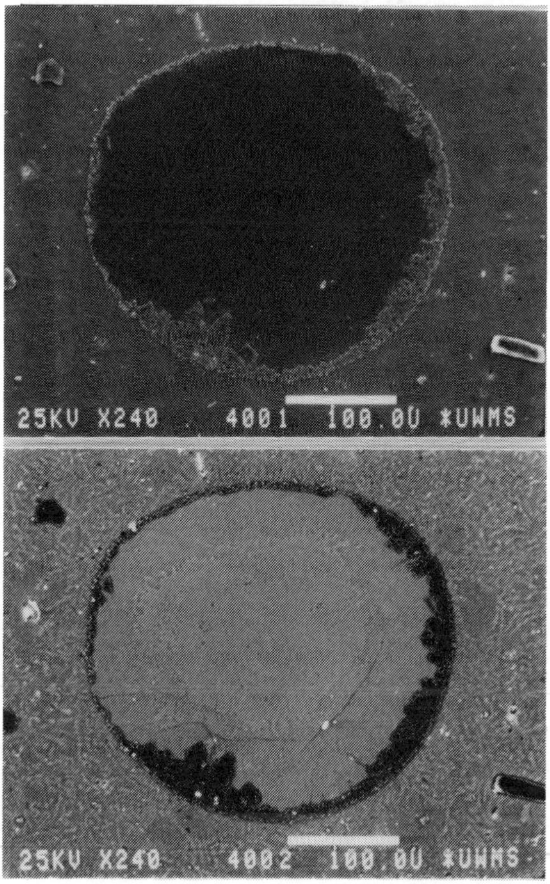

(a)

(b)

Fig. 2. The SEM secondary electron image (SEI) (a) and backscattered electron image (BEI) (b) of IMS CA fiber reinforced Al-Si alloy composite processed at 927 °C for 2h.

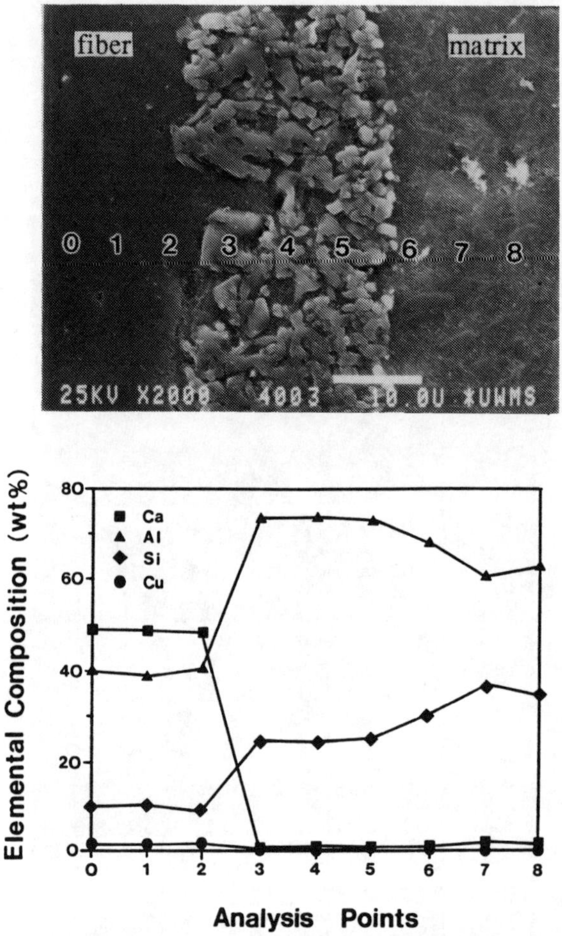

(a)

(b)

Fig. 3. The blown-up SEM secondary electron image (SEI) of the interface of IMS CA fiber reinforced Al-Si alloy composite processed at 927 °C for 2h (a) and compositional scan for the corresponding interface (b).

XRD was performed on the powdered CA fibers separated from the composites. Fig. 4 shows the XRD pattern for the powdered CA fiber separated from Al-Si composite processed at 927 °C. The broad background comes from the vitreous CA fiber. The peaks at 26.26 (1), 25.95 (2), 40.79 (3),

35.27, 16.43 (4), 33.22 (5), 60.71 (6), 42.59 (7), 57.56, 30.95 (8), 64.57 (9), 53.88, 36.99 ° (10) of mullite were identified by comparing the data in JCPDS card.

The mechanism of mullite formation at the interface of the composite can be expressed as following.[25,26]

$$Si + O_2 \text{------------------>} SiO_2 \qquad \Delta G° = -690 \text{ kJ/mol at } 927\ °C \quad (5)$$

$$3Al_2O_3 + 2SiO_2 \text{------>} 3Al_2O_3 \cdot 2SiO_2 \quad \Delta G° = -5292 \text{ kJ/mol at } 927°C \quad (6)$$

This implies that Si in the Al-Si matrix is first oxidized to SiO_2 even at the high vacuum atmosphere and the SiO_2 at the interface region reacted with Al_2O_3 in the CA fiber to form the mullite phase. The oxygen partial pressure to obtain Si oxidation reaction at 927 °C is only 6.93 x 10^{-31} atm. Thus, even the high vacuum (~10^{-6} torr: 1.3 x 10^{-9} atm) of the hot press has a high enough oxygen partial pressure to oxidize Si to SiO_2. The negative ΔG° values for the formation of SiO_2 and $Al_2O_3 \cdot 2SiO_2$ indicate that these are stable phases at 927°C.

Although only ~ 15 vol% of IMS CA fibers was transformed to mullite phase, this would affect the overall mechanical properties of the composite, especially the creep properties. Furthermore, if redrawn IMS CA fibers of ~10μm diameter were used, the whole fibers would be transformed to a mullite phase during processing in this composite system. Mechanical properties of this composite system are presently being studied and will be reported later.

4. CONCLUSIONS

An interphase of mullite was formed between the IMS CA fiber and the Al-Si matrix for the 927 °C melt infiltration. The thickness of mullite phase region was approximately 15 μm. Excellent interfacial wetting was observed for these composites.

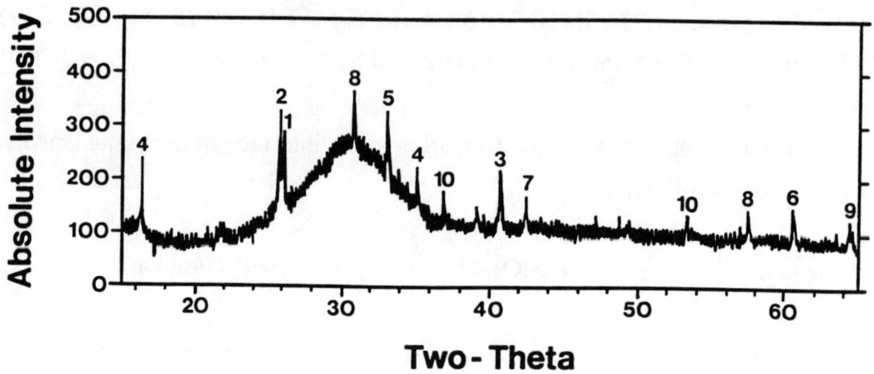

Fig. 4. The X-ray diffraction (XRD) pattern for the powdered CA fiber separated from IMS CA fiber reinforced Al-Si alloy composites processed at 927 °C for 2h. The numbers on the top of the peaks correspond to the peak numbers listed in section 2.

ACKNOWLEDGEMENT

The authors wish to thank Dr. Douglas A. Granger in Alcoa Technical Center (Alcoa Center, PA) for kindly providing the aluminum alloys.

REFERENCES

1. J. M. Anglin, in Engineered Materials Handbook, Vol. 1 Composites, edited by ASM Int'l (Metals Park, OH, 1987) p. 801.
2. T. T. Seafani, ibid., p. 810.
3. H. B. Dexter, ibid., p.823.
4. M. E. Buck, R. J. Supinkas, ibid., p. 851.
5. J. A. McCelman, ibid., p. 858.
6. J. C. Romine, ibid., p. 874.
7. D. M. Essock, ibid., p. 878.

8. W. C. Harrigan, Jr., ibid., p. 889.
9. J. L. Cook, W. R. Mohn, ibid., p. 896.
10. M. W. Toaz, ibid., p. 903.
11. R. E. Cunningham, L. F. Rakestraw, and S. A. Dunn, in Spinning Wire From Molten Metal, AIChE Symposium Series, Vol. 74, edited by J. W. Mottern and W. J. Privott. (American Institute of Chemical Engineers, NY, 1978) 20.
12. B. S. Mitchell, K.-Y. Yon, S. A. Dunn, and J. A. Koutsky, *Mater. Lett.*. **10** (1990) 71.
13. F. T. Wallenberger, N. E. Weston, and S. A. Dunn, *J. Non-Cryst. Sol.* **124** (1990) 116.
14. F. T. Wallenberger, N. E. Weston, and S. A. Dunn, *SAMPE Quart.*, **9** (1990) 121.
15. F. T. Wallenberger, *Ceram. Bull.*, **69** (1990) 1646.
16. B. S. Mitchell, K.-Y. Yon, S. A. Dunn, and J. A. Koutsky, *Chem. Eng. Comm.*, **106** (1991) 87.
17. F. T. Wallenberger, N. E. Weston, K. Motzfelt, and D. G. Swartzfager, *J. Am. Ceram. Soc.* **75** (1992) 629.
18. B. S. Mitchell, K.-Y. Yon, S. A. Dunn, and J. A. Koutsky, *J. Non-Cryst. Sol.* **152** (1993) 143.
19. B. S. Mitchell, Ph. D. Thesis (University of Wisconsin, 1991).
20. Y.-M. Sung, S. A. Dunn, and J. A. Koutsky, accepted by *Ceram. Int.* (1994).
21. I. A. Aksay, D. M. Dabbs, and M. Sarikaya, *J. Am. Ceram. Soc.* **74** (1991) 2343.
22. C. G. Levi, G. J. Abbaschian, and R. Mehrabian, *Met. Trans.* A **9A** (1978) 697.
23. U. Madeleno, H. Liu, T. Shinoda, Y. Mishima, and T. Suzuki, *J. Mater. Sci.* **25** (1990) 3273.
24. K.-Y. Yon, Ph. D. Thesis (University of Wisconsin-Madison, 1993).
25. D. R. Gaskell, Introduction to Metallurgical Thermodynamics 2nd edition (McGraw-Hill Book Comp.).
26. JANAF Termodynamical Data, Army-Navy-Air Force Thermodynamical Panel, (The Dow Chemical Co., MI, 1962-63).

FIBER-MATRIX INTERPHASE DEVELOPMENT IN CARBON/CARBON/COMPOSITES

G. S. Rellick, R. J. Zaldivar and P. M. Adams
Mechanics and Materials Technology Center
The Aerospace Corporation
El Segundo CA 90245

INTRODUCTION

In carbon/carbon (C/C) composites—i.e., a composite in which a carbon matrix is reinforced with carbon fiber—it is found that when the matrix is derived from a thermosetting resin, we always observe a distinct, highly graphitizable and well-oriented matrix interphase structure adjacent to the fibers.[1,2] Qualitatively, the orientation of the interphase is the same as the fiber. It is important to note that thermosetting resins are nongraphitizing when heated in bulk; they form isotropic, amorphous "glassy" carbon. The structure of this interphase becomes more prominent, i.e., more graphitic, as the heat treatment exceeds about 2200°C. We have postulated[2] that the basis for this graphite interphase development is molecular orientation induced in the degradation of the polymer matrix to carbon as a consequence of restraint of pyrolysis shrinkage at the fiber-matrix interface. More specifically, we have hypothesized that the critical factor for development of lamellar graphite (by subsequent high-temperature heat treatment) in this interphase, rather than amorphous glassy carbon, is a state of multiaxial tensile deformation during pyrolysis.[2]

An example of this interphase structure is shown in Figure 1. The matrix is derived from polyarylacetylene (PAA) resin.[3,4] The fiber is the Amoco PX-7, open-wedge type. The photograph is from a scanning electron microscope (SEM), in which the specimen has been argon-ion etched following polishing. This ion-etching procedure has the effect of revealing the graphite layer-plane microstructure, particularly when the graphite layer planes are predominantly perpendicular to the plane of section. Using this approach, we can effectively distinguish the graphitic interphase zone from the matrix proper—i.e., that portion of matrix removed some distance from the fiber and which is not transformed to graphite. Figure 2 shows a longitudinal section of a PAA-based composite heat

To the extent authorized under the laws of the United States of America, all copyright interests in this publication are the property of The American Ceramic Society. Any duplication, reproduction, or republication of this publication or any part thereof, without the express written consent of The American Ceramic Society or fee paid to the Copyright Clearance Center, is prohibited.

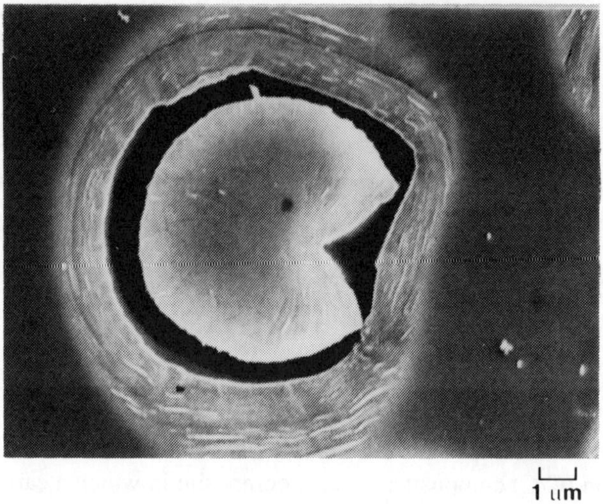

Figure 1. SEM showing "stress-graphitized" matrix around Amoco pitch-based PX-7 fiber following 2750°C HTT. Resin precursor was PAA. (Reprinted from *Carbon*, 29, R. J. Zaldivar and G. S. Rellick, "Some Observations on Stress Graphitization in Carbon-Carbon Composites," 1155–1163, Copyright 1991, with permission from Elsevier Sciences Ltd., Pergamon Imprint.)

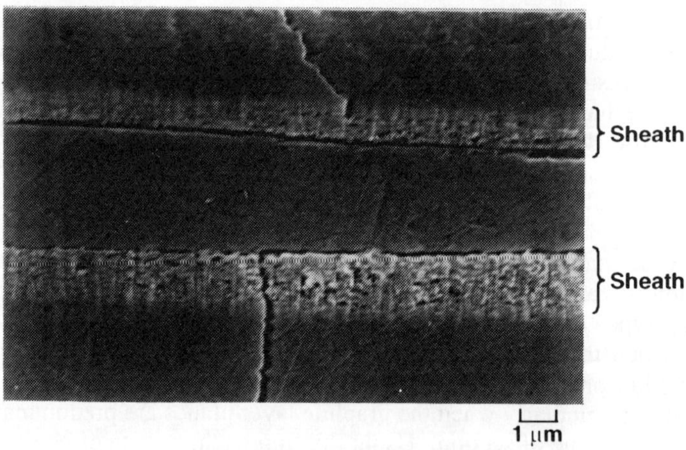

Figure 2. Longitudinal fiber-matrix interface of PAA/T50 composite heat treated to 2750°C, showing two strength-enhancing mechanisms: crack deflection along fiber-matrix interface (bottom), and blunting of crack tip within well-ordered sheath (top). (From "Processing Effects on the Mechanical Behavior of PAA-Derived C/C Composites," R. J. Zaldivar, G. S. Rellick, and J. M. Yang, *SAMPE J.* Reprinted by permission of the Society for the Advancement of Material and Process Engineering.)

treated to 2750°C that was strained to failure in tension; the fiber is the PAN-based T50 from Amoco. What we see in this very interesting photograph is both crack blunting and deflection in the interphase zone. The crack at the top of the photo is blunted at the interphase/glassy-carbon interface, whereas the bottom crack is deflected along the fiber/interphase boundary.

Our objective is to study the structural details of the interphase region in more detail using transmission electron microscopy in conjunction with selected area (electron) diffraction (SAD). We also wish to explore the influence of matrix microstructure on fracture behavior by using an in-situ SEM flexure stage that permits us to follow the crack tip as it advances through the different microstructural regions.

RESULTS

Figure 3 reveals the matrix orientation between two fibers for a low HTT of 1100°C. The composite is unidirectional, with the fiber being mesophase-based E75 (modulus of 75Mpsi ≅ 520 GPa) from DuPont, and the matrix is PAA. The SAD pattern is consistent with the layer planes oriented parallel to the fibers, as expected based on the previous discussion and evidence from optical microscopy

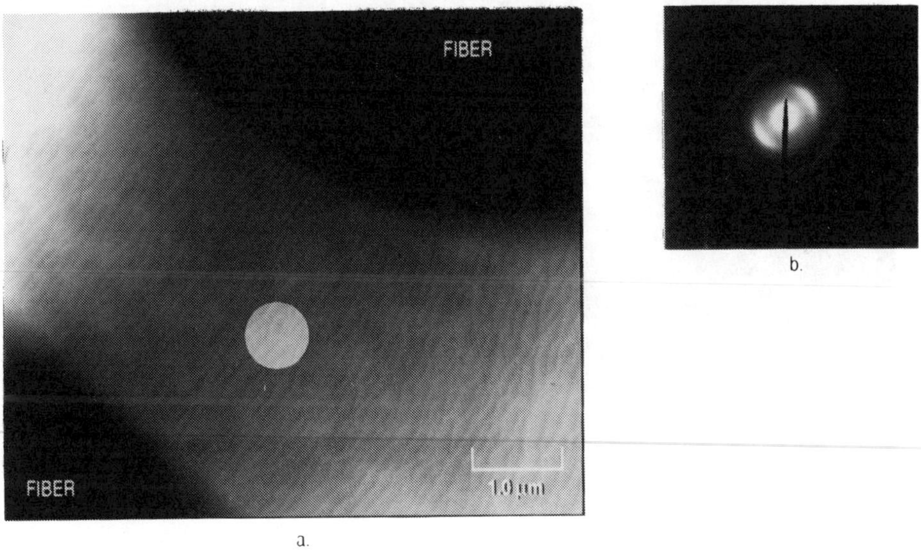

Figure 3. Transverse sections of carbonized unidirectional composites: (a) BF image of E75/PAA/1100°C; (b) matrix SAD of aperture zone in (a). (Reprinted from *Carbon* 32(1), G. S. Rellick and P. M. Adams, TEM Studies of Resin-Based Matrix Microstructure in Carbon/Carbon Composites," 127–44, Copyright 1994, with permission from Elsevier Sciences Ltd, Pergamon Imprint.)

and SEM.[1,2] Note the rippling or striation effect in the matrix. These striations run perpendicular to the layer-plane orientations and are believed to be the result of the interaction of the ion beam with the oriented layer planes.[5] Following higher HTTs, the matrix reveals a highly lamellar texture, brought out by the ion milling (Figure 4); SADs of this region confirmed its highly graphitized nature.

One immediate consequence of this strain-induced, oriented, and graphitized matrix interphase in C/C is that the matrix becomes a two-phase system, the second phase being the nongraphitized, glassy carbon. We would expect, therefore, that the *average* density and shrinkage of a thermoset-resin-derived matrix will depend on the relative proportion of the two different carbon structures and their respective densities. Furthermore, we would expect the amount of graphitized interphase to increase with the fiber volume fraction (ignoring the complicating

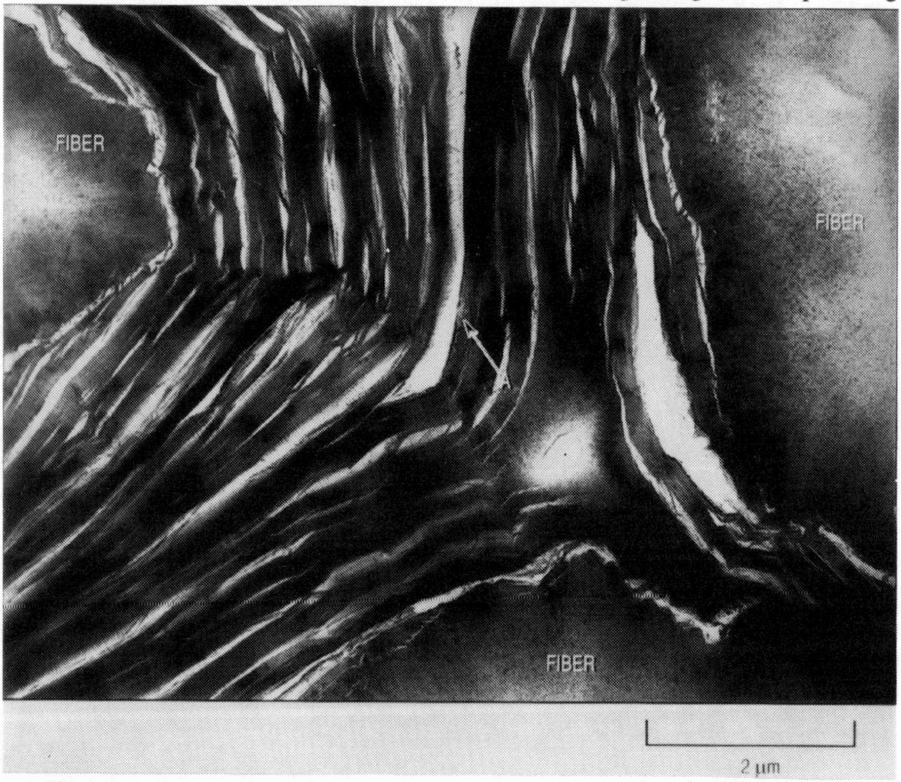

Figure 4. Kink bands in transverse section of T50/SC1008/2900°C, BF image. (Reprinted from *Carbon* 32(1), G. S. Rellick and P. M. Adams, TEM Studies of Resin-Based Matrix Microstructure in Carbon/Carbon Composites," 127–44, Copyright 1994, with permission from Elsevier Sciences Ltd, Pergamon Imprint.)

factor of fiber distribution). We recently attempted a measurement of the in situ matrix shrinkage and in situ density as a function of HTT for a series of unidirectionals fabricated from T50 fiber and both PAA resin and phenolic resin.[6] The results confirmed our suspicions—for both resin-matrix precursors, the average matrix density for each HTT from 1800 to 2750°C increased with fiber volume fraction. In addition, there were significant density differences between the two matrices. For example, the density of the phenolic-derived matrix after 2750°C HTT was 1.90 g/cm^3, while for the PAA-derived matrix, it was only 1.77 g/cm^3. The respective fiber volume fractions were 0.45 and 0.32.

It was a characteristic feature of the phenolic resin to consolidate more efficiently and, thereby, yield a higher fiber volume composite. However, the higher density may also have had a contribution from a greater "intrinsic" graphitizability of the phenolic resin. It was not possible to separate this effect from the fiber-volume effect. However, to repeat, both matrices showed a positive dependence of density on fiber volume fraction, thereby establishing this as a real effect.

A much more interesting, but certainly more complicated, issue is the role of the matrix (and the interphase region) in C/C fracture behavior. We recently studied this phenomenon by using an in situ SEM flexure stage in order to observe the interactions between the advancing crack tip and the microstructural features of the composite in the frontal process zone. The composite was a single-tow unidirectional made with DuPont E130 fiber and PAA-resin precursor. A sketch of the experiment is shown in Figure 5.

A number of interesting phenomena were observed for HTTs of 1100 to 2400°C. Following the lowest HTT of 1100°C, failure was dominated by the well-bonded brittle matrix; a tortuous crack path in the E130 fibers (Figure 6) appeared to contribute to a relatively high utilization of fiber strength in spite of this brittle matrix

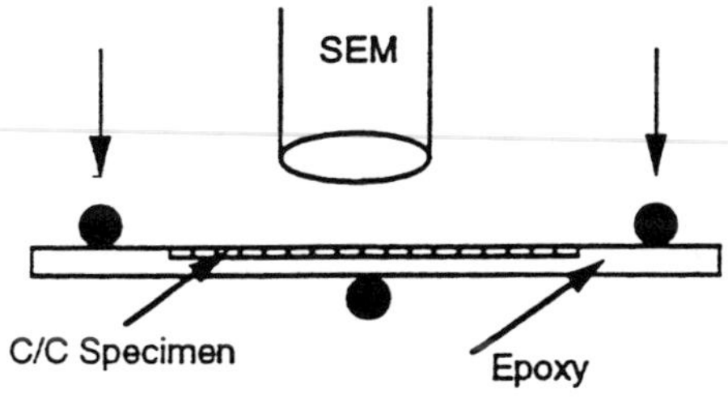

Figure 5. Schematic diagram of experimental arrangement for in situ flexure testing

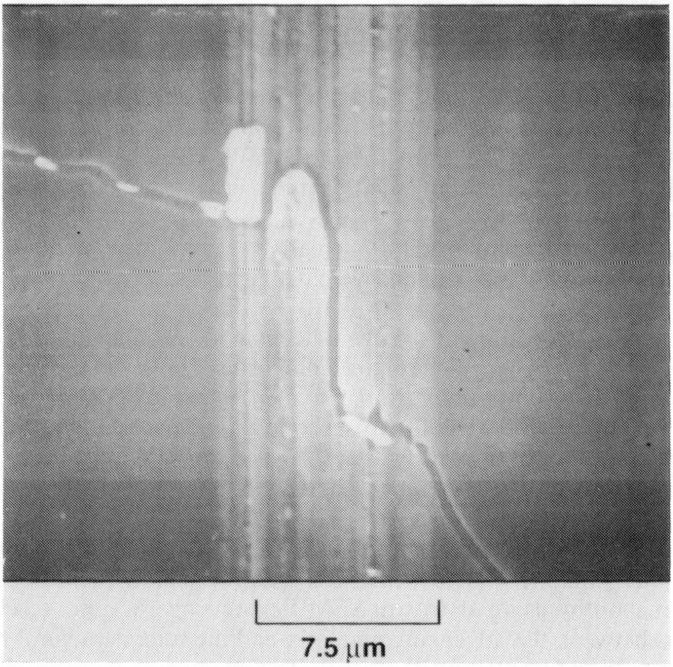

Figure 6. View of crack propagation in E130 composite prior to crack opening revealing intrafilament crack deflection.

failure. We attempted to calculate the interfacial shear stress (IFSS) that might be generated by matrix shrinkage during pyrolysis of the polymer to carbon. These values were compared to approximate calculations of crack-tip interfacial shear stresses using the Cook-Gordon approach. The results suggested that the strong bonding in the 1100°C HTT composite and the consequent absence of crack deflection cannot be accounted for by friction alone, and, therefore, chemical bonding or some type of fiber-matrix mechanical interlocking must be involved.

Higher HTTs lead to progressive weakening of the fiber-matrix interface, thereby reducing the likelihood of matrix-dominated failure. For a HTT of 1600°C, we observe fiber bridging across matrix cracks; this bridging contributes to crack-tip shielding by the filaments. However, there are still well-bonded regions that result in brittle crack propagation. The result typically is a mixed-mode type of failure. With heat treatment to 2150°C, multiple matrix cracking (MMC) is observed. Using the crack-spacing model of Aveston, Cooper, and Kelly (ACK), an IFSS of 6 MPa was estimated for the MMC case. Attempts to calculate the matrix failure strain using the ACK formulation led to a large overprediction of the failure strain, although a number of the parameters used in the calculation are known only very approximately. In addition, the onset of matrix interphase (sheath) structure development is also first evident at the 2150°C HTT. In Figure 7 the interphase microstructure for the 2150°C HTT is seen in the SEM photo to cause blunting of the crack, but no deflection of the crack along the sheath (interphase) length. Something approaching an optimum strength utilization is realized from the 2150°C HTT, based on independent uniaxial tensile tests.[8] As discussed previously,[2] this interphase is a region of higher preferred orientation and enhanced crystalline development[5] as a consequence of matrix deformation during pyrolysis and carbonization. For the case of the SEM in Figure 7, the greater brightness of the sheath is most likely the result of its higher density,[6] which results in a higher electron emission. Also, because of the relatively low HTT of 2150°C, this matrix interphase sheath structure is not highly graphitized and, therefore, in contrast, does not have the lamellar graphitic texture seen in the highly graphitized E130 fiber adjacent to it.

With heat-treatment to 2400°C, the oriented matrix interphase now becomes well graphitized, as evidenced by its lamellar texture, and is almost indistinguishable from the graphitic E130 fiber (Figure 8 compared to Figure 7). Crack blunting at

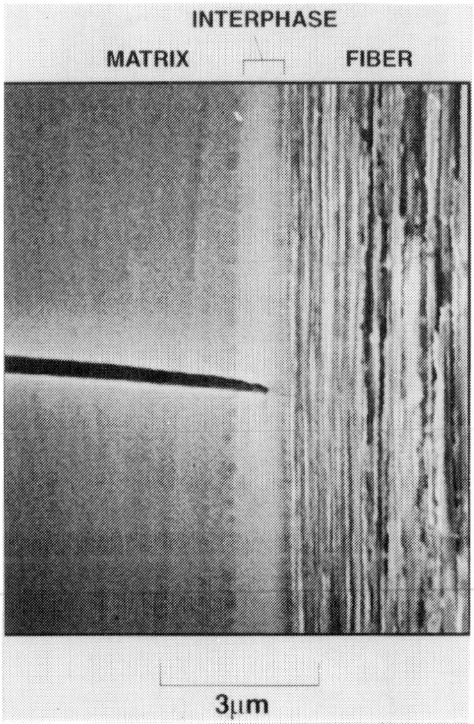

Figure 7. High-magnification micrograph showing crack blunting in interfacial region of 2150°C heat-treated composite.

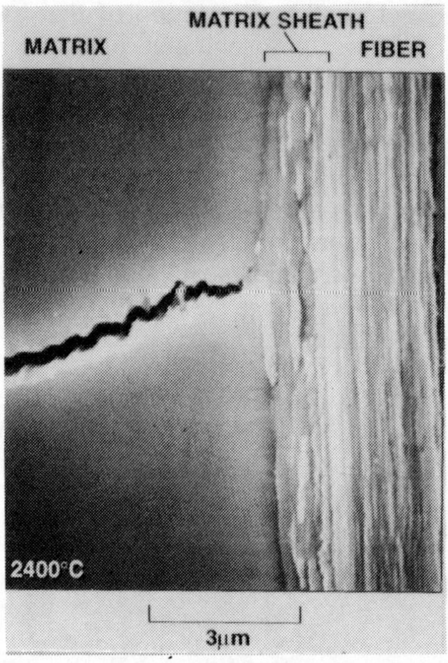

Figure 8. High-magnification micrograph showing crack blunting at graphitized matrix sheath structure in 2400°C heat-treated composite.

Figure 9. Short-range longitudinal intramatrix splitting in 2400°C heat-treated composite.

the graphitized matrix sheath is also shown in Figure 8. The high degree of orientation and lamellar graphitic structure of the sheath leads to significant delamination cracking in the sheath along the direction of the filament (Figure 9). However, this intramatrix cohesive failure also reduces the load-carrying capabilities of the matrix. Figure 10 is a higher-magnification view of a crack zone showing well-defined regions of nongraphitized matrix, graphitized matrix sheath, and fiber. Note the extensive debonding between fiber and matrix.

Following the highest HTT of 2750°C, the development of the graphitized sheath interphase is very extensive along the length of the interfacial region. Figure 11 shows a sequence of two micrographs of the composite as it is strained. As the crack propagates across the width of the composite, it is first deflected around a zone of sheath (left side of Figure 11a). The fiber associated with this sheath has been removed by polishing. The crack then becomes blunted at fiber F after it

Figure 10. Crack zone showing well-defined regions of non graphitized matrix, graphitized sheath, and fiber in 2400°C heat-treated composite.

passes through a second sheath. Further straining (Figure 11) leads to partial fracture of filament F and extensive longitudinal splitting of the matrix sheath region labeled S. Note the saw-toothed structure of the fractured sheath. At this point, the intra-matrix cohesive failure of the sheath is very extensive, leading to a

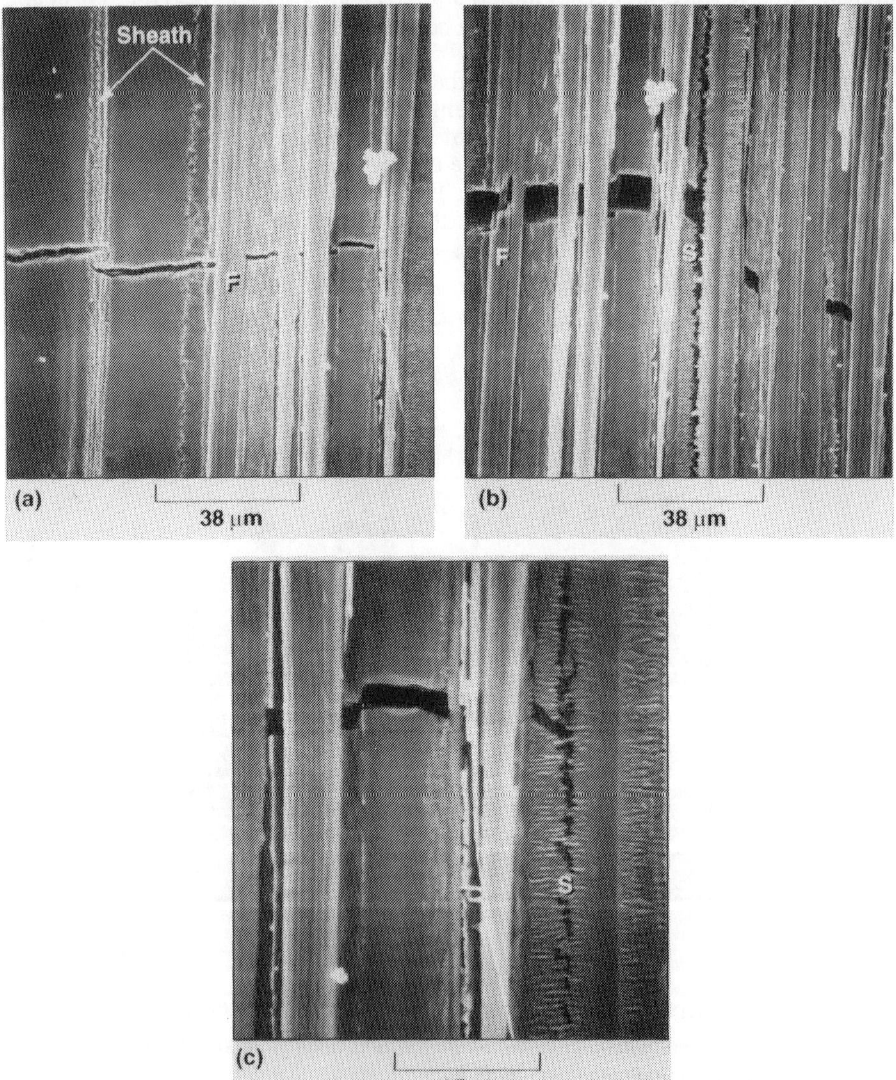

Figure 11. Crack propagation through a 2750°C heat-treated composite at two different degrees of strain.

decoupling of the surrounding matrix from the fibers. In addition to the very weak coupling between fiber and matrix for this HTT, in a previous study, we also found evidence of fiber degradation resulting from the C/C processing.[8]

CONCLUSIONS

Carbon/Carbon composites that are produced with a thermosetting-resin matrix precursor form a distinct interphase of graphitized matrix adjacent to the fibers. This interphase "sheath" can be as large as 1–2 μm in thickness. This structure differs significantly from carbon matrix, which is at the same distance from the fiber and which does not experience the orienting stresses at the interface, which are believed responsible for the development of the stress-oriented and graphitizable structure. Rather, this matrix, although usually fairly well oriented, remains glassy, i.e., nongraphitic in structure. In many ways, it resembles the structure of PAN-based carbon fibers. The structure of this interphase can be studied most effectively by SEM and TEM. The development of a graphitized-carbon interphase can have major effects on properties. Two are discussed in this publication. The first is the effect on matrix density and associated matrix shrinkage. The other is the effect on mechanical properties. The latter is much more complicated and involves phenomena such as crack-tip blunting in the interphase in the early stages of its development and longer-range crack deflection as well as intramatrix cohesive failure in the later stages of development (HTTs of 2400 and 2750°C) where the interphase becomes highly graphitized.

REFERENCES

1. R. J. Zaldivar and G. S. Rellick, "Some Observations on Stress Graphitization in Carbon-Carbon Composites," *Carbon 29*, 1155–1163 (1991).

2. G. S. Rellick, D. J. Chang, and R. J. Zaldivar, "Mechanisms of Orientation and Graphitization of Hard-Carbon Matrices in Carbon/Carbon Composites," *J. Materials Research 7*, 2798–2809 (1992).

3. R. J. Zaldivar, G. S. Rellick, and J. M. Yang, "Processing Effects on the Mechanical Behavior of Polyarylacetylene-Derived Carbon-Carbon Composites," *SAMPE J. 27*, 29–36 (Sept/Oct 1991).

4. R. J. Zaldivar, R. W. Kobayashi, G. S. Rellick, and J. M. Yang, "Carborane-Catalyzed Graphitization in Polyarylacetylene-Derived Carbon-Carbon Composites," *Carbon 29*, 1145–1153 (1991).

5. G. S. Rellick and P. M. Adams, "TEM Studies of Resin-Based Matrix Microstructure in Carbon/Carbon Composites," *Carbon* 32(1), 127–44 (1994).

6. G. S. Rellick and R. J. Zaldivar, "In Situ Matrix Properties in Unidirectional Carbon/Carbon Composites," *Ext. Abst., 21st Biennial Conf. on Carbon*, Buffalo, New York, pp. 68–69 (1993).

7. R. J. Zaldivar, G. S. Rellick, and J. M. Yang, "Failure Modes of the Unidirectional Ultra-High Modulus Carbon-Fiber/Carbon-Matrix Composite," *J. Amer. Ceram. Soc.*, submitted for publication.

8. R. J. Zaldivar, G. S. Rellick, and J. M. Yang, "Fiber Strength Utilization in Carbon-Carbon Composites," *J. Materials Research*, 8(3), 501–511 (1993).

SILICON OXYCARBIDE REACTION WITH GLASS MATRIX COMPOSITIONS

Gang Qi and Carlo G. Pantano
Department of Materials Science and Engineering
The Pennsylvania State University
University Park, PA16802

ABSTRACT

A residual silicon oxycarbide phase has been observed in Nicalon and other polymer derived SiC fibers. This phase is believed to control the thermal stability and interface reaction of these fibers in ceramic composites. In this study, silicon oxycarbide was synthesized in bulk to study, directly, its reaction with a barium magnesium aluminosilicate glass (BMAS) using reaction couples. A strong relationship between decomposition of the silicon oxycarbide and the transport of glass (matrix) constituents (especially Al) was observed. In many cases, a carbon rich layer was observed to form within the decomposed surface of the oxycarbide at the oxycarbide/glass interface. The absence of a reaction layer at the interface of SiC and BMAS glass indicates that it is the oxycarbide phase in Nicalon fiber that provides the carbon-rich interphase.

INTRODUCTION

The fiber-matrix interface plays an important role in SiC fiber reinforced glass-ceramic matrix composites. It determines the mechanical properties of the composites, especially toughness, which is thought to be critical for applications of glass-ceramic matrix composites. Normally, the interface consists of an interphase that may form as a result of interface chemical reactions, or a compositionally-modified zone which is caused by interdiffusion of reactants. Thus, chemical reactions and diffusion govern the features of the interface, and thereby, the mechanical performance of the composite.

To the extent authorized under the laws of the United States of America, all copyright interests in this publication are the property of The American Ceramic Society. Any duplication, reproduction, or republication of this publication or any part thereof, without the express written consent of The American Ceramic Society or fee paid to the Copyright Clearance Center, is prohibited.

Many papers discuss the interface in ceramic composites [1-17]. These studies basically fall into two categories- (a), the observation of the interface, and (b), the mechanism of interface formation. Brennan [1] first reported the interfacial carbon layer in Nicalon fiber reinforced lithium aluminum silicate (LAS) glass matrix composites. This has been confirmed in many other systems including calcium aluminum silicate (CAS) [2-4], magnesium aluminum silicate (MAS) [2,5,6] and related compositions with several additives such as Nb_2O_5 and As_2O_3 [2,7,8]. Cooper and Chyung [2] characterized the interface by using electron microscopy (TEM); a graphite interfacial layer was observed in Nicalon reinforced CAS matrix composites. Chen et al. [6] reported the presence of carbon rich layers using scanning Auger microscopy in MAS composites with several additives. In addition, carbide species [6,7,8] have been observed at interfaces when specific additives are present.

There are several papers on the mechanisms of the interface reaction, especially carbon layer formation. Bonney and Cooper [3] proposed a mechanism based mainly on the morphology of the interface. They reasoned that the interphase of carbon was the result of SiC oxidation, and that the planar carbon layer required the rapid diffusion of silicon across the carbon layer. Homeny, et al. [10] measured the time-dependent thickness of carbon rich layers from the concentration profile across the interface using scanning Auger microscopy and observed parabolic growth behavior. Therefore, they suggested that the carbon layer formation was diffusion controlled, but from their study, the specific rate-controlling reaction was not clear. Qi , Spear and Pantano [14,15] modeled the effect of additives, especially As_2O_3, on the interface reaction as well as the carbon layer growth. They proposed that SiC and carbon oxidation could possibly occur in matrices with As_2O_3 or other variable valence oxides. The competition between SiC oxidation (which produced carbon) and carbon oxidation (which consumed carbon) controlled the extent of interfacial carbon formation. Therefore, carbon interphase formation, in-situ, is transient in nature, and depends on processing parameters such as temperature, time and thermal history.

Most fiber/matrix reactions have been observed in Nicalon fiber reinforced glass-ceramic matrix composites. Nicalon is a well known non-stoichiometric SiC fiber and contains excess oxygen and carbon. The excess oxygen is believed to be in the form of silicon oxycarbide. NMR experiments have confirmed that this residual phase exits in Nicalon fiber.[11,16] However, few studies have been conducted on the effect of the silicon oxycarbide phase, itself, on the formation of a carbon interphase. This is not at all surprising since the silicon oxycarbide phase is metastable, and thereby, difficult to synthesize in bulk.

In this work, silicon oxycarbide glass was made using a sol/gel technique [18,19], and was used to simulate the Nicalon fiber and the effect of glass composition on the interface reaction. Aluminum was of particular interest because it is the most prevalent species observed in Nicalon fibers after composite processing [3,6].

EXPERIMENT

1. Materials

Silicon oxycarbide glass is a key material for this study. It was synthesized using the sol-gel process. The starting material was methyltrimethoxysilane. The details of preparation of the solution and gel are described by Zhang and Pantano [18]. The dried gels were pyrolyzed at 1200°C under an argon atmosphere. The glass frits were then hot pressed at 1500°C. The pyrolysis and hot-pressing have been previously described by Hammond and Pantano[19]. The hot-pressed silicon oxycarbide samples were cut into blocks and were polished on both sides.

In this study, barium magnesium aluminum silicate (BMAS) glass frit, commercially available from Corning (Corning Code SP 1089), was chosen as the matrix material because it is a very useful glass-ceramic matrix material. The composition of BMAS glass frit is 50 wt%SiO_2, 28.3 wt%Al_2O_3, 14.2 wt% BaO, and 7.5 wt% MgO. In order to study the effect of Al_2O_3 migration, an Al_2O_3 matrix was used for comparison. In addition, polycrystalline CVD SiC (Morton International) was used for comparison to silicon oxycarbide glass.

2. Preparation of reaction couples

The polished silicon oxycarbide pieces were embedded in the BMAS glass frit. They were then cold-pressed into pellets. The pellets were heated to 1400°C, 1500°C and 1600°C, respectively. The heat-treatments were performed in an induction carbon furnace under ultra-high-purity argon. The heating rate was about 60°C/min and the cooling rate was about 100°C/min from the target temperature to 1000°C, and 50°C/min below 1000°C. Subsequently, the samples were cut and the cross sections were polished to a 1 micron finish.

3. Electron microprobe analysis

The microprobe technique was chosen as a main characterization tool in this work. It can easily provide compositional information on a microscopic scale for a heterogeneous specimen. The sample preparation for analysis was described above. A carbon coating was used to avoid charging problems during analysis.

The carbon concentration was indirectly determined using the difference of the others species concentrations. The compositional analysis started at the interface, then was extended to both sides at two micron intervals until the concentration reached a constant value. Since the spatial resolution of this technique is 1 micron, the interface resolution is no better than 1 micron.

RESULTS

Figure 1 is an example of the concentration profile across an interface. The profile shows the existence of a reaction layer between the BMAS glass and silicon oxycarbide glass. It contains, primarily, carbon, aluminum, silicon and oxygen. In Figure 2, an SEM image clearly shows the reaction layer at the interface. Similar layers were formed on the other samples treated for different times and temperatures. The cross sectional profiles indicated that the chemical compositions of these reaction layers were almost the same, although the thicknesses were different. Figure 3 summarizes the change in reaction layer thickness with time. Since interaction and diffusion could occur before the target temperature was reached, a time correction was made to account for this variation. The correction was determined experimentally to be 0.5 min. The

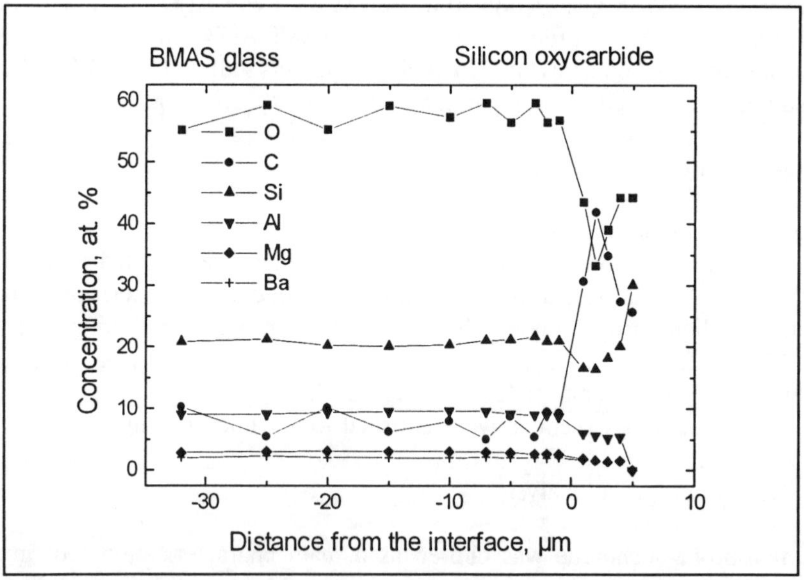

Figure 1 Electron microprobe cross-sectional concentration profile between BMAS glass and silicon oxycarbide glass after heat treatment, at 1500°C for 5 min

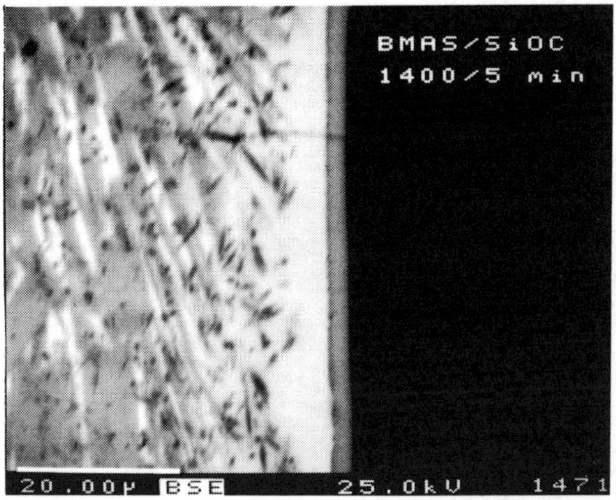

Figure 2 Backscattered electron image of a BMAS/silicon oxycarbide glass interface, at 1400°C for 5 min

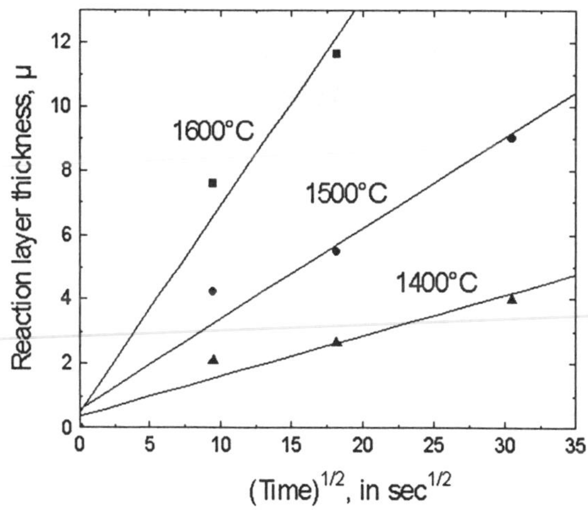

Figure 3 Reaction layer thickness vs. square root of time for BMAS glass/silicon oxycarbide glass reaction couples

linear relation in the plot of thickness against the square root of time indicates that growth of the layer follows the well-known parabolic law. Since aluminum was the main species in the reaction layer, the growth of the layer is thought to be related to the mass transport of Al_2O_3.

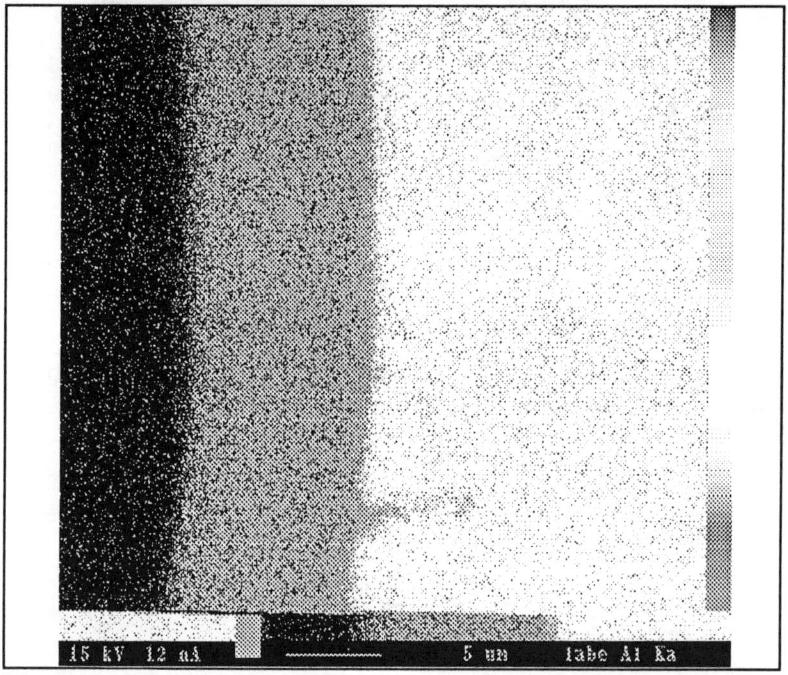

Figure 4 Electron microprobe x-ray map for Al in Al_2O_3 (right), silicon oxycarbide glass (left), and interphase (center)

In order to investigate the effect of Al_2O_3, BMAS glass was replaced by an Al_2O_3 matrix. The Al x-ray map in Figure 4 shows a thicker diffusion layer, when compared to BMAS glass reacted with silicon oxycarbide glass under the same conditions. Obviously, the Al_2O_3 matrix enhanced the mass transport of Al_2O_3 by increasing the activity of Al_2O_3.

No reaction layer was observed in the BMAS/SiC couple under the same conditions as BMAS/silicon oxycarbide couple (Fig.5). The main difference between these two reaction couples is the presence of oxide phase in the silicon oxycarbide glass. This suggests that the formation of alumino-silicate glass provides a mechanism for Al_2O_3 transport.

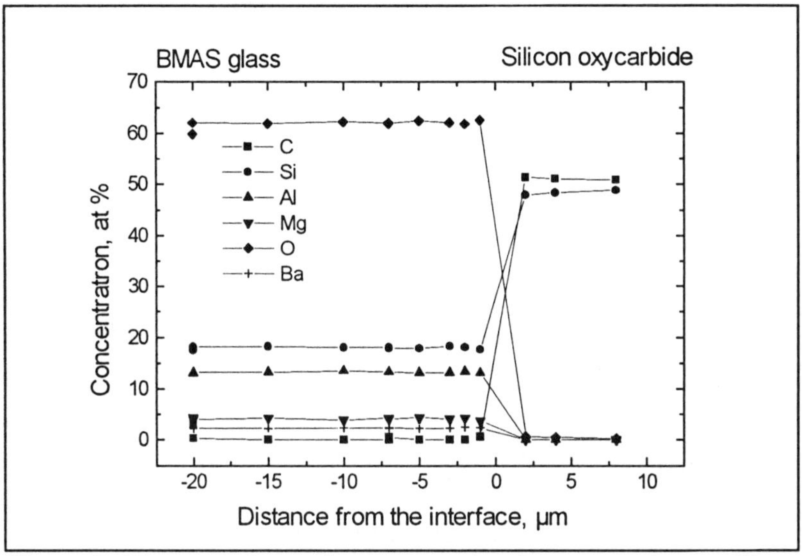

Figure 5 Electron microprobe cross-sectional concentration profile between BMAS glass and SiC after 5 min at 1600°C

The tendency for carbon enrichment at the interface of BMAS/silicon oxycarbide is similar to that observed in Nicalon fiber reinforced composites. Figure 6 is a plot of the carbon concentration in the interface against reaction layer thickness. It is interesting to note that the carbon concentration reached a maximum value as the thickness increased, then decreased with increasing thickness. This is consistent with the behavior predicted by the model of Qi, Spear and Pantano[14,15].

In summary, the experiments produced the following observations:

1. Silicon oxycarbide glass has a greater reactivity than polycrystalline SiC.
2. Carbon enrichment is related to interfacial reactions and Al_2O_3 mass transport.
3. The reaction layer thickness corresponds to Al_2O_3 mass transport.
4. The cross sectional profile and carbon behavior are similar to that observed in Nicalon reinforced glass matrix composites using scanning Auger microscopy and high resolution TEM.

DISCUSSION

Because there is shortage of thermodynamic data for silicon oxycarbide, the common model used to represent silicon oxycarbide is a mixture of SiO_2, SiC and

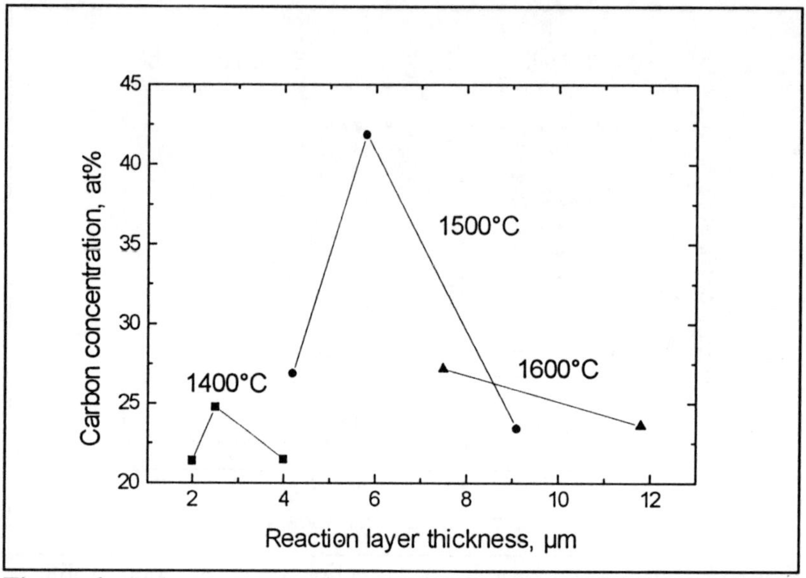

Figure 6 Carbon concentration vs. reaction layer thickness in BMAS glass and silicon oxycarbide glass reaction couples

carbon. In terms of thermodynamics, there is an equilibrium condition in silicon oxycarbide. The following reactions represent this equilibrium:

$$SiC + 2\ SiO_2 = 3\ SiO + CO \quad (1)$$
$$3\ C + SiO_2 = SiC + 2\ CO \quad (2)$$

When the silicon oxycarbide glass contacts the BMAS glass, the equilibrium state at the interface is changed because gradients exist in SiO_2 activity and CO partial pressure. The SiO_2 activity and CO partial pressure are related to each other through the above reactions. We assume that the activity of SiO_2 in silicon oxycarbide glass is unity. Obviously, the SiO_2 activity in BMAS glass is much less than unity. Thus, an activity gradient is set up at the interface. This activity gradient in SiO_2 is probably the driving force for the interface reaction. Here, the interface reaction may be the formation of glass, such as an aluminum silicate with a low SiO_2 activity. The result of interface reaction minimizes the gradient and smoothes the activity of SiO_2. In addition, the change in SiO_2 activity will cause a change in CO partial pressure.

A gradient in CO pressure will drive CO molecular diffusion which can also change the SiO_2 activity. Therefore, there are two ways to decrease the gradient in SiO_2 activity. One way is through Al_2O_3 mass transport. Al_2O_3 in contact with the SiO_2 phase can lead to the formation of an alumino-silicate glass which decreases the activity of SiO_2. The other way is the outward diffusion of CO from the silicon oxycarbide glass. From reactions 1 and 2, a decreasing CO pressure causes a decrease in SiO_2 activity. Figure 7 shows the change of SiO_2 activity with Al_2O_3 and CO diffusion calculated using Chemsage[20]. The input composition is that of Nicalon fiber[3]. In the calculation, an ideal mixture model was used for the glass phase. The free SiO_2 concentration was treated as the activity of SiO_2. The numbers on the curve represent the moles of Al_2O_3 transferred between the matrix and fiber. The curves indicate the change in SiO_2 activity as CO is removed from the system. From the plot, the capability of Nicalon to produce CO molecules increases as the activity of SiO_2 is reduced, and will decrease as the Al_2O_3 content increases. The left lines represent the change in SiO_2 activity in the glass matrix when SiO_2 from Nicalon moves into the glass matrix while Al_2O_3 moves into the fiber. The intersection of the lines corresponds to the final equilibrium states where both sides of the interface have the same activity of SiO_2. These points show that the presence of Al_2O_3 in the

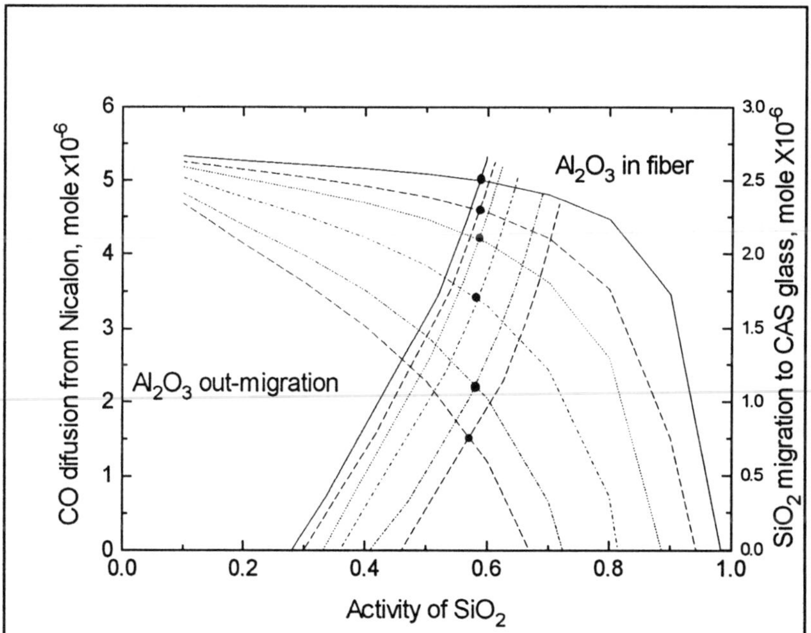

Figure 7 The Calculation of change in SiO_2 activity with Al_2O_3 and CO transport between CAS glass and Nicalon fiber

Nicalon(i.e. in the oxycarbide phase) lowers the driving force for CO production. It is worth noting that the mass transport of CO molecules and SiO_2 also represent the migration of oxygen from the Nicalon fiber. Therefore, in order to compare these two effects, both vertical axes used the equivalent oxygen scale, i.e., both vertical axes utilize the same oxygen content.

When CO molecules reach the interface, the following reactions probably occur:

$$CO + SiO = C + SiO_2 \quad (3)$$
$$2\ CO + SiC = SiO_2 + 3\ C \quad (4)$$

Obviously, the CO molecules act as a vehicle for carbon transport into the interface from the bulk of the silicon oxycarbide glass.

When the processing is carried out for a longer time, more Al_2O_3 will migrate into the silicon oxycarbide. Therefore, the capability for CO generation will be reduced, and at the same time, the CO partial pressure gradient will decrease as the reaction layer grows. Finally, the carbon enrichment will slow down, or stop.

Since the sample was heated in a carbon furnace under ultra high purity Ar, the oxygen partial pressure was very low (about 10^{-4} atm). Therefore, the CO partial pressure was of the same order of magnitude as the oxygen partial pressure. However in the SiO_2, SiC and C system, the equilibrium partial pressure of CO is about 10^{-1} atm. Therefore, there exists a CO pressure gradient from the interface to the surface of sample. If we adopt the concept proposed by Qi, Spear and Pantano [14,15], the outward diffusion of CO will reduce the enrichment of carbon. The outward-diffusion of CO can explain the reduction in carbon enrichment as the reaction layer grows.

In general, CO acts as a vehicle for carbon mass transport. The Al_2O_3 transport decreases the SiO_2 activity, and furthermore, reduces the capability for CO generation by silicon oxycarbide. The maximum value of carbon enrichment is the result of competition between CO diffusion to the interface and CO diffusion out from the interface.

SUMMARY

Silicon oxycarbide glass is more reactive toward oxide glass-matrices, compared to polycrystalline SiC, because it contains a silicate phase. Carbon enrichment was found to exhibit a time dependence. The transient nature of carbon enrichment was very similar to the behavior of the carbon layer in Nicalon fiber reinforced composites. In this sense, silicon oxycarbide is a good model for the chemical behavior of Nicalon. Altogether, this study shows that the silicon oxycarbide phase provides a mechanism for carbon layer formation under the non-oxidizing conditions used in composite fabrication.

ACKNOWLEDGEMENTS

The authors gratefully acknowledge the financial support of the Office of Naval Research (N00014-90-J-1392, Steve Fishman, Contract Monitor). The authors also thank M. Angelone for the microprobe analyses, and M. Hammond for the silicon oxycarbide glass samples.

REFERENCES

1. J. J. Brennan, "Interfacial characterization of glass and glass-ceramic matrix Nicalon SiC fiber composites", *Tailoring multiphase and composite ceramics, Mater. Sci. Res.* R. E. Tressler, G. L. Messing, C. G. Pantano, and R. E. Newham, editors, 20(1986)549-60
2. R. F. Cooper, and K. Chyung, "Structure and chemistry of fibre-matrix interfaces in silicon carbide fibre-reinforced glass-ceramic composites: an electron microscopy study", J. of materials Science 22(1987)3148-3160
3. L. A. Bonney, and R. F. Cooper, "Reaction-layer interfaces in SiC-fiber reinforced glass-ceramics: a high-resolution scanning transmission electron microscopy analysis", J. Am. Ceram. Soc., 73[10](1990)2916-21
4. S. M. Bleay, V. D. Scott, B. Harris, R. G. Cooke, and F. A. Habib, "Interface characterization and fracture of calcium aluminosilicate glass-ceramic reinforced with Nicalon fibers", J. of Materials Science 27(1992)2811-2822
5. R. Chaim, and A. H. Heuer, "Carbon interfacial layer formed by oxidation of SiC in SiC/Ba-stuffed cordierite glass-ceramic reaction couples", J. Am. Ceram. Soc., 74[7](1991)1663-67
6. M. Y. Chen, J.M. Battison and T. I. Mah, "Interface analysis of Si-C-O fibre/magnesium aluminosilicate matrix composites", J. of Materials Science 24(1989)3213-3220
7. E. Bischoff, M. Ruhle, O. Sbaizero, and A.G.Evans, "Microstructural studies of the interfacial zone of a SiC-fiber-reinforced lithium aluminum silicate glass-ceramic", J. Am. Ceram. Soc., 72[5](1989)741-45
8 H.C. Cao, E. Bischoff, O. Sbaizero, M. Ruhle, A.G. Evans, D.B.Marshall and J. J. Brennan, "Effect of interfaces on the properties of

fiber-reinforced ceramics", J. Am. Ceram. Soc., 73[6](1990)1691-99

9. J. Y. Hau, and R.F. Speyer, "Interfacial phenomenology of SiC fibre reinforced $Li_2O.Al_2O_3.SiO_2$ glass-ceramic composites", J. of Materials Science, 27(1992)374-380
10. J. Homeny, J. R. VanValzah, and M. A. Kelly, " Interfacial characterization of silicon carbide fiber/lithia-alumina-silica glass matrix composites", J. Am. Ceram. Soc., 73[7](1990)2054-59
11. V.S.R. Murthy, M. W. Pharaoh, and M. H. Lewis, "Interface microstructure and matrix crystallization in SiC-borosilicate (pyrex) composites", Materials Letters, 10[4-5](1990)161-164
12. V. S. R. Murthy, L. Jie, and M. H. Lewis, "Interfacial microstructure and crystallization in SiC-glass ceramic composites", Ceram. Eng. Sci Proc. 10[7-8](1989)938-951
13. L. Porte and A. Sartre, "Evidence for a silicon oxycarbide phase in the Nicalon silicon carbide fibre", J. of Materials Science, 24(1989)271-275
14. G. Qi, K. E. Spear and C. G. Pantano, "Carbon-layer formation at silicon-carbide-glass interfaces", Materials Science and Engineering, A162(1993)45-52
15. C. G. Pantano, K. E. Spear, G. Qi, and D. M. Beall, "Thermochemical modeling of interface reactions in glass matrix composites", *Advances in ceramic-matrix composites, Am. Ceram. Soc,* N. Bansal, Editor, 38(1993)173-198
16. C. Laffon, A. M. Flank, P. Lagarde, M. Laridjani, R. Hagege and P. Olry, J. Cotteret, J. Dixmier, J. L. Miquel, H. Hommel, and A. P. Legrand, "Study of Nicalon-based ceramic fibers and powders by EXAFS spectrometry,X-ray diffractometry and some additional methods", J. of Materials Science 24(1989)1503-1512
17. P. M. Benson, K. E. Spear, and C. G. Pantano, "Interfacial characterization of glass matrix/Nicalon SiC fiber composites: a thermodynamic approach", Ceram. Eng. Sci. Proc., 9[7-8](1988)664-670
18. H. Zhang, and C. G. Pantano, "Synthesis and Characterization of Silicon Oxycarbide Glasses", J. Am. Ceram. Soc., 73[4](1990)958-65
19. M. Hammond, E. Breval, and C. G. Pantano, "Microstructure and viscosity of hot pressed silicon oxycarbide glasses", Ceramic Engineering & Science proceedings, Editor J. B. Wachtman, 14[9-10](1993)947-954
20. G. Eriksson, and K. Hack, *ChemSage, Edition 2.1.1*, October 1991 (GTT MbH, Mies-Van Der Rohe Str. 25, 5110 Aachen, Germany)

ELECTRON MICROSCOPY AND THERMOGRAVIMETRY STUDIES OF NICALON™ FIBER-REINFORCED BLACKGLAS™ MATRIX COMPOSITES

Sandeep R. Koppikar and Honghua Du
Department of Materials Science and Engineering and
Advanced Technology Center for Surface-Engineered Materials
Stevens Institute of Technology, Hoboken, NJ 07040

ABSTRACT

Nicalon fiber-reinforced Blackglas matrix composites were investigated by transmission electron microscopy and thermogravimetry to study the structural and phase characteristics of the fiber, matrix, and interphase and the oxidation stability. It was shown that the Nicalon fiber retained its nanocrystallinity and the silicon oxycarbide matrix was electron-diffraction amorphous. The fiber and the matrix were separated by a 10-30 nm thick graphite interphase. Appreciable weight loss occurred when the composites were exposed to air above 500°C due to rapid oxidative removal of the interphase. The oxidation process resulted in direct bonding between the fiber and the matrix with the interface region rich in oxygen as compared with the matrix phase. Oxidation embrittlement of the composites and strength reduction are expected in high-temperature oxidation environments.

INTRODUCTION

Fiber-reinforced ceramic matrix composites are excellent candidate materials for replacing metals and alloys in advanced turbine engines and hypersonic vehicles with significantly improved performance. Much interest exists in the development of Nicalon fiber-reinforced Blackglas matrix composites (NFRBMC) for aero-structural applications.[1] The NFRBMC's offer, potentially, a combination of low density, high strength and toughness, good oxidation resistance and property retention to high temperature, reduced probability of failure at relatively low load, and a more controlled failure mode instead of the catastrophic fracture characteristic of monolithic ceramics.

To the extent authorized under the laws of the United States of America, all copyright interests in this publication are the property of The American Ceramic Society. Any duplication, reproduction, or republication of this publication or any part thereof, without the express written consent of The American Ceramic Society or fee paid to the Copyright Clearance Center, is prohibited.

The performance and reliability of the NFRBMC's depends critically on the nature of the fiber-matrix interface, which controls such toughening mechanisms as crack deflection, fiber-matrix debonding, and fiber pull-out. The interface should be strong enough to ensure proper load transfer from the matrix to the fiber yet weak enough to avoid catastrophic failure by promoting various toughening processes.[2,3] Any chemical reactions and/or interdiffusion which lead to strong fiber-matrix bonding must be inhibited during composite fabrication and use in high-temperature oxidation environments.

The potential of the NFRBMC's in aerospace applications will not be realized unless their structural and phase characteristics, particularly in the interface region, and their oxidation stability can be well understood and controlled. The purpose of this research was to investigate the structural and phase nature of the fiber, matrix, and interphase of the NFRBMC's and to study the effect of oxidative exposure on the composite, and especially the fiber-matrix interphase using electron microscopic and thermogravimetric techniques.

EXPERIMENTAL

The NFRBMC's used in this investigation were fabricated at Grumman Corporate Research Center (Bethpage, NY) following a proprietary process. They are essentially 0°/90° arrays of carbon-coated ceramic-grade Nicalon fibers (~15 μm in diameter) embedded in a Blackglas matrix phase pyrolyzed from an organosilicon copolymer precursor supplied by Allied Signal, Inc. (Morristown, NJ). The volume fraction of the fibers was estimated to be around 70% in the composites.

Thermogravimetric analysis (TGA) of the NFRBMC samples was conducted in air using a Perkin-Elmer TGS-2 system sensitive to 1 μg weight change. The TGA experiments were performed at a constant heating rate of 3.3°C/min in the temperature range of 300°-900°C or under isothermal conditions at 500°C and 700°C for over 300 min. Transmission electron microscopy (TEM) of the as-fabricated and oxidized NFRBMC's was carried out using a Philips EM400T (120 kV) and a Topcon EM002B microscope (200 kV). The latter is equipped with a Gatan EELS system for microchemical analysis. The TEM specimen preparation (perhaps the most challenging part of this investigation) was done first by lapping the composite down to about 50 μm thick with the fibers running parallel to the surface and then by thinning in a Gatan 600 dual-gun ion-milling system. The as-fabricated and oxidized NFRBMC's were also studied by scanning electron microscopy (SEM) with a JEOL JSM840 microscope (15 kV).

RESULTS AND DISCUSSION

Shown in Figure 1 is a TEM micrograph of a fiber in the as-fabricated NFRBMC sample. The inset is a selected-area electron diffraction (SAD) pattern from the fiber, illustrating the presence of distinct diffraction rings. Indexing analysis of the first, second, and third rings in the SAD pattern yielded d-spacings of 0.266 nm, 0.164 nm, and 0.139 nm, respectively. These values correspond to that of ß-SiC.[4] A close examination of the TEM micrograph reveals the presence of ß-SiC lattice fringes (such as those indicated by the arrow) with dimensions in the order of 2 nm. The TEM results offer a strong evidence that the Nicalon fiber is nanocrystalline ß-SiC. Documented studies of the as-synthesized ceramic-grade Nicalon fibers showed similar grain dimensions.[5] These results indicate that grain growth did not take place in the fiber during composite processing. The preservation of the nanocrystalline structure after composite fabrication is critical in the property retention of the fibers.

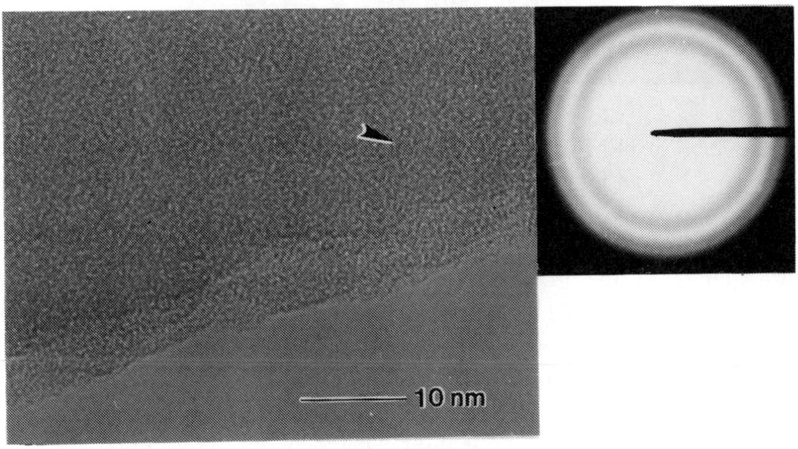

Figure 1. TEM micrograph of the Nicalon fiber in the as-received NFRBMC. The inset is a SAD pattern from the fiber region.

Depicted in Figure 2 is a TEM micrograph of the as-fabricated NFRBMC sample with the matrix in region **a**, an interphase **b**, and the fiber **c**. The inset is a selected-area electron diffraction (SAD) pattern from region **a**, which indicates the amorphous nature of the silicon oxycarbide matrix phase. No crystalline phase was detected in the matrix in the TEM analysis. Documented studies of silicon oxycarbide derived from various polymeric precursors have also revealed the amorphous nature of such phase.[6,7] SEM analysis indicated that the matrix material contains some microcracks as in monolithic Blackglas material due to appreciable sample shrinkage during pyrolysis.

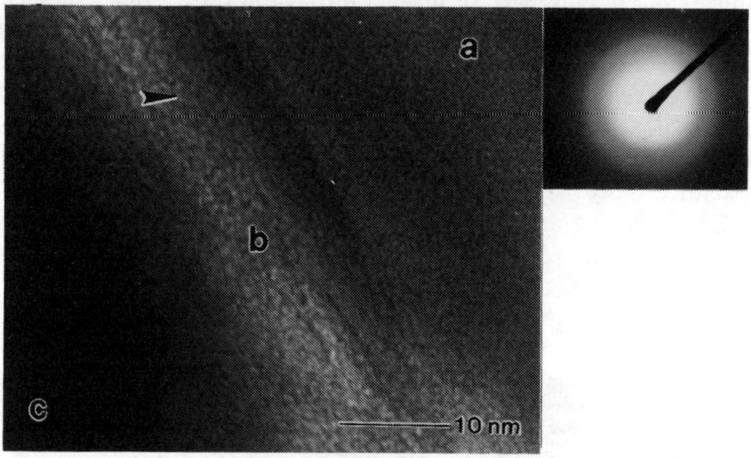

Figure 2. TEM micrograph of the as-fabricated NFRBMC with the matrix **a**, an interface **b**, and the fiber **c**. The inset is a SAD pattern from **a**.

The interphase shown in region **b** of Figure 2 is 10-30 nm thick. This phase corresponds to the graphite coating applied onto the Nicalon fibers before composite fabrication to offer protection to the fibers from surface damage and importantly to promote fiber pull-out during fracture. Graphite fringes are identifiable in this region (indicated by the arrow).

The graphite fringes are even more visible in the TEM micrograph shown in Figure 3, where **a** corresponds to the interphase and **b** to the fiber. The presence of the graphite layer is responsible for the high tensile strength, high strain to fracture, and excellent fracture toughness of the NFRBMC's at room temperature and in high-temperature inert environments.[1] The extensive fiber pull-out and matrix-fiber debonding assisted by graphite (an excellent solid lubricant) during composite fracture is illustrated in Figure 4, the SEM micrograph of the as-fabricated NFRBMC fractured in bending.

Literature studies of Nicalon fibers and silicon oxycarbide glasses indicated that these materials exhibit excellent oxidation resistance at temperatures up to

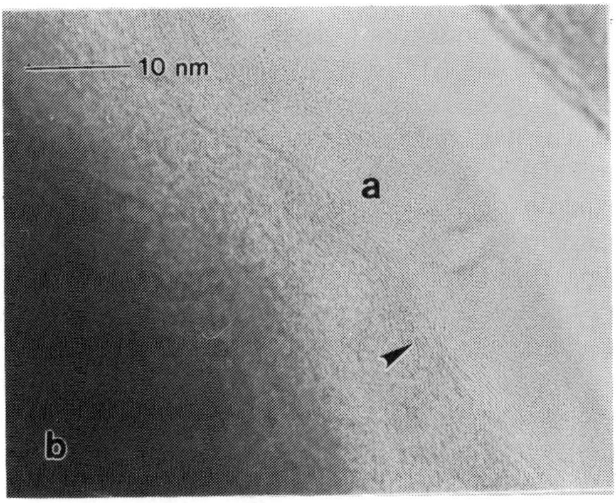

Figure 3. TEM micrograph of the as-fabricated NFRBMC with **a** being the interphase and **b** the fiber.

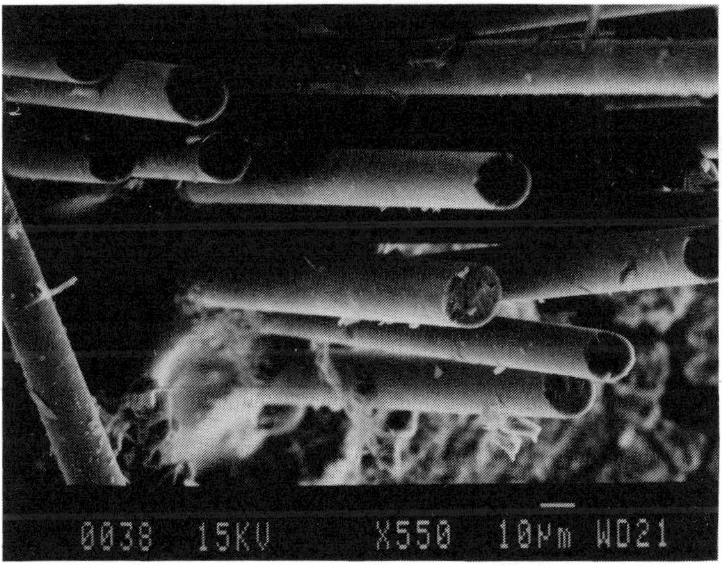

Figure 4. SEM micrograph of the as-fabricated NFRBMC after fracture in bending.

1100°C.[6,8,9] Due to the presence of the graphite interphase, the same conclusion may not be drawn to the NFRBMC's. Shown in Figure 5 is the TGA curve obtained by heating an NFRBMC sample (14.10 mg, 5x5x1.5 mm^3) in flowing air at 3.3°C/min from 300°-900°C. The fibers were in general encased in the Blackglas matrix and were not directly exposed to the environment except some surface fibers and fiber-ends. The seemingly periodic oscillation associated with the weight change is not characteristic of the composite. It can be attributed, instead, to slight fluctuation in the velocity of the air flow regulated by a rotameter. This fluctuation in turn results in variation in the buoyancy the sample pan experiences during the TGA test. As indicated in Figure 5, the weight change of the composite was small below around 500°C. Above this temperature, however, it underwent rapid weight loss. A total of 3% in weight reduction was observed after this experiment.

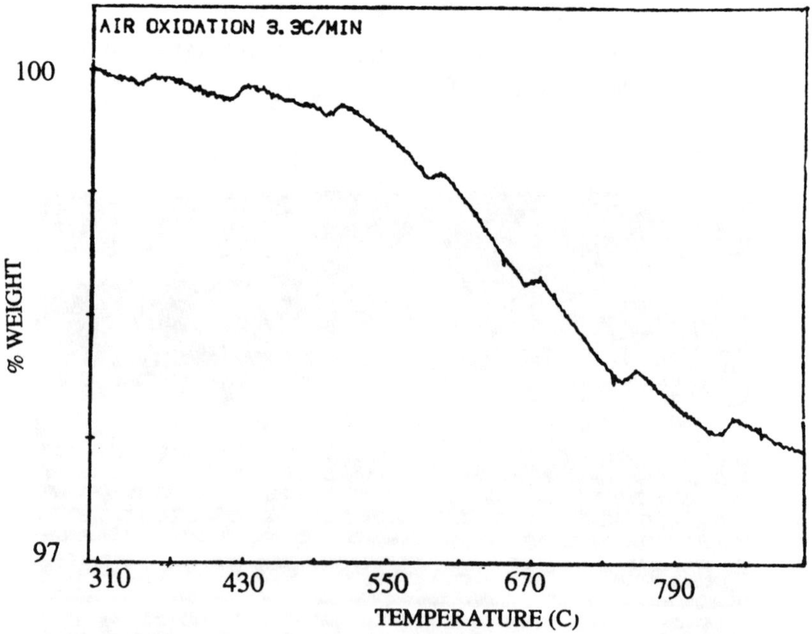

Figure 5. TGA of the NFRBMC in air during heating from 300°C to 900°C at a rate of 3.3°C/min.

The NFRBMC's were further investigated using TGA in flowing air under isothermal conditions at 500°C and 700°C for over 300 min. The results are presented in Figure 6. The weight loss of the NFRBMC sample was gradual throughout the entire oxidation duration at 500°C and a total of about 1.3% weight change occurred after 300 min of the oxidation exposure. In contrast, the weight reduction was very dramatic at 700°C, most of which took place within the initial 60 minutes or so of oxidation. Beyond this period, the rate of weight change was similar to that at 500°C.

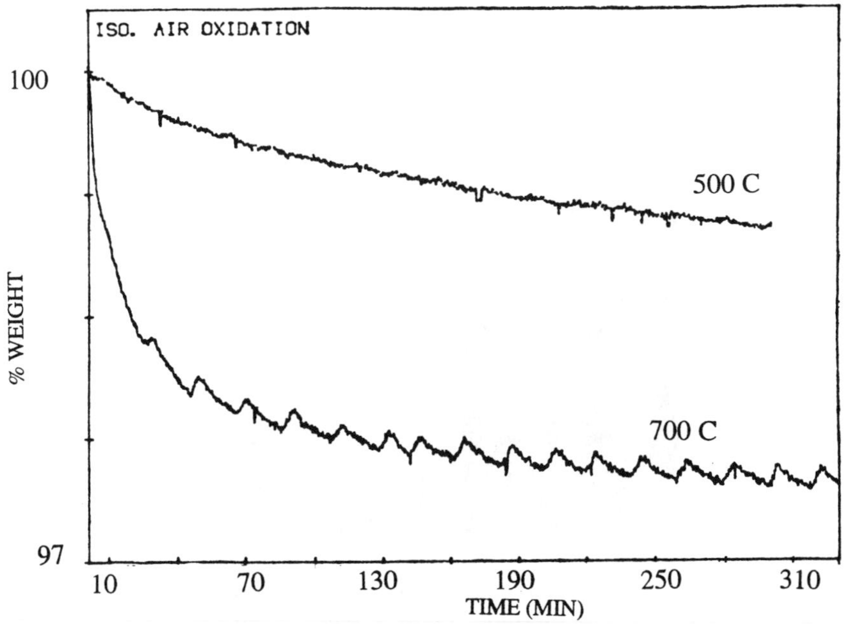

Figure 6. TGA of the NFRBMC in air at 500°C and 700°C with a respective initial sample weight of 12.09 mg and 9.60 mg.

The NFRBMC composites after the oxidation treatment were also studied using SEM and TEM. Despite the appreciable sample weight loss, oxidation under the current conditions resulted in no apparent change in the morphology of polished NFRBMC composite surfaces according to the SEM analysis. However, changes in the interface characteristics of the composites are significant as revealed by TEM.

Shown in Figure 7 is a TEM micrograph of an NFRBMC sample after oxidation in air at 500°C for 240 min. Region **a** in the figure corresponds to the fiber and

c to the matrix based upon SAD analysis. Region **b** is also amorphous based upon SAD analysis. Microchemical analysis using electron energy loss spectroscopy (EELS) indicated apparent variation in chemistry in the fiber (region **a** and the matrix (region **c**). The interphase (region **b**) appears to have a higher oxygen to carbon ratio than both the matrix and the fiber components. Although a definitive conclusion requires further investigation, it is probable that region **b** resulted from partial oxygen substitution of carbon in the silicon oxycarbide network structure during the oxidation exposure.

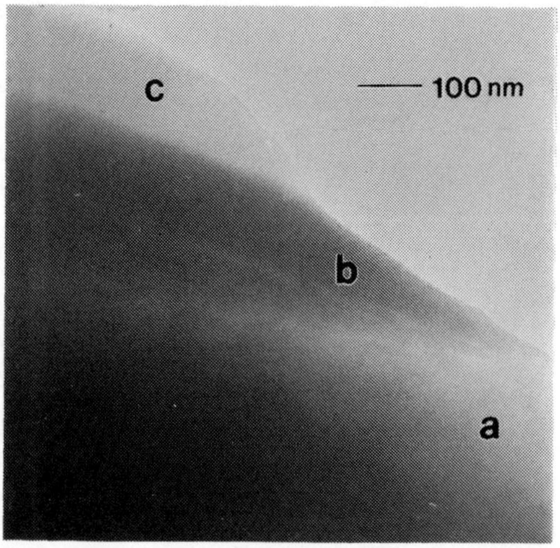

Figure 7. TEM micrograph of the NFRBMC after oxidation in air at 500°C for 260 min. Region **a** corresponds to the fiber, **c** to the matrix, and **b** an interphase.

Most strikingly, the graphite interphase disappeared after the oxidation treatment of the NFRBMC, as evidenced by the absence of lattice fringes in the TEM analysis. The weight loss the composites experienced during the TGA analysis can therefore be attributed mainly to oxidative removal of the graphite interphase since the weight gain resulting from passive oxidation of the fiber and the matrix is expected to be small in the temperature range chosen in the current investigation.[6,9]

The weight loss of the composite was insignificant when exposed in air below ~500°C and became appreciable above this temperature (Figure 5). This may be attributed to the strong temperature dependence of the chemical reaction and diffusion processes from the kinetic standpoint. The oxidation process of the NFRBMC's is expected to be quite complex. On one hand, the microcracks present in the matrix may accelerate gaseous diffusion. On the other the bonding between the fiber and matrix interface due to oxidation may slow down the diffusion step. Detailed investigation is needed to reveal mechanistic information on the oxidation process.

The oxidative removal of the graphite interphase and consequently the formation of strong fiber-matrix bonding will undoubtedly result in degradation of the mechanical properties, and worse yet, the change in failure mode of the NFRBMC's. The latter is illustrated in Figure 8, the SEM micrograph of a fractured NFRBMC sample after oxidation exposure in air at 500°C for 240 min. Oxidation resulted in significant reduction in the fiber pull-out process. The detailed effect of oxidation exposure on the mechanical properties of the NFRBMC's is being investigated at Grumman Corporate Research Center.

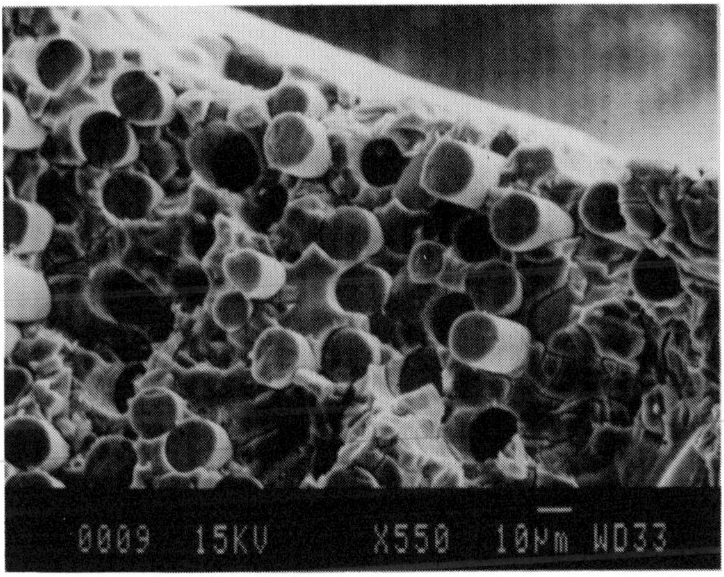

Figure 8. SEM micrograph of a bend-fractured NFRBMC sample after oxidation in air at 500°C for 240 min.

CONCLUSIONS

The as-fabricated Nicalon fiber-reinforced Blackglas matrix composite is arrays of 0°/90° nanocrystalline silicon carbide fibers embedded in an electron-diffraction amorphous silicon oxycarbide matrix. The fiber and the matrix are separated by a 10-30 nm thick graphite interphase. The absence of grain growth in the fiber reinforcement during fabrication should prove critical in retaining the superior mechanical properties of the fibers in the composite. The composite is expected to exhibit excellent strength and toughness and high strain to failure below 500°C in oxidizing environments and up to ~1200°C in inert environments.

However, there is sufficient evidence that the NFRBMC's will experience rapid oxidation in the form of oxidative removal of the graphite interphase above 500°C in air. Consequently, strong bonding between the fiber and the matrix occurs. Oxidation and the associated processes are expected to degrade the mechanical properties and the reliability of the composites. Until the introduction of an effective means to improve the oxidation stability of a functional fiber-matrix interphase (or interface), the application of the otherwise promising NFRBMC's in advanced aero-structures which must undergo oxidation exposure will be limited.

ACKNOWLEDGMENTS

The authors would like to acknowledge the support from NASA and the Grumman Corporation via the New Jersey Associated Institutions for Materials Science (AIMS). Dr. David Smith, Director of AIMS, is the program manager. The valuable input from Drs. P. Adler, T. Donnellan, and A. Tobin of Grumman is greatly appreciated.

REFERENCES

[1]P. Adler, T. Donnellan, and A. Tobin (Grumman Corporate Research Center), Private Communication.

[2]R.J. Kerans, R.S. Hay, N.J. Pagano, T.A. Parthasarthy, "The Role of the Fiber-Matrix Interface in Ceramic Composites," *Ceramic Bulletin*, **68** [2], 429-442 (1989).

[3]M. Ruhle and A.G. Evans, "High Toughness Ceramics and Ceramic Composites," *Progress in Mater. Sci.*, **33**, 85-167 (1989).

[4]Powder Diffraction File (Inorganic Compounds), Card No. 1-1119. JCPDS, Swarthmore, PA 1977.

[5]L.C. Sawyer, R. Arons, F. Haimbach IV, M. Jaffee, and K.D. Rappaport, "Characterization of Nicalon: Strength, Structure, and Fractography," *Ceram. Eng. & Sci. Proc.*, Vol. 6, pp. 567-75 (1985).

[6]H. Zhang and C.G. Pantano, "Synthesis and Characterization of Silicon Oxycarbide Glasses," *J. Am. Ceram. Soc.*, **73** [4], 958-63 (1990).

[7]G.M. Renlund, S. Prochazk, and R.H. Doremus, "Silicon Oxycarbide Glasses: Part II. Structure and Properties," *J. Mater. Res.*, **6** [12], 2723-34 (1991).

[8]T. Mah, N.L. Hecht, D.E. McLullum, J.R. Hoemigman, H.M. Kim, A.P. Katz, and H.A. Lipsitt, "Thermal Stability of SiC Fibers (Nicalon™), *J. Mater. Sci.*, **19** [4], 1191-1201 (1984).

[9]M. Huger, S. Souchard, and C. Gault, "Oxidation of Nicalon SiC Fibers," *J. Mater. Sci. Lett.*, **12**, 414-16 (1993).

CHARACTERIZATION OF THE INTERFACE IN BORON-DOPED, OXIDATION RESISTANT NICALON REINFORCED GLASS-CERAMIC MATRIX COMPOSITES

Douglas M. Beall and Carlo G. Pantano
Department of Materials Science and Engineering
The Pennsylvania State University
University Park, PA 16802

ABSTRACT

It is well established that an in situ carbon interphase develops during processing of certain Nicalon reinforced glass-ceramic matrix composites. This interlayer is responsible for the debonding and fiber pullout that results in the exceptional room temperature toughness observed in these composites. However, exposure to a high temperature, oxygen-rich environment results in oxidation of the interfacial layer and embrittlement of the composite. It has been found that doping of a Nicalon reinforced barium magnesium alumino-silicate (BaMAS) glass-ceramic matrix with a small amount of borosilicate glass results in increased oxidation resistance of the composite at high temperatures. In this paper, the chemistry and structure of the interface was characterized and used to describe the mechanism of increased oxidation resistance in this material.

INTRODUCTION

The performance of fiber-reinforced glass-ceramic matrix composites is largely determined by the nature of the fiber/matrix interface. Several interface requirements must be met in order to realize the full potential of this type of composite material. First, in order to achieve high strength and toughness, the interface is required to be strong enough to provide load transfer from matrix to fibers, yet weak enough to allow debonding and frictional sliding during

To the extent authorized under the laws of the United States of America, all copyright interests in this publication are the property of The American Ceramic Society. Any duplication, reproduction, or republication of this publication or any part thereof, without the express written consent of The American Ceramic Society or fee paid to the Copyright Clearance Center, is prohibited.

fiber pullout. A second requirement of the composite interface is that it serve as a reaction barrier against extensive chemical interaction between the fiber and matrix which could result in degradation of the fiber strength and/or strong bonding between fiber and matrix. A third criterion for the interface is thermochemical stability. The intended use of these composite materials is in high temperature structural component applications. Thus, the interface must be able to meet the first two requirements over a wide range of temperature in oxygen containing environments.

Often, a chemically distinct interphase material is necessary to meet the mechanical and chemical requirements outlined above. One interphase material of interest has been carbon. A carbon interphase can be produced either through coating methods or, as in the case of Nicalon fibers, can form in-situ during composite fabrication. Composites having a carbon interphase have been found to exhibit exceptional room temperature strength and toughness. However, application of these materials in aerospace and turbine structural components has been restricted by the inherent reactivity of carbon toward oxygen at temperatures in excess of about 600°C. Exposure of the composite to these conditions leads to oxidation of the carbon, and a subsequent increase in interfacial shear strength between fiber and matrix resulting in embrittlement of the composite[1]. Thus, carbon alone does not meet the thermal stability criterion and is therefore not a suitable interphase material. Research on substitute interphase materials is ongoing[2,3]; however, to date, none have matched the mechanical requirements as well as carbon.

An alternative approach to replacing the carbon interphase is to protect it against oxidation through the use of fiber coatings. One coating material, which has been extensively researched as an oxidation inhibitor for carbon, is boron oxide (B2O3). Borate glass is one of the few materials known to wet the surface of carbon. It can form a continuous oxide barrier to molecular diffusion of oxygen to, and carbon monoxide away from, the carbon. Due to the viscous nature of the glass at elevated temperature, it has the additional advantage of being able to flow to and seal regions exposed by matrix cracks . The main drawback to the use of a borate glass coating is its volatility at temperatures above about 1000°C, particularly in atmospheres containing water vapor. When the coated carbon is exposed to air at temperatures in excess of 1000°C, rapid oxidation of the carbon is observed. Further, the coating process itself adds an extra step to the fabrication process which increases cost and complexity to the composite manufacture.

An alternate method for oxidation protection of the carbon interphase is the addition of boron dopants to the matrix in hopes of getting improved oxidative stability without the use of coatings. This was first attempted by Brennan[4] who added B2O3 to an LAS matrix composition which was reinforced with Nicalon fibers. It was found that the borate phase significantly increased the oxidative stability. Unfortunately, the borate phase segregated to the grain boundaries and excessive creep resulted, lowering the practical use temperature of the composite to less than 800°C. Larsen and coworkers[5] added small amounts of a borosilicate glass to a barium-stuffed cordierite matrix (BaMAS) reinforced with Nicalon fibers. In this case, the glass separated to isolated pockets within the matrix and creep resistance was maintained. More importantly, exceptional oxidation resistance of the carbon interphase was observed at high temperature in air for very long times (5000 hrs. at 1100°C).

The objective of this study was to characterize the structure and chemistry of the interface region in the borosilicate glass (BSG) doped matrix composites[5] and to relate the interface modifications to the observed improvement in thermal durability.

EXPERIMENTAL

For this study, the matrix composition chosen was a barium-stuffed cordierite doped with 5% potassium borosilicate glass. The compositions of the doped and undoped matrices along with the composition of the potassium borosilicate dopant are given in Table I.

Table I. Matrix and Dopant Oxide Compositions by Weight (%)

	BaMAS	doped-BaMAS	BSG dopant
SiO_2	43.4	45.3	77.0
Al_2O_3	36.9	35.0	
MgO	12.9	12.2	
BaO	6.1	5.8	
As_2O_3	0.7	0.6	
B_2O_3		1.0	20.0
K_2O		0.2	3.0

The glasses were melted at 1650°C for 16 hours, poured and ball milled to produce a glass frit. The frit was then used to fabricate two different sample types. First, composite samples reinforced with Nicalon ceramic grade fiber were produced. Prepregs were made via the slurry dip technique. The prepregs were stacked and hot-pressed in a graphite die at 1250°C for 20 minutes at a pressure of about 10.3 MPa. The composite sections were cut and polished normal to the fiber direction for microscopic analysis.

In addition to the composite samples, reaction couples between the BSG-doped matrix composite and substrates of polycrystalline silicon carbide, highly oriented pyrolytic graphite (HOPG) and vitreous carbon were made. The SiC was obtained from Morton International. It was made by a CVD process and polished to 1 micron. The vitreous carbon was obtained from EMC Industries and polished to 1 micron. The HOPG graphite was obtained from Union Carbide with a basal plane orientation at the surface. These substrates were compacted with the frit glass in a graphite die and hot pressed in an induction furnace at 1250°C for 20 minutes at a pressure of about 10.3 MPa.

After cooling, the reaction couples were cut into cross-section and were also split along the interface so that both the matrix and substrate surfaces of the interface could be characterized. By using reaction couples, a large interfacial area could be obtained and surface analyses could be carried out which were not possible on composite samples due to the small diameter of the Nicalon fibers. Silicon carbide and carbon were chosen as substrate materials to represent the interface materials that would be encountered by the matrix in a Nicalon reinforced composite.

The analytical techniques used to characterize the interfaces were: SEM, TEM, atomic force microscopy (AFM), x-ray photoelectron spectroscopy (XPS), and wavelength dispersive electron microprobe analysis.

For TEM analysis, cross sections of the reaction couple samples, and composite samples were prepared for microscopy by polishing to 1 micron and treating with a weak HF solution to bring out the microstructure. Replicas of the samples were prepared by making a casting of the surface sample with acetylcellulose replicating film and evaporating platinum at an angle to the surface. A carbon film was then applied and the casting dissolved. The replica was placed on a copper grid and then viewed in a transmission electron microscope.

XPS spectra were measured with a 1253.6 eV Mg Kα radiation source.

Binding energy spectra were collected to obtain core photoelectron intensities for B 1s, Si 2p, Al 2s, Mg 2s, Ba 4d, As 3s, and the O 1s peak. The binding energies were obtained with respect to the C 1s core level peak at 285.0 eV (due to hydrocarbon contamination on the surface).

RESULTS

Figures 1 and 2 are TEM replicas showing the interface microstructures for the BSG-doped BaMAS and graphite and silicon carbide, respectively. In both cases a distinct interphase has developed which is of the order of 3-5 μm in thickness. Also in both cases, the interphase has been preferentially etched relative to the matrix as evidenced by the lack of polishing scratches in the interphase zone.

Atomic force microscopy was performed on the SiC reaction couple sample to verify that the preferential etching was within the interphase zone. Figure 3 shows an AFM three-dimensional image of the surface. The vertical scale in this micrograph is 1μm per division. As can be seen by the micrograph, preferential polishing and / or etching has taken place in this region. This evidence was an indication that the interphase material was glassy.

Figure 4 is a replica micrograph of the polished and etched composite. The fiber and matrix are clearly visible in the micrograph. In addition, some glass phase regions within the matrix are indicated. The glass phase is isolated in pockets from the rest of the matrix. More important is the interfacial zone between fiber and matrix. A thin interphase region which has been etched preferentially is visible. This region is much thinner than those seen in the reaction couples (about 0.2μm), but otherwise is quite similar.

To identify the composition of the interphase, x-ray photoelectron spectroscopy and electron microprobe analysis were employed. For XPS, the reaction couple samples were sectioned after hot pressing, and then split along the interface between matrix and substrate. These surfaces were then introduced into the vacuum chamber and analyzed to obtain chemical information concerning the glass constituents. The interface scans were compared to fracture surfaces far from the interface to obtain a comparison between bulk composition and composition at the interface.

Figure 5a shows an XPS spectra comparison between the bulk glass-ceramic composition and the composition at the silicon carbide interface. The spectra

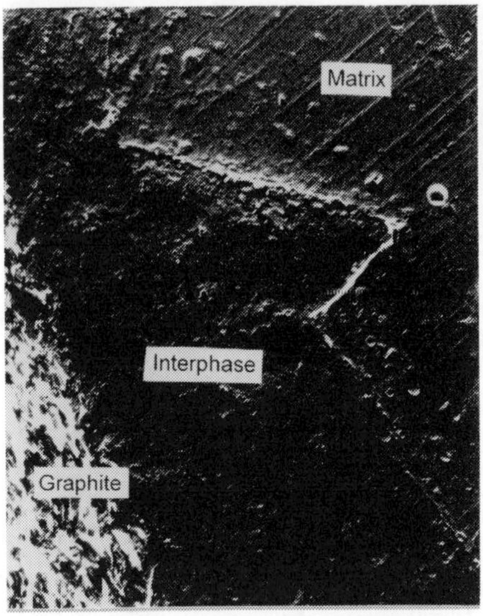

Fig. 1. TEM replica micrograph of BSG-doped matrix / HOPG

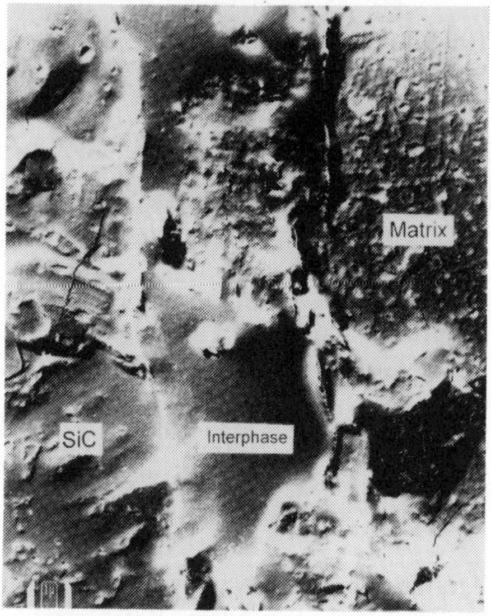

Fig. 2. TEM replica micrograph of BSG-doped matrix / SiC substrate

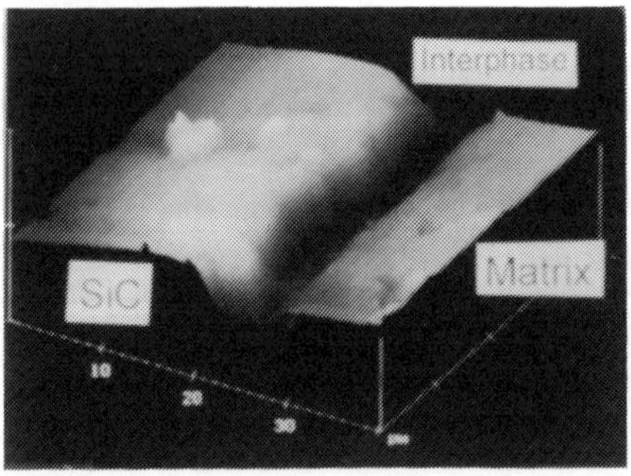

Fig. 3. AFM Micrograph of BSG-doped Matrix/SiC substrate

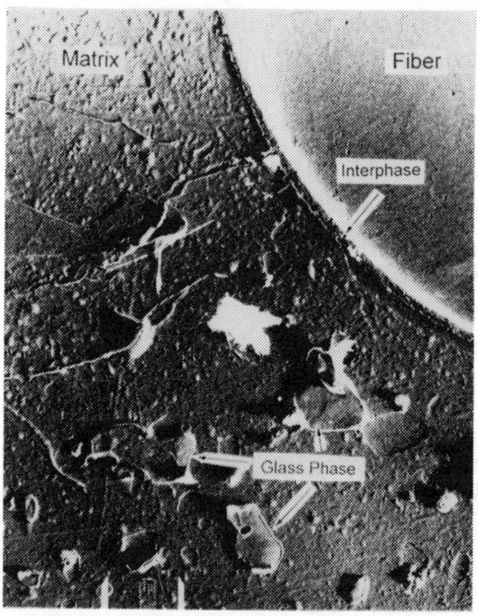

Fig. 4 TEM replica of Nicalon fiber /BSG-doped matrix composite

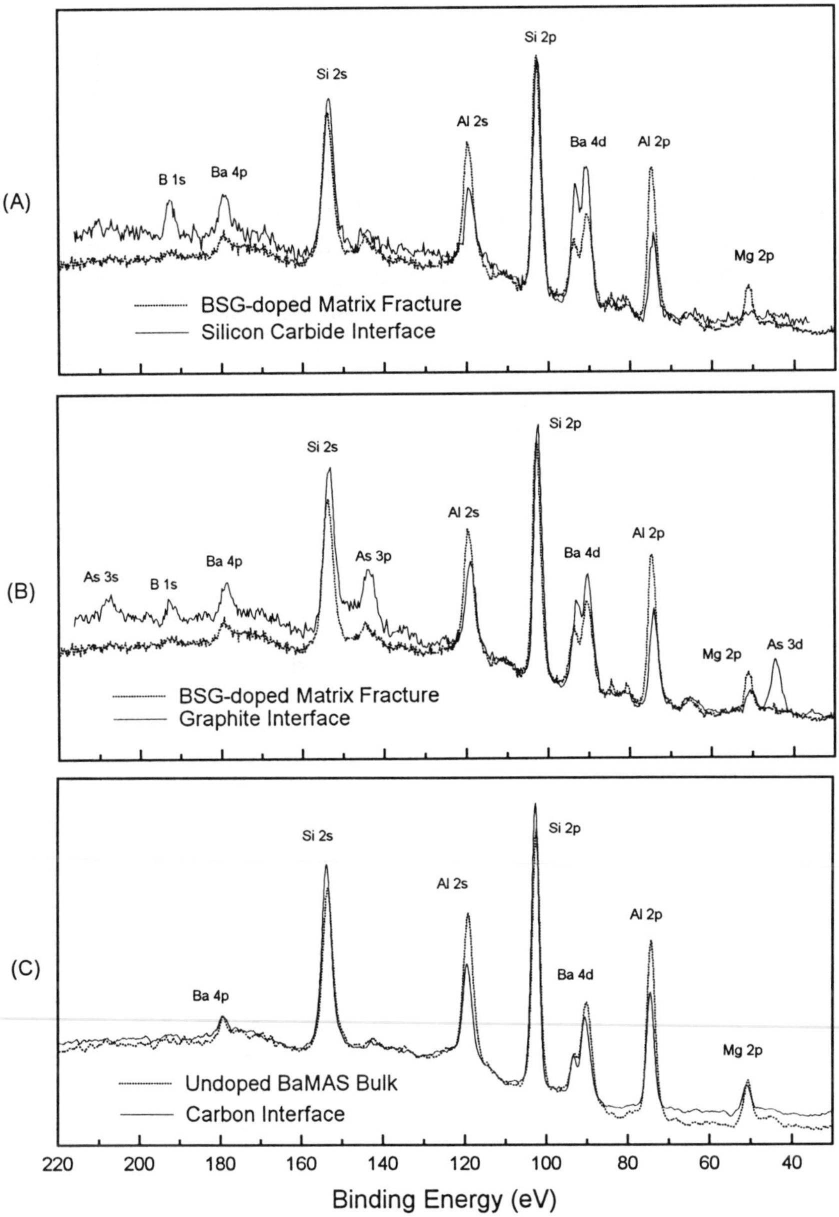

Figure 5. XPS Spectra: Comparison between bulk and interfacial compositions. (A) BSG-doped matrix / SiC, (B) BSG-doped matrix / HOPG, (C) Undoped BaMAS matrix / amorphous carbon substrate

show peaks for all the cation species in the glass. As shown by the figure, the peak intensities of boron and barium have increased at the interface with respect to the bulk composition. In addition, the intensities of magnesium and aluminum have been depleted.

Figure 5b shows a comparison between the graphite interface and bulk composition. The same trend of increased levels of boron and barium at the expense of aluminum and magnesium are observed. However, at the graphite interface a set of peaks corresponding to arsenic are also seen. Arsenic is added to the glass composition as a fining agent and appears to segregate to the graphitic interface.

Figure 5c is a comparison between the carbon interface and bulk composition in the undoped BaMAS matrix. The data show that the compositions are similar. This indicates that an interphase is not developing in the undoped samples and that the BSG-doping is responsible for the modification of the interface compositions in figures 5a and b. Compositions measured by XPS for the interphase materials as well as the bulk matrix are given in Table II.

Table II Interfacial and Bulk Compositions Measured by XPS (wt. %)

	BSG-doped matrix (batched)	BSG-doped matrix (measured)	SiC interphase	Graphite interphase	Vitreous C interphase
Ba	5.6	6.8	13.9	11.3	11.6
Mg	7.8	7.6		4.0	2.1
Al	18.5	18.9	12.6	13.0	10.5
Si	20.5	23.6	24.9	21.9	21.9
B	0.3		2.8	1.8	1.1
As	0.8			2.4	
O	47.3	43.1	45.8	45.6	52.8

The interface compositions were also evaluated with the electron microprobe. For this analysis, a polished cross section of the BSG-doped matrix and silicon carbide interface was examined. Figure 6 is a backscattered SEM image of this interface. Contrast is provided by differences in average atomic number, thus differences in composition are readily discernable.

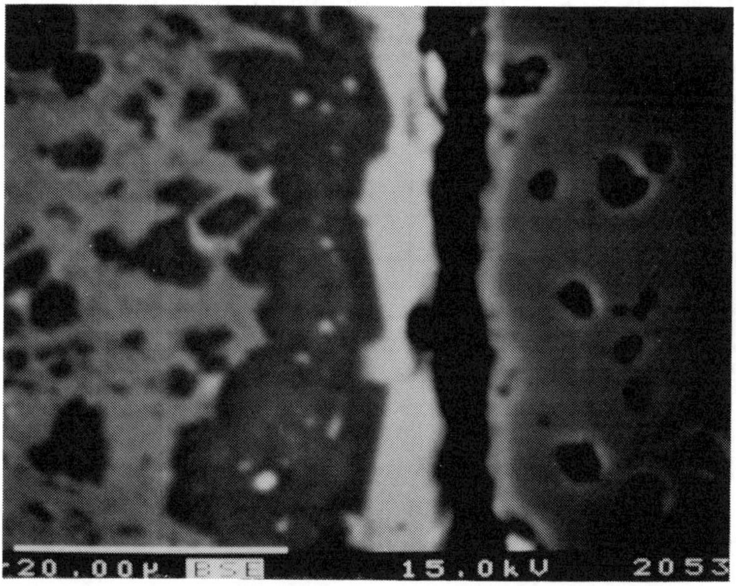

Figure 6.. SEM Backscatter Image of SiC/BSG-doped Matrix Interface. The SiC substrate is on the right and the matrix on the left separated by a crack.

A continuous bright layer is visible at the interface and its composition is easily distinguished from that of the bulk. Barium enrichment in this layer is responsible for the brightness; the chemical composition of the interphase as measured by microprobe analysis, is similar to that found in XPS.

A dark region is found as a continuous layer immediately adjacent to the bright interphase. It was found that barium is almost completely depleted in this area. This phase has a composition which is very close to that for stoichiometric cordierite (2MgO $2Al_2O_3$ $5SiO_2$). It appears that almost all of the barium has segregated from this region into the interfacial zone. The next brightest region farther from the interface corresponds to the batched Ba-stuffed cordierite composition. Compositions, as measured quantitatively by the microprobe, are given in Table III for the three different phases found in the matrix.

Table III. Matrix Phase Compositions Measured by Electron Microprobe (wt %) and Comparison to Known Cordierite Compositions

	Interphase	Matrix Phase	Cordierite 2:2:5 MAS	Matrix Phase	Ba-stuffed Cordierite
Ba	16.0	1.5		8.2	5.6
Mg	2.8	7.9	8.2	7.4	7.8
Al	11.1	18.3	18.5	19.2	19.7
Si	23.0	23.4	24.0	19.6	20.4
B	1.5				
O	45.5	48.7	49.3	45.4	46.6

DISCUSSION

In the BaMAS matrix doped with borosilicate glass, it seems clear that there is a segregation of certain glass species toward interfaces with SiC and carbon. The result is the formation of a continuous glassy interphase. This interphase was observed in both the reaction couples and composites reinforced with Nicalon fibers. The glassy layer seems to form as a result of the BSG-doping since no interlayer was observed in the undoped samples. However, the composition of the interphase is distinctly different from that of either the dopant or the matrix.

The segregation of boron to the interface is of primary interest with respect to composites reinforced with Nicalon fibers since boron is known to improve the oxidation resistance of carbon[6]. Pantano et al. observed segregation of boron to the Nicalon interface in composites with a sodium borosilicate glass matrix[7]. In that work, scanning Auger microscopy found a higher level of boron at the matrix / Nicalon interface than in the bulk glass. The driving force for this segregation is not completely understood presently, however, it is known that borate glass and some borosilicate glasses are able to wet the surface of carbon[8]. Barium was also observed to segregate to the interface in the reaction couples. Both barium and boron are also known to decrease the surface tension in glasses. Therefore a lower interfacial surface energy condition may result when the substrate is in contact with the glass which could provide a driving force for such segregation. Likewise, the phase separation observed throughout the BSG-doped matrix may be enhanced at the interface, and thereby could lead to a continuous interphase.

Arsenic also was seen to segregate to the interface in the case of the carbon substrates; however, it was not observed in the case of the silicon carbide. Arsenic is a critical batch addition for composites reinforced with Nicalon fibers. It has been found that arsenic oxide and other mixed valence oxides can control the formation of the *in-situ* carbon layer[9]. The oxygen potential at the interface determines the extent of the reaction which generates the carbon interlayer. At high temperature As_2O_5 can be converted to As_2O_3 and release oxygen. There is a driving force for this reaction in the vicinity of the carbon substrates due to a local oxygen activity gradient. The fact that the arsenic is enriched at the interface confirms the role of arsenic in the interface reaction. The observation that the arsenic is not enriched at the silicon carbide interface is likely due to the fact that the processing temperature is too low for oxidation of the SiC.

The composition of the glass phase, both within the matrix bulk and at the interface, is not that of the potassium borosilicate dopant. Rather, it contains larger percentages of baria, and alumina and a lower silica content. The approximate oxide composition as calculated from both the XPS and microprobe data is given below (wt %):

SiO_2	52
Al_2O_3	22
BaO	16
B_2O_3	7
MgO	3

This composition is expected to form a stable glass which would be viscous at high temperatures and be able to flow to areas of the carbon interface which were exposed by cracks in the matrix. This explains the excellent high temperature mechanical properties found in this type of material. A distinct advantage of this type of protective layer over that of a borate coating is that this glass would be quite stable at temperatures where the B_2O_3 would volatilize and oxidation embrittlement would occur. Another advantage is that a coating process is not necessary. The glassy protective coating as well as the carbon layer both form *in-situ* during composite manufacture.

The problematic condition for this material is extended times at intermediate temperatures (about 600°C). At these temperatures, oxidation embrittlement is found to occur. This is most likely due to the fact that the viscosity of the glass is too high at the lower temperatures to flow and heal areas exposed by cracks. However, it should be possible to tailor the glass composition to have a longer glass transition range by adjusting either or both the matrix and dopant compositions to help alleviate this problem.

SUMMARY

It has been shown that when a small amount of potassium borosilicate glass is added to a BaMAS matrix composition, a continuous glassy layer is formed at the interface of the matrix and substrates of SiC and carbon. This layer is also observed in composites reinforced with Nicalon fibers. Boron and barium are found to segregate into the glassy interphase and the composition is nominally a barium boro-aluminosilicate. This glass is stable at high temperatures, it wets carbon, and is able to flow to heal areas of the carbon interlayer exposed by matrix cracks.

ACKNOWLEDGEMENTS:

The authors gratefully acknowledge Ken Chyung and Dave Larsen for provision of samples and helpful discussion during the course of this work. Thanks are also extended to Temel Buyuklimanli for the XPS studies, Mark Angelone for the electron microprobe analysis, and Bryan Wheaton for assistance with the TEM and AFM work. The authors also acknowledge with appreciation, the financial support of the Office of Naval Research (NOOO14-90-J-1392, Steve Fishman , Contract Monitor).

REFERENCES:

1. R.L. Stewart, K. Chyung, M.P. Taylor, R.F. Cooper, "Fracture of SiC-Fiber/Glass-Ceramic Composites as a Function of Temperature": pp 33-51 in Fracture Mechanics of Ceramics, Vol.7, *Composites, Impact, Statistics and High Temperature Phenomena*, Ed. by R.C. Brandt, A.G. Evans, D.P.H. Hasselman, and F.F. Lange, Plenum Press, New York, 1986

2. R.F. Cooper, P.C. Hall, "Reactions between Synthetic Mica and Simple Oxide Compounds with Application to Oxidation-Resistant Ceramic Composites", *J. Am. Ceram. Soc.* **76** [5] 1276-73, 1993

3. J.B. Davis, J.P.A. Lofvander, A.G. Evans, E. Bischoff, M.L. Emiliani, "Fiber Coating Concepts for Brittle Matrix Composites", *J. Am. Ceram. Soc.* **76** [5] 1249-57, 1993

4. J.J. Brennan, "Interfacial Studies of SiC Fiber Reinforced Glass-Ceramic Matrix Composites", U.T.R.C. Final Report Contract N00014-82-C-0096 for O.N.R. October 15, 1987

5. D. Larsen, H.T. Godard, R.L. Stewart, K. Chyung, G.D. Linsey, and D.S. Murphy, "BSG-doped CMC's for Thermal Durability", to be published in NASA Conference Publication: 17th Annual Conference on Composites, Materials and Structures. Proceedings of a Conference held in Cocoa Beach FL, Jan. 10-15 1993

6. D.W. McKee, "Borate Treatment of Carbon Fibers and Carbon/Carbon Composites for Improved Oxidation Resistance", *Carbon* **24** [5] 737-41 1986

7. W.K. Treadway, K.M. Prweo, C.G. Pantano, "Fiber/ Matrix Interfacial Effects in Carbon Fiber-Reinforced Glass Matrix Composites", *Carbon* **27** [5] 717-27 1989

8. C.G. Pantano, G. Chen, D. Qi, "Interface Reactions and Wetting in Carbon-Fiber-Reinforced Glass Matrix Composites", *Mater. Sci. and Eng. A.* **126** 192-201 1990

9. C.G. Pantano, K.E. Spear, G. Qi, D.M. Beall, "Thermochemical Modelling of Interface Reactions in Glass Matrix Composites", pp 173-198 in Advances in Ceramic Matrix Composites, Ed. by N. Bansal, Am. Ceram. Soc., v. 38 1993

COATING OF CERAMIC FIBERS WITH YTTRIA FILMS

Jay P. Daniel, Lutgard C. DeJonghe
Center for Advanced Materials, Materials and Chemical Sciences Division, Lawrence Berkeley Laboratory, Berkeley, California, 94720

ABSTRACT

Two aqueous solution methods have been developed for coating ceramic fibers with a film of yttria. The first utilizes the decomposition of urea in the presence of yttrium nitrate to yield yttrium basic carbonate(YBC). YBC precipitates heterogeneously forming a film. In the second method, a concentrated aqueous solution of yttrium nitrate, polyethylene oxide, and a commercially available wetting agent is prepared. Ceramic fibers are drawn through the solution, wetting the fiber with a film of the solution. The films are converted to yttria by heating in an oxidizing atmosphere. The film morphologies are characterized using scanning electron microscopy. The film constituencies are characterized using thin film x-ray diffraction spectroscopy. Limits to film thickness are explored.

INTRODUCTION

The search for structural materials with desirable mechanical properties at high temperatures has led to the effort to develop composites incorporating ceramic fibers into a ceramic or metal matrix. Development of ceramic fibers that can resist degradation during processing into a ceramic/ceramic or metal matrix composite has been a significant impediment though, to the commercial use of these composites. During sintering of ceramic/ceramic composites in which the ceramic fiber consists of the same material as the matrix phase, grain growth into the fiber or the sintering together of the matrix and the fiber can degrade the toughness of the sintered product. In metal matrix composites, the heated metal is often highly corrosive and can reacte with the reinforcing phase during incorporation producing undesirable reaction products and degraded mechanical properties. This is particularly true in the case of nickel and titanium alloys, candidates for high temperature service.

It is thus desirable to develop a reinforcement coating in ceramic composites that will prevent grain growth into the reinforcing phase and will prevent the matrix and reinforcing phase from sintering together. In metal matrix composites, it is desirable to develop a coating that will resist hot metal corrosion. Yttrium oxide, Y_2O_3, is a good candidate to constitute this coating. It has a very high free energy of formation [1]. Thus, it resists corrosion very well and is very stable.

To the extent authorized under the laws of the United States of America, all copyright interests in this publication are the property of The American Ceramic Society. Any duplication, reproduction, or republication of this publication or any part thereof, without the express written consent of The American Ceramic Society or fee paid to the Copyright Clearance Center, is prohibited.

Previous investigators have deposited yttria films to study metal/yttria film/substrate interactions. A variety of techniques to deposit yttria films were employed. Mehan, Jackson, and McConnell [2] used sputtering to directly deposit yttria films. Larkin and Interrante [3] developed a technique depositing a yttria layer by CVD. Kieschke, Somekh, and Clyne [4,5,6] also deposited yttria layers. They formed an yttria film by sputtering yttria or by sputtering yttrium followed by oxidation. These investigators reported success in achieving protection of the substrate from attack by use of a yttria interlayer. Development of a relatively inexpensive technique to coat fibers with yttria would thus be desirable. Aqueous techniques for film deposition were considered since they generally require significantly less capital expense and cheaper starting materials.

Garg and DeJonghe [7] investigated deposition of yttria precursor films out of aqueous solution utilizing a technique involving the decomposition of urea in the presence of yttrium nitrate. The reactions and rate equations leading to YBC are as follows.

Chemistry

The process of coating a substrate with an oxide precursor, in our case yttrium basic carbonate(YBC), consists of nucleation and growth of the matrix phase (from solution) on the surface of the substrate. Precipitation of this yttria precursor has been investigated, specifically the homogeneous precipitation of spherical powders [8]. After calcination, yttria powder has been produced using this method.

YBC is formed via the hydrolysis of yttrium nitrate in an aqueous solution of decomposing urea. The urea decomposes at a significant rate above 50°C in moderately acidic solutions. The decomposition of urea in solution is a well known reaction studied by Shaw and Bordeaux [9,10]. Cyanate and ammonium are produced by first order kinetics according to (1):

$$H_2NCONH_2 \dashrightarrow NH_4^+ + OCN^- \tag{1}$$

The rate expression cited by Shaw and Bordeaux[9,10] is:

$$r_{dec} = - k_o [H_2NCONH_2] \tag{2}$$

$$k_o = 4 \times 10^{14} e^{(-32.4/RT)} \tag{3}$$

with k_o in sec^{-1} and E_a in kcal. This reaction proceeds in the reverse direction, although the reverse rate for urea formation is not significant until approximately 5% of the urea has decomposed. These levels of urea decomposition are not reached in the systems studied.

The decompostion of cyanate in queous solution was studied by I.A. Kemp and G. Kohnstam [11]. The equilibrium reaction (4) leading to carbonate is the dominant route for cyanate decomposition at the conditions studied.

$$OCN^- + 2H_2O \longleftrightarrow CO_3^{2-} + NH_4^+ \tag{4}$$

From the data cited by Kemp and Kohnstam, the dominant rate expression at moderate to low pH is:

$$r_{dec} = - k_1 [OCN^-] [H3O^+] \tag{5}$$

$$k_1 = 1.5x10^{13} e^{(-14.6/RT)} \tag{6}$$

with k_1 in l/mol.min and E_a in kcal. The reverse rate of (4) is insignificant compared to (5) for this work. Yttrium ions are weakly hydrolyzed in water to $YOH(H2O)_n{}^{2+}$ [12]. These complexes react with carbonate to form YBC. The formation of YBC was cited by Her [13]:

$$[YOH(H_2O)_n]^{2+} + H_2CO_3 \text{ ------> } Y(OH)CO_3.H_2O + 2H^+ + (n-1)H_2O \tag{7}$$

Garg and DeJonghe were interested in coating discrete particles in aqueous solution to improve the structural properties of composites. This investigator initially focused on extending this technique to alumina, niobium, and silicon carbide substrates.

Dip Coating

While the urea decomposition technique offered a superior technique for coating discrete particles that could be dispersed in solution, such as platelets or whiskers, the technique was less applicable to large quantities of continuous fibers. Due to relatively slow film deposition kinetics, this investigator refocused attention on alternative methods. A technique wherein filaments would be drawn through a bath and coated would lend itself better to industrial production for continuous fiber morphology. Dip coating fibers in concentrated solutions of yttrium salt and wetting agent was thus attempted.

EXPERIMENTAL

Urea Decomposition

Alumina fibers and niobium coated alumina fibers were utilized as received from 3M Corporation. Niobium was sputter coated on the alumina fibers to a thickness of 30-50 nm. The fibers were cut into 2 cm lengths. Approximately 1000 fibers were prepared for a run. A pH meter with digital readout(Cole Palmer Model 5654-16) was calibrated using buffers of pH 4.01 and 7.00 to $\pm$ 0.02. The pH probe was a glass electrode(Cole Palmer 5654 80) with temperature compensation. A 300 ml three neck distilling flash was prepared with the pH probe in one neck, a water cooled condensing column on the middle neck, and a K-type thermocouple in the third neck.

A solution of yttrium nitrate (NOAH Technologies Corporation) and urea (Mallinckrodt, Inc.) was prepared at ambient temperature with concentrations of 0.106 g·mol/l and 1.5 g·mol/l respectively. This solution was transferred to the distilling flask. The alumina fibers were transferred to the solution. A Eurotherm temperature controller was used to modulate current to a heating mantle based on a temperature setpoint for the reaction solution. The solution was stirred by a teflon coated magnetic stir bar.

A temperature setpoint was entered into the temperature controller and the reaction solution would ramp to setpoint. Data was gathered on solution temperature and pH over the course of the reaction. A portion of fibers was retrieved at intervals over the reaction run. The fibers were washed with water and subsequently washed with acetone. The fibers were then allowed to dry at ambient temperatures.

Sapphire substrate, 20 mm in diameter by 2 mm height, for use in thin film XRD analysis, was also coated using an identical procedure.

After the washing procedure, the coated alumina fibers and sapphire substrate were subsequently heated to 725°C over three hours in an oxidizing environment and allowed to cool to ambient.

Wetting Solution Of Surfactant, PEO, and Yttrium Nitrate

Aqueous solutions of polyethylene oxide,(Aldrich 18,945-6 MW 9e5) were prepared to yield solutions from 0.0 - 1.0 wt%. Yttrium nitrate(NOAH Technologies Corporation 99.99%) was added to yield final yttrium nitrate concentrations of 47% (saturated) to 23%. Merpol SH™(Dupont), a commercially available wetting agent, was used as received. The wetting agent was added to the solutions of yttrium nitrate in strengths varying from 0.0% to 0.5%. Alumina filaments of 12-14 μm diameter were utilized as received from 3M. The filaments were supplied in a toe that was wound around a reel. The toe consisted of 122 filaments.

A 5 cm length of a toe was separated from the reel. Individual alumina filaments were manually separated out from toe of alumina filaments. Care was taken to avoid contact with areas of filaments that would later be exposed to solution coating. Each filament was weighted at one end with a drop of epoxy (Hardman Extra Fast Setting Double Bubble™). The epoxy was used to weight the filaments at one end so they could be dip coated. The epoxy was dried at ambient for one day.

The filaments were then dipped manually in a solution of yttrium nitrate and wetting agent. The filaments were transferred to an alumina crucible. The filaments were ramped to 725°C from ambient over three hours in air in a box furnace. The furnace was then cooled to near ambient. The filaments were retrieved from the crucible and approximately 2 cm of filament on the epoxy end was cut off and discarded. The filaments were characterized using SEM.

A cylindrical sapphire substrate was dipped in a aqueous solution of 35% yttrium nitrate and 0.3% wetting agent. The substrate was transferred to a alumina crucible and heated to 725°C over three hours. The substrate was then cooled to near ambient. Thin film XRD was performed to characterize the surface of the sapphire.

RESULTS

Urea Decomposition

Based upon the reaction kinetics determined by Shaw and Bordeaux [9,10], and the rate expressions for the decomposition of cyanate ion given by Kemp and Kohnstam [11], a model for the production rate of yttrium basic carbonate may be generated. Figure 1 shows the time dependency of the concentration of carbonate in solution during a typical reaction run. The concentration of carbonate was calculated using the rate expressions listed in the introduction as well as the temperature and pH profiles of the reaction run [Fig. 2-3]. Concentrations on hydronium ion were obtained from continuous pH measurement of the solution during the reaction run.

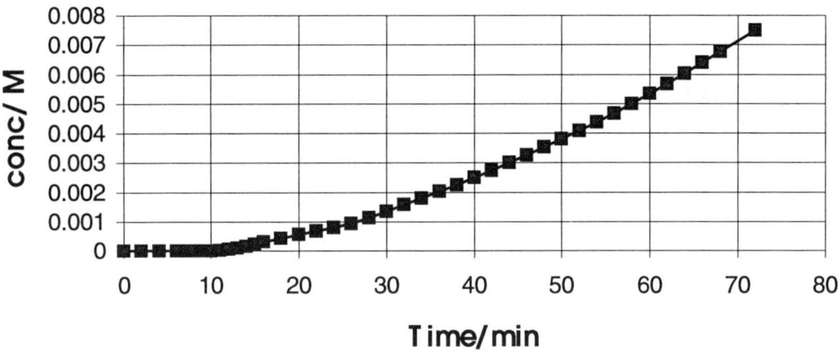

Figure 1: Cyanate ion concentration during reaction run.

During coating of alumina fibers, small groups of fibers were removed from solution during various points in the coating run. Using the SEM, the thickness and morphology of the YBC coating from various points in the reaction was determined. The time of initial heterogeneous precipitation in the reaction run was established in this way. Based on this time, and the kinetic model of the formation of YBC, a heterogeneous precipitation concentration may be calculated.

The heterogeneous precipitation for this system occurs at a carbonate concentration of approximately 0.004 g.mol/l. Upon reaching this concentration, in this chemical system, yttrium basic carbonate will coat out on alumina surface. Based upon the kinetic model for carbonate formation and visual observation of the solution, a homogeneous precipitation can also be calculated. In work done by A. Garg, initial nucleation at the surface occurred with a solid phase as little as 5 nm in diameter. Thus, the homogeneous precipitation limit calculated based upon the time which opacity is first observed is limited by the ability of the instrument to detect light scattering in the reaction solution. The homogeneous precipitation concentration, based on visual observation, occurs at a carbonate concentration of approximately 0.007 g.mol/l.

The thicknesses of the YBC films were controlled by varying time of exposure of the fiber to solution at different times in the reaction run. The progress of the batch reaction in the region where precipitation occurred could be quantified by monitoring the pH of the solution with time. See figures 2 and 3 for a plot of the solution temperature and pH during a reaction run. It is believed that the initial drop in pH during heating is due to dissociation of the yttrium complexes. The rising pH for the next 40 minutes of the reaction is due to the generation of ammonium by the urea and cyanate decompositions. The drop off in pH was associated with precipitation and is believed to occur due to the removal from solution of hydroxide and carbonate resulting in more free hydronium. Film growth began at the peak of the reaction pH curve.

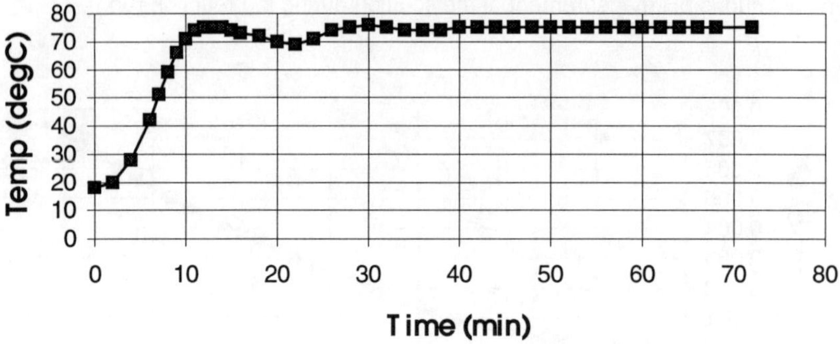

Figure 2: Temperature profile during reaction run

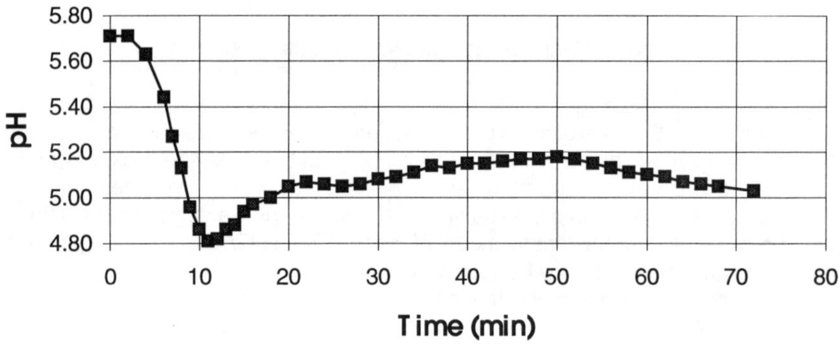

Figure 3: pH profile during a reaction run.

The thickness of the films were controlled by controlling the timing of the exposure of the fibers to the reaction solution. The benchmark that was used to remove fibers from the solution was the peak in pH that occurred approximately 50 minutes into the reaction run. Film thicknesses up to 1 μm were generated. Films were successfully deposited on alumina, niobium [Fig. 4] and silicon carbide fibers.

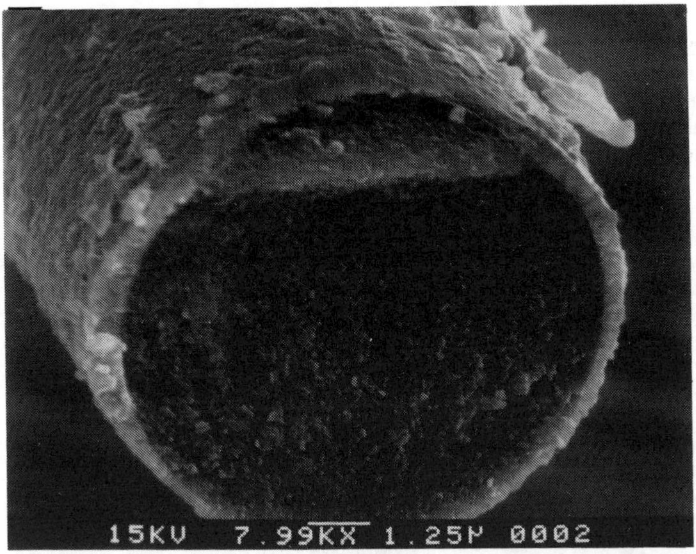

Figure 4: Alumina fiber coated by sputtering with thin(3-50 nm) layer of niobium and subsequently coated with yttria by urea decomposition process.

A reaction run was also conducted to coat a sapphire substrate. Thin film x-ray spectroscopy was used to characterize the film after calcination. The material analyzed as yttria although two peaks, at 2θ of 61.2 and 74.4, were unidentified [Fig. 5].

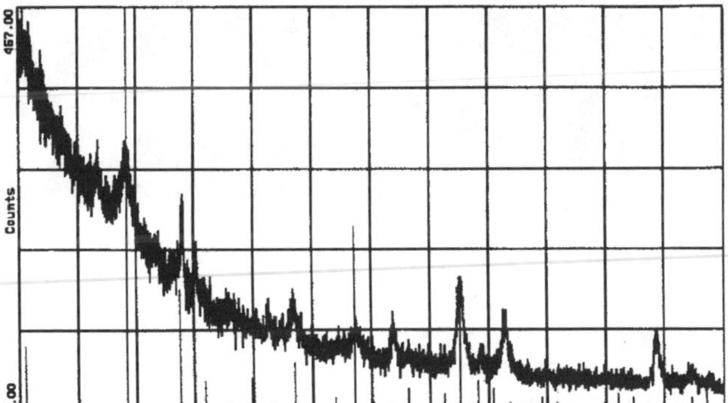

Figure 5: Thin film XRD pattern for yttria film deposited on single crystal alumina.

Wetting Solution Of Surfactant, PEO, and Yttrium Nitrate

Deposition of a film of yttria upon alumina filaments was attempted using an aqueous solution of 47% $Y(NO_3)_3$ without any wetting agent. After annealing, the alumina filaments were characterized by SEM. Islands of yttria were seen on the filaments, although these islands were intermittent. There were no contiguous films on the filaments.

When Merpol SH™ was used in conjunction with the $Y(NO_3)_3$, films of yttria were achieved. The minimum concentration of wetting agent that was used was 0.25%, although lower concentrations may be effective. Varying the concentration of yttrium nitrate produced varying thickness of films as characterized by SEM. Figure 6 shows the film morphology obtained after annealing of filaments when 47% yttrium nitrate and 0.5% wetting agent was used. The yttria coatings exhibited a tendency towards periodic axial cracking of the yttria film. Circumferential cracks occur approximately every 100 μm along the filament length. Film thickness was approximately 0.4-0.5 μm.

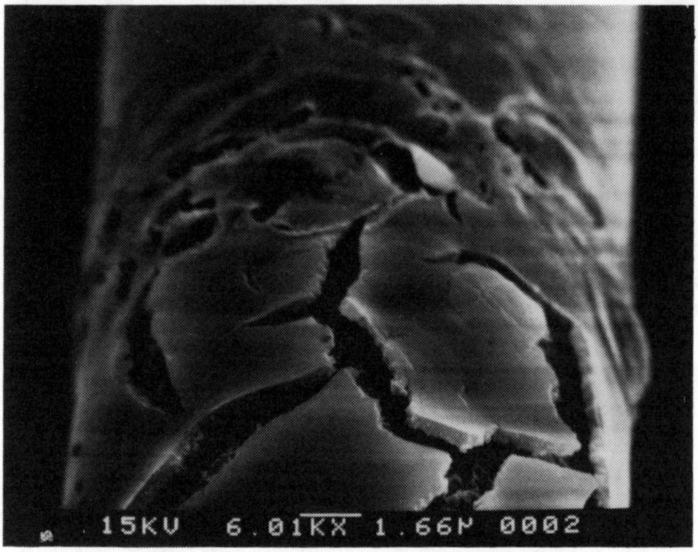

Figure 6: Cracked film of yttria on alumina fiber.

Filaments were also coated at lower concentrations of yttrium nitrate and wetting agent. Using solutions of yttrium nitrate concentration and wetting agent concentration of 39% and 0.3% respectively, yttria films of 0.10-0.15 μm were produced. These films exhibited cracking similar to that seen in figure 6, although the crack widths seen for these films were much smaller than that seen in figure 6 (0.1 as opposed to 0.5 μm).

Contiguous, uncracked films were realized when coating filaments with solution containing 23% yttrium nitrate and 0.25% wetting agent. Film thicknesses of 0.05 - 0.10 μm were achieved [Fig. 7].

Figure 7 is a SEM image of a filament end that was coated with yttria and broken to visually characterize the coating. The image was taken at approximately 40° with horizontal. Higher resolution images of filaments coated with these solutions show the yttria coating to be approximately 0.05 μm thick.

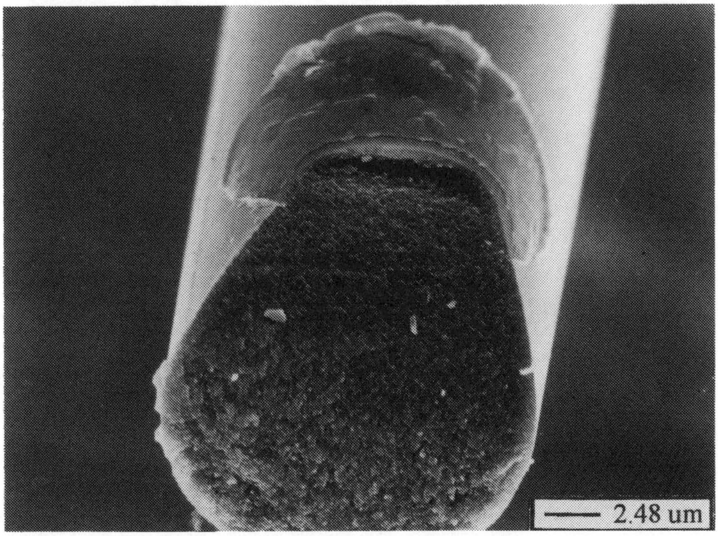

Figure 7: Yttria films obtained from 23% $Y(NO_3)_3$ and 0.25% surfactant.

Filaments dipped in solutions containing PEO allowed thicker contiguous films to be deposited. PEO of 9xE5 molecular weight yielded the best films. Higher molecular weight PEO yielded solutions with very high viscosities that produced poor coating results during dipping. Solutions with concentrations of 9xe5 PEO greater than 0.5% produced difficulty in dipping due to high viscosity. Optimal films were obtained from a solution of 30% $Y(NO_3)_3$, 0.25% PEO and 0.5% surfactant[Fig. 8]. Micrographs taken at higher resolution showed the yttria films from this solution were approximately 0.15-0.20 μm thick.

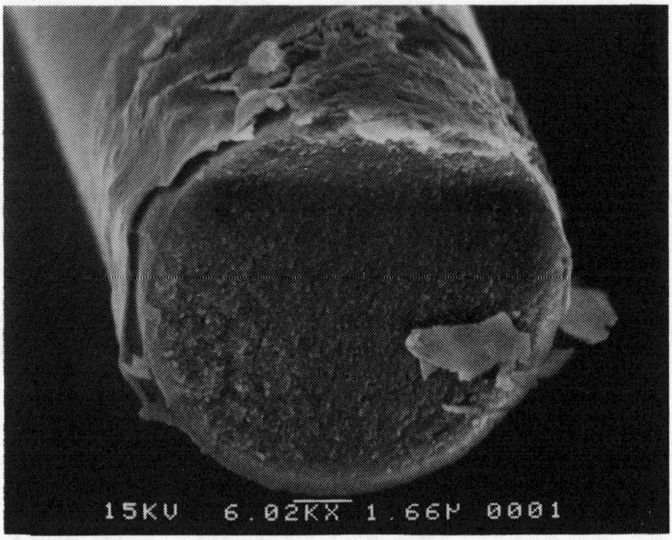

Figure 8: Fracture surface of alumina fiber with yttria film obtained from 30% $Y(NO_3)_3$, 0.5% surfactant, and 0.25% PEO.

The wetting agent utilized to improve wetting of the yttrium nitrate solution to the alumina substrate is a nonionic, hydrocarbon surfactant. It has the structure, R - (C_2H_4 - O)$_n$ - CH_3OH. While the compound contains solely hydrogen, carbon, and oxygen, there was some concern that residual carbon would contaminate the yttria film upon annealing. To examine this, thin film XRD was also utilized to characterize the surface constituency of the film. The profile [Fig. 9] matched yttria very well inferring that very little contamination remained following annealing to 725°C. The literature reports conversion of yttrium nitrate, $Y(NO_3)_3$, to yttria, Y_2O_3, at 650°C[14].

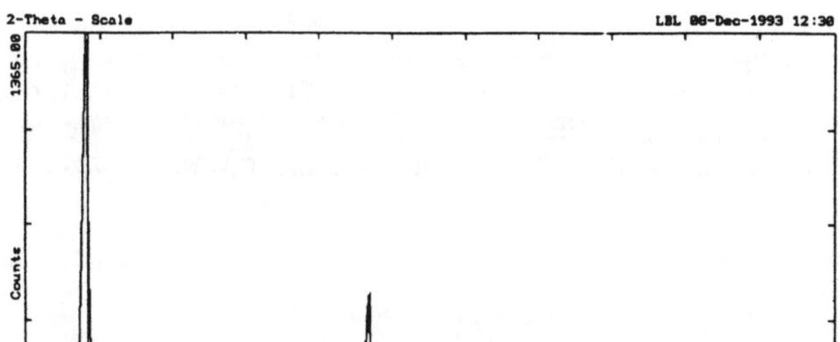

Figure 9: Thin film XRD pattern from yttria film on sapphire substrate.

DISCUSSION

At pH values below 9, alumina is known to adsorb hydroxyl groups out of water to its surface generating a net release of hydronium to solution. It is also known that nonionic surfactants, such as that used in this study, will adsorb onto the surface of alumina, lowering the alumina solution surface energy [15]. Recalling Young's equation (1),

$$\gamma^{sv} = \gamma^{sl} + \cos(\theta)\gamma^{lv} \tag{8}$$

this surfactant adsorbs to the solution alumina interface, lowering the value of γ^{sl}. Assuming constant γ^{sv} and γ^{lv}, $\cos(\theta)$ would be forced to increase to retain equivalency driving θ towards zero. The wetting agents used in this study also segregate at the water air interface, lowering the surface energy from 73 to 27 mN/m [16]. The wetting agent decreases γ^{lv} significantly, again forcing $\cos(\theta)$ to increase and thus lowering θ towards zero.

From the geometry of the cracks seen in the thicker films in this study, it seems likely that the cracks formed during the drying of the aqueous solution. The significant opening of the cracks, without any peeling or sloughing of the film, indicates that the film was relatively free to move along the substrate during shrinkage. This ability to move along the substrate is more probable for the hydrated yttrium nitrate salt with a surfactant layer between the salt and the fiber, than for the annealed yttria layer. Previous investigators [17] have had success adding soluble long chain organic consituents, such as polyethylene oxide, to solutions being used for spin coating to prevent cracks from forming during drying of the films. This technique proved useful here. Such constituents might even allow quicker annealing of the films since they provide fuel to the abundant oxidizer concentration provided by the nitrates. The glycine nitrate process [18] is an example of film deposition occurring via combustion of nitrates and a fuel. Previous investigators have cited the need to produce yttria films of 0.8 μm thickness to fully protect substrates from attack by nickel aluminide. Films of 0.3 to 0.5 μm were adequate to protect substrates from attack by titanium. Ti 6-Al 4-V alloys are the probable candidate for inclusion in a titanium alloy matrix, ceramic fiber MMC due to its superior structural properties. Ti 6-Al 4-V is less corrosive to ceramics, and should require a thinner yttria interlayer to provide protection than that needed for titanium.

It is interesting to note that the thickness of the yttria films did not scale with the concentration of the used in solution. Aqueous solution of 47% $Y(NO_3)_3$ produced films of 0.5 μm thickness while solution of 23% $Y(NO_3)_3$ produced films of 0.05 μm thickness. This was most likely due to the higher thickness of aqueous solution that adhered to the alumina filaments due to the higher viscosity of the 47% $Y(NO_3)_3$ solution. This result was indicated by Scriven[19]. He derived the relation of film thickness to fiber pulling velocity and coating solution viscosity where h is the film thickness, c_1 is a coefficient close to 0.8 for Newtonian fluids, η is the coating solution viscosity, U is the fiber velocity, and ρ is the solution density.

$$h = c_1(\eta U/\rho g)^{1/2} \tag{9}$$

Landau and Levich [20] extended this for systems with low coating viscosity and substrate speeds. γ^{lv} is the liquid vapor surface energy.

$$h = 0.94(\eta U/\gamma^{lv})^{1/6}(\eta U/\rho g)^{1/2} \quad (10)$$

In both cases, the film thickness is near proportional to the root of the coating solution viscosity.

Conclusions

A method for producing yttria films via urea decomposition in aqueous solution has been explored. The method would be appropriate for industrial application in coating discrete particles, such as platelets or whiskers. A method for producing thin yttria films on alumina filaments via dip coating using a simple aqueous solution method has also been elucidated. The method is easily adaptable to continuous processing. Limits to film thickness have been explored.

Acknowledgements: This work was funded by Advanced Research Projects Agency. Dr. David White and Dr. Robert Kieshke of 3M corporation graciously provided alumina fibers and niobium coated alumina fibers.

References

[1] K.A. Gschneider, L. Eyring, 1979, Handbook of the Physics and Chemistry of the Rare Earths, Volume 3 - Non-Metallic Compounds, North-Holland Publishing, New York
[2] R.L. Mehan, M.R. Jackson, M.D. McConnell, J. Mat. Sci, **18**, 3195 (1983)
[3] D.J. Larkin, L.V. Interrante, J. Mater. Res., **5**, 2706 (1990)
[4] R.E. Kieschke, Somekh, T.W. Clyne, Acta Met. Mater., **39**, 427 (1991)
[5] R.E. Kieschke, Somekh, T.W. Clyne, Acta Met. Mater., **39**, 437 (1991)
[6] R.E. Kieschke, Somekh, T.W. Clyne, Acta Met. Mater., **39**, 445 (1991)
[7] A.K. Garg, L.C. DeJonghe, J. Mater. Res., **5**(1), 136 (1990)
[8] Her, Y-S., Matijevic, E., Wilcox, W.R., J. Mater. Res. **7**(8), 2269 (1992)
[9] W.H.R. Shaw , J.J. Bordeaux, J. Am. Chem. Soc. **77**, 4729 (1955).
[10] W.H.R. Shaw , J.J. Bordeaux, Anal. Chem. **27**, 138, (1955)
[11] I.A. Kemp, G. Kohnstam, J. Chem. Soc. 900-11 (1956)
[12] C.F. Baes and R.E. Mesmer, 1976, The Hydrolysis of Cations, John Wiley and Sons, New York
[13] Y-S Her., E. Matijevic, and W.R. Wilcox , J. Mater. Res., **78**, 2269 (1992)
[14] L. Domingues, R. Wilfong, L. Furlong, Bureau of Mines Report of Investigation, No. 6029, (1962)
[15] A. Bayman, P.K. Hansma, Surface Science, **125**, (1983), 613-623.
[16] Dupont Nonionic Hydrocarbon Surfactant Sales Literature
[17] M.E. Khavari, F. F. Lange, P. Smoth, D. B. Marshall, Mater. Res. Soc. Proc. Symp. Ultrastr. Proc., **121**(3), 617-621 (1988)
[18] L.A. Chick, L.R. Pederson, G.D. Maupin, J.L. Bates, L.E. Thomas, G.J. Exarhos, Materials Letters, **10**, 6-12, (1990).
[19] L.E. Scriven, Better Ceramics Through Chemistry III eds. C. J. Brinker, D.E. Clark, D.R. Ulrich, Mat. Res. Soc., Pittsburg Pa., 1988, pp. 717-729
[20] L.D. Landau, B.G. Levich, Acta Physiochim, U.R.S.S., **17** (1942) 42-54.

DEPOSITION OF Si-B-C MATERIALS FROM THE VAPOR PHASE FOR APPLICATIONS IN CERAMIC MATRIX COMPOSITES

S. Goujard* and L. Vandenbulcke
Laboratoire de Combustion et Systèmes Réactifs - CNRS, 45071 Orléans Cedex 2, France.

ABSTRACT

The codeposition from the vapor phase of silicon-boron-carbon ternary solids is carried out in a hot-wall reactor as a function of various parameters, including the initial gaseous mixture composed of methyltrichlorosilane, boron trichloride and hydrogen. The results are compared to thermodynamic predictions and the influence of different deposition kinetics of boron and silicon from their chloride precursors is pointed out. Experimental conditions far from equilibrium are shown to allow the deposition of solids which present both good thickness and chemical composition uniformity. Various solid compositions can be obtained as a function of the deposition parameters. These solids are employed with other materials like SiC to protect oxidizible components of ceramic matrix composites.

INTRODUCTION

Advanced ceramic fiber-reinforced composites have being developed for use in high temperature corrosive environments. However at present all their components do not present a sufficient oxidation resistance and a protection must be performed. External sealant materials can be used [1-4] but such surface coatings would only be one components of more complex systems. Improved oxidation resistant non-oxide matrix have also been developed which employ boron-containing materials associated to pure silicon carbide [5-6]. The combination of the Si-B-C elements in one material appeared promising for both type of protection if an accurate control of the composition could be obtained. A chemical vapor deposition (CVD) study of the Si-B-C ternary system was carried out with that aim and the potentiality of such a material was first tested as an external coating.

A thermodynamic approach was undertaken for the purpose of predicting the composition of the Si-B-C-H-Cl system at equilibrium, and to determine the

**Present address: Société Européenne de Propulsion, Les Cinq Chemins - Le Haillan BP 37, F-33165 Saint-Médard-en-Jalles, France.*

To the extent authorized under the laws of the United States of America, all copyright interests in this publication are the property of The American Ceramic Society. Any duplication, reproduction, or republication of this publication or any part thereof, without the express written consent of The American Ceramic Society or fee paid to the Copyright Clearance Center, is prohibited.

most interesting domain of the experimental parameters, especially the inlet composition. A relatively high temperature of 1400 K was chosen to minimize the amount of departure from equilibrium at the gas-solid interface. High temperatures are known to favor the control by mass transport which is the rate-limiting step, but a classical direct comparison of the global thermodynamic calculations and experiments will be reported using this approach. First the influence of mass transfer on a local thermochemical equilibrium was neglected at the gas-solid interface where the material balance is changed by the element segregation transport phenomena [7,8] or by the well-known reactant consumption and product formation at downstream positions inside the reactor, especially a hot-wall. The influence of fluid mechanical and chemical kinetics effects will be deduced from the experimental results and the deposition conditions will be extended to higher supersaturations, at lower temperature and higher mass flow rate, in order to obtain uniform coatings. The protection of an oxidizible part will be finally given as an example.

THERMODYNAMIC STUDY

The hypotheses necessary to carry out the thermodynamic calculations concern the choice of the solid phases, the choice of the thermodynamic data and the model used to describe the solid solutions. They were described in detail in a previous paper [9]. To take into consideration the "boron carbide" solid solution, which extend from about $B_{10}C$ to B_4C, the sublattice model developed by Hillert and Staffansson [10] and generalized by Sundman and Aagren [11] was used. The following solid phases were therefore considered: Si, B, C, SiC, SiB_3, SiB_6, SiB_{14} and B_xC. All the thermodynamic data were taken from JANAF [12,13], except those concerning the silicon borides and the "boron carbide" solid solution which required a prior critical assessment of the thermodynamic data [14]. The calculations were carried out by the "Melange" program [15].

Fig. 1 shows the deposition diagram of the three elements (Si-B-C) as a function of the initial concentrations of methyltrichlorosilane (CH_3SiCl_3 or MTS), BCl_3 and H_2 at T = 1400 K and P = 40 kPa (300 torrs). The nature of the deposits is changed through the following ratios:

$$\alpha = [H_2]_0/[MTS]_0 \quad \text{and } \beta = [MTS]_0/[BCl_3]_0$$

where $[X]_0$ represents the initial number of X mole. For values of α between 0.1 and 15, the thermodynamic calculations predict the appearance of free carbon in the deposits. For $\alpha > 15$, in the main part of the domain, two phases are obtained: SiC and the rhombohedral boron carbide phase which becomes richer in boron as α increases or, in other words, as the proportion of hydrogen in the gaseous mixture is higher. Boron carbide is formed in preference to the silicon borides in the main part of the domain.
The equilibrium yields of the different gaseous and solid species were also obtained. They allow especially to calculate the solid composition at equilibrium which will be compared to the experimental results.

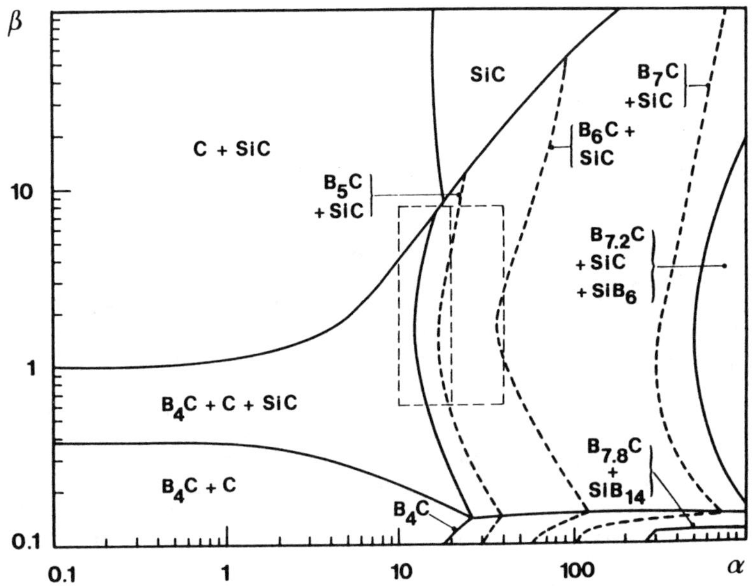

Fig. 1 : Deposition diagram as function of $\alpha = [H_2]_0/[MTS]_0$ and $\beta = [MTS]_0/[BCl_3]_0$ (T = 1400 K, P = 40 kPa).

For protection from oxidation, it is advisable to obtain deposits having no free carbon. Boron and silicon carbides are the most worthwhile phases because they should be deposited in a large range where the α and ß values are easy to control. From Fig. 1 the experimental study area can be chosen within the α = 10-40 and ß = 0.6-8 range.

EXPERIMENTAL TECHNIQUES

The deposition apparatus was described in detail previously [9]. The samples were hung inside a graphite susceptor which was heated by a high frequency generator regulated by a thermocouple fitted at the top of the susceptor. The substrate temperature was calibrated with a small thermocouple inserted in one substrate; it allows also to verify the temperature uniformity along the substrate (±5 K). The gas mixture was introduced by a nozzle from the bottom. The desired flow rates of H_2, MTS or BCl_3 were adjusted by mass flowmeters. The hydrogen was of 99.995% purity and boron trichloride and MTS of 99.9% purity. The substrates were initially fine-textured, prepolished graphite having an average density of 1.75, cleaned in alcohol in an ultrasonic device, and dried before weighing. Samples size was (18mm x 10mm x 2mm). Then various C/C, C/SiC and SiC/SiC composites were coated.

The thickness measurements were obtained on polished cross-sections with the help of an optical microscope. The chemical composition of the deposits was obtained by the γ prompt photon spectrometry method. The details of this method are given elsewhere [16]. X-ray diffraction was also used to attempt to determine the crystalline phases of some coatings.

COMPARISON BETWEEN THE THERMODYNAMIC CALCULATIONS AND THE EXPERIMENTS AT T = 1400 K.

To carry out this comparison, the theoretical and experimental chemical compositions have been reported as a function of the ß ratio = $[MTS]_0/[BCl_3]_0$ for a temperature of 1400 K, a pressure of 40 kPa, a mass flow rate of 0.320 g/min, and a fixed α ratio of 20 (the coatings were generally badly-crystallized and a comparison with the predicted phases cannot be done from X-ray diffraction results). The experimental chemical compositions were measured at the bottom of the samples, at a distance h=3 mm from the lower end. This position was chosen to minimize the reactant depletion and maintain the overall α and ß ratios as close as possible to the inlet ones. The curves in continuous lines represent the calculated compositions while the dashed lines, the experimental ones (Fig. 2). The experimental and the thermodynamic values vary in the same way, but the incorporation of boron into the deposits was always greater than the predictions of the thermodynamic calculations while the carbon incorporation was lower.

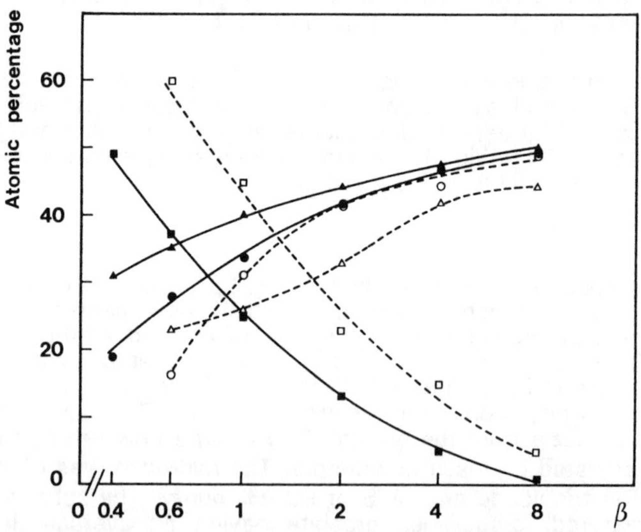

Fig. 2 : Comparison between the thermodynamic calculations and the experiments at α = 20, ß variable, T = 1400 K, P = 40 kPa, ——— thermodynamic, — — — experiment. ■ □ Boron, ● ○ Silicon, ▲ △ Carbon.

This discrepancy between the global equilibrium predictions and the experimental results may be due to the influence of fluid mechanical and chemical kinetics effects [9].

The influence of mass transport, consumption of reactants and product formation at downstream positions has been studied in the same deposition conditions as those reported in figure 2. The variations of the deposition rate and the solid composition are given in figure 3 as a function of the distance h along the samples. Neither the thickness nor the chemical composition were uniform for all the deposition conditions reported here. The decrease in total thickness was higher when the ß ratio increased, that is for conditions which led to the greatest deposition rate and obviously to the greatest reactant depletion. The deposit compositions were not uniform, with an especially significant decrease in the boron incorporation as a function of the position along the sample, in the gas flow direction.

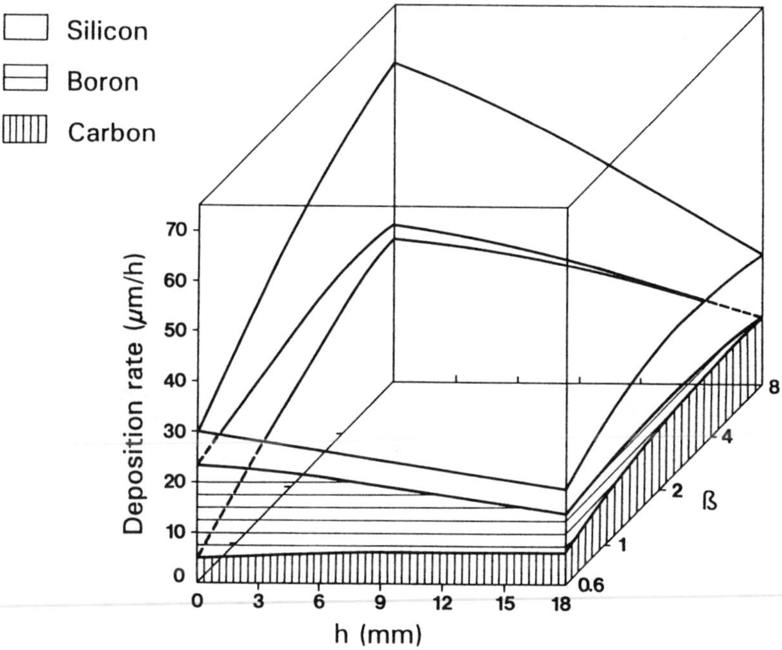

Fig. 3 : Thickness and chemical composition (in at%) uniformities as a function of ß (T = 1400 K, P = 40 kPa, α = 20, F = 0.320 g/min).

DEPOSITION IN CONDITIONS OF HIGHER SUPERSATURATION

It is well-known that the departure from equilibrium at the gas solid interface of a CVD system increases when the temperature decreases and the mass flow rate increases as a result of a deposition process which is first controlled by the mass transfer, then by a combination of mass transfer and chemical (reaction) kinetics and finally which is almost only controlled by chemical kinetics alone. When the supersaturation varies, it influences both the thickness and composition in a multicomponent system, hence their uniformity along the surface. This evolution was observed here for the codeposition of the three elements (Si-B-C).

Figure 4 presents, for an inlet composition α = 20 and ß = 1, the variations of the deposition rate and the solid composition as a function of the position on the sample for different temperatures (the mass flow rate was increased here to 0.6 g/min). At lower temperatures and higher mass flow rates, it is clear that the deposition is more controlled by the surface kinetics which is the slowest step and a uniform thickness and composition is favored. However the kinetic control may vary in such a ternary system as a function of the inlet composition.

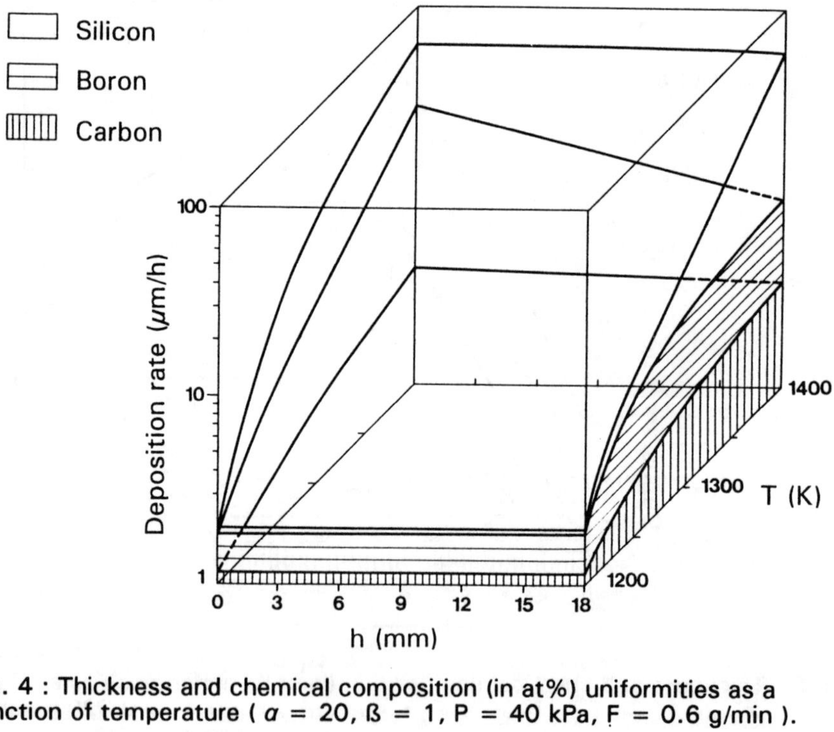

Fig. 4 : Thickness and chemical composition (in at%) uniformities as a function of temperature (α = 20, ß = 1, P = 40 kPa, F = 0.6 g/min).

For α = 20, T = 1200 K and a moderate mass flow rate of 0.32 g/min, the variations of the deposition rate and the composition along the samples are reported in figure 5 for different values of ß, as in figure 3 at 1400 K.

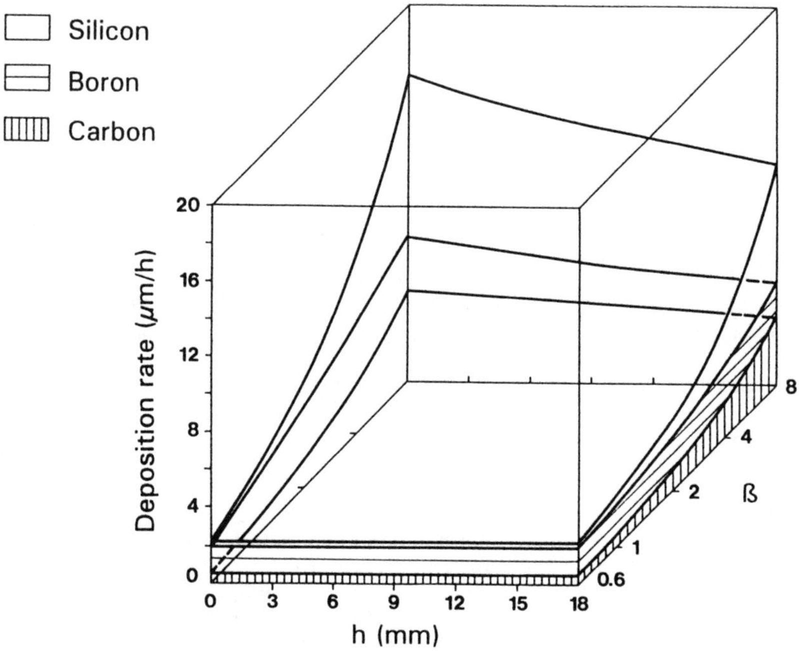

Fig. 5 : Thickness and chemical composition (in at%) uniformities as a function of ß (T = 1200K, P = 40 kPa, α = 20, F = 0.320 g/min).

When the temperature decreased from 1400 to 1200 K, the incorporation of boron and/or silicon increased while the atomic percentage of carbon decreased. This influence is well-known in carbide deposition systems where free element incorporation is kinetically-favored in comparison to carbide formation [7]. High supersaturation in CH_x species should therefore be employed to deposit C-rich coatings.

On the other hand, the chemical kinetics control, which increases at a lower temperature with a high boron or silicon content, leads to greater uniformity of composition. This control was particularly good when T = 1200 K and ß was low (0.6 to 1). Under these conditions the departure from equilibrium was high and the depletion in the gas phase and increase concentration of the gas products were reduced along the samples, in accordance with a lower deposition rate obtained for low ß values (even at 1400 K the results were better for low ß values). This kinetic limitation at low temperature and low ß values led, accordingly, to the best results for thickness uniformity as the deposition rate was reduced from about 30 μm/h at 1400 K to 3 μm/h at 1200 K when α = 20 and ß = 1.

It can be therefore concluded that under a low supersaturation in CH_x species produced by a relatively high α value of 20, which leads to a low carbon content in the solid, the kinetics limitation which is encountered at the lowest temperature of 1200 K depends on the ß value. The process is more controled, with the best uniformity for composition and thickness, when the ß value is low. When the ß value increases (for inlet concentration of MTS higher than BCl_3), the process is less kinetically controled by the silicon incorporation in accordance with a higher silicon deposition rate (at 1200 K the global deposition rate varied from 2.2 μm/h for boron-rich coatings deposited at ß = 0.6 to about 15 μm/h for silicon-rich coatings deposited at ß = 8). However the variations of α' ($\alpha' = [H_2]_0/[MTS]_0 + [BCl_3]_0$) with ß (Table I) should not be overlooked in interpreting these chemical kinetics and final uniformity differences.

Table I : Variations of α' as a function of ß for α = 20.

ß	0.6	1	2	4	8
α'	7.5	10	13.3	16	17.8

It may be noted that fairly uniform coatings were obtained here on small substrates at relatively high deposition rates that could not be employed in a large-scale reactor without leading to high inhomogeneity. The influence of the total pressure was not studied here, but it is well-known that lower pressures allow to increase the surface kinetic control of the deposition process. This point is especially interesting for depositing coating in industrial reactors. It is also important to extend the process to chemical vapor infiltration (CVI).

APPLICATIONS OF Si-B-C AS PROTECTIVE MATERIALS

In a first attempt, single-layers containing the three elements (Si-B-C) were tested in an oxidizing atmosphere to protect 2D C/SiC or 3D C/C composites provided by the "Société Européenne de Propulsion" (S.E.P.). Then, a solution based on a multilayer : SiC internal layer, (Si-B-C) intermediate layer and SiC external layer has been developed and tested up to temperatures of about 1600°C [17]. Figure 6 presents for example the results of cyclic oxidation tests carried out on 3D C/C at 1500°C (samples were put inside and taken out a furnace in about 30s for 6 cycles of one hour followed by 4 cycles of 6 hours). For this temperature of 1500°C, a single-layer Si-B-C coating with a low boron content (10 at%) was chosen. Better results were obtained with a multilayer coating which include an intermediate Si-B-C layer which contains more boron (40 at%) and an intrinsically more resistant pure SiC external layer. Such multilayer coatings can operate in a large range of temperature if the boron content is sufficiently high to allow sealing of the cracks by internal glass formation while retaining the intrinsic oxidation resistance of the external SiC layer.

The process for the manufacture of a refractory composite material with a Si-B-C matrix type has been patented by S.E.P. [18].

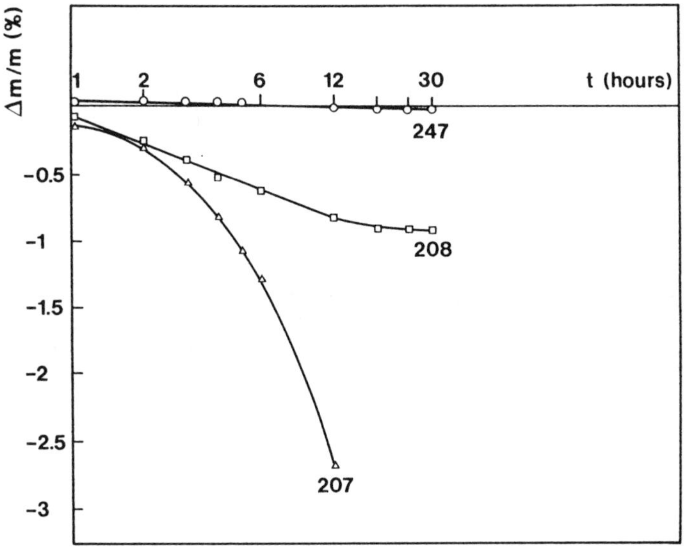

Fig. 6 : 3D C/C cyclic oxidation tests at T = 1500°C in air. (mass variations $\delta m/m$ in wt%).
208. [(Si(50%) B(10%) C(40%)] (180 μm)
247. SiC (180 μm) + [(Si(30 %) B(40 %) C(30 %)] (60 μm) + SiC (60 μm)
207. SiC (160 μm)
(Percentages expressed in at%).

CONCLUSION

The comparison between the thermodynamic and the experimental study at T = 1400 K showed the possibility for predicting the tendencies of the composition obtained at the bottom of the samples. However, a good correlation between the experimental results and a global thermodynamic approach can be fortuitous in a hot-wall reactor because classical fluid dynamic and chemical kinetics effects can change the solid composition as a result of the variations of the local gas phase composition.

It has been shown that a high supersaturation at the gas-solid interface leads to a good uniformity of the thickness and the composition of (Si-B-C) coatings. A higher supersaturation is obviously obtained at lower temperatures and higher mass flow rates, but its amount is also a function of the gaseous composition since different kinetic limitations can arise, especially here as a function of the relative amount of the boron and silicon precursors. The best control of both thickness and composition is clearly obtained at the lowest deposition rate, when boron-rich coatings are deposited at 1200 K and $[MTS]_0/[BCl_3]_0 \leq 1$.

This C.V.D. study made it possible to choose experimental conditions with high deposition rates on small substrates, even when the process control was not fully optimized, and to test the oxidation resistance of these coatings as either a (Si-B-C) monolayer or when associated with other compounds in multilayer coatings. More detailed results are published elsewhere [17].

ACKNOWLEDGMENTS :

The authors gratefully acknowledge the financial support of the "Société Européenne de Propulsion" (SEP) under contract N° 430 190. This work was part of the thesis of S. Goujard (Orléans, 1990) granted by the SEP and CNRS.

REFERENCES :

1. D.M. Shuford, "Composition and method for forming a protective coating on C/C substrates", US Patent N° 4,465,777, (Aug. 14, 1984).
2. R.A. Holzl,"Self protecting carbon bodies and method for making same", US Patent N° 4,515,860, (May 7, 1985).
3. G. Rousseau and N. Pastureau, "Matériau composite à matrice et fibres de renforcement carbonées et son procédé de fabrication", Fr Patent N° 8702496, (25 février 1987).
4. D.W. Mc Kee, Oxidation behavior and protection of C/C composites", *Carbon*, 25(4), 551-557, (1987).
5. T. Vasilos, "Oxidation resistant C/C composites", Fr Patent N° 8315411, (Sept. 28, 1983).
6. T. Strangman and R.J. Keiser, "Interstitially protected oxidation resistant C/C composite", US Patent N° 4,668,579 (May 26, 1987).
7. L. Vandenbulcke, "Mass transfer, equilibrium and kinetics in the CVD of a polycomponent system: application to boron - carbon", Proc. 7th Int. Conf. on C.V.D., T.O.Sedgwick and H. Lydtin Eds., The Electrochem. Soc., Pennington, N.J., (1979), 315-331.
8. D.E. Rosner, R. Nagarayan, S.A. Gokoglu, M. Kori, "Maximum effect of vapor phase chemical reactions on CVD-rates and deposition onset conditions in the absence of interfacial kinetic barriers" Proc. 10th Int. Conf. on C.V.D., G.W. Cullen, J.M. Blocher Jr. Eds., The Electrochem. Soc., Pennington, N.J., (1987), 61-80.
9. S. Goujard, L. Vandenbulcke, C. Bernard, G. Blondiaux and J.L. Debrun, "Thermodynamic and experimental study of the chemical vapor deposition in the silicon-boron-carbon system at 1400 K", *The Electrochem. Soc.*, 141-2, 452-461 (1994).
10. M. Hillert, L.I. Staffansson, "Regular solutions for stoichiometric phases and ionic melt solution model", *Acta Chem. Scand.*, 24, 3618-3626, (1970).
11. B. Sundman, J. Aagren, "A regular solution model for phases with several components and sublattices, suitable for computer applications", *J. Phys. Chem. Solids*, 42, 297-301, (1981).

12. Janaf Thermochemical Tables, 2[d] ed. Edited by D.R. Strull and H. Prophet, N.B.S. Plu., N° NBS 537, (1971). 13. Janaf Thermochemical Tables, 3rd ed, Edited by M.W. Chase,Jr., C.A. Davies, J.R. Downey,Jr., D.J. Frurip, R.A. McDonald and A.N. Syverud, N.B.S., M.D. 20899, (1986).
14. S. Goujard, L. Vandenbulcke, C. Bernard, "Thermodynamic study on the CVD in the Si-B-C-H-Cl system", *Calphad*, 18 (4), (1994) in print.
15. J.N. Barbier and C. Bernard, "Melange, versatile computer program for all kinds of complex equilibria", Proc. 15th Calphad Meeting, L. Kaufman ed., *Calphad*, 10 (3-4), 206-207, (1986).
16. G. Blondiaux, G. Ducouret, J.L. Debrun, S. Goujard and L. Vandenbulcke, "Development of a nuclear analytical technique for the analysis of ternary coatings (B+C+Si) and application to the optimization of a CVD process", *NIM-B*, B79, 521-523, (1993).
17. S. Goujard, L. Vandenbulcke and H. Tawil, "Oxidation tests of carbonaceous composite materials protected by Si-B-C CVD coatings", *Thin Solid Films*, in press.
18. S. Goujard, L. Vandenbulcke, J. Rey, J.L. Charvet and H. Tawil, "Process for the manufacture of a refractory composite material protected against corrosion" Fr Patent N° 9013323, (Sept. 26, 1990).

SURFACE MODIFICATION OF ZIRCONIA-ALUMINA COMPOSITE CERAMICS

Syamal K. Ghosh and Dilip K. Chatterjee
Eastman Kodak Company
Rochester, New York 14652-4347

ABSTRACT

Surface modification of tetragonal zirconia and α-alumina composite was accomplished by diffusion of MgO during sintering at 1500°C. α-Al_2O_3 powder, 2 to 60 wt%, was thoroughly mixed with 5 wt% yttria-alloyed zirconia powder, cold pressed and sintered in contact with fine MgO powder. In the diffusion zone, α-Al_2O_3 was transformed to spinel, $MgAl_2O_4$, and transformation of tetragonal zirconia to cubic phase depended primarily on alumina concentration. The modified surface was, therefore, a composite of spinel, tetragonal (t) and/or cubic zirconia as distinguished from the core, comprised of α-Al_2O_3 and t-ZrO_2. Tribological characteristics of the modified outer surface are also reported.

INTRODUCTION

Yttria stabilized, tetragonal zirconia polycrystals (Y-TZP) has high flexural strength and high fracture toughness. The improved fracture toughness is due to the stress-induced martensitic transformation from the metastable tetragonal to the stable monoclinic phases (1). Recently, it was reported that Y-TZP containing 20 wt% Al_2O_3 showed flexural strength in excess of 2 GPa (2) and fracture toughness of 15 $MPa{\cdot}m^{1/2}$ (3). The addition of fine alumina particles in a matrix of fine-grained Y-TZP is believed to serve two purposes: (1) an increase in elastic modulus which has a minimal effect on strength, and (2) act as grain boundary pinning points preventing grain growth of the tetragonal phase (t-ZrO_2); thus, lowering the probability of auto-transformation to monoclinic phase (4,5).

To the extent authorized under the laws of the United States of America, all copyright interests in this publication are the property of The American Ceramic Society. Any duplication, reproduction, or republication of this publication or any part thereof, without the express written consent of The American Ceramic Society or fee paid to the Copyright Clearance Center, is prohibited.

In this study, we modified the surfaces of Y-TZP - alumina composites to improve their tribological properties. The specific purpose of this paper is to report the process developed to alter the crystalline structure of the zirconia-alumina composite ceramic surfaces by in-situ doping with MgO during sintering, similar to what we reported for Y-TZP (6). The high fracture toughness of zirconia-alumina composites was maintained, and at the same time, the surface characteristics were modified. This process, analogous to diffusion controlled surface hardening of steel, transformed zirconia-alumina composite surfaces to $MgAl_2O_4$, a spinel crystalline phase. This transformation is believed to have occurred by a chemical reaction between MgO and Al_2O_3 species at relatively high temperatures. We reported in our earlier observation (6) that magnesia diffusion during sintering transforms the tetragonal phase of Y-TZP to a cubic phase which makes the surface more resistant to wear and abrasion than the tetragonal phase. MgO dopant which causes this phase transformation to occur has a partitioning effect in Y-TZP-alumina composites in that the cubic zirconia phase is formed only when alumina concentration is very low.

In-situ surface modification of composite ceramics allowed tailoring surfaces to specific applications. Sliding and abrasion characteristics of the modified zirconia-alumina composite surfaces with different crystal structures are also presented.

EXPERIMENTAL

Material: Alloy Composition

Yttria-alloyed zirconia powders (containing 5 wt% yttria) mixed with 10, 20, 30, 40 and 60 wt% high purity alumina were obtained from Tosoh Corporation. Composites containing 2 and 5 wt% alumina were made in our laboratory, as they were not available commercially.

Sintering

The zirconia-alumina composite pellets were prepared by uniaxially cold pressing the powder at a pressure of about 100 - 110 MPa, followed by sintering at 1500°C for 2 h. The heating and cooling cycles were carefully controlled as reported earlier (7). In addition, the conventional sintering operation was modified; wherein, the "green" pellets, after pressing, were placed in contact with fine magnesia powder and subjected to the sintering schedule reported earlier (8). Glancing angle x-ray diffraction revealed the presence of a surface cubic zirconia and spinel layer in these samples; whereas, the bulk of the material had a composite t-ZrO_2 and α-Al_2O_3 crystal structures.

X-ray Diffraction

Both coupled (θ-2θ) and glancing angle diffraction work was carried out using a Rigaku (RU300) diffractometer with CuK_α radiation. In glancing angle x-ray diffraction, the x-rays were made incident at a glancing angle to the sample surface for enhanced surface sensitivity. For this work, the typical glancing angle, θ, was varied between 0.5° and 2.0°. Sampling depth for 1° glancing angle is estimated to be about 570 nm. Identification of phases was mainly based on the following: (i) the presence of diffraction peaks at 2θ = 28.2° and 31.5° for ($\bar{1}11$) and (111) reflections from the monoclinic zirconia phase, respectively; (ii) the presence of a diffraction peak at 2θ = 35.2° due to the (200) reflection from the cubic zirconia phase; (iii) splitting of the diffraction peak at 2θ = 34.7°-35.2° due to the (002) and (200) reflections from the tetragonal zirconia phase; (iv) the presence of diffraction peaks at 2θ = 31.27° and 36.85° due to (220) and (311) reflections from cubic spinel phase, respectively; and (v) the presence of diffraction peaks at 2θ = 25.58°, 35.14°, 37.79°, 43.36° and 52.55° due to (012), (104), (110), (113) and (024) reflections, respectively of α-Al_2O_3 (9).

Microhardness

Microhardness measurements were carried out using a Tukon micro-hardness tester, Model 300 FM/DF. All measurements were made on sintered flat surfaces using Knoop indenter for reproducible results. The indentation load for all tests was 500 g. Samples containing 5 and 20 wt% Al_2O_3 were used only.

Wear and Abrasion

Ceramic-to-ceramic wear tests were carried out using a reciprocating wear tester as shown in Figure 1. Flat samples (counterface) were placed on a reciprocating platform and subjected to wear by a rider ball, nominally 12 mm diameter, mounted on a weighted arm. The riders (balls) used in this study were made from Y-TZP containing 5 wt% yttria and 99.9% Al_2O_3. The counterfaces (flat samples) were made from ZrO_2-20 wt% Al_2O_3 composite ceramics only. The normal load on the ball was 500 g. For all wear tests, the reciprocating frequency was 60 cycles min^{-1}, the stroke length was 25 mm and the test duration was 2 h. All tests were done in triplicate to obtain a statistical average. Sliding wear volumes for the rider and the counterface were computed from the profilometric traces of the wear scar.

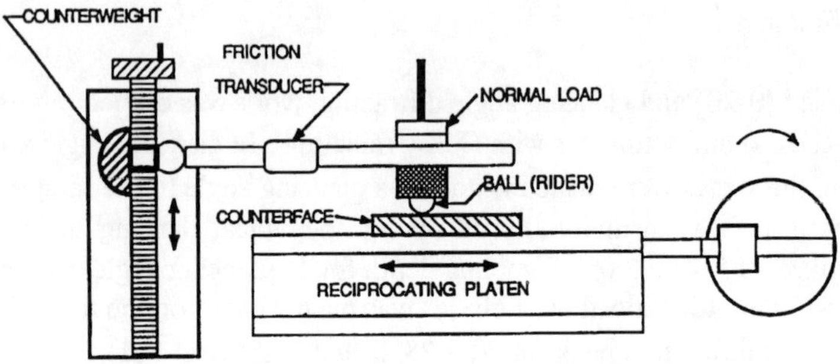

Figure 1. Schematic diagram of the sliding wear tester

Abrasion tests were carried out using a loop abrasion tester as shown in Figure 2. This tester was built to simulate abrasive wear that is encountered in web handling equipment. An endless running abrasive belt was run against the flat surfaces of ceramic samples. The abrasive particles on the belt were silica, SiO_2, average particle size being 24 μm. The belt speed, normal load and test duration for each test were 0.5 m s^{-1}, 200 g and 8 h, respectively. These tests were also done in triplicates. Abrasive wear volumes were computed similarly from profilometric trace measurements.

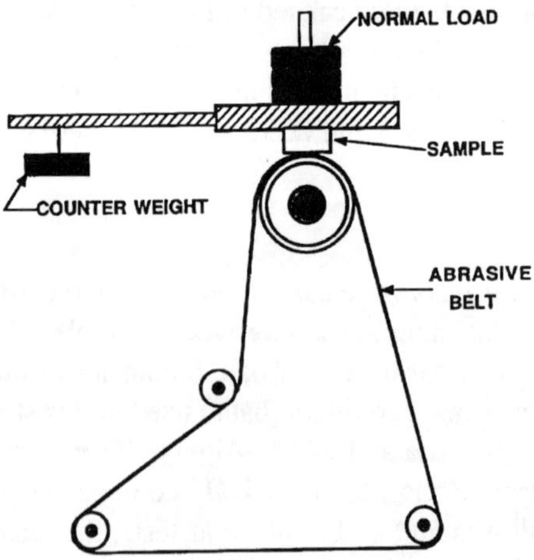

Figure 2. Schematic diagram of the loop abrasion tester

RESULTS AND DISCUSSIONS

Surface Modification in Zirconia-Alumina Composites

The bulk crystalline structure of zirconia in zirconia-alumina composites is tetragonal. Surfaces in contact with MgO powders were modified. Two distinct characteristic features were observed in the modified surfaces. For alumina content greater than 5 wt%, it transformed to spinel, $MgAl_2O_4$, and t-ZrO_2 remained unchanged in the modified surfaces; a trace amount of monoclinic zirconia was also present. However, when alumina content was reduced to less than 5 wt%, not only Al_2O_3 transformed to spinel but also t-ZrO_2 transformed to cubic phase. Transformation to spinel is believed to have occurred by chemical reaction between MgO and Al_2O_3 species at high temperatures as thermodynamically favored by the large negative free energy of formation (ΔG = -523 kcal/mole) of spinel. We reported in our earlier observation (6) that MgO diffusion during sintering transforms the tetragonal phase of Y-TZP at the surface to cubic phase and that renders the surface more wear and abrasion resistant. In zirconia-alumina composites, phase transformation of zirconia from tetragonal to cubic phase occurs only when alumina concentration is below 5 wt%. MgO dopant, which causes these phase transformations, has a partitioning effect in Y-TZP-alumina composites. It is believed that chemical reaction between MgO and Al_2O_3 takes precedence over phase transformation of zirconia. t-ZrO_2 on the modified surfaces transforms to cubic phase only, if alumina concentration is below 5 wt% because excess MgO is available after spinel formation.

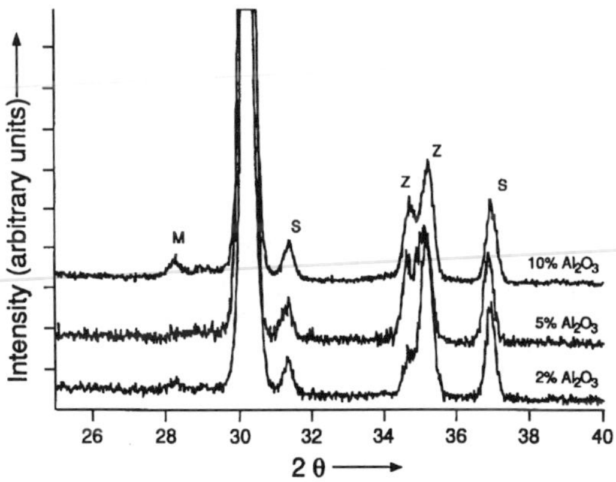

Figure 3. Typical glancing angle x-ray diffraction patterns on modified zirconia-alumina composite surfaces

Typical glancing angle x-ray diffraction patterns taken at 0.5° on modified composite surfaces containing 2, 5 and 10 weight% alumina are shown in Figure 3. A typical x-ray diffraction pattern taken on unmodified surfaces is shown in Figure 4, which shows the presence of α-Al_2O_3 and t-ZrO_2 phases. Diffraction peaks due to different phases are identified in Figures 3 and 4 as follows: cubic and tetragonal zirconia as Z, monoclinic as M, spinel as S, and α-Al_2O_3 as A.

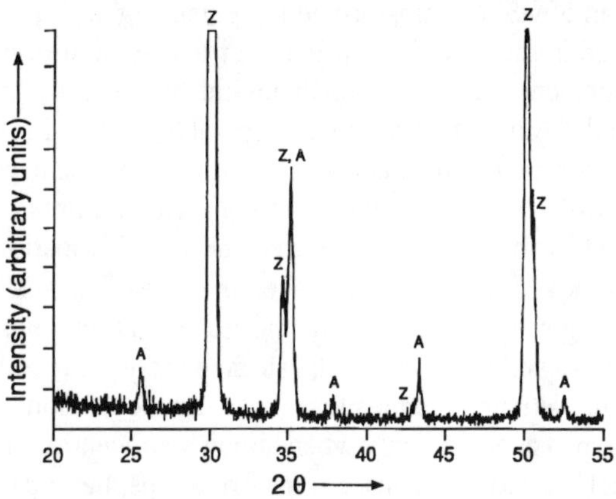

Figure 4. Typical glancing angle x-ray diffraction pattern on as-sintered zirconia-alumina composite surface

Surface Hardness

Surface hardness measurements of zirconia-2% alumina composites also verified the above observations. Average hardness of the modified surface, consisting of cubic zirconia and spinel phase, was 2085 KHN whereas that of the unmodified surface, consisting of t-ZrO_2 and α-Al_2O_3, was 1593 KHN. The modified surface was significantly harder. The hard cubic phase contributed to the increase in hardness. On the other hand, there was no significant difference in surface hardness between the modified and unmodified surfaces for zirconia-20 wt% alumina composites, being 2104 and 2015 KHN, respectively. The increase in hardness from that of zirconia-2 wt% alumina composite is primarily due to the presence of higher volume fraction of hard α-alumina particles.

The volume losses for the riders (Y-TZP and alumina) and the counterfaces (zirconia-20 wt% alumina) for various tribocouples in sliding wear tests are shown in Figure 5. These results reveal that the wear resistance is least for counterfaces when the rider is alumina. This is significantly different when the rider is Y-TZP. The modified surface (t-ZrO_2 and spinel) is more wear resistant than the unmodified surface. As reported earlier, wear of the riders does not follow a specific pattern (6). The modified and unmodified surfaces have identical hardness and that may explain why their tribological characteristics are similar. An order of magnitude greater volume loss of the unmodified counterface occurred when the rider is Y-TZP. From these experimental results, it is apparent that the wear mechanism for a given tribocouple is determined by the boundary conditions of the system. Sliding wear is a form of adhesive wear that occurs when two smooth bodies are slid over each other, and fragments are pulled off one surface to adhere to the other. Later, these fragments may come off the surface on which they adhere and be transferred back to the original surface, or else form loose wear particles. Adhesive wear arises from the strong adhesive forces set up when atoms of different species come into intimate contact (10). Appreciable wear of either or both surfaces is expected if the adhesive force between the tribocouple is significantly higher than the cohesive forces among atoms in each mating surface. We do not have enough information on the cohesive forces among atoms at the various crystallographic surfaces studied in this work. Owing to the complex nature of adhesive wear, it is difficult to offer any definitive explanation of our experimental results at this point in time.

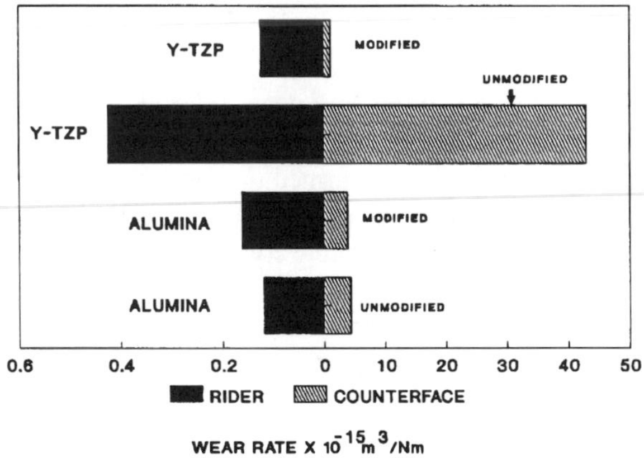

Figure 5. Sliding wear results from various tribocouples

Abrasive wear is a form of wear that occurs when a rough and hard surface, or a soft surface containing hard particles, slides on a softer surface and ploughs a series of grooves in it (10). The material from the grooves is displaced in the form of wear particles that are generally loose. The results of the abrasion tests of zirconia-alumina composites against SiO_2 are summarized in Figure 6. It is clear that the modified, as well as unmodified, composite surfaces have similar abrasion resistance as contrasted to what was observed in sliding wear. Since the surface hardness of both surfaces was practically identical, it was not surprising to observe that their abrasion properties were also very similar.

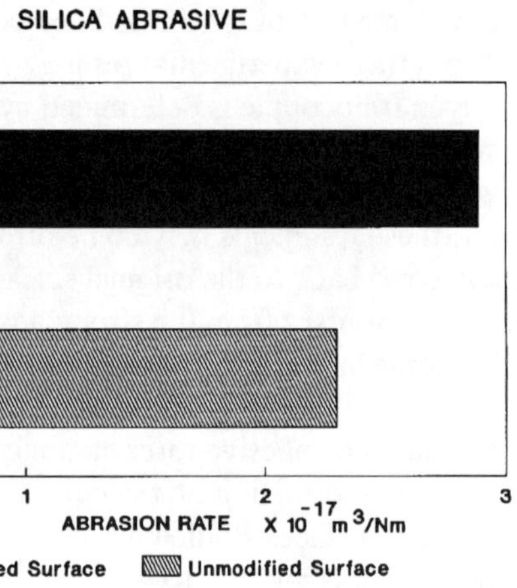

Figure 6. Abrasive wear results of zirconia-alumina composites

CONCLUSION

In this study, we have shown that the crystal structure of the outer surfaces of zirconia-alumina composites can be modified by in-situ doping with MgO during sintering. X-ray diffraction studies confirmed the formation of spinel along with cubic and/or t-ZrO_2 phases depending on alumina concentration in the composites. The t-ZrO_2 phase was transformed to cubic crystal structure only when the alumina content was less than 5% and that is believed to be due to the partitioning effect of MgO. Tribological properties were not significantly affected by the surface modification as contrasted to what we observed in Y-TZP.

ACKNOWLEDGMENTS

The authors wish to thank Mark Kohler for wear and abrasion experiments, and Thomas Blanton for x-ray diffraction work.

REFERENCES

1. T. K. Gupta, F. F. Lange, and J. H. Beshtold, "Effect of Stress-induced Phase Transformation on the Properties of Polycrystalline Zirconia Containing Metastable Tetragonal Phase," J. Mater. Sci., 13, 1467-70 (1978).
2. K. Tsukuma, K. Ueda, and M. Shimada, "Strength and Fracture Toughness of Isostatically Hot-Pressed Composites of Al_2O_3-Partially Stabilized Zirconia," J. Am. Ceram. Soc., 68 [1] C-4 - C-5 (1985).
3. T. S. Yen and J. K. Gao, "The Role of ZrO_2 in TZP and ZTA," p. 581, in Advances in Ceramics, 24B: Science and Technology of Zirconia III, 1st Edited by S. Somiya, N. Yamamoto, and H. Yangida, American Ceramic Society, Westerville, Ohio, 1988.
4. R. Stevens, "Fabrication of Toughened Ceramic - Alumina Toughened Zirconia," pp.24-26, in Zirconia and Zirconia Ceramics, Publication No. 113, Magnesium Elektron Inc., Flemington, NJ, 1986.
5. H. Tsubakino, R. Nozato, and M. Hamamoto, "Effect of Alumina Addition on the Tetragonal-to-Monoclinic Phase Transformation in Zirconia-3 mol% Yttria," J. Am. Ceram. Soc., 74 (2), 440-43 (1991).
6. D. Chatterjee, D. Majumdar, and S. Ghosh, "Surface Modification and Wear Characteristics of Yttria-doped Zirconia," Surf. and Coatings Tech., 54/55, 516-520 (1992).
7. S. K. Ghosh, D. K. Chatterjee, D. R. Koziol, and D. Majumdar, "Net-Shape Fabrication of Y-TZP Ceramic Through a Statistically Designed Experiment," Ceram. Eng. Sci. Proc., 14[11-12], 80-86 (1993).
8. D. K. Chatterjee, D. Majumdar, S. K. Ghosh, and J. M. Mir, "Ceramic Articles and Methods for Preparing Ceramic Articles and for Sintering," Patent # 5 290 332, USA, 1994.
9. JCPDS Diffraction File, Nos. 14-534, 27-997, 21-1152, 10-173, and 36-420, International Center for Diffraction Data, Swarthmore, PA, 1991.
10. E. Rabinowicz, "Friction and Wear of Materials," Wiley, New York, 1965.

COMBUSTION SYNTHESIS OF $MoSi_2$ - SiC OXIDATION RESISTANT COATINGS

K.A. Monroe, S. Govindarajan, J.J. Moore, B. Mishra, and D.L. Olson
Department of Metallurgical and Materials Engineering
Colorado School of Mines
Golden, CO 80401

ABSTRACT

Several high-temperature, oxidation resistant materials were investigated as possible coatings for molybdenum furnace elements. Molybdenum disilicide ($MoSi_2$) composites are potential candidates for these high temperature coating applications. Composite materials were selected based on several factors, including oxidation resistance, coefficients of thermal expansion, compatibility at the coating/substrate interface, thermodynamic stability over a large temperature range and adhesion. The synthesis of dense $MoSi_2$-SiC composites through simultaneous combustion and hot-pressing is currently being investigated. Functionally graded materials (FGM's) are being studied to overcome some of the property differences between a coating and a substrate. Oxidation testing was used to determine the protective qualities of the FGM's.

INTRODUCTION

In recent years, intermetallics have been studied because of their potential use for high temperature structural applications. Amongst these, molybdenum disilicide has emerged as one of the leading candidates. The excellent resistance to high temperature oxidation of $MoSi_2$ was established by Watts [1] in 1906 and it was first used as a high temperature corrosion-resistant coating material by Hoenigschmid [2]. The subsequent history of the development of $MoSi_2$ and $MoSi_2$-based composites has been the subject of an excellent review by Vasudevan and Petrovic [3]. The potential for using $MoSi_2$ and $MoSi_2$-based

To the extent authorized under the laws of the United States of America, all copyright interests in this publication are the property of The American Ceramic Society. Any duplication, reproduction, or republication of this publication or any part thereof, without the express written consent of The American Ceramic Society or fee paid to the Copyright Clearance Center, is prohibited.

composites as high temperature coating materials was the subject of a detailed review by Schlichting [4].

The primary factors that have contributed to the choice of $MoSi_2$ as a matrix material for high temperature structural composites or coatings are: (1) a high melting point of around 2300K; (2) low density (6.24g/cc) when compared with nickel-based superalloys; (3) superior oxidation resistance; (4) high temperature strength; and (5) high thermal and electrical conductivity. The excellent oxidation resistance of $MoSi_2$ is due to the formation of a protective layer of silica. This has resulted in the use of $MoSi_2$ as high temperature furnace heating elements, for use in air at temperatures of up to 2000K.

However, $MoSi_2$ suffers from a serious limitation for high temperature structural applications - i.e. room temperature brittleness ($MoSi_2$ has a ductile to brittle transition temperature (DBTT) of around 1273K). The low toughness at room temperature prevents the use of monolithic $MoSi_2$. for structural applications involving stresses at room temperature. In addition, $MoSi_2$ has poor strength and creep resistance and can suffer the phenomenon of "pesting"[5-7].

During the last two decades, there has been a strong research effort directed towards the development of $MoSi_2$-based composites [3,4,8-13]. The majority of studies have focused on the improvement of mechanical properties, and/or oxidation resistance, by compositing $MoSi_2$ with ZrO_2, TiB_2, SiC, Al_2O_3, TiC, MoB, Mo_2B_5 and Nb [3,12,13]. Amongst these composites, $MoSi_2$-SiC appears to offer the best combination of oxidation resistance and superior mechanical properties while the lower CTE of SiC can be used to decrease the overall expansion of the system.

The significant CTE mismatch between Mo (8.4×10^{-6} K^{-1} at 1800K)and the $MoSi_2$ (10.0×10^{-6} K^{-1} at 1800K) coating causes cracking and spalling of the coating during high temperature cyclic operation. A functionally graded material (FGM) coating can be designed to give a smooth CTE transition from the Mo substrate to the $MoSi_2$ coating by designing layers in the coating with increasing amounts of $MoSi_2$ to SiC.

The synthesis of $MoSi_2$ has been examined by a number of researchers [14-17]. Bhattacharya [15] developed a model, simulating the synthesis of $MoSi_2$ and $MoSi_2$ + SiC by SHS. In agreement with the model, the steady state combustion velocity was observed to decrease with increasing SiC content, eventually leading to possible extinction at SiC levels above 45 volume percent.

Based on these considerations, a program for assessing different processing routes and the effects of different processing parameters on the production of $MoSi_2$, $MoSi_2$-SiC composites and functionally-graded composite materials was initiated. In order to achieve a dense product, hot pressing has been used in conjunction with simultaneous combustion synthesis. The preliminary results are discussed below.

EXPERIMENTAL PROCEDURE

Molybdenum (<10μm, 99.5%), silicon (<7μm, 99.9%) and lampblack carbon (<0.09μm, 99.6%) powders were used to make the $MoSi_2$ and $MoSi_2$-SiC composites. The green density of the samples varied from 53% to 63%, depending on the applied load. In the case of the functionally-graded material (FGM) samples, individual layers of varying stoichiometry were mixed to provide varying ratios of $MoSi_2$:SiC and manually stacked, one above the other, prior to compaction.

A combination of simultaneous combustion synthesis and hot pressing was carried out in the system schematically presented in Figure 1. In simultaneous combustion the whole sample is uniformly heated to the ignition temperature in an inert atmosphere. On detection of the reaction exotherm, the sample was pressed under an applied pressure of up to 4680 psi (32.3 MPa).

The density of the reacted samples was determined by Archimedes Principle (i.e. sample weighed in air, saturated with water, and saturated and immersed in water). Scanning electron microscopy (SEM) was used to observe the microstructure of each material. Energy dispersive spectroscopy (EDS) and X-ray diffraction (XRD) were used to verify the phases present in the reacted sample.

Oxidation testing was performed in air at 1873K for 10, 50, 100 and 200 hours. The samples were weighed before and after oxidation to determine the weight change, and examined using optical and scanning electron microscopy.

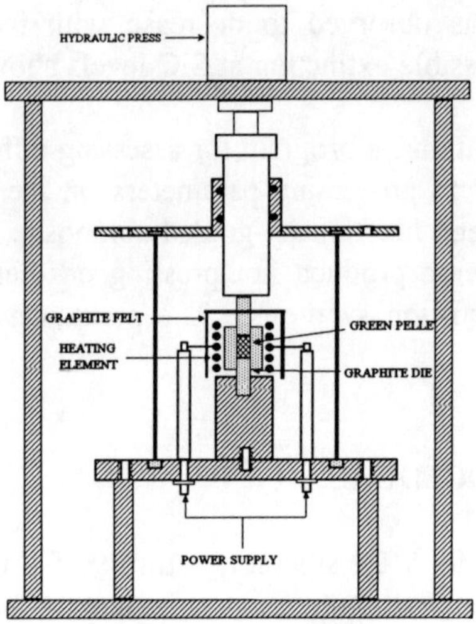

Figure 1. Setup for the simultaneous combustion and hot pressing chamber.

RESULTS AND DISCUSSION

Simultaneous Combustion (SC) + Hot Pressing (HP) of $MoSi_2$ + X SiC in Argon

Four reactions were studied in the synthesis of $MoSi_2$ + X SiC composites:

$$Mo + 2\ Si + X\ SiC \rightarrow MoSi_2 + X\ SiC \quad (1)$$
$$Mo + (2+X)\ Si + X\ C \rightarrow MoSi_2 + X\ SiC \quad (2)$$
$$1/2\ Mo_2C + 5/2\ Si + X/2\ SiC \rightarrow MoSi_2 + (1+X)/2\ SiC \quad (3)$$
$$1/2\ Mo_2C + (4+X)/2\ Si + (X-1)/2\ C \rightarrow MoSi_2 + X/2\ SiC \quad (4)$$

The most exothermic reactions are (2) and (4), with both reactions fully reacting in the simultaneous combustion and pressing mode when X was less than 1. A pressure of 32.3 MPa was applied on observing the exotherm and reapplied once a minute for 5 minutes, then held for 5 minutes, at 1873K. Reactions (1) and (3) contained SiC as a diluent and were incompletely reacted, with $MoSi_2$, SiC, and varying amounts of Si and C observed in the X-ray diffraction data. Examination of the photomicrographs (Figure 2 (a,b)) and XRD analysis of the products from reaction (2) show both the center (a) and surface (b) to be fully reacted for values of $X \leq 0.5$. The average density was determined to be 89% of the theoretical density, but it is evident from the photomicrographs that the surface was very porous while the center was extremely dense.

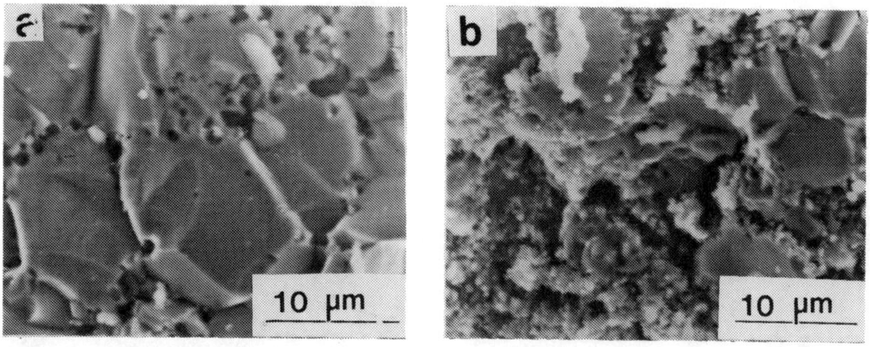

Figure 2. Center (a) and surface (b) regions of the reaction Mo + 2.5Si + 0.5C→ $MoSi_2$ + 0.5 SiC synthesized by SC and HP in argon.

Synthesis of FGM's by Simultaneous Combustion (SC) and Hot Pressing(HP).

The following FGM's were synthesized through the SC and HP route:

* $MoSi_2$ + x SiC - 6 layers with x varying from 0 to 2
* $MoSi_2$ + x SiC - 3 layers with x varying from 0 to 0.5

The ignition temperature for all the FGM's was close to the melting point of silicon (1685K). The strongest exotherm (i.e. a temperature rise of approximately 280K) was observed for the 3-layer FGM with a maximum value of 0.5 moles SiC. This is due to the reaction enthalpy for the formation of $MoSi_2$ being significantly higher than for SiC.

SEM analysis of the FGM's showed an interesting variation in morphology in the different layers of the sample. For comparison, the SEM photomicrographs for the individual layers of the FGM of $MoSi_2$ + x SiC, with x = 0 to 1.934, are shown in Figure 3. The composition where x = 1.934 was chosen because a ratio of 1 mole $MoSi_2$ to 1.934 moles SiC has a CTE similar to Mo when calculated by the rule of mixtures.

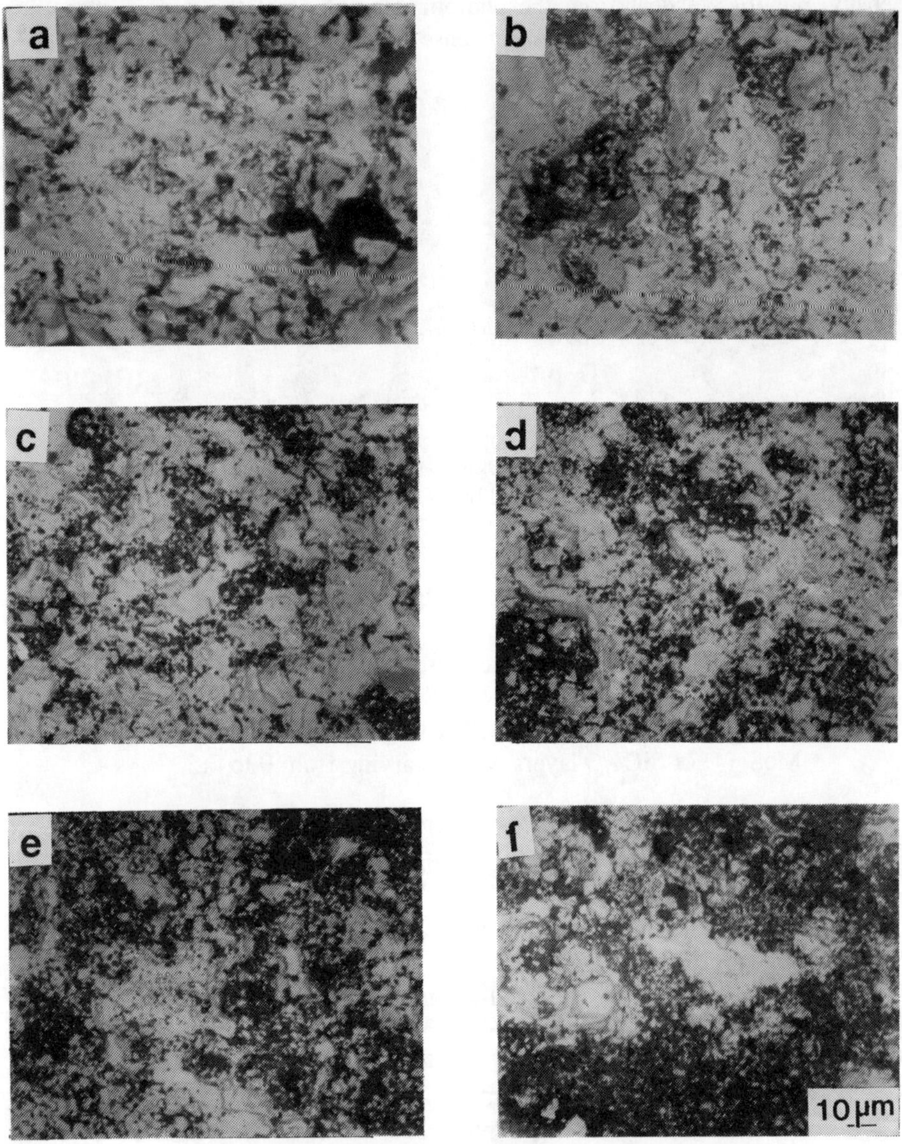

Figure 3. Backscattered electron imaging (BEI) photomicrographs of individual layers of FGM (vol% $MoSi_2$ to vol% SiC). (a) 100% $MoSi_2$, (b) 90/10, (c) 80/20, (d) 70/30, (e) 60/40, (f) 50/50.

Oxidation of $MoSi_2$ and $MoSi_2$-SiC Composites

The need for a functionally graded material is demonstrated in Figure 4 in that thermal cracking has occurred in the $MoSi_2$ coating initiating at the Mo-$MoSi_2$ interface. The thermal expansion mismatch is confirmed by the severe cracking throughout the coating.

Figure 4. $MoSi_2$ coating on molybdenum substrate showing cracking due to CTE mismatch..

The weight change as a function of time at 1873K for $MoSi_2$, 90% $MoSi_2$:10% SiC and 50% $MoSi_2$:50% SiC has been plotted in Figure 5. At longer times, the weight change of $MoSi_2$ remains relatively constant once the SiO_2 scale layer has formed. The 90% $MoSi_2$:10% SiC and 50% $MoSi_2$:50% SiC materials have considerably higher weight changes, with the latter composition having the greatest increase.

The photomicrographs of the FGM after oxidation for 10 hours are shown in Figure 6 and can be compared with the same sample, before oxidation, shown in Figure 3. The amount of Mo_5Si_3 in the oxidized sample has increased near the edge of the $MoSi_2$ due to the silicon diffusion to the surface to form the SiO_2 layer, hence the depletion of $MoSi_2$ to Mo_5Si_3. This is verified by EDS and is plotted in Figure 7 as the atomic percent of molybdenum and silicon as a function of distance in the sample.

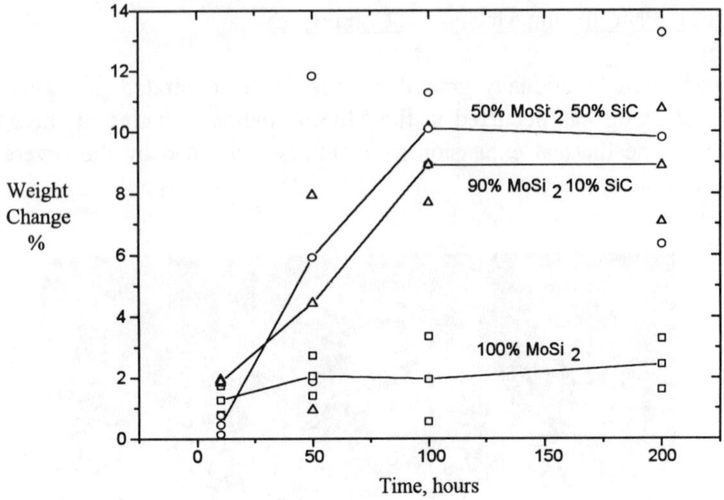

Figure 5. Weight change as a function of time for oxidation at 1873K.

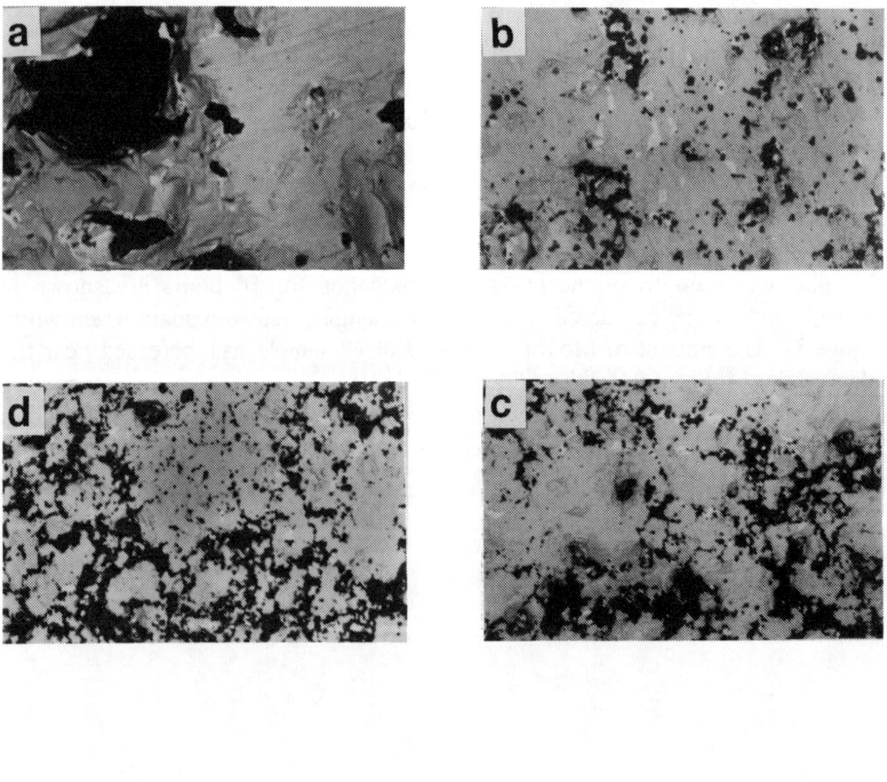

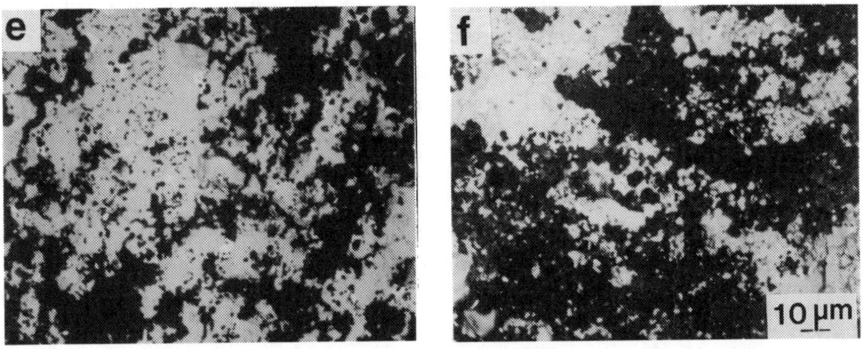

Figure 6. BEI photomicrographs of individual layers of FGM, oxidized for 10 hours (vol% $MoSi_2$ to vol% SiC). (a) 100% $MoSi_2$, (b) 90/10, (c) 80/20, (d) 70/30, (e) 60/40, (f) 50/50.

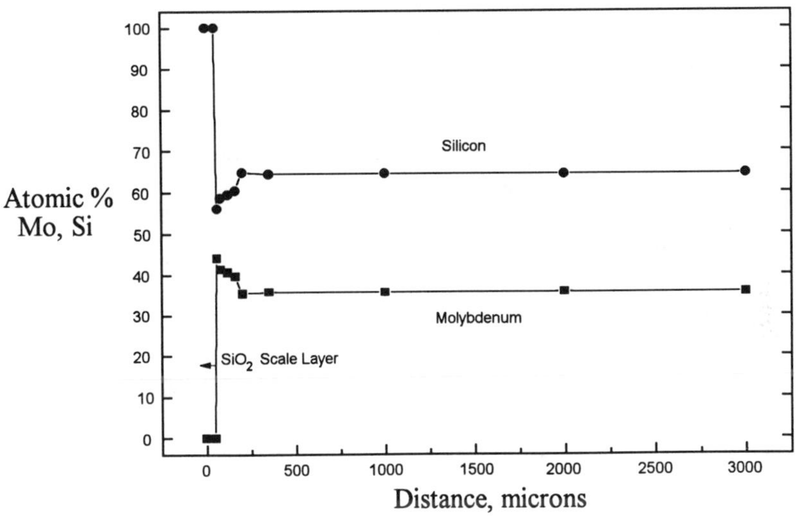

Figure 7. Atomic percent Mo, Si as a function of depth into $MoSi_2$ layer.

Discussion

The enthalpy-temperature diagram for the synthesis of $MoSi_2$ + X SiC is shown in Figure 8. The SiC additives (X SiC) to the SHS reactants act as "diluents" and reduce the adiabatic temperature. The resulting microstructure of the product phases has been shown to be controlled by T_c, in that lowering T_c will refine the particle size of the phases and, therefore, affects the mechanical properties [24] and protective coating capabilities. Examination of the enthalpy-temperature data and the criteria for a reaction to be self-sustaining suggest that the formation of $MoSi_2$ by SHS is possible, as shown experimentally. However, addition of X SiC reduces the exothermicity of the reaction and at a certain level, the reaction will quench out. In the case of FGM's, the presence of 100% $MoSi_2$ layers in the top of the samples provides sufficient heat to allow the reaction to propagate past the layers where the reaction should have theoretically been quenched out.

The molybdenum metal which was covered with $MoSi_2$ (prepared by simultaneous combustion and hot pressing) exhibited severe cracking during isothermal oxidation, thus proving the need for a coating with similar thermal expansion to the substrate. The FGM which was tested in an oxidizing environment was affected only at the surface of the sample. Silicon diffusion to the surface and the formation of the SiO_2 layer resulted in protection of the inside of the sample. The functionally graded material approach appears to offer excellent oxidation protection while providing a coating which is less likely to crack due to the CTE gradient built through the coating.

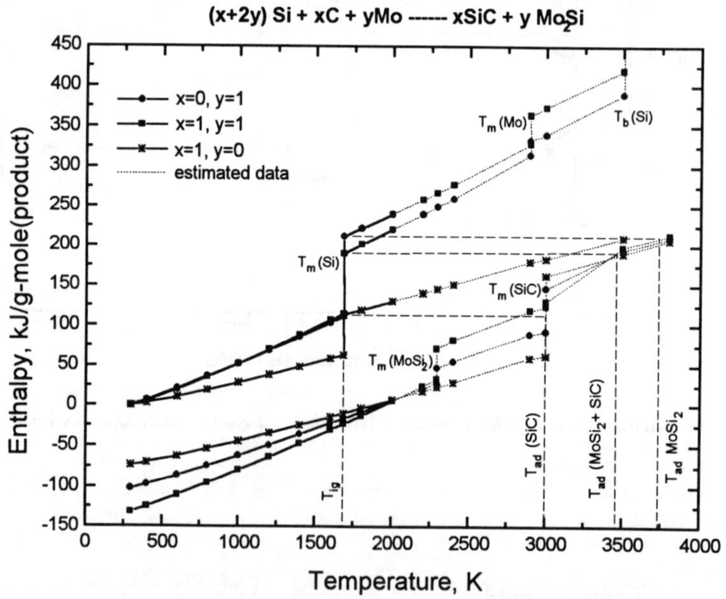

Figure 8. Enthalpy-Temperature diagram for the x SiC + y $MoSi_2$ system.

Future investigators may want to consider optimizing the adhesion of the coating to the substrate while preventing diffusion of silicon from the coating into the substrate with a barrier layer, such as Ta_5Si_3 or Ti_5Si_3 [4,12,13]. Finite element analysis may also be considered for designing the coating layer compositions and thicknesses.

CONCLUSIONS

(1) $MoSi_2$ + X SiC was produced by an affordable, one step simultaneous combustion and hot pressing process. The most exothermic reactions appear to be the following:

$$Mo + (2+X)\ Si + X\ C \rightarrow MoSi_2 + X\ SiC$$
$$1/2\ Mo_2C + (4+X)/2\ Si + (X-1)/2\ C \rightarrow MoSi_2 + X/2\ SiC$$

Photomicrographs of the reacted samples showed uneven porosity throughout the sample. Product phases were primarily $MoSi_2$ and SiC as found by X-ray diffraction.

(2) The 6 layer FGM, with x = 0 to 1.934 moles SiC, produced by SC and HP exhibited very low exothermicity (temperature rise of around 80K above the ignition temperature). However, the 3 layer FGM, with x > 0.5 exhibited a much higher exothermicity (temperature rise of around 280K).

(3) Oxidation of the $MoSi_2$ and $MoSi_2$-SiC composites resulted in a SiO_2 oxidation layer followed by a zone of depleted silicon, and thus, the formation of Mo_5Si_3 in a matrix of $MoSi_2$ or $MoSi_2$-SiC.

REFERENCES

1. O.P. Watts, Trans. Am. Electrochem. Soc., 9 (1906), 105.
2. O. Hoenigschmid, Monatsh. Chem., 28 (1907), 1017.
3. A.K. Vasudevan and J.J. Petrovic, "A comparitive overview of molybdenum disilicide composites," Materials Science and Engineering, A155 (1992), 1-17.

4. J. Schlichting, "Molybdenum disilicide as a component of modern composites for application at high temperatures," High Temperatures-High Pressures, 10 (1978), 241.
5. T.C. Chou and T.G. Nieh, "New observations of $MoSi_2$ pest at 500 degrees C," Scripta Metallurgica et Materialia, 26 (10) (1992), 1637.
6. C.G. McKamey, P.F. Tortorelli, J.H. DeVan and CA Carmichael, "A study of pest oxidation in polycrystalline $MoSi_2$," J. Mater. Res., 7 (10) (Oct. 1992), 2747.
7. P.J. Meschter, "Low temperature oxidation of molybdenum disilicide," Met. Trans. A, 23A (June 1992), 1763-1772.
8. J. Cook, R. Mahapatro, E.W. Lee, A. Khan and J. Waldman, "Oxidation behaviour of $MoSi_2$ composites," Ceram. Eng. and Sci. Proc., 12 (9-10) (1991), 1656-1670.
9. M.J. Maloney, "Development of fiber reinforced $MoSi_2$-base composites," Mat. Sci. and Eng., A155 (1992), 19-32.
10. A. Mueller, G. Wang, R.A. Rapp and E.L. Courtwright, "Deposition and cyclic oxidation behavior of a protective (Mo, W) Si, $Ge)_2$ coating on Nb-base alloys," J. Electrochem. Soc., 139 (5) (May 1992), 1266-1275.
11. T.A. Kircher and E.L. Courtwright, "Engineering limitations of $MoSi_2$ coatings," Mater. Sci. and Eng., A155 (1992), 67-74.
12. D.H. Carter, J.J. Petrovic, R.E. Honnell and W.S. Gibbs, "SiC-$MoSi_2$ composites," Ceram. Eng. Sci. Proc., 10 (9/10) (1989), 1121-29.
13. M. Singh and A. Bose, "Molybdenum disilicide matrix composites," Processing and Fabrication of Advanced Materials for High Temperature Applications, V.A. Ravi and T.S. Srivatsan, editors, TMS (1992), 95-109.
14. S. Zhang and Z.A. Munir, "Synthesis of molybdenum silicides by the self propagating combustion method," J. Mater. Sci., 26 (1991), 3685-88.
15. A.K. Bhattacharya, "Modeling of combustion synthesis of $MoSi_2$ and it's composites," Ceram. Eng. and Sci. Proc., 12 (9/10) (1991), 1697-1722.
16. D.A. Hardwick, P.L. Marting and R.J. Moores, "Reaction synthesis of $MoSi_2$ from high purity elemental powders," Scripta Metallurgica et Materiallia, 27 (1992), 391-394.
17. S. Jandhyala, "Thermochemical evaluation of combustion synthesis of $MoSi_2$-SiC composites," J. Amer Ceram. Soc., 76 (1) (1993), 226.
18. H.J. Feng, J.J. Moore and D.G. Wirth, "Combustion Synthesis of Dense Ceramic-Metal Composites" in Proceedings of the Symposium on Ceramic Matrix Composites, 1993, Annual American Ceramic Society Conference, Cincinatti, April 1993, to be published by American Ceramic Society.

DIAMOND-LIKE NANOCOMPOSITE FILMS: SOME CHARACTERISTICS AND APPLICATIONS

Benjamin Dorfman, Arvind Goel and Michael Abraizov
Advanced Refractory Technologies, Inc.
699 Hertel Avenue
Buffalo, NY 14207

ABSTRACT

Diamond-Like Nanocomposite (DLN) coatings are a novel class of thin film materials. The structure of these films is analogous to composite materials, engineered at the atomic scale. The films can be doped with a wide variety of metal and ceramic species, allowing properties to be tailored to achieve desired performance. Key characteristics of DLN films are high thermal stability, versatility, and excellent adhesion to a wide range of substrates. Research has yielded attractive property data, and significant advantages over other hard thin films currently being developed. Properties of DLN are described and applications discussed briefly in this review paper.

BACKGROUND

During the past decade several kinds of novel protective coatings have been proposed, discovered and developed, including hard carbides, nitrides, borides, oxides and carbo-oxy-nitrides of transitions metals, diamond-like carbon (DLC), diamond, cubic boron nitride, and lately carbon nitride[1,2]. A variety of deposition techniques are used, including chemical vapor deposition, plasma deposition, ablation and ion beam deposition. DLC films possess many of the technologically important properties of diamond, such as high hardness[3], high electrochemical resistance, good optical and infrared transmission[4], and high wear resistance[5]. The amorphous nature of DLC, unlike diamond, allows the low temperature deposition of uniform, smooth, pore-free, thin films with thicknesses as low as 10 nm and low coefficient of friction. DLC can be synthesized at close to room temperature, compared to the ~800°C required to deposit adherent diamond films and is formed via a variety of low pressure processes such as DC, RF or microwave plasma decomposition of hydrocarbons, laser ablation of graphite, and low energy carbon ion beam deposition.

However, DLC films suffer from several disadvantages, including low adherence to a number of useful substrates (although some adhesion problems have recently been addressed by means of interlayers), high residual stress levels especially at the interface[5], and low threshold of fatigue. The high electrical resistivity and low dielectric strength of DLC films further limit their field of application. Another key problem is the relatively low thermal stability of DLC films - complete graphitization occurs at temperatures above 600°C[6], accompanied by a severe degradation of

To the extent authorized under the laws of the United States of America, all copyright interests in this publication are the property of The American Ceramic Society. Any duplication, reproduction, or republication of this publication or any part thereof, without the express written consent of The American Ceramic Society or fee paid to the Copyright Clearance Center, is prohibited.

properties. Extensive work has been devoted to carbon-based microcomposites containing metallic inclusions in various concentrations, but each composition has severe property limitations. Research is also ongoing in the area of doped DLC and diamond to achieve specific properties (e.g., diamond semiconductors)[7,8,9]. The issue of the stability of doped-DLC structures has not been adequately addressed, however. Dual ion beam synthesis of DLC has been proposed as a solution to some of DLC's drawbacks[10]. While this technique is believed to result in highly adherent DLC coatings on a variety of substrates including plastics, it is not clear whether this leads to significant changes in residual stress levels and fatigue behavior of DLC films.

A sharp decrease in adhesion at elevated temperature occurs for many of the coatings listed above. To solve this problem certain techniques have been developed, such as roughening of the substrate surface, or other alteration of the surface or sub-surface region of the substrate to promote reaction or diffusion between the coating and the substrate. Another solution to the adhesion problem is the use of adhesion enhancing interlayers, or 'bond coats'. The most effective bond coat is a functionally gradient interlayer. Interlayers can be produced via chemical vapor deposition, sputtering or other techniques. Since the interlayer must be compatible with both the coating and the substrate, it must be specifically chosen for each system. This approach is therefore very limited in application. Coefficients of thermal expansion, electrical and other properties of the coatings can be controlled within narrow ranges only. Thus, although specialized applications may exist for some of these materials, their use has so far been limited.

The novel diamond-like nanocomposite (DLN) films described in this paper not only overcome some of the adhesion problems common with hard thin films while eliminating the need for interlayers, but also provide an unprecedented level of versatility in allowing control over film properties by modifying composition and deposition conditions.

DIAMOND-LIKE NANOCOMPOSITE THIN FILMS

The fundamental structure of these composites consists of two atomic-scale random networks, each stabilized chemically by additional atomic species. The a-C:H and a-Si:O networks do not exhibit mutual bonding (e.g. to form silicon carbide), and also stabilize each other. A schematic of this structure is shown in Figure 1. The evidence for this structure includes

- advanced imaging techniques, including STM, high resolution SEM, cross-sectional STEM, and low angle electron and X-ray diffraction, that show a uniform clusterless structure,
- advanced spectroscopy techniques, including EXAFS, XPS, AES, FTIR and Raman that show no evidence of SiC or other metal carbides, and
- chemical techniques in which one network component, either carbon or silicon, can be removed by oxidation or etching of very thin films, leaving the other component intact.

This two network structure imparts unusual thermal stability to the amorphous films, as described in a subsequent section. Atomic-scale composites such as DLN represent a fundamental limit of resolution in composite structures. These films have also been described in detail elsewhere[11].

The pure DLN film can be doped with metallic phases, that may exist as random dispersions, a third random network, or even constitute the bulk of the film. Dopants can also include ceramic species, such as TiN, BN and ZrN. These dopant species can be preserved in the amorphous state at elevated temperatures as a result of the structural stabilization of the various networks. Carbide formation is prevented at metal concentrations as high as 45%. If the metal constitutes the bulk of the film, the result is an amorphous metal film stabilized by the DLN film structure.

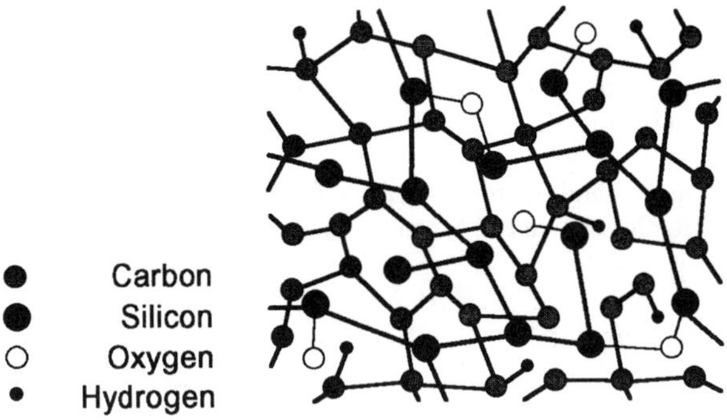

$a\text{-}(C{:}H_{0.15})_{0.7}\ a\text{-}(Si{:}O_{0.3})_{0.3}$

Figure 1. Atomic structure of DLN - a schematic representation

The mutual stabilization of interpenetrating atomic-scale filaments and the random network structure prevents the growth of graphitic carbon even at high temperature. The high adhesion and lower internal stress can also be attributed to this structure. At the temperature of crystallization, glassy metals are usually crystallized within one hour. In Me-DLN however, amorphous metallic inclusions are stable to long-term annealing at over 1000°C in oxygen-free atmospheres.

The characteristics of these diamond-like composite films demonstrate that they are unlike any other diamond-like or other protective films currently in use or being developed worldwide, and provide performance and cost superiority over competitive approaches. Properties of DLN and Me-DLN coatings are briefly described below.

ADHESION

Due to the atomic scale homogeneity of DLN films, problems of network-matrix misalignment and non-uniform adhesion or delamination do not occur. The silicon-oxygen and the carbon-hydrogen components of the films allow at least one of these components to bond well to substrates containing metals, ceramics, or organics, resulting in high adhesion. DLN films have excellent adhesion to a variety of metals, semiconductors, quartz, alumina, ferromagnetic ceramics, polycrystalline diamond-based ceramics, zirconia, metal ceramics, synthetics, plastics[12] and even composite materials.

THERMAL STABILITY

Although thermal stability is normally a key problem for materials in the amorphous state, DLN and Me-DLN structures actually exhibit unusual stability at elevated temperatures. At temperatures below 400°C, the lifetime of DLN films is practically unlimited. Although oxidation occurs under extended exposure at temperatures above 500-600°C (depending on composition) in air, DLN films are stable to much higher temperatures for short times. For example, the properties of tungsten-doped (W-DLN) films did not change after 10^7 pulses with maximum temperatures of 1000-1200°C in air. The films are also highly stable under severe thermal shock conditions (10^8 K/s from 1200°C). Schottky diodes - W-DLN on GaAs - are formed by annealing at 850°C, while preserving the amorphous state of the film. High adhesion strength and excellent thermal shock resistance prevents spalling under high thermal shock.

FTIR studies indicate the high thermal stability of DLN films. Figures 2 and 3 show typical spectra before and after heat treatment. No structural changes are observed.

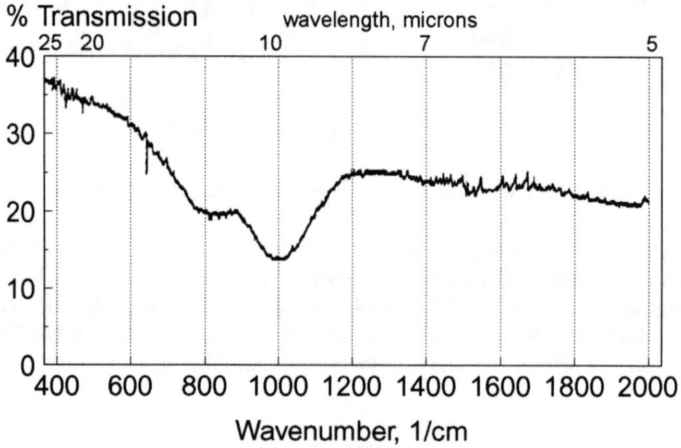

Figure 2. FTIR spectrum on as-deposited DLN

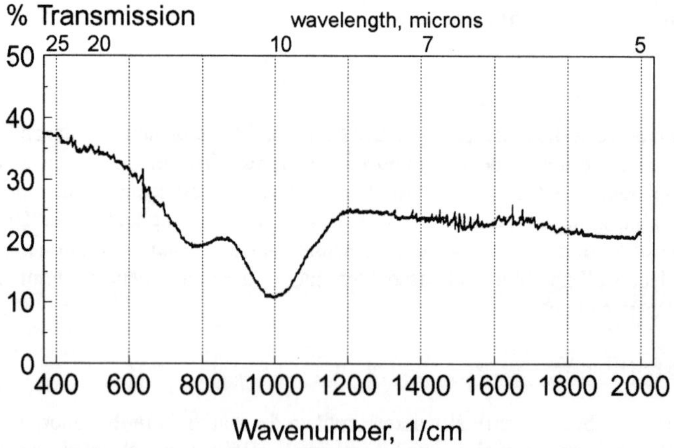

Figure 3. FTIR spectrum on DLN heat treated for 9 hr at 400°C in air, followed by 1 hr at 600°C in a vacuum

CORROSION RESISTANCE

Me-DLN form stable corrosion resistant coatings against most organic and inorganic agents at thicknesses of 15 to 20 nm. Coatings of specific compositions showed no degradation after 2 hours at 1250°C in anhydrous HCl flow, over 400 hr in 10% hydrochloric acid, and in tests with various other inorganic and organic acids and salt solutions[12]. Coatings on stainless steel (410) and aluminum (6066) were unaffected in a 48 hr ASTM salt fog test, while the substrates exhibited noticeable corrosion. DLN films also exhibit good diffusion barrier properties as a result of their pore-free structure. In 10% H_2SO_4/HNO_3 however, most Me-DLN films dissolve at a rate of ~1 nm/hr. HF and aqueous HCl can also corrode DLN films of certain compositions.

MECHANICAL PROPERTIES

DLN films possess high microhardness (1000-2000 kg/mm^2) combined with high elasticity, extremely low coefficients of friction (dynamic: 0.045-0.1 in air, static: <0.5), minimal intrinsic strain and stress, and high wear resistance similar to that of the diamond-like state. In comparative tests of DLC and DLN protective coatings for magnetic hard disks, DLN performed as well as the best DLC, while the friction coefficient of DLN was two times lower than that for DLC. DLN also performed well in tests for other applications such as protection of thermal printer heads, thermoresistors for ink-printer heads (2×10^8 thermal shock cycles), and electromechanical devices (10^7 cycles).

DLN coatings on plastic films exhibit low stress and high flexibility. Flexibility can again be attributed to the films' atomic-scale composite structure. Additional flexibility is provided by the presence of nanopores with diameters of about 3 Å.

OPTICAL PROPERTIES

DLN films are opaque to ultraviolet radiation - the transmission threshold is ~250 nm. Transmission increases monotonically in the visible and near IR range up to a wavelength of about 1 μm, and remains at over 80% in the 1-9 μm range. DLN films possess high transparency

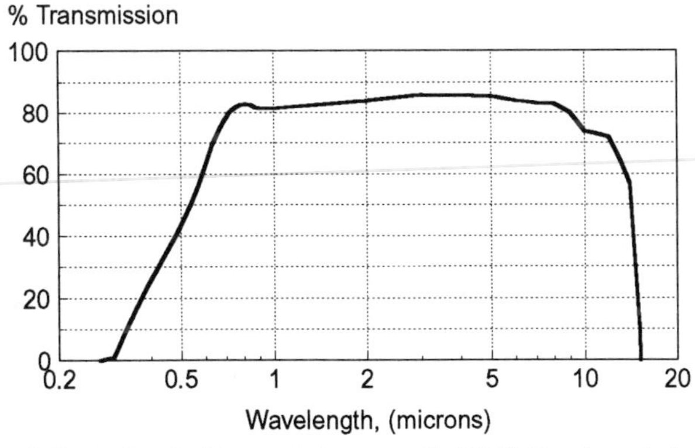

Figure 4. Typical optical transmission curve for DLN film deposited on ZnS.

Substrates as large as 760 mm in diameter can be accommodated in existing equipment, although there is no theoretical limit to the size of the part to be coated. Additional design and further scale-up work is currently ongoing.

Hurdles: Although DLN films have been shown to be extremely versatile, possessing many attractive properties tailorable over wide ranges, they can not always be used directly for all applications where these properties are required. The films must be carefully customized for each new application, to achieve the combination of desired properties on the specific substrate for the specific shape of substrate. In some cases tradeoffs may exist among different properties, and focused development is required to achieve the combination of properties desired. Such development programs are currently underway at ART for several areas of use.

APPLICATIONS

The low friction, high wear resistance and temperature stability of DLN/Me-DLN makes these coatings very attractive for a wide range of applications. These may include wear surfaces for moving parts in chemical equipment, automotive engine assemblies, transportation systems, and areas where energy savings and longer life may result due to low friction. Properties can be tailored for individual applications, making this a uniquely versatile technology. The films have already shown promising performance in the protection of magnetic and optical media.

Pilot scale applications have included protection of the following devices: integrated circuits, thermal printing heads, orthopedic implants, photomasks, visible optics, and infrared germanium optics. These materials have been used or successfully tested as active layers in thermoresistors for ink jet printing heads, electromechanical devices, Schottky diodes (Me-DLN/Si and Me-DLN/GaAs), and high sensitivity sensors (bolometer, tensiometer, photomasks, x-ray masks). They also have the potential for use as anti-electrostatic protective coatings. The high chemical and wear resistance also serves as a basis for creating new coatings for biomedical applications. Preliminary indications are that the films are biocompatible.

Me-DLN containing tungsten and other hard-melting transition metals can provide extremely stable thermal resistors (e.g., for deicing and defogging). Amorphous multilayered structures could be useful for X-ray masks and X-ray focusing optics. DLN coatings containing catalytically active elements (V, Ni, Pt, Pd, Hf, etc.) can serve as inexpensive catalysts, since high surface exposure is possible using very small amounts of precious metals. Multifunctional, cost-effective Me-DLN coatings can be used for implanted electronic crystals and packages, for creation of improved orthopedic devices and biomaterials, and on surgical tools and artificial implants. Other applications continue to be discovered as this novel material technology is studied further.

CONCLUSIONS

DLN/Me-DLN thin film technology offers unique solutions to some traditional problems across a wide range of industries. The properties of these coatings can be controlled to fit specific requirements. The unique atomic-scale composite structure of these films, and the ability to dope the films with a range of metallic and ceramic materials, impart to the films attractive characteristics, properties and performance.

up to at least 40 μm. A characteristic transmission spectrum for DLN film on zinc sulfide in the visible-IR range is shown in Figure 4. This spectral response depends on deposition conditions and can be modified. Refractive index is in the range of 1.7-2.5 for undoped DLN, and up to 3.1 for Me-DLN. Refractive index can also be controlled based on deposition conditions.

The films are transparent to X-rays and free-standing DLN films have been successfully evaluated for use as X-ray and electron windows. Protection of infrared windows is another application where DLN films show promise.

ABLATION RESISTANCE

DLN and Me-DLN films are highly resistant to damage by laser radiation[13]. In a comparative evaluation, DLN, Me-DLN and DLC coated samples were exposed to a KrF excimer laser radiation (20 ns pulses, wavelength 248 nm, fluence ~900 mJ/cm^2). Under these laser irradiation conditions DLC graphitized or was completely etched in less than 5 min., while the three types of DLN and Me-DLN films underwent no observable changes in structure or thickness. Because of this property, DLN films are expected to find uses in protection of military sensors, protective goggles, laser optics, insulators used in plasma devices, and other related applications.

ELECTRICAL PROPERTIES

Undoped DLN films exhibit high dielectric breakdown strength[13]. For any thickness and constant electrical loading the threshold is 2.5 - 5.0 x 10^6 V/cm (DLN on silicon). After annealing for over 100 min. at 450°C, this dielectric strength is preserved, or even increased in some cases. In contrast, DLC films graphitize and become conductive at these temperatures. Even at 400°C, the electrical conductivity of DLC grows by over four orders of magnitude[6]. For short pulses (< 1 sec), the electrical strength of the films exceeds 10^7 V/cm - some films have exhibited this strength level for extended loading as well. DLN films on silicon have exhibited breakdown in excess of 10^8 V/cm. Due to these high dielectric strength properties, DLN coated insulators are being evaluated for use in high energy plasma devices.

In Me-DLN, unique combinations of electrical properties can be attained. All the electron transport mechanisms ranging from (low-temperature) superconductivity, dielectric behavior, metallic behavior, and intermediate electron percolation phenomena are observed, depending on conditions of film growth and film composition. Electrical conductivity can be varied over 18 orders of magnitude (10^{14}-10^{-4} Ω.cm) by controlling metal concentration. Temperature coefficients of resistance of 10^{-5}/K have been observed, implying stable resistivity. Alternating conducting-dielectric layers can also be prepared in one process. Such multilayer superstructures have been shown to be highly stable, and have potential uses in electrical, optical and electro-optical devices.

TECHNOLOGY STATUS

Films are deposited at low temperature (25-200°C), and substrate size is limited only by chamber diameter. Although the thickness of DLN films has no theoretical limit, the existing technology and available equipment allow thicknesses typically in the range of 1-5 μm, up to a maximum of 6-10 μm. Deposition rates range from 0.5-3 μm/hr.

ACKNOWLEDGEMENTS

Part of this work was supported by the Department of Defense in the form of SBIR contracts (Kirtland Air Force Base, Contract number F29601-93-0160; Naval Surface Warfare Center, Contract number N60921-93-C-A356).

REFERENCES

[1]Chem. Engg., April 1994, pg. 35-39.
[2]Machine Design, April 4, 1994, pg. 60-64.
[3]C.J. McHargue, Applications of Diamond Films and Related Materials, Ed. Y. Tzeng *et al*, Elsevier Science Publishers, 1991, pg. 113-120.
[4]S. Mitura, A. Sokolowska, J. Szmidt, and A. Balasinski, *ibid*, pg. 795-797.
[5]A. Grill, B. Patel and B.S. Myerson, Diamond and Diamond-Like Films and Coatings, Ed. R.E. Clausing et al., Plenum Press, New York, 1991, pg. 417-426.
[6]P. Koidl, C. Wild, R. Locher and R.E. Sah, *ibid*, pg. 243-265.
[7]L. Martino, Diamond and Diamond-Like Films and Coatings, Ed. R.E. Clausing *et al*, Plenum Press, NY, 1991, pg. 467.
[8]H. Biederman, Vacuum, Vol. 39, 1, 1989, pg. 13.
[9]H. Biederman, Vacuum, Vol. 40, 1990, pg. 251.
[10]M. Mirtich et al, Thin Solid Films, Vol. 131, no. 3-4, Sept. 1985, pg. 245-254.
[11]V.F. Dorfman, Thin Solid Films, vol. 212, 1992, pg. 267-273.
[12]Final project report, DOD SBIR Contract number N60921-93-C-A356, March, 1994.
[13]Final project report, DOD SBIR Contract number F29601-93-0160, Feb. 1994.

Environmental Effects

OXIDATION KINETICS OF Al-DOPED SiC SINGLE CRYSTALS

S.A.Newcomb, R.E. Tressler and K.E. Spear
Center for Advanced Materials
The Pennsylvania State University, University Park, PA USA 16802

ABSTRACT

Several studies have stated that when SiC is used as a reinforcement phase in oxide matrices, its oxidation rate constant is over three orders of magnitude greater than for bare SiC. The common explanation given for this behavior is increased oxygen permeation through the oxidized silica product layer which has been modified by the oxide matrix cation. However, in the alumina matrix case, Al_2O_3-SiO_2 diffusion couple experiments have shown the modification process is much slower than the oxide growth rate on bare SiC at temperatures less than 1800°C. SiC single crystals implanted with approximately 15 a/o Al were oxidized between 1000° and 1400°C to clarify the role of Al. Although Al increased the oxidation rate at all temperatures and had a very strong effect at 1400°C, the observed increases were not enough to explain the reported results for the oxidation of Al_2O_3-SiC composites.

INTRODUCTION

Although silicon carbide reinforced alumina composites have found applications in the cutting tools industry[1], their sustained use at elevated temperatures has not been successful due, in part, to the internal oxidation of the reinforcement phase and consequent degradation in mechanical properties[2-7]. Park and Luthra have reported oxidation rate constants for these composites which are several orders of magnitude greater than bare SiC (see Table I)[2]. The common explanation given for this behavior is increased oxygen permeation through the reaction product (alumina, mullite and aluminosilicate glass) relative to molecular oxygen permeation through pure amorphous silica[2]. Results from alumina/silica diffusion couples[9], however,

To the extent authorized under the laws of the United States of America, all copyright interests in this publication are the property of The American Ceramic Society. Any duplication, reproduction, or republication of this publication or any part thereof, without the express written consent of The American Ceramic Society or fee paid to the Copyright Clearance Center, is prohibited.

indicate that the interdiffusion of the two oxides (i.e., the migration of Al into the silica and vice versa) is much slower than the oxide growth rate on bare SiC. To clarify the role that Al plays in the oxidation process, an experiment was designed to separate the effect of Al on oxygen permeation through silica from the transport of Al to the reaction interface. This was accomplished by oxidizing SiC single crystals which had been implanted with Al, and, thus, the Al was present at the interface at the beginning of the oxidation process.

Table I. Oxidation Constants for SiC Oxidation

		K_{par} (cm^2/s)		
Temperature °C	Al_2O_3-SiC [2]	polycrystal sintered[8]	single crystal[10] (0001)	single crystal[10] $(000\bar{1})$
1200	-	2.9×10^{-14}	6.3×10^{-15}	2.9×10^{-14}
1400	1.5×10^{-10}	1.6×10^{-13}	8.0×10^{-14}	9.6×10^{-14}
1500	1.1×10^{-09}	-	2.1×10^{-13}	2.0×10^{-13}

EXPERIMENTAL PROCEDURE

Material Preparation

Single crystal SiC made from the Acheson process was used for the experiment. The crystals were of the same type used by Zheng et al.[10] and their major impurity was reported to be nitrogen. All of the crystals were polished on both sides parallel to the basal planes to create uniform starting surfaces. Diamond wheels and pastes were used to first grind the specimens flat and then polish the surfaces. Successive diamond grit sizes were used down to 3μm paste. After polishing the crystals were RCA cleaned[11], placed in 1 v/o HF for 30 s, and rinsed with distilled water.

The SiC crystals were oxidized at 1000°C (dry O_2, 342 min) before ion implantation to provide an initial oxide film. The purpose of the film was to prevent the formation of an alumina layer at the beginning of the oxidation experiments of the doped crystals. The average oxide thickness on the 12 crystals was 1599 ± 18 Å, as determined by ellipsometry.

Ion Implantation

Multiple ion implantation was used to create a constant Al concentration in the implant region. The $(000\bar{1})$ faces of the crystals were implanted with Al by Implant Sciences (Wakefield, MA) at 800°C to insure in-situ recrystallization [12]. The parameters of the implant are given in Fig. 1 along with the predicted Al concentration profile calculated with the PROFILE computer code [13]. Although the goal was to have approximately 1000Å of silica remaining on the crystals after implantation, the implantation process sputtered all of the oxide from the surface.

Oxide Thickness Determination

Ellipsometry could not be used to measure the thermally grown oxide thickness due to the presence of the Al. Therefore an etchant was developed to remove part of the Al-doped silica layer. A wide range of acids and acid combinations were tried with a mixture of results. Only a 3:1 mixture of concentrated HNO_3 and 49 v/o HF could remove all of the scale throughout the oxide growth. Unfortunately it also attacked the Al-doped SiC equally well and was not used for this reason.

The second etch developed was 10 v/o HF. This etch was strong enough to remove Al-doped silicates, but too weak to significantly etch the Al-doped SiC. The etchant was also too weak to completely remove crystallites that formed on the surface of samples oxidized for long periods of time or at high temperature. To insure that the etchant was completely removing the oxide, microprobe analysis showed that the oxygen content of etched surfaces was lower than that measured on SiC crystals with only a native oxide.

The etchant was used to remove part of the oxide growth layer to form a step. This was achieved by partially dipping each oxidized crystal into liquid wax,

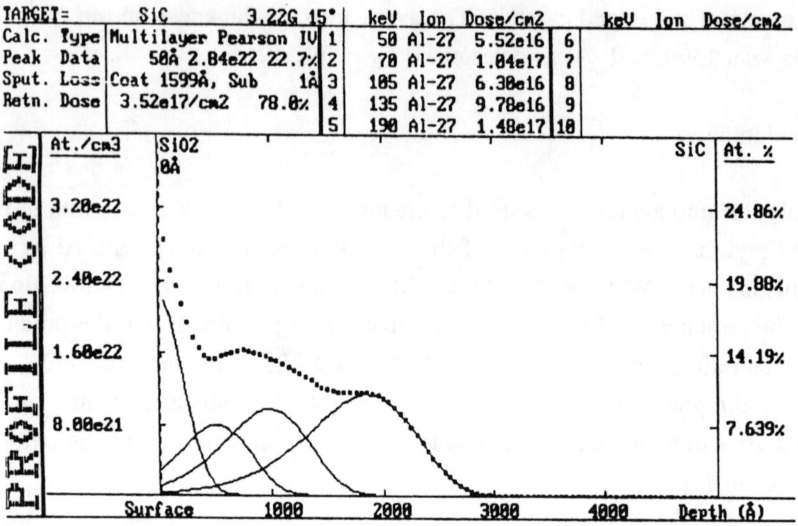

Figure 1. Simulation of multiple implantation of Al in SiC using PROFILE computer program.

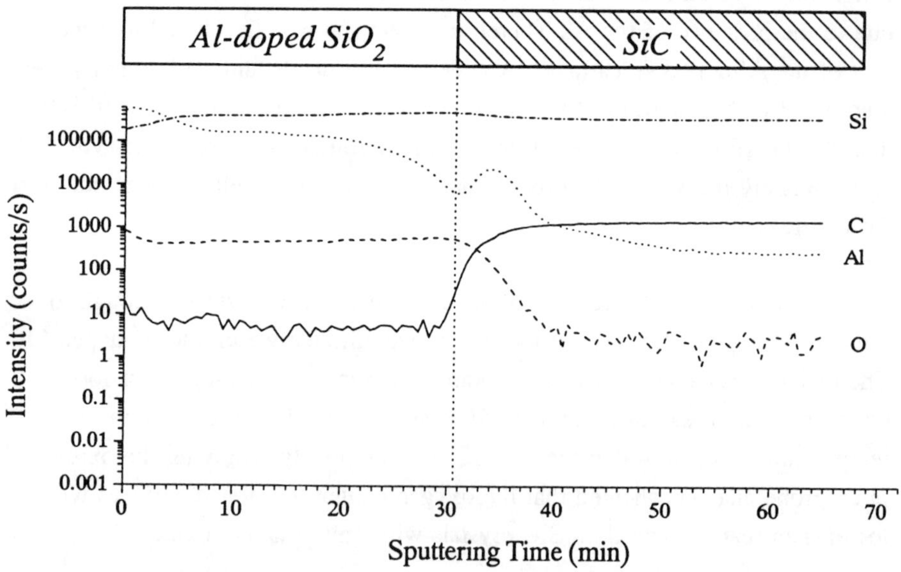

Figure 2. Elemental depth profiles from SIMS analysis of sample oxidized for 24 hrs at 1200°C.

and then etching them from 5 to 60 min in 10 v/o HF at room temperature. The wax was removed by heating the crystal on a hot-plate and wiping off the excess wax. The crystal was cleaned by sonicating for 1 min each in toluene, isopropyl alcohol and acetone. A profilometer was then used to measure the step height.

Oxidation Trials

Samples were oxidized in a mullite tube furnace at 1000°, 1200° and 1400°C with dry oxygen at 1 atm (~150 sccm). The crystals were placed on top of a fused silica slide which was placed on a small alumina boat. This was necessary to prevent the fused silica slide from sticking to the mullite tube during oxidation. Runs were started by sliding the alumina boat into the furnace under 400 sccm of Ar. The procedure lasted about 15 minutes. Temperatures were determined using a S-type (Pt-Pt10%Rh) thermocouple placed next to the alumina boat. The oxygen was turned on once the furnace returned to its temperature set-point (~ 3 min). Removal of the sample was similar to insertion except 1 min intervals were used.

RESULTS AND DISCUSSION

Oxide Film Characterization

The oxide films grown in this study were generally smooth, although they were not always homogeneous. Small crystals nucleated on the surface of the SiC crystals at all three temperatures. One would expect the crystallites to be mullite based on the SiO_2-Al_2O_3 phase diagram. Mullite has been seen both in the reaction product of Al_2O_3-SiC composites[2] and the oxidation of Al-doped SiC crystals in an experiment similar to the present one [14]. The precipitation of the crystals was not surprising since the Al concentration was above the solubility limit for silica at the temperatures studied [9].

During oxidation the Al segregated to both the oxide surface and the internal interface between silica and SiC. A SIMS profile of a crystal oxidized 24 hrs at

1200°C shows elevated Al concentration at both interfaces (see Fig. 2). The surface segregation phenomena has been reported in previous investigations of the oxidation of Si wafers with a lower concentration of Al doping [15,16]. Other than near the interfaces the Al concentration was approximately constant. The SIMS profile in Fig. 2 also shows that the SiC/oxide interface remained sharp throughout the oxidation of the sample as evidenced by the quick increase in the carbon signal after 30 min of sputtering.

Oxide Growth Rate

A plot of the square of the thickness versus time is shown for all three temperatures in Fig. 3. The lines show the least-squared regression of the square of the oxide thicknesses on time, with the y-intercept forced to go through the origin. The parabolic rate constants were 1.3 to 19 times greater than the rate constants for the $(000\overline{1})$ face of SiC at the same conditions (see Table II). Although it is not possible to confirm with the present data, there appears to be a change in the activation energy from 1200° to 1400°C. The rate constant for the 1400°C test was much larger than the value which would have been expected from the tests at lower temperatures. Although the oxidation rates were accelerated, the parabolic rate constants were still much lower than the rates reported for Al_2O_3-SiC composites.

Table II. Oxidation Constants for Al-doped SiC

Temperature °C	K_{par} (cm^2/s) Al-doped	Ratio Kpar(Al-doped) / Kpar(bare SCSC)	Ratio Kpar(Al_2O_3-SiC) / Kpar(bare SCSC) [2]
1000	$1.7x10^{-14}$	-	-
1200	$3.8x10^{-14}$	1.3	-
1400	$1.8x10^{-12}$	18.7	1563
1500	-	-	5500

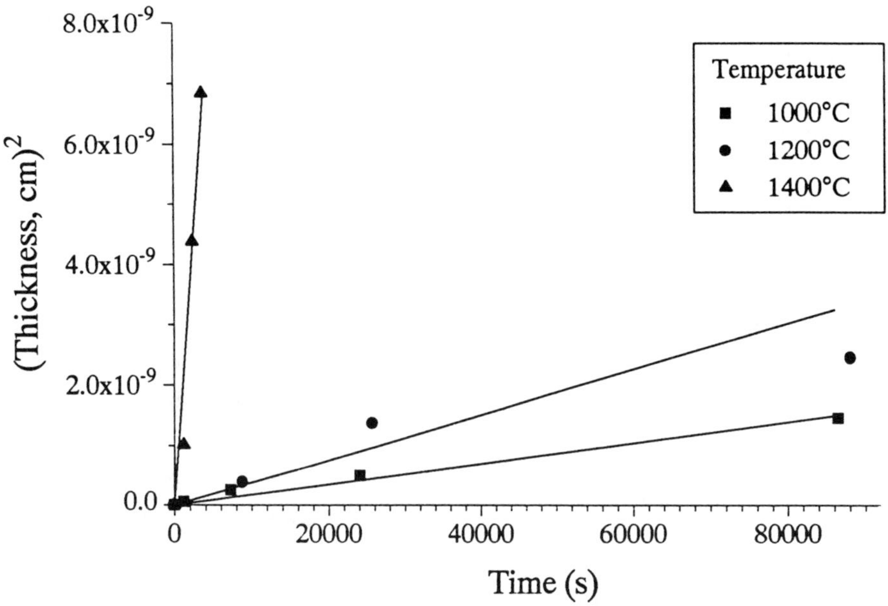

Figure 3. Plot of (oxide thickness)^2 vs. time for the oxidation of Al-doped SiC single crystals.

It is difficult to understand how the composites can oxidize so much more rapidly than the Al-doped samples. If Al plays the key role, as has been suggested and is likely, why does a sample in which the Al is already present at the interface oxidize more slowly than a sample in which the Al has to migrate from the alumina matrix?

A possible explanation is that there could be additional routes formed during oxidation for the oxygen to reach the SiC in the composite case. The internal stresses caused by the volume increase for the transformation of SiC to SiO_2 could cause cracks or mass flow (as opposed to diffusion). The films observed in the present study were always coherent and would not have shown this effect.

A second possibility is that the rate constants derived from the thermo-gravimetric data of the oxidation of composites may not be related in a simple manner to the rate constants determined from 1-dimensional thickness changes (which a majority of the SiC oxidation experiments have used). Many assumptions have to be made to derive and compare the rate constants for these different types of experiments. The apparent discrepancies indicate further improvement in the formation and analysis of these models is needed.

CONCLUSIONS

The following conclusions were made:

1. Al increases the oxidation rate of SiC, particularly above 1200°C.

2. The increases are not large enough to completely explain the parabolic rate constants reported for Al_2O_3-SiC composites.

3. Methods used for evaluating the oxidation rate constants of composites need to be further developed.

ACKNOWLEDGEMENTS

The authors would like to thank H. Du of the Stevens Institute of Technology for his assistance in calculating the implant profiles. We would also like to thank the Gas Research Institute for funding this work under contract No. 5084-238-1302.

REFERENCES

1. E.R. Billman, P.K. Mehrotra, A.F. Shuster and C.W. Beeghly, "Machining with Al_2O_3-SiC-Whisker Cutting Tools," *Am. Ceram. Soc. Bull.*,**67** 1016-19 (1988).

2. K.L. Luthra and H.D. Park, "Oxidation of Silicon Carbide-Reinforced Oxide-Matrix Composites at 1375° to 1575°C," *J. Am. Ceram. Soc.*, **73** [4] 1014-23 (1990).

3. S. Karunanithy, "Chemical Processes that Degrade Composites of Alumina with SiC Whiskers," *Mat. Sci. Eng.*, **A112** 225-31 (1989).

4. P.F. Becher, P. Angelini, W.H. Warwick and T.N. Tiegs, "Elevated-Temperature-Delayed Failure of Alumina Reinforced with 20 vol% Silicon Carbide Whiskers," *J. Am. Ceram. Soc.*, **73** 91-96 (1990).

5. H.Kim and A.J. Moorhead, "Corrosion and Strength of SiC-Whisker-Reinforced Alumina Exposed at High Temperatures to H_2-H_2O Atmospheres," *J. Am. Ceram. Soc.*, **74** 1354-59 (1991).

6. M. Backhaus-Ricoult, "Oxidation Behavior of SiC-Whisker-Reinforced Alumina-Zirconia Composites," *J. Am. Ceram. Soc.*, **74** 1793-802 (1991).

7. W. Deqing and H.F. Lopez, "Interfacial Cracking during Thermal Oxidation in a 30 vol% SiC-Whisker Reinforced Alumina," *Mater. Let.*, **16** 313-316 (1993).

8. J.A. Costello and R.E. Tressler, "Oxidation Kinetics of Silicon Carbide Crystals and Ceramics: I, In Dry Oxygen," *J. Am. Ceram. Soc.*, **69** 674-81 (1986).

9. R.F. Davis and J.A. Park, "Diffusion and Reaction Studies in the System Al_2O_3-SiO_2," *J. Am. Ceram. Soc.*, **55** [10] 525-31 (1972).

10. Z. Zheng, R.E. Tressler and K.E. Spear, "Oxidation of Single Crystal Silicon Carbide, Part II. Kinetic Model," *J. Electrochem. Soc.*, **137** [9] 2812-16 (1990).

11. W. Kern and D.A. Puotinen, "Cleaning Solutions Based on Hydrogen Peroxide for Use in Silicon Semiconductor Technology," *RCA Rev.*, **31** 187-206 (1970).

12. H. Du, Z. Yang, M. Libera, D.C. Jacobson, Y.C. Wang and R.F. Davis, "Chemistry and Structure of Beta Silicon Carbide Implanted with High-Dose Aluminum," *J. Am. Ceram. Soc.*, **76** 330-35 (1993).

13. A.J. Armini, Implant Sciences, private communication.

14. Z. Yang, H. Du, M. Libera, S. Withrow, "Oxidation Behavior of SiC Implanted with Aluminum to High Doses at Multiple Energies," presented at the 20th Fall Materials Research Society Meeting, Boston, Nov.29, 1993.

15. J.W. Palmour, R.F. Davis, H.S. Kong, S.F. Corcoran and D.P. Griffis, "Dopant Redistribution during Thermal Oxidation of Monocrystalline Beta-SiC Thin Films," *J. Electrochem. Soc.*, **136** 502-07 (1989).

16. B.E. Deal, A.S. Grove, E.H. Snow and C.T. Sah, "Observation of Impurity Redistribution During Thermal Oxidation of Silicon Using the MOS Structure," *J. Electrochem. Soc.*, **112** [3] 308-314 (1990).

OXIDATION KINETICS OF COATED SILICON CARBIDE FIBER-REINFORCED SILICON CARBIDE (SiC/SiC)

Dennis S. Fox, National Aeronautics and Space Administration,
Lewis Research Center, Cleveland, Ohio

Silicon carbide fiber-reinforced silicon carbide (SiC/SiC) was exposed for 100 hours to dry, flowing oxygen. Oxidation kinetics were determined via thermogravimetric analysis at 981°, 1204° and 1316°C (1800°, 2200° and 2400°F). The effectiveness of three external coating systems applied for oxidation protection is discussed. In all cases, weight gains were observed, and the pyrolytic carbon interface layer remained intact. A CVD SiC external coating is the most promising due to low oxidation kinetics resulting from solid silica formation. A borosilicate glass was observed on the surface of two of the materials that have boron-containing coatings.

Introduction

The low density and high temperature capability of continuous fiber-reinforced ceramic matrix composites (CMCs) makes this class of materials appealing for advanced aerospace engine applications. Continuous fiber reinforcement of Si-based matrices allows graceful failure if a component is damaged. An interface material is needed to prevent fiber/matrix bonding during fabrication. This material must also debond under high applied stress to dissipate matrix crack energy. In many composite systems, carbon is the interface material of choice. The major drawback of carbon is that it oxidizes at temperatures much lower than the intended use temperatures of CMCs. An extensive literature review of carbon oxidation in CMCs is beyond the scope of this paper. References 1-3 provide analysis of the oxidation of carbon in SiC/SiC systems, as well as appropriate literature reviews.

One approach used to protect the exposed carbon interface in these materials is

To the extent authorized under the laws of the United States of America, all copyright interests in this publication are the property of The American Ceramic Society. Any duplication, reproduction, or republication of this publication or any part thereof, without the express written consent of The American Ceramic Society or fee paid to the Copyright Clearance Center, is prohibited.

to apply an oxidation-resistant external coating. One such coating is SiC deposited via chemical vapor deposition (CVD). If crack-free, this coating would provide satisfactory protection of the carbon. Oxidation would be further impeded by the solid silica layer formed on SiC at high temperature. However, if cracks do occur in this coating, oxidation will proceed at a rapid rate nearly independent of the crack opening displacement.[3] An alternative coating system comes directly from technology used to protect carbon/carbon (C/C). A SiC coating applied to C/C will be cracked due to thermal expansion mismatch. A glass-forming particulate such as boron can be added to the matrix[4], or a glass-forming sealant layer can be applied to the C/C.[5] If the outer CVD SiC becomes cracked, the particulate or sealant can react with oxygen to produce a glass that seals the crack and inhibits oxidation. This technology is currently being applied to SiC/SiC composites.

Reliable CMCs are now being developed that possess good mechanical properties and oxidation resistance. The response of CMCs to high temperature environments can therefore be determined. The purpose of this paper is to expose one such silicon carbide fiber-reinforced silicon carbide (SiC/SiC) to a dry, flowing oxygen environment. The oxidation kinetics are determined via thermogravimetric analysis. The effectiveness of three different external coating systems in protecting the carbon interface is presented.

Experimental Procedure

The material used in this study (BFGoodrich Aerospace) is based on a 2D plain-weave 0/90° Nicalon™ (Nippon Carbon Co., Tokyo, Japan) fiber fabric layup. Material properties are listed in Table I. The fiber/matrix interface material is pyrolytic carbon 0.3 microns thick. The 114 × 254 mm preform is densified with silicon carbide via chemical vapor infiltration (CVI). Specimens were machined from the large panel into their final geometry of 25.4 × 13.0 × 3.5 mm. Because these samples were machined after densification was completed, the fiber edges and carbon interface were not sealed by SiC. These uncoated samples will be referred to as Type U. Three different coating systems were

Table I. Material Description & Properties

Fiber Content	45 volume %
Density	2.54 g/cc
Porosity	10 volume %
Tensile Strength	215 MPa
Tensile Modulus	183 GPa
CTE	$3.0 \times 10^{-6}/°C$

applied on additional SiC/SiC samples: Type A, a SiC via chemical vapor deposition; Type B, a particulate-based sealant with an outer layer of CVD SiC; and Type C, a boron-rich inner layer with an outer layer of CVD SiC. Polished cross sections (to 20 microns) show details of the three different types of coatings (Figure 1). During the CVI technique, infiltration of the fiber weave becomes more difficult with time as porosity decreases. This often results in the formation of a layer of SiC on the surface of the composite. This layer can be seen in the upper portion of the Fig. 1 photos between the respective coatings and the SiC/SiC substrate. Cracks were observed in this layer in the as-received material in all three cases.

Oxidation experiments were conducted in dry oxygen flowing at 100 cm^3/min in a vertical alumina tube furnace. Three temperatures were investigated: 981°, 1204° and 1316°C (1800°, 2200° and 2400°F). One must be aware that Nicalon fibers contain excess carbon and SiO_2. At temperatures above 1200°C, Nicalon decomposes to form CO.[6,7] The oxidation kinetics were obtained by thermogravimetric analyses. Weight change was continuously monitored with a recording microbalance (C-1000, Cahn Instruments, Inc., Cerritios, CA). Samples were held in an alumina cup with an inner sleeve of fused silica. Exposure time was 100 hours.

Sample morphologies were studied via scanning electron microscopy (SEM, JEOL JSM-840A, Tokyo, Japan). The microscope was equipped with thin-window, energy dispersive spectrometry (EDS, KEVEX 3600-0178, Foster City, CA) that allowed elemental dot mapping of polished cross sections. Polished cross sections were prepared by sawing each sample in two, and polishing to a 20 micron finish with high-purity kerosene. Samples were then cleaned with trichloroethane, acetone and alcohol. This non-aqueous preparation technique was required to prevent hydration of the borosilicate glasses formed during exposure of Type B and Type C-coated SiC/SiC. Due to the experimental nature of these coatings, only one sample of each type was available for oxidation at each of the three temperatures.

Results and Discussion

Figure 2 contains the kinetic curves for the Type U (uncoated) material. The initial weight loss is due to oxidation of the exposed carbon interface material. The subsequent weight gains are due to SiC and Nicalon fiber oxidation. X-ray maps of the interior of the exposed samples confirm that the carbon interface is completely oxidized and silica (SiO_2) has formed in its place.

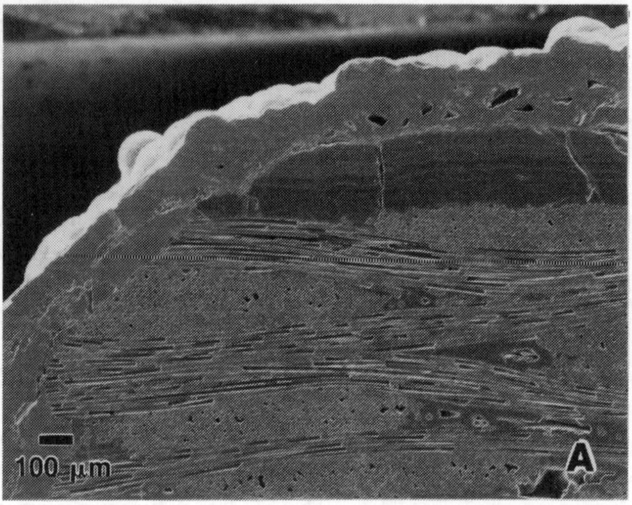

Figure 1.a. Cross section of Type A SiC/SiC (CVD SiC coating).

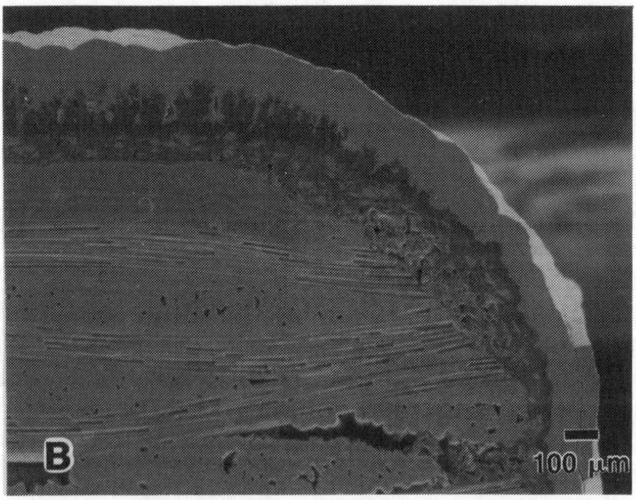

Figure 1.b. Cross section of Type B SiC/SiC (particulate-based inner sealant layer and outer CVD SiC coating).

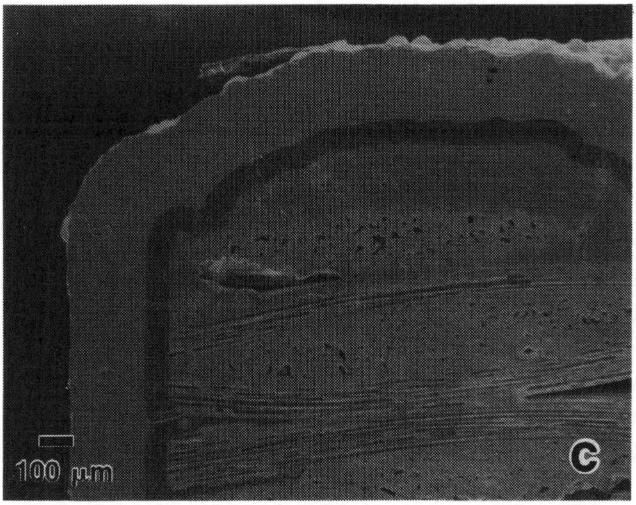

Figure 1.c. Cross section of Type C SiC/SiC (boron-rich inner layer and outer CVD SiC coating).

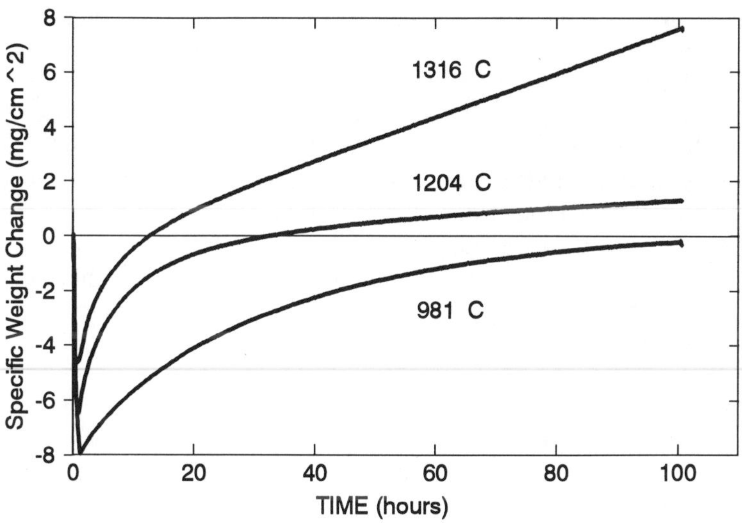

Figure 2. Oxidation kinetics of Type U (uncoated) SiC/SiC.

The Type A material exhibits parabolic oxidation kinetics at the two lower temperatures (Fig. 3). Thin silica scales were observed on the samples after oxidation. The parabolic rate constant (K_p) measured at 1204°C was 3.0×10^{-4} mg^2/cm^4 h. By comparison, the K_p measured on a bulk monolithic CVD SiC under these same conditions in a previous study was 1.5×10^{-4} mg^2/cm^4 h. The kinetics at 1316°C deviate from parabolic. This is likely due to decomposition of the Nicalon fibers that results in the formation of a non-protective, porous oxide scale. No weight loss was observed for any of the Type A samples. X-ray maps of polished cross sections after exposure confirm that carbon remains at the matrix/fiber interface.

Figure 4 contains the kinetic curves for the Type B material. Note that the measured weight gains are an order of magnitude higher than that observed for Type A SiC/SiC. No weight loss is noted. Low parabolic kinetics are observed at 981°C. However, large weight gains are observed at the two higher temperatures. At 1204°C, one corner of the sample reacted with the sample cup. A crack in the outer SiC layer of the duplex coating evolved from this reaction site. A glass is observed in the vicinity of the crack. As discussed above, the purpose of this glass is to seal the crack and inhibit oxidation. A polished cross-section of Type B SiC/SiC after 100 h at 1204°C in dry oxygen is shown in Fig. 5. The crack in the outer CVD SiC layer of the duplex coating is denoted by Arrow 1. A small crack observed in the substrate SiC matrix is denoted by Arrow 2. X-ray maps of the area just below this crack (Fig. 6), an area where one would expect enhanced oxidation, reveal that carbon is still present at the fiber/matrix interface. The surface oxide of the 1316°C sample is quite porous, which leads to the enhanced kinetics. This porous nature could result from gas evolution from decomposition of the Nicalon fibers or volatilization of a boron-containing surface glass. SEM/EDS shows carbon to be present in the sample interior. From the kinetic curves and interior X-ray analysis, it is clear that coating Type B protects the interface from oxygen under the conditions of this study.

The Type C coating also contains boron, and borosilicate surface glass is observed after exposure at the two higher temperatures. Oxidation kinetics are enhanced (Fig. 7) because the oxide is liquid at the test temperature. A cross section of Type C coated SiC/SiC after 100 h at 1204°C in dry oxygen is shown in Figure 8. A crack in the outer SiC layer allows oxygen penetration to the boron-containing inner layer of the coating. A borosilicate glass forms and acts to seal the crack. As shown in Fig. 8, some of the glass also travels down cracks in the inner CVI SiC of the composite to the near-surface fiber bundle. Only minor oxidation of the interface is observed. X-ray dot maps of polished

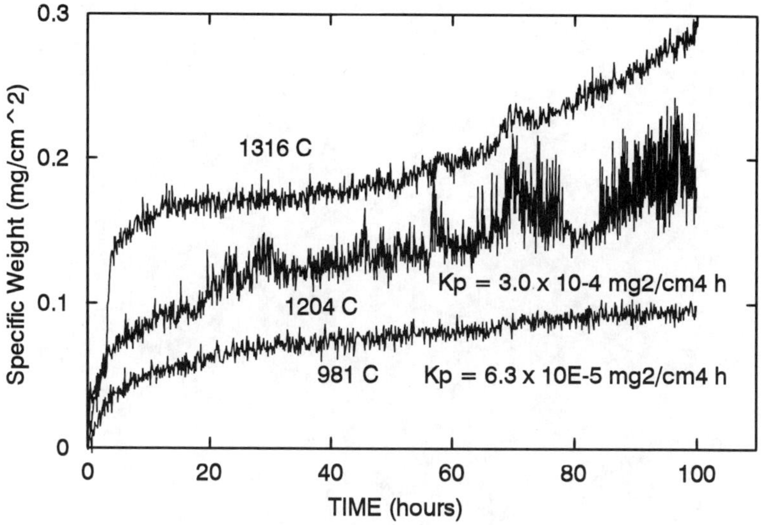

Figure 3. Oxidation kinetics of Type A SiC/SiC.

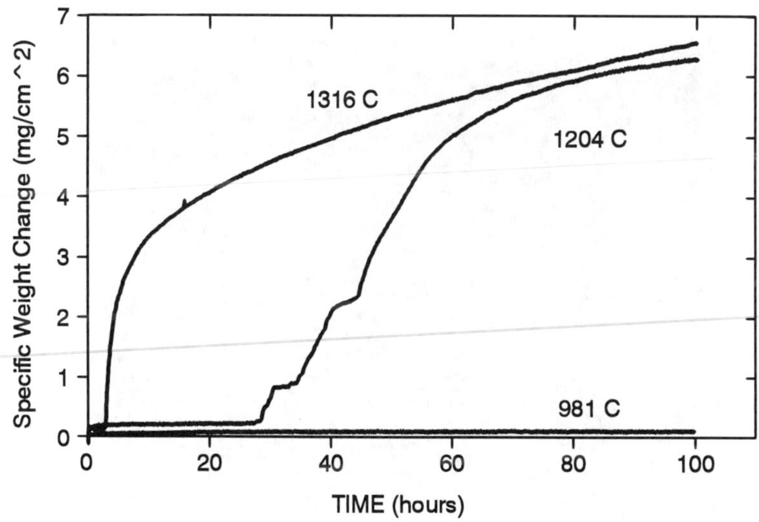

Figure 4. Oxidation kinetics of Type B-coated SiC/SiC.

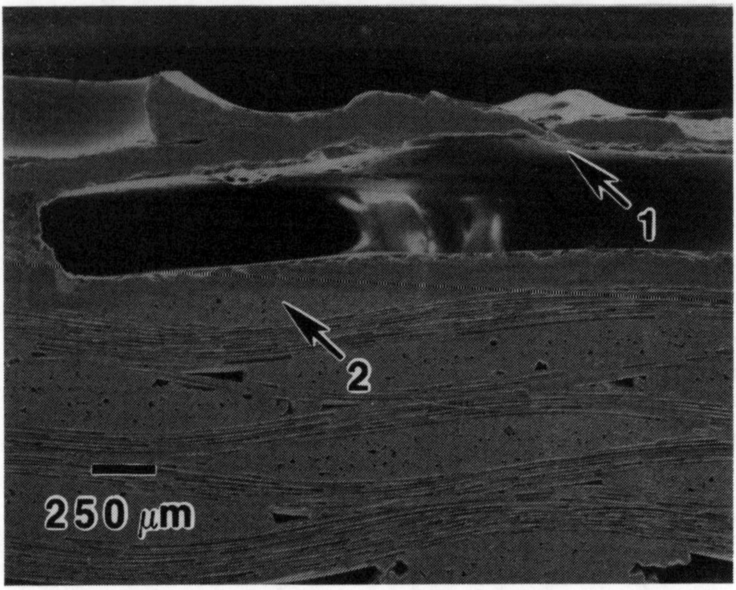

Figure 5. A cross-section of Type B SiC/SiC after 100 h at 1204°C in dry oxygen. Arrow 1 points to the crack in the outer SiC layer of the duplex coating. Arrow 2 points to the area just below a crack in the substrate analyzed for carbon using SEM/EDS.

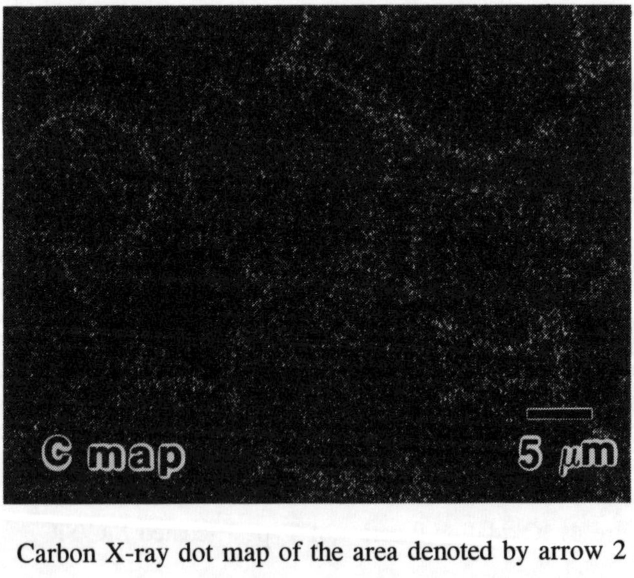

Figure 6. Carbon X-ray dot map of the area denoted by arrow 2 in Fig. 5.

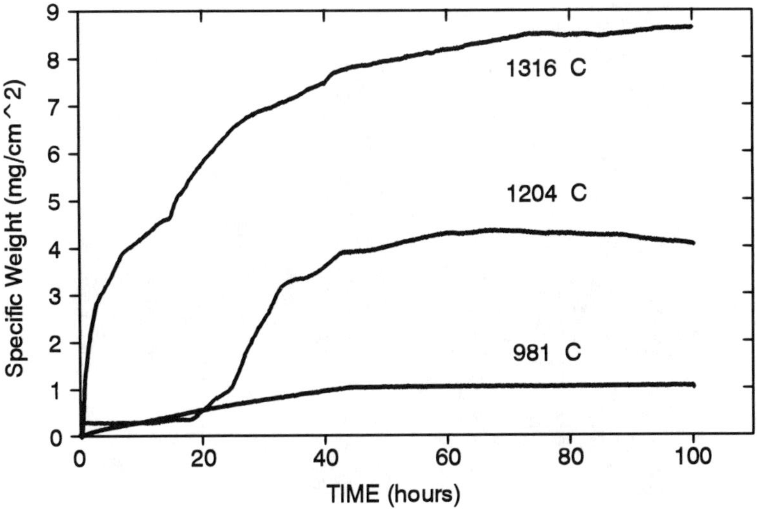

Figure 7. Oxidation kinetics of Type C-coated SiC/SiC.

cross sections at all three temperatures show carbon at the interface. As with the other two coatings, coating Type C adequately protects the interface from oxidation under the conditions of this study.

One must be aware of the possible problems associated with the surface borosilicate glass observed after exposure of Type B SiC/SiC at 1204°C and Type C SiC/SiC at 1204° and 1316°C. Surface oxidation will be enhanced because the boron-containing oxide is liquid at temperature, as compared to the solid silica formed on Type A material. Carbon oxidation will be inhibited as long as the glass is present. However, the glass may volatilize or actually be driven off of the sample by the high velocities typical of actual engine operating environments. If the glass is no longer present, oxidation of carbon at the interface will occur rapidly. Another concern is hydration of the glass that could cause its rapid depletion. Burner rig oxidation experiments are currently underway in these laboratories on Type C coated SiC/SiC. A burner rig closely models the high velocity conditions encountered in a gas turbine. Utilizing kerosene as the fuel, the oxidizing gases will include not only oxygen, but also water vapor and carbon dioxide. In 100-hour tests wherein weight gains are measured, exposed Type C SiC/SiC tensile samples retain their as-fabricated room-temperature strengths.

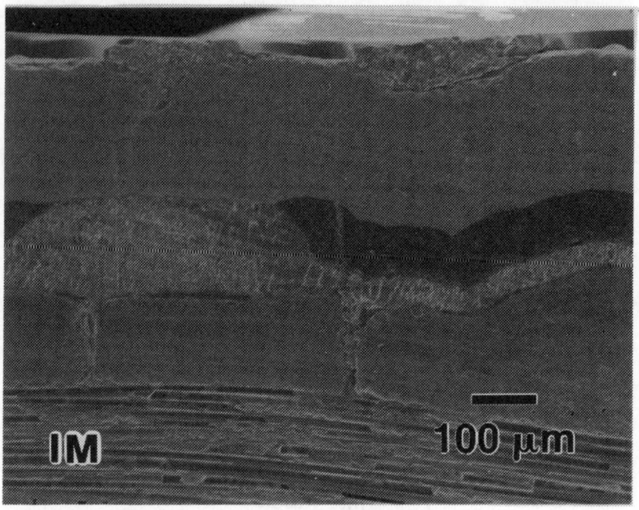

Figure 8.a. Cross-section of Type C SiC/SiC after 100 h in dry oxygen.

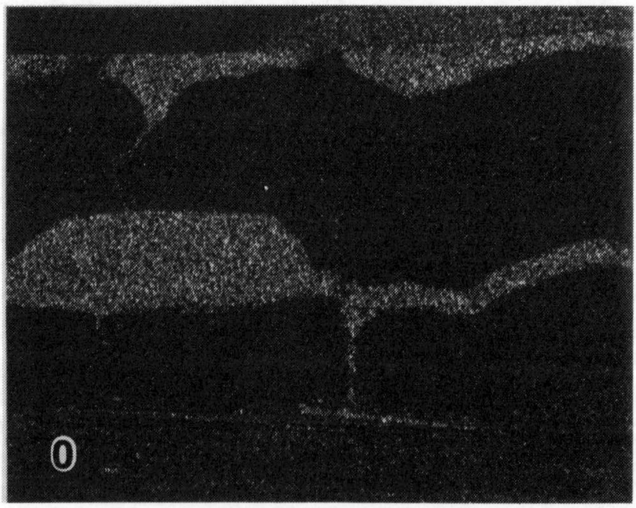

Figure 8.b. Type C SiC/SiC after 100 h in dry oxygen; Oxygen map.

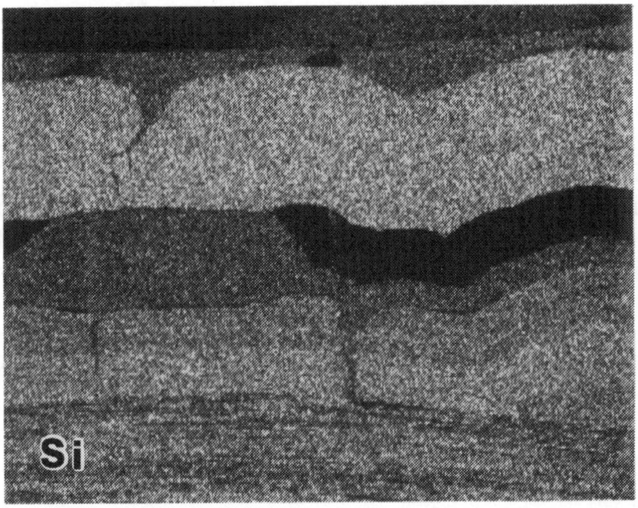

Figure 8.c. Type C SiC/SiC after 100 h in dry oxygen; Silicon map.

Figure 8.d. Type C SiC/SiC after 100 h in dry oxygen; Carbon map.

SUMMARY AND CONCLUSION

A continuous silicon carbide fiber-reinforced silicon carbide matrix composite (SiC/SiC) was exposed to a dry, flowing oxygen environment at 981°, 1204° and 1316°C. The oxidation kinetics were determined via thermogravimetric analysis. Three different external coating systems were effective in protecting the carbon interface from oxidation for the 100 hours of exposure time. Coating Type A (CVD SiC) is the most promising due to low oxidation kinetics resulting from solid silica formation. Oxidation is enhanced due to the formation of borosilicate surface glass at 1204° and 1316°C in both Type B and Type C coatings. The enhanced oxygen diffusion through the glass, as compared to pure SiO_2, results in accelerated oxidation kinetics. The glass seals surface cracks and inhibits oxygen penetration. In conclusion, the carbon interface in SiC/SiC is adequately protected by all three coatings under low flow conditions in dry oxygen. Hydration of the surface borosilicate glass that was formed during exposure of Type B and Type C materials can result in sealant depletion. This may limit the useful lifetime of those materials in environments where this effect occurs.

REFERENCES

1) L. Filipuzzi, G. Camus, R. Naslain and J. Thebault, "Oxidation Mechanisms and Kinetics of 1D-SiC/C/SiC Composite Materials: I, An Experimental Approach," J. Am. Ceram. Soc., **77** [2] 459-66 (1994).

2) L. Filipuzzi and R. Naslain, "Oxidation Mechanisms and Kinetics of 1D-SiC/C/SiC Composite Materials: II, Modeling," J. Am. Ceram. Soc., **77** [2] 467-80 (1994).

3) J. D. Cawley, A. J. Eckel and T. A. Parthasarathy, "Oxidation Kinetics of a Continuous Carbon Phase in a Non-Reactive Matrix," submitted to J. Am. Ceram. Soc.

4) T. D. Nixon and J. D. Cawley, "Oxidation Inhibition Mechanisms in Coated Carbon-Carbon Composites," J. Am. Ceram. Soc., **75** [3] 703-708 (1992).

5) H. Dietrich, "Carbon/Carbon, Protected/Protected," Materials Engineering, [8] 34-35 (1991).

6) T. Mah, N. L. Hecht, D. E. McCullum, J. R. Hoenigman, H. M. Kim, A. P. Katz and H. A. Lipsitt, "Thermal Stability of SiC Fibres (Nicalon™)," J. Mater. Sci., **19** [4] 1191-1201 (1984).

7) J. A. DiCarlo, "Fibers for Structurally Reliable Metal and Ceramic Composites," J. Metals, **37** [6] 44-49 (1985).

OXIDATION OF SiC WHISKER AND PLATELET REINFORCED Al_2O_3 COMPOSITES AND RESULTANT MECHANICAL PROPERTIES,

Lori A. Leaskey, Raouf O. Loutfy, and James C. Withers, Materials and Electrochemical Research (MER) Corporation, 7960 South Kolb Road, Tucson, AZ 85706

ABSTRACT

The effect of oxidation of SiC whisker- or platelet-reinforced Al_2O_3 composites on microstructure and mechanical properties has been investigated. It was found that short-term, high temperature oxidation conditions promote the formation of a low porosity reaction layer which results in significant increases in both strength (up to 23%) and fracture toughness (up to 176%). The extent of improvement was found to be dependent upon the thickness and porosity content of the reaction layer, starting composition, and bulk characteristics of the unoxidized composite.

INTRODUCTION

Perhaps the major obstacle prohibiting the use of ceramic materials in numerous applications is their poor reliability, which is commonly a result of inhomogeneities or defects introduced during processing or handling which degrade the materials' mechanical properties. Because fracture origins are frequently associated with surface defects, the elimination or stabilization of these defects should result in an improvement in both mechanical properties and reliability. Different approaches have been utilized to eliminate or minimize any defects present in the surface or bulk of the ceramic, or to improve the mechanical performance of the ceramic in general. One approach which has resulted in an enhancement of strength for numerous materials is the introduction of surface compressive stresses.[1-13] Although there are many effective ways to produce compressive surface layers, including low temperature quenching and surface coatings, most have

To the extent authorized under the laws of the United States of America, all copyright interests in this publication are the property of The American Ceramic Society. Any duplication, reproduction, or republication of this publication or any part thereof, without the express written consent of The American Ceramic Society or fee paid to the Copyright Clearance Center, is prohibited.

produced only marginal improvements in mechanical properties.

Alumina (Al_2O_3) is a ceramic oxide material whose superior corrosion resistance would make it an ideal candidate for various applications, including heat exchangers or components exposed to corrosive flue gases such as in coal gasification environments. However, an Al_2O_3 monolith has a high probability of failure as a result of its low strength, poor thermal properties, and thus poor thermal shock resistance. On the other hand, silicon carbide whisker (SiC_w)-reinforced Al_2O_3 composites have received considerable attention during the past few years as candidates for structural applications because of their potential for high strength and fracture toughness and good resistance to thermal shock, slow crack growth, and creep. Similarly, SiC platelets (SiC_{pl}) have recently been investigated as reinforcements for Al_2O_3, particularly since these are benign, considerably less expensive than whiskers, and when incorporated into composites by pressureless sintering, higher densities than those containing comparable amounts of whiskers are achieved. However, the strength of the SiC_{pl} containing composites is limited by the platelet sizes, which at the time of this study were available only in the range of 5-500 μm. Since the critical flaw size is usually controlled by the maximum platelet size, the strength is limited by the size of the platelets.

The purpose of this work was to improve the strength and fracture toughness of SiC-reinforced Al_2O_3 composites by producing a compressive surface layer to negate the effect of surface defects in degrading these properties. The compressive surface layer was achieved by creating a higher volume surface layer through controlled oxidation of the samples to convert the SiC reinforcement in the surface region to SiO_2. Few studies have been performed on the oxidation of SiC-reinforced Al_2O_3 composites which directly pertain to effect on mechanical properties. Kriven et al.[14] investigated the effect of relatively low temperature (≤1200°C) and extended time (100-1000h) oxidation on the room temperature flexural strength and under all conditions observed a reduction in strength. This investigation focused on the oxidation of Al_2O_3-SiC composites at higher temperatures (up to 1600°C) for short time periods (≤10h).

EXPERIMENTAL PROCEDURE

Oxidation studies were conducted on both hot pressed and pressureless sintered specimens of Al_2O_3 reinforced with SiC whiskers or platelets. The Al_2O_3 powder (Ceralox Corporation, Tucson, AZ) had an average particle size of 0.5 μm, the SiC whiskers (Advanced Composite Materials Corporation, Greer, SC) had an average diameter of 0.5 μm, and the SiC platelets (Third Millennium Technologies, Knoxville, TN) had an average diameter of 10 μm. For hot pressed SiC_w-reinforced samples, mixtures of Al_2O_3 powder and SiC_w (28 vol%) were homogenized by turbomilling in isopropanol, sieving, drying by rotary evaporation, and grinding with a mortar and pestle. Consolidation to full densification was accomplished by hot pressing at 34.5 MPa and a temperature of 1650°C for 20 min. A hot pressed Al_2O_3-SiC_{pl} composite was supplied by Third Millennium Technologies, Inc., and this sample contained 10 vol% SiC_{pl} and an unspecified amount of MgO. This composite was hot pressed at 1600°C with 13.8 MPa of pressure.

Pressureless sintered SiC_w and SiC_{pl}-reinforced Al_2O_3 composites were also fabricated. The SiC_w composites contained 15 vol% SiC_w and 2-6 vol% Y_2O_3, while the SiC_{pl} composites contained 10-30 vol% SiC_{pl} and 0.5-6 vol% Y_2O_3. Green bodies were produced by slip casting and drying pressing for SiC_w- and SiC_{pl}-reinforced composites, respectively. Sintering conditions of 1825°C for 1.5h were utilized for the SiC_w-reinforced materials and 1650-1850°C for ≤2h for SiC_{pl}-reinforced composites.

Oxidation of the samples was conducted in a molydisilicide furnace in air at temperatures up to 1600°C for periods of time up to 10h. To ensure uniform oxidation of all surfaces the specimens were placed on knife edged Al_2O_3 supports. The hot pressed Al_2O_3-SiC_w samples were placed in the furnace at the oxidation temperature, while the pressureless sintered Al_2O_3-SiC_w and all Al_2O_3-SiC_{pl} samples were inserted into the furnace at 700°C and heated to 1600°C at a rate of 25°/min since these samples were susceptible to thermal shock. Prior to oxidation, each sample (which usually contained a notch for fracture toughness measurement after oxidation) was weighed on a 5 digit balance. The materials were heat treated at the desired oxidation temperature, allowed to equilibrate to room temperature, and measured for changes in weight, strength, and fracture toughness. Strength was measured using the conventional three point flexure (MOR) test, and fracture toughness was determined using the single edge notched beam method. SEM analyses of polished and fracture surfaces were conducted to examine microstructure and determine reaction layer thickness.

RESULTS AND DISCUSSION

The first step in the oxidation of Al_2O_3-SiC composites is that SiC is oxidized to form SiO_2, which can then react with Al_2O_3 to form mullite ($3Al_2O_3 \cdot 2SiO_2$) or an aluminosilicate glass.[15] Numerous studies[16-27] have been performed on the oxidation of SiC monoliths, and it has been found that on exposure to oxygen atmospheres at high temperatures, SiC will oxidize either by passive or active oxidation. Passive oxidation occurs at lower temperatures and higher partial pressures of oxygen to form a protective SiO_2 layer which protects the material from further oxidation due to the low diffusion of oxygen through silica. Passive oxidation results in a net mass increase as shown in equation (1) below, where two oxygen species (weight = 2 moles x 16 g/mole = 32g) replace one carbon species (weight = 1 mole x 12 g/mole = 12g).

$$SiC(s) + 3/2\ O_2(g) \rightleftharpoons SiO_2(s) + CO(g) \qquad (1)$$

Active oxidation occurs at high temperatures and low partial pressures of oxygen, and results in a net mass decrease as shown in equation (2) since only gaseous products are produced.

$$SiC(s) + O_2(g) \rightleftharpoons SiO(g) + CO(g) \qquad (2)$$

Thus, active oxidation commonly results in erosion of the surface layer and degradation of mechanical properties. There is a broad range of data concerning the precise point at which the transition from active to passive oxidation occurs, all of which indicate that the oxidation conditions utilized in this study (i.e. $\leq 1600°C$, ambient pressure) are in the passive oxidation region for SiC. This was verified by an increase in weight for all samples after oxidation.

The mechanical properties of the composites as-sintered and after oxidation at 1600°C are shown in Figure 1. Fracture toughness values for samples oxidized for 1 and 1.5 h were not measured due to an insufficient number of samples. As shown, significant increases in both strength and toughness were observed for oxidation times up to 1h. The highest strength and toughness of 1040 MPa and 17 $MPa{\cdot}m^{1/2}$ were obtained in only 15 minutes, after which a gradual decrease in these values was observed. Examination of the microstructures of the cross section of each sample following oxidation for 0.25, 1 and 1.5 h as shown in Figures 2(a) through (c), respectively, reveals that as oxidation time at 1600°C increases, the number of strength limiting flaws or pores also increases within the reaction layer. In addition, as the oxidation time increases, the oxidation layer thickness increases. An increase in oxidation layer thickness may contribute to the retention of

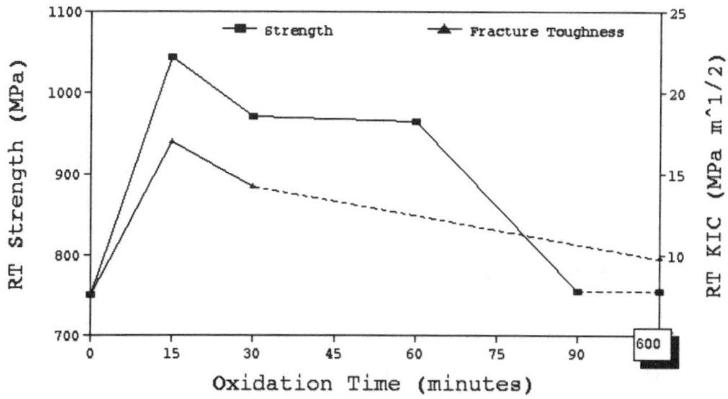

Figure 1. Mechanical properties as a function of oxidation time at 1600°C.

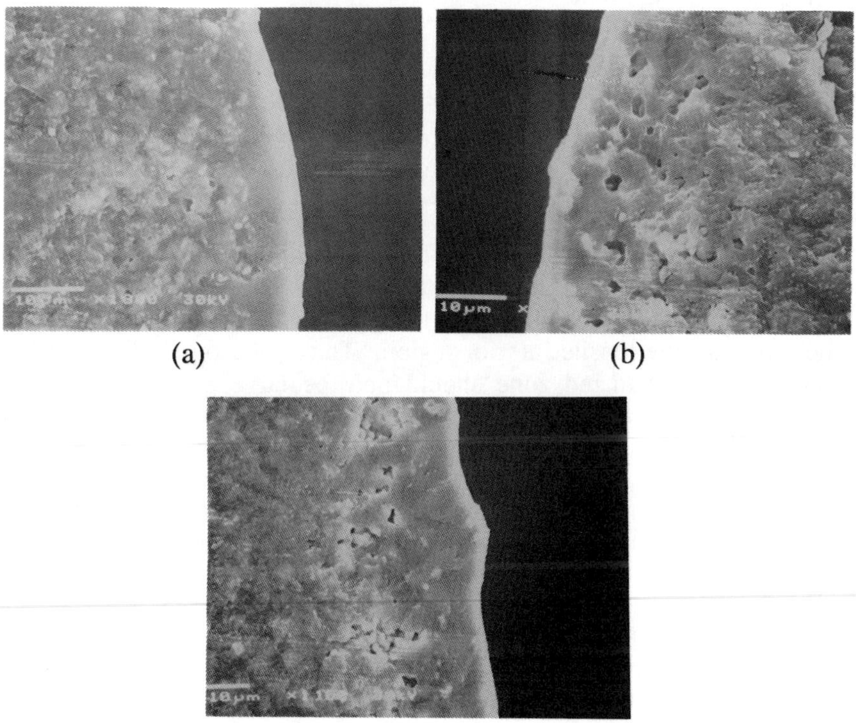

Figure 2. SEM micrographs showing cross-section of reaction layer after oxidation of Al_2O_3-SiC_w composites at 1600°C for (a) 0.25h,(b) 1h, and (c)1.5h.

porosity by increasing the distance the CO(g) product must travel before escaping the sample. The amount of retained porosity is also a function of oxygen and carbon monoxide diffusion rates and reaction products. It is logical that higher diffusion rates (as effected by increased temperature oxidation) and smaller thickness oxidation layers (as effected by decreased time at temperature) could promote the development of low porosity oxidation layers which are apparently critical to achieving superior mechanical properties.

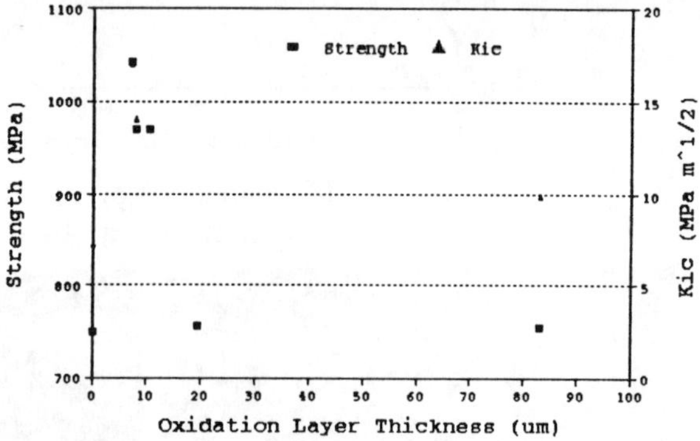

Figure 3. Mechanical properties vs. oxidation layer thickness.

The effect of layer thickness on mechanical properties is depicted in Figure 3. It is clear that there is a critical thickness at which mechanical properties are optimized. As the thickness of oxidized zone increases, the structure initially becomes stronger and subsequently weaker. From a fracture mechanics point of view, a thin compressive surface zone results in surface crack prevention benefits that can outweigh the continuum loss of strength associated with the tensile interior region. Thus, it becomes clear that the thickness of the oxidized zone should only be large enough to prevent surface cracks. In this study the best mechanical properties for the hot pressed Al_2O_3-SiC_w composites for the oxidation conditions employed were achieved with a thin layer of ≈ 8 μm. Thus, the attainment of high strength and fracture toughness composites through a post-oxidation treatment is dependent on both the oxidation layer thickness and characteristics, where a low porosity oxidation layer of a critical thickness is desired.

It has been shown that the oxidation treatment can be utilized to improve the mechanical properties of high density structural materials. The investigation was continued to determine whether the same effect could be incurred for weaker materials such as the hot pressed SiC_{pl} reinforced composite. The far superior mechanical properties of the SiC_w- reinforced composite in comparison to the SiC_{pl} composite can be partially attributed to the higher SiC_w content and also to the detrimental effect of the large SiC_{pl} on flexural strength in particular.

Oxidation of the hot pressed Al_2O_3-10 vol% SiC_{pl} composite was performed at 1600°C, and the resultant mechanical properties as a function of oxidation time are shown in Figure 4. A significant increase in toughness was achieved to 7.4 MPa·m$^{\frac{1}{2}}$, which is double that of the unoxidized sample. A slight improvement is strength was also achieved, although not of the magnitude expected. This is a result of the porous and degraded microstructures which were produced for all investigated oxidation times up to 2h. Figure 5(a) shows the microstructure of the sample prior to oxidation, and Figure 5(b) shows the microstructure of the same sample after oxidation at 1600°C, revealing coarsening and crack development. Figure 6 shows a cross sectional view of the oxidation layer, which contains large pores. It is likely that the SiC_{pl} content of this composite may be too low to create a sufficient volume of reaction products to uniformly and completely encase the material with a dense oxidation layer. However, the improvement in toughness in particular indicates that the post oxidation treatment could result in superior platelet-reinforced composites if the composition and microstructure of the base material were improved.

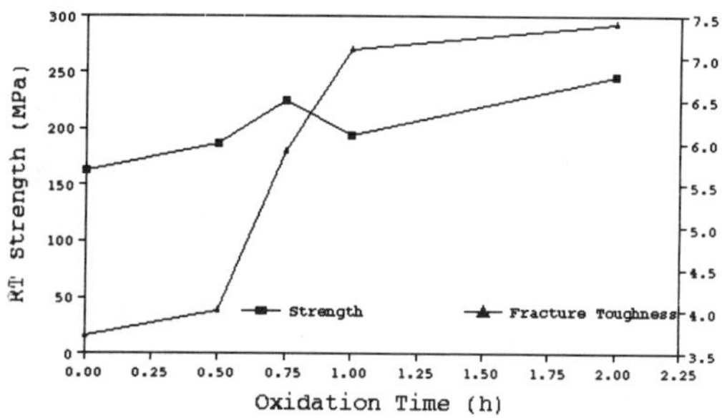

Figure 4. Mechanical properties of SiCpl composite after oxidation at 1600C.

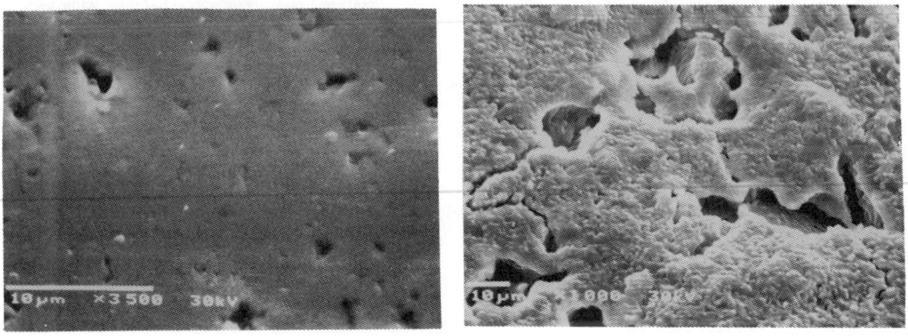

(a) (b)

Figure 5. SEM micrographs of Al_2O_3-SiC_{pl} composite (a) before and (b) after oxidation at 1600°C.

Similar results were observed for the pressureless sintered Al_2O_3-SiC_w composites, which also had some porosity in the as-sintered samples. Figures 7 and 8 show the toughness and strength, respectively, of the composites before and after oxidation at 1600°C for periods of time up to 1h. All measured toughness values showed an increase up to ≈8 MPa·m$^{1/2}$ over the toughness of the base material of ≈5 MPa·m$^{1/2}$. However, in most cases little improvement in the strength was observed. Examination of the fracture surfaces of these materials revealed that the failure origins were commonly in the bulk ceramic and not the surface as shown in Figure 9, where the fracture origin is indicated by the arrow. Other common defects found in the bulk material can be exemplified by that shown in Figure 10, which illustrates porous regions possibly caused by whisker clumping. In addition, the lower whisker content of these composites (15 vol%) resulted in a disruptive oxidized surface layer as shown in Figure 11 after oxidation for 1h. These results suggest that

Figure 6. SEM micrograph of cross-section of SiCp-reinforced composite.

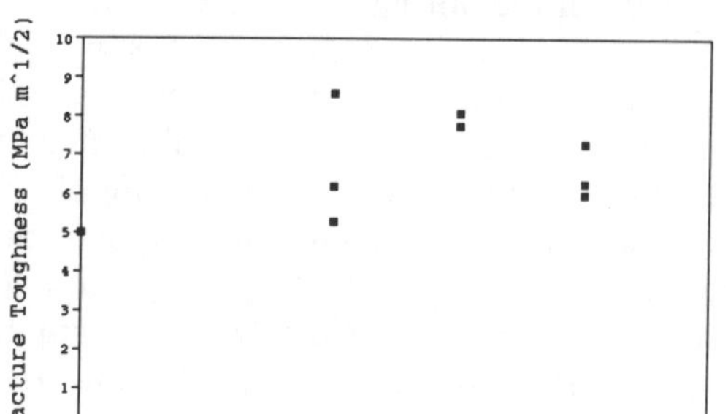

Figure 7. Fracture toughness of pressureless sintered composite after 1600°C oxidation.

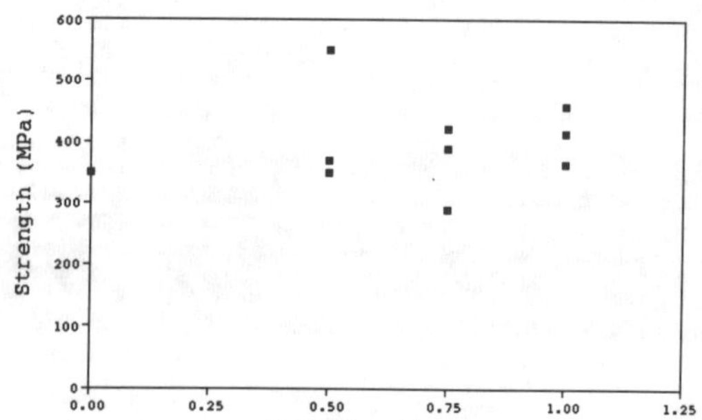

Figure 8. Strength of pressureless sintered composite after 1600°C oxidation.

processing must be optimized to eliminate any large pre-oxidation defects, and that the SiC content must be great enough to create a mainly continuous oxide surface layer or sufficient compressive stress.

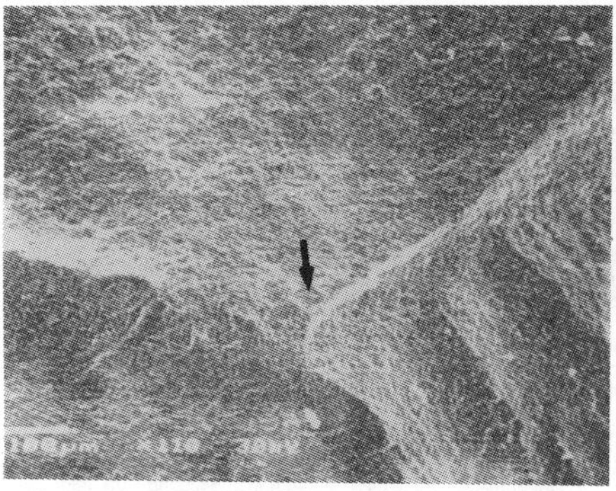

Figure 9. Failure origin in bulk of pressureless sintered composite.

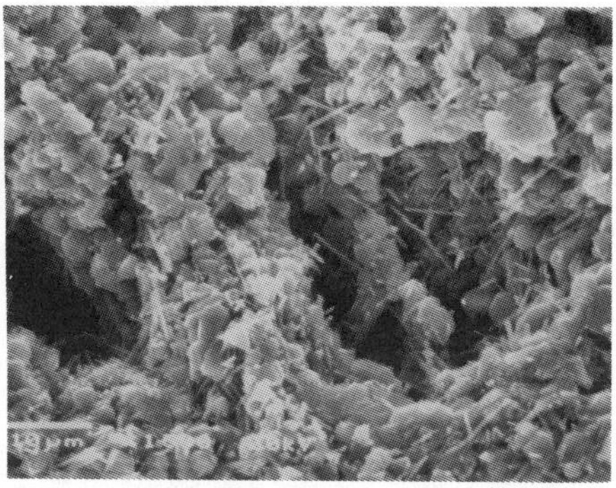

Figure 10. Defect observed in bulk of pressureless sintered composite.

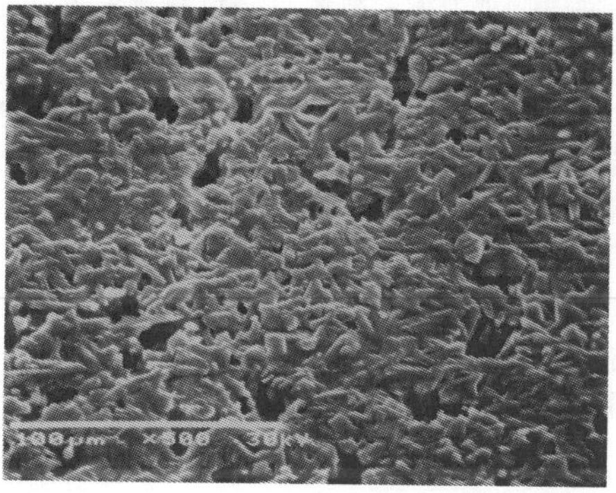

Figure 11. Oxidized surface of pressureless sintered SiC_w-reinforced composite.

Finally, the major objective of this effort was to eliminate the dependence of mechanical properties of structural ceramics on surface defects which are commonly introduced merely as a result of simple handling and use. Figures 12(a) and (b) show the surface of a sample oxidized for 0.25h and 1.5h, respectively. The shorter term oxidation resulted in surface flaws in the form of pores which were as large as 50 μm in diameter. Increasing the oxidation time to >1h resulted in the elimination of these pores, likely through the flow of the glassy oxidation layer which formed. Interestingly, the strength of the sample with the porous surface was 970 MPa (much higher than would be associated with 50 μm strength limiting flaws) while that of the sample with the nonporous surface was only 756 MPa. This again supports the theory that the mechanical properties of the samples are governed by the characteristics of the reaction layer itself and not the surface or bulk characteristics of the material. Thus, optimization of the oxidation cycle to optimize the reaction layer characteristics of the material can result in a tremendous improvement in mechanical properties. At the same time, these properties will not be later degraded by the introduction of surface flaws.

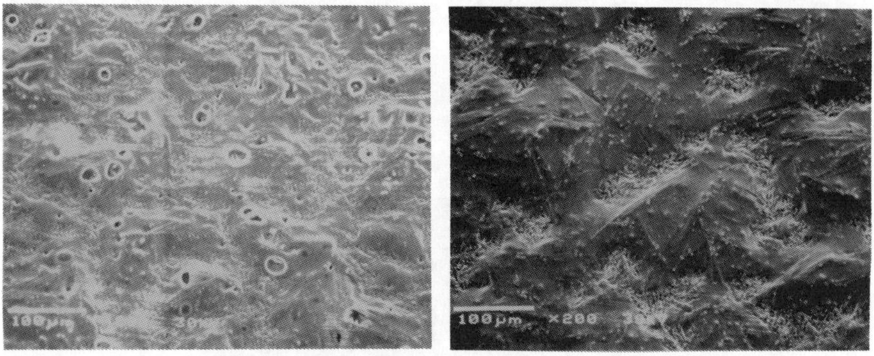

(a) (b)

Figure 12. Surfaces of hot pressed SiC_w-reinforced composites after oxidation at 1600°C for (a) 0.25h and (b) 1.5h.

CONCLUSIONS

Oxidation of the SiC component in SiC-reinforced Al_2O_3 composites to create a compressive surface layer is a valid technique for improvement of room temperature strength and fracture toughness if the following conditions apply: (1) the SiC content is large enough to create a continuous oxide surface layer, (2) the oxidation conditions are optimized to produce a low porosity oxidation layer of some critical thickness, and (3) large size flaws in the bulk of the composites are eliminated prior to oxidation.

ACKNOWLEDGEMENTS

We gratefully acknowledge DoD (DARPA) support under Contract DAAH01-92-C-R277.

REFERENCES

[1]D.G. Holloway, The Physical Properties of Glass; pp. 190-96. Wykeham Publications Ltd., London, 1973.

[2]D.A. Duke, J.E. Megles, J.E. MacDowell, and H.E. Bopp, "Strengthening Glass-Ceramics by Application of Compressive Glazes," J. Am. Ceram. Soc. 51 [2]98-102 (1968).

[3]R.A. Cutler and A.V. Virkar, "The Effect of Binder Thickness and Residual Stresses on the Fracture Toughness of Cemented Carbides," J. Mater. Sci. 20, 3557-73 (1985).

[4]M.L. Torti, Jr. and D.W. Richerson, "High Strength Composite Ceramic Structure," U.S. Pat. No. 3 911 188, Oct. 7, 1975.

[5]R.T. Pascoe and R.C. Garvie, "Surface Strengthening of Transformation Toughened Zirconia," pp.774-84 in Ceramic Microstructures '76. Edited by R.M. Fulrath and J.A. Pask. Westview Press, Boulder, CO, 1977.

[6]J.S. Reed and A. Lejus, "Effect of Grinding and Polishing on Near-Surface Phase Transformations in Zirconia," Mater. Res. Bull. 12 [10]949-4 (1977).

[7]N. Claussen, "Stress-Induced Transformation of Tetragonal ZrO_2 Particles in Ceramic Matrices," Acta Metall. 61 [1-2]85-86 (1978).

[8]T.K. Gupta, "Strengthening by Surface Damage in Metastable Tetragonal Zirconia," J. Am. Ceram. Soc. 63 [1-2]117 (1980).

[9]F.F. Lange and A.G. Evans, "Erosive Damage Depth in Ceramics: A Study on Metastable, Tetragonal Zirconia," J. Am. Ceram. Soc. 62 [102]62-65 (1979).

[10]D.J. Green, F.F. Lange, and M.R. James, "Factors Influencing Residual Surface Stresses due to a Stress-Induced Phase Transformation," J. Am. Ceram. Soc. 66 [9]623-29 (1983).

[11]D.J. Green, "A Technique for Introducing Surface Compression into Zirconia Ceramics," J. Am. Ceram. Soc. 66 [10]C-178-C-179 (1983).

[12]A.V. Virkar, J.L. Huang, and R.A. Cutler, "Strengthening of Oxide Ceramics by Transformation-Induced Stresses," J. Am. Ceram. Soc. 70 [3]164-70 (1987).

[13]R.A. Cutler, J.D. Bright, A.V. Virkar, and D.K. Shetty, "Strength Improvement in Transformation-Toughened Alumina by Selective Phase Transformation," J. Am. Ceram. Soc. 70 [10]714-18 (1987).

[14]W.M. Kriven, G. Van Tendeloo, T.N. Tiegs, and P.F. Becher, "Effect of

High Temperature Oxidation on the Microstructure and Mechanical Properties of Whisker-Reinforced Ceramics," pp. 939-947 in Ceramic Microstructures '86, Edited by Pask and Evans, Plenum Publishing Corporation, 1988.

[15]K.L. Luthra and H.D. Park, "Oxidation of Silicon Carbide-Reinforced Oxide-Matrix Composites at 1375° to 1575°C," J. Am. Ceram. Soc. 73 [4] 1014-23 (1990).

[16]M. Maeda, K. Nakamura, and M. Yamada, "Oxidation Resistance Evaluation of Silicon Carbide Ceramics with Various Additives," J. Am. Ceram. Soc. 72 [3] 512-14 (1989).

[17]B. Berton et al., "High-Temperature Oxidation of Silicon Carbide in Simulated Atmospheric Re-Entry Conditions," J. Mat. Sci. 27, 3206-3210 (1992).

[18]M. Balat et al., "Active to Passive Transition in the Oxidation of Silicon Carbide at High Temperature and Low Pressure in Molecular and Atomic Oxygen," J. Mat. Sci. 27, 697-703 (1992).

[19]W. Vaughn and H.G. Maahs, "Active-to-Passive Transition in the Oxidation of Silicon Carbide and Silicon Nitride in Air," J. Am. Ceram. Soc. 73 [6] 1540-43 (1990).

[20]H. Kim and A.J. Moorhead, "Effects of Active Oxidation on the Flexural Strength of α-Silicon Carbide," J. Am. Ceram. Soc. 73 [7] 1868-72 (1990).

[21]N. Frety et al., "Oxidizing Ageing Effects on SiC-SiC Composites," J. Mat. Sci. 27, 5084-5090 (1992).

[22]K.L. Luthra, "Some New Perspectives on Oxidation of Silicon Carbide and Silicon Nitride," J. Am. Ceram. Soc. 74 [5] 1095-1103 (1991).

[23]J. Dongliang, S. Jihong, and T. Shouhong, "Strengthening of Silicon Carbide Ceramics by Surface Nitridation during Hot Isostatic Pressing," J. Am. Ceram. Soc. 75 [9] 2586-89 (1992).

[24]J.W. Hinze and H.C. Graham, "The Active Oxidation of Si and SiC in the Viscous Gas-Flow Regime," J. Electrochem. Soc. 123 [7] 1066-1073 (1976).

[25]C. Wagner, "Passivity During the Oxidation of Silicon at Elevated Temperatures," J. Appl. Phys. 29 [9] 1295-97 (1958).

[26]Y. Takeda and N. Nakamura, "Oxidation Behavior of Silicon Carbide Ceramics at 1500°C," pp. 459-60 in Proceedings of the Annual Meeting of the Ceramic Society of Japan, Okayama, May 22-24, 1985. The Ceramic Society of Japan, Tokyo, Japan, 1985.

[27]W.D. Carruthers, "Durability Testing of Ceramic Materials"; pp. 427-43 in Proceedings of the Automotive Technology Development Contractor Coordination Meeting, Dearborn, MI, Oct. 26-29, 1981.

OXIDATION MECHANISMS OF MULLITE/ZrO_2/SiC-PLATELET COMPOSITES

Avigdor Zangvil, Youren Xu and Gaoling Fu
University of Illinois at Urbana-Champaign, Urbana, IL 61801

INTRODUCTION

Reinforcement of ceramics with SiC whiskers or platelets has been receiving more and more attention because of the ability of these reinforcements to increase the fracture toughness and strength of composite materials [1-9]. The stability of SiC whiskers or platelets within the ceramic matrix as well as in the environment is extremely important for high temperature applications. Alumina and mullite have been selected as oxide matrix materials for their excellent high-temperature stability and low oxygen permeabilities. Very comprehensive work has been performed on the oxidation of SiC-reinforced alumina and mullite matrix composites [10] and mullite/zirconia matrix composites [11]. Two oxidation modes were proposed in the literature [11-12]. Mode I is defined as the case where oxygen can react with the whole SiC particle before it diffuses further into the matrix, resulting in a clear boundary between a layer with completely oxidized SiC particles and the underlying composite with unoxidized SiC particles, as shown in Fig. 1a. Mode II is the case where oxygen can deeply penetrate into the matrix before SiC particles in the outer region are completely oxidized, leaving a long range of partially oxidized SiC particles behind, as shown in Fig. 1b. Mode I was found in mullite/SiC-whisker and mullite/Mg-PSZ/SiC-whisker composites while mode II was observed in mullite/Y-PSZ/SiC-whisker composites. These cases were qualitatively interpreted on the basis of the relationship between oxygen diffusivities in the matrix and in the oxide layer [11].

Recently, a simple model has been developed for the purpose of the rationalization of the oxidation modes concept [13]. The molecular oxygen flux is given by:

$$\frac{1}{A}\frac{dn}{dt} = \frac{D_m C_m}{z} \frac{1}{1+\dfrac{2D_m h}{D' z}} \tag{1}$$

Where A is the surface area of the platelets; n is the number of diffusing O_2

To the extent authorized under the laws of the United States of America, all copyright interests in this publication are the property of The American Ceramic Society. Any duplication, reproduction, or republication of this publication or any part thereof, without the express written consent of The American Ceramic Society or fee paid to the Copyright Clearance Center, is prohibited.

molecules; t is the time; D_m and D' are the oxygen diffusivities in the matrix and SiO_2 layer, respectively; C_m is the oxygen solubility in the matrix; h and z are the SiO_2 layer thickness and the depth of the platelet below the surface, respectively. This enabled the derivation of an approximate predicted relationship between the thickness of the oxide layer on the surface of an embedded platelet and the distance of that platelet from the outer surface of the composite [13]:

$$2h = -z\frac{D'}{D_m} + \sqrt{z^2\left(\frac{D'}{D_m}\right)^2 + \frac{C_m V_0 D' t}{\sigma}} \qquad (2)$$

Where V_0 is the molar volume of SiO_2, and σ is a stoichiometry factor (the number of moles O_2 reacting with one mole SiC). The equation indicates that the thickness of the silica layer around each oxidizing SiC particle is not only a function of oxygen diffusivities in the matrix and silica, but also of the depth of the SiC particle in the matrix. Therefore, microscopic measurements of h vs. z can be compared with predictions of the model and enable better understanding of oxidation modes in composites. The calculated thickness of silica layer vs. depth curve showed a good agreement with the experimental results [11] of oxidation of mullite/Y-PSZ/SiC-whisker composites, as shown in Fig. 2 [13].

To improve the understanding of oxidation modes in composites, we conducted systematic studies of mullite/zirconia/SiC-platelet composites as a model system. This was based on the following considerations: (1) mullite has very low oxygen permeability which can lead to a matrix diffusion-controlled oxidation process; (2) zirconia has a high oxygen diffusivity, so that control of the volume fraction of zirconia can adjust oxygen diffusion rate in the matrix. This will help in obtaining silica layer diffusion-controlled or mix-controlled regimes. (3) SiC platelets are single crystals with flat (111) surfaces, which should enable easy detection and measurement of oxide layers on them.

EXPERIMENTAL PROCEDURE

A coarse mullite powder (particle size: 5 - 25 μm, Aremco Products, Inc., Ossining, NY), fine Y_2O_3-stabilized zirconia powder (HSY-3.0U, particle size: ~ 50 nm, Zirconia Sales (America), Inc., Marietta, GA) and two kinds of α-SiC platelets (-100,+200 mesh and -400 mesh, Third Millennium Technologies, Inc., Knoxville, TN) were used in this study. The thicker platelets have a thickness of 5-15 μm, and the thinner ones have a thickness of 0.5-3 μm. Three compositions were selected for this study: mullite-10% SiC platelet, mullite-20%ZrO_2-10%SiC platelet and mullite-40%ZrO_2-10%SiC platelet (all compositions are given in volume%). Thicker and thinner SiC platelets were used for each composition. Samples without zirconia were hot-pressed at 1750°C, and samples

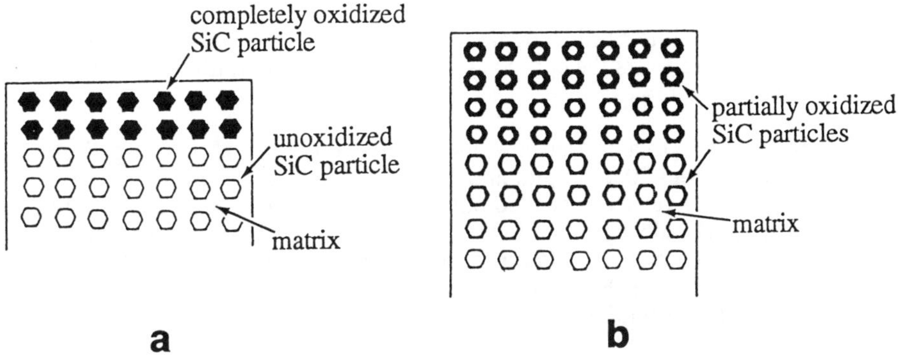

Fig. 1 Schematic drawings of oxidation mode I (a), and mode II (b) [11].

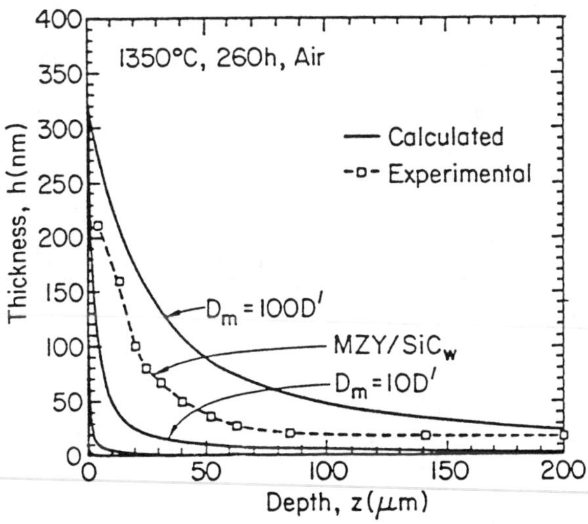

Fig. 2 Calculated (solid) and experimental curves of oxide layer thickness vs. depth into the composites, showing good agreement for SiC whisker oxidation in a mullite/Y-PSZ matrix. D' and D_m are oxygen diffusivities in the SiO_2 layer and the matrix, respectively [13].

with zirconia were hot-pressed at 1600°C, both for 20 min. Hot-pressing was conducted in a graphite heating element furnace in vacuum and an axial pressure of 33.3 MPa. Table I lists the compositions, processing conditions and densities of all composites samples.

Specimens were cut to a desired size, then ground on diamond grinding discs with a sequence from 70 to 40 μm, and finally polished to a 3-μm-diamond finish. The exposure test was carried out in an alumina tube furnace with a SiC heating element (Lindberg, Watertown, WI), where samples were exposed to flowing oxygen at a rate of ~3 *l/min* at 1430°C. Exposure times for samples with zirconia were selected as 8, 16, 32, and 64 h, and 8, 24, 56, and 80 h exposure times were used for samples without zirconia.

Table I. Compositions, Processing Conditions and Densities of Mullite/Zirconia/SiC-Platelet Composites

Sample Code*	Composition**	Processing Condition	Relative Density
40L	mullite-40%ZrO_2-10%SiC	1600°C-20 min	99.9% TD
40S	mullite-40%ZrO_2-10%SiC	1600°C-20 min	99.9%TD
20L	mullite-20%ZrO_2-10%SiC	1600°C-20 min	97% TD
20S	mullite-20%ZrO_2-10%SiC	1600°C-20 min	97%TD
L	mullite-10%SiC	1750°C-20 min	97%TD
S	mullite-10%SiC	1750°C-20 min	97% TD

* L = thick SiC platelets; S = thin SiC platelets
** ZrO_2 content is given as vol.% of the *matrix*

As-hot-pressed and exposed samples were examined in a polarizing optical microscope (Carl Zeiss Ultraphot II) and a high resolution scanning electron microscope (SEM) (S-800, Hitachi Ltd., Japan). A thin layer of Au/Pd was applied to specimen surfaces prior to SEM observation. The oxidation depth and microstructures in oxidized layers can best be viewed in the optical microscope and SEM by observing polished cross-sections. Samples for this purpose were prepared by cutting the oxidized sample in half perpendicular to one of the oxidized surfaces, and then gluing together using a 5-Minute® epoxy (Devcon Corp., Danvers, MA) with the two oxidized surfaces facing each other. The sandwiched samples were ground and polished to a 3-μm-diamond finish.

RESULTS AND DISCUSSION

1 Microstructure of as-hot-pressed samples

Fig. 3 shows the optical micrographs of the typical microstructures of as-hot-pressed 40L and 40S samples, parallel to the direction of hot-pressing. The brightest image represents the SiC platelet, the gray is the image of zirconia phase regions, and the dark particles are mullite. It is clear that both thicker and thinner platelets are well aligned with their axis parallel to the hot-pressing direction. This preferential orientation was found in all samples. Fig. 4 consists of the SEM images showing the microstructures parallel to the hot-pressing axis of as-prepared 40L and 20L samples. It is difficult to distinguish SiC platelets from mullite grains in SEM micrographs; however, the zirconia phase is still distinguishable. It appears that zirconia is the continuous phase in the 40% zirconia sample. Because the particle size of mullite is much larger than the particle size of zirconia, it is easier for zirconia to form a continuous phase. This tendency could also be found in 20% zirconia sample, although the degree of phase continuation is greatly reduced. In comparison to the dense 40L sample, a few percent of porosity can be seen in the low zirconia content sample. This agrees with the measured densities in Table I.

2. Characteristics of exposed samples

Fig. 5 shows the polarized light micrographs of the cross section images of 40L and 40S samples which were each exposed at 1430°C for 8, 16, and 32 h. There are two main features in these micrographs: (1) At the same exposure time, the thickness of the layer with extensively oxidized platelets (exhibiting brighter contrast in polarized light) in 40L samples is considerably larger than that in 40S samples. This is probably due to the fact that the total surface area which thinner platelets provided was significantly larger than the same volume of thicker platelets provided. Therefore, at the same depth, small platelets consume more oxygen than the same volume of large platelets, lowering oxidation rate with depth. (2) A fraction of unoxidized SiC platelets can be found in 40L, meaning that the white layer is not a completely oxidized layer. This phenomenon was not observed in the sample with thinner platelets. For samples without zirconia, no significant difference in layer thickness could be found. This is because the mullite matrix offers a very low oxygen permeation and controls the oxidation process, leading to a mode I oxidation.

The cross sections of the microstructures of all exposed samples were observed by SEM. Fig. 6 shows the typical images for exposed samples 20L and 40L. The white layer, which was distinguished in the polarizing microscope, was no longer observed in SEM images. These images seem to be close to the Mode II microstructure although considerable deviations were seen in some images. Fig. 7 shows the relationship between the thickness (h) of the silica layer around the partially oxidized platelet and the depth (z) from the outer surface to that platelet,

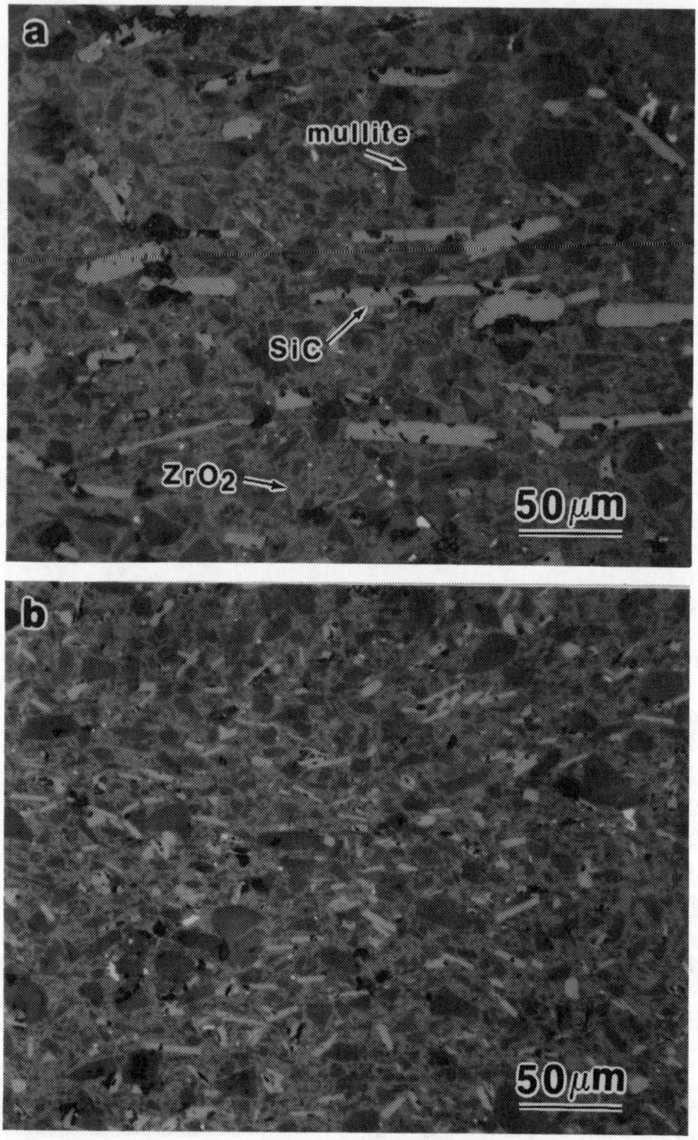

Fig. 3 Optical micrographs of as-hot-pressed mullite/zirconia/SiC-platelet composites: (a) sample 40L, and (b) sample 40S. Platelets tend to align parallel to the direction of hot-pressing.

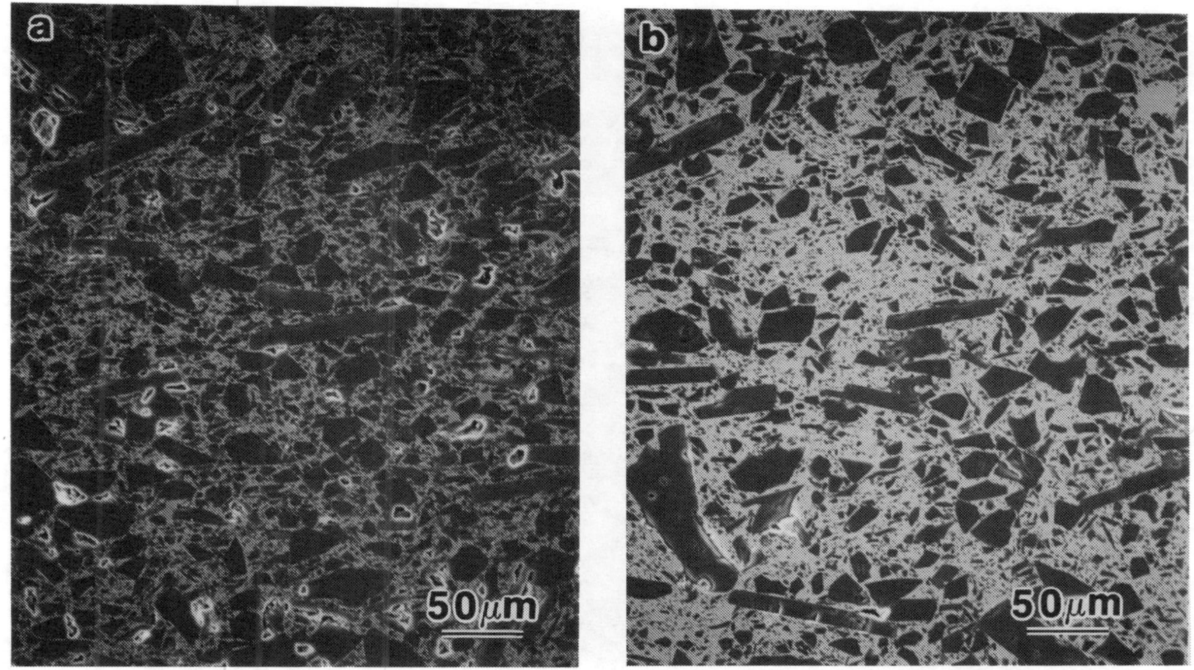

Fig. 4 SEM images of as-hot-pressed composites: (a) 20L, and (b) 40L. The hot-pressing axis is in the vertical direction.

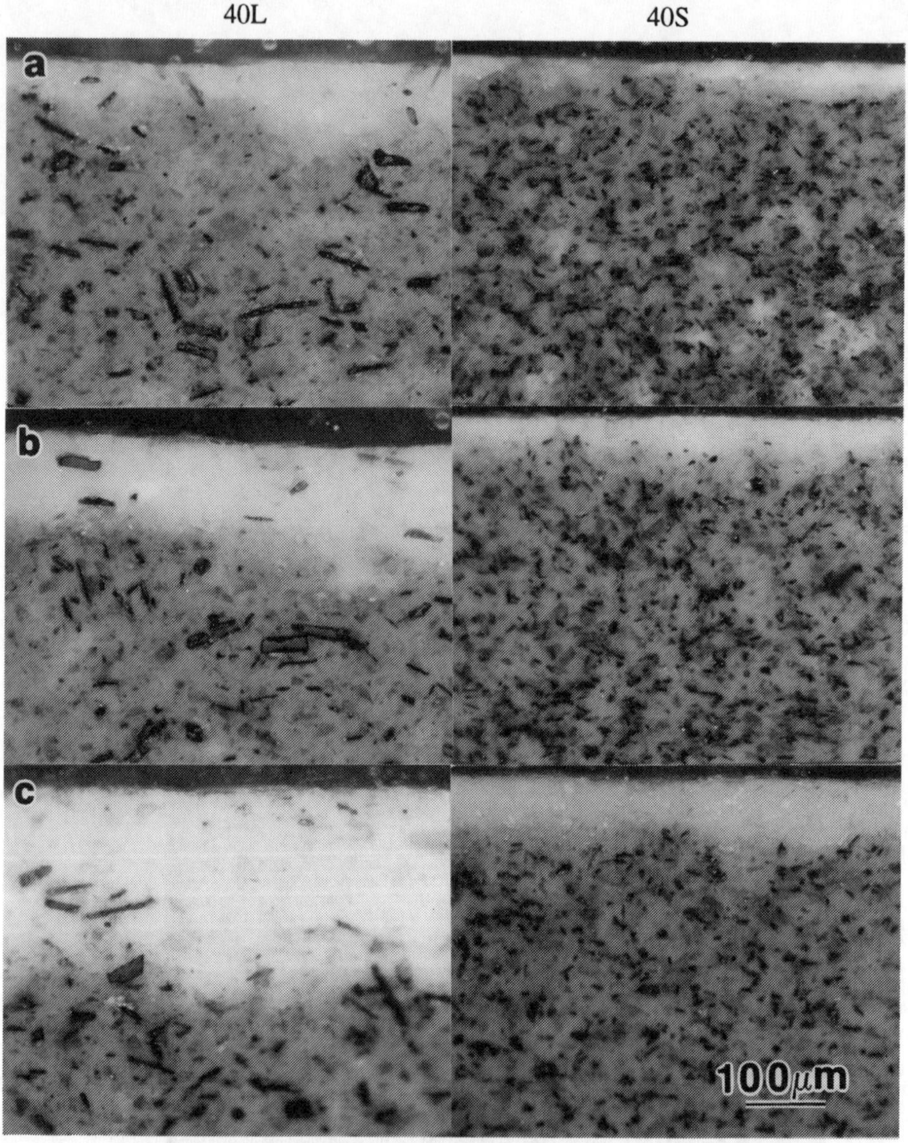

Fig. 5 Polarized light micrographs of exposed 40L and 40S samples, showing the cross section images of the composites after exposure at 1430°C for different times: (a) 8 h; (b) 16 h; and (c) 32 h.

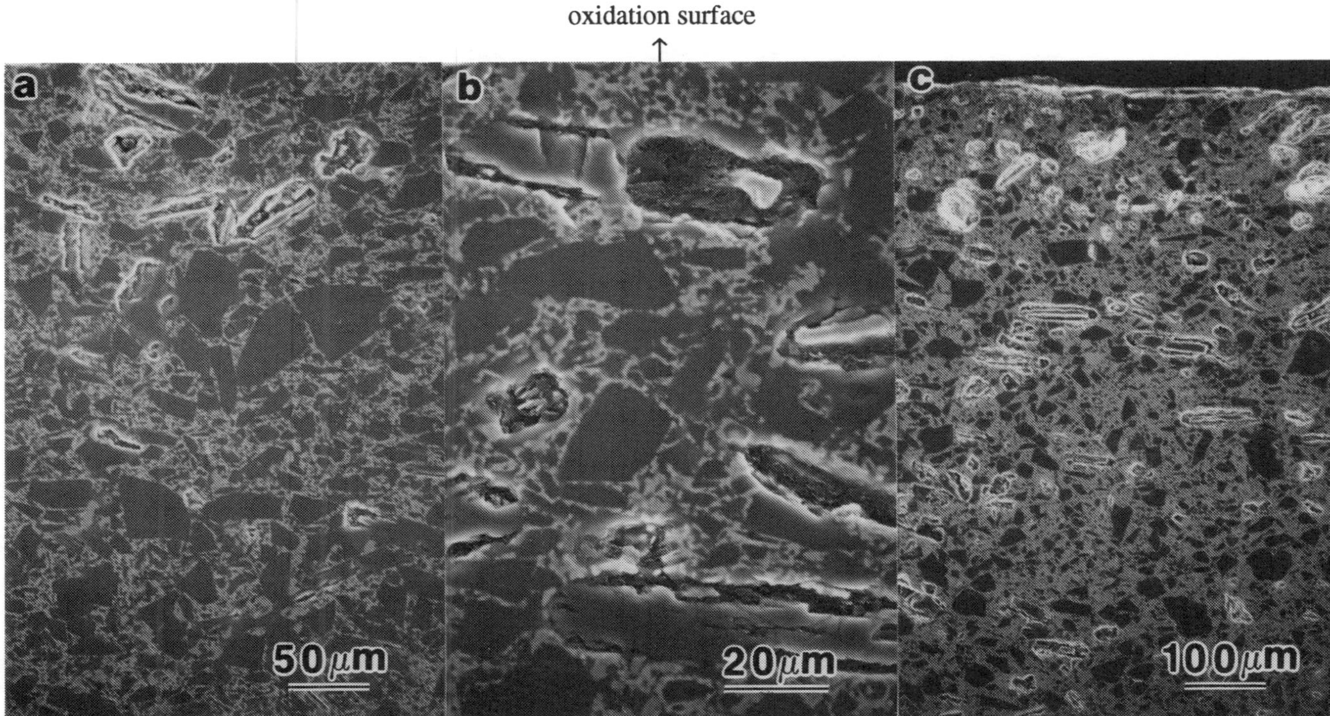

Fig. 6 SEM cross section images of exposed samples. (a) Low magnification image showing the change of the thickness of the silica layer around SiC platelet with depth for 32 h exposed 20L sample; (b) Medium magnification image of partially oxidized SiC platelets in 32 h exposed 20L sample; (c) Low magnification image of an 8 h exposed 40L sample.

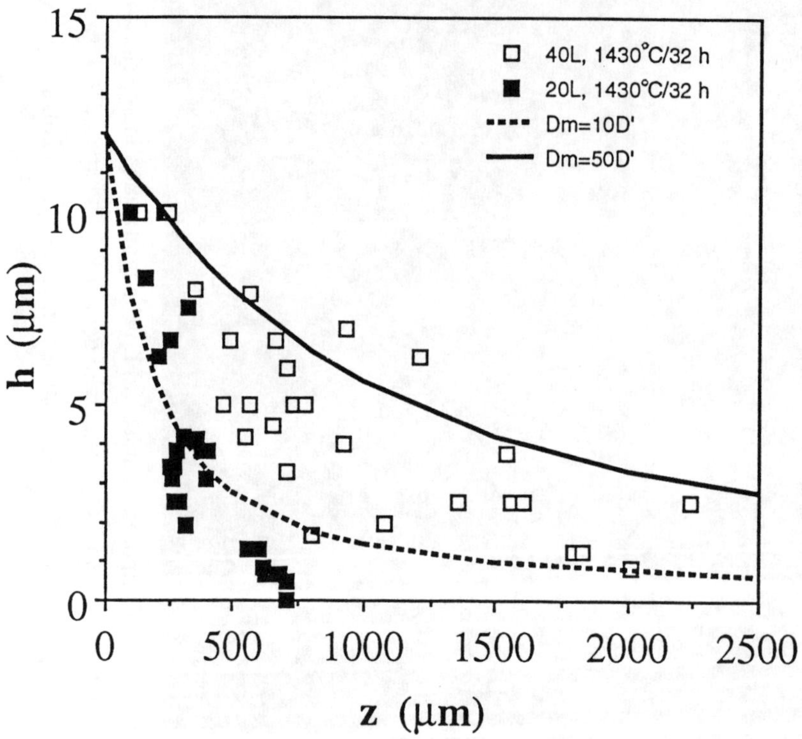

Fig. 7 Microscopically measured data and calculated curves of h vs. z. The values used for calculation were: $C_m = 2.87 \times 10^{-6}$ mole/cm^3; $D' = 10^{-6}$ cm^2/sec; $V_0 = 26.1$ cm^3/mole, and $\sigma = 1.5$.

for two exposed samples. Evidently, partial oxidation depth (z) is much larger for the 40L sample, due to the higher oxygen diffusivity in its mullite + 40% ZrO_2 matrix. These data are rather scattered, yet show reasonable agreement with the calculated curves. The simulated curves suggest that the O_2 diffusivity in 40L is 5 times larger than in 20L. It should be pointed out that the calculated curve was based on the assumption that no fully oxidized layer was formed on the surface. The reason that the present results generally fit the calculated curves is probably due to the oxygen permeability remaining essentially unchanged from the light oxidized layer to the underlying materials (with partially oxidized platelets), and the oxygen diffusion in the whole sample still being dominated by the zirconia phase.

SUMMARY

1. The oxidation behavior of mullite/zirconia/SiC-platelet composites could be microscopically described as "mode II", where a large range of partially oxidized SiC platelets were present.

2. Experimental results of the silica layer thickness (h) vs. the depth (z) are in reasonably good agreement with calculated curves based on a simplified model. It was found that the oxidation of SiC platelets in a mullite/zirconia ceramic matrix is not simply a function of diffusivity in silica or matrix, but also of the depth of the SiC platelet. This results from the mode II behavior which, in turn, results from the high oxygen diffusivity of the matrix.

ACKNOWLEDGMENTS

The authors would like to thank Third Millennium Technologies, Inc. for providing SiC platelets. This work is supported by the US Department of Energy, Division of Materials Sciences, under Contract No. DE-FG02-91ER45439.

REFERENCES

1. P.F. Becher and G.C. Wei, "Toughening Behavior in SiC-Whisker-Reinforced Alumina," J. Am. Ceram. Soc., 67[12] C-267-C-269 (1984).

2. G.C. Wei and P.F. Becher, "Development of SiC-Whisker-Reinforced Ceramics," Am. Ceram. Soc. Bull., 64[2] 298-304 (1985).

3. T.N. Tiegs and P.F. Becher, "Whisker Reinforced Ceramic Composites"; pp. 639-47 in Tailoring Multiphase and Composite Ceramics, Edited by R.E. Tressler, G.L. Messing, C.G. Pantano, and R.E. Newnham. Plenum Press, New York, 1986.

4. N. Claussen and G. Petzow, "Whisker-Reinforced Zirconia-Toughened Ceramics," pp. 649-62 in Tailoring Multiphase and Composite Ceramics, Edited by R.E. Tressler, G.L. Messing, C.G. Pantano, and R.E. Newnham. Plenum Press, New York, 1986.

5. N. Claussen, "Ceramic Platelet Composites"; in Proceedings of the 11th RISO International Symposium on Metallurgy and Materials Science 1990, Structural Ceramics Processing, Microstructure and Properties. Edited by J.J. Bentzen, J.B. Bilde-Sorensen, N. Christiansen, A. Horsewell, and B. Ralph. RISO National Laboratory, Roskidle, Denmark, 1990.

6. C. Nischik, M.M. Seibold, N.A. Travitzky, and N. Claussen, "Effect of Processing on Mechanical Properties of Platelet-Reinforced Mullite Composites," J. Am. Ceram. Soc., 74[10] 2464-68 (1991).

7. Y.-S. Chou and D. J. Green, "Silicon Carbide Platelet/Alumina Composites: I, Effect of Forming Technique on Platelet Orientation," J. Am. Ceram. Soc., 75[12] 3346-52 (1992).

8. Y.-S. Chou and D. J. Green, "Silicon Carbide Platelet/Alumina Composites: II, Mechanical Properties" J. Am. Ceram. Soc., 76[6] 1452-58 (1993).

9. Y.-S. Chou and D. J. Green, "Silicon Carbide Platelet/Alumina Composites: III, Toughening Mechanisms" J. Am. Ceram. Soc., 76[8] 1985-92 (1993).

10. K.L. Luthra and H.D. Park, "Oxidation of Silicon Carbide-Reinforced Oxide-Matrix Composites at 1375° to 1575°C," J. Am. Ceram. Soc., 73[4] 1014-23 (1990).

11. C.C. Lin, "Microstructural Development, Phase Stability, and Toughening in Mullite Matrix Composites"; Ph.D. thesis, University of Illinois at Urbana-Champaign, 1991.

12. C.C. Lin, A. Zangvil and R. Ruh, "Oxidation of SiC Whisker Reinforced Mullite/Zirconia Matrix Composites: II, Oxidation Modes," submitted to J. Am. Ceram. Soc.

13. A. Zangvil, "Modeling of the Oxidation Behavior of SiC-Reinforced Ceramic Matrix Composites," to be published

STRESS-OXIDATION TESTS IN SiC_f/SiC COMPOSITES

S. Raghuraman[a], M.K. Ferber[b], J.F. Stubbins[a], A.A. Wereszczak[b]
[a]Department of Nuclear Engineering, University of Illinois at Urbana-Champaign, Urbana, IL 61801
[b]High Temperature Materials Laboratory, Oak Ridge National Laboratory, Oak Ridge, TN 37830

ABSTRACT

This research focusses on the stress assisted high temperature oxidation of a 2-D SiC_f/SiC woven composite. 3-point flexure specimens were subjected to static stresses at 950°C in air for time durations of 1-100 hours. At the end of the test, the surviving specimens were ruptured in fast fracture, in order to measure their remanent strengths. At low stresses (below matrix cracking stress), the composite exhibited no reduction in strength even after 100 hours of exposure to oxidizing atmosphere. Above the matrix cracking stress, the composite shows enhanced susceptibility to oxygen induced embrittlement, and strength degradation. The formation of matrix crack allows oxygen ingress to the fibers and interphases, and results in embrittlement. This is substantiated by SEM, AEM, and TEM observations.

1. INTRODUCTION

SiC_f/SiC material is being actively investigated with respect to its mechanical properties at room and elevated temperatures [1-10]. It is clear that oxidation effects in this material are very important. The CVI-SiC matrix is known to be stable beyond 2000°C [11], but the Nicalon-SiC fibers begin to degrade above 1200°C, although the degradation is slower when it is incorporated in the composite [12]. The carbon interphase layer is of the greatest interest because of its important role in determining the fracture behavior. The interfacial properties that are most beneficial for high fracture toughness and crack-resistant characteristics, can be greatly affected by exposure to oxidation above 700°C. The effect of oxidation on mechanical properties will depend on several factors such as interphase composition and thickness, temperature, and exposure time. The effect of stress assisted oxidation has not been studied, although it might be present in all of the structural applications of this material. The intent of this study is to examine stress-oxidation and ageing in a SiC_f/SiC material, using a combination of stresses and time durations in oxidizing environment at 950°C.

2. EXPERIMENTAL PROCEDURE

2.1. Material

The SiC_f/SiC composite used in this study is a commercially available material made by Du Pont CVI composites. It is a two-dimensional composite of Nicalon^{TM}-SiC fibers in a chemical vapor infiltrated SiC matrix. Individual fibers were bundled into tows which were woven into a plain weave. The fibers have a thin coating (50-100 nm) of C on their surface. Table 1 lists pertinent properties

To the extent authorized under the laws of the United States of America, all copyright interests in this publication are the property of The American Ceramic Society. Any duplication, reproduction, or republication of this publication or any part thereof, without the express written consent of The American Ceramic Society or fee paid to the Copyright Clearance Center, is prohibited.

of the material [13]. The woven material has a very porous structure, with small intrabundle pores and large interbundle cavities, arising from the CVI process.

2.2. Experiments

Rectangular bars (dimensions: 25 x 4.2 x 3.2 mm) were loaded in a 3-point flexure fixture of 20 mm span, with the laminates parallel to the loading direction. In order to distinguish high-temperature ageing effects from stress-oxidation effects, the specimens were prestressed at various static loads for time durations of 1-100 hours at 950°C. The temperature was selected based on published oxidation data [14], where it was shown that rapid oxidation of C interphase occurs. The prestresses were selected using the matrix cracking stress (σ_o) as a guideline. Using brittle matrix composite theory [15], and interphase strength of 20 MPa [16], $\sigma_o \cong 120$ MPa. At the end of the hold period, the surviving specimens were fractured at a fast loading rate (22 N/s). The specimens were then characterized using SEM, AEM, and TEM.

3. RESULTS

3.1. Stress-Oxidation Tests

The remnant strengths are plotted as a function of stress-oxidation exposure time in Figure 1, for all of the prestress conditions. It can be seen that when the holding stress is lower than the matrix cracking stress ($\sigma_o \cong 120$ MPa), the strength is unaffected by the stress-oxidation holding period. Above σ_o, there is drastic reduction in strength, even after short exposure times. Some of the specimens prestressed above σ_o did not survive the test. Those specimens were incorporated in the static fatigue data-set, and plotted as stress vs time to failure.

3.2. Scanning Electron Microscopy

The effect of stress assisted oxidation clearly depends on the ability of the applied prestress to initiate a matrix crack. Similar to the static load tests [8] only the cracked portion of the matrix is embrittled. For prestress below σ_o, (Figure 2) where no crack is formed, the entire fracture surface shows fiber pullout indicative of substantial strength. Fibers in unaffected zones exhibit rough features and fracture mirrors characteristic of high strengths. For prestress above σ_o, (Figure 3) the fracture surface shows an embrittled zone devoid of fiber pullout and an unaffected zone with pullout. Fibers in the embrittled zone have smooth fracture surfaces which are indicative of low fracture strengths.

3.3. Auger Electron Microscopy

High-Resolution AEM of the matrix, interphase and fiber regions of fractured specimens was carried out using a Perkin Elmer PHI 660 instrument at

10 keV. Two control specimens and one stress-oxidation specimen were analyzed. The two control specimens were cut from as-received material in the form of rectangular bars (dimensions: 16 x 2 x 2 mm). One of them was exposed to 950°C for 11 hours in a laboratory furnace. In order to avoid contamination from atmosphere, the control specimens were fractured *in-situ* in the AEM prior to analysis. Auger Electron Spectra (AES) was obtained at three points (1: fiber surface, 2: trough surface, 3: matrix). The samples were then intermittently sputtered with Argon ions for 20-30 seconds interval and AES data was obtained after each interval. This yielded depth profiles of interphase regions in the samples.

(a) control specimen 1: as-received, fractured *in-situ*

The initial AES data, and depth profiles are shown in Figure 4. Presence of C interphase is evident from AES data of the fiber and trough surfaces. The interphase is 60-100 nm thick. The depth profiles across the interphase shows the transition between the C layer to the fiber or matrix respectively.

(b) control specimen 2: oxidized for 11 hours at 950°C, fractured *in-situ*

The initial AES data obtained from interior of the specimen after *in-situ* fracture shows the same spectra as that of the as-received material (Figure 5). Although the specimen was less than 2 mm thick, exposure to 950°C oxidation did not affect the composition of the constituent phases. The depth profiles resembled that of the as-received material, so they are not shown here.

(c) stress-oxidation specimen: 140 MPa at 950°C, fractured after 2.6 hours

The initial AES data, depth profiles and final AES data are shown in Figure 6. There was a contamination layer on the surface due to exposure (consisting of C, O). Presence of silica was detected in all three regions. The fiber depth profile shows a drop in C concentration after only 5 nm, indicating that this was a contamination product and not the original interphase C. Comparison with the depth profile for the as-received material indicates that the C interphase is replaced with silica.

3.4. Transmission Electron Microscopy

Specimens were analyzed using a Philips EM-400 TEM operating at 120 keV. Both as-recieved and oxidized foils were analyzed. As the SEM and AEM showed that oxidation takes place only on the crack surfaces, we tried to make a TEM foil from the fracture surface of a stress-oxidation specimen. Due to the uneveness of the fracture surface, this attempt was not successful. Therefore it was decided to analyze a foil that was oxidized at 950°C for 11 hours in a laboratory furnace. Because of the small thickness of the foil, it is assumed that the foil would be oxidized in the same manner as the fracture surface.

Considerable care and time is required in preparing foil specimens for TEM due to high porosity of the material. Thin areas suitable for TEM examination are very small so that only a few fibers can be analyzed. Foils were prepared for TEM by first cutting 300 μm slices using a slow-speed diamond saw, followed by rough grinding (40 μm diamond disc), rough and fine polishing (15 μm and 6 μm diamond compound). This procedure yielded 100-125 μm thick foils. The foils were then further thinned to 45-50 μm in a dimpler with a flatting tool using 3 μm and 1 μm diamond slurry. Finally, the foils were mounted on copper washers and sputter-thinned using 4 keV, 5 mA argon ion beam in an ion miller.

Figure 7 shows TEM micrographs of the fiber (F), interphase (I) and matrix (M). The interphase C layer is seen to be continuous. Figure 8 shows TEM micrographs of oxidized specimens. The annulus left behind by interphase oxidation is seen clearly (Figure 8 (a)). The fiber shows an affected zone of ~ 250 nm thickness as indicated by a dark band along the entire peripheral surface, as indicated by arrows. In some cases a crack was observed to form in this region (Figure 8 (b)), associated with spalling of the layer. No discernable differences could be made between the peripheral and interior regions of the fiber other than in contrast and brightness. Auger electron microscopy of the peripheral zone showed increase in oxygen content, indicating formation of oxides. High resolution electron microscopy or convergent beam diffraction analysis will be required to accurately identify the affected zone.

4. DISCUSSION

Oxidation phenomenon in this material is a diverse process, depending on temperature, fiber architecture, oxygen partial pressure, interphase thickness, and exposure time. When exposed to 950°C in air, the interphases that are closest to the surface (specimen faces and crack faces) are consumed rapidly. Thermogravimetric analyses at 900-1000°C indicate that weight loss occurs initially corresponding to C interphase burnup [12, 14]. An annular opening is created between the fiber and matrix surfaces, through which oxygen can diffuse to the C interphase present at the end. Simultaneously, the fiber and the matrix surfaces also oxidize to form silica, which gradually fills the opening. TGA measurements record a weight gain during this phase [12, 14]. Competition from fiber/matrix oxidation, and also reduction in the annulus cross-section, is effective in reducing or preventing further interphase consumption [12]. This explanation is consistent with the fact that no oxidation products were found in the AES analysis of the interior of control specimen 2 (§ 3.3. *case b*, Figure 5). TEM analysis of an oxidized sample shows that the fiber surfaces have a different morphology. Fiber and matrix surfaces oxidize and spread into the annulus because of positive volume change associated with silica formation [12]. The relatively short time (11 hours) allowed for oxidation was not enough to completely fill the annular gap with silica. It is possible that oxidation rates are different for fiber and matrix and depend on the partial oxygen pressure at the annulus, such that fiber oxidizes more than the matrix. This is also seen in the AEM depth profiles of fiber and matrix of stress-oxidation specimen (§ 3.3. *case c*, Figure 6).

Thus two effects of oxidation on strength and fracture properties can be possible. First, the formation of silica in the annulus will result in a stronger and chemically bonded interfacial region. Such a strong bond will prevent debonding, increase stress in a crack-bridging fiber, and expose crack-tip fiber to stress concentration. When fracture of crack-bridging fiber occurs, pullout will be limited. Secondly, the fiber may be degraded due to formation of affected zone.

5. CONCLUSIONS

Stress-oxidation tests were performed on 2-D SiC_f/SiC composite at 950°C. Based on test results, SEM, AEM and TEM, the following conclusions are drawn:

1. Stress-assisted oxidation is a significant phenomenon due to its role in determining mechanical properties in actual applications.
2. Matrix cracking stress is the limiting stress for long term intermediate temperature applications.
3. Bulk oxidation does not take place despite the high porosity.
4. No ageing induced degradation was seen at 950°C upto 100 hours.
5. Oxidation of interphase can reduce strength and fracture toughness both by chemical bonding of fiber and matrix and by degrading fiber.

REFERENCES

1. N. Frety, M. Boussuge, Composi. Sci. Tech., **37** (1990) 177-89.
2. N. Frety, R. Molins, M. Boussuge, J. Mater. Sci., **27** (1990) 5084-90.
3. M. Gomina, P. Fourvel, M.-H. Rouillon, J. Mater. Sci., **26** (1991) 1891-8.
4. J.-M. Yang, W. Lin, C.J. Shih, W. Kai, S.M. Jeng, C.V. Burkland, J. Mater. Sci., **26** (1991) 2954-60.
5. L. Filipuzzi, G. Camus, J. Thebault, R. Naslain, Proc. of 11th RISΦ Intl. Symp. on Metall. and Mater. Sci., (1990) 283-9.
6. G. Grathwohl, A. Hahnel, B. Meier, E. Pippel, G. Richter, J. Woltersdorf, J. Eur. Ceram. Soc., **10** (1992) 1-12.
7. C.H. Henager, Jr., R.H. Jones, Ceram. Engr. Sci. Proc., **14** [7-8] (1993) 408-15.
8. S. Raghuraman, J.F. Stubbins, M.K. Ferber, A.A. Wereszczak, J. Nucl. Mater., in press.
9. S. Raghuraman, J.F. Stubbins, M.K. Ferber, A.A. Wereszczak, Ceram. Engr. Sci. Proc., in press.

10. D.A. Woodford, D.R. Van Steele, J.A. Brehm, L.A. Timms, J.E. Palko, J. Materials, **45** [5] (1993) 57-63.
11. D. Cojean and M. Monthioux, Br. Ceram. Trans. J., **91** (1992) 188-95.
12. L. Filipuzzi, G. Camus, R. Naslain, J. Thebault, J. Am. Ceram. Soc., **77** [2] (1994) 459-66.
13. Preliminary Engineering Data Sheet, Du Pont CVI Ceramic Matrix Composites.
14. P.F. Tortorelli, S. Nijhawan, L. Riester, R.A. Lowden, Ceram. Engr. Sci. Proc., **14** [7-8] (1993) 358-16.
15. D.B. Marshall, B.N. Cox, A.G. Evans, Acta Metall., **33** [11] (1985) 2013-21.
16. F. Abbe', J.-L. Chermant, J. Am. Ceram. Soc., **73** [8] (1990) 2573-5.

ACKNOWLEDGMENTS

This work was supported by the Basic Energy Sciences - U.S. Department of Energy, through the Materials Research Laboratory at University of Illinois at Urbana-Champaign, under contract no. DE-AC02-76ER01198. Mechanical testing was performed at High Temperature Materials Laboratory at Oak Ridge National Laboratory. Microscopy was performed at the Center for Microanalysis of Materials at MRL, UIUC. Assistance of John McIntosh, Nancy Finnegan, and Laura Reister with microscopy is acknowledged.

Table 1. Properties of SiC_f/SiC material [13].

Fiber Volume	Porosity	Fiber Lay-up	Fiber Diameter	Modulus
40%	10%	0°/90° Plain Weave	10-15 μm	211 GPa

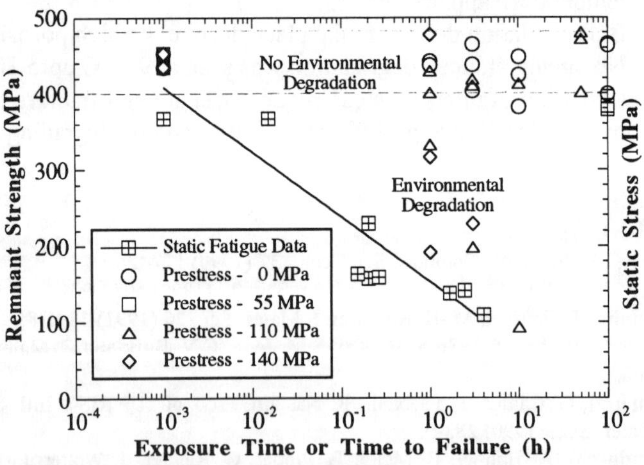

Figure 1. Stress-oxidation test results: strength after exposure to oxidation at 950°C at the indicated prestress conditions. Static stress vs time to failure is shown for prematurely fractured specimens.

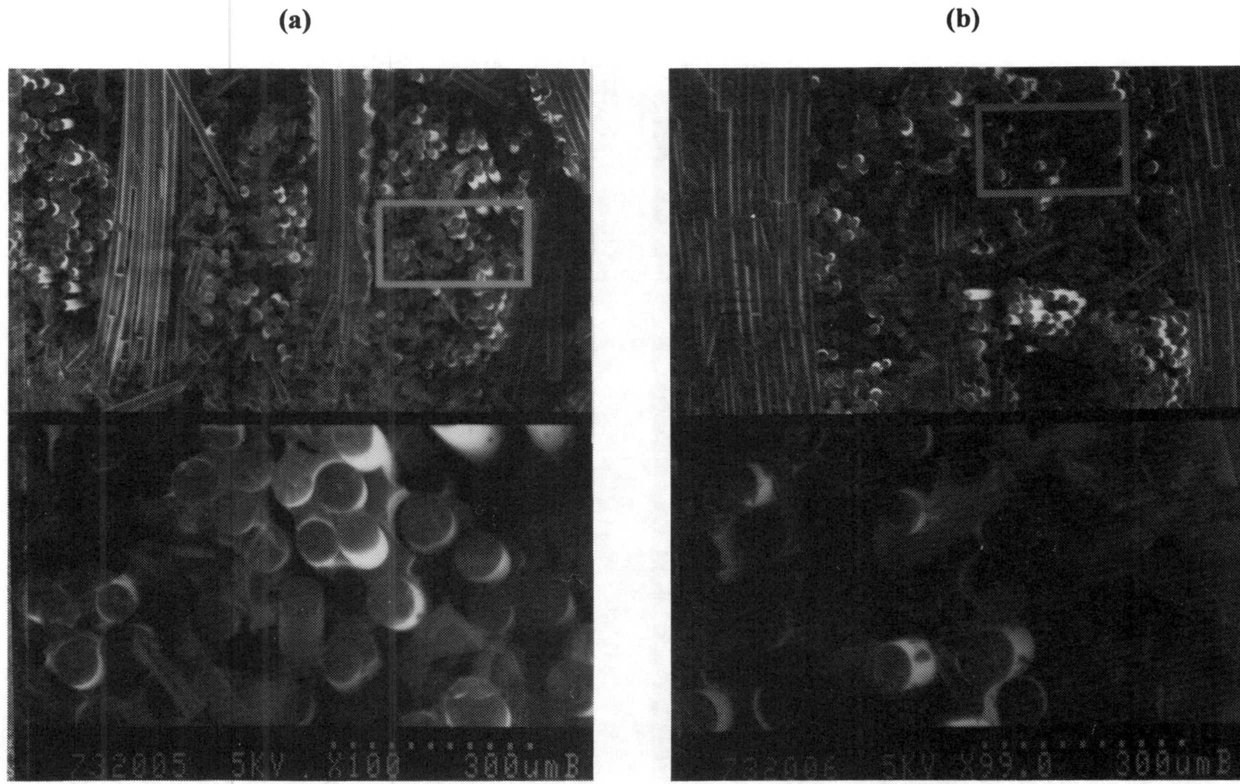

Figure 2. Fracture surface of stress-oxidation specimen for prestress < σ_o: (a) tensile region (b) compressive region.

(a) (b)

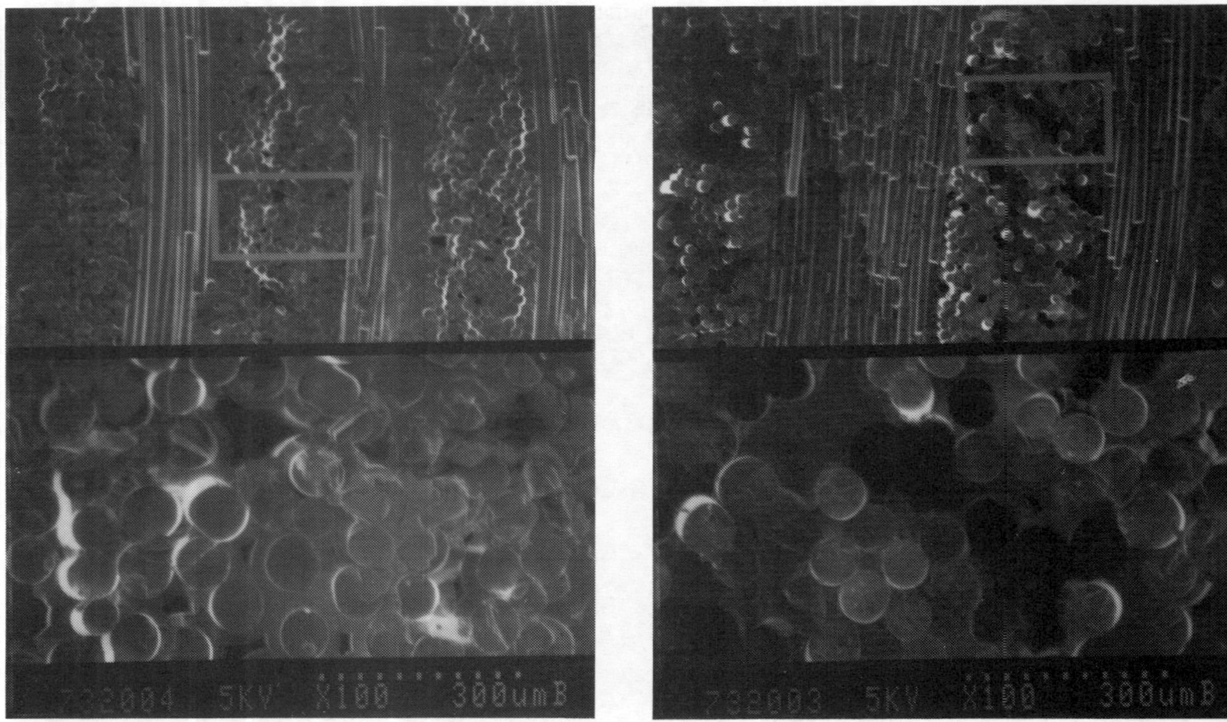

Figure 3. Fracture surface of stress-oxidation specimen for prestress > σ_o:
(a) tensile region (b) compressive region.

Control Specimen 1: As-Received

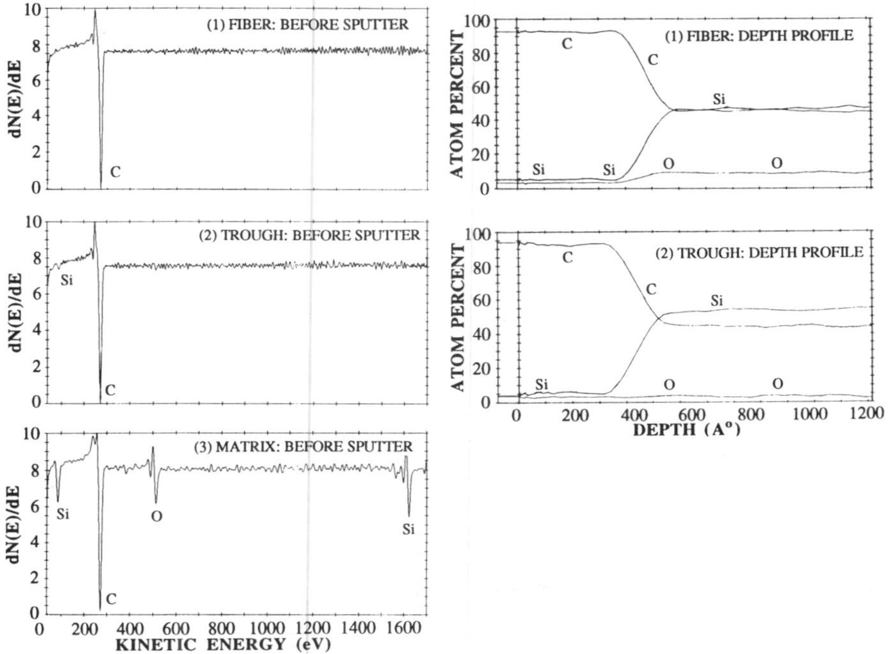

Figure 4. Initial AES data and depth profile for control specimen 1.

Control Specimen 2: Heat-Treated

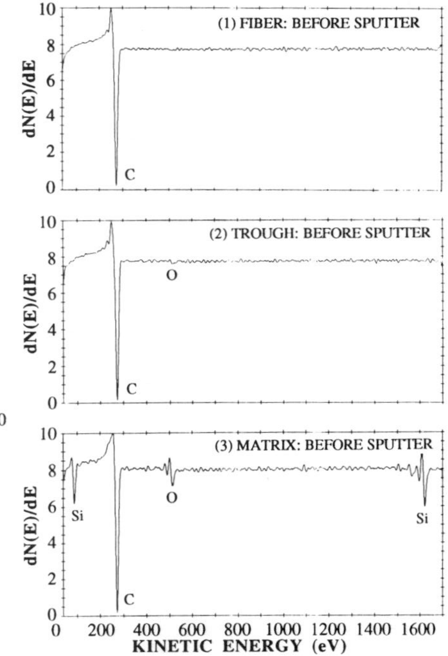

Figure 5. Initial AES data for control specimen 2.

Stress-Oxidation Specimen (Prestress = 140 MPa, 950°C)

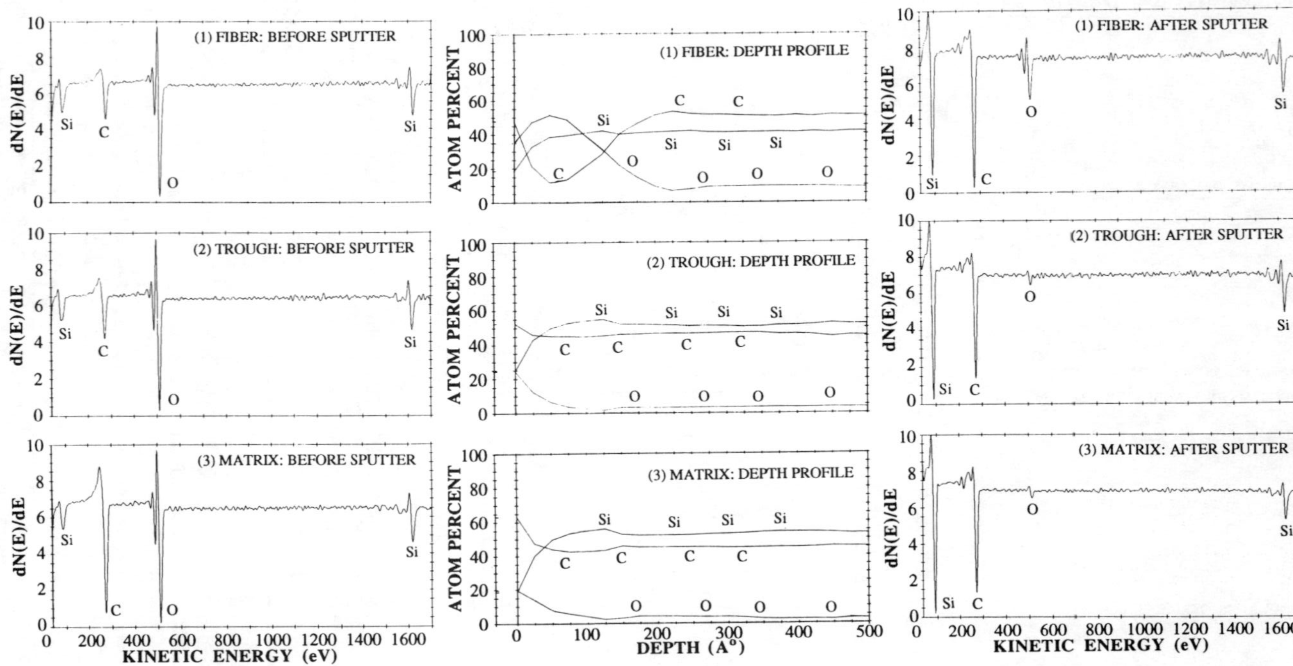

Figure 6. Initial AES data, depth profile, and final AES data for stress-oxidation specimen.

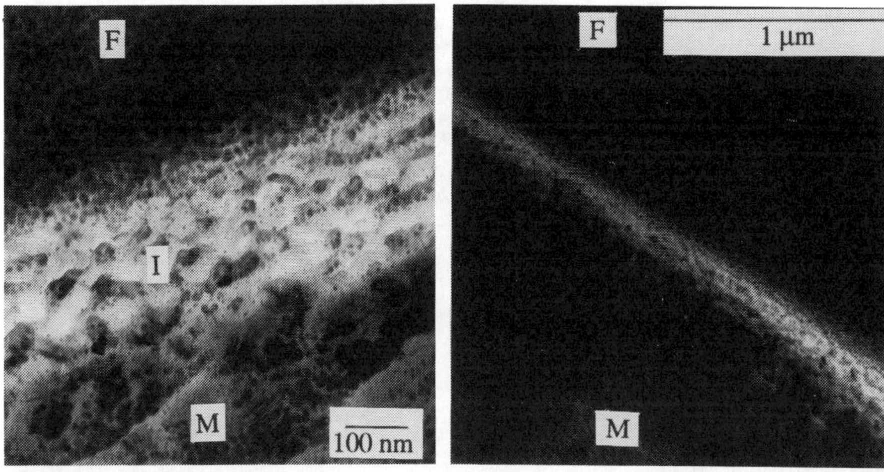

Figure 7. TEM images of as-received specimen.

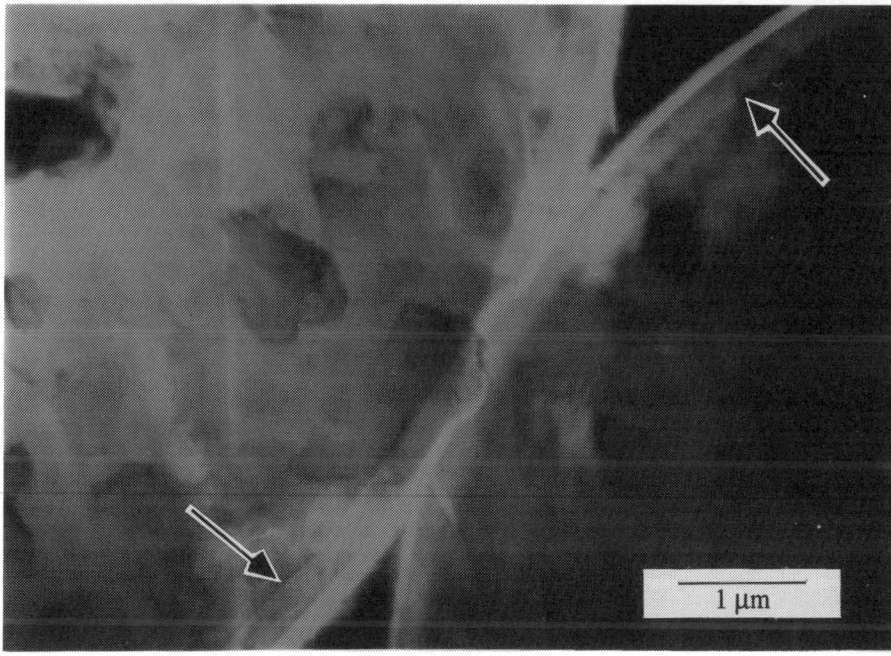

Figure 8. TEM images of oxidized specimen:
(a) annular interfacial opening, and affected zone in fiber (at arrow).

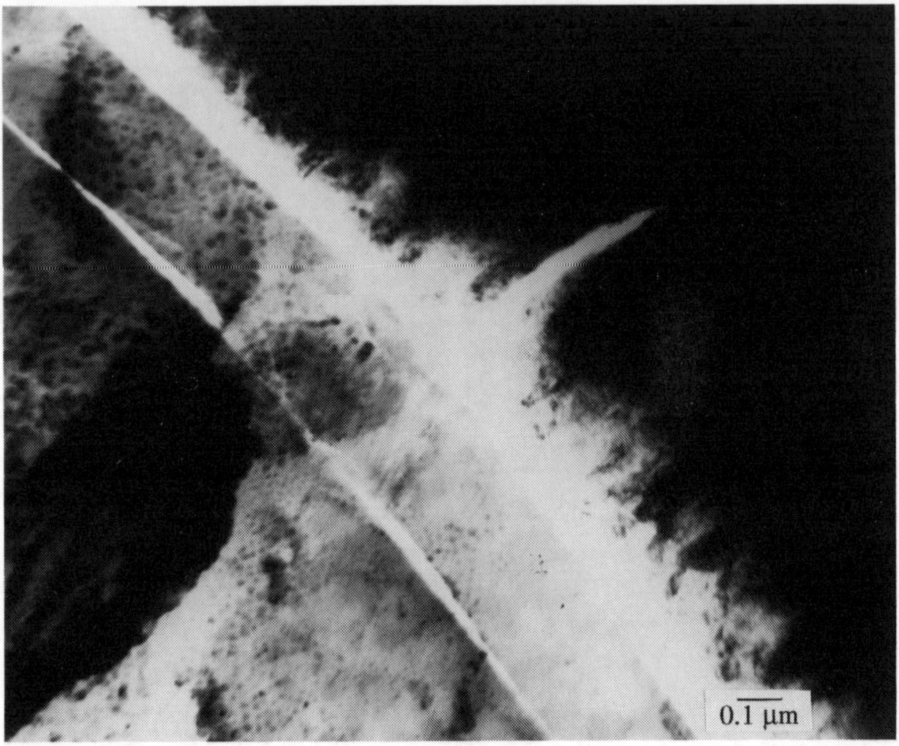

Figure 8. TEM images of oxidized specimen:
(b) crack in affected zone of fiber.

ENVIRONMENTAL EFFECTS ON FIBER REINFORCED CERAMIC MATRIX COMPOSITES BASED ON BLACKGLAS™

S. S. Campbell and S. T. Gonczy, AlliedSignal, Inc. P.O. Box 5016, Des Plaines, IL 60017-5016

Y.S. Park and M.J. McNallan, 2095 ERF, University of Illinois at Chicago, 842 W. Taylor St., Chicago, IL 60607

Abstract

Physical and mechanical properties of ceramic matrix composites intended for use at elevated temperatures may be degraded by chemical interactions between the environment and either the fibers, the matrix, or the interface compounds. Property degradation is characterized by measurement of mass and room temperature strength changes as well as by microstructural examination of the material after exposure. The effects of 700°C exposure on ceramic matrix composites prepared from Nextel 312™ fibers and AlliedSignal Blackglas™ matrices in dry air, humid air, and humid air containing KCl vapor are compared.

Introduction:

Ceramic materials have advantages over metals and alloys for high temperature applications because of their lower density and superior high temperature strength and creep resistance. However, the brittle nature of most monolithic ceramics makes them unsuitable for many applications in which they would otherwise outperform competing metals. Ceramic fiber reinforced ceramic matrix composites have the potential to overcome the toughness limitations of high temperature ceramics when fabricated with a suitable interface layer.

To the extent authorized under the laws of the United States of America, all copyright interests in this publication are the property of The American Ceramic Society. Any duplication, reproduction, or republication of this publication or any part thereof, without the express written consent of The American Ceramic Society or fee paid to the Copyright Clearance Center, is prohibited.

As discussed in an earlier work[1], the presence of an interface on continuous ceramic fibers is required for the fabrication of strong, damage tolerant ceramic fiber/ceramic matrix composites (CFCMC's). A suitable interface can be generated on Nextel 312™ fibers by treating the fibers in an ammonia containing atmosphere at elevated temperatures (>1200°C).[2,3] Equally important is the stability of the system (fiber, matrix, and interface coating) in application environments. While ceramic fibers and matrices are generally considered to be more oxidation resistant than metals, the degradation (by oxidation, corrosion, or evaporation) of the interface coating in service leads to one of two mechanisms of component degradation; 1) matrix/fiber reactions causing low strength brittle failure or 2) gap formation between the fiber and matrix causing loss of strength and modulus but the retention of crack deflection and fibrous failure. Both of these mechanisms have been observed in oxidation studies of carbon coated Nicalon™ and carbon coated Nextel 440™ CFCMC's[4].

A study of the stability of nitrided Nextel 312™ fiber/Blackglas™ ceramic matrix composites in air at temperatures between 600° and 800°C with an emphasis on the effect of oxidation on the room temperature mechanical properties and failure mechanisms indicated that the principal effects of air oxidation on the composites could be attributed to mechanism (2) above.[5] After oxidation in the temperature range of 600°C to 800°C for periods of up to 100 hours, the composites retained a fibrous fracture mode at room temperature with no sign of brittle failure. Oxidation at the higher temperatures and for longer times produced high failure strains (+50%), improved fiber pull-out, lower moduli (-50%), and a slow, gradual reduction in flexure strength(-30%). SEM analysis of the fiber interfaces in the oxidized composites did not provide sufficient information to identify the mechanism of oxidation in the interface region or to explain the effects on the mechanical properties.

While oxidation resistance in air represents a minimum environmental stability criterion for any commercially useful ceramic matrix composite, more severe environments are likely to be encountered in many important high temperature processes. Experience has shown that the presence of water vapor accelerates the high temperature oxidation of many non-oxide ceramics,[6,7] and alkali containing environments can be particularly damaging.[8,9] The simultaneous presence of both alkali halide vapors and water vapor has a synergistic effect on the corrosion rate which causes more severe attack than either component alone.[10] This paper describes the results of experiments performed on Blackglas™ matrix composites containing nitrided Nextel 312™

fibers in high temperature environments containing water vapor and alkali chloride vapors.

Experimental Procedures:

Materials:

Nextel 312™/Blackglas™ ceramic matrix were prepared in the same manner as described in the previous studies[1,5]. In this case, five 10.2 cm (4") x 91 cm (36") strips of 5 harness satin weave fabric were nitrided simultaneously to provide adequate material for the preparation of 45 CFCMC sample bars 7.6 cm (3") by 0.64 cm (0.25") x 0.17 cm (0.065").

In the fabrication of composites, AlliedSignal Blackglas™ pre-ceramic polymer products were used. Blackglas™ 489C B-stage resin was used to pre-preg fabric samples and Blackglas™ 489A/489B neat/catalyst solutions were used for densification.

Fabrication:

Blackglas™ CMC's were prepared with the treated fabric using a lamination/pyrolysis/infiltration procedure described in detail elsewhere[4]. Sample bars 8.9 cm (3.5") by 0.64 cm (0.25") by 0.17 cm (0.065") were cut from polymer preform panels with a diamond saw prior to pyrolysis and densification. Pyrolysis was performed under flowing nitrogen at temperatures below 900°C. Re-infiltration was achieved by evacuation and backfilling with resin followed by a 3-hour cure at 55°C.

Six samples bars were retained to provide baseline mechanical property information. Three samples each were exposed to controlled oxidizing conditions for fixed periods of time. The test atmospheres consisted of dry air, air containing 3% H_2O vapor by volume (saturated with liquid H_2O at 25°C), dry air containing 80ppm KCl vapor, and air containing both 3%H_2O and 80 ppm KCl. Exposures were performed at 700°C for 5, 20, and 100 hours and at 900°C for 5 hours. The test exposures without KCl (g) were performed in a Thermolyne laboratory box furnace in flowing air. The test exposures in atmospheres containing KCl(g) were performed in the two zone furnace shown schematically in Figure 1. Test exposures were performed in the upper zone of the furnace, while temperature of the lower zone fixed the KCl content of the environment. Air was metered and purified and passed downward through an alumina tube in the center of the furnace where it discharged into a crucible packed with solid KCl. KCl (g) was added to the gas by vaporization from the crucible. The vapor pressures of KCl were checked by measuring the weight changes of the crucible before and after each run. The low KCl content of the test gases was required by the low temperatures of some of the tests and the need to avoid condensation of KCl within the furnace. All samples were weighed before and after oxidation to provide weight change information.

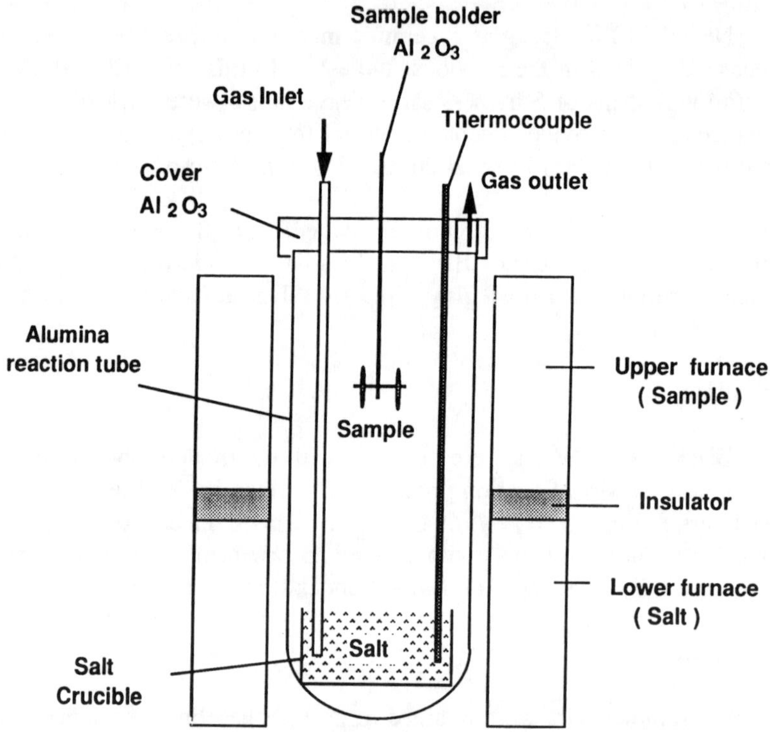

Figure 1: Schematic diagram of furnace used for alkali chloride vapor exposures.

Composite mechanical properties were evaluated at room temperature using an Instron 4507 universal testing machine. For 3-point flexure testing a crosshead speed of 0.050 cm/min and a span of 7.1 cm were used. Subsequently, the fractured composites were examined by using a JEOL T-200 Scanning electron microscope equipped for elemental microanalysis using energy dispersive X-ray spectroscopy.

Results and Discussion:

After five infiltrations, the composite bars had a nominal density of 2.2 g/cc and an estimated porosity of 10%.

Weight Change:

Figure 2 compares the mean weight change results for three composite specimens from each test at 700°C for different periods of time in dry air and in air containing H_2O and KCl(g). In dry air, the specimens continuously lose mass during the exposure as a result of oxidation of carbon from the Blackglas composite and possibly by oxidation and volatilization of the BN interface. The results are consistent with the results of earlier studies of similar composites.[5] In the KCl containing environment, there is a smaller degree of weight loss, and there is a slight increase in mass after the longest time exposure. This result was not unexpected on the basis of earlier experiments on silicon based ceramics which indicated that the corrosion products formed in these environments consisted of alkali silicates.[10]

SEM examination of Blackglas™ surfaces after exposure to the KCl containing environments indicated a slightly different mechanism of attack than has been observed for SiC or Si_3N_4. Figure 3 compares the surfaces of monolithic BlackglasTM coupons before and after exposure to Air-3%H_2O-80 ppm KCl at 700°C. The unexposed specimen is severely microcracked as a result of pyrolysis without the presence of reinforcing fibers. The exposed specimen has essentially the same structure. EDS analysis of the exposed specimen detected potassium but not chlorine, indicating that the potassium was absorbed into the Blackglas™ matrix as an oxide. This absorption did not result in the formation of a separate solid or liquid corrosion product layer even in the tests performed at 900°C.

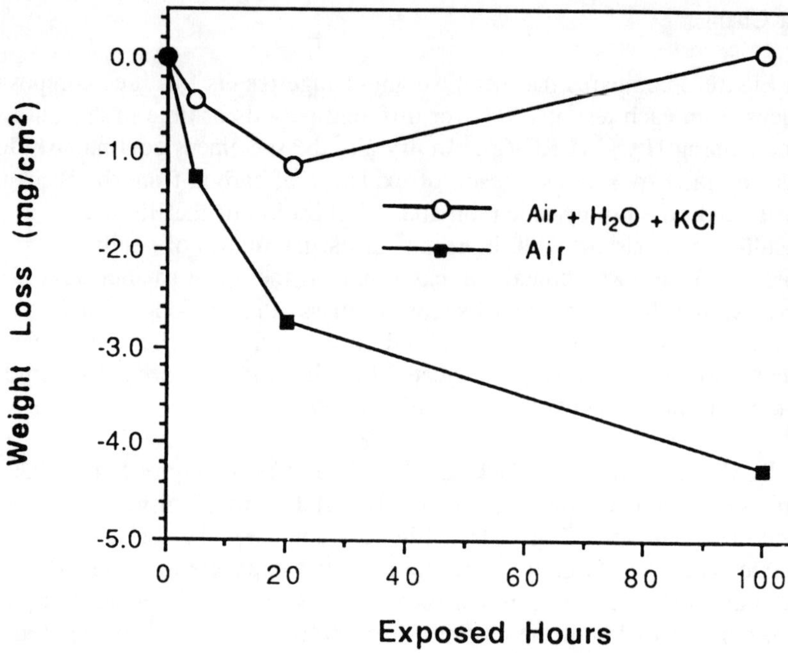

Figure 2: Mass change per unit of exposed area for composite specimens after exposure to air or air-3%H_2O-80 ppm KCl at 700°C,

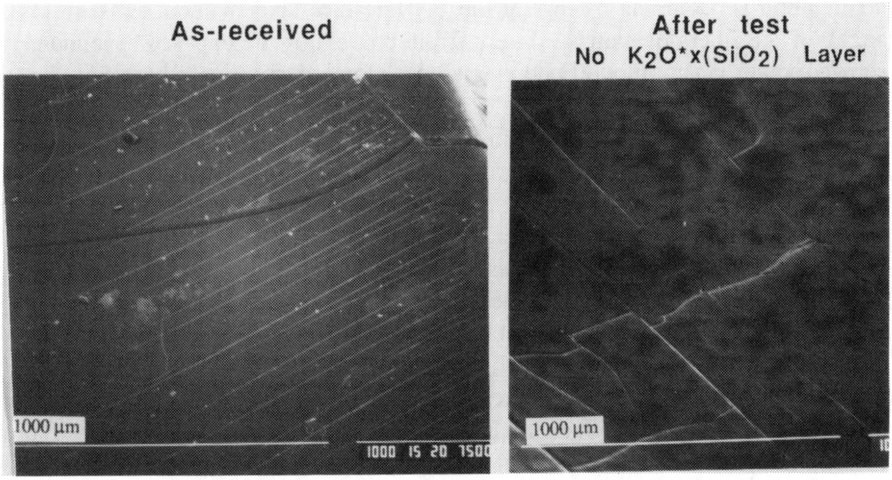

Figure 3: Scanning electron micrographs of Blackglas™ surfaces before and after exposure to air-3%H_2O-80 ppm KCl.

Mechanical Properties:

Figure 4 presents the mean modulus of rupture for three specimens exposed to each environment and tested at room temperature in comparison to the mean of six specimens tested in the as fabricated state. In general, the flexure strengths of the composites decrease with increasing time of exposure at 700°C, and are lower after five hours of exposure at 900°C than after a comparable period at the lower temperature. The loss in strength appeared to be associated with oxidation effects. The addition of KCl(g) to the environments did not result in lower strengths compared with bars tested in uncontaminated environments. The specimens tested in air-3%H_2O-80 ppm KCl at 900°C lost less strength than the specimens tested in the other environments.

Figure 5 presents the mean strain at failure for the specimens exposed to each environment in a format similar to that of Figure 4. In general, the strains at failure were unaffected or slightly increased by the exposures. This is consistent with the results of earlier oxidation tests in the same temperature range.[5] All of the failures were fibrous with substantial amounts of fiber pull-out. Brittle failure was not observed in any of the specimens. This, together with the reduced strength and modulus of the specimens, indicates that the degradation is occurring by loss of the BN interface layer without substantial attack of the fibers or matrix. Stress-strain curves for the exposed specimens showed lower elastic moduli and extended regions of non-linear behavior, consistent with a loss of load transfer between the matrix and fibers. Even the exposures at 900°C did not result in the formation of a bond between the fibers and matrix which could produce brittle fracture.

SEM examination of the fracture surfaces gave results which were consistent with the mechanical measurements. Extended fiber pull-out was observed. EDS microanalysis indicated that potassium was absorbed near the outside surfaces of the bend bars, but did not penetrate into the centers of the bars. Boron and nitrogen are too light to detect by EDS, and there was no sign of interfacial phases after the exposures to either the potassium containing or uncontaminated environments. Further investigations are underway to determine the fate of the interfacial BN layer.

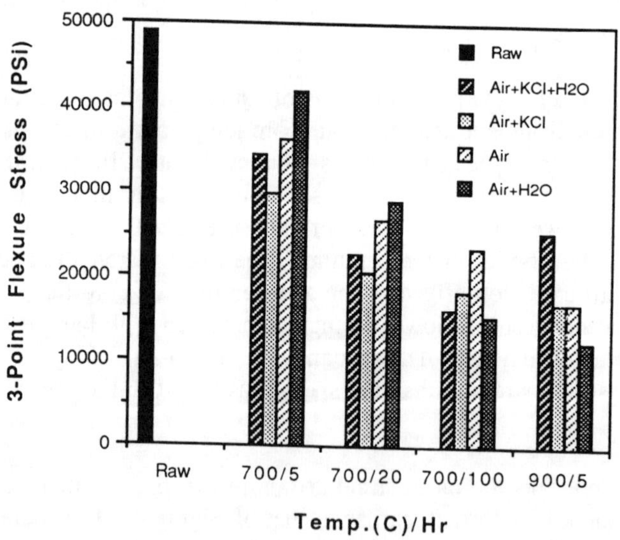

Figure 4: Three point bend modulus of rupture for specimens after exposure to oxidizing environments for various times.

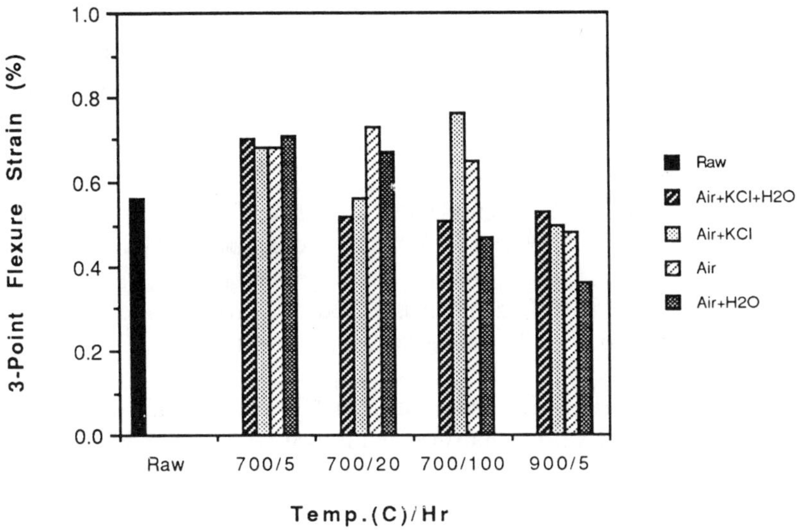

Figure 5: Mean strain at failure in three point bend tests for specimens after exposure to oxidizing environments for various times.

Conclusions:

Blackglas™ composites fabricated with nitrided Nextel 312™ fabric in a 2-D configuration exhibit fibrous fracture at room temperature. Oxidation in air and in gas mixtures of air-3%H_2O, air-80 ppm KCl and air-3%H_2O-80 ppm KCl did not change this characteristic. The strain to failure was slightly increased by oxidation at 700°C.

The modulus of rupture decreases with increasing exposure time at 700°C and is lower after 5 hours of exposure at 900°C than after the same exposure at 700°C.

Potassium penetrates into the Blackglas™ matrix, probably as K_2O, but does not form a distinct corrosion product layer or degrade the mechanical properties of the composites under the conditions of these tests.

References:

1. Campbell, S. and Gonczy, S.; "In-situ Formation of Boron Nitride Interface coatings on Nextel 312™ Continuous Fiber I: Nitriding Process and Blackglas Ceramic Matrix Composite Properties"; Ceramic Engineering and Science Proceedings, 1994, in press.

2. Simpson, F.H. and Verzemnieks, J., "Barrier Coated Ceramic Fiber and Coating Method," U.S. Patent #4,605,588, August 12, 1986.

3. Simpson, F.H. and Verzemnieks, J., Boron Nitride Coated Ceramic Fiber and Coating Method," U.S. Patent #4,948,662, August 12, 1986.

4. Gonczy, S.; Leung, R.; and Evans, J.; "Oxidation Effects on Fibrous Fracture in Blackglas™ Composites Reinforced with 3-D and 2-D Nextel™ 440 (Carbon Coated)"; The 16th Conference on Metal Matrix, Carbon, and Ceramic Matrix Composites; NASA Conference Publication 3175 Part 2, 1992, p. 561.

5. Campbell, S.S. and Gonczy, S.T., "In-Situ Formation of Boron Nitride Interfaces on Nextel 312[312] Continuous Ceramic Fiber II: Oxidation of Blackglas™ Ceramic Matrix Composites," Ceramic Engineering and Science Proceedings, 1994, in press.

6. Singhal, S.C., "Effect of Water Vapor on the Oxidation of Hot Pressed Silicon Nitride and Silicon Carbide," **J. Amer. Ceram. Soc., 59**, 1976, p 81.

7. Irene, E.A., and Ghez, R., "Silicon Oxidation Studies: The Role of H_2O," **J. Electrochem. Soc., 124**, 1977 p 1757.

8. V. Pareek and D. Shores, "Oxidation of Silicon Carbide in Environments Containing Potassium Salt Vapors," **J. Amer. Ceram. Soc., 74**, 1991, p 556.

9. Z. Zheng, R.E. Tressler, K.E. Spear, "The Effect of Sodium Contamination on the Oxidation of Single Crystal Silicon Carbide," **Corros. Sci., 33**, 1992, p 545.

10. M.J. McNallan, P.P. Hsu, S.Y. Lee, "Oxidation of Si Based Ceramics in the Presence of Alkali Halide Vapors," **JOM, 45**, No. 12, 1993, p22.

THE EVOLUTION OF A NON-PLANAR OXIDATION FRONT IN RECEDING CARBON FIBERS

W. H. Glime and J. D. Cawley
Case Western Reserve University, Cleveland, OH

A. J. Eckel
NASA Lewis Research Center, Cleveland, OH

ABSTRACT

Carbon fibers have been observed to develop a non-planar surface during oxidative recession into a non-reactive matrix. The origin and development of the non-planar morphology was experimentally investigated through controlled oxidation of carbon cores in SCS-6® SiC fibers. Also, a numerical simulation of carbon fiber recession during oxidation was developed. Results of the simulation for two idealized microstructures were compared to experimental observations.

INTRODUCTION

Continuous fiber ceramic-matrix composites (CFCC's) containing carbon constituents (e.g., fibers or interfacial layers) are potentially useful for many high temperature structural applications. The susceptibility of carbon to oxidation is a severe limitation to these material systems. In order to accurately predict service life it is necessary to make quantitative measurements of recession rates. In addition, careful measurements also have the potential to shed light on the aspects of microstructure that influence the rate of recession.

Recession of an oxidizing constituent into a non-reactive matrix is a function of diffusive oxygen transport and the rate of chemical reaction at the oxidizing surface (Figure 1). The coupling of diffusive transport and chemical reaction rate results in linear-parabolic recession kinetics described by Equation 1. Experimental methods which indirectly measure the oxidative recession rate (e.g., TGA or electrical resistance techniques) typically assume the oxidizing carbon surface is planar. The actual morphology of the receding carbon surface is, however, seldom verified. In studies where the oxidative recession distance was measured directly, a highly non-planar carbon surface morphology was observed.

In a study of the rate of recession of carbon rods, 3 mm in diameter, into an alumina sheath below 900°C in air, a non-planar "conic" tip morphology was observed to develop[1] (Figure 2a).

To the extent authorized under the laws of the United States of America, all copyright interests in this publication are the property of The American Ceramic Society. Any duplication, reproduction, or republication of this publication or any part thereof, without the express written consent of The American Ceramic Society or fee paid to the Copyright Clearance Center, is prohibited.

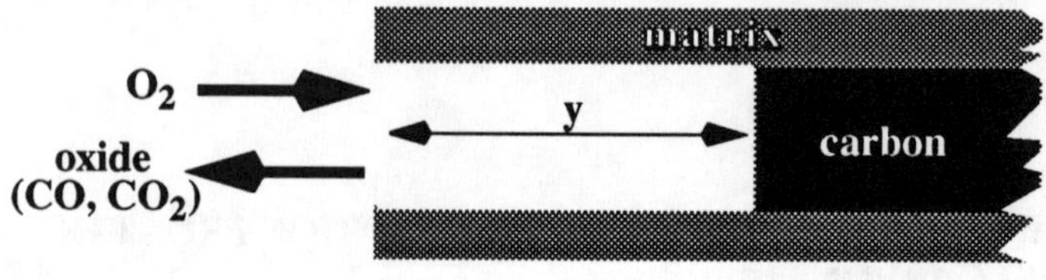

Figure 1. Schematic of carbon fiber recession into a non-reactive matrix during oxidation. Recession kinetics described by Equation 1.

$$\frac{y}{k_l} + \frac{y^2}{k_p} = t \qquad (1)$$

where: $k_l = \frac{k\,C_g}{N}$ $\quad k_p = \frac{D\,C_g}{N}$

t = oxidation time (s)
y = recession distance (m)
k_l = linear rate constant (m/s)
k_p = parabolic rate constant (m^2/s)
k = chemical reaction rate (m/s)
C_g = O_2 concentration in ambient ($mols/m^3$)
D = gas phase diffusivity (m^2/s)
N = molar density of carbon ($mols/m^3$)

A similar observation was made in a study of the recession of SCS-6 cores during oxidation below 1000°C [2]. Recession distance was determined by grinding the SiC fiber parallel to the cylinder axis, exposing the pitch-based carbon core. A non-planar morphology in the receding carbon surface was noted (Figure 2b).

It is striking that the non-planar tip is observed in both cases, since the diameters differ by a factor of 100. This suggests that understanding "tip" formation is of general relevance. The work reported here focused on understanding the underlying mechanisms for development of a non-planar oxidation front during recession of SCS-6 cores.

a) carbon rod in an alumina sheath

b) carbon core of SCS-6 SiC fiber

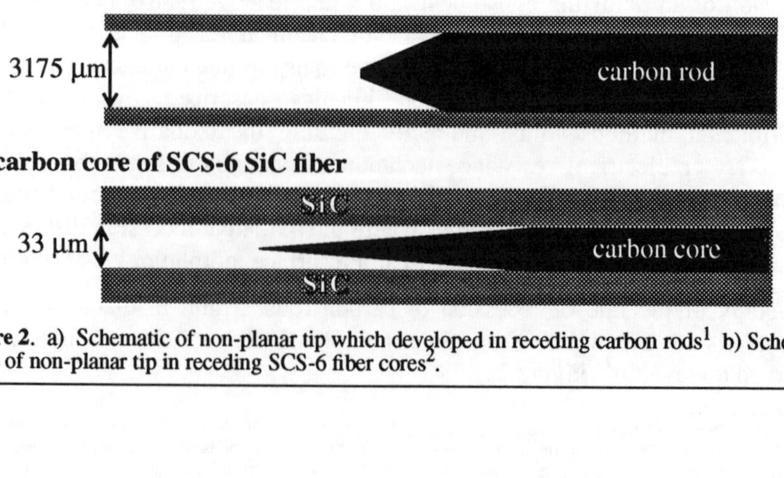

Figure 2. a) Schematic of non-planar tip which developed in receding carbon rods[1] b) Schematic of non-planar tip in receding SCS-6 fiber cores[2].

EXPERIMENTAL ANALYSIS

SCS-6 fibers are produced by CVD of SiC (50 μm) around a carbon monofilament (r=16 μm). Before SiC deposition, a 1.5 μm layer of pyrolytic carbon is deposited on the carbon monofilament. SEM micrographs of an SCS-6 fiber cross-section are given in Figure 3a and 3b.

Exposure of the SCS-6 cross-section to air at 900°C produced microstructures depicted in Figure 3c and 3d. These possess the following features of interest:

- Pitting develops on the surface of the carbon core.
- A crevice opens up around the core perimeter.

Further oxidation results in the core morphologies depicted in Figures 4a and 4b. Figure 4a is a SEM micrograph obtained by grinding the fiber along the axial direction, exposing the carbon core. Figure 4b was obtained by sliding the SiC sheath away from a crack, exposing the core and pyrocarbon layer. The micrographs indicate that the non-planar "conic" tip morphology develops in the core at the onset of oxidation and remains stable over large recession distances ($y > 1$ mm). Based on comparison of dimensions in Figures 3b and 3d, the pyrocarbon layer appears more resistant to oxidation than the core. The increased resistance to oxidation of the pyrocarbon is also evident in Figure 4b, where the "conic" core tip is estimated by geometry to extend over 100 μm into a free-standing sheath of pyrocarbon.

Two possible mechanisms for the observed development of a non-planar morphology are:

- The presence of a debond between the pyrocarbon layer and the core allowing the ingress of oxygen which results in radial oxidation of the core and pyrocarbon layer.
- The existence of a highly oxidizable layer in the region of the core/pyrocarbon interface.

The presence of a debond is feasible based on large core pullout lengths (up to 1 mm) after fiber fracture. The existence of a highly oxidizable layer in the region of the pyrocarbon/core interface is supported by observations of Lamouroux *et al.* on PAN based (T-300) fibers with a pyrocarbon coating[3]. Preferential oxidation occurred in the region of the pyrocarbon/fiber interface and was attributed to an increased pore density observed in TEM analysis of the fiber radial perimeter.

NUMERICAL MODELING OF OXIDATION

A computer simulation of recession during oxidation of a carbon fiber into a non-reactive matrix was generated to evaluate possible mechanisms for non-planar

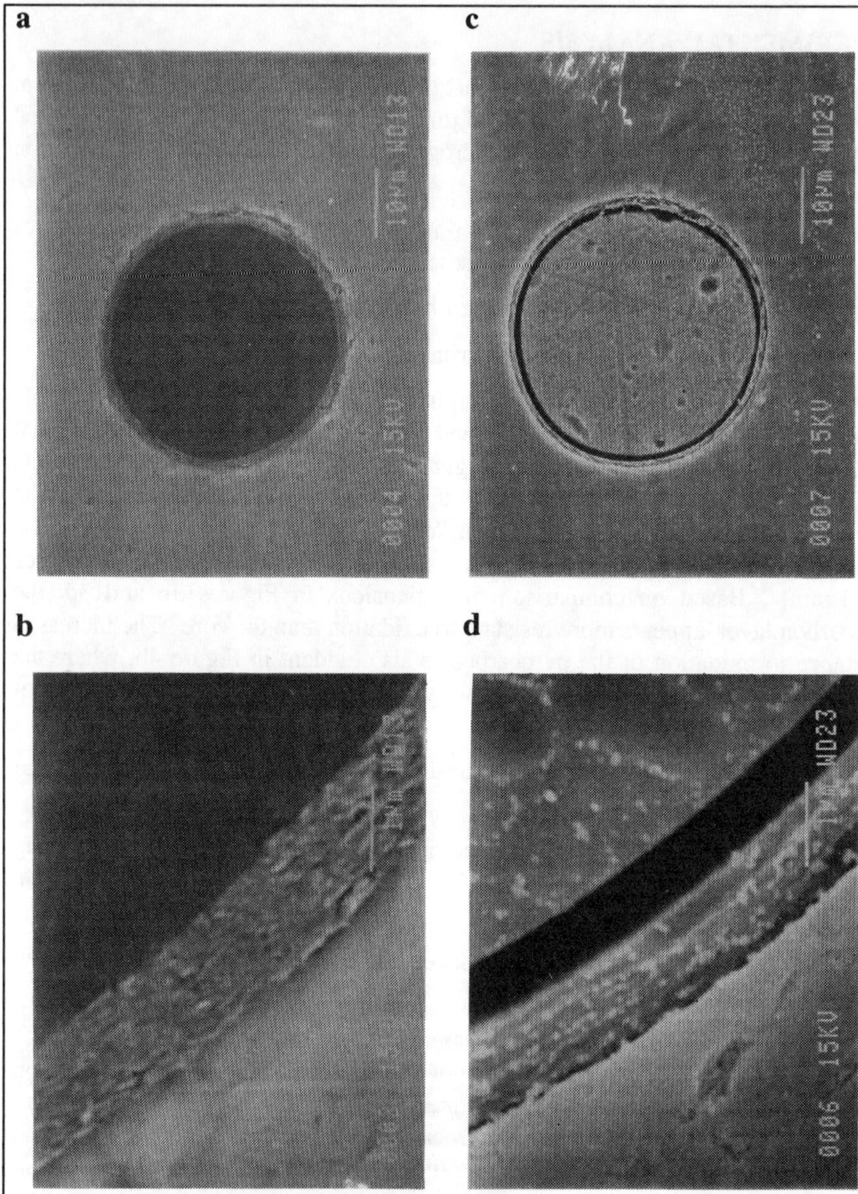

Figure 3. SEM micrographs of SCS-6 fiber cross-section a) carbon core b) region of pyrocarbon layer c) core after 900°C anneal in air (t < 5 minutes) d) region of pyrocarbon layer after 900°C anneal in air (t < 5 minutes).

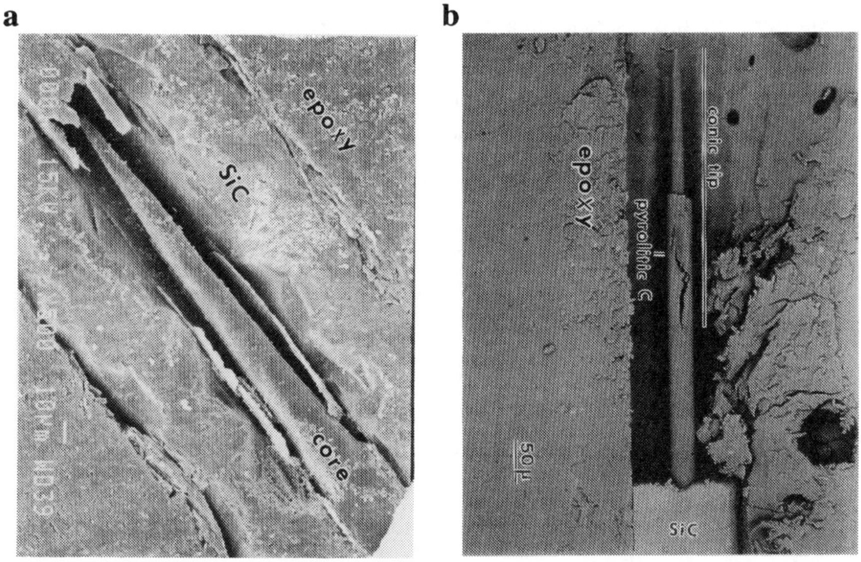

Figure 4. Micrographs of non-planar "conic" morphology which develops in SCS-6 fiber core during oxidation. a) 10 min in flowing O_2, prepared by grinding in axial direction b) 1 hour in flowing O_2, prepared by sliding off SiC coating. Note the presence of the free standing pyrocarbon layer extending well beyond the oxidative penetration depth.

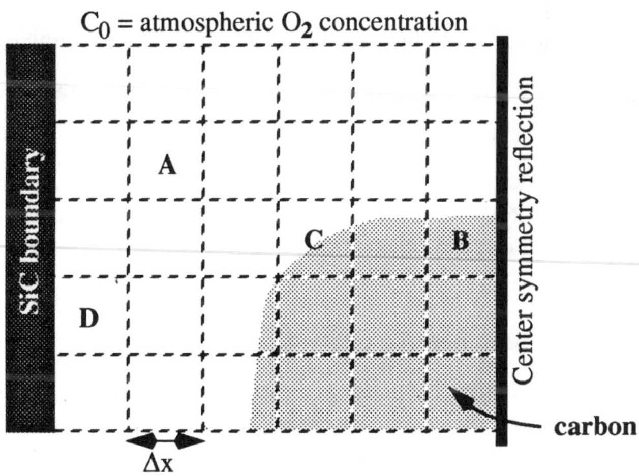

Figure 5. Schematic of mesh used in computer simulation of carbon fiber oxidation.

tip evolution. Figure 5 schematically represents the basis of the simulation which is briefly described as follows:

- The SiC layer and symmetry along the center provide no-flux boundary conditions. (i.e., J = 0 across these boundaries)
- The concentration at the end of the pore is given by the atmospheric oxygen concentration (C_o).
- The oxygen concentration [C(i,j)] of cells that do not contain carbon (e.g., A & D) is defined by diffusive flux balance with adjacent cells and evaluated using the following expression;

  $$C(i,j) = (C(i,j-1)+C(i,j+1)+C(i-1,j)+C(i+1,j)) / n_a$$

  i, j = cell position (col, row)
  n_a = # of adjacent cells

- The oxygen concentration of cells representing the surface of the carbon fiber which contain both O_2 and carbon (e.g., B & C) is defined by balancing the chemical reaction rate with the diffusive flux, yielding

  $$C(i,j) = (\Sigma C_{ex})/(n_{ex}(1- k\,\Delta x/D))$$

  n_{ex} = # of adjacent cells devoid of carbon (n_{ex} = 2 for cell C and 1 for cell B)
  k = chemical reaction rate constant (m/s)
  D = gas phase diffusivity (m^2/s)
  Δx = cell dimension (1 μm)

- Cell O_2 concentrations are determined by iterating until steady state is obtained.
- The time step (Δt) for carbon removal is determined by calculating the time required to empty each individual carbon surface cell and then choosing the shortest of the calculated times.
- Carbon is removed from the surface cells using the following expression;

  $$M(i,j,p+1) = M(i,j,p) - k\ C(i,j)\,\Delta t$$

  M(i,j,p) = carbon in cell i,j at time-step p

- Following carbon removal the entire process is repeated.

A VAX 2200 workstation was used to execute simulations. Graphic output was created on a Macintosh PC with filled pixels indicating cells containing carbon and unfilled containing oxygen only.

RESULTS

Oxidation in the presence of a debond

A debond was simulated by placing a gap of finite length (300 μm) between the pyrocarbon layer and the core. A value of 0.02 cm/s was chosen for the chemical reaction rate (k) of the core, based on the experimental value of k measured by Eckel *et al.*[1] at 800°C in flowing oxygen. For the pyrocarbon layer, which was observed to exhibit a higher resistance to oxidation than the carbon core, a value of $k = k_{core}/2$ = 0.01 cm/s was used. The value of gas phase diffusivity (D) used in all simulations was 2.0 cm^2/s, based on tabulated values[4]. Figure 6a graphically depicts the initial conditions used to simulate oxidation in the presence of a debond.

During the early stages of oxidation (Figure 6b), the gap around the perimeter of the core widens, consistent with experimental observations. As oxidation progresses (Figure 6c) a pointed morphology begins to form but rapidly disappears (Figure 6d) due to the surface to volume ratio of the point being much larger than that of the relatively planar crevice bottom. Further recession of the core then proceeds with a relatively planar front. Thus, the results calculated using this model are not consistent with the observation that the tip remains self-similar even after large recession distances.

Figure 6. Simulated oxidation of SCS-6 core with a finite debond.

Highly oxidizable layer near pyrocarbon/core interface

Figure 7 schematically represents the procedure for simulating a highly oxidizable layer in the region of the pyrocarbon/core interface. Values for the reaction rate are the same as those of the simulated debond (k_{core}=0.02 cm/s, k_{pyc}=0.01 cm/s), but a highly oxidizable layer (thickness=1 μm, k=20k_{core}=0.4 cm/s) was placed between

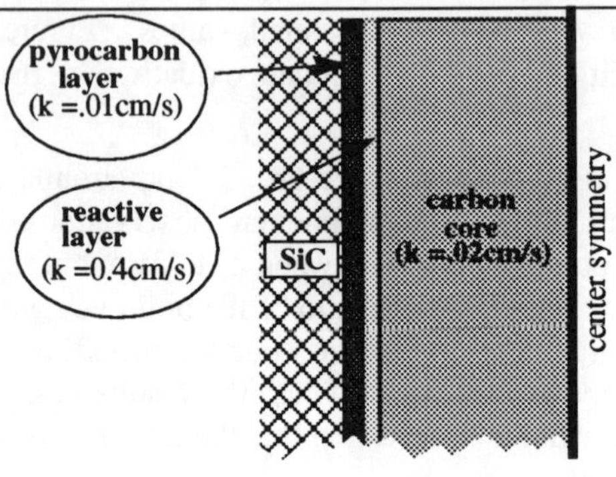

Figure 7. Diagram of initial conditions for simulation of fiber oxidation in the presence of a highly oxidizable layer.

the pyrocarbon layer and the core in the simulation.

Figures 8a-8d depict morphologies obtained during simulated oxidation in the presence of a highly oxidizable layer. At the onset of oxidation, the highly oxidizable layer quickly recedes leaving a gap in its wake (Figure 8a). The

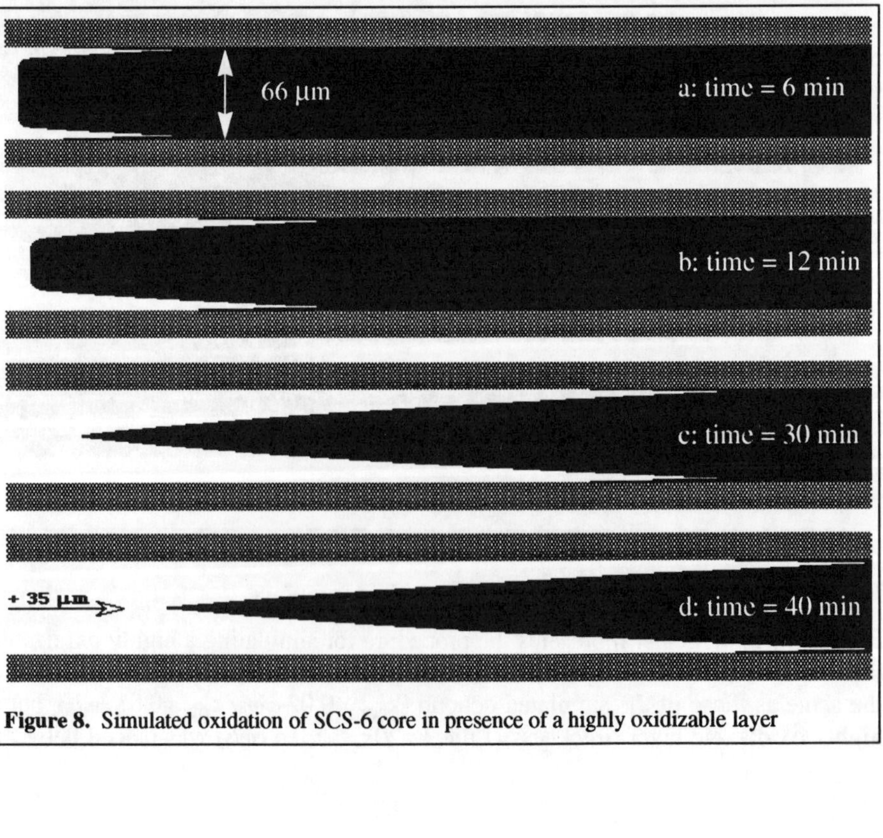

Figure 8. Simulated oxidation of SCS-6 core in presence of a highly oxidizable layer

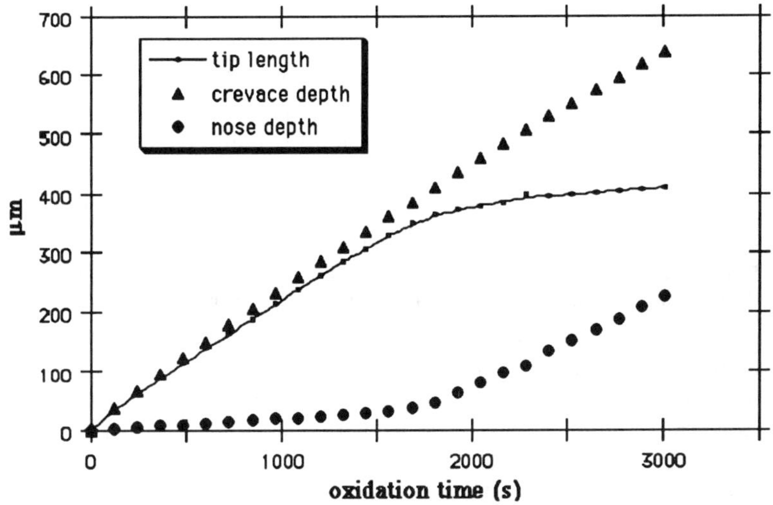

Figure 9. Plot of recession distance vs. time for the highly oxidizable layer (crevice depth) and the tip of the carbon core (nose depth). The "tip height" is the difference between the nose depth and the crevice depth.

crevice left by the rapidly receding layer allows radial oxidation of the core and pyrocarbon and the core begins to form a pointed morphology (Figure 8b). As oxidation continues, the pointed morphology of the core continues to evolve (Figure 8c) until the rate of recession of the core "tip" approaches that of the highly oxidizable layer due to the increase in relative surface area. When the recession rate of the tip matches that of the highly oxidizable layer, oxidation proceeds with a steady state morphology (Figure 8d).

Plotting the recession of the highly oxidizable layer and the core "tip" with time (Figure 9) shows the evolution of a steady-state tip morphology. Initially, the highly oxidizable layer recedes at a rate ~20 times that of the core center (tip) as defined by the initial conditions. The recession of the highly oxidizable layer slows modestly with time as the oxygen concentration at the bottom of the crevice is reduced. The core tip, which begins with a planar morphology, recedes in a linear fashion, defined by the chemical reaction rate, until a pointed morphology develops. At this point, the recession of the tip abruptly increases until it matches that of the rapidly oxidizing layer. The fact that the recession rates are matched means that steady-state has been reached. The tip height in Figure 9 represents the difference in recession distance between the rapidly oxidizing layer and the tip of the core. Plotting the tip height as a function of time depicts the development of a steady state morphology (stable tip height).

CONCLUSIONS

During the oxidative recession of the carbon core in SCS-6 SiC fibers, a highly non-planar "pointed" oxidation front has been experimentally observed. The pointed morphology develops at the onset of oxidation and remains stable over large recession distances. Numerical simulation of carbon fiber oxidation was performed and has proven useful in understanding the development of non-planar morphologies in the receding carbon cores. Results from the simulation indicate that the presence of a finite debond will induce a pointed tip morphology but it will not be maintained over large recession distances. The presence of a highly oxidizable layer (i.e., containing porosity, impurities which catalyze the reaction, a weak interface, etc.) produced simulated results in agreement with experimental observations. Results of the simulation indicate that the rate of carbon core recession is governed by that of the highly oxidizable component. The increase in the recession rate of the carbon core is caused by an increase in the surface exposed to oxidation. The exposed surface area of the constituents less susceptible to oxidation will increase until the recession rate of each component is identical. In the case of the SCS-6 core, the development of a non-planar oxidation front can produce an order of magnitude increase in the oxidative recession rate as compared to recession with a planar morphology.

REFERENCES

1. A. J. Eckel, J. D. Cawley, and T. A. Parthasarathy, "Oxidation of a Continuous Carbon Phase in a Non-Reactive Matrix", submitted to *Journal of the American Ceramic Society,* April 1994.
2. J. D. Cawley, O. Unal, and A. J. Eckel, "Oxidation of Carbon in Continuous Fiber Reinforced Ceramic Matrix Composites", *Ceramic Transactions, Advances in Ceramic Matrix Composites*, Vol. 38, ed. by N. P. Bansal, p 541-552, (1994).
3. F. Lamouroux, X. Bourrat, R. Naslain, and J. Sevely, "Structure Oxidation Behavior Relationship in the Carbonaceous Constituents of 2D-C/PyC/SiC Composites", *Carbon,* Vol. 31, pp. 1273-1299, (1993).
4. G. H. Geiger and D. R. Poirier, *Transport Phenomena in Metallurgy,* Addison-Wesley, 1973, pp. 468.

ACKNOWLEDGMENTS

This research was jointly funded by an Ohio Aerospace Collaborative Core Research Program (CCRP 93-1-027) and a NASA/CWRU Cooperative Agreement on Ceramic Processing (NCC3-139).

® SCS-6 fibers are manufactured by Textron, Lowell, MA.

INDEX